A TEXTBOOK OF
PLANT PHYSIOLOGY, BIOCHEMISTRY AND BIOTECHNOLOGY

[For B.Sc. and M.Sc. Students of All Indian Universities]

S.K. VERMA

M.Sc., Ph.D., F.B.S.

Reader
Department of Postagraduate Studies in Botany
St. Andrew's College
(Deen Dayal Upadhyay University)
GORAKHPUR–273001

and

MOHIT VERMA

M.Sc. (Botany)

S. CHAND
PUBLISHING

S Chand And Company Limited
(ISO 9001 Certified Company)
RAM NAGAR, NEW DELHI - 110 055

S Chand And Company Limited
(ISO 9001 Certified Company)

Head Office: 7361, RAM NAGAR, QUTAB ROAD, NEW DELHI - 110 055
Phone: 23672080-81-82, 66672000 Fax: 91-11-23677446
www.schandpublishing.com; e-mail: info@schandpublishing.com

Branches:

Ahmedabad	:	Ph: 27541965, 27542369, ahmedabad@schandpublishing.com
Bengaluru	:	Ph: 22268048, 22354008, bangalore@schandpublishing.com
Bhopal	:	Ph: 4209587, bhopal@schandpublishing.com
Chandigarh	:	Ph: 2625356, 2625546, 4025418, chandigarh@schandpublishing.com
Chennai	:	Ph: 28410027, 28410058, chennai@schandpublishing.com
Coimbatore	:	Ph: 2323620, 4217136, coimbatore@schandpublishing.com (Marketing Office)
Cuttack	:	Ph: 2332580, 2332581, cuttack@schandpublishing.com
Dehradun	:	Ph: 2711101, 2710861, dehradun@schandpublishing.com
Guwahati	:	Ph: 2738811, 2735640, guwahati@schandpublishing.com
Hyderabad	:	Ph: 27550194, 27550195, hyderabad@schandpublishing.com
Jaipur	:	Ph: 2219175, 2219176, jaipur@schandpublishing.com
Jalandhar	:	Ph: 2401630, jalandhar@schandpublishing.com
Kochi	:	Ph: 2809208, 2808207, cochin@schandpublishing.com
Kolkata	:	Ph: 23353914, 23357458, kolkata@schandpublishing.com
Lucknow	:	Ph: 4065646, lucknow@schandpublishing.com
Mumbai	:	Ph: 22690881, 22610885, 22610886, mumbai@schandpublishing.com
Nagpur	:	Ph: 2720523, 2777666, nagpur@schandpublishing.com
Patna	:	Ph: 2300489, 2260011, patna@schandpublishing.com
Pune	:	Ph: 64017298, pune@schandpublishing.com
Raipur	:	Ph: 2443142, raipur@schandpublishing.com (Marketing Office)
Ranchi	:	Ph: 2361178, ranchi@schandpublishing.com
Sahibabad	:	Ph: 2771235, 2771238, delhibr-sahibabad@schandpublishing.com

© 1995, S.K. Verma & Mohit Verma

All rights reserved. No part of this publication may be reproduced or copied in any material form (including photocopying or storing it in any medium in form of graphics, electronic or mechanical means and whether or not transient or incidental to some other use of this publication) without written permission of the copyright owner. Any breach of this will entail legal action and prosecution without further notice.

Jurisdiction: *All disputes with respect to this publication shall be subject to the jurisdiction of the Courts, Tribunals and Forums of New Delhi, India only.*

First Edition 1995
Subsequent Editions and Reprints 1997, 99, 2000, 2002, 2003, 2005, 2007 (Twice), 2008, 2010, 2011, 2012, 2013, 2014 (Twice), 2016 (Twice), 2017, 2018
Reprint 2019

ISBN : 978-81-219-0627-2   **Code :** 1003C 202

PRINTED IN INDIA

By Vikas Publishing House Pvt. Ltd., Plot 20/4, Site-IV, Industrial Area Sahibabad, Ghaziabad-201010 and Published by S Chand And Company Limited, 7361, Ram Nagar, New Delhi-110 055.

PREFACE TO THE SIXTH EDITION

Present edition is thoroughly revised and is based on the suggestions of students and teachers from various universities. To update and meet the needs of B.Sc. students, who have offered unified U.G.C. prescribed syllabus, a new section, "Part III- BIOTECHNOLOGY" is added. This part deals with the things only prescribed in the syllabus and a few important things needed in the genetic engineering part for students appearing in NET and other competitive examinations. Now, the book is entitled as "A text book of Plant Physiology, Biochemistry and Biotechnology".

Hope, this edition will be more useful and will fulfil the needs of students of Indian Universities.

S.K. Verma
Mohit Verma

PREFACE TO THE THIRD EDITION

As the second edition of "*A Textbook of Plant Physiology and Biochemistry*" was warmly received by the readers and it exhausted within no time. Several useful suggestions were received from learned teachers and students, which necessitated the printing of this third revised edition. During the preparation of this edition the general plan of the book was retained as such except the additional information added in almost every chapter. Besides, a few new chapters have been added. The entire text has been thoroughly checked, revised and updated. To update the book following additional information has been added in respective chapters:

Biochemistry

Chapter
1. Experiments related to chromatography
2. Some important tests for carbohydrates and related experiments.
3. Protein content of the plant
4. Derived lipids (Steroids)
5. Isolation and purification of enzymes, Models for explaining enzyme action, Differences between enzymes and hormones, Tests for some enzymes.
7. Differences between Vitamins and hormones.
8. Differences between auxins and gibberellins.
9. Forms of double stranded DNA, Biosynthesis of nucleotides — Purine nucleotides: IMP, GMP, AMP and Uric acid, Biosynthesis of Pyrimidine ribonucleotides — CMP and TMP, Mechanism of DNA replication, Test for the presence of DNA and RNA in Plants.

(New) 12. *Seconary metabolites — Terpenoids —* Structure, Types and sources, Biosynthesis of various terpenoids, Functions of terpenoids. *Alkaloids —* Types, Distribution and Localization, Biosynthesis of Various alkaloids, Biological functions of alkaloids. *Flavonoids.*

Plant Physiology

1. Differences between 70S and 80S ribosomes, Cell components and their main functions."
2. Demonstration of diffusion, Preparation of artificial semipermeable membrane, Factors affecting permeability; Water potential concept, Differences between DPD and water potential (ψ), Water potential changes in plasmolysis and deplasmolysis; Significance of imbibition, Demonstration of imbibition; Other experiments.
3. Importance of water to the plants, Soil-waters, Forms of underground water, Hygroscopic coefficient, Moisture equivalent, Wilting and Wilting Coefficient,

      Path of water in roots, Theories regarding development of root pressure, Apoplast and symplast concept.
4. Evidences in support of Cohesion Theory, its criticism.
5. Experiments related to transpiration.
6. Differences between trace and tracer elements, Hydroponics, Precautions during preparation of nutrient solution, Differences between chlorosis and etiolation.
10. Photosynthetic apparatus, Photosynthetic Pigments — Chlorophylls, Phycobilins, Carotenoids, Distribution of Plant Pigments, Biosynthesis of chlorophyll; Nature of light — Theories, Mechanism of light absorption, components of electromagnetic spectrum; Absorption spectrum; Action spectrum; PS I and PS II, ET chain components; Significance of CAM, Bacterial Photosynthesis — Non-cyclic and Cyclic photophos-phorylation, CO_2 fixation, Differences between plant and bacterial photosynthesis; Chemosynthesis, Differences between Photosynthesis and Chemosynthesis, Other experiments.
11. Terminal oxidation of NADH (extramitochondrial), M-O-A shuttle and G-3-P DHAP shuttle, Oxidative phosphorylation — site, ATP synthetase — Structure and Chemistry, Structure and Function of Electron carriers present in ET chain, Mechanism — Various theories; Respiratory inhibitors, Cyanide resistant respiration, Types and determination of RQ, Differences between RQ and PQ.
(New) 13. *Nitrogen metabolism and Nitrogen cycle:* Significance of N_2, N_2 in soil, Nitrate reduction in plants; Nitrogen fixation — Non-biological and biological — Non-symbiotic, Symbiotic and Associative, Symbiotic N_2 fixation in Leguminous plants — Formation of root nodules; Biochemistry of N_2 fixation — Requirements, Nitrogenase enzyme, Steps in Asymiotic and symbiotic N_2 fixation, Source of N_2 in insectivorous plants; Denitrification, N_2 Cycle.
New 14. *Nucleic Acid Metabolism: Introduction;* Nitrogenous Bases — Pyrimidines and Purines; Phosphoric acid; Nucleosides and Nucleotides — Structure; Biosynthesis of purines — IMP, GMP, AMP and uric acid; Pyrimidines — CMP, UMP and TMP; DNA Structure, Watson-crick model, Duplication — Methods and Mechanism, Enzymes of DNA synthesis, DNA unwinding proteins; RNA, Differences between RNA and DNA, RNA types, Microchemical test for DNA and RNA.
15. *Metabolism of Amino-acids (New) and Proteins :* Biosynthesis of Amino acids — By reductive amination and Transamination; Biosynthesis of amino acids derived from pyruvate, a-ketoglutarate, oxaloacetate and Aromatic amino-acids; Amphoteric nature of amino acids; Isoelectric point.
18. Phytochrome — Differences between PR and PFR forms, Physioco-chemical nature, Chromophore Structure.
19. Dormancy of Seeds and Seeds Germination (New)
Changes during seed germination; Light dependent germination, Changes in food reserves; Factors affecting germination.
21. *Growth and Metabolism of Growth Hormones* (New)
22. *Phototropism* — Diaphototropism, Photoreceptor pigment, Location of photoreceptor, Role of Auxins in Phototrapism, First and Second positive response, Curry Hypothesis, Demonstration of phototropism; Geotropism — Types, Effects of conditions, Steps in Geotropic response, Role of Auxins in Geotropism, Role of other hormones, Theories of Geotropism; Difference between Tropic and Nastic movements.

With above information I hope this edition will be more beneficial to the readers and will continue to serve the purpose effectively.

Criticisms and constructive suggestions from the readers for further improvement of the book are cordially invited.

                                                                                       S.K. VERMA

PREFACE TO THE FIRST EDITION

The present book entitled "A Textbook of Plant Physiology and Biochemistry" is basically written to meet the needs of B.Sc. II and III year students of various Indian universities. However, the students of M.Sc. will also benefited from this book.

The book is divided into two parts. Part I deals with eleven chapters or Biochemistry and Part II deals with twenty-one chapters on Physiology. Effort have been made to provide the students with a lucid, simple, authentic and recent account of each and every topic. Hope, the book will be beneficial for the students.

I wish to thank all my friends and colleagues and specially to Prof. M.S. Khan for going through the manuscript and giving valuable suggestions during the preparation of the book. I am also thankful to S. Chand & Co. Ltd. for taking keen interest in the publication of this book.

Constructive suggestions from readers will always be welcomed and entertained wholeheartedly.

Shankerpuri
Canal Road
Gorakhpur - 273014
Tele. 0551-2283470

S.K. VERMA

By the Same Author
A TEXTBOOK OF PLANT PHYSIOLOGY

CONTENTS
PART I : PLANT PHYSIOLOGY

1. **THE CELL**   1 — 53
   Introduction; Definition of Cell; Brief History of Cell; Types of Cells; Structure of Cell under Electron Microscope — Cell Wall; Plasma membrane; Protoplasm; Centrosome; Cilia and Flagella; Endoplasmic Reticulum; Golgi Complex; Lysosomes; Microbodies; Mitochondria; Plastids; Ribosomes; Differences between 70S and 80S ribosomes; Metabolically Inactive Cell Inclusions; Vacuoles; Nucleus; Cell Components and their main functions.

2. **OSMOTIC RELATIONS OF PLANT CELLS**   54 — 70
   Diffusion; Diffusion Pressure of Liquids; Factors affecting the Rate of Diffusion; Importance of Diffusion in Plants; Demonstration of diffusion; Permeability; Preparation of Artificial Semipermeable membrane; Factors affecting Permeability; Osmosis; Demonstration of Osmosis by Simple Ometer and U-Tube; Diffusion through Artificial and Natural Membranes; Types of Osmosis — Endosmosis and Exosmosis; Difference between Diffusion and Osmosis; Osmotic Pressure; Factors affecting Osmotic Pressure; Significance of Osmosis in Plants; Plant Cell as Osmatic System and Relationship among Turgor Pressure, Wall Pressure and Osmotic Pressure; Inter-relationship among DPD, S.P., O.P., T.P. and W.P.; Water Potential Concept; Differences between DPD and ψ; Plasmolysis and Deplasmolysis; Water Potential Changes in Plasmolysis and Deplasmolysis; Advantages of Plasmolysis; Difference between Osmosis and Plasmolysis; Imbibition; Significance of Imbibition, Demonstration of Imbibition; Other Experiments.

3. **ABSORPTION OF WATER**   71 — 80
   Importance of water to the plants; Soil water; Forms of Underground Water; Hygroscopic coefficient; Moisture Equivalent; Wilting and Wilting Coefficient; Water absorbing parts of the Plants; Root System in Plants; Characteristics of Root; Regions of the Root; Internal Structure of Root; Structure of Root Hair; Mechanism of Water Absorpiton; Active Absorption and Passive Absorption; Path of Water in Roots; Root Pressure; Theories Regarding Development of Root Pressure; Demonstration of Root Pressure; Non-osmotive Active Absorption; Passive Absorption; Difference between Active and Passive Absorption; Apoplast and Symplast Concept; Factors affecting the Rate of Water Absorption.

4. **ASCENT OF SAP**   81 — 84
   Introduction; Path of Ascent of Sap; Experiments; Theories of Ascent of Sap — Vital Theories; Root Pressure Theory; Godlewski Theory; Vital Force Theory; Physical Theories — Imbibition theory, Capillary Force Theory, Cohesion Theory of Dixon and Jolly; Evidences in Support of Cohesion Theory, Criticism; Demonstration of Transpiration Pull.

5. **TRANSPIRATION**   85 — 102
   Introduction; Differences between Transpiration and Evaporation; Kinds of Transpiration — Cuticular, Lenticular and Stomatal; Structure of Stomata; Distribution of Stomata on Leaf; Types of Stomata; Daily Periodicity of Stomatal Movement; Mechanism of Transpiration; Mechanisms of Opening and Closing of Stomata; Theories of Opening and Closing of

Stomata, Theory of Photosynthesis in Guard Cells; Theory of Starch — Sugar Inter-conversion; Theory of Starch - Glucose Inter-conversion, Theory of Glycolate Metabolism; Theory of Proton Transport; Opening and Closing of Stomata in Succulent Plants; Plant Anti-transpirations; Factors Affecting Transpiration; Special Features in Plants to Reduce Transpiration; Benefits of Transpiration to Plants or Importance of Transpiration; Demonstration of Suction Pressure due to Transpiration; Other Methods of Water-loss; Differences between Transpiration and Guttation; Experimental Demonstration of Guttation; Bledding; Demonstration of Transpiration — Experiments.

6. **MINERAL NUTRITION IN PLANTS** 103 — 117

   Introduction; Composition of Plant - ash; Essential and Non-essential Elements; Macro-nutrients and Micro-nutrients; Differences between Trace and Tracer Elements; Hydroponics; To Study the Importance of Mineral Elements in Plants— Solution Culture Method and Sand Culture Method; Nutrient Solutions; Precautions during Preparation and use of Nutrient Solution; Inorganic Salts in Soil; Inorganic Salts in Soil-water; General Roles of Mineral Elements in Plants; Differences between Chlorosis and Etiolation; Source; Occurrence and Functions of Essential Elements — Carbon, Hydrogen, Oxygen, Nitrogen, Sulphur, Phosphorus, Potassium, Calcium, Magnesium, Iron, Manganese, Zinc, Boron, Copper, Molybdeunum, Chlorine, Vanadium; Methods to Overcome Mineral Deficiency — Soil Application, Foliar Application and Injection Method.

7. **MINERAL SALT ABSORPTION** 118 — 123

   Introduction; Passive Absorption — Outer and Apparent Free Space Theory; Ion Exchange; Donnan Equilibrium; Mass Flow; Active Absorption; Mechanism of Active Salt Absorption- Carrier Concept, Cytochrome Pump Theory; Carrier Mechanism Involving ATP; Factors affecting Salt Absorption; Translocation of Mineral Salts.

8. **TRANSLOCATION AND STORAGE OF FOOD IN PLANTS** 124 — 128

   Introduction; Translocation of Food; Direction of Translocation; Path of Translocation; Mechanism of Translocation; Diffusion Hypothesis; Protoplasmic Streaming Theory; Munch's 'Mass Flow' Hypothesis; Storage of Food; Food Storage Organs; Forms of Stored Food.

9. **SPECIAL MODES OF NUTRITION IN PLANTS** 129 — 136

   Introduction; Types of Plants, Depending upon the Mode of Nutrition; Autotrophic; Heterotrophic; Special Type — Insectivorous (Utricularia, Drosera, Nepenthes and Dionaea).

10. **PHOTOSYNTHESIS** 137 — 183

    Brief History; Photosynthetic Apparatus; Photosynthetic Pigments — Chlorophylls, Phycobilins, Carotenoids, Distribution of Pigments; Biosynthesis of Chlorophyll; Definition of Photosynthesis; Photosynthesis as a Chemical Process; Evidences in support of Light and Dark Reaction; Nature of Light — Theories; Mechanism of Absorption of Light; Components of Electromagnetic Spectrum; Absorption Spectrum; Action Spectrum; Mechanism of Photosynthesis; Red Drop and Emerson's Enhancement Effect; Two Pigment Systems — PS I and PS II; Differences Between Pigment System I and Pigment System II; Photo-Oxidation of Water; Production of Assimilatory Powers; ET Chain Components; Non-cyclic Electron Transport System and Non-cyclic

Photophosphorylation; Cyclic Electron Transport System and Cyclic Photophosphorylation; Differences between Non-cyclic and Cyclic Photophosphorylation; Dark Reaction (Calvin Cycle); Hatch-Slack Cycle; Characteristics of C_4 Plants; Significance of C_4 Cycle; Differences Between Calvin Cycle and Hatch-Slack Cycle; Crassulacean Acid Metabolism (CAM Cycle); Significance of CAM; Bacterial Photosynthesis — Non-cyclic Photophosphorylation; Cyclic Photophosphorylation, CO_2 Fixation; Differences between Plant and Bacterial Photosynthesis; Chemosynthesis; Differences between Photosynthesis and Chemosynthesis; Factors Affecting Rate of Photosynthesis; External Factors — Light, CO_2, Temperature, Water, Internal Factors — Chlorophyll, Protoplasm, Significance of Photosynthesis; Experiments Relating Photosynthesis.

11. RESPIRATION 184 — 228

Introduction; History of Respiration; Differences between Respiration and Combustion; Changes Associated with Respiration; Types of Respiration — Aerobic Respiration, Anaerobic Respiration; Differences between Aerobic and Anaerobic Respiration; Classification of living beings based on Respiration; Respiratory Substrate; Mechanism of Respiration; *Glycolysis*; Summary of Glucolysis; *Aerobic Oxidation of Pyruvic Acid*; *Kreb's Cycle*; Summary of Aerobic Respiration of Glucose; ATP production during Aerobic Respiration of one molecule of Glucose; *Electron Transport System*; Terminal Oxidation of reduced Coenzymes; Oxidation of Extramitochondrial NADH — MOA, G-3-P DHAP Shuttles; *Oxidative Phosphorylation;* Its site, ATP Synthetase — Structure and Chemistry; Structure and Function of Electron Carriers Present in ET Chain; Mechanism of Oxidative Phosphorylation — Various Theories; Respiratory inhibitors; Cyanide Resistant Respiration; Adenosine Triphosphate; Causes Differences between Oxidative and Photophosphorylation; Fermentation; Differences between Fermentation and Anaerobic Respiration; Differences between Respiration and Fermentation; Types of Fermentation; Alcoholic Fermentation; Lactic Acid Fermentation; Acetic Acid Fermentation; Butyric Acid Fermentation; Relation between Anaerobic Respiration and Fermentation; Respiratory Quotient; Types and Determination; Photosynthetic Quotient; Differences between RQ and PQ; Factors Affecting the Rate of Respiration; Experiments related with Respiration; Differences between Respiration and Photosynthesis; Demonstration of Fermentation.

12. PHOTORESPIRATION 229 — 232

Introduction; Site of Photorespiration; Biochemistry of Photorespiration; Evidences in support of Photorespiration; Significance of Photorespiration; Difference between Photorespiration and Dark Respiration.

13. NITROGEN METABOLISM AND NITROGEN CYCLE 233 — 245

Significance of Nitrogen; Nitrogen in Soil; Nitrate reduction in Plants; Nitrogen Fixation — Non-biological and Biological — Non-symbiotic, Symbiotic, Associative; Symbiotic N_2 Fixation in Leguminous plants — Formation of Root Nodules; Biochemistry of N_2 Fixation — Requirements, Nitrogenase Enzyme; Steps in Asymbiotic N_2 Fixation and Symbiotic N_2 Fixation; Sources of N_2 in Insectivorous Plants; Denitrification; N_2 Cycle.

14. NUCLEIC ACIDS METABOLISM 246 — 270

Introduction; Nitrogenous bases — Pyrimidines and Purines; Phosphoric Acid; Nucleosides; Nucleotides — Structure, Biosynthesis of Purines —

IMP, GMP, AMP uric acid; Pyrimidines — CMP, UMP and TMP; DNA — Structure; Watson — Crick Model, Duplication — Methods and Mechanism; Enzymes for DNA Synthesis; DNA unwinding Proteins; RNA, Differences between RNA and DNA, RNA Types — rRNA, mRNA, tRNA, Microchemical for DNA and RNA.

15. **METABOLISM OF AMINO-ACIDS AND PROTEINS**   271 — 300

Biosynthesis of Amino Acids — By Reductive Amination, By Transamination; Biosynthesis of Amino Acids Derived from Pyruvate, a-Ketoglutarate, Oxaloacetate and Aromatic Amino Acids; Amphoteric Nature of Amino Acids; Isoelectric Point; Biosynthesis of Proteins; Mechanism of Protein Synthesis; Transcription; Maturation of mRNA from HnRNA in Eukaryotes; Translation — Various Steps; Initiation steps in *E. coli* and Eukaryotes; Regulation of Protein Synthesis; Enzyme Induction and Repression; Operon Model; Mechanism of Operon Model — In Inducible System, In Repressible System; Mutations in Controlling Genes and Production of Constitutive Strains; The differences between Regulator Mutant and Operator Mutant; Promoter Gene and Promoter Region; Genetic Code : Properties of Genetic Code : New Genetic Codes In Mitochondria And Ciliate Protozoans, Rule Regarding Codon-anticodon pairing, methods of cracking genetic code; oxidation of proteins.

16. **LIPID METABOLISM**   301 — 314

Introduction; Biosynthesis of Fats — Synthesis of Fatty Acids, Synthesis of Glycerol, Condensation of Fatty Acids and Glycerol; Biosynthesis of Lecithim; Biosynthesis of Cholesterol; Biosynthesis of Sphingosine and Sphingomyelins; Fat Oxiation — Hydrolysis of Fat (Triglyceride) by Lipase; Metabolism of Glucerol; Oxidation of Fatty Acids; β-oxidation; α-oxidation; ω-oxidation; Conversion of Fats into Carbohydrates (Glyoxylate Cycle).

17. **CARBOHYDRATE METABOLISM**   315 — 327

Introduction; Biosynthesis of Carbohydrates; Oxidation (Catabolism) of Carbohydrates; Direct Oxidation Pathway; Inter-conversions of Monosaccharides; Gluconeogenesis.

18. **PHOTOPERIODISM AND PHOTOMORPHOGENESIS**   328 — 335

Introduction; Critical Day Length; Classification of Plants according to Photoperiodic Reaction; Photoperiodic Induction; CO_2 Supply and Photoperiodic Induction; Perception and Transmission of Stimulus; Importance of Dark Period; Importance of Photoperiod; Flowering Hormone—Florigen; Action Spectrum of Light; Phytochrome—Differences between P_R and P_{FR} Forms; Occurrence, Physico-Chemical Nature, Chromophore, Properties, Mode of Action, Isolation.

19. **DORMANCY OF SEEDS AND SEED GERMINATION**   336 — 343

Introduction; Seed Dormancy; Causes of the Seed Dormancy; Dormancy due to specific light requirement; Dormancy due to germination inhibitors; Methods of breaking Seed Dormancy; Advantages of Seed Dormancy. Changes during Seeds Germination; Light Dependent Germination, Changes in Food Reserves; Factors Affecting Germination.

20. **VERNALIZATION**   344 — 347

Introduction; Perception of the Cold Stimulus; Morphological changes associated with Seed Vernalization; Induced State; Presence of a Floral Harmone; Other conditions necessary for Vernalization; Mechanism of

Vernalization; Devernalization; Vernalization and Gibberellins; Importance of Vernalization.

21. **GROWTH AND METABOLISM OF GROWTH HORMONES**   348 — 370

    Introduction; Course of Growth; Measurement of Growth; Factors affecting the Growth of Plants; Growth Hormones — Auxins, Experiments, Isolation, Extraction, Bioassays, Biosynthesis of Indole Auxins — Indole Pyruvic Acid Pathway, Tryptamine Pathway, Indoleacetaldoxime Pathway and Tryptophol — Pathway. Biosynthesis of Non-indole Auxins, Mechanism of Auxin Action and Affected Processes, Functions of Auxins; Gibberellins — Discovery, Structure, Extraction, Bioassays Methods, Biosynthesis, Functions; Differences between Auxins and Gibberellins; Cytokinins — Structure, Bioassay and Effects; Florigen; Growth inhibitors — Ethylene, Abscisic Acid (ABA) and Morphactins.

22. **PLANT MOVEMENTS**   371 — 386

    Introduction; Law of Summation of Stimuli; Classification of Movements: Movements of Locomotion — Autonomous Movements — Ciliary Movement, Amoeboid Movement, Excretory Movement, Cyclosis; Induced Movements — Phototaxis, Thermotaxis, Chemotaxis, Rheotaxis, Galvanotaxis; Movement of Curvature — Vital Movements: Autonomous, Movements of Variation, Movements of Growth, Nutation, Hyponasty, Ephemeral Movements; Induced Movements — Tropic Movements. Phototropism — Diaphototropism, Photoreceptor Pigment, Location of Photoreceptor, Role of Auxin in Phototropism, First and Second Positive Response, Curry-hypothesis for Phototrapism, Demonstration of Phototropism; Geotropism — Types, Effects of Conditions, Steps in Geotropic response, Role of Auxinisin Geotropism, Role of Other Hormones, Theories of Geotropism; Hydrotropism, Thigmotropism, Chemotropism, Thermotropism, Aerotropism; Traumatotropism, Rheotropism, Galvanotropism, Osmotropism; Nastic Movements, Seismonastic Movement, Nyctinastic, Photonastic, Thermonastic, Haptonastic, Hydronasty, Chemonasty; Differences between Tropic and Nastic Movements.

23. **STRESS PHYSIOLOGY**   387 — 398

    Physiological Responses in Plants Growing under Stress Conditions (Stress Physiology)—Introduction; Stress and Stress physiology; Types of Stress; Stress resistaaance; Effects of Stress; Drought Stress; Salt Stress; Temperature Stress; Pollution stress

PART II : BIOCHEMISTRY

1. **METHODS OF BIOCHEMICAL ANALYSIS**   1 – 23

    Chromatography; Introduction; Kinds of Chromatography; Outline Steps in Different Kinds of Chromatography; Chromatographic Methods of Popular Use; Adsorption Column Chromatography; Ion Exchange Chromatography; Partition Chromatography; Paper Chromatography; Thin Layer Chromatography; Laws of Absorption; Colorimetry; Spectrophotometry; Electrophoresis; Centrifugation and Ultra-centrifugation; X-Ray Diffraction; Tracer Technique; Auto-radiography; Experiments related to Chromatography.

2. **CARBOHYDRATES**   24 – 48

    Introduction; Classification of Carbohydrates; Chemistry of Monosaccharides; Isomerism; Ring Structure; Classification of

Monosaccharides; Some Important Reactions of Monosaccharides; Other Sugar Derivatives — Sugar Phosphates; Amino-Sugars; Deoxy sugars; Ascorbic Acid; Reducing and Non-Reducing Sugars; Chemistry of Oligo saccharides; Sucrose; Maltose; Lactose; Cellobiose; Raffinose; Gentianose; Melezitose; Classification and Chemistry of Polysaccharides; Starch; Inulin; Glycogen; Chitin; Cellulose; Agar; Gum ArAbic; Pectins; Mucopoly-Saccharides; Distinction Between Mono-; Oligo- and Poly-saccharides; Significance of Carbohydrates; Some important tests for Carbohydrates; Experiments.

3. **AMINO ACIDS AND PROTEINS** 49 – 64

What are Amino-acids; Classification of Amino-acids; Essential and Non-Essential Amino-acids; Separation of Amino-acids; Proteins; Characteristic Features; Protein Content of the Plant; Classification of Proteins; Chemical Tests of Amino-acids and Proteins; Protein Structure; Peptide Bond; Desulphide Bond; Hydrogen Bonds; Hydrophobic Bonds; Structure of Proteins — Primary; Secondary; Fine Structure, Tertiary and Quarternary Structures; Denaturation and Renaturation of Proteins.

4. **THE LIPIDS** 65 – 77

Introduction; Classification of Lipids — Simple Lipids; Compound Lipids, Classification of Fatty Acids; Properties of Fatty Acids and Fats; Waxes; Phosph- Lipids; Glycoproteins; Lipo-Proteins; Derived lipids (steroids); Importance of Lipids, Tests for Lipids.

5. **THE ENZYMES** 78 – 100

Introduction Occurrence; Nomenclature and Classification; Major Clases of Enzymes; Isoenzymes; Isolation and Purification of Enzymes; Chemical Nature of Enzymes; Mode of Action of Enzymes; Derivation of Michaelis Constant; Models for Explaining Enzyme Action; Energy of Activation in the Mechanism of Enzyme Action; Properties of Enzymes; Enzyme Inhibition; Factors Affecting Enzyme Activity, Key to Numbering and Classification of Enzymes; Differences between Enzymes and Hormones; Tests for Some Enzymes.

6. **COENZYMES** 101 – 118

Introduction; Structure and Classification; Action of Coenzymes; Some Important Coenzymes — NAD and NADP, Riboflabin Coenzymes, Coenzyme - A; Lipoic Acid; Thiamine Pyrophosphate, Cytochromes, Biotin, Pyridoxal Phosphate; Ascorbic Acid; Tetrahydrofolic Acid; Cytidine Di-Phosphate; Uridine-Di-Phosphate, Cyanocobalamine, Coenzyme - Q.

7. **VITAMINS** 119 – 133

Introduction ; General Characteristics of Vitamins; Vitamins and other related Compounds; Vitamins and Hormones; Differences between Vitamins and Hormones; Nomenclature and Classification; Fat Soluble Vitamins — Vitamin-A, Vitamin-D, Vitamin-E, Vitamin-K; Water Soluble Vitamins — Vitami- C, Vitamin-B_1, Vitamin-B_2, Niacin, Vitamin-B_6; Vitamin-B_3, Vitamin-H, Vitamin-B_{12}; Lipoic Acid; Folic Acid; Summary of Vitamins.

8. **PLANT GROWTH SUBSTANCES** 134 – 154

Introduction; Auxins; Gibberellins; Differences between Auxins and Gibberellines; Cytokinis; Abscisic Acid; Ethylene; Morphactins; Other Hormones.

9. NUCLEIC ACIDS $\quad$ 155 – 200

Introduction; Nitrogenous Bases – Pyrimidine Bases, Purine Bases; Chemistry of Structure of Bases; Free Purines and Pyrimidines; Pentose Sugars; Phosphoric Acid; Nucleosides; Inter-conversions of the Nucleosides; The Nucleotides; Free Nucleotides; The Nucleoside di-and tri-Phosphates; Coenzyme Nucleotides; Properties of Nucleotides; Biosynthesis of Nucleotides – Purine nucleotides, Biosynthesis of TMP, GMP, AMP and Uric acid, Biosynthesis of Pyrimidine riporucleotides – CMP and TMP; DNA – Structure of DNA, Classes of DNA - Molecular Structure of DNA, Watson and Crick Model of DNA, Forms of double stranded DNA; DNA Duplication – Watson and Crick Model and Meselson and Stahl's Theory, Mechanism of ONA Replication; Mitochondrial- DNA, Chloroplast- DNA, Single Stranded DNA and Circular DNA, Distinguishing Features between Native DNA and Single Stranded (SS) DNA, Circular DNA, Single Stranded Circular DNA, Double Stranded circular DNA Molecular Weight of DNA, Denaturation and Renaturation of DNA, Hydrolysis of DNA, RNA - Differences between RNA and DNA; Types of RNA, Hydrolysis of RNA, Test for the presence of DNA and RNA in Plants.

10. INTRODUCTION TO BIO-ENERGETICS $\quad$ 201 – 208

Energy; Free Energy; Energetic Coupling; Energy Rich Compounds; Cases of Energy Richness of ATP; Other Energy Rich Compounds; Laws of Thermodynamics; First Law of Thermodynamics and its Applications; Entropy; Physical Significance of Entropy; Concept of Entropy in Living Systems; Chemical Equilibrium; Thermodynzmic Equilibrium; Dynamic Equilibrium and Steady State.

11. BIOLOGICAL OXIDATION AND REDUCTION $\quad$ 209 – 218

Introduction; Oxidation and Reduction; Redox Reactions in Biological Systems; Oxidation - Reduction Potential and Its Measurements; Biologically Important Redox Systems.

12. SECONDARY METABOLITES $\quad$ 219 – 238

Terperoids — Structure, Types and Sources, Biosynthesis of various terpenoids, Functions of Terperoids; Alkaloids — Types, Distribution and Localization, Biosynthesis of various alkaloids, Biological functions of Alkaloids; Flavoroids.

Part III – BIOTECHNOLOGY

1. GENETIC ENGINEERING $\quad$ 1 – 64

Introduction; steps involved in gene transfer; RECOMBINANT DNA TECHNOLOGY-Tools involved in R-DNA technology: ENZYMES-Exonucleases, Endo-nucleases, Restriction enzymes, modification by methylation, Nomenclature of RE, OTHER ENZMES-DNA ligases, Alkaline phosphatase, S_1 nuclease, DNA polymerase, Reverse transcriptase; FOREIGN DNA; VECTORS: Natural-Plasmids, experimental procedure for formation of hybrid plasmid, pBR322; PHAGES as vectors, Insertion and Replacement phage vectors; VIRUSES as vectors-Simian virus 40(SV40); Reconstituted Vectors–Cosmids, Phasmids. Techniques involved in R-DNA technology – Palindromes and staggered cut method, Addition of poly dA at 3'end of the vector and poly dT at 3'end of DNA clone, Blunt end ligation by T_4 DNA ligase. Cloning vectors. cDNA libraries, GENOMIC LIBRARIES; Methods to pick up correct desired clone from a library– colony hybridization, DNA Probes, Single plaque hybridization, by antibodies directed against the gene encoded protein;

Gene cloning technique–in bacteria and eukaryotes. SOME IMPORTANT TECHNOLOGIES IN GENETIC ENGINEERING: Hybridoma technology and production of monoclonal antibodies; Blotting techniques–Southern, Northern and Western, DNA finger Printing; PCR technique; Protoplast fusion technique; Techniques of gene transfer–Microprojectile bombardment, Microinjection, Liposome mediated gene transfer, Electroporation, Ultrasonication, Coprecipitation of DNA with calcium phosphate, using DNA complexes, Laser microbeam technique. Gene transfer in cultured cells, plants, animals and production of transgenic individuals. TRANSPOSABLE ELEMENTS–IS elements; Transposons–in prokaryotes and eukaryotes, Classes on the basis of mecnanism of transposition, Transposable elements in *Drosophila*, Yeast and corn; Retroelements, Mechanism of transposition. TECHNIQUES OF DNA SEQUENCING: Maxam and Gilbert procedure, Sanger' procedure; Technique of genetic mapping–Chromosome Walking.

BIOTECHNOLOGY: 65 – 100

Introduction and functional definition; Uses of biotechnology; Some foreign biotechnology companies; CETUS; Biotechnology Boards, Institutes and Centres in India; NBT; Techniques of Biotechnology. BASIC ASPECTS OF PLANT TISSUE CULTURE: Brief history; Requirements for *in vitro* cultures–Tissue culture laboratory, maintenance of aseptic environment, Nutrient media; Callus and suspension cultures: Methods of plant tissue culture (Basic steps); Some important plant material cultures–Explant culture, Callus cultures, Root culture, Shoot or meristem culture, Cell and suspension cultures, Anther and pollen culture, Embryo culture, Ovary culture; Cryopreservation. CELLULAR TOTIPOTENCY, DIFFERENTIATION AND MORPHOGENESIS–Micropropagation. BIOLOGY OF AGROBACTERIUM: *Agrobacterium Ti* and *Ri* plasmids or vectors for gene delivery, *Ti* plasmids, *Ri* plasmids, their properties, Gene transfer in explants through *Agrobacterium*, Mechanism of T-DNA transfer. MARKER GENES–TK, DHFR, XGPRT, NPT, CDA protein. SALIENT ACHIEVEMENTS IN CROP BIOTECHNOLOGY–Genetic engineering in cloning of *nif* genes; Transfer of *nif* genes and production of biofertilizers; Gene transfer in dicots, Gene transfer in monocots, Transgenic plants–herbicide resistant, Insect resistant, Viral resistant, Bacterial and Fungal resistant. Disadvantages or Potential hazards of Genetic engineering and Biotechnology.

Index 101 – 112

PART I
PLANT PHYSIOLOGY

1
The Cell

(Latin, Cellula = a small compartment)

1.1. INTRODUCTION

"Cell is the structural and functional unit of living organisms." Actually it is true also because plants and animals both are made up of these units. On the basis of number of these units, living organisms have been divided into two groups: (*i*) Unicellular (*ii*) Multicelluar.

Unicellular organisms are those which start their life from a single cell or unit and till death continue it in the same manner, i.e., their body is made up of a single cell. The examples of unicellular organisms are yeast, diatoms, bacteria and protozoans. Multicellular organisms are those which are made up of more than one cells. These organisms also start their lives from a single cell but later on it undergoes continued divisions and develop into a multicellular body. All higher plants and animals are multicellular. The cells in multicellular organisms get arranged in a definite manner to form different parts of the body. These parts perform different functions and all are governed by the common activity of each cell present in that part of the body.

Electron microscopic studies have revealed that each cell contains in it several living sub-cellular particles called *cell-organelles* which perform different functions. The chief cell-organelles are mitochondria, endoplasmic reticulum, ribosomes, plastids, lysosomes, golgi body and nucleus etc. Thus, it can be said that despite being the unit of living organisms, cell has a much complex structure.

1.2. DEFINITION OF CELL

Although it is very difficult to define a cell, many scientists have defined it in their own way. Some of the definitions are as follows :

(*i*) "Cell is the structural and functional unit of living organisms."
(*ii*) "Cell is the smallest living unit capable of independent existence."
(*iii*) "Cell is the smallest mass of living matter (protoplasm) containing nucleus or nuclear material."
(*iv*) "Cell is a piece of nucleated cytoplasm surrounded by cell-wall or plasmamembrane."

None of the above definitions is true because exceptions of each definition are available. *Loewy and Siekevitz* (1963) gave the definition of cell which has been accepted by majority of the scientists. The definition is as follows :

"Cell is the unit of biological activity delimited by a semipermeable membrane and capable of self reproduction in a medium free from other living systems."

1.3. BRIEF HISTORY OF CELL

The history of cell starts with the invention of microscope. The simplest microscope was prepared for the first time by *Jenssen* (1590) and *Galileo* (1610). Later on, *Robert Hooke* (1665) made the first compound microscope and examined thin slices of cork. He observed that the cork was composed of small box-like compartments similar to honey-comb. He called these compartments, cells (cell : GK. kytos; L; cellula = hollow space).

In seventeenth century *Grew* and *Malpighi* repeated the observations of *Robert Hooke* in different plants and recognised in them minute cavities in the midst of homogenous mass which they called *bladders* or *utricles*. Later on, *Antony van Leeuwenhock*, a Dutch draper, in 1723 made a special type of microscope and observed unicellular organisms in a drop of pond water. He also studied several micro-organisms, blood-cells, spermatozoans and protozoans. It was Leeuwenhock who for the first time observed nuclei of living cells and RBC of Salmon blood although he could not explained the significance of what he saw.

Fig. 1.1: Cork-cells as seen by Robert Hooke.

Till the beginning of nineteenth century, the knowledge of cell remained stationary but in the last quarter of nineteenth century most of the important discoveries were made. For this reason, this period is known as 'classical period' of cytology.

After 1933 when *Knoll* and *Ruska* invented the electron microscope, the actual detailed structure of cell came into view.

1.4. TYPES OF CELLS

The cells are generally classified into two main groups :

1. Prokaryotic cells
2. Eukaryotic cells.

1. Prokaryotic cells (GK. Pro = primitive, karyon = nucleus). They are characterized by the absence of nuclear membrane, nucleus, nucleolus and most of the well developed cytoplasmic organelles. They possess 70S ribosomes. They have been considered the most primitive type of cells and in them nuclear material is found freely distributed in the cytoplasm, i.e., the nuclear materials like DNA, RNA and proteins are in direct contact with protoplasm. The chromosomes of prokaryotic cells lack histone proteins. Bacteria, viruses, rickettsia, bacteriophages, PPLO and blue green algae are the examples of prokaryotic cells.

2. Eukaryotic cell (GK. eu = good or well; karyon = nucleus) Eukaryotic cells possess all the characters which prokaryotic cells lack. Thus, they are characterized by the presence of definitely organised nucleus with a nuclear membrane and nucleolus and presence of well organized cytoplasmic-organelles like mitochondria, plastids, ribosomes, endoplasmic reticulum, lysosomes, golgi body etc. They possess 80S ribosomes. Nuclear materials like DNA, RNA and proteins in these cells are in contact with karyoplasm. Their chromosomes contain histone proteins. Examples of eukaryotic cell are algae (except blue green algae) and all higher plant and animal cells.

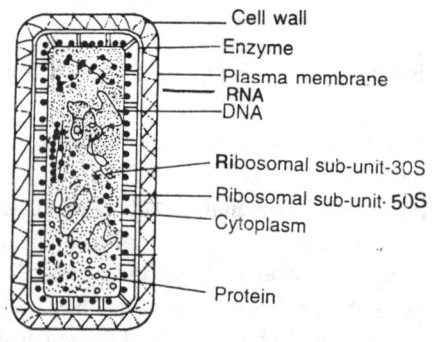

Fig. 1.2: A prokaryotic (bacteria) cell.

1.5. STRUCTURE OF CELL UNDER ELECTRON MICROSCOPE

The structure of cell reveals that a typical cell contains many sub-cellular structures (Fig. 1.4). Each sub-cellular particle has a characteristic structure and functions of its own. The study of cell under electron microscope can be divided into many sub-heads

The Cell

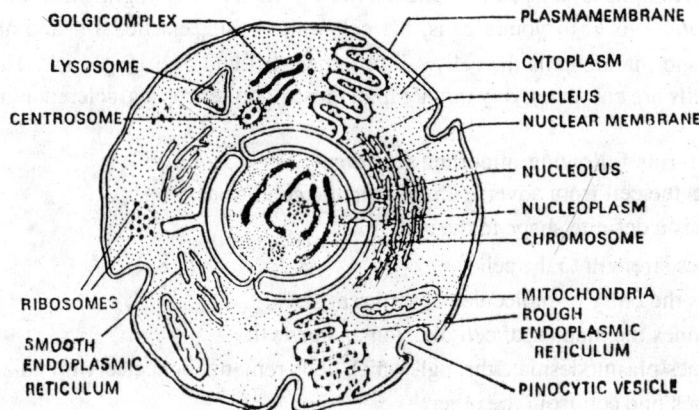

Fig. 1.3: An eukaryotic cell (animal cell).

Cell-Wall

Cell-wall is the characteristic feature of all the plant-cells. It is the outermost layer and covering of the plasmamembrane. The cell-wall is entirely lacking in animal cells. The cell-wall is rather rigid, strong, thick, porous and non-living structure and is secreted by the living matter of the cell.

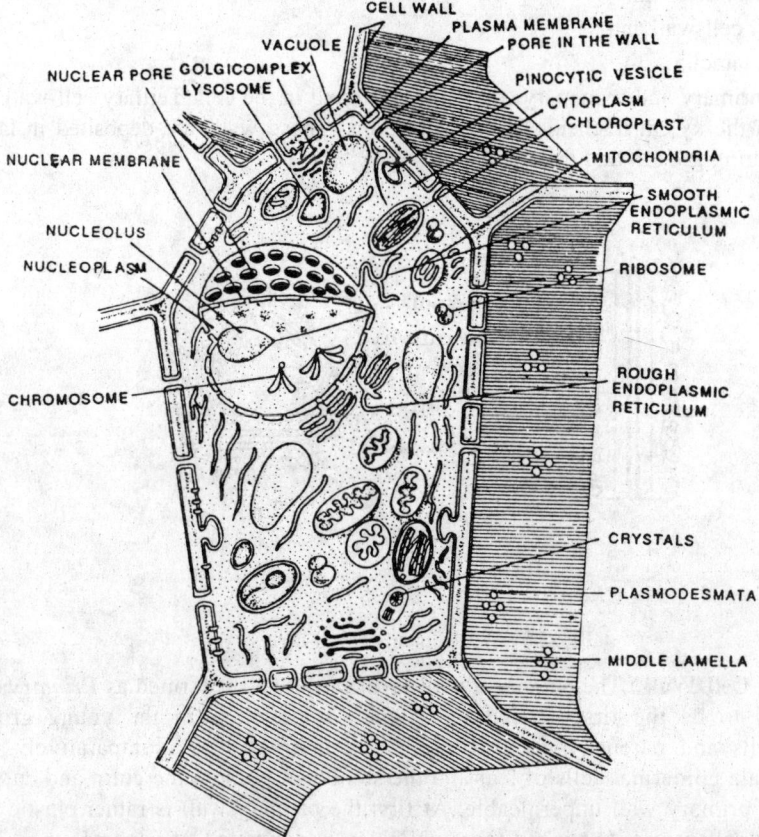

Fig.1.4: Structure of cell under electron microscope (three-dimensional).

It is laid down at the telophase stage and is believed to be formed by the fragments of the endoplasmic reticulum (*Northcote*, 1968). In young cells, the cell-wall is thin, elastic, soft and about 1 to 3 μ thick while in old and mature cells the cell-wall becomes stiff and 5 to 10 μ thick. The cell-wall of parenchymatous cells are comparatively thinner than those of collenchyma, sclerenchyma and xylem vessels.

Cell-wall performs following important functions :
(*i*) It protects the cell from adverse environmental conditions.
(*ii*) It provides a definite shape to the cell.
(*iii*) It provides strength to the cell.
(*iv*) It permits the entry of molecules of different sizes.
(*v*) It determines the manner of cell division and growth.
(*vi*) It possesses plasmodesmata through which cells remain connected with adjacent cells.
(*vii*) It separates one cell from the other.

Structure of Cell-Wall

The cell-walls are complex and highly differentiated in some tissues. They also possess a special sequence of their arrangement. The wall of a solitary cell can be differentiated into following layers :
(*i*) Primary cell-wall
(*ii*) Secondary cell-wall
(*iii*) Tertiary cell-wall and
(*iv*) Middle lamella

Generally, primary and secondary cell-walls are found in the cell. Tertiary cell-walls are rare and are found in the xylem tracheids of Gymnosperms. These walls are deposited in layers one after the other during growth and differentiation.

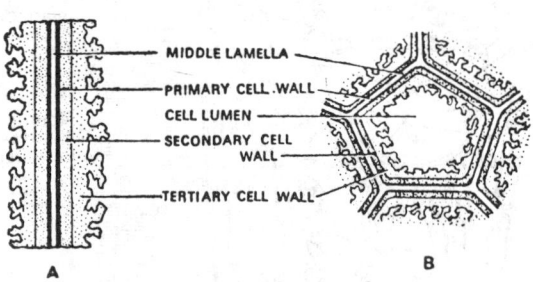

Fig. 1.5: Arrangement of wall-layers.

1. Primary Cell-Wall : The outermost wall layer of the cell is termed as *Primary cell-wall* and is regarded to be the first deposition product. It develops in the young growing or meristematic cells and parenchymatous cells. The primary wall is comparatively thin and permeable. Certain epidermal cells of leaf and the stem also possess the cutin and cutin-waxes which make the primary wall impermeable. At first the primary wall is rather elastic and able to extend as the cell grows, but when more cellulose is deposited, it becomes more rigid. In many fungi the cell-wall is composed of chitin.

The Cell

2. Secondary Cell-Wall : Next to the primary cell-wall is another layer known as *secondary cell-wall*. It is comparatively thicker than the primary wall. The secondary cell-wall is generally found in the mature, permanent or non-growing cells. The outer surfaces of epidermal cells with secondary wall of some leaves and the stem may have a cuticle rich in fats and waxes which tends to limit water loss. The primary cell-wall is composed of cellulose but the secondary wall in addition to cellulose contains pectins, non-cellulose polysaccharides, lignin and a phenolic polymer, which imparts hardness and mechanical rigidity to the wood. In certain cells it is further differentiated into outer, middle and inner layers.

3. Tertiary Cell-Wall : Rarely in certain cells a third layer is added inside the secondary wall which is known as *tertiary cell-wall*. The presence of tertiary cell-wall was described by *Bucher* (1953) in the xylem tracheids of Gymnosperms where it is quite thin and produces many warty outgrowths. The tertiary wall differs from the primary and secondary walls in its morphology, chemical composition and staining properties. Tertiary cell-wall is mainly composed of xylan instead of cellulose.

4. Middle Lamella : The primary walls of the two adjacent cells are often separated by a layer or a structure known as *middle-lamella*. During the development of cell-walls, the middle lamella is formed first. It is composed of calcium and magnesium pectates. The pectates are viscous and gelatinous. Middle lamella binds the adjoining cells firmly. In mature and aged cells the middle lamella is dissolved and consequently the cells are loosened. During the maturation of fruits the pectic substances of the middle lamella become soluble due to the action of pectolytic enzymes. The ripe fruits are, therefore, soft.

Table 1.1: Showing differences among primary, secondary and tertiary cell-walls

	Primary cell-wall	*Secondary cell-wall*	*Tertiary cell-wall*
1.	It is the outermost layer of the cell-wall.	It is found situated below primary cell-wall.	It is situated below secondary cell-wall.
2.	It is comparatively thin, elastic, delicate and permeable.	It is comparatively thick hard and impermeable.	It is thicker than secondary cell-wall.
3.	It is found in meristematic, young and parenchymatous cells.	It is found in mature and permanent cells.	It is found in the tracheids of mature and gymospermous plants.
4.	It is made up of only cellulose.	It is made up of cellulose, pectin, lignin and other substances.	It is made up of xylan substance.
5.	It is simple and smooth.	It is also simple.	It forms finger like processes by envagination of the wall.

Chemical Composition of Cell-Walls

The primary cell-wall consists of intertwined cellulose fibres. It may have a deposition of pectin, lignin, hemicellulose etc. The cellulose molecules are polymers of disaccharide cellobiose having approximately 3000 glucose units. The glucose molecules are arranged in the form of chain which are joined by β-1, 4-linkages. The cellulose fibrils are about 0.25 μ wide and up to 1 μ long. They are woven into an irregular net with a mesh of about 0.3 μ.

Many chains of cellulose molecules lie parallel to each other to form the bundles. A bundle of 100 cellulose molecular chains forms the elementary fibril known as *micelle*. The 20 parallely arranged micelles form another bundle known as *microfibril*. It is about 250Å thick. Similarly, 250 microfibrils form the large-sized bundle known as *macrofibril* (Fig. 1.7). The macrofibrils ultimately form the main framework of the cell-wall.

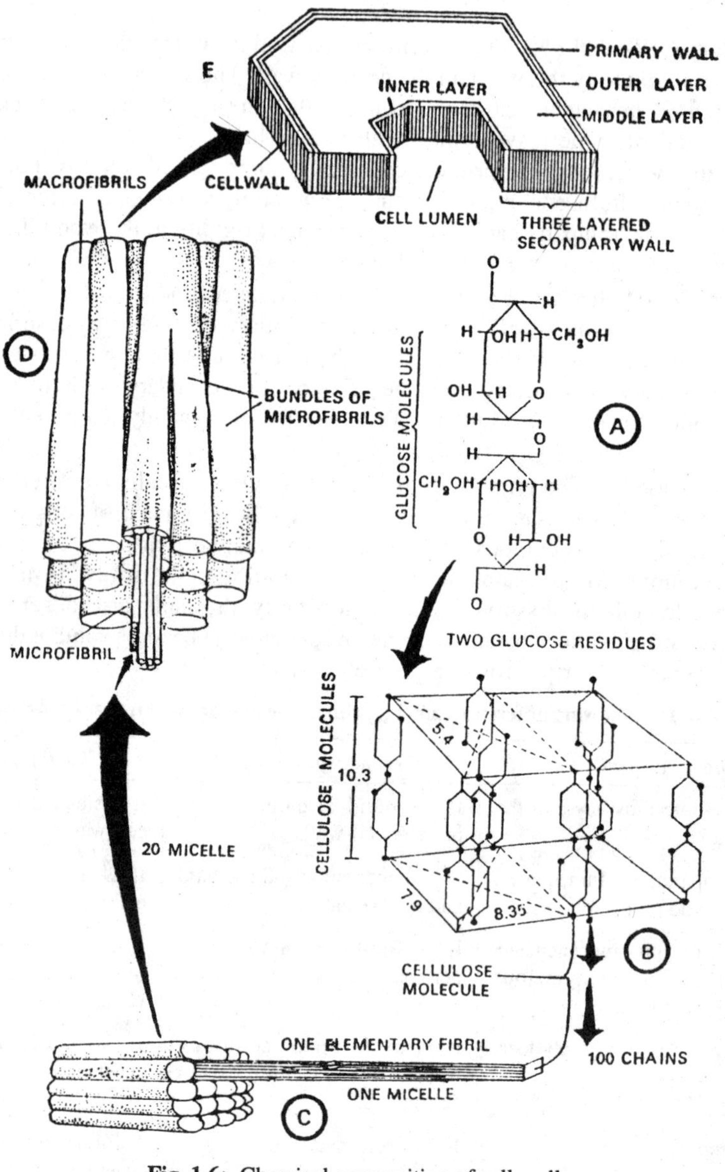

Fig. 1.6: Chemical composition of cell-wall.

The hemicelluloses are the polysaccharides of pentose as well as hexose sugars like arabinose,

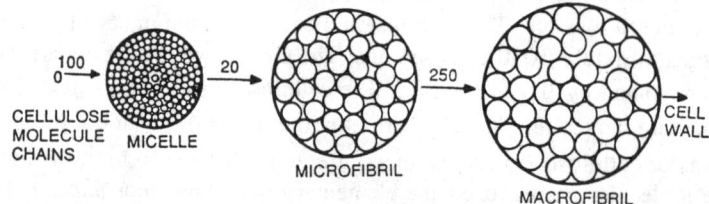

Fig. 1.7: Structure of Macrofibril.

xylose, mannose and galactose. The pectic substances are the long chains of repeating units of uronic acid derivatives of hexoses, glucuronic and galacturonic acids. Ligin is composed of coniferyl

The Cell

alcohols and cutin of many fatty acids. Chitin is a polymer of glucosamine.

The secondary wall is also composed of cellulose and contains lignin. The formation and structure of microfibrils and macrofibrils are the same as in case of primary cell-wall. The only difference is in their arrangement. In the secondary walls the cellulose microfibrils in a macrofibril are comparatively more compactly arranged.

The tertiary wall is composed of *xylan* instead of cellulose and the middle lamella is composed of Ca and Mg pectates.

The relative concentration of cellulose, hemicelluloses, and pectic substances in the primary cell-wall is not constant in all the cells. It varies from one cell-type to another cell-type. *Bishop* (1958) reported the presence of high concentration of hemicelluloses. *Jensen* (1960) found that the provascular cells had high concentration of pectic substances and hemicelluloses in their cell-walls. The cells of the cortex and protoderm had lower concentrations. *Setterfield* and *Bayley* (1961) reported the analysis of primary cell-walls as shown in Table 1.2.

Table 1.2: Showing average analysis of primary cell-wall.

Sl. No.	Substance	Per cent of fresh weight
1.	Water	60
2.	Hemicelluloses	5–15
3.	Cellulose	10–15
4.	Pectic substances	2–8
5.	Proteins	1–2
6.	Lipids	0.5—30

Growth of Cell-Wall

In primary stages, the cell-wall is thin (about 1-3 µ) elastic and delicate. At the time of cell-growth, it is stretched and new cell-wall forming materials are deposited upon it. As soon as the secondary growth is completed and the cell becomes mature, its thinckness is increased to about 5-10 µ. The growth of cell-wall is of following two types :

1. *Growth by Intussusception* : This type of growth results in the increase in volume of the cell-wall. During cell-growth, at first the wall is stretched and a tension is created. Now, the new cell-wall material secreted from the protoplasm is filled in the intercellular spaces of the stretched primary wall. This type of growth takes place during the growth period of the cell.

2. *Growth by Opposition* : During this type of growth, the cell-wall forming material secreted from the protoplasm is deposited towards the inner side of the primary cell-wall in the form of layers. Thus, the cell-wall becomes comparatively more thicker. Such layers are formed uniformly at all places of the cell-wall except at the pits.

Plasmodesmata

Different types of pores or pits are found on the cell-walls. When the cell-wall forming material is deposited on primary cell-wall, certain places are left as such. These are known as *primary pit fields*. Addition materials like cutin, suberin, calcium carbonate and calcium oxalate are also not deposited on these pits. The pores on the wall of living cells are usually very small and rounded. Through these pores the cytoplasm of adjacent cells remains connected with each other. Such cytoplasmic connections are known as *Plasmodesmata*. Through these connections, various substances are transported.

The plasmodesmata are found in multicellular plants with thick cell-walls and also in animal cells such as bone cells. Within the plasmodesmata, tubules have been observed in continuity with vacuoles or cisternae of endoplasmic reticulum.

Formation of plasmodesmata is related to the formation of cell plate or *Phragmoplast* (*Porter* and *Machado*, 1960). *Frey-Wyssling* and *Muhlethaler* (1964) and *Hepler* and *Newcomb* (1967)

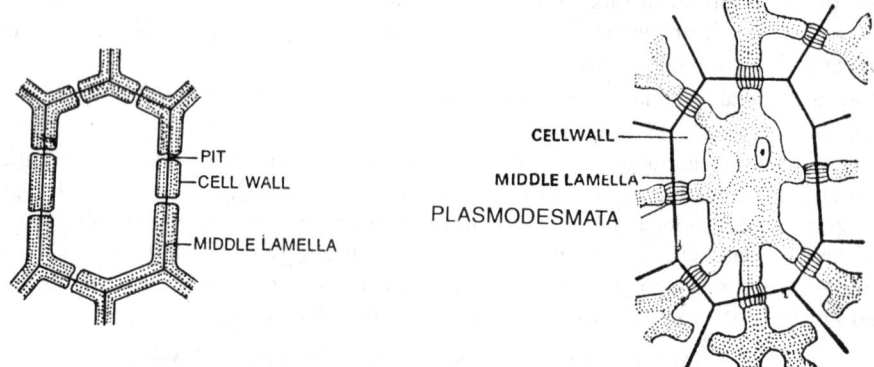

Fig. 1.8: Primary pit on cell-wall.

Fig. 1.9: Plasmodesmata in the cells of palm endosperm.

reported that at telophase of cell division the cell plate is crossed by vesicles and tubules of the endoplasmic reticulum that determine the location of the plasmodesmata.

Thickening of Cell-Wall

The cell-wall of many cells like xylem etc. become very thick and hard during secondary growth. The reason behind this is that the lignin and other materials are deposited upon it. As soon as the lignin starts depositing on the primary wall, the protoplasm of the cell starts decreasing. After deposition of lignin in great amount, the entire protoplasm of the cell is lost and the cell becomes dead. The deposition of lignin takes place in a definite sequence. It is first deposited on middle lamella, then on primary wall and then finally on secondary wall. Similary, other materials are also deposited on the wall. Due to this deposition of lignin and other materials the cell-wall becomes very thick and the thickening may take following shapes :

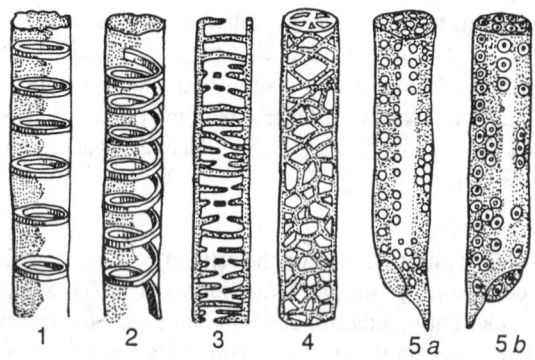

Fig. 1.10: Different types of thickenings on cell-wall.

(*i*) *Annular thickening* : When the lignin is deposited in the form of rings e.g., thickening in the cells of protoxylem.

(*ii*) *Spiral thickening* : When the lignin and other materials are deposited in the form of screw like strip.

The Cell

(iii) *Scalariform or ladder like thickening* : When the deposition of materials takes place in the form of transverse strips.

(iv) *Reticulate thickening* : When deposition takes place in the form of a net.

(v) *Pitted thickening* : When the deposition of materials takes place on entire cell-wall except at certain small places.

The pits are formed in pairs on opposite walls. The part of cell-wall which is found in between the opposite pits is known as *closing* membrane. In bordered pits, the closing membrane becomes slightly swollen and is called *torus*.

Types of Pits : They are of following two types:

(i) *Simple pits* : When the pitted portion contains only middle lamella and primary wall, the pit is called *simple pit* and these structures (middle lamella and primary wall) are called *pit membranes*. Simple pits are found in thick walled parenchymatous cells.

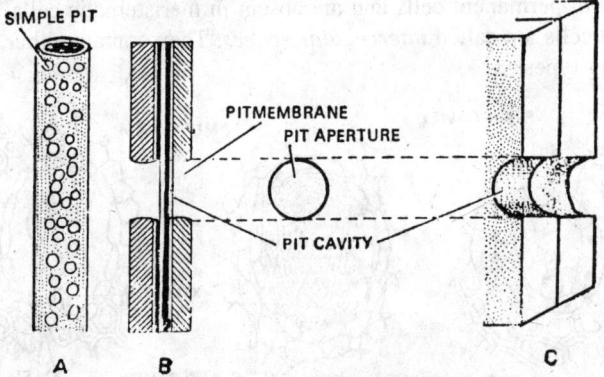

Fig. 1.11: Structure of simple pit. (A) cell-wall (B) L.S. of vessel, (C) Simple pit in L.S.

(ii) *Bordered pits* : When the depositing material is deposited around the primary wall and forms a funnel like structure, the pit is called *bordered pit*. It is usually represented in the form of two circles. The smaller circle represents the pit-aperture and the larger the area of the border pit. The closing membrane, here, swells and forms the *torus* which functions as a valve and controls the diffusion of liquids. When the pressure of the liquid is same on both the sides of torus, the pit remains open. If the pressure increases at any side, it becomes closed. Bordered pits are found in xylem tracheids and vessels.

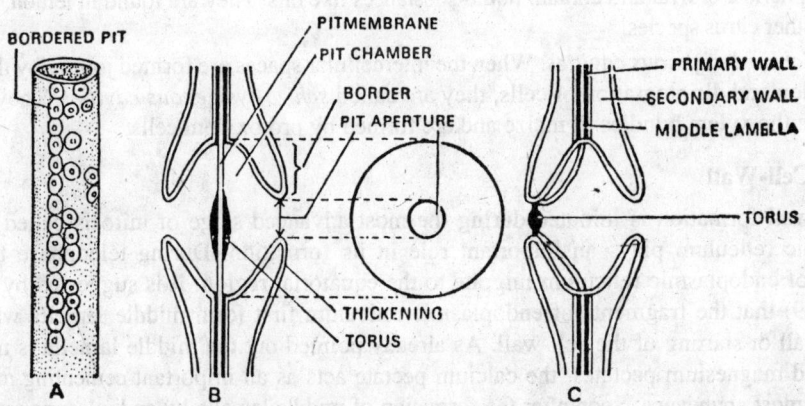

Fig. 1.12: Structure of Bordered pit. (A) front view, (B) L.S., (C) section of bordered pit.

Chemical Changes in the Cell-Wall

Following chemical changes may occur during the growth of the cell-wall :

(i) *Lignification* : The deposition of lignin on cell-wall or the conversion of cellulose into lignin is called *lignification*.

(ii) *Cutinization* : The conversion of cellulose into cutin in the cell-wall is called *cutinization*.

(iii) *Suberization* : The deposition of suberin on the cork cell-wall is known as *suberization*.

(iv) *Mucilaginous changes* : Sometimes, the cellulose of cell-wall is converted into mucilage. The phenonenon is known as *mucilaginous change*.

(v) *Mineralization* : The deposition of mineral substances like silica, calcium carbonate and calcium oxalate on the cell-wall is called as *mineralization*.

Intercellular Spaces

They are found in permanent cells and are absent in meristematic cells. The spaces found in between the adjacent cells are called *intercellular spaces*. They contain either air or water in them. They are of following types :

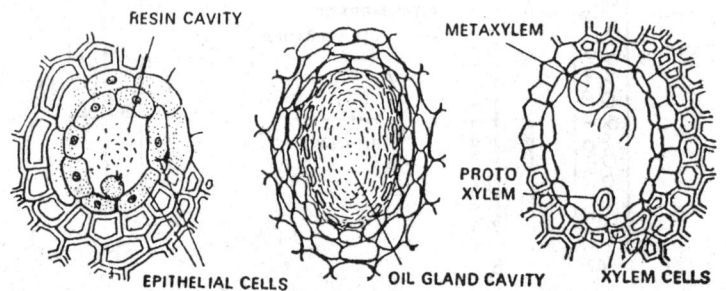

Fig. 1.13: Intercellular spaces. (A) Schizogenous cavity, (B) Lysigenous cavity in orange, (C) Shizo-lysigenous cavity in xylen tissue.

(i) *Schizogenous cavities*: When the intercelular spaces are formed by the contraction of the cell-walls of two adjacent cells, they are called *Schizogenous cavities*. They are small, narrow and remain filled with gases or liquids. They help in the diffusion of gases and osmosis of liquids. Resin ducts in the *Pinus* represents this type of schizogenous cavities.

(ii) *Lysigenous cavities*: When the intercellular spaces are formed by the disintegration or degeneration of cells, they are called *lysigenous cavities*. They are comparatively larger, spherical or oval and contain liquid substances like oils. They are found in lemon, orange and other citrus species.

(iii) *Schizo-lysigenous cavities*: When the intercellular spaces are formed jointly by the contraction and disintegration of cells, they are called *schizo-lysigenous cavities*. They are found in the xylem bundles of maize and are formed by protoxylem cells.

Origin of Cell-Wall

Cell-wall formation is initiated during the most advanced stage of mitosis called telophase. Endoplasmic reticulum plays an important role in its formation. During teleophase the tubular fragments of endoplasmic reticulum migrate to the equatorial region. It is suggested by *Northcote* (1968, 1969) that the fragments of endoplasmic reticulum first form middle lamella which is the new cell-wall or starting of the cell-wall. As already pointed out the middle lamella is made up of calcium and magnesium pectates, the calcium pectate acts as an important cementing material and is found in most abundance. Soon after the formation of middle lamella by endoplasmic reticulum, on either sides of it cellulose fibrils accumulate in concentric rows which finally form the primary cell-wall.

The Cell

The cytoplasm forms the plasmamembrane on the inner side of primary cell-wall. Later on, the deposition of cellulose, lignin, saccharides etc. takes place on the primary cell-wall which ultimately form the secondary cell-wall.

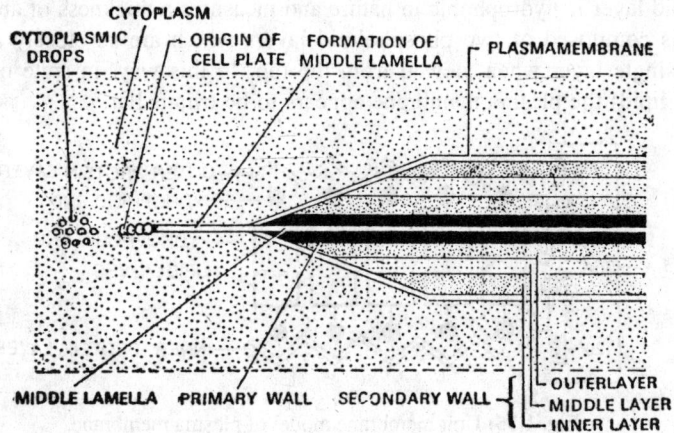

Fig. 1.14 : Origin and development of plant cell-wall showing sequence of formation of different layers.

Plasma Membrane

The plasmamembrane is very thin, elastic and semipermeable. It has been discribed under a variety of names like cell-membrane (*C. Nagh* and *C. Cramer,* 1855) biological membrane or plasmalemma (*J.Q. Plowe*, 1931). Microdissection shows that the plasma membrane cannot be removed or badly damaged even for a short time without death of the cell.

The plasmamembrane is generally considered as the outermost living part of the cell. The plant cells differ from animal cells in possessing an additional non-living cell-wall external to the plasmamembrane.

Definition

On the basis of structure, position and function, is can be defined as: "The outermost boundary of the cytoplasm of the cell which controls the entry and exit of molecules and ions and thus, helps in maintaining the difference in the ionic concentrations of the cytoplasm and the surrounding medium".

Structure of Plasma Membrane

Plasma membrane is a living, ultrathin, flexible, porous and semipermeable structure. Many scientists gave various theories and models to explain its structure. The names of some important models with their discoverers are given below:

   (*i*) *Paucimolecular theory* : It was given by *Danielli* and *Davson* (1930).

  (*ii*) *Unit membrane model* : It was proposed by *Robertson* (1959).

 (*iii*) *Kavanau's lipid-pillars model* : It was given by *Kavanau* (1965).

 (*iv*) *Hydrophobic binding model* : proposed by *Benson* (1966) and *Korn* (1966) separately.

  (*v*) *Greater membrane model* : proposed by *Lehninger* (1968).

 (*vi*) *Composite model* : proposed by *Scheide and Lin* (1970).

(*vii*) *Fluid mosaic model* : proposed by *Singer* and *Nicolson* (1972).

Out of the above proposed models, the unit membrane model proposed by *Robertson* (1959) is most accepted and widly used one.

Unit Membrane Model

According to this model, plasma membrane is a trilamellar structure having a thickness of about 75Å. The outer layers on both the sides are made up of protein. They are osmophilic in nature and each one measures a thickness of about 20Å. A layer of lipid is present between two layers of protein. This lipid layer is hydrophobic in nature and measures a thickness of about 35Å. Actually the lipid layer is composed of two phospholipid layers which are so closely arranged that they appear to be a single layer when viewed from the top. In this way, in spite of four layers (two protein and two lipid) the plasma membrane appears to be trilamellar.

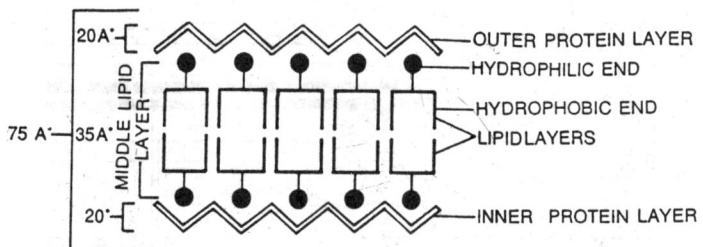

Fig. 1.15: Unit membrane model of plasma membrane.

The membranes of almost all the cell-organelles have similar trilamellar structure as described above. Due to this reason Robertson named this model as the *unit membrane* model. The thickness of membrane varies in different cells, e.g., in epithelium cells of intestine it is about 195Å while in R.B.C. about 215Å.

Fluid Mosaic Model

This model was proposed by *Singer* and *Nicolson* (1972). It also shows two layers of lipids and proteins each. But according to it the protein layers have two types of molecules. Some molecules enter deeply into the lipid layers and are called *intrinsic proteins*. They cannot be easily separated. Second type of protein molecules which do not enter deeply into the lipid layers but remain attached to their outer surface only, are called *extrinsic proteins*. They are easily separable.

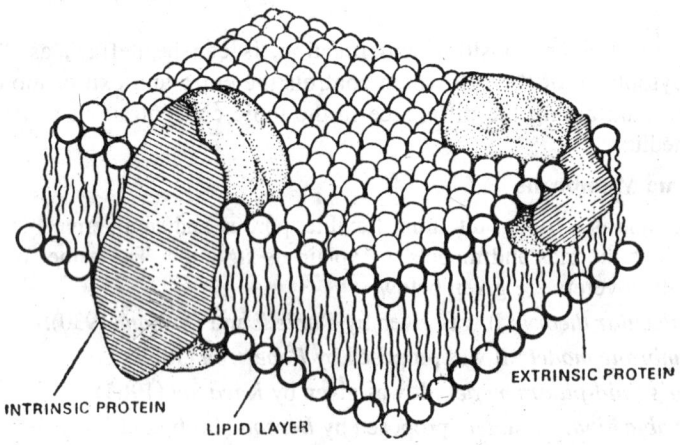

Fig. 1.16: Fluid mosaic model of plasma membrane.

Proteins have two important functions :
  (*i*) The help in movement of molecules of substances across the membrane.
  (*ii*) They help in different enzymatic activities of the membrane.

The Cell

Protein molecules are also effected by light. In presence of light they enter about half into the lipid layer while in absence of light, their only 1/3 portion remains inside.

Chemical Composition

Chemically the plasma membrane consists of lipids and proteins but a small amount of carbohydrates has also been reported from certain cell membranes.

(*i*) *Lipid fraction of plasma membrane* : The lipids constitute 20 to 40 per cent of the total dry weight in plasmamembrane. The main lipid constituents are phospholipids (phosphatidylcholine or lecithin) and cholesterol. Other important lipids are sphingolipids, glycolipids, glycophosphatides.

The phospholipids contain phosphate groups, glycerol, fatty acids and a nitrogenous base. They possess long hydrogen chains. It has been suggested that plasma membrane contains more cholesterol than cytoplasmic membranes. Its higher amount provides strength and stability to plasma membrane. Of the total lipids phospholipids represent about 55 to 57 per cent.

The glycolipids of plasma membrane are compounds containing sugar and lipid components. The glycolipid fraction in human erythrocyte contains hexa-amine and no sialic acid.

Like lipids, the plasma membrane of mammalian erythrocytes also contains a number of fatty acids such as palmitic, stearic, oleic linoleic, and arachidonic acids etc. Variations in compound lipid levels between species have also been reported.

(*ii*) *Protein fraction of plasma membrane* : The proteins which constitute the plasma membrane are 60 to 80 per cent of the total dry weight. *Maddy* (1964) and *Rega* (1967) reported that the red cell ghosts contain 90 per cent soluble proteins. It has been reported that the proteins of plasma-membrane are high molecular weight substances but their exact size is still unknown. The proteins isolated from thered cell membranes have been named as *taktins* by *Mazia and Ruby* (1968) because they resembly in aminoacid composition and other properties with *action* protein of muscle and actin like protein of microtubles.

The proteins are made up of number of aminoacids united by peptide linkages. The proteins may be glycoproteins as in blood group substances. Here they are known as mucopolysaccharide peptides. The liver plasma membranes consist of a phospholipoglycoprotein to which soluble proteins are attached.

The membrane associated proteins usually contain bounded enzymes but some of them are involved in purely structural and non-catalytic role. Glycoproteins on the outside may be responsible for adherence of cells.

(*iii*) *Carbohydrates in plasma membrane* : The presence of carbohydrates in plasma membrane was suggested by *Bell* (1962). The red cell ghosts contain about 5 per cent carbohydrates, while the liver plasma membrane contains less than one per cent of the total dry weight. The most common carbohydrates of these cells are hexose, hexosamine, fucose, and sialic acid. The sialic acid occurs in small amount in the form of gangliosides or glycolipids. Gangliosides in the plasma membrane of nerve endings have been reported by *Lapetina, Soto* and *De Robertis* (1967). Sialic acid is sensitive to neuraminidase and is attached to proteins by N-acetylgalactosamine on the outer surface of the membrane.

(*iv*) *Enzymes of plasma membrane* : About thirty enzymes have been detected from isolated plasma membranes. The most common and important enzymes are 5-nucleotidase, Mg^{++} ATPase, Na^+ - K^+ activated Mg^{++} ATPase, alkaline phosphatase, acid phosphomonoesterase, Glucose-6-phosphatase, $NADH_2$ and Cytochrome-c reductase. The plasma membrane lacks the respiratory chain and glycolytic activity.

Functional Activity

Following important functions have been suggested for the plasma membrane :

Permeability : The plasma membrane is often inaccurately described as semipermeable or differentially permeable but the best term for it is selectively permeable. Permeability determines which substances can enter the cell. The plasma membrane allows the movement of small ions and molecules of various substances and regulates the outflow of excretory material and water from the cell. The permeability of the plasma membrane is changed from time to time due to change in K^+ ion concentration. Depending upon the permeability, the membranes can be classified into following categories :

(i) *Impermeable membranes* : These membranes do not allow to pass anything through them except the gases. They are found in the unfertilized eggs of certain fishes.

(ii) *Semipermeable membranes* : These are those membranes which allow only water and no solute particles through them.

(iii) *Selective permeable membranes* : Selective permeable membranes are those which allow only selected ions and small molecules to pass through them. The plasma membrane and other intracellular unit membranes are selective permeable in nature.

(iv) *Dialysing membranes* : The membranes having certain extraneous coats are known as *dialysing membranes*. The water molecules and crystalloids are forced through them by the hydrostatic pressure forces. The basement membranes of endothelial cells are dialysing membranes.

Osmosis and Osmotic Pressure : The plasma membrane also performs the phenomenon of osmosis. Osmosis is a special type of diffusion of liquids. "When two solutions of different concentrations are separated by means of a semipermeable membrane, the diffusion of water or the solvent from the solution of lower concentration to the solution of higher concentration is known as Osmosis.".

Diffusion of the solvent takes place both ways across the membrane. When the diffusion of water molecules or solvent molecules (of lower concentration) takes place into the living cells (of higher concentration), it is known as *endosmosis*, while the reverse process which involves the exit of the water or solvent molecules from the cell is known as *exosmosis*. Thus, when a plant cell is placed in an isotonic solution (solution having similar concentration and osmotic pressure), the cytoplasm remains adherent to the cellulose wall and is not changed. When the solution of the medium is more concentrated (hypertonic solution), the cell loses water and the cytoplasm retracts from the rigid cell-wall showing exosmosis. This phenomenon is known as *plasmolysis*. On the other hand, when the solution of the medium is less concentrated than the intracellular fluid (hypotonic solution), the cell swells and eventually bursts showing endosmosis. Since plasma membrane of the cell is permeable to water and certain solutes, the osmotic pressure, caused by hydrostatic pressure is maintained by a mechanism that regulates the concentration of the dissolved substances within the cell. The hydrostatic pressure is caused by the amount of water inside the cell. 0.66 per cent solution of sodium chloride acts as a hypotonic for mammalian cells while the same solution acts as isotonic for amphibian erythrocytes.

Diffusion or passive transport : When the particles or molecules of various substances pass through the plasma membrane without consuming any energy, the phenomenon is known as *diffusion*. The diffusion of ions through the plasma membrane depends on the concentration and electrical gradients.

Active transport : This process is quite similar to that of passive transport except that here the movement of ions or molecules requires energy. This shows that certain amount of work must be done in order to penetrate the molecules or ions. The energy, in the form of ATP, is mainly produced by the oxidative phosphorylation in mitochondria. For this reason, active transport is generally related or coupled to the cell respiration.

Certain enzymes like ribonuclease are found to assist in this active transport. The active transport of ribonuclease through plasma membrane has been reported in the living plant cells, eggs, flagellates and ascitic tumors. Certain chemical compounds like urea, formamide, and glycerol also pass through plasma membrane by active transport. *Brachet* (1957) reported the transport of large molecules of certain proteins through this process.

The Cell

Endocytosis : Endocytosis includes pinocytosis and phagocytosis through which solid or fluid materials enter the cell. Both these processes are related to the activity of plasma membrane.

(*a*) *Pinocytosis* : Pinocytosis helps in the entry of high molecular weight substances such as proteins. The presence of proteins, certain aminoacids and ions seems to act as a stimulus to pinocytosis.

In all cultures, it has been observed that the plasma membrane first invaginates and then buds off internally to form a vesicle. It encloses the droplet of medium in the cytoplasm. The process is known as *Pinocytosis* and the vesicle so formed is called a *Pinocytotic vesicle*. Pinocytosis was first observed by *Edwards* and *Lewis* (1937) in amoeba and cell cultures. The pinocytotic vesicles were named as *Pinosomes* by Lewis.

Micropinocytosis : The pinocytosis which occurs at submicroscopic level is known as *Micropinocytosis*. During this process the micropinocytic vesicles of about 650Å are formed. These vesicles have opening on both outer and inner surfaces which suggest the possible transfer of fluid through these vesicles to the cell. Micropinocytotic vesicles were first observed in endothelial cells and later on in *Schwann and Satellite* cells, muscle cells, retiuclar cells and macrophages.

(*b*) *Phagocytosis* : The ingestion of solid particles by the living cell has been termed as *Phagocytosis* and the vocuole formed by this process is called *phagocytotic vesicle* or *phagosome* or *food vacuole*. The phagosomes migrates to the interior of the cell where they come in contact with the primary lysosome containing many enzymes. The membranes of the two fuse to form digestive vacuole. Here, the large substances of phagosome are progressively digested by the hydrolytic enzymes and the products diffuse through the membrane into the cell. When the digested food of the phagosome diffuses into the cytoplasm, the vacuole is left with indigestible material only inside it. This vacuole is now known as *residual body*. It is ejected into the external environment by a process similar to phagocytosis in reverse called *ephagy* or *egestion*.

The process of phagocytosis is found in most protozoans, metazoans and certain multi-cellular organisms.

When the colloidal particles are ingested by the process of phagocytosis, it is known as *ultraphagocytosis* or *colloidopexy* and if the absorbed colloid is a chromogen, it is known as *chromopexy*.

Emeiocytosis or Cell Vomiting : This phenomenon is the reverse of endocytosis and is also known as *exocytosis*. Through this process the cells which have secretory functions pass out their secretions outside the cell. During the process there is a step at which the secretory vacuoles fuse with the plasmamembrane to discharge their contents outside.

Protoplasm (Prots = first, plasm = substance)

The substance found inside the plasma membrane of each cell of living organisms is called *protoplasm*. The vital activities occurring in living organisms are brought up by the protoplasm. This is the reason why protoplasm is called "physical basis of life". *Hugo Von Mohl* (1846) explained the importance of protoplasm while describing the protoplasm of plant cells. *Max Schultze* (1861) propounded the "theory of protoplasm" according to which, the body of a living organism is a group of units made up of protoplasm.

Physical Properties

Protoplasm is clear, colourless, semi-transparent and semi-fluid substance. It consists of about 60 to 80 per cent water, in which different organic and inorganic substances remain dissolved.

Different substances like sugar, salt and mineral salts etc. remain dissolved in protoplasm as true solution, i.e., the solute molecules are dissolved in the solvent medium to form a homogeneous system. This type of solution is called *true solution*. The size of solute particles in true solutions is smaller than 0.001 μ. The solute particles are usually found in ionic or molecular form. Sometimes the solute particles present in the solvent attains a size more than 0.1 μ. They can be seen with naked eyes. Such particles are called *suspension*, e.g., the solution of soil particles in water.

The particles of many organic substances, e.g., gum, starch, albumin, egg etc. dissolved in protoplasm, have a size more than ions or molecules but smaller than suspension i.e., the size of molecules ranges between 0.1 μ and .001 μ. These particles can be seen with the help of ultramicroscope—Such types of solutions are called *colloidal solutions* and the dissolved substances as *colloids*.

Thus, it is clear that the protoplasm exhibits the properties of true solution, suspension and complex colloidal solution. Due to colloidal nature, the protoplasm exhibits the properties like elasticity, plasticity, contration and expansion.

Besides, it also possesses the following properties :

Streaming Movement

It may be of following types :

(A) **Cyclosis** : The movement of protoplasm inside the cell is called *cyclosis*. It is of two types :

> (i) *Rotation* : When the protoplasm moves around the large vacuole in a cell in only one direction, it is called *rotation*. It can be observed in the leaves of *Hydrilla* and *Vallisneria*.

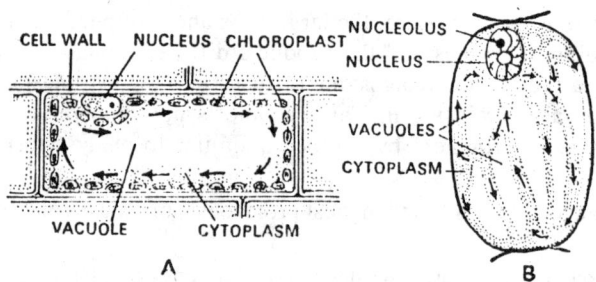

Fig. 1.17: Streaming movement of Protoplasm. (A) Rotation in a cell of Hydrilla leaf, (B) Circulation in a cell of staminal hair of Tradescantia.

> (ii) *Circulation* : When the protoplasm moves in different directions around several vacuole within a single cell, it is called *circulation*. It can be observed in the cells of staminal hairs of *Tradescantia* and *Rhoeodiscolor*.

(B) **Amoeboid movement** : The movement of protoplasm through pseudopodia is called *amoeboid movement*. It can be observed in Amoeba, leucocytes, naked gametes of *Spirogyra* and slime fungi.

(C) **Ciliary movement** : The movement of protoplasm with the help of cilia is called *ciliary movement*, e.g., Chlamydomonas.

Coagulation

Due to certain physical factors like temperature etc. the colloidal particles of protoplasm combine together to form large particles. It is called *coagulation*. The larger particles again break up into smaller ones under certain limits.

Two Forms of Protoplasm

Like other colloidal solutions, the protoplasm is also found in two states called *sol* and *gel*. The protoplasm shows reversible sol-gel transformation. Sol is thin and dilute form in which the molecules of various dissolved organic substances are found scattered. The molecules of organic substances in gel are found nearer and attached to each other. Therefore, it is thick and viscous than sol.

The Cell

Brownian Movement : In sol, the suspended particles or molecules perform zig-zag movement by colliding with each other. It is called *Brownian movement*. It was for the first time observed by

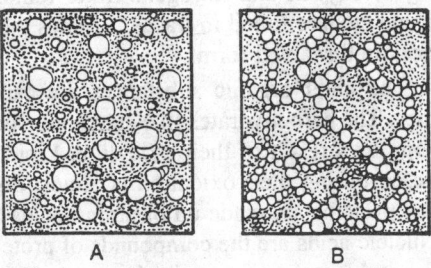

Fig. 1.18: Protoplasm : (A) Sol state (B) Gel state.

Brown (1829). Due to this movement the colloidal particles of protoplasm always remain in suspended state and do not settle down because the gravitational pull is counteracted. This type of movement is not found in gel state of protoplasm.

Nature of Protoplasm

Various scientists have given different theories regarding the nature of protoplasm. On the basis of nature of suspended particles of protoplasm, following theories have been proposed for the structure of protoplasm:

(A) *Granular theory*: It was proposed by Altmann (1893). According to this theory, the protoplasm is made up of minute granules or particles which remain scattered in a homogeneous liquid medium.

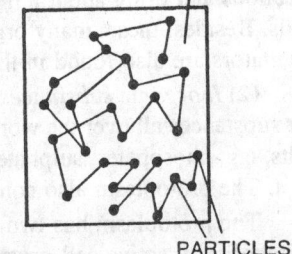

Fig. 1.19: Brownian movement.

(B) *Alveolar theory*: It was proposed by Butschilli (1892). According to it, protoplasm is a lather like frothy emulsion made up of two substances of different densities. The substance with higher density forms the ground substance or the matrix while the other with lower density remain suspended in the matrix.

(C) *Fibrillar theory*: It was proposed by Fleming. According to it, the protoplasm is made up of fine threads or fibrils of protein which are called *micelles*.

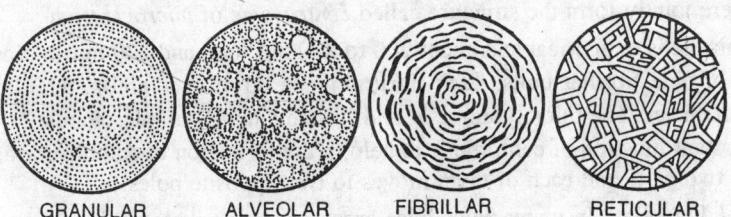

Fig. 1.20: Physical forms of protoplasm.

(D) *Reticular theory*: It was proposed by Klein and Carnoy. According to this theory the protoplasm is made up of a fine net of fibrils. Now, it has been confirmed that fibrils found in the protoplasm are the chains of aminoacids. Besides these fibrils, the presence of enodplasmic reticulum also provides a reticulate appearance to the protoplasm.

(E) *Colloidal theory*: It was proposed by Wilson (1925). According to this theory, the protoplasm is partially a true solution and partially a colloidal system.

Chemical Properties

Protoplasm is a complex mixture containing about 30 elements. The chief elements of the protoplasm are carbon, hydrogen, oxygen and nitrogen. Other remaining elements like sodium, potassium, calcium, iron, sulphate etc. are found in traces. Besides them, protoplasm also contains organic and inorganic substances in dissolved form.

(1) *Organic substances*: The chief organic substances are carbohydrates, lipids, proteins, nucleic acids and a few enzymes. The carbohydrates like starch are composed of carbon, hydrogen and oxygen. They release energy as a result of their oxidation. Lipids are made up of fatty acids and glycerol. The energy released during their oxidation is almost double to that released during oxidation of carbohydrates. The proteins are made up of amino-acids and they help in growth and repair of broken body parts. Nucleic acids are the compounds of proteins and carbohydrates. They help in the synthesis of proteins and transfer of genetic characters from one generation to the next. The protoplasm contains following tow types of nucleic acids.

(*i*) *Ribonucleic Acid* (RNA) : Major percentage of RNA is found in the cytoplasm.

(*ii*) *Deoxyribonucleic Acid* (DNA) : Its major percentage is found in the nucleus.

The enzymes found in the protoplasm are also made up of proteins. They catalyse most of the reactions and bring about a major change in the velocity of biochemical reactions occurring in the cells. Besides, these, many organic substances like pigments, latex, vitamins, alkaloids and growth regulators are also found in the protoplasm which perform specialized functions.

(2) *Inorganic substances*: They include water, salts and some gases. Water is the best solvent for substances all over the world. Protoplasm comprises about 80 per cent of water. Some important salts, e.g, phosphates, sulphates, chlorides and carbonates of Na, K, Ca and Fe are found dissolved in it. The protoplasm also contains gases like oxygen and CO_2 in dissolved form.

The protoplasm has two parts: (*i*) Cytoplasm and (*ii*) Nucleus. The cytoplasm contains many metabolically active cell-organelles an inactive cell-inclusions.

Metabolically, Active Cell-Organelles

Centrosome, Cilia and Flagella

Centrosome: It is a spherical structure situated near the external surface of the nucleus and occupies almost a central position. It is mainly found in animal cells and in a few plant cells like algae and fungi. The centrosome is usually made up of two spindle-shaped centrioles which remain surrounded by a zone of hyaline cytoplasm known as *centrosphere* or *idiozome*. Thus, centrioles and centrosphere jointly form the structure called *centrosome* of *microcentrum*.

Each centriole is cylindrical and about 200 to 200 mμ long and 200 mμ in diameter. Further, it is made up of nine fibrillar units. Each fibril is made up of *a*, *b*, and *c*, three microtubules of about 20 mμ diameter. The microtubules '*b*' and '*c*' are simple but microtubule '*a*' is armed. The chief and important function of centriole is to help in the formation of spindle during cell division. It divides into two parts and each of them moves to two opposite poles.

Cilia and Flagella : In many unicellular organisms and ciliated epithelium of multicellular animals, whip-like appendages are produced which are called *cilia* and *flagella*. They help in the locomotion of cell and emerge from the *basal bodies*, also known as *kinetosomes*.

Basal bodies, cilia and flagella have almost similar internal organization and resemble with that of centrosome. They are also made up of fibrillar units but their number varies from centrosome. In cilia and flagella eleven fibrils (filaments) are present, of which nine remain arranged in an outer ring and two in the centre. Each fibril of the ring consists of two microtubules—*a* and *b*. They are also knwon as sub-filament *a* and *b*. The microtubule—*a* is armed. The two central fibrils are surrounded by inner sheath. All eleven fibrils remain connected with each other with the help of fibres. The basal bodies lack two central fibrils and resemble centrioles and can multiply like them.

The Cell

Although, cilia and flagella are morphologically and physiologically similar structures but they differ from each other in their number, size and functions. The differences between cilia and flagella are given in Table 1.3.

Sometimes, cytoplasm forms immobile structures called *sterocilia* which are quite different from true cilia. The true cilia are called kinocilia. Chemically, cilia and flagella are made up of 70-84 per cent proteins, 13-26 per cent lipids, 1-6 per cent carbohydrates and 0.2-4 per cent nucleotides.

Table 1.3 : Showing differences between flagella and cilia

Flagella	*Cilia*
1. Their number in an organism may be one or two.	1. Their number in an organism is quite high (3000-14000).
2. They are comparatively larger (about 150 μ) than Cilia	2. They are comparatively smaller than flagella (5-10 μ).
3. They are found at one end of the cell.	3. They are found on entire surface of the cell.
4. They show undulating movement.	4. They show sweeping movement.
5. They move freely.	5. They move in a co-ordinated rhythm.

Endoplasmic Reticulum

It was discoverd by *Granier* (1897). But the name endoplasmic reticulum was first used by *Porter et al.*, (1945) to describe certain structures observed under electron microscope in the cytoplasm of fibroblast—like cells in cultures of chick embryonic tissues. It is now more generally applied to describe similar structures found in almost all the cells. According to Porter (1961) *"the endoplasmic reticulum is a complex, finely divided vacuolar system of interconnected vesicles and tubules extending in the cytoplasm between the nuclear membrane and the plasma membrane."*

The name *ergastoplasm* was suggested by *Granier* to the ER due to its basophilic nature but later on certain objections were raised by different workers because the basophila like structures were not associated with ER. They called them *acidophila* showing its acidophilic nature. The studies of *Caspersson* (1955) and *Brachet* (1957) have demonstrated that the basophilic nature of the ergastoplasm is due to the high RNA content.

Sjostrand (1956) criticized the lamellate structure and proposed the name '*double membrane*' or '*double edge membrane*' on the basis of cytoplasmic membranes. They are of three types:

(*i*) α-cytomembranes: They are double contour and always associated with some granules.

(*ii*) β-cytomembranes: They are also double contour but do not have granules.

(*iii*) γ-cytomembranes: All the membranes of golgi-complex are γ-cytomembranes.

However, the name endoplasmic reticulum has been used most commonly by different workers. *Fawcett* and *Ito* (1958), *Thirery* (1958), *Rose* and *Pomerat* (1960) and *Christensen* (1960) have made important contributions in the field of structure and distribution of endoplasmic reticulum.

Definition

A number of definitions have been proposed by different workers. Some of them are :

"The system of ER represents the complex arrangement of membrane bound spaces which have a distinct inside and outside." According to *Ambrose* and *Easty* (1970, 71), "The endoplasmic reticulum is a complex branching network of membrane-bound cavities (or cisternae) and is more concentrated in the inner, endoplasmic region of the cell than in the peripheral or ectoplasmic region." They further stressed that the branches of the network are intercommunicating and divide the cytoplasm into two main regions, the first being enclosed within the membrane system, the second forming the outer region or cytoplasmic matrix.

Robert M. Dowben (1971) defined the Endoplasmic Reticulum as "An intracellular membrane system of branching and intercommunicating tubules which is involved in many cellular functions. The outer surfaces of these membranous systems are often covered with dense granules of about 250Å in diameter."

Occurrence: The endoplasmic reticulum is of universal occurrence, present in almost every cell except the mammalian erythrocytes. It occurs abundantly in the cytoplasm of mature differentiated cells. It is often absent in eggs, sperms and in embryonic or undifferentiated cells but increases with differentiation. In spermatocytes only a few vacuoles can be observed. The cells which are active in lipid metabolism such as adipose cells, brown fat cell and adrenocortical cells contain simple type of endoplasmic reticulum. *Christensen* (1960) reported that the interstitial cells of Opposum's testis contain only smooth or agranular type of ER. On the contrary, the cells which are actively engaged in protein synthesis, such as pancreatic cells and goblet cells contain rough or granular type of ER in the form of cisternae. The liver cells possess both the types of endoplasmic reticulum.

In meristematic cells the number is quite large. In prokaryotic cells, it is generally absent while in eukaryotic cells it is present in large number.

Structure or morphology : Endoplasmic strands appear as a double membrane and can widen out locally into vesicles and frequently chains of vesicles connected with one another by canalicules or large ventricles and caverns. ER exhibits great variation in morphology depending upon the metabolic state of the cell. Reticulum usually occurs in three patterns : (*i*) Cisternae, (*ii*) Vesicles, (*iii*) Tubules.

1. **Cisternae :** The cisternae are long, flattened, parallely arranged, and usually unbranched tubules of about 40-50 µ diameter. They form successive layers around the nucleus. Cisternae are found in the cells active in protein-synthesis like the cells of liver, pancreas, brain and notochord. These are basophilic in nature.

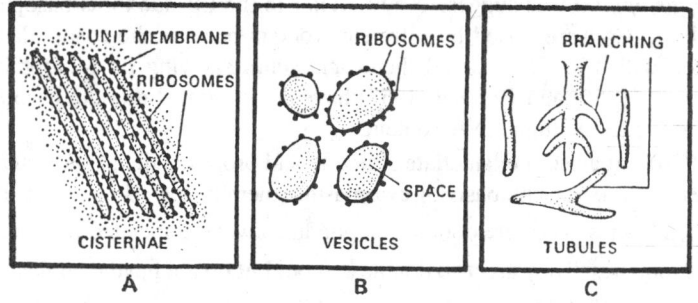

Fig 1.21: Three different structures of Endoplasmic Reticulum.

2. **Vesicles :** The vesicles are almost rounded, spherical or ovoidal spaces with 25 to 500mµ in diameter. Vesicles are found in abundance in synthetically active cells, such as, in liver cells, pancreatic cells and developing spermatocytes of mammals, where they form a short of network along the peripheral region.

3. **Tubules:** The tubules are small, smooth walled, branches, tubular spaces having different forms. They have a diameter of about 50-100mµ. The tubules are usually found in non-secretory cells and also in the cells that synthesise steroids like cholesterol, glycerides and hormones. They are irregularly distributed in the cytoplasm of developing spermatids of giunea pigs and muscle cells.

Above three forms of reticulum may occur in a single cell at the same time or may appear at different times during the cell cycle. The pattern of arrangement of endoplasmic reticulum also differs in different cells. For example, mammalian liver cells show parallel lamellar elements (cisternae) of more or less uniform size during metabolic activity whereas pancreatic cells show haphazard arrangement. In striated muscles, the ER forms network of tubules which is commonly called as *Sarcoplasmic reticulum*.

Ultrastructure of Membranes

The membrane of ER is semipermeable and showed the 'unit membrane' structure of *Robertson* (1959). Thus, the membrane of ER is trilamellar composed of lipo-proteins. The outer and inner dense layers are of proteins and the middle layer is of phospholipids. The middle layer is formed by the close association of two thin and transparent layers of phospholipids, thus showing a single layer. The characteristic 'double membranes' which occur frequently must be regarded as being in fact pairs of unit membranes separated by a narrow space of about 50-60Å. There is a mass deficient. So, they must be filled with serum which is known as *enchylema*. *Palade* (1956) has observed secretory granules in the cavity of ER.

Nature of Endoplasmic Reticulum

The endoplasmic reticulum occurs in two forms depending upon the presence or absence of ribosomes : (1) Smooth surfaced reticulum (Agranular) and (2) Rough surfaced reticulum (Granular).

Smooth Surfaced or Agranular Reticulum

The ER which does not possess ribosomes is known as *smooth surfaced ER*. It appears in the form of smooth vesicles of 300-700Å in diameter and bounded by unit membranes. The smooth surfaced ER occurs in different shapes and form especially in those cells which are almost inactive in protein synthesis. It is found in the cytoplasm of mature leucocytes, spermatocytes, retinal cells, adipose cells, interstitial cells, glycogen storing cells of the liver and steroid hormone synthesising cells. In certain immature oocytes, it represents annular appearances and is therefore, known as *annulated reticulum*. *Ward and Ward* (1968) reported that the annulated lamellae generally are arranged in groups. They may be associated either close to the nucleus or in the cytoplasm but are perforated by numerous pores. In pigmented cells it exists in the form of tightly packed vesicles and tubes known as *Myeloid-bodies*. The muscle cells are also rich in smooth type of ER and here it is known as *Sarcoplasmic reticulum*. Smooth walled reticulum can be converted into the rough form and *vice-versa*, depending upon the protein requirements of the cell.

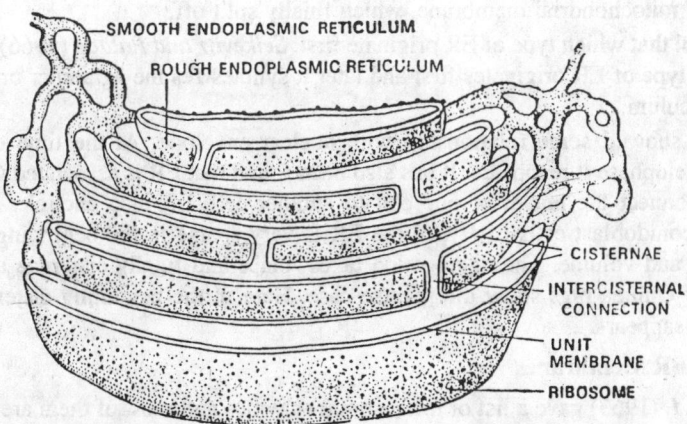

Fig. 1.22: Three-dimensional structure of Endoplasmic Reticulum showing smooth and rough surfaced ER.

Rough Surfaced or Granular Reticulum

It is characterized by the attachment of ribosomes to its outer surface. Ribosomes were first discovered by *Palade* (1955). They are made up of ribonucleoproteins and play a vital role in the process of protein synthesis. They serve as a 'marker' to identify the rough ER. The rough ER is found in abundance in those cells which are actively busy in the protein synthesis, such as plasma cells, goblet cells, pancreatic cells and liver cells. Rough ER is basophilic in nature because of higher RNA contents. In cross-sections the rough ER appears as a system of parallel double membranes while in tengential section as whorls of saccules (*Kessel*, 1963; *Ward and Ward*, 1968; *Maul*; 1968). Due to basophilic nature, the region of matrix containing granular ER has been named as ergastoplasm, basophilic bodies, chromophilic substances or Nissl bodies by earlier cytologists.

Irrespective of the differences, the two types of ER are formed of double unit membranes and it is regarded that there is no sharp morphological discontinuity between smooth and rough type of ER (*Fawcett*, 1963).

Origin and Relationship of ER

The origin of ER is a matter of controversy as different workers have suggested its possible origin from the membranes of different cell organelles. Some of the common views regarding the origin of ER are as follows :

1. **De novo Origin :** De novo origin has been confirmed in many cases but in case of ER it is doubtful.

2. **From Plasma membrane :** Palade had suggested the origin of ER by infoldings of the cell membrane. This view was supported by *Buvat* (1963).

3. **From Nuclear Envelope :** It has been suggested by *Barer, Joseph and Meek* (1959) and *Essner and Novikoff* (1962). The membranes of ER resemble with the nuclear membrane and plasma membranes and at telophase stage the ER membranes form the nuclear envelope. Therefore, it is normally assumed that the ER might have originated by the evagination of the nuclear membranes.

4. **From Mitochondria:** There are number of authors who told that the ER is associated with mitochondria and the former (ER) is being formed on the surface of mitochondrial membrane, which finally split off.

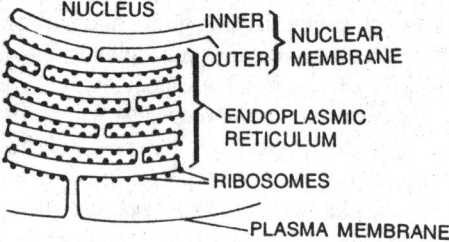

Fig. 1.23: Relationship among plasma membrane, Endoplasmic reticulum and Nuclear Membrane.

It is doubtful that which type of ER originate first. *Seikevitz and Palade* (1966) have suggested that the granular type of ER originates first and later it synthesizes the agranular or smooth type of endoplasmic reticulum.

ER always showed some relationship with nuclear envelope. At the time of prophase ER dissolves and at telophase it reappears. It has also been found that ER is associated with plasmodesmata as it can connect the nuclei of one cell to another cell. ER also encircles the plastids. In ectodermal and conidoblast cells of *Hydra* the ER is very sparse in the beginning and later on it increases in size and volume. The aggregation takes place and finally cisternae are formed. The spermatocytes of Guinea pigs show the reverse case. Here in the beginning cisternae are present which later on disappears.

Enzymes of the ER Membranes

Rothschild. J. (1963) gave a list of microsomal enzymes and most of them are associated with the membranes of ER. The membranes provide a large inner surface and participate in the different metabolic reactions by means of the attached enzymes. Stereases and NADH-cytochrome reductase were first identified and used as a marker of the microsomal membranes. Other important enzymes

The Cell

of the ER membranes are NADH-diaphorase, glucose-6-phosphatase and Mg^{++} activated ATPase. Some of the enzymes of ER membranes and then functions are shown in Table 1.5.

Functions of ER

The endoplasmic reticulum functions as circulatory, secretory, storage, and nervous system for the cell. The functions of the two types of ER are being discussed separately.

Common Functions of Smooth and Rough ER

(i) ER provides an ultrastructural skeleton framework to the cell and gives mechanical support to the colloidal cytoplasmic matrix.

(ii) ER in plants, plays a special role in the inter-connection of cells through plasmodesmata.

(iii) The membranes of ER possess osmotic properties and thus helps in the regulation of entry and exist of materials.

(iv) The ER acts as a circulatory or transporting system for various substances (*Palade*, 1956). During the transport of various products the different parts of the vacuolar system interact and show the following directional flow :

Granular ER → Agranular ER → Golgi membrane → Lysosomes (*Essner and Novikoff*, 1962).

(v) The ER membranes are found to conduct intracellular impulses e.g., the sarcoplasmic reticulum transmits impulses from the surface membranes into the deep region of muscle fibres.

(vi) It determines the plane of cell division (*Northcote*, 1963).

(vii) It plays an important role in the formation of nuclear envelope after each nuclear division (*Northcote*, 1963).

(viii) It determines the formation of exine pattern in pollen grains and sites of germ pores in exine (*Heslop-Harrison*, 1966).

(ix) It also determines the development of pores in nuclear membranes.

(x) The development of pores in seive-plates is attributed to the prior disposition of ER at such places (*Esau*, 1962).

(xi) The ER protects the cell by the toxic effects of various substances by a process of "detoxification" (*Jones* and *Fawcett*, 1966).

(xii) ER provides increased surface for various enzymatic reactions.

(xiii) It helps in the transportation of genetic material.

(xiv) ER membranes provide a site for ATP synthesis in the cell.

Function of Granular ER

The granular ER plays a major role in protein synthesis because of the presence of ribosomes (*Palade*, 1958; *Seikevitz*, 1959). The ribosomes are the centres of protein synthesis. The polyribosomes on the ER are said to accelerate the activity of m-RNA and protein synthesis. The proteins are transported by the ER to the exterior of the cell via smooth ER, golgi complex and secretory granules (*Palade*, 1969).

Functions of the Agranular or Smooth ER

1. **Synthesis of Lipids :** The agranular type of ER is abundant in those cells which are involved in the synthesis of lipids (*Christensen and Fawcett*, 1961). The synthesis of triglycerides and lipo-protein complexes is performed by the agranular ER and the golgi complex (Claude, 1968).

2. **Polysaccharide Metabolism:** Smooth ER in plant cells develops at the surface where the cellulose wall of the cell is being formed, which implies a relationship with polysaccharide metabolism. In liver cells, too, the smooth form is present in regions of the cell which are rich in glycogen. It is important, however, for the metabolism of lipids and steroids.

3. Synthesis of Glycogen: *Porter and Poruni* (1960) reported that in fasted animals, the residual glycogen remained associated with the tubules and vesicles of the ER. *Porter and Mochado,* (1960) reported that in plant cells the agranular ER develops along the surface where the cellulose walls are being formed. *Porter* (1961), *Luck* (1961) and *Peters Kelly* and *Dembitzen* (1963) suggested that the agranular reticulum is related to glycogenolysis (digestion of glycogen) and not to glycogenesis (synthesis of glycogen).

4. Detoxification: *Jones and Fawcett* (1966) reported the detoxification effect of ER which is due to the increased activity of enzymes related to detoxification and a considerable hypertrophy of the agranular ER.

ER also performs the function of intracellular impulse conduction.

Table 1.4: Enzymes of ER Membranes and their functions

(*i*)	Synthesis of glycerides
	Triglycerides
	Phosphatides
	Glycolipids and plasmalogens
(*ii*)	Metabolism of plasmalogens
(*iii*)	Fatty acid synthesis
(*iv*)	Steroid biosynthesis:
	Cholesterol biosynthesis
	Steroid hydrogenation of unsaturated bonds
(*v*)	$NADPH_2 + O_2$ — requiring steroid transformations:
	Aromatization
	Hydroxylation
(*vi*)	$NADPH_2 + O_2$ — requiring drug detoxification:
	Aromatic hydroxylations
	Side-chain oxidation
	Deamination
	Thio-ether oxidation
	Desulphuration
(*vii*)	L-Ascorbic acid biosynthesis
(*viii*)	UDP-uronic acid metabolism
(*ix*)	UDP-glucose dephosphorylation
(*x*)	Aryl- and steroid-sulphatase.

Modified from *Rothschild, J.* (1963) and after *DeRobertis, Nowinski* and *Saez* (1970).

Golgi Complex

"The Golgi complex, like endoplasmic reticulum (ER) is a canalicular system with sacs, but unlike the ER it has parallely arranged flattened membrane bounded vesicles."

The most characteristic aspect of this cell organelle is its ability to deposit heavy metals like osmium and rubidium when the cell is impregnated with the oxides of these metals. Thus, they are stainable with osmium tetraoxide and silver salts. *Golgi* observed a network of anastomosing, dense filaments near the nuclei of neural cells of the barn owl and the cat by these staining reactions. After the name of its discoverer Golgi complex was variously named as Golgi region, Golgi bodies, Golgi apparatus, Golgiosome, Golgi material and Golgi membranes.

History: Golgi complex was first observed by *George* (1867) and was described by an Italian cytologist, *Camillo Golgi* (1898). *Gatenby* (1917) and *Hirschler* (1918) have shown that irrespective of morphological differences the golgi complex of vertebrates is homologous to the dictyosome of invertebrate cells. *Parat* and *Painleve* (1924) put forward the vacuole hypothesis for the structure of Golgi complex, according to which the vacuoles stained with neutral red were the living representation of Golgi reticulum. *Worley* (1943) and *Baker* (1944-49), using methylene blue as a

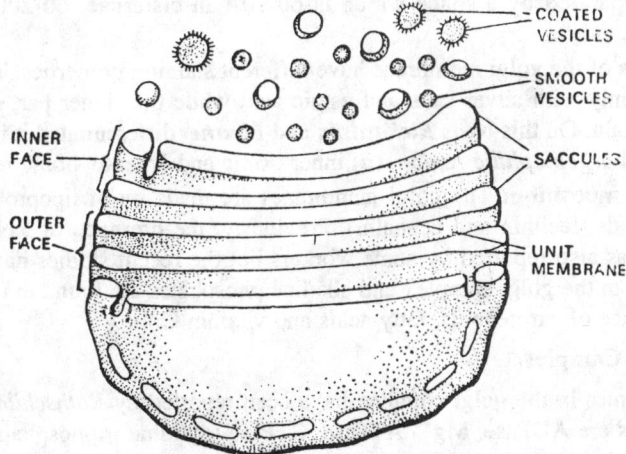

Fig. 1.24: Three-dimensional structure of Golgi body.

staining reagent, concluded that the droplets stained with methylene blue are the precursors of Golgi bodies. *Baker* (1957) used the term *Lipochondria* for them and emphasized the variability in their chemical constitution. *Walker* and *Allen* (1927) observed the golgi complex in fixed cells and considered it as an artifact. *Bowen* (1929) established the similarity between the golgi complex and developing acrosome of the spermatid.

Morphology: Golgi complex exhibits great variation in its morphology from cell to cell and from time to time depending upon the functional stages. However, the shape is characteristic for each type. The Golgi complex is generally a disc-shaped organelle and consists of following components:

1. **Flattened Sacs (Cisternae):** The cisternae are flat or curved and elongated tubes arranged in parallel bundles. The cisternae are separated from each other by distance of about 200Å to 300Å. The enclosing space of each cisternae is about 60-90Å.

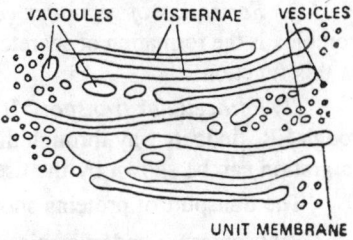

Fig. 1.25: Golgi body in cross-section.

Morrie (1969) and others have shown that the stacks of cisternae have got the polarity. They are polarized in such a way that the proximal pole of each cisternae is associated with the endoplasmic reticulum or the nuclear envelop and the distal pole with the formation of secretory vesicles. The membranes of proximal pole are thinner and morphologically similar to the endoplasmic reticulum. The membranes of distal pole are thicker and show similarities with the plasma membrane.

2. **Vacuoles :** The vacuoles are large, spaceous and rounded sac like structures. They are generally present at the edge of the golgi complex. The formation of vacuoles takes place by the expansion of flattened sacs (cisternae) in which the two membranes of sacs get widely separated.

3. **Vesicles:** The vesicles are small droplet–like structures. Their diameter ranges from 400 to 800Å. They are found in clusters around the ends and outer surface of the cisternae. The vesicles are formed from the cisternae either by budding or by the constriction of the ends of the sacs.

Ultrastructure: The ultrastructure of golgi complex shows that its membranes are the unit membranes. They are differentiated into outer and inner membranes and remain separated by an intra-membranous space. The golgi membranes are about 60-70Å thick. The adjacent membranes are separated by a space which is 60-90Å in cisternae, 60-200Å in vacuoles and 30-40Å in vesicles.

The two faces of the golgi membrane have different staining properties. The outer part of the membrane reacts only with silver salts and osmic acid while the inner part of the membrane in contrast takes no stain. On this basis *Richardson* and *Bourne* differentiated it into the two regions: (*i*) outer osmic and *argentophilic region*, (*ii*) inner osmic and argentophobic region.

Chemical Composition: The golgi membranes are made up of lipoproteins. The lipids are rich in phospholipids (lecithin and cephalin). Previously, the presence of RNA and alkaline and acid phosphates was also reported by some workers but the recent studies have shown that RNA is totally absent from the golgi complex and alkaline phosphates are found in traces. *Cain* has also reported the presence of carotenoids, fatty acids and vitamin C.

Enzymes of Golgi Complex

Various enzymes in the golgi complex have been reported by *Rothschild* (1963) and others. The important ones are ADPase, Mg^{++} ATPase, CTPase (cytidine triphosphatase), thiamine pyrophosphatase, acid phosphatase, UDP-N-acetylglycosamine transferase, galactosyl transferase and glucose-6-phosphatase. Glucose-6-phosphatase, a marker of the endoplasmic reticulum, occurs in low concentrations while others occurs in high concentrations.

Functions of Golgi Complex

There has been a great controversy regarding the functions of golgi complex. However, its secretory, storage and biogenetic functions have been suggested by many workers. The various functions of golgi complex are as follows:

(1) *Formation of secretory vesicles or primary lysosomes:* The primary function of the golgi complex is the formation of secretory vesicles or primary lysosomes. Golgi membranes are involved in this function.

(2) *Intracellular transport*: It has been shown that a portion of the protein, synthesized by the rough ER, finds its way through the lumen of ER into the golgi apparatus. The sequence of protein migration can be shown by the use tritium-labelled aminoacids.

The transport of proteins shows the following sequence :

Ribosomes → endoplasmic reticulum → golgi complex → zymogen granules → lumen.

(3) *Formation of the acrosome*: M.H. Burgos, and D.W. Fawcett (1955) discovered that golgi complex is related to the formation of acrosome during the maturation of the sperm (*spermatoleosis*).

(4) *Production of hormones*: Cowdry has established that the golgi complex helps the endocrine cells in the secretion of hormones. He observed when the golgi complex of endocrine cells were injured, the secretion of the hormones was ceased.

(5) *Cell-plate formation during mitosis:* Northcote (1968, 1969) has shown that during mitosis, material for cell plate formation is carried by vesicles of golgi complex.

(6) *Storage of proteins and lipids* : The main function of the golgi complex is the storage of proteins and lipids. The proteins are glycoproteins which are first stored in the cavity of ER and later on transmitted to the golgi complex where they are stored for the longer time.

(7) *Activation of mitochondria* : Golgi complex activates the mitochondria to produce the ATP which is utilized in various mechanical, physiological and synthetic processes.

(8) *Synthesis of carbohydrates* : Pierre Favard and Juniper and Roberts (1966) demonstrated that the golgi complex is concerned with the synthesis and transport of polysaccharides.

(9) *Regeneration of membranes* : Golgi complex also regenerates the membrane system

The Cell

throughout the cell. These membranes absorb various materials for the synthetic functions of the cell.

(10) To *Check the oxidation of synthetic materials* : Vitamin C is found associated in good concentration within the golgi complex. Therefore, it is suggested that this storage of Vitamin C is due to the fact that the golgi material provides the cytoplasm a segregated area of high reducing capacity that checks the oxidation of synthetic materials during increased metabolism.

(11) *Secretion of enzymes and hormones* : De Robertis and *Sabatini* (1960) have confirmed the relationship between the golgi complex and secretion, which was postulated by *Cajal* (1914). Here, the sacs of golgi complex are related to the ER and to the secretion of droplet. These droplets may be of lipids, bile, enzymes, hormones, yolk etc. *Bown* has suggested that the golgi complex are the centres of enzyme formation.

(12) *Modification of proteins* : Golgi complex is concerned the modification of proteins by the addition of polysaccharides and prosthetic groups.

Dictyosomes

Dictysomes are the structures similar to golgi complex. Due to similarities in between the two structures, most of the cytologists have used the two terms (Golgi complex and dictyosome) as synonymous. *Ambrose and Easty* (1970) used the term dictyosome strictly and only in case of plants. According to them, structures similar to those of golgi complex which have been observed in plant cells are known as *dictyosomes*. The dictyosomes are generally found rendomly distributed in the cytoplasm. *De Robertis, Nowinski and Saez* (1970) used the term dictyosome in case of plant cells and invertebrate tissues.

The dictyosomes are dispersed throughout the cytoplasm without any definite polarization. A single dictyosome has a plate like arched shape and the size is similar to that of mitochondrion or may be smaller. The ultrastrucuture of the dictyosome is typical. It consists of a stack of flattened vasicles (*cisternae*) surrounded by the vesicles. The cisternae are slightly dilated at the edges and the vesicles are formed by localized dilatations. The vesicles are probably produced by the activity of dictyosomes.

Dictyosomes contain some specific enzymes such as *thiamine pyrophopsphate* and *inosinic-disphosphatase*.

Dictyosomes and associated vesicles are involved in the synthesis of mucilage which is produced at the expense of the starch bodies present in plastids (*Northcote* and *Pickett Heaps*, 1966). They also help in the formation of cell plate during cell-division and secretion of substances.

Lysosomes

The lysosomes (lyso = digestive; soma = body) were described as a new group of cytoplasmic particles from the liver tissue by *de Duve* (1949). For long time they were called as 'Pericanalicular dense bodies', suggesting their location but not their function. It was *Christian de Duve* who in 1955 renamed these bodies as *lysosomes* because they contain digestive enzymes. This name is based upon their functional activity.

(1) *Definition:* The lysosomes are generally defined as tiny sac-like granules containing enzymes in solution for intracellular digestion and surrounded by a membrane which is impervious to several substances, particularly to the substances of the enzymes.

(2) *Occurrence and Distribution*: Lysosomes occur in animal cells and a few plant cells. They abundant in the cells of those organs which are related with the enzymatic actions and secretions. They are found in the cells of pancreas, liver, spleen, brain, thyroid and kidney etc. W.B.C., some protozoans and the meristematic cells of the plants also contain lysosomes. The lysosomes often remain distributed in the cytoplasm. *Matile* (1965) described them from the young seedling of *Zea maize* and tobacco. They have also been reported from the aleurone cells of wheat, germinating onion and pollen tubes etc.

(3) *Morphology*: Lysosomes are polymorphic and heterogeneous. They do not have any typical shape and size. The lysosomes are generally spherical bodies but may be irregular in shape as in certain root meristems. *C. de Duve* (1965) reported that the size of lysosomal particles ranges from 0.2 to 0.8 μ. In the cells of mammalian kidney, they are about 5 μ.

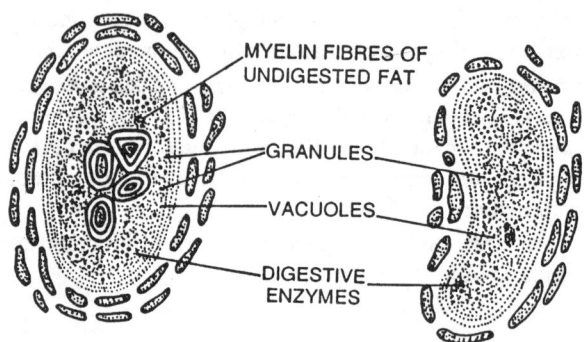

Fig. 1.26: Structure of Lysosomes.

(4) *Ultrastructure*: The ultrastructure of lysosome shows that it is basically a vacuole containing a high concentration of various enzymes which are used in digestive processes inside the cell. Each lysosome is surrounded by a single outer unit membrane of *Robertson* (1959). The internal organisation is quite variable due to their different functional activities. Some lysosomes are uniformly solid, others have a very dense outer zone with a less dense core and still others possess cavities (or vacuoles) with granular material.

(5) *Chemical composition*: Chemically, the unit membrane of lysosomes is made up of lipoproteins. The ratio of the lipids and proteins may be 1:1 or it may vary. The unit membrane encloses a number of enzymes. The dense material of lysosomes is rich in acid phosphatases (tissue dissolving enzymes). The other enzymes are arylsulphatase A and B, acid lipase, phospholipase A, phosphatidic acid phosphatase, hyaluronidase, amino peptidase A, dextranase, saccharase and lysozyme, Mg^{++} activated ATPase, indoxylacetate, esterase and plasminogen activator. The membrane of the lysosome can be ruptured easily by the progesterone, endotoxins, vitamin A and E, proteaes, lecithinase, X-radiation, ultraviolet irradiation and bile salts. In living cells the lysosomes are stable and this property is due to certain substances present in the membrane of lysosome e.g., cholesterol, cortisone, cortisol and chloroquine etc.

(6) *Polymorphism of the lysosome*: When a cell organelle possesses many shapes, the phenomenon is known as *polymorphism*. It has been reported in lysosomes. The lysosome particles in different cell types or even within a single cell are of many types. Their shape varies due to their contents at various stages of digestion. The recent literature regarding the polymorphism of lysosome suggests that it is due to association of ribosomes with the materials phagocytized by the cell. It may also be due to association of primary lysosomes with the different materials that are phagocytized by the cell.

Following four types of lysosomes have been reported from different cells or even within a single cell.

(1) **Primary lysosome :** Primary lysosomes are also known as storage granules. They are small sac like bodies containing many enzymes. These enzymatic contents are synthesized by the ribosomes and accomulated in the endoplasmic reticulum. From there it pentrates into the golgi region.

Primary lysosomes originate either from the endoplasmic reticulum (*Novikoff*, 1965) or from the cisterane of golgi complex.

(2) **Secondary lysosome :** The secondary lysosomes are also known as *heterophagosomes* or *digestive vacuoles*. They enclose the foreign material and the enzymes within the membrane.

The Cell

They are formed by the fusion of primary lysosome and pino- or phagosome. The cell membrane first invaginates and then buds off enternally to form a vesicle which encloses the droplet

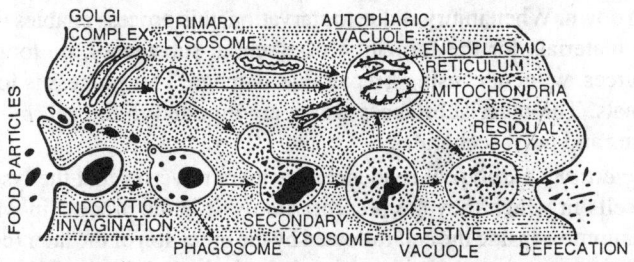

Fig. 1.27: Digestion of foreign particles and intracellular substances by a Lysosome.

of medium in the cytoplasm. This process is known as *Pinocytosis* and the vesicle so formed is called Pinosome or *Pinocytotic vesice*. When the solid particles are ingested through this way, the process is known as *phagocytosis* and the vacuole thus formed is called *phagosome*.

When the pinosome of phagosome contacts a primary lysosome, the membranes of the two fuse and form secondary lysosome (digestive vacuole). Here, the large particles of pinosome or phagosome are progressively digested by the hydrolytic enzymes and the products diffuse through the membrane into the cell.

(3) **Residual body :** When the digested food of the secondary lysosome diffuses into the cytoplasm, the vacuole is left with indigestible material only inside it. This is now known as *residual body*. Thus, the residual bodies are formed if the digestion is incomplete. In some cells, e.g., amoeba and other protozoa, these residual bodies are eliminated by defecation (a process reverse of pino- or phagocytosis). In other cells they may remain for a long time and play an important role in the aging process.

(4) **Autophagic vacuole :** The autophagic vacuoles are also known as cytolysosomes or autophagosomes. These are those lysosomes which contain certain cell organelles like mitochondrion or portion of endoplasmic reticulum. Cell injury involves autodigestion or autophagy. It is the process in which the cell feeds on its intracellular organelles. The autophagic vacuoles are formed in certain physiologic and pathologic processes. *de Duve* (1957) and *Allison* (1967) have observed that during starvation of the organism many authophagic vacuoles develop in the liver cells which feed on the cellular components.

(7) **Origin of lysosomes:** The exact origin of lysosome has not yet been clearly known. It has been suggested that they have a multiple origin, depending upon the tissue in which they are located or their function in a cell. Some workers are of the opinion that the lysosomes have originated from the pinocytotic vacuoles which in their turn originated from the plasma membrane. Thus, lysosomes show *extracellular origin*.

Novikoff (1965) has shown the origin of lysosomes directly from granular endoplasmic reticulum.

There are also evidences to suggest that it has originated from the golgi complex.

(8) **Functions of lysosomes:** Lysosomes perform following important functions :

(*i*) *Digestion of external particles* (*lysosomal digestion*): Foreign proteins and other larger particles of food are digested by lysosomes. The cell engulfs the particles by forming an invagination which becomes pinched off from the cell membrane in the form of food vacuole. Depending upon the physical state of food, the process is known as *pinocytosis* and *phagocytosis* and collectively as *endocytosis* and the food vacuoles as *pinosomes* and *phagosomes*. These vacuoles migrate towards the interior of the cytoplasm where they fuse with the lysosomes. The contents of the two bodies are mixed together so that the enzymes of lysosome come in contact with the ingested molecules and digest them.

(*ii*) *Digestion of intracellular substances (Autophagy)* : Cell injury involves another aspect of lysosomal function known as *autophagy* (autodigestion) in which some of the cell-organelles like mitochondria, ribosomes or endoplasmic reticulum are engulfed by the lysosome and broken down. When injury is due to starvation, this process enables the cell to use some of its own materials for the formation of essential substances, no longer available from outside sources, without causing irrepairable damage. This process is found in protozoans and mammals. The other substances which are digested in the cell by the lysosomes are proteins, fats and polysaccharides.

(*iii*) *Cellular digestion (Autolysis)*: Evidences are there to verify that the lysosomes can digest the entire cells in which they are present. *Autolysis* is the process in which the lysosome membrane ruptures inside the cell followed by the digestion of the later (cell) by the released enzymes. This process occurs either in pathological conditions or when the cells die. The cells are continuously being replaced, and the activity of lysosomes in them provides a mechanism by which the dead cells are removed. One of the most striking example is the regression of the tadpole tail during the later stages of tadpole's development.

de Duve (1963) coined the term 'suicide bag' for lysosome as it releases its complete enzymes within a normal cell and this results in autolysis or cell death.

(*iv*) *Extracellular digestion*: A cell can discharge lysosomal enzymes outside to destroy the surrounding structures, perhaps by reverse phagocytosis. The enzymes from a lysosome are released outside the cell where they digest the structure concerned. Thus, it has been suggested that lysosomes play a part in the fertilization of ova. Similarly, the cartilage and bone cells are digested by the chondrioblast cells and osteoblast cells respectively by the process of extracellular digestion.

(*v*) *As trigger of cell-division*: Allison (1967) reported that the dividing cells have comparatively fewer lysosomes and that too lie on the periphery of the cell instead of near the nucleus. This suggests that the break down of lysosomes may act as a trigger for mitosis in cells.

(*vi*) *As causative agent of certain diseases*: There are number of evidences that the lysosomes are involved in causing some diseases. *Allison and Mollucci* (1964) suggested that the release of lysosomal enzymes may be involved in carcinogenesis (the changing of normal cells to cancer cells). They have also shown that three of the most effective co-carcinogens (substances which promote the effect of carcinogens) are croton oil, a certain class of detergent, and a high oxygen level which affect the permeability of lysosomal membranes.

According to *Allison* (1967), the chromosomal changes, because of liberation of enzyme DNAase by the lysosomal rupture, may help to explain the origin of certain type of cancers.

Microbodies

(1) *Glyoxysomes*: Glyoxysome particles were first discovered by Beevers in 1961 in crude mitochondrial preparations from castor bean endosperm. The sucrose density gradient of these particles in 1.25. These are minute spherical particles having a diameter of 0.5 to 1.0 μm. They remain enclosed by a single unit membrane. The membrane encloses a fine granular ground substance called *stroma* which contains two chief enzymes of glyoxalate cycle viz., *isocitrate lyase* and *malate synthetase*. Besides these, other enzymes of glyoxalate cycle like citrate synthetase, aconitase, malate dehydrogenase, glycolate oxidase and catalase are also found. Enzymes like succinic dehydrogenase, fumarase and NADH oxidase are not found in them.

Glyoxysomes are found only in those plant tissues where conversion of fat into carbohydrate takes place or where glyoxylate cycle is found, e.g., seedings of peanuts and watermelon.

(2) *Peroxisomes*: They were first isolated from liver cells by *Beaufay* and *Berther* and (1963). Later on, in plant kingdom they were discovered by Tolbert and his coworkers (1968) from spinach

leaves. Since then they have been reported frequently from a number of plants that show photorespiration and are responsible for metabolism of glycolic acid, e.g., tea leaves, Phleum proteuse, sun flower, Pea, wheat and tobacco etc. A close relationship between chloroplast and peroxisome has been suggested by *Frederick* and *Newcomb* (1969).

The number of peroxisomes is approximately five times to that of lysosomes. Peroxisomes are minute, somewhat spherical, single unit membraned particles with a diameter varying from 0.5 to 1.0 μ. The stroma inside them is quite dense. *Tolbert et al.* (1968) discovered the different enzymes present in them. The main enzymes of peroxisomes are glycolate oxidase, glyoxylate reductase, calalase, glyoxylate transminase and malate dehydrogenase.

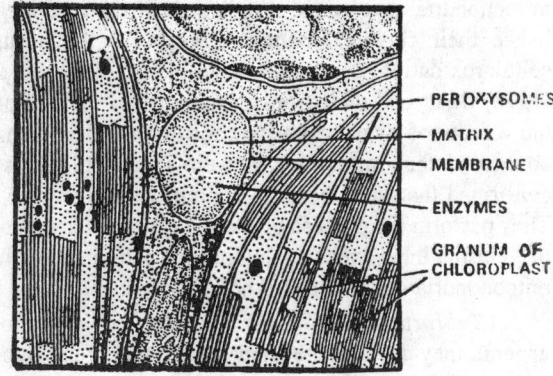

Fig. 1.28: Peroxisomes in a portion of *Phleum proteuse* leaf as seen under electron microscope.

The chief function of peroxisomes is to help in the metabolism of glycolate substance synthesized by chloroplast during photosynthesis. Peroxisomes have the ability to reduce oxygen to water by a two step mechanism involing H_2O_2 as an intermediate. Like lysosomes, peroxisomes also originate from endoplasmic reticulum (ER).

(3) *Sphaerosome*: They are minute spherical particles which were first described by *Perner* (1953). *Semadeni* (1967) isolated them from the seedings of maize. They possess high hydrolase activity and a relative density about 1.105 or 1.138 gm cm^3. They are single unit membraned structures originated from endoplasmic reticulum. Sphaerosomes contain enzymes like acid hydrolase, acid ribonuclease, acid phosphatase and acid asterase but lack the important enzymes of lysosomes, e.g., β-Glucuronidase, phosphatidase-C, lipase, and aryl sulphate A and B. Sphaerosomes are found only in plant cells and are similar to lysosomes found in animal cells but both differ on the basis of number of enzymes found in them. The chief functions of sphaerosomes are the synthesis, storage and translocation of fat.

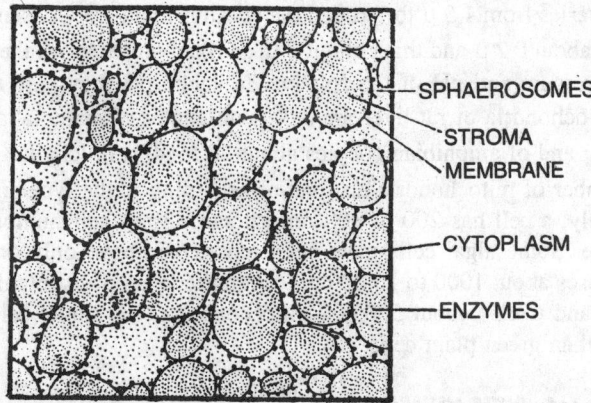

Fig. 1.29: Sphaerosomes in the cells of groundnut under electron microscope.

Mitochondria

Mitochondria (Gr: mito = thread; chondrion = granule) (Singular: mitochondrion) are most essential cell-organelles found distributed in the cytoplasm of plant and animal cells and protozoa. However, they are not found in bacteria and red blood cells of multicellular animals. Mitochondria are characterized by a number of morphological, biochemical and functional properties like, size, shape, special staining properties, the specific structural organization, lipoprotein composition, and presence of certain specific enzymes and coenzymes.

From the physiological viewpoint, mitochondria are known as "*Biochemical machines*" because they recover the energy contained in the food stuffs through Kerb's Cycle and the respiratory chain. The energy is produced in the form of ATP (adenosine-triphosphate) by the process of phosphorylation. ATP possesses the high energy phosphate bond. Most commonly mitochondria are also known as *Power-House* of the cell because they produce the energy necessary for many cellular functions.

(1) *History* : Mitochondria were first observed by *Kollicker* (1850). *Flemming* (1892) named them 'fila'. Later on *Altmann* (1894) described them as bioblasts. *Benda* (1897) named them as mitochondria. *Michaelis* (1900) stained them supravitally with janus green. *Regard* (1908) established their chemical nature. *Kingsbury* (1912) suggested that mitochondria were the sites of cellularoxidation. *Lewis and Lewis* (1914) observed their sensitivity to metabolic conditions. *Meves* (1918) described their transformation into various other type of cells. *Keilin* (1923) demonstrated the effects of cyanide and carbon monoxide on respiration. *Hogeboom, et al.* in (1948) finally confirmed that mitochondria were indeed the sites of cellular respiration. Very recently it has confirmed that mitochondria contain a specific type of DNA different from that of nuclear DNA. They perform the process of protein synthesis in them and also take part in the genetical phenomenon of "Plasma inheritance". This plasma-inheritance where the characters are transmitted through the mitochondria of female, is known as *mitochondrial inheritance*.

(2) *Morphology–Shape*: The shape of mitochondria varies according to the type of cell. In general, they are sausage shaped or filamentous. Depending upon the physiological conditions they may be club-shaped, tennis recket-shaped, rod-shaped or vescular in shape. Such shapes are only for a short period and after 48 hours these changes cease and the mitochondria regain their original form.

Size : The size of mitochondria is also variable and chiefly depends upon the functional stage of the cell. The length of mitochondria varies from 1.5 μ to 7 μ and the width is relatively constant being 0.5 μ. Very thin mitochondria are about 0.2 μ and thick about 2 μ. In fixed preparations their size and shape depend upon osmotic pressure and pH of the fixative. In acidic pH they tend to fragment and to become vescicular. Mitochondria of rat-liver are 3.3. μ in length; of mammalian exocrine pancrease about 10 μ in length; and of amphibian cocytes about 20 to 40 μ in length.

Number: Like shape and size, number of mitochondria also varies according to the type and functional activities of the cell. Generally, a cell has 200 to 300 mitochondria but their number may reach from a few to 1000 or more. Some algal cells have been reported to have only one mitochondria. A normal liver cell possesses about 1000 to 16,000 mitochondria; eggs of sea urchin 14,000 to 150,000; ovocytes 300,000; and *Chaos* about 500,000 mitochondria. The animal cells posses greater number of mitochondria than green plant cells.

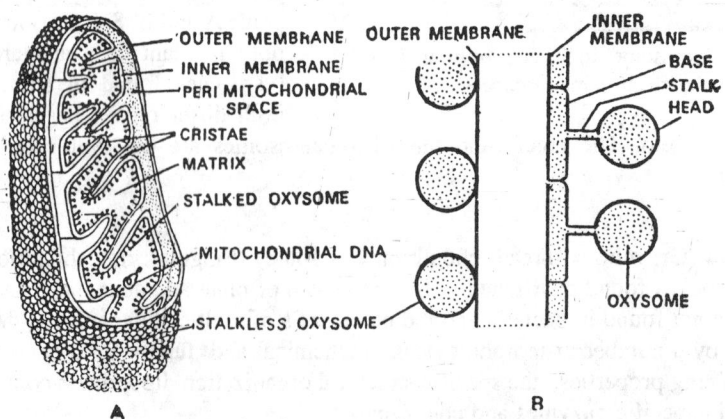

Fig. 1.30: (A) Three-dimensional structure of a mitochondrion; (B) Enlarged view of a part of mitochondrion with stalkless and stalked Oxysomes.

The Cell

(3) *Location*: In general the mitochondria are found uniformly distributed in the cytoplasm. However, they may exceptionally found around the nucleus and towards the periphery of the cytoplasm. Such exceptions are frequent in pathological conditions but sometimes may be overloading with inclusions like glycogen and fat.

Mitochondria are usually observed in large number in that area of the cell where the metabolic activities are at their maximum. During mitosis, they are found concentrated near the spindle. Mitochondria either move freely in the cytoplasm or are permanently placed near the regions of the cell which need more energy.

(4) *Orientation*: Mitochondria possess a more of less definite orientation in some cells. For example, in cylindrical cells, their orientation is in the basal-apical direction, parallel to the main axis. In leukocytes, their arrangement is redical with respect to the centrioles. The orientations of mitochondria depend upon the direction of the diffusion currents within cells (*Pollister*, 1941) and are also related to the sub-microscopic organisation of the cytoplasmic matrix and vacuolar system.

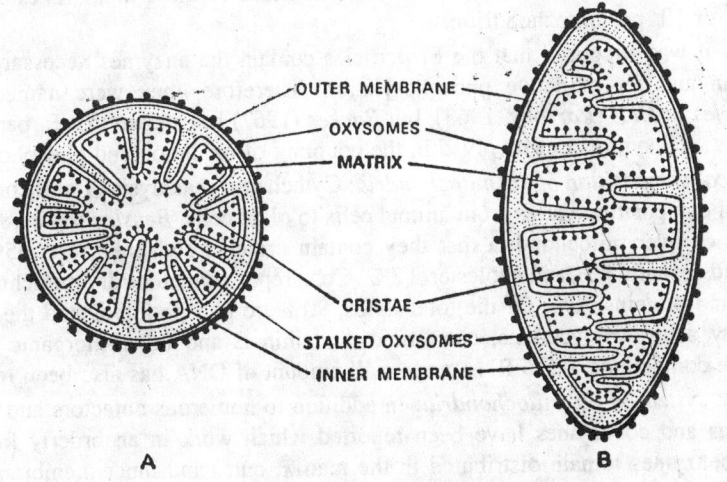

Fig. 1.31: Mitochondrion. (A) Transverse section, (B) Longitudinal section.

(5) *Ultrastructure*: Ultrastructure of the mitochondria reveals that they consist of two membranes (out and inner) and two compartments. The larger compartment contains the mitochondrial matrix. Each mitochondrion is surrounded by an outer limiting membrane of about 60Å thickness. It encloses an inner membrane of about same thickness. The inner membrane projects into mitochondrial cavity certain complex infoldings called *mitochondrial crests* or *cristae*. The space between outer and inner membrane is about 60 to 80Å. The inner membrane divides the mitochondrion into two chambers :

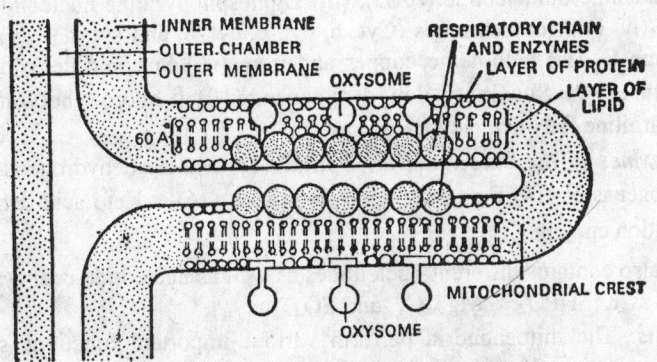

Fig 1.32: Molecular organisation of Cristae.

(i) *The outer chamber*: it is situated in between the two membranes ;

(ii) *The inner chamber*: it is bounded by the inner membrane. The inner chamber is filled with a relatively dense and homogenous material called **mitochondrial** matrix. Small and highly dense granules have also been observed in certain mitochondrial matrix. These granules are considered as sites for binding bivalent cations, particularly Mg^{++} and Ca^{++}. The mitochondrial cristae are in the form of incomplete septa or ridges which do not interrupt the continuity of the inner membrane. Thus, within mitochondrion its matrix remains continuous.

Further studies by electron microscope have shown that mitochondrial membranes (outer and inner) are trilamellar and about 60Å in thickness. The two extreme outer layers are of proteins and about 20-25Å in thickness. In between the protein layers is a lipid layer of about 25Å thickness. More recently, by the use of negative staining techniques it has been shown that the inner membrane and the crests are covered by particles of about 80 to 100Å. These particles have a stem like structure which links them with the membrane. The stemed particles are known as *elmentary* or F_1 *particles*. They are placed at the interval of 100Å on the inner surface of these membranes. Their number varies from 10^4 or 10^5 per mitochondrion.

Previously it was supposed that the F_1 particles contain the enzymes necessary for electron transport system and for oxidative phosphorylation. Therefore, they were named as *electron transport particles* or *ETP* (Parsons, 1963), but Racker (1967) has shown that F_1 particles contain special ATPase (ATP synthetase) involved in the coupling of oxidation and phosphorylation.

(6) *Chemical composition of the mitochondria*: Cytochemical analysis of mitochondria reveals that its chemical composition varies from animal cells to plant cells. *Bensley* reported from the dry weight analysis of liver mitochondria that they contain proteins and unknowns 65%, glycerides 29%, lecithin and cephalin 4%, and cholesterol 2%. Cohn reported that the dry mitochondria contain lipids 25-30% and proteins 70%. Of the total lipids, 90% are phospholipids and the rest 10% are cholesterol, fatty acids, triglycerides, carotenoids, vitamin E and other inorganic materials. In addition to these compounds, 0.5% RNA and small amount of DNA has also been reported.

(7) *Enzyme system of the mitochondria*: In addition to numerous cofactors and metals, more than 70 enzymes and coenzymes have been reported which work in an orderly fashion. These enzymes and coenzymes remain distributed in the matrix, outer and inner membrane, and in the space between outer and inner membrane. A.L. Lehninger (1969) reported the following list of important mitochondrial enzymes.

(A) *The enzymes of outer mitochondrial membrane*: (1) Monoamine oxidase, (2) Rotenone-insensitive NADH-cytochrome reductase, (3) Kymurenine hydroxylase, (4) Fatty acid CoA ligase.

(B) *Enzymes of the space between outer and inner membranes (enzymes of outer mitochondrial chamber):* (5) Adenylate kinase, (6) Nucleoside diphosphokinase.

(C) *The enzymes of inner mitochondrial membrane*: (7) Respiratory chain enzymes (enzymes of electron transport pathways) viz., (*i*) Nicotinamide adenine dinucleotide (NAD). (*ii*) Flavin adenine dinucleotide (FAD), (*iii*) Diphosphopyridine nucleotide (DPN) dehydrogenase, (*iv*) Four cytochromes (Cyt. b, cyt. c, cyt. a, and cyt. a3), (*v*) Ubiquinone Q or coenzyme Q, (*vi*) Non-heme copper and iron, (*vii*) Succinic dehydrogenase, 98) ATP synthetase, (9) Succinate dehydrogenase, (10) β-hydroxybutyrate dehydrogenase, (11) Carnitine fatty acid acyltransferase.

(D) *The enzymes of the mitochondrial m*atrix: (12) Malate dehydrogenases (13) Isocitrate dehydrogenases, (14) Fumarase, (15) Aconitase, (16) α-Keto acid dehydrogenases, (17) β-oxidation enzymes.

The matrix also contains different nucleotides as well as nucleotide-coenzymes and inorganic electrolytes, such as K^+, HPO_4^{++}, Mg^-, Cl^- and SO_4^-.

(8) *Functions:* The mitochondria perform various important functions such as oxidation, dehydrogenation, oxidative phosphorylation and electron transfer which are closely related with the

structural organisation and chiefly depend upon the enzymes present in the mitochondria. The main function of mitochondria is to produce and store the energy in the form of ATP and to release it for various important functions. The release of energy takes place as a result of biological oxidation of food stuffs like carbohydrates and fats. After complete oxidation the different materials are converted into CO_2 and H_2O. Oxygen is utilized during oxidation and hydrogen is released. The released hydrogen is broken down into electrons and protons. The electrons are passed through respiratory chain components (cytochromes) for the conversion of adenosine diphosphate (ADP) into adenosine triphosphate (ATP).

Oxidation of Carbohydrates

All the green plants possess chloroplasts which contain chlorophyll in them. It absorbs quanta of red light and converts the CO_2 and H_2O into carbohydrates (glucose). These are stored in the form of starch molecules. At the time of energy requirement, the cells utilize the enzymes of cytoplasm and mitochondria and break the glucose molecules. The process involves a number of steps. The steps occurring in the cytoplasm comprise glycolysis and those occurring in the mitochondria comprise the Kreb's cycle (in the outer part) and oxidative phosphorylation (in the inner part). Various pathways through which oxidation is completed are as follows:

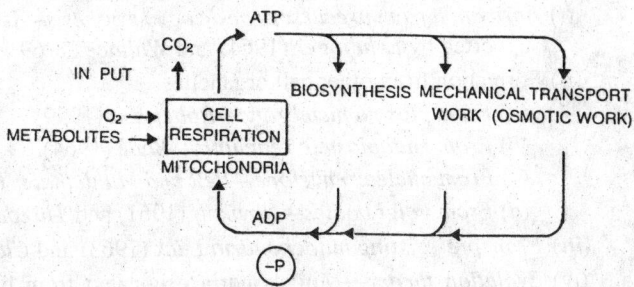

Fig. 1.33: Generation of ATP and energy transformation.

(i) Glycolysis or Embden-Meyerhof Paranas Pathways.

(ii) Oxidative decarboxylation.

(iii) Kreb's cycle or Citric acid cycle.

(iv) Respiratory chain.

(v) Oxidative phosphorylation.

(9) *Mitochondrial DNA*: The presence of mitochondrial DNA (*m*-DNA) was reported by *M.M.K. Nass* and *S. Nass* in (1963) and it was confirmed later on by *Kleinschmidt*. A single mitochondria may contain one or more molecules of DNA according to its size. The common length which has been reported from many mitochondria is 5 μ. The molecules of DNA are highly twisted, double stranded and circular in shape, a feature similar to bacterial DNA. The mitochondrial DNA (m-DNA) differs from nuclear DNA (*n*-DNA) in following points:

(i) Mitochondrial DNA is circular in shape whereas *n*-DNA is ladder like.

(ii) Mitochondrial DNA contains more guanine and cytosine (G-C) contents then the *n*-DNA (Rabinowitz, 1968).

(iii) M-DNA has higher denaturation temperature than the *n*-DNA.

(iv) The m-DNA is shorter and contains few coded informations than the *n*-DNA.

(v) DNA polymerase of the mitochondria is also different from the nuclear-DNA polymerase.

Recently, it has been established that the mitochondria have the ability to conduct protein synthesis. They have the enzyme DNA polymerase for the self duplication of *m*-DNA. During duplication, the *m*-DNA behaves as if it is a mitochondrial chromosome. *Suzam* and *Boawer* (1966) and *Breidenbach et al.* (1967) assumed that the mitochondria are self-replicating systems and show continuity through generations as they contain DNA and RNA in them. Thus, they play an important role in cytoplasmic inheritance and particularly in mitochondrial inheritance. *Nass* (1969) reported that *m*-DNA codes for some structural proteins of the mitochondria and the structural genes for the *m*-enzymes are essentially localized on the nuclear genome.

(10) *Mitochondrial ribosomes* : Only in recent years, it has been established that isolated mitochondria have ribosomes and may synthesize proteins from amino-acids. The mitochondrial ribosomes are smaller than cytoplasmic ribosomes and are almost similar to bacterial ribosomes. Protein synthesis in mitochondria is inhibited by chloramphenicol, while the *cytoplasmic* protein synthesis is not affected. The ribosomes of mitochondria are 73S.

(11) *Origin or biogenesis of mitochondria:* The origin of mitochondria has been discussed and reviewed by several workers (*Lehninger*, 1964). There has been a great controversy regarding its origin. Some of the important views are as follows :

   (*i*) *Formation from precursors or de novo synthesis*—It was proposed by *Novikoff* (1961) and supported by *Lehninger* (1964) and *Wallace* (1969).

   (*ii*) Formation from other cell organelles.

   (*a*) *From plasma membrane—Robertson* (1959).

   (*b*) *From endoplasmic reticulum—Bade* (1964).

   (*c*) *From nuclear envelope—Bell and Muhlethaler* (1962).

   (*d*) *From golgi bodies—Novikoff* (1961) and *Thread Gold* (1967).

   (*iii*) *From pre-existing mitochondria Luck* (1963) and *Claude* (1965).

   (*iv*) *Evolution theory*—Mitochondria originated from bacteria like organisms (*Nass*, 1965; *Sagar*, 1967).

   (*v*) *Symbiont theory—Sagar* (1967)

Plastids

Plastids are characteristic structures of almost all the plant cells and a few animal cells like *Euglena*. They represent the largest cytoplasmic cell organelles and are intimately related with the synthesis and storage of carbohydrates, proteins and lipids.

The 'plastid' is a common name given to a group of cytoplasmic organelles which possess naturally occurring coloured substances (pigments) or contain oils, proteins and carbohydrates as storage material. The term 'chromatophere' is often used for the plastid like colouring bodies of blue green algae, fungi and certain flagellate protozoans.

The information about shape, size, number, ultrastructure and chemical composition of plastids have been obtained through electronmicroscopic, cytochemical and histochemical studies.

(1) *Classification of plastids* : The plastids are generally classified into three main groups depending upon the presence or absence of colour (pigment) contents.

(A) *Leucoplasts* : These plastids are devoid of pigments. They are mainly meant for storing the food materials like starch (carbohydrates), lipids and proteins. The leucoplasts vary in shape

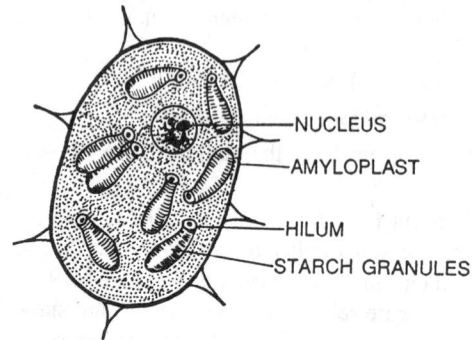

Fig. 1.34: Amyloplast containing starch grains.

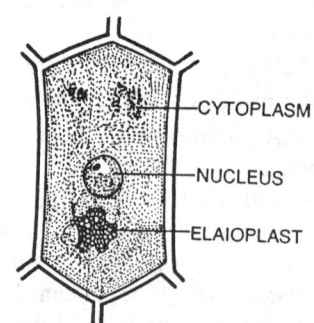

1.35: Elaioplast in the epidermal cell of Polyanthes tuberosa.

The Cell

and size and are usually rod-like or spheroid. They are of common occurrence in embryonic cells, sex cells and meristematic cells. Depending upon the presence of storage material, they are further classified into following heads :

   (i) *Amyloplasts* : The starch storing leucoplasts are known as *amyloplasts*. They are found in those cells which store the starch and particularly in storage tubers, cotyledons and endosperm.

   (ii) *Elaioplasts* : The oil storing leucoplasts are known as *elaioplasts*. They are found in the seeds of both monocotyledonous and dicotyledonous plants. It has been observed that in most of the mono cotyledonous plants the chloroplasts after losing their chlorophyll start storing oil. In the epidermal cells of Orchidaceae and Liliaceae the disorganised plastids fuse to form oil droplets.

   (iii) *Proteinoplasts* : The protein storing plastids are known as *proteinoplasts* (or aleurone-plasts). They have been reported in the epidermal cells of *Helleborus* and seeds of *Ricinus* and *Brazil nut*.

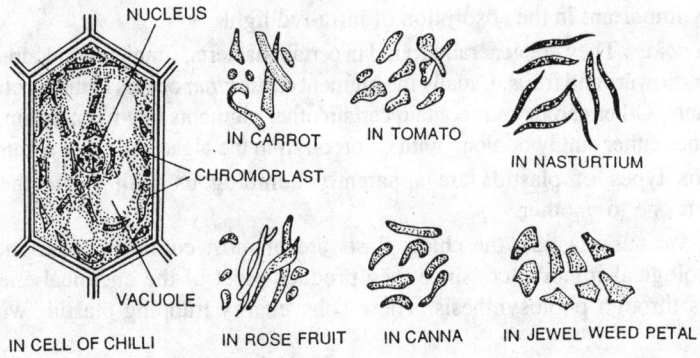

Fig. 1.36: Chromoplasts in plant cells.

(B) *Chromoplasts* : The coloured plastids are known as *Chromoplasts*. They contain variety of pigments and most of them synthesize food through the process of photosynthesis. The chromoplasts are found in the cells of leaves, many flowers and fruits. Some of the common chromoplasts of plant cells are as follows :

   (i) *Chloroplast* : The green plastids are known as *Chloroplasts*. They contain the pigments chlorophyll A and chlorophyll-B, DNA and RNA. The chloroplasts are found in green algae and higher plants where they play a role in the photosynthesis. *Rhoades* and *Carvalho* (1944) reported that the specialized chloroplasts of parenchymatous sheath cells of the maize leaf and other members of the sub-family Pani-coideae of the grasses do not function in photosynthesis.

   (ii) *Phaeoplast* : The brownish colour plastids are known as *phaeoplasts*. They are found in the brown algae (Phaeophy ceae), diatoms and dinoflagellates. These appear brown due to masking effect of brown carotenoids. In brown algae, it is the pigment *fucoxanthin* which is responsible for the absorption of light and transfer of energy to the chlorophyll.

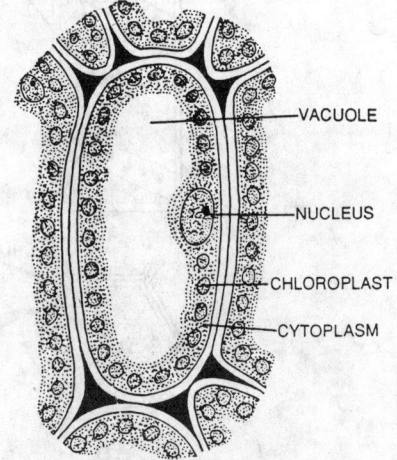

Fig. 1.37: Chloroplasts in a mesophyll cell of a leaf.

(*iii*) *Rhodoplasts* : They are red-colour plastids and contain the pigment **phycoerythrin.** This pigment plays the role similar to fucoxanthin. Rhodoplasts are found in the members of Rhodophyceae, the red algae.

(*C*) *Chromatophores* : In the cells of blue green algae, fungi and bacteria, the pigments are not organized within a discrete plastid body but are often arranged on lamellar structures in concentric rings or plates. As such here the term chromatophore is used instead of plastids. Various types of chromatophores are as follows :

(*i*) *Blue green chromatophores* : They are found in blue green algae which contain the pigments c-phycocyanin and c-phycoerythrin along with chlorophyll A and carotenoids. These accessory pigments impart bluish hue in addition to green colour and perform photosynthesis.

(*ii*) *Bacterial chromatophores* : Bacterial chromatophores are found in certain bacteria, e.g., purple sulphur bacteria, and contain the pigment *bacteriochlorophyll*. The green sulphur bacteria contain the pigment *bacterio-viridin* in their chromatophores. Bacterio-chlorophyll is very important in the absorption of infra-red light.

(*iii*) *Carotenoids* : They are generally found in certain bacteria, fungi, red-coloured ripe tomato and pepper, flowers and fruits. Usually the pigment *capsanthin* occurs in the carotenoids of bacteria and fungi. Other carotenoids contain certain other pigments like fucoxanthin, xanthophyll and carotenes either singly or along with chlorophyll in the algae and other plants.

The various types of plastids are apparently homologous with each other. They can be transformed from one to another.

Among of various plastids, the chloroplasts are of most common occurrence in plants and have greatest biological importance, since they produce most of the chemical energy used by the living organisms through photosynthesis. These solar energy trapping plastids will be considered here in detail.

Chloroplasts

(1) *Morphology*: The shape, size and number of chloroplasts may vary in different cells within a species, but they are relatively constant within cells of the same tissue.

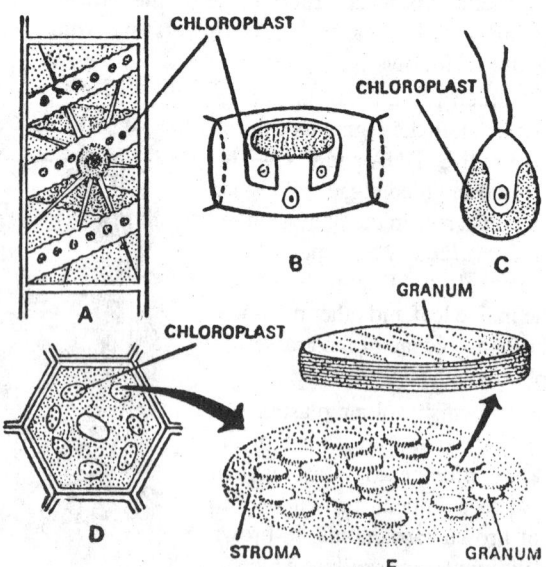

Fig. 1.38: Different shapes of chloroplasts. (A) Spiral in *Spirogyra*, (B) Collar-shaped in *Ulothrix*, (C) Cup-shaped in Chlamydomones, (D) Oval-shaped in a plant cell, (E) Enlarged view of a chloroplast.

The Cell

Shape : The various shapes of the chloroplast include vesicular, spheroid, ovoid, discoid, club and saucer. Most algal cells have single huge chloroplast which may be reticulate, spiral or stellate.

Size : The size of the chloroplasts is also variable. In higher plants, the average-sized chloroplasts are 4 to 6μ in diameter. The size of these plastids also depend upon the chromosome number and habitat of the plant. For example, the polyploid cells and the plants grown in the shade have larger chloroplasts than the diploid and sunny plants respectively.

Number : The number of chloroplasts is relatively constant in different plants. The algal cells usually have a single and huge chloroplast. The cells of the higher plants contain 20 to 40 chloroplasts.

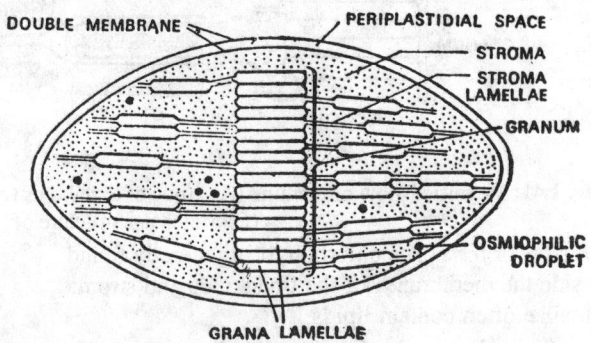

Fig. 1.39: Chloroplast in section (under Electron Microscope).

(2) *Ultrastructure of chloroplast*: Mature chloroplasts of higher plants have a complex structure. Each chloroplast is enclosed by two concentric unit membranes (outer and inner). Each membrane is lipoproteinous, trilamellar and about 50Å thick. These membranes are smooth, continuous and differentially permeable. The membranes enclose two distinct systems within it : (*i*) a granular matrix or stroma, composed of ribosomes and soluble proteins and (*ii*) A lamellar system which is differentiated in the midst of this matrix into (*a*) grana lamellae or thylakoids and (*b*) intergrana lamellae or stroma lamellae or fret.

1. *The matrix (= stroma)* : The outer and inner membranes of the chloroplast enclose the granular, transparent substance known as *matrix* or *stroma*. The lamellar system (grana lamellae and intergrana lamellae) remain embedded in the matrix. In addition to the lamellar system, granules, lipid droplets, starch grains, vesicles, ribosomes and proteins are also found in the matrix but not in all the cases. In the matrix of the algal cells eye spots and pyrenoids are often found.

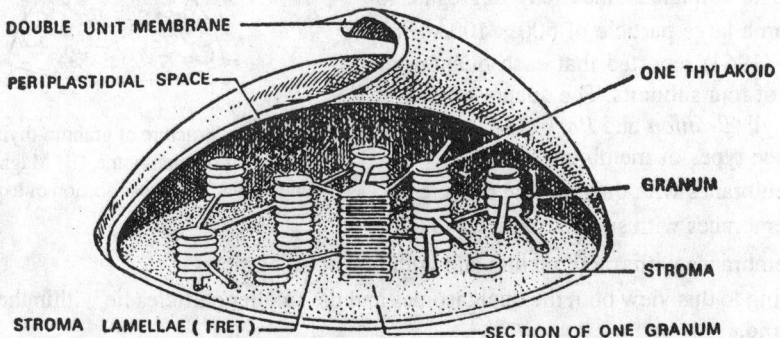

Fig. 1.40: Three-dimensional structure of chloroplast.

2. *The grana (= lamellar system)* : The photosynthetically active pigments are confined to the lameller system of the chloroplasts. In higher forms, the pigments are restricted to certain areas of the lamellae. These areas are usually found layered on top of each other, the stacks being called *grana*. The size of the grana may range from 0.3 to 1.7 μ and the number of grana per chloroplast may vary from 40 to 60.

Each granum of the chloroplast is formed by superimposed closed compartments called *thylakoids*. The number of thylakoids per granum may vary from a few to 50 or more. Each thylakoid remains separated from the matrix of the chloroplast by its unit membranes. The adjacent grana are interconnected by a network of flexuous, anastomsing tubules which join certain compartments but not others. The interconnecting membranes of the grana are known as *stroma lamellae* or *intergrana*

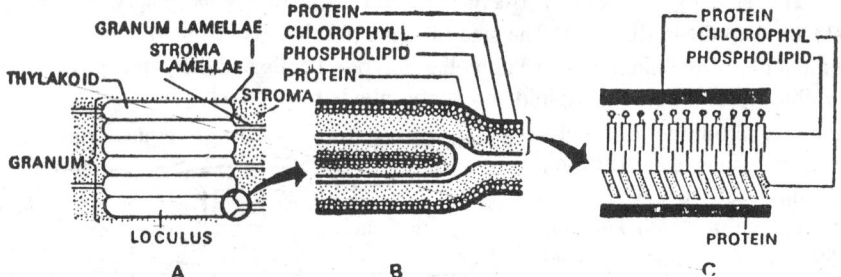

Fig. 1.41: Magnified view of granum. (A) group of thylakoids (granum). (B) Structure of granum-thylakoids (C) Molecular structure of lamellae.

lamellae or *frets*. The chlorophyll is generally found in side the membranes of the thylakoids and stroma lamellae often contain lipids.

The chloroplasts of the mesophyll cells typically possess grana while they are absent in the chloroplasts of cells associated with vascular elements.

(3) *The quantasome concept* : Park and *Pon* (1963) have shown that the membranes of the thylakoids have integral globular subunits. They are about 185Å long, 155Å wide and 100Å thick and usually randomly scattered. According to them, these isolated particles with a part of the membrane represent the smallest morphological photosynthetic unit called *quantasome*. Saucer and *Calvin* (1962) have shown that 3 to 6 quantasomes may aggregate together to form a large particle of 500 × 100Å. *Park* and *Beggins* (1964) reported that each quantasome is composed of four subunits. The quantasomes contain chlorophyll. *Branton* and *Park* (1967) observed following three types of membranes :

(*i*) Membranes with quantasome particles

(*ii*) Membranes with smaller particles (110Å)

(*iii*) Membranes with rough texture and with few or no particles

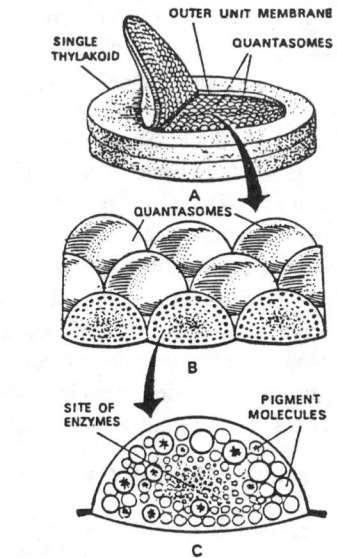

Fig 1.42. Structure of granum-thylakoid. (A) Quantasomes on membrane, (B) Magnified view of quantasomes, (C) Cross-section of a quantasome.

According to this view both the quantasomes and the smaller particles lie within the membrane of the thylakoid.

Howell and *Moundrianakis* (1967) gave the concept of quantasome as a photosynthetic unit. They suggested that the chloroplast is one of the most elaborated biochemical machine producing energy transformation at a molecular level.

(4) *Chemical composition*: The dry weight analysis of the chloroplasts of higher plants showed that chemically the chloroplasts are composed of following substances :

(*i*) Proteins	35 to 55%	(*ii*) Lipids	25 to 35%
(*iii*) Chlorophyll A and B	5 to 10%	(*iv*) Carotenoids	1 to 4.5%
(*v*) RNA	2 to 3%	(*vi*) DNA	0.5%
(*vii*) Vitamin E	0.089%	(*viii*) Vitamin K	0.004%
(*ix*) Carbohydrates	Variable	(*x*) Mg, Fe, Cu, Mn, Zn and P	Traces

The Cell

About 80 per cent of the protein is insoluble and intimately bound to lipids to form lipoproteins. The lipids are mainly phospholipids and contribute to the formation of lamellae and plastid wall. The lipid fraction comprises neutral fats, steroids, waxes and phospholipids. The common alcohols of the lipids are choline, inositol, glycerol, ethanolamine, and serine.

The chlorophyll was discovered by Willstatler (1915). The chlorophyll is an asymmetric molecule having a hydrophilic head made up of four pyrrolic nuclei located around a magnesium atom, and a long tail formed by a hydrophobic chain (phytol tail). The chlorophyll consists of 75% chlorophyll-A and 25 per cent chlorophyll-B.

The chemical formula of chlorophyll A is $C_{55}H_{72}O_5N_4Mg$ and chlorophyll B is $C_{55}H_{70}O_6N_4Mg$. The chlorophyll A contains methyl (= CH_3) group while chlorophyll B contains aldehyde (=CHO) group in place of $-CH_3$.

Carotenoids are hydrocarbons and may be of two types: (i) Carotenes, $C_{40}H_{56}$ and (ii) Xanthophylls = $C_{40}H_{56}O_2$. Carotenes are orange coloured while xanthophylls are yellow-coloured. Certain plants like *Spirogyra*, *Anthoceros* and *Selaginella* etc. contain pyrenoids in their chloroplasts which are made up of protein. They remain surrounded by starch grains.

(5) *DNA in chloroplasts* : A characteristic DNA occurs in the chloroplasts of algae and higher plants. *Ris* and *Plaut* (1962) reported the presence of fine filaments of DNA from the chloroplasts of *Chlamydomonas*. The presence of DNA has been confirmed in several other algae and higher plants. Segments of DNA as long as 150μ have been separated from the chloroplasts by *Woodcock* and *Fernandez-Morgan* (1968). DNA of the chloroplast resembles closely with the bacterial DNA and differs with the nuclear DNA. No circular DNA molecules have been found in chloroplasts. The algal chloroplast DNAs from several species have lower G-C content than the corresponding nuclear DNA but chloroplast DNAs from angiosperms possess higher G-C ratios than nuclear DNA. The DNA of the chloroplasts has been related to the presence of a special non-chromosomal genetic system of cytoplasmic heredity by *Rhoades* (1955).

(6) *Ribosomes in chloroplasts* : The ribosomes of chloroplasts are smaller than ribosomes of the cytoplasm and resemble bacterial ribosomes in size. Polysomes have also been separated from chloroplasts by Lyttleton (1962) and *Clark et al.* (1964). 70S ribosomes containing 23S and 16Sr-RNA were isolated from chloroplasts by *Stutz* and *Noll* (1967) and *Bager* and *Hamilton* (1967). They also contain the two characteristic sub-units. Chloroplast ribosomes of pea seedlings show a sedimentation coefficient of 62S that can dissociate into 46S and 32S subunits in contrast to 76S cytoplasmic ribosomes with 54S and 38S subunits. RNA of the chloroplast ribosomes is homologous to regions of chloroplast DNA. The isolated chloroplasts are able to synthesize DNA, RNA and proteins.

(7) *Functions of the Chloroplast*:

(1) *Photosynthesis* : The most important function of chloroplast is photosynthesis. During the process of photosynthesis, the chlorophyll contained in the chloroplasts traps the energy of sunlight emitted as photons (quanta) and transforms it into chemical energy. Because this process takes place inside the chloroplast, they are considered as the centres of photosynthetic activity. The chemical energy is stored in the chemical bonds that are produced during the synthesis of various food stuffs like carbohydrates, lipids and proteins.

The photosynthesis can be defined with the help of following equation :

$$nCO_2 + nH_2O \xrightarrow[\text{Chlorophyll}]{\text{light}} (CH_2O)_n + nO_2$$

The equation shows that photosynthesis is the process in which carbon dioxide and water are converted into carbohydrates in presence of light and chlorophyll. The O_2 is evolved during the process.

The photosynthesis consists of two stages, the light or photochemical reaction and a dark or thermochemical reaction. The photochemical reaction is light sensitive and thermochemical (dark reaction) reaction is temperature sensitive. The photochemical reaction takes place in the following steps :

(*i*) Photophosphorylation, in which ATP is formed from ADP through a chain of electron carriers.

(*ii*) Hydrolysis and ionization of water, in which NADP is reduced to $NADPH_2$.

In dark reaction CO_2 is fixed and reduced by thermochemical mechanisms. This mechanism also involves a number of reactions in which the conversions of one product into another takes place in presence of certain specific enzymes. The initial enzyme, carboxydismutase, is responsible for the formation of phosphoglyceric acid molecules from ribulose diphosphate and CO_2.

2. *Protein synthesis:* Brawerman and *Eisenstadt* (1968) reported the special ribosomes associated with chloroplasts and suggested that chloroplasts contain a specific protein synthesizing system. The chloroplasts contain sufficient amounts of mRNA of its own. In the presence of CO_2, as the sole source of carbon, chloroplasts actively incorporate amino acids into proteins.

3. *Cytoplasmic heredity* : The presence of DNA in chloroplasts has been related to the presence of a special non-chromosomal genetic system. Thus, the chloroplasts act as carrier for genetic material.

4. *Role in Kreb's cycle and fatty acid synthesis* : It is assumed that chloroplasts contain enzymes necessary for Kerb's cycle and for synthesis of fatty acids.

5. *Mutation* : Some of the plastids even chloroplasts, undergo mutation which is known as *plastid-mutation*. After mutation the plastids may perform altered functions. The genes of the plastids are known as *plastogenes*.

(11) *Ribosomes*: The present knowledge about ribosomes is the outcome of electron microscopy, ultracentrifugation and radioisotopic techniques. They are dumb-bell shaped or almost spherical, dense, minute particles having a diameter of 140 to 160Å. They are made up of ribonucleic acid (RNA) and protein, due to which they are also called **ribonucleoprotein granules** or particles (RNP granules = RNP paticles).

(1) *History:* Ribosomes were for the first time observed under electron microscope by *Palade* (1955). *Hanstein* (1890) named them as *sphaerosomes*, Claude (1940) as *microsomes* and *Hoflet* (1957) as *meiosomes*. Due to presence of excess RNA content in these particles, most of the scientists have supported the name *ribosome* which is being most widely used now a days. *Lakes* (1976) proposed asymmetrical model while *Stoffler* and *Wittmann* (1977) proposed quasi-symmetrical model for structure of ribosomes.

(2) *Occurrence*: Ribosomes occur in all the living cells. In eukaryotic cells, the ribosomes are found either freely distributed in the cytoplasm or attached to the surface of the endoplasmic reticulum. The ribosomes have also been reported in the chloroplasts, mitochondria and nucleolus. They are found abundantly in protein synthesizing cells.

(3) *Number and concentration*: The number and concentration of ribosomes are directly related to the RNA content of the cell and to the basophilic properties of the cytoplasm. Groups of ribosomes are found in those cells which contain ergastoplasm (basophilic substance) e.g., gland cells, plasma and liver cells, Nissl bodies of nerve cells, all rapidly growing plant and animal cells and bacteria. *De Vries* (1965) correlated the concentration of ribosome with the staining properties and RNA content of the cell.

In *E. coli* growing at maximum rate ribosomes account for 25 to 30 per cent of the cell mass and their number is about 20,000 to 30,000 per cell (*Watson*, 1970). However, if the rate of protein synthesis is slowed down by unfavourable nutritional conditions, the ribosomal content also drops considerably. A coli bacillus and yeast cells possess about 8,000 ribosomes.

(4) *Classes of Ribosomes*: Ribosomes are remarkably uniform in size, structure and composition in different cells from which they have been separated. On the basis of size and sedimentation coefficient (S) they have been classified into two main categories :

1. *70S Ribosomes* : They are comparatively smaller in size and are found in prokaryotic cells (bacteria), mitochondria and chloroplast of eukaryotic cells. The molecular weight of each 70S ribosome is about 2.7×10^6 daltons.

2. *80S Ribosomes* : They are found in eukaryotic cells. Each 80S ribosome possesses a molecular weight of about 4×10^6 daltons.

(5) *Structure and subunits of Ribosomes*: Electron microscopy indicates that ribosomes are composed of two rounded sub-units fitted together to give a complete unit of about 200Å in diameter. One sub-unit is approximately twice the size of other. *Huxley and Zubay* (1960) reported that in *E. coli* the larger subunit is dome-shaped (14 to 16Å) and smaller subunit forms a cap on the flat surface of the other. *Nanninga* (1968) reported that the larger subunit have a regular icosahedral structure. Its rounded appearance is due to collapse of structure during fixation.

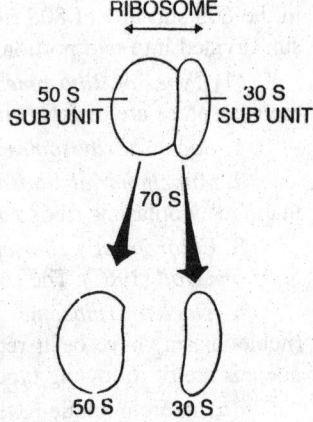

Fig. 1.43: Two sub-units of a 70S Ribosome.

The sedimentation coefficient of intact prokaryotic and eukaryotic ribosome is 70S and 80S respectively. The 70S ribosomes possess a larger subunit of 50S and smaller of 30S. The 80S ribosomes possess a larger subunit of 60S and cohesion smaller of 40S. Ribosomes require low concentration of Mg^{2+} (0.001M) for structural cohesion. If the Mg^{2+} concentration is increased ten fold, two ribosomes combine and form a structure known as *dimer*. It has the molecular weight twice of individual ribosomes. If the Mg^{2+} concentration is lowered (.000lM), the single intact ribosome dissociates into its sub-units. At high concentration of Mg^{2+} many ribosomes unite on a strand of *m*RNA and form a structure known as *Polyribosomes* (Polysomes).

Fig. 1.44: Formation of Ribosome, Dimer and Polyribosomes.

(6) *Ultrastructure*: Fine structure of the ribosome is very complex. The preparations with Uranyl ions show that each ribosome is star-shaped with 4 to 6 arms implanted on a dense axis.

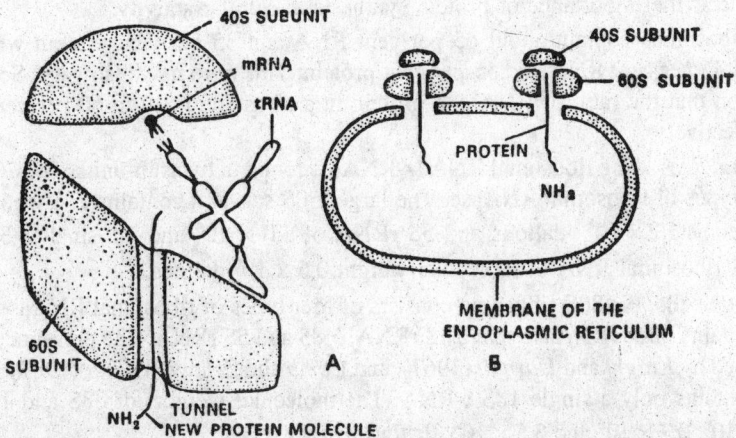

Fig. 1.45: Ultrastructure of Ribosome (A) Two subunits of 80S Ribosome and relationship with other RNA molecules, (B) Ribosomes on ER and entrance of polypeptide chain in ER.

Nanninga (1967) reported that the 50S sub-unit of *Bacillus subtilis* is pentagonal in shape and is a compact particle of 160 to 180Å. He also showed that the centre of this subunit possesses a rounded area of 40 to 60Å. *Florendo* (1968) has described an electron transparent core in the larger sub-unit. It has been confirmed that the 50S sub-unit has a kind of hole in the centre through which neither proteolytic enzymes nor ribonucleases may enter. The presence of such holes has also been reported in the 60S sub-unit of 80S ribosomes. The smaller sub-units possess irregular shape and tend to be sub-divided into two portions that are interconnected by a strand of about 30 to 60Å thickness.

(7) *Types of Ribosomes*: Depending upon the occurrence in the cytoplasm or its organelles, the ribosomes are of following types :

1. *Bacterial ribosomes* : They occur in bacteria and are of about 70S.

2. *Mitochondrial ribosomes* : They occur in mitochondria. They are very much smaller in size than the cytoplasmic ribosomes and are more similar to bacterial ribosomes. They are of about 73S.

3. *Chloroplast's ribosomes* : The presence of ribosomes in chlorpasts has been reported by *Stutz and Noll* (1967). The coefficient of these ribosomes is 67S. They may also synthesize proteins.

4. *Nuclear ribosomes* : They are persent in the nucleus. RNP granules in the nucleus (nucleoplasm) have been reported by *Simard* (1966) and *Bernhard* (1968). The granules in the nucleus are of following types :

(*a*) Interchromatic RNP fibrils : situated near the chromatin,

(*b*) Inter-chromatic RNP granules : They are about 250Å in diameter with the same localization.

(*c*) Perichromic RNP granules : They are about 450Å in diameter.

(*d*) RNP coiled bodies.

5. *Nucleolar ribosomes* : Fibrillar and granular RNP granules have been reported in the nucleolus. Most of the nucleoli possess following type of granules :

(*a*) Dense RNP granules of about 150 to 200Å in diameter.

(*b*) RNP fibrils of about 50 to 80Å length.

Both are precursors of the cytoplasmic ribosomes. The different aspects of nucleolar particles have been studied by *Marinozzi* (1963); *Perry* (1964) and *Simard* and *Bernhard* (1967).

6. *Cytoplasmic Ribosomes* : They are large-sized ribosomes of the cytoplasm. The largest ribosome has been reported in pea seedlings which is about 230Å.

(8) *Chemical Composition of Ribosomes*: Chemically ribosomes are composed of RNA and proteins. In many organisms the ratio of RNA and protein in ribosomes is approximately 1 to 1, but in *E. coli* is 2 to 1 respectively. In *E. coli*, ribosomes comprise about a quarter of the total cell mass, which reflects the importance of protein synthesis in cellular activity.

The 70S ribosomes contain about 65 per cent RNA and 35 per cent protein while the 80S ribosomes about 45 per cent RNA and 55 per cent protein. This ratio may also vary. Some workers are of the opinion that the ratio of RNA and protein in a ribosome is 40 to 60 per cent and 60 to 40 per cent respectively.

1. *Ribosomal RNA* : The ribosomal RNA (*r*-RNA) is found in two sub-units. The 70S ribosome consists of two types of ribosomal RNAase. The larger 50S sub-unit contains 23S ribosomal RNA of molecular weight 1.2×10^6 daltons and 5S rRNA of 3.0×10^4 and the smaller 30S sub-unit contains the 16S ribosomal RNA of molecular weight 0.6×10^6 daltons.

In eukaryotic cells, the 80S ribosome consists of four types of ribosomal RNAase. The larger 60S sub-unit contains three rRNAase viz., 28S rRNA, 5.8S and 5S rRNA. The presence of 5S rRNA has been described by *Knight* and *Darnell* (1967), and *Comb* and *Zehavi-Willner* (1967). The smaller 40S sub-unit contains only a single 18S r-RNA. The molecular weights of 28S and 18S and 5.8S rRNA are 1.7×10^6, 0.7×10^6 and 3.5×10^4 daltons respectively.

The 5S rRNA has a clover leaf shape. Its length is equal to 120 nucleotides (*Forget* and

Weissman, 1968). The extent of base pairing in 5S r-RNA has been determined by *Canter,* (1968). Its function is still unknown.

2. *Ribosomal Proteins* : Fractional studies of the ribosomes have demonstrated that their protein content is highly complex. About 50 ribosomal proteins have been identified so far.

(9) *Synthesis of Ribosomes*: Nomura (1968) has synthesized in vitro a unit, simila to 30S ribosomal subunit from 16S r-RNA and proteins of a variety of bacteria.

(10) *Biogenesis of Ribosomes*: Ribosomes may originate by any of the following methods :
   (*i*) de novo formation.
   (*ii*) Autoreplication.
   (*iii*) Nucleolar origin.

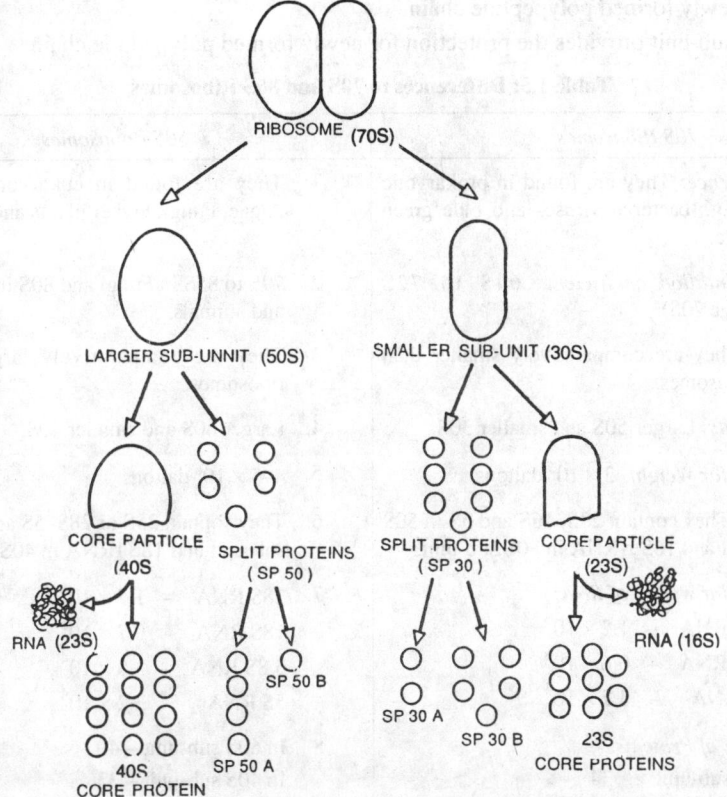

Fig. 1.46: Ribosomal proteins.

Out of these, the nucledor origin is most widely accepted one. It has been demonstrated that the nucleolus plays a key role in the biogenesis of ribosomes.

(11) *Function of Ribosomes*: A number of scientists have demonstrated that the ribosomes are concerned chiefly with the process of protein synthesis.

(12) *Role of different subunits of Ribosome* : (*a*) *Role of 50S sub-unit* : The 50S sub-unit contains the enzyme peptidyl transferase (or peptide synthetase). The subunit performs the following two functions :
   (*i*) The enzyme present in the 50S sub-unit is involved in the formation of peptide bond (*Monro,* 1967).
   (*ii*) The 50S subunit provides two binding sites for the two tRNA molecules. These two sites should be next to each other in order to permit the formation of the peptide bond.

Role of 30S sub-unit : The following roles are played by the 30S sub-unit :
  (i) 30S sub-unit is used for the reading of genetic message of mRNA.
  (ii) It forms initiation complex by binding the small 30S sub-unit to the first codon of the mRNA.

Role of 60S sub-unit :
  (i) 60S sub-unit is used in the attachment with endoplasmic reticulum membrane. The attachment may be related to the transfer of the newly synthesized protein into the cisternae of the endoplasmic reticulum. It has been shown in vivo by *Halliman, Murty* and *Grant* (1968) that attached ribosomes are more efficient in protein synthesis than free ribosomes lying in the cell-matrix.
  (ii) The tunnel in the centre of 60S sub-unit provides the space for the unidirectional flow of the newly formed polypeptide chain.
  (iii) 60S sub-unit provides the protection for newly formed polypeptide chain.

Table 1.5: Differences in 70S and 80S Ribosomes

70S Ribosomes	80S Robosomes
1. *Occurrence:* They are found in prokaryotic cells, e.g., bacteria, viruses and blue green algae etc.	1. They are found in eukaryotic cells, e.g., Algae, Fungi, higher plants and animals.
2. *Sedimentation coefficient:* 64S to 72S (Average 70S).	2. 79S to 85S in Fungi and 80S in higher plants and animals.
3. *Size:* They are comparatively smaller than 80S ribosomes.	3. They are comparatively larger than 70S ribosomes.
4. *Sub-units:* Larger 50S and smaller 30S.	4. Larger 60S and smaller 40S.
5. *Molecular Weight:* 3×10^6 daltons.	5. $4-5 \times 10^6$ daltons.
6. *RNAs:* They contain 23S, 16S and 5S in 50S sub-unit and 16S r-RNA in 30S sub-unit.	6. They contain 25S or 28S, 5S and 5.8S in 60S sub-unit and 18S rRNA in 40S sub-unit.
7. *Molecular weight of RNA:* 23S RNA $- 1.2 \times 10^6$ 16S RNA $- 0.6 \times 10^6$ 5S RNA $- 3.2 \times 10^4$	7. 28S RNA $- 1.7 \times 10^6$ 18S RNA $- 0.7 \times 10^6$ 5.8S RNA $- 3.5 \times 10^4$ 5S RNA $- 3.2 \times 10^4$
8. *Number of Proteins:* In 50S sub-unit $- 34$ In 30S sub-unit $- 21$	8. In 60S sub-unit $- 49$ In 40S sub-unit $- 33$
9. *Average molecular weight of Protein:* 1.8×10^4	9. 2.1×10^4
10. *RNA, Protein ratio* : 2 : 1	10. 1 : 1
11. *Mol. Wt. of Sub-units :* 50S $- 1.8 \times 10^6$ 30S $- 1.0 \times 10^6$	11. 60S $- 2.7 \times 10^6$ 40S $- 1.3 \times 10^6$

(13) *Polyribosomes (Polysomes)*: When ribosomes exist in groups on the strand of mRNA, they are called as *polyribosomes* or *polysomes*. They are formed during protein synthesis. The function of polysomes in protein synthesis could not be discovered until 1962. In *E. coli*, polyri-

bosomes are formed by 50 units. Polyribosomes may be free in the cytoplasmic matrix or bound to the membranes of ER. It is assumed that in polysomes the mRNA is situated in between the two subunits of the ribosome. The polysomes in the sections show helical structure. In the helical structures the smaller sub-units are arranged around the central axis and the larger sub-units are disposed at the periphery.

The polyribosomes can be liberated by treatment with detergent such as deoxycholate. Treatement of polyribosomes with low concentrations of RNase yields single 70S ribosomes. Intact single ribosomes are relatively resistant to RNase.

1.6. METABOLICALLY INACTIVE CELL INCLUSIONS

In living cells, many non-living materials are produced due to various types of metabolisms. Some of them are found in the soluble form and others in the solid form as reserve food materials. Such materials of living cells can be divided into following three groups :

    A. Reserve materials

    B. Secretory materials

    C. Excretory materials

1. *Reserve materials*: They are synthesized in the living cells of the plants. They provide nutrition to them and are stored mainly in underground stems, roots, buds and seeds. Reserve materials may be of following types :

Carbohydrates: Carbohydrates are made up of carbon, hydrogen and oxygen. They release energy as a result of oxidation. Among them some are soluble and others insoluble in water. Main carbohydrates are as follows :

(*a*) *Sugars* : They have been divided into following groups :

1. *Monosaccharides* : They are simple sugars and cannot be hydrolysed. Depending upon the number of carbon atoms present in the monosaccharides, they have been further divided into following types :

Trioses : They contain 3 carbon atoms, e.g., glyceraldehyde and dihydroxyacetone.

Tetroses : They contain 4 carbon atoms, e.g., erythrose.

Pentoses : They contain 5 carbon atoms, e.g., ribose and deoxyribose.

Hexoses : They contain 6 carbon atoms, e.g., glucose, fructose, galactose and mannose etc.

Septoses : They contain 7 carbon atoms, e.g., sedoheptulose.

2. *Disaccharides* : These are sugars which upon hydrolysis yield two molecules of monosaccharides. The common sugar which we use in our daily lives is a disaccharide and is known as *sucrose* ($C_{12}H_{22}O_{11}$). It is extracted from sugarcane and sugarbeet. Other disaccharide sugars are *maltose*, *lactose* and *cellobiose*. Upon hydrolysis, sucrose yields one molecule of glucose and fructose each and maltose yields two molecules of glucose.

1. $$\underset{\text{sucrose}}{C_{12}H_{22}O_{11}} \xrightarrow{+H_2O} C_6H_{12}O_6 + C_6H_{12}O_6$$
                                               glucose   fructose

2. $$\underset{\text{maltose}}{C_{12}H_{22}O_{11}} \xrightarrow{+H_2O} C_6H_{12}O_6 + C_6H_{12}O_6$$
                                               glucose   glucose

Similarly, upon hydrolysis trisaccharides yield 3 molecules of monosaccharides; tetrasaccharides 4 molecules and pentasaccharides 5 molecules of monosaccharides. The examples of tri—and tetra—saccharides are *raffinose* and *stachyose* respectively.

3. *Polysaccharides* : They are non-sugars of high molecular weight and upon hydolysis yield many molecules of monosaccharides. Main polysaccharides are *pentosans* e.g., *araban* and *xylan*; and *hexosans* e.g., *starch, cellulose, glycogen* and *inulin*.

Starch : It is insoluble in water and is formed by sugars. Usually, it is found in the cells of green plants. Its chemical formula is $(C_6H_{10}O_5)n$. At the time of necessity it is converted into sugars. In each starch grain, there is a deep point known as *hilum*. Starch layers are formed around this hilum. Such layers are known as *stratifications*. The starch grains of maize, wheat and rice contain the hilum in the centre and are known as *concentric* starch grains, whereas the starch grains of potato contain this hilum at one end side and are known as *eccentric* starch grains. Sometimes, a single starch grain possesses 2 or more hilum. Such grains are known as *compound* starch grains. Potato starch grains are eccentric as well as compound types. The hydrolysis of starch yields glucose which reacts with iodine solution and form blue-violet coloured starch iodide.

Inulin : It is a water soluble carbohydrate and can be converted into sugars. It is found in the cells of *Dahlia*.

Cellulose : It is water insoluble carbohydrate and is found deposited on the cell-walls of plants.

Fats, Lipids and Oils

They are also synthesized by carbon, hydrogen and oxygen like carbohydrates but contain comparatively very low amount of oxygen. They are insoluble in water but soluble in organic solvents like chloroform, ether and acetone. In cell-cytoplasm, they are found in the form of drops. The oxidation of these substances produces a very high amount of energy as compared to carbohydrates.

At ordinary temperature, the fats are liquids and are called as *Oils*. The different parts of plants like seeds, dry fruits and endosperms contain the fats in the form of oil drops.

Nitrogenous Organic Materials

(a) *Proteins* : They are very complex colloidal organic substances of high molecular weight but indefinite melting point. Thye are made up of carbon, hydrogen, oxygen, nitrogen and sulphur. In addition, they also contain iron and phosphorus. The percentage of various elements is given below:

$C = 50 - 55\%$; $H = 6.5 - 7.3\%$; $N = 15 - 19\%$;

$C = 21.9 - 24\%$ and $S = 0 - 2.8\%$

Each protein molecule is made up of several aminoacids joined by amide bonds. Such bonds are known as *peptide bonds*. In addition, they also possess disulphide, hydrogen and other types of bonds.

Proteins are usually insoluble in water, alcohol and ether but soluble in dilute salt solutions. Upon heating all proteins coagulate but upon hydrolysis they produce a mixture of α-aminoacids. They are found in all the living cells specially in seeds, chlorophyll and cytochromes.

(b) *Amides* : The salts of aminoacids are called *amides*. *Asparagine* and *glutamine* are common plant amides. They are found in the form of *aleurone grains* in oil cells of inulin containing seeds. In the endosperm of *Ricinus* seeds, amides are found in the form of layers. The protein matrix contains one large crystalline regions known as *crystalloid* and another small rounded region known as *globoid*. The crystalloid region possesses nitrogen and globoid region possesses calcium and magnesium phosphates. They are more complex molecules. The amide *gliadin* ($C_{685}H_{1068}N_{196}O_{211}S_5$) is found in wheat protein.

Secretory Materials

A few plants possess special type of cells which secrete useful substances like colour producing substances, enzymes, nectar and hormones. They are useful to the plants as well as human beings. A brief description about them is given below :

Colouring matters : The flower fruits and leaves of plants are coloured due to presence of special substances. In them, the green colour is due to chlorophyll a and b, yellow due to xanthophylls, red and orange due to carotenes and blue violet or pink due to various type of anthocyanins. Anthocyanins are usually found in the vacuolar sap of fruits in the dissolved form.

Enzymes : They are proteinaceous, colloidal, organic catalysts which help in digestion and other metabolic processes of the plants. Each enzyme (holoenzyme) is made up of two parts (*i*) protein part, (*ii*) prosthetic group. The protein part is called *apoenzyme* and the prosthetic group may either be organic known as *coenzyme* or inorganic known as *cofactor*. Thus, apoenzyme and prosthetic group form complete enzyme called *holoenzyme*.

Nectar : It may contain sucrose, glucose and fructose. Due to presence of such sugars it is sweet in taste. It is secreted in the nectaries (glands) mostly present below the gynoecium of the flowers and helps in insect pollination.

Hormones : They are important organic compounds required for the growth and elongation of plant cells.

Excretory Materials

Several excretory materials are produced due to various activities of the plant cells. Unlike animals, the plants do not possess excretory organs. So, they are found in different parts of the plant in the form of dry substance. Excretory materials are of two types : (*a*) organic, (*b*) inorganic.

Organic Materials

(1) *Alkaloids* : They are nitrogenous compounds which are produced by the fragmentation of proteins inside the cells. They are bitter in taste and are usually found in the storage regions of the plants like seeds, bark and leaves etc. Alkaloids are very useful to the human beings. Now a days, they are being used in the preparation of medicines. The alkaloid *nicotine* is extracted from the leaves of *Nicotiana*. Similarly, *quinine* is extracted from the bark of *Cinchona*, *Caffine* from the seeds of coffee and *morphine* from the fruits of poppy.

(2) *Tannins* : Tannins are found in the bark of plants, immature fruits and tea leaves in greater amounts. It remains dissolved in the cell-cytoplasm. They provide strength to the wood. They are being used in the preparation of ink, medicines and hard and good leather (tanning industry).

(3) *Essential Oils* : They are volatile in nature and are found in special oil secreting glands. Such glands are found in the rinds of orange and lemon, leaves of *Eucalyptus* and flowers of *Jasmine*, *Rose*, *Michelia* and others. Essential oils are used in the preparation of medicines and soaps. In plants, they help in insect pollination.

(4) *Latex* : It is white, yellow or grey coloured milky substance and is produced in the latex cells or laticiferous vessels. It helps in the healing of wounds, digestion and production of rubber.

(5) *Resins* : They are insouble in water but soluble in ether and alcohol. Resins are yellow coloured compounds and are produced by the oxidation of essential oils. They are used in the preparation of paints, varnishes, terpine oil, leather and medicines.

(6) *Glucosides* : They are produced by the hydrolysis of carbohydrates and are used in the preparation of medicines.

(7) *Gums* : They are formed by the hydrolysis of cellulose cell-wall. They are soluble in water and are used in pasting, medicines and oral eating (*Gum arabica*).

(8) *Organic acids* : They are soluble in water and sour in taste. They are found in the leaves and fruits of the plants. *Tartaric acid* is found in the fruits of *Tamarindus* and grapes. Similarly, *citric acid* is found in the fruits of orange, lemon and other Citrus sp.; *Oxalic acid* in the leaves of *oxalis* and sour spinach and *malic acid* in apple fruits.

Inorganic Materials

(1) *Calcium carbonate* ($CaCO_3$) : It is usually found in the cells of leaves of *Ficus* sp. The crystals of $CaCO_3$ remain hanged in the form of a bunch and are attached to the epidermal cells through a stalk. Such a structure is known as *cystolith*.

(2) *Silica* : Silica grains are found in the form of layers on the cell-walls of certain plants.

In abundance silica grains are present in the leaves of onion, straw of wheat and stem of *Equisetum*.

(3) *Calcium oxalate* : The crystals of calcium oxalate are found in different shapes likes rod-shape, prismatic, pyramidal, rhomboidal and needle shape. Needle like crystals are found in the leaves of *Pothas, Eichhornia* and *Colocasia* sp. and are called *raphides*. Star shaped or rounded crystals are called *sphaeraphides*. They are found in *Papaya* and *Opuntia*.

1.7. VACUOLES

They are found usually in plants cells. In young cells, they are very small in size and indefinite in number but in mature cells their size increases and number decreases. Sometimes only one vacuole is found in a cell. The shape of vaculoes is also variable. Each vacuole is surrounded by a thin layer of cytoplasm, known as *tonoplast* or *vacuolar membrane*. The tonoplast encloses the liquid substance known as *vacuolar sap*. The vacuolar sap contains carbohydrates, amides, amino acids, proteins, organic acids, anthocyanins, waste products and mineral salts like nitrates, chlorides and phosphates.

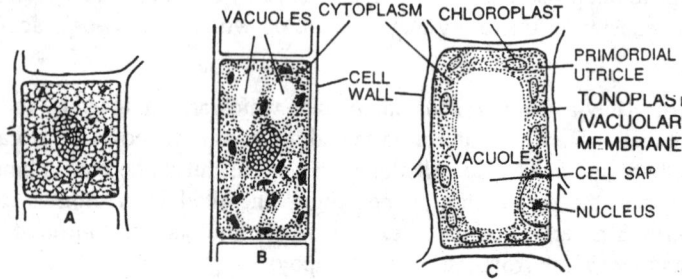

Fig. 1.47: Different stages of growth of a plant cell and relationship between size and number of vacuoles in them.

The vacuolar sap helps in maintaining the turgor pressure of the cell. They also act as storage for various substances. The vacuoles of flower petals and fruits contain various types of anthocyanins in the dissolved form and provide various colours to them.

The formation of vacuoles depends upon the increase in cell size and change in the amount of cytoplasm during cell growth. Initially all vacuoles are smaller in size but at later stages of growth, they become larger. Large-sized vacuoles are formed by the fusion of several small vacuoles.

1.8. NUCLEUS

Nucleus is the most important cell-organelle which directs and controls all its cellular activities. *Robert Brown* (1831) was the first to recognize the nucleus as a permanent organelle of the cell. It occurs in all the eukaryotic cells while the prokaryotic cells lack well defined nucleus. The morphology, especially, shape and size, of nucleus varies from cell to cell and organism to organism. The number of nuclei in a cell also varies. Usually single nucleus is found in the cells of higher plants and animals. But some-

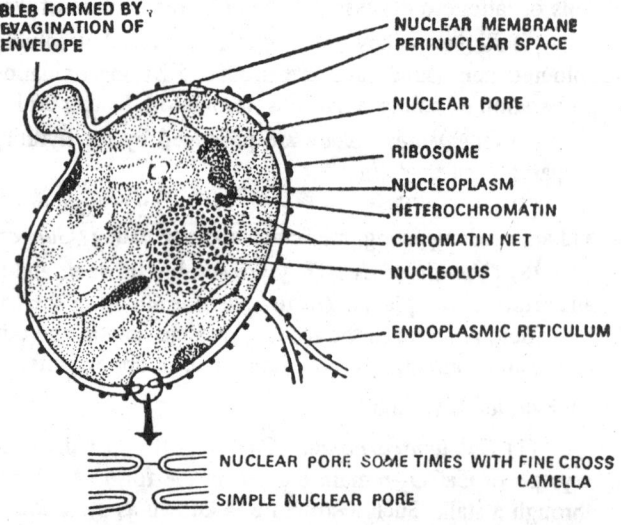

Fig. 1.48: Structure of nucleus under electron microscope.

times two nuclei, e.g., in binucleate cells of *Paramecium*, or more nuclei, e.g., polynucleate cells of *Vaucheria*, are also found.

Ultrastructure of nucleus reveals that it is made up of (*i*) Nuclear envelope or nuclear membrane, (*ii*) Nucleoplasm or nuclear sap or karyoplasm, (*iii*) Chromatin material and (*iv*) Nucleolus. The last three structures do not possess any limiting membrane. The nuclear envelope is again made up of inner and outer membrane, perinuclear space, pores, annuli material and inner dense lamella. The nuclear membrane encloses the clear, transparent, homogeneous and colloidal liquid of variable consistency called *nucleoplasm*. Nucleoplasm is mainly composed of nucleoproteins and a small amount of orgnic and inorganic substances like nucleic acids, dissolved phosphorus, proteins, enzymes, minerals, ribose sugars and nucleotides. The nuclear reticulum and nucleous remain suspended in the nucleoplasm.

Chromatin is basophilic in nature and most of the chromatin material is transformed into a specific number of chromosomes during cell-division. The chromatin material and chromosomes lie embedded in the nucleoplasm. In the interphase nucleus, the chromatin forms a system of irregularly tangled threads. These thread like structures are called *chromonemata*. The chromonemata anastomose to form a fine net work known as *nuclear reticulum*. The chromatin threads show beaded appearance being deposited with minute dark stained chromatin granules, the *chromocentres*. They are composed of chromatin proper. The chromatin material may be heterochromatin, euchromatin and sex-chromatin.

The nucleus reveals a spherical, often eccentrically placed body, the *nucleolus*. In a single nucleus, the number of nucleoli may be one or more. It was discovered by *Fontana* (1781). The ultrastructure reveals that it is made up of following regions (*i*) Pars granulosa region, (*ii*) Pars fibrosa region, (*iii*) Pars amorpha region and (*iv*) Pars chromosoma region.

The nucleolus lacks membrane. The importance of nucleus was demonstrated by *J. Hammerling* (1934) in *Acetabularia*.

Table 1.6: Cell components and their main functions.

Cell component	Functions
1. Cell Wall	1. Provides a definite shape and strength to the cell.
	2. Protects plasma membrane and protoplasm.
(*a*) Pits	1. Provide passage for the movement of water and minerals in thick-walled cells.
	2. Maintain protoplasmic connections between the cells.
(*b*) Plasmodesmata	1. Maintain continuity with neighbouring cells and help in transporting materials from cell to cell.
2. Plasma membrane	1. Protect the cytoplasm.
	2. Maintains selective permeability.
	3. Regulates the movement of water and minerals.
	4. Helps in protein synthesis and excretion of waste materials.
3. Cytoplasm	1. Provides site for enzyme activity.
	2. Regulates different plant metabolisms like glycolysis, pentose phosphate shunt and fat hydrolysis etc.
4. *Cell-organelles* (*a*) Mitochondria	1. Help in oxidation of food materials.
	2. Work as power house of the cell.

(*Contd.*)

	3. Act as site of Kerb's cycle, oxidative phosphorylation and produce ATP.
(b) Endoplasmic reticulum	1. Provides mechanical strength to the cell and helps in transportation of materials.
	2. Provides surface area for several chemical reactions.
	3. Helps in metabolism of cholesterol and steroid hormones.
	4. Provides site for protein synthesis.
	5. Granular ER helps in protein synthesis and smooth ER in fat synthesis.
(c) Golgi bodies (dictyosomes)	1. Help in the formation of primary lysosomes, vacuoles and cell-plate.
	2. Help in production of hormone.
	3. Help in intercellular transport.
	4. Store protein and fat.
	5. Help in membrane transformation and secretion.
(d) Ribosomes	Act as site and help in protein synthesis.
(e) *Plastids*	
(i) Chloroplast	1. Photosynthesis.
	2. Synthesis of starch (in green cells), protein and fatty acids.
	3. Heredity and mutation.
	4. Stroma is the site for dark reaction of photosynthesis.
	5. Grana is the site for light reaction of photosynthesis.
(ii) Amyloplast	1. Storage and synthesis of starch in non-green cells.
(iii) Lipoplasts	1. Storage and synthesis of lipids.
(iv) Proteinoplast	1. Storage of proteins.
(v) Chromoplast	1. Synthesis of coloured substances.
(f) Lysosomes	1. Digest external and intercellular materials.
	2. Autolysis
	3. Supports cancer formation.
(g) *Microbodies*	
(i) Glyoxysome	1. Conversion of fat into carbohydrates.
(ii) Peroxysome	1. Helps in photorespiration.
	2. Helps in glycolate metabolism.
(iii) Sphaerosomes	1. Synthesis of fat.
	2. Storage and translocation of fat.
(iv) Lomasomes	1. Synthesis of cell-wall.
(h) Vacuoles	1. Store cell sap.
	2. Maintain osmotic relations of cell.

(*Contd.*)

The Cell

	3. Provide colour to the flowers and fruits.
(i) Nucleus	1. Apparatus of heredity.
	2. Control heredity characters.
(i) Nuclear membrane	1. Regulates movement of materials entering and coming out of nucleus.
	2. Separates cytoplasm from nucleoplasm.
	3. Helps in protein synthesis.
(ii) Nucleoplasm	1. Acts as reservoir of various enzymes and nucleotides.
(iii) Nucleous	1. Formation of ribosomes
	2. Helps in cell-division.
	3. Synthesis of RNA and protein.
	4. Reservoir of RNA.
(j) Centrosome	1. Formation of spindle during cell-division in animal cells.

2

Osmotic Relations of Plant Cells

2.1. DIFFUSION

The movement of free molecules of gases, liquids and solids from the region of higher concentration to the lower concentration due to internal or external forces is called *diffusion*. Blowing of wind, dispersal of good smell of agarbattis in a room, dissolution of sugar in water, intake of CO_2 and liberation of O_2 in photosynthesis, intake of O_2 and liberation of CO_2 during respiration and dissolution of $KMnO_4$ particles in water are all the examples of diffusion. Such a diffusion of molecules in any region continues till it spreads throughout the available area. The movement of molecules depends upon the internal kinetic energy. The molecules in the region of higher concentration contain more kinetic energy and that is why they show fast movement. During the movement all the molecules collide themselves and produce a pressure in the medium called *diffusion pressure*

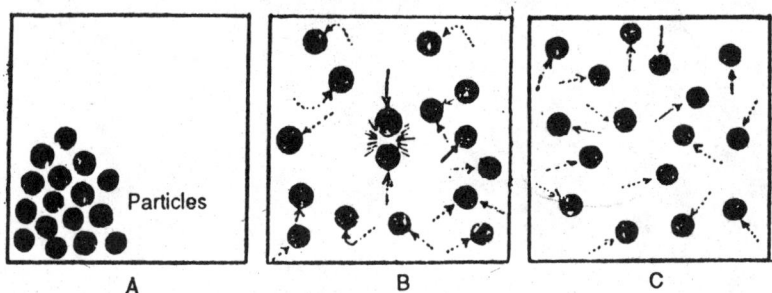

Fig. 2.1: Demonstration of diffusion.

which is directly proportional to the concentration and temperature of diffusing molecules. It means if the temperature and concentration of diffusing molecules will be more, the diffusion pressure will also be more and if the temperature and concentration is less, the diffusion pressure will also be less.

(*i*) Diffusion pressure α concentration of molecules.

(*ii*) Diffusion pressure α temperature.

Due to this reason, the diffusion always takes place from the region of higher diffusion pressure (conc.) to the region of lower diffusion pressure.

Mostly gases, liquids and solutes diffuse at different rates in different direction at the same place without affecting each other. This is the reason that O_2 and CO_2 both diffuse in different directions from a single stoma of leaf during respiration. The rate of diffusion is maximum

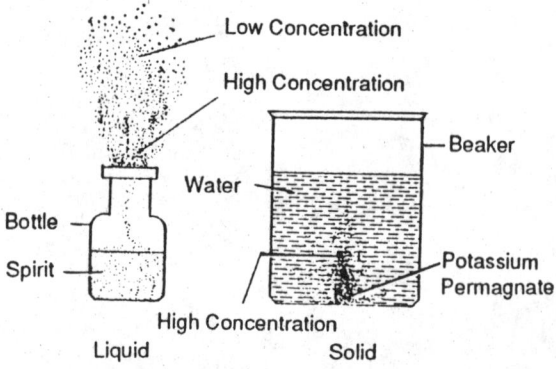

Fig. 2.2: Demonstration of diffusion in liquids and solids.

in gases and minimum in solids when dissolved in solvent. Liquids stand on intermediate position in this regard.

Diffusion Pressure of Liquids

Like gases, solvents and liquids also possess diffusion pressure. The diffusion pressure of a pure solvent is always maximum and when solute particles are added in it, the diffusion pressure of the solution is reduced. The difference between the diffusion pressure of a solvent and its solution is called *diffusion pressure deficit* (DPD). When a deficiency in diffusion pressure of the solution is created, it starts absorbing more solvent particles to overcome this deficiency. Thus, can be found out the absorbing capacity of a solution through DPD and DPD of a solution gives an idea about the absorbing capacity of that solution. That is why DPD is also called *suction pressure* (S.P.).

Factors Affecting the Rate of Diffusion

(1) *Temperature*: It is directly proportional to the rate of diffusion. On increasing the temperature, the rate of diffusing particles is also increased because of increase in the velocity of these particles.

(2) *Concentration of the medium*: It is inversely proportional to the rate of diffusion. On increasing the concentration of the medium, the rate of diffusion is reduced and on decreasing the concentration, it is increased.

(3) *The size and mass of the diffusing particles*: If the size and mass of the diffusing particles is smaller, the rate of their diffusion will always be faster.

(4) *Solubility of solutes*: The rate of diffusion increases with the rate of dissolution of solute in solution. Thus, more the solubility of any solute in solution, faster will be the rate of diffusion.

(5) *Diffusion pressure gradient* (DPG): The rate of diffusion of molecules of gases and liquids also depends upon the diffusion pressure gradient (DPG). When the difference in the diffusion pressures is more, faster is the rate of diffusion. In other words, steeper the rate of diffusion pressure gradient, the faster is the rate of diffusion.

(6) *Density of the diffusing particles*: The rate of diffusion of gases is related with the density of diffusing particles. According to Grahm's law of diffusion, "The rate of diffusion of any gas particle is inversely proportional to the square root of its density."

$$r = \frac{1}{\sqrt{d}}$$ where; r = rate of diffusion of a gas

d = density of a gas

The gases having higher densities show slower rate of diffusion whereas those possessing lower densities show faster rate of diffusion. For example, the density of hydrogen (H) is one and of oxygen(O)16. Therefore, according to Grahm's law of diffusion:

$$\frac{rH}{rO} = \frac{\sqrt{dO}}{\sqrt{dH}} = \frac{\sqrt{16}}{\sqrt{1}} = \frac{4}{1}.$$

Thus, the hydrogen will diffuse four times faster than oxygen.

Importance of Diffusion in Plants

(i) The exchange of O_2 and CO_2 gases in the atmosphere through stomata of leaves takes place by the process of diffusion. O_2 gas participates in respiration whereas CO_2 in photosynthesis.

(ii) During stomatal transpiration the water vapours from intercellular spaces diffuse in the atmosphere through stomata by the process of diffusion.

(iii) The diffusion of ions of mineral salts during passive absorption also takes place by this process.

(iv) The absorption of water through roots is also performed by diffusion.

Experiment : To demonstrate the phenomenon of diffusion: Take a beaker containing about 200 ml clean water. Place a small crystal of $KMnO_4$ at the bottom of the beaker. Leave if for some time and observe. The $KMnO_4$ crystal being soluble in water starts dissolving and the red colour spreads slowly throughout the water. After sometime, the complete crystal is dissolved and the entire water becomes red. It indicates the diffusion of $KMnO_4$ in water.

2.2. PERMEABILITY

The entrance or exit of any substance in the cell depends upon the permeability of the plasma membrane. The permeability of plasma membrane is usually changed from time to time due to change in either K^+ ions concentration or nature of other particles surrounding it. Depending upon the permeability properties the membranes can be classified into following categories:

(a) *Impermeable* : These are those membranes through which only exchange of gases takes place.

(b) *Semi-permeable* : These are the membranes through which the exchange of only water or solvent molecules takes place. Solute particles neither can enter nor leave through such membranes. Parchment membrane, urinary bladder of goat, and egg membranes are the examples of semipermeable membranes.

(c) *Selective permeable* : These are the membranes through which the exchange of only selected ions or small molecules takes place. Plasma membrane is mainly selective permeable and not only semipermeable.

(d) *Dialysing membranes* : These membranes possess other layers in the form of outer covering.

Plasma membrane is usually permeable for gases like CO_2, O_2, N_2; solvents like alcohol, ether, chloroform and organic substances like mono- and disaccharides, fatty acids, amino acids and glycerol whereas it is normally impermeable for polysaccharides, phospholipids and proteins. The permeability of plasma membrane for gases and solvents is more than organic substances.

Factors Affecting Permeability

A number of external and internal factors affect the permeability of plasma membrane.

External factors

1. *Physical agents:* The physical agents like high temperature, heat, low pressure of O_2 and CO_2, radiations etc. increase the permeability of plasma membrane.

2. *Chemical agents:* A number of chemicals like ether, benzene, chloroform, acetone, toxic substances etc. when added in the external medium or solution, increase the permeability.

3. *Ageing of cells:* The permeability varies at different ages of the cells. In young cells, it is low but at the time of senescence it increases. The older cells approaching death also show increase in permeability.

Internal factors

4. *Membrane constitution:* The permeability of plasma membrane depends upon its constitution, i.e. percentage of lipids, proteins, carbohydrates and phosphate etc., thickness, porocity and degree of hydration.

Experiments

1. *To prepare artificial chemical semipermeable membrane:* Take *copper sulphate* ($CuSO_4$) solution in a porous cylindrial and place it in a beaker containing *potassium ferrocyanide* solution $CuSO_4$ solution will diffuse outward while the potassium ferrocyanide inward into the porous pot. Both the solution will meet and react in the porous wall to form a thin layer of *copper ferrocyanide* which act as semipermeable membrane. Take out the porous pot carefully and drain its extra $CuSO_4$

solution. Now, it can be used as semipermeable membrane (Fig. 2.3).

2. *To demonstrate the effect of temperature and alcohol on the permeability of plasma membrane:* Cut small and equal-sized cylindrical pieces of beet root with the help of cork-borer and wash them thoroughly in water. Place one piece each in separate seven test tubes containing equal amount of water at different temperatures i.e., 0°, 10°, 20°, 30°, 40°, 60° and 70°C. Take another fresh test tube, pour alcohol and place one piece in it at room temperature. Leave as such all these test tubes for sometime.

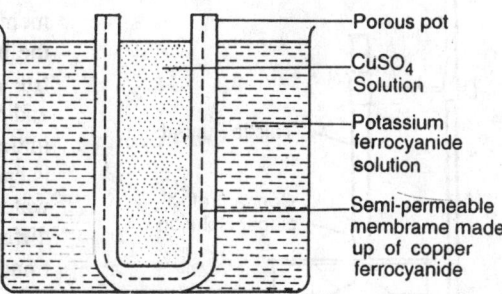

Fig. 2.3: Preparation of chemical semipermeable membrane.

It is observed that the water in the test tubes kept at low temperatures (0°, 10°, 20° and 30°C) remain unchanged (colourless), while the water in the test tubes kept at high temperatures (40°, 60° and 70°C) becomes red coloured. The alcohol in the test tube also becomes red coloured. It is because the anthocyanin pigment present in the cell sap diffuses out in the water of test tubes kept at higher temperatures. This diffusion increases towards higher temperature increasing the intensity of colour while decreases at lower temperatures. At very low temperatures, the diffusion of pigment does not take place. The diffusion increases due to gradual loss in permeability of the plasma membrane. At high temperature the permeability is completely lost.

The diffusion of pigment is also rapid in case of alcohol because it kills the cells of beet root and destroys their plasmamembrane.

2.3. OSMOSIS

When two solutions of different concentrations are separated by a semipermeable membrane, the diffusion of water or solvent molecules takes place from the solution of lower concentration towards the solution of higher concentration. This process is called *osmosis*. (Fig. 2.4).

Actually the diffusion of solvent molecules takes place across the membrane on both the sides but it is faster from the lower concentration side than the higher concentration side. This is according to the principle of diffusion because the solution of lower concentration possesses higher concentration of solvent molecules whereas the solution of higher concentration possesses comparatively lower concentration of solvent molecules. Thus, it can be said that osmosis is a type of diffusion through semipermeable membrane.

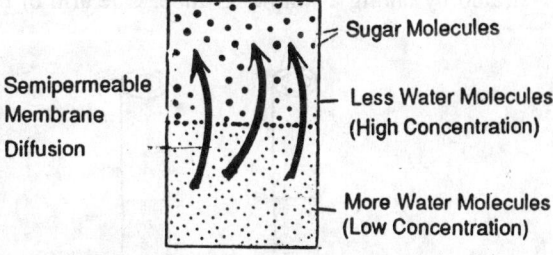

Fig. 2.4: Demonstration of Osmosis.

From biological point of view the solutions may be of following three types :

1. *Hypertonic solutions* : Those solutions whose concentration and osmotic pressure (OP) are more than the concentration and OP of cell-sap, are called *hypertonic solutions*.

2. *Hypotonic solutions* : Those solutions whose concentration and OP are less than the concentration and OP of the cell-sap, are called *hypotonic solutions*.

3. *Isotonic solutions* : Those solutions whose concentration and OP are equal to the concentration and OP of cell-sap, are called *isotonic solutions*.

Demonstration of Osmosis

Osmosis can be demonstrated by following experiments :

(1) *By Simple Osmometer* : It is also called thistle funnel experiment (Fig 2.5).

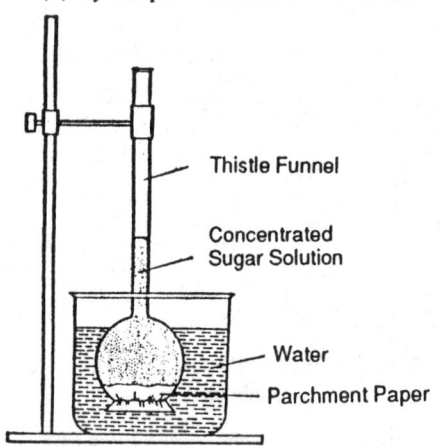

Fig. 2.5: Thistle funnel experiment for osmosis.

Take a thistle funnel and bind a semipermeable membrane on its wider mouth. Hang this funnel with the help of stand in a beaker filled with water. Now, fill up the concentrated sugar solution in the funnel and mark it. After sometime the level of sugar solution in the funnel increases to a certain extent. This process is called *osmosis*. This process takes place in between water and sugar solution or any two solutions of different concentrations. This process of osmosis continues till the concentration of solutions on both the sides of semipermeable membrane becomes equal.

Actually the semipermeable membrane contains many minute pores. The molecules of water or solvent are so small that they reach into the sugar solution of thistle funnel through these pores of semipermeable membrane. The sugar molecules are comparatively larger in size and they cannot reach into the water of beaker through the pores of membrane.

See Fig. 2.4 and note the number of water molecules present on both the sides of semipermeable membrane. On the upper side of membrane the number of water molecules in the sugar solution is quite less in comparison to the number of water molecules on the lower side of membrane. Due to this reason the water molecules from outside (higher concentration) diffuse into sugar solution (lower concentration) through semipermeable membrane.

(2) *Demonstration by U-Tube*: Like thistle funnel experiment osmosis can also be demonstrated by taking a simple U-tube. One arm of the tube is filled with 5 per cent sugar solution

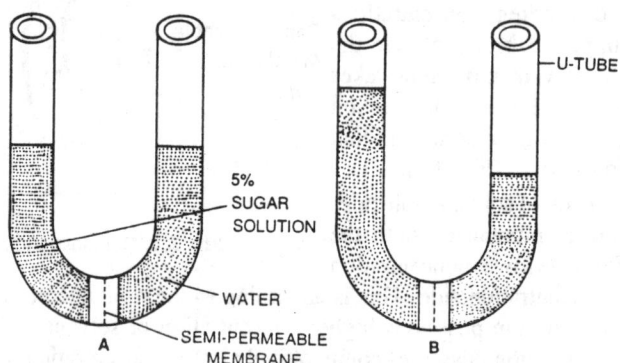

Fig. 2.6: Demonstration of Osmosis by U-tube (A) Before experiment (B) After expriment.

and other arm with pure water and both are separated by a semipermeable membrane. The levels on both the sides are marked (Fig. 2.6). This experiment is left for sometime. Now, it is observed that the level of left side (sugar solution side) increases and right side (water side) decreases. It happens due to phenomenon of osmosis and this increase in level continues till the concentrations of both the arms become equal. The explanation for this can be given just like thistle funnel experiment.

Duffusion Through Artificial and Natural Semipermeable Membranes

Artificial semipermeable membranes have been prepared by several scientists (*Maurice Traube*, 1867; *Pfeffer*, 1877) to study the phenomenon of osmosis. The first artificial semipermeable membrane was prepared by *M. Traube* (1867). The artificial membranes were analogous to the plasma membrane. The common artificial membrane is of copper ferrocyanide. Osmosis was first demonstrated by *Nollet* in 1748 using pig's bladder. He filled this bladder (semipermeable membrane) with alcohol and immersed in water. The increase in the level of alcohol was noted showing osmosis phenomenon. *Dutrochet* and *Vierdot* (1827-48) demonstrated osmosis with the help of other animal membranes but they filled those membranes with salt solutions and immersed in water. The rise in the level of salt solution was noticed by them under hydrostatic pressure called osmotic pressure.

Types of Osmosis : Osmosis in Plants is of Two Types

(1) *Endosmosis* : When water or solvent molecules enter into the cell through plasmamembrane from the outer medium, it is called *endosmosis*. It can be demonstrated by potato osmometer (Fig 2.7). Take a potato tuber and make a cavity inside it with the help of a knife. Now, fill the cavity with concentrated sugar solution and mark the level with the help of the alpin. Place this solution filled tuber in a bowl containing water. After sometime the level of sugar solution in the potato cavity rises. Here the plasma membranes of potato cells work like a semipermeable membrane. In this experiment the movement of water molecules is taking place from outer medium into the potato cavity. Thus, it is the phenomenon of endosmosis. Similarly, when dry raisins are placed in the water, they swell up due to endosmosis (Fig. 2.8).

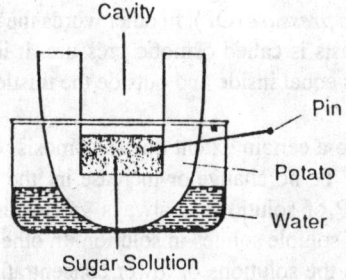

Fig. 2.7: Demonstration of endosmosis by potato osmometer.

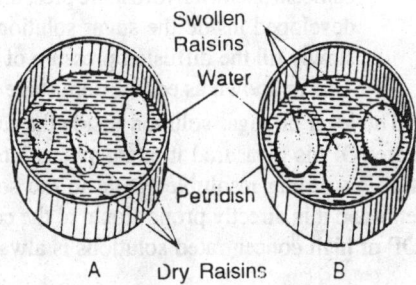

Fig. 2.8. Endosmosis in raisins.

(2) *Exosmosis*: When a plant cell is placed in concentrated solution, the water molecules move from cell into the outer conc. medium through plasma membrane. It is called *exosmosis*. Thus, in exosmosis the water or solvent molecules leave the cell into the outer medium through plasma membrane. It can be demonstrated by the following experiment.

Experiment : Place a few grapes into the petridish containing 30 per cent sugar solution. After sometime you will see that the swollen grapes are collapsed (Fig. 2.9). It happens because the grapes contain sugar solution of lower concentration in comparison to the outer more concentrated solution. Thus, the concentration of outer medium is more than the concentration of sugar solution of grape cells. Due to this reason the water molecules from the cells of grapes (solution of lower concentration) move into the sugar solution of outer medium (solution of higher concentration) and grapes are collapsed. Thus, it explains the phenomenon of exosmosis.

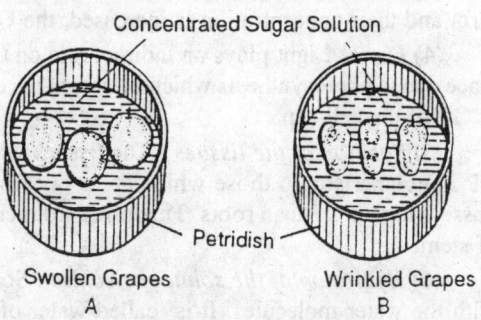

Fig. 2.9: Exosmosis in grapes.

Table 2.1: Differences Between Diffusion and Osmosis

Diffusion	Osmosis
1. Diffusion occurs in solids, liquids and gases.	Osmosis occurs only in liquids.
2. Diffusion does not require a semipermeable membrane.	Osmosis requires a semipermeable membrane.
3. In this process the movement of molecules takes place from higher concentration to lower concentration.	In osmosis, the movement of water or solvent molecules takes place from the solution of lower concentration to the solution of higher concentration.

Osmotic Pressure (*OP*)

In the thistle funnel experiment of osmosis you have studied that the water molecules diffuse into sugar solution through semipermeable membrane. It is called *osmosis*. Why this diffusion of water molecules takes place? There may be two following reasons for this :

1. The solute (sugar) particles being larger in size cannot cross the semipermeable membrane. They collide continuously with the membrane and exert a pressure on it due to which water molecules go inside.
2. The water molecules after continuously reaching inside the sugar solution of funnel through osmosis exerts a hydrostatic pressure called *osmotic pressure* (OP). In other words the pressure developed inside the sugar solution through osmosis is called osmotic pressure. It increases slowly till the diffusion pressure of water becomes equal inside and outside the thistle funnel. This in known as equilibrium stage of osmosis.

The level of sugar solution inside the funnel rises up to a certain extent due to osmosis. Osmotic pressure (OP) is measured in atmospheres units. There will be no change or increase in the OP of a solution on adding insoluble solute in the solution. The OP of solutions is always greater than pure solvents and it is directly proportional to the concentration of soluble solutes in solution. In other words, the OP of high concentrated solutions is always greater than the solutions of lower concentration.

Factors Affecting Osmotic Pressure

(1) *Concentration of solute particles*: The concentration of solute particles in solution is directly proportional to the OP. The osmotic pressure of concentrated solutions is always greater than the OP of solutions of low concentration. If the concentration of solute particles is increased in any solution, its OP will also increase.

(2) *Temperature*: The temperature is also directly related with the OP. If the temperature of the solution is increased, its OP will also increase.

(3) *Ionization of the solute molecules*: If solute particles in the solution are present in ionic form and their concentration is increased, the OP of the solution is also increased.

(4) *Light* : Light plays an indirect role on OP. In sunlight the synthesis of carbohydrates takes place due to photosynthesis which increases the concentration of the protoplasm resulting in increase in OP of the cell-sap.

(5) *Position of the tissues* : The tissues which are nearer to soil or water supply possess low OP in comparison to those which are away from the water supply. Due to this reason leaves possess higher OP than roots. This pressure increases simultaneously from roots to upper portion of stem.

(6) *Hydration of the solute molecules* : Sometimes a few molecules of solutes remain bound with the water molecules. It is called water of hydration. The solution which possesses water of hydration shows high OP because these water molecules which remain bound with the solute molecules are not the part of solution.

Significance of Osmosis in Plants

1. The water is most important factor for life because all the biological activities are directly or indirectly controlled by water. Osmosis helps the plants in the absorption of water from soil through roots.
2. The turgidity in plant cells is maintained by the absorption of water through osmosis.
3. The diffusion of water from one cell to another and its distribution in different plants parts is also performed by osmosis.
4. The opening of stomata depends upon the turgidity of guard cells which is maintained by osmosis.
5. Osmotic pressure and turgor pressure help in the growth of young cells.
6. High osmotic concentration in plants increase the resistance against freezing temperature and dehydration.
7. Osmosis helps in the dehiscence of fruits and sporangia.

Plant Cell as Osmotic System and Relationship among Turgor Pressure (TP), Wall Pressure (WP) and Osmotic Pressure (OP)

When a plant cell is placed in water, the water starts entering into the cell through osmosis because the cell possesses an outer cell-wall of cellulose enclosing a semipermeable plasma membrane. In addition the osmotic concentration of cell-sap is also greater than water. As soon as

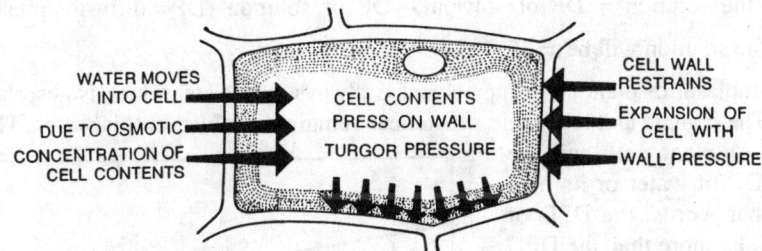

Fig. 2.10: Demonstration of osmotic concentration, turgor pressure and wall pressure.

the water enters into the cell, it exerts a pressure on cell inclusions and organelles. The pressure increases slowly by slowly with the entry of water molecules and a stage comes when large of water enters into the cell and this quantity of water exerts a pressure on the cell-wall. The pressure exerted on the cell-wall from the inner side by the cell inclusions and cell-organelles through the entry of water eq. is called *turgor pressure* (TP). Such swollen cells are called *turgid cells* and the stage of cells is called *turgidity*.

Due to increase in turgor pressure the cell-wall stretches towards outside. Since the cell-wall provides a definite shape to the cell and is elastic in nature, it also exerts a pressure on cell-sap from outside *i.e.*, in opposite direction, equal to TP. The pres-

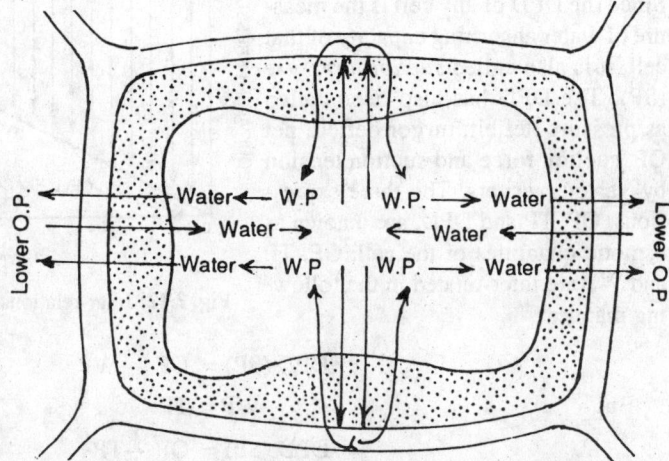

Fig. 2.11: Osmotic system in a plant cell. Relationship among OP, TP and WP at equilibrium stage.

sure exerted by the cell-wall on the cell-sap (cell inclusions and organelles) is called wall pressure (WP). Due to this WP the cell-wall stretched through TP regains its orginial position.

As more and more water enters into the cell, the cell-wall stretches slowly by slowly towards outside and the TP also increases. After sometime a stage of equilibrium comes when the tendency of water to enter the cell due to osmotic concentrations of cell contents is balanced by the tendency of water to leave the cell due to TP. At this stage, the turgor pressure balances the osmotic pressure. In other words, both TP and OP become equal.

$$OP = TP$$

or $$OP - TP = 0 \; (OC = TP \; \text{or} \; OC - TP = 0)$$

Inter-relationship among Diffusion Pressure Deficit (DPD) or Suction Pressure (SP), Osmotic Pressure (OP), Turgor Pressure (TP) and Wall Pressure (WP)

It has been discussed earlier that the diffusion pressure of pure solvents is always greater than the diffusion pressure of solutions. The quantitative amount of diffusion pressure of solution in atmospheres which is less than the diffusion pressure of its own pure solvent, is called diffusion pressure deficit (DPD). In other words, the difference between the diffusion pressure of any solution and its pure solvent in atmospheres is called DPD.

Suppose, the diffusion pressure of solvent = 25 atm.

and diffusion pressure of solution = 15 atm.

∴ DPD of the solution = DP of solvent − DP of solution (DP = diffusion pressure)

∴ DPD of the solution will be = 25 − 15 = 10 atm.

In the cytoplasm of plant cells, the solvent is always pure water which is absorbed from the soil by roots. The organic and inorganic substances remain dissolved in this water. Thus, the DP of cytoplasm of plant cells will be less than the DP of water or its own solvent. In other words, the DPD of cytoplasm will be more than the DPD of water. When such cells are placed in pure water, the water starts entering into the cell due to lack of water in the cytoplasm or high DPD of cytoplasm. Since the DPD of any cell is the measure of water absorbing capacity of that cell, it is also called *suction pressure* (SP). The DPD has also been called as pressure deficit, turgor deficit, net OP, suction force and suction tension by various writers. The three expressions, OP, TP and DPD, are known as osmotic quantities of the cell. OP, TP and WP are inter-related in the following manner:

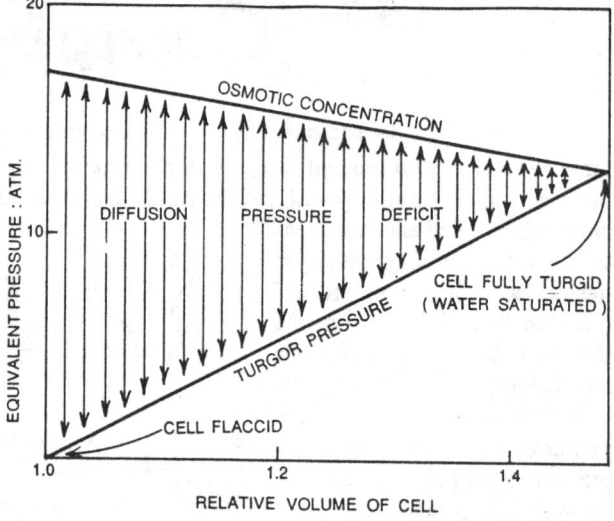

Fig. 2.12: Inter-relationship among DPD, OP and TP in a cell.

$$DPD \; (SP) = OP - WP$$

But $$WP = TP$$

∴ $$DPD \; (SP) = OP - TP$$

Due to continuous entry of water in the cell, the OP of cell-cytoplasm decreases and simultaneously the TP increases. After sometime a stage of equilibrium comes when the cell becomes

Osmotic Relations of Plant Cells

fully turgid and the TP balances the OP. In other words the TP becomes equal to OP at this stage.

$$DPD\ (SP) = OP - TP$$
$$\therefore \quad OP = TP \quad \text{(when the cell is fully turgid)}$$
$$\therefore \quad DPD\ (SP) = OP - OP$$
or $\quad DPD\ (SP) = TP - TP$
or $\quad DPD\ (SP) = 0$

When the cell is placed in hypertonic solution, the water from cell-cytoplasm goes out by the process of exosmosis resulting in increase in the OP and decrease in TP of cell-cytoplasm. As a result of exosmosis the cytoplasm contracts and it leaves the cell-wall. This phenomenon of cell is called *plasmolysis*. Such cell is called plasmolysed cell and this stage as flaccid stage. When the cell becomes completely plasmolysed, at that time the TP of the cell is reduced to zero.

$$TP = O \quad \text{(when the cell is completely plasmolysed)}$$
$$\therefore \quad DPD\ (SP) = OP - TP$$
Therefore, $\quad DPD\ (SP) = OP - 0$
or $\quad DPD\ (SP) = OP$

If the cell is placed in some other hypotonic solution instead of water, its DPD or SP will be as follows :

$$DPD\ (SP) = (OP - OP_1) - OP\ (\text{where } OP_1 = OP \text{ of other hypotonic solution}).$$

Thus, the DPD or SP of plant cell is not directly proportional to the OP or Osmotic concentration (OC) of that cell. It depends upon OP and TP. When the OP the cell-sap is more, the DPD of cell will also be more but the TP will be less. Under such condition the water enters into the cell. Sometimes it happens that the OP and TP of one of the adjacent cells are more than the OP and TP of other cell, even then the water does not enter into it. The reason is that the DPD of that cell is less and there is no effect of high OP on the entry of water.

Water Potential Concept

The water is a good solvent. The water molecules possess free energy and "the difference between free energy of water molecules in pure and in any other system is called *water potential*." Any other system means the water in a solution or in a plant cell or tissue. It is designated by the symbol Ψ (psi, a Greek letter). The water potential is measured in terms of a pressure e.g., bars or atmospheres. One bar is equal to 14.5 lb/m^2, 750 mm Hg or 0.987 atm.

The movement of water, as in osmosis, cannot be accurately explained in terms of differences in concentration or any other linear expressions but the spontaneous movement of water from one region or system to another can be best expressed in terms of differences in free energy of water between the two regions or systems. The water moves from the system having higher free energy towards the system having lower free energy. Usually the free energy of pure water is greater than the free energy of other solutions prepared in water.

If the two regions in an aquous system have water potentials ΨA and ΨB respectively, the difference in water potential will be:

$$\Delta \Psi = \Psi A - \Psi B \qquad (\Delta \Psi = \text{difference in water potentials})$$

If ΨA is greater than ΨB, $\Delta \Psi$ will be positive and the water will move from A to B region. If ΨB is greater than ΨA, $\Delta \Psi$ will be negative and water will move from B to A region.

The water potential of protoplasm is equal but opposite in sign to the diffusion pressure deficit (DPD) or suction pressure (SP). The water potential of a solution is usually measured using pure water as the standard. The water potential of pure water at atmospheric pressure is zero (0). When the solute particles are added in water and solution is prepared, they reduce the free energy of water. Thus, the water potential of solutions is decreased and comes in negative values i.e., less than zero. Due to this reason the water potential of a solution is always less than zero.

If the two systems having different water potentials are separated by a semipermeable membrane, the movement of water molecules always takes place from the system having higher water potential (less concentrated solution) towards the system having lower water potential (more concentrated solution). The movement of water will continue till the water potentials of the two systems become equal and a stage of equilibrium is reached. At this stage, the net transfer of water will cease.

The Components of Water Potential

When a typical plant cell containing cell-wall, vacuole and cytoplasm is subjected to movement of water, a number of factors are involved which determine the water potential of cell-sap. The water potential of a living cell is determined by three major components (potentials). (1) matric potential (Ψm), (2) solute potential (Ψs) and (3) pressure potential (Ψp). The water potential (Ψ) is actually the sum of all the above three potentials.

$$\Psi = \Psi m + \Psi s + \Psi p$$

(1) *Matric Potential* (Ψm): The term *matric* is used for such surfaces which can adsorb water molecules, e.g., cell-walls, protoplasms and soil particles etc. Matric potential (Ψm) is the component of water potential influenced by the presence of a matric and possesses negative value. The Ψm in case of plant tissues and cells is often neglected because it is insignificant in osmosis. The above equation may be written in simplified form as follows:

$$\Psi = \Psi s + \Psi p$$

The equation indicates that water potential of plant cells and tissues is usually the sum of solute potential and pressure potential.

(2) *Solute Potential* (Ψs): It is also a component of water potential and is also known as *osmotic potential*. The amount of solute present in water is called *solute potential* (Ψs). The presence of solute in water reduces the value of water potential and reduction values of Ψ are directly proportional to the amount of solute present in water. Since the potential of pure water is zero, the values of solute potential (Ψs) or ($\Psi 0$) are always negative i.e., less than zero. The solute potential is expressed in bars with negative sign and the term Ψs has replaced the osmotic pressure, the term used earlier.

(3) *Pressure potential* (Ψp): The cell-wall of a plant cell, which is made up of cellulose, provides a definite shape to the cell and is elastic in nature. It also exerts a presure on the cellular contents inwards (inward wall presure) resulting into development of hydrostatic pressure in the vacuole called *turgor pressure*. The values of Ψp in plant cells are positive and Ψp is equivalent to wall pressure or turgor pressure.

Osmotic Relations in three Physical States (Ψ, Ψs and Ψp) according to Water Potential

(a) *In fully turgid cell:*

1. Net movement of water into cell stops.
2. The cell will be at equilibrium state with water of outer medium.

The water potential (Ψ) will be zero.

The cell at full turgor possesses Ψs equal to Ψp but with opposite sign.

∴ its Ψ will be zero (0).

If Ψp of a cell is 5 bars.

Osmotic Relations of Plant Cells

Then $\therefore \Psi = \Psi s + \Psi p$

$\therefore \Psi = (-5) + (+5)$

or $\Psi = 0$

(b) *In flaccid cell:*
  1. The turgor becomes zero.
  2. At zero turgor, the cell shows osmotic potential (Ψs) equal to its water potential (Ψ), i.e., $\Psi s = \Psi$.

If a flaccid cell has an osmotic potential (Ψs) of –5 bars and pressure potential (Ψp) of 0 bars, i.e., $\Psi s = -5$ and $\Psi p = 0$, then,

$\therefore \Psi = \Psi s + \Psi p$

$\therefore \Psi = -5 \text{ bars} + 0 \text{ bars}$

or $\Psi = -5 \text{ bars}.$

Thus, the water potential (Ψ) of the cell will be –5 bars. This value is less than the Ψ of pure water i.e., zero.

(c) *In plasmolysed cell:*
  1. The pressure potential (Ψp) has negative value, *i.e.*, the cell shows negative turgor pressure.
  2. The water potential in such a cell will be more negative. If in a plasmolysed cell $\Psi s = -5$ bars and $\Psi p = -1$ bars,

Then $\therefore \Psi = \Psi s + \Psi p$

$\therefore \Psi = (-5) + (-1)$

or $\Psi = -6 \text{ bars}$

Problems Regarding Osmotic entry of Water

Problem 1: There are two adjacent living cell A and B. Cell A has an osmotic potential (Ψs) of –10 bars and pressure potential (Ψp) of 5 bars, whereas cell B has an osmotic potential of –5 bars and pressure potential of 2 bars. What will be the direction of water flow in the cells?

Solution:

Cell A	Cell B
$\Psi s = -10$	$\Psi s = -5$
$\Psi p = 5$	$\Psi p = 2$
$\because \Psi = \Psi s + \Psi p$	$\because \Psi = \Psi s + \Psi p$
$\therefore \Psi = -10 + 5 = -5$	$\therefore \Psi = -5 + 2 = -3$

Fig. 2.13

In the problem, the cell A has $\Psi s = -10$ and $\Psi p = 5$. Therefore, the water potential (Ψ) of cell A will be –5 bars.

$\therefore$ Water potential (Ψ) = Osmotic potential (Ψs) + Pressure potential (Ψp)

$= -10 + 5 = -5$

bars. The cell B has $\Psi s = -5$ and $\Psi p = 2$. Therefore, the water potential (Ψ) of cell B will be -3 bars.

$$\Psi = \Psi s + \Psi p$$
$$= -5 + 2 = -3$$

Since the osmotic entry of water always takes place from higher water potential to lower water potential, the direction of water flow will be from cell B (-3 bars) to cell A (-5 bars).

Problem 2: There are two adjacent living cell A and B. Cell A has an osmotic potential (Ψs) of -7 bars and pressure potential (Ψp) of 4 bars, whereas cell B has an osmotic potential of -8 bars and pressure potential of 3 bars. What will be the direction of water flow in the cells?

Solution:

Cell A	Cell B
$\Psi s = -7$ $\Psi p = 4$	$\Psi s = -8$ $\Psi p = 3$
$\therefore \Psi = \Psi s + \Psi p$ $\therefore \Psi = -7 + 4 = -3$	$\therefore \Psi = \Psi s + \Psi p$ $\therefore \Psi = -8 + 3 = -5$

Fig. 2.14

In the problem cell A has $\Psi s = -7$ and $\Psi p = 4$. Therefore, the water potential (Ψ) of cell A will be -3 bars.

$$\therefore \Psi = \Psi s + \Psi p$$
$$= -7 + 4 = -3 \text{ bars.}$$

The cell B has $\Psi s = -8$ and $\Psi p = 3$. Therefore, the water potential (Ψ) of cell B will be -5 bars.

$$\therefore \Psi = \Psi s + \Psi p$$
$$= -8 + 3 = -5 \text{ bars.}$$

Since osmotic entry of water always takes place from higher water potential to lower water potential, the direction of water flow will be from cell A (-3 bars) to cell B (-5 bars).

Table 2.2. Differences between diffusion pressure deficit (DPD) and water potential (Ψ).

Diffusion pressure deficit (DPD)	*Water potential (Ψ)*
1. The difference between the diffusion pressure of any solution and its pure solvent in atmospheres is called DPD. Since DPD of any cell is the measure of water absorbing capacity of that cell, it is also called suction pressure (SP), DPD has also been called pressure deficit, turgor deficit, net OP, suction force and suction tension etc.	1. Water potential is the chemical potential of water. It is equivalent to DPD with negative sign. It is the difference between free energy of water molecules in pure water and in a solution.
2. Diffusion pressure deficit is denoted as DPD.	2. Water potential is denoted by a Greek letter psi (Ψ)
3. DPD is measured in *atmospheres* (atm).	3. It is measured in *bars*, a pressure unit.

(Contd.)

Osmotic Relations of Plant Cells

4. The direction of water movement is from lower DPD to higher DPD.	4. The direction of water movement is from higher water potential to lower water potential i.e., in energetically down hill direction.
5. DPD = OP – TP (where OP = osmotic pressure and TP = Turgor pressure).	5. $\Psi = \Psi m + \Psi s + \Psi p$. (where Ψm = matric potential, Ψs = solute potential and Ψp = pressure potential).

Plasmolysis and Deplasmolysis (Fig. 2.15)

When a plant cell is placed in hypertonic solution, the process of exosmosis starts where the water from cell-sap starts diffusing into the solution of external medium through the plasma membrane. It reduces the tension of cell-wall and results in contraction of the protoplasm due to continuous loss of water. Now the protoplasm starts separating from the cell-wall and after some time complete protoplasm of the cell becomes rounded in shape due to contraction. This phenomenon of protoplasm in the cell is called *plasmolysis* and such a cell is called *plasmolysed cell*. The space between cell-wall and contracted protoplasm is filled with the external solution of the medium. The initial stage of plasmolysis where the protoplasm of the cells just starts leaving the cell-wall is called *incipient plasmolysis*.

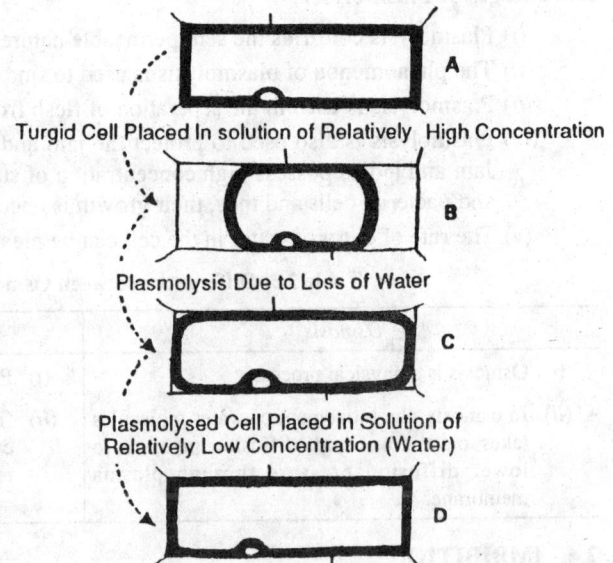

If completely plasmolysed cell is again placed in water or hypotonic solution, the process of endosmosis starts and the cell-protoplasm regains its original stage or shape. Thus, the protoplasm again starts touching the cell-wall and the cell becomes turgid. This phenomenon is just reverse of plasmolysis and is called *deplasmolysis*.

Fig. 2.15: Plasmolysis and Deplasmolysis.

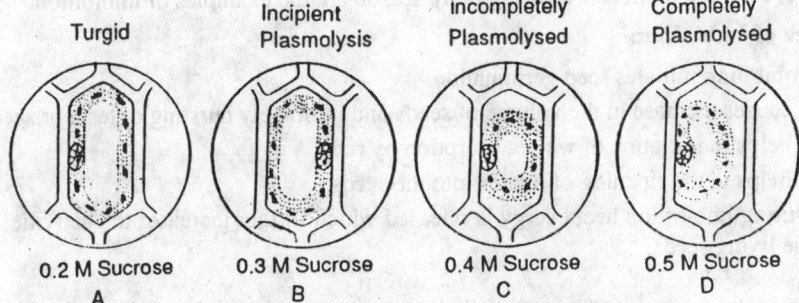

Fig. 2.16: Different stages of plasmolysis. (A) Normal cell, (B) Incipient plasmolysis, (C) Late stage of plasmolysis, (D) Completely plasmolysed cell.

When the cell is placed in isotonic solution, no change except a slight movement in the plasma membrane, could be noticed.

Water Potential Changes in Plasmolysis and Deplasmolysis

In a plasmolysed cell the pressure potential becomes zero and the cell-walls do not remain in touch with the protoplasm. Further, the extremely plasmolysed cells, the protoplasmshrinks and forms a small spherical ball like structure. In extremely plasmolysed cells, however, the negative pressure potential does not develop because there is no inward pull on the cell-wall. Deplasmolysis always occurs when the plasmolysed cell is placed in the solution of higher water potential (water).

The *hypertonic solution* is that which has lower water potential or higher osmotic potential than that of the cell, whereas the *hypotonic solution* is that which has higher water potential or lower osmotic potential. The isotonic solution has the same osmotic potential as that of cell.

Advantages of Plasmolysis

(*i*) Plasmolysis confirms the semipermeable nature of plasma membrane.

(*ii*) The phenomenon of plasmolysis is used to find out the OP of cell-sap.

(*iii*) Plasmolysis is used in the separation of flesh from bones.

(*iv*) Plasmolysis is also used to protect the jam and jelly from fungal and bacterial infection. Jam and jellies possess high concentration of sugar which plasmolyses the fungal hyphae and bacterial cells and thus, their growth is checked.

(*v*) The rate of entry of water in the cell can be measured by plasmolysis.

Table 2.3: Differences between Osmosis and Plasmolysis

Osmosis	Plasmolysis
(*i*) Osmosis is a physical process.	(*i*) Plasmolysis is a biological process.
(*ii*) In osmosis, the movement of water molecules takes place from high diffusion pressure to lower diffusion pressure through plasma membrane.	(*ii*) In plasmolysis, the diffusion of water from cell cytoplasm takes place through exosmosis resulting in contraction of cytoplasm.

2.4. IMBIBITION

Some materials like gum, starch, cellulose, gelatin, agar and protein etc. when placed in water, they absorb it and swell up. This phenomenon is called *imbibition*. This is also a type of diffusion which is performed like osmosis. In this process, there is a great difference between diffusion pressure of the liquid of imbibing cell and the liquid of external medium. Due to this reason the process of imbibition takes place and it continues till the pressures of cell-liquid and external medium becomes equal or balanced. The swelling of dry seeds when placed in water, the swelling of wooden windows, tables, doors, etc., due to moisture during rainy season are the examples of imbibition.

Significance of Imbibition

1. Imbibition initiates seed germination.
2. It causes increase in the volume of seeds and ultimately bursting of testa or seed coat.
3. It helps in initiation of water absorption by roots.
4. It helps in the ripening of ovules into the seeds.
5. During imbibition heat energy is released which further increases the activities of cells of the living seeds.

Experiment

To demonstrate the phenomenon of imbibition : Take a beaker and fill it up about 1/3 with dry gram seeds. Now pour the water in beaker and leave it for overnight. In the morning you will see that the seeds become swollen and almost fill the beaker. The seeds swell due to imbibition of water. The difference in increase in volume can be noticed.

Osmotic Relations of Plant Cells

2.5. OTHER EXPERIMENTS

1. *To compare two solutions (hypertonic and hypotonic) by plasmolytic method* (Fig. 2.17): For this experiment take the leaf of Rhoeodiscolor or Tradescantia. Now, peel off its lower coloured epidermis, mount in water and observe under high power of microscope. Note that all the cells are completely filled with red-coloured cell-sap. Now take another two peels and dip one peel in one solution and other in second solution. The two solutions should be of different concentrations (one weak and other strong). Leave the peels in both the solutions for sometime and then observe separately under microscope.

The peel dipped in strong solution shows the cells with plasmolysis while the peel dipped in weak solution shows no plasmolysis or only a small amount of plasmolysis. It indicates that the solution which causes plasmolysis of cells is physiologically concentrated or hypertonic while the solution which does not or slightly cause the plasmolysis of cells is physiologically weak or less concentrated or hypotonic. Thus, out of the two solutions, one will be hypertonic and other hypotonic. It is also observed that in isotonic solution, about 50 per cent cells show incipient plasmolysis.

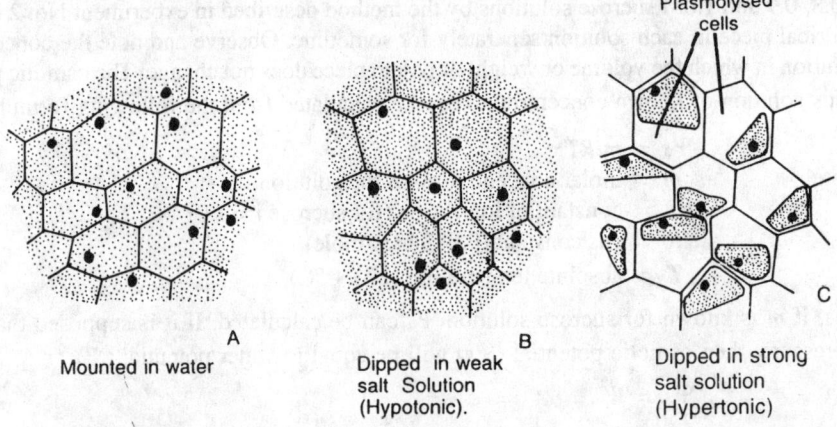

Fig. 2.17: Comparison of hypertonic and hypotonic solution by plasmolytic method.

2. *To determine the osmotic pressure of the cell sap by plasmolytic method:* For this experiment take the leaf of Rhoeodiscolor and prepare solutions of different concentrations. To prepare solutions of different concentrations, first prepare standard molar solution (SS) of sucrose by dissolving 85.5. gm of sucrose in 100 ml water and now add 150 ml water in it to make its final volume 250 ml. From this standard solution, prepare 0.1, 0.2, 0.3, 0.4, 0.5 and 0.6 M solutions by further dilutions as follows:

1. 1 ml of SS + 9 ml water = 0.1 M solution   (SS = standard solution)
2. 2 ml of SS + 8 ml water = 0.2 M solution
3. 3 ml of SS + 7 ml water = 0.3 M solution
4. 4 ml of SS + 6 ml water = 0.4 M solution
5. 5 ml of SS + 5 ml water = 0.5 M solution
6. 6 ml of SS + 4 ml water = 0.6 M solution.

Now, peel off the lower coloured epidermis of the leaf of Rhoeodiscolor and cut into six small pieces. Dip one piece in each solution for sometime and observe separately under microscope after mounting on the slide for plasmolysis. Note the concentrations of solutions where the cells do not show plasmolysis and the cells show plasmolysis. Suppose, the plasmolysis does not occur in 0.1, 0.2 and 0.3M solutions and occurs in 0.4, 0.5 and 0.6 M solutions. It indicates that the isotonic solution should be in between 0.3 and 0.4M sucrose solution. Now, prepare further dilutions of the solution between 0.3 to 0.4 i.e., 0.32, 0.34, 0.36 and 0.38M. Repeat the experiment with these dilutions and find out the concentration where 50 per cent of the cells show incipient plasmolysis. This concentration of sucrose solution will be isotonic with the cell-sap.

Osmotic pressure (O.P.) of the cell-sap can be calculated by the following formula:

$$OP = CRT$$

(where C = molar concentration of solution at incipient plasmolysis)
R = gas constant (0.083) and
T = absolute temperature (273 + °C)

The above formula is suitable only for non-electrolytes e.g., sucrose solution. If the experimental solution is an electrolyte e.g., Nacl, its osmotic pressure is calculated first and then multiplied by its degree of dissociation.

3. To determine the water potential of a plant cell or tissue by measuring changes in the volume or weight of the tissue : For the above experiment potato tubers are taken and peeled off. Now, cut uniform small cylindrical pieces from the peeled potato. Prepare 0.1, 0.2, 0.3, 0.4, 0.5, 0.6, 0.7, 0.8, 0.9 and 1.0M sucrose solutions by the method described in experiment No. 2 Immerse one cylindrical piece in each solution separately for sometime. Observe and note the concentration of that solution in which the volume or weight of potato piece does not change. The osmotic potential (Ψs) of this solution of known concentration can be calculated from the following formula:

$$-\Psi s = miRT$$

where m = molar concentration of the solution
i = constant of ionization (for sucrose i = 1.0)
R = gas constant (0.083 bars/mole)
T = absolute temperature (273 + °C).

Thus, if m is known for sucrose solution Ψs can be calculated. If it is supposed that there is no wall pressure, then osmotic potential (Ψs) will be equal to water potential (Ψ).

$$\Psi s = \Psi$$

4. To determine osmotic potential (Ψs) of the cell sap by plasmolytic method. To determine the osmotic potential, the same method is used as for experiment No. 3 Through this method the concentration of isotonic sucrose solution is determined and the osmotic potential is calculated by the following formula:

$$-\Psi s = miRT$$

Through plasmolytic method, the osmotic potential of only coloured tissues can be determined and not of colourless tissues because it is difficult to see plasmolysis in colourless tissues.

5. To determine pressure potential of the cell-sap : Once we get the water potential and osmotic potential through the methods described above, the pressure potential can be calculated by the equation:

$$\Psi = \Psi s + \Psi p$$

where Ψ = Water potential
Ψs = Osmotic potential
Ψp = Pressure potential

It can also be determined by a osmometer directly.

3

Absorption of Water

3.1. IMPORTANCE OF WATER TO THE PLANTS

Although water plays many roles in plants but some of the important ones are:

1. The amount of water present in the soil changes the morphology and anatomy of the plants. Accordingly the plants are mesophytes, hydrophytes and xerophytes.
2. Water is a good solvent. It also acts as a reactant in various chemical reactions in plant cells.
3. Water helps in the formation of protoplasm.
4. The absorption and translocation of mineral salts and dissolved substances take place through water.
5. During photosynthesis, water is oxidized and O_2 is produced.
6. Water effects transpiration, seed germination, respiration, dispersal of fruits and seeds, activation of enzymes, hydrolysis of ATP growth and all other metabolic processes.
7. Water maintains the temperature in plant tissues.
8. Water maintains the turgidity of plant body.

3.2. SOIL WATER

Soil water is the most important constituent of soil because it provides the medium for the absorption of minerals and organic matter by the roots and also activates various enzymes and metabolic processes. It affects morphological, anatomical and physiological processes directly or indirectly. Soil-water later on reaches to the various other parts of the plant through roots. The vegetation of an area depends upon the quantity of soil water. The water present in the soil may be of following types:

1. *Gravitational water:* The water which reaches deeply into the soil after rains due to gravitation is called *gravitational water*. The plants cannot absorb this water through their roots.

2. *Capillary water:* The water which remains present in the intercellular spaces of soil particles is called *capillary water*. It is the only available water which is absorbed by plant roots.

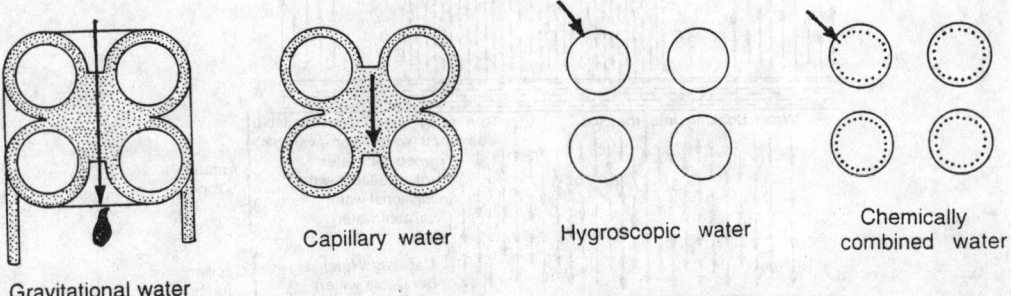

Fig. 3.1: Different types of soil water and their position in soil particles.

3. *Hygroscopic water:* The water which is present around the soil particles in the form of thin vapour layer is called *hygroscopic water*. It is also non-available water and is not absorbed by the plant roots.

4. *Crystalline water or chemically combined water:* The water which remain chemically bound to the soil particles is called *crystalline water*. It is also non-available water, *e.g.*, $CuSO_4.5H_2O$, $FeSO_4.7H_2O$ etc.

5. *Running water:* The non-available water which after rains flows down through the slopes is called *running water*.

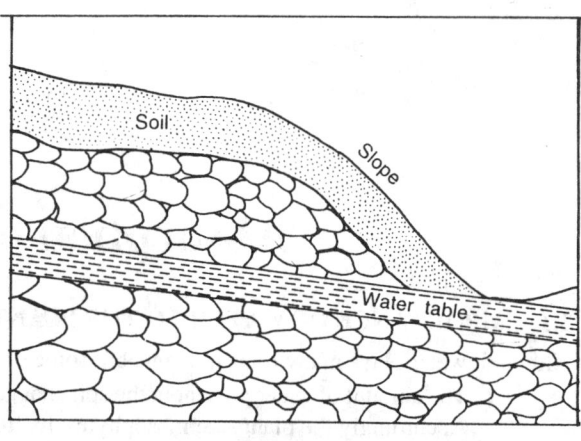

Fig. 3.2: Steepness of water-table in soil and its relation with the slope.

3.3. FORMS OF UNDERGROUND WATER

Underground water is found in following two forms:

1. *Water-table:* It is found quite deep in the soil near rocky layer. It contains free water which comes down into the soil through gravitation. The water from water-table ascends towards soil-surface through *capillary forces*.

2. *Capillary fringe* (Fig. 3.3): It is present just above the water-table and represents that part of soil where the water ascends from water-table by capillary forces. Capillary fringe is absent in gravel soil. The soil portion which is present above the capillary fringe is called *aeration zone*. It contains more field-capacity. The field capacity of any soil is the percentage of water present in that soil which remains after the gravitational water is drained away. Thus, field capacity of any soil is the sum of capillary water, hygroscopic water and crystalline water.

Field capacity = Capillary water + hygroscopic water + Crystalline water.

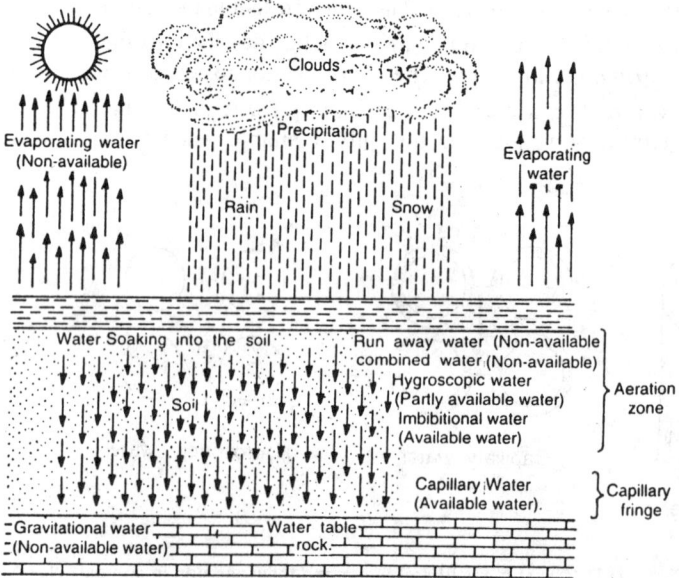

Fig. 3.3: Different types of soil water and capillary fringe.

The maximum quantity of capillary water, hygroscopic water and crystalline water in any soil represents the *water-holding capacity* of that soil.

3.4. WATER CONTENT OR HYGROSCOPIC COEFFICIENT

The percentage of water present in terms of the dry weight of the soil is called *hygroscopic coefficient*.

It can be calculated by collecting soil samples from the field, weighing them in balance and drying in an oven at 105°C to obtain constant weight. The loss in weight in the soil sample will be water content or hygroscopic coefficient of that soil.

3.5. MOISTURE EQUIVALENT

The percentage of water a soil can hold in opposition to pull 1,000 times that of gravity is called *moisture equivalent*.

To determine it, soil samples are placed in especially designed cup with a perforated bottom and whirled in a centrifuge for half an hour. The loosely held water by whirling is displaced. Now, the water retained by soil is estimated by the usual procedure of weighing and oven-drying. The result is expressed in percentage of dry weight. This value represents the moisture equivalent.

3.6. WILTING

In dry atmosphere, sometimes the plants show scaricity of water in them and their leaves become wilted. It is called *wilting*. There are two reasons for it:
1. High rate of transpiration by plant in comparison to less absorption of water by roots. The wilting due to this reason is called *temporary wilting* because the leaves regain their natural condition during night due to low transpiration rate.
2. Less quantity of water in soil. When the leaves become wilted due to less quantity of water in soil, it is called *permanent wilting*. In such stage, the leaves do not regain natural condition at night. The permanent wilting can be overcome by adding water in the soil.

(1) *Permanent Wilting Point:* The soil moisture content present at the time of permanent wilting is called *permanent wilting point*.

(2) *Wilting Coefficient:* The percentage of water present in soil after permanent wilting is attained is called *wilting coefficient*. For its determination, the seedlings are grown in glass containers containing soil and mixture of paraffin and vaseline. When wilting is attained and it does not recover overnight, the moisture content of the soil is estimated by drying at 105°C and percentage dry weight is calculated. The wilting coefficient thus obtained is the *observed wilting coefficient*. This method is applicable only in case of seedlings and it is not suited to mature plants or plants with advanced vegetative phase.

If hygroscopic coefficient, water-holding capacity and moisture equivalent are ascertained, moisture retentiveness of a soil sample, called *calculated wilting coefficient*, can be calculated with the help of following formulae.

$$W = \frac{\text{Hygroscopic coefficient}}{0.68 \pm 0.012}$$

$$W = \frac{\text{Water–holding capacity}}{2.9 \pm 0.016}$$

$$W = \frac{\text{Moisture equivalent}}{1.84 (1 \pm 1.007)}$$

In all the above formulae W stands for wilting coefficient. This value is of great importance to ecologists, agronomists and physiologists.

3.7. WATER ABSORBING PARTS OF THE PLANTS

The absorption of water in different types of plants is performed by some of their special organs. In lower plants like aquatic fungi and algae, the water is absorbed by their all cells of plant body as they are found submerged in water. In other aquatic plants, the absorption of water takes place by the cells of external surface of different organs. Such plants do not have any special organs for water absorption. In liverworts and mosses the water absorption takes place through unicellular or multicellular rhizoids which are similar to those of root hairs persent in higher plants and also perform the function of roots. Liverworts possess unicellular rhizoids whereas mosses possess multicellular rhizoids. In pteridophytes, gymnosperms and angiosperms the water is mainly absorbed by the roots present in the soil. These roots remain spread in the soil water. The roots possess many unicellular root hairs which absorb only capillary water of the soil. The epiphytes like orchids, possess aerial roots which contain velamen layer. The cells of this layer are hygroscopic in nature and absorb water present in the atmosphere. In some plants like *Vitis, Lycopersicum, Beta, Sorghum,* and *Phaseolus,* the water is absorbed by leaves. The following factors affect water absorption by leaves :

(*i*) Dry or wet surface of leaf.
(*ii*) Presence of hairs on the leaf-surface.
(*iii*) The structure of epidermis and cuticle and their permeability.
(*iv*) Lack of water in the cells of cortex.

3.8. ROOT SYSTEM IN PLANTS

The part of the plant which develops from the radicle of germinated seed is called *root*. At the time of seed germination, first radicle emerges which forms primary root. The secondary and tertiary roots are formed from primary and secondary roots respectively. Primary, secondary and tertiary roots collectively form the root system. In some plants the special roots are formed which do not develop from radicle.

3.9. CHARACTERISTICS OF ROOT

(*i*) Root is the descending part of the plant axis which develops from the radicle of embryo.
(*ii*) Roots are negatively phototropic.
(*iii*) They are usually brown, white or ash coloured.
(*iv*) Roots lack nodes, internodes, leaves, fruits and flowers.
(*v*) Roots lack buds. In some plants the roots possess vegetative buds which develop into new plants.
(*vi*) Roots possess a root cap at their tips which protects the root from injury at the time of growth.
(*vii*) Roots possess many unicellular root hairs which help in the absorption of water and mineral salts.

3.10. REGIONS OF THE ROOT (Fig. 3.4)

A root possesses following distinct five regions :

(A) *Root Cap region* : It is found at the tip of root. It protects the root from injury at the time of growth. During growth when old cells become dead, new cells are formed.

(B) *Meristematic* region : It is also known as *region of cell division* and *region of growth*. It is found just above the root cap region in about 1 mm in length. It remains covered by many root hairs. The cells of this region are active, thin cell-walled and with large protoplasm. They divide and redivide to form new cells.

(C) *Region of elongation* : It is found just above the meristematic region in about 2–5 mm in length. In this region, the cells elongate resulting in increase in length.

(D) *Region of root hair* : It is found just above the region of elongation and possesses many unicellular root hairs. Here, newly-formed meristematic cells are converted into permanent tissues.

Absorption of Water

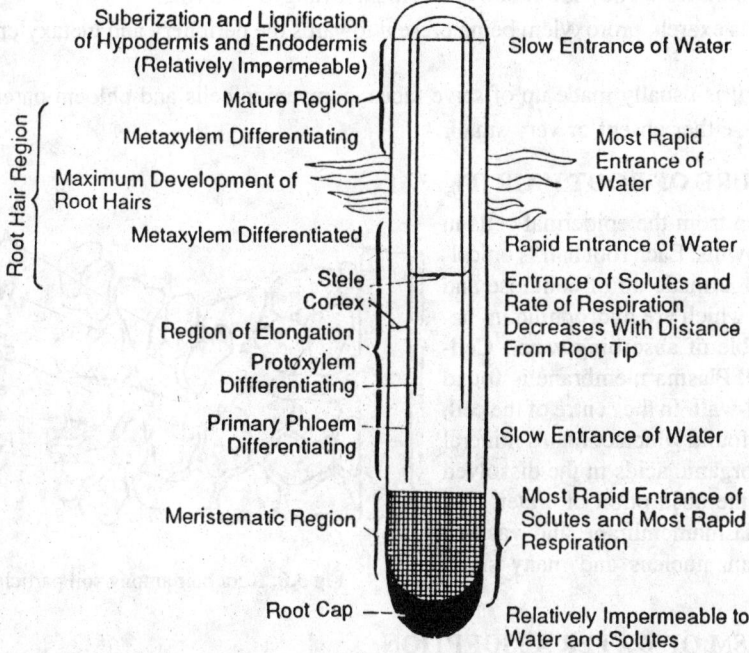

Fig 3.4 : Various regions of the root.

Root hairs are important organs of plants because they help in the absorption of water and mineral salts. One millimetre area possesses about 2000 root hairs. *Utricularia, Hydrilla* and *Spirogyra* etc. do not possess root hairs.

(E) *Mature region* : This region is present above the region of root hair. It produces many additional roots which provide anchorage to the plant.

3.11. INTERNAL STRUCTURE OF ROOT (Fig. 3.5)

Following structures are seen in a TS of root:

1 *Epidermis* : It is single-layered, hairy and without stomata and cuticle.

2. *Cortex*: It is multiallular, parenchymatous and with many well developed intercellular air spaces.

3. *Endodermis* : It is well developed, single layered and made up of barrel shaped cells. Endodermis possesses passage cells and casparian strips.

4. *Pericycle*: It is single-layered, parenchymatous and without intercellular spaces.

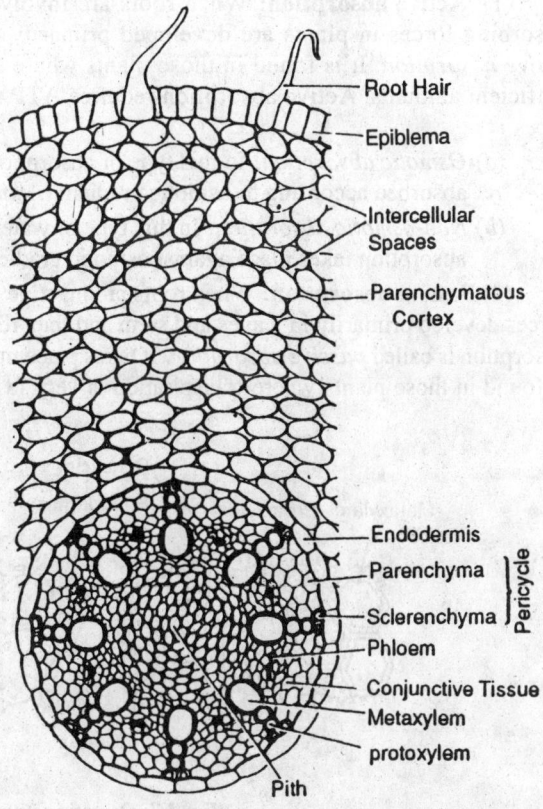

Fig. 3.5 : Internal structure of root (T.S. of root).

5. *Vascular bundles* : They are radial and remain arranged in a ring.

6. *Xylem* : It is exarch, protoxylem being present towards the periphery and metaxylem towards the centre.

7. *Phloem* : It is usually made up of seive tubes, companion cells and phloem parenchyma.

8. *Pith* : It is either absent or very small.

3.12. STRUCTURE OF ROOT HAIR (Fig. 3.6)

Root hairs develop from the epidermal cells in the form of outgrowths. Each root hair is unicellular. Its cell-wall is made up of cellulose and pectic substances which are hydrophilic in nature and are capable of absorbing water. Cell-wall is permeable. Plasma membrane is found just below the cell-wall. In the centre of the cell, a large vacuole is found which contains mineral salts, sugars and organic acids in the dissolved form. It controls the absorption of water. The space between plasmamembrane and vacuole contains cytoplasm, nucleus and many small vacuoles.

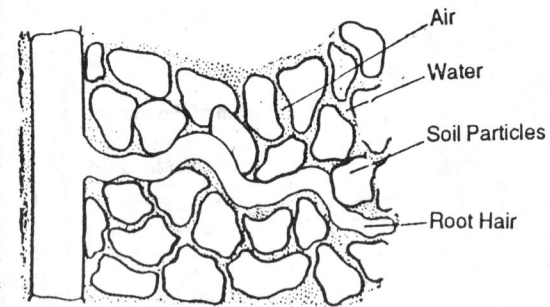

Fig 3.6: Root hair among soil particles.

3.7. MECHANISM OF WATER ABSORPTION

According to *Renner* (1912 and 1915), the mechanism of water absorption is of following two types:

(1) **Active absorption:** When roots are involved actively in water absorption and water absorbing forces in plants are developed primarily in roots, such type of absorption is called *active absorption*. It is found in those plants where transpiration is less and water is present in sufficient amounts. Active absorption requires ATP released during respiration. It is also of two types :

(a) *Osmotic absorption:* In this type of absorption the roots act like a osmometer and water is absorbed according to osmotic gradient (*Atkins*, 1961 and *Priestley*, 1922).

(b) *Non-osmotic absorption*: In this type of water absorption more ATP is required and water absorption takes place against osmotic gradient (*Thimann*, 1951 and *Kramer,* 1959).

(2) **Passive absorption:** When roots are inactive in water absorption and the water absorption forces develop primarily in leaves and stem and then reach to root through xylem, this type of water absorption is called *passive absorption*. It takes place mainly due to transpiration. Passive absorption is found in those plants where transpiration is very fast and it does not require ATP.

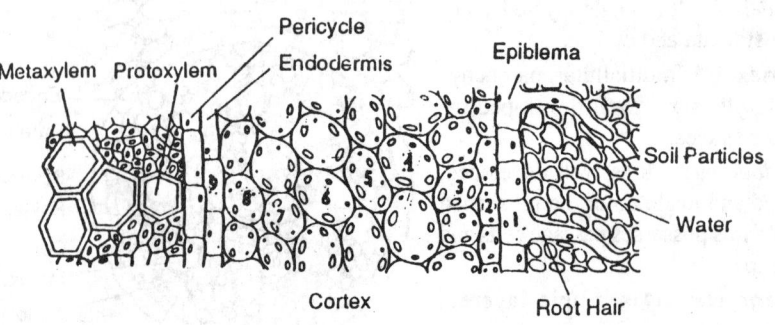

Fig 3.7: Osmotic active absorption.

(a) *Osmotic Active Absorption*: Soil is made up of irregularly arranged soil particles. Well developed air spaces are found among these particles. The spaces contain capillary water where air and mineral salts remain dissolved. The root hairs remain spread in this water. It has already been described earlier that each root hair contains a vacuole which remains filled with mineral salts, sugar and organic acids. The cell-wall of root hair is permeable and it does not create any hindrance in the entry and exit of the liquids in the cell. Plasma membrane is semipermeable and it allows only the diffusion of water and important dissolved salts into the cytoplasm. The cell-wall of root hair being hydrophilic in nature first absorbs soil water through imbibition. The cytoplasm of root hair is usually concentrated than the capillary water of soil. The OP of cell-sap of root is also greater (OP = 3 – 8 atm.) than the OP of capillary water of soil (OP = 1 atm.). Thus, the DPD and SP become more in root hairs resulting in osmotic diffusion or endosmosis of water and its dissolved substances into the root hair.

After a definite time, the cells of root hairs become turgid and their OP, DPD and SP are reduced and TP is increased due to continuous absorption of water into the root hair. At this time, the cytoplasm of root hair becomes thin and its OP, DPD and SP are reduced in comparison to the cytoplasm of its adjacent first cortex cell—resulting in osmotic diffusion of water and its dissolved substances from root hair into the first cortex cell. Now, the OP, DPD and SP of this first cortex cell are also reduced and TP is increased. Similarly the osmotic diffusion of water and its dissolved substances from first cortex cell to second adjacent cell takes because the second cortex cell possesses greater OP, DPD and SP than the first cortex cell. This process continues till the water reaches to pericycle through all cortex cells and endodermal cell. The endodermis possesses passage cells and casparian striped cells. The passage cells are thin walled and found against protoxylems. Through these passage cells the water diffuses into cells of pericycle and ultimately into xylem vessels. As the water diffuses from root hair to first cortex cell, the cytoplasm of root hair cell again becomes concentrated and its OP, DPD and SP are also increased resulting again in osmotic diffusion of capillary water of soil into the root hair. This phenomenon continues for long time and thus, osmotic active water absorption takes place.

Pathway of Water in Root

The pathway of water in root cells is as follows:

Root hair ⟶ Epidermal cell ⟶ Various successive cortex cells ⟶ Endodermal cell (passage cell) ⟶ Cells of pericycle ⟶ Xylem cells ⟶ Xylem duct ⟶ Upward movement of water.

Root Pressure: When the water enters from the turgid pericycle cells into xylem vessels, a pressure is created in the xylem of roots due to which the water rises to a certain height in the xylem. This pressure is called *root pressure*. Actually the root pressure is a type of hydrostatic pressure which is produced in the cell-sap of xylem vessels.

Demonstration of Root Pressure (Fig. 3.8): For the demonstration of root pressure, take a potted plant of tomato, *Bryophyllum* or *Zenia* and cut the main stem in water with the help of knife at about 4-5 cm above the soil. Now connect the cut end of stem with a manometer glass tube through rubber-ring as shown in Fig. 3.8 and fill the tube first with some water and then with mercury. Leave the experiment for sometime and note the rise in mercury level in manometer which is mainly due to root pressure. The water absorbed by the root hairs is also pushed up in xylem vessels due to root pressure. If the upper portion of stem of any plant is cut, some liquid is oosed out due to root pressure.

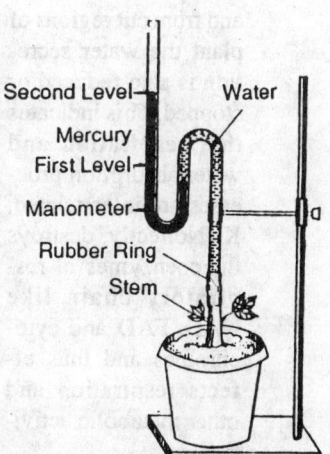

Fig 3.8: Demonstration of Root Pressure.

The well-known example of demonstration of root pressure is the production of toddy juice from the cuts of upper portion of stem.

Theories Regarding Development of Root Pressure

Following theories have been proposed to explain how the root pressure develops in plants:

1. *Secretion theory:* According to this theory the root pressure develops by the secretion of water into the xylem. The secretion of water results due to difference in the permeabilities of water in two sides of the root cells. It is higher towards inner side and lower towards outer side. Some energy may be required for this work which is provided by respiration through root cells.

2. *Electro-osmotic theory:* According to this theory the root pressure develops due to electro-osmotic movement of water into the xylem. It has been proved experimentally that when an electric current is passed into the membrane, water moves across this membrane. However, this theory has been discarded.

3. *Osmotic theory:* According to this theory the root pressure develops due to osmosis. The root behaves like an osmometer due to accumulation of solutes. The soil solution remains less concentrated than xylem sap. Thus, water moves from soil solution to xylem through various tissues of root developing root pressure. The details of this theory will be discussed under the mechanism of water absorption.

Non-Osmotic Active Absorption

According to *Thimann* (1951) and *Kramer* (1959) the water absorption is an active process which takes place against the osmotic gradient. In other words, sometimes the water absorption also takes place when the osmotic pressure of soil water, is greater than the OP of cytoplasm of root hair. This is against osmosis. Such type of water absorption is called *non-osmotic active absorption* and it requires ATP produced during respiration of root cells. How this energy is utilized, it is not well worked out. The energy may be utilized directly, or indirectly. The theory of non-osmotic active absorption is supported by the following facts :

(i) *Correlation between water absorption and rate of respiration* : Usually in presence of respiration reducing factors the rate of water absorption is also reduced. Thus, both the processes are inter-related.

(ii) *Reduction of water absorption in presence of respiratory inhibitors* : It has been observed when a plant is placed in the solution of respiratory inhibitors like KCN, the rate of water absorption is reduced and from cut regions of plant the water secretion is also reduced or stopped. This indicates that respiration and water absorption processes are inter-related. KCN directly, destroys the coenzymes of respiratory chain like NAD, FAD and cytochromes and thus effects respiration and other metabolic activities.

(iii) *Wilting of plants in oxygen deficient soil* :

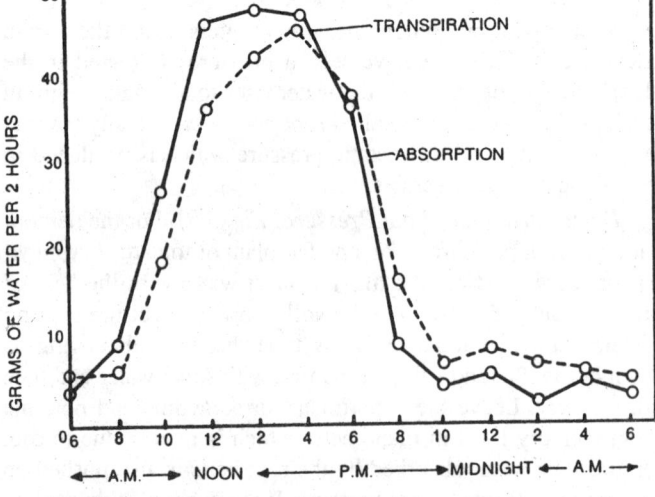

Fig. 3.9: Relationship between transpiration and passive absorptions.

In O_2 deficient soil, the plants do not get sufficient O_2 for respiration and ultimately wilt.

(iv) *Effects of Auxins* : In presence of auxins, the rates of metabolisms along with water absorption are increased.

(v) *The occurrence of water absorption and respiration only in living cells* : Both these processes are found only in living plants and not in dead plant.

Passive Absorption

It takes place mainly due to transpiration. In passive absorption, the roots remain inactive and the water absorbing forces are first produced in the cells of leaves. When DPD increases in the cells of leaves due to transpiration, the water diffuses from the xylem cells of leaves to all mesophyll cells. When the rate of transpiration is high, a tension is created in the water column of xylem which increases the DPD of water. This tension is like negative pressure and it moves from leaves to roots. At this stage, the DPD of peripheral cells of young roots becomes very high. The DPD in roots increases from xylem cells to epidermal root hairs. The rate of passive water absorption is directly proportional to the rate of transpiration and it requires sufficient amount of water in the soil.

Table 3.1: Difference Between Active and Passive Absorption

Active absorption	*Passive absorption*
1. It requires energy or ATP.	It does not require energy.
2. Active absorption creates root-pressure.	Root pressure is not created due to passive absorption.
3. It requires oxygen.	It does not require oxygen.
4. The movement of water takes place from the solution of higher concentration to the solution of lower concentration i.e., against the concentration gradient.	The movement of water takes place from the solution of lower concentration to the solution of higher concentration i.e., according to osmotic gradient.

3.14. APOPLAST AND SYMPLAST CONCEPT

The two terms, apoplast and symplast, were used by *Munch* (1930) to explain the flow of water and minerals in plants. He used the term apoplast for the *non-living or dead portions* of the plants like interconnecting cell-walls, intercellular spaces, cell-walls of endodermal cells excluding casparian strips, cell-walls of pericycle, xylem tracheids and vessels. All these non-living portions constitute a single system called *apoplast*. This system is continuous throughout the plant and through this system the water moves chiefly due to capillary action or free diffusion along the gradient. The term *symplast* was used by Munch for the *living portions* of the plants like cytoplasm of all living cells connected through plasmodesmata present in the cell-walls. Since only the living system is included under symplast, here the water moves chiefly due to osmosis.

3.15. FACTORS AFFECTING THE RATE OF WATER ABSORPTION

(1) *Available soil-water* : It has been discussed earlier that soil water is found in many forms (gravitational, capillary, hygroscopic and chemically bound water) but out of these only capillary water is availabale for the absorption of plants. The rate of water absorption in plants remains constant the same in between field capacity and permanent wilting percentage. If the amount of water is increased than the field capacity, it creates a bad effect on soil aeration and also reduces the rate of water absorption. Similarly, on shortage of soil water or reduction in permanent wilting percentage, the water absorption rate is also reduced.

(2) *Soil aeration* : The rate of water absorption increases when the aeration in soil is high. In other words, the rate of water absorption is decreased in absence of oxygen. The reduction in rate

of water absorption is due to lack of O_2 and accumulation of CO_2 which effects the plants in following manner:

(i) CO_2 reduces metabolic activities and respiration rates.

(ii) It reduces the size and growth of roots.

(iii) It increases the concentration of protoplasm which results in the reduction of permeability.

Due to above effects of CO_2, the rate of water absorption is decreased. This situation is created in water-logged soil where the aeration is quite poor. Such soil is usually physiologically dry and is not fit for water absorption.

(3) *Concentration* of the *soil solution* : The osmotic pressure of any solution is directly proportional to its concentration. In other words the OP of more concentrated solutions is always greater than the OP of less concentrated solutions. If the soil water contains more salts, its conc. and OP will also be more. The rate of water absorption will be reduced if the OP of soil solution will be more than the OP of cell sap of roots. Due to this reason, the plants growing in alkaline and marsh soil show a very little or no water absorption.

(4) *Soil temperature*: Generally, the rate of water absorption is maximum between 20–30°C. It is reduced if the temperature is less than 20°C or more than 30°C. At very low temperature or at 0°C, the water absorption is almost stopped and its rate becomes zero. On very much reduction in temperature, the rate of water absorption is affected directly or indirectly in the following manners:

(i) It increases the concentration of protoplasm.

(ii) It reduces the permeability of plasma membrane.

(iii) The growth of roots is reduced which results in reduction of their length.

(iv) The rate of diffusion of soil-water into roots is also reduced.

(v) Metabolic activities of root cells are also reduced.

(5) *Root system*: Root system also affects the rate of water-absorption. Those plants which possess hairy and well developed root system show higher rates of water absorption in comparison to those which possess very small roots and less number of root hairs.

4
Ascent of Sap

4.1. INTRODUCTION

The water and soluble mineral salts absorbed by the roots reach to the leaves through roots, stem and branches of the plant. The phenomenon of ascending of absorbed water against gravitation through the vessels and tracheids of xylem is called *ascent of sap*.

4.2. PATH OF ASCENT OF SAP

The water ascends through the vessels and tracheids of xylem. In other words, the path of ascent of sap is xylem. It can be demonstrated by the following experiments.

(1) *Experiment No. 1*: It can be studied easily in the white flowered Lupin plant or Rose plant. Cut a flowering twig of either Lupin plant or Rose plant in water to avoid the entry of air bubbles and place it in a beaker containing a solution of eosine in water. After a few hours, the minute veins of the petals of flowers become redish. Now, cut the TS of stem and observe that xylem vessels and tracheids are red. It proves that ascent of sap in plants takes place through vessels and tracheids of xylem.

(2) *Experiment No. 2*: Cut the lower portion of white Balsam stem carefully in water so that the air bubbles could not enter into it. Place this portion vertically in eosine solution as shown in Fig. 4.1 and observe after a few hours that the stem possesses many minute vertical red linings due to ascent of eosine solution. Now, cut the TS of this stem and observe that only the vessels and tracheids of xylem become red whereas other tissues have no effect of eosine and remain white.

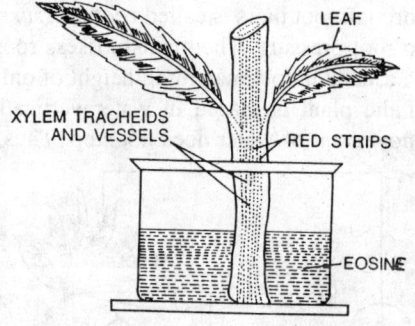

Fig. 4.1: Demonstration of ascent of sap.

It again proves that the absorbed water in the stem ascends only through the vessels and tracheids of xylem.

(3) *Experiment No. 3– Ringing Experiment*: This experiment was performed for the first time by *Stephen Hales*. It also demonstrates the ascent of sap in the stem through vessels and tracheids of xylem. For this experiment, take a potted plant and remove in circular fashion at some place of stem, all the outer tissues (epidermis, cortex, endodermis, pericycle and phloem) except vessels and tracheids of xylem and pith. Leave the experiment for a few days and observe that the leaves of the plant are still green. It proves that the water and soluble salts ascend through xylem

Fig 4.2: Ringing experiment.

tissues. The question may arise that the water may also ascend through pith cells instead of xylem cells. To confirm that water and dissolved salts ascend through xylem tissues and not through pith cells, destroy the pith cells also with the help of a large needle or wire and set the experiment. After a few days you will observe again that the leaves of this plant are still green. This confirms that the water ascends only through the xylem tissues and not through any other tissue.

From all the above experiments, it becomes clear that the water and dissolved mineral salts ascends in the stem through vessels and tracheids of xylem (xylem tissues). Now, the question arises, how the water ascends in the stem. In very tall plants like **Sequoia gigantia**, the water ascends to a height of about 115 metres. How this water ascends to this height? What is that force which helps the water to reach such a height against gravitation? In this context, there are many theories called *"Theories of ascent of sap."*

4.3. THEORIES OF ASCENT OF SAP

They are divided into two main groups : (a) Vital theories, and (b) Physical theories.

Vital Theories

(1) *Root pressure theory*: The pressure developed in the xylem cells by the absorption of water through root hairs is called *root pressure*. The water ascends in the xylem vessels and tracheids due to root pressure. The plants possess root pressure of only two atmospheres which helps the water to ascend upto a maximum height of only 20 metres. It has been confirmed by the experiments that if the plant is placed in water vertically after removing its root portion, even then the upward movement of water does not stop. Thus, the root pressure theory proves to be wrong.

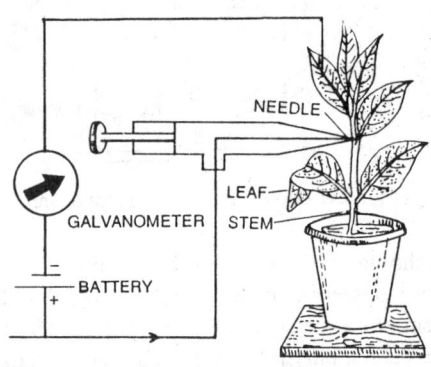

Fig 4.3: The electric probe of Sir J.C. Bose.

(2) *Godlewski theory* : Godlewski (1884) proposed that the water ascends in the plant due to pumping action of the cells of xylem parenchyma and medullary rays.

(3) *Vital force theory*: Well-known Indian scientist Sir J.C. Bose (1923) advocated that the cells of inner layer of cortex possess pulsating action like heart. When these cells expand, the water of adjacent cortex cells diffuses into them and they (cells of inner layer of cortex) become turgid. When these cells contract, the water ascends in the plant forcibly. Bose confirmed his observations with the help of an apparatus, electric probe, prepared by himself. He took the electric probe and connected it with a galvanometer through battery (Fig. 4.3). When he inserted the needle of the probe into the stem slowly by slowly, he observed the movement in the needle of galvanometer as the needle of probe entered into the inner layer of cortex but the needle of galvanometer did not show any movement during the entry of prob's needle into various outer layers of cortex. Most of the modern scientists are against this view.

Strasburger (1893) proved all the vital theories wrong. He observed the continuous ascending movement of water even if the cells of parenchyma and medullary rays were destroyed with the help of poisonous chemical substances like picric acid.

Physical Theories

(1) *Imbibition theory* : It was first proposed by *Unger*

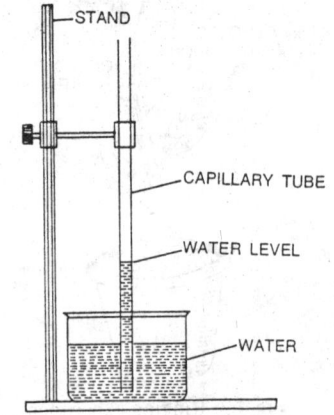

Fig 4.4: Ascent of water in capillary tube.

(1868) and then supported by *Sachs* (1878). According to this theory, the water ascends due to imbibitional forces through the walls of xylem tracheids and vessels and their lumen or cavities have no relation with the ascent of sap, but modern scientists have proved that the ascent of sap stops and the leaves wilt if the lumens of xylem vessels and tracheids are blocked with the wax. Thus, imbibition theory is also proved to be wrong.

(2) *Capillary force theory* : It was proposed by *Boehm* (1899). He was of the opinion that the water ascends in the xylem vessels through capillary action. If we place vertically a capillary tube in the water, the water ascends automatically in the tube up to some distance. It happens due to capillary force. The xylem vessels are also quite thin like capillary tube. Thus, the water ascends in them like capillary tube. It has been proved that through capillary force the water can ascend only up to a very small distance. Thus, this theory is wrong and will not be applicable for very tall plants.

(3) *Cohesion theory of Dixon and Jolly*: It was proposed by *Dixon and Jolly* (1894). It is most important and widely accepted view. They explained the theory in the following manner.

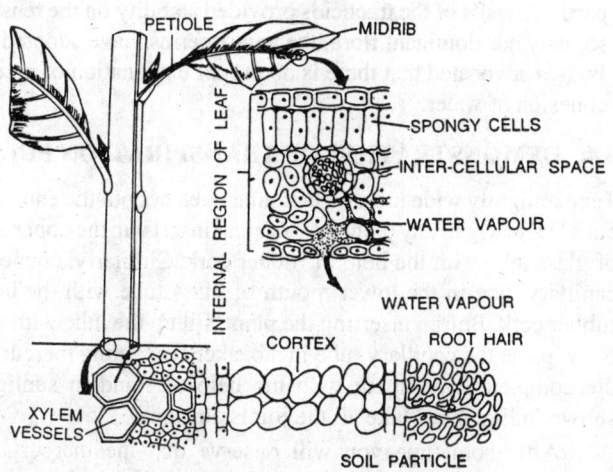

Fig 4.5: Demonstration of ascent of sap.

The water molecules remain attracted by a force called *cohesive force*. Cohesion is the phenomenon of attraction between similar molecules. The cohesive force maintain the continuity of water column in the xylem vessels. In other words, the water remains filled in the vessels continuously from roots to the leaves. The water evaporates continuously from the leaves due to transpiration. The mesophyll cells of leaves possess intercellular spaces filled with water vapours which go out in the atmosphere through stomata continuously. To overcome this deficiency the water vapours diffuse from the neighbouring mesophyll cells into intercellular spaces. Due to loss of water vapours the cytoplasm of mesophyll cells becomes concentrated and its osmotic pressure and suction pressure are increased resulting in the absorption of water from the adjacent cells. This process continues slowly by slowly upto the cells of xylem vessels. The attraction of water molecules from xylem vessels towards mesophyll cells also starts. The xylem vessels of roots, stem and leaves are connected with each other due to which the water absorbed through root hairs forms a water-column. Due to this cohesive force produced among water molecules, the water, ascends in the stem at a very height. This pulling force or suction force is called *transpiration pull* and the water column in the xylem vessels as *transpiration stream* because the force is created due to transpiration. The ascent of sap and water is directly proportional to the rate of transpiration.

According to cohesive force theory the suction force in leaves is created due to transpiration which pushes the water-column of xylem vessels towards leaves. The continuity of water column does not break due to cohesive force among water molecules. Due to both these forces the water column reaches to the upper apical leaves of very tall plants.

Evidences in Support of Cohesion Theory

1. With the help of several experiments Scholander provided evidences in favour of continuous freely mobile sap column and absence of metabolic pump.
2. In tallest trees during ascent of sap the combined all forces are only 50 atm whereas the

cohesive force of water is up to 350 atm. The hinderance created by combined forces of a tree is thus counteracted by cohesive force.

3. *McDermott* (1914) observed that if a vessel with water under tension was punctured, water snapped apart from the punctured portion.
4. *Thut* (1932) demonstrated that a leafy twig when cut under water and the cut end of twig is sealed to the top of mercury manometer, the mercury level ascends.
5. *Kramer and Kozlowski* (1960) demonstrated through dendrographic measurements the diameter variations in tree trunks. They observed that the diameter of the stem decreased during the day when transpiration was high. The shrinkage in the diameter of the cells was obviously due to strain caused on the water column by tension.

Criticism: Cohesion theory assumes tracheids to be more efficient than vessels. Dixon believed that partition walls of the tracheids provided stability on the tensilely stressed transpiration stream. If it is so, why our dominant flora, the angiosperms, have adopted vessels in the place of tracheids. *Barton Wright* advocated that there is no proper explanation of ascent of sap but they supported the view of cohesion of water.

4.4. DEMONSTRATION OF TRANSPIRATION PULL

Take a slightly wide mouth glass tube open at both the ends. Now, cut a leafy twig of any plant in water and insert it in the upper mouth of glass tube with the help of rubber cork. Similarly, connect the capillary tube in the lower mouth of glass tube with the help of rubber cork. Before inserting the plant, fill up the tube with water. Now, place the capillary tube in a beaker containing mercury. Set the complete experiment with the help of stand in sunlight as shown in Fig. 4.6. Make all the joints air tight.

After sometime you will observe that the mercury starts ascending in the capillary tube. Why does it happen? The reason is that in sunlight the leaves transpire due to which a suction is created in the xylem vessels, called *transpiration pull* resulting in increase in the mercury column of capillary tube. In other words the mercury ascends in the capillary tube due to transpiration pull. The plant twig should be cut in the water to avoid the entry of air bubbles and blockage of xylem vessels.

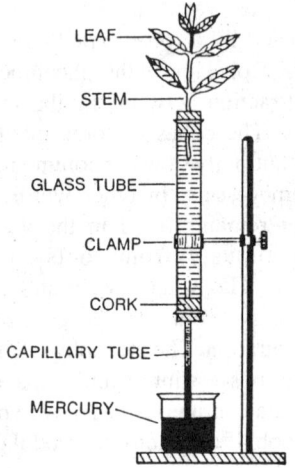

Fig. 4.6: Demonstration of transpiration pull.

5
Transpiration

5.1. INTRODUCTION

The roots of plants absorb water and mineral salts continuously. The total amount of water absorbed is not retained in the plants but only a very small amount necessary for various activities and composition of various organs of plants is retained. The excess amount is transpired through the aerial parts (stem, leaves, flowers and buds) of the plant. Thus, only 5 per cent of the absorbed H_2O is retained in the plants and remaining 95 per cent is lost through aerial parts. The leaves are most important for transpiration. *"The loss of excess water in the form of vapours from the various aerial parts of the plants is called transpiration."* The transpiration is like evaporation but both differ from each other in following points.

Table 5.1. Differences between Transpiration and Evaporation

Transpiration	Evaporation
(i) Transpiration is a biological phenomenon.	(i) Evaporation is a simple physical phenomenon.
(ii) It is controlled by guard cells.	(ii) It is not controlled by guard cells.
(iii) It takes place through the surface of leaves.	(iii) It takes place through the surface of various open water bodies and does not require the living organs of plants like leaves.
(iv) It takes place due to osmotic pressure and suction pressure.	(iv) Suction pressure or osmotic pressures are not involved in evaporation.
(v) It takes place in living cells.	(v) Plant cells are not required for the evaporation.
(vi) The temperature of the plant is maintained due to transpiration.	(vi) It has no relation with plant temperature.

5.2. KINDS OF TRANSPIRATION

The process of transpiration takes place through all the aerial parts of the plant but through leaves it is maximum. Transpiration may be of following three types :

Cuticular Transpiration

The loss of water in the form of vapours from the cuticle of various aerial parts is called *cuticular transpiration*. The water vapours reach to the cuticle through the internal tissues of aerial parts by diffusion and from cuticle it diffuses in the atmosphere. Of the total transpired water only about 5 to 10 per cent is lost through cuticle. The plants growing in shady places transpire up to 15 per cent of the total transpired water. The amount of water loss in xerophytic plants is quite little or negligible. In flowers, stems, and fruits mainly cuticular transpiration is found.

Lenticular Transpiration

The loss of water in the form of vapours through lenticels is called *Lenticular transpiration*. Lenticels are the small pores present below the bark of old trees.

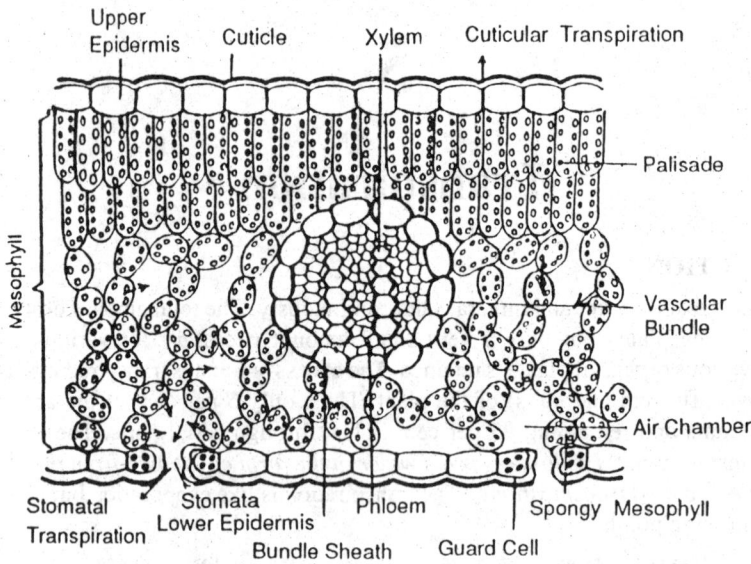

Fig. 5.1: Representation of cuticular and stomatal transpiration in T.S. of leaf.

Stomatal Transpiration

The loss of water in the form of vapours through the stomata of leaves is called *stomatal transpiration*. The maximum amount of (80-90 per cent) absorbed water is transpired through stomatal transpiration. Stomatal transpiration is commonly found in the leaves and stems of young plants. The crop plants like wheat, gram, maize, mustard and potato etc. also transpire through stomata. The water loss of 250 tonnes/day/acre has been recorded from wheat crop.

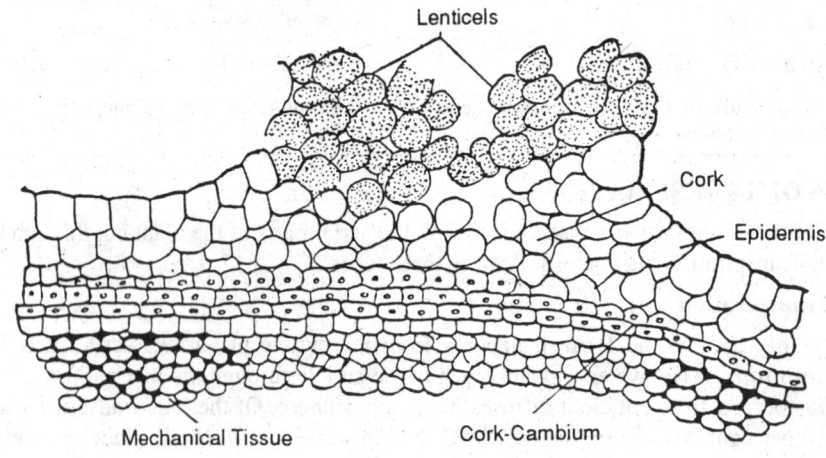

Fig. 5.2: TS of old stem showing lenticular transpiration.

5.3. STRUCTURE OF STOMATA

The epidermis of leaves and green stems possess many small pores called *stomata* (singular : stoma). The length and breadth of open stomata is about 10-40 μ and 3-10 μ respectively. Each pore remains surrounded by special semi-lunar or kidney shaped living epidermal cells called *guard cells*. The pore and guard cells jointly called **stoma**. The stoma opens to the interior into a cavity called *substomatal cavity* which remains connected with the intercellular spaces.

The guard cells possess a nucleus, cytoplasm and several chloroplasts. Their inner walls are thicker than the outer walls. Each guard cell is surrounded by two or more living cells called *subsidiary* or *accessory cells*.

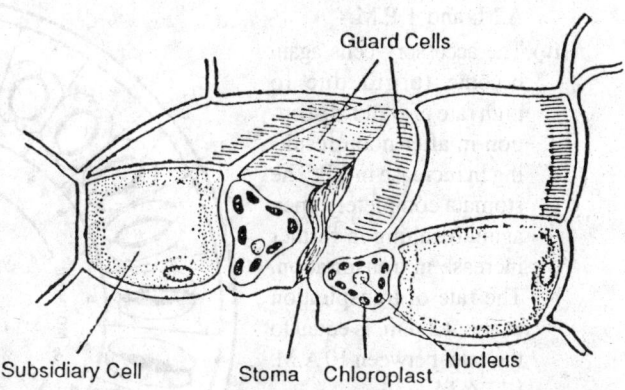

Fig. 5.3: Three-dimensional view of structure of stoma.

5.4. DISTRIBUTION OF STOMATA ON LEAF

The stomata are found distributed on the upper and lower surfaces of the leaves. In dorsiventral leaves, their number is more on the lower side. In isobilateral leaves *e.g.*, Monocotyledonous leaves, their number is equal on lower and upper both the surfaces. In floating leaves *e.g.*, water-lily, the stomata are found only on the upper surface. In submerged plants like *Hydrilla* and *Potamogeton* non-functional stomata are found. In xerophytes like *Capparis decidua, Casuarina* and *Pluchia lanceolata*, sunken stomata are found. In *Nerium* and *Pinus*, the stomata are found sunken in the cavities.

In dicotyledonous leave usually the stomata are found scattered where as in monocotyledonous leaves they are arranged in parallel rows. The number of stomata in a leaf varies from 1000 to 6000 per square centimetre. All the stomata of a leaf cover about 1-2 per cent of total area of leaf.

5.5. TYPES OF STOMATA

The stomata may be classified into following 5 types according to their distribution on leaves :
   (*i*) *Apple type*: When the stomata are found only on the lower surface of the leaf, the condition is called *hypostomatous*. This condition is found in Apple, Peach Mulberry and Walnut.
   (*ii*) *Potato type*: When the stomata are found more on the lower surface than the upper surface. *e.g.*, in Potato, Tomato and Pea.
   (*iii*) *Oat-type*: When the stomata are found equally on both the surfaces *e.g.*, Oat, Maize and Grasses.
   (*iv*) *Water-lily type*: When the stomata are found only on the upper surface *e.g.*, water lily.
   (*v*) *Potamogeton type* : When the stomata are absent or non-functional *e.g.*, *Potamogeton*.

5.6. DAILY PERIODICITY OF STOMATAL MOVEMENT

The rate of stomatal transpiration does not remain the same all times (24 hours) but it varies from morning to night in every hour. This variation in the rates of stomatal transpiration during each hour depends upon the opening and closing of stomata, which ultimately depends upon the intensity of light. Daily periodicity of stomatal movement is as follows :
   (*i*) When the sun rises in the morning, the stomata starts opening in sun-light. The transpiration also starts with the opening of stomata. Although its rate is negligibe or very low.
   (*ii*) The intensity of light increases with the increase in time resulting in more opening of stomata and increase in rate of transpiration. The stomata completely open in between 10 and 11 A.M. At this time the rate of transpiration is maximum.
   (*iii*) The high rate of transpiration decreases the turgidity of leaf cells and creates the internal water deficit. This results in more transpiration by leaves and less absorption of water by

roots. The TP of accessory cells decreases with the decrease in the turgidity resulting in flaccid condition of accessory cells and partially closing of stomata. Thus, the rate of transpiration is decreased. It happens between 11 A.M. and 1 P.M.

(iv) The accessory cells again become turgid due to high rate of water absorption in after noon resulting in increase in TP. The stomata completely open again resulting in further increase in transpiration. The rate of transpiration at about 3 p.m. is equal to the rate between 10 A.M. - 11 A.M.

(v) At about 4 P.M., the stomata start closing with the decrease in quantity and intensity of light. Thus, the rate of transpiration also starts decreasing.

(vi) At sunset the stomata become completely closed and continue during whole night. The transpiration rate is nil during night. The stomata again open with the sun rise. Thus, the stomata remain open during the day hours and closed during night hours.

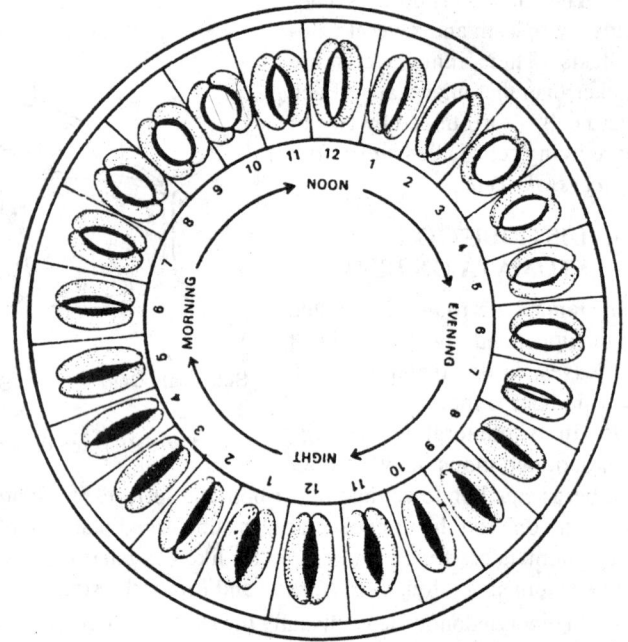

Fig. 5.4: Diagrammatic representation of daily periodicity of stomatal movement.

Although the stomata remain closed during night, the exchange of gases due to respiration continues.

5.7. MECHANISM OF TRANSPIRATION

The mechanism of transpiration is completed in two stages:

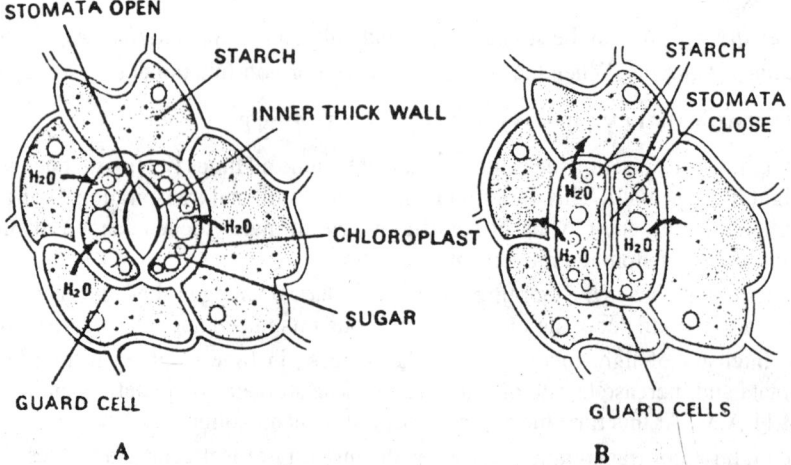

Fig. 5.5: Representation of mechanism of transpiration.

1. The diffusion of water of mesophyll cells into intercellular spaces.
2. The diffusion of water vapours of intercellular spaces into the outer dry atmosphere.

The roots of plants continuously absorb water and mineral salts from the soil which ascend through the xylem vessels and reach to the leaves. The mesophyll cells of leaves become turgid due to excess amount of this absorbed water. At this stage the TP of cells is increased and DPD is decreased which results in diffusion of water from mesophyll cells into intercellular spaces. The intercellular spaces now become saturated with water and their water vapour pressure becomes greater than the water vapour pressure of atmosphere. Simultaneosly DPD of intercellular spaces is decreased much than the DPD of water vapours present in the atmosphere. Thus, the water from the intercellular spaces diffuses into the atmosphere in the form of vapours through stomata, lenticels and cuticle.

About 95 per cent of the absorbed water is transpired through stomata. The intercellular spaces of mesophyll cells remain connected with the sub-stomatal chambers or respiratory chambers. Thus, the water vapours continuously diffuse from intercellular spaces into substomatal chambers and the air of these cavities remain always saturated with water vapours. When the air of outer atmosphere is unsaturated, the water diffuses from sub-stomatal chambers into the atmosphere, thus, the process of transpiration continues.

5.8. MECHANISM OF OPENING AND CLOSING OF STOMATA

The observations on open and closed stomata under microscope indicate that the guard cells of open stomata always remain in turgid condition, while those of closed stomata remain in flaccid condition. It indicates that the opening and closing of stomata depend upon the turgid and flaccid condition of guard cells respectively.

When the guard cells become turgid, stomata open and they become closed when the guard cells become flaccid. The size of stomatal opening depends upon the degree of turgidity of guard cells. It has been explained earlier that the outer wall of guard cells is thin and elastic whereas inner wall is quite thick and non-elastic. When the guard cells absorb water from the surrounding cells, they become turgid and their TP is also increased. The TP exert a pressure on the outer wall of guard cells resulting in stretching or spreading of outer walls due to its thin and elastic nature, but the inner wall being thick and non-elastic in nature does not spread much. Due to stretching of its outer wall towards outside, the inner wall also stretches towards outside resulting in opening of stomata. When the turgidity and TP of guard cells start decreasing, both the walls also start regaining their original state and the stomata start closing. When the guard cells become flaccid and their TP becomes zero, the outer walls regain their original position and the stomata become closed.

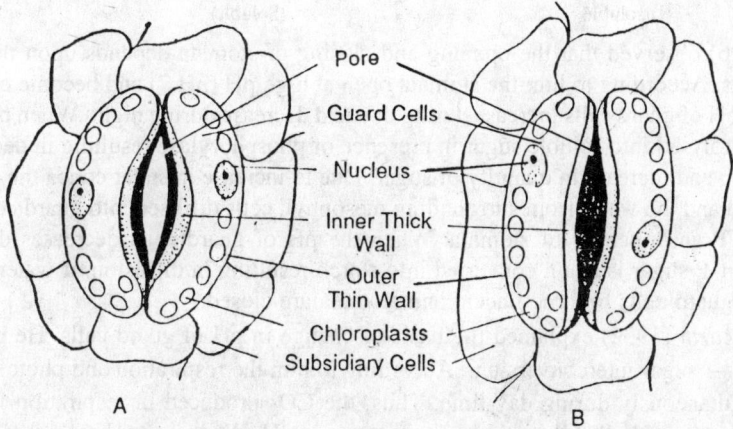

Fig. 5.6: Structure of stomata in open and closed condition.

Different theories were proposed by various physiologists to explain the reasons of change in turgidity and TP of guard cells. Some of the important theories are as follows:

(1) *Theory of Photosynthesis in Guard Cells:* Von Mohl (1856) observed that the stomata remain open in light or day time and closed in dark or during night. On this basis he proposed the theory of photosynthesis in guard cells and told that the chloroplasts present in the guard cells photosynthesis in light-resulting in the formation of carbohydrates and increase in OP of guard cells due to which the water enters into the guard cells by osmosis from the surrounding cells and the guard cells become turgid. With the turgidity of cells, TP also increases which results in opening of stomata. During night the OP and TP of guard cells are very much reduced or become zero due to lack of photosynthesis and the stomata become closed.

Following are the objections regarding this theory:

(i) The chloroplasts of guard cells perform insufficient photosynthesis.

(ii) The chloroplasts of stomatal guard cells of certain plants are completely unable to photosynthesis carbohydrates.

(iii) The concentration of cell sap of stomatal guard cells increases 2 to 3 times in 5 to 30 minutes while the amount of chlorophyll present in the guard cells is quite less.

(iv) The guard cells already possess much amount of stored sugar.

(v) Sometimes the guard cells of stomata of young leaves also possess starch grains before the opening and formation of buds.

(vi) Some plants when are kept in dark, their leaves still possess starch.

(vii) The leaves of some plants which are without chlorophyll also possess starch grains in the guard cells of stomata.

Summary of "Theory of Photosynthesis in guard cells"

Light → Photosynthesis by guard cells → Formation of Sugar → Increase in osmotic pressure of cell-sap → Endosmosis → Entrance of water from mesophyll cells → Increase in turgidity of guard cells → Stomata open.

(2) *Theory of Starch ⇌ Sugar Interconversion:* According to *Lloyd* (1905), *Loftfield* (1921) and *Sayre* (1926), the amount of starch in the guard cells increases during night and decreases during day time. The insoluble starch present in the guard cells is hydrolysed into soluble glucose-I-P in presence of phosphorylase enzyme during day time and soluble glucose I-P is converted into insoluble starch during night. Thus, both are reversible reaction.

$$\text{Starch} + \text{Phosphorylase} \underset{\text{Night}}{\overset{\text{Day}}{\rightleftharpoons}} \text{Glucose} - 1 - PO_4$$
$$(\text{Insoluble}) \qquad\qquad\qquad\qquad (\text{Soluble})$$

Sayre (1926) observed that the opening and closing of stomata depends upon the change in pH of guard cells. According to him the stomata open at high pH (pH 7) and become closed at low pH (pH 5). The pH of guard cells increase during day and decrease during night. When pH increases, the starch is hydrolysed into soluble sugar in presence of phosphorylase resulting in decrease in the quantity of starch and increase in quantity of sugar. Due to increase in sugar conc., the OP of guard cells is increased and the water from surrounding mesophyll cells diffused into guard cells resulting in increase of TP and opening of stomata. When the pH of guard cells decreases during night, soluble glucose 1-P sugar is again converted into starch resulting in diffusion of water from guard cells. Thus, the guard cells become flaccid and stomata are closed.

Later on *Scarth* (1932) explained the basis of change in pH of guard cells. He proposed the theory of starch → sugar interconversion. According to him the respiration and photosynthesis are carried out simultaneously during day time. Thus, the CO_2 produced in respiration is utilized in photosynthesis by mesophyll cells resulting in increase in pH. Whenever pH increases, the enzyme phosphorylase becomes active and hydrolyses starch into soluble sugars. During night the CO_2 produced in respiration instead of being utilized in photosynthesis accumulated in the intercellular spaces of mesophyll cells resulting in decrease in pH of guard cells. Whenever pH decreases, the soluble sugars are converted into starch, resulting in decrease in OP of guard cells and diffusion of water from guard cells. Thus, the guard cells become flaccid and stomata are closed.

Changes taking place during closing of stomata
- (i) The CO_2 becomes accumulated in intercellular spaces.
- (ii) The pH of cell-sap of guard cells is reduced to 5.0.
- (iii) Soluble sugars are converted into starch.
- (iv) OP of guard cells is reduced.
- (v) The water of guard cells diffuses outside due to exosmosis resulting in flaccid stage of guard cells and closing of stomata.

Changes taking place during opening of stomata:
- (i) The concentration of CO_2 in leaves is reduced much due to photosynthesis.
- (ii) The pH of cell-sap of guard cells is increased (pH 7).
- (iii) The starch is converted into soluble sugars due to action of enzyme phosphorylase.
- (iv) OP of guard cells is increased.
- (v) The water from mesophyll cells diffuses into guard cells due to endosmosis and the guard cells become turgid resulting in opening of stomata.

$$\underset{\substack{\text{(Insoluble and} \\ \text{osmotically} \\ \text{inactive)}}}{\text{Starch} + PO_4} \underset{\substack{\text{Dark, pH 5.0} \\ \text{(Synthesis)}}}{\overset{\substack{\text{pH7} \\ \text{Phosphorylase} \\ \text{(Hydrolysis) Light}}}{\rightleftharpoons}} \underset{\substack{\text{(Insoluble and} \\ \text{osmotically} \\ \text{active)}}}{\text{Glucose–1–Phosphate (G-1-P)}}$$

(3) *Theory of Starch $\rightleftharpoons$ Glucose Interconversion* : Steward (1964) criticised the theory of starch sugar interconversion and proposed another modified theory for opening and closing of stomata called theory of *starch glucose interconversion*. The scheme given by Steward is shown in Fig. 5.7. According to this theory the stomata open when starch is hydrolysed into glucose-1-P in presence of phosphorylase enzyme and glucose-1-P is converted into glucose and iP in presence of phosphoglucomutase and phosphatase enzymes. The stomata become closed when glucose and inorganic phosphate are again converted into glucose- 1-P in presence of ATP and hexokinase enzyme and glucose-1-P into starch. When starch is converted into glucose and inorganic phosphate, the osmotic pressure of guard cells increases along with the pH of the cell-sap which becomes seven (pH 7). When glucose and iP are converted into starch, the osmotic pressure of guard cells is reduced and the pH of cell-sap becomes five (pH5).

The summary of Opening and Closing of stomata is as follows :

A. *Opening of Stomata* :

(i) Starch + phosphate $\underset{}{\overset{\text{Phosphorylase}}{\rightleftharpoons}}$ Glucose -1-P

(ii) Glucose-1-P $\underset{}{\overset{\text{Phosphoglucomutase}}{\rightleftharpoons}}$ Glucose-6-P.

(iii) Glucose-6-P $\xrightarrow{\text{Phosphatase}}$ Glucose + iP.

B. *Closing of Stomata*

(i) Glucose + iP + ATP $\xrightarrow{\text{Hexokinase}}$ Glucose-1-P

(ii) Glucose-1-P $\xrightleftharpoons{\text{Phosphorylase}}$ Starch + Phosphate

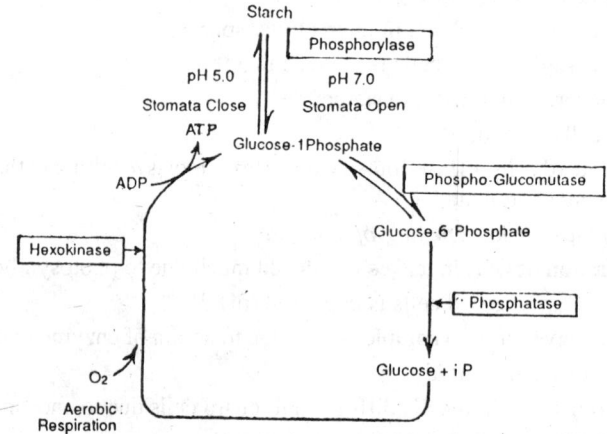

Fig. 5.7: Steward's Scheme (1964) for opening and closing of stomata.

Many physiologists do not agree with the theories of starch sugar interconversion and starch glucose interconversion due to following reasons:

(i) The guard cells of monocotyledonous plants do not contain starch even then they function like other dicotyledonous plants.

(ii) The stomata become closed in noon without any change in the quantity of starch.

(iii) The rate of interconversion of starch and sugar and starch and glucose is insufficient for opening and closing of stomata.

(iv) The concentration of CO_2 is insufficient for activation of guard cells and change in the pH of cell-sap.

(v) Sometimes starch is converted into malic acid in place of sugars.

(4) *Theory of Glycolate Metabolism*: This theory was proposed by Zelitch (1963) according to which glycolic acid plays an important role in the opening of stomata. Glycolic acid is formed in the guard cells when conc. of CO_2 is reduced. Glucolate also synthesizes carbohydrates. The OP of guard cells increased with the formation of glycolate which requires ATP for its synthesis. The whole process may be shown as below:

(i) $NADP^+ + ADP + iP + H_2O \xrightarrow{\text{Light}} NADPH + H^+ + ATP + \frac{1}{2}O_2$

(Non-cyclic phosphorylation)

(ii) Glyoxylate + NADPH + H^+ $\xrightarrow{\text{Glyoxylate reductase}}$ Glycolate + $NADP^+$

(iii) Glycolate + $\frac{1}{2}O_2$ $\xrightarrow{\text{Glycolate oxidase}}$ Glyoxylate + H_2O_2

(Hydrogen peroxide)

(iv) $H_2O_2 \xrightarrow{\text{Catalase}} H_2O + \frac{1}{2}O_2$

(5) *Theory of Proton Transport*: This theory was proposed by Levitt (1974). According to

Transpiration

this theory the opening and closing of stomata depend upon the entry and exit of potassium ions in the guard cells. At first malic acid is formed from starch in the guard cells which dissociates into cations and anions.

$$[R(COOH)_2 \rightleftharpoons R(COO^-)_2 + 2H^+].$$
$$H^+ \rightleftharpoons K^+$$

The anions exit out from the guard cells and to replace them, K^+ ions enter into guard cells from surrounding mesophyll cells. The K^+ ions react with malic acid to form potassium malate which is transported into cell vacuoles. It increases the OP of guard cells resulting in diffusion of water from mesophyll cells into guard cells. Thus, the guard cells become turgid and due to increase in TP the stomata open.

Noggle and *Fritz* (1976), supporters of Proton transport theory, summarized and proposed following scheme of opening and closing of stomata.

Opening of Stomata : Light → starch → Production of malic acid in guard cells → Dissociation into hydrogen and malate ions → Influx of K^+ and efflux of H^+ ions → Formation of potassium malate → Transport of potassium malate into the vacuoles → Osmotic entrance of water into guard cells → Increase of turgor pressure →Stomata open. During closing of stomata reverse reactions take place.

5.9. OPENING AND CLOSING OF STOMATA IN SUCCULENT PLANTS

In succulent plants like opuntia the stomata open during night and remain closed during day time. Why does it happen ? In these plants incomplete oxidation of carbohydrates takes place during night resulting in the formation of organic acids which accumulate in the guard cells. Thus, CO_2 is not released outside and the stomata remain open during night.

During day time the organic acids accumulated during night are oxidised resulting in liberation of large amount of CO_2 which is utilized in photosynthesis. Thus, the stomata remain closed during day time.

During night (In Dark)

$$2 C_6H_{12}O_6 + 3O_2 \longrightarrow 3C_4H_6O_5 + 3H_2O$$
Carbohydrate   Oxygen   Malic acid   Water
                                 (Organic acid)

During day (In Light)

$$C_4H_6O_5 + 3O_2 \longrightarrow 4CO_2 + 3H_2O$$
Malic acid   Oxygen   Carbon dioxide   Water

5.10. PLANT ANTITRANSPIRANTS

Those substances which reduce the rate of transpiration in plants are called *plant antitranspirants*. Some of the main antitranspirants are colourless plastics, Polythene bags, Silicone oil, low viscosity waxes, Phenyl mercuric acetate, abscisic acid, CO_2, oxy-ethylene, and decenyl succinic acid. They retard the transpiration without affecting CO_2, fixation in photosynthesis or plant growth.

5.11. FACTORS AFFECTING TRANSPIRATION

Following factors effect the transpiration : (*a*) External or Environmental factors, and (*b*) Internal or Structural factors.

External or Environmental Factors

(1) *Humidity in air* : Wet and dry air affects the transpiration. The **dry air absorbs moisture**. If the air of the atmosphere is dry and the air present inside the intercellular spaces of leaves is wet,

the diffusion of water-vapours will take place from intercellular spaces of leaves into the atmospheric air. Thus, less humidity in the air increases the rate of transpiration. In contrast high humidity in the air decreases the rate of transpiration.

(2) *Temperature*: The rate of transpiration is directly proportional to the temperature of atmosphere. Temperature also affects the humidity of air. The rate of transpiration increases with the increase in temperature and decreases with the decrease in temperature of the atmosphere. The high temperature of atmosphere lowers and low temperature increases the humidity.

(3) *Light*: Transpiration increases in high light intensity and decreases in low intensity. In high intensity of light, the water present in mesophyll cells diffuses rapidly resulting in increase in humidity of internal air. This increases the rate of transpiration. The stomata remain open during day hours and closed during night hours.

(4) *Wind velocity* : High wind velocity increases the rate of transpiration and low velocity decreases the rate of transpiration. In high wind velocity, the humid air surrounding the leaves moves all the side away from the plant, resulting in decrease in the humidity around the leaves and increase in rate of transpiration.

(5) *Water content in soil*: If excess water is present in the soil, more absorption of water will take place resulting in high rate of transpiration. In water deficient soil, the rate of transpiration decreases.

Internal Factors

(1) *Presence of cuticle on epidermis*: Some leaves possess thick cuticle on the epidermis. It reduces the rate of transpiration.

(2) *Hairy leaves*: Some leaves possess epidermal hairs which retain moisture or humid air resulting retardation in transpiration.

(3) *Presence of sunken stomata*: The xerophytic leaves possess sunken stomata which also reduce the transpiration.

(4) *Number of stomata on leaves*: The rate of transpiration is related with the number of stomata. More number of stomata present on the leaves, more will be the transpiration rate and less number of stomata show lower transpiration rates.

5.12. SPECIAL FEATURES IN PLANTS TO REDUCE TRANSPIRATION

Following three types of features are found in plants to reduce transpiration :

(A) Morphological features.

(B) Anatomical features.

(C) Physiological features.

Morphological Features

(i) Presence of dry, hard and cylindrical stem *e.g.*, *Salvadora* and *Leptadaenia*. Some stem possess ridges and furrows *e.g.*, *Casuarina*.

(ii) Presence of bark and cork on stems *e.g.*, *Acacia*.

(iii) Reduction in number of branches and size of stems. Some stems become flat and fleshy *e.g.*, *Opuntia*.

(iv) Decrease in length and number of branches in roots. They reach deeply in the soil. *e.g.*, *Alfalfa*.

(v) Presence of fleshy roots for storage of water *e.g.*, *Asparagus*.

(vi) Presence of small scaly and reduced leaves e.g., *Ruscus* and *Tamarix*.

(vii) Presence of quite small leaves which soon fall off *e.g.*, *Capparis* and *Euphorbia*.

(viii) Division of leaves into many small leaflets or lamina *e.g.*, *Acacia*.

Transpiration

    (ix) Presence of long narrow and needle like leaves e.g., *Pinus*.
    (x) Presence of thick leaves covered by thick waxy layer or cuticle. e.g., *Calotropis, Ficus* and *Salvadora*.
    (xi) The leaves of some plants become thick, fleshy and mucilagenous or possess resin or gum e.g., *Aloe* and *Agave*.
    (xii) Modification of stipules into thorn e.g., *Ziziphus* and *Acacia*.
    (xiii) Presence of excess hairs or hairy covering on the leaves, flowers and stems e.g., *Gnephalium, Aerva, Calotropis*.
    (xiv) Rolling and folding of leaves e.g., *Marrum grass*.
    (xv) Presence of smooth and shining leaves e.g., *Nerium*.
    (xvi) Production of gum, resin and mucilage etc. in the cells of plants e.g., *Pinus, Opuntia* and *Cycas*.

Anatomical Features

    (i) Presence of thick cuticle e.g., *Agave, Dianthus, Ficus* and *Nerium* etc.
    (ii) Presence of another waxy layer on the cuticle e.g., *Salix glaucophylla*
    (iii) Presence of multilayered epidermis e.g., *Nerium, Ficus* and *Pepromia*.
    (iv) Presence of hairs and scales on the epidermis.
    (v) Presence of sunken and less number of stomata. e.g., most of the xerophytes and sometimes in the depressions of epidermis e.g., *Nerium*.
    (vi) Presence of mucilage, gum, resin, latex and tanin in the cells of hypodermis and cortex e.g., *Opuntia, Pinus, Ficus, Euphorbia* and *cycas*.
    (vii) Presence of exess amount of sclerenchyma e.g., *Salvadora*
    (viii) Presence of compactly arranged palisade cells e.g., *Cycas* and *Nerium*.
    (ix) Formation of cork and bark e.g., *Acacia, Ziziphus* and *Prosopis* etc.
    (x) Formation of cork layer in the stem.

Physiological Features

    (i) Presence of high OP in the cell-sap of leaves.
    (ii) Presence of hydrophylic substances in the cells of stems and leaves which retain water and become fleshy.
    (iii) Presence of excessive growth in the roots due to which they become quite long and reach at a greater depth to absorb the water e.g., *Alhagi cameloran*.
    (iv) Closing of stomata during adverse conditions of environment.

5.13. BENEFITS OF TRANSPIRATION TO PLANTS OR IMPORTANCE OF TRANSPIRATION

    (1) The water and minerals absorbed by the roots from the soil reach continuously in different parts of the plant through transpiration.
    (2) Transpiration maintains the conc. of mineral salts.
    (3) Transpiration maintains the temperature of plants.
    (4) Transpiration increases the SP of mesophyll cells of the leaves which results in increase in the absorption power of roots.

5.14. EXPERIMENT : TO DEMONSTRATE THE PRODUCTION OF SUCTION PRESSURE DUE TO TRANSPIRATION

Take "4" shaped tube as shown in Fig. 5.8 and fill it with water. Now insert a plant twig previously cut in water in the lateral arm and capillary tube in the lower mouth of long arm with the help of cork.

Place the lower end of capillary tube in a Hg filled beaker. Make all the joints air-tight and set the apparatus on the stand. After sometime you will observe that the level of Hg rises in the capillary tube. It happens because the leaves of the twig transpire and a suction pressure is created due to which the H_2O of the tube ascends and to fill this gap Hg ascends in the capillary tube. It shows that SP is created due to transpiration.

5.15. OTHER METHODS OF WATER LOSS

(1) *Guttation*: The herbaceous flowering plants growing in moist places posses special type of structure at the vein ending margins of leaf called *hydathodes*. The water exudates through hydathodes in the form of small liquid droplets. This phenomenon is called *guttation*. It was discovered by *DeBary* in 1869. The liquid coming out of *hydathodes* is not pure water but a solution containing a number of disolved substances like P, K, Na, Fe, Mg, glucose, fructose, sucrose, thiamine and riboflavin.

It has been pointed out earlier that the stomata become closed on sunset and remains such during night. In contrast, the hydathodes remain open whole day and night (24 hrs). The *hydathodes* remain connected with the veins. Each hydathode open to the exterior by means of a pore called *stoma*. The stoma is surrounded by two guard cells and opens internally in to an air chamber. Just below the air chamber there is a group of parenchymatous cells with intercellular spaces called *epithem*. The xylem vessels and tracheids open just behind the epithem. The exudation of water in the form of liquid drops takes place through stoma of *hydathodes*. The liquid moves from xylem elements into cells of epithem and then exudes through stoma. The phenomenon of *guttation* takes place due to development of positive root pressure.

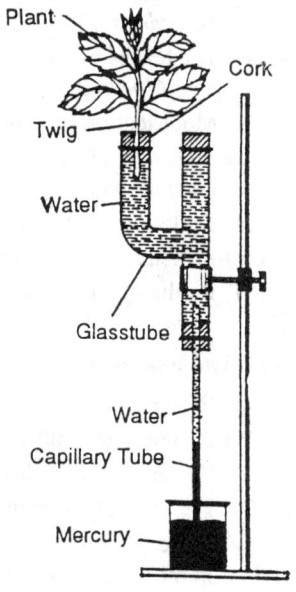

Fig 5.8: Representation of suction pressure in transpiration.

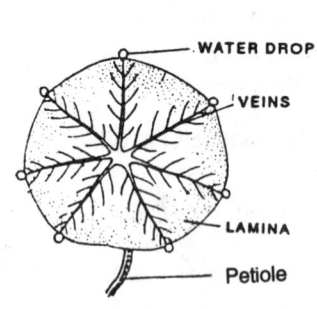

Fig 5.9: Representation of guttation.

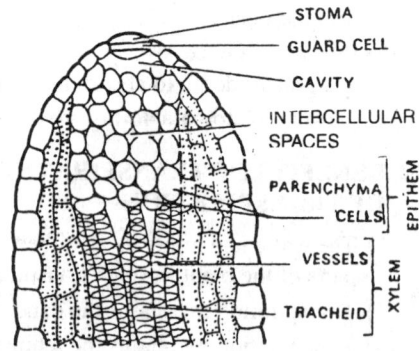

Fig 5.10: Structure of epithem (in section).

When soil condition favour rapid water absorption and there is high humidity in the atmosphere, at that time the transpiration takes place slowly but the root pressure increases continuosly due to active absorption which results in guttation. It can be observed early in the morning after a moist and warm night in the leaves of herbaceous plants like grasses, *Nasturtium*, tomato, potato, brinjal, and Alocasia.

Transpiration

Table 5.2. Differences between Transpiration and Guttation

	Transpiration		Guttation
(i)	In transpiration, the water loss take place in the form of water vapours.	(i)	In guttation, water loss takes place in the form of liquid drops.
(ii)	Transpired water is pure water.	(ii)	The liquid coming out in guttation is not pure water but it contains a number of dissolved substances like minerals, salts, sugars, and coenzymes.
(iii)	Transpiration take place through stomata, lenticels and cuticle.	(iii)	Guttation take place through hydathodes which are found at the margins of the leaves.
(iv)	The stomata of leaves become closed or open according to the need, usually they remain open during day and closed during night.	(iv)	The hydathodes remain open whole day and night.
(v)	It takes place during day hours.	(v)	It takes place either in the morning or during the night.
(vi)	It is controlled by the guard cells.	(vi)	It is not controlled by guard cells.
(vii)	Root pressure is not involved in transpiration.	(vii)	Guttation takes place due to development of root pressure.
(viii)	Transpiration maintains the temperature of the plant.	(viii)	It has no relation with the temperature.
(ix)	It takes place in all the higher terrestrial plants.	(ix)	It takes place mainly in herbaceous plants like *Nasturtium*, tomato, potato, brinjal, and grasses.

Experimental Demonstration of Guttation

Take a J-shaped tube and fill it with some mercury (Hg). Now pour some water above the Hg in the smaller arm and fit a single hole cork at this mouth. Insert a petiole of *Nasturtium* leaf in the hole of cork. The petiole should be cut in water to avoid the entry of air bubbles. Now pour some more Hg in the long arm of J-tube and observe.

The exudation of water at the vein ending margins of leaf take place. The reason behind this is that the Hg presses the water like root pressure and the water ascends into petiole through xylem elements and comes out of leaf through veins in the form of *guttation* fluid drop.

5.16. BLEEDING

The exudation of water and cell-sap through the cuts or wounds of plants is called *bleeding*. It also happens due to positive root pressure. In certain plants the pressure is developed either in the phloem elements or in the cells surrounding the cut or wounds. The exudation of toddy juice from the wounded stem of *Borassus*. Todday and latex from *Haevea* (rubber) stem are the examples of bleeding. It has been utilized in the production of various economically important products like, Rubber, Sugar, and Alcoholic drinks.

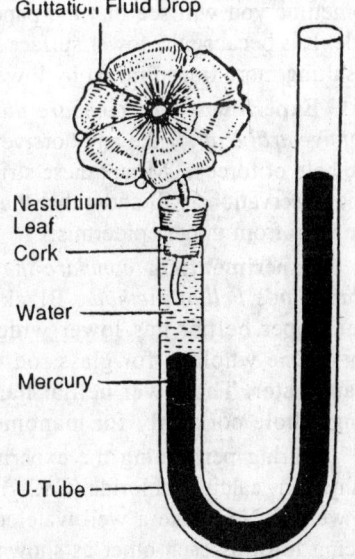

Fig 5.11: Experimental representation of Guttation.

5.17. EXPERIMENTS RELATED TO TRANSPIRATION

(1) *Experiment*: To demonstrate that transpiration takes place through the aerial parts of the plants like stems, leaves, and flowers.

Belljar Method

Take a potted plant and cover the pot of soil with help of thick paper or rubber sheet to avoid the evaporation of soil water. Now cover the complete plant with the help of belljar and make it air-tight. (see Fig. 5.12). After sometime you will see many drops on the wall of belljar. Test this liquid with the help of cobalt chloride paper which turns pink. It confirms that these drops are of water because it is the property of cobalt chloride paper that when it comes in contact with water, it turns pink. From where these water drops in the jar have appeared? They appeared in the jar due to transpiration by aerial part of the plant because the pot was covered with rubber sheet and the belljar was also neat and clean. It confirms that the transpiration takes place through aerial part of the plants.

(2) *Experiment*: To demonstrate that more transpiration takes place through the lower surface of leaf than the upper surface.

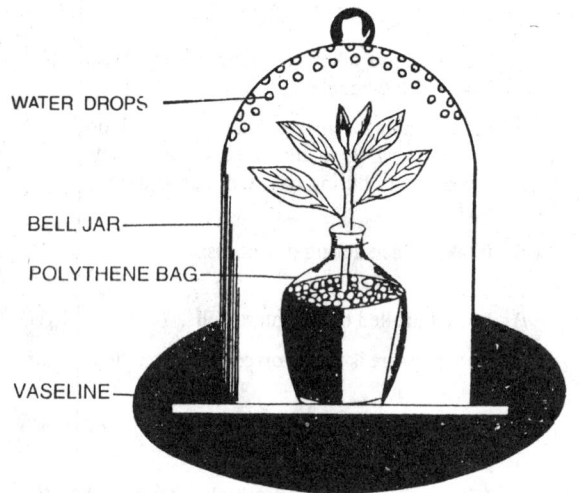

Fig. 5.12: Demonstration of transpiration.

Cobalt Chloride Method

Take a potted plant and cover one leaves from both the sides with previously prepared cobalt chloride paper by dipping ordinary blotting paper in aquous solution of cobalt chloride. After sometime you will see that the paper of lower side becomes pink earlier than the paper of upper side. It is because the lower surface of leaf possesses more number of stomata, than the upper side resulting more transpiration by lower surface than the upper surface.

Experiment : *To compare the number of stomata (frequency) present on two surfaces of a dorsiventral leaf :* Take any dorsiventral leaf and peel of its lower and upper epidermal strips with the help of forceps. Mount these strips in water separately on slides and observe under microscope. The observations will show that the strip from lower epidermis has more number of stomata than the strip from upper epidermis.

Experiment: *To compare the rate of transpiration of upper and lower surfaces of leaf by Blackman's belljar method :* Blackman's belljar apparatus actually consists of two belljars. The upper belljar has lower wide opening and upper narrow opening fitted with two hole cork. One whole is for glass rod to hang $CaCl_2$ tube with the help of a thread and other for manometer. The lower belljar has the same openings but the narrow opening is fitted with single hole cork only for manometer.

During performing the experiment first two small ignition tubes are taken and filled with anhydrous calcium chloride ($CaCl_2$). They are weighed in such a way that both the tubes are equal in weight. Now take a well watered plant and place its one healthy leaf in between two belljars facing towards each other as shown in Fig. 5.13. Clamp these belljars to the stand tightly. Now, hang one $CaCl_2$ tube in the upper belljar through a thread and glass rod fitted in one hole of the cork and place second $CaCl_2$ tube at the bottom of second lower belljar. Connect one manometer in the whole of the cork for manometer in the upper belljar and the second manometer in the hole

Transpiration

of cork of lower belljar. Finally make the apparatus air tight with the help of grease and keep it in sunlight for about an hour.

Now take out $CaCl_2$ tubes from both the belljars and weigh them separately. It is observed that the weight of both the tubes has increased. It is because the anhydrous $CaCl_2$ has the properly to absorb water vapours which comes through transpiration by both the surfaces of leaf. Further, the weight of the tube placed in lower belljar will be more than the tube placed in upper belljar. It indicates that the rateof transpiration in lower epidermis is greater than the upper epidermis because the lower epidermis has more number of stomata than the upper epidermis.

Experiment : *To measure the rate of transpiration by Ganong's potometer* : Ganong's potometer is made up of a vertical wide glasstube whose lower end remains connected with a long and narrow horizontal tube. The long end of the horizontal tube is graduated and bends downward at 90°. The opening of this end terminates in the form of a small hole situated laterally. At one place the horizontal tube is connected with a larger water reservoir through a glass stopper. The complete apparatus remains fixed on a wooden stand present opposite to water reservoir.

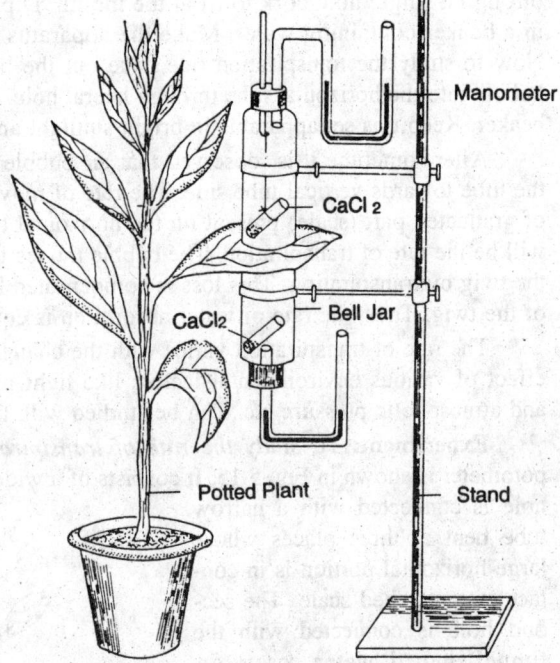

Fig. 5.13: Comparison of transpiration from two surfaces of leaf by Blackman's belljar.

To measure the rate of transpiration, first of all fill the apparatus with water through water reservoir and close the stopper. Now select one healthy leafy plant twig and cut it under water

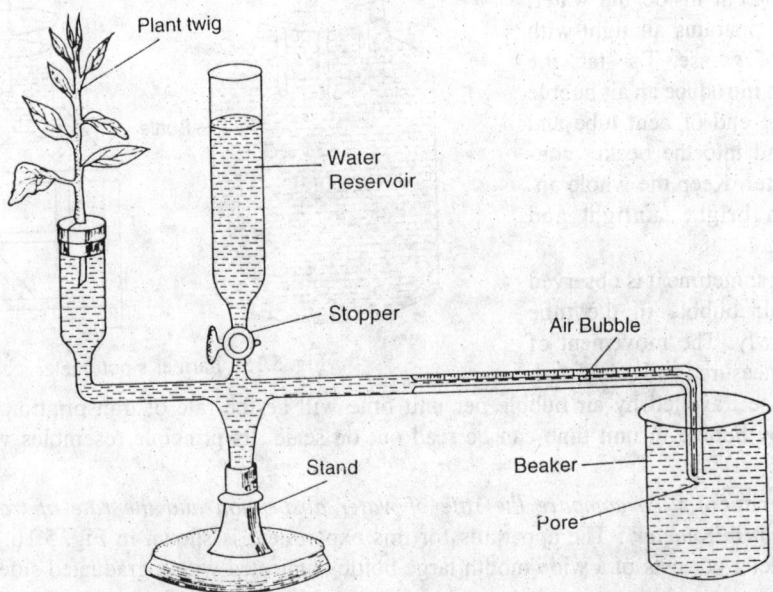

Fig. 5.14: Measurement of transpiration by Ganong's potometer.

to avoid the entry of air bubble in the xylem vessels. Insert this twig into the wider vertical tube through a single hole cork to fit at the mouth. Dip the bent end of the graduated horizonal tube in a beaker containing water. Make the apparatus air tight by applying grease around the cork. Now to study the transpiration rate, take out the bent end from the beaker and introduce an air bubble into the horizontal tube through lateral hole. Again place the terminal end of the tube in the beaker. Keep this set apparatus in bright sunlight and observe.

After sometime it is observed that air bubble moves, slowly along the graduated portion of the tube towards vertical tube side. The rate of movement of air bubble is measured with the help of graduated part (scale) present on the horizontal tube. The movement of air bubble in unit time will be the rate of transpiration. The bubble moves forward due to water loss through the leaves of the twig by transpiration. This loss is compensated by water absorption through the xylem portion of the twig. Thus, the rate of water absorption is equal to the rate of transpiration.

The rate of transpiration varies with the changes in the environmental conditions. Therefore, effect of various environmental factors like light intensity, humidity, temperature, wind velocity and atmospheric pressure etc., can be studied with the help of this apparatus.

Experiment: *To study the rate of transipiration by Farmer's potometer :* The Farmer's potometer is shown in Fig. 5.15. It consists of a wide mouth bottle fitted with three hole cork. First hole is connected with a narrow tube bent at three places whose large horizontal portion is in contact with attached scale. The second hole is connected with the funnel-shaped water reservoir. The third central hole is used for inserting the plant twig.

To perform the experiment, set the apparatus according to Figure 5.15 i.e., fill the apparatus with water, insert the twig in the central hole priorly cut inside the water. Make the apparatus air tight with the help of grease. To start the experiment introduce an air bubble through the end of bent tube and dip this end into the beaker containing water. Keep the whole apparatus in bright sunlight and observe.

After sometime it is observed that the air bubble in the tube moves slowly. The movement of bubble is measured and calculated.

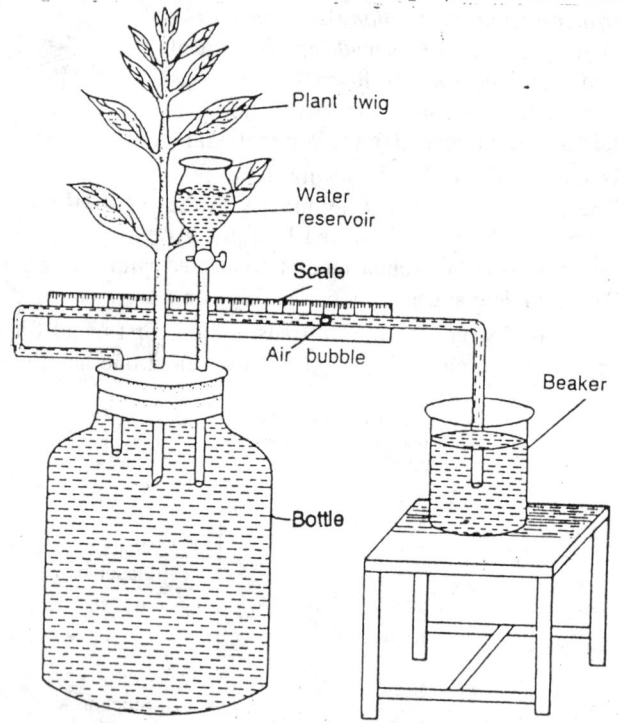

Fig. 5.15: Farmer's potometer.

The distance travelled by air bubble per unit time will be the rate of transpiration. The distance travelled by bubble in unit time can be read out on scale. Its principle resembles with Ganong's potometer.

Experiment : *To compare the rate of water absorption and the rate of transpiration by absorbo-transpirometer :* The apparatus for this experiment is shown in Fig. 5.16. The absorbo-transpirometer consists of a wide mouth large bottle connected with a graduated side vertical tube. Fill the apparatus with water and insert a small rooted herbaceous plant in the mouth of the bottle through a hole in the cork. Make the apparatus air tight with the help of grease and put a few drops

Transpiration

of oil in the open end of the vertical side tube to check the evaporation of water. Weigh the whole apparatus and note the level of water inside the graduated side tube. Keep the apparatus in bright sunlight for observation.

After a few hours, it is observed that the level of water in the graduated tube falls down. This reading is also noticed. The difference between initial and final reading will be the amount of water absorbed by the plant roots. Now reweigh the whole apparatus. The difference between initial weight and final weight will be the amount of water transpired. Thus, the amount of water absorbed by roots and amount of transpired water through the shoot can be compared. Mostly in plants the absorption of water is almost equal to the amount of water transpired.

Experiment : *To compare the rate of stomatal and cuticular transpiration by four leaves method :* Take four fresh dorsiventral leaves of almost same age and size. Hang them in a series by tieing their petioles with a thread. The two ends of the thread to two separate stands so that the leaves could hang freely in a straight line.

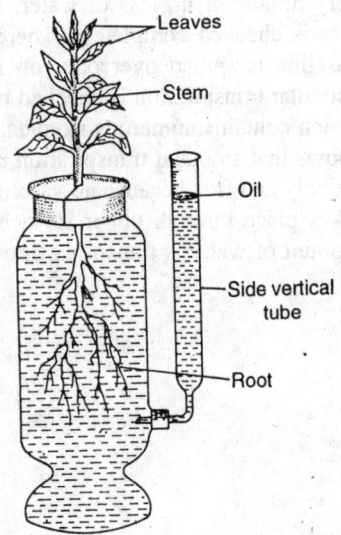

Now spread the vaseline or grease on the surfaces of leaves as follows:

1. Spread the vaseline on both the surfaces of leaf A.
2. Spread the vaseline only on the lower surface of leaf B.
3. Spread the vaseline only on the upper surface of leaf C.
4. Do not spread the vaseline on any leaf surface of leaf D.

Fig. 5.16: Comparison of rates of water absorption and transpiration by absorbo-transpirometer.

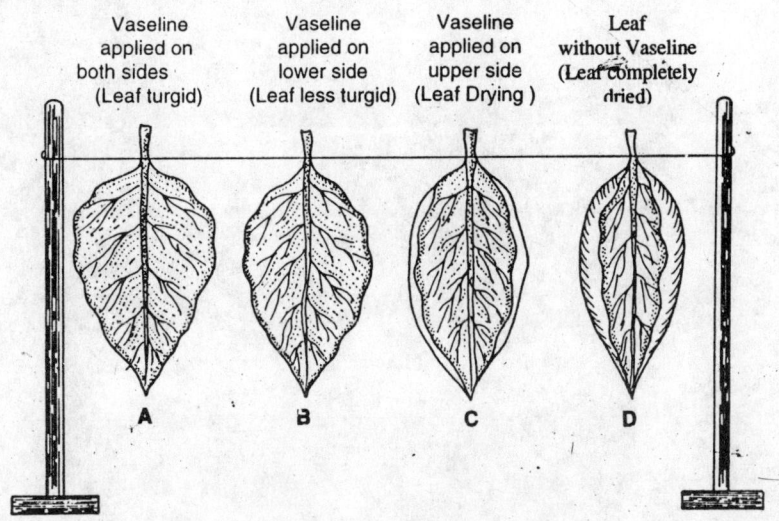

Fig. 5.17: Four leaves experiment.

Keep the above processed leaves in sunlight and observe. After a few hours it is observed that leaf D wilts first followed by C and B. The leaf A does not wilt and remains turgid. The leaf C becomes less turgid as compared to A and B. The reason behind is that the stomata remain distributed mainly on the lower surface of dorsiventral leaves. When the vaseline is spread over this surface, the stomata are blocked and water loss is checked. Thus, the leaf B remains turgid. In this leaf only cuticular transpiration takes place through the upper surface which is almost insignificant as it leads very minute or no loss of water. By spreading vaseline on both the surfaces of leaf A, the water loss is checked completely. Therefore, leaf A remains maximum turgid. In leaf, C, because the vaseline is spread over only on the upper surface which contains no stomata, therefore, only cuticular transpiration is checked but the stomatal transpiration continues through the lower surface which contains numerous stomata. Thus, leaf C starts wilting due to loss of water and turgor. This shows that stomatal transpiration rate is much higher than cuticular transpiration. The leaf D wilts quickly and first because no vaseline was spread over on any surface. Hence, cuticular transpiration takes place through upper surface and stomatal transpiration through lower surface. Thus, great amount of water is transpired through both the surfaces showing complete loss of turgor.

6

Mineral Nutrition in Plants

6.1. INTRODUCTION

The importance of mineral nutrition is known since ancient times. *Woodward* (1699) for the first time observed that plants grow better in muddy water than in rain water. *de Saussure* (1804) confirmed that inorganic mineral elements of the plants ash are obtained from the soil. He also recognised that mineral elements present in the soil are important for the growth and development of plants. *Liebig* (1840) for the first time gave evidences about the functions of these plant elements.

6.2. COMPOSITION OF PLANT ASH

From the analysis of plants and their organs it became evident that they are made up of water and some solid matters. The plants contain more amount of water which is calculated by drying these plants. The amount of water present in any plant is called *moisture content*. To know moisture content of any plant, first it is taken out from the soil, washed thoroughly and then surface water is dried with blotting paper. This plant is weighed directly or its parts are cut first and then weighed. This weight is called *fresh weight*. Now the plant is oven dried at 100°C so that the moisture or water present is evaporated. The dried plant is again weighed carefully. This weight is called *dry weight*. Now the dry weight is subtracted from the fresh weight to get moisture content. If the weight is calculated in percentage, it is called *moisture percentage*.

The moisture content and percentage is variable in different plants and their parts in different conditions. In aquatic plants it is about 95 to 98 per cent; in herbaceous parts it is 70 to 80 per cent and in woody parts 40 to 50 per cent, in succulent parts 25 to 90 per cent and in dormant seeds only 10 to 20 per cent.

Oven dried plant (dry matter) is ignited at 400-600°C which results in the oxidation of organic materials present in the plant. Thus, carbon, hydrogen, nitrogen and sulphur elements are converted into CO_2 ; H_2O; CH_4 ; N_2 ; NH_3 and SO_2 respectively and are liberated. At the end incombustible matter remains which is called **plant ash**. The weight of plant ash is called *ash weight*. The percentage of plant ash varies in different plants. The ash contains about 40 elements in variable amounts. All these elements are not essential for the nutrition of plants. After absorption these elements are not found in free form in plants but they are found usually in the form of ions or components of organic compounds. The main elements present in the plant ash are C, H, O, N, S, P, K, Ca, Mg, Fe, Mn, B, Zn, Al, Cu, Mo and Cl. Out of these C, H, O, N, S, and P elements are eliminated in the form of gases and other elements remain as oxide salts in the plant ash.

6.3. ESSENTIAL AND NON-ESSENTIAL ELEMENTS

All the 40 elements obtained by the analysis of plant ash are not essential for the nutrition of plants but a few are essential for the growth and development of plants. Other elements are non-essential. It is known since ancient times that C, H and O are essential elements for plants. It has been confirmed by water-culture and sand-culture experiments that N, S, P, K, Ca, Mg and Fe are also essential elements in addition to C, H, O, elements. Their deficiencies produce many diseased symptoms. B, Zn, Mn, Cu, Mo and chlorine are also essential elements although they are required in very small amounts in plants. C, H, O, N, S, P, K, Ca, Mg, Fe, B, Mn, Zn, Cu, Mo and Cl are sixteen essential elements discovered so far and others which are not required for the growth and development of plants are called *non-essential elements*.

Essential elements are divided into following two groups depending upon their requirement.

(1) *Macronutrients = Macroelements = Major elements*

Those elements which are required in large quantities (more than 100 mg/litre of water) are called *macronutrients* or *macro-elements*. They usually participate in body construction and are ten in number (C, H, O, N, S, P, K, Mg, Ca and Fe).

(2) *Micronutrients = Micro elements = Minor elements = Trace elements*

Those elements which are required in smaller quantities (100 or less mg/litre of water) are called *microelements* or *trace elements*. They usually participate in various metabolisms and are six in number (B, Mn, Zn, Cu, Mo and Cl).

Table 6.1. Differences between trace and tracer elements

Trace elements	*Tracer elements*
1. These are nutrient elements required for plants growth and their various metabolisms.	1. These are radioisotopic elements required for detecting various metabolic pathways.
2. They may be supplied through nutrient solution during deficiency in plants.	2. They are produced through radioactivity and introduced into the plants artificially through different plant parts.
3. They are required for plant growth in very minute amounts (100 or less mg/litre of water) and therefore, they are called trace elements or micro-elements.	3. They can be detected by Geiger-Muller counter or by any other methods. Thus, they are traceable in plant systems.
4. The main trace elements are boron (B) manganese (Mn), Zinc (Zn), Copper (Cu), Molybdenum (Mo) and Chloride (Cl).	4. The main tracer elements used in detecting various metabolic pathways are Carbon (^{14}C), Nitrogen (^{13}N and ^{15}N), Sulphur (^{35}S), Cobalt (^{60}Co), Magnesium (^{28}Mg), Phosphorus (^{32}P), Oxygen (^{18}O), Potassium (^{42}K) etc.
5. They are supplied through liquid medium (nutrient solutions.)	5. They may be incubated in plants as gaseous form or liquid form. Deuterium and tritium, isotopes of hydrogen and nitrate ($^{15}NO_3$) etc., are given as water or in water but $^{14}CO_2$ is given in gas form.

6.4. HYDROPONICS

The term **hydroponics** has been used for growth of plants in water and sand cultures. The terms soil less agriculture, bath-tube farming, test-tube farming, tank farming and chemical gardening have also been used for such work. The technique of growing plants without soil has been known for a long back (about 100 years) but it became popular only during the past few decades.

Commercially hydroponic cultures are maintained in large shallow tanks made up of concrete, cement, wood or metal-sheets. In such containers the gravel and solutions are kept. The tanks are generally equipped with pumps and empty auxiliary tanks to pump out and circulate the growth solution and to maintain satisfactory aeration of the nutrient solution.

Commercially, the hydroponics have been used for the production of horticultural and floricultural crops. This practice provided better yield than in soil in gladioli, carrots, radish, potatoes, roses, snapdragons, cucumbers, green peppers, tomatoes and lettuce etc. Such large yield of various plants led to the speculation that hydroponic agriculture might some day replace soil agriculture. The gardeners are utilizing hydroponic methods in such localities where good soil is not available. Such techniques were employed during World War II in remote oceanic islands, particularly in

Mineral Nutrition in Plants

Pacific theatre, as an emergency means of raising supplementary greens and not the staple food.

Advantages of Hydroponics

(1) It is possible to provide whatever nutrient environment is desirable. (2) The acid-base balance (pH) of the nutrient solution can be easily set and maintained in the range most adequate for a given crop. (3) There are no soil colloids present to immobilise any of the nutrients through adsorption. (4) Frequent replacement of the nutrient solution prevents the accumulation of toxic organic decomposition products such as often occurring in soils. (5) Conditiose relatively unfavourable for soil-borne bacteria, fungi and insects provide a means of growing plants in an area where soil is lacking (e.g., coal reef islands) or where the soil is present and fertile but is contaminated with pathogens or some toxic principles (7) Through pumping devices, the solutions may be circulated, thus enabling a regularity of aeration not possible in soils. (8) The equipment can be made automatic, avoiding the labour and expense of watering the plants. (9) No tilling is necessary. (10) Mulching, changing of soil and weeding are eliminated. (11) There is general uniformity of plant growth and plant products. (12) Many experienced growers have observed this method to be a cheaper and more favourable medium than soil for growing plants or, at least, certain of their crops. (13) The nutrition can be altered at any time to coordinate with fluctuations in the weather, especially changes in light and temperature. (14) Since the onset as well as degree of flowering and fruiting are affected by nitrogen supply, the nutrition can be changed at any time in order to regulate the vegetative or reproductive phases of the plant.

Disadvantages of Hydroponics

According to Gauch (1972), disadvantages are as follows: (1) As compared with field production, production by hydroponics is limited. (2) In the economic sense, its use under greenhouse conditions is generally limited to high-value speciality crops. (3) Considerable technical skill is required to design equipment, plan routine procedures and handle special problems which may arise. (4) If a disease appears, it may affect all plants in the container; here circulation of the nutrient solution tends to spread the pathogenic organisms to the roots of all plants. (5) Special modifications of technique are necessary for certain crops. Carrots, parsnips and potatoes may be undesirably changed in shape with coarse gravel but not when sand is employed.

To Study the Importance of Mineral Elements in Plants through Hydroponics

Artificial methods like soil culture and sand culture are employed to know which mineral element is essential for the growth and development of the plant. Both these methods are soilless.

(1) *Solution Culture Method*: In this method the roots of the plant are placed in liquid nutrient solution. The solution which contains those essential minerals which the plants absorb from the soil by roots, is called *nutrient solution*. When all the mineral elements are present in a solution, it is called *normal nutrient solution* and when one of the essential elements is not added in the nutrient solution, it is called *deficient nutrient solution*.

If we want to study the importance of any mineral, it is not added in the nutrient solution. It means the solution is deficient nutrient solution for that particular element. Now, the plants are grown in this deficient solution and are compared with those grown in normal solution. If there are deviations in morphological, anatomical and physiological characteristics of plants from the normal ones, these deviations are due to deficiency of that elements. By this method we can find out the importance and functions of various elements.

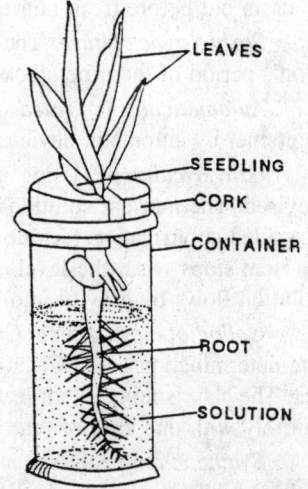

Fig. 6.1 : Study of plants by solution culture method.

Following conditions are essential for a normal nutrient solution :
   (i) It should contain all the essential elements in soluble state.
   (ii) The solution should be changed from time to time.
   (iii) The solution should be balanced so that the absorption of essential elements could be done.
   (iv) The solution should get normal temperature, light and air.
   (v) pH of the solution should be kept according to requirement.

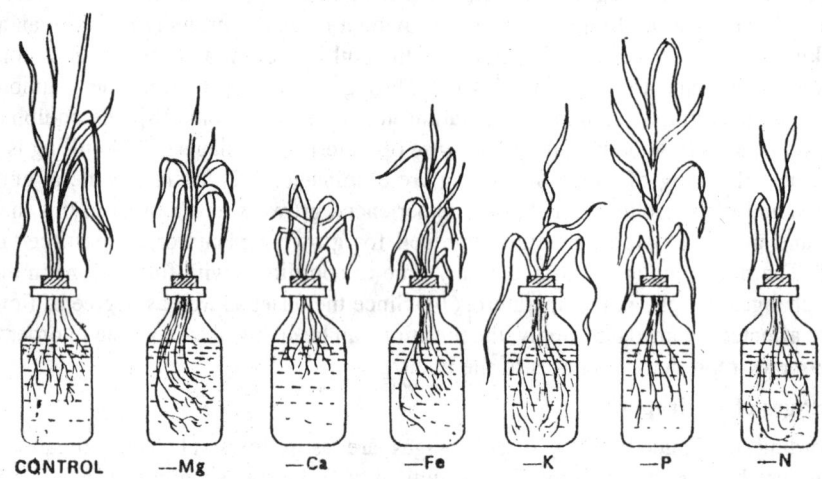

Fig. 6.2: Culture of plants in different deficient nutrient solutions.

(2) *Sand Culture Method*: In this method the plants are grown on sand (solid media) and the nutrient solutions are added to sand culture by slope culture (pouring over the surface) or drip culture (dripping on the surface) or sub-irrigation (forcing solution up from the bottom of the container). For sand culture, following methods are also used:

Drip culture: The sand is kept moist by nutrient solution which siphonos and drips from a capillary glass tube, one end of which passes into the solution contained in an inverted jar placed next to it.

Slope culture: A nutrient solution is periodically supplied to the surface of the sand and allowed to drain out before fresh nutrient is added.

Soxhlet-type culture: The sand is alternately flooded and drained several times during 15-min "on" period of the time clock.

Automatically irrigated culture: The nutrient solution is forced up from the bottom of the container by automatic devices several times daily.

Gravel culture : Using gravel as an inert medium, automatic irrigated systems have been devised. The nutrient solution is delivered into the bottom of a crock or greenhouse bench by the so-called subirrigation technique and the pumping time is adjusted so that further delivery of the nutrient stops when the level of solution nears or reaches the surface of the gravel. The excess of solution flows by gravity into the reservoir.

Collodial clay culture: Colloidal clay is prepared from a subsoil high in beidellite clay. After the determination of BEC, nutrients are added in the amount and ratio desired and the pH value set. The clay is mixed with leached white sand to provide a suitable culture medium. It is saturated entirely with one kind of ion or with definite quantities of several types of ions.

Synthetic ion-exchange materials : Plants can be grown successfully in sand culture or gravel culture with cations adsorbed on Permutit (an artifical zeolite) and anions on De-ocidite (an aniline material). Amberlites are also suitable for sources of adsorbed cations and anions.

Mineral Nutrition in Plants

Nutrient vapour bath : In this type, a nutrient solution supplied by means of a common vapouriser bathes the roots in vapour form.

Sterile root medium : An apparatus for growing a plant with its roots in a sterile medium has been developed by Blanchard and Diller (1950). Here contamination occurred in only about 5 per cent of the cultures.

This method is supposed to be better than solution culture method because of the following reasons:

(*i*) Soil provides better aeration to the plants.
(*ii*) Soil provides better support to plants. The plants can stand erect in the soil.
(*iii*) Roots spread well and easily in the soil.

6.5. NUTRIENT SOLUTIONS

Different scientists like Sachs, Knop, Pfeffer, Shive and Hoagland prepared their own nutrient solutions with minor variations. The composition of their solutions is shown in Table 6.2.

Table 6.2: Chemical compositions of some plant nutrient solutions prepared by Sachs, Knop, Pfeffer, Shive and Hoagland (gm/litre)

Salts	Sachs solution (1860) gm/litre		Knop's solution (1865) gm/litre	Pfeffer's solution (1900) gm/litre	Shive's solution (1915) gm//litre	Hoagland's solution (1920) gm/litre
$Ca(NO_3)_2 \cdot 4H_2O$	...	0.8	...	I	II	Sol. I	Sol. II
$Ca_3(PO_4)_2$	0.5	...	...	1.228	1.842	1.18	0.95
$Ca(NO_3)_2$	...	...	0.8	....	....	....	....
$CaSO_4$	0.5	...	...	....	....	....	....
KNO_3	0.1	0.2	0.2	....	...	...	...
KCl	...	...	0.2	....	...	0.51	0.61
$MgSO_4 \cdot 7H_2O$	0.5	2.0	...	....	...	...	...
$MgSO_4$	...	....	0.2	3.700	0.49	0.49	0.49
KH_2PO_4	...	0.2	0.2	....	....	...	...
$NaCl$	0.25	...	...	2.45	1.47	0.14	...
$FeSO_4$	(trace)	(trace)	...	....	...	...	...
$FePO_4$	...	...	....	.....	...	...	...
$FeCl_3$	...	...	(trace)	...	...	...	...
Fe	...	...	...	1 cc of 0.5% ferric phosphate	1 cc of 0.5% of ferric phosphate	...	...
$NH_4H_2PO_4$	...	...	...	...	...	...	0.12
Ferric tartarate	...	...	....	...	...	0.005	000.5

Some other important solutions are as follows:

1. Corne's, 1904

	gm/l
KNO_3	1.0
$Ca_3(PO_4)_2$	0.25
$CaSO_4$	0.25
$Fe_3(PO_4)_2, 7H_2O$	0.25
$MgSO_4, 7H_2O$	0.25

2. Duggar's 1920

	gm/l
$Ca_3(PO_4)_2$	1.0
$CaSO_4, 2H_2O$	0.5
"Soluble ferric phosphate"	1.0
$Mg_3PO_4.8H_2O$	0.5
KNO_3	1.5

3. Knudson's, 1922

	gm/l
$Ca(NO_3)_2.4H_2O$	1.00
KH_2PO_4	0.25
$MgSO_4, 7H_2O$	0.25
$(NH_4)_2SO_4$	0.50
$FePO_4$ (insoluble)	0.05

4. Arnon and Hoagland's, 1940

	gm/l
KNO_3	1.02
$Ca(NO_3)_2$	0.492
$NH_4H_2PO_4$	0.23
$MgSO_4, 7H_2O$	0.49

	Mg/l
H_3BO_3	2.86
$MnCl_2, 4H_2O$	1.81
$CuSO_4, 5H_2O$	0.08
$ZnSO_4, 7H_2O$	0.22
$H_2M_0O_4, H_2O$	0.09
$FeSO_4, 7H_2O$ — 0.5% ⎫	0.6 ml/l
Tartaric acid — 0.4% ⎭	3x weekly

Precautions During Preparation and Use of Nutrient Solutions

While preparing and using nutrient solutions, following precautions must be taken into consideration:

1. The solution must contain an ample supply of all essential elements.
2. All solutions must possess the same osmotic pressure.
3. Solutions must be changed at definite intervals to maintain their concentration.
4. A constant pH of solutions must be maintained.
5. Aeration of roots in solutions must be proper.

Mineral Nutrition in Plants

6. If salts tend to crystallise and stems of plants become wilt, they must be washed down with water occasionally.
7. If plants are grown in only pure solutions, they must be supported by stakes.
8. The growth controlling factors like light and temperature etc. must be adequate.

6.6. INORGANIC SALTS IN SOIL

The soil contains following inorganic salts:
   (*i*) Particles of lime-stone ($CaCO_3$).
   (*ii*) Particles of sand stone.
   (*iii*) Crystals of silicate salts. They contain K, Na, Ca, Mg, Fe and Al elements.
   (*iv*) Other mineral salts.

According to volume, the garden soil usually contains 40 per cent mineral matter, 10 per cent organic matter, 25 per cent water and 25 per cent air.

6.7. INORGANIC SALTS IN SOIL WATER

The water present in the spaces of soil particles is called *capillary water*. It contains many ions in the dissolved form. This water is also called *soil solution*. It is quite dilute and contains following inorganic ions:

(1) *Inorganic cations*: H^+, K^+, NH_4^+, Na^+, Ca^{++}, Mg^{++}, Fe^{++}, Fe^{+++}, Zn^{++}, Cu^{++}, Mo^{++}, Mn^{++} and Al^{+++} etc.

(2) *Inorganic anions*— Sulphate (SO_4^-), Nitrate (NO_3^-), Nitrite (NO_2^-), Bicarbonate (HCO_3^-), Phosphates ($H_2PO_4^-$, HPO_4^-, PO_4^{--}) Chlorine (Cl), Borate (BO_3^-) etc.

6.8. GENERAL ROLES OF MINERAL ELEMENTS IN PLANTS

(1) *Formation of Protoplasm and Cell-wall*: C, H, O, N, S and P elements etc. participate in the formation of protoplasm and cell-wall. In addition, N element is also involved in protein and nucleic acid synthesis, S in protein synthesis, P in nucleic acid synthesis, Mg in chlorophyll synthesis and Ca in the formation of middle lamella of cell-wall.

(2) *Influence on the Osmotic Pressure of Plant-Cell*: The mineral salts and organic matter present in the cell-sap produce osmotic pressure in the cell.

(3) *Influence on the pH*: The mineral elements absorbed from the soil affect the pH of the cell-sap.

(4) *Catalytic Function*: Fe, Mn, Zn, Cu, elements etc. act as catalysts in various enzymatic processes.

(5) *Antagonistic or Balancing Function*: Ca, Mg, Na, K elements etc. or their salts neutralise the toxic effects of other elements in the cell. Thus, the ions in the cell remain balanced.

(6) *Toxic Effects of Mineral Elements*: As, Cu, Hg, etc. show toxic effects in certain stages on plants.

(7) Certain ions like K^+ and Ca^{++} by deposition on cell-membranes change their permeability.

Table 6.3. Differences between Chlorosis and etiolation.

Chlorosis	Etiolation
1. It is a physiological disease.	1. It is a physiological phenomenon.
2. Chlorosis is caused due to deficiency of certain elements like Mg^{+2}, iron, sulphur, nitrogen, potassium, manganese etc., when the plants are grown in light.	2. Etiolation is caused in green plants when they are grown in dark. Mineral deficiency is not involved in such plants.

(*Contd.*)

3. During chlorosis, the leaves become non-green even in presence of all other accessory pigments like xanthophylls and carotenoids etc.	3. During etiolation the stem becomes elongated, the leaves become colour less or yellow, young leaves remain unexpanded. Pigments involved in photosynthesis are not synthesized.
4. It may be complete or inter-veinal chlorosis.	4. Absence of light is the only factor in etiolation and mostly the entire leaf becomes colourless or yellow.
5. Chlorosis can be overcome by supplying the deficient element through any method.	5. Etiolation can be avoided if the plant is kept in proper sunlight.

6.9. OCCURRENCE AND FUNCTIONS OF ESSENTIAL ELEMENTS

(1) **Carbon (C):** It is obtained from the air in the form of CO_2.

(2) **Hydrogen (H):** It is obtained from the soil in the form of H_2O.

(3) **Oxygen (O):** It is obtained from air and soil in the form of O_2 and H_2O respectively. C, H and O mainly participate in the formation of protoplasm and cell-wall of plants. Without them, the life is impossible.

(4) **Nitrogen (N):** It is obtained from the soil in the form of nitrate (NO_3^-), nitrite (NO_2^-) and ammonium (NH_4^+) salts. Insectivorous plants use molecular nitrogen.

  (*i*) **Occurrence:** It is found in entire plant body, seeds and food storing regions in both organic and inorganic forms.

  (*ii*) **Functions:** (*i*) Protein synthesis, (*ii*) Formation of protoplasm, nucleic acids, purine and pyrimidine bases, chlorophyll, alkaloids, NAD and NADP coenzymes etc.

Deficiency Symptoms

  (A) *External Symptoms*:

  (*i*) The leaves become yellow. This phenomenon is called *chlorosis*.

  (*ii*) Lowering of respiratory rates and reduction in growth.

  (*iii*) Poor development of foliage and roots.

  (*iv*) Late and less or no flowering.

  (*v*) Development of anthocyanin pigment in the stems and petiole and veins of leaves due to which they become red.

  (*vi*) Production of erect leaves in grasses, potato, tomato cereals and *Linum* etc.

  (*vii*) Necrosis (death) of tissues at advanced stage.

  (B) *Anatomical Symptoms*:

  (*i*) Deposition of excess amount of lignin on the cell-walls.

  (*ii*) Increase in starch contents and decrease in protein contents.

  (*iii*) Cells become smaller and thick walled.

  (*iv*) Reduction in the size of nuclei and cell-lumen.

(5) **Sulphur (S)**

(*a*) *Source*: It is obtained either from soil solution in the form of sulphate (SO_4^{--}) ions or through the activity of micro-organisms by biological oxidation.

(*b*) *Occurrence*: Sulphur is found in all the parts of plants in both organic and inorganic forms. In organic forms it is present in amino-acids like *Cysteine, Cystine* and *Methionine*, in respiratory pigment as *glutathione*; in *glucosides* and also in mustard oil. In inorganic form, it is present in the form of sulphate ions.

(*c*) *Functions of Sulphur*

  (*i*) Formation of sulphur-containing amino-acids like cystine, cysteine and methionine.

Mineral Nutrition in Plants

   (ii) Synthesis of sulpur-containing vitamins like *biotin thiamine* and *coenzyme-A*.
   (iii) It imparts distinctive odour and flavour to garlic, onion and mustard oil.
   (iv) Increases growth, cell-division and fruiting.
(1) *External deficiency Symptoms*: These symptoms resemble to that of nitrogen. The main among them are :
   (i) Creates hindrance in chlorophyll formation resulting in chlorosis.
   (ii) Retards cell division and growth.
   (iii) Development of anthocyanin pigment in the stem and leaves.
   (iv) Suppression of fruit formation and delaying in ripening.
(2) *Anatomical and Chemical Changes Symptoms*
   (i) Reduction in the number of stroma lamellae and increase in number of grana lamellae in chloroplasts (*Hall et. al.*, 1972).
   (ii) Accumulation of starch, sucrose and soluble nitrogen in tomato and sunflower plants (*Eaton*, 1942, 51).
Other symptoms resemble with nitrogen.

(6) Phosphorus (P)

(1) *Source*: It is obtained from soil solution in the form of phosphate ions ($H_2PO_4^-$ and HPO_4^-).

Occurrence: Phosphorus is more abundantly found in the meristematic tissues and storage regions as fruits and seeds. In inorganic form, it is found in the form of $H_2PO_4^-$ and HPO_4^- while in organic forms, it is present in nucleic acids (RNA, DNA), in phospholipids, hexose phosphates, coenzymes NADP, NAD and ATP, GTP, CTP etc.

Functions: (i) Synthesis of nucleoproteins.
   (ii) Formation of ATP, NAD and NADP which are involved in photosynthesis, respiration, synthesis of fatty acids and proteins etc.

External Deficiency Symptoms: The deficiency symptoms of phosphorus resemble with nitrogen but they are comparatively less developed. The main deficiency symptoms of phosphorus are :
   (i) Plants show stunted growth due to abnormal cell-division.
   (ii) Favours healthy root growth by helping the translocation of solutes.
   (iii) Development of dead necrotic spots on the leaves, petioles and fruits which ultimately cause falling of the leaves.
   (iv) Promotes fruit ripening.
   (v) Reduction of nitrates is checked resulting in slow down of protein synthesis.
   (vi) Development of anthocyanin pigmentation.

Internal Deficiency Symptoms
   (i) Hinders cambial activity.
   (ii) Lignification.
   (iii) More development of pith and reduction of vascular tissues.
   (iv) Elements of xylem and phloem become thin-walled.
   (v) Accumulation of carbohydrates.
   (vi) Production of large-sized intercellular spaces in the stem due to necrosis of pith cells.

(7) Potassium (K)

(1) *Source*: It is found in soil solution in non-exchangeable or fixed form and in the cell in free ionic form.
(2) *Occurrence*: It is found mostly in all cells except cork-cells. It is most common in

cytoplasm and vacuoles, more abundant in apical meristems of roots and shoots, and absent from nucleolus and nucleus.

(3) *Functions*: It is supposed that potassium does not participate directly in the formation of organic compounds present in plants but plays important roles in respiration, photosynthesis, development of chlorophyll and acts as inorganic catalyst in various processes. It plays following important roles:

    (*i*) Maintains water-balance and hydration of protoplasm and controls permeability of cytoplasm.

    (*ii*) Controls enzymatic activities of various enzymes like diastase, catalase, reductase and invertase.

    (*iii*) Antagonises the toxic effects of calcium.

    (*iv*) Plants become fleshy and succulent in its proper supply.

    (*v*) Affects the synthesis of sugars, starches, fats and proteins.

    (*vi*) Adds in enzymatic hydrolysis of starch and hence essential for manufacture and translocation of carbohydrates (*White*, 1936).

    (*vii*) Neutralises the effects of organic acids.

    (*viii*) Enhance meristematic activity.

(4) *External Deficiency Symptoms*

    (*i*) Foliage leaves become yellow at the margins and roll inward and their tips curve downward (mottled chlorosis).

    (*ii*) Plants become less resistant to pathological diseases.

    (*iii*) Growth become stunted and internodes become very much short.

    (*iv*) Occurrence of leaf wilting and abscission.

    (*v*) At early stage, the leaves become slender and light-green.

(5) *Internal Deficiency symptoms*

    (*i*) Mechanical tissues remain poorly developed.

    (*ii*) Phloem elements degenerate.

    (*iii*) More lignification of cells.

    (*iv*) Reduction in the activity of secondary meristematic and cambial meristematic tissues.

    (*v*) Cell walls of pericycle and collenchyma become thick.

    (*vi*) Accumulation of starch grains in cortex, phloem, medullary rays and pith as in tomato stem.

    (*vii*) Exessive proteolysis (*Malavolte et al.*, 1955).

    (*viii*) Xylem becomes parenchymatous.

(8) Calcium (Ca)

(1) *Source*: In soil, it is present in the form of cations or in mineral salts like anorthite ($CaAl_2Si_2O_8$) and calcite ($CaCO_3$).

(2) *Occurrence*: It is abundantly found in leaves, fruits and seed-coats. In middle lamella of cell-wall, it is found in the form of calcium pectate and in the leaves of **Ficus** sp. and other plants it is present in the form of calcium carbonate and calcium oxalate respectively. Leguminous plants contain higher percentage of calcium while grasses contain comparatively lower percentage.

(3) *Functions*

    (*i*) It is the main constituent of middle lamella of cell-wall in the form of calcium pectate.

    (*ii*) Maintains the semipermeable nature of plasma membrane.

    (*iii*) Initiates the development of root hairs.

    (*iv*) Neutralises the organic acids like oxalic acid.

Mineral Nutrition in Plants

(v) Ca antagonises the toxic effect of certain mineral salts like NaCl in acidic soil.
(vi) Formation of lipids and cell-membrane.
(vii) Helps in the binding of nucleic acids with protein.
(viii) Activate phospholipase, arginine kinase and ATP.
(ix) Helps in the metabolism of fats.
(x) Transportation of carbohydrates and amino-acids.
(xi) Helps in the formation of chromosomes.

(4) *External Deficiency Symptoms*
(i) Poor development of root hairs.
(ii) Chlorosis occurs at the leaf tips and necrosis at the margins.
(iii) The roots become short, stubby and brown.
(iv) Absorption of nitrate is reduced and as such the rate of protein synthesis is retarded.
(v) More absorption of Mg takes place resulting in toxic effect and dying of plant.
(vi) Cell-walls remain weaker due to decrease in the cell-cytoplasm.
(vii) Hooking of leaf tip and leaf-apex killed.

(5) *Internal Deficiency Symptoms*
(i) Incomplete separation and lumping of chromosomes during cell-division.
(ii) Cell-walls become rigid or brittle with the break down of middle lamella.
(iii) Decrease in the cytoplasm of cells occurs resulting in weakening of cell-wall.
(iv) Protoplasm becomes more granulated.
(v) Cells of the shoot apex become enlarged and vacuolated.

(a) Magnesium (Mg)

(1) *Source*: It is present in the soil solution in the form of magnesite ($MgCO_3$), dolomite ($MgCO_3 \cdot CaCO_3$), and olivine [$(MgFe)_2 SiO_4$].

(2) *Occurrence*: It is mainly found in the leaves, seeds, fruits and meristematic cells. The oily seeds (castor) contain more Mg as compared to starchy seeds (wheat and barley). Leguminous plants contain a higher percentage of Mg than grasses. In organic form, it is present in chlorophyll while in inorganic form it is present in cytoplasm.

(3) *Functions*
(i) Chief role is the formation of chlorophyll.
(ii) Activates the enzymes of carbohydrate metabolism.
(iii) Helps in the binding of ribosomal subunits during protein synthesis.
(iv) Helps in fat synthesis and nucleoprotein synthesis.
(v) Activates the enzymes participating in phosphate transfer processes.

(4) *External Deficiency Symptoms*
(i) *Interveinal chlorosis*: The chlorosis develops from the base of the leaves and then proceeds upward.
(ii) Development of anthocyanin pigment and necrotic spots in the leaves.
(iii) Reduction or ceasing of protein synthesis.
(iv) Reduction in carbohydrate and fat synthesis.

(5) *Internal Deficiency Symptoms*
(i) Cells size is reduced.
(ii) Number of chloroplasts in cells is very much increased.
(iii) Pith cells become smaller.

(10) Iron (Fe)

(1) *Source*: It is absorbed in the form of ferrous ions (Fe^{++}) from the soil solution. In neutral and alkaline soil, it is present in insoluble form while in acidic soil in soluble form.

(2) *Occurrence*: It is found in all the parts of plant and protoplasm but in very small quantities. Its greatest concentration is found inside the vacuoles.

(3) *Functions*

    (i) Acts as catalytic agent in chlorophyll synthesis.

    (ii) Helps in the formation of important respiratory enzymes and coenzymes like flavoprotein, iron porphyrin, cytochrome peroxidase and catalase etc.

    (iii) Formation of ferridoxin which plays important role in biological nitrogen fixation and primary photochemical reaction.

(4) *External Deficiency Symptoms*

    (i) Occurrence of excessive interveinal chlorosis.

    (ii) Reduces respiration rate.

(5) *Internal Deficiency Symptoms*

    (i) Chloroplast formation is checked.

    (ii) Protein synthesis is stopped.

    (iii) Accumulation of large quantities of free amino acids and amides.

(11) Manganese (Mn)

(1) *Source*: It is found in the soil in the form of bi-, tri-, or tetravalent ions. Of these ions only bivalent ions are found in soluble state in soil solution.

(2) *Occurrence*: It is found in the plant ash especially in the leaves.

(3) *Functions*

    (i) It helps in the formation of chlorophyll (*Eyster et al.*, 1958).

    (ii) Oxidises indole 3-acetic acid (IAA) (*Goldacre*, 1961).

    (iii) Activates the enzymes of Krebs cycle like malic dehydrogenase and oxalosuccinic decarboxylase and the enzymes of nitrogen metabolism like nitrate reductase and hydroxylamine reductase (*Nason*, 1956; *McElroy et al.*, 1957).

    (iv) Participates in primary photochemical reaction of photosynthesis (*Arnon*, 1954).

    (v) Acts as a cofactor in oxidative phosphorylation (*Lindberg* and *Ernster*, 1954).

(4) *External Deficiency Symptoms*

    (i) Retardation in chlorophyll formation (*Hopkins*, 1934).

    (ii) Causes interveinal chlorosis and necrosis of leaves.

    (iii) Retardation in growth as in Pea (*Piper*, 1941).

    (iv) Formation of seeds slowed down (*Piper*, 1941).

    (v) Respiratory rate is lowered due to reduction in oxygen carrying power of oxidase.

    (vi) Retardation in nitrogen assimilation.

(5) *Internal Deficiency Symptoms*

    (i) Chloroplasts lose chlorophyll (*Eltinge*, 1941).

    (ii) Disintegration of starch grains as in tomato (*Eltinge*, 1941).

(6) *Effect of Mn when present in excess*

    (i) Browning of roots and leaves as in barley, wheat, lettuce and tomato (*Williams* and *Velamis*, 1957).

    (ii) Causes Chlorosis in Pine apple and Citrus (*Haas*, 1932).

(12) Zinc (Zn)

(1) *Source*: It is present in the soil in the form of divalent ions which are released by the weathering of minerals like magnetite, biotite and hornblende.

(2) *Occurrence*: It is usually found in the seeds.

(3) *Functions*

   (i) Participates in the synthesis of tryptophan and auxins-IAA (*Tsui*, 1948).

   (ii) Activates the metabolism of enzyme alcohol dehydrogenase (*Vassel*, 1951).

   (iii) Enhances the production of cytochrome a and b and cytochrome oxidase (*Grim* and *Allen*, 1953).

   (iv) Involved in the formation of enzyme carbonic anhydrase (*Keilin* and *Mann*, 1940).

   (v) Participates in protein synthesis (*Possingham*, 1956).

(4) *External Deficiency Symptoms*

   (i) Depression in chlorophyll formation and development of interveinal chlorosis followed by necrotic spots as in cotton (*Brown* and *Wilson*, 1952).

   (ii) Results in an increase in the amide and free amino acids, particularly asparagine and glutamine (*Possingham*, 1956, 57).

   (iii) Decrease in protein synthesis.

   (iv) Creates hindrance in seed formation.

   (v) Causes diseases like rosette disease of walnut, mottled leaf disease of walnut and apple, and white bud disease of maize.

   (vi) Checks growth of vegetative parts particularly size of inter-nodes and leaves is reduced.

   (vii) Interveinal chlorosis which starts from tips and margins of leaves.

(5) *Symptoms when Zn present in excess*

   (i) Promotes antibiotic production in **Fusarium** (*Kalyansundarum* and *Saraswathi Devi*, 1955).

   (ii) Promotes nicotine production in tobacco plants (*Steinberg* and *Jeffery*, 1956).

(13) Boron (B)

(1) *Source*: It is present in the soil in the form of boric acid, calcium and manganese borate and silicates. It is always absorbed in the form of borate anions or tetraborate ions.

(2) *Occurrence*: The percentage of boron is higher in woody plants than herbs.

(3) *Functions*

   (i) Boron helps in easy transport of sugar in phloem which ultimately implicates in various metabolic activities of plants.

   (ii) Changes the concentration of vitamin C.

   (iii) Helps in the metabolism of nitrogen, phosphorus, fats and hormones.

   (iv) Helps in the absorption of salts and in photosynthesis.

(4) *External Deficiency Symptoms*

   (i) Causes death of shoot and root apex.

   (ii) Checks flowering.

   (iii) Leaf veins become copper coloured.

   (iv) Accumulates carbohydrates and amino acids in leaves.

   (v) Causes shortening of roots.

   (vi) Causes "*top sickness*" disease in tabacco.

   (vii) Retards formation of root nodules in leguminous plants.

   (viii) Occurrence of brittleness in stems and leaf petioles as in *Vicia faba*.

(5) *Internal Deficiency Symptoms*
  (*i*) Causes disintegration and browning of internal tissues in sugarbeet, called *"heart rot disease of sugarbeet"*.
  (*ii*) Similar disease of cauliflower is called *"browning disease"* where watery areas are produced.
  (*iii*) Results in browning of cork cambium in apples, called internal cork of apples.
  (*iv*) Results is breaking down of conducting tissues in tomato.
  (*v*) Size of root cells is reduced.

(14) Copper (Cu)

(1) *Source*: In soil, the copper is mainly present as chalcopyrite ($CuFeS_2$) and copper sulphide. It is absorbed in the form of divalent copper cations.

(2) *Functions*
  (*i*) Participates in the formation of phenolase lactase and ascorbic acid oxidase enzymes.
  (*ii*) Helps in the biosynthesis of chlorophyll (*Stiles*, 1951).
  (*iii*) Helps in absorption of CO_2 and photosynthesis.
  (*iv*) Acts as catalyst in oxidation processes.
  (*v*) Imparts black pigmentation to the spores of *Aspergillus niger*.
  (*vi*) Acts as fungicide to prevent various diseases as 'late blight of potato'.

(3) *External Deficiency Symptoms*
  (*i*) Produces necrotic spots at the tips of young leaves.
  (*ii*) Causes 'exanthema' disease where erruptions are produced on the stem and branches.
  (*iii*) Causes 'reclamation' or 'Moor sickness' disease in fruit trees, cereals and leguminous plants. Here the growth of plants in spring is vigour and leaves become abnormally large but later on become chlorotic.

(15) Molybdenum (Mo)

(1) *Source*: It is absorbed from the soil solution in the form of molybdate ions (MoO_4^- or $HMoO_4^-$). In exchangeable form it is found adsorbed on soil particles while in non-exchangeable form it is present as constituent of soil minerals and organic matter.

(2) *Functions*
  (*i*) The chief role of molybdenum is to activate nitrate reductase enzyme during nitrogen metabolism, (*Possingham*, 1956, 57).
  (*ii*) Controls ascorbic acid synthesis (*Hewitt et al.*, 1950).
  (*iii*) Increases carbohydrate metabolism (*Agarwal* and *Hewitt*, 1955).
  (*iv*) Helps in the formation of pectic substances (*Agarwal*, 1952).

(3) *External Deficiency Symptoms*
  (*i*) Chlorotic interveinal mottling of lower leaves followed by a marginal necrosis and infolding of the leaves.
  (*ii*) Inhibition of flowering.
  (*iii*) Causes "whiptail" disease of cauliflower.
  (*iv*) Under more severe conditions, mottled areas of lower leaves become necrotic resulting in wilting of leaves.

(4) *Internal Deficiency Symptoms*
  (*i*) Depression of ascorbic acid (*Hewitt et al.*, 1950).
  (*ii*) Reduction in sugar content of plants (*Agarwal* and *Hewitt*, 1955).
  (*iii*) Formation of poorly developed middle lamella as in cauliflower (*Agarwal*, 1952).

(iv) Marked depression in soluble nitrogenous compounds (*Hewitt et al.*, 1957 ; *Possingham*, 1956 and 57).

(v) Accumulation of amino acids and protein synthesis is stopped.

(16) **Chlorine:** It is absorbed from the soil solution in the form chloride ions (Cl^-). Chlorine is an essential factor in photophosphorylation (*Arnon*, 1959). It helps in transfer of electrons during photosynthesis (*Vernon* and *Ke*, 1966).

(17) **Vanadium:** *Arnon* (1959) emphasized the importance of this element in *Scenedesmus*. In its deficiency, chlorophyll synthesis is depressed and photosynthetic oxygen evolution is inhibited. These processes can be restored by supplying vanadium element to the deficient cells.

6.10. METHODS TO OVERCOME MINERAL DEFICIENCY

Following methods are generally used to overcome the mineral deficiency :

(1) *Soil application*: In this method, the mineral elements are added in the soil. Usually, the deficiency of N, P and K elements is produced in the soil which is overcome by adding fertilizers available in the market. These three elements are called *critical elements*.

(2) *Foliar application*: In this method, first the dilute solution of soil deficient minerals is prepared and then sprayed over the young leaves of the plants. The minerals are absorbed by the surface of leaves.

(3) *Injection method*: In this method also, first the dilute solution of minerals which are deficient in soil is prepared by dissolving their salts and then injected through different parts and organs of the plants like apex, petiole and veins of leaves; apex of stem and branches; and roots or any part of the plant.

7

Mineral Salt Absorption

7.1. INTRODUCTION

Early scientists are of the opinion that inorganic salts are passively carried into plants with the absorption of water, and the absorbed salts are translocated to the different parts of the plant through transpiration stream. They considered that osmotically active substances diffuse from soil solution into the plant along the concentration gradient. Thus, initially the salt absorption and translocation were considered only a physical mechanism neglecting the role of metabolic energy in these processes. Now a days, it has been established that mineral salt absorption is an active process rather than a passive process as it was considered earlier.

The contributions in the field of mineral salt absorption are those of *Steward* (1932), *Vanden Honert* (1937), *Overstreet* (1939, 51), *Overstreet et al.* (1940), *Hober* (1945), *Robertson* and *Turner* (1945), *Lundegardh* (1950, 55), *Hope* (1953), *Hylmo* (1953, 55), *Bennet-clark* (1956), *Hopkins* (1956), *Kramer* (1956), *Kylin* and *Hylmo* (1957), *Devlin* (1966), etc. The mineral salt absorption may be of two types—Passive absorption and active absorption.

7.2. PASSIVE ABSORPTION

When the absorption of mineral salts takes place without any expenditure of metabolic energy (ATP) and simply by diffusion into the plant cells, it is called *passive absorption*.

(1) *Outer and apparent free space theory*: Salt absorption takes place through the intimate contact of the root system with the soil colloids or soil solution. Passive or non-metabolic absorption of ions has been demonstrated by a number of scientists. It has been found frequently that when a plant cell or tissue is transferred from a medium of low salt concentration to the medium of relatively higher salt concentration, there is an initial rapid uptake of ions which is followed by a slow and steady uptake. During rapid initial uptake of ions the metabolic energy is not involved. If the above plant tissue is returned to the lower salt solution or to the pure water, some of the ions taken up will diffuse out into the external medium. In other words, a part of the cell or tissues immersed in the salt solution is open to free diffusion of ions. Since free diffusion implies, that ions can move freely in or out of the tissue, the part of the tissue opened to free diffusion will reach an equilibrium with the external medium and the ion concentration of this part will be the same as found in the external medium. The part of the plant cell or tissue which allows diffusion is called an *outer space*.

Hope and *Stevens* (1952) believes that the cell wall and a part of the cytoplasm are included in the outer apparent free space. The term apparent free space was introduced to describe the apparent volume allowing for the free diffusion of ions.

In certain cases accumulation of salts does not involve metabolic energy and is obtained against the concentration gradient. It cannot be explained by simple process of diffusion. How ions are accumulated against a concentration gradient without the participation of metabolic energy can be explained through ion exchange mechanism and Donnan equilibrium.

(2) *Ion Exchange*: The mineral elements are absorbed by plants in the form of ions. The anions or cations of the plant cells are exchanged from the anions or cations of the equivalent charge

from the external medium in which tissue is immersed. It has been experimentally confirmed in excised barley roots. When the roots were placed in the solution of radioactive K^+ions, the exchange of these ions takes place with non-radioactive K^+ions of the tissue. A similar exchange mechanism operates between soil solution and clay micelles.

The ion exchange mechanism has been explained by following two theories:

(a) *The contact exchange theory* (Fig. 7.1) : It was proposed by *Jenny and Overstreet* (1939). According to this theory the ions are transferred from soil particles to the root or *vice versa* without passing into free solutions. It is a well-known fact that the ions adsorbed electrostatically to the surface of root cells or clay particles are not held tightly but due to thermal agitation each of them oscillates within a small volume of the space. It is termed as *oscillation volume*. When the oscillation volumes of two ions with same charge overlap, one ion is exchanged from the other. This process has been called *contact exchange*. The contact exchange of ions takes place not only between the soil particles but also between soil particles and the root surface.

Suppose H^+ is adsorbed on the root cell surface and K^+ is adsorbed on the clay micelle and both oscillate in such a way that the oscillation volume of H^+ overlaps with that of K^+. It will result in the transfer of H^+ to the clay micelle and K^+ to the root surface.

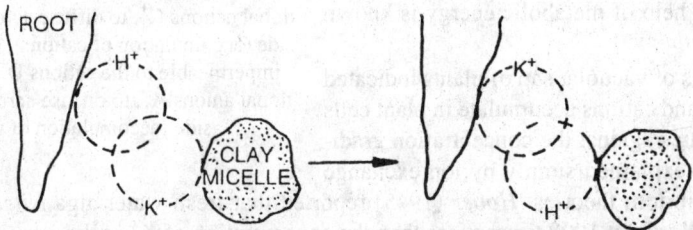

Fig. 7.1: Diagrammatic representation of the contact exchange theory.

(b) *Carbonic acid exchange theory* (Fig. 7.2): According to this theory the soil solution plays an important role in exchange of ions by providing a medium. The theory explains that the CO_2 released during respiration of root cells combines with water to form *carbonic acid* (H_2CO_3) in the soil solution. Carbonic acid dissociates into H^+ and HCO_3^- (anion). A cation *e.g.* K^+, adsorbed on the clay micelle may be exchanged with H^+ of the soil solution. This cation (K^+) may diffuse to the root surface in exchange for H^+ ion. The cations may also be absorbed on root cells in exchange as ion pairs with bicarbonate.

(3) *Donnan equilibrium* (Fig. 7.3) : In Donnan equilibrium the fixed or indiffusible ions play an important role. According

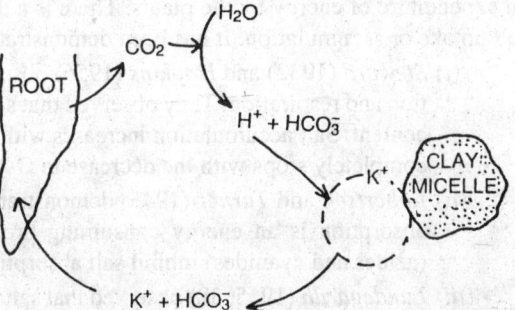

Fig. 7.2: Diagrammatic representation of carbonic acid exchange theory.

to this theory there are certain pre-existing ions inside the cells which cannot diffuse outside through membrane. Such ions are called *fixed* or *indiffusible ions*. The outer membrane is impermeable to fixed ions. However, the cell membrane is permeable to both anions and cations present in the external medium. If the cell is immersed in an external salt solution and on the inner side of the memberane there are fixed anions (—), the movememt of equal number of anions and cations takes place until an equilibrium (Donnan equilibrium) between cell-sap and external medium is reached. Here, additional cations will also move from the external medium to balance the negative changes of the fixed anions on the inner side of the membrane. Thus, the cation concentration in the cell-sap would be greater than in the external medium. This balance or equilibrium is called *Donnan equilibrium*.

In the same way if the membrane is impermeable to *fixed cations*, additional anions will accumulate from the external medium.

The Donnan equilibrium theory explains the accumulation of ions against a concentration gradient without the participation of metabolic energy. (Fig. 7.3).

(4) *Mass Flow:* Hylmo (1953, 55), Kramer (1956), Kylin and Hylmo (1957) and others believe that ions can move through roots along with the mass flow of water. According to this theory an increase in transpiration increases in the absorption of ions. This view was supported by Lopushinsky (1964) through his experiments on detopped tomato plants.

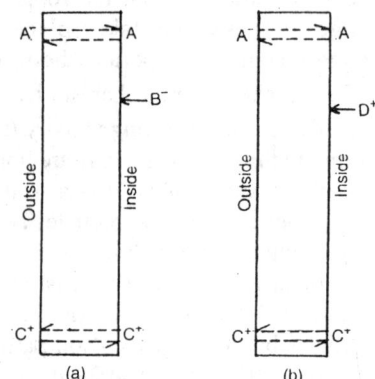

Fig. 7.3: Diagram showing diffusion of ions across membrane. (a) Membrane is impermeable to the anions B^- (fixed anions), causing additional cations C^+, to diffuse across from the outside (accumulation of cations). (b) Membrane is impermeable to the cations D^+, causing additional anions A^-, to diffuse across from the outside (accumulation of anions).

7.3. ACTIVE ABSORPTION

The absorption of ions against the concentration gradient or with the help of metabolic energy is known as *active absorption*.

The analysis of vacuolar sap of plants indicated that both anions and cations accumulate in plant cells in a large quantities against the concentration gradient. It cannot be explained simply by ion exchange or Donnan equilibrium theories. *Hober* (1945) reported that fresh water alga *Nitella* accumulated K^+ ions in the cells about 1000 times more than the concentration of K^+ in the surrounding medium. Such observations were also confirmed in marine alga *Valonia* by Hoagland and in many other higher plants.

The absorption of ions and their retention within the cells at the highest concentration require an expenditure of energy by the plants. There is a direct relationship between metabolic energy and salt uptake or accumulation. It has been demonstrated experimentally by a number of workers, e.g.:

(i) *Steward* (1932) and *Hopkins* (1956) observed a close relationship between salt accumulation and respiration. They observed that salt accumulation is directly proportional to the O_2 content. Salt accumulation increases with the increase in O_2 content and decreases or even completely stops with the decrease in O_2 content of the nutrient medium.

(ii) *Robertson* and *Turner* (1945) demonstrated with the help of metabolic inhibitors that salt absorption is an energy consuming process. They observed that respiratory inhibitors (azides and cyanides) inhibit salt absorption.

(iii) *Lundegardh* (1955, 58) observed that salt uptake or absorption increases and stimulates the rate of respiration.

(1) **Mechanism of active salt absorption**: Following theories have been proposed to explain the phenomenon of active absorption.

(a) *The Carrier Concept* (Fig. 7.4): This theory was proposed for the first time by *Vanden Honert* (1937). He explained the role of carrier compounds in the transport of ions across the membrane. The space in the tissue or cell can be divided into three parts: (i) *Outer free space*—The space which permits free diffusion of ions, (ii) *Inner space*—The space in which penetration of ions takes place with the aid of metabolic energy, and (iii) *The intermediate space between outer free space and inner space*—It is not clearly defined. However, it is thought to occur in the middle of the cytoplasm. This space is found to be impermeable to free ions. Through this space, ions can be carried only be *specific carriers* which are thought to exist in this membrane space.

Mineral Salt Absorption

The carrier concept explains that :

(i) Some carrier molecules act as carrier of ions.

(ii) The carrier molecule combines with the ion in the outer free space to form *carrier-ion complex*.

(iii) The carrier-ion complex moves through intermediate space into the inner space where it releases ions.

(iv) The ions once taken in into inner space can not go back to outer space.

(v) The carrier compound returns back to outer space to pick up fresh ions.

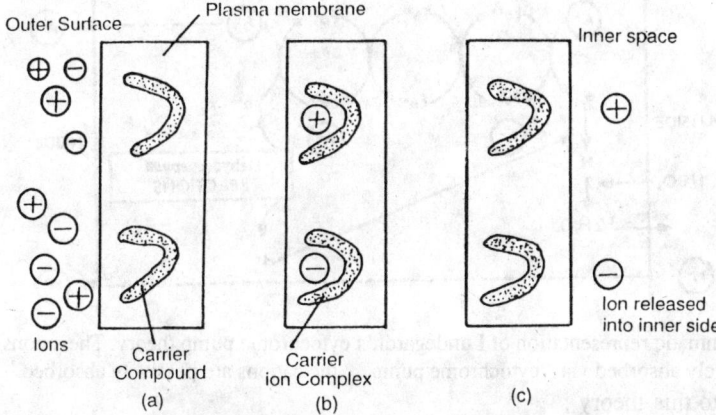

Fig 7.4: Diagram showing the carrier concept.

Evidences in Support of Carrier Concept

The following three observations *viz.* isotopic exchange, saturation effect and specificity, greatly support the carrier concept of active absorption of mineral salts :

1. *Isotopic exchange*: The use of radioactive ions by a number of physiologists demonstrated that the ions failed to diffuse through the cell membrance. This led to the discovery that certain carriers intervein in the diffusion of ions through the impermeable membrane.

2. *Saturation effect*: It has been observed that continuous increase in the concentration of salts in the external medium does not increase the rate of active salt absorption beyond a certain limit. This is because all the active sites of carriers are occupied with the ions. The maximum rate becomes at the time of complete saturation of active sites of carrier compound. After this, there will be no effect of increase in salt concentration in the external medium. It is called *saturation effect*. The carrier compounds are either enzymes or enzyme like molecules.

3. *Specificity*: The carriers are specific in nature and they absorb specific ions. This also explains the selective and unequal absorption of ions by roots.

Theories of Salt Absorption

Based on the carrier concept there are two possible mechanisms for salt absorption. In one cytochromes are involved while in another ATP is used.

(1) **Cytochrome pump theory** (Fig. 7.5): Earlier workers believed that salt accumulation depends upon metabolic energy but there is no quantitative relationship between salt absorption and respiration. However, *Lundegardh* and *Burstrom* (1933) observed such relationship between anion absorption and respiration. They observed that the rate of respiration increases when a plant is transferred from water to a salt solution. This increase in rate of respiration over the normal or ground respiration due to anion absorption is called *anion respiration* or *salt respiration*.

Later on *Lundegardh* (1950, 1954) proposed the cytochrome pump theory based on the following assumptions:

(i) The absorption of anions is independent of absorption of cations.
(ii) The mechanism of anion and cation absorption is different.
(iii) An oxygen concentration gradient exists from outer surface to the inner surface of the membrane which favours oxidation at the outer surface and reduction at the inner surface.
(iv) The anions are absorbed through cytochrome oxidase system by an active process.
(v) The cations are absorbed passively.

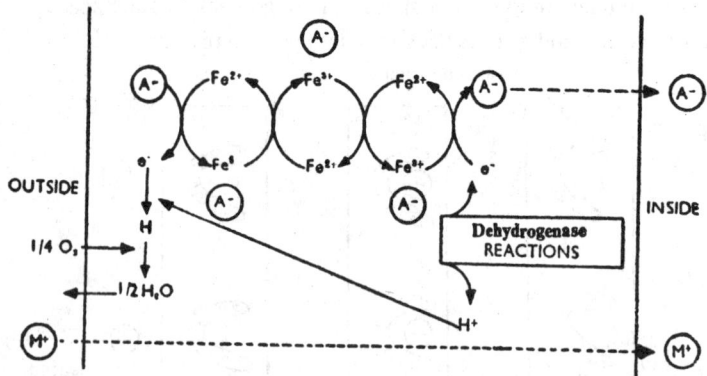

Fig 7.5: Diagrammatic representation of Lundegardh's cytochrome pump theory. The anions (A^-) are actively absorbed via "cytochrome pump"; the cations are passively absorbed.

According to this theory :
(i) Dehydrogenase reactions on the inner side of the membrane produce protons (H^+) and electrons (e^-).
(ii) The electrons move towards outside via a cytochrome chain, which finally reduce terminal cytochrome on the outer surface of membrane. The ion plays an important role in this mechanism.
(iii) At the outer surface, the reduced iron of cytochrome is oxidized with the help of oxygen by releasing the electron (e^-) and picking up an anion.
(iv) The released electron unites with a proton (H^+) and oxygen to form water.
(v) The anion (A^-) moves towards inner side over the cytochrome chain.
(vi) When anion reaches the terminal cytochrome towards inner side, the oxidized cytochrome becomes reduced by taking an electron produced through the dehydrogenase reactions, and the anion (A^-) is released towards inner side through cytochrome.
(vii) As a result of anion absorption, a cation (M^+) moves passively from outside to inside to balance the anion.

Criticism of Cytochrome Pump Theory

This theory has been criticised by a number of workers on the following points:
(i) It does not explain the involvement of metabolic energy in salt absorption.
(ii) It explains only the active absorption of anions.
(iii) It does not explain selective uptake of ions.
(iv) Overstreet (1955) observed that cations also stimulate salt respiration.
(v) 2,4-Dinitrophenol (DNP), an inhibitor of oxidative phosphorylation, increases respiration but inhibits salt absorption (*Robertson et al.*, 1951).

Carrier Mechanism Involving ATP (Fig. 7.6)

The observation of *Robertson et al.* (1951) that 2,4-DNP inhibited salt absorption, presents a strong case for the participation of ATP in active salt absorption.

Mineral Salt Absorption

Bennet-clark (1956) proposed a carrier mechanism involving phospholipids (lecithin) and ATP. According to this mechanism lecithin is synthesized and hydrolysed in a cyclic manner, picking up ions on the outer surface and releasing them on hydrolysis into inner space. The synthesis of at least one of the components of this lecithin cycle requires ATP. Here lecithin acts as a carrier which helps in the movement of ions inside through the cell membrane which otherwise is impermeable to ions.

The compound lecithin is composed of phosphatide and choline. The phosphate group in the phosphatide acts as a active centre for binding the cation and choline group acts as active centre for binding the anion. The lecithin molecule when combines with cations and anions, moves inside and decomposes into phosphatidic acid and choline into inner surface by the action of enzyme *lecithinase*. Here, both the ions are released and lecithin is regenerated from phosphatidic acid and choline by the action of enzymes choline acetylase and choline esterase in presence of ATP.

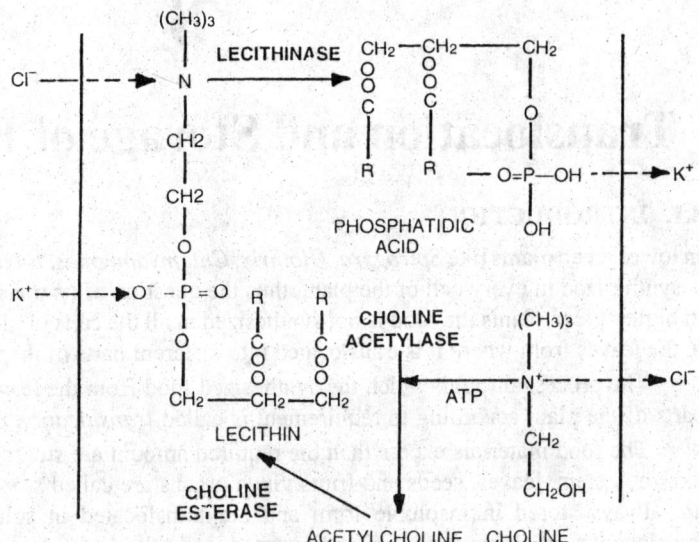

Fig 7.6: Diagrammatic representation of the Bennet-Clark's protein lecithin theory.

7.4. FACTORS AFFECTING SALT ABSORPTION

1. *Temperature*: The absorption of salt increases with an increase in temperature, but in confined to very narrow range. At very high temperature the salt absorption is inhibited most probably due to denaturation of enzymes (proteins) involved.

2. *Hydrogen ion concentration* (pH): The decrease in the pH of soil solution accelerates the absorption of anions while increase in pH favours the absorption of cations. The pH values across the physiological range may inhibit salt absorption.

3. *Light*: It indirectly effects the transpiration and photosynthesis showing its influence in salt absorption.

4. *Oxygen*: It greatly influences active salt absorption. In absence of oxygen, active salt absorption is inhibited.

5. *Interaction of ions*: The absorption of one ion is influenced by the presence of other ions in the medium, *e.g.*, the absorption of potassium is found to be affected by the presence of calcium, magnesium and other cations in the external medium.

6. *Growth*: Active cell division, elongation and developmental processes promote absorption of salt.

With the ageing of plant parts, such as roots, the absorbing capacity is decreased and the roots become more suberized and cuticularized.

7.5. TRANSLOCATION OF MINERAL SALTS

The translocation of mineral salts takes place both by xylem and phloem. The upward movement (from root to shoot) usually occurs through xylem while bidirectional movement (upward and downward from mature leaves) occurs through phloem.

8

Translocation and Storage of Food in Plants

8.1. INTRODUCTION

In lower green plants like *Spirogyra, Ulothrix, Chlamydomonas, Euglena* and all other algae, the food is synthesized in every cell of the plant, thus the question of its translocation does not arise in them. In higher green plants the food is not synthesized in all the cells of plants but it is mainly synthesized in the leaves from where it is translocated into different parts of the plant.

The process through which the synthesized food from the leaves is translocated into different parts of the plant according to requirement is called *translocation of food*.

The food materials excess than the required amount are stored in different organs of the plant like root, stem, leaves, seeds and fruits. Such organs are called *storage organs*. The food materials are always stored in insoluble form and are translocated in solution or soluble form. During translocation insoluble food is first converted into soluble form.

8.2. TRANSLOCATION OF FOOD

(1) *Direction of translocation*: The translocation of food in plants takes place in downward, upward and lateral directions.

(2) *Downward translocation*: According to this hypothesis different organic food materials synthesized in the leaves are translocated downward in the soluble form due to which they are stored in stems and roots of the plants. In plants some food is utilised in the formation of new cells, some part in the nutrition of old cells and the remaining food is converted into insoluble form and is stored in different storage organs like stems *e.g.*, in potato tubers, and in roots *e.g.*, Radish, Carrot and Beta root etc.

(3) *Upward translocation*: Certain stages of the plants are found when the food materials are translocated upwardly. Some stages are as follows:

    (*i*) Germination of seeds.
    (*ii*) The formation of new stems and leaves from the underground storage organs.
    (*iii*) Formation and development of new buds.
    (*iv*) Development of fruits.

(4) *Lateral translocation*: In certain parts of stems and roots, the food is translocated in lateral direction. This is mainly performed by medullary rays.

Path of Translocation

In higher plants, the food is translocated in different organs of the plants through vascular system which is made up of vascular bundles. The vascular bundles are made up of complex tissues, xylem and phloem. The xylem possesses tracheids, vessels, fibres and paren-

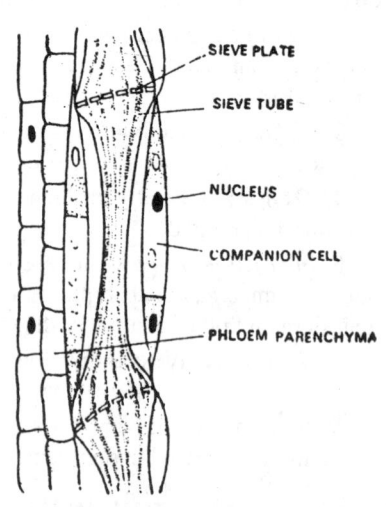

Fig. 8.1 : Phloem tissue of stem.

chyma cells whereas the phloem possesses sieve tubes, companion cells, phloem fibres and phloem parenchyma cells. The xylem elements are mainly involved in ascent of sap and phloem elements (sieve tubes) in translocation of food.

(1) *Path of downward translocation of foods*: The food materials synthesized in the leaves are translocated in the soluble form through the sieve tubes of phloem elements downward. It has been confirmed by various experiments and can be demonstrated by following experiments.

Ringing experiment: Take a plant and remove its all the tissues (phloem and bark) except xylem and pith in the form of a ring at any place of the stem. The stem part from where the tissues have been removed is closed or sealed with the help of melted wax. After 7-8 days you will see that the epidermis and cortex of upper portion of the ring become very much swollen and from this swollen part the adventitious roots

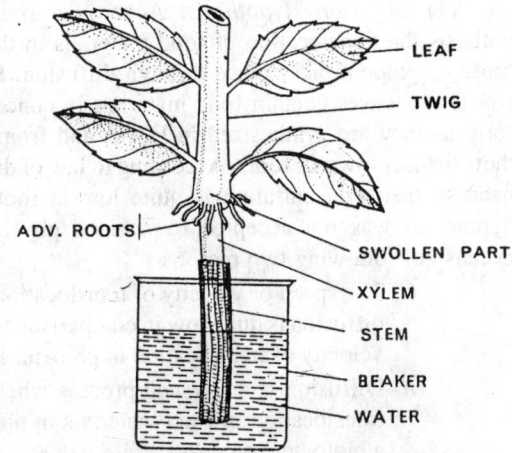

Fig. 8.2: Ringing experiment.

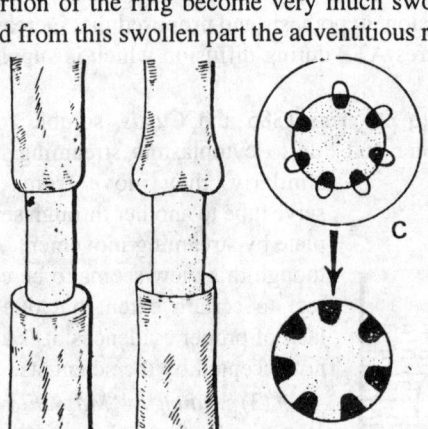

Fig. 8.3: Ringed stem and its TS.
A. Removal of bark in the form of ring.
B. Swelling in upper portion of the ring.
C. TS of barked stems.
D. TS of ringed portion.

emerged out. It happens because the food material translocated from the leaves does not pass through the ring and is stored in upper portion of the ring resulting in its swelling. After sometime the lower portion of stem becomes weak and due to starvation or non-supply of the food materials. If this experiment of the phloem is not removed from the ring, the swelling in upper portion does not occur. This indicates that the food materials synthesized in the leaves are translocated through the phloem in downward direction.

(2) *Path of upward translocation of food materials*: In the beginning, *Dixon* proposed that upward movement of food materials in plants takes place through xylem. Later on *Curtis* confirmed through certain experiments that the upward movement of food material takes place through phloem only, and xylem is not involved in this phenomenon. Curtis made following alterations in three different young woody plants.

(*i*) From first plant, he removed phloem tissues and no xylem in a ring. He also removed the leaves from the upper portion of ring.

(*ii*) From second plant, he removed xylem from particular portion of a stem, the leaves were also removed.

(*iii*) In third plant, the xylem and phloem were left in normal condition but the leaves were removed.

He observed that the dry weight of the first plant is the lowest in comparison to others two and the growth take place in the upper portion of the ring. It clearly indicates that the translocation of food materials in upward direction is ceased due to removal of phloem tissues. The question does not arise of reaching newly formed food materials of leaves into stem because the leaves were already removed from this portion of stem.

8.3. MECHANISM OF TRANSLOCATION

Following theories were proposed to explain the mechanism of translocaion.

(1) *Diffusion Hypothesis*: According to this hypothesis the translocation of food materials in the seive tubes of phloem takes place through diffusion. In other words the leaves contain food materials in concentrated form as they are synthesized in leaves and from leaves they diffuses towards roots. According to law of diffusion because their concentration is quite low in roots. This hypothesis was not accepted by various physiologists because of following two reasons :

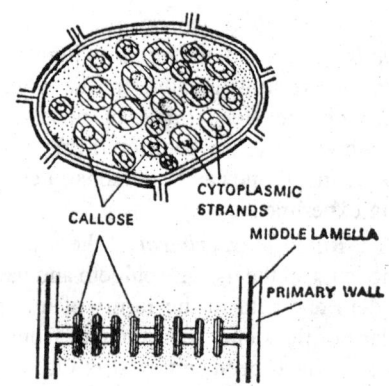

Fig. 8.4: Callose formation in Sieve plate.

(i) The speed or velocity of translocation due to diffusion is quite low in comparison to actual velocity of translocation in phloem.

(ii) Diffusion is a physical process whereas the translocation of food materials in phloem is a biological process.

Masson and *Phillis* (1936) slightly modified diffusion hypothesis and proposed that increased in the velocity of translocation of food materials requires ATP during diffusion which is supplied from the protoplasm of seive tubes of phloem.

(2) *Protoplasmic streaming theory*: According to *De-vries* 1885 and *Curtis*, soluble food materials in seive tubes move from one end to another end due to cytoplasmic streaming and similarly, they move from one seive tube to another through seive plate by streaming movement. Although this view seems to be correct to certain extent but due to lack of proper evidences, it is also not accepted by the scientists.

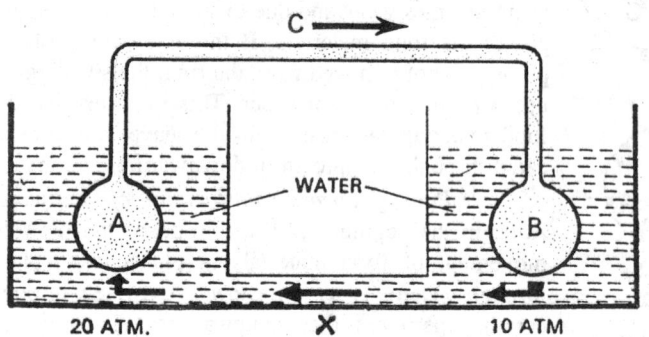

Fig. 8.5: Experimental demonstration of Munch's "Mass Flow" hypothesis.

(3) *Munch's "Mass Flow" Hypothesis*: Some scientists are of the opinion that the soluble food material in the phloem move just like the blood which moves in the blood vessels.

Munch (1930) proposed a hypothesis, according to which the soluble food materials in the phloem show mass flow. The main reason behind this is that the sugars are synthesized in the mesophyll cells of leaves due to which the OP of mesophyll cells is increased resulting in absorption of water through xylem by endosmosis.

Now, the turgor pressure of mesophyll cells increases towards upper side which affects and produces mass flow in the protoplasm of seive tube towards lower side. Thus, the sugars move downwards into the roots where they are utilised during respiration and growth and their concentration is also reduced. Thus, the food materials always move from higher concentration towards lower concentration continuously.

The main drawback or defect of Munch-hypothesis is that it explains only unidirectional downward flow of soluble food materials.

8.4. STORAGE OF FOOD

The carbohydrates and other food materials are synthesized in excess than the required amount inside the leaves. This excess food is translocated from leaves and stored in various storage organs like roots, stems and leaves, seeds of fruits. The storage mainly takes place in parenchyma cells, meduallary rays and xylem parenchyma cells. These cells contain certain enzymes which convert soluble food materials into insoluble food and other enzymes again convert insoluble food into soluble food when required. In carrot, radish, and turnip the food is stored inside the roots whereas in others it is stored in underground stems, leaves, seeds, flowers and fruits.

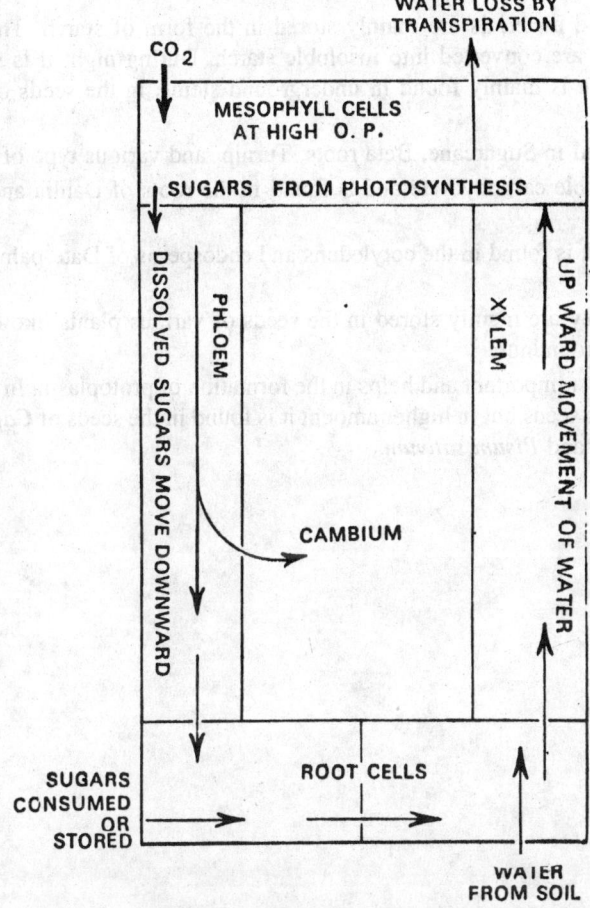

Fig. 8.6: Diagrammatic representation of Munch's Hypothesis.

(a) *Food storage organs*: Food storage organs are as follows:

   (i) *Leaves:* In certain plants the food is stored inside the leaves e.g., Onion, Garlic, Cabbage, and *Bryophyllum* etc.

   (ii) *Roots:* In certain plants food is stored inside the roots. e.g., Carrot, Radish, Turnip, *Beta vulgaris, Asparagus* and Dahlia etc.

   (iii) *Stems:* In certain plants, the food remain stored inside the stems only. The examples are as follows:

Phylloclade	:	*Opuntia*
Rhizome	:	*Turmeric, Zingiber,* Banana
Corm	:	*Aulocacia*

Tuber	:	Potato
Bulbil	:	*Agave*, Lily

(iv) Flowers: e.g., Cauliflower

(v) Seeds: Normally the seeds of all flowering plants contain stored food either inside the cotyledons or inside the endosperm. This food is utilized during germination of seed and formation of seedling.

(b) Forms of stored food: The food may be stored inside the plants in the following forms:

(1) Carbohydrates

(A) Starch: The food materials are mainly stored in the form of starch. The soluble sugars synthesized in the leaves are converted into insoluble starch. During night it is stored in various storage organs. The starch is mainly found in underground stems, in the seeds of Wheat, Maize, Rice and in fleshy roots.

(B) Sugar: It is found in Sugarcane, Beta roots, Turnip and various type of fruits.

(C) Inulin: It is soluble carbohydrate and is found in the roots of Dahlia and in members of Compositae.

(D) Hemicellulose: It is found in the cotyledons and endosperns of Date-palm, Lupin, Coffee, etc.

(2) Oil and Fat: They are mainly stored in the seeds of various plants like Linum, Mustard, Almond, Coconut, and Groundnut.

(3) Protein: It is most important and helps in the formation of protoplasm. In smaller amounts it is found in almost all the seeds but in higher amount it is found in the seeds of *Cajanus, Phaseolus radiatus, Cicer arietinum* and *Pisum sativum*.

9

Special Modes of Nutrition in Plants

9.1. INTRODUCTION

Like animals, plants also need food. Green plants can synthesise their own food but others cannot synthesise and they depend upon other sources like plants or organic and inorganic substances for the food. All the plants can be divided into following groups depending upon the mode of nutrition.

(*i*) Autotrophic
(*ii*) Heterotrophic
(*iii*) Special, Insectivorous

9.2. AUTOTROPHIC

Those plants which can synthesise their own food are included under this group. The leaves of such plants are green and synthesise their food from CO_2 and H_2O in presence of sunlight by the process of photosynthesis. Some plants possess green and soft branches which also photosynthesize and participate in food formation.

Epiphytes: The plants growing on other tree branches and capable of manufacturing their own food are called *epiphytes*. In other words, the autotrophic plants growing on the branches of other trees are called *epiphytes*. They possess two types of roots :

(*a*) *Aerial roots*: Their ends are spongy and they absorb moisture from the air.

(*b*) *Clinging roots*: They absorb mineral substances by entering into the bark of trees and provide stability to the epiphytic plant. The well-known examples of epiphytes are *orchids* and some ferns.

9.3. HETEROTROPHIC

Those plants which are unable to synthesise their own food are included under this group. They are usually achlorophyllous and depend on other sources for their food. Heterotrophic plants may be of the following types :

A. Parasite
B. Saprophyte

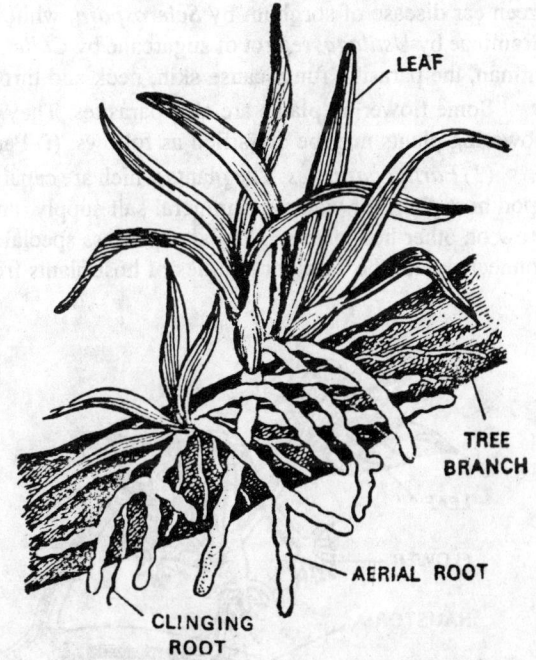

Fig. 9.1: An orchid growing on a tree.

C. Symbiotic

Parasites

The plants growing on other living trees or animals are called *parasites e.g.*, Bacteria and fungi.

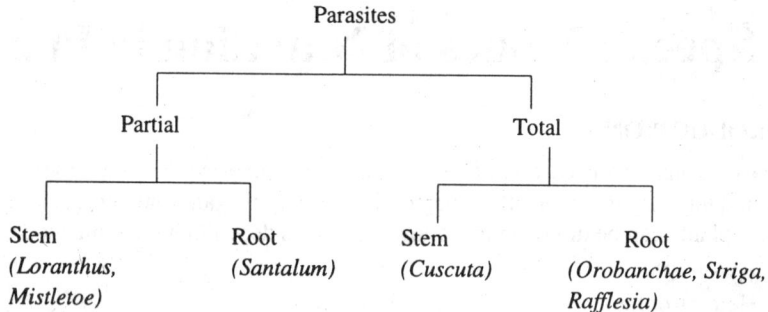

(1) *Parasitic bacteria*: They cause several diseases in plants and animals. For example, the fire disease of apple and pear is caused by *Irwinia amylovera*; black rot disease of radish turnips and cabbage by *Xanthomonas campestris*. The common diseases caused by bacteria in human beings are diarrhoea, pneumonia, tetanus, leprosy, gonoerrhoea and plague.

(2) *Parasitic fungi*: They also cause various diseases in plants and animals. For example, rust disease of wheat is caused by *Puccinia graminis*, late blight disease of potato by *Phytophthora*, green ear disease of sorghum by *Sclerospora*, white rust of crucifers by *Cystopus* smut disease of Graminae by *Ustilago*, red rot of sugarcane by *Colletorichum* and early blight disease by *Alternaria*. In man, the parasitic fungi cause skin, neck and throat diseases.

Some flowering plants are also parasites. They may be partial or total parasites. The parasitic flowering plants may be classified as follows. (*i*) Partial parasites, and (*ii*) Total parasites

(3) *Partial parasites*: The plants which are capable of manufacturing their own food but depend upon host plant for water and mineral salt supply, are called *partial parasites*. They are green and grow on other host plants. They also possess special type of organs called *haustoria* which remain connected with the vascular bundles of host plants from where they absorb water and mineral salts.

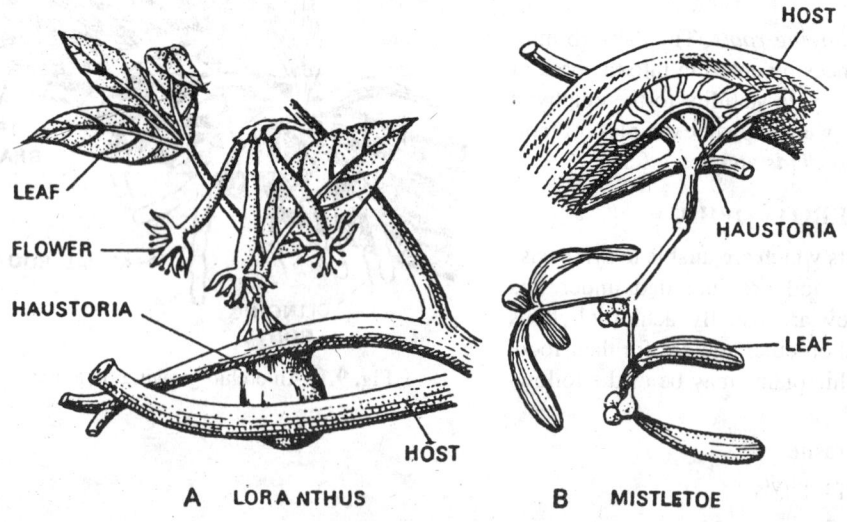

Fig. 9.2: Partial stem parasites. (A) Loranthus, (B) Mistletoe.

Special Modes of Nutrition in Plants

Partial parasites may be of two types : 1. Partial stem parasites, and 2. Partial root parasites.

1. *Partial Stem Parasites*: The partial parasites growing on the stems of other plants are called *partial stem parasites*. They depend for water and mineral supply on the stem of host plant. *Loranthus, Mistletoe* and *Viscum* are common *partial stem parasites*. Loranthus usually grows on mango trees whereas *Mistletoe* commonly grows on the stems of *Citrus, Artocarpus* and *Mangifera* plants. The seeds of *Mistletoe* are sticky and dispersed by birds (Ornithophilous). Such seeds germinate on host plant and get water and mineral supply from the host. After sometime the host plant dries up. (Fig. 9.2).

2. *Partial Root Parasites*: The partial parasites growing on the roots of host plants are called *partial root parasites*. They produce several thin roots through which they absorb water and mineral salts. The roots of *Santalum* are partial root parasites.

(4) *Total Parasites*: Those plants which completely depend on other host plants for their food supply, are called *total parasites*. They are achlorophyllous and unable to synthesise their food. Total parasites are of two types : 1. Total stem parasites, 2. Total root parasites

1. *Total Stem Parasites*: Total parasites growing on the stem of host plants are called *total stem parasites*. *Cuscuta* is common and important total stem parasite. The plants of *Cuscuta* possess only small flowers and no leaves. Its seeds germinate in the soil and develop into twinner. After sometime it becomes attached with the neighbouring plant and dissociates with the soil. Now, the haustoria are produced from the stems of *Cuscuta* plant which get associated with the vascular bundles of host plant and absorb food materials (Fig. 9.3).

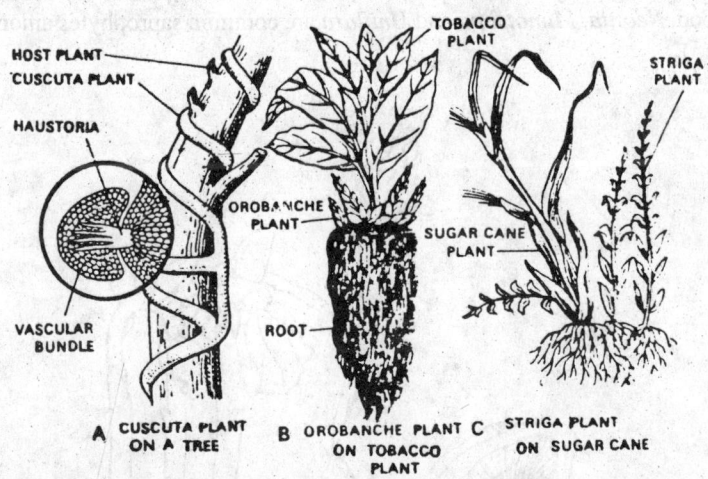

Fig. 9.3: Total stem parasites. (A) Cuscuta, (B) Orobanche, (C) Striga.

2. *Total Root Parasites*: The total parasites growing on the roots of host plants are called *total root parasites*. Following are the common and important total root parasites.

(*i*) **Striga**: It grows on the roots of sugarcane and sorghum plants.

(*ii*) **Rafflesia**: It is also total root parasite. It grows on the roots of Lianas plants in the forests of Jawa and Sumatra. The body of Rafflesia is made up of very small fibres which enter into the soil and get attached with the roots of other plant for the absorption of food materials. Only the flowers of this plant are seen on the soil which are largest among the flowers of all angiospermous plants in the world. The diameter of each flower is about one metre and its weight about 15 kg. (Fig. 9.4).

(*iii*) **Orobanche**: It grows on the roots of potato, tomato, brinjal and tobacco etc. It also possesses haustoria for the absorption of food materials.

(2) *Saprophytes*: The plants growing on the dead organic materials are called *saprophytes*. They are without chlorophyll and unable to synthesise their own food. The dead organic materials provide nourishment to these soprophytes.

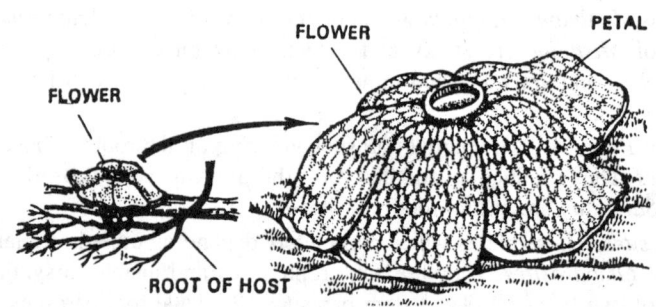

Fig. 9.4: Rafflesia on the roots of Lianas.

Most of the bacteria and fungi are saprophytes. They decompose the dead bodies of plants and animals into simpler compounds and clean the environment. Alcohols, curd and cheese are prepared from such saprophytic bacteria. *Yeast, Mucor, Rhizopus, Penicillium, Morchella, Agaricus* etc. are saprophytes. Of these *Mucor, Rhizopus* and *Penicillium* spoil various foodstuffs, cloths and leather products during rainy season. Species of *Morchella* and *Agaricus* are used in the preparation of delicious food. *Neottia, Monotropa* and *Uniflora* are common saprophytes among angiospermous plants (Fig. 9.5).

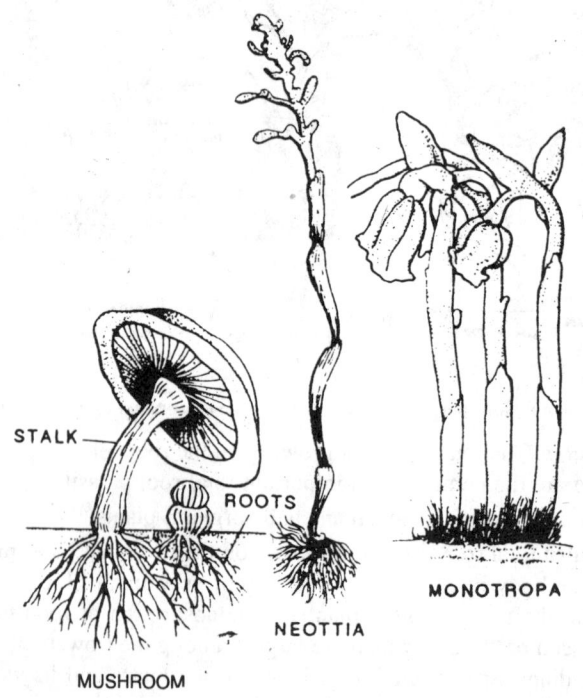

Fig. 9.5: Saprophytic Plants.

Special Modes of Nutrition in Plants

(*i*) *Mushrooms*: They grow on the dead bark and wood of plants and on the soil.

(*ii*) *Neottia*: This is a type of orchid which grows in humus rich soil. Its stem is fleshy and yellow and bears only a few or none very small leaves. Its roots contain fungal hyphae which form mycorrhiza. *Neottia* plants absorb food materials through their fungal hyphae.

(*iii*) *Monotropa*: The plants of *Monotropa* are also fleshy and bear brown scale leaves at the base of stem. They also absorb food materials through the fungal hyphae present in the roots.

Symbionts

When two plants grow together and are mutually benefited, the plants are known as *symbiotic plants* and the phenomenon is called *symbiosis*. Following are the examples which explain the phenomenon of symbiosis.

(1) *Lichens*: Lichens form the group of those plants which are formed by the association of algae and fungi. The fungus absorbs water for algae and the algae photosynthesise the food materials for fungus. Thus, both are mutually benefited. The lives of both become impossible without the help of any one of them. It is called *symbiosis*. Lichens grow on the barks of trees or on the stones. They are capable enough of facing unfavourable conditions (Fig. 9.6).

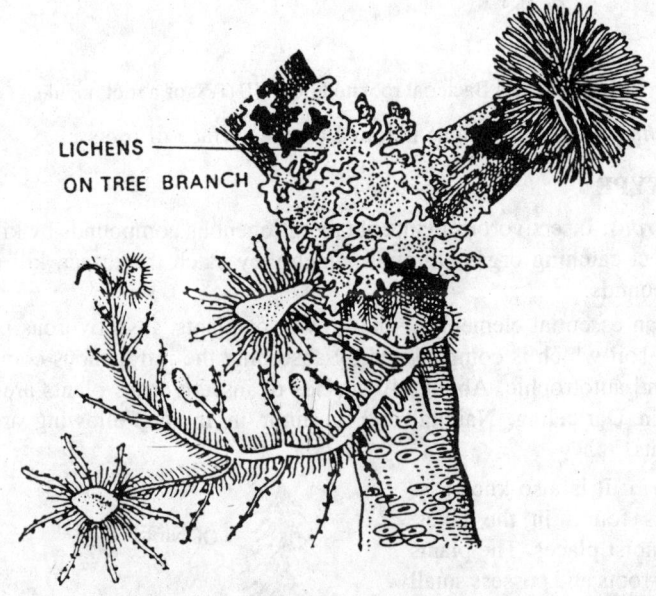

Fig. 9.6: The lichen on a tree branch.

(2) *Bacteria of Root nodules*: Leguminous plants like *Cajanus, Cicer, Pisum, Dolichos* and *Lens esculentus* etc. possess different-sized glands in their roots, called *root nodules* or *bacterial nodules*. They contain the bacteria called *Rhizobium leguminoserum* which help in the fixation of free nitrogen of soil air into nitrates. The nitrates are absorbed in the form of manure by the roots of these plants. In exchange, the plants provide food and living space to the bacteria. Thus, both are benefited mutually and explain the symbiotic relationship (Fig. 9.7).

(3) *Mycorrhiza*: The symbiotic relationship of fungus with the roots of any plant is called *mycorrhiza*. It is observed in the roots of forest plants especially on the surface of *Pinus* roots. The fungus forms the mantle on the roots and behaves like root hairs, performing the function of absorption of water and mineral salts from the soil. *Pinus* plant provides living space and food material to the fungus in exchange. *Mycorrhiza* is of two types :

(1) *Endotrophic*: When the fungus is found in the cells of root cortex.

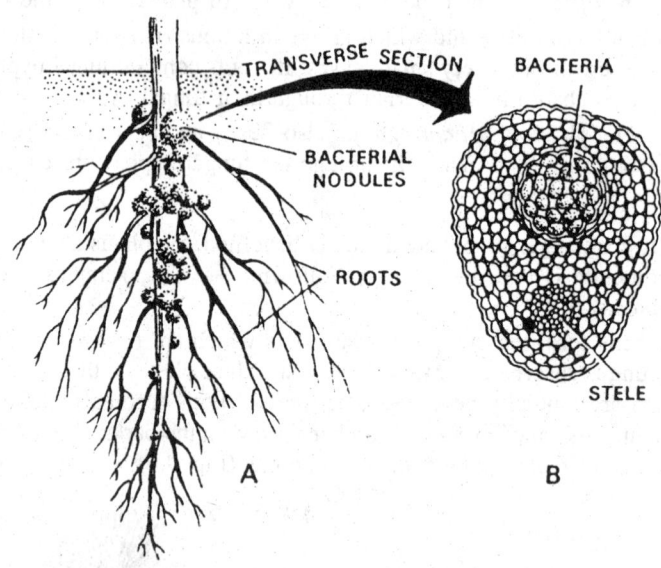

Fig. 9.7: (A) Bacterial root nodulues, (B)T.S. of a root nodule.

(2) *Ectotrophic*: When the fungus is found on the surface of roots.

9.4. SPECIAL TYPE

(1) *Insectivorous*: Insectivorous plants obtain nitrogenous compounds by killing small insects. They possess insect catching organs through which they catch the insect, kill it and then absorb nitrogenous compounds.

Nitrogen is an essential element for the growth of plants. Insectivorous plants are found in nitrogen deficient soil which is compensated by absorbing the nitrogenous compounds of insects. They are green and autotrophic. About 440 species of insectivorous plants are known. A few of them are found in Darjeeling, Nainital and Kashmir in India. Following are a few common insectivorous plants.

(a) *Utricularia*: It is also known as bladderwort. It is found in the lakes, ponds and other moist places. The plants of *utricularia* lack roots and possess small dissected leaves and coloured attractive flowers. A few leaves of plants form bladders which possess trap door with valves for insects on the mouth and many digestive glands on the walls. Many insects enter into the bladder through trap door along with water stream. Now, the valves of trap door become closed and the digestive glands produce digestive juice for the digestion of insects. Finally, the digested materials along with nitrogenous compounds are absorbed by the plant.

(b) *Drosera*: It is also known as

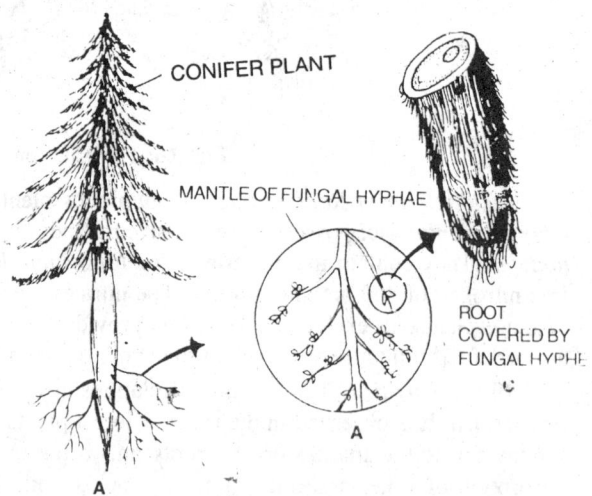

Fig. 9.8: (A) Fungal hyphae on the roots of Pinus tree. (B) Enlarged view. (C) Mantle of fungl hyphae on root.

Special Modes of Nutrition in Plants

sundew plant. It is herbaceous, about 8-20 cm long and found in the marshy places and hills of Kashmir and Nainital. The leaves of this plant are flat and round and produce many pin head like glandular hairs called *tentacles*. The tantacles are sensitive and produce a sticky liquid which helps in catching the insects. The digestive glands produce the juice for the digestion of insects.

(*c*) *Nepenthes*: It is also known as pitcher plant. It is found in the Assam hills. Its stem is herbaceous and climbing and the leaves are modified into pitcher like structure. The mouth of the pitcher possesses a lid and its walls possess many digestive glands which produce scented digestive juice. The juice helps in attraction and digestion of insects. When any insect sits on the mouth of pitcher, it slips into pitcher through the lid and the hard hairs present on the walls of pitcher check the insect to go out. The digestive juice oozes out from the glands and digest the insect. Finally, the walls absorb the nitrogenous compounds present in the insect (Fig. 9.9).

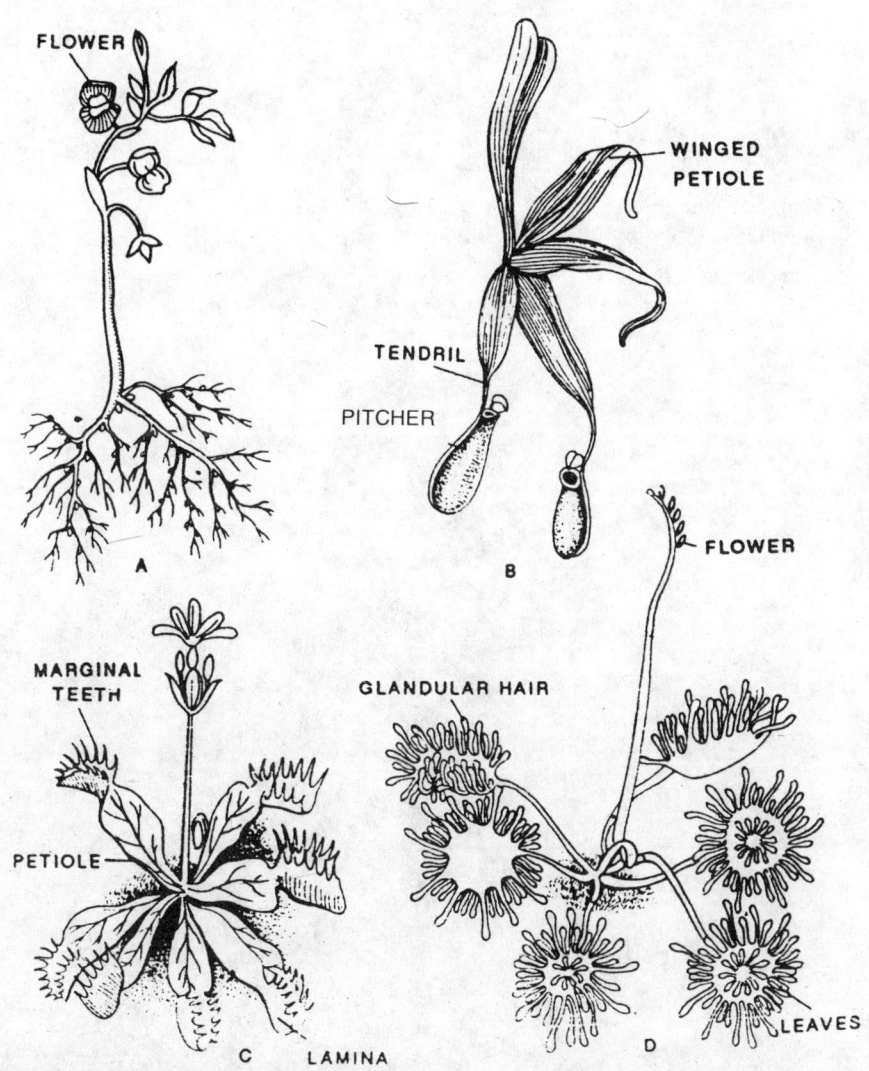

Fig. 9.9: (A) —Utricularia, (B)—Nepenthes, (C) —Dionaea, (D) —Drosera.

(*d*) *Dionaea*: It is also known as Venus fly Trap. The complete plant is spherical and looks like a large flower. It is herbaceous and found on moist places. The potiole of the leaf looks like a leaf and its lamina become rounded and bears 10-20 marginal teeth and three pairs of sensitive hairs in the centre. When any insect touches the sensitive hairs, the lamina from both the sides turns and closes the insect. The marginal teeth check the insect from going out. Digestive glands are found near the sensitive hairs which produce the juice for the digestion of insect (Fig. 9.9).

10

Photosynthesis

10.1. BRIEF HISTORY

1. In 1771-72, *Joseph Priestly* performed the first experiment on photosynthesis and concluded that the plants restore used air. This experiment was the first demonstration that the plants produce oxygen.
2. In 1779, *Jan Ingen Housz* repeated the experiment of Priestly and made following important conclusions:
   (a) The evolution of O_2 took place only within a few hours and not weeks.
   (b) The evolution of O_2 took place only during day time *i.e.* in presence of sunlight.
   (c) Only green parts of plants could restore used air. Ingen Housz was the first to give an equation for photosynthesis.

$$CO_2 + H_2O \xrightarrow[\text{Green Plants}]{\text{Light}} O_2 + \text{Organic matter.}$$

3. In 1782, *Jean Senebier* demonstrated that fixed air (CO_2) is neccesary for the production of O_2 by green plants.
4. In 1804, *Nicholas Theadore de Saussure* observed that water is essential for photosynthesis. He also correlated the importance of light during intake of CO_2 and evolution of O_2.
5. After 1780s, *Antonie Lavoisier* told that photosynthesis is a chemical process.
6. In 1837, *Dutrochet* experimentally demonstrated that chlorophyll is necessary for photosynthesis.
7. In 1845, *Liebig* told that CO_2 is the main source of all organic compounds synthesised by green plants.
8. By the early *nineteenth* century, the necessary requirements for photosynthesis were known. These can be summarized by the following equation:

$$\underset{\text{Raw materials}}{\text{Carbon dioxide}} + \underset{\text{}}{\text{Water}} \xrightarrow[\text{Green Plants}]{\text{Necessary conditions} \atop \text{Light}} \underset{\text{Products}}{\text{Organic matter}} + O_2$$

(Senebier)   (de Saussure)   (Ingen-Hausz)   (Ingen-Hausz)   (Priestly)

9. In 1864, *Sachs* first told that the organic matters produced by plants were carbohydrates.
10. In 1923, after Warburg's work the green alga *Chlorella vulgaris* was extensively used in photosynthetic studies.
11. In 1941, *Ruben, Randall* and *Kamen* confirmed by the use isotopic studies that the O_2 evolved during photosynthesis comes from H_2O and not from CO_2.
12. In 1949, *Calvin, Benson* and co-workers studied the different steps of dark reaction in an unicellular alga *Chlorella* through tracer technique.
13. In 1950, *F.F. Blackman* proposed the *'Law of limiting factor'*.
14. In 1960, *Hill* and *Bendall* suggested Z-scheme.

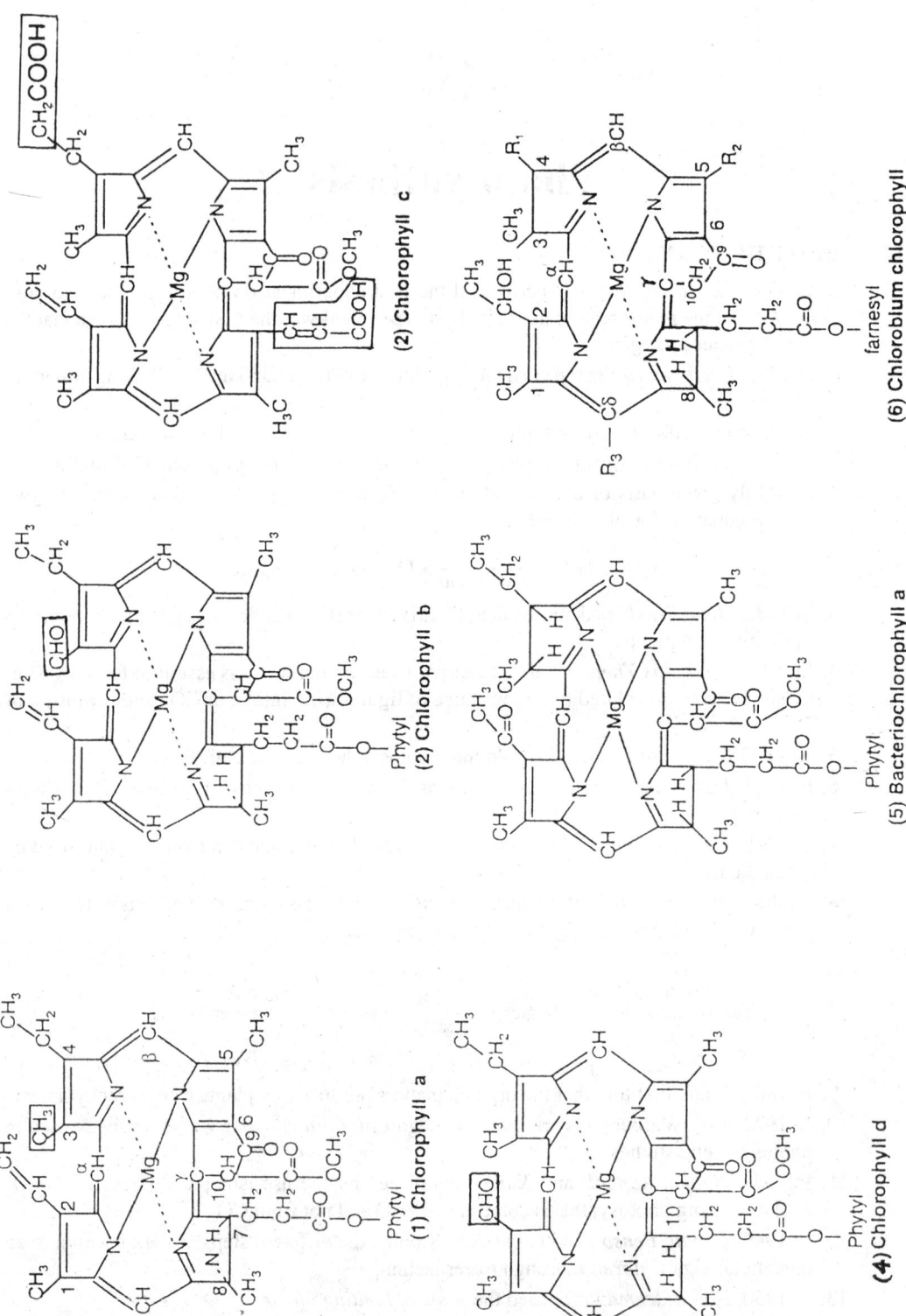

Fig. 10.1: *Structures*. Chl *a, b, c, d*, Bacteriochlorophyll *a* and Chlorobium chlorophyll.

15. In 1961, *Peter Mitchell* proposed chemi-osmotic hypothesis for the production of ATP in photosynthesis.
16. In 1966, *M.D. Hatch* and *C.R. Slack* proposed C_4 cycle for the CO_2 fixation in higher plants.

10.2. PHOTOSYNTHETIC APPARATUS

(Consult Chapter "The Cell").

10.3. PHOTOSYNTHETIC PIGMENTS

There are three major classes of photosynthetic pigments:

(1) Chlorophylls (2) Phycobilins and (3) Carotenoids.

Chlorophylls: They are insoluble in water but soluble in organic solvents. Various type of chlorophylls like chlorophyll *a, b, c, d, e* bacteriochlorophyll *a, b, c* and *d* chlorobium chlorophylls and bacterioviridin etc. have been isolated either from higher plants or algae or bacteria. The members of higher plants or algae or bacteria may have single or more than one type of chlorophylls. The chlorophylls absorb light in the blue 450 nm and in red of 650-700 nm regions of the visible spectrum. The distribution of various plant pigments is shown in Table 10.1.

Bacterio chlorophyll *b* has been isolated from *Rhodopseudomonas* spp, but its structure is not yet know. Chlorobium chlorophyll exists in two forms (650 and 660). They differ from chlorophyll *a* in following points:

1. They are esterified with farnesol rather than phytol.
2. They lack the $CH_3.O.C$ group at C_{10}
3. They possess a $CH_3.CHOH$-group at C_2
4. They have different substituent at C_4, C_5 and the δ methane carbon atom.

A number of chlorophylls with distinct spectra have been isolated which perform different roles in photosynthesis (see photophosphorylation). Depending upon their absorption in the region of visible light they are called Chl $a_{670-673}$, Chl $a_{680-683}$, Chl $a_{695-705}$ and P_{690} and P_{700}.

All chlorophylls possess almost similar structure with slight variation. The chlorophyll is an

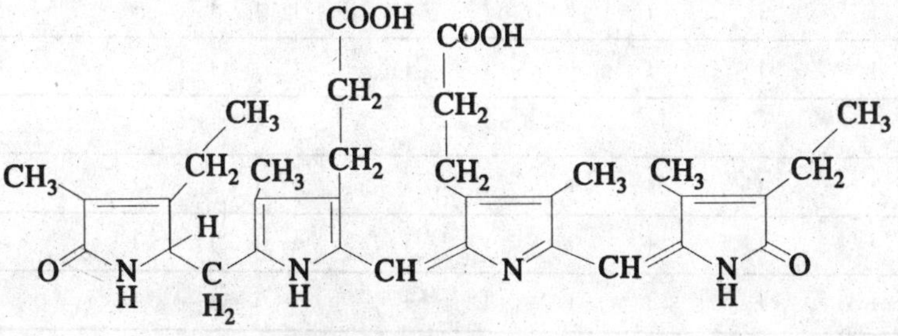

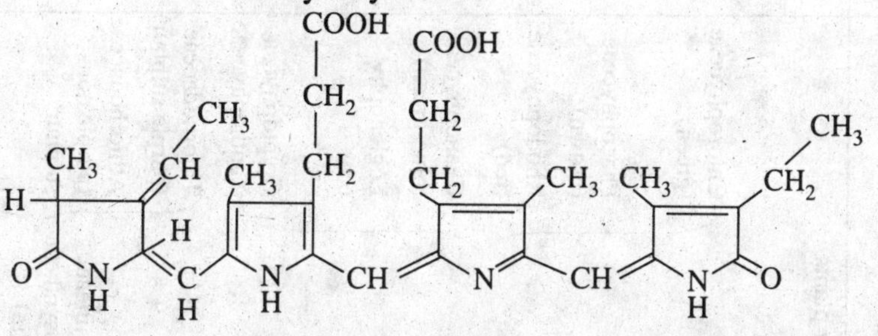

Fig. 10.2: Structure of A. Phycoerythrobilin B. Phycocyanobilin.

Table 10.1. Distribution of Major Photosynthetic Pigments.

Organism		Chlorophylls						Bacterio-chl			CN onobi chl	Bacterio virldin	Phycobilins			Major Carotenoids	
		a	b	c	d	e	a	b	c	d			Phyco-erythrin	Phyco-cyannin		Carotenes	Xanthophylls
Higher plants		+	+	–	–		–	–	–	–		–	–	–	β-carotene	lutein violaxanthin neoxanthin
Algae	Chlorophyceae (green)	+	+	–	–		–	–	–	–		–	–	–	β-carotene	lutein violaxanthin neoxanthin
	Phaeophyceae (brown)	+	–	+	–		–	–	–	–		–	–	–	β-carotene	fucoxanthin
	Rhodophyceae (red)	+	–	–	+		–	–	–	–		–	+++	+	α-carotene β-carotene	lutein zeaxanthin
	Cyanophyceae (blue-green)	+	–	–	–		–	–	–	–		–	+	+++	β-carotene	echinenone myxoxanthophyll
	Euglenophyceae	+	+	–	–		–	–	–	–		–	–	–	β-carotene	violaxanthin diadinox anthin neoxanthin
	Cryptophycea Xanthophyceae	+	–	+	–	+	–	–	–	–		–	+	+	α-carotene	alloxanthin
Rhodo-bacteriineae (photosynthetic bacteria)	Thiorhodaceae (purple sulphur)	–	–	–	–		+	+	+	–	–	–	–	–	–	Spirilloxanthin[a]
	Athiorhodaceae (purple non-sulphur)	–	–	–	–		+	+	+	+	–	–	–	–	lycopene (traces)[b]	Spirilloxanthin[b] spheroldene[c] spheroldenone[c]
	Chlorobacteriaceae (green sulphur)	–	–	–	–		+	–	+	–	+	+	–	–	chlorobactene γ-carotene (occasionally)	hydroxychloro-bactene (traces)

Chlor-770 / 650 or 660

[a] In Chromatium [b] In Rhodospirillum rubrum [c] In Rhodopseudomonas spheroides

asymmetric molecule having a hydrophilic head (porphyrin) made up of four substituted pyrrolic rings located around a divalent magnesium (Mg^{++}) which is complexed with the four nitrogen atoms of pyrrole rings (one nitrogen atom from each ring) and a long tail formed by a hydrophobic isoprenoid chain (phytol tail) with a phytol alcohol and 20 carbon atoms. The phytol tail remains linked with IVth pyrrole ring with ester linkage (except in Chl. c). The fifth non-pyrrole made up of only carbon atoms, is also found attached with IIIrd pyrrole ring.

The empirical formula of Chl. a is $C_{55}H_{72}O_5N_4Mg$ and Chl. b is $C_{55}H_{70}O_6N_4Mg$. The chlorophyll b occurs in two forms Chl b_{640} and Chl b_{650}. PSII contains major amount of Chl b. The chlorophyll b differs from chlorophyll a in having an aldehyde (–CHO) group at C_3 in IInd pyrrole ring. Chlorophyll a contains methyl group (—CH_3) in place of aldehyde at the same position.

Phycobilins: Another class of photosynthetic pigments includes phycobilins. They are soluble in water and in structure, they resemble with chlorophylls in having tetrapyrroles. Phycobilins are however, open chain tetrapyrrole compounds without Mg^{2+} and phytol chain and are of two types. The phycocyanins (blue) and phycoerythrins (red). Phycobilins have been isolated only from Rhodophyceae, cyanophyceae and cryptophyceae. Phycobilins absorb light of visible spectrum between 500-600 nm. There distribution is shown in Table 10.1. The structure of phycoerythrobilin and phycocyanobilin are as follows:

They constitute the major photosynthetic pigments of blue green and red algae and impart a distinctive colour due to presence of bile pigments in their cells. Bile pigments impart red colour and phycocyanins impart blue colour to the algae.

In vivo, phycobilins exist as prosthetic group of conjugated proteins known as *biliproteins*. The linkage between phycobilin and the apoprotein is a covalent type whose cleavage requires vigorous hydrolysing conditions. It is observed that the location of the main absorption of these pigments varies with the protein to which the chromophoric-prosthetic group is attached and with the mode of attachment. The absorption maxima of biliproteins varies between 500-600 nm.

Carotenoids: They form the third class of photosynthetic pigments and are of two types— Carotenes (orange) and Xanthophylls (yellow). Carotenes are usually found in PSI and Xanthophylls in PS II. They are insoluble in water and soluble in organic solvents. Various types of carotenoids have been isolated from higher plants, green algae, brown algae, red algae, blue green algae, cryptomonads and photosynthetic bacteria. In addition to light absorption, carotenoids perform other unrelated functions like protection of chlorophyll molecules from photo-oxidation. They absorb light of visible spectrum between 450-500 nm.

Carotenoids are always found associated with chlorophylls and in thylakoids are present as chromoproteins. Carotenes are isoprenoid compounds with unsaturated hydrocarbons. They absorb blue and green lights and transmit red and yellow lights. The xanthophylls are oxygen derivatives of carotenes containing 1-8 oxygen atoms. For structure of carotenes consult chapter on 'vitamins' of this book.

10.4. DISTRIBUTION OF PIGMENTS
The distribution of different pigments is shown in Table 10.1.

10.5. BIOSYNTHESIS OF CHLOROPHYLL
Various steps in biosynthesis of chlorophyll are as follows:
1. The succinyl CoA, which is an intermediate compound of Kreb's cycle, condenses with the amino-acid glycine to form an unstable compound α-*amino* β-*ketoadipic acid* which after decarboxylation yields δ-aminolevulinic acid. The reaction is catalysed by enzyme δ-aminolevulinic acid synthetase in presence of pyridoxal phosphate, iron and light.

$$\text{Succinyl CoA + glycine} \longrightarrow \text{α-amino β-ketoadipic acid (unstable)}$$

$$\xrightarrow[\text{Fe}^{2+}, \text{ Light}]{\text{Pyridoxal phosphate}} \text{δ-aminolevulinic acid.}$$
(with CO$_2$ released)

2. Two molecules of δ-aminolevulinic acid condense and fuse to form porphobilinogen in presence of enzyme δ-aminolevulinic acid dehydrase.

$$\text{2 mols δ-aminolevulinic acid} \xrightarrow{\text{δ-aminolevulinic acid dehydrogenase}} \text{Porphobilinogen.}$$

3. Four molecules of porphobilinogen condense to form uroporphyrinogen III in presence of enzyme uroporphyrinogen I synthetase and uroporphyrinogen III cosynthetase. During the condensation 4 ammonium ions are released.

$$\text{4 mols Porphibilinogen} \longrightarrow \text{Uroporphyrinogen III} + 4NH_4^+$$

4. Four molecules of uroporphyrinogen III combine with 4-acetic acid molecules to produce coproporphyrinogen III in presence of enzyme uroporphyrinogen decarboxylase. 4CO$_2$ molecules are released in this reaction.

$$\text{4 mols. Uroprophyrinogen III + 4 mols. Acetic acid} \longrightarrow \text{coproporphyrinogen III} + 4CO_2$$

5. Coproporphyrinogen III combines with O$_2$ in presence of coproporphyrinogen oxidative decarboxylase to produce protoporphyrinogen IX.

$$\text{Coproporphyrinogen III} + 2O_2 \longrightarrow \text{Protoporphyrinogen IX} + 2CO_2 + 4H^+$$

6. Protoporphyrinogen IX undergoes oxidation to form protoporphyrin IX.

$$\text{Protoporphyrinogen IX} + O_2 \longrightarrow \text{Protoporphyrin IX} + 6H^+$$

7. Protoporphyrin IX combines with Mg^{2+} to form Mg protoporphyrin IX.

$$\text{Protoporphyrin IX} + Mg^{2+} \longrightarrow \text{Mg protoporphyrin IX.}$$

8. Mg protoporphyrin IX is methylated with 5-adenosyl methionine to form Mg protoporphyrin IX monomethyl ester. The transfer of methyl group from 5-adenosyl methionine to Mg protoporphyrin IX is catalysed in presence of Mg protoprophyrin methyl esterase.

9. Mg protoporphyrin IX monomethyl ester is converted to protochlorophyllide.

10. Protochlorophyllide now combines with a phytol group to form protochlorophyll which in presence of light is reduced to chlorophyll *a*.

$$\text{Protochlorophyllide + Phytol} \longrightarrow \text{Protochlorophyll.}$$

$$\text{Protochlorophyll} + 2H^+ \xrightarrow{\text{Light}} \text{Chlorophyll } a.$$

Recent researches have shown that chlorophyllide a is a precursor of chlorophyll *a*. When etiolated seedlings are subjected to light, protochlorophyllide is reduced to chlorophyllide *a*.

$$\text{Protochlorophyllide} + 2H^+ \xrightarrow{\text{Light}} \text{Chlorophyllide } a.$$

11. Finally chlorophyllide *a* is esterified with phytol group to form chlorophyll *a* in presence of enzyme chlorophyllase.

$$\text{Chlorophyllide } a + \text{Phytol} \xrightarrow{\text{Chlorophyllase}} \text{Chlorophyll } a.$$

10.6. DEFINITION OF PHOTOSYNTHESIS

The synthesis of organic compounds like carbohydrates or glucose by the cells of green plants in presence of sunlight with the help of CO_2 and H_2O is called *photosynthesis*. During this process the radiant energy of sunlight is stored in carbohydrates. Because the carbon of used CO_2 in this process is assimilated, it is also called *carbon assimilation* process. The mechanism of photosynthesis can be represented by the following simple equation :

$$6CO_2 + 6H_2O \xrightarrow[\text{chlorophyll}]{\text{Light}} C_6H_{12}O_6 + 6O_2$$

Ruben and *Kamen* (1941) demonstrated that the evolution of O_2 in this process takes place by the oxidation or dissociation of water molecules. Therefore, the above equation can be correctly written as :

$$6CO_2 + 12H_2O \xrightarrow[\text{chlorophyll}]{\text{Light}} C_6H_{12}O_6 + 6H_2O + 6O_2\uparrow.$$

Because six molecules of water are insufficient for the evolution of 6 molecules of O_2, therefore, more molecules of water have to be shown in the equation (at least 12 molecules of water for the formation of one hexose sugar).

Now, the question arises that from where the raw materials are derived by plants for photosynthesis. The terrestrial plants obtain CO_2 from the atmosphere (where it is only 0.03%) through the stomata of leaves; water from the soil through roots; and light from the sun. The chlorophyll is found in the chloroplasts of mesophyll cells of leaves. Aquatic plants obtain CO_2 from water, where it is found in the dissolved form in about 0.3 or more per cent, through their entire suface; water from the aquatic medium by absorption through the surfaces of their organs; and light from the sun. The chlorophyll is found in the chloroplasts of leaves like terrestrial plants. Aquatic plants photosynthesise more and produce larger amount of carbohydrates than terrestrial plants. The reason behind this is that the water-bodies, where aquatic plants grow, contain more amount of CO_2 in water which alongwith water is absorbed through the entire surface of plants.

10.7. PHOTOSYNTHESIS IS A CHEMICAL PROCESS

Photosynthesis is a chemical, oxidation-reduction, process in which water molecules are oxidized to form O_2, and CO_2 molecules are reduced to form carbohydrates (glucose). The reduction of CO_2 into carbohydrate requires assimilatory powers like ATP and NADPH + H^+. The reduction of CO_2 takes place in dark conditions while the production of assimilatory powers takes place in presence of light. Therefore, two phases occur during photosynthesis.

(1) *Light dependent phase*—It requires light and is also called *photochemical reaction* or *light reaction* or *Hill-reaction*.

(2) *Light independent phase*—It does not require light and is also called *chemical dark reaction* or *dark-reaction* or *Blackman's reaction*. (Fig. 10.3).

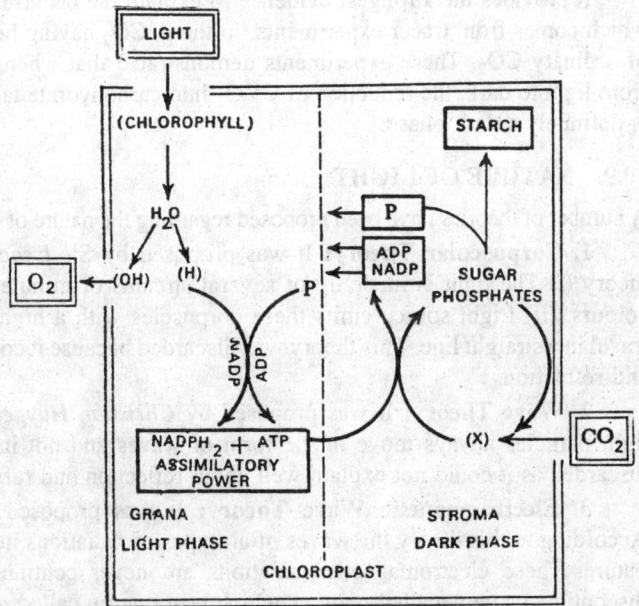

Fig. 10.3: Demonstration of light dependent and light independent phases during photosynthesis.

10.4. EVIDENCES IN SUPPORT OF LIGHT AND DARK REACTIONS

1. Evidence from Intermittent Light

Warburg (1919) observed in *Chlorella vulgaris* and *Scenedesmus obliques* that when intermittent light was given to them, the rate of photosynthesis per second was higher as compared to the continuous supply of the same intensity of light. This indicates that one phase of photosynthesis is light independent. It may also be possible that the rate of dark reaction is reduced due to continuous supply of light.

2. From Temperature Coefficient

The temperature coefficient of any reaction is the ratio of the rate of reaction at a given temperature and at a temperature of 10°C lower than this. It is represented by symbol Q_{10}. The value of Q_{10} is almost unity *i.e.*, $Q_{10}=1$, for a photochemical or light reaction because its rate does not increase with the rise of temperature. The value of Q_{10} is always equal to 2 or 3 *i.e.*, $Q_{10}=2$ or 3, for the dark reaction. Blackman for the first time demonstrated that if a plant is given more light and sufficient quantity of CO_2, the value of Q_{10} always comes 2 or more. It indicates that in photosynthesis at least one phase or reaction is influenced by temperature. This may be light independent phase (dark reaction). In contrast, the value of Q_{10} for photosynthesis in low intensity of light comes almost unity which indicates that in photosynthesis at least one phase is controlled by light. It may be light dependent phase (light reaction). In low intensity of light, the light reaction becomes limiting, which also limits dark reaction. Now, if the temperature is increased 10°C, the rate of dark reaction increases but the rate of light reaction becomes limited. Thus, ultimately the rate of photosynthesis does not increase. It also indicates that light and dark reactions are although independent but are interlinked.

3. From CO_2 Reduction in Dark

It provides the strongest evidence to explain the occurrence of two phases in photosynthesis which comes from tracer experiments. In them, CO_2 having heavy carbon ($C^{14}O_2$) is used in place of ordinary CO_2. These experiments demonstrated that when chlorophyllous cells are transferred from light to dark, the reduction of $C^{14}O_2$ into carbohydrate takes place. It indicates that this phase is definitely a dark phase.

10.9. NATURE OF LIGHT

A number of theories have been proposed regarding the nature of light. Some of the important ones are :

1. Corpuscular Theory: It was proposed by Sir *Issac Newton* (1666). According to this theory (*i*) The light is made up of several streams of minute particles or corpuscles of different colours. (*ii*) Light source emits these corpuscles with a high speed. (*iii*) The corpuscles always travel in a straight line. This theory was discarded because it could not explain the laws of reflection and refraction.

2. Wave Theory: It was proposed by *Christian Huygens* (1678). According to this theory light particles always move in the form of waves and not in straight line. This theory was also discarded as it could not explain well about reflection and refraction.

3. Electromagnetic Wave Theory: It was proposed by *James Clark Maxwell* (1860). According to this theory the waves of all types of radiations including light are electro-magnetic in nature. These electromagnetic radiations are never continuous and are emitted by matter as discontinuous units called *photons* which bear energy called *quantum*.

4. Quantum Theory: It was proposed by *Max Planck* (1900). According to this theory the radiant energy including light is made up of discrete energy particles (corpuscles) called *quanta*. This concept was called *quantum theory*. Planck also stated that the size of a quantum of energy is directly proportional to the frequency of radiation and the intensity of light depends upon the number of photons. The energy of quantum (photon) can be calculated by the following Planck's equation:

$E_{photon} = hv$ (where E = energy; h = Planck's constant and v = frequency of radiation)
Planck's constant = 1.5×10^{-37} Kcal sec/quantum or 6.62×10^{-27} erg sec.

$v = c/\lambda$ (where c = velocity of light λ = wavelength of light in cm
= 2.9977×10^{10} cm/sec.)

$\therefore E_{photon} = h \times \dfrac{c}{\lambda} = \dfrac{hc}{\lambda}.\text{erg.} = \dfrac{hc}{4.184 \times 10^7 \lambda}$ calories (4.184×10^7 ergs = 1 calorie)

The light quanta are called *photons*. They have been considered as the physical unit of light while the quantum as the unit of energy of photons. The blue light photons have more energy than red light photons because the energy of photons depends upon the frequency of wavelength of radiation and the blue light had more wavelength than red light. It should be noted that although blue light photon possesses more energy than red light photon but the red light is more effective in promoting photosynthesis than blue light. The reason being behind that chlorophyll molecules undergo differential changes in their energy states in blue and red light.

5. Einstein Law of Photochemical Equivalence: According to this law a molecule in a photochemical reaction can react only after absorbing one photon. It follows, therefore, that 1 mole can react only after absorbing 6.023×10^{23} photons (6.023×10^{23} is Avogadro's number which is the number of molecules in a mole of any substance). Now the energy of 6.023×10^{23} photons is (6.023×10^{23}) hv or (6.023×10^{23}) hc/λ. This quantity of energy is called an *Einstein*. Thus, an Einstein is the amount of radiant energy which must be absorbed by 1 mole of a substance before it can react in a photochemical reaction. The energy of an Einstein varies directly with frequency (v) and thus inversely with the wavelength (λ); the greater the frequency, the shorter the wavelength and the larger the energy of the Einstein. The energy of an Einstein can thus be calculated as follows:

Suppose the wavelength of light is 700 nm.

1. Einstein 700 nm = $\dfrac{Nhc}{4.184 \times 10^7 \lambda}$ calories

where
N = Avogadro's number = 6.023×10^{23}
h = Planck's constant = 6.62×10^{-27} erg sec.
c = Velocity of light = 2.9977×10^{10} cm/sec.
λ = Wavelength of light in cm (1 nm = 10Å = 10^{-7} cm)

$= \dfrac{6.023 \times 10^{23} \times 6.62 \times 10^{-27} \times 2.9977 \times 10^{10}}{4.184 \times 10^7 \times 0.70 \times 10^{-4}}$

$= \dfrac{2.856}{0.70 \times 10^{-4}}$ [note $\dfrac{Nhc}{4.184 \times 10^7}$ is a constant = 2.856]

= 40,800 calories.

10.10. MECHANISM OF ABSORPTION OF LIGHT

When a visible light photon with wavelength (λ) between 380 nm and 780 nm strikes a chlorophyll molecule, it releases an electron from chlorophyll to an outer molecular orbital. This state of chlorophyll molecule is called *excited* or *activated state*. The normal state of an atom or a molecule is called *ground state* or *singlet state* (S_0) in which the *electrons are present in even number and paired condition*. These paired electrons always possess spins in opposite direction. They possess lowest energy in the molecular orbital and the molecule shows all spins paired and no magnetic moment.

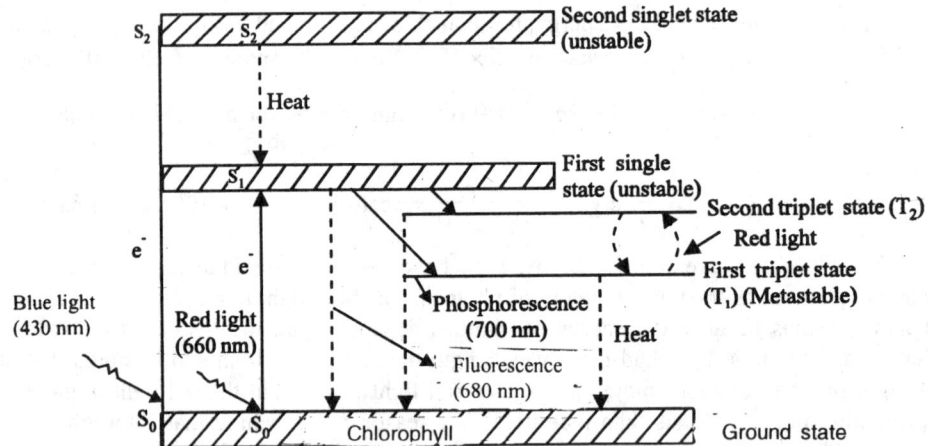

Fig. 10.4: Various energy states of chlorophyll molecule during light absorption.

The excited chlorophyll molecules may show at least four states: (*i*) first singlet state (S_1), (*ii*) second singlet state (S_2); (*iii*) first triplet state (T_1) and (*iv*) second triplet state (T_2).

When a red light photon strikes a chlorophyll molecule, it becomes photoexcited and an electron is released from its ground level molecular orbital to an outer molecular orbital. It is called *first excited singlet state* (S_1). It is unstable state having a half life period of only 10^{-9} seconds and the molecule contains high energy provided by the visible red light photon. Thus, in this state two molecular orbitals, each having one electron, are produced.

When blue light photons strike a chlorophyll molecule, its one electron from ground level molecular orbital (paired unit) is released to an outer molecular orbital. The chlorophyll molecules become photoexcited and the electron is raised more high than S_1 condition. It is because blue light photons possess more energy than red light photons. It is called *second excited singlet state* (S_2). This state is also unstable having half life period of less than 10^{-9} seconds. In S_2 state also two molecular orbitals are produced.

The S_1 and S_2 both states being unstable are converted into ground (S_0) by releasing energy through some processes like heat, phosphorescence, fluorescence or chemical energy. The S_1 state is converted to S_0 state by the release of radiation energy (light) in the red region. It is called *fluorescence*. It should be noted that the total energy is not lost as fluorescence but a very small amount of energy still remains present which is utilised in driving photosynthetic reactions. The S_2 state is first converted into S_1 state by the release of some energy in the form of heat. The S_1 state when releases only small amount of radiation energy, it is converted into T_1 and T_2 states. Both are interconvertible. T_1 is converted to T_2 with the absorption of second quantum (red light) by the pigment in T_1 state. T_1 state is converted to So state by the release of only small amount of radiation energy. This delaying effect of light emission is called *phosphorescence*. The remaining major amount of energy of T_1 is utilized in photochemical reaction. T_1 is the metastable state having a half life period of about 10^{-2} seconds. The singlet and triplet states of excitation differ in the spinning of electrons in their orbitals.

10.11. COMPONENTS OF ELECTROMAGNETIC SPECTRUM

Sunlight as the main source of energy for photosynthesis. Although it appears white, it is actually made up of different colours. The *electromagnetic spectrum* consists of radiations of different wavelengths including cosmic rays, gamma rays, X-rays, ultraviolet (UV) rays, the visible spectrum,

Photosynthesis

infrared rays (IR) and radio waves (Fig. 10.5). The waves of each of these types have a characteristic range of wavelengths. The visible spectrum represents only a small region in electromagnetic spectrum and consists of different coloured bands. It ranges from 3800-7600Å. Of the total visible spectrum only a small part is used in photosynthesis.

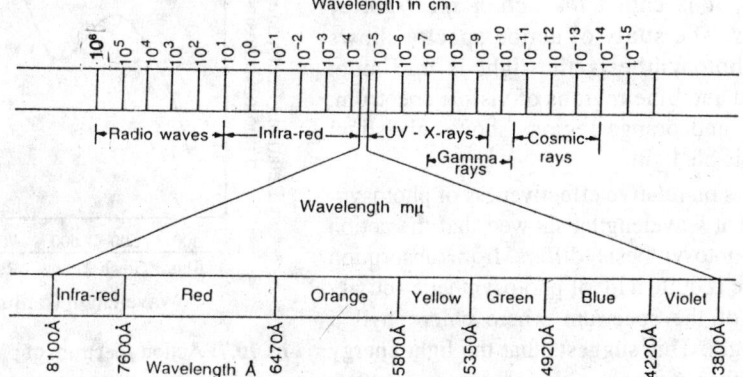

Fig. 10.5: Details of electromagnetic spectrum.

10.12. ABSORPTION SPECTRUM

The amount of absorption of light at different wavelengths by its object is called *absorption spectrum*. If chlorophyll is extracted and the light of different wavelengths is passed through it, the absorption at each wavelength can be measured by a spectrophotometer. Absorption of different wavelengths of light by a particular pigment is plotted and called the *absorption spectra* of that pigment. Different pigment absorb different wavelength of light. An absorption spectrum of chlorophyll *a* shows that light is mainly absorbed in the blue and red regions. Other lights like green, yellow and orange are absorbed only slightly. It is because the chlorophyll being green coloured does not absorb green light but reflects it. The exact position of peaks depends upon the solvent used in the extraction of pigments. In ether solution, the chlorophyll *a* shows maximum absorption at 662 mμ in red region and 430 mμ in blue region. In living plants, the maximum absorption peak of chlorophyll *a* is obtained at 683 mμ. The chlorophyll *b* in ether solution shows maximum absorption

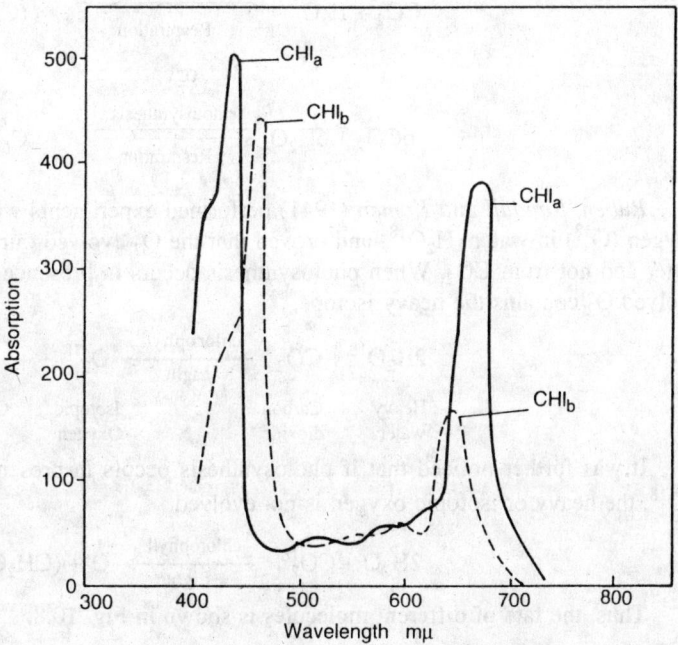

Fig. 10.6: Absorption spectra of Chl_a and Chl_b.

peak at 644 mμ in red region and 455 mμ in blue region. The absorption peaks of chlorophylls in blue-violet region are called *soret bands*.

Bacteriochlorophyll shows maximum absorption in infrared and blue-violet regions.

10.13. ACTION SPECTRUM

The effectiveness of different wavelengths of light utilized in the action of a process is called *action spectrum*. In case of photosynthesis, it is utilized in CO_2 fixation, O_2 production and $NADP^+$ reduction etc. Therefore, it is called *the action spectrum of photosynthesis*. The study of action spectra shows that during photosynthesis the light is maximum absorbed in red and blue regions of visible spectrum. Green, yellow and orange regions show only slight absorption of visible light.

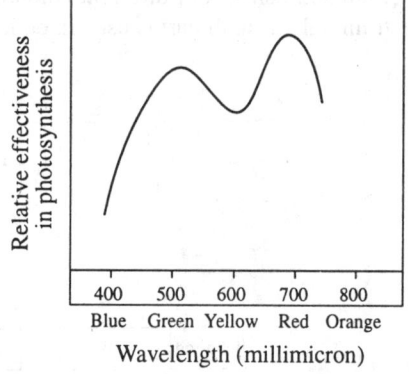

Fig. 10.7: Action spectrum of photosynthesis.

The studies on relative effectiveness of photosynthesis at different wavelengths showed that the action spectrum of photosynthesis differs from absorption spectrum. There is quite a lot of photosynthetic activity even in parts of the spectrum where chlorophyll *a* absorbs little light. This suggests that the light energy absorbed by other pigments like carotenes, xanthophylls and other forms of chlorophyll is transferred to chlorophyll *a*.

10.14. MECHANISM OF PHOTOSYNTHESIS

(1) Light Reaction: Earlier scientists upto 1930 believed that the process of photosynthesis is just reverse of respiration and the O_2 evolved during photosynthesis comes from CO_2. The water and CO_2 combines together to form carbohydrates.

$$CO_2 + H_2O \underset{\text{Respiration}}{\overset{\text{photosynthesis}}{\rightleftharpoons}} (CH_2O) + O_2$$

or

$$6CO_2 + 6H_2O \underset{\text{Respiration}}{\overset{\text{photosynthesis}}{\rightleftharpoons}} C_6H_{12}O_6 + 6O_2$$

Ruben, Randall and *Kamen* (1941) performed experiments with the help of heavy isotope of oxygen (O^{18}) in water (H_2O^{18}) and proved that the O_2 evolved during photosynthesis comes from water and not from CO_2. When photosynthesis occurs in presence of H_2O^{18} and normal CO_2, the evolved O_2 contains the heavy isotope.

$$2H_2O^{18} + CO_2 \xrightarrow[\text{Light}]{\text{chlorophyll}} O_2^{18} + (CH_2O) + H_2O$$

Heavy water — Carbon dioxide — Isotopic Oxygen — Carbohydrate — water

It was further noticed that if photosynthesis occurs in presence of normal water (H_2O) and CO_2^{18}, the heavy or isotopic oxygen is not evolved.

$$2H_2O + CO_2^{18} \xrightarrow[\text{Light}]{\text{chlorophyll}} O_2 + (CH_2O^{18}) + H_2O^{18}$$

Thus, the fate of different molecules is shown in Fig. 10.8:

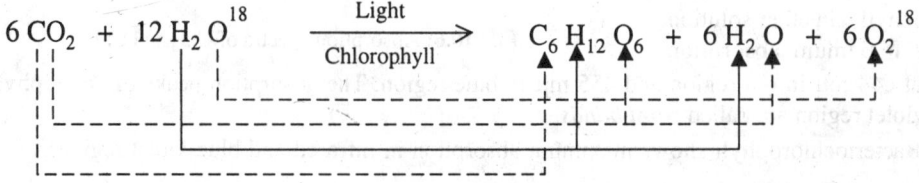

Fig. 10.8: Fate of different molecules during photosynthesis.

Light reaction was studied by *Robert Hill* (1941), therefore, it is called *Hill's Reaction*. He demonstrated that when isolated chlorophyll from green plants is illuminated in presence of hydrogen acceptors in a beaker, evolution of O_2 takes place. This phenomenon was named after Hill as Hill's reaction. Light reaction always takes place in the grana portion of chloroplast in presence of light.

The light reaction can be studied under following sub-heads:
I. Red drop, Emerson's enhancement effect and two pigment systems.
   (*a*) Red drop and Emerson's enhancement effect.
   (*b*) Two pigment systems.
II. Photo-oxidation of Water.
III. Production of assimilatory powers.
   (*a*) Non-cyclic electron transport system and non-cyclic photophosphorylation.
   (*b*) Cyclic electron transport system and cyclic photophosporylation.

(1) *Red Drop and Emerson's Enhancement Effect, and Two Pigment Systems.*

(*a*) *Red Drop and Emerson's Enhancement Effect*: Robert Emerson while determining the quantum yield of photosynthesis in *Chlorella* by using monochromatic light of different wavelengths observed a sharp decrease in quantum yield at wavelengths greater than 680 mμ. This decrease in quantum yield took place in the red part of the spectrum. Due to this reason, this phenomenon was called *red drop*. The number of oxygen molecules released per light quanta absorbed is called quantum yield of photosynthesis.

Later on, *Emerson* and his co-workers observed that if *Chlorella* plants are give the inefficient far-red light of 680 mμ or more wavelength along with light of shorter wavelength in alternate fashion, the quantum yield of photosynthesis becomes greater than the sum of yields when two types of beams of light are used separately. Thus, inefficient far-red light becomes efficient with the light of shorter wavelength. This enhancement of quantum yield of photosynthesis is referred to as *Emerson's enhancement effect*.

(*b*) *Two Pigment Systems or Photosystems:* The discovery of red drop and Emerson's enhancement effect has clearly shown that light reaction of photosynthesis is driven by two distinct photochemical processes, one driven by wavelengths exceeding 680 mμ and other by shorter wavelengths. These two processes are associated with two different systems of pigments called *Pigment System I* (PS I) and *Pigment System II* (PS II). Wavelengths of light exceeding 680 mμ affect only pigment system I while wavelengths of light shorter than 680 mμ affect both the pigment systems.

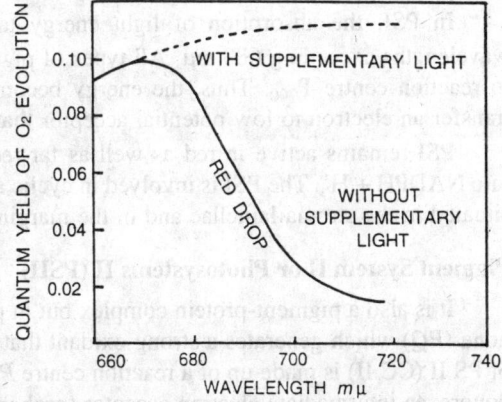

Fig. 10.9: Graphic representation of red drop and Emerson's enhancement effect in Chlorella.

A partial physical separation of two pigment systems has been achieved recently from thylakoid membranes. This separation was achieved by ultracentrifuging thylakoid membranes which resulted into two fractions—one lighter and other heavier. The lighter fraction gave the characteristic reactions of PS I while the heavier fraction gave the characteristic reactions of PS II. Later on it was noticed that the chlorophyll molecules are generally associated with hydrophobic protein forming complexes called *pigment-protein complexes (PPC)*.

The two pigment systems or photosystems (PS I and PS II) are found within thylakoid lamellae as structural entities and form separate pigment-protein complexes. Each pigment system is made up of:

(i) **Core Complex (CC):** It is a polypeptide containing 40-60 chlorophyll molecules, a reaction centre, electron donors and electron acceptors.
   (ii) **Light Harvesting Complex (LHC) or Antenna Complex (AC):** It is made up of many other remaining chlorophyll molecules and accessory pigments.

The pigment molecules of antenna complex absorb different wavelengths of light and transfer their absorbed energy to its reaction centre situated in the core complex. The core complex containing reaction centre is actually involved in photo-chemical act.

Pigment System I or Photosystems I (PSI)

The PSI is the pigment-protein complex which in presence of light can induce weak oxidant that oxidizes plastocyanin and also induce a strong reductant which transfers electrons to ferredoxin (Fd). PSI is made up of Chl a (absorbing higher wavelength of light). Chl b and carotenoids carotenes and xanthophylls. The carotenes in PSI are in dominating position as compared to xanthophylls while in PS II xanthophylls are more as compared to carotenes. The *core complex* of PSI (CCI) is made up of special type of pigment molecules similar to Chl$a$ – P_{700} (Probably Chl a_{700}) which absorb wavelength of 700 mμ, two iron containing proteins similar to ferredoxin, called *Fe-S* proteins. The *Fe-S* proteins are primary electron acceptors of PSI. Thus, core complex of PSI contains P_{700} which constitutes the reaction centre.

The light harvesting complex or antenna complex of PSI (LHCI or ACI) is made up of pigments absorbing at higher wavelengths of light. The definite number of pigment molecules in PSI is not known but *Muller et al.* (1980) estimated about 110 pigment molecules per reaction centre.

In PSI, the absorption of light energy takes place by pigment molecules which absorb wavelengths exceeding 680 mμ. All types of pigments after absorbing light energy finally transfer to reaction centre–P_{700}. Thus, the energy becomes stored in P_{700} molecules which is utilized to transfer an electron to low potential acceptor that via intermediate reduces NaDP$^+$.

PSI remains active in red as well as far-red light both and carries out reduction of NADP$^+$ into NADPH + H$^+$. The PSI is involved in cyclic and non-cyclic electron transport both and is found situated in the stroma lamellae and in the margins of grana.

Pigment System II or Photosystems II (PSII)

It is also a pigment-protein complex but in presence of light it induces reduction of plastoquinone (PQ) which generates a strong oxidant that recovers electrons from water. The *core complex* of PS II (CC II) is made up of a reaction centre P_{680} (most probably chl a_{680}), two or more electron donors, an intermediate electron acceptor (probably phaeophytin) and two bound quinones (Q_a and Q_b) which act as primary and secondary electron acceptors of PS II respectively. The antenna complex of PS II (AC II or LHC II) is made up of about 200 chlorophyll molecules absorbing shorter wavelengths of light, chlorophyll b and more xanthophylls than carotenes.

In PS II, the absorption of energy takes place by chlorophyll. a molecules absorbing wavelength of 680 mμ (Chl a_{680} or P_{680}) chlorophyll b and carotenoids. The reaction centre in this system is P_{680} and the energy is stored in these molecules. All energy absorbing pigments except chlorophyll.a are called *accessory pigments*.

The quantity of chlorophyll.b differs in both the pigment systems. Only chlorophyll.a molecules in both the pigment systems participate in photochemical reaction.

PS II remains active in red light only but becomes inactive in far-red light, *i.e.*, beyond 680 mμ. It carries out photo-oxidation of water *i.e.*, it is involved in evolution of molecular oxygen and release of protons (H$^+$) and electrons (e^-), and Hill reaction in presence of hill oxidants. PS II is involved in only non-cyclic electron transport and photophosphorylation and reduces the reaction centre of PSI (P_{700}).

Table 10.2: Differences between Pigment System I and Pigment System II

Pigment System I (PS I)	Pigment System II (PS II)
(i) PS I is made up of about 200 to 400 chlorophyll molecules; 50 carotenoid molecules; one molecule of reaction centre P_{700} (Chl. a_{700}), Plastocyanin, Cytochrome f each; two molecules of Cyt. b_{563} (Cyt b_6) and one or 2 molecules of ferredoxin.	(i) PS II is made up about 200 Chlorophyll molecules; 50 carotenoid molecules; one molecule of reaction centre P_{680} (Chl. a_{680}), unknown substance Z, Mn^{++}, Cl^-, Quinone, plastoquinone, Cytochrome b_6 (b_{559}), Cytochrome f and Plastocyanin each.
(ii) Usually the number of Chl. b. molecules is less.	(ii) Usually the number of Chl. b molecules is more.
(iii) Its reaction centre is P_{700}.	(iii) Its reaction centre is P_{680}.
(iv) It is located on the outer surface of thylakoid.	(iv) It is located on the inner surface of thylakoid.
(v) In this system, molecular O_2 is not evolved.	(v) In this system, molecular O_2 is evolved as a result of break down of water.
(vi) $NADPH + H^+$ is formed in this system.	(vi) $NADPH + H^+$ is not formed in PS II.
(vii) It participates in both cyclic and non-cyclic photophosphorylation.	(vii) It participates only in non-cyclic photophosphorylation. It gives electron to PS I when NADP is reduced to $NADPH + H^+$ in PS I.
(viii) When chloroplasts are ultracentrifuged, PS I being lighter comes up. Thus, PS I comes under lighter fraction.	(viii) On ultracentrifugation PS II being heavier settles down. Thus, PS II comes under heavier fraction.

(2) Photo-Oxidation of Water

Van Niel (1941) considered that first photochemical reaction in photosynthesis is the photo-oxidation of water and evolution of molecular oxygen. It is mainly associated with PS II. The light reaction starts when the light (photons) fall on pigment molecules of PS II. The light energy is first harvested by *antenna complex* of PS II and then transferred to its *reaction centre* – P_{680} (Chl a_{680}). The P_{680} gets excited, *i.e.*, its outer electron passes into high energy state called *excited state*. The excited P_{680} which has very short life (approximately 5×10^{-9} seconds) immediately transfer its electron to adjacent primary acceptor (most probably to *pheophytin*). The transfer of electron from P_{680} creates a gap which is filled by electron produced by the photo-oxidation of water. After receiving electron, the reaction centre-P_{680} comes to the ground state.

The water molecules are oxidized to H^+ and OH^- ions as follows :

$$H_2O \xrightarrow[chlorophyll]{Light} [H] + [OH]$$

$$[H] \longrightarrow H^+ + e^-$$

$$e^- + [OH] \longrightarrow OH^-$$

The transfer of electron to P_{680} takes place via an unknown compound Z. The OH^- molecules now reunite to form water and evolve molecular oxygen. This step depends upon some unknown enzyme because it requires the presence of Mn^{++} and Cl^- ions and Ca^{++} ions. Manganese occurs in abundance (about 5-8 atoms per 400 chlorophyll molecules). The protons (H^+) are carried by some carrier, possibly plastoquinone *(PQ)*.

The observations of *Kok et al.* (1970) also confirmed the requirement of unknown enzyme during photo-oxidation of water and evolution of molecular oxygen. According to them this enzyme is a *water oxidizing enzyme complex* which exists in five states (S_0, S_1, S_2, S_3 and S_4). This complex is called in *S state*. *The removal of O_2* from two molecules of water is as follows :

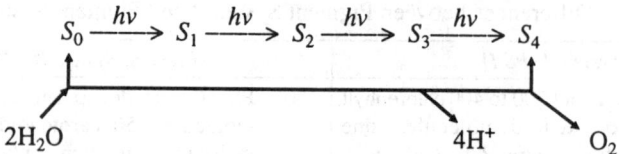

The complete process can be best studied if enzyme complex is considered as (M.Z. $P_{680}Q$) where:

M = An intermediate which after reaction produces oxygen.

Z = An unknown electron donor which donates electron to P_{680}.

P_{680} = Reaction centre of PS II.

Q = An electron acceptor.

The complete state of enzyme complex is regarded as S i.e., M.Z.P_{680}.Q. The intermediate M can gain up to four oxidizing charges, i.e., M^+, M^{2+}, M^{3+} and M^{4+}. The complete state of enzyme changes with the change of M. The complete S state is divisible into 5 different states—S_0, S_1, S_2, S_3, S_4 and S cycle runs through these five states in presence of light and each transition requires one quantum of light per electron transport. *The evolution* of O_2 is complete in 4 steps. *Molecular oxygen is released when S_4 state is converted into S_0 state* in presence of some unknown compound Z which accepts electrons from water and transfers them to reaction centre-P_{680} in presence of above enzyme, Mn^{+2}, Cl^- and Ca^{+2} ions. The transition into different states of S showing changes in the state of M intermediate with oxidizing charges in each state are shown below:

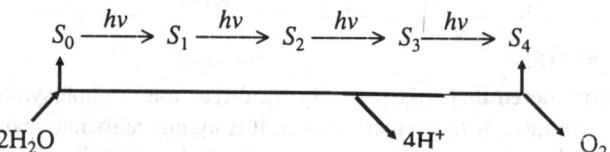

where:

$S_0 = M.Z.P_{680}.Q$; $\quad S_1 = M^+.Z.P_{680}.Q$; $\quad S_2 = M^{2+}.Z.P_{680}.Q$

$S_3 = M^{3+}.Z.P_{680}.Q.$ $\quad S_4 = M^{4+}.Z.P_{680}.Q.$

Because during formation of one molecule of O_2, 4 molecules of water dissociate and 4 electrons are liberated, therefore, for the formation of 6 molecules of O_2, $4 \times 6 = 24$ water molecules will dissociate and 24 electrons will be liberated. These 24 electrons and 24 protons (H^+) will combine with 12 molecules of NADP to form 12 NADPH + H^+ molecules.

$$4 H_2O \longrightarrow 4(OH^-) + 4H^+$$
$$4(OH^-) \longrightarrow 2H_2O + O_2 + 4e^-$$

Or

$$24 H_2O \longrightarrow 24(OH^-) + 24H^+$$
$$24(OH^-) \longrightarrow 12 H_2O + 6O_2 + 24 e^-$$
$$12 NADP + 24 H^+ + 24e^- \longrightarrow 12 NADPH + H^+$$

(3) Production of assimilatory powers (ATP and NADPH + H^+)

The term assimilatory powers were used by *Arnon* (1956) for ATP and $NADPH_2$. The mechanism of reduction of NADP into NADPH + H^+ may be called as *electron transport system in photosynthesis* while the mechanism of production of ATP from ADP and inorganic phosphate (iP) with the help of light energy is called *photophosphorylation*. During photosynthesis, some amount of light energy absorbed by chlorophyll is used in the production of ATP while the remaining light energy is used in the reduction of NADP.

(4) Electron Transport Chain (ET Chain)

The light reaction of photosynthesis involves two separate photosystems or pigment systems (PS I and PS II). Which remain connected in series with each other by the components of ET chain. The ET chain in each photosystem involves following four complexes:

1. Antenna Complex (AC) or Light Harvesting Complex (LHC): It is a pigment protein complex made up of chlorophyll molecules and other accessory pigment molecules. This complex primarily performs the function of light harvesting. The light harvesting complex of PSI is called *LHC* I and of PS II is called *LHC* II.

2. Core Complex (CC): It is also a pigment protein complex made up of polypeptides containing 40-60 chlorophyll molecules, a reaction centre, electron donors and electron acceptors. The core complex of PSI is called *CC* I and of PSII is called *CC* II.

3. Cyt b_6.–f complex: This complex was first isolated and characterized by *Nelson* and *Neumann*, 1972. It connects PS I and PS II. It is a non-pigmented complex which shows plastoquinone (PQ) plastocyanin (PC) oxidoreductase activity. The PQ and PC components act as shuttle in ET chain. PQ acts as shuttle between PS II and Cyt b_6–F complex and PC acts in between Cyt.b_6-f and PSI complexes.

4. ATPase Complex or Coupling factor: It is found localized within and upon the surface of chloroplast membrane. This complex is made up of CF_1 and CF_0 factors. This complex utilize energy derived from ET and converts ADP and inorganic phosphate (iP) into ATP.

Electron Transport System (ETS) and Photophosphorylation

There are two types of electron transport systems (ETS) and photophosphorylations (Fig. 10.3). Both phenomena can be jointly discussed as follows:

(4) Non-cyclic Electron Transport System and Non-Cyclic Photophosphorylation (Fig. 10.11)

PS I and PS II both are involved in non-cyclic ETS and photophosphorylation. They include following steps:

In PS I Complex

(1) The chlorophyll and accessory pigments of light harvesting antenna complex of PSI (LHCI) absorb light of higher wavelengths in the form of photons and transfer their absorberd energy to the reaction centre – P_{700} (Chl a_{700}). The P_{700} gets photoexcited and loses its outer valance electron to intermediate electron acceptor of PSI–(A_1). Although the nature of A_1 is not well-known but according to *Fajer et.al* (1980) it is probably a monomeric Chl *a* anion radical. When an electron is released from P_{700}, it becomes oxidized. The oxidized P_{700} is reduced by the electron donated by plastocyanin (PC).

(2) The reduced A_1 (Chl a^+) then passes electrons to primary electron acceptor of PSI–A_2 (probably a Fe–S protein in nature). The reduced A_2 then transfers its electron to secondary electron acceptor of PSI–A_3 (most probably P_{430}). The sequence of electron transfer is as follows:

$$P_{700} \xrightarrow{e^-} A_1 \xrightarrow{e^-} A_2 \xrightarrow{e^-} A_3$$
$$\text{(Chl } a^+\text{)} \quad \text{(Fe-S protein)} \quad (P_{430})$$

(3) The reduced A_3 (P_{430}) then passes its electron to ferredoxin (Fd) present at outer surface of thylakoid membrane. Thus, Fd is reduced and A_3 is oxidized.

$$A_3 \xrightarrow{e^-} Fd$$

$$Fd + A_3 \longrightarrow A_3 + Fd$$

(Oxi) (red.) (Oxi) (red.)

(4) The electrons from reduced Fd finally passes to $NADP^+$ to reduce it. The protons (H^+) are directly taken from the medium. The reduction of $NADP^+$ to $NADPH + H^+$ involves an enzyme called *ferredoxin–NADP–reductase*.

$$Fd + NADP^+ + 2H^+ \xrightarrow[2e^-]{Fd-NADP-Reductase} Fd + NADPH + H^+$$

(Red) (Oxl)

(5) The reduced $NADP^+$ ($NADPH + H^+$) further participates in CO_2 assimilation.

In PSII Complex

(6) Simultaneously the chlorophyll and accessory pigments of light harvesting antenna complex of PS II (LHC II) also absorb photons (light) of specific wavelengths and transfer their absorbed energy to the reaction centre–P_{680} (Chl a_{680}). The P_{680} gets photo-excited and loses its outer valance electron creating an electron deficiency or a 'hole' in P_{680} molecule.

(7) The emitted electron from reaction centre of PS II (P_{680}) is trapped by an intermediate acceptor called *phaeophytin* (Pheo). Thus, phaeophytin is reduced and P_{680} is oxidized (P_{680}^+).

(8) The oxidized P_{680} is reduced by trapping electrons from H_2O. This step is known as *photo-oxidation of water* and results in evolution of molecular oxygen. It requires an enzyme known as water oxidizing enzyme or *S* complex and an unknown compound *Z* which is most probably a quinonoid compound (*Q*) as suggested by *Warden etal.* (1976). For details of photo-oxidation of water consult the mechanism described earlier.

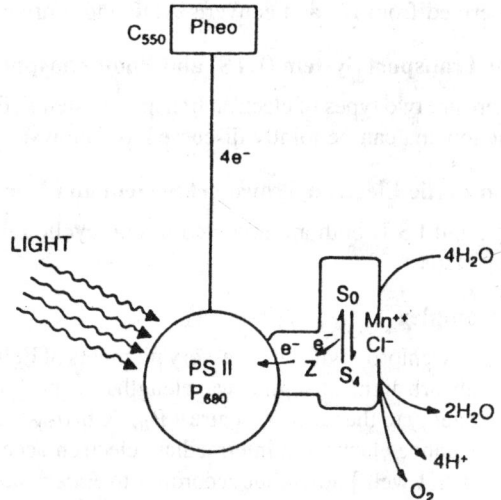

Fig. 10.10: Oxidation of H_2O and transfer of electron to PS II via unknown compound Z.

The water molecules are photo-oxidized as follows :

The oxygen evolving system is called *S* which is made up of M, Z, P_{680} and Q. where:

   M = an intermediate which after reaction produces O_2.

   Z = an electron donor,

   P_{680} = Reaction centre of PS II, and

   Q = An electron acceptor

The oxygen is evolved when S_4 is converted so state.

$$4H_2O \xrightarrow[Mn^{2+}, Cl^-]{\overset{S_4 \quad S_0}{\overset{\curvearrowright}{Z}}} 4H^+ + 4e^- + 2H_2O + O_2\uparrow$$

The oxidation of water into $4e^-$ and $4H^+$ takes place in presence of Mn^{2+} and Cl^-, water oxidizing enzyme and light falling on pigment molecules. The electrons are carried over to unknown compound Z which transfers them to the reaction centre of PS II (P_{680}). As stated earlier, actually the evolution of one molecule of O_2 requires $4H_2O$ molecules. The molecular oxygen and $4H^+$ are released inside the thylakoid membrane.

(9) The reduced phaeophytin (Pheo) then transfers its electron to Q_a, an unknown primary acceptor of PS II. The Q_a is probably a bound quinone in thylakoid membrane.

$$Pheo \xrightarrow{e^-} Q_a$$

(10) The Q_a then transfers electron to Q_b to reduce it. Q_b is an unknown secondary electron acceptor of PSII and probably it is membrane bound quinone which accepts two electrons. The Q_b is reduced in following manner—

(i) $\quad Q_a^- \xrightarrow{e^-} Q_b \longrightarrow Q_b^-$

(ii) $\quad Q_a^- \xrightarrow{e^-} Q_b^- \longrightarrow Q_b^{2-}$

Thus, Q_b is reduced with the help of two electrons and Q_a is oxidized.

In Cyt_{b6} – f complex

(11) The reduced Q_b^{2-} now transfers its electrons to plastoquinone (PQ) which is a hydrogen carrier. The reduction of PQ requires electrons from Q_b^{2-} and protons (H^+) from the medium. The reduced PQ is called *plastohydroquinone* (PQH_2).

$$Q_b^{2-} \xrightarrow{2e^-} PQ \xrightarrow{\overset{2H^+ (from\ medium)}{\downarrow}} PQH_2$$

(12) The PQH_2 then transfers electrons to cytochrome f (Cyt.f) of Cyt b_6–f complex and releases protons (H^+) to the inner side of thylakoid membrane. Thus, PQH_2 is oxidized to PQ and Cyt.f is reduced. In reduction of Cyt.f actually $4e^-$ are needed.

$$\begin{array}{c} ADP + ip \quad ATP \\ \underset{(Oxi)}{Cyt.f} + \underset{(Red)}{PQH_2} \xrightarrow{2e^-} \underset{(Red)}{Cyt.f} + \underset{(Oxi)}{PQ} \\ 2H^+ \\ (in\ medium) \end{array}$$

At this step, ADP combines with inorganic phosphate (ip) to form one molecule of ATP (it is photophosphorylation step) in presence of *ATPase complex*. It may be simply written as follows:

$$\underset{(Oxi)}{Cyt.f} + PQH_2 + ADP + iP \xrightarrow{4e^-} \underset{(Red)}{Cyt.f} + \underset{(Oxi)}{PQ} + ATP$$

(13) The reduced Cyt.f transfers electrons to plastocyanin (PC). The Cyt.f is oxidized and PC is reduced at this step. PC is mobile and acts as shuttle between Cyt.f and PSI complex.

$$\text{Red. Cyt.f} \xrightarrow{4e^-} \text{PC} \longrightarrow \text{PSI complex.}$$

(14) The reduced PC now transfer the electron to the PS I reaction centre of (P_{700}) so that its hole created by emission of electron gets again filled and reaction centre in PS I (P_{700}) is again reduced. Thus, PC is oxidized and P_{700} is reduced.

$$\text{PC} \xrightarrow{4e^-} P_{700}.$$

In the above scheme of electron transport, the electron ejected from PS II did not return to its place of origin, instead it was passed on to PS I. Similarly, the electron emitted from PS I did not cycle back but was used to reduce $NADP^+$ into $NADPH + H^+$. Therefore, this electron transport has been called as *non-cyclic electron transport* and the phosphorylation as *non-cyclic photophosphorylation*. Thus, in this system the PS I is reduced by electrons coming from PS II and the PS II is reduced by electrons coming from water. This non-cyclic electron transport is associated with photo-oxidation of water releasing molecular O_2.

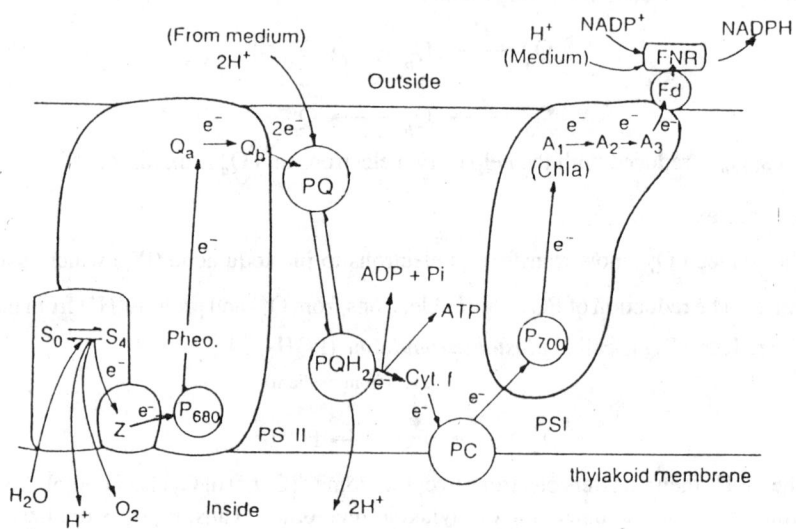

Fig. 10.11: Non-cyclic Electron Transport System and photophosphorylation. S_0—S_4 = states of unknown photo-oxidizing substance, Z = unknown electron donor of PS II, Pheo = pheophytin, Qa and Qb = quinonoid acceptors, PQ = plastoquinone. PQH_2 = plastohydroquinone. Cyt. = cytochrome, PC = plastocyanin, A_1, A_2 and A_3 = first, second and third unknown electron acceptors of PS I, Fd = ferredoxin, FNR = ferredoxin NADP reductase.

(5) Cyclic Electron Transport System and Cyclic Photophosphorylation (Fig. 10.12)

In cyclic electron transport only PS I is involved and the formation of $NADPH_2$ does not take place. It takes place under conditions which exclude non-cyclic photoposphorylation. The cyclic electron transport and photophosphorylation occur when the activity of PS II stops and the process of non-cyclic photophosphorylation and electron transport is ceased due to certain reasons. The activity of PS II stops due to following two reasons :

(i) When specific inhibitors like 3 –(4^1–chlorophenyl)–1, 1—dimethyl urea (CMU) and 3–($3'$ $4'$—dichlorophenyl)—1, 1— dimethylurea (DCMU) are used.

Photosynthesis

   (ii) When wavelengths of light greater than 680 mµ are used.
   Under above conditions the chlorophyll a_{673} (P_{673}) molecules of PS II stop their functioning due to which following changes occur :
   (i) Only PS I remains active.
   (ii) Photolysis of water does not take place.
   (iii) Blockage of non-cyclic ATP formation causes a drop in CO_2 assimilation in dark reaction.
   (iv) The amount of oxidized $NADP^+$ is also reduced.

Following steps occur during cyclic electron transport and cyclic photophosphorylation :

(1) The chlorophylls and accessory pigments of light harvesting antenna complex of PSI absorb photons of specific wavelengths of light and transfer their energy to the reaction centre of PSI–P_{700} (or Chl a_{700}). The P_{700} gets phoexcited and emits its outer valance electron resulting into oxidation of P_{700} molecule. Thus, a gap is created in the reaction centre (P_{700}).

(2) The emitted electron is trapped by an intermediate primary acceptor of PSI called A_1 (or Chl a). The A_1 reduces to primary acceptor called A_2, a Fe–S protein, and A_2 reduces to A_3 (or P_{430}) similar to that of non-cyclic E.T.S.

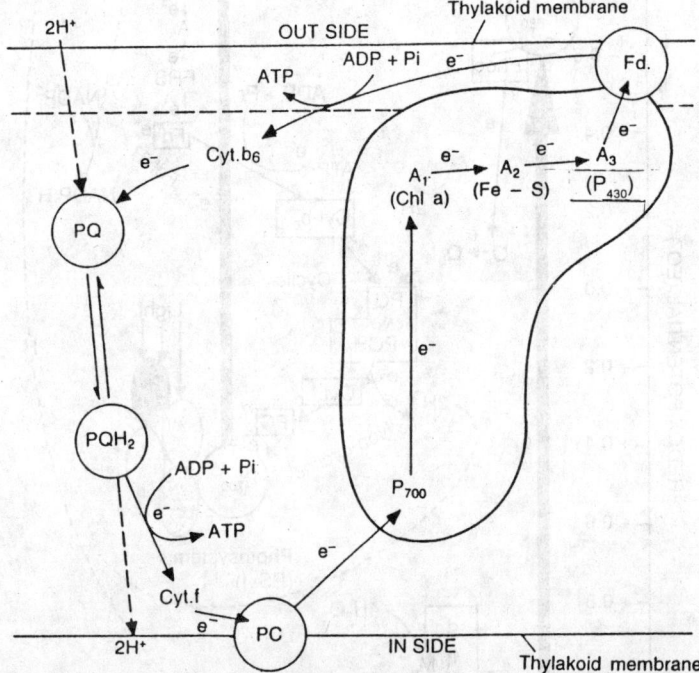

Fig. 10.12: Cyclic electron transport system and photophosphorylation. Abbreviations: PQ = plastoquinone, PQH_2 = plastohydroquinone, Cyt = cytochrome, PC = plastocyanin, A_1, A_2 and A_3 = first, second and third unknown electron acceptors of PS I, Fd = ferredoxin.

(3) The reduced A_3 (P_{430}) then passes electron to ferredoxin (Fd) via some ferredoxin reducing substance (FRS). Thus A_3 is oxidized and Fd is reduced.

$$\text{Red. } A_3 \xrightarrow{e^-} \text{FRS} \xrightarrow{e^-} \text{Fd}$$
$$(P_{430})$$

(4) The reduced Fd is incapable to reduce $NADP^+$, thus, passes its electron to cytochrome b_6 (Cyt. b_{563}). At this step one molecule of ATP is synthesized with the help of ADP and inorganic phosphate (Pi).

$$\text{Fd} + \text{Cyt. b}_6 + \text{ADP} + \text{Pi} \longrightarrow \text{Fd} + \text{Cyt.b}_6 + \text{ATP}$$
(red.) (oxi) (oxi) (red.)

(5) The reduced cytochrome b_6 (Cyt b_{563}) may then transfer its electron to hydrogen carrier plastoquinone (PQ) or to plastocyanin (PC). This step is controversial and not very much clear but the evidences given by *Gimmler* and *Avron* (1972) suggested that plastoquinone is the acceptor of electron coming from Cyt b_6. The protons (H^+) are supplied from the outer medium. Thus, PQ is reduced to plastohydroquinone (PQH_2).

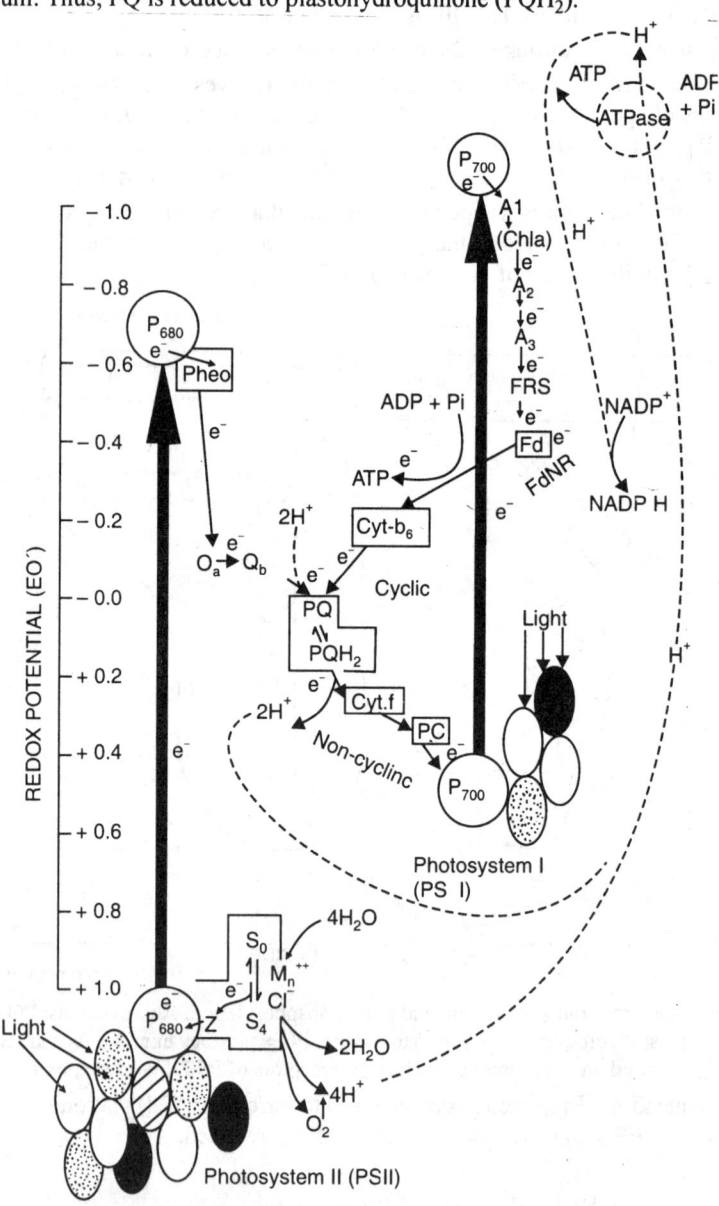

Fig. 10.13: A schematic diagram of the electron transport system nd photophosphonylation in photosynthesis involving two pigment systems. Abbreviations: Pheo = pheophytin, PQ = plastoquinone, PQH_2 = plastohydroquinone, Cyt = cytochrome, PC = plastocyanin, A_1, A_2 and A_3 = first, second and third unknown electron acceptors of PS I, Fd = ferredoxin, Fd NR = ferredoxin $NADP^+$ reductase, S_0—S_4 = states of unknown photo oxidizing substance, Z = unknown electron donor of PS II, Qa and Qb = quinone like substances.

Photosynthesis

$$\text{Cyt b}_6 \xrightarrow{e^-} \text{PQ} \xrightarrow{\text{2H+ (from outer medium)}} \text{PQH}_2$$

(6) PQH_2 then passes its electron to Cyt f and $2H^+$ to the inner side of thylakoid membrane. At this step one molecule of ATP is synthesized from ADP and Pi.

$$\underset{\text{(oxi)}}{PQH_2} + \underset{}{\text{Cyt } f} + ADP + Pi \xrightarrow{e^-} \underset{\text{(oxi)}}{PQ} + \underset{\text{(red)}}{\text{Cyt } f} + ATP$$

(7) Reduced Cyt f then transfers its electron to plastocyanin (PC). Thus, Cyt f is oxidized and PC is reduced.

$$\underset{\text{(Red.)}}{\text{Cyt } f} \xrightarrow{e^-} PC$$

(8) The reduced PC finally migrates and transfers its electron to photo-oxidized reaction centre of PS I (P_{700}). Thus, the same electron which was emitted from P_{700} molecule of PS I returns back to its original position.

Because in the above electron transport system the electron which was emitted from P_{700} molecule is cycled back, the process has been called as *cyclic electron transport* and the phosphorylation as *cyclic photophosphorylation*. The electron path way in this system is not clearly known. Some workers suggest that electron from Fd falls to PQ via Cyt b_6 while others suggest that the electron falls from Fd to PC or Cyt f. Cyclic phosphorylation provides essential ATP for dark reaction which is non-sufficiently produced in non-cyclic process. Cyclic transport is uncommon.

Table 10.3: Differences between Non-cyclic and Cyclic Photophosphorylation.

	Non-cyclic Photophosphorylation	*Cyclic Photophosphorylation*
(i)	In this PS I and PS II both participate.	In this only PS I participates.
(ii)	It requires free electron donor. The water acts as the source of electrons. Here, $NADP^+$ is the final electron acceptor. The electron emitted from the reaction centre does not come back to its original place.	Here, the system works like a closed loop and the electron emitted from the reaction centre cycles back to its original place.
(iii)	$NADP^+$ is reduced to $NADPH + H^+$ which is utilized in carbon assimilation.	$NADP^+$ is not reduced due to which the rate of CO_2 assimilation is reduced.
(iv)	O_2 evolution takes place.	O_2 is not evolved.
(v)	It occurs in green plants and not in bacteria.	It occurs mostly in bacteria.
(vi)	It stops in presence of certain inhibitors like CMU and DCMU.	DCMU and antimycine etc. have no effect on the process of cyclic photophosphorylation.
(vii)	The transport of electrons in this phosphorylation is as follows : PS II (P_{680}) → Pheo → Q_a → Q_b → PQ → Cyt. f → PC → PSI (P_{700}) → A_1 → A_2 → A_3 → Fd → $NADP^+$.	The transport of electrons in cyclic system is as follows : PS I(P_{700}) → A_1 → A_2 → A_3 → FRS → Fd → Cyt b_6 → PQ → Cyt f → PC → PSI
(viii)	Only one ATP molecule is synthesized i.e., during reduction of Cyt. f from PQ.	Two ATP molecules are synthesized. (i) Fd $\xrightarrow{e^-}$ Cyt. b_6. (ii) PQ $\xrightarrow{e^-}$ Cyt. f.
(ix)	Photo-oxidation of water takes place.	Photo-oxidation of water does not take place.

Dark Reaction

It is also known as *Blackman's reaction* or *thermochemical reaction* or *path of carbon in photosynthesis*. It does not require light i.e., it takes place in presence and absence both the conditions of light. It mainly occurs in the stroma portion of chloroplast. The different steps of dark reaction were studied by *Calvin, Benson* and their colleagues in 1949 in an unicellular green alga, *Chlorella*, with the help of tracer technique by using $C^{14}O_2$. Calvin represented different steps of dark reaction in the form of a cycle. Therefore, it is called as *Calvin cycle*.

Photosynthetic Carbon Reduction Cycle or Calvin Cycle (Fig. 10.14)

Through this cycle (Fig. 10.14) the reduced pyridine nucleotide and ATP produced in the light phase are used to convert CO_2 into carbohydrate. Various steps of the Calvin cycle are as follows :

1. Six molecules of CO_2 react with 6 molecules of ribulose 1, 5-diphosphate to form 12 molecules of 3-phosphoglyceric acid. The reaction is catalysed by the enzyme Ru DP carboxylase. This reaction is rather complex because at first CO_2 is added to a 5C sugar to form 6C sugar which then splits into two molecules of a 3C compound as follows.

$$
\begin{array}{c}
CH_2O(P) \\
| \\
C=O \\
| \\
H-C-OH \\
| \\
H-C-OH \\
| \\
CH_2O(P)
\end{array}
\xrightleftharpoons[\text{Isomerization}]{}
\left[
\begin{array}{c}
CH_2O(P) \\
| \\
C-OH \\
\| \\
C-OH \\
| \\
H-C-OH \\
| \\
CH_2O(P)
\end{array}
\right]
\xrightarrow{CO_2}
\left[
\begin{array}{c}
CH_2O(P) \\
| \\
HOOC-C-OH \\
| \\
C=O \\
| \\
H-C-OH \\
| \\
CH_2O(P)
\end{array}
\right]
\xrightarrow{H_2O}
\begin{array}{c}
CH_2O(P) \\
| \\
HOOC-C-OH \\
| \\
H \\
+ \\
HO-C=O \\
| \\
H-C-OH \\
| \\
CH_2O(P)
\end{array}
$$

Ribulose 1, 5 di (P)   Enediol of ribulose 1, 5- di(p)   β-Keto acid Intermediate   3-Phosphoglyceric acid

The cleavage of the β-Ketoacid intermediate takes place by water between C_2 and C_3 of ribulose chain (Reaction 1, Fig 10.14).

$$6RuDP + 6CO_2 \xrightarrow{6H_2O} \beta\text{-Ketoacid} \longrightarrow 12 \text{ mols. 3 – phosphoglycericacid (PGA)}$$

(5C) Carbon acceptor   (6C) Intermediate   (3C)   ... (Reaction-1)

2. The 12 molecules of 3-phosphoglyceric acid thus formed by cleavage of β-ketoacid, are phosphorylated in presence of 12 ATP molecules to 1, 3-diphosphoglyceric acid. The reaction is catalysed by the enzyme phosphoglycerokinase (Reaction 2, Fig 10.14).

12 mols, 3 -Phosphoglyceric acid + 12 ATP

$$\xrightarrow{\text{Phosphoglycerokinase}}$$

12 Mols. 1,3 diphosphoglycericacid + 12 ADP   ...(Reaction-2)

3. The 12 molecules of 1,3-diphosphoglyceric acid are now reduced to form 12 molecules of 3-phosphoglycer aldehyde by 12 molecules of reduced pyridine nucleotide (NADPH), produced in the light phase of photosynthesis. The reaction is catalysed by the enzyme 3-phosphoglyceraldehyde dehydrogenase. (Reaction 3, Fig. 10.14).

12 mols, 1, 3-diphosphoglyceric acid + $12NADPH_2$

$$\xrightarrow{\text{3–Phosphoglycer aldehyde dehydrogenase}}$$

12, mols. 3 Phosphoglycer aldehyde 12NADP.   ... (Reaction -3)

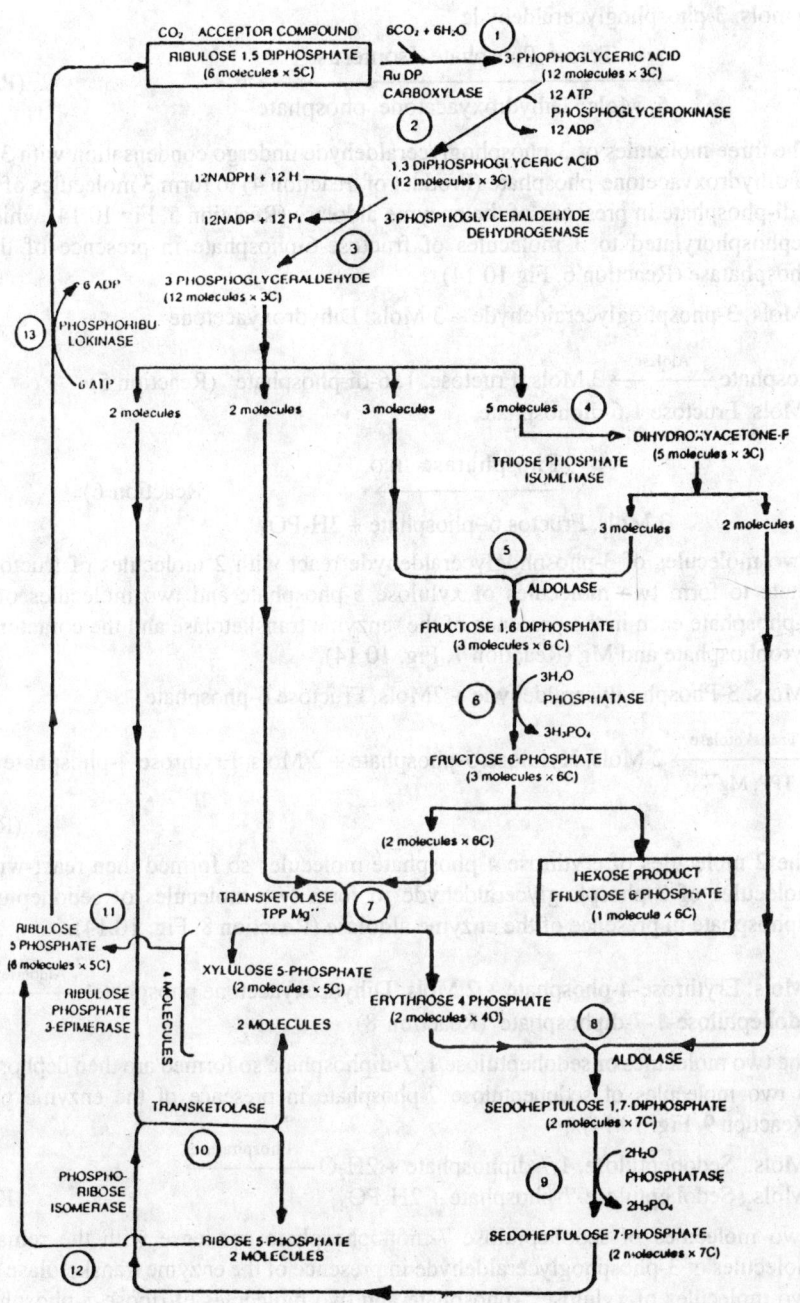

Fig. 10.14: Calvin cycle.

4. The 3-phosphoglyceraldehyde molecules can be utilized in four different ways.
   (i) Five molecules out of 12, are isomerized to 5 molecules of di-hydroxyacetone phosphate (5 molecules) in presence of enzyme triose phosphate isomerase (Reaction 4, Fig 10.14).

   $$\text{5 mols, 3-phosphoglyceraldehyde} \xrightarrow{\text{Triose Phosphate Isomerase}} \text{5, Mols. Dihydroxyacetone phosphate} \quad \text{... (Reaction -4)}$$

   (ii) The three molecules of 3-phosphoglyceraldehyde undergo condensation with 3 molecules of dihydroxyacetone-phosphate (Product of reaction 4) to form 3 molecules of fructose 1, 6 di-phosphate in presence of the enzyme aldolase (Reaction 5, Fig 10.14) which are then dephosphorylated to 3 molecules of fructose-6-phosphate in presence of the enzyme phosphatase (Reaction 6, Fig 10.14).

   $$\text{3 Mols, 3-phosphoglyceraidehyde + 3 Mols. Dihydroxyacetone phosphate} \xrightarrow{\text{Alolase}} \text{3 Mols, Fructose. 1, 6-di-phosphate} \quad \text{(Reaction 5)}$$

   $$\text{3 Mols, Fructose 1,6-diphosphate} \xrightarrow{\text{Phosphatase 3H}_2\text{O}} \text{3 Mols, Fructos 6-phosphate} + 3H_3PO_4 \quad \text{(Reaction 6)}$$

   (iii) Two molecules of 3-phosphoglyceraldehyde react with 2 molecules of fructose-6-phosphate to form two molecules of xylulose 5-phosphate and two molecules of erythrose 4-phosphate each in the presence of the enzyme transketolase and the cofactors thiamine pyrophosphate and Mg (Reaction 7, Fig. 10.14).

   $$\text{2 Mols. 3-Phosphoglyceraldehyde + 2Mols. Fructose 6-phosphate} \xrightarrow[\text{TPP, Mg}^{++}]{\text{Transketolase}} \text{2 Mols, Xylulose-5-phosphate + 2 Mols, Erythrose-4-phosphate}$$

   ...(Reaction-7)

   The 2 molecules of erythrose 4-phosphate molecules so formed then react with the two molecules of 3-phosphoglyceraldehyde to form two molecules of sedoheptulose 1, 7-diphosphate in presence of the enzyme aldolase (Reaction 8, Fig. 10.14).

   $$\text{2 Mols, Erythrose-4-phosphate + 2 Mols. Dihydroxyacetone phosphate} \xrightarrow{\text{Aldolase}} \text{2 Mols, Sedoheptulose-1-7-diphosphate} \quad \text{(Reaction-8)}$$

   The two molecules of sedoheptulose 1, 7-diphosphate so formed are then dephosphorylated to two molecules of sedoheptulose 7-phosphate in presence of the enzyme phosphatase (Reaction 9, Fig. 10.14).

   $$\text{2 Mols., Sedoheptulose, 1,7-diphosphate} + 2H_2O \xrightarrow{\text{Phosphatase}} \text{2 Mols., Sedoheptulose-7-phosphate} + 2H_3PO_4 \quad \text{(Reaction-9)}$$

   (iv) Two molecules of sedoheptulose 7-monophosphate condense with the remaining two molecules of 3-phosphoglyceraldehyde in presence of the enzyme transketolase to produce two molecules of xylulose 5-phosphate and two molecules of ribose-5-phosphate (Reaction-10, Fig 10.14).

   $$\text{2 Mols., Sedoheptulose 7-phospaate + 2 Mols., 3-phosphoglyceraldehyde} \xrightarrow{\text{Transketolase}} \text{2 Mols., Xylulose-5-phosphate + 2 Mols., Ribose 5-Phosphate} \quad \text{...(Reaction-10)}$$

   (v) All the 4 molecules of xylulose-5-phosphate (2 formed by reaction 7, and 2 formed by

Photosynthesis

reaction-8) now undergo epimerization in presence of the enzyme ribulose phosphate 3-epimerase to form 4-molecules of ribulose 5-phosphate (Reaction-11, Fig. 10.14).

$$\text{4. Mols., Xylulose 5-Phosphate} \xrightarrow{\text{Ribulose Phosphate-3-Epimerase}} \text{4 Mols., Ribulose 5-Phosphate}$$
...(Reaction-11)

(vi) The two molecules of ribose-5-phosphate, formed by reaction-10, undergo isomerization in presence of the enzyme phosphoribose isomerase to form 2 molecules of ribulose-5-phosphate (Reaction-12, Fig. 10.14).

$$\text{2. Mols., Ribose 5-Phosphate} \xrightarrow{\text{Phosphoriboseisomerase}} \text{2, Mols., Ribulose-5-Phosphate}$$
(Reaction-12)

Thus six molecules of ribulose-5-phosphate are formed up to this stage (two by reaction 12 and 4 molecules by reaction 11).

(vii) In the final step, all the 6 molecules of ribulose-5-phosphate are phosphorylated at the expense of 6 molecules of ATP in presence of the enzyme phosphoribulokinase to form 6 molecules of the carbon acceptor, ribulose 1, 5-diphosphate. Thus, all the 6 molecules of ribulose 1, 5-diphosphate are regenerated and then again re-enter into the cycle (Reaction-13, Fig . 10.14).

$$\text{6. Mols., Ribulose 5-Phosphate} + 6 ATP \xrightarrow{\text{Phosphoribulokinase}} \text{6 Mols., Ribulose 1,5-diphosphate} + 6 ADP$$
(Reaction-13) (carbon acceptor)

Thus, in this cycle 6 molecules of CO_2 are used and one molecule of fructose-6-phosphate is formed as a by-product after reaction 6 of Fig. 10.14, at the expense of 12 molecules of NADPH and 18 molecules of ATP. The overall reaction can be shown as follows:

$$6CO_2 + 12\,NADPH + 12H^+ + 18\,ATP + 11\,H_2O \longrightarrow$$
$$= 1\,Mol., F-6-P + 12\,NADP^+ + 18ADP + 17Pi$$

10.15. Hatch and Slack Cycle (C_4 – Dicarboxylic acid path way or C_4 Cycle) (Fig. 10.15)

Up to 1965, it was believed that the fixation of CO_2 during photosynthesis of higher plants and algae takes place only by Benson Calvin Cycle, But *Kortschak, Hartt* and *Burr* (1965) demonstrated with the use of $C^{14}O_2$ that in sugarcane leaves the chief synthesized labelled products are C_4-dicarboxylic acids like malate, aspartate etc. Their observations were confirmed by *M.D. Hatch* and *C.R. Slack* (1966). They told that during photosynthesis in sugarcane leaves 4-carbon substances like *Oxaloacetate*, *malate* and *aspartate* are synthesised within a very short time. Later on, these observations have been confirmed in other monocotyledonous plants like *Zea mays, Sorghum, Panicum maximum* and *Cyperus rotundus* and some dicotyledonous plants like *Amaranthus* and *Atriplex* etc. Thus, this cycle occurs in the members of Cyperaceae and some dicotyledonous plants in addition to members of Graminae. This cycle was named after the discoverers as *Hatch-Slack Cycle*. It is also called β-*carboxylation path way* and *co-operative photosynthesis*. The first stable compound of Hatch-Slack Cycle is **4-carbon oxaloacetic acid.** Therefore, it is called C_4-*Cycle*. Such plants which possess C_4 Cycle are called C_4 *plants*.

Reactions of Hatch and Slack Cycle

Hatch and Slack cycle is completed in the chloroplasts of mesophyll cells and bundle sheath cells. Following reactions occur during this cycle:

Reactions Occurring in the Chloroplast of Mesophyll Cells (Fig. 10.15)

1. *Formation of Oxalo-acetic acid*—The primary acceptor of CO_2 in this cycle is a 3C–compound-phosphoenol pyruvic acid. In mesophyll cells, the atmopspheric CO_2 first combines with water to form bicarbonate ion (HCO_3^-) in presence of enzyme *carbonic anhydrase*.

$$CO_2 + H_2O \xrightarrow{\text{Carbonic anhydrase}} HCO_3^- + H^+$$

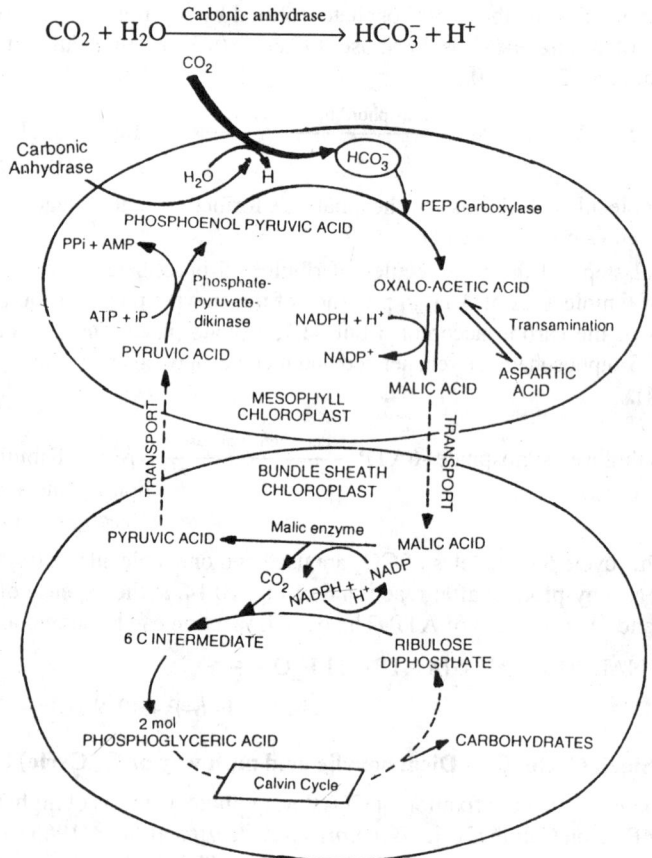

Fig. 10.15: Hatch and Slack Cycle.

The CO_2 acceptor, phosphoenol pyruvic acid (PEP), combines with CO_2 and forms a 4C acid—oxaloacetic acid in presence of enzyme **PEP carboxylase**. The enzyme remain present in large amounts in mesophyll cells.

$$\underset{\text{(PEP)}}{\begin{array}{c}COOH\\|\\C-O.P\\||\\CH_2\end{array}} + CO_2 + H_2O \xrightarrow{\text{PEP carboxylase}} \underset{\text{(Oxaloacetic acid)}}{\begin{array}{c}COOH\\|\\C=O\\|\\CH_2\\|\\COOH\end{array}} + H_3PO_4$$

2. *Formation of malic acid and aspartic acid*—Oxaloacetic acid is quite unstable and is converted either into malic acid or aspartic acid. The oxaloacetic acid is reduced to malic acid by using light-generated $NADPH+H^+$. This reaction is catalysed by enzyme **malic dehydrogenase**.

Photosynthesis

$$\underset{\text{Oxaloacetic acid}}{\begin{array}{c}\text{COOH}\\|\\\text{C}=\text{O}\\|\\\text{CH}_2\\|\\\text{COOH}\end{array}} + \text{NADPH} + \text{H}^+ \xrightarrow{\text{malic dehydrogenase}} \underset{\text{Malic acid}}{\begin{array}{c}\text{COOH}\\|\\\text{H}-\text{C}-\text{OH}\\|\\\text{CH}_2\\|\\\text{COOH}\end{array}} + \text{NADP}^+$$

The oxaloacetic acid can also be converted into aspartic acid in presence of enzyme **aspartic transaminase** (transamination reaction).

$$\text{Oxaloacetic acid} \xrightleftharpoons{\text{Aspartic transaminase}} \text{Aspartic acid.}$$

The C_4 acids i.e., malic acid and aspartic acid are then transported to the chloroplasts of bundle sheath.

Reactions Occurring in bundle Sheath Chloroplast

(3) *Formation of pyruvic acid*—In bundle sheath chloroplast, the malic acid undergoes oxidative decarboxylation to yield pyruvic acid and CO_2 in presence of **malic enzyme**.

$$\underset{\text{Malic acid}}{\begin{array}{c}\text{COOH}\\|\\\text{H}-\text{C}-\text{OH}\\|\\\text{CH}_2\\|\\\text{COOH}\end{array}} + \text{NADP}^+ \xrightarrow{\text{malic enzyme}} \underset{\text{Pyruvic acid}}{\begin{array}{c}\text{COOH}\\|\\\text{C}=\text{O}\\|\\\text{CH}_3\end{array}} + \text{NADPH} + \text{H}^+ + CO_2$$

(4) The CO_2 and NADPH+H$^+$, produced by oxidative decarboxylation of malic acid enter into Calvin Cycle. The CO_2 combines with ribulose diphosphate (Ru DP) to yield 2 molecules of phosphoglyceric acid (PGA).

$$CO_2 + \text{RuDP} \longrightarrow 2 \text{ Mols. PGA.}$$

(5) In a few C_4 plants the aspartic acid undergoes transamination to form oxaloacetic acid which is then decarboxylated to pyruvic acid. This reaction is catalysed by **aspartate transaminase**.

$$\text{L-Aspartic acid} \xrightarrow{\text{Transamination}} \text{Oxaloacetic acid} \xrightarrow{\text{Decarboxylation}} \text{Pyruvic acid.}$$

In Mesophyll Cells

(6) *Formation of Posphoenol Pyruvic acid (PEP)*—The pyruvic acid produced by oxidative decarboxylation (reaction 3) is transported back to the mesophyll cells where it is phosphorylated to phosphoenol pyruvic acid in presence of enzyme **pyruvate phosphate dikinase**. This enzyme is unusual because it splits one molecule of ATP, (synthesized in photosynthetic light reaction), into AMP and PPi. PPi is then degraded to Pi.

$$\underset{\text{(Pyruvic acid)}}{\begin{array}{c}\text{COOH}\\|\\\text{C}=\text{O}\\|\\\text{CH}_3\end{array}} + \text{ATP} + H_3PO_4 \xrightarrow{\text{Pyruvate dikinase}} \underset{\text{(PEP)}}{\begin{array}{c}\text{COOH}\\|\\\text{C}-\text{O.PO}_3H_2\\||\\\text{CH}_2\end{array}} + \text{AMP} + \text{P} - \text{Pi}$$

10.7. CHARACTERISTICS OF C_4 PLANTS

(i) The leaves of C_4 plants possess special anatomy called *kranz type*. The vascular elements (xylem and phloem) in C_4 leaves remain surrounded by a layer of bundle sheath cells containing chloroplasts in abundance. The bundle sheath is surrounded by one to three layers of mesophyll cells which possess very small intercellular spaces.

(ii) The chloroplasts in C_4 leaves are *dimorphic i.e.*, distinctly of two types—(a) The chloroplasts of mesophyll cells are of normal type, (b) The chloroplasts of bundle sheath cells are comparatively quite **larger in size, without grana or PS II but contain starch grains** and arranged **centripetally.**

(iii) PEP carboxylase enzyme occurs in mesophyll cells.

(iv) C_4 cycle is performed in mesophyll cells while C_3 cycle is performed in the cells of bundle sheath.

(v) They possess two types of CO_2 acceptor—(i) Phosphoenol pyruvate which occurs in mesophyll cells, (ii) Ribulose diphosphate (RuDP) which occurs in the bundle sheath cells.

(vi) In them, the first stable compound formed is *Oxaloacetic acid*.

(vii) C_4 plants are found in tropical (dry and hot) and subtropical regions.

(viii) They grow fast at high temperature and in more light intensities. Therefore, C_4 plants are called *efficient plants*. The optimum temperature for their growth varies from 30 to 45°C.

(ix) In C_4 plants, the O_2 has no inhibitory effect.

(x) They lack photo-respiration.

10.8. SIGNIFICANCE OF C_4 CYCLE

(i) In C_4 plants, it increases the photosynthetic yield two to three times more than C_3 plants.

(ii) In C_4 plants, it performs a high rate of photosynthesis even when the stomata are nearly closed.

(iii) It increases the adaptability of C_4 plant to high temperature and light intensities.

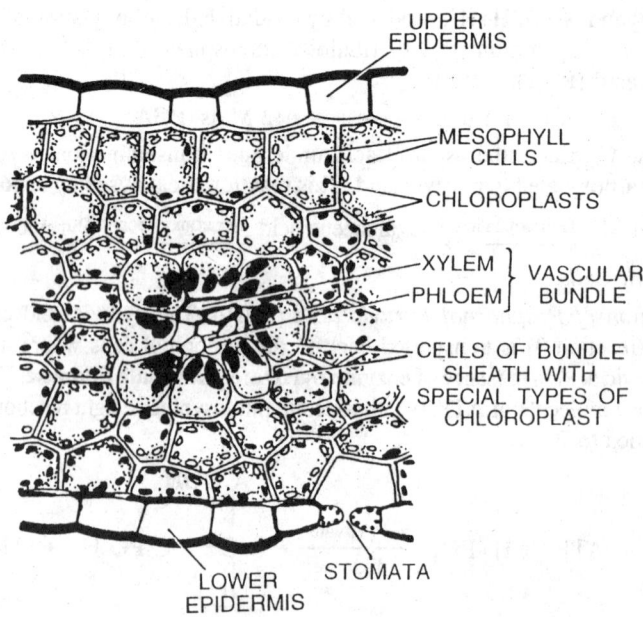

Fig. 10.16: Cross-section of leaf showing 'Kranz' type anatomy.

Photosynthesis

(iv) It increases the rate of CO_2 fixation at 25-30°C in C_4 plants as compared to C_3 plants.

(v) It reduces the rate of photorespiration at 25-30°C.

Table 10.3: Differences between Calvin Cycle and Hatch-Slack Cycle

	Calvin Cycle (C_3 Cycle)	Hatch-Slack Cycle (C_4 Cycle)
1.	The primary CO_2 acceptor is a 5C compound—**ribulose diphosphate** (RuDP).	The primary CO_2 acceptor is a 3C compound—**phosphoenol pyruvic acid** (PEP).
2.	The first stable compound formed is **phosphoglyceric acid** (PGA) which contains 3 C atoms.	The first stable compound is a 4C **Oxaloacetic acid**.
3.	C_3 cycle is completed in only one type of chloroplast present in mesophyll cells.	C_4 cycle is completed in two types of chloroplasts, one occurring in mesophyll cells and other in bundle sheath cells.
4.	It takes place at comparatively low temperature.	It takes place at high temperature and more light intensities.
5.	Photorespiration occurs in C_3 Plants.	Photorespiration does not occur in C_4 plants.
6.	The rate of photosynthesis is comparatively lower.	The rate of photosynthesis is comparatively higher.
7.	It occurs in C_3 plants which show normal anatomy.	It occurs in C_4 plants which show Kranz anatomy.

10.18. CRASSULACEAN ACID METABOLISM OR CAM CYCLE (Fig. 10.17)

It occurs mostly in succulent plants which grow under semi-arid conditions. Since the cycle was first observed in the plants belonging to family crassulaceae e.g., *Bryophyllum, Sedium calycinum* and *Kalanchoe* etc., it was named as Crassulacean Acid Metabolism (CAM). Similar metabolism has been reported in other plants belonging to cactus (Opuntia), orchid and pine apple families. CAM plants possess following characteristics:

(i) They fix atmospheric CO_2 in dark and accumulate large amount of malic acid.

(ii) They show diurnal pattern of organic acid formation, i.e., they accumulate organic acids in the leaves at night and decrease during the day. In such plants, the pH of cell sap substantially decreases with the accumulation of organic acids.

(iii) They are usually succulents. In CAM plants, the vacuoles normally function as a site of accumulation of organic acid (malic acid).

(iv) They possess xerophytic characters like thick cuticle, sunken stomata, thorns and reduced leaves.

(v) The stomata remain closed during the day (light) and open at night (dark).

(vi) They show maximum gaseous exchange at night because of nocturnal opening of stomata.

(vii) They show decrease in starch content during night and increase during the day.

(viii) In them, the chlorophyllous cells contain large storage vacuoles.

(ix) They possess high level of phosphoenol pyruvate and an active decarboxylase.

CAM cycle is completed in two parts (a) Acidification (b) Deacidification. Acidification takes place in dark while deacidification occurs during day time.

(1) Acidification

Various steps during acidification are as follows:

(i) The stored carbohydrates are converted into phosphoenol pyruvic acid (PEP) through glycolysis. The CO_2 diffuses freely into the leaf through open stomata in night.

(ii) The PEP is carboxylated into oxaloacetic acid (OAA) in presence of enzyme **PEP caboxylase**.

Phosphoenol pyruvic acid (PEP) + CO_2 + H_2O ⟶ Oxaloacetic acid (OAA) + H_3PO_4

(*iii*) The Oxaloacetic acid is now reduced to malic acid in presence of enzyme **malic dehydrogenase**. This reaction is facilitated in presence of reduced $NADP^+$ ($NADPH + H^+$) formed during glycolysis.

Oxaloacetic acid + $NADPH + H^+$ ⟶ Malic acid + $NADP^+$

The malic acid, thus, produced in dark as a result of acidification is stored in the vacuole. The oxaloacetic acid may also be interconverted into aspartic acid.

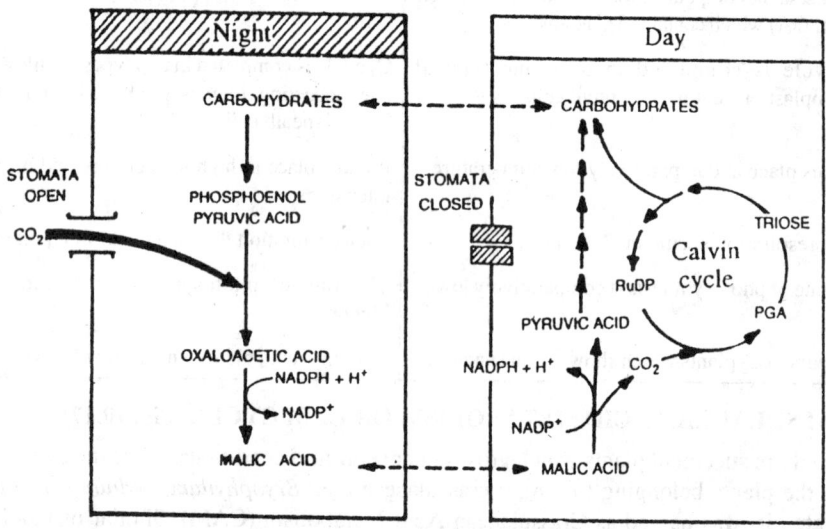

Fig. 10.17: Crassulacean acid metabolism in succulents

(2) **Deacidification**

The decarboxylation of malic acid into pyruvic acid and CO_2 in presence of light is called *deacidification*. In light, during deacidification the malic acid formed during night is converted into pyruvic acid and CO_2 in presence of **malic enzyme**. In certain plants, this reaction is catalysed by *PEP carboxykinase*. One molecule of $NADP^+$ is also reduced in this reaction.

Malic acid + $NADP^+$ ⟶ Pyruvic acid + $NADPH + H^+ + CO_2$

The pyruvic acid formed in this reaction is either oxidized to CO_2 through Kreb's cycle or reconverted to PEP or phosphoglyceric acid to synthesise sugar through C_3 Cycle.

The CO_2 liberated by deacidification of malic acid is accepted by ribulose diphosphate (RuDP) to fix it into carbohydrate through C_3 Cycle (Calvin cycle). However, the fate of pyruvic acid is still not clearly known.

(3) **Significance of CAM**

1. As CAM plants are able to fix CO_2 in dark, they can survive for longer periods in light withoug CO_2 uptake.
2. The stomata of leaves remain closed during the day and open at night. This is an adaptation to conserve water, since succulents exhibiting CAM are found in dry habitat.
3. During the night CO_2 is taken into the leaves through open stomata. This limits the photosynthesis. It is also limited by stored organic acid and carbohydrates causing slow growth of the plants. Thus CAM plants are generally slow growing.
4. They are drought resistant and possess xerophytic adaptations like thick fleshy leaves and

if the leaves are absent, a swollen photosynthetic stem. Other characters include sunken-stomata, thick cuticle, thorns, well developed xylem etc.

10.19. BACTERIAL PHOTOSYNTHESIS

During photosynthesis the common green plants reduce CO_2 into organic constituents in presence of light using H_2O as reductant and evolve O_2. Certain micro-organisms like bacteria during photosynthesis are able to reduce CO_2 into essential organic constituents in presence of light using H_2S, hydrogen and other inorganic and organic reductants instead of water. In such bacterial photosynthesis O_2 is not evolved and the bacteria inhabit anaerobic environment. Such bacteria are called *photosynthetic bacteria* and the photosynthesis as *bacterial photosynthesis*. The photosynthetic bacteria usually grow in such places where the compounds serving as a hydrogen donor occurs. Depending upon the pigmentation and substrate used for reduction of CO_2, following groups of bacteria have been distinguished:

1. *Green Sulphur Bacteria*: They belong to family chlorobacteriaceae and contain the pigment bacterioviridin or bacteriopurpurin which almost resemble in structure and composition with chlorophyll *a*. The important bacteria of this group are *Chlorobacterium* and *Chlorobium*. The members of chlorobacteriaceae possess a special type of chlorophyll called *Chlorobium chlorophyll* which shows a absorption maxima at 750 nm. These bacteria utilise H_2S for the reduction of CO_2. The element sulphur is produced from H_2S by removing hydrogen.

$$2H_2S + CO_2 \xrightarrow{Light} [CH_2O] + 2S + H_2O + Energy$$

2. *Purple Sulphur Bacteria*. They belong to family Thiorhodaceae and contain the pigment bacteriochlorophyll *a* or *b* which shows an absorption maxima at 590 nm. They also contain large amount of carotenoids. The important bacteria of this group is *chromatium* which can utilize a variety of sulphur compounds (thiosulphates), sometimes selenium compounds and even molecular hydrogen. The reactions of chromatium photosynthesis are as follows:

$$2CO_2 + 5H_2O + Na_2S_2O_3 \xrightarrow{Light} 2(CH_2O) + 2H_2O + 2NaHSO_4 + Energy$$

$$3CO_2 + 8H_2O + Energy \longrightarrow 3(CH_2O) + 3H_2O + 2H_2SO_4 + Energy$$

3. *Purple non-sulphur Bacteria:* They belong to family Athiorhodaceae and contain the pigment bacteriochlorophyll *a* and *b*. They can live in presence of alcohol, organic acids etc. They are heterotrophic. The photoreduction of CO_2 by purple non-sulphur bacteria may be represented as follows :

$$2CH_3CHOHCH_3 + CO_2 \xrightarrow{Light} CH_2O + 2CH_3COCH_3 + H_2O$$

(1) Non-cyclic Photophosphorylation in Photosynthetic Bacteria (Fig. 10.18)

The mechanism of non-cyclic photophosphorylation in photosynthetic bacteria differs from that of green plants in following points:

1. In photosynthetic bacteria there is only one pigment system and only one photochemical act takes place.
2. The electron donor is an inorganic compound like H_2S instead of water.
3. Each species of bacterium utilizes and confines itself to one electron donor (*e.g.*, fatty acids, 4C-dicarboxylic acid or aliphatic alcohols or other inorganic compounds).
4. In second electron transport (ET) chain either ubiquinone and/or vitamin K_2 is found instead of plastoquinone.
5. The cytochromes *b* and *c* in photosynthetic bacteria are quite different from those of higher plants and chromatium is devoid of Cyt. *b*.

The mechanism of non-cyclic photophosphorylation in *chromatium* bacterium using H_2S as electron donor is outlined in Fig. 10.18, which shows:

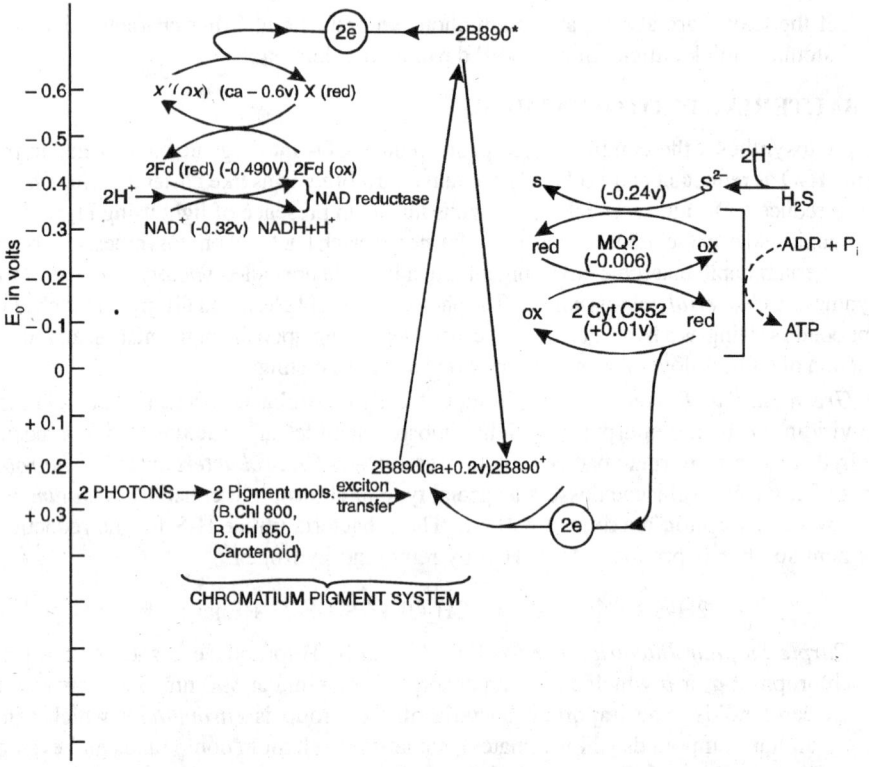

Fig. 10.18: Non-cyclic photophosphorylation in Chromatium.

(1) The involvement of one pigment system and two ET chains.
(2) First ET chain, which receives electrons from the pigment system, comprises of components: (i) an unknown oxidation-reduction system X ($E_0' = -0.6$ V), (ii) bacterial ferredoxin ($E_0' = -0.49$ V) and (iii) NAD ($E_0' = -0.32$ V).
(3) Second ET chain receives electrons from the electron donor H_2S and passes them to the pigment system. Its components are : (i) Bacterial cytochrome C_{552}, also called C_{553} and $C_{423.5}$ ($E_0' = +0.01$ V) and (ii) Menaquinone or vitamin K_2 ($E_0' = -0.006$ V).

Steps: Various steps of non-cyclic photophosphorylation in *Chromatium* bacterium are as follows:

1. Two photons of appropriate wavelength photoexcite two pigment molecules (B Chl_{800}, B Chl_{850} and carotenoids) and the energy is transferred to reaction centre, the $BChl_{890}$ molecules which attain an excited singlet state and lose their outer valence electrons to primary acceptor unknown compound X. The oxidized B^+_{890} is reduced by receiving two electrons from reduced Cyt. C_{552} of second ET chain.

2. These electrons released by B_{890} reduce $X/\overline{X}$ oxidation reduction system.
3. The reduced system X then reduces ferredoxin (Fd).
4. The electrons from Fd then drop down an electrochemical gradient and reduce one molecule of NAD^+. The protons (H^+) are directly taken from the medium. The reduction of NAD^+ to $NADH + H^+$ involves an enzyme called Fd-NAD reductase.
5. Cyt C_{552} molecules receive electrons from H_2S via menaquinone, a two electron oxidation-reduction system.

The H_2S can ionize to produce $2H^+$ and a sulphide ion S^{2-}. The S^{2-} forms the reduced component of sulphide/sulphur (S^{2-}/S) oxidation-reduction system ($E_0' = -0.24V$). This is more electronegative than Cyt C_{552} and the menaquinone oxidation-reduction system. The sulphide is, therefore, a sufficiently good reducing system to reduce these systems. Thus, one sulphide ion donates two electrons to two molecules of Cyt C_{552} through a single molecule of menaquinone and is itself converted to an atom of sulphur. Thus, two electrons required to reduce a molecule of NAD^+ ultimately come from a molecule of H_2S. The free sulphur gets deposited in *chromatium*.

One ATP molecule is synthesized from ADP and Pi when elecrons pass from menaquinone to Cyt C_{552}. In the above system, the electrons are not cycled back but they start from a donor and end up at pyridine nucleotide. Due to this reason, it is called non-cyclic.

(2) Cyclic Photophosphorylation in Photosynthetic Bacteria

In cyclic photophosphorylation the production of reducing power ($NADH + H^+$) is independent of light. However, light is required to produce ATP which is needed for CO_2 reduction. Various steps of cyclic phosphorylation are as follows :

1. Two protons are absorbed by the pigment system ($Bchl_{800}$, $Bchl_{850}$ and carotenoids) and

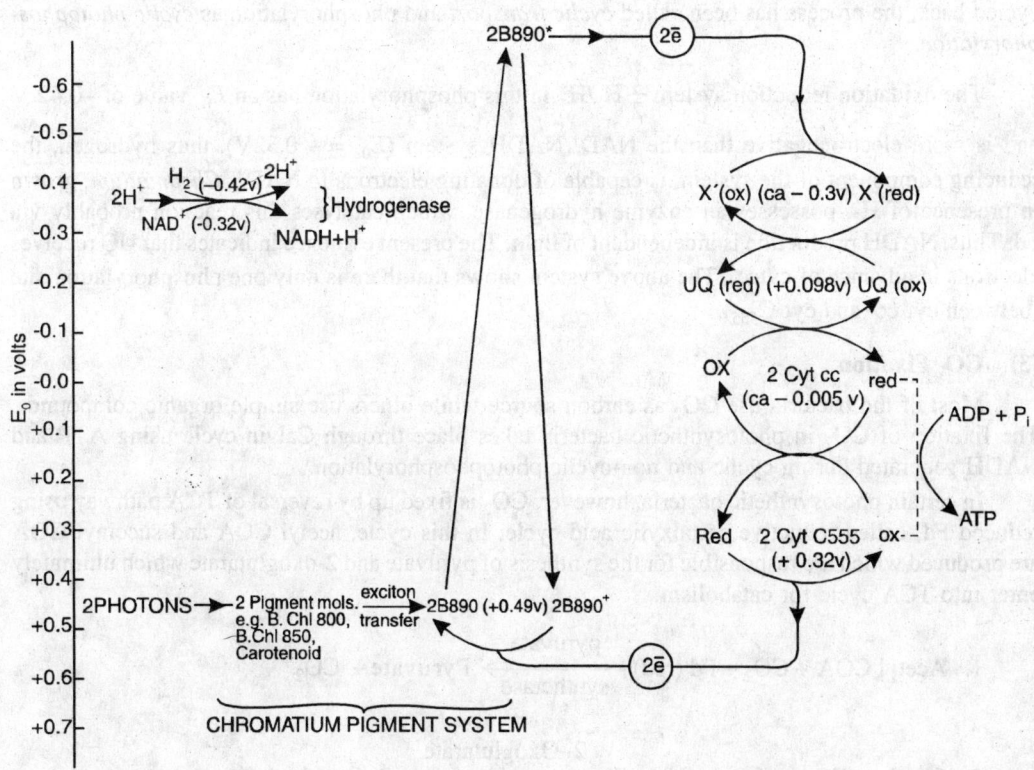

Fig. 10.19: Cyclic photophosphorylation in Chromatium. Key: UQ—Ubiquinone, Cyt—Cytochrome.

their energy is transferred to $Bchl_{890}$ pigment molecules (the reaction centre). $Bchl_{890}$ molecules become photoexcited. These, molecules emit their outer valance electrons resulting into oxidation of $Bchl_{890}$ molecules. This creates gaps in the reaction centre.

2. The emitted electrons are trapped by the unknown oxidation-reduction system (X' or P_{430}) which becomes reduced (Ca–0.3V). This system X' of cyclic photophosphorylation with $E_0' = -0.6V$ differs from X system of non-cyclic photophosphorylation in having E_0' values of $-0.3V$. The two $BChl_{890}$ molecules become positively charged ions ($Bchl_{890}^+$).

3. The reduced X' substance then passes electrons to ubiquinone (UQ). Thus X' is oxidized and UQ is reduced (+ 0.098V).

4. Reduced UQ passes electrons to Cyt. cc' also called RHP and C_{430} ($E_0' = -0.005$ V). Thus, UQ is oxidized and Cyt. cc' is reduced.

5. Reduced Cyt. cc' then passes electrons to Cyt. C_{555} (also called C_{556} and C_{422}). Thus, Cyt. cc' is oxidized and Cyt. C_{555} is reduced (+0.32 V). During this electron transfer one molecule of ATP is synthesized.

6. Finally reduced Cyt. C_{555} ($E_0' = +0.32$ V) passes electrons to $Bchl_{890}^+$ molecules ($E_0' = +0.49$). Thus, the gaps created during emission of electrons from $Bchl_{890}$ molecules are compensated by electrons receiving from Cyt. C_{555}.

Because in above electron transport system (ETS), the electrons emitted from $Bchl_{890}$ are cycled back, the process has been called *cyclic transport* and phosphorylation as *cyclic photophosphorylation*.

The oxidation reduction system $\frac{1}{2} H_2/H^+$ in this phosphorylation has an E_0' value of -0.42 V and is more electronegative than the $NAD^+/NADH$ system ($E_0' = -0.32V$), thus hydrogen, the reducing component of the system, is capable of donating electrons to NAD^+. *Chromatium*, grown in presence of H_2, possesses an enzyme hydrogenase, which catalyses this reaction probably via Fd. Thus, NADH production is independent of light. The present evidence indicates that UQ receives electrons in advance of cytcc'. The above system shows that there is only one phosphorylation site. (between cyt cc' and cyt C_{555}).

(3) CO_2 Fixation

Most of the bacteria use CO_2 as carbon source while others use simple organic compounds. The fixation of CO_2 in photosynthetic bacteria takes place through Calvin cycle using ATP and NADH generated during cyclic and non-cyclic photophosphorylation.

In certain photosynthetic bacteria, however, CO_2 is fixed up by reversal of TCA pathway using reduced Fd, called reductive carboxylic acid cycle. In this cycle, acetyl COA and succinyl COA are produced which are responsible for the synthesis of pyruvate and 2-oxoglutarate which ultimately enter into TCA cycle for catabolism.

1. Acetyl COA + CO_2 + Fd (red) $\xrightarrow{\text{pyruvate synthetase}}$ Pyruvate + CO_2

2. Succinyl COA + CO_2 + Fd (red) $\xrightarrow{\text{2-Oxoglutarate synthetase}}$ 2-Oxoglutarate.

Table 10.5: Differences between Plant Photosynthesis and Bacterial Photosynthesis.

Plant Photosynthesis	Bacterial Photosynthesis
1. Plants (green) contain definite chloroplasts.	1. Bacteria lack definite chloroplasts.
2. In plant photosynthesis, the pigments involved are chlorophylls, carotenoids and phycobilins.	2. In bacterial photosynthesis, the pigments involved are bacterio-chlorophylls, bacterioviridin and open chain aliphatic carotenoids.
3. It takes place at wavelengths between 400 to 700 mμ.	3. It takes place at wavelengths above 700 mμ.
4. The CO_2 reductant is $NADPH + H^+$.	4. The reductant is $NADH + H^+$.
5. The electron donor in this is only H_2O.	5. The electron donors in this are H_2S, inorganic compounds and reduced organic compounds.
6. Oxygen is evolved.	6. O_2 is not evolved.
7. Two pigment systems are involved.	7. It involves only one pigment systmen.
8. The reaction centre of PS I is P_{700} and of PS II is P_{673} or P_{680}.	8. The reaction centre is only P_{890}.
9. Non-cyclic photophosphorylation is dominant.	9. Cyclic photophosphorylation is dominant.
10. Emerson's enhancement effect occurs.	10. It is not reported.

10.20. CHEMOSYNTHESIS

There are certain bacteria which do not carry any photosynthesis and do not depend on food produced by other organisms through photosynthesis. They obtain energy for food synthesis by oxidizing such inorganic materials as H_2S, sulphur, CO, NH_3, molecular oxygen or ferrous carbonates. These bacteria produce organic matter from CO_2 and use energy by oxidizing one of the above inorganic compounds. This process of food synthesis where bacteria utilize energy by oxidizing inorganic compounds and reduce CO_2 is called *chemosynthesis*. Various-types of chemosynthesizing bacteria are as follows:

1. **Colourless sulphur bacteria:** *Beggiatoa, Thiobacillus,* and *Thiothrix etc.* are colourless sulphur bacteria. They obtain energy by oxidizing H_2S and other compounds with reduced sulphur. They first oxidize H_2S to sulphur and then to sulphuric acid. The energy produced is utilized in the decomposition of CO_2.

$$2H_2S + O_2 \longrightarrow 2H_2O + 2S + 65 \text{ cal.}$$
$$2S + 3O_2 + 2H_2O \longrightarrow 2H_2SO_4 + 282 \text{ cal.}$$

2. **Carbon bacteria:** They obtain energy by oxidizing CO to CO_2. The released energy is utilized for the synthesis of carbohydrates and other organic compounds.

$$2CO + O_2 \longrightarrow 2CO_2 + \text{energy}$$

3. **Nitrogen fixer:** They are colour bacteria and are of two types:

(*i*) Those which oxidize ammonia to nitrous acid *e.g.,* **Nitrosomonas, Nitrosocystis** and **Nitrosospira**.

$$2NH_3 + 3O_2 \rightleftharpoons 2HNO_2 + 2H_2O + 158 \text{ cal.}$$

(*ii*) Those which oxidize nitrous acid into nitric acid, *e.g.,* **Nitrobacter** and **Bactoderma**.

$$2HNO_2 + O_2 \rightleftharpoons 2HNO_3 + 38 \text{ cal}$$

The released energy is utilized for the synthesis of organic substances and to maintain various vital activities like growth and movement etc.

4. **Iron bacteria:** The important iron bacteria are **Ferrobacillus, Leptothrix Ochracea** and **Spirophyllum ferrugineum**. They live in water and obtain energy for carbohydrate synthesis by oxidizing ferrous salts into ferric salts.

$$4FeCO_3 + O_2 + 6H_2O \longrightarrow 4Fe(OH)_2 + 4CO_2 + 81 \text{ cal}$$

5. **Hydrogen bacteria:** The hydrogen bacterium, Bacillus **pantotrophus**, live saprophytically in soil and obtain H_2 from it. They can also be grown in artificial medium containing H_2, O_2 and CO_2. They obtain energy for carbohydrate synthesis by oxidizing hydrogen as follows:

$$O_2 + 2H_2 \longrightarrow 2H_2O + 137 \text{ cal}$$
$$2H_2 + 2CO_2 \longrightarrow (CH_2O) + H_2O + 112 \text{ cal}.$$

6. **Methane fixers:** Methane fixer bacteria like **Methanomonas** live in aerobic conditions and obtain energy and carbon for carbohydrate synthesis by oxidizing only methane as follows:

$$CH_4 + 2O_2 \longrightarrow CO_2 + 2H_2O + \text{Energy}$$

Table 10.6: Differences between photosynthesis and Chemosynthesis.

Photosynthesis	*Chemosynthesis*
1. It takes place in all green plants including algae, green bacteria and higher plants.	1. It takes place only in colourless aerobic bacteria like *Nitrosomonas* and Nitrobacter etc.
2. It involves the conversion of CO_2 and H_2O into carbohydrates in presence of light and chlorophyll. Here H_2O is oxidized evolving O_2 and CO_2 is reduced.	2. It invovles the reduction of CO_2 into carbohydrates in absence of light and chlorophyll.
3. Molecular oxygen is evolved.	3. Molecular O_2 is not evolved.
4. In this process light energy (radiant energy) is converted into chemical energy (carbohydrates).	4. Here, the carbohydrates are synthesized by utilizing chemical energy produced as a result of oxidation of inorganic substances.
5. It involves two pigment systems for the absorption of light energy and to perform photochemical act.	5. It does not involve any pigment system or molecules and here photochemical act does not take place.
6. Photophosphorylation takes place.	6. Photophosphorylation does not take place.

10.21. FACTORS AFFECTING THE RATE OF PHOTOSYNTHESIS

The rate of photosynthesis is affected by several factors which have been divided chiefly into two main groups:

(1) **(A) External Factors**

  (*i*) Light

  (*ii*) Carbon dioxide

  (*iii*) Temperature

  (*iv*) Water

(2) **(B) Internal Factors**

  (*i*) Chlorophyll

  (*ii*) Protoplasm

  (*iii*) Accumulation of the end-products of photosynthesis

Before discussing the factors, it is essential to know in brief about *Blackman's law of limiting factors*.

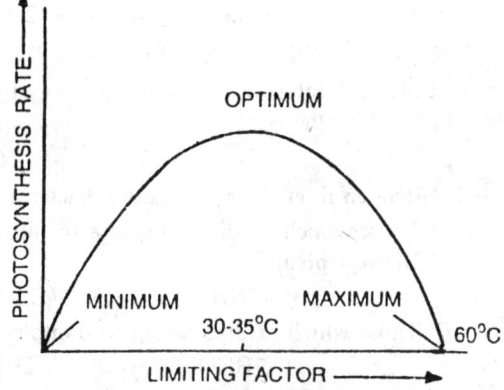

Fig. 10.20: Graphic representation of the idea of three cordinal points.

(3) Blackman's Law of Limiting Factors

Earlier scientists believed that the quantity or intensity of any factor at which the process of photosynthesis starts is called *minimum point* while the quantity or intensity at which the photosynthesis takes place at its maximum rate is called *Optimum point*. The quantity or intensity at which the photosynthesis does not take place is called *maximum point*. All the above three points are called *cordinal points*.

Leibig (1883) advocated the *law of minimum*. According to this law "the rate of any process is limited by the pace of the slowest factor". F.F. *Blackman* (1905) extended this law to formulate the principle of limiting factors. This was later known as "*Blackman's Law of Limiting Factors*". According to him the law states that "**When a process is conditioned as to its rapidity by a number of separate factors, the rate of the process is limited by the pace of slowest factor.**"

The principle was explained by *Blackman* as follows:

Suppose a leaf is exposed to such light intensity which allows the leaf to utilize 5 mg of CO_2 per hour in photosynthesis. The photosynthesis will not occur if the CO_2 is totally absent from the atmosphere. If one mg of CO_2 enters the leaf in an hour, the rate of photosynthesis is limited due to CO_2 factor. If the CO_2 concentration is increased in the atmosphere from 1 to 5 mg/hour, the rate of photosynthesis also increases along the line AB. Thus, the increase in photosynthetic rate will be proportionate with the increase in CO_2 concentration up to 5 mg. Any further increase in the CO_2 concentration will have no effect on the rate of photosynthesis which has become constant along the line BC. This is because the light factor (low intensity) has now become the limiting factor. Now the rate of photosynthesis will increase further along the line BD only if the light intensity is also increased from low to medium. The medium light intensity again becomes a limiting factor at point D and any addition of CO_2 will not increase the rate of photosynthesis. The rate becomes constant along the line DE. In the same way, when light intensity is increased from medium to higher, an increase in photosynthetic rate will take place along the line DF by adding CO_2. When the rate attains maximum at F, further increase in CO_2 will not increase the rate of photosynthesis which becomes constant along the line FG. Here, higher light intensity again becomes a limiting factor.

From the Fig. 10.21 it become clear that the rate of photosynthesis cannot be increased by increasing only one factor. The other factors should also be increased in proper proportion for favourable effect. Besides CO_2 and light, other factors such as temperature, water etc. may also become limiting under certain conditions.

Blackman's law of limiting factors has been criticised by many workers like *Jensen* (1918), *Harder* (1921) and *Hoover* (1933). According to them, the limiting effect may be because the chloroplasts in a leaf are not under same environmental conditions at a given time. Thus, the concept of limiting effect is not absolute but it is relative one. Due to this reason some scientists call it '*relatively limiting factor*'.

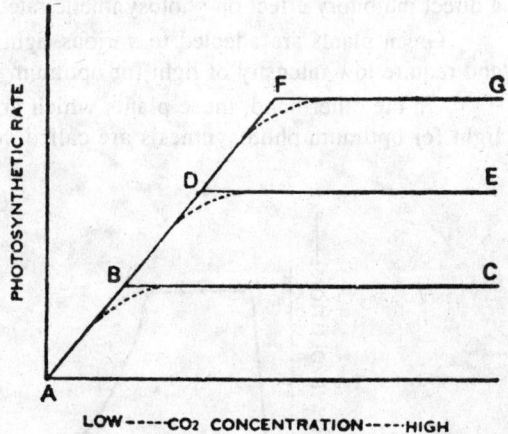

Fig. 10.21: Illustration of Blackman's law of limiting factor.

External Factors

(1) *Light*: The green plants require light for the synthesis of carbohydrates (photosynthesis) and chlorophyll. Normally, the plants obtain light from the sun for such processes but sometimes artificial light (through electricity) is also given for this purpose.

(a) Quality of light: The photosynthesis takes place only in visible part of the spectrum of light in between 3600 Å and 7600 Å or from 350 to 750 mµ wavelengths of light. It does not take place in ultraviolet, green and infra-red light. The maximum photosynthesis occurs in red light and slightly less to it in blue light. In green light it is minimum or almost nil. The rate of photosynthesis is higher in white light than in the light rays of a particular wavelength (monochromatic). The effect of different wavelengths of light on the photosynthetic rate is called *action spectrum*.

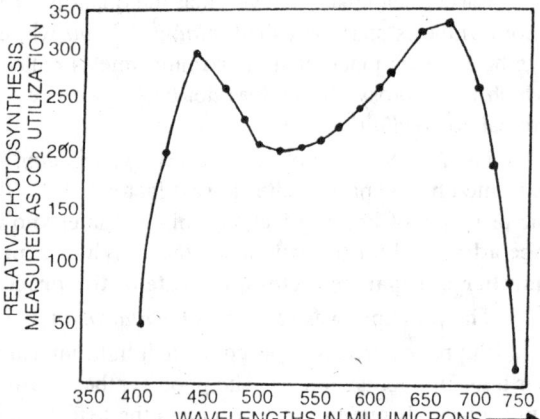

Fig. 10.22: Action spectrum of photosynthesis.

(b) Light Intensity: The intensity of light has favourable effect on the rate of photosynthesis. The total light perceived by a plant depends on its general form and arrangement of leaves. Of the light falling on the leaf, about 80 percent is absorbed, 10 percent is reflected and 10 percent is transmitted.

According to *Wolkoff* (1866) the rate of photosynthesis is directly proportional to the intensity of light. Normally, the chlorophyll pigments absorb only 3 percent of the total light absorbed by leaves. The rate of photosynthesis is usually low under low light intensity and it increases with the increase in light intensity until some other factor becomes limiting. It has been observed through various experiments that the rate of photosynthesis increases if the light intensity is increased gradually from 100 foot candle to 3000 foot candle and other factors are available in sufficient amount. At very high light intensity beyond a certain point the temperature of cells increases resulting into *photo-oxidation* of its constituents. The phenomenon is called *solarization*. It shows a direct inhibitory effect on photosynthetic rate.

Green plants are adapted to various light intensities. Those plants which are shade loving and require low intensity of light for optimum photosynthesis are called *sciophytes*.

On the other hand, those plants which grow in sunny places and require high intensity of light for optimum photosynthesis are called *heliophytes*.

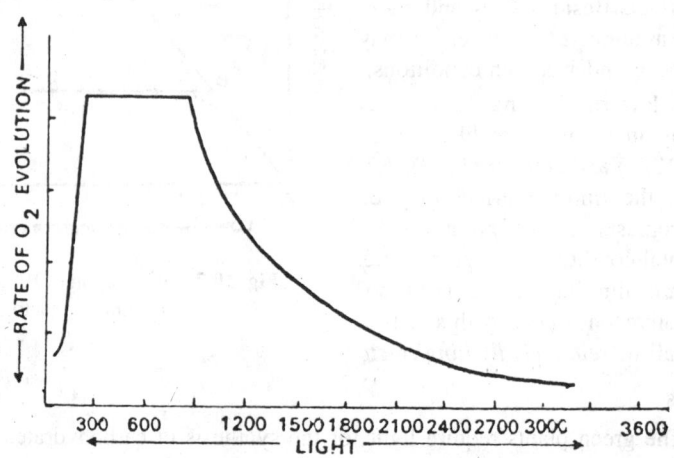

Fig. 10.23: Effect of increasing light intensity on photosynthetic rate in terms of rate of O_2 evolution.

The light intensity at which the amount of CO_2 used in photosynthesis and amount of CO_2 liberated in respiration become volumetrically equal is known as *compensation point*.

(c) *Duration of light*: Even a brief flash of light is enough for photosynthesis to occur. However, the rate of photosynthesis is greater in intermittent light than in continuous light because in continuous light the assimilatory power accumulates and is not consumed in the dark reaction at the same rate at which it is produced in light reaction. Longer duration of light period favours photosynthesis in leaves without being damaged. Good photosynthetic yield occurs if a plant gets 10 to 12 hours light per day.

(2) *Effect of CO_2 concentration*: The atmospheric air contains only 0.03% CO_2 concentration by volume. As the CO_2 concentration in the atmosphere increases, the rate of photosynthesis also increases but after a definite concentration (0.9%) of CO_2, the rate of photosynthesis does not increase. In this case, the light acts as a limiting factor. It has been experimentally demonstrated that the rate of photosynthesis varies in different plants and depends upon the CO_2 concentration. For example, in hydrophytes the rate of photosynthesis increases upto 1.1% CO_2 concentration while in *Triticum aestivum* maximum photosynthesis occurs at 0.15% CO_2.

(3) *Temperature*: Temperature shows a little effect on photosynthetic rate as compared to other processes. The variation in temperature affects only calvin cycle of photosynthesis and not light reaction. It has been observed during certain experiments that in most of the plants the rate of photosynthesis increases from 10°C to 30°C. An increase of each 10°C temperature up to 30°C doubles the photosynthetic rate. It is denoted by Q_{10}. Thus, in case of photosynthesis $Q_{10} = 2$. The ratio of photosynthetic rates at t°C and $t-10$°C is called *temperature coefficient* ($Q_{10}C$). The temperature beyond 30°C initially increases the rate of photosynthesis but after sometime reduces the rate of photosynthesis. The photosynthesis will stop in many plants at about freezing point but in certain conifers it takes place even at −35°C. In certain species of algae indigenous to hot spring, it can take place even at 75°C. Usually, the temperatures beyond 40°C-50°C retard photosynthesis in most of the plants because most of the enzymes present in the chloroplast become inactive and stop functioning.

(4) *Water*: Water is an essential raw material in photosynthesis. This rarely acts as a limiting factor because less than 1 percent of the water absorbed by a plant is used in photosynthesis. However, the rate of photosynthesis may decrease if the plants are inadequately supplied with water. According to some scientists, the rate of photosynthesis decreases up to 87 percent in water deficient soil. The reasons behind this are (i) closure of stomata (ii) stopage of CO_2 absorption and (iii) reduction in the activity of photosynthetic enzymes.

Internal Factors

Chlorophyll: Chlorophyll is a necessary factor for photosynthesis. Due to this reason, the photosynthesis does not take place in etiolated and achlorophyllous plants. In variegated leaves, photosynthesis occurs only in those parts of leaves which possess chlorophyll. When the chlorophyll formation does not take place at certain portions in the leaves due to deficiency of certain minerals such as N, P, K, Mg, Fe etc., the leaves become variegated. This phenomenon is called *chlorosis*.

Willstater used the term *assimilation number* to find out the importance of chlorophyll. This number decreases with the age of the plants. The quantity of chlorophyll present in the cell is directly related with the photosynthetic-rate.

Protoplasm: In cells there are certain unknown factors which are catalytic in nature and affect the rate of photosynthesis. These factors have been known as *protoplasmic factors*. They effect the dark reaction and are enzymatic in nature. Proper hydration of protoplasm is also essential for photosynthesis.

Accumulation of the end products of photosynthesis: If the photosynthetic products are not translocated, there is a retarding effect upon photosynthesis. Quick translocation of the carbohydrates or the end products of photosynthesis will have a favourable effect on the photosynthetic rate.

10.22. SIGNIFICANCE OF PHOTOSYNTHESIS

The phenomenon of photosynthesis is a boon to the nature and human beings. Without this the life is impossible on the earth. Photosynthesis is a process through which solar energy is converted into chemical energy which is ultimately utilized by the living beings for various activities. It helps in the following:

Production of Food Material

Green plants synthesise various food materials like carbohydrates, proteins and fats etc. through photosynthesis in light and interconversions of the products. These food materials are used in the formation of plant-body and its constituents and some of them are used as food for animals. Thus, green plants are chief producers in nature and all living organisms depend upon them for their nutrition. The green alga *Chlorella* is being used as food in space travel.

Atmospheric Control and Air Purification

Each living organism produces CO_2 and energy as a result of oxidation of carbohydrates, fats and proteins during respiration. On the other hand, CO_2 is also released in the atmosphere by burning coal, petrol and diesel etc. If this CO_2 is accumulated continuously in the atmosphere, the life of men and other living organisms will be impossible. Green plants utilize this extra CO_2 of atmosphere for photosynthesis, due to which CO_2 balance in the atmosphere (0.03%) is maintained.

Photosynthesis and respiration both are complementary processes to each other. Thus, they will not take place in absence of any one of them. In photosynthesis CO_2 is utilized to synthesise carbohydrates (food materials) and O_2 is released. These carbohydrates or food materials and O_2 are used in respiration resulting into oxidation of food materials and liberation of CO_2 and energy. The CO_2 released during respiration is again utilized in photosynthesis. Thus, both the processes continued in the nature and the atmoshpheric air is purified.

10.23. SOME EXPERIMENTS RELATED WITH PHOTOSYNTHESIS

Experiment 1: *To demonstrate that starch formation takes place inside green leaves during photosynthesis.*

Take a potted plant and keep it for 38 hours to destarch its leaves. To know that earlier synthesised starch is lost, boil single leaf of the plant in ethyl alcohol on the bath. The chlorophyll of the leaf will be dissolved in the alcohol and the leaf will be colourless.

Dry this leaf with the help of blotting paper and pour a few drops of iodine solution on it. You will see no change in the leaf colour.

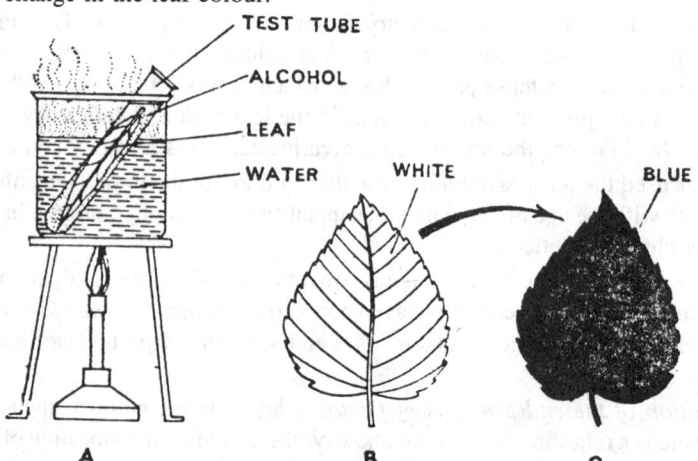

Fig. 10.24: Starch test: (A) Separation of chlorophyll from leaf, (B) Destarched white leaf, (C) Blue coloured leaf.

Photosynthesis

Now keep the plant in sunlight for 6-7 hours and then test its one leaf by the above method. You will observe that this time the leaf turns blue. This blue colour is due to formation of starch iodide. The starch present in the leaf reacts with iodine solution and starch iodide is formed. It proves that the leaves synthesise starch in presence of sunlight.

Experiment 2: *To demonstrate that O_2 is evolved during photosynthesis.*

Take a beaker filled with water. Now, insert the twigs of *Hydrilla* in the funnel in such a way that the cut ends are firmly introduced in the tube of funnel but spreading in its wide mouth. Place this funnel fully immersed in water and then invert a test tube filled with water over the stem of the funnel. Keep this apparatus in sunlight for a few hours and observe.

The air bubbles will come out through the stem of the funnel indicating the evolution of a gas. The gas will be collected in the tube after replacing water. When the test tube is completely filled with the gas, remove it from the funnel and test the gas with the help of either by burning match stick or by pyrogallol solution. The burning match stick will burn more quickly and the gas will be dissolved in the pyrogallol solution which will not give test with burning match stick further. Both the tests indicate the presence of O_2 in the test tube which comes from the plant. The O_2 is released as a result of photosynthesis. The evolution of O_2 takes place due to photolysis of water.

In the above experiment if the following modification are done, the photosynthesis will be effected as under:

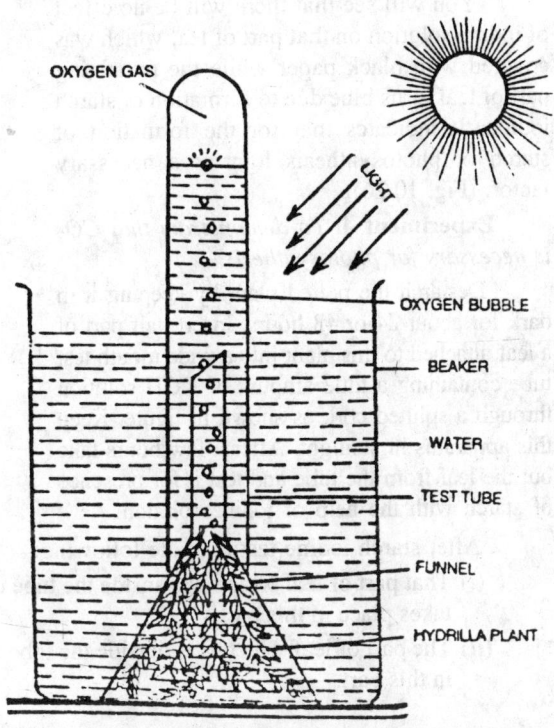

Fig. 10.25: Experiment to demonstrate evolution of O_2 during photosynthesis.

(1) *If boiled water is used instead of pond water*: If boiled water is used instead of pond water, neither photosynthesis will occur nor evolution of O_2 bubbles will take place because

(i) CO_2 will be removed from water and in absence of CO_2 photosynthesis will not take place.

(ii) Temperature of water will be increased which will effect metabolisms of the plants. Due to decrease or cease of metabolisms, the plants will die.

(2) *If terrestrial plant is used in place of Hydrilla*: The leaves of terrestrial plant (mesophytic) cannot absorb the CO_2 present in water from their surface. The leaf stomata will be clogged and closed due to which the exchange of gases will stop resulting into non-occurrence of photosynthesis.

(3) *If Sodium bicarbonate is added to the water*: By adding $NaHCO_3$ in water, the rate of O_2 bubbles production will immediately increase resulting into increase in the rate of photosynthesis. $NaHCO_3$ will react with water molecules to produce CO_2 which will increase CO_2 concentration in water. With this increase in CO_2 concentration in water, the rate of photosynthesis will also increase. After sometime CO_2 will act as a limiting factor.

$$2NaHCO_3 + H_2O \longrightarrow Na_2CO_3 + H_2CO_3$$
$$H_2CO_3 \longrightarrow H_2O + CO_2\uparrow$$

(4) *If some toxic or harmful or anesthetic substance is added to water*: When such substances are added to water, they enter into plant and kill the enzymes and coenzymes present in the plant cells. It will result into death of protoplasm. Thus, the rate of photosynthesis will first decrease and ultimately stop.

Experiment 3. *To demonstrate that light is necessary for photosynthesis.*

Destarch the leaf of a potted plant by keeping it in dark for about 24 to 48 hours. Now, test the leaf for starch presence. When it is confirmed that the leaf is devoid of starch, cover the central part of a leaf with the help of a black paper strip and clips. Keep this apparatus in sunlight for a few hours. After 6-7 hours take out the leaf and test for starch by starch iodide test method as described earlier.

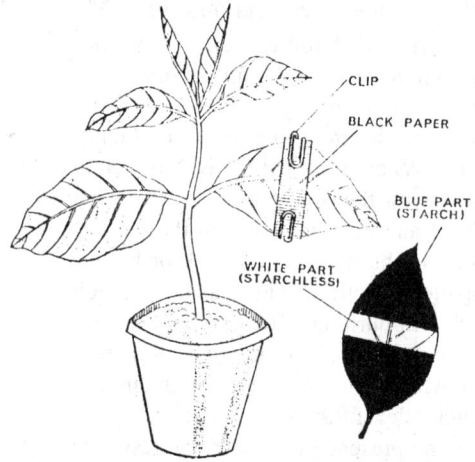

You will see that there will be no effect of iodine solution on that part of leaf which was covered with black paper while the remaining part of leaf turns blue due to formation of starch iodide. It indicates that for the formation of starch or photosynthesis light is a necessary factor. (Fig. 10.26).

Experiment 4. *To demonstrate that CO_2 is necessary for photosynthesis.*

Destarch the potted plant by keeping it in dark for about 24 or 48 hours. Insert half part of a leaf attached to this plant into a wide mouth test tube containing a little amount of KOH solution through a splitted cork as shown in figure. Keep this apparatus in sunlight. After a few hours take out the leaf from the tube and test it for presence of starch with the help of iodine solution.

Fig. 10.26: Demonstration of light necessity in photosynthesis.

After starch iodine test note the following:

(*i*) That part of leaf which was inside the tube does not turn blue because no starch formation takes place in this part.

(*ii*) The part of leaf which was outside the tube turns blue because starch formation took place in this part.

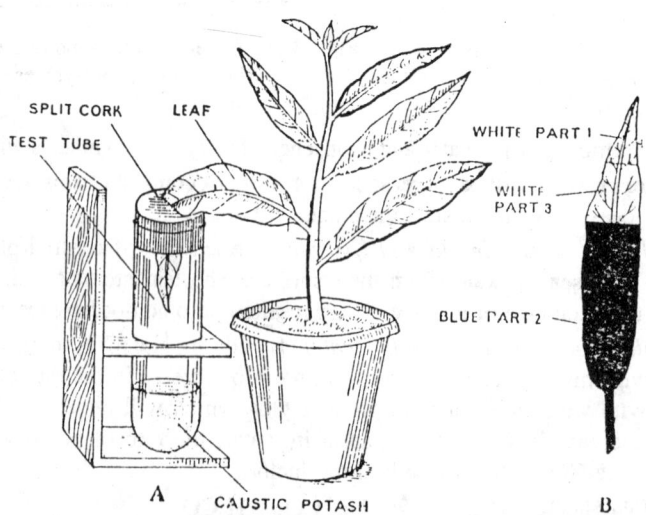

Fig. 10.27: Demonstration of CO_2 necessity in photosynthesis.

(*iii*) The part of leaf which was inside the splitted cork also shows no effect of iodine solution as no starch formation took place here.

In case of (*a*) starch formation could not take place because of non-availability of CO_2 which was absorbed by KOH. Thus, this part received all the requirements like chlorophyll, water and light except CO_2. In case of (*b*) starch formation took place because this part received all the requirements including CO_2. In case of (*c*) also starch formation could not take place because this part could not receive sunlight. (Fig. 10.27).

Experiment 5. *To demonstrate that Chlorophyll is necessary for photosynthesis.*

Pluck a variegated leaf of *Coleus* or *Croton* plant kept in sunlight. Test this leaf for starch by iodine solution (starch iodide test) method described earlier. You will see that only those parts of leaf turn blue which were originally green. The yellow parts or white parts of leaf show negative test. It indicates that chlorophyll is necessary for photosynthesis, (Fig. 10.28).

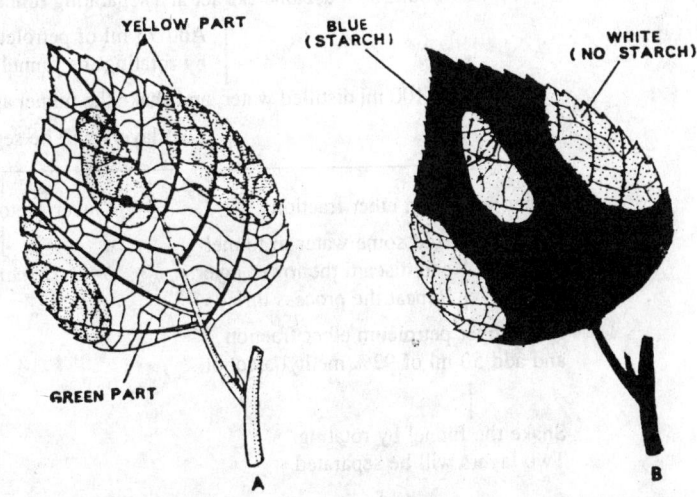

Fig. 10.28: To demonstrate that chlorophyll is necessary for photosynthesis.

Experiment 6: *To determine the rate of photosynthesis by Wilmott's bubbler.*

The Wilmotts bubbler consists of a wide mouth bottle fitted with single hole cork, a glass tube with lower end having wider opening to insert plant twigs and upper end with a narrow bent nozzle, and a water reservoir (Fig. 10.29).

Fill the bottle of Wilmott's bubble with water and insert a few branched twigs of *Hydrilla* (aquatic plant) in the lower wider end of the tube fixed in the cork of bottle. The twigs of Hydrilla should be cut inside the water to avoid the entry of air bubbles in the xylem. Now immerse the plant twigs in the bottle and fix the cork in the mouth of the bottle. Add some more water in the reservoir so that the nozzle of the tube dips in the water. Keep the apparatus in sunlight and observe.

After sometime it is observed that the bubbles start coming out from the cut ends of the plant twigs which finally escape through the nozzle of the tube. When bubbles of uniform size escape continuously, they are counted in a definite time. The number of bubbles coming out per unit time will be the rate of photosynthesis. The effect of various factors affecting photosynthesis can be studied through this apparatus.

Experiment 7 : *Extraction and separation of plant pig-*

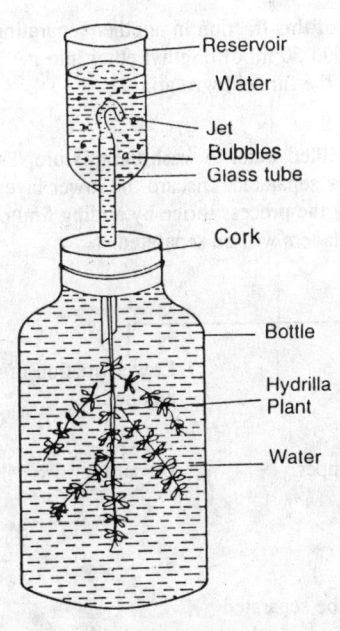

Fig. 10.29: Wilmott's bubbler

ments *(Chlorophylls and carotenoids) by chemical method through separating funnel.*

Grind about 120 gms of fresh spinach leaves in 50 ml of 80% acetone in a pestle and morter containing a pinch of $CaCO_3$. Homogenise it and filter the homogenate with the help of Buchner funnel. The deep green coloured filtrate containing chlorophylls and carotenoids is called *acetone extract*. Now separate the plant pigments from this extract through separating funnel (Fig. 10.30) with the help of following chart :

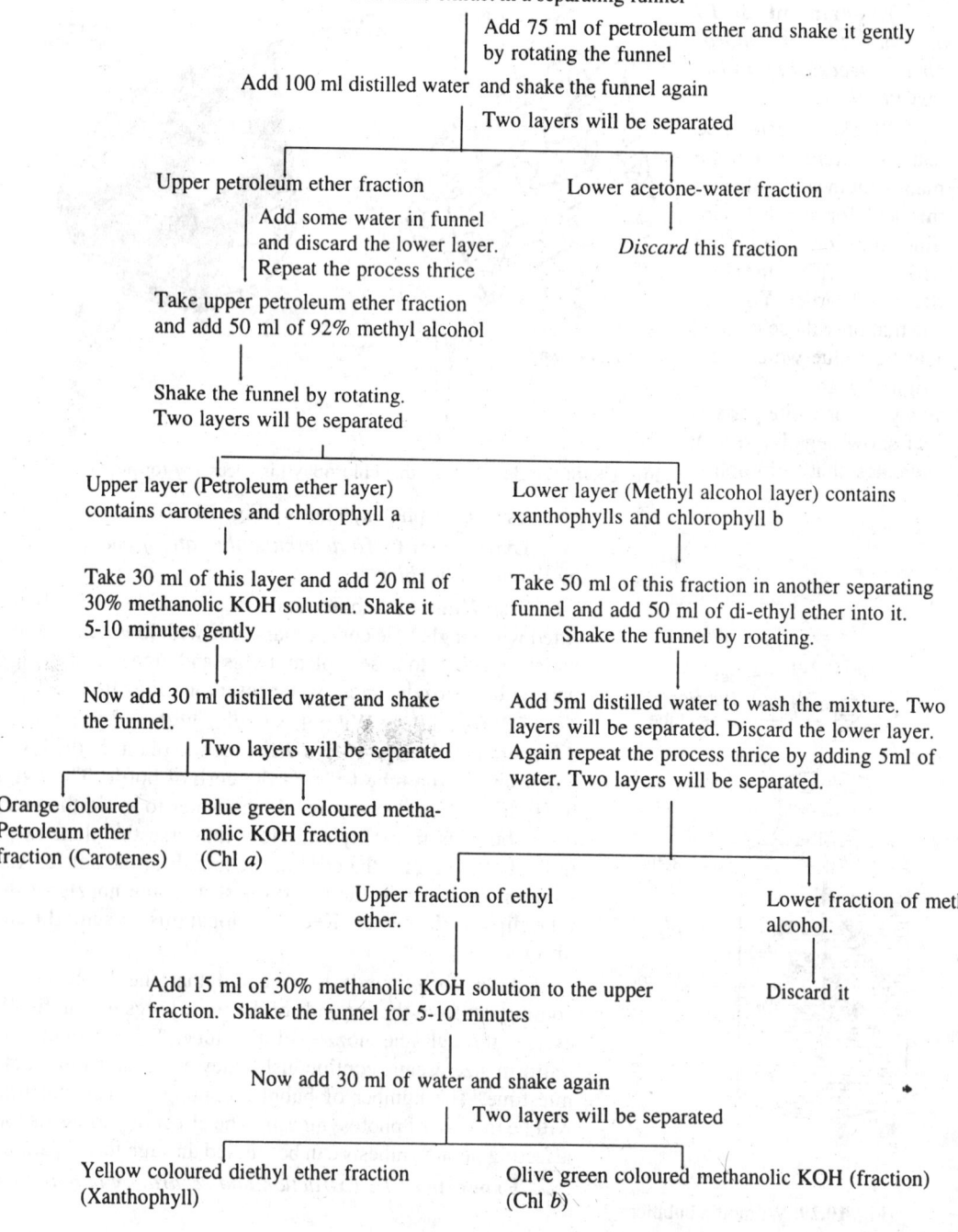

Photosynthesis

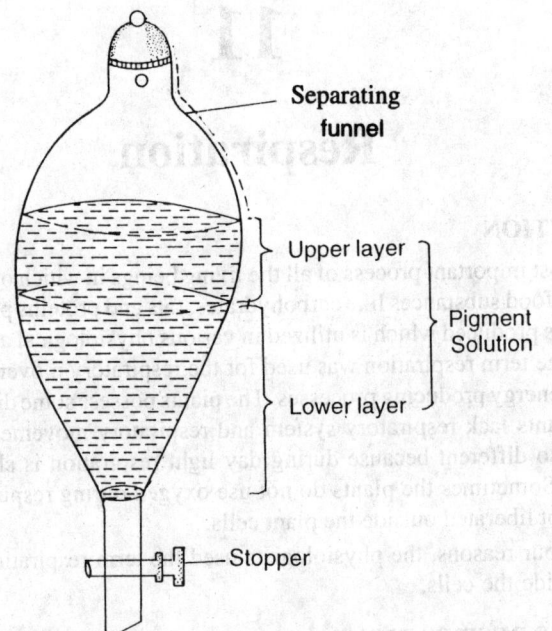

Fig. 10.30: Separating funnel showing two layers containing pigments.

11

Respiration

11.1. INTRODUCTION

Respiration is the most important process of all the living beings in which oxygen enters into the cells and oxidises various food substances like carbohydrates, fats and proteins present in them. As a result of oxidation energy is produced which is utilized in various physiological and synthetic processes. In the very beginning the term respiration was used for the respiratory movements of animals but later on it included all the energy producing processes. The plants possess some different type of respiration because—(*i*) The plants lack respiratory system and respiratory movements, (*ii*) The exchange of gases in plants is also different because during day light respiration is slightly suppressed due to photosynthesis, (*iii*) Sometimes the plants do not use oxygen during respiration (anaerobic), (*iv*) In some cases CO_2 is not liberated outside the plant cells.

Due to above four reasons, the physiologists used the term respiration only for the oxidation of food materials inside the cells.

11.2. HISTORY OF RESPIRATION

The term "respiration" was used in the beginning of fifteenth century but its importance was worked out by *Crooke* (1615). He told that the life is impossible without respiration. At that time people believed that respiration is found only in animals and not in plants. *Malpighi* (1619) demonstrated that oxygen is required in high amount during germination of seeds. *Sheele* (1777) told that during respiration seeds take O_2 inside and liberate CO_2. *Lavoisier* demonstrated that respiration is a combustion process in which the energy is produced in the same way as is produced by burning the coal but both the processes differ in following points.

Table 11.1: Differences between Respiration and Combustion

	Respiration	*Combustion*
(*i*)	It is a biological process.	It is a chemical process.
(*ii*)	It takes place at normal temperature.	It takes place at high temperature.
(*iii*)	Respiration is a slow process completed in different steps. Thus, the energy is also liberated in several steps and remain stored in the form of ATP.	Combustion is a fast process in which the energy is liberated only in one step resulting in increase in temperature and production of fire.

Ingen-Housz (1779) concluded on the basis of his experiments that all plants liberate CO_2 in dark. *De Saussure* (1804 and 1822) measured the amount of O_2 taken and CO_2 liberated by plants. *Sachs* (1864) told that the growth is directly proportional to the rate of respiration. *Sachs* (1865) also told that exchange of gases is related with two main processes, of which one is now called photosynthesis and other as respiration. *Pasteur* (1870) studied the fermentation by yeast cells and demonstrated that alcohol is formed as a result of fermentation and anaerobic respiration. *Mac Munn* (1886) discovered histohaematin (cytochromes). *Bach* (1901) confirmed that plant extracts can oxidise phenols in presence of O_2. *Pfeffer* (1900) proved that the first stage in aerobic and anaerobic respiration both is similar.

Respiration

Respiration is the process of gaseous exchange where O_2 *is taken in and CO_2 is liberated outside*. *Hackett* (1959) called respiration 'life with air'. According to him, respiration is a complex process which includes following processes:

(*i*) Oxidation and fragmentation of organic compounds.

(*ii*) Transfer of electrons which ultimately form water by the union of hydrogen and oxygen.

(*iii*) Liberation of energy which is utilised in various physiological processes.

According to *Stiles* and *Leach* (1960), *the respiration is a complex process where complex compounds are oxidised into simpler compounds and energy is released*. Respiration is completed into two parts. The first part where carbohydrates etc. are converted into pyruvic acid is called *glycolysis* and second part where pyruvic acid is oxidised into CO_2 and H_2O is called *Kreb's Cycle*. The glycolysis was discovered by three scientists namely *Embden, Meyerhof* and *Parnas*. On their names, the glycolysis is also called E.M.P. pathway. Kreb's Cycle was discovered by *Sir Hans Kreb* (1937). *Millard et. al.* (1951) separated mitochondria from plant cells. *Davis* and *Majelis Young* (1953-60) demonstrated that all the enzymes of Kerb's Cycle are found in mitochondria.

11.3. CHANGES ASSOCIATED WITH RESPIRATION

Following changes occur during respiration:

(*i*) O_2 is taken in.

(*ii*) This O_2 is utilized or exhausted during the oxidation of food materials.

(*iii*) Energy is released as a result of oxidation.

(*iv*) Intermediate products are formed in the cells due to oxidation.

(*v*) CO_2 and H_2O are formed as the end products which are released later on outside the plant.

(*vi*) The dry weight of the plants is reduced due to liberation of CO_2 and H_2O.

In respiration, mainly hexose sugars are oxidised and the energy is liberated. One molecule of glucose after oxidation liberates about 673k. calories of energy.

$$C_6H_{12}O_6 + 6O_2 \longrightarrow 6CO_2 + 6H_2O + 673 \text{ K. Cal.}$$
Glucose   Oxygen   Carbon   Water   Energy
                          dioxide

11.4. TYPES OF RESPIRATION

Respiration is mainly of two types : (*i*) Aerobic respiration, (*ii*) Anaerobic respiration.

(1) **Aerobic Respiration:** Respiration occurring in presence of oxygen is called *aerobic respiration*. In most of the plants usually aerobic respiration is found which is stopped in absence of O_2. During aerobic respiration, the food materials like carbohydrates, fats and proteins are completely oxidised into CO_2 and H_2O and large amount of energy is released. The oxidation of one molecule of glucose produces about 673 k. calories of energy. Of the various end products, CO_2 is liberated into atmosphere, water is retained inside the plant cells and energy is utilised in various biological and synthetic processes. Aerobic respiration can be shown by following equation.

$$C_6H_{12}O_6 + 6O_2 \longrightarrow 6CO_2 + 6H_2O + 673 \text{ k.cal.}$$

The produced energy is absorbed by ADP and is stored in the form of ATP. The energy is again produced by the hydrolysis of ATP, when required. Although, aerobic respiration has been shown above by a single equation but it is a very complex process and is completed in several steps. The whole process may be divided into two parts:

(*a*) Conversion of glucose into pyruvic acid (glycolysis).

(*b*) Aerobic oxidation of pyruvic acid into CO_2 and H_2O (oxidative decarboxylation and Kreb's cycle).

Kostychev demonstrated in *Aspergillus niger* that during aerobic respiration the oxidation of lactic acid, glycerol and mannitol takes place. During oxidation of peptones, amino-acids are formed first which are then oxidised into CO_2 and H_2O.

(2) **Anaerobic Respiration:** Respiration occurring in absence of oxygen is called *anaerobic respiration*. It is found only in certain plants. During anaerobic respiration, the food materials like carbohydrates, fats and proteins are imcompletely oxidised resulting in the formation of alcohol and CO_2. The energy released during anaerobic respiration is quite less (about 21 k.cal. per molecule) in comparison to aerobic respiration (about 673 k.cal./mol.). Of the various end products, CO_2 is liberated in the atmosphere, alochol is retained inside the cells and energy is utilised only in a few processes. The alcohol may be reoxidised by aerobic respiration.

$$C_6H_{12}O_6 \longrightarrow 2C_2H_5OH + 2CO_2 + 21 \text{ k. cal.}$$
glucose   alcohol   energy

Anaerobic respiration occurs during resting stage of seeds, initial stages of seed germination, in the cells of fruit walls, several micro-organisms and Ascaris etc.

Table 11.2. Differences between Aerobic and Anaerobic Respiration.

	Aerobic	Anaerobic
(i)	It occurs in all living cells of higher plants.	It occurs in bacteria, certain fungi, germinating seeds and fleshy fruits.
(ii)	It requires oxygen.	Oxygen is not required.
(iii)	The end products are CO_2 and H_2O.	The end products are alcohol and CO_2.
(iv)	The oxidation of one molecule of glucose produces 38 ATP molecules.	The number of ATP molecules produced is only 2.
(v)	All the reactions except the reactions of glycolysis take place inside mitochondria.	All the reactions take place in cytoplasm.
(vi)	Organic compounds are completely oxidised and high energy is released.	Organic compounds are incompletely oxidised and very small amount of energy is released.
(vii)	Not toxic to plants.	Toxic to higher plants.

11.5. CLASSIFICATION OF LIVING BEINGS BASED ON RESPIRATION

Based on respiration, the living beings may be classified into following four classes:

(1) *Aerobes*: The living beings respiring in presence of oxygen are called *aerobes*.

(2) *Anaerobes*: The living beings respiring in absence of oxygen are called *anaerobes*.

(3) *Obligate aerobes*: Those living beings or plants which normally respire and produce energy under aerobic conditions (presence of O_2) but cannot survive under anaerobic conditions (absence of O_2) are called *obligate aerobes*.

(4) *Facultative anaerobes*: Those living beings or plants which normally respire and produce energy uder aerobic conditions but can also survive and respire under anaerobic conditions are called *facultative anaerobes*.

11.6. RESPIRATORY SUBSTRATES

Those substrates which are used as fuel in respiration are called *respiratory substrates*. The main respiratory substrates are carbohydrates and fats but proteins may also be used under special circumstances. Cabrohydrates include mono-, di- and polysaccharides. Among monosaccharides glucose and fructose, among disaccharides sucrose, and among polysaccharides starch, inulin and hemicelluloses are commonly used as respiratory substrates. The complex carbohydrates are first hydrolysed into simple hexose sugars in presence of special enzyme systems before entering into the process. Similarly, the fats are hydrolysed in presence of lipase enzyme into glycerol and fatty acids which are converted into hexose sugars for respiratory substrate. The proteins are also used as substrate in absence of carbohydrates and fats. Based upon the type of substrate, Blackman used the term *floating respiration* where the respiratory substrate was a carbohydrate and *protoplasmic respiration* where the substrate was a protein.

11.7. MECHANISM OF RESPIRATION

The mechanism of respiration is quite complex. The chemical bonds present inside the sugars are broken and hydrogen is liberated during respiration. The energy is also released which is absorbed by ADP and is stored in the form of ATP. At this stage hydrogen ions are broken into protons and electrons which are transferred through various coenzymes and cytochromes.

The first part in aerobic and anaerobic respiration *i.e.*, formation of pyruvic acid from carbohydrates through glycolysis, is the same and common but later on the fate of pyruvic acid depends upon the presence or absence of oxygen. In presence of oxygen, pyruvic acid is oxidised into CO_2 and H_2O whereas in absence of oxygen, pyruvic acid is oxidised into CO_2 and alcohol. (Fig. 11.1).

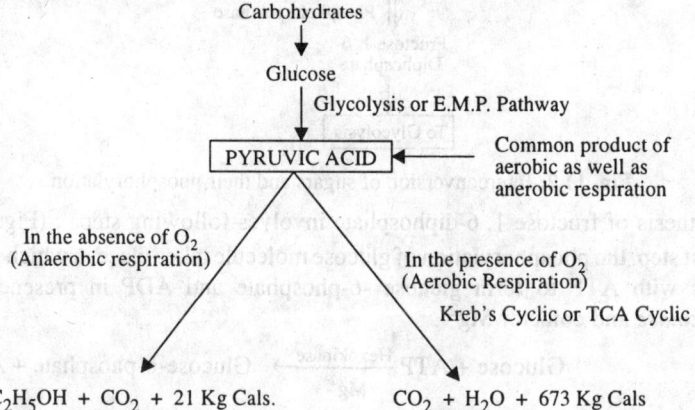

Fig. 11.1: Scheme showing relationship between aerobic and anaerobic respiration.

Glycolysis *(GK : glycos* = sugar *: lysis* = dissolution)

The steps of glycolysis were first studied by three German scientists, *Embden, Meyerhof* and *Parnas*. On their names, the glycolysis is also called E.M.P. pathway. The steps of glycolysis take place in the cytoplasm and through glycolysis glucose is oxidised into pyruvic acid in presence of many enzymes and coenzymes present in the cytoplasm. The oxygen is not required at any step during glycolysis.

Glycolysis can be divided into following four stages:

(1) *First Stage of glycolysis* = *Synthesis of Fructose 1, 6-diphosphate*

The synthesis of fructose 1, 6-diphosphate takes place by the phosphorylation of hexose sugars which combine with phosphate to form different types of hexose phosphates. The attachment of phosphate with hexose sugars is called *phosphorylation*. If the food material is a disaccharide, *e.g.* sucrose or a polysaccharide, *e.g.* starch or hemicellulose etc., they are first hydrolysed into hexose sugars and later on these sugars are phosphorylated. During phosphorylation, ATP plays an important role by acting as phosphate donor and it is converted into ADP. All these reactions are catalysed by a group of phosphorylases and transphosphorylases enzymes.

During synthesis of fructose 1, 6-diphosphate, glucose first reacts with ATP to form glucose-6-phosphate and ADP. This reaction is catalysed by an enzyme *hexokinase*. In addition to glucose other hexose sugars like mannose, fructose, galactose etc. also form glucose-6-phosphate. During this process, initally all sugars form their phosphates which later on are converted into glucose-6-

phosphate with the help of enzymes **isomerases**. Glucose-6-phosphate undergoes isomerisation to form fructose-6-phosphate. (Fig. 11.2).

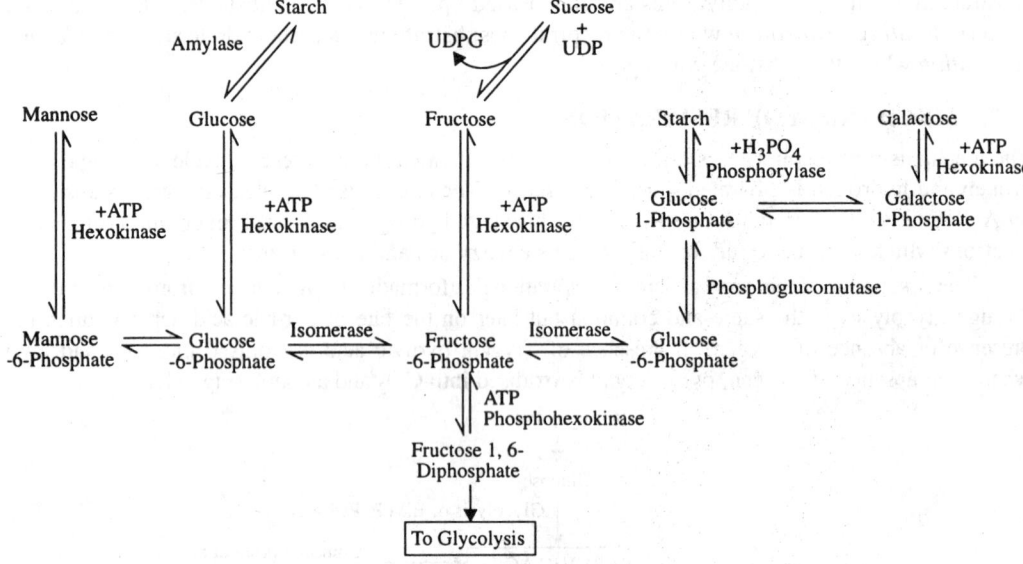

Fig. 11.2: Interconversion of sugars and their phosphorylation.

Thus, synthesis of fructose 1, 6-diphosphate involves following steps : (Fig. 11.3)

1. In first step, the phosphorylation of glucose molecule takes place in which glucose molecule reacts with ATP to form glucose -6-phosphate and ADP in presence of an enzyme *hexokinase* and cofactor Mg^{2+}.

$$\text{Glucose + ATP} \xrightarrow[Mg^{2+}]{\text{Hexokinase}} \text{Glucose-6-phosphate + ADP}$$

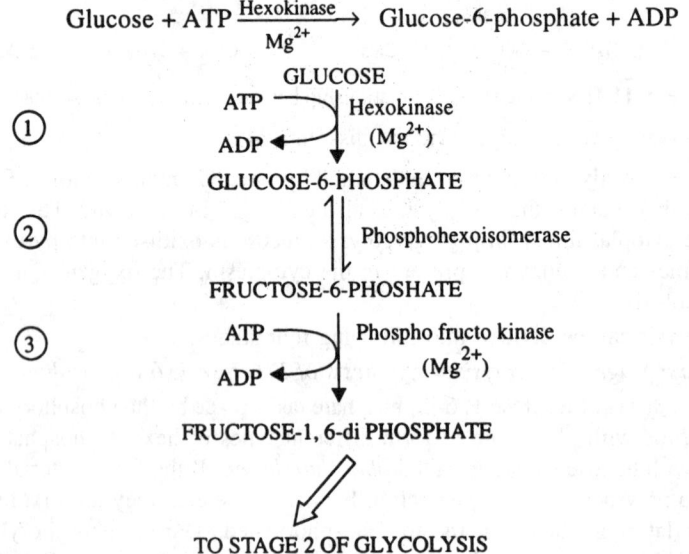

Fig. 11.3: Glycolysis—First stage : Synthesis of fructose 1,6-diphosphate.

2. Glucose-6-phosphate undergoes isomerisation in presence of enzyme *phosphohexoisomerase* to form fructose-6-phosphate. It is a reversible reaction.

$$\text{Glucose-6-phosphate} \xrightleftharpoons{\text{Phosphohexoisomerase}} \text{Fructose-6-phosphate.}$$

Respiration

3. In this step, further phosphorylation of fructose-6-phosphate takes place with the help of ATP resulting into formation of fructose 1, 6-diphosphate in presence of enzyme **phosphofructo kinase** and magnesium as cofactor. This reaction is not reversible.

$$\text{Fructose 6-phosphate} + \text{ATP} \xrightarrow[\text{Mg}^{2+}]{\text{Phosphofructokinase}} \text{Fructose 1, 6-diphosphate} + \text{ADP}.$$

(2) *Stage 2 of glycolysis = Cleavage of fructose 1, 6-diphosphate* (Fig. 11.4).

4. Fructose 1, 6-diphosphate formed at step 3 of stage 1 now splits into two 3-carbon compounds, 3-Phosphoglyceraldehyde and dihydroxyacetone phosphate with the help of enzyme aldolase. Both these compounds are called *triose phosphates* and are interconverted into each other in presence of an enzyme **phosphotriose isomerase** so that a balance is maintained. In further reactions two molecules of 3-carbon will be involved.

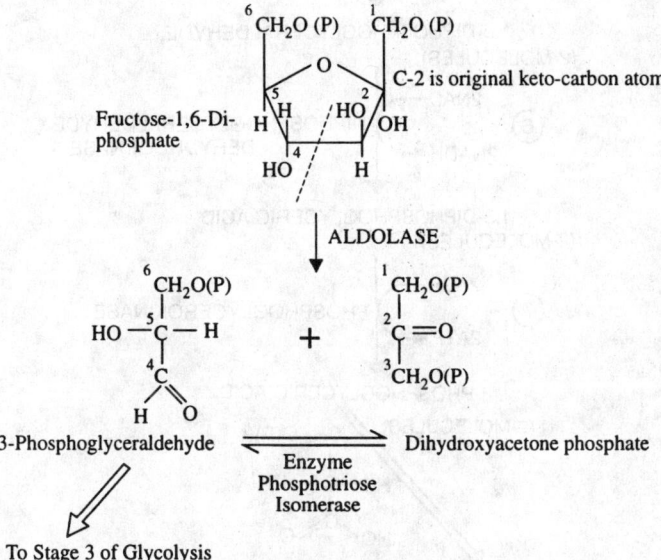

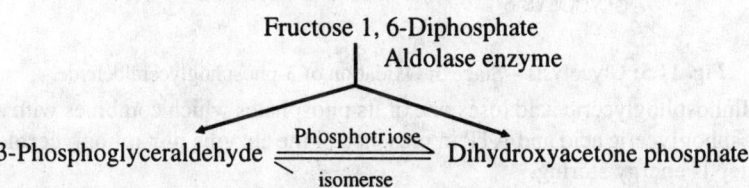

Fig. 11.4: Glycolysis—Stage 2: Splitting of fructose 1, 6—diphosphate.

(3) *Stage 3 of glycolysis = Oxidation of 3-phosphoglyceraldehyde and the formation of 3-phosphoglyceric acid.* (Fig. 11.5).

3-Phosphoglyceraldehyde formed in stage 2 now enters into stage 3 which involves following steps:

5. In this step, 3-phosphoglyceraldehyde reacts with phosphoric acid (H_3PO_4) resulting into formation of 1, 3-diphosphoglyceraldehyde.

3-phosphoglyceraldehyde + $2H_3PO_4$ ⟶ 1,3-diphosphoglyceraldehyde
(2 molecules) (2 molecules)

6. Now 1, 3-diphosphoglyceraldehyde is oxidised into 1, 3-diphosphoglyceric acid in presence of an enzyme **diphosphoglyceraldehyde dehydrogenase** and a coenzyme NAD. During

this reaction two molecules of hydrogen are released from diphosphoglyceraldehyde which reduces NAD to $NADH_2$.

$$1,3\text{–diphosphoglyceraldehyde} + 2NAD \xrightarrow{\text{Diphosphoglyceraldehyde dehydrogenase}} 1,3\text{,-diphosphoglyceric acid} + 2NADH_2$$
(2 molecules) → (2 molecules)

3-PHOSPHO GLYCERALDEHYDE
(2-MOLECULES)

⑤ ↓ —H_3PO_4

1,3-DIPHOSPHOGLYCERALDEHYDE
(2-MOLECULES)

⑥ 2NAD ⤵ DIPHOSPHOGLYCERALDEHYDE DEHYDROGENASE
    2NADH ⤴

1,3-DIPHOSPHOGLYCERIC ACID
(2-MOLECULES)

⑦ 2ADP ⤵ PHOSPHOGLYCEROKINASE
    2ATP ⤴

3-PHOSPHOGLYCERIC ACID
(2-MOLECULES)

$CH_2\,O(P)$
|
$HO-C-H$
|
$HO-C=O$

TO STAGE 4 OF GLYCOLYSIS

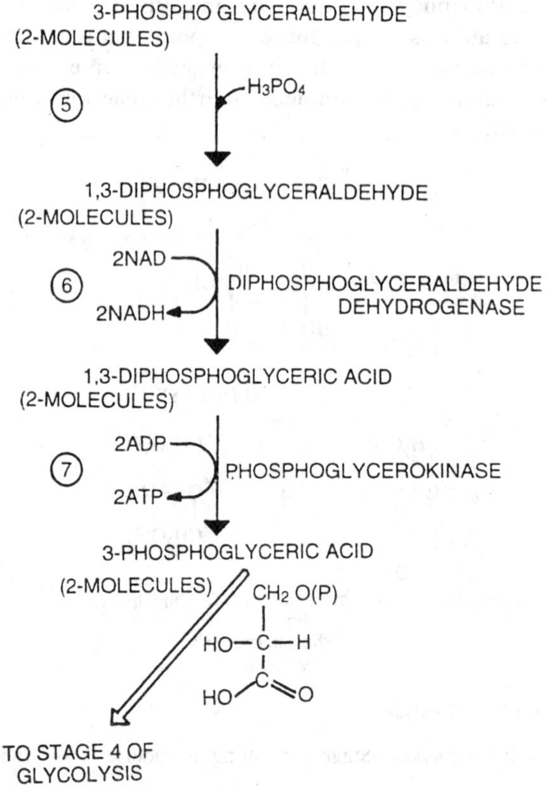

Fig. 11.5: Glycolysis—Stage 3: Oxidation of 3-phosphoglyceraldehyde.

7. The diphosphoglyceric acid loses one of its phosphates which combines with ADP to form 3-phosphoglyceric acid and ATP in presence of the enzyme **phosphoglycerokinase**. Thus, this step is energy storing.

$$1,3\text{–diphosphoglyceric acid} + 2ADP \xrightarrow{\text{Phosphoglycerokinase}} 3\text{-phosphoglyceric acid} + 2\ ATP$$
(2 molecules) → (2 mols)

(4) *Stage 4 of glycolysis = Synthesis of pyruic acid* (Fig. 11.6)

All the steps after the formation of 3-phosphoglyceric acid come under stage 4. It includes following steps:

8. In this step, an isomeric change takes place in 3-phosphoglyceric acid resulting into transfer of phosphate from Carbon 3 (C_3) to Carbon 2 (C_2) and formation of 2-phosphoglyceric acid in presence of enzyme **phosphoglyceromutase**.

3–Phosphoglyceric acid $\xrightarrow{\text{Phosphoglycero mutase}}$ 2–Phosphoglyceric acid
(2 molecules)   (2 motecules)

9. 2-phosphoglyceric acid loses two molecules of water in presence of enzyme **enolase** and is converted into phosphoenol-pyruvic acid.

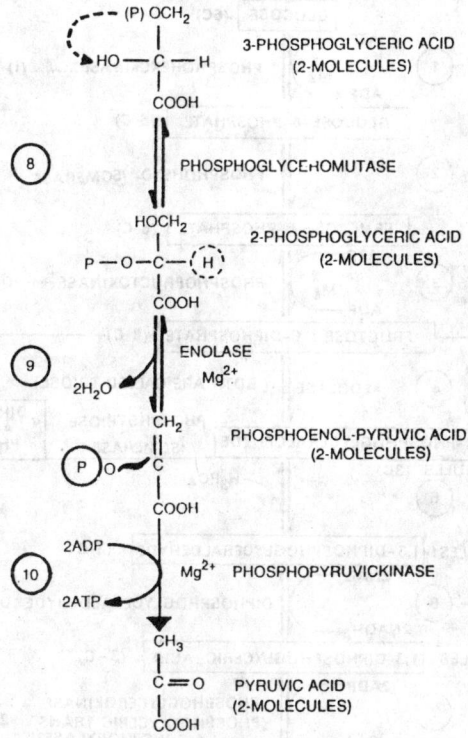

Fig. 11.6 : Synthesis of Pyruvic acid.

2-Phosphoglyceric acid $\xrightarrow[\text{Mg}^{+2}]{\text{Enolase}}$ Phosphoenol-pyruvic acid + $2H_2O$
(2 molecules)   (2 molecules)

10. Now, one phosphate group is removed from phospho-enol-pyruvic acid with the production of pyruvic acid under the influence of an enzyme **phosphopyruvic kinase** and cofactor Mg^{2+}. The removed phosphate group is accepted by ADP to form ATP.

Phosphoenol pyruvic acid + 2ADP $\xrightarrow[\text{kinase Mg}^{+2}]{\text{Phosphopyruvic}}$ Pyruvic acid + 2ATP
(2 molecules)   (2 molecule)

All the four stages of glycolysis with various steps are shown in Fig. 11.7.

(5) *Summary of Glycolysis*

(i) In glycolysis, from one molecule of glucose, *two molecules* of *pyruvic are formed*.

(ii) In this, four molecules of ATP are formed (2 ATP at stage 3, step 7 + 2ATP at stage 4, step 10). Because two molecules of ATP are consumed in phosphorylation reactions (1 ATP at stage 1, step 1 + 1 ATP at stage 1, step 3 = 2 ATP molecules), therefore, during glycolysis there is a *net gain* of 2 ATP *molecules* (4 ATP — 2ATP = 2 ATP).

(iii) Two molecules of NAD are reduced to two molecules of $NADH_2$ (stage 3, step 6) which later on oxidised aerobically to yield six molecules of ATP (one NAD molecule after oxidation produces 3 molecules of ATP). Thus, the total gain of ATP molecules during glycolysis in presence of oxygen will be increased to eight instead of two.

(iv) Thus, the energy of glucose become stored partly in ATP molecules and partly in $NADH_2$ molecules.
(v) Oxygen is not required during glycolysis.
(vi) In glycolysis, CO_2 is also not produced.

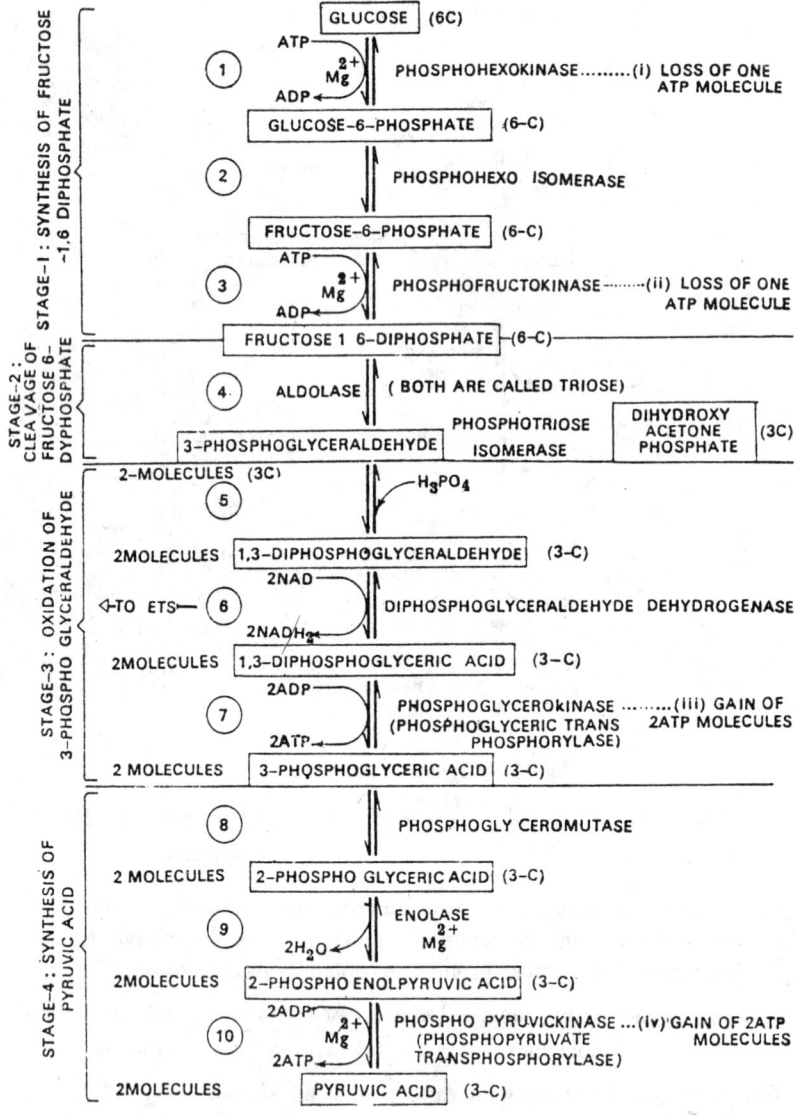

Fig. 11.7: Glycolysis: Different stages with various steps in a sequence.

Aerobic Oxidation of Pyruvic Acid

The oxidation of pyruvic acid (3-carbon compound) in aerobic conditions takes place through tricarboxylic acid cycle (TCA Cycle) discovered by Sir Hans Kreb. The pyruvic acid before entering into Kreb's Cycle is first converted into 2-carbon compound, acetyl coenzyme-A from 3-carbon compound with the help of several enzymes and coenzymes. During this process, one molecule of CO_2 is liberated from pyruvic acid. This mechanism or process is called *oxidative decarboxylation*. It takes place in the mitochondria of aerobic cells.

(1) *Reactions of Oxidative Decarboxylation*: Oxidative decarboxylation is completed in several

Respiration

steps catalysed by a enzyme complex **pyruvate dehydrogenase**. The enzyme complex includes following enzymes and coenzymes.

Enzymes: Pyruvic acid decarboxylase, Dihydroxy Lipoyl transacetylase, Dihydrolipoyl dehydrogenase.

Coenzymes: Thiamine pyrophosphate (TPP), Lipoic acid, Coenzyme A, and NAD.

Different steps of oxidative decarboxylation are given below and the enzymes catalysing particular steps are given in the brackets.

1. In first step, one molecule of CO_2 is evolved from pyruvic acid to form acetaldehyde. (Enzyme—Pyruvic acid decarboxylase).

$$CH_3.CO.COOH \xrightarrow{Pyruvic\ acid\ decarboxylase} CH_3.CHO + CO_2$$
Pyruvic acid → Acetaldehyde

2. Now, acetaldehyde (acetyl group) reacts with coenzyme thiamine pyrophosphate (TPP) to form active acetaldehyde.

$$CH_3.CHO + TPP \xrightarrow{Mg^{+2}} CH_3.CHO.TPP$$
Acetaldehyde + Thiamine pyrophosphate → Active acetaldehyde

3. Active acetaldehyde after combining with lipoic acid forms acetyl lipoic acid (Enzyme—Dihydroxylipoyl transacetylase).

$$CH_3CHO.TPP + Lipoic\ Acid \xrightarrow{Dihydroxylipoyl\ Transacetylase} Acetyl\text{-}lipoic\ acid + TPP$$
Active acetaldehyde

4. Acetyl-lipoic acid reacts with Coenzyme A resulting into formation of acetyl CoA and reduced lipoic acid. (Enzyme—Dihydrolipoyl dehydrogenase).

Acetyl Lipoic acid + coenzyme A $\xrightarrow{Dihydroxylipoyl\ dehydrogenase}$ Reduced Lipoic acid + Acetyl CoA.

5. Reduced lipoic acid after reacting with NAD is oxidized and NAD is reduced to $NADH_2$. (Enzyme-Lipoate dehydrogenase).

Reduced lipoic acid + NAD^+ $\xrightarrow{Lipoate\ dehydrogenase}$ Oxidised lipoic acid + $NADH_2$

The sum of all the above five reactions gives following single equation:

$$CH_3.CO.COOH + Co.A.SH + NAD^+ \longrightarrow CH_3CO.S.CoA + CO_2 + NADH_2$$
Pyruvic acid   Coenzyme-A   Acetyl conzyme A.

The $NADH_2$ formed in the above equation combines with half molecule of oxygen to form one molecule of water. In this process, three molecules of ADP are also oxidized resulting into formation of three molecules of ATP.

$$NADH_2 + 1/2\ O_2 + 3ADP \longrightarrow NAD^+ + H_2O + 3ATP.$$

(2) Summary of Oxidative Decarboxylation

(i) Two molecules of pyruvic acid form two molecules of acetyl CoA, CO_2 and $NADH_2$ each.

(ii) Two molecules of $NADH_2$ after being oxidized produce 6 molecules of ATP.

(iii) This process acts as a link between glycolysis and Kreb's Cycle.

(3) Kreb's Cycle or TCA Cycle or Citric Acid Cycle

An English biochemist, Sir H.A. Kreb's (1937), discovered this cycle for the first time in nematodes. After his name, the cycle was called Kreb's cycle. He was awarded Nobel Prize for this discovery. All the reactions of Kreb's cycle takes place inside the mitochondria and through this pyruvic acid, fatty acids, fats and amino acids are oxidized into CO_2 and water. Thus, Kreb's cycle represents the common path for the metabolism of carbohydrates, fats and proteins (Fig. 11.8).

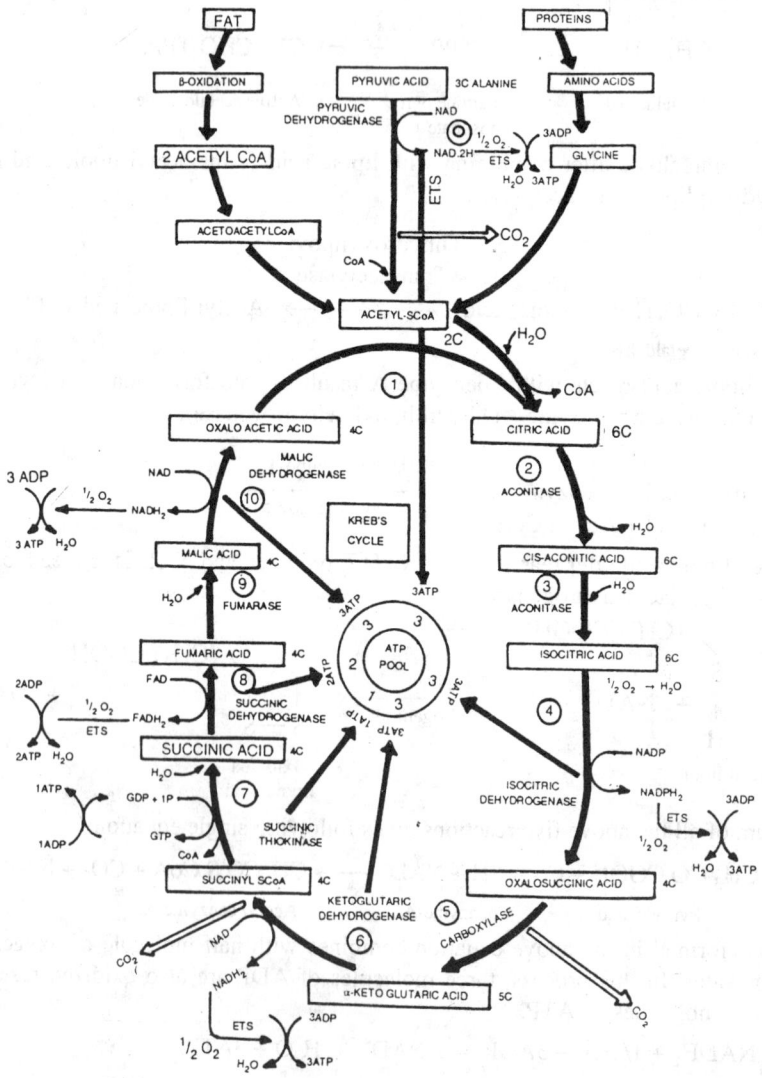

Fig. 11.8: Kreb's Cycle.

Respiration

Like glycolysis, Kreb's cycle is also completed in several steps which are catalysed by various enzymes and coenzymes.

Acetyl CoA formed during oxidative decarboxylation enters into the Kreb's cycle for further oxidation. Acetyl CoA acts as a connecting link between glycolysis and Kreb's cycle. Its complete oxidation takes place through Kreb's cycle and Electron Transport System (ETS).

The different steps of Kreb's cycle (Fig. 11.8). The enzymes catalysing each step are given in the bracket.

Reactions

1. Two carbon compound acetyl CoA reacts with one molecule of four carbon compound oxaloacetic acid and one molecule of water to form 6-carbon compound citric acid and CoA. Citric acid is degraded step-wise until oxaloacetic acid is regenerated. (Enzyme: Condensing enzymes—**Citrate synthetase**).

$$\begin{array}{c}CH_3\\|\\C=O\\|\\Co\sim A\end{array} + \begin{array}{c}COOH\\|\\CH_2\\|\\C=O\\|\\COOH\end{array} + H_2O \xrightarrow{\text{Citrate synthetase}} \begin{array}{c}COOH\\|\\CH_2\\|\\HO-C-COOH\\|\\CH_2\\|\\COOH\end{array} + Co\sim A$$

Acetyl CoA + oxaloacetic acid → Citric acid + Coenzyme-A

2. Citric acid undergoes dehydration to form cis-aconitic acid. (Enzyme—**Aconitase**).

$$\begin{array}{c}COOH\\|\\CH_2\\|\\HO-C-COOH\\|\\CH_2\\|\\COOH\end{array} \xrightarrow{\text{Aconitase}} \begin{array}{c}COOH\\|\\CH_2\\|\\C-COOH\\||\\CH\\|\\COOH\end{array} + H_2O$$

Cis-aconitic acid → Cis-aconitic acid

3. Cis-aconitic acid accepts one molecule of water (hydration process) to form isocitric acid (Enzymes — **Aconitase**).

$$\begin{array}{c}COOH\\|\\CH_2\\|\\C-COOH\\||\\CH\\|\\COOH\end{array} + H_2O \xrightarrow{\text{Aconitase}} \begin{array}{c}COOH\\|\\CH_2\\|\\H-C-COOH\\|\\HO-C-H\\|\\COOH\end{array}$$

Cis-aconitic acid → Isocitric acid

4. Isocitric acid is oxidised in presence of NADP into oxalosuccinic acid. Two molecules of hydrogen are released from the isocitric acid which reduce NADP into $NADPH_2$. (Enzyme—**Isocitric dehydrogenase**)

$$\underset{\text{Isocitric acid}}{\begin{array}{c} \text{COOH} \\ | \\ \text{CH}_2 \\ | \\ \text{H}-\text{C}-\text{COOH} \\ | \\ \text{HO}-\text{C}-\text{H} \\ | \\ \text{COOH} \end{array}} + \text{NADP}^+ \xrightarrow{\underset{\text{dehydrogenase}}{\text{Isocitric}}} \underset{\text{Oxalosuccinic acid}}{\begin{array}{c} \text{COOH} \\ | \\ \text{CH}_2 \\ | \\ \text{H}-\text{C}-\text{COOH} \\ | \\ \text{O}=\text{C}-\text{COOH} \end{array}} + \text{NADPH} + \text{H}^+$$

5. The 6-carbon oxalosuccinic acid undergoes decarboxylation to form 5-carbon α-ketoglutaric acid. One molecule of CO_2 is released in this reaction. (Enzyme—**Oxalosuccinic decarboxylase**).

$$\underset{\text{Oxalosuccinic acid}}{\begin{array}{c} \text{COOH} \\ | \\ \text{CH}_2 \\ | \\ \text{H}-\text{C}-\text{COOH} \\ | \\ \text{O}=\text{C}-\text{COOH} \end{array}} \xrightarrow{\underset{\text{decarboxylase}}{\text{Oxalosuccinic}}} \underset{\text{α-Ketoglutaric acid}}{\begin{array}{c} \text{COOH} \\ | \\ \text{CH}_2 \\ | \\ \text{CH}_2 \\ | \\ \text{C}=\text{O} \\ | \\ \text{COOH} \end{array}} + CO_2$$

6. α-Ketoglutaric acid undergoes oxidative decarboxylation like pyruvic acid to form a 4-carbon, succinyl coenzyme-A. During this reaction, one molecule of Co~A is used up, one molecule of CO_2 is released and NAD^+ is reduced to $NADH + H^+$. (Enzyme—**α-Ketoglutaric dehydrogenase**).

$$\underset{\text{α-Ketoglutaric acid}}{\begin{array}{c} \text{COOH} \\ | \\ \text{CH}_2 \\ | \\ \text{CH}_2 \\ | \\ \text{C}=\text{O} \\ | \\ \text{COOH} \end{array}} + \text{NAD}^+ + \text{CoA~SH} \xrightarrow{\underset{\text{dehydrogenase}}{\text{α-Ketoglutaric}}} \underset{\text{Succinyl CoA.}}{\begin{array}{c} \text{COOH} \\ | \\ \text{CH}_2 \\ | \\ \text{CH}_2 \\ | \\ \text{O}=\text{C}-\text{SCo~A} \end{array}} + \text{NADH} + \text{H}^+ + CO_2$$

7. Succinyl Co~A is hydrolysed to succinic acid in presence of **succinic thiokinase**. In this reaction, CoA~SH is regenerated and guanosine di-phosphate (GDP) undergoes substrate phosphorylation to produce GTP. GTP again reacts with ADP to form ATP and GDP. (Enzyme—**Succinic thiokinase**)

$$\underset{\text{Succinyl CoA}}{\begin{array}{c} \text{COOH} \\ | \\ \text{CH}_2 \\ | \\ \text{CH}_2 \\ | \\ \text{O}=\text{C}-\text{S.Co~A} \end{array}} + H_2O \xrightarrow{\text{Succinic thiokinase}} \underset{\text{Succinic acid}}{\begin{array}{c} \text{COOH} \\ | \\ \text{CH}_2 \\ | \\ \text{CH}_2 \\ | \\ \text{COOH} \end{array}} + \text{CoA~SH}.$$

Respiration

$$GDP + iP \longrightarrow GTP$$
$$GTP + ADP \longrightarrow GDP + ATP.$$

8. Succinic acid is oxidised to fumaric acid. During this reaction, two molecules of hydrogen are released which reduce the coenzyme flavin adenine dinucleotide (FAD) to $FADH_2$. (Enzyme—**Succinic dehydrogenase**).

```
COOH                                          COOH
 |                                             |
CH₂                                            CH
 |    + FAD   succinic dehydrogenase           ||    + FADH₂
CH₂         ─────────────────────→             CH
 |                                             |
COOH                                          COOH
Succinic acid                                 Fumaric acid.
```

9. Fumaric acid reacts with one molecule of water to form malic acid. (Enzyme—**Fumarase**)

```
COOH                                          COOH
 |                                             |
CH                                             CH₂
||    + H₂O    Fumarase                        |
CH            ─────────→                      HO—C—H
 |                                             |
COOH                                          COOH
Fumaric acid                                  Malic acid
```

10. Finally, Malic acid is oxidized to oxaloacetic acid—which again reacts with acetyl CoA and enters into Kreb's cycle to be oxidized. In this reaction, two hydrogen molecules are released from malic acid which reduce NAD^+ to $NADH + H^+$. (Enzyme—**Malic dehydrogenase**).

```
COOH                                          COOH
 |                                             |
CH₂                                            CH₂
 |     + NAD⁺   Malic dehydrogenase            |     + NADH+H⁺
HO—C—H         ─────────────────→             C=O
 |                                             |
COOH                                          COOH
Malic acid                                    Oxaloacetic acid.
```

Thus, as a result of oxidation of pyruvic acid one molecule of CO_2 in oxidative decarboxylation and two molecules of CO_2 in Kreb's cycle (steps 5 and 6) are liberated. The total number of CO_2 evolved become three which indicate that the 3-carbon pyruvic acid has been completely oxidized. Now, because two molecules of pyruvic acid which are formed by one molecule of glucose in glycolysis, enter into Kreb's cycle for oxidation, a total of 6 CO_2 molecules will be evolved (2 molecules of pyruvic acid × 3 molecules of CO_2 = 6 molecules of CO_2).

All the $NADH_2$ and $FADH_2$ molecules synthesized during glycolysis and Kreb's cycle enter into electron transport system for oxidation where they are oxidized in presence of oxygen to produce ATP molecules. One molecule of $NADH_2$ after oxidation produces three molecules of ATP where as $FADH_2$ produces only two ATP molecules. The complete oxidation of one molecule of glucose produces 38 ATP molecules of which 8 ATP molecules are produced in glycolysis, 6 ATP molecules in oxidative decarboxylation and 24 ATP molecules in Kreb's cycle. The summary of aerobic oxidation of glucose is as follows :

(1) *Summary of Aerobic Oxidation of Glucose*

	Reaction	
(*i*) Glucose $\longrightarrow$ 2 Pyruvic acid + 2H$_2$ (Glycolysis)	... 8 ATP	...(*i*)
(*ii*) 2 Pyruvic acid + 2 CoA. SH $\longrightarrow$ 2 Acetyl. CoA + 2H$_2$ + 2CO$_2$	... 6 ATP	...(*ii*)
(oxidative decarboxylation)	(in 2 turns)	
(*iii*) 2 Acetyl. CoA + 6 H$_2$O $\longrightarrow$ 2CoA.SH + 8H$_2$ + 4 CO$_2$	... 24 ATP	...(*iii*)
(Kreb's cycle)	(in 2 Turns)	

Sum of reactions (*i*), (*ii*) & (*iii*): Glucose + 6H$_2$O $\longrightarrow$ 6CO$_2$ + 12H$_2$ + 38ATP ...(*iv*)

$$12H_2 + 6O_2 \longrightarrow 12H_2O \quad ...(v)$$

Sum of reactions (4) and (5): Glucose + 6O$_2$ → 6CO$_2$ + 6H$_2$O + 38ATP.

(2) *ATP Production during aerobic oxidation or respiration of one mole of glucose.*

Table 11.3 : ATP producing or consuming steps during glycolysis and Kreb's cycle.

	Gain or loss of ATP molecules	Step No.
In Fig. 11.7		
Glucose → Glucose-6-(P)	– 1	1
Fructose 6-(P) → fructose 1, 6 di–(P)	– 1	3
2 mol. [3-Phosphoglyceraldehyle] → 2 [1, 3-diphosphoglyceric acid] [NADH → NAD$^+$]	+6	5 & 6
2 [1, 3-diphosphoglyceric acid] → 2 [3- phosphoglyceric acid] [~p]	+2	7
2 [Phosphoenol pyruvic acid] [→ 2 [Pyruvic acid] [~p]	+2	10
In Fig. 11.8		
2 [Pyruvic acid] → 2 [Acetyl CoA] [NADH → NAD$^+$]	+6	0
2 [Isocitric acid] → 2 [α–ketoglutaric acid] [NADH → NAD$^+$]	+6	4 & 5
2 [α -ketoglutaric acid] → 2 Succinyl CoA] [NADH → NAD$^+$]	+6	6
2 [Succinyl CoA] → 2 [Succinic acid] [~p]	+2	7
2 [Succinic acid] → 2 [Fumaric acid] [FADH$_2$ → FAD$^+$]	+4	8
2 [Malic acid] → 2 [Oxalo acetic acid] [NADH → NAD$^+$]	+6	10
Total Net gain	+38	ATP

(4) *Electron Transport System or Terminal Oxidation of Reduced Coenzymes*: The respiratory break down of glucose in presence of oxygen is an oxidative process. During aerobic respiration simple carbohydrates and intermediates like phosphoglyceraldehyde, pyruvic acid, isocitric acid, α-ketoglutaric acid, succinic acid and malic acid are oxidized. Each oxidative step involves release of a pair of hydrogen atoms (2H) which dissociate into two *protons* (2H$^+$) and two *electrons* (2e$^-$).

$$2H \longrightarrow 2H^+ + 2e^-$$

The pairs of hydrogen atoms (2H$^+$ + 2e$^-$) released in each oxidative step of Kreb's cycle do not combine directly with oxygen but pass through a series of coenzymes and cytochromes, which form electron transport system, before reacting with O$_2$ to form H$_2$O. The electron transport system is made up of following coenzymes and proteins:

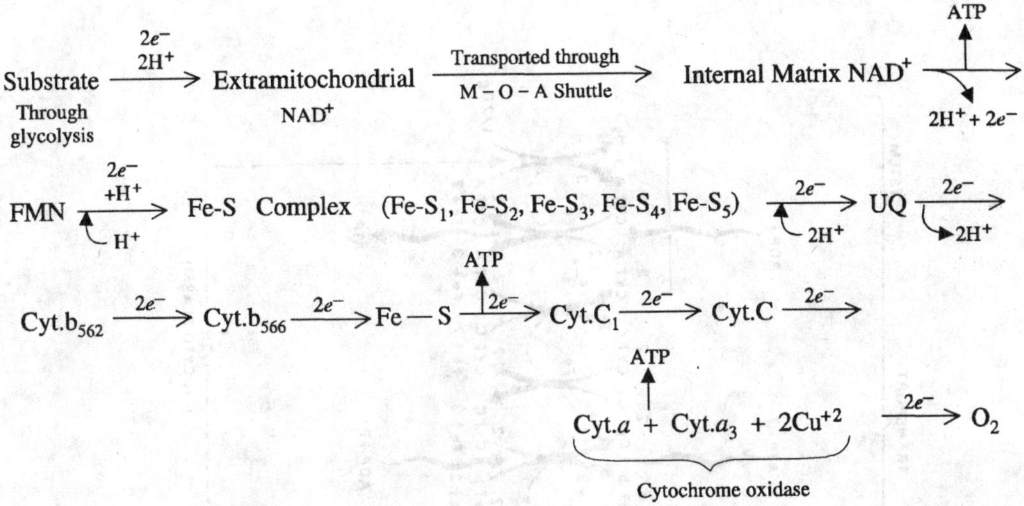

Respiration

$$\text{Substrate} \xrightarrow[\text{glycolysis}]{\substack{2e^- \\ 2H^+}} \text{Extramitochondrial NAD}^+ \xrightarrow{\text{Transported through M-O-A Shuttle}} \text{Internal Matrix NAD}^+ \xrightarrow{\substack{\uparrow \text{ATP} \\ \downarrow \\ 2H^+ + 2e^-}}$$

$$\text{FMN} \xrightarrow[H^+]{\substack{2e^- \\ +H^+}} \text{Fe-S Complex } (Fe\text{-}S_1, Fe\text{-}S_2, Fe\text{-}S_3, Fe\text{-}S_4, Fe\text{-}S_5) \xrightarrow[2H^+]{2e^-} UQ \xrightarrow[2H^+]{2e^-}$$

$$Cyt.b_{562} \xrightarrow{2e^-} Cyt.b_{566} \xrightarrow{2e^-} Fe-S \xrightarrow[\uparrow \text{ATP}]{2e^-} Cyt.C_1 \xrightarrow{2e^-} Cyt.C \xrightarrow{2e^-}$$

$$\underbrace{Cyt.a + Cyt.a_3 + 2Cu^{+2}}_{\text{Cytochrome oxidase}} \xrightarrow{2e^-} O_2$$

$\uparrow \text{ATP}$

Fig. 11.9: A complete set of electron carriers in plants aerobic respiration showing the transfer of electrons and protons both or only electrons as in case of cytochromes.

1. Nicotinamide adenine dinucleotide (NAD).
2. Flavo proteins (FMN) or FAD
3. Fe–S protein complex
4. Coenzyme-Q (Co. Q) or ubiquinone (UQ)
5. Cytochrome b (Cyt. b)— Cyt b_{562} and Cyt b_{566}
6. Fe–S, protein
7. Cytochrome C_1 (Cyt. C_1)
8. Cytochrome C (Cyt. C)
9. Cytochrome a (Cyt. a)
10. Cytochrome a_3 (Cyt. a_3)

All the above coenzymes are found in F_1 particles of mitochondria.

During the transfer of hydrogen atoms from one conenzyme to another conezyme, a large amount of energy is released which is picked up by ADP to form ATP with the help of inorganic phosphate (iP).

During respiration, the electron pairs liberated from respiratory compounds are picked up by coenzymes like NAD^+ or $NADP^+$ and FMN etc. (Fig. 11.10). The transfer of electrons in all compounds except succinic acid takes place first in NAD^+ or $NADP^+$ and later on in FMN. The transfer of electrons from succinic acid takes place directly to the FAD and not through NAD^+ or $NADP^+$. Due to this reason only two molecules of ATP are produced in the formation of fumaric acid from succinic acid whereas in case of other compounds three ATP molecules are produced because in these cases the electrons are first picked up by NAD.

The different steps of electron transport system are as follows (Fig. 11.10).

1. The hydrogen pairs released from different substrates of Kreb's cycle except succinic acid reacts with matrix NAD^+. Two electrons and one proton (H^+) are transferred to NAD^+ causing its reduction and one proton is released in the medium.

$$2H \longrightarrow \underset{\text{(protons)}}{2H^+} + \underset{\text{(electrons)}}{2e^-}$$

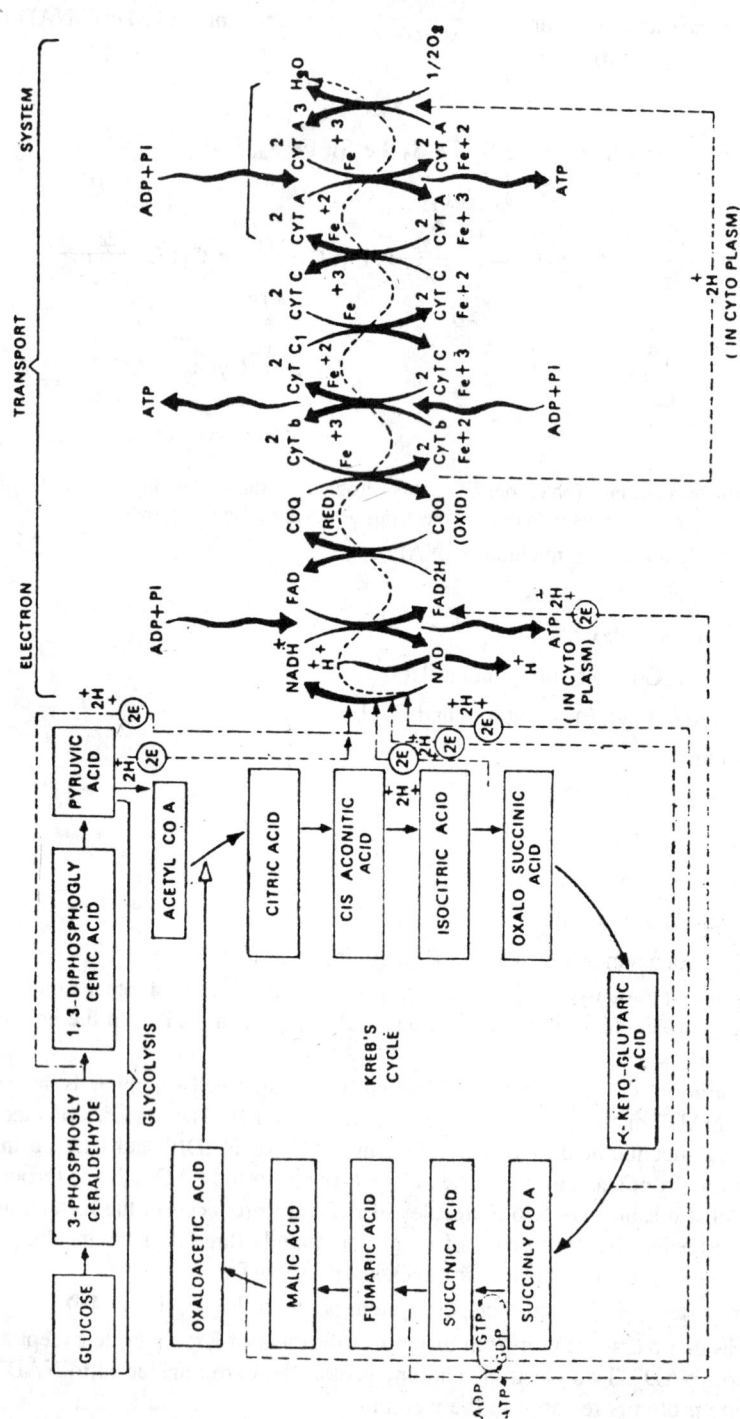

Fig. 11.10: Electron transport system and its relation with Kreb's Cycle.

Respiration

$$NAD^+ + 2H^+ + 2e^- \longrightarrow \underset{\text{(reduced)}}{NADH} + \underset{\text{(ion pool)}}{H^+}$$

2. Now, two electrons and one proton are transferred from NADH to flavoprotein-flavin mononucleotide (FMN) causing oxidation of NADH into NAD^+ and reduction of FMN into $FMNH_2$. One hydrogen ion (H^+) is picked up from hydrogen ion pool to complete this reaction.

$$\underset{\text{(reduced)}}{NADH} + H^+ + \underset{\text{(oxidized)}}{FMN} + 2e^- \longrightarrow \underset{\text{(oxidized)}}{NAD^+} + \underset{\text{(reduced)}}{FMNH_2}$$

The free energy released at this step is stored during oxidative phosphorylation and one molecule of ATP is synthesised from ADP and inorganic phosphate.

The hydrogen pair from succinic acid is first transferred to FAD to form $FADH_2$. The $FADH_2$ transfers electrons to Ubiquinon (UQ) through Fe-S and from UQ the electrons pass to cytochromes arranged in normal series on the basis of redox potentials. Thus, during oxidation of succinic acid only 2ATP molecules are generated.

3. The oxidation of $FMNH_2$ takes place by transferring electrons to Fe-S protein to form reduced Fe-S and oxidized FMN. The protons ($2H^+$) are released in the space.

4. The reduced Fe-S then transfers its 2 electrons to ubiquinone (UQ) or CoQ one by one. The two protons ($2H^+$) are picked up from the matrix (medium). The UQ is reduced to UQH_2.

5. The reduced ubiquinone (UQH_2) then transfers a pair of electrons ($2e^-$) (one at a time) to cytochrome b (Cyt. b) while two hydrogen ions are released in the medium. Thus, UQH_2 is oxidized and Cyt b is reduced ($Fe^{+3} \rightarrow Fe^{+2}$).

6. The reduced Cyt. b transfers its electron ($2e^-$) to Fe–S protein causing oxidation of Cyt. b ($Fe^{2+} \rightarrow Fe^{3+}$) and reduction of Fe–S.

7. The reduced Fe–S protein transfers electrons to Cyt c_1 to reduce it. The energy released at this step is coupled to form ATP from ADP and iP.

8. Reduced Cyt. c_1 transfers its electrons to Cyt.c causing reduction of Cyt.c and oxidation of Cyt.c_1.

9. Reduced Cyt.c transfers a pair of electrons to Cyt. a causing reduction of Cyt. a.

10. Pair of electrons are then transferred from reduced Cyt. a to Cyt. a_3. Thus, Cyt. a_3 is reduced. The energy released at this step is coupled to form ATP from ADP and iP.

11. Reduced Cyt. a_3 loses a pair of electrons which are accepted by molecular oxygen along with a pair of protons ($2H^+$) from the medium (hydrogen ion pool) to form one molecule of water.

It should be noted that four electrons and four protons (two pairs of hydrogen) will be needed during the real reduction of one oxygen molecule to form water.

$$O_2 + 4e^- \longrightarrow 2(O^{--})$$
$$2(O^{--}) + 4H^+ \longrightarrow 2H_2O.$$

The reduction of various cytochromes requires only electrons and no protons. Each cytochrome possesses an iron element in the centre which functions for accepting ($Fe^{3+} \xrightarrow{+e^-} Fe^{2+}$) or donating ($Fe^{2+} \xrightarrow{-e^-} Fe^{3+}$) electrons. When a cytochrome accepts electrons, it is reduced and if a cytochrome donates electrons, it is oxidized.

Flavin mononucleotide (FMN) is a metalloprotein and Fe–S protein is iron-sulphur protein. The last step (11) is called *terminal oxidation* which is catalysed by enzyme *cytochrome oxidase*. This enzyme contains inseparable Cyt.a and Cyt.a_3 components and a polypeptide containing two copper ions ($2Cu^{2+}$). Both iron and copper undergo reversible changes in their oxidized states (*i.e.*, $Fe^{2+} \rightleftharpoons Fe^{3+} + e^-$; $Cu^+ \rightleftharpoons Cu^{2+} + e^-$) during electron transport by cytochrome oxidase.

Summary of Electron Transport System

1. It is made up of coenzymes NAD^+ or $NADP^+$, FAD and coenzyme Q and cytochromes b, C_1, C, a and a_3.
2. The transfer of electrons in all compounds except succinic acid takes place first in NAD^+ or $NADP^+$ and later on in FAD.
3. The transfer of electrons from succinic acid takes place directly to the FAD.
4. 3ATP molecules are produced for each $NADH + H^+$ or $NADPH + H^+$ molecule.
5. Only 2ATP molecules are produced for each $FADH_2$ molecule.
6. The reduction of various cytochromes requires only electrons and no protons.
7. The formation of one molecule of water requires $\frac{1}{2}O_2 + 2e^- + 2H^+$ while reduction of one molecule of oxygen (O_2) requires $4e^- + 4H^+$.
8. The reduction and oxidation of coenzymes and cytochromes take place in a sequence and stepwise because in electron transport chain they are arranged in a series according to their redox potential. The first coenzyme (NAD^+) possesses low redox potential while last cytochrome (Cyt. a_3) highest. Thus, the transfer of electrons proceeds from compounds with low redox potential to those with high redox potential.

Oxidation of Extramitochondrial NADH or NADH Shuttle Systems

Normally NADH does not penetrate the inner mitochondrial membrane but it is continuously produced and accumulated in cytosol through glycolytic enzyme 3-*phosphoglyceraldehyde dehydrogenase*. The NADH produced in the cytosol is called *extramitochondrial NADH*. This, however, under aerobic conditions does not accumulate and is oxidized by mitochondrial respiratory chain or electron transport chain. It is facilitated through *special shuttle systems* where the electrons from cytosolic NADH are carried across the inner mitochondrial membrane by an indirect route. Following two important shuttle systems explaining such penetration of NADH into inner mitochondrial membrane are described here:

1. **Malate-oxaloacetate-aspartate shuttle (Fig. 11.11)**

   This shuttle is most common and of universal occurrence* in plants, in heart, liver and kidney of animal mitochondria. It involves three membrane carriers and four enzymes It is a readily reversible shuttle, *i.e.*, can be operated in both directions either into or out of the mitochondria and its complexity is due to impermeability of mitochondrial membrane to oxaloacetate. When 2NADH are transported through this shuttle, 6ATP molecules will be produced. A net gain of ATP per glucose molecule oxidized will be 38. This type of shuttle mechanism can be explained through following steps:

   (1) The cytosolic NADH first transfers its electrons to cytosolic oxaloacetate to form malate in presence of enzyme *cytosolic malate dehydrogenase*.

   (2) The malate, carrying electrons, easily passes through inner mitochondrial membrane into the matrix by a dicarboxylate transport system (A).

   (3) This malate in matrix transfers its electron to the matrix NAD^+ in presence of *matrix malate dehydrogenase*. This results into reduction of NAD^+ into NADH and formation of oxaloacetate.

   (4) The NADH formed in matrix now passes its electrons directly to the respiratory chain (= Electron transport chain) and 3 ATP molecules are generated when this pair of electron passes to O_2.

---

* Can consult books (1) *Outlines of Biochemistry* – by Conn & Stumpf, (2) *Introduction to Plant Biochemistry* – by Goodwin & Mercer.

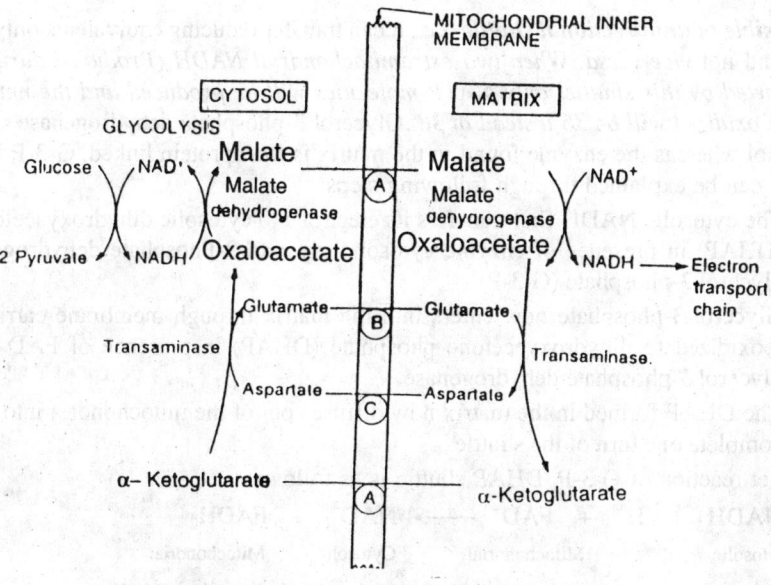

Fig. 11.11: Malate-Oxalate-Aspartate (M–O–A) shuttle.

(5) The oxaloacetate formed in matrix cannot pass back into the cytosol from the matrix through inner membrane. So it is converted into *aspartate* and α-*ketoglutarate* by the action of enzyme *transaminase*. The aspartate can pass via the amino-acid transport system (C).

(6) The transport system (B) regenerates oxaloacetate into the cytosol and helps in exchange of aspartate to glutamate.

(7) The α-ketoglutarate is carried out to cytosol through dicarboxylate transport system (A) in exchange of malate which has passed inward.

The net reaction of malate-aspartate shuttle is as follows:

$$NADH + NAD^+ \rightleftharpoons NAD^+ + NADH$$
Cytosolic   Mitochondrial   Cytosolic Mitochondrial

2. Glycerophosphate-dihydroxyacetone phosphate shuttle or G–3–P–DHAP shuttle (Fig. 11.12)

This shuttle is not very common and is found mainly in brain and in the insect muscles and other eukaryotic animal cells. It involves membrane carriers and two enzymes: (*i*) Cytosolic glycerol 3-Phosphate dehydrogenase and (*ii*) Mitochondrial (matrix) glycerol 3-phosphate dehydrogenase.

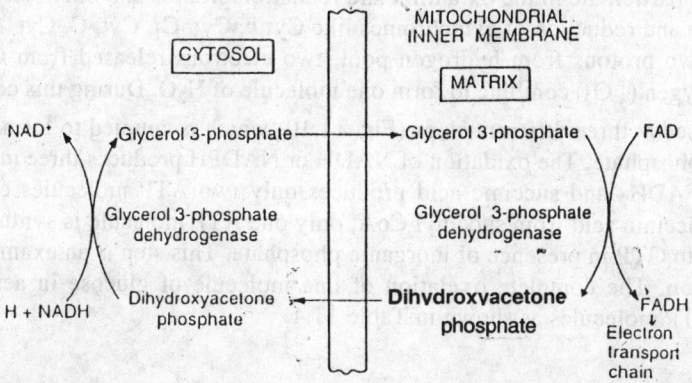

Fig. 11.12: Glycerophosphate-Dihydroxyacetone phosphate (G-3-P-DHAP) shuttle.

It *is irreversible* or **unidirectional shuttle**, *i.e.*, it can transfer reducing equivalents only from cytosol to matrix and not *vice-versa*. When two extramitochondrial NADH (Produced during glycolysis) are transported by this shuttle, only 4 ATP molecules will be produced and the net yield of ATP per glucose oxidized will be 36 instead of 38. Glycerol 3-phosphate dehydrogenase is NAD-linked in the cytosol whereas the enzyme found in the matrix is flavoprotein linked. G-3-P-DHAP shuttle mechanism can be explained through following steps:

(1) The cytosolic NADH first transfers its electrons to cytosolic dihydroxyacetone phosphate (DHAP) in presence of enzyme cytosolic glycerol 3-Phosphate dehydrogenase to form glycerol 3-phosphate (G-3-P).

(2) Glycerol-3-phosphate now enters into the matrix through membrane carrier where it is reoxidized to dihydroxyacetone phosphate (DHAP) in presence of FAD-bound matrix glycerol 3-phosphate dehydrogenase.

(3) The DHAP formed in the matrix now diffuses out of the mitochondria into the cytosol to complete one turn of the shuttle.

The net reaction of G-3-P, DHAP shuttle is as follows:

$$NADH + H^+ + FAD^+ \longrightarrow NAD^+ + FADH_2$$

Cytosolic — Mitochondrial — Cytosolic — Mitochondrial

(3) ATP molecules Produced Per Glucose Molecule Oxidized under Aerobic Conditions (Respiration)

(1) *Through Malate-Oxaloacetate-aspartate shuttle*

$$Glucose + 38\, Pi\, (H_3PO_4) + 38 ADP + 6O_2 \longrightarrow 6CO_2 + 38 ATP + 44 H_2O$$

(2) *Through glycerophosphate shuttle*

$$Glucose + 36\, Pi\, (H_3PO_4) + 36 ADP + 6O_2 \longrightarrow 6CO_2 + 36 ATP + 42\, H_2O$$

11.8. OXIDATIVE PHOSPHORYLATION

Synthesis of ATP during oxidation of coenzymes in electron transport system of aerobic respiration is called *oxidative phosphorylation*. In this process, the substrate is first oxidized by releasing a pair of hydrogen atoms which dissociate into protons ($2H^+$) and electrons $2e^-$. They are picked up by NAD^+ due to which NAD^+ is reduced to $NADH + H^+$. Reduced NAD (NADH) is oxidized to NAD^+ by transferring one proton and two electrons to FMN. Thus, FMN is reduced to $FMNH_2$ which needs one more proton from the medium. $FMNH_2$ transfers two electrons to ubiquinone (UQ) through Fe-S protein causing reduction of UQ to UQH_2. The UQH_2 transfers its electrons to cytochromes b and protons ($2H^+$) to the medium. Thus UQH_2 is oxidized to UQ and Cyt b is reduced. The reduced Cyt b transfers electrons to Cyt c_1 through Fe—S protein. The protons after reduction of UQ now do not participate in the oxidation and reduction process and only electrons participate in the oxidation and reduction of cytochromes like Cyt b, Cyt. C_1, Cyt. C, Cyt. a and Cyt. a_3. In the last step, two protons from hydrogen pool, two electrons released from Cyt. a_3 and half molecule of oxygen ($\frac{1}{2} O_2$) combine to form one molecule of H_2O. During this complete process, energy is released at three different steps (Fig. 11.10) which is coupled to form ATP from ADP and inorganic phosphate. The oxidation of NADH or NADPH produces three molecules of ATP while that of $FADH_2$ and succinic acid produces only two ATP molecules each. During the formation of succinic acid from succinyl CoA, only one ATP molecule is synthesized and GDP is converted into GTP in presence of inorganic phosphate. This step is an example of **substrate phosphorylation**. The complete oxidation of one molecule of glucose in aerobic conditions produces 38 ATP molecules as shown in Table 11.4.

Respiration

Table 11.4: Demonstration of ATP molecules produced during various steps of respiration.

Substrates to be oxidized	Compounds after oxidation	Coenzyme (acceptor)	No. of ATP produceed
2 Pyruvic acid	2 Acetyl CoA	NAD^+	$2 \times 3 = 6$
2 Isocitric acid	2 Oxalosuccinic acid	NAD^+	$2 \times 3 = 6$
2 α-Kertoglutaric acid	2 Succinyl CoA	NAD^+	$2 \times 3 = 6$
2 Succinyl CoA	2 Succinic acid	GTP	$2 \times 1 = 2$
2 Succinic acid	2 Fumaric acid	FAD	$2 \times 2 = 4$
2 Malic acid	2 Oxalosuccinic acid	NAD^+	$2 \times 3 = 6$
Total			= 30(Thirty)
Net gain of ATP in Glycolysis			= 8 (eight)
	Total = 30 + 8 = 38 ATP molecules		

11.9. SITE OF OXIDATIVE PHOSPHORYLATION

For structure of mitochondria consult chapter on "*Cell*" of this book.

Mitochondria have been considered the site of oxidative phosphorylation as they contain coenzymes of respiratory chain arranged in cristae, ATP synthetase molecules, the enzymes of citric acid cycle (Kreb's cycle) and enzymes of fatty acid oxidation. Each mitochondrion possesses two membranes—outer and inner, central matrix and an intermembrane space between outer and inner membranes. The inner membrane forms numerous cristae towards central matrix.

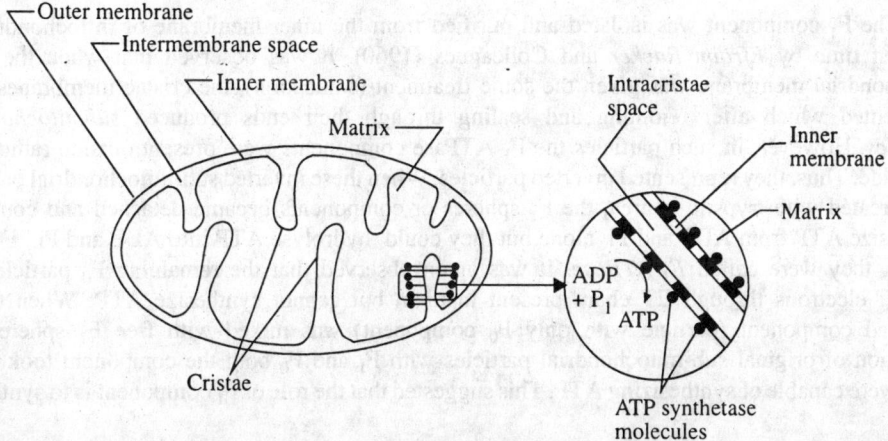

Fig. 11.13: Biochemical anatomy of a mitochondrion. The base pieces of ATP synthetase molecules are located within the inner membrane. ATP is made in the matrix as shown.

The outer membrane contains a few enzymes and is permeable to many small molecules and ions. The inner membrane is impermeable to all ions and uncharged molecules and possesses ET chain, succinate dehydrogenase and ATP synthesizing enzymes. The number of ET chains and enzymes in a single mitochondrion vary from its plain surface to cristae. The inner membrane surface contain more than 10,000 ET chains and ATP synthesizing enzymes as in liver mitochondrion while the cristae of heart mitochondrion contain more than 30,000 such ET chains and enzymes. The intermembrane space contains enzymes *adenylate kinase* and a few other enzymes. The matrix contains most of the enzymes of citric acid cycle, fatty acid oxidation and pyruvate/ dehydrogenase system. Coenzymes ATP, ADP, AMP, NAD, NADP, CoA, Pi and several ions like K^+, Mg^{2+} and Ca^{2+} etc., are also found in the matrix.

11.10. ATP SYNTHETASE (=$F_0 F_1$ AT Pase)

ATP synthetase or $F_0 F_1$ ATPase is a ATP synthesizing enzyme complex which is found in the inner membrane of mitochondria. It is made up of two components (factors) F_0 and F_1. F_1 component is a knob like structure protruding from the inner membrane and is present towards the matrix while F_0 component is a rectangular piece like structure found embedded in the inner membrane. F_1 and F_0 components remain connected with the help of a stalk.

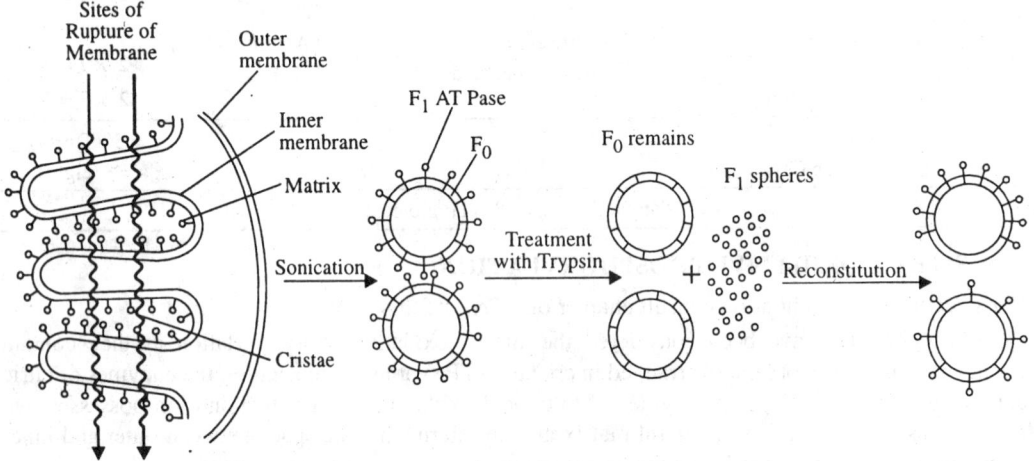

Fig. 11.14: Sonication of mitochondrial inner membrane.

The F_1 component was isolated and purified from the inner membrane of mitochondria for the first time by *Efraim Racker* and Colleagues (1960). It was observed that when the inner mitochondrial membrane was given the sonic treatment (sonication), the cristae membranes were fragmented which after rejoining and sealing through their ends produced *submitochondrial particles*. However, in such particles the F_1 ATPase components were present outside rather than the inside. Thus, they represented inverted particles. When these inverted sub-mitochondrial particles were treated with trypsin or urea, the F_1 spheres or components became detached and could not synthesize ATP from ADP and Pi alone but they could hydrolyse ATP into ADP and Pi. For this reason, they were called *F_1 ATPase*. It was again observed that the remaining F_0 particles can transfer electrons through ET chain present in them but cannot synthesize ATP. When the F_1 detached component (particle with only F_0 component) was mixed with free F_1 spheres, the formation of original sub-mitochondrial particles with F_1 and F_0 both the component took place. They were capable of synthesizing ATP. This suggested that the role of F_1 component is to synthesize ATP.

11.11. CHEMISTRY OF F_1 AND F_0 COMPONENT OF ATPase OR ATP SYNTHETASE

The knob like F_1 component of ATPase is made up of 9 polypeptide chains arranged in clusters and many ATP and ADP binding sites. Each polypeptide chain is further composed of five sub-units designated as $\alpha, \beta, \gamma, \delta, \epsilon$. The molecular weight of F_1 component is about 360 k.dal. It acts as a head.

The rectangular F_0 component is made up of only four polypeptide chains of hydrophobic nature. It acts as a base piece and normally extends across the inner membrane. It contains ET chain and represents the *proton channel of the enzyme complex*. The molecular weights of four polypeptide chains are 29, 22, 12 and 8 k.dal. respectively.

The cylindrical stalk acts as a link between F_1 and F_0 components because it joins the two components. It contains several other proteins. One of them provides sensitivity to the enzyme complex for oligomycin which is an antibiotic that block ATP synthesis by interferring with the utilization of proton gradient. The *stalk acts as a communicating portion of the enzyme complex.*

F_0 and F_1 ATPase is called an ATPase because in dissociated state, it hydrolyses ATP to ADP and Pi. Since the intact particle on inner mitochondrial membrane catalyses the synthesis of ATP from ADP and Pi, it will be more appropriate to call it ATP synthetase.

Table 11.5: Components of the mitochondrial ATP synthetase

Subunits	Mass (in kcal)	Role	Location
F_1	360	Contains catalytic site for ATP synthesis.	Spherical headpiece on matrix side
α	53		
β	50		
γ	33		
δ	17		
ε	7		
F_0	29	Contains proton channel	Transmembrane
	22		
	12		
	8		
F_1 inhibitor	10	Regulates proton flow and ATP synthesis	Stalk between F_0 and F_1
Oligomycin-sensitivity conferring protein (OSCP)	18		
$Fe_2(F_6)$	6		

11.12. STRUCTURE AND FUNCTIONS OF ELECTRON CARRIERS (COENZYMES) PRESENT IN ET CHAIN

Consult chapter on 'Coenzymes'.

11.13. MECHANISM OF OXIDATIVE PHOSPHORYLATION

Three important theories have been proposed to explain the mechanism of oxidative phosphorylation. These theories explain how the energy transfer between electron transport and ATP synthesis takes place.

(1) Chemical Coupling Hypothesis

It was first proposed by *Slater* in 1953 and is based on the principles of substrate level phosphorylation. The hypothesis postulates that a high energy intermediate is produced as electrons are passed from one carrier to the next. However, no such high energy intermediates have been shown to exist and the need for intact mitochondrial membranes for effective oxidative phosphorylation is not explained by this hypothesis.

For the explanation of hypothesis it is proposed that two hypothetical coupling factors (enzymes), called X and E, are involved at each ATP generating step (Fig. 11.15). It is further proposed that coupling factors required at three ATP generating steps are different. They are designated at X_1 and E_1, X_2 and E_2 and X_3 and E_3. Various steps of mechanisms are as follows:

1. The coupling factor X first combines with the respiratory enzyme like Cyt. *b* to form a high energy intermediate complex ($Fe^{3+} \sim X$).

$$Fe^{2+} + X \longrightarrow Fe^{3+} \sim X$$

2. The high energy intermediate complex combines with PO_4^- to form phosphorylated intermediate $(X \sim P)$ containing a high energy phosphate group. At this step the respiratory enzyme is removed.

$$Fe^{3+} \sim X + PO_4^- \longrightarrow X \sim P + Fe^{2+}$$

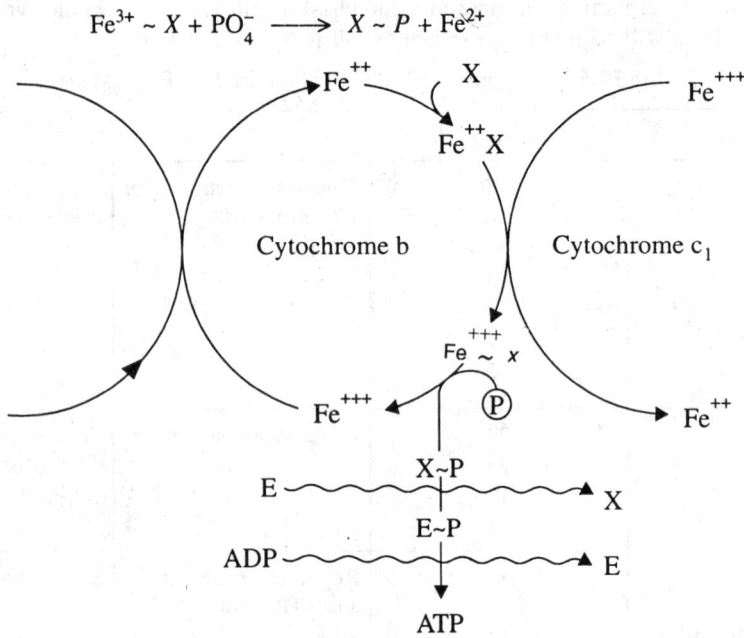

Fig. 11.15: Chemical coupling hypothesis of oxidative phosphorylation.

3. The phosphorylated intermediate $(X \sim P)$ now combines with another coupling factor E to replace first coupling factor X and to form energy rich phosphorylated complex $(E \sim P)$ which catalyses the synthesis of ATP from ADP and regeneration of the coupling factor (enzyme) E.

$$X \sim P + E \longrightarrow E \sim P + X$$
$$E \sim P + ADP \longrightarrow ATP + E$$

Thus, it is presumed that the function of coupling factors X and E is to transfer the energy released in redox reaction for ATP synthesis.

(2) Conformational coupling hypothesis

It was first proposed by *Boyer* in 1964. According to this hypothesis the energy produced during electron transfer is conserved by conformational changes in the molecules comprising the

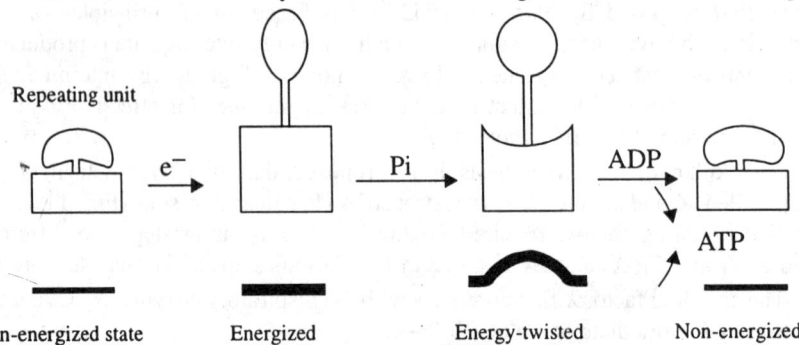

Fig. 11.16: Conformational coupling hypothesis.

Respiration

mitochondrial membrane (protein component of respiratory electron carriers) and matrix which may be the driving force for ATP formation.

Main conformational changes have been observed in ATP synthetase particles and cristae of mitorhondria. The cristae assume different forms during different functional states of the mitochondrion. Similarly the ATP synthetase particles also assume different shapes like disc, dumb-bell, or spherical. When there is lack of energy supply and the mitochondrion is in non-energized state, the cristae are in the form of straight flattened sacs and the stalk of ATP synthetase particles, also called repeating units, becomes contracted and its head piece or F_1 part becomes flattened to form a disc. When isolated mitochondria are supplied the substrates or kept in solution containing ATP, they are converted into energized state. At this stage, the cristae become more organized and assume vesicular form whereas the stems of ATP synthetase particles assume an extended form and F_1 part becomes dumb-bell shaped. When only inorganic phosphate and no ADP is supplied to mitochondria, the membranes of the cristae become convoluted assuming zig-zag shape (energy twisted) whereas the stem of ATP synthetase particles becomes elongated and F_1 part becomes spherical. When ADP is added, the energy-twisted state is changed to energized state which can be converted to non-energized state by addition of such cations or uncouplers that can be transported across the membrane

Of the above conformational changes, the twisted form stores the energy released during electron transport. When energy is needed for ATP synthesis, the energy twisted or energized state returns to stable non-energized state releasing energy which is utilized for the synthesis of ATP from ADP and inorganic phosphate.

Boyer (1965) proposed that there is a direct communication between electron transfer catalysts and ATP synthesizing components through polypeptide polypeptide interaction.

Boyer and Slater (1974) proposed a *modified conformational coupling hypothesis* which postulates that electron transfer induces conformational changes leading to translocation of protons. Conformational changes in electron transfer proteins induce changes in ATP synthesizing protein components. They believe that passage of protons through F_1 can change the conformation of its protein and such proton induced conformational changes near the active site can synthesize ATP.

ADP and inorganic phosphate can combine spontaneously to form ATP in the active site of F_1 of ATPase without requiring free energy (Figs. 11.17 & 11.18). ATP formed is tightly bound to

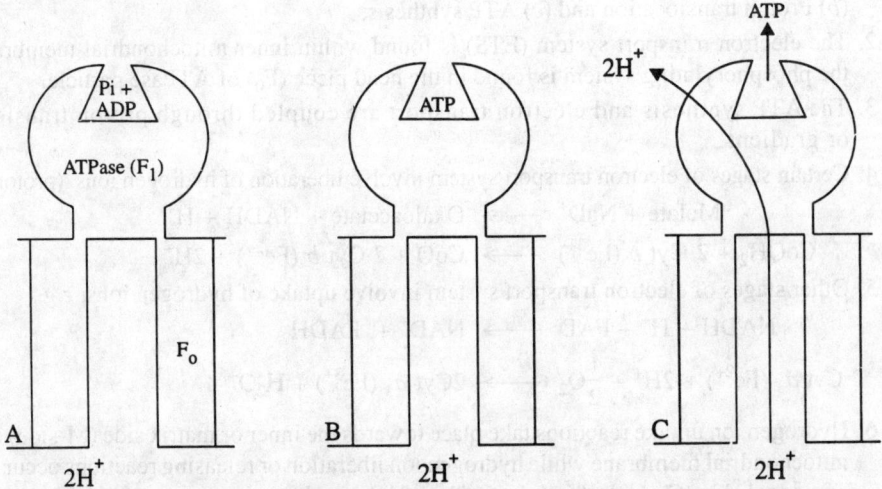

Fig. 11.17: Conformational change mechanism (modified Boyer). (A) ADP and Pi combine spontaneously in active site of ATPase, (B) ATP formed is tightly bound to ATPase, (C) Energy supplied by protons causes conformational change of ATPase, releasing ATP.

ATPase. The energy is, however, required to release tightly bound ATP molecules from ATPase. The protons when bind elsewhere other than the active site, can cause conformational changes in F_1 part of ATPase resulting into release of ATP. The protons are released into the solution on M-side of the membrane.

ADP + Pi ⟶ ATP (at the active site of F_1 of ATPase)

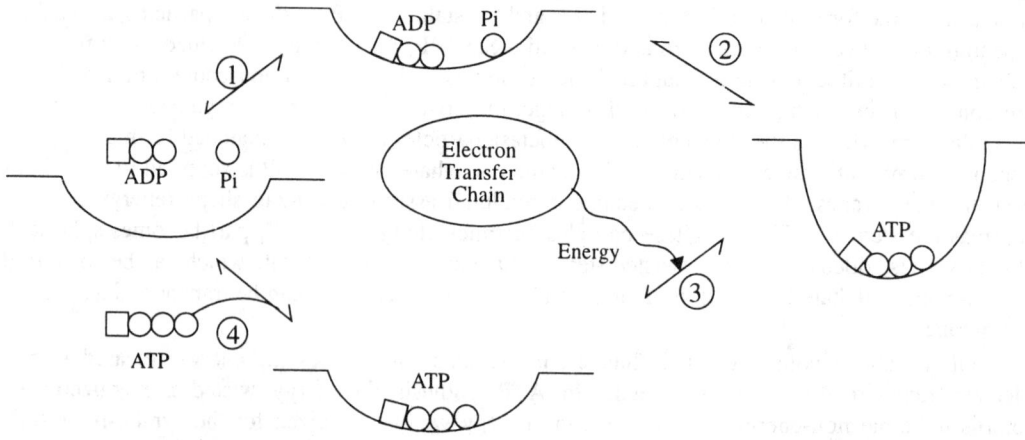

Fig. 11.18: Oxidative phosphorylation, modified Boyer model. (1) ADP and Pi attach to active site of ATPase, (2) ADP + Pi combine spontaneously to form ATP, (3) Energy-supplied proton translocation by the electron transfer chain causes conformational change of ATPase, (4) ATP released as a result of conformational change.

Proton binding to F_1 of ATPase ⟶ conformational changes in F_1 of ATPase ⟶ Release of ATP.

(3) The Chemiosmotic hypothesis

It was proposed by *Peter Mitchell*, a British biochemist, in 1961. This theory is most convincing and acceptable to date. The hypothesis can be explained through following points:

1. The inner mitochondrial membrane possesses three kinds of flow: (*a*) Electron transport, (*b*) Proton translocation and (*c*) ATP synthesis.
2. The electron transport system (ETS) is found within inner mitochondrial membrane and the phosphorylating system is found in the head piece (F_0) of ATPase particle.
3. The ATP synthesis and electron transport are coupled through proton translocation or gradient
4. Certain stages of electron transport system involve liberation of hydrogen ions (protons) e.g.,

    Malate + NaD^+ ⟶ Oxaloacetate + NADH + H^+

    $CoQH_2$ + 2 Cyt b (Fe^{3+}) ⟶ CoQ + 2 Cyt b (Fe^{2+}) + $2H^+$

5. Other stages of electron transport system involve uptake of hydrogen ions. e.g.,

    NADH + H^+ + FAD ⟶ NAD^+ + $FADH_2$

    2 Cyt a_3 (Fe^{2+}) + $2H^+$ + $\frac{1}{2}O_2$ ⟶ 2Cyt a_3 (Fe^{3+}) + H_2O

6. Hydrogen ion uptake reactions take place towards the inner or matrix side (M-side) of inner mitochondrial membrane while hydrogen ion liberation or releasing reactions occur outside or cytosol side (C-side) of inner mitochondrial membrane.
7. The electron transfer carriers in the membrane of mitochondrion are arranged in such a way that the transfer of electrons takes place from the carriers of low redox potential to the

Respiration

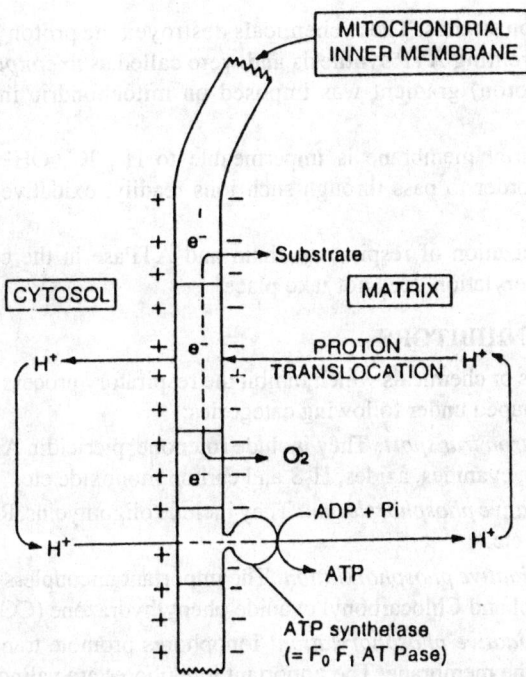

Fig. 11.19: Diagram illustrating principles of the chemiosmotic coupling hypothesis.

carriers of high redox potential and couple with transport of protons across the membrane. In other words, electron transport system operates a *proton pump* which transports protons (H^+) only from M-side to C-side of of mitochondrial membrane at high electro-chemical potential or proton motive force (pmf). This movement of protons is called *proton translocation*. Six protons are generated per electron pair transported.

8. The transport of protons to the C-side of membrane causes positive charges on the outer surface of membrane creating a *proton concentration gradient* or *membrane potential* across the inner mitochondrial membrane.

9. The proton concentration gradient forces the protons from C-side to M-side. The proton gradient across the membrane generates sufficient electrochemical energy which helps in driving the process of ATP synthesis (oxidative phosphorylation).

10. It is proposed that inner membrane of mitochondria is impermeable to hydrogen ions (H^+) and also to K^+, OH^- and Cl^- ions. Due to this reason, the protons do not flow back *i.e.*, from C-side to M-side, directly through the membrane but they flow back into the matrix through a specific region called *proton channel* or *pore* present in the F_0 portion of $F_0 - F_1$ ATPase (ATP synthetase) molecule. F_1 portion acts as an active site of ATP synthesis.

11. The back flow of protons through proton channel helps in the synthesis of ATP from ADP and Pi. One ATP molecule is produced for every two protons passing through $F_0 - F_1$ complex.

12. When hydrogen ions move to C-side, the H_2O dissociates into OH^- and H^+ ions to replace them. The OH^- ions combine with free H^+ to form water. Thus, OH^- efflux is essentially similar to H^+ influx.

(1) Evidences in support of Chemiosmotic hypothesis

There are a number of evidences in the support of chemiosmotic hypothesis for oxidative phosphorylation but the most important one came by the use of chemicals, like 2, 4-dinitrophenol, during phosphorylation studies. These chemicals destroyed the proton gradient across mitochondrial membranes preventing ATP synthesis and were called as *uncouplers*. ATP was further synthesized when pH (proton) gradient was imposed on mitochondria in absence of electron transport.

The inner mitochondrial membrane is impermeable to H^+, K^+, OH^- and Cl^- ions. If the membrane is damaged in order to pass through such ions readily, oxidative phosphorylation will not take place.

If the vectorial organization of respiratory chain and ATPase in the coupling membrane is changed, oxidative phosphorylation does not take place.

11.14. RESPIRATORY INHIBITORS

The substances, compounds or chemicals which inhibit the respiratory process are called *respiratory inhibitors*. They can be grouped under following categories:

1. *Inhibitors of electron transport:* They include rotenone, piericidin A, barbiturates, antimycins, dimercaprol, cyanides, azides, H_2S and carbon monoxide etc.
2. *Inhibitors of oxidative phosphorylation:* They include oligomycins, Rutamycin, Atractylate and Bongkrekate etc.
3. *Uncouplers of oxidative phosphorylation:* The important uncouplers are 2, 4-dinitrophenol (DNP), Dicumarol and Chlocarbonyl cyanide phenylhydrazone (CCCP).
4. *Ionophores of oxidative phosphorylation:* Ionophores promote transport of cations other than H^+ through the membrane. The important ionophores are valinomycin, Gramicidin A and Nigericin.

11.15. CYANIDE RESISTANT RESPIRATION (CRR) IN PLANTS

Studies with animal mitochondria indicate that respiration is *cyanide-sensitive* because *cytochrome oxidase* (a_3) which catalyses the electron transfer between cytochrome a_3 and O_2 is rapidly inactivated by reaction with cyanide leading to inhibited O_2 uptake and ATP generation. However, the respiration of a number of higher plants (potato, Arum, Suaromatium etc.) or plant organs, several algae, fungi and a few bacteria has been reported to be *cyanide-resistant*. A full explanation of this phenomenon has not yet been given but it is proposed that these cyanide-resistant plants and micro-organisms have an alternate insensitive *non-cytochrome terminal pathway* which allows the transfer of electrons. In this pathway the electrons are transported from CoQ to oxygen through *non-cytochrome oxidase* enzyme. However, in this transfer only heat is evolved instead of ATP formation. The ATP is formed only in complex I prior to CoQ.

```
        ADP + Pi  ATP
NADH  ─────┴────→ Flavoprotein ────→ CoQ ────→ Non-cytochrome ────→ O₂
                                              oxidase                ↘
                                                                      H₂O
```

Fig. 11.20: ETS in cyanide resistant respiratioin.

Arum-spadix shows highly specialized structure and high rate of respiration which increases inner temperature of the spadix from 10 to 15°C indicating production of high amount of heat. This heat causes evaporation of certain odoriferous compounds like ammonia, amines and indole etc. Thus, cyanide resistant respiration through non-cytochrome oxidase is an adaptation for the survival of plants at high temperature. The evaporation of odoriferous compounds helps in insect pollination.

Following points must be considered before one assumes that the existence of cyanide-resistant respiration involves a non-cytochrome terminal pathway:

Respiration

1. The inhibitor must be acting on a rate-limiting step in the reaction.
2. If intact tissues are used, it must be certain that inhibitor penetrates into the cell.
3. Plants metabolize small amounts of cyanide quickly.
4. Plants may contain a special cytochrome oxidase (Ferrocytochrome $C:O_2$ oxidoreductase) which is cyanide-resistant.

It was reported that the mitochondria of **Arum** spadix contained large amount of $Cyt.b_7$ and their cyanide resistant was due to the passage of electrons through $Cyt.\ b_7$. $10^{-3}M$ cyanide concentration hardly affected oxidation of components of TCA cycle but inhibited 80 per cent oxidation of NADH. Considerable inhibition of cytochrome oxidase occurs in both the cases but the low rate of oxidation of TCA cycle acids allow the transfer of electrons through to oxygen without significant inhibition. When electrons pass through from NADH at a much faster rate, the reduced activity of cytochrome oxidase is revealed.

11.16. ADENOSINE TRIPHOSPHATE (ATP)

ATP is an energy rich compound which acts as a link between cellular exergonic (energy releasing) and endergonic (energy requiring) reactions. It is a triphosphate ester compound of adenine ribonucleoside which is formed by the union of one molecule of purine base—adenine, one molecule of pentose sugar—ribose and three molecules of phosphoric acid. Adenine and ribose combine to form a nucleoside called *adenine ribonucleoside*. Three phosphate esters of adenine ribonucleoside are found which are called *adenosine monophosphate* (AMP), *adenosine diphosphate* (ADP) and *adenosine triphosphate* (ATP). In all these three compounds (ATP, ADP and AMP) the CH_2OH group of ribose forms an ester link with the phosphate group of phosphoric acid, H_3PO_4 (Fig. 11.21).

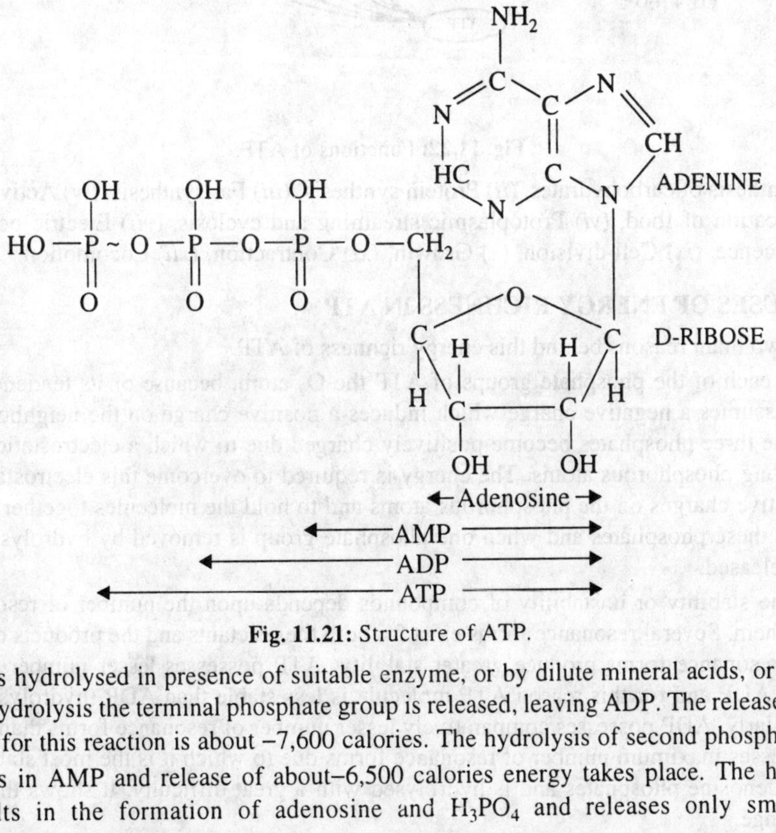

Fig. 11.21: Structure of ATP.

ATP is hydrolysed in presence of suitable enzyme, or by dilute mineral acids, or alkalies. As a result of hydrolysis the terminal phosphate group is released, leaving ADP. The release of standard free energy for this reaction is about $-7,600$ calories. The hydrolysis of second phosphate group in ADP results in AMP and release of about $-6,500$ calories energy takes place. The hydrolysis of AMP results in the formation of adenosine and H_3PO_4 and releases only small amount, about $-2,200$ calories, of energy.

$$\text{ATP} + \text{H}_2\text{O} \longrightarrow \text{ADP} + \text{H}_3\text{PO}_4 \quad (\Delta G' = -7{,}600 \text{ cal./mol.})$$
$$\text{ADP} + \text{H}_2\text{O} \longrightarrow \text{AMP} + \text{H}_3\text{PO}_4 \quad (\Delta G' = -6{,}500 \text{ cal./mol.})$$
$$\text{AMP} + \text{H}_2\text{O} \longrightarrow \text{Adenosine} + \text{H}_3\text{PO}_4 \quad (\Delta G' = -2{,}200 \text{ cal./mol.})$$

$\Delta G'$ represents free energy change. The energy released by the hydrolysis of ATP is utilized in following functions (Fig. 11.22).

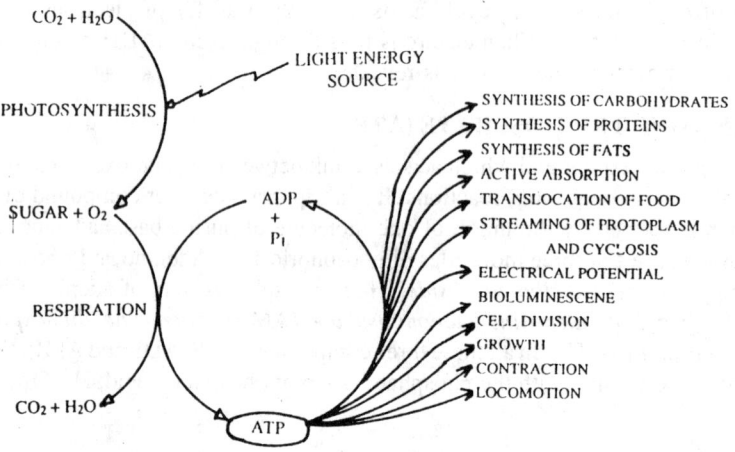

Fig. 11.22: Functions of ATP.

(*i*) Synthesis of carbohydrates, (*ii*) Protein synthesis, (*iii*) Fat-synthesis, (*iv*) Active absorption, (*v*) Translocation of food, (*vi*) Protoplasmic streaming and cyclosis, (*vii*) Electric potential, *(viii)* Bioluminescence, (*ix*) Cell-division, (*x*) Growth, (*xi*) Contraction, (*xii*) Locomotion.

11.17. CAUSES OF ENERGY RICHNESS IN ATP

There are two main reasons behind this energy richness of ATP.

1. In each of the phosphate groups of ATP the O_2 atom, because of its tendency to acquire electrons, assumes a negative charge which induces a positive charge on the neighbouring atoms. Thus, all the three phosphates become positively charged due to which a electrostatic repulsion is created among phosphorous atoms. The energy is required to overcome this electrostatic repulsion of like positive charges on the phosphorous atoms and to hold the molecules together. This energy is stored in these phosphates and when one phosphate group is removed by hydrolysis, the stored energy is released.

2. The stability or unstability of compounds depends upon the number of resonance forms present in them. Several resonance forms exist for both the reactants and the products of hydrolysis. The more resonance forms produce greater stability. ATP possesses lesser number of resonance forms than ADP and for this reason ATP molecule is less stable than ADP (hydrolysis product of ATP). Similarly, ADP possesses comparatively lesser number of resonance forms than AMP. Thus, AMP possesses maximum number of resonance forms due to which it is the most stable among all the three adenosine phosphates and is hydrolysed with a great difficulty. It shows the lowest free energy change.

Respiration

Table 11.6: Differences between Oxidative Phosphorylation and Photophosphorylation.

	Oxidative Phosphorylation	Photophosphorylation
(i)	It occurs during respiration.	It occurs during photosynthesis.
(ii)	It takes place inside mitochondria.	It takes place inside the chloroplast.
(iii)	It occurs inside the F_1 particles present on the inner membrane of cristae of mitochondria.	It occurs in the thylakoid membrane of chloroplast.
(iv)	It requires molecular oxygen for terminal oxidation.	Molecular oxygen is not required.
(v)	Phosphorylation occurs during electron transport system.	It occurs during cyclic and non-cyclic electron transport (system).
(vi)	In this process, pigment systems are not involved.	Pigment system I (PS I) and PS II, both are involved.
(vii)	In this, during electron transport energy is released as a result of oxidation and reduction which is coupled with ADP + iP to form ATP.	In photophosphorylation, the chief source of energy is sun-light.
(viii)	The ATP molecules are released into the cytoplasm which are used for various metabolic processes.	The ATP molecules synthesised in this process are utilised in dark reaction for CO_2 assimilation.

11.18. FERMENTATION

Fermentation is a type of anaerobic respiration where the substrates are incompletely oxidized in absence of oxygen. The glucose solution is incompletely oxidized into CO_2 and ethyl alcohol (C_2H_5OH) during fermentation. Sometimes instead of alcohol, organic acids like lactic acid, acetic acid, butyric acid, oxalic acid or citric acid are formed.

Table 11.7: Differences between Fermentation and Anaerobic Respiration.

	Fermentation	Anaerobic Respiration
(i)	Fermentation takes place in presence of yeast or bacterial cells.	It takes place in absence of yeast or bacterial cells.
(ii)	In fermentation, the substrate remain present outside the plant cells in the liquid medium.	In anaerobic respiration, the substrate is found inside the plant cells.

Table 11.8: Differences between Aerobic Respiration and Fermentation

	Aerobic Respiration	Fermentation
(i)	It occurs in all the living cells throughout the day and night.	In occurs outside the plant cells and in certain micro-organisms.
(ii)	It occurs in presence of O_2.	It occurs in absence of O_2.
(iii)	The end products are CO_2 and H_2O.	End products are CO_2 and alcohol or other organic acids.
(iv)	Not toxic to plants.	It is toxic to higher plants.
(v)	Food materials are completely oxidized.	Food materials are incompletely oxidized.
(vi)	In this process, large amount of energy (about 673 k. cal./glucose molecule) is released.	In this process, very small amount of energy (about 21 k. cal./glucose molecule) is released.
(vii)	The complete oxidation of one molecule of glucose produces 38 ATP molecules.	In fermentation, only two molecules of ATP are produced.
(viii)	Zymase enzyme is not required but many other enzymes and coenzymes are required.	The fermentation requires enzyme **Zymase** particularly in case of carbohydrates.
(ix)	The reaction of respiration is as follows : $$C_6H_{12}O_6 + 6O_2 \rightarrow 6CO_2 + 6H_2O + 673 \text{ k. cal.}$$	The reaction of fermentation is as follows : $$C_6H_{12}O_6 \xrightarrow[\text{(Yeast)}]{\text{Zymase}} 2C_2H_5OH + 2CO_2 + \text{Energy.}$$

Types of Fermentation

Based upon the type of end products produced during incomplete oxidation of glucose, following types of fermentation are commonly found.

(1) **Alcoholic Fermentation:** When the end product is ethyl alcohol, the fermentation is called *alcoholic fermentation*. In other words, the fermentation where glucose is converted into C_2H_5OH and CO_2, is called alcoholic fermentation. During this process, at first glucose in converted into pyruvic acid and then pyruvic acid into C_2H_5OH and CO_2. The intermediate product of alcoholic fermentation is acetaldehyde. The formation of alcohol from pyruvic acid is completed in following two steps:

(i) In first step, the pyruvic acid is converted into acetaldehyde in presence of enzyme **pyruvic decarboxylase**. One molecule of CO_2 is also liberated.

$$\underset{\text{(Pyruvic acid)}}{CH_3.CO.COOH} \xrightarrow{\text{Pyruvic decarboxylase}} \underset{\text{(Acetaldehyde)}}{CH_3.CHO} + CO_2 \quad \ldots(i)$$

(ii) Acetaldehyde is reduced to ethyl alcohol in presence of enzyme **alcohol dehydrogenase**. At this step, one molecule of NADH is oxidized in presence of H^+ into NAD.

$$\underset{\text{(Acetaldehyde)}}{CH_3CHO} + NADH + H^+ \xrightarrow{\text{Alcohol dehydrogenase}} \underset{\text{(Ethyl alcohol)}}{C_2H_5OH} + NAD^+ \quad \ldots(ii)$$

After completion of alcoholic fermentation process, two molecules of C_2H_5OH and 2 molecules of CO_2 are produced from each glucose molecule because two pyruvic acid molecules are produced from one molecule of glucose through glycolysis.

$$\underset{\text{(glucose)}}{C_6H_{12}O_6} \xrightarrow[\text{(yeast)}]{\text{Zymase Complex}} \underset{\text{(ethyl alcohol)}}{2\ C_2H_5OH} + 2CO_2 + 21 \text{ k. cal.}$$

The process of alcoholic fermentation stops when the concentration of alcohol increases more than 12-15 per cent because with the much increase in concentration of alcohol, the growth of yeast cells stops and they die.

Alcoholic fermentation may occur in any sugar solution. The fruit juices show alcoholic fermentation when yeast powder is added or the juice is left as such open in the air.

(2) *Lactic acid fermentation*: The fermentation where sugar is converted into lactic acid is called *lactic acid fermentation*. It takes place in presence of **Bacterium lactic acidi** and **Bacterium acidi lactici** which convert the milk sugar into lactic acid. It is completed in following steps :

$$\underset{\text{(Lactose)}}{C_{12}H_{22}O_{11}} + H_2O \longrightarrow \underset{\text{(glucose)}}{C_6H_{12}O_6} + \underset{\text{(Galactose)}}{C_6H_{12}O_6}$$

$$\underset{\text{(Hexose)}}{C_6H_{12}O_6} \xrightarrow{\text{Bacterium acidi lactic}} \underset{\text{(Lactic acid)}}{2C_3H_6O_3}$$

The hexose sugar, glucose, is converted into lactic acid and ethyl alcohol in presence of **Bacterium lactic acidi**. CO_2 is also liberated.

$$\underset{\text{(glucose)}}{C_6H_{12}O_6} \xrightarrow{\text{Bacterium lactici acidi}} \underset{\text{(Lactic acid)}}{CH_3.CHOH.COOH} + \underset{\text{(Ethyl alcohal)}}{C_2H_5OH} + CO_2$$

(3) *Acetic acid fermentation*: The fermentation where gulcose is converted into acetic acid is called *acetic acid fermentation*. It is quite different from other types of fermentation because in this fermentation atmospheric O_2 is used. It takes place in presence of **Acetobacter aceti**

and atmospheric O_2. It is completed in two steps. In first step, glucose is converted into ethyl alcohol and in second step, ethyl alcohol is oxidised into acetic acid in presence of oxygen.

(i) $C_6H_{12}O_6 \longrightarrow 2C_2H_5OH + 2CO_2$.

(ii) $C_2H_5OH + O_2 \xrightarrow{\text{Acetobacter aceti}} CH_3COOH + H_2O + 118.2$ k. cal.
(Ethyl alcohol)   (Acetic acid)

(4) *Butyric acid fermentation*: When the end product of fermentation is butyric acid, it is called *butyric acid fermentation*. It takes place in presence of **Clostridium butyricum** and **Bacillus butyricus** bacteria which convert hexose sugars and lactic acid into butyric acid. Such type of fermentation is normally found in rotten butter due to which it gives fowl smell.

$$C_6H_{12}O_6 \longrightarrow C_4H_8O_2 + 2H_2 + 2CO_2$$
(Hexose)   (Butyric acid)

$$2C_3H_6O_3 \longrightarrow C_4H_8O_2 + 2H_2 + 2CO_2$$
(Lactic acid)   (Butyric acid)

11.19. RELATION BETWEEN ANAEROBIC RESPIRATION AND FERMENTATION

According to most of the scientists, anaerobic respiration and alcoholic fermentation are the similar processes and are related to each other. In both the processes, at first glucose is converted into pyruvic acid through glycolysis and then pyruvic acid is converted into ethyl alcohol and CO_2 through an intermediate product acetaldehyde. The energy is also liberated in both the processes. Alcoholic fermatation and anaerobic respiration are shown by the following equations :

(i) $C_6H_{12}O_6 \xrightarrow[\text{(Yeast)}]{\text{Zymase}} 2C_2H_5OH + 2CO_2 + \text{Energy}$ (Alcoholic fermentation)

(ii) $C_6H_{12}O_6 \longrightarrow 2C_2H_5OH + 2CO_2 + 21$ k. cal. (Anaerobic respiration)

Both the above equations show following facts :

(i) Glucose is incompletely oxidized in both the processes, *i.e.*, all carbon atoms of glucose are not oxidized into CO_2.

(ii) Both processes produce the same organic product and liberate CO_2.

(iii) Both processes do not utilise the atmospheric O_2 for the oxidation of glucose.

(iv) A very small amount of energy is produced in both.

Fermentation and anaerobic respiration are supposed to be similar processes due to following reasons :

(i) In both the processes, the substrate and the end products are the same, *i.e.*, glucose is converted into C_2H_5OH and CO_2.

(ii) In both the processes, similar type of chemical reactions take place and **zymase** enzyme is used.

(iii) They require similar phosphate salts and due to this phosphate, the rate of CO_2 production is increased.

(iv) The intermediate product in both the processes is acetaldehyde.

(v) Energy is produced and ATP is synthesised in both the processes.

11.20. RESPIRATORY QUOTIENT OR R.Q.

The ratio of volumes of CO_2 liberated and O_2 used during respiration is called *respiratory quotient*. It is denoted by R.Q.

$$\text{Respiratory Quotient (R.Q.)} = \frac{\text{The volume of } CO_2 \text{ liberated}}{\text{The volume of } O_2 \text{ used.}}$$

It can be find out through respiratory quotient that which food material is being oxidized during respiration because the different food materials like corbohydrates, fats, proteins and organic acids etc. possess different R.Q. values. The value of R.Q. may be unity or one, less than one or more than one depending upon the substrate used. The R.Q. values in different respiratory substrates will be as follows :

(1) *Carbohydrates*: When the respiratory substrate is a carbohydrate or hexose sugars, the R.Q. value will be one or unity because one molecule of CO_2 is produced for each molecule of O_2. In other words, the volume of CO_2 liberated will be equal to the volume of O_2 consumed. It can be shown by the following equation.

$$C_6H_{12}O_6 + 6O_2 \longrightarrow 6CO_2 + 6H_2O + 673 \text{ k.cal. Energy}$$
(glucose)

$$\text{Respiratory Quotient (R.Q.)} = \frac{\text{Volume of } CO_2 \text{ evolved}}{\text{Volume of } O_2 \text{ used}} = \frac{6}{6} = 1 \text{ or unity}$$

In germinated seeds like wheat, oat, barley or paddy etc., the value of R.Q. is always one because in them the respiratory substrate is a carbohydrate.

(2) *Fats:* If the respiratory substrate is a fat as in case of germinated seeds of mustard, castor, linseed etc., the R.Q. of respiring cells will be less than one because the volume of CO_2 liberated is quite less in comparison to volume of O_2 consumed. The fats always require more amount of O_2 for their oxidation. The oxidation of fat can be shown by taking an example of tripalmitin fat. It takes place usually at the time of seed germination.

$$2C_{51}H_{98}O_6 + 145 O_2 \longrightarrow 102\, CO_2 + 98\, H_2O.$$
(Tripalmitin)

$$\text{R.Q.} = \frac{\text{Volume of } CO_2 \text{ evolved}}{\text{Volume of } O_2 \text{ used}} = \frac{102}{145} = 0.7 \text{ (less than one).}$$

(3) *Organic acids:* If the respiratory substrate is organic acid, the R.Q. of respiring cells will be more than one because the volume of CO_2 liberated is more than the volume of O_2 consumed. The acids already contain more O_2, so they further need only a small amount of O_2. The oxidation of malic acid can be taken as an example for this purpose.

$$C_4H_6O_5 + 3O_2 \longrightarrow 4CO_2 + 3H_2O$$
(Malic acid)

$$\text{R.Q.} = \frac{\text{Volume of } CO_2 \text{ evolved}}{\text{Volume of } O_2 \text{ used}} = \frac{4}{3} = 1.33 \text{ (More than one).}$$

(4) *Succulents:* The R.Q. value varies under different conditions in succulents like **Opuntia** and **Bryophyllum**.

(a) *Dark fixation* or *Acidification*—It occurs in dark particularly at night when the stomata remain open in succulents. The carbohydrates are incompletely oxidized to organic acids. The incomplete oxidation of glucose molecule results in the formation and storage of malic acid. The CO_2 is also evolved but in very small amount which is again taken back for dark fixation. Thus, ultimately there will be no production of CO_2.

$$2C_6H_{12}O_6 + 6O_2 \longrightarrow 2C_4H_6O_5 + 4CO_2 + 6H_2O.$$

$$\text{R.Q.} = \frac{\text{Volume of } CO_2 \text{ evolved}}{\text{Volume of } O_2 \text{ consumed}} = \frac{4}{6} = 0.67 \text{ (less than unity)}$$

Respiration

In dark fixation

$$2C_6H_{12}O_6 + 3O_2 \longrightarrow 2C_4H_6O_5 + 3H_2O.$$
(glucose) (malic acid)

$$R.Q. = \frac{\text{Volume of } CO_2 \text{ evolved}}{\text{Volume of } O_2 \text{ used}} = \frac{0}{3} = 0 \text{ (Zero)}.$$

(b) *Deacidification:* It takes place in succulents in either light (day time) or prolong darkness. Here, the organic acids act as respiratory substrate. Therefore a good amount of CO_2 is given out but due to closure of stomata at day time this gas (CO_2) does not come out. The trapped gas is fixed by the photosynthesis process in presence of light. In this condition on CO_2 will come out from the plant and R.Q. will be zero at day time.

(5) *Anaerobic respiration:* During this process, the R.Q. value will infinite (∞) because O_2 is not used while 2 molecules of CO_2 are evolved.

$$C_6H_{12}O_6 \longrightarrow 2CO_2 + 2C_2H_5OH$$

$$R.Q. = \frac{\text{Volume of } CO_2 \text{ evolved}}{\text{Volume of } O_2 \text{ used}} = \frac{2}{0} = 2 \text{ (Infinite, } \infty\text{)}$$

The R.Q. values of different substrates can be measured by Ganong's respirometer. It is made up of two big glass tubes connected with a rubber tube. The upper portion of one glass tube is bulb-like in which the material is kept and it is also graduated. The reading of change in volume is directly read out from the scale.

(6) *R.Q in red coloured organs of the plant*: The red colour in various parts of the plants like petals and leaves etc. is due to presence of anthocyanin, the synthesis of which requires O_2. Such parts of the plant also require O_2 for respiration and there is evolution of CO_2.

It means due to anthocyanin synthesis and respiration processes there is need of more amount of O_2 than the CO_2 evolved. Therefore, R.Q. value will be less than one. For example, red petals of flowers, red leaves and other red parts of the plant.

(7) *R.Q. in germinating seeds (Fig. 11.23)*: In germinating seeds, some part of embryo (radical) is exposed to atmosphere and other part remains concealed within the seed coat. Therefore, aerobic respiration runs in exposed part and anaerobic respiration runs in concealed part. In both the types of respiration CO_2 is evolved but only in aerobic respiration O_2 is used. It means germinating seeds use less O_2 than CO_2 evolved. Therefore, R.Q. will be more than one.

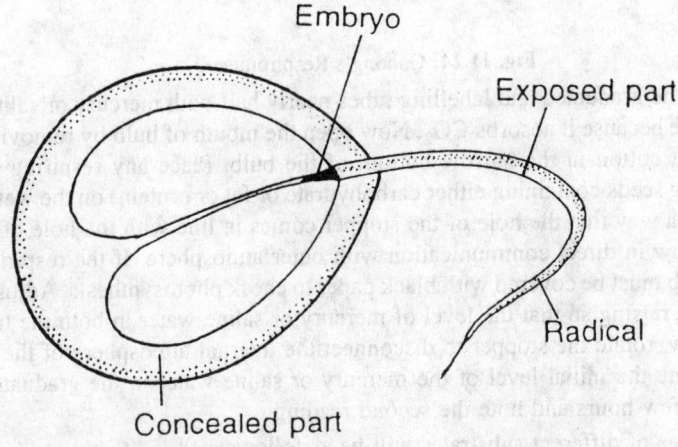

Fig. 11.23: Respiration in germinating seed.

11.21. EXPERIMENT

To determine the value of respiratory quotient (R.Q.) of different substrates by Ganong's respirometer

The apparatus: Ganong's respirometer (Fig. 11.24) consists of (*i*) a glass bulb connected with a graduated glass tube and (*ii*) a labelling glass tube. Graduated glass tube and labelling glass tube both are vertically placed and fixed on a stand. The lower narrow ends of both the tubes are connected with a rubber tube. The neck of the bulb has a small hole. A similar hole is found in the side wall of the glass stopper which fits in the opening of glass bulb. The holes of stopper and neck remain in a line with one another.

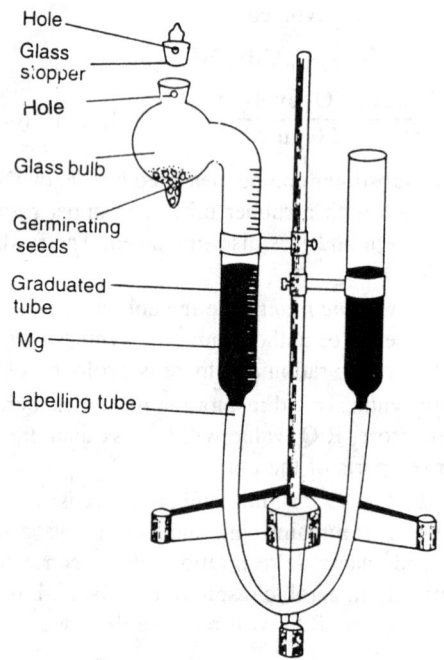

Fig. 11.24: Ganong's Respirometer.

Method: Fill the graduated and labelling tubes nearly half with mercury or saline water. Pure water should not use because it absorbs CO_2. Now open the mouth of bulb by removing the stopper and place some wet cotton in the narrow bottom of the bulb. Place any respiring plant material (usually germinating seeds containing either carbohydrate or fat or protein) on the wet cotton. Close the stopper in such a way that the hole of the stopper comes in line with the hole of the bulb. The air of the bulb is now in direct communication with outer atmosphere. If the respiring material is green tissue, the bulb must be covered with black paper to check photosynthesis. Adjust the labelling tube by lowering or raising so that the level of mercury or saline water in both the tubes comes to the same level. Now, rotate the stopper to disconnect the internal atmosphere of the bulb with the atmospheric air. Note the initial level of the mercury or saline water in the graduated tube. Keep the apparatus for a few hours and note the second reading.

The R.Q. values of different substrates will be as follows:

(1) *In case of carbohydrates:* If the respiring plant material is a carbohydrate (germinating

Respiration

seeds of wheat, barley, Oat or paddy etc.), the level of graduated tube will not change because in such a case the amount of CO_2 released during respiration will be equal to the amount of O_2 absorbed from the air of the bulb. Thus, the R.Q. is unity.

$$C_6H_{12}O_6 + 6O_2 \longrightarrow 6CO_2 + 6H_2O + 673 \text{ K cal}$$
(Glucose)

$$RQ = \frac{\text{Volume of } CO_2}{\text{Volume of } O_2} = \frac{6}{6} = 1 \text{ or unity.}$$

In case of green leaves also the value of RQ remains unity.

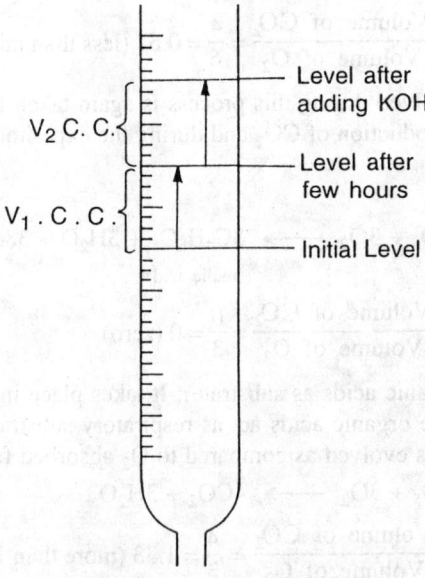

Fig. 11.25: Graduated tube with Hg.

(2) *In case of fats:* If the respiring plant material is a fat (germinating seeds of castor, linseed, til or mustard), *the level of merucury in the graduated tube will rise up* (Fig. 11.25) because in such cases the amount of CO_2 released will be less than the amount of O_2 absorbed. Suppose the second reading (after rising the mercury or saline water level) is V_1 cc. It represents the excess volume of O_2 absorbed. Now add KOH crystals into the graduated tube through the open end of levelling tube which will absorb CO_2 and the level of mercury or saline water will further rise. Suppose this reading is V_2 cc. The value of R.Q. can be calculated as follows:

$$R.Q. = \frac{\text{Volume of } CO_2}{\text{Volume of } O_2} = \frac{V_2 \text{ c.c.}}{V_1 \text{ c.c.} + V_2 \text{ c.c.}}$$

It can also be shown by the following equation:

$$2C_{51}H_{98}O_6 + 145 O_2 \longrightarrow 102 CO_2 + 98 H_2O$$
(tripalmitin fat)

$$R.Q. = \frac{\text{Volume of } CO_2}{\text{Volume of } O_2} = \frac{102}{145} = 0.7 \text{ (less than 1 or unity).}$$

(3) *In case of Proteins:* If the respiring plant material contains protein (germinating pea, gram, bean or moong seeds), the level of mercury or saline water in the graduated tube will rise up because in such cases also like fat containing seeds the amount of CO_2 released will be less than the amount of O_2 absorbed. Thus, the value of R.Q. in this case will be calculated exactly like that of fat.

(4) *In case of succulents:* If the respiratory plant material is a succulent (*Opuntia, Bryophyllum* etc.), R.Q. values vary under different conditions.

(a) Dark fixation or Acidification: It occurs in dark particularly at night and this condition may be created if the apparatus is kept in dark or the bulb is covered with black cloth or paper. During night the stomata remain open in succulents and the carbohydrates are incompletely oxidized to organic acids. The incomplete oxidation of glucose molecule results in the production and storage of malic acid. The CO_2 is also evolved but in very small amount.

$$2C_6H_{12}O_6 + 6O_2 \longrightarrow 2C_4H_6O_5 + 4CO_2 + 6H_2O$$
(glucose) (malic acid)

$$R.Q. = \frac{\text{Volume of } CO_2}{\text{Volume of } O_2} = \frac{4}{6} = 0.67 \text{ (less than unity)}$$

The amount of CO_2 evolved during this process is again taken back for dark fixation. Thus, ultimately there will be no production of CO_2 and during the experiment the level in the graduated tube will continue to rise.

In dark fixation

$$2C_6H_{12}O_6 + 3O_2 \longrightarrow 3C_4H_6O_5 + 3H_2O + 386 \text{ k.cal.}$$
(glucose) (malic acid)

$$R.Q. = \frac{\text{Volume of } CO_2}{\text{Volume of } O_2} = \frac{0}{3} = 0 \text{ (zero).}$$

(b) Deacidification (organic acids as substrate): It takes place in succulents in either light or prolonged darkness. Here, the organic acids act as respiratory substrate. Organic acids are rich in O_2, so more amount of CO_2 is evolved as compared to O_2 absorbed from the air of the bulb.

$$C_4H_6O_5 + 3O_2 \longrightarrow 4CO_2 + 3H_2O$$

$$R.Q. = \frac{\text{Volume of } CO_2}{\text{Volume of } O_2} = \frac{4}{3} = 1.33 \text{ (more than 1 or unity)}$$

If organic acid acts as a substrate, *the level of mercury or saline water will fall* (Fig. 11.26).

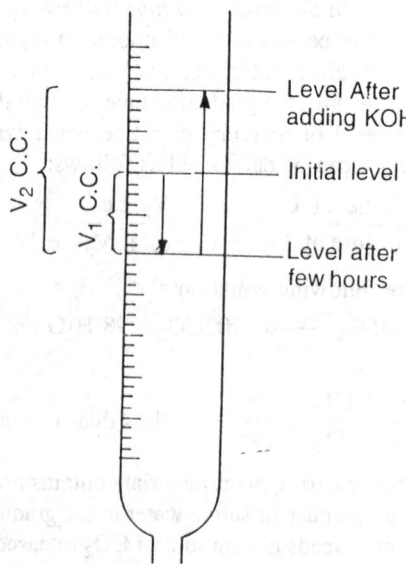

Fig. 11.26: Graduated tube with Hg.

Respiration

Suppose this reading is V_1 cc. It represents the excess volume of CO_2 given off. Add KOH crystals into the graduated tube through the open end of levelling tube. KOH will absorb CO_2 and the level of mercury or saline water will rise. Suppose this reading in V_2 cc. Then the value of R.Q. can be calculated as follows:

$$R.Q. = \frac{\text{Volume of } CO_2}{\text{Volume of } O_2} = \frac{V_2 \text{ c.c.}}{V_2 + V_1 \text{ c.c.}}$$

Note: (1) The amount of V_1 = first reading – initial reading.

(2) The amount of V_2 = second reading – first reading.

(3) The initial reading is the reading which is taken at the time of starting the experiment.

(4) The first reading is that which is taken after a few hours starting the experiment.

(5) The second reading is that which is taken after addition of KOH.

11.22. PHOTOSYNTHETIC QUOTIENT

It is the ratio of the volumes of O_2 evolved and CO_2 absorbed during photosynthesis. It is denoted by P.Q.

$$P.Q. = \frac{\text{Volume of } O_2 \text{ evolved}}{\text{Volume of } CO_2 \text{ absorbed}}$$

Table 11.9: Differences between RQ and PQ.

Respiratory Quotient (R.Q.)	Photosynthetic Quotient (P.Q.)
1. It is the ratio of the volumes of CO_2 liberated and O_2 used during respiration in a given period of time. $$R.Q. = \frac{\text{Volume of } CO_2 \text{ liberated}}{\text{Volume of } O_2 \text{ used}}$$	1. It is the ratio of the volumes of O_2 evolved and CO_2 absorbed during photosynthesis in a given period of time. $$P.Q. = \frac{\text{Volume of } O_2 \text{ evolved}}{\text{Volume of } CO_2 \text{ absorbed}}$$
2. It gives an idea that which food material is being oxidized during respiration because different food materials like carbohydrates, fats, proteins and organic acids possess different R.Q. values.	2. It gives an idea that which food material is being synthesized during photosynthesis.
3. The value of R.Q. varies with varying respiratory substrates. The value of R.Q. may be unity or one (*e.g.*, in carbohydrates), less than one (*e.g.*, proteins and fats) or more than one (*e.g.*, organic acids).	3. The value of P.Q. is nearly always unity or one. In CAM plants in light, the value of P.Q. becomes more than one. Such plants contain high acid contents and when they get highly illuminated, deacidification starts. They liberate O_2 without absorption of CO_2 from the atmosphere. In leaves containing low acid content, the value of P.Q. becomes less than one.
4. It is measured by the apparatus called *Ganong's respirometer*.	4. It is measured by the apparatus called *Ganong's photosynthometer*.

11.23. FACTORS AFFECTING THE RATE OF RESPIRATION

(A) Internal Factors

(1) *Protoplasm*: The rate of respiration depends upon the quantity of protoplasm. Usually the meristematic cells possess more protoplasm than mature permanent cells. Therefore, the rate of respiration and division capacity in meristematic cells is maximum. Thus, rate of respiration is indirectly related with the protoplasm and it increases or decreases according to the quantity of protoplasm.

In addition to amount of protoplasm, its component like amount and type of enzymes, quantity of water etc., also affect the rate of respiration. All the above components of protoplasm jointly affect the rate of respiration.

(2) *Food Materials* : The rate of respiration increases when the organic food materials are sufficient in quantity. It is seen that the rate of respiration is directly proportional to the concentration of glucose in a plant.

(B) External Factors

(1) *Temperature* : Temperature plays an important role in respiration. It takes place in plants even at below 0°C and increases with the increase in temperature. The rate of respiration is maximum between 35 and 40°C but above this the rate decreases simultaneously.

(2) *Oxygen* : Aerobic respiration does not take place in absence of O_2. The air contains about 20.8% oxygen. Normal respiration takes place in between 2-4 per cent O_2.

(3) *Light* : The stomata in leaves remain open in presence of light which help in maximum exchange of gases resulting in increase of rate of respiration. Light also affects temperature.

(4) *Water* : It is an important part of protoplasm. The protoplasm becomes inactive in absence of water. The activity of enzymes requires presence of water. Thus, the rate of respiration increases in excess of water.

(5) *Carbon-di-oxide* : The rate of respiration decreases in excess of CO_2 around the plant.

(6) *Wound* or *Injury* : Wounds or injury cause increase in the rate of respiration but it again becomes normal after about 24 hours. The reason behind this is that the healing of wound requires the formation of several new cells. This process needs a large amount of energy which is supplied through respiration. Thus, the rate of respiration is increased.

(7) *Effects of Chemicals*: Some special type of enzyme inhibiting chemical substances like cyanide, carbon-mono-oxide, iodo-acetate etc., reduce the rate of respiration. They produce the same effect whether their quantity is high or low.

11.24. EXPERIMENTS RELATED WITH RESPIRATION

Experiment 1: *To demonstrate that heat energy is released during respiration.*

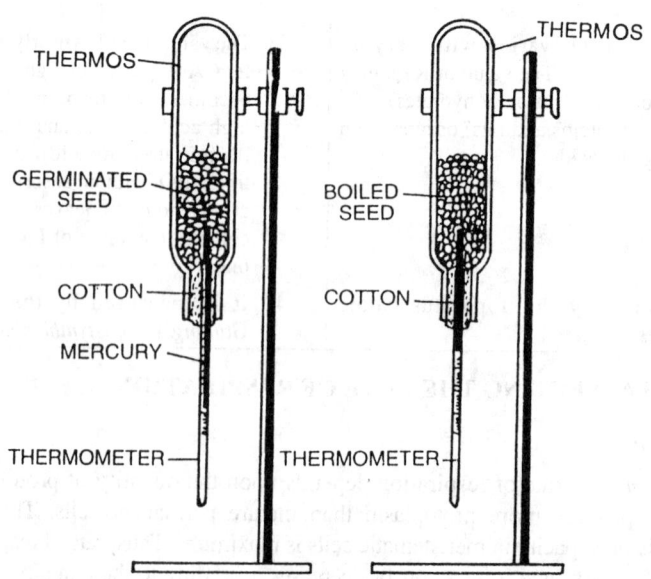

Fig. 11.27: Demonstration of liberation of heat energy during respiration.

Respiration

Keep some germinated seeds in thermos bottle and cover its mouth with moist cotton. Insert single hole cork at its mouth and a thermometer in the hole. Now, invert the bottle and clamp it with the stand according to Fig. 11.27. Similarly take another thermos bottle and keep boiled (dead) seeds and set it according to the first thermos. Notedown the temperatures of both the thermometer and leave the experiment for 2-3 days. Again note the temperatures. You will find that the temperature of first bottle's thermometer is increased while that of second thermometer remains unchanged. It is because the seeds of first bottle are living and they respire to produce heat energy while the seeds of second bottle are dead and they do not respire. Thus, the temperature of the thermometer of first thermos bottle is increased.

Experiment 2: *To demonstrate that the air taken through stomata of leaf reaches to different parts of the plant.*

Take a wide mouth bottle and fill its 3/4 part with water. Now, fit a two hole cork in it and insert a freshly cut plant twig in one hole in such a way that it should be inside the water while in second hole a tube bent at 90° angle in such a way that it should not touch the water. Make the apparatus air tight with the help of grease. Now, connect the tube with suction pump and observe the bubbles coming out through leaf-petiole. It is the same air which has entered through the stomata of leaves.

Repeat the experiment by rubbing grease on both the surfaces of leaves. After sometime you will observe that there is no evolution of air-bubbles. It is because the stomata become closed due to grease and the entry of air through stomata is checked.

Experiment 3: *To demonstrate that the plants take oxygen and evolve carbon-di-oxide during respiration.*

Set the apparatus according to Fig. 11.29. by taking KOH in U-tube, lime-water in two wide mouth bottles, one potted plant, bell-jar and black-cloth. During day time the potted plants is covered with black-cloth to check photosynthesis. Make the apparatus air-tight and start the aspirator. After sometime you will find that the lime water of second bottle turns milky. The explanation for this

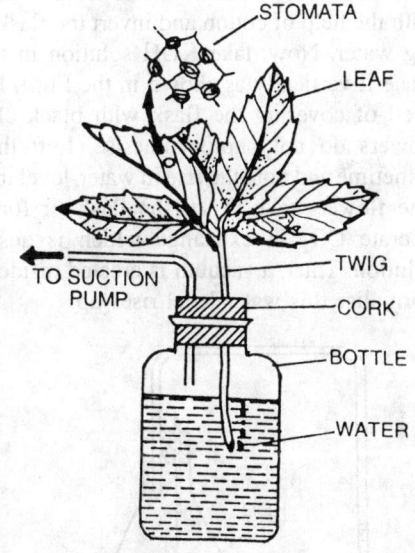

Fig. 11.28: Relationship of stomata with atmosphere.

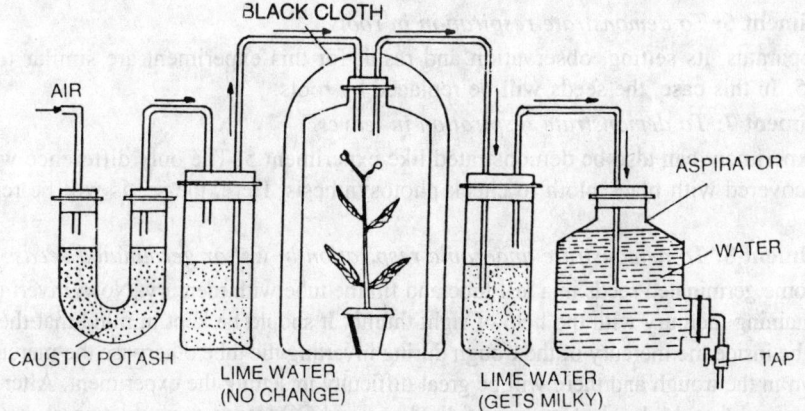

Fig. 11.29: CO_2 is evolved during respiration.

is that when the water comes out from aspirator, the atmospheric air enters into the apparatus through the second end and passes through the U-tube containing caustic potash into the tube containing lime water. The caustic potash absorbs the CO_2 of the air. Thus, CO_2 free air reaches into the lime water so it does not turn milky. It indicates the air now does not contain even trace of CO_2. When this air reaches into the lime water of second tube through bell jar having potted plant covered with black cloth to check photosynthesis, it turns milky. It proves that CO_2 is evolved during respiration.

Experiment 4: *To demonstrate that respiration also occurs in flowers.*

Place some fresh flowers in the bottom of the flask with the help of cotton and invert the flask in dish containing water. Now, take KOH solution in the test-tube and place it vertically as shown in the Fig. 11.30. There is no need of covering the flask with black cloth because the flowers do not photosynthesise. Left the apparatus for sometime and note the rise in water-level in the flask. Why? The flowers take O_2 of the flask-air for respiration and liberate CO_2 in exhcange which is absorbed by KOH solution. Thus, a vacuum is created inside the flask and to normalise this water-level rises up.

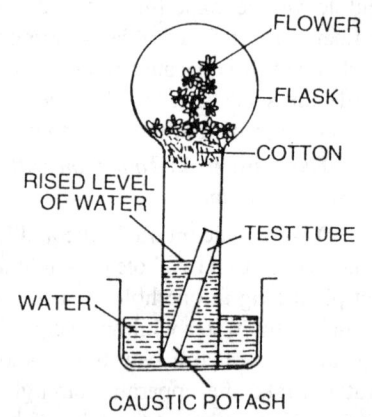

Fig. 11.30: Demonstration of respiration of flowers.

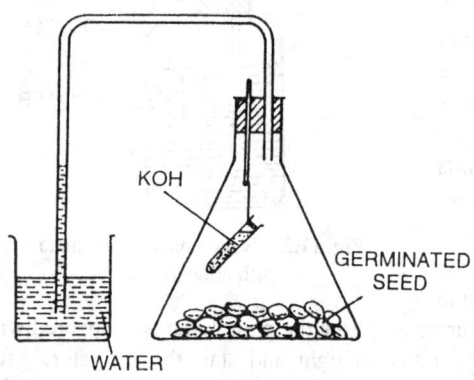

Fig. 11.31: Demonstration of respiration in seeds.

Experiment 5: *To demonstrate respiration in germinated seeds.*

Place some germinated seeds in a conical flask and tight it with the help of two-holed cork. Hang a small tube containing KOH at the base fitted in one hole and connect a glass tube bent at 90° at two places through the second hole of the cork to a beaker containing water or mercury, as shown in Fig. 11.31. After sometime the rise in mercury level will be observed. Why? The germinated seeds respire. They take the O_2 of the flask and liberate CO_2 which is absorbed by KOH. Thus, a vacuum will be created which will be replaced by water showing rise in its level.

Experiment 6: *To demonstrate respiration in roots.*

The apparatus, its setting, observation and result for this experiment are similar to those of experiment 5. In this case, the seeds will be replaced by roots.

Experiment 7: *To demonstrate respiration in leaves.*

This experiment can also be demonstrated like experiment 5. The only difference will be that the flask is covered with black cloth to check photosynthesis. Here, the seeds will be replaced by leaves.

Experiment 8: *To demonstrate anaerobic respiration in wet or germinated seeds.*

Take some germinated seeds in a test-tube and fill the tube with mercury. Now, invert the tube in a trough containing mercury with the help of right thumb. It should be kept in mind that the mouth of the test-tube be inside the mercury of the trough during inverting the tube otherwise the mercury of tube will fall down in the trough and there will be great difficulty in setting the experiment. After sometime it will be observed that the level of mercury falls down and CO_2 starts accumulating at the upper end of the test-tube. It indicates that the seeds also respire in absence of O_2.

Respiration

To test that the liberated gas is actually CO_2, introduce the pellets of KOH through the mouth of test-tube with the help of forceps. After sometime it will be observed that the mercury level in the test-tube again rises. It is because the KOH absorbs the CO_2 liberated during anaerobic respiration. It proves that anaerobic respiration takes place in wet or germinated seeds in absence of O_2 and CO_2 is liberated outside (Fig. 11.32).

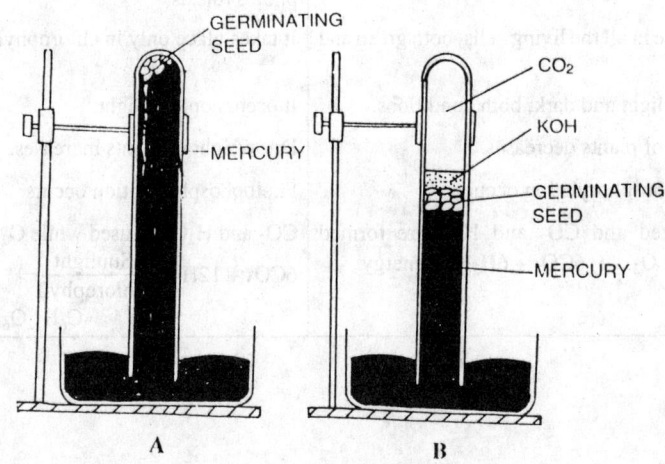

Fig. 11.32: Demonstration of anaerobic respiration in seeds.

Experiment 9: *To demonstrate fermentation.*

The process of fermentation is demonstrated by *Kuhne' tube* (Fig. 11.33). It is made up of a vertical and closed glass tube which is slight oblique or laterally turned towards base and connected with a side bulb open at the mouth.

Fill the apparatus with 10 per cent glucose solution through the mouth of the bulb and add baker's yeast into it. Half of the bulb should be empty. Leave the apparatus for sometime after closing the mouth of the bulb with the help of cork or cotton plug. It is observed that fermentation takes place resulting into production of ethyl alcohol and CO_2. The CO_2 starts accumulating at the upper end of straight tube and the level of solution falls down. It is because the yeast contains the enzyme zymase which catalyses the fermentation of glucose into C_2H_5OH and CO_2. The liberated CO_2 exerts a pressure on the solution so that the solution-level falls down in the vertical tube but the volume or quantity of solution in the bulb increases. A stage comes when the bulb becomes completely filled with the solution. Now, introduce the KOH pellets through the mouth of bulb. They will absorb the CO_2 resulting into again increase in the solution level in the vertical tube. It proves that CO_2 gas is released during fermentation and C_2H_5OH is formed which gives a smell.

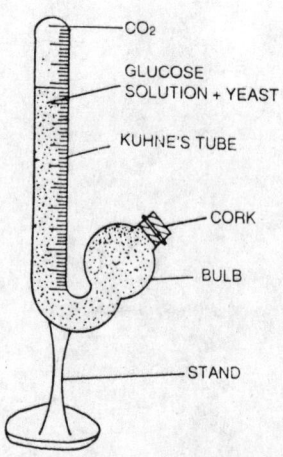

Fig. 11.33: Demonstration of fermentation by Kuhne's tube.

Table 11.10: Differences between Respiration and Photosynthesis.

	Respiration	Photosynthesis
(i)	It is a catabolic process.	It is an anabolic process.
(ii)	Carbohydrates are oxidized.	Carbohydrates are synthesised.
(iii)	Energy is liberated in the form of ATP.	Light energy is stored in the form of glucose or chemical energy.
(iv)	The amount of CO_2 in the air increases during respiration.	The amount of CO_2 in the air decreases during photosythesis.
(v)	It takes place in all the living cells, both green and non-green.	It takes place only in chlorophyllous cells.
(vi)	It occurs in light and dark, both conditions.	It occurs only in light.
(vii)	Dry weight of plants decreases.	Dry weight of plants increases.
(viii)	Oxidative phosphorylation occurs.	Photophosphorylation occurs.
(ix)	O_2 is utilized and CO_2 and H_2O are formed $C_6H_{12}O_6 + 6O_2 \rightarrow 6CO_2 + 6H_2O + energy.$	CO_2 and H_2O are used while O_2 is evolved. $6CO_2 + 12H_2O \xrightarrow[\text{Chlorophyll}]{\text{Sunlight}} C_6H_{12}O_6 + 6O_2 + 6H_2O.$

12

Photorespiration

12.1. INTRODUCTION

Earlier it was believed that the rate of respiration in light was almost equal to the respiration in darkness. Recently it has been observed that light affects respiration and the rate of respiration in light is about 3 to 5 times more than the respiration in darkness. This led to the discovery of photorespiration. It is also known as C_2 cycle.

The existence of photorespiration was first demonstrated by *Decker* in the year 1955 and 1959. He and his associates were the first to use the term photorespiration.

The photorespiration may be defined as "the respiration that occurs in green cells in presence of light resulting into excess evolution of CO_2" or it may be simply defined as "the release of CO_2 in respiration in presence of light".

The photorespiration has been reported in green cells of different plants like *Nicotiana, Phaseolus, Pisum, Petunia, Gossypium, Capsicum, Antirrhinum, Oryza, Glycine, Helianthus, Chlorella* and *Nitella* etc. It has rarely been reported in tropical grasses.

Photorespiration always requires light and its rate is maximum between 25°C and 35°C. It also depends on oxygen concentration. Photorespiration is quite different from that of normal or ground or dark respiration.

12.2 SITE OF PHOTORESPIRATION

Earlier it was supposed that the site of photorespiration is chloroplast (*Hew* and *Krotkov*, 1968). The discovery of peroxisomes, containing enzymes of glycolate metabolism, suggests a correlation between the peroxisome and photorespiration (*Tolbert et al.*, 1968). Thus, peroxisome may be the site of photorespiration. *Kisaki* and *Tolbert* (1969) showed that peroxisome is not the actual site of CO_2 evolution during photorespiration. The peroxisome simply provides a substrate for CO_2 evolution. *Tolbert* (1971) proposed that CO_2 evolution during photorespiration takes place in mitochondria. The above findings of various workers indicate a close relationship among chloroplast, peroxisome and mitochondria. At present most of the physiologists believe that chloroplast, peroxisome and mitochondria—all the three cell-organelles participate in photorespiration. Thus, they jointly form the site of photorespiration.

12.3. BIOCHEMISTRY OF PHOTORESPIRATION

In photorespiration, chloroplast, mitochondria and peroxisome are involved. They are found in close association with each other. In chloroplast when photosynthesis occurs, an early product glycolate is produced which is used as a primary substrate for photorespiration. Actually glycolic acid is produced as a result of oxidation of ribulose diphosphate when the concentration of CO_2 in the external atmosphere is less than 1 per cent. At first ribulose diphosphate is oxidized into 3-phosphoglyceric acid (PGA) and 2-phosphoglycolic acid in presence of enzyme **RuDP carboxylase** (also known as **RuDP oxygenase**) (Fig. 12.1).

$$\text{Ribulose 1,5-diphosphate} \xrightarrow[\text{RuDP carboxylase}]{+O_2} \text{3 phosphoglyceric acid}$$
$$+ \text{ 2-phosphoglycolic acid.}$$

The 2-phosphoglycolic acid loses phosphate group to form glycolic acid.

$$\text{2-phosphoglycolic acid} + H_2O \longrightarrow \text{Glycolic acid} + \text{Phosphoric acid.}$$

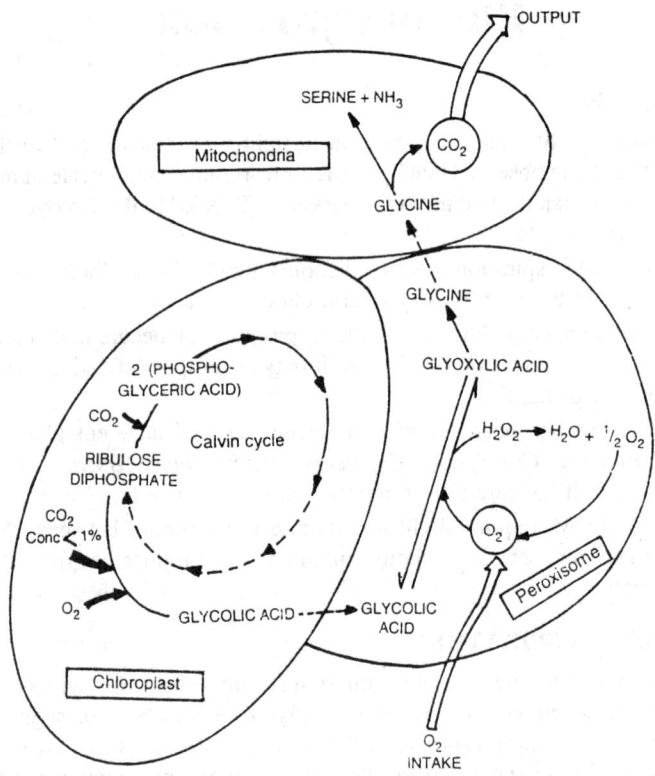

Fig. 12.1: Diagrammatic representation of various steps of photorespiration.

The glycolic acid, thus, synthesised in the chloroplast is transferred to peroxisome where it is oxidized into glyoxylic acid and hydrogen peroxide (H_2O_2) in presence of enzyme **glycolic acid oxidase**. This reaction was worked out by *Zelitch* (1966).

$$\text{Glycolic acid} + O_2 \xrightarrow[\text{light}]{\text{glycolic acid oxidase}} \text{Glyoxylic acid} + H_2O_2.$$

The hydrogen peroxide is catalysed into water and O_2 in presence of enzyme **catalase**.

$$2H_2O_2 \xrightarrow{\text{Catalase}} 2H_2O + O_2.$$

The glyoxylic acid formed as a result of oxidation of glycolic acid is converted into an amino-acid-*glycine* in presence of enzyme **glutamate-glyoxylate transaminase.**

$$\text{Glyoxylate} \xrightarrow{\text{glutamate–glyoxylate transaminase}} \text{Glycine.}$$

Now, the glycine amino-acid moves into the mitochondria where two molecules of glycine interact to form one molecule of serine amino acid and one molecule of CO_2. The CO_2 is then released in photorespiration.

$$2 \text{ Mols. Glycine} \longrightarrow 1 \text{ Mol. Serine} + 1 \text{ Mol. } CO_2.$$

The serine is then transported out from mitochondria to the cytoplasm where it is converted into hydroxypyruvate and then into phosphoglycerate which enters in Calvin-cycle and forms glucose (carbohydrate).

12.4. EVIDENCES IN SUPPORT OF PHOTORESPIRATION

Following evidences support the occurrence of photorespiration:

1. *Use of isotopes* : Isotopes have been used to find out that from where CO_2 comes during photorespiration. The leaves of tobacco were given labelled carbon through solution and the metabolism revealed that the first carbon atom of glycolate is liberated as CO_2. During photorespiration the amount of CO_2 liberated remains almost the same during the rise in temperature from 25°C to 35°C. It indicated that carboxylic group of glycolate acts as the donor of CO_2 to photorespiratory CO_2 release. Zelitch (1966) also demonstrated the release of CO_2 from C_1 of glycolate during photorespiration.

2. *Use of Photosynthetic inhibitors* : El-Sharkawy et al. (1967) demonstrated that when illuminated corn leaves are treated with photosynthetic inhibitor like DCMU [3-(3,4-dichlorophenyl)-1-dimethyl urea], they release CO_2. These leaves normally do not release CO_2 in CO_2 free atmosphere. It indicated that this inhibitor only checks, photosynthesis and not dark respiration.

3. *Use of glycolate oxidase inhibitor* : The glycolate oxidase inhibitor is α-*hydroxy-2-pyridine methane sulphonic acid*. When this inhibitor is used, it checks the formation of glyoxylate and H_2O_2. Thus, photorespiration is also inhibited.

4. *Burst of CO_2 after illumination* : When green plants are shifted from continuous light to darkness, a burst of CO_2 is observed. This burst of CO_2 occurs only in photorespiring plants while it is absent in non-photorespiring plants.

5. *Chlorophyll deficient leaves* : Green leaves perform photorespiration. The chlorophyll deficient leaves do not perform photo-respiration. The rate of CO_2 evolution in non-green leaves remains the same in dark as well as in light conditions.

12.5. EFFECT OF O_2

Usually O_2 concentration has no effect on dark respiration but photorespiration depends upon O_2 concentration and is directly proportional to it.

It indicates that respiration in dark is quite different from respiration in light.

12.6. SIGNIFICANCE OF PHOTORESPIRATION

(i) Photorespiration helps in classifying the plants into two groups : (a) Plants with photorespiration, (b) Plants without photorespiration. The plants of both the groups have different characteristics.

(ii) It helps plants to photosynthesise rapidly by increasing the concentration of CO_2 in intercellular spaces of leaf tissue.

(iii) Because CO_2 is evolved during photorespiration, it prevents the total depletion of CO_2 in the area of chloroplasts.

(iv) It also gives an idea about the evolution of plants. In earlier days, the photorespiration was most common because the concentration of CO_2 at that time was quite low. Later on, the rate of photorespiration decreased with the increase of CO_2 in the atmosphere.

(v) Photorespiration can take place even in very low concentration of oxygen.

Table 12.1: Differences between Photorespiration and Dark Respiration

	Photorespiration	Dark Respiration
(i)	It takes place only in presence of light.	It takes place in presence of light and dark both conditions.
(ii)	It occurs only in green cells.	It occurs in all the living cells.
(iii)	The respiratory substrate in photorespiration is 2-carbon glycolic acid.	The respiratory substrate in dark respiration may be carbohydrates, fats or proteins.
(iv)	Respiratory substrate is immediately formed.	Respiratory substrate may be previously stored or immediately formed.
(v)	It occurs among chloroplast, cytoplasm, peroxisome and mitochondria.	It occurs in cytoplasm and mitochondria.
(vi)	During photorespiration, hydrogen peroxide (H_2O_2) is formed.	H_2O_2 is not formed.
(vii)	ATP molecules are not synthesised.	ATP molecules are synthesised.
(viii)	NADH is oxidized to NAD^+.	NAD^+ is reduced to $NADH^+ H^+$.
(ix)	The process of transamination occurs during photorespiration.	This process does not occur during dark respiration.
(x)	It depends upon O_2 concentration and is directly proportional to it.	O_2 concentration has no affect on this process.
(xi)	Its maximum rate is in between 25°C-35°C.	Its maximum rate is in between 35°C-40°C.

13

Nitrogen Metabolism and Nitrogen Cycle

13.1. SIGNIFICANCE OF NITROGEN

N_2 is an important element for all the living organisms. It is obtained from the soil in the form of nitrate (NO_3^-), nitrite (NO_2^-) and ammonium (NH_4^+) salts. Insectivorous plants use molecular N_2. The main source of N_2 is air which contains about 78 per cent N_2 by volume. From the atmosphere (air) it reaches into the soil in the form of N_2 compounds through different methods. Its deficiency in soil is created due to continuous absorption by plants. To overcome this deficiency different blue green algae and bacteria present in the soil help in N_2 fixation.

N_2 participates in protein synthesis and in the formation of protoplasm, nucleic acids, purine and pyrimidine bases, chlorophyll, alkaloids and many coenzymes, in which the N_2 is found in the form of organic combinations. Due to deficiency of N_2 the leaves of plants become yellow, and this phenomenon is called *chlorosis*. Other effects due to deficiency of N_2 are reduction in cell-division, cell growth and plant growth, late flowering and development of anthocyanin pigment in various organs and leaf petiole.

13.2. NITROGEN IN SOIL

N_2 is found in the form of organic and inorganic compounds in the soil.

(A) *In the form of Organic Substances*

Organic compounds are mainly found in the humus. The formation of the humus takes place by the decay of different organs of dead plants and animals. It is quite rich in protein which cannot be absorbed directly by the plants. Therefore, it is essential to break the proteins into many other simpler compounds, like aminoacids and ammonium. This function is performed by some special type of bacteria in the soil which are called *ammonifying bacteria*. The process in which organic compounds decompose into ammonia is called *ammonification*. The ammonia liberated by the decomposition of organic compounds is converted into NH_4^+ ions with the help of these ammonifying bacteria. The ammonium ions are directly absorbed by some green plants. The nitrifying bacteria are found in the soil and natural water which convert ammonium ions into nitrites (HNO_2) and water. This conversion is performed by *Nitrosomonas* and *Nitrococcus* becteria in presence of oxygen.

$$2NH_3 + 3O_2 \xrightarrow{\text{Nitrosomonas/Nitrococcus}} 2HNO_2 + 2H_2O + 158 \text{ k.cal.}$$

Other type of nitrifying bacteria like **Nitrobacter** oxidise the nitrites into nitrates in presence of oxygen.

$$2HNO_2 + O_2 \xrightarrow{\text{Nitrobacter}} 2HNO_3 + 38 \text{ k.cal.}$$

Thus, the NH_3 present in the soil is converted into nitrate. This process where NH_3 is converted into nitrate, is called *nitrification*. The nitrates are soluble and absorbed by the plants through roots in the form of manure.

(B) In the form of Inorganic Substances

In inorganic substances, the nitrogen is found in the form of ammonium (NH_4^+), nitrite (NO_2^-) and nitrates (NO_3^-).

13.3. NITRATE REDUCTION IN PLANTS

The nitrates are absorbed by the roots of plants and are directly transported to the leaves through transpiration stream. The nitrates are chiefly reduced in leaves but their reduction has also been reported in roots and shoots. The nitrates are not directly utilized by the plants. They are first reduced to ammonia and water and then converted into organic form.

The nitrogen in nitrate (NO_3^-) remains present in highly oxidized state but in ammonia it remains present in reduced form. Thus, the conversion of nitrate to ammonia is a reductive process.

The reduction of nitrate to ammonia is completed in several steps mediated by specific enzymes. At each step *two electrons* are added and ultimately NO_3 in which nitrogen has *five positive charges* is converted into NH_3 in which nitrogen has *three negative charges*. In such steps, the electrons are supplied by NADH and NADPH.

$$\underset{\text{Nitrate}}{\overset{+5}{NO_3}} \xrightarrow{2e^-} \underset{\text{Nitrite}}{\overset{+3}{NO_2}} \xrightarrow{2e^-} \underset{\text{Hyponitrite}}{\overset{+1}{H_2N_2O_2}} (?) \xrightarrow{2e^-} \underset{\substack{\text{Hydroxyl}\\\text{amine}}}{\overset{-1}{NH_2OH}} (?) \xrightarrow{2e^-} \underset{\text{Ammonia}}{\overset{-3}{NH_3}}$$

The overall reaction may be written as follows :

$$HNO_3 + 8(H) \longrightarrow NH_3 + 3H_2O$$

The entire process of nitrate reduction is completed in following two steps :

(1) Reduction of Nitrate to Nitrite

The reduction of nitrate to nitrite is catalysed by the enzyme *nitrate reductase (Nitrate NADH oxidoreductase)* which requires coenzyme $NADH + H^+$ or $NADPH + H^+$ as an electron donor.

$$\underset{\substack{\text{Nitrate}}}{NO_3} + \underset{\substack{\text{reduced}\\\text{coenzyme}}}{NADH} + H^+ \xrightarrow{\text{Nitrate reductase}} \underset{\text{Nitrite}}{NO_2} + \underset{\substack{\text{oxidized}\\\text{coenzyme}}}{NAD^+} + H_2O$$

The enzyme nitrate reductase is a *molybdoflavo protein* with an operative sulfhydryl group. It contains FAD as prosthetic group and molybdenum (Mo) as an activator of enzyme. The enzyme was first isolated by *Evans* and *Nason* in 1953 from a fungus, *Neurospora* and leaves of Soyabean. The enzyme isolated from *Neurospora* could utilise only NADPH while the enzyme isolated from soyabean leaves could utilise both NADH and NADPH. The enzyme nitrate reductase is an *inducible* type of enzyme whose synthesis is inducted in many plant tissues by NO_3^-. The enzyme is found in cytoplasm.

The actual reduction of nitrate to nitrite and path of electron transfer is shown in Fig. 13.1.

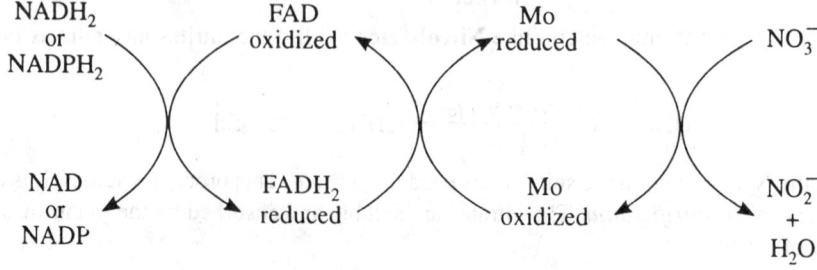

Fig. 13.1: Reduction of nitrate to nitrite.

Nitrogen Metabolism and Nitrogen Cycle

The electrons are first donated by NADH + H⁺ or NADPH + H⁺ to FAD⁺. Thus, FAD⁺ is reduced to $FADH_2$. From reduced $FADH_2$ the electrons are transferred to oxidized molybdenum causing its reduction. From reduced molybdenum the electrons are finally transferred to NO_3^-. Thus, NO_3^- is reduced to NO_2^- and H_2O and Mo is Oxidized.

The recent studies have indicated that the enzyme nitrate reductase is a complex enzyme present in higher plants and micro-organisms. This complex is an 8S unit in which NADPH–*Cytochrome b_{557} reductase* is found as a separate (3.7 – 4.5 S) subunit. The latter is bound to the nitrate reductase by molybdenum (Mo) binding protein. The electrons transfer from NADH + H⁺ or NADPH + H⁺ to nitrate takes place in two steps (Fig. 13.2).

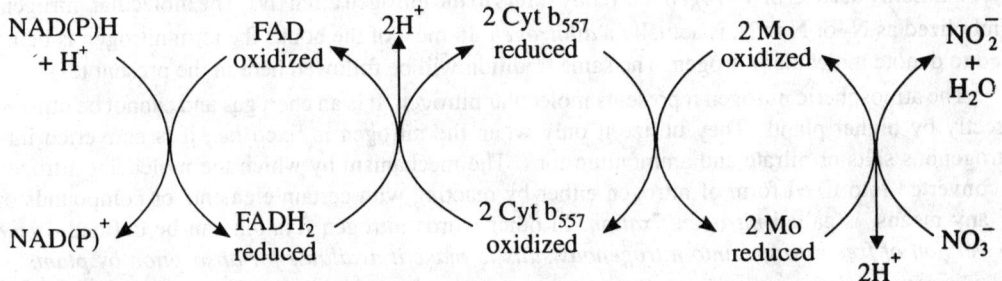

Fig. 13.2: Recent view of reduction of nitrate to nitrite.

(*i*) In first step, electrons are transferred from NADH + H⁺ or NADPH + H⁺ to Cyt b_{557} through FAD⁺. This reaction is catalysed by NADH/NADPH– *Cytochrome b_{557} reductase*.

(*ii*) In second step, electrons are transferred from Cyt.b_{557} to NO_3^- through Mo. This step is catalysed by enzyme *nitrate reductase* of *nitrate reductase complex*.

Two molecules of each, Cyt b_{557} and Mo, are required to transfer $2e^-$ through them as both are *one electron carriers i.e.*, each Cyt b_{557} and Mo can carry only one electron. Two protons are released during electron transfer from $FADH_2$ to Cyt b_{557} and two protons are needed during electron transfer from reduced Mo to NO_3^-.

The source of electrons during reduction of nitrates in blue green algae is reduced ferredoxin.

(2) Reduction of Nitrite to Ammonia

This step takes place in presence of enzyme *nitrite reductase* which catalyses the $6e^-$ reduction from NO_2^- to NH_4^+. During this reduction the intermediates, nitroxyl (NOH) or hyponitrite ($H_2N_2O_2$) and hydroxylamine (NH_2OH), are formed as follows :

$$NO_2^- \xrightarrow[-OH^-]{2e^- + 2H^+} \underset{Nitroxyl}{NOH} \xrightarrow{2e^- + 2H^+} \underset{Hydroxyl-amine}{NH_2OH} \xrightarrow[-H_2O]{2e^- + 2H^+} \underset{Ammonia}{NH_3}$$

(Nitrite)

The most probable electron donor in this reaction is *reduced ferredoxin* (an iron containing, non-heme protein of low molecular weight). The reduction of ferredoxin takes place during light reaction of photosynthesis in green leaves.

The electron donor in non-green tissues like roots is probably $FADH_2$ or NADPH + H⁺ which are generated in respiratory metabolism.

The enzyme *nitrite reductase* was first isolated by *Nason* and his colleagues from *Neurospora* and Soyabean leaves. In green tissues, this enzymes is found in chloroplasts. Previously it was believed that this enzyme contains Mn as an activator but now it is believed that *sirohaem*, an iron porphyrin, is found associated with this enzyme which mediates six electrons transfer.

The formation of intermediate compounds, hyponitrite and hydroxylamine is doubtful because:

(1) Hyponitrite is quite unstable.

(2) Hydroxylamine is toxic.

(3) These intermediates have never been observed in free state in the cells.

Now it is believed that these intermediates are formed at the surface of the enzyme. They leave the surface as they are completely reduced to further intermediate or ammonia.

The overall reaction of nitrite reduction may be written as follows :

$$NO_2^- + 6e^- + 8H^+ \longrightarrow NH_4^+ + 2H_2O$$

13.4 NITROGEN FIXATION

The commonly used term *'nitrogen'* correctly refers to the nitrogen atom (N). The molecular nitrogen, symbolized as N_2 or $N\equiv N$, is actually a *dinitrogen*. In most of the books the term nitrogen is being used to denote molecular nitrogen. The same tradition will be followed here in the present text.

The atmospheric nitrogen represents molecular nitrogen. It is an enert gas and cannot be utilized directly by higher plants. They utilize it only when the nitrogen is fixed *i.e.*, it is converted into nitrogenous salts or nitrate and ammonium ions. The mechanism by which the molecular nitrogen is converted into fixed form of nitrogen either by reacting with certain elements or compounds or by any means, is called *nitrogen fixation*. In other words nitrogen fixation can be defined *as the conversion of free nitrogen into nitrogenous salts to make it available for absorption by plants.*

In nature, nitrogen fixation takes place by both non-biological (physico-chemical) and biological means. The biological nitrogen fixation is more significant as compared to non-biological because more amount of nitrogen is fixed by biological means (140 to 700 mg/m²/year) than non-biological (about 35 mg/m²/year). Thus, nitrogen fixation in nature is of following two types:

(i) Non-biological (physical), (ii) Biological.

(1) *Non-biological (Physical)*: In non-biological nitrogen fixation the micro-organisms like algae, fungi and bacteria etc., do not participate. Non-biological nitrogen fixation is usually found in rainy season during lightning, thunder storms and atmospheric pollution. During lightning the free nitrogen of atmosphere combines with the oxygen to form nitric oxide. Nitric oxide is oxidised into nitrogen per oxide (NO_2) in presence of excess oxygen. Now, NO_2 may react with only water to form HNO_2 and HNO_3 or may react with the atmospheric oxygen and rain water to form nitric acid which reaches to the soil with rain water and reacts with the alkaline substances like calcium and ammonia present in the soil to form calcium and ammonium nitrates. Thus, free nitrogen of atmosphere is fixed. Such fixed nitrogen (nitrogenous substances in the form of calcium and ammonium nitrates) is absorbed by the plants as nitrates are soluble in water.

(i) $\quad N_2 + O_2 \xrightarrow{\text{Lightning}}_{\text{Thunder}} 2NO$

(Nitric oxide)

(ii) $\quad 2NO + O_2 \xrightarrow{\text{oxidation}} 2NO_2$

(Nitrogen per oxide)

(iii) $\quad 2NO_2 + H_2O \longrightarrow HNO_2 + HNO_3$

(iv) $\quad 4NO_2 + 2H_2O + O_2 \longrightarrow 4HNO_3$

(Nitric acid)

(v) $\quad CaO + 2HNO_3 \longrightarrow Ca(NO_3)_2 + H_2O.$

(Calcium nitrate)

(vi) $\quad HNO_3 + NH_3 \longrightarrow NH_4NO_3$

(Ammonium nitrate)

(vii) $\quad HNO_2 + NH_3 \longrightarrow NH_4NO_2$

(Ammonium nitrite)

Nitrogen Metabolism and Nitrogen Cycle

(2) *Biological*: The fixation of atmospheric nitrogen into nitrogenous salts with the help of micro-organisms like bacteria, fungi and algae is called **Biological nitrogen fixation**. Biological nitrogen fixation is mainly carried by two types of micro-organisms (1) free living or *non-symbiotic* and (2) *symbiotic*. A third category of micro-organisms which fix nitrogen in association with the roots of cereals and grasses has also been recognised. It is called *associative symbiotic nitrogen fixation.*

Non-symbiotic

The fixation of free nitrogen of the soil by all the micro-organisms living freely or outside the plant cell is called *non-symbiotic* biological N_2 fixation. It is performed by the following living organisms like aerobic and anaerobic bacteria and blue green algae. Studies with radioactive isotopes of nitrogen (^{15}N) indicated that several other organisms also fix atmospheric nitrogen. Non-symbiotic nitrogen fixation is performed by bacteria, fungi and blue green algae. Non-symbiotic free living N_2 fixers are quite primitive. They fix N_2 more actively under poor aeration, provided no H_2 gas is being produced.

(*i*) *By Bacteria* : The soil contains a special type of bacteria called **nitrogen fixing bacteria**. They convert free N_2 of the soil into soluble compounds which are absorbed from the soil by plants. *Azotobacter, Beijerenckia, Derxia, Clostridium, Aerobacter, Methanobacterium, Chlorobium, Rhodopseudomonas* colourless sulphur bacteria, *Rhodospirillum Desulfovibrio* are all nitrogen fixing bacteria. They may be classified into following four categories :

1. Free living aerobic nitrogen fixing bacteria : *e.g., Azotobacter, Beijerenckia, Derxia.*
2. Free living anaerobic nitrogen fixing bacteria : *e.g., Clostridium.*
3. Free living photosynthetic bacteria : *e.g., Chlorobium, Rhodospirellum, Rhodopseudomonas.*
4. Free living chemosynthetic bacteria : *e.g., Desulfovibrio.*

(*ii*) *By Free living fungi* : *e.g., yeasts* and *uollularia*. The yeast present in the soil is called *soil yeast*. A few species of soil yeasts are involved in N_2 fixation. *Pullularia* fungus is also involved in nitrogen fixation.

(*iii*) *By Blue Green Algae* : About 40 species of blue green algae including *Nostoc, Anabaena, Calothrix, Oscillatoria, Tolypothrix, Cylindrospermum* and *Scytonema* are found freely in the soil where they fix free N_2 into nitrogenous and ammonium compounds. Most of them bear thick-walled *heterocysts* which are the sites of nitrogen fixation.

Symbiotic

The fixation of free nitrogen of the soil by micro-organisms living symbiotically inside the plants, is called *symbiotic biological nitrogen fixation.*

The term "*symbiosis*" was coined by **DeBary** in 1879. The term "*mutualism*" has also been introduced for such type of symbiotic relationship. The symbiotic biological N_2 fixatin may be grouped under following three categories :

1. *Nitrogen fixation through nodule formation in Leguminous plants* : About 2500 species belonging to family Leguminosae including common plants like *Cicer arietinum, Pisum, Cajanus, Dolichos, Glycine soja* and *Arachis hypogea* produce root nodules containing *Rhizobium* spp. These bacteria can fix nitrogen only when they are present inside the nodules. If these bacteria are present outside the nodules in rhizosphere, they will not fix nitrogen. The association of bacteria inside the nodules is symbiotic because the host plant (especially roots) provides space to live and carbohydrates as food for bacteria and in return bacteria supply fixed nitrogen to the host plant.

When the crop is harvested, the root nodules remain burried inside the soil. The bacteria present in the root nodules convert the free nitrogen of the air into nitrates. Thus, the fertility of the soil is increased. When the plants of the next crop are grown in this soil, they absorb these soil nitrates in the form of manure through roots. Thus, the soil where leguminous plants were grown remain always fertile and it needs no manuring.

2. Nitrogen fixation through nodule formation in non-leguminous plants : In addition to Leguminous plants there are many plants specially shrubs and trees belonging to families other than Leguminosae which produce root-nodules. Some of the non-leguminous plants producing root nodules for nitrogen fixation are :

(i) *Casuarina equisetifolia* – It contains bacterium *Frankia*, a member of actinomycete.

(ii) *Alnus* – It also contains *Frankia*.

(iii) *Myrica gale* – It also contains *Frankia*.

(iv) *Parasponia* – It contains bacterium *Rhizobium*.

Sometimes nodules are also produced in the roots of certain gymnosperms *e.g.*, *Podocarpus* and in the leaves *e.g.*, *Pavetta zinumermanniana* and *Chomelia*.

3. Nitrogen fixation through non-nodulation: It includes those plants where root nodules are not formed but symbiotic nitrogen fixation takes place. Some of the examples are as follows :

(i) *Lichens* : They show symbiotic association of fungi and algae. The algae may be blue green (cyanobacteria) or green chlorophyllous.

(ii) *Anthoceros* : It is a bryophyte which contains blue green alga *Nostoc* inside mucilage cavities present on ventral side.

(iii) *Azolla* : It is a fern which contains blue green alga *Anabaena azollae*.

(iv) *Cycas* : It is a gymnosperm which contains *Anabaena* or *Nostoc* in the algal zone of coralloid roots.

(v) *Gunnera macrophylla* : It is an angiospermous plant whose stem contains *Nostoc*.

(vi) *Digitaria, Maize and Sorghum* : Their roots show symbiotic association with *Spirillum notatum*.

(vii) *Paspalum notatum* : It is an angiosperm whose roots show symbiotic association with *Azotobacter paspali*.

(viii) *Phyllosphere Association* : Ruinen (1954, 1961) reported that a variety of micro-organisms are found abundantly on the surface of leaves of many trees and shrubs growing under tropical humid conditions and fix atmospheric nitrogen. The environment provided by the wet leaf surface for growth of micro-organisms is called *phyllosphere* and this association as *phyllosphere association*.

Associative Symbiotic Nitrogen Fixation

When the bacteria live in close association with the roots of cereals and grasses and fix nitrogen, the association is of loose mutualism type and is called *associative symbiosis* whereas this nitrogen fixation is called *associative symbiotic nitrogen fixation*. In such association the bacteria live in *rhizosphere*, a transition zone between soil and root. The bacteria may remain in rhizosphere or may enter into the roots. The bacteria fix nitrogen and supply to the roots and in return the roots provide carbohydrates for the nourishment of bacteria. Some of the examples of associative symbiotism are as follows :

(i) *Azotobacter paspali*, a nitrogen fixing bacteria living in the rhizosphere of a tropical grass – *Paspalum notatum*.

(ii) *Azospirillum brasilense*, a bacteria living in the rhizosphere of cereal roots.

(iii) *Beijerinckia*, a bacteria living in the rhizosphere of sugarcane.

13.5. SYMBIOTIC NITROGEN FIXATION IN LEGUMINOUS PLANTS

In Leguminous plants the symbiotic nitrogen fixing bacteria are found in small, knob-like protuberances called *root-nodules* on the roots of these plants. The root-nodules vary in their size and shape. They may be spherical, flat, finger like or elongated in shape and from pin-head to one centimeter in diameter in size. The genus *Rhizobium*, formerly called *Bacillus radicicola*, is mainly found in the nodules. Various species under genus *Rhizobium* have been reported. Most of the species have been

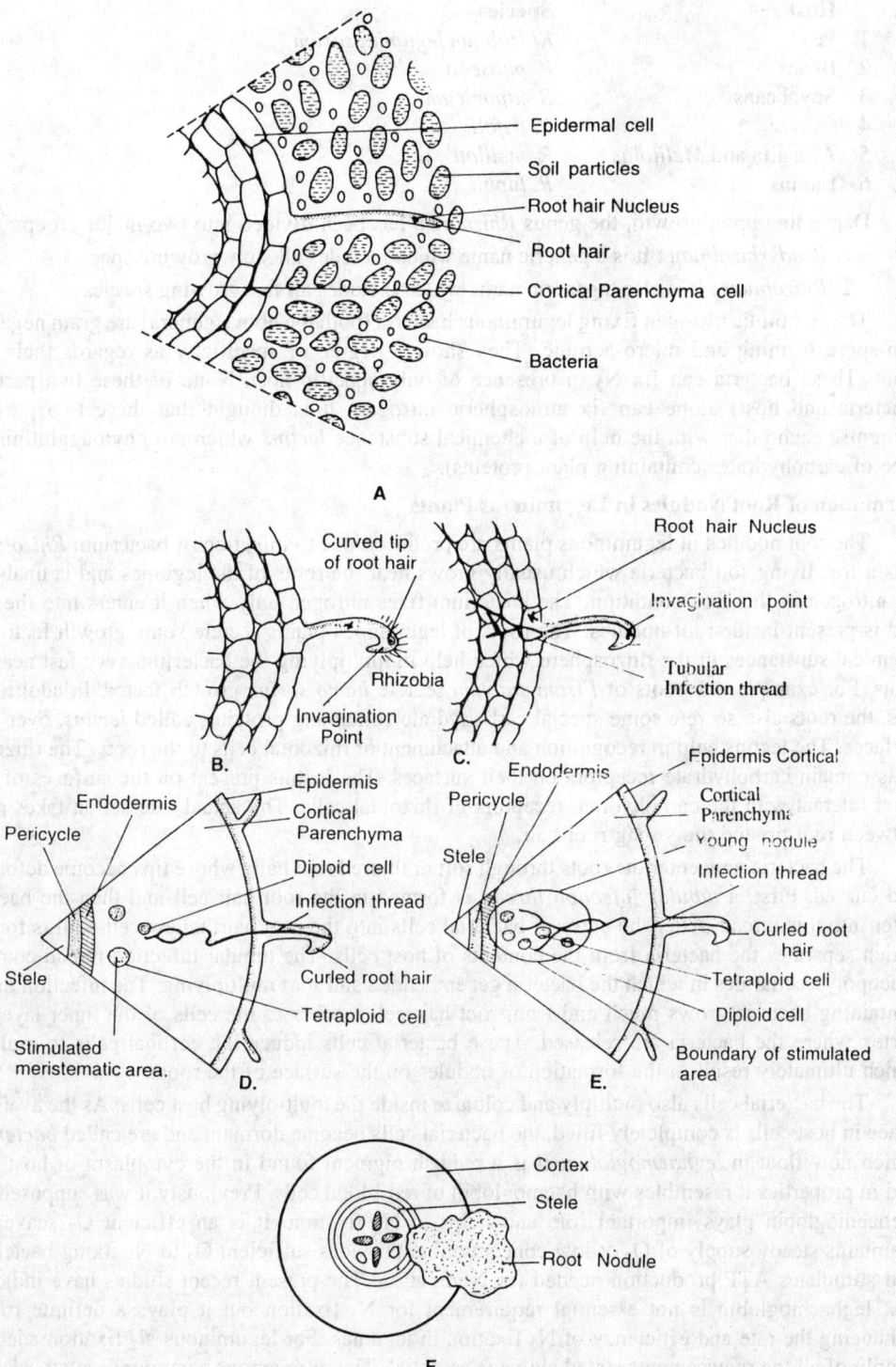

Fig. 13.3: A–F. Penetration of Rhizobia into the root hair and formation of root nodule in Leguminous plant.

named depending upon which plant they occur. Some of the species and hosts are as follows :

Host	Species
1. Pea	*Rhizobium leguminosarum*
2. Beans	*R. phaseoli*
3. Soyabeans	*R. japonicum*
4. Clover	*R. trifolii*
5. Alfa alfa and *Melilolus*	*R. meliloti*
6. Lupins	*R. lupini*

Depending upon growth, the genus *Rhizobium* has been divided into two major groups:
 1. *Bradyrhizobium* : It is a generic name which includes all slow growing species.
 2. *Rhizobium* : It is also a generic name which includes all fast growing species.

The symbiotic nitrogen fixing leguminous bacteria (both the above genera) are gram negative, non-spore forming and micro-aerobic. They show a degree of specificity as regards their host plant. These bacteria can fix N_2 in presence of only specific host. None of these two partners (bacteria and host) alone can fix atmospheric nitrogen. It is thought that these two partners recognise each other with the help of a chemical substance *lectins* which are phytoagglutinins (a type of carbohydrates containing plant proteins).

Formation of Root Nodules in Leguminous Plants

The root nodules in leguminous plants are produced due to infection of bacterium *Rhizobium*. It is a free living soil bacteria which usually grows near the roots of the legumes and is unable to fix nitrogen in this free condition. The bacterium fixes nitrogen only when it enters into the root and is present inside root-nodules. The roots of leguminous plants secrete some growth factors or chemical substances in the rhizosphere which help in multiplying the bacterium very fast near the roots. For example, the roots of *Pisum sativum* secrete *homo serine* growth factor. In addition to this, the roots also secrete some special carbohydrate containing proteins, called *lectins*, over their surfaces. The lectins help in recognition and attachment of rhizobial cells to the roots. The rhizobial cells contain carbohydrate receptors on their surfaces. The lectins present on the surfaces of root hairs interact with the carbohydrate receptors of rhizobial cells. The actual interaction takes place between root tip and the young root-hair.

The bacteria now enter the roots through soft or injured root hairs whose tips become deformed and curved. First, a *tubular infection thread* is formed in the root hair cell and then the bacteria enter into this thread. After the entry of bacterial cells into the root hair, a new cell-wall is formed which separates the bacteria from the contents of host cells. The tubular infection thread contains mucopolysaccharides in which the bacteria get embedded and start multiplying. The infection thread containing bacteria grows much and from root hair cell reaches to the cells of the inner layers of cortex where the bacteria are released. These bacterial cells induce the cortical cells to multiply which ultimately result in the formation of nodules on the surface of the roots.

The bacterial cells also multiply and colonize inside the multiplying host cells. As the available space in host cells is completely filled, the bacterial cells become dormant and are called *bacteroids* which now float in *leghaemoglobin*. It is a reddish pigment found in the cytoplasm of host cells and in properties it resembles with haemoglobin of red-blood cells. Previously it was supposed that leghaemoglobin plays important role and helps in N_2 fixation. It is an efficient O_2 scavenger, maintains steady supply of O_2 at low concentration, provides sufficient O_2 to N_2 fixing bacteroids and stimulates ATP production needed for N_2 fixation. The present recent studies have indicated that leghaemoglobin is not essential requirement for N_2 fixation but it plays a definite role in enhancing the rate and efficiency of N_2 fixation in legumes. For leguminous N_2 fixation adequate supply of all the requirements stated above is essential. The nitrogenous compounds synthesized in root nodules are translocated to different parts of the plants through vascular tissues. The groups of rhizobia remain surrounded by double membranes. The latter originate from the host cell-wall. The bacteroids lack a firm wall and are osmotically labile.

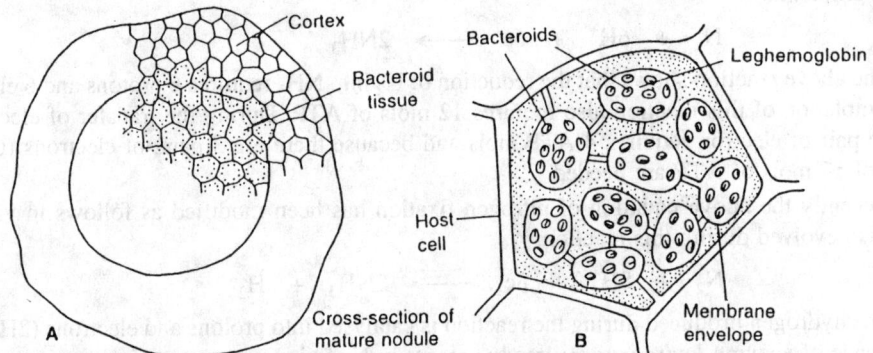

Fig. 13.4: Structure of root nodule and bacteroids. A. Cross section of root nodule.
B. Host cell filled with bacteroids.

The studies have indicated that nodule forming cortical cells possess double the number of chromosomes present in other somatic cells. Although this view has been criticized by some workers and according to them nodules may be formed by diploid and tetraploid cells both.

Biotechnological studies with free living bacterium *klebsiella pneumoniae* and certain free living species of *Rhizobium* have demonstrated that nitrogen fixation is controlled by a set of nitrogen fixing genes called *Nif genes*. In free living bacteria, the nif genes are repressed but in *Rhizobia* associated with root nodules these genes become functional even in presence of NH_4^+.

13.6 BIOCHEMISTRY OF NITROGEN FIXATION

Although the biochemistry of nitrogen fixation is not fully known but the use of isotopic nitrogen *i.e.*, ^{15}N labelled dinitrogen by *Schneider et al.* (1960) and *Carnahan et al.* (1960) in nitrogen fixing cell free preparations have confirmed the conversion of nitrogen into ammonia. The intermediates produced during this conversion could not be detected so far.

Basic Requirements for Nitrogen Fixation

Following important things are required for nitrogen fixation :
1. Presence of *nitrogenase* and *hydrogenase* enzymes in the nitrogen fixing cells or organisms.
2. A mechanism which could protect enzyme nitrogenase against oxygen.
3. *Ferredoxin*, a non-heme iron protein acting as electron carriers.
4. The hydrogen releasing system or substance or electron donor which is usually pyruvic acid but may also be sucrose of glucose etc.
5. A constant supply of ATP which is produced through the conversion of acetyl phosphate into acetate.
6. The coenzymes and cofactors like TPP, CoA, inorganic phosphate (iP) and Mg^{+2}.
7. Cobalt (Co) and molybdenum (Mo).
8. A carbon compound for trapping released ammonia.

Nitrogenase Enzyme : This enzyme plays a key role in nitrogen fixation which is usually found in N_2 fixing organisms (non-symbiotic) and in bacteroids of root nodules of leguminous plants (symbiotic N_2 fixation). The enzyme remains active only during anaerobic condition. The enzyme nitrogenase actually is made up of two protein subunits and both are required for the activity of enzyme. Out of the two subunits, one is *non-heme iron protein*. It is commonly called *Fe-Protein* or *dinitrogen reductase*. The second subunit is an *iron molybdenum protein*. It is commonly called *Mo Fe-Protein* or *dinitrogenase*.

The first subunit (Fe-Protein) reacts with ATP (which is produced during the conversion of acetyl phosphate into acetate) and reduces second subunit (Mo Fe Protein) which ultimately reduces N_2 into ammonia.

$$N_2 + 6H^+ + 6e^- \longrightarrow 2NH_3$$

The above reaction shows that the reduction of N_2 into NH_3 requires 6 protons and 6 electrons. The completion of this reaction also requires 12 mols of ATP. During the transfer of electrons to N_2, one pair of electron requires 4 ATP mols and because there are 3 pairs of electrons (6 e^-), so a total of 12 mols of ATP are needed.

Recently the above equation of nitrogen fixation has been modified as follows in view that H_2 is also evolved during this reaction.

$$N_2 + 8H^+ + 8e^- \longrightarrow 2NH_3 + H_2$$

The hydrogen produced during the reaction is catalysed into protons and electrons ($2H^+ + 2e^-$) in presence of enzyme *hydrogenase* already present in N_2 fixing micro-organisms.

$$H_2 \xrightleftharpoons{\text{Hydrogenase}} 2H^+ + 2e^-$$

Steps in Pathway of N_2 Fixation in Asymbiotic Organisms (Fig. 13.5)

1. First phosphoroclastic breakdown of pyruvic acid (pyruvate), an end product of glycolysis, takes place resulting into production of acetyl phosphate, CO_2 and H_2. This reaction is completed in two steps. First step requires enzyme *pyruvic acid dehydrogenase* and second step requires enzyme *phosphotransacetylase*. Hydrogen is broken into protons and electrons in presence of enzyme *hydrogenase*.

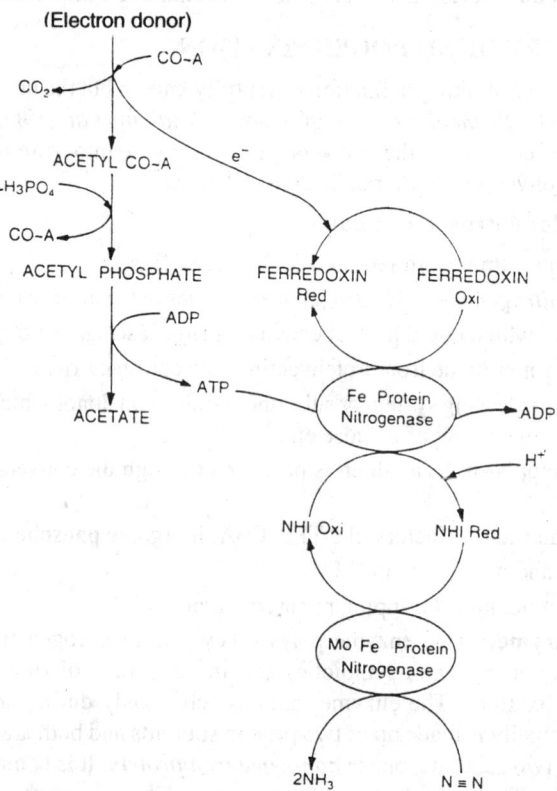

Fig. 13.5: Pathway of N_2 fixation in asymbiotic organisms.

Nitrogen Metabolism and Nitrogen Cycle

(i) Pyruvic acid + Coenzyme-A $\longrightarrow$ Acetyl Co-A + CO_2 + H_2

$$H_2 \underset{}{\overset{Hydrogenase}{\rightleftharpoons}} 2H^+ + 2e^-$$

(ii) Acetyl Co-A + H_2PO_4 $\longrightarrow$ Acetyl phosphate + Co-A

2. Acetyl phosphate reacts with ADP and is coverted into *acetate*. One molecule of ATP is produced which acts as source of energy in nitrogen fixation.

Acetyl phosphate + ADP $\longrightarrow$ Acetate + ATP

The ATP produced during oxidative or photophosphorylations as in certain photosynthetic bacteria like **Chromatium** and in blue green algae, can also be used as source of energy for nitrogen fixation.

3. The electrons released during breakdown of pyruvic acid are accepted by electron carrier, *ferredoxin*. Thus, ferredoxin is reduced.

4. The reduced ferredoxin donates electrons to Fe-Protein component (dinitrogenase reductase) of enzyme nitrogenase. Thus, first subunit (Fe-Protein) of enzyme nitrogenase is reduced.

5. The reduced Fe-Protein combines with ATP in presence of Mg^{+2} to become activated and to reduce second subunit (Mo Fe-Protein) of enzyme nitrogenase.

6. The reduced Mo Fe-Protein subunit donates electrons to N_2 stepwise to reduce N_2 into NH_3. The enzyme is set free only when N_2 has been completely reduced to NH_3.

7. The NH_3 is mostly transformed into glutamic acid and translocated through vascular tissues to the various parts of the plant.

A scheme of sequential reduction of N_2 of NH_3 by enzyme nitrogenase as proposed by *Chatt et al.* (1978) and *Throneley et al.* (1980) is given below :

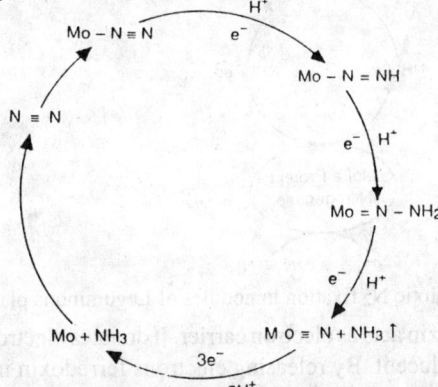

Fig. 13.6: A scheme of sequential reduction of N_2 to NH_3 by enzyme nitrogenase (after Chatt *et al.* 1978 and Thorneley *et al.* 1980).

Steps in Pathway of Symbiotic N_2 Fixation in Root Nodules of Leguminous Plants

The steps in pathway of symbiotic N_2 fixation in the root nodules of Leguminous plants are almost similar to those of asymbiotic nitrogen fixation described earlier except a few following initial steps :

1. Here *glucose-6-phosphate* acts as an electron donor instead of pyruvic acid in N_2 fixation. The glucose-6-phosphate is produced as follows :

Sucrose $\xrightarrow{\text{Translocated to roots}}$ Sucrose $\xrightarrow{\text{Invertase}}$ glucose and fructose $\longrightarrow$
(synthesized (in roots)
in leaves)

$\xrightarrow{\text{glycolytic enzymes in bacteroids}}$ glucose-6-phosphate.

2. Glucose-6-Phosphate is converted into *6-phosphogluconic acid* in presence of Co-enzyme NADP⁺ and enzyme glucose-6-phosphate dehydrogenase. The electrons released by glucose-6-phosphate are accepted by NADP⁺ and it is reduced to NADPH + H⁺.

Glucose 6-phosphate + NADP⁺ + H_2O ⟶ 6-Phosphogluconic acid + NADPH + H⁺

3. NADPH donates electrons to ferredoxin in presence of enzyme NADP-Fd-oxidoreductase. The protons (H⁺) are released in the medium and ferredoxin is reduced.

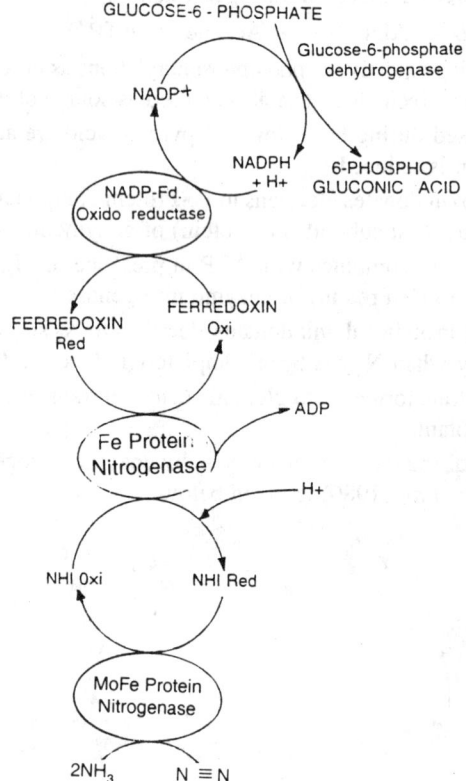

Fig. 13.7: Pathway of symbiotic N_2 fixation in nodules of Leguminous plants. NHI=Non heme iron.

4. The reduced ferredoxin acts as electron carrier. It donates electrons to Fe-Protein component of nitrogenase to reduce it. By releasing electrons ferredoxin in oxidized.

Further steps of nitrogen fixation (N_2 → NH_3) through Mo-Fe-Protein component are exactly similar to those of asymbiotic nitrogen fixation described above.

13.7. SOURCE OF NITROGEN IN INSECTIVOROUS PLANTS

Some plants get nitrogenous compounds by killing small insects. Such plants are called *insectivorous* plants. These plants possess insect catching organs through which they kill and digest the insects and ultimately absorb the nitrogenous compounds present in them. Nitrogen is an important element for the growth of the plants. The insectivorous plants are usually found in nitrogen deficient soil and this deficiency of N_2 is overcome by killing the insects and getting nitrogenous compounds from their body. The insectivorous plants are green and autotrophic. *Drosera, Utricularia, Nepenthes* and *Dionia* etc. are the examples of insectivorous plants.

13.8. DENITRIFICATION

From the above facts it is clear that the plants take the free nitrogen of the atmosphere or soil and synthesise protein and other nitrogenous organic compounds. The animal get N_2 by eating the plants. If this process continues, a day will come when all the N_2 of atmosphere will be exhausted. But in nature it never happens because denitrifying bacteria are found in the soil which convert nitrates of the soil into nitrites and nitrites into ammonia. Ultimately from ammonia free N_2 is released in the atmosphere continuously. This formation of free nitrogen from the nitrates of the soil is called *denitrification*. It is performed by some special denitrifying bacteria like *Bacterium denitrificans, Pseudomonas denitrificans, Bacillus subtilis* and *Micrococcus denitrificans*.

13.9. NITROGEN CYCLE (13.8)

The nitrogen is found freely in the atmosphere which is fixed by above methods resulting into formation of soluble nitrates. The nitrates are absorbed from the soil by roots of the plants which help in the formation of plant amino-acids, proteins, nucleic acids and other organic compounds where the N_2 remain present in bound form. In animals, the proteins decompose into amino-acids and help in the formation of protoplasm. The nitrogenous excretory products like ammonia, urea and uric acid are formed in the animal body by the dissociation of amino-acids which are discharged outside in the soil along with the animal urine. Insectivorous plants get N_2 by eating insects. The excretory waste products of animals and dead remains of the plants and animals body are decomposed by certain decomposers like bacteria and fungi and release free N_2 in the form of ammonia in the atmosphere. This process is completed by denitrifying bacteria. Thus, the nitrogen cycle continues in the atmosphere among green plants, animals and bacteria which are producers, consumers and decomposers respectively (Fig. 13.8).

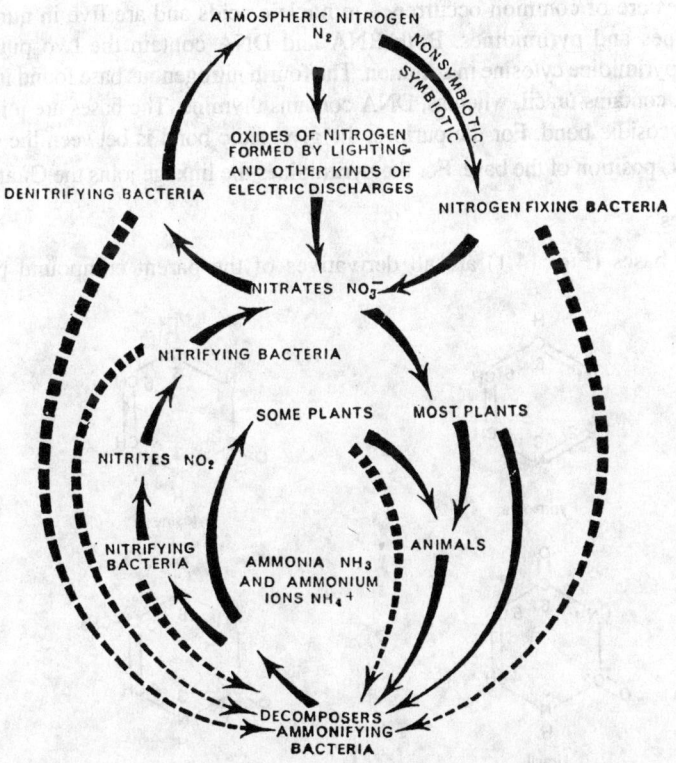

Fig. 13.8: Nitrogen cycle.

14

Nucleic Acids Metabolism

14.1. INTRODUCTION

The nucleic acids are of considerable importance in biological systems. They are of two types : (1) Ribose nucleic acid, and (2) Deoxyribose nucleic acid. The basic chemical subunits of the nucleic acids are nucleotides. The nucleotides are made up of three components : (*i*) A heterocyclic ring containing nitrogen, known as a nitrogenous base, (*ii*) A five carbon pentose sugar, and (*iii*) A phosphate group. The bases found in nucleic acids are of two kinds—purines and pyrimidines. Adenine and guanine are purine and cytosine, uracil and thymine are pyrimidine bases.

Ribose nucleic acid (RNA) is also of common occurrence in plants as well as animals. It is of three types: (*i*) ribosomal RNA (r-RNA); (*ii*) soluble RNA or transfer RNA (t-RNA) and (*iii*) messenger RNA (m-RNA). Ribosomal-RNA is found in small sub-cellular particles, the ribosomes. RNAs with sendimentation Coefficient value, 5S, 16S and 23S have been reported from 70S ribosomes, while 18S, 28S, 5.8S and 5S r-RNAs have been reported from 80S ribosomes. t-RNA is found in free form in the cytoplasm. m-RNA is found in small quantities in association with ribosomes.

14.2. NITROGENOUS BASES

Nitrogenous bases are of common occurrence in nucleic acids and are five in number. They are of two kinds—purines and pyrimidines. Both RNA and DNA contain the two purines adenine and guanine and one pyrimidine cytosine in common. The fourth nitrogenous base found in two nucleic acids is different. RNA contains uracil, whereas, DNA contains thymine. The bases are joined to the pentose sugar by N-C glycosidic bond. For the purines, the glycosidic bond is between the C_1 position of the pentose and the N_9 position of the base. For the pyrimidines the linkage joins the C_1 and N_3 positions.

Pyrimidine bases

Pyrimidine bases (Fig. 14.1) are all derivatives of the parent compound pyrimidine which

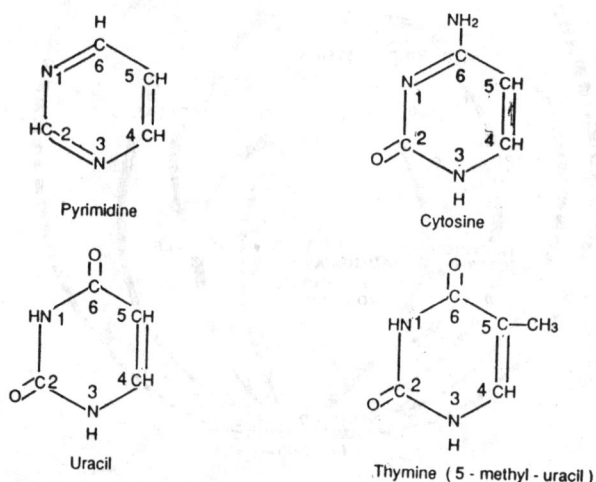

Fig. 14.1: Pyrimidines.

Nucleic Acids Metabolism

shows a six-membered ring. The derivatives are found in the nucleic acids. Cytosine is found in both types of nucleic acids, uracil in RNA, and thymine in DNA.

The thymine differs from uracil only in having a methyl substituent in the C_5 position.

Purine Bases

Adenine and guanine comprise the purine bases and occur in both DNA and RNA. In RNA a large number of bases are methylated and occur as methyladenine, methylguanine or methylcytosine.

A purine ring is formed by the fusion of a 6-membered pyrimidine ring and a 5-membered imidazole ring. In a purine the positions are designated by the numbers 1 through 9 (Fig. 14.2).

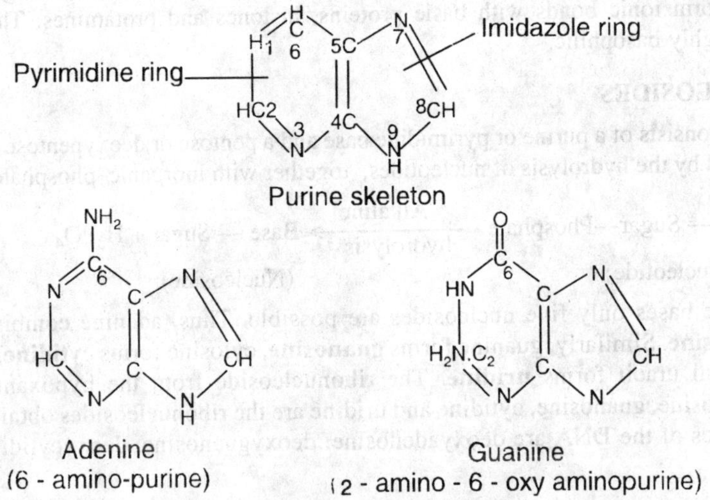

Fig. 14.2: Purine.

14.3. PENTOSE SUGARS

They contain five carbon atoms and are found in a combined state in the plants. They reduce the fehling solution and are not fermentable. Two naturally occurring pentoses ribose and deoxyribose found in RNA and DNA respectively have a pentagonal ring with 5 carbons, two of which (3' and 5') are linked to phosphoric acid and a third one (Carbon C_1) to the base (Fig. 14.3).

The pentose sugar in DNA, as the name implies, is deoxyribose. It is always the OH on the C_1 which is the point of attachment of the base. In deoxyribose sugar the oxygen on the second

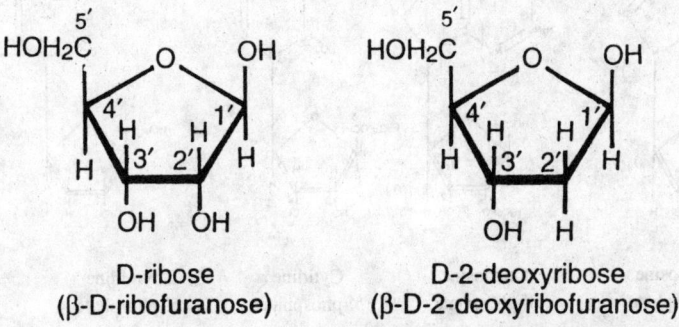

Fig. 14.3: Showing the strucuture of different pentose sugars.

carbon is lacking. This sugar is responsible for the Feulgen reaction, a very specific reaction for DNA.

14.4. PHOSPHORIC ACID

Nucleic acids are formed by the union of many nucleotides or polynucleotides. The union between two nucleosides is made through phosphoric acid. The nucleoside becomes a nucleotide, with the attachment of a phosphate group. During attachment of the two consecutive nucleosides the phosphate forms a phosphate ester bond between the pentoses of the nucleosides. These bonds link carbon 3' of one nucleoside with carbon 5' of the next.

In this way the phosphoric acid uses two of its acid groups. The 3rd acid group enables the molecule to form ionic bonds with basic proteins, histones and protamines. This group makes nucleotides highly basophilic.

14.5. NUCLEOSIDES

A nucleoside consists of a purine or pyrimidine base and a pentose or deoxypentose component only and is obtained by the hydrolysis of nucleotides, together with inorganic phosphate.

$$\text{Base — Sugar — Phosphate} \xrightarrow{\text{Alkaline hydrolysis}} \text{Base — Sugar} + H_3PO_4$$

(Nucleotide) (Nucleoside)

With five bases only five nucleosides are possible. Thus, adenine combines with ribose to form **adenosine**. Similarly, guanine forms **guanosine**, cytosine forms **cytidine**, thymine forms **thymidine**, and uracil forms **uridine**. The ribonucleoside from the hypoxanthine is named **inosine**. Adenosine, guanosine, cytidine and uridine are the ribonucleosides obtained from RNA. The nucleosides of the DNA are deoxyadenosine, deoxyguanosine, deoxycytidine and thymidine.

14.6. THE NUCLEOTIDES

The nucleotides are the phosphate esters of the nucleosides. There, phosphoric acid (H_3PO_4) forms an ester linkage with one of the free hydroxyl groups on the ribose or deoxyribose component. Those nucleotides derived from ribose nucleosides are usually called as **ribonucleotides** and those from deoxyribose nucleosides as **deoxyribonucleotides**.

The ribose portion of a ribonucleoside has three possible positions where the phosphate could be esterified, the 2'-hydroxyl, 3'-hydroxyl, and 5'-hydroxyl, whereas, the deoxyribonu-

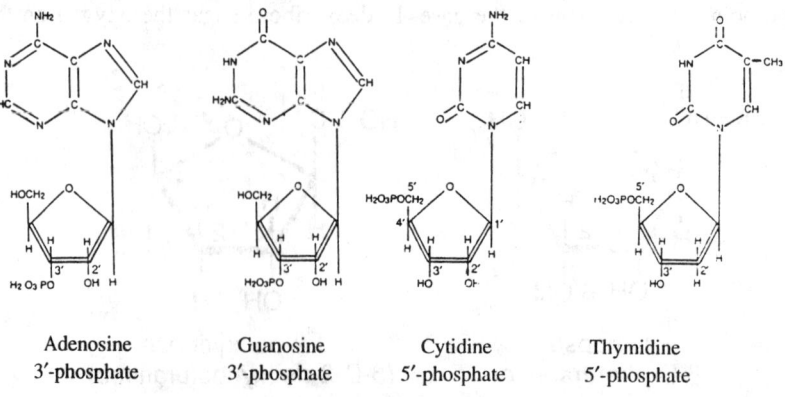

Adenosine 3'-phosphate   Guanosine 3'-phosphate   Cytidine 5'-phosphate   Thymidine 5'-phosphate

Fig. 14.4: Showing structure of different nucleotides.

cleoside has only the 3'-and 5'-positions available. Since there are three positions available in the ribose of ribonucleosides, three possible nucleoside monophosphates can be formed. Adenosine, for example, can give rise to three monophosphates (adenylic acids), adenosine 5'-phosphate, adenosine 3'-phosphate, and adenosine 2'-phosphate. Nucleotides of all three types have been isolated and identified.

In the same way guanosine, cytidine, and uridine can give rise to three guanosine monophosphates (guanylic acids), three cytidine monophosphates (cytidylic acids) and three uridine monophosphates (uridylic acids) respectively. The structure of nucleotides or nucleoside monophosphates are shown in Fig. 14.4.

14.7 BIOSYNTHESIS OF NUCLEOTIDES

Synthesis of Purines

Purines are synthesized in the cell in the form of their *nucleotide monophosphates*. The synthesis of purines has been discussed by a number of workers including Hartman and Buckmann (1959), Warren (1961), Schulman (1961) and Grav (1967).

A purine ring (skeleton) is made of four nitrogen atoms and five carbon atoms (Fig. 14.5). It is proposed that these atoms of the purine ring originated from the following molecules.

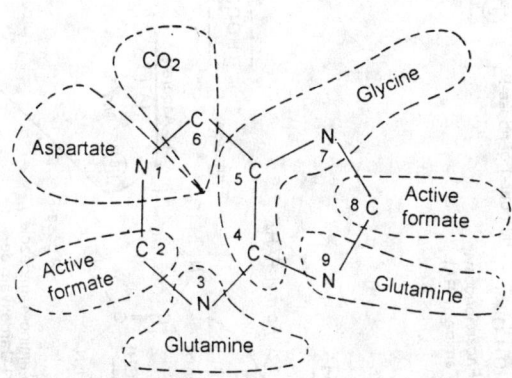

Fig. 14.5: Origin of carbon and nitrogen atoms of purine skeleton.

1. N_3 and N_9 – from NH_3 after its conversion into amide group of *glutamine*.
2. N_1 – Also from NH_3 but after its conversion in to *aspartate*.
3. N_7 – from *glycine* molecule.
4. C_4 and C_5 – from *glycine* molecule.
5. C_6 – from CO_2 molecule.
6. C_2 and C_8 – from *active formate*. Free formate cannot be incorporated as such.

Biosynthesis of Inosine 5-monophosphate (IMP)

During the biosynthesis of purine bases e.g., guanine and adenine, first inosine 5-monophosphate (IMP) is to be formed. The formation of IMP is completed in following steps (Fig. 14.6).

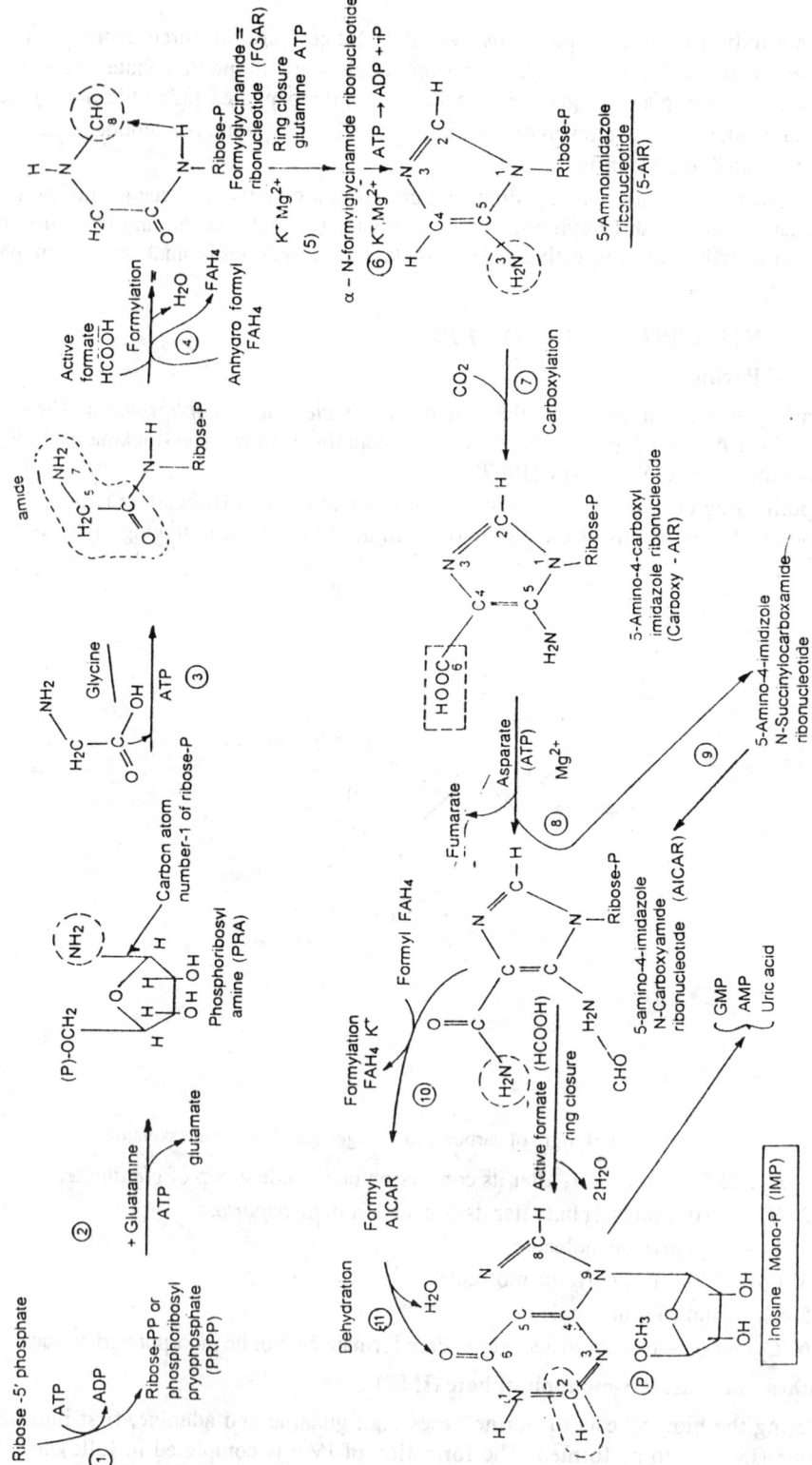

Fig. 14.6: Outline of purine synthesis (Numbers refer to atoms in the final purine molecule.)

Nucleic Acids Metabolism

1. Ribose 5-phosphate is phosphorylated by ATP to produce 5-phosphoribosyl 1-pyrophosphate (PRPP) in presence of enzyme *phosphorylase*.
2. PRPP combines with glutamine to form 5-phosphoribosyl amine (PRA), glutamate and pyrophosphate group in presence of enzyme *amidophosphoribosyl transferase*. This is an amination reaction where glutamine acts as most efficient donor. NH_2 becomes attached at C_1 of ribose.
3. PRA combines with glycine to produce glycinamide ribonucleotide (GAR). During the reaction amino group (NH_2) of PRA molecule reacts with the carboxyl group of glycine to form an amide linkage between glycine and PRA. Ultimately Glycinamide robonucleotide is produced with the liberation of one molecule of water.
4. GAR is formylated by the transfer of formyl group from the N_5, N_{10} anhydroformyltetrahydrofolic acid and formyl glycinamide ribonucleotide (FGAR) is formed. Here Anhydroformyl FAH_4 is converted into FAH_4 and one molecule of H_2O is eliminated.
5. FGAR reacts with glutamine in presence of ATP, K^+, Mg^{2+} to produce α-N formylglycinamide ribonucleotide, glutamate and ADP. Here, FGAR undergoes amination in which amino group is transferred from glutamine to FGAR. Now the ring become closed.
6. α-NFGAR in presence of K^+, Mg^{2+} and ATP is converted into 5-aminoimidazole ribonucleotide (5-AIR) and ADP. 5-AIR contains a 5-membered imidazole ring portion of a purine-molecule.

   Subsequent reactions will synthesise the pyrimidine ring of the purine molecules.
7. 5-AIR undergoes carboxylation reaction with CO_2 to produce 5-amino-4-carboxyimidazole ribonucleohide (Carboxy AIR).
8. Carboxy AIR reacts with aspartate in presence of Mg^{2+} and ATP and forms 5-amino-4-imidazole N-succinylocarboxamide ribonucleotide.
9. 5-Amino 4-imidazole N-succinylocarboxamide ribonucleotide is converted into 5-amino-4-imidazole N-Carboxyamide ribonucleotide (AICAR) with the liberation of fumarate.
10. AICAR undergoes formylation reaction with formyl FAH_4 to form formyl AICAR and FAH_4.
11. Formyl AICAR undergoes dehydration reaction resulting into ring closure and production of inosine 5-monophosphate (IMP) which is a precursor ribonucleotide for the synthesis of AMP, GMP and uric acid.

Reactions of Biosynthesis of IMP

(1) Ribose-5 phosphate + ATP $\longrightarrow$ 5-Phosphoribosyl 1-pyrophosphate (PRPP)

(2) 5-Phosphoribosyl 1-Pyrophosphate (PRPP) + Glutamine $\xrightarrow{\text{Transamination}}$ 5′ Phosphoribosyl amine (PRA) + glutamate + Pp.

(3) 5′ Phosphoribsylamine (PRA) + Glycine + ATP $\xrightarrow{Mg^{2+}}$ Glycinamide ribonuclcotide (GAR) + H_2O + ADP + ip

(4) Glycinamide ribonuclcotide (GAR) + active formate (HCOOH) $\xrightarrow[\text{Formylation}]{\text{Anhydro formyl } FAH_4 \quad FAH_4 \quad H_2O}$ Formylglycinamide Ribonucleotide (FGAR)

(5) Formyl glycinamide ribonucleotide (FGAR) + Glutamine + ATP $\xrightarrow[\text{amination}]{K^+, Mg^{2+}}$ 5-Aminoimidazole ribonuclcotide (5-AIR) + ADP

(6) α-N-Formylglycinamide ribonucleotide + ATP (α-N-FGAR) $\xrightarrow{K^+, Mg^{2+}}$ 5-Aminoimidazole ribonuclcotide (5-AIR) + ADP

(7) 5-Aminoimidazole ribonucleotide + CO2 (5-AIR) $\xrightarrow{carboxylation}$ 5-Amino-4-carboxy imidazole ribonuclcotide (Carboxy AIR)

(8) 5 Amino-4-carboxyimidazole ribonucleotide + Aspartate + ATP (carboxy AIR) $\xrightarrow{Mg^{2+}}$ 5-Amino-4-imidazole N-Succinylo Carbox amide ribonucleotide + ADP

(9) 5-Amino-4-imidazole N-Succinylocarboxamide ribonucleotide. $\longrightarrow$ 5-Amino-4-imidazole N-Carboxyamide ribonuclotide (AICAR) + Fumarate

(10) 5-Amino-4-imidazole N-carboxyamide ribonucleotide (AICAR) $\xrightarrow[K^+]{N^{10}\text{-Formyl FAH}_4 \quad FAH_4}$ Formyl AICAR

(11) Formyl AICAR $\xrightarrow[H_2O]{dehydration}$ Inosine 5'-Phosphate (IMP)

Synthesis of Guanosine monophosphate (GMP) from IMP

The synthesis of GMP from IMP takes place through following steps (Fig. 14.7).

(1) IMP undergoes oxidation via NAD^+ to produce xanthosine mono-phosphate (XMP) or Xanthylic acid. The purine base of XMP is called *Xanthine*.

(2) XMP undergoes amination reaction in which XMP reacts with glutamine in presence of ATP and Mg^{2+}. During the reaction amino group ($-NH_2$) of glutamine is transferred to XMP at C_2 position resulting into synthesis of GMP. The glutamine and ATP are converted into glutamic acid and AMP respectively.

Xanthylic acid (XMP) + ATP + glutamine $\xrightarrow{Mg^{2+}}$ Guanylic acid (GMP) + glutamic acid + AMP + PPi

Synthesis of Adenosine monophosphate (AMP) from IMP

The synthesis of AMP from IMP is outlined in Fig. 14.7. The steps are as follows :

(1) IMP undergoes condensation with aspartic acid in presence of GTP to form adenylosuccinate (nic-acid) which contains a nitrogen bridge.

(2) Adenylosuccinic acid subsequently cleaves into adenosine monophosphate (AMP = adenylic acid) and fumaric acid.

Synthesis of uric acid from IMP

The synthesis of uric acid from IMP takes place in following steps (Fig. 14.8)

(1) IMP first breaks into hypoxanthine and ribose-5-phosphate. Hypoxanthine is a monohydroxy purine.

(2) Hypoxanthine undergoes oxidation to produce xanthine or dihydroxy purine.

(3) Xanthine undergoes further oxidation to produce tri-hydroxy purine or uric acid.

(4) Trihydroxy purine form of uric acid may undergo tautomeric shift to produce triketo purine form of uric acid.

Nucleic Acids Metabolism

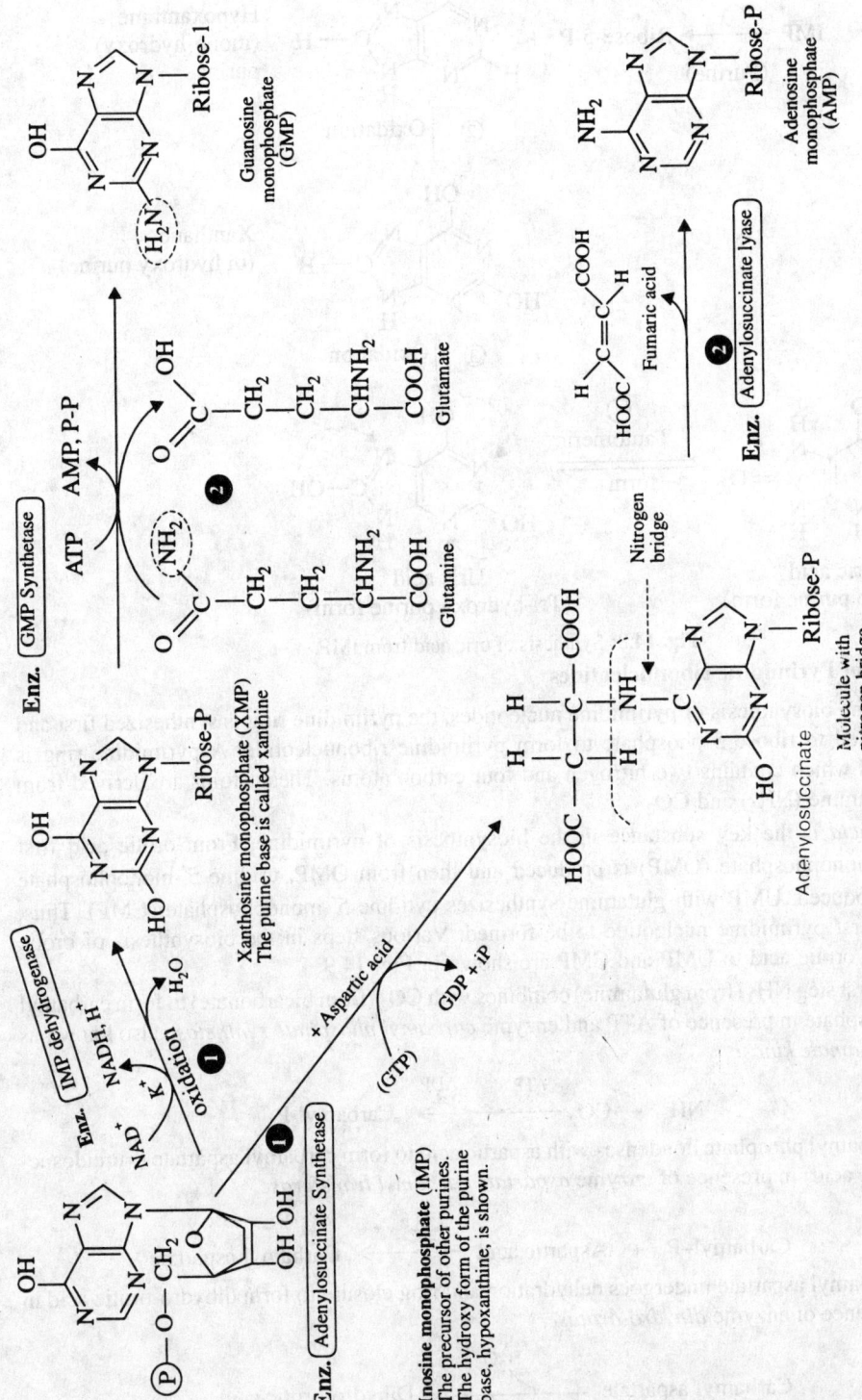

Fig. 14.7: Outline of the synthesis of GMP and AMP from inosine monophosphate (IMP).

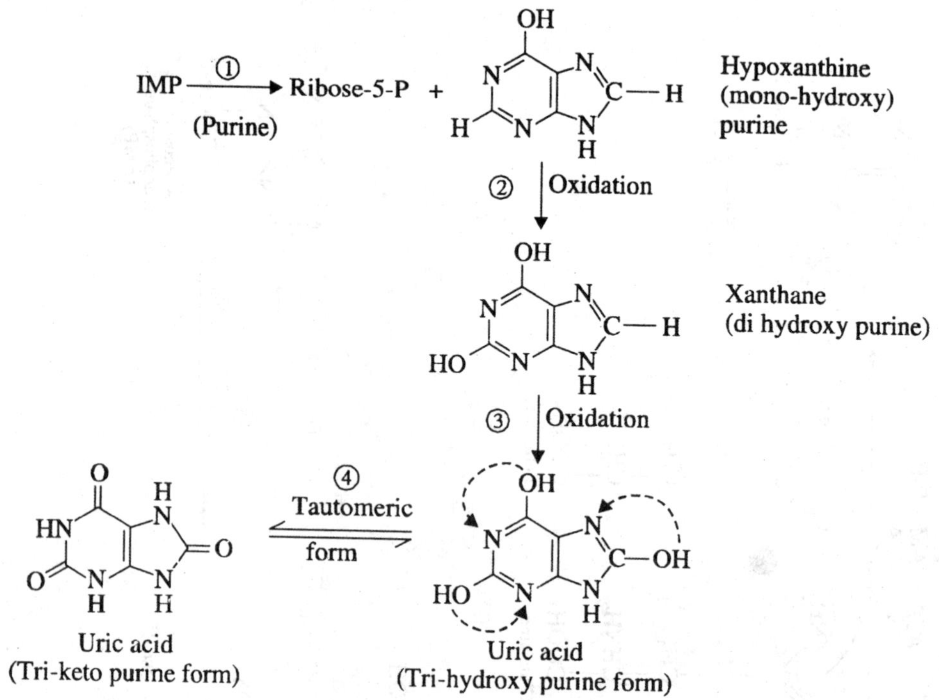

Fig. 14.8: Synthesis of uric acid from IMP.

Biosynthesis of Pyrimidine ribonucleotides

During the biosynthesis of pyrimidine nucleotides, the pyrimidine ring is synthesized first and then it is linked to ribose 5-phosphate to form pyrimidine ribonucleotide. A pyrimidine ring is six-membered which contains two nitrogen and four carbon atoms. These atoms are derived from aspatate, glutamine (NH_3) and CO_2.

Orotic acid is the key substance in the biosynthesis of pyrimidine. From orotic acid first orotidine 5'-monophosphate (OMP) is produced and then from OMP, uridine 5'-monophosphate (UMP) is produced. UMP with glutamine synthesizes cytidine 5'-monophosphate (CMP). Thus, *UMP* is the first pyrimidine nucleotide to be formed. Various steps in the biosynthesis of orotic acid and from orotic acid to UMP and CMP are shown in Fig. 14.9.

(1) In first step NH_3 (from glutamine) combines with CO_2 (from bicarbonate) to form carbamyl phosphate in presence of ATP and enzyme *carbamyl phosphate synthetase* also known as *carbamate kinase*.

$$NH_3 + CO_2 \xrightarrow{ATP \quad ADP} Carbamyl\text{-}P$$

(2) Carbamyl phosphate dondenses with aspartic acid to form carbamyl aspartate (=ureidosuccinic acid) in presence of enzyme *aspartate carbamyl transferase*.

$$Carbamyl\text{-}P + Aspartic\ acid \xrightarrow{Pi} Carbamyl\ aspartate$$

(3) Carbamyl aspartate undergoes dehydration and ring closure to form dihydro-orotic acid in presence of enzyme *dihydro-orotase*.

$$Carbamyl\ aspartate \xrightarrow[ring\ closure]{H_2O} Dihydro\text{-}orotic\ acid$$

Nucleic Acids Metabolism

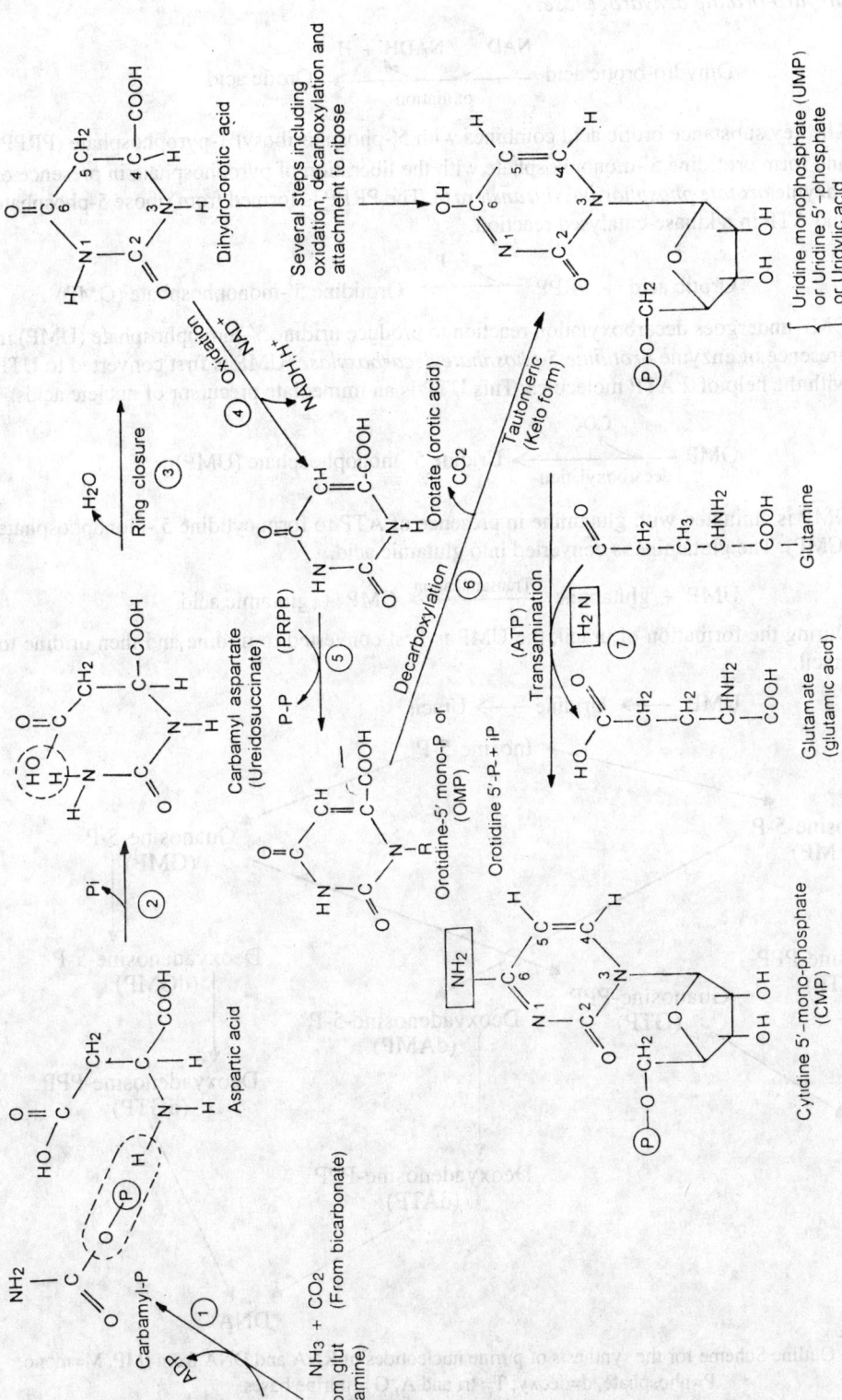

Fig. 14.9: Outline of pyrimidine synthesis.

(4) Dihydro-orotic acid is oxidized with NAD⁺ coenzyme to orotic acid in presence of enzyme *dihydro-orotate dehydrogenase*.

$$\text{Dihydro-orotic acid} \xrightarrow[\text{oxidation}]{\text{NAD}^+ \quad \text{NADH + H}^+} \text{Orotic acid}$$

(5) The key substance orotic acid combines with 5'-phosphoribosyl 1-pyrophosphate (PRPP) and form orotidine 5'-monophosphate with the liberation of pyrophosphate in presence of enzyme *orotate phosphoribosyl transferase*. The PRPP is formed from ribose 5-phosphate and ATP in a kinase-catalysed reaction.

$$\text{Orotic acid + PRPP} \xrightarrow{\text{P-P}} \text{Orotidine 5'-monophosphate (OMP)}$$

(6) OMP undergoes decarboxylation reaction to produce uridine 5'-monophosphate (UMP) in presence of enzyme *orotidine 5-phosphate decarboxylase*. UMP is first converted to UTP with the help of 2 ATP molecules. This UTP is an immediate precursor of nucleic acids.

$$\text{OMP} \xrightarrow[\text{decarboxylation}]{\text{CO}_2} \text{Uridine 5'-monophosphate (UMP)}$$

(7) UMP is aminated with glutamine in presence of ATP to form cytidine 5'-monophosphate (CMP). The glutamine is converted into glutamic acid.

$$\text{UMP + glutamine} \xrightarrow{\text{Transamination}} \text{CMP + glutamic acid.}$$

(8) During the formation of uracil, the UMP is first converted to uridine and then uridine to uracil.

$$\text{UMP} \longrightarrow \text{Uridine} \longrightarrow \text{Uracil}$$

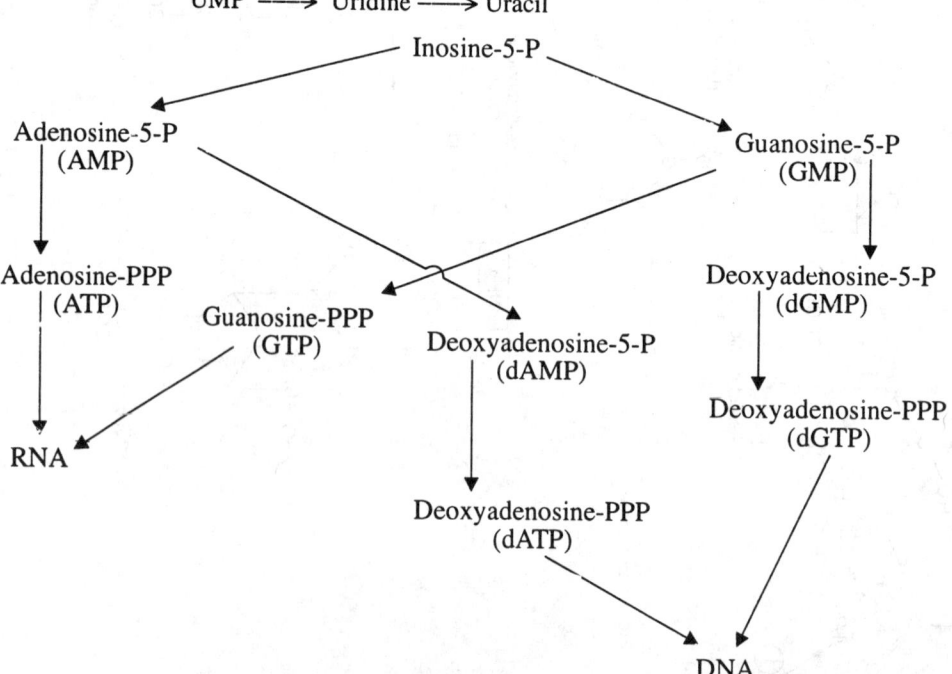

Fig. 14.10: Outline Scheme for the synthesis of purine nucleotides of RNA and DNA from IMP. M=mono, P=phosphate, d=deoxy; T= tri and A, G = purine bases.

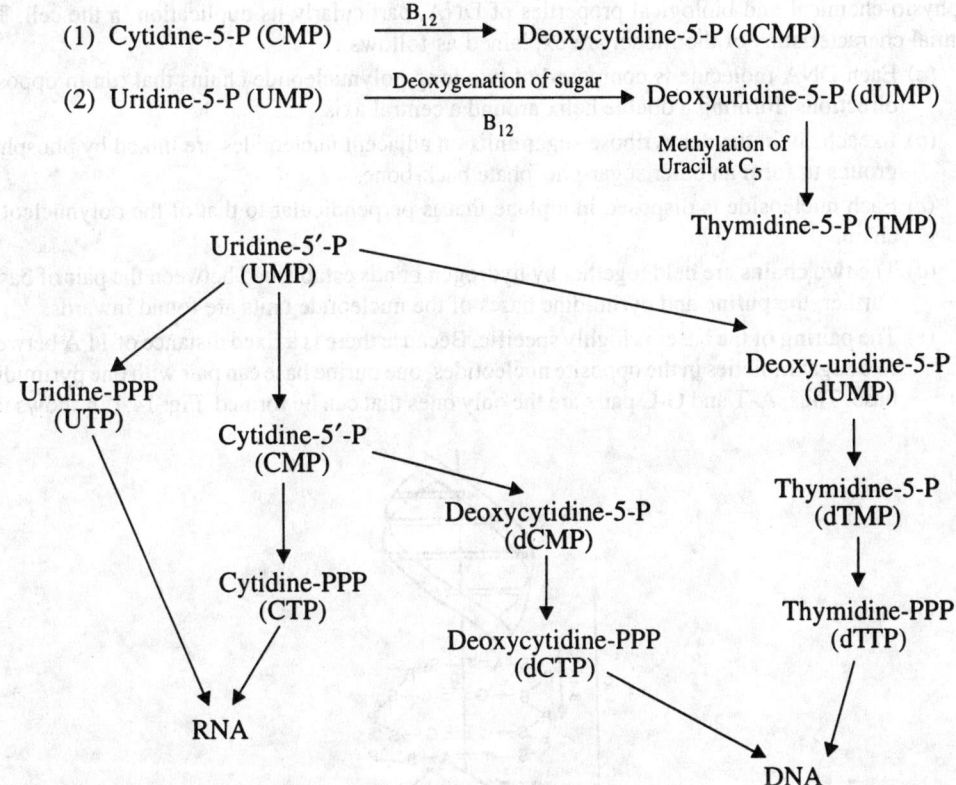

Fig. 14.11: Outline Scheme for the synthesis of pyrimidine nucleotides of RNA and DNA.

14.8. THE DEOXYRIBONUCLEIC ACID (DNA)

Studies since 1940s have indicated that DNA is the universal genetic material of all forms of life except certain viruses, which have RNA as their genetic material. In the nuclei of plant and animal cells the DNA is present in the chromosomes in association with protein molecules.

The structure of DNA

The building units of DNA are nucleotides, each one of which is made up of a nitrogenous base, a deoxyribose sugar and a phosphoric acid molecule. The nucleotide units are joined together to form a polynucleotide chain.

Chargaff studied the chemical structure of DNA and showed that the DNA contained equal proportions of the large purine bases and the smaller pyrimidines, and even more interesting that adenine and thymine were present in equimolecular proportions and so are cytosine and guanine. This equivalence of 'A' and 'T' and of 'G' and 'C' was of the utmost importance in relation to the formation of the DNA helix. This is referred to as "Chargaff's rule", There are two deviations from the rule :

(a) In wheat germ DNA 'G' and 'C' are not present in equimolar amounts.

(b) In φ × 174 DNA the adenine is not equimolar with thymine nor is guanine with cytosine.

This is because φ × 174 DNA is single stranded.

Watson and Crick Model of DNA

Watson and *Crick* (1953) constructed their famous "double helix" model of DNA, which explained all the evidences then available, and for which they and Wilkins were later awarded the Noble-Prize. The model is illustrated in Fig. 14.12 a, b. This model is important because it explains the physio-chemical and biological properties of DNA, particularly its duplication in the cell. The essential characteristics of the model are explained as follows :

(a) Each DNA molecule is composed of two long polynucleotide chains that run in opposite directions, forming a double helix around a central axis.

(b) In each chain the deoxyribose sugar units on adjacent nucleotides are linked by phosphate groups to form an outer sugar-phosphate back-bone.

(c) Each nucleoside is disposed in a plane that is perpendicular to that of the polynucleotide chain.

(d) The two chains are held together by hydrogen bonds established between the pair of bases. Further, the purine and pyrimidine bases of the nucleotide units are found inwards.

(e) The pairing of the bases is highly specific. Because there is a fixed distance of 11 Å between two sugar moieties in the opposite nucleotides, one purine base can pair with one pyrimidine base. Thus, A-T and G-C pairs are the only ones that can be formed. Fig. 14.12b shows that

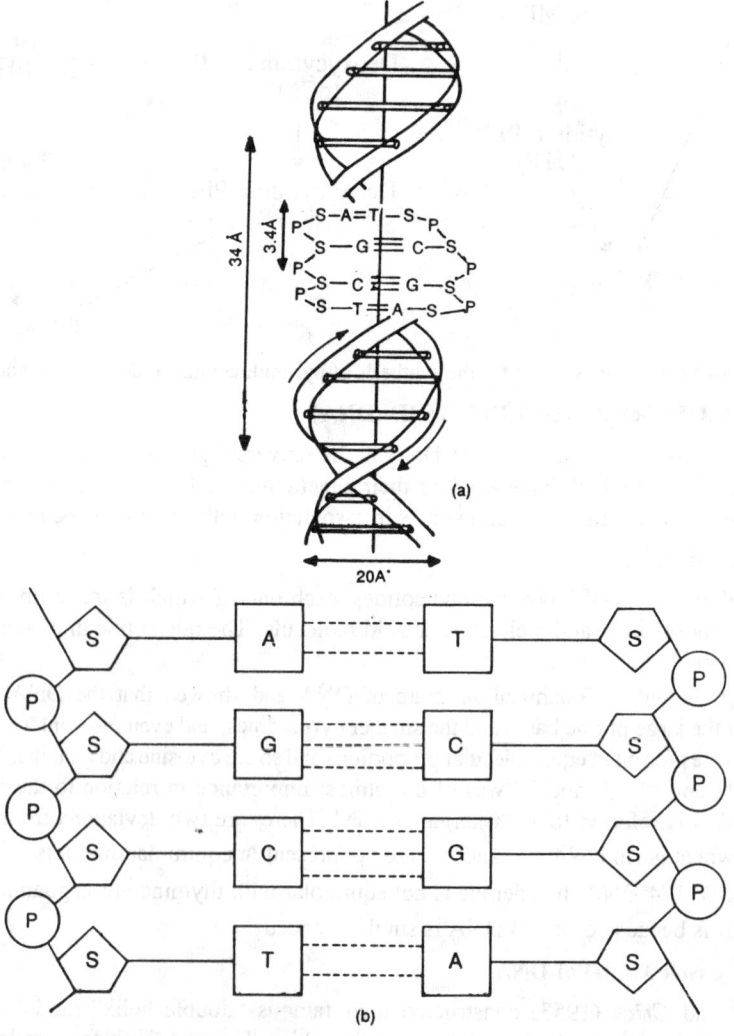

Fig. 14.12: (a) Diagrammatic representation of DNA molecule as proposed by Watson & Crick.
(b) Diagrammatic representation of part of a hypothetical polynucleotide chain in DNA.

Nucleic Acids Metabolism

two hydrogen bonds are formed between A and T, and three hydrogen bonds are formed, between C and G.

Because of this type of base pairing, the model satisfies *Chargaff's* chemical observations that the DNA molecule contains equal number of adenine and thymine bases, and of cytosine and guanine bases. It also satisfies the Wilkins' X-ray diffraction observations.

(f) The two polynucleotide chains are complementary to each other, in that there is complementary relationship between their sequence of bases. Thus, if one chain has a region which goes -adenine -guanine-cytosine-thymine-guanine, then the corresponding region of the complementary chain will go— thymine—cytosine—guanine—adenine—cytosine. This can be shown as follows :

```
1st Chain   A,  G,  C,  T,  G
            ||  ||| ||| ||  |||
2nd Chain   T   C   G   A   C
```

Thus, the molecule of DNA consists of two complementary strands and not indentical strands. The axial sequence of bases along one polypeptide chain may vary considerably. During duplication, each chain of the DNA molecule serves as a model or template on which its complements are built up. The number and sequence of bases in a DNA molecule form the basis of the genetic code.

(g) The distance between the bases is 3.4 Å, the diameter of the molecule is about (20 Å), and the two twisted chains form a molecule with alternate wide and narrow grooves. A complete turn of chain occurs every (34 Å), and 10 nucleotide units are present in each turn.

14.9. METHODS OF DNA DUPLICATION OR REPLICATION

There are two common methods of replication :

Watson and Crick Model

Watson and *Crick* (1953) also proposed the method of DNA replication. Their model itself suggests the manner in which DNA molecules are replicated (Fig. 14.13).

According to their hypothesis, during replication the nucleotides of a DNA molecule split by separation of their hydrogen bonds. As a result, the two strands of the DNA molecule unwind and separate completely. The strands, thus, separated are complements of one another. When the separation is completed, the nucleotides of the separated strands attract other complementary nucleotides which are already present in the cell nucleus as nucleotides pool. During the formation of complementary strand of each separated strand, first purine attaches with the pyrimidine and pyrimidine with purine ('A' couples with 'T' and 'G' with 'C' or *vice-versa*.) The attachment of bases takes place through hydrogen bonds. After the attachment of bases, the sugar molecules unite with one another by their phosphate components. Thus, the formation of nucleotide and polynucleotide chain takes place. In the final stage, the complementary units link up to form two DNA molecules, each of them is an exact copy of the original molecule. This is known as DNA replication or DNA duplication.

Meselson and Stahl's theory

This theory was proposed by *Meselson* and *Stahl* in 1958. According to them, during replication the two strands do not separate completely and synthesize their replica, but instead they start unzipping at one end and simultaneously the unzipped segments start attracting their respective nucleotide pairs. Thus, the unzipping of the original DNA strands and synthesis of fresh DNA strands go side by side and at the end of replication two daugher DNA molecules, having one parental and one synthesized strand, are formed. The two daughter DNA molecules thus produced are exact copies of the original DNA molecule. This type of DNA replication is called as "Semiconservative type", since each of the daughter DNA molecule consists of one

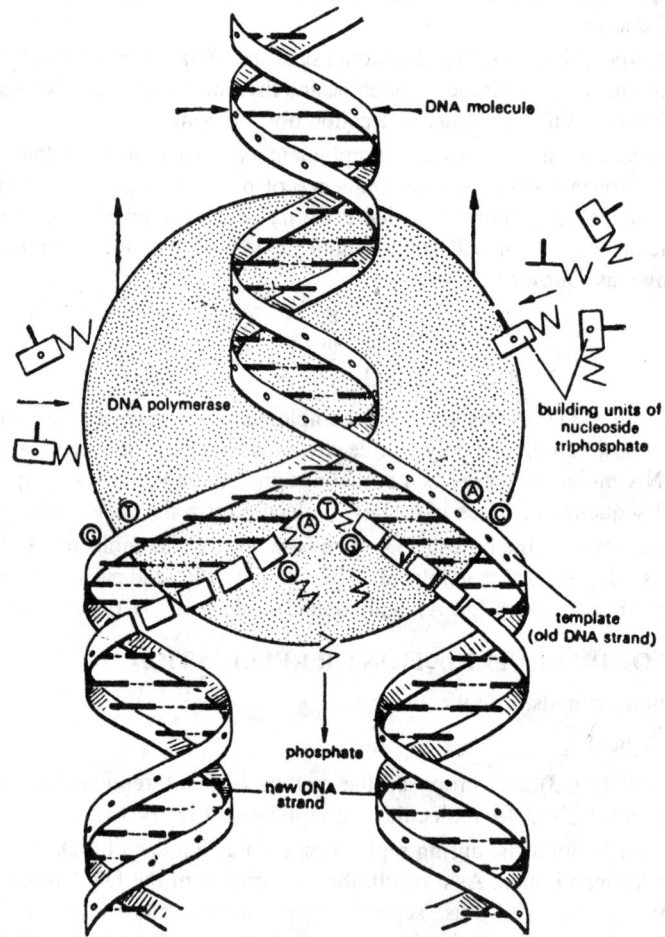

Fig. 14.13: Diagrammatic representation of the replication of DNA molecule.

'Old' polynucleotide strand from the parent molecule, and one newly synthesized strand.

The over-all process is summarised in the following steps:

(a) During replication the DNA molecule first unzips (the two strands move in the opposite direction and start to open).

(b) During unzipping process the hydrogen bonds between the organic bases are broken. As a result two strands start to open.

(c) Attachment of new complementary nucleotides, which are present in the nucleus, takes place.

(d) Two similar DNA molecules are formed. The semi-conservative type of DNA replication was shown by Meselson and Stahl in *E. coli* by the use of ^{15}N isotope of nitrogen which was heavier than ^{14}N. The experiment is shown in Fig. 14.14.

14.10. MECHANISM OF DNA REPLICATION

The complete process of DNA replication involves following steps in *E.coli*.

1. Recognition of the initiation point : DNA replication starts at a specific point, called **initiation point** or **origin** where replication fork begins. This is a nucleotide sequence of 100 to

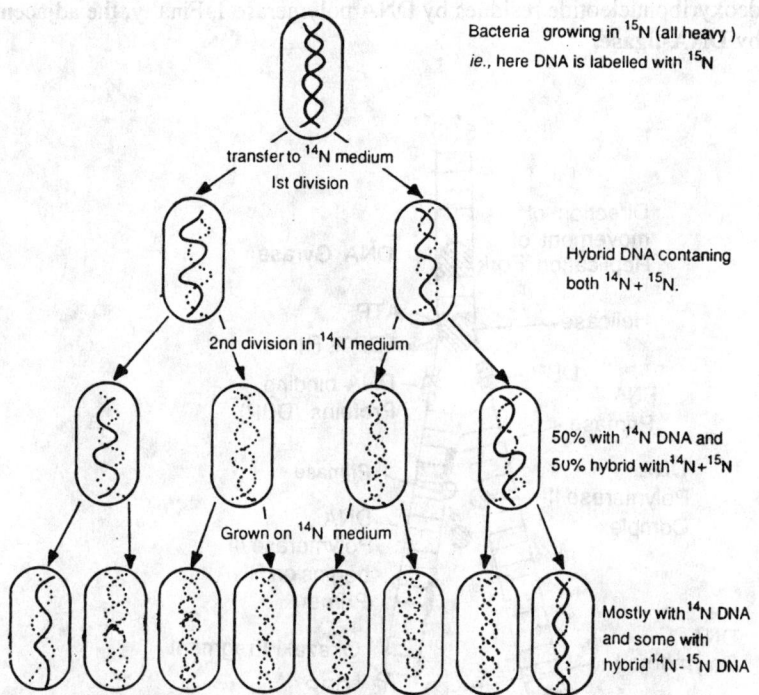

Fig. 14.14: Meselson and Stahl's experiment with E. Coli bacteria, demonstrating that DNA replication is a semi-conservative process.

200 base pairs. **Specific initiator proteins** recognise the initiation point on DNA. Such proteins along with DNA directed RNA polymerase initiate the synthesis of RNA primer for the formation of DNA chain. Prokaryotic chromosomes usually prossess one initiation point or replication fork whereas eukaryotic chromosomes may possess several (about a thousand) replication forks.

2. **Unwinding of DNA :** When the DNA duplex molecule is cut open (nicked) to form a bubble or fork, the unwinding protein gets attached at the point of nick which helps in the separation of two strands of DNA duplex.

3. **RNA priming :** Now RNA primers are synthesized from the DNA directed RNA polymerase. The priming RNA strands (RNA primers) are complementary to the two strands of DNA and are made up of 50 to 100 nucleotides.

4. **Formation of DNA or RNA primers :** The new strands of DNA are synthesized in the 5' – 3' direction from the 3'–5' template DNA by the addition of deoxyribonucleotides to the 3' end of the primer RNA. This addition is achieved by DNA polymerase III (poly III-copol III) in presence of ATP. Once the synthesis of DNA strand has been initiated, copol III detaches and poly III carries out replication of the DNA strand. The unwinding proteins separate the DNA duplex strands ahead of the replication fork.

The **leading strand of DNA** is synthesized in 5' – 3' direction as one piece. The **lagging strand of DNA** is synthesized in its opposite direction in short segments consisting of 1,000 to 2,000 nucleotides. These segments are called **Okazaki fragments**.

5. **Excision of RNA primers :** Once a small segment of an Okazaki fragment has been formed, the nucleotides of RNA primer are removed from the 5' end one by one by the action 5' – 3' exonuclease activity of DNA polymerase-I.

6. **Joining of Okazaki fragments:** The gaps left between Okazaki fragments are filled with

complementary deoxyribonucleotide residues by DNA polymerase-I. Finally, the adjacent 5' and 3' ends are joined by **DNA-ligase.**

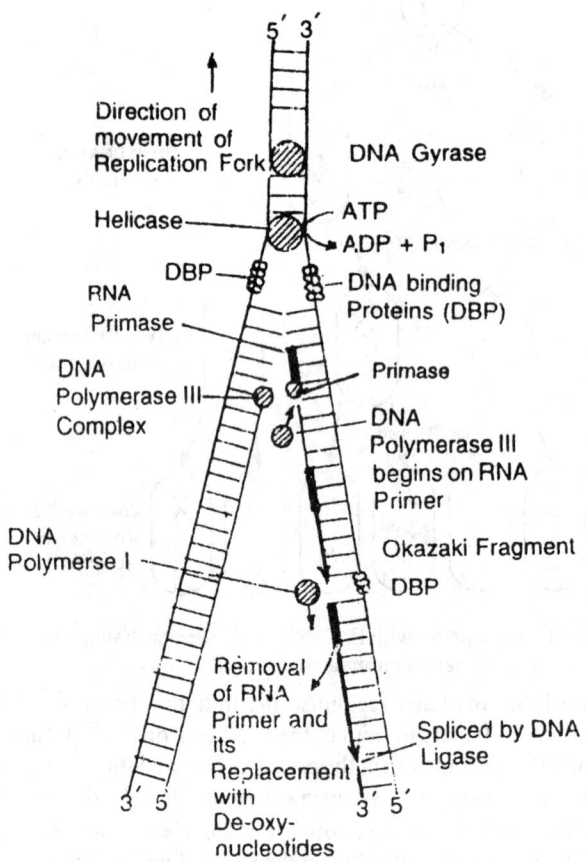

Fig. 14.15: Summary of major steps in DNA replication.

In eukaryotic cells, the mechanism of DNA replication is expected to be more complex than in prokaryotes. The replication of eukaryotic DNA begins at multiple points of origin. Their RNA primer is formed of about 10 nucleotides and the Okazaki fragments are much shorter (about 100 to 150 nucleotides).

14.11. ENZYMES OF DNA SYNTHESIS

About 20 or more different proteins and enzymes are required during DNA replication. These collectively form a **DNA replicase system** or **replisome.** The enzymes fall into two types—(*i*) **DNA polymerase** and (*ii*) **polynucleotide ligase**. DNA polymerase has three sites for attachment. One of them attaches to the template DNA, the second to the triphosphate nucleotide and the third one to the 3'—OH end of the DNa primer. Thus, **DNA polymerase** adds triphosphate nucleotides to primer DNA from 5' end to the 3' end of the polynucleotide chain. The new strands are synthesized in fragments and these fragments are then added up by the enzyme, **polynucleotide ligase.**

I. DNA Polymerase Enzymes

There are three DNA polymerase enzymes that participate in the process of DNA replication:

(i) DNA polymerase-I (Pol. I)
(ii) DNA polymerase-II (Pol. II)
(iii) DNA polymerase-III (Pol. III)

(a) **DNA Polymerase-I** : This enzyme has been studied in *E. coli* in detail. It is roughly spherical with a diameter of about 6.5 mm. It has a molecular weight of 1,90,000 and is formed of a single polynucleotide chain of about 1,000 amino acid residues. It possesses a sulphydryl group, single interchain disulphide and one zinc molecule at the active site.

A DNA polymerase-I molecule in reality is a complex structure being formed of :

(i) DNA polymerase 3' – 5' exonuclease
(ii) 5' – 3' exonuclease.

There are five specific binding sites on the spherical molecule of DNA polymerase-I :

(i) **Template site** for binding the template DNA.
(ii) **Primer site** for binding primer strand of DNA.
(iii) **Primer terminus site** for 3' – hydroxyl terminus of primer.
(iv) **5'– triphosphate site** – a locus for incoming deoxyribonucleotide 5'—triphosphate group.
(v) 5' – 3' **exonuclease site,** a locus for 5' – 3' exonuclease activity situated in the path of growing chain.

DNA polymerase-I was discovered by KORNBERG and his colleagues in 1955, then it was considered to carry out DNA replication. It is now known that DNA polymerase-I performs varied functions (**multifunctional enzyme**).

(1) It catalyses the addition of mononucletide units (the deoxyribonucleotides) to the free 3'–OH end of DNA strand, thus, participating in the repair of DNA molecule. A pure DNA polymerase-I can add about 1000 nucleotide units per minute per molecule of enzyme at 37°C.

(2) It catalyses 3' – 5' exonuclease activity and removes nucleotide residues of primer RNA at 3' end.

(3) It also catalyses 2' – 3' exonuclease activity.

(b) **DNA Polymerase-II** : Its role is not yet fully understood. It is effective only on DNA duplex with gaps and it cannot replicate long strands of DNA.

(c) **DNA Polymerase-III** : It was discovered by **T.Kornberg** and **M.L. Gefter** in 1972.

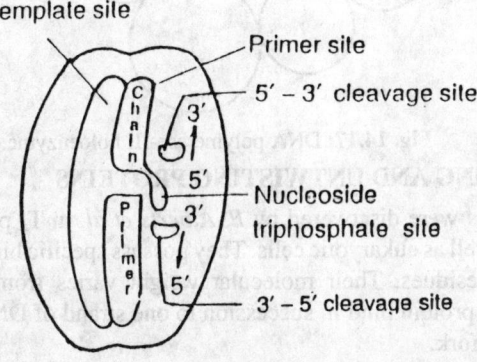

Fig. 14.16: A model of DNA polymerase-I enzyme with different sites.

It is most active enzyme among all the three polymerases. It is made up of α, β, θ, δ and ε sub-units.

DNA polymerase-III is chiefly responsible for DNA chain elongation. It is complex molecule with molecular weight (MW) about 550,000, the sub-unit β is also known as **copolymerase-III**. It recognizes and binds to the primer strand of parental DNA. The copolymerase III is released just after the polymerase-III binds to the correct initiation point. DNA polymease-III now helps in elongating the DNA stand in 5'–3' direction by addition of new nucleotide residues to the 3' end of primer strand. Thus, DNA polymerase-III does not initiate replication but helps in elongation of chain.

It also acts as 5'–3' exonuclease and 3'–5' exonuclease. It can hydrolyse terminal nucleotides from either end of a DNA strand.

II. DNA-Ligase or Polynucleotide Ligase

DNA-ligase enzyme helps in joining or sealing the joints of the two DNA fragments by catalysing the synthesis of a phosphodiester bond between a 3'–OH group at the end of one chain and 5'–phosphate group at the end of other chain. This enzyme was extracted from *E. coli*. It is made up of single polynucleotide chain with a molecular weight about 77,000. Each *E. coli* cell contains about 2000 to 4000 copies of this enzyme. DNA ligase enzyme performs following important functions:

1. It helps in joining the DNA fragments during DNA replication.
2. It helps in repairing single-stranded nicks in duplex DNA molecule.
3. It helps in linking the ends of linear DNA duplexes to form circular DNA.
4. It helps in joining the segments of DNA during recombination which takes place during meiosis, genetic transformation or transduction.

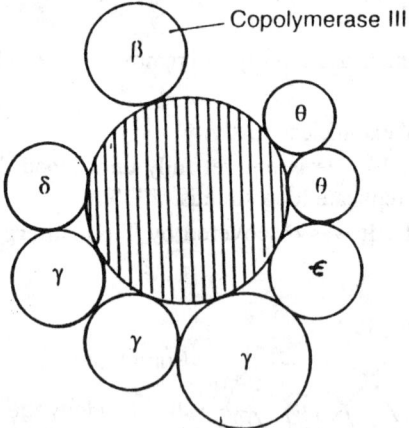

Fig. 14.17: DNA polymerase-III holoenzyme.

14.12. DNA UNWINDING AND UNTWISTING PROTEINS

DNA unwiding proteins were discovered by *B. Alberts et al.* in T_4 phage. A number of them are found in prokaryotic as well as eukaryotic cells. They possess specific binding sites for short segments of about 8 nucleotide residues. Their molecular weight varies from 10,000 to 75,000. Several molecules of unwinding protein bind in succession to one strand of DNA duplex in advance for the formation of replication fork.

The untwisting proteins cause nicks in one of the two strands of super-coiled DNA. This allows some unwiding of DNA molecules which relieves to torsional stress. The nick is released afterwards completing the strand.

Nucleic Acids Metabolism

14.13. RIBONUCLEIC ACID (RNA)

RNA exists chiefy in the cytoplasm of animal and plant cells, but nucleus also contains a very small amount of RNA. The naturally occurring RNA does not show the regularity of base composition as is found in DNA. They also do not show the equivalence of adenine to uracil, or of guanine to cytosine. It is, therefore, not to be expected that the structure, of RNA will correspond to a perfect two-stranded helix of the *Watson-Crick* type. There are a number of evidences which show that the RNA is single-stranded and it forms partially helical structure. The helical regions remain separated by the non-helical regions. Thus, the RNA shows loop like structures.

The structure of RNA is closely related with the DNA; like DNA, RNA is also made up of a pentose sugar, organic bases (Purine and Pyrimidine) and a phosphoric acid molecule. The pentose sugar in RNA is ribose *instead of deoxyribose* found in DNA. The pyrimidines in RNA are cytosine and uracil instead of cytosine and thymine which are found in DNA. In this way we can say that RNA is made up of :

  (*i*) Ribose sugar
  (*ii*) Purines-Adenine, Guanine; Pyrimidines-Cytosine and Uracil.
  (*iii*) Phosphoric acid molecule.

14.14. DIFFERENCES BETWEEN RNA AND DNA

  (*i*) RNA molecule is much shorter than DNA.
  (*ii*) Mostly RNA is single-stranded, while the DNA is double-stranded.
  (*iii*) The sugar in RNA is ribose, while in DNA it is deoxyribose.
  (*iv*) DNA contains the pyrimidine-thymine, but RNA contains uracil.
  (*v*) RNA is generally confined to the cytoplasm and its inclusions, while DNA is found always in nucleus. In recent years the presence of DNA in mitochondria and chloroplast and of RNA in chromosomes has been reported by many workers.
  (*vi*) The two types of nucleic acids are distinguishable by their stainable reactions. DNA is stainable by methyl green and RNA is stainable by pyromine. Differences no. 3 and 4 are most important.

14.15. TYPES OF RNA

In animal and plant cells the following three types of RNAs have been reported:

  (*i*) Ribosomal RNA (r-RNA)
  (*ii*) Messenger RNA (m-RNA)
  (*iii*) Transfer RNA (t-RNA)

Robosomal RNA (r-RNA)

Ribosomal RNA is also known as insoluble RNA. It is confined mainly in the cytoplasm in the minute ribonucleoprotein granules known as *ribosomes*. It accounts up to 80 per cent of the total cell RNA. The ribosomes which are minute particles, are made up of two subunits, one smaller and other larger. The smaller subunit attaches at the top of larger subunit to form a cap like structure. Two types of ribosomes have been reported, one having sedimentation coefficient value of 70 S (in Prokaryotes) and other of 80S (in Eukaryotes). The 70 S ribosomes are again made up of 50S and 30S subunits, while 80 S are made up of 60S and 40S subunits. Both subunits of the two types of ribosomes contain different types of ribosomal RNAs. 70S ribosomes contain RNAs with sedimentation Coefficient 5S, 16S and 23S, of which 16S r-RNA is found in 30S subunit and 5S, 23S r-RNA in 50S subunit. The 80S ribosomes contain 18S, 28S, 5.8S and 5S r-RNAs, of which 28S, 5.8S and 5S r-RNAs are found in 60S subunit and 18S r-RNA is found in 40S smaller subunit.

The 28S r-RNA and 23S r-RNA have the molecular weight of about 1.7×10^6 daltons and

1.1×10^6 daltons respectively, while 16S and 18S r-RNAs about 6×10^5 and 7×10^5 daltons respectively. 5S r-RNA with mol. wt. 3.2×10^4 was discovered by *Knight* and *Darnell* (1967) and by *Forget & Weissman* (1968). Forget and Weissman reported that 5S r-RNA has a length to 120 nucleotides and it has a clover leaf shape, similar to 4S t-RNA 5.8S rRNA has mol. wt. 3.5×10^4 daltons.

The role of r-RNS in the process of protein synthesis is still little understood. However, it is supposed that they definitely play a significant role in the process of protein synthesis. The function of 5S r-RNA is still quite unknown.

Processing of r-RNA

The r-RNA is formed first in the nucleolus with the help of nucleolar organizer which contains ribosomal DNA (r-DNA) cistrons. The r-DNA cistrons first form a precursor molecule of 45 S r-RNA which after a number of events is broken into two molecules of 28S and 18S r-RNA and ultimately they reach the cytoplasm from the nucleolus (Fig. 14.18). The 18S r-RNA enters the smaller subunit of ribosome (40S), whereas, 28S rRNA enters the larger subunit (60S) of ribosome.

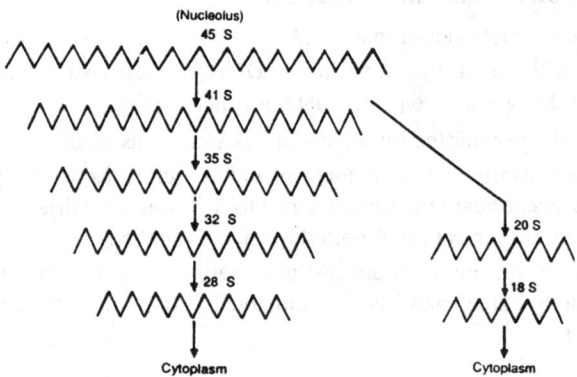

Fig. 14.18: Processing of 45S nucleolar RNA into 28S and 18S r-RNA.

The 45S r-RNA precursor molecule is cleaved into 32S and 18S r-RNAs through several intermediate steps. The 18S r-RNA is immediately exported to cytoplasm, but 32S r-RNA is again cleaved to form 28S r-RNA. The formation of 28S rRNA from 32S rRNA takes about 40 minutes. After its formation the 28S rRNA remains in the nucleolus for 30 minutes and then it is transferred to the cytoplasm.

Miller (1970) after doing a number of experiments on amphibian nucleoli suggested that each r-DNA cistron coding for a 45S molecule is separated by segments of non-transcribed DNA. It is also believed that there are about 100 RNA polymerases which act on each r-DNA cistron at the same time. Each polymerase transcribes a single 45S RNA.

The 5S r-RNA is transcribed outside the nucleolus because the genes forming it are not linked to the nucleolar organizer and are present as anucleolate as shown in Fig. 14.19. Those regions of the DNA that code for messenger RNA are called the **structural genes** and other regions which code for the different ribosomal and transfer RNAs are frequently called the **determinants** for RNA. The 5S rRNA, which has been formed anucleolate, finally enters the 60S ribosomal subunit. Fig. 14.19 summarizes, at the cellular level, the transcription and transport of nuclear RNAs, in the eukaryotic cell and also m-RNA participation in protein synthesis. It also indicates the transcription sites in DNA for 4S and 18-180S m-RNAs.

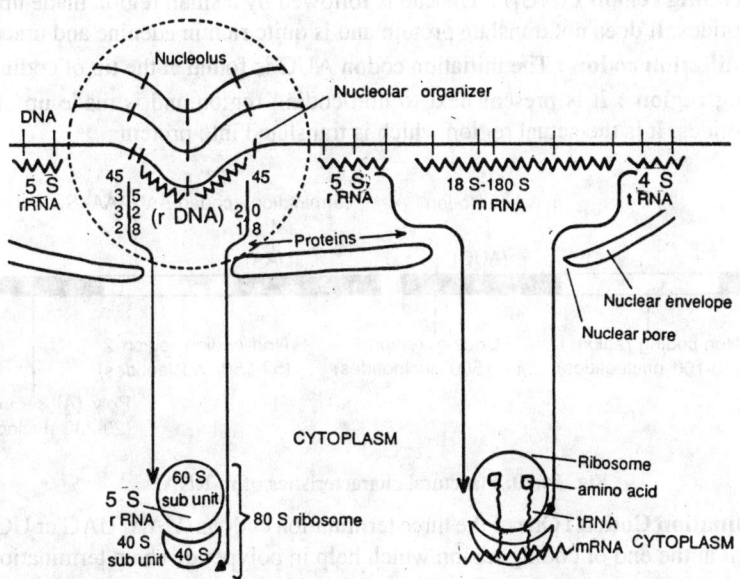

Fig. 14.19: Diagram showing transcription and transport of a nuclear RNA's in a eukaryotic cell and participation of tRNA and mRNA in protein synthesis.

Messenger RNA (m-RNA)

The RNA which carries the genetic informations from the chromosomal DNA to the cytoplasm (particularly to the ribosomes) for protein synthesis, is known as **messenger RNA** or m-RNA. The name messenger RNA was proposed by *Jacob* and *Monod* in 1961. It is also known as 'informational', 'Complementary', 'translational' and 'transcript' RNA. It is formed in the nucleus under the influence of DNA and frequently has the same base ratio as the DNA except that in the formation of m-RNA — the thymine from DNA segment is replaced by uracil and so m-RNA will have uracil bases in place of thymine. The m-RNA has high molecular weight (perhaps up to several millions) and represents only about one per cent of the total cellular RNA. The main function of the m-RNA is to direct the process of protein synthesis. After the formation of m-RNA in the nucleus, it moves to the cytoplasm where it attaches to the 30S subunit of the ribosome. The t-RNA and aminoacids also help in this complex process of protein synthesis. The m-RNA has a rapid turnover and heterogeneity with respect to both size and base content. Usually the m-RNA is also single-stranded like r-RNA, but it has got different sizes. The size of m-RNA is directly related with the size of codons for the protein molecules. Two main types of m-RNA have been discovered so far.

(*a*) *Monocistronic m-RNA* : These *m*-RNA are those which carry the codes of single cistron of the DNA i.e, codes for one complete protein molecule.

(*b*) *Polycistronic m-RNA* : It is also known as polygenic m-RNA. These m-RNA are those which carry the codes from several adjacent DNA cistrons and become much longer in size. This type of m-RNA has been recorded during the metabolism of the histidine.

Characteristics of m-RNA: m-RNA possesses following characteristic features :

1. **Cap :** The m-RNA of eukaryotic cells and animal viruses possess a 'cap' at 5' end. It is a blocked methylated structure containing several methylated bases like m^7Gpp Cmp Amp or m^7 Gpp Cmp Cmp Amp. Cmp or Amp contains 2'O methyl ribose. The rate of protein synthesis depends upon the presence of the cap which helps the m-RNA in binding with ribosomes to form polyribosome. In absence of the cap, the m-RNA remains loosely attached with ribosomes.

2. **Non-coding region I (NC$_1$)** : The cap is followed by a small region made up of 10 to 100 nucleotides. It does not translate protein and is quite rich in adenine and uracil residues.
3. **The initiation codon** : The initiation codon AUG is found at the tip of coding region.
4. **Coding region** : It is present next to non-coding region and is made up of about 1500 nucleotides. It is the actual region which is translated into protein.

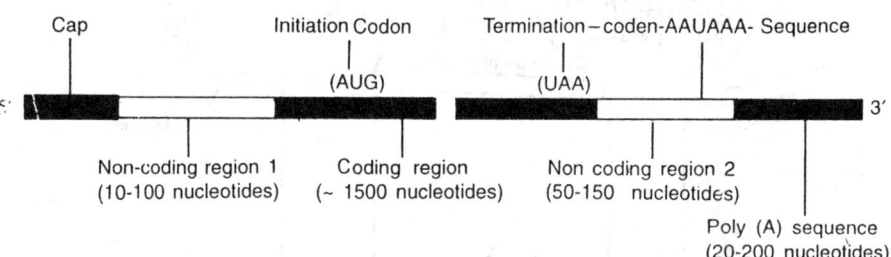

Fig 14.20: Structural characteristics of m-RNA.

5. **Termination Codon** : One of the three termination codons (UAA, UAG or UGA) is always present at the end of coding region which help in polypetide chain termination.
6. **Non-coding Region II (NC$_2$)** : It is present next to termination code. It is made up of 50–150 nucleotides and is not involved in translation of protein. NC$_2$ contains AAUAAA sequence.
7. **Poly (A) sequence** : The 3' end of m-RNA contains a polyadenylate or Poly (A) sequence which is added in the nucleus before its entry into the cytoplasm. Initially this sequence contains 200-250 nucleotides but with the lapse of time nucleotide number is reduced.

Transfer RNA (t-RNA)

Transfer RNA is also known as soluble RNA (s-RNA) and acceptor RNA. It is found free in the cytoplasm and accounts for 15 to 20 per cent of the total cellular RNA. These contain 75 to 80 nucleotides and have a molecular weight of about 25,000 daltons-which is much lower than r RNA. The t-RNA functions in the transportation of amino acids from the cytoplasm to the ribosomes. Each tRNA is specific for each amino acid. Thus, there are about twenty or twenty-two different types of t-RNAs which help in the transfer of the same number of aminoacids from the cytoplasm to the ribosomes. The t-RNA differs from other RNAs (r-RNA or m-RNA) in having unusual bases which are mostly methylated derivatives of more common bases, e.g., methylated purines, pseudo-uridines, methyl cytosine, methylaminopurine etc.

All t-RNAs possess almost similar molecular weights (25,000), but different nucleotide sequences. The sedimentation constant of t-RNA is 4S. The nucleotide sequence within the RNA molecule is known as primary structure and when within the single stranded RNA the base-pairing takes places, it gives rise to the secondary structure of the RNA. Previously, it was supposed that the single strand of RNA twists to form a hair pin like structure (helical structure) which has mostly paired bases. The unpaired bases which are generally found in the centre are known as **anti-codons**. The terminal end has also exposed unpaired bases (trinucleotide sequence) which consists of two cytidylic residues followed by an adenosine in all t-RNAs. Later on, Lake and *Beeman* (1967) suggested that each t-RNA consists of three folds of double helix to give it the shape of the clover leaf. This clover leaf pattern is the characteristic of all t-RNA's.

Clover leaf model of t-RNA : (Two-dimensional structure)

The detailed structure of t-RNA is most important because, (1) It recognises the amino acid through the enzyme amino acid RNA synthetase (AA-RNA synthetase) (2) It recognises the appropriate base sequences on m-RNA.

Nucleic Acids Metabolism

The detailed study of t-RNA was easily worked out because they are comparatively smaller than other types of RNAs. The first t-RNA which was worked out in detail is alanyl t-RNA of yeast. The studies were made by *Holley* and colleagues for which he received Nobel Prize along with *Khorana* and *Nirenberg* in 1968. This information was later used by Khorana and his colleagues for the synthesis of the gene for this t-RNA. The detailed structure of yeast alanyl t-RNA as worked out by *Robert Holley* is shown in Fig. 14.21.

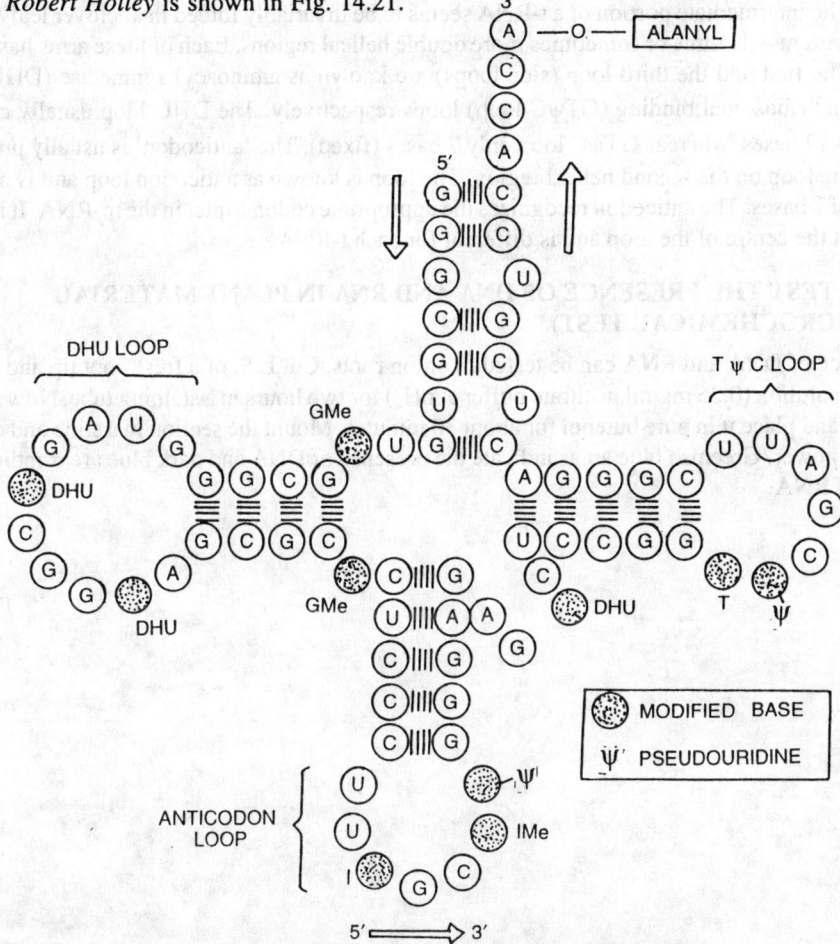

Fig. 14.21: Structure of yeast alanyl t-RNA.

On the basis of the detailed structures of several t-RNAs worked out, the following general conclusions may be drawn :

(*i*) Each t-RNA has the two ends, 3'-end and 5'-end.

(*ii*) The 3'-end has the base sequence 'CCA' and 5'-end has 'G'-nucleotide (guanine residue).

(*iii*) The amino acid is always attached at 3'-end only and the sequence 'CCA' remains unpaired.

(*iv*) Each t-RNA contains a large number of unusual nucleosides e.g. pseudouridine (ψ), Inosine (I), dihydrouridine (DHU) etc. The unusual bases are formed by the process of methylation of normal bases. The phenylalanine t-RNA of wheat germ shows the methylated bases like A^{1M}, $G7^{Me}$, $G^{2O'Me}$, G^{2Me}, G^{DiMe}, $C^{2,OMe}$ etc.

(*v*) The ratios of A:U and G:C are near unity which, suggests the formation of DNA like double helical segments (secondary structure).

(vi) In these double helical regions of RNA, G:C base pairs are more common than A:U.

(vii) Each t-RNA possesses a tertiary structure and Mg^{++} ion concentration plays an important role in its stabilization. Therefore, although each t-RNA is a polynucleotide, it resembles with enzymes in its structure and function. The details of the tertiary structure of t-RNA are still unknown.

(viii) The intermediate portion of a t-RNA seems to be invariably folded in a 'clover leaf' pattern with mostly three or sometimes more double helical regions. Each of these arms has a loop. The first and the third loop (side loops) are known as aminoacyl synthetase (DHU loop) and ribosomal binding (GTψC loop) loops respectively. The DHU loop usually contains 8-12 bases, whereas GTψC loop only 7 bases (fixed). The 'anticodon' is usually present in the loop on the second helical region. The loop is known as anticodon loop and is made up of 7 bases. The anticodon recognizes the appropriate codon triplet in the m-RNA. It is found at the centre of the loop and is different for each t-RNA.

14.16. TO TEST THE PRESENCE OF DNA AND RNA IN PLANT MATERIAL (MICROCHEMICAL TEST)

The presence of DNA and RNA can be tested in onion roots. Cut L.S. of a fresh root tip and place it in Azure B solution (0.25 mg/ml in citrate buffer at pH_4) for two hours in a staining tube. Now remove the section and place it in pure butenol for about 30 minutes. Mount the section in a slide and observe under high power. Greenish blue areas indicate the presence of DNA and dark blue areas indicate the presence of RNA.

15
Metabolism of Amino Acids and Protein

15.1 BIOSYNTHESIS OF AMINO ACIDS

Aminoacids are important as they are precursors of proteins. In plants, they are synthesized from organic acids (produced during glycolysis and Kreb's cycle) and inorganic nitrogen (NH_3). The inorganic nitrogen enters into plants either in the form of nitrate or ammonium. The nitrate is first reduced to ammonium and then is incorporated into amino acids. The leguminous plants possess symbiotic nitrogen fixing bacteria in their root nodules and they use atmospheric nitrogen to synthesize aminoacids. The micro-organisms differ in their capacity to synthesize amino acids. There are following two basic mechanisms of amino acid biosynthesis :

By Reductive Amination

The conversion of inorganic nitrogen (NH_3) into organic nitrogen (amino acid) through organic acids, where amination and reduction at keto group of the organic acid takes place, is called *reductive amination*. It is the primary pathway of amino acid biosynthesis. For example the amino acid *glutamic acid* is synthesized from α-ketoglutaric acid (an organic acid produced as an intermediate of Kerb's cycle) in presence of enzyme *glutamic dehydrogenase* and $NADPH_2$.

$$\alpha\text{-ketoglutaric acid} + NH_3 + NADPH + H^+ \xrightarrow{\text{Glutamic dehydrogenase}} \text{glutamic acid} + NADP^+ + H_2O$$

The amino acid glutamic acid serves as a precursor for several other amino acids like proline, hydroxyproline, ornithine, citrulline, arginine, r-aminobutyric acid etc. These amino acids are formed by transamination reaction of glutamic acid.

The glutamic acid can also be synthesized by an alternate path way called *glutamine synthetase α-ketoglutaric acid aminotransferase* (GS-α-K-GAT) *pathway*. Here, glutamic acid is synthesized through the amide *glutamine*, which itself is synthesized from glutamic acid and ammonia.

$$\text{Glutamic acid} + NH_3 \xrightarrow[\text{ATP} \quad \text{ADP + Pi}]{\text{Glutamine synthetase}} \text{Glutamine}$$
(One molecule)

α-ketoglutaric acid combines with glutamine and undergoes amination reaction to produce 2 molecules of glutamic acid in presence of enzyme glutamic acid synthetase.

$$\alpha\text{-ketoglutaric acid} + \text{glutamine} \longrightarrow \text{glutamic acid}$$
(Two molecules)

Thus, one glutamic acid molecule synthesize two molecules of glutamic acid showing net production of one glutamic acid molecule.

The aspartic acid and alanine amino acids can also be synthesized directly by the incorporation of ammonia into oxaloacetic acid and pyruvic acid respectively as follows.

$$\text{Oxaloacetic acid} + NH_3 \rightleftharpoons \text{Aspartic acid}$$
$$\text{Pyruvic acid} + NH_3 \rightleftharpoons \text{Alanine}$$

1.2 BY TRANSAMINATION

The transfer of amino group (–NH$_2$) of an amino acid to carbonyl group of a keto acid is called *transamination*. About 17 aminoacids are synthesized from glutamic acid by transamination. During the synthesis of various amino acids, first glutamic acid is produced by reductive amination and then other aminoacids are synthesized by the transfer of its amino group to various keto acids. The transamination reaction is catalysed in presence of enzyme *transaminase*. Each aminoacid synthesis requires a separate transaminase enzyme. The coenzyme *pyridoxal phosphate* is also required in all transamination reations.

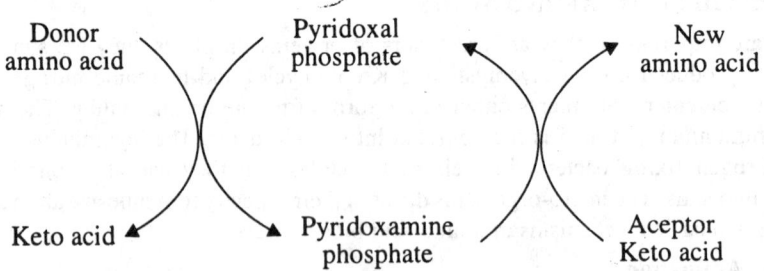

Fig. 15.1 : Transamination reaction.

In the above reaction the coenzyme pyridoxal phosphate acts as a carrier of amino group. It picks up amino group from the donor amino acid and is converted into *pyridoxamine* phosphate which ultimately transfers the containing amino group to the acceptor ketoacid to form a new aminoacid and to regenerate pyridoxal phosphate. The formation of aspartic acid and α-alanine aminoacids by transamination reactions is as follows :

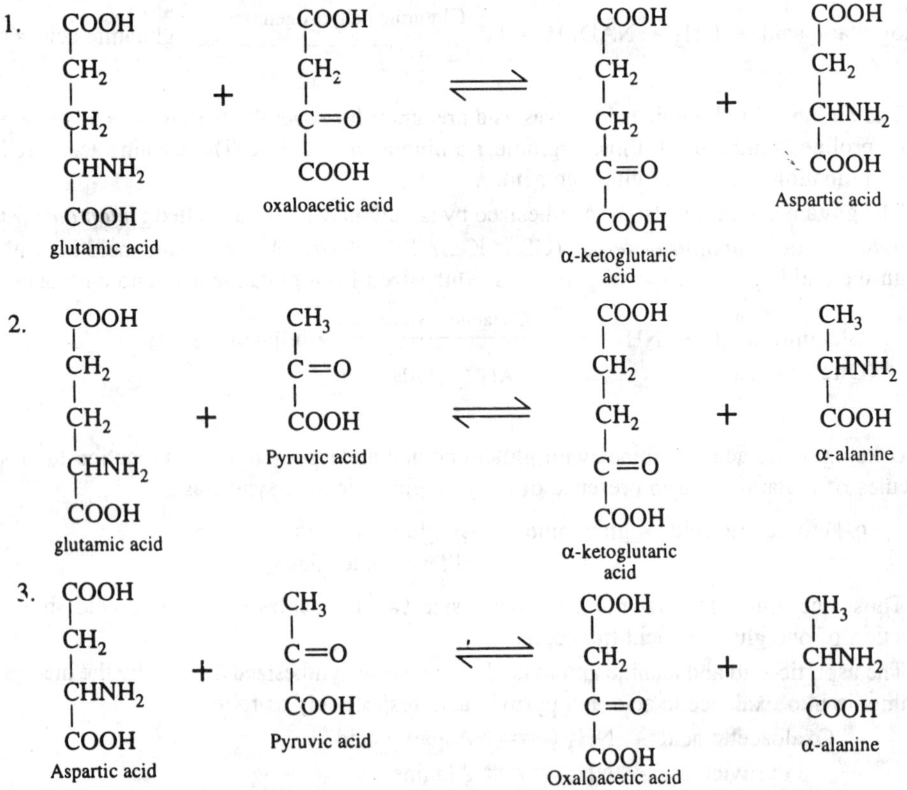

1. glutamic acid + oxaloacetic acid ⇌ α-ketoglutaric acid + Aspartic acid

$$\begin{array}{c}\text{COOH}\\|\\\text{CH}_2\\|\\\text{CH}_2\\|\\\text{CHNH}_2\\|\\\text{COOH}\end{array} + \begin{array}{c}\text{COOH}\\|\\\text{CH}_2\\|\\\text{C=O}\\|\\\text{COOH}\end{array} \rightleftharpoons \begin{array}{c}\text{COOH}\\|\\\text{CH}_2\\|\\\text{CH}_2\\|\\\text{C=O}\\|\\\text{COOH}\end{array} + \begin{array}{c}\text{COOH}\\|\\\text{CH}_2\\|\\\text{CHNH}_2\\|\\\text{COOH}\end{array}$$

2. glutamic acid + Pyruvic acid ⇌ α-ketoglutaric acid + α-alanine

$$\begin{array}{c}\text{COOH}\\|\\\text{CH}_2\\|\\\text{CH}_2\\|\\\text{CHNH}_2\\|\\\text{COOH}\end{array} + \begin{array}{c}\text{CH}_3\\|\\\text{C=O}\\|\\\text{COOH}\end{array} \rightleftharpoons \begin{array}{c}\text{COOH}\\|\\\text{CH}_2\\|\\\text{CH}_2\\|\\\text{C=O}\\|\\\text{COOH}\end{array} + \begin{array}{c}\text{CH}_3\\|\\\text{CHNH}_2\\|\\\text{COOH}\end{array}$$

3. Aspartic acid + Pyruvic acid ⇌ Oxaloacetic acid + α-alanine

$$\begin{array}{c}\text{COOH}\\|\\\text{CH}_2\\|\\\text{CHNH}_2\\|\\\text{COOH}\end{array} + \begin{array}{c}\text{CH}_3\\|\\\text{C=O}\\|\\\text{COOH}\end{array} \rightleftharpoons \begin{array}{c}\text{COOH}\\|\\\text{CH}_2\\|\\\text{C=O}\\|\\\text{COOH}\end{array} + \begin{array}{c}\text{CH}_3\\|\\\text{CHNH}_2\\|\\\text{COOH}\end{array}$$

Fig. 15.2 : Reactions to show formation of aspartic acid and α-alanine amino-acids by transamination.

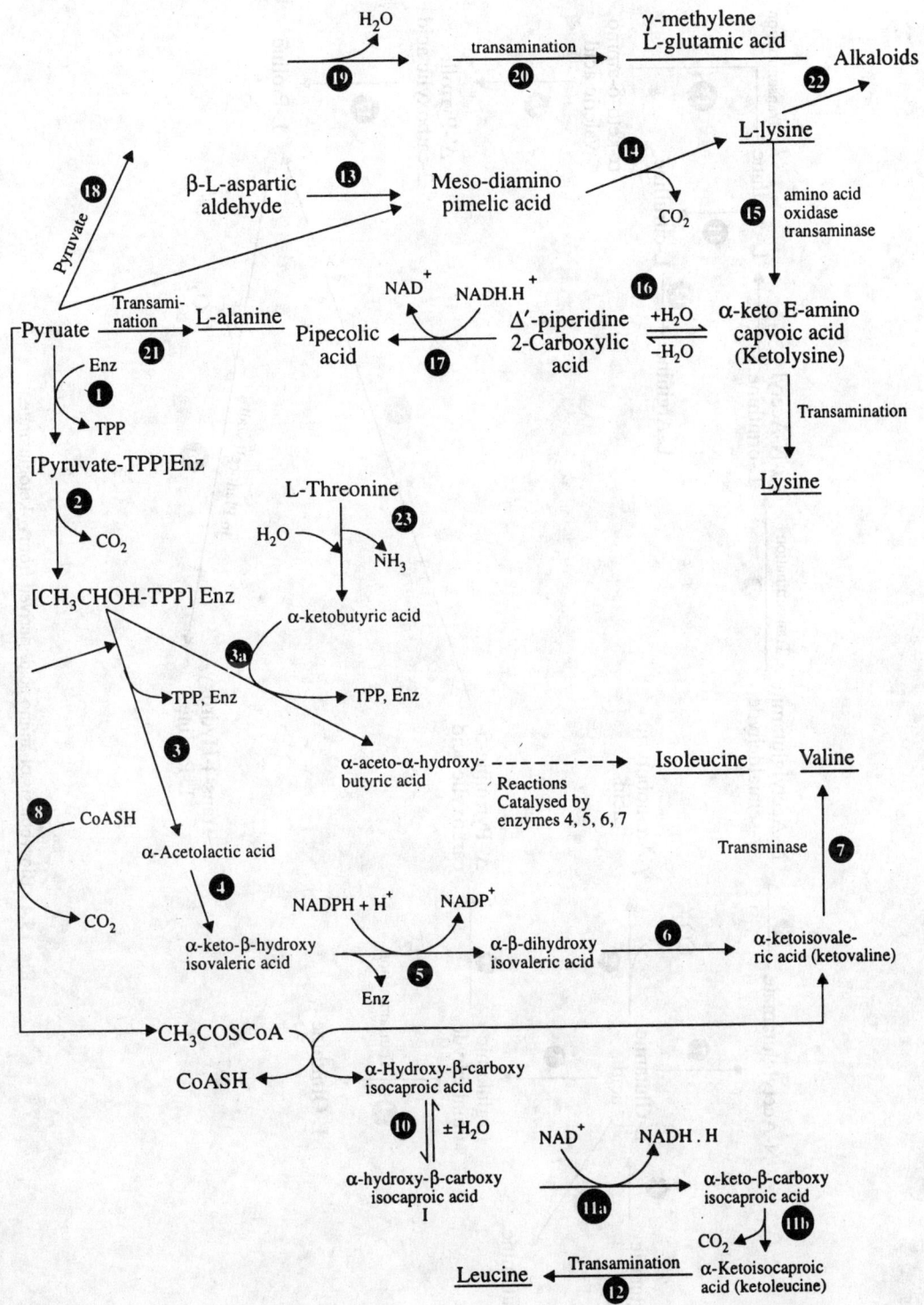

Fig. 15.3 : Biosynthesis of amino acids derived from pyruvate.

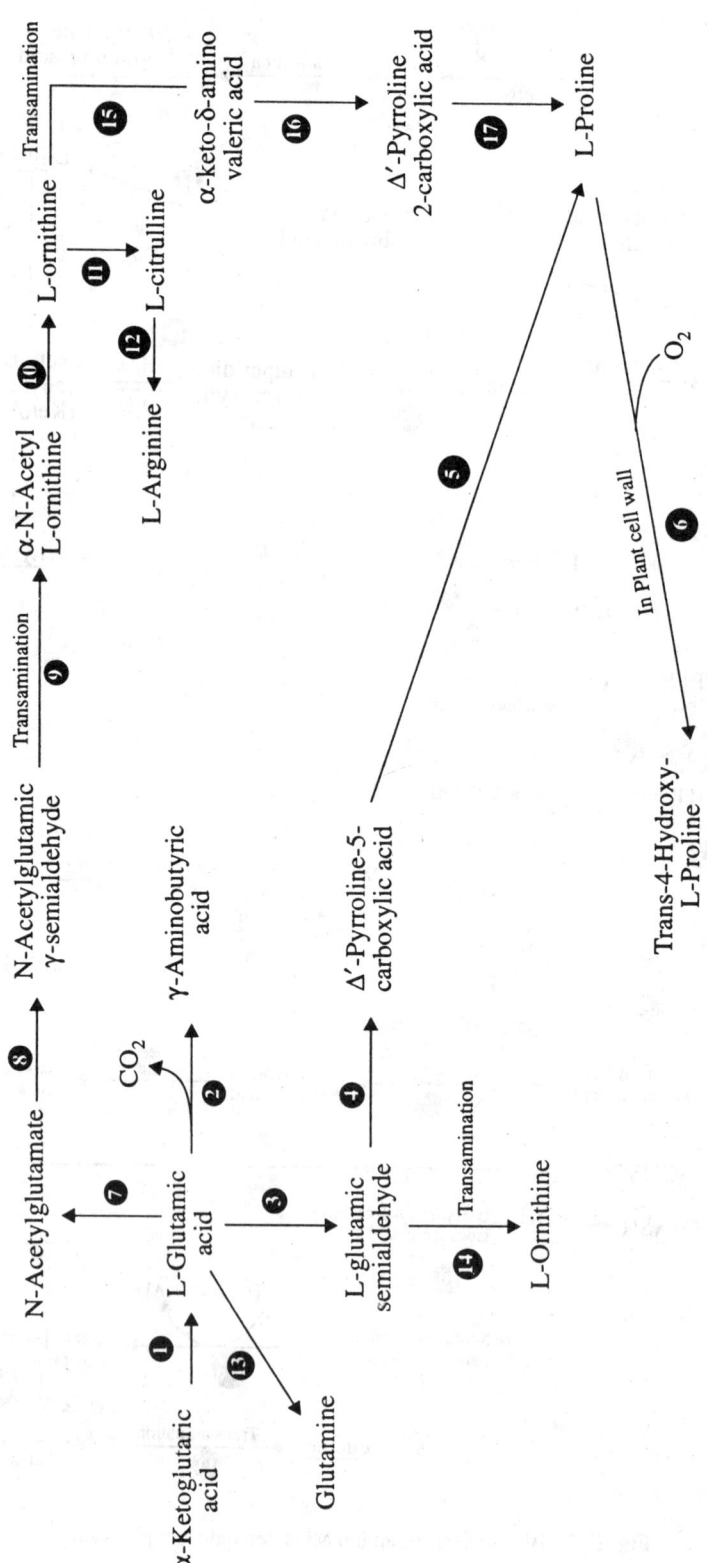

Fig. 15.4 : Biosynthesis of amino acids derived from α-ketoglutarate.

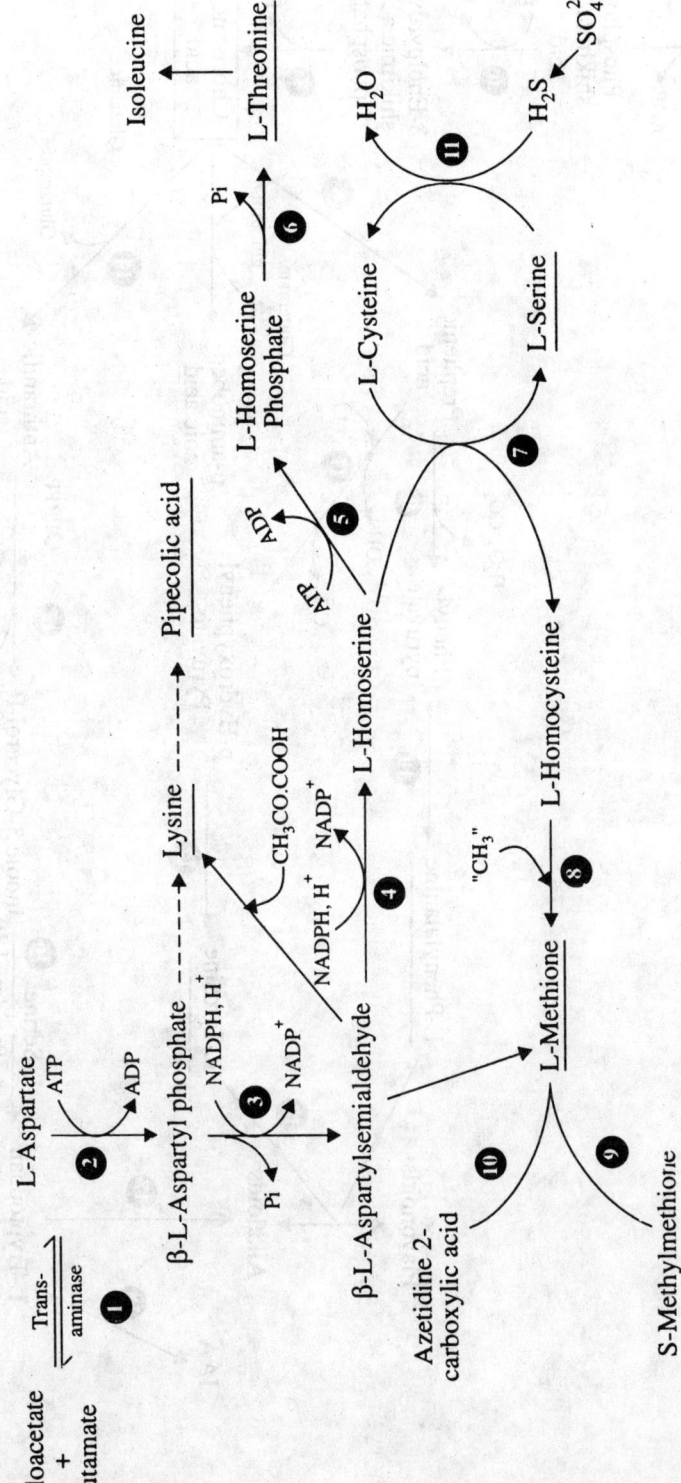

Fig. 15.5 : Biosynthesis of aminoacids derived from oxaloacetate.

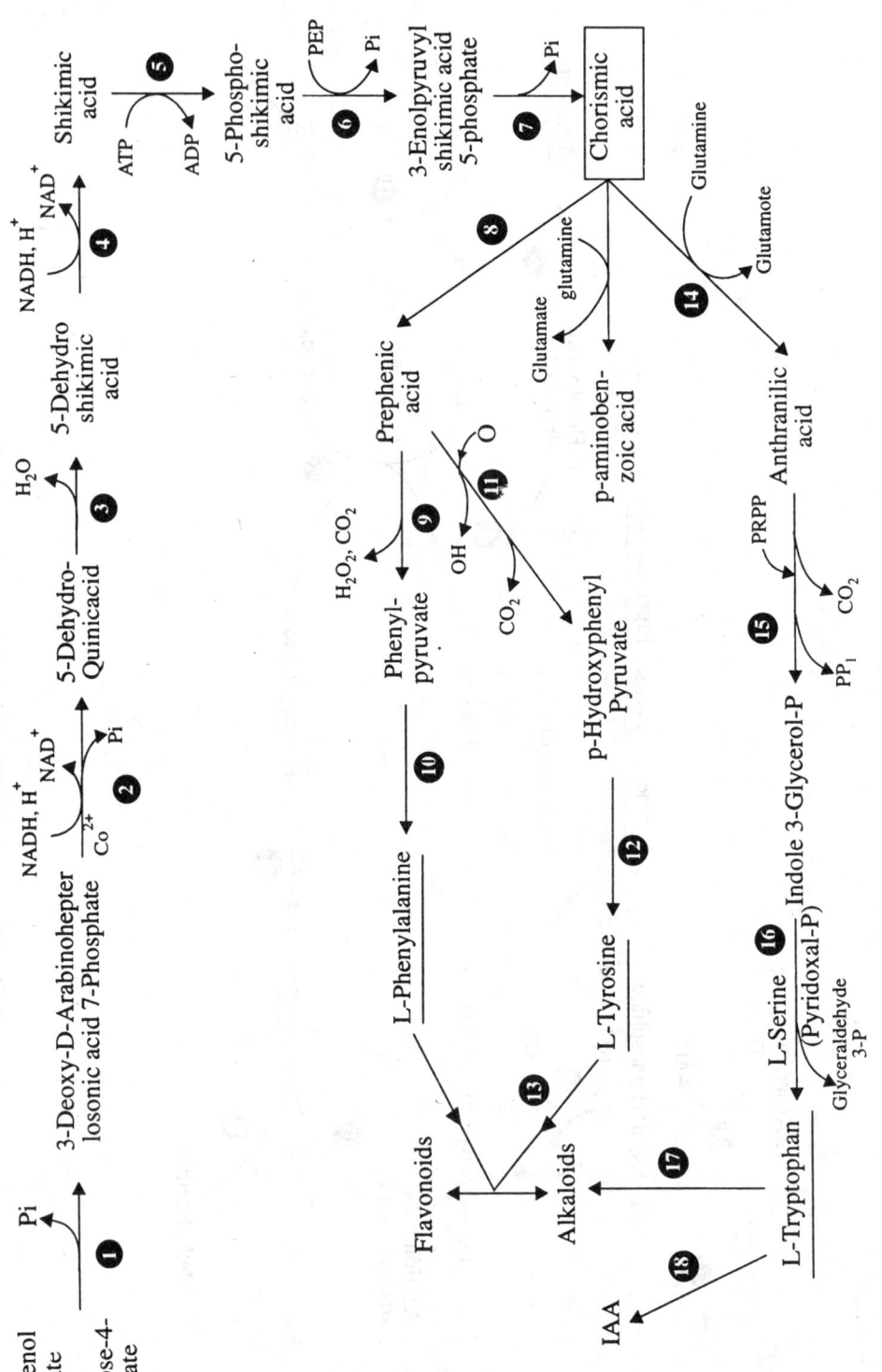

Fig. 15.6 : Biosynthesis of aromatic amino acids.

Transamination reactions are also involved in the synthesis of glycine, leucine, isoleucine, valine, serine and the aromatic aminoacids. The biosynthesis of different type of amino acids derived from pyruvate, α-keto glutarate, oxaloacetate and aromatic amino acids is given in Fig. 15.3, 15.4, 15.5 and 15.6 respectively.

15.2. AMPHOTERIC NATURE OF AMINO ACIDS

Amino acids are **amphoteric** in nature because they show **basic** as well as *acidic* properties due to the presence of ionizable **amino** and **carboxylic** groups.

ZWITTER-ION

In amphoteric compounds *e.g.*, amino acids, there are two ionizable groups.

$$-NH_2 \rightarrow -NH_3^+$$
$$-COOH \rightarrow -COO^-$$

If there is simultaneous ionization at both the groups, there will be two opposite charges on the ion. Such an ion in called as a **Zwitter-ion** (Fig. 15.7).

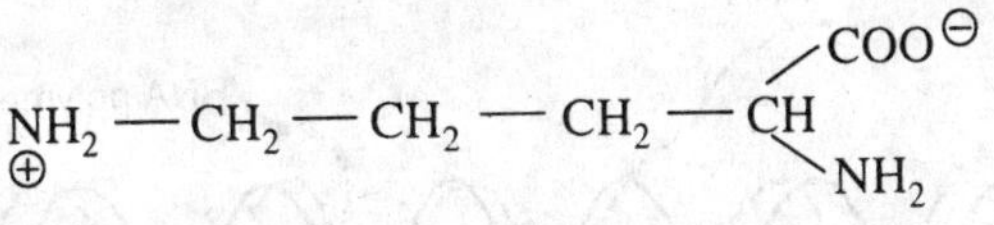

$$\overset{\oplus}{NH_2}-CH_2-CH_2-CH_2-CH\overset{COO^{\ominus}}{\underset{NH_2}{\diagdown}}$$

Fig. 15.7 : Zwitter ion of ornithine.

15.3. ISOELECTRIC POINT

As discussed earlier the amphoteric compounds like amino acids occur in Zwitter-ion state in solution, At lower pH (acidic values here is excess of H$^+$ in the solution the amino acids will be positively charged, since they will have excess $-NH_3^+$. On the other hand, at higher pH (alkaline) value the amino acids will be negatively charged due to excess $-COO^-$ groups, since the deficit of H$^+$ ions in the solution will be made up by the OH$^-$ ions in the sulution drawing H$^+$ ions from the $-COOH$ groups of the amino acids.

There will be a certain intermediate pH value where the number of positive and negative charges will be equal and the amino acids will become **neutral** or **uncharged**. This is called as the **isoelectric point**.

15.4. BIOSYNTHESIS OF PROTEINS

In most cases, biosynthesis of protein is under direct control of DNA which contains the necessary information for protein synthesis. This information is first transcribed into mRNA and then is translated into protein. There are a few organisms such as viruses where DNA is absent and the genetic information for protein synthesis is contained in the RNA. There are three ways in which the genetic information flows from DNA to protein:

(1) Unidirectional flow :

$$DNA \xrightarrow{Transcription} RNA \xrightarrow{Translation} Protein$$

(2) Circular flow :

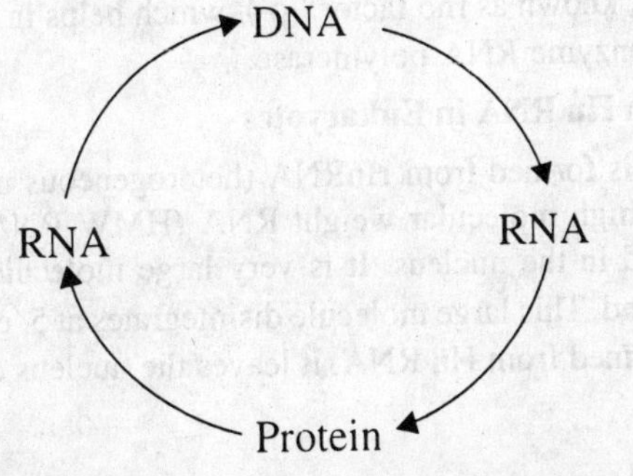

$$DNA \rightarrow RNA \rightarrow Protein \rightarrow DNA$$

(3) Reverse flow through Reverse Transcriptase :

$$RNA \xrightarrow{\text{Reverse Transcriptase}} DNA \xrightarrow{\text{Transcription}} RNA \xrightarrow{\text{Translation}} Protein$$

Unidirectional flow of information is explained by the Central dogma in which information flows from DNA to RNA and then to protein. The circular flow of information was proposed by *Barry Commoner* (1968). The reverse flow of information from RNA to DNA was demonstrated by *H. Temin* and *D. Baltimore* (1970) in RNA tumour viruses in which double-stranded circular DNA is formed from RNA with reverse transcriptase. The information from circular DNA to RNA flows as in other cases. For this discovery *Temin* and *Baltimore* were awarded Nobel Prize for 1975 along with *R. Dulbecco* in medicine.

15.5. MECHANISM OF PROTEIN SYNTHESIS

The mechanism of protein synthesis involves the following steps as shown in Figs 15.8. and 15.9.

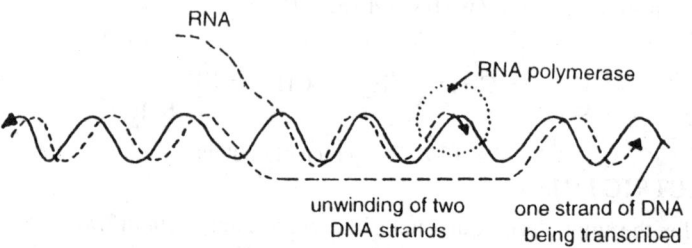

Fig. 15.8 : Synthesis of RNA on DNA template (Transcription).

Transcription

The process of formation of mRNA from DNA is known as *transcription* (Fig. 15.8). When protein synthesis begins, the genetic information contained in DNA is transferred to mRNA. The two strands of DNA molecule uncoil. One of the strands acts as a template for the formation of mRNA. The mRNA is formed according to the triplet codes of DNA by the copying process. Transcription process is completed with the help of RNA polymerase. In bacteria, RNA polymerase is made up of 4 different polypeptide chains (σ, β', β, α_2). The β', β, α_2 are jointly called as *core enzyme* whereas σ is known as *sigma factor*. Sigma factor is found usually loosely attached and, therefore, core enzyme (β', β, α_2) can be easily separated.

The sigma factor helps in the recognition of start signals on DNA molecule and directs RNA polymerase in selecting the initiation sites. Sometimes, sigma factor is not found in RNA polymerases. In that case core enzyme helps in the selection of initiation sites. As soon as the RNA synthesis is initiated, sigma factor dissociates and the core enzyme now helps in the elongation of m-RNA. The dissociated sigma factor can again combine with the core enzyme to form RNA polymerase.

There is another factor, known as rho factor (ρ), which helps in termination of mRNA chain. This factor is not a part of enzyme RNA polymerase.

Maturation of mRNA from Hn RNA in Eukaryotes

In eukaryotes, mRNA is formed from HnRNA (heterogeneous nuclear RNA). The precursor Hn RNA is also known as high molecular weight RNA (HMW RNA) and DNA like RNA. The Hn RNA is first synthesized in the nucleus. It is very large molecule which, contains about 200 adenylic acid residues at 3' end. This large molecule disintegrates at 5' end to form mRNA molecule. As soon as the mRNA is formed from Hn RNA, it leaves the nucleus and reaches in the cytoplasm

Metabolism of Amino Acids and Protein

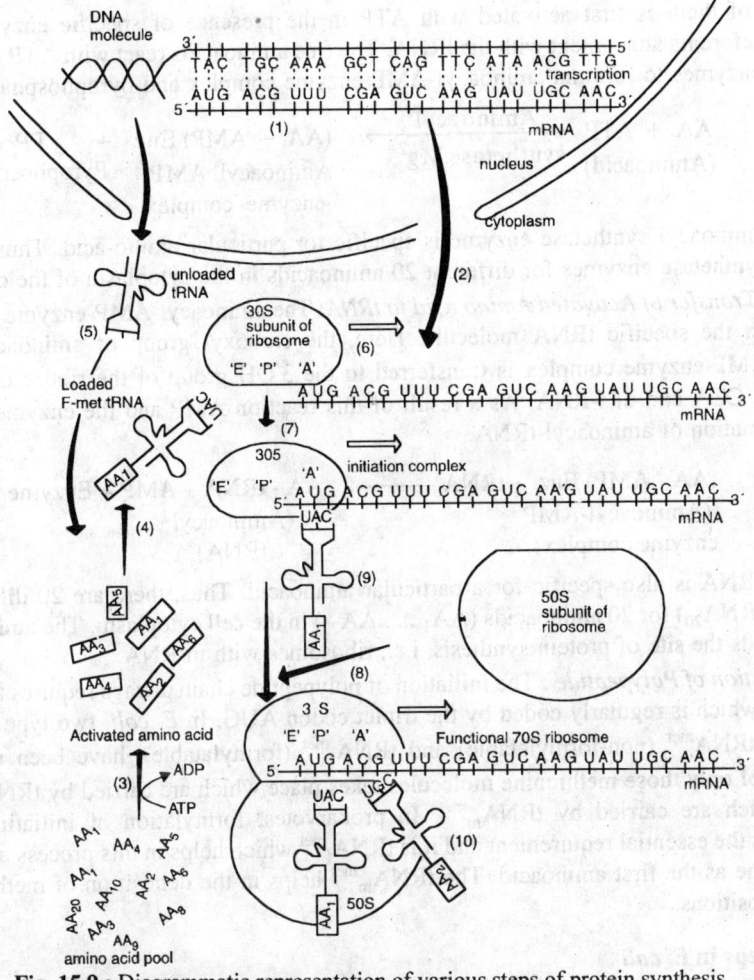

Fig. 15.9 : Diagrammatic representation of various steps of protein synthesis.

Step.
1. Synthesis of mRNA (transcription)
2. Maturation and movement of mRNA to cytoplasm through nuclear pore.
3. Activation of aminoacids.
4. Transfer of activated aminoacid to tRNA.
5. Formylation of AA_1 (methionine)—tRNA.
6. Attachment of mRNA (5' end) with 30S ribosomal subunit.
7. Attachment of f-AA-tRNA^{met} with 30S-mRNA complex to form initiation complex.
8. Association of 50S subunit to form 70S ribosome.
9. Binding of AA_1-tRNA at "A" site of ribosome.
10. Elongation of polypeptide chain by addition of amino acids.

through nuclear pores of the nuclear membrane where it attaches with the 40S subunit of the ribosome for protein synthesis. Recent studies have shown that m-RNA of eukaryotes possesses polyadenylic acid sequence at 3' end and 7-methyl guanosine at 5 'end.

Translation

The process of formation of protein from the language available in the form of mRNA is known as *translation*. It is very complex process and involves the following steps:

1. *Activation of Amino Acids* : In the cytoplasm, all the 20 amino acids occur in an inactive

state. Each of them is first activated with ATP in the presence of specific enzyme *aminoacyl synthetase* before its attachment with the tRNA. The free aminoacids react with ATP in the presence of specific enzymes to form an aminoacyl-AMP-enzyme complex and pyrophosphate.

$$AA + ATP \xrightarrow{\text{Aminoacyl synthetase, Mg}^{2+}} (AA - AMP) \text{Enz} + PP$$
(Aminoacid)   Aminoacyl AMP– Pyrophosphate enzyme complex

Each aminoacyl synthetase enzyme is specific for particular amino-acid. Thus, there are 20 aminoacyl synthetase enzymes for different 20 aminoacids in the cytoplasm of the cell.

2. *The Transfer of Activated Amino acid to tRNA*: The aminoacyl-AMP-enzyme complex, now collides with the specific tRNA molecule. Here, the carboxyl group of aminoacid residue of aminoacyl-AMP-enzyme complex is transferred to the 3'OH group of the ribose of the terminal adenosine at CCA end of t-RNA. As a result of this reaction AMP and the enzyme are liberated with the formation of aminoacyl-tRNA.

$$(AA - AMP) \text{Enz.} + tRNA \longrightarrow AA\text{-}tRNA + AMP + Enzyme$$
(Aminoacyl-AMP enzyme complex)   (Aminoacyl tRNA)

Each t-RNA is also specific for a particular aminoacid. Thus, there are 20 different tRNAs ($tRNA_1$ ......$tRNA_{20}$) for 20 aminoacids (AA_1 ......AA_{20}) in the cell cytoplasm. The aminoacyl-tRNA moves towards the site of protein synthesis, i.e., ribosomes with m-RNA.

3. *Initiation of Polypeptide* : The initiation of polypeptide chain always requires the aminoacid methionine, which is regularly coded by the triplet codon AUG. In *E. coli*, two type of $tRNA_s$ for methionine, $tRNA_m^{met}$ (non-formylatable) and $tRNA_f^{met}$ (formylatable), have been reported. The formylation of only those methionine molecules takes place which are carried by $tRNA_f^{met}$ and not of those which are carried by $tRNA_m^{met}$. In prokaryotes, formylation of initiating aminoacid methionine is the essential requirement and it is $tRNA_f^{met}$ which helps in this process and it deposits the methionine as the first aminoacid. The $tRNA_m^{met}$ helps in the deposition of methionine at the intercalary positions.

Initiation Steps in *E. coli*

(i) *Formation of met-$tRNA_f^{met}$*: The formation of met-$tRNA_f^{met}$ takes place first by activation of methionine and then its transfer to $tRNA_f^{met}$.

$$\text{Methionine} + ATP \xrightarrow{\text{Methionine acyl synthetase, Mg}^{2+}} (\text{Met–AMP})Enz_1 + PP$$

$$(\text{Met-AMP})Enz_1 + tRNA_f^{met} \rightarrow \text{Met-}tRNA_f^{met} + AMP + Enz_1$$

(ii) *Formation of formyl-methionyl $tRNA_f^{met}$ (f -met-$tRNA_f^{met}$)*: The formylation of met-$tRNA_f^{met}$ is brought about by enzyme transformylase in presence of formyltetrahydrofolic acid (f-THFA).

$$\text{Met} + tRNA_f^{met} + \text{Formyl-tetrahydrofolic acid} \xrightarrow{\text{transformylase}} \text{f-met-}tRNA_f^{met}$$

(iii) *The attachment of mRNA (5'end) with 30S Subunit of ribosome*: It has been observed that prior to protein synthesis the ribosomes occur in dissociated (50S and 30S) and inactive state. The 5'end of the mRNA carrying AUG triplet codon, binds with the 30S ribosomal subunit in presence of initiation F_3 protein factor.

$$mRNA + 30S \xrightarrow{F_3} 30S\text{-}mRNA$$

Metabolism of Amino Acids and Protein

(iv) *The Attachment of f-met-tRNA$_f^{met}$ with 30S-mRNA to form initiation complex*: The f-met-tRNA$_f^{met}$, formed in the cytoplasm by steps (i) and (ii), now attaches with the first triplet codon (AUG) of 30S-mRNA complex, formed by step (iii), to initiate the process of protein synthesis and to form the initiation complex, 30S-mRNA-f-met-tRNA$_f^{met}$. The formation of initiation complex is facilitated by the GTP (guanosine triphosphate) and three protein initiation factors (F_1, F_2 and F_3).

$$30S\text{-mRNA} + \text{f-met-tRNA}_f^{met} \xrightarrow[F_1, F_2 \text{ and } F_3]{GTP} 30S\text{-mRNA-f-met-tRNA}_f^{met} \text{ (Initiation complex)}$$

(v) *Association of Ribosomal Subunits*: Soon after the formation of initiation complex, 30S-mRNA-f-met-tRNA$_f^{met}$, the 30S ribosomal subunit associates with 50S ribosomal subunit to form the 70S ribosome. This union of ribosomal subunits requires Mg^{+2} ion of .001M concentration. When the Mg^{+2} concentration is increased, many ribosomes unite to form a structure known as *polyribosome* or polysome on a common mRNA.

$$30S\text{-mRNA-f-met-tRNA}_f^{met} + 50S \xrightarrow{0.001M\ Mg^{2+}} 70S\text{-mRNA-f-met-tRNA}_f^{met}$$

Initiation of Polypeptide in Eukaryotes

In eukaryotes the formylation of methionine does not occur due to absence of tRNA$_f^{met}$ in plants and enzyme transformylase in animals. Thus, initiation in higher organisms takes place without formylation. In mammals, the initiation is, however, completed in the following steps :

$$\text{Methionine} + \text{ATP} \xrightarrow[\text{acyl synthetase, } Mg^{2+}]{\text{Methionine}} (\text{Met-AMP})\ ENZ_1 + PP$$

$$(\text{Met-AMP})\ ENZ_1 + \text{tRNA}_f^{met} \rightarrow \text{Met-tRNA}_f^{met} + AMP + ENZ_1$$

$$40S + \text{Met-tRNA}_f^{met} \rightarrow 40S\text{-met tRNA}_f^{met}$$

$$40S\text{-met tRNA}_f^{met} + \text{mRNA} \rightarrow 40S\text{-mRNA-met-tRNA}_f^{met}$$

$$40S\text{-mRNA-met-tRNA}_f^{met} + 60S \rightarrow 80S\text{-mRNA-met-tRNA}_f^{met}$$

Thus, in eukaryotes (mammals) the initiation does not require formylation of methionine and here the smaller 40S ribosomal subunit directly associates with met-tRNA$_f^{met}$ without the help of mRNA.

4. *Elongation of Polypeptide*: As soon as the formation of 70S-mRNA-f-met-tRNA$_f^{met}$ complex takes place, the polypeptide chain elongates and it is done by regular addition of aminoacids. The various steps in elongation of polypetide are:

(i) *Binding of AA-tRNA at aminoacid site of ribosome*: The formation of polypetide chain occurs in the 50S ribosomal subunit. The 50S subunit contains two binding sites, viz., aminoacid tRNA site ('A' site) and peptide tRNA site ('P' site), for the tRNA. It is still unknown whether f-met. tRNA$_f^{met}$ comes on 'P' site or on 'A' site, but ultimately it has to be present on 'P' site, to make 'A' site available for the next aminoacyl tRNA (AA-tRNA).

The tRNAs which carry activated aminoacids along them enter in the 50S subunit from one side and attach with the 'A' site. After leaving the aminoacids to 'P' site, the tRNAs exit from the other side. The peptide tRNA site contains the tRNA along with growing polypeptide chain. The entrance of AA-tRNA to 'A' site requires energy rich compound GTP (guanosine triphosphate) and a transfer factor TF. They form a complex AA-tRNA-GTP-TF which deposits AA-tRNA at 'A' site with the liberation of GDP (guanosine diphosphate), transfer factor (TF) and phosphate group. The binding of AA-tRNA to 'A' site takes place only when the tRNA has the complementary codon to that of m-RNA.

(ii) *Formation of peptide (-CO-NH) bond*: The formation of peptide bond takes place between the free carboxylic group (—COOH) of the peptidyl tRNA at the 'P' site and the free amino group (—NH$_2$) of aminoacyl tRNA at the 'A' site. This reaction is catalysed by the enzyme *peptidyl transferase*, present in 50S ribosomal subunit, with the liberation of one molecule of water. The first peptide bond is formed between—COOH group of formyl methionine (AA$_1$) and the amino group of AA$_2$. After the formation of peptide bond, the tRNA at 'P' site is deacylated and the tRNA at 'A' site now carries the polypeptide. The enzymes which cause the transfer of aminoacids are termed as *transferase I, transferas II*---and so on.

(*iii*) *Translocation of peptidyl tRNA from 'A' to 'P' site*: The enzyme transferase I kicks off tRNA from formyl methionine (AA$_1$) and flips the AA$_1$ to the aminoacyl-tRNA (AA$_2$-tRNA) bound at 'A' site. The aminoacyl tRNA (AA$_1$-AA$_2$-tRNA) also known as peptidyl-tRNA present at 'A' site is now translocated to 'P' site with the help of another enzyme known as transferase II or translocase or G factor. Associated with this translocating activity by transferase II there is relative movement of ribosome over the mRNA, so that ribosome reaches to the next triplet code. The translocating activity, thus, opens the 'A' site for the third aminoacid (AA$_3$) coming in as AA$_3$-tRNA. Each aminoacid follows the cycle of aminoacyl-tRNA binding, formation of peptide bond by peptidyl transferase and translocation by transferase II. As the cycle is completed once, one aminoacid is added to polypetide chain.

(5) *Chain termination*: When the synthesis of polypeptide chain is completed according to the codons of mRNA, the process of chain termination occurs. It is brought about due to the presence of one of the three terminating codons, namely UAA, UAG and UGA. These terminating codons are present at the end of mRNA and act as terminating signals. These codons are also known as *non-sense codons* because they do not code for any aminoacid. At the end, polypeptide chain is released with the help of two releasing protein factors known as R$_1$ and R$_2$ by spliting of the carboxyl end of polypeptide and the last tRNA carrying the chain. Now, the ribosome dissociates into two subunits (30S and 50S) with the help of F$_3$ dissociation factor.

The polypeptide chain, thus, formed may be modified by many enzymes like deformylase, exopeptidases etc. The deformylase causes the removal of formyl group of methionine and exopeptidases of other amino acids either from N-terminal end or C-terminal end or both. Sometimes, the same chain undergoes folding to give it the secondary, tertiary or quarternary structure and sometimes it unites with other polypeptide chains to give a complex protein. The union takes place with the help of different type of bondings described in chapter 3 of Part I.

15.6. REGULATION OF PROTEIN SYNTHESIS

Protein synthesis is a very complex process and it is regulated by a system of many genes. In the body of an organism many proteins and enzymes are formed, but all of them are not required at one time. However, at all times in the life-cycle, every cell contains the same set of genes. It would be necessary, therefore, to have a machanism which would allow only the desired genes to function at a particular time. Some of these mechanisms for the regulation and control of protein synthesis have been demonstrated in bacteria, but it is quite difficult to demonstrate them in higher organisms.

Enzyme Induction and Repression

Certain micro-organisms could change their enzymic machinery under the influence of specific substrates (food stuffs added to the medium). This phenomenon is designated as "enzyme induction". The substrates, whose introduction into the growth medium specifically increases the amount of an Enzyme, are known as "inducers". Their corresponding enzymes are called "inducible enzymes" and the genetic system responsible for the synthesis or increase of such an enzyme are known as "inducible systems".

One of the best studied systems is β-*galactosidase*, the enzyme that catalyses the hydrolysis of lactose into galactose and glucose in *E. coli*. The enzyme is produced only when lactose (inducer) is present in the medium. In the absence of lactose, enzyme synthesis virtually stops. It has been demonstrated that inducer of the enzyme and the substrate on which it acts may or may not be identical. In fact, not all inducers are substrates and not all substrates are inducers. For example, methyl-β-D-thiogalactoside is a substrate but not an inducer. Neither of these substances supports growth alone, but if they are put together, growth takes place. Because of these findings, the concept of enzyme induction has been superseded by the concept of induced enzyme synthesis, (Fig. 15.10).

The opposite phenomenon is called *enzyme repression* where the end product of a biosynthetic sequence represses the synthesis of enzymes involved. The end products, whose presence in the medium checks the synthesis of an enzyme, are known as "Co-repressors", their corresponding enzymes are called "repressible enzymes" and these genetic systems are known as "repressible systems".

The example of this type of system was also studied in *E. coli* where the enzymes involved in histidine synthesis are produced only when the amount of free histidine in the cell is drastically reduced. In the presence of free histidine, enzyme production stops. It has been demonstrated that the cells growing in a medium without any aminoacid contain all the enzymes necessary for the biosynthesis of the 20 necessary aminoacids, (Fig. 15.11).

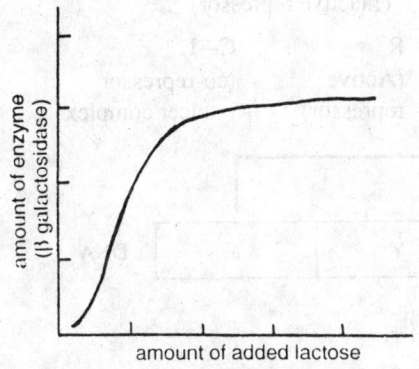

Fig. 15.10 : Induction of β-galactosidase by lactose.

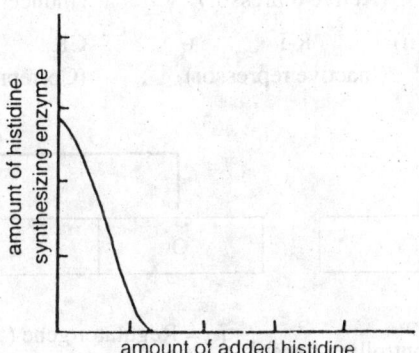

Fig. 15.11 : Repression of histidine synthesizing enzymes by histidine.

The Operon Model

The famous "Operon model" was proposed by *F. Jacob* and *J. Monod in* 1961, in order to explain the induction or repression of β-galactosidase enzyme synthesis in *E. coli*. The model proposed by Jacob and Monod assumes that there are two general types of genetic elements: Controlling genes and structural genes. The controlling genes are of two types: regulators and

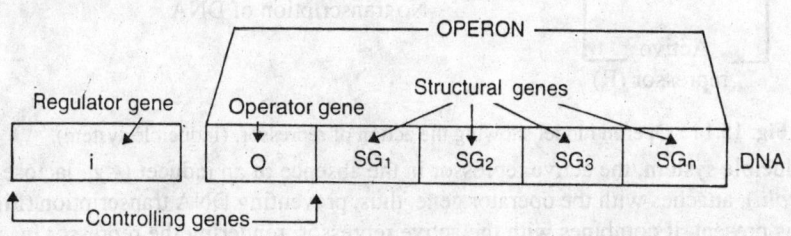

Fig. 15.12 : "Operon model" with various genes on DNA segment.

operators. Regulators may or may not be located close to the genes which they regulate. Operators are always located close to and, in fact, may be a part of one of the structural genes they control. The operator and its structural genes constitute an "Operon". Thus, Operon, in general can be represented as follows: (Fig. 15.12)

Mechanism of Operon Model

In Inducible System : In inducible systems, the presence of inducers (1) is an essential feature for the synthesis of inducible enzymes. In the absence of an inducer, the system does not synthesize the enzyme. In case of *E. coli*, 'Lac' operon model regulates the synthesis of β-galactosidase enzyme, for which the presence of β-galactoside (lactose) is essential. It means that in the absence of an "inducer" the genes responsible for the synthesis of β-glactosidase do not function, and because of the presence of repressors (R) in the cell the activity of genes is checked. The repressors (R) are synthesized by the activity of regulator gene (*i*). When active repressors come in contact with inducers, they are converted into inactive repressors. The inactive repressors may again get converted into active repressors by the addition of co-repressors. The co-repressors are simply the end products which check the synthesis of enzymes.

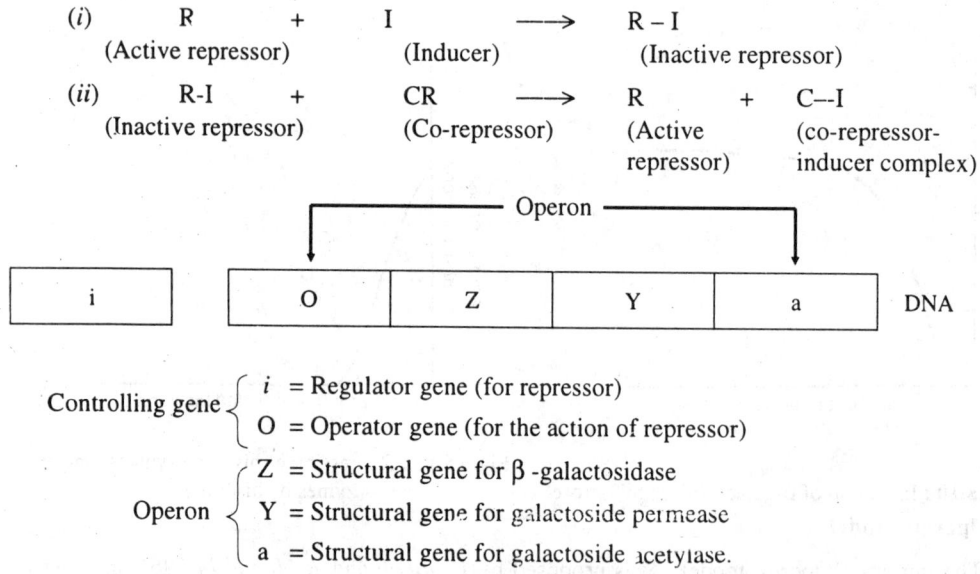

Fig. 15.13 : The lactose operon of E. coli and its regulatory gene.

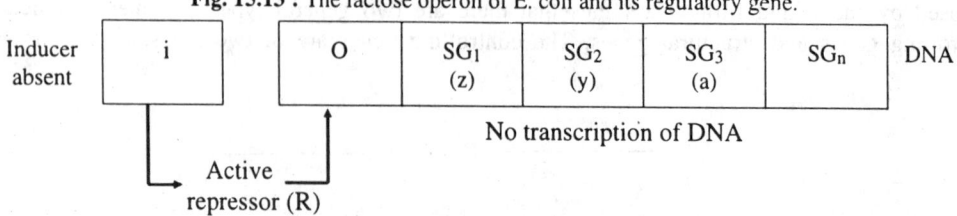

Fig. 15.14 : Operon model showing the action of repressor. (Inducible system).

In an inducible system, the active repressor in the absence of an inducer (*e.g.*, lactose in "lac" operon of *E. coli.*), attaches with the operator gene, thus, preventing DNA transcription (Fig. 15.14). If an inducer is present, it combines with the active repressor, rendering the repressor inactive. The inactive repressor is unable to attach with operator gene. This permits the structural genes to be transcribed (Fig. 15.15).

Metabolism of Amino Acids and Protein

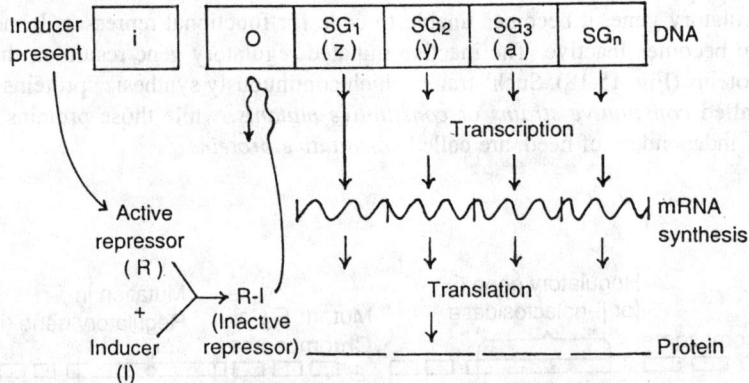

Fig. 15.15 : Operon model showing induction by inducer (Inducible system).

In Repressible System: In a repressible system, the repressor (R), in absence of a corepressor (CR), (e.g., histidine in *E. coli*), does not attach with the operator gene, thus permitting DNA transcription (Fig. 15.16). If a co-repressor (CR) is present, the repressor substance combines with

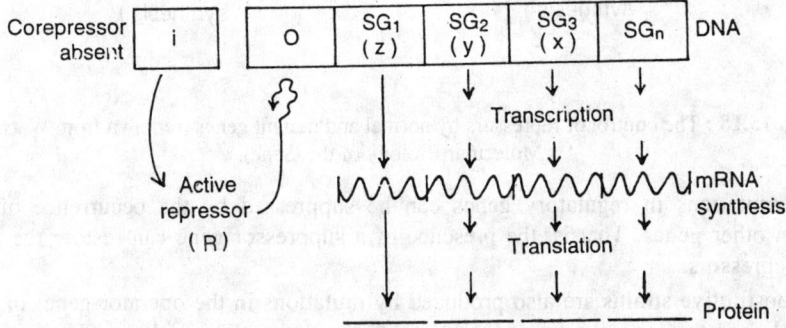

Fig. 15.16 : Operon model showing the action of repressor in absence of corepressor (Repressible system).

it to form a repressor-corepressor complex (R-CR) which blocks the operator gene. Thus, the DNA transcription does not take place (Fig. 15.17.).

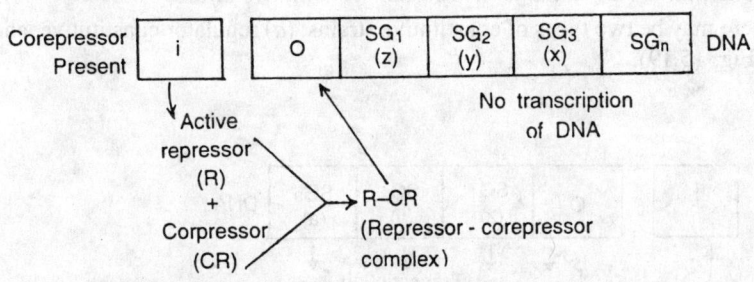

Fig. 15.17 : Operon model showing the action of repressor when co-repressor is also present (Repressible system).

$$\text{Repressor (R)} + \text{Corepressor (CR)} \longrightarrow \text{Repressor-Corepressor Complex (R-CR)}$$

Mutations in Controlling genes and production of Constitutive Strains: The regulator and operator genes are known as *controlling genes*. It is now a well-known fact that each repressor blocks the synthesis of one or more proteins, and like all other proteins, repressors are coded by chromosomal DNA. The genes which code for them are called *regulatory genes*. When a mutation occurs in a regulatory gene, it becomes unable to code for functional repressor. It means that the regulatory gene becomes inactive. The inactive mutated regulatory gene results in the unchecked synthesis of proteins (Fig. 15.18). Such strains which continuously synthesize proteins independent of need, are called *constitutive strains* or *constitutive mutants*, while those proteins produced in fixed amounts, independent of need, are called *constitutive proteins*.

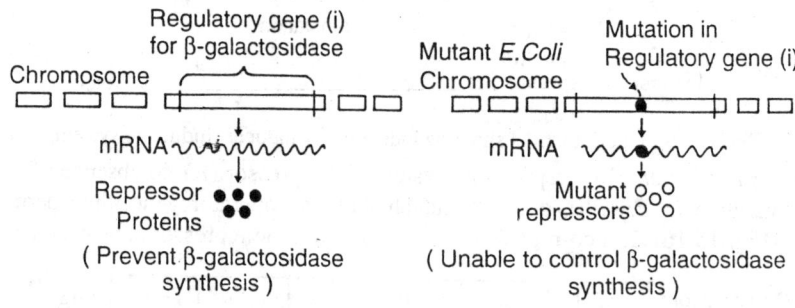

Fig. 15.18 : The control of repressors by normal and mutant genes (redrawn from Watson: Molecular Biology of the Gene).

Some mutations in regulatory genes can be suppressed by the occurrence of suppressor mutations in other genes. That is, the presence of a suppressor gene can restore the synthesis of functional repressors.

The constitutive strains are also produced by mutations in the operator gene. In the absence of functional operator gene, the corresponding repressor cannot check the synthesis of the specific mRNA, and as a result, their is a constitutive synthesis of its corresponding protein. The normal operator gene undergoes mutation to form an inactive operator which prevents the working of repressors. When this happens, constitutive enzyme synthesis takes place. These mutants are known as *operator constitutive* (O^c) *mutants* or *Operator constitutive strains*.

Thus, there may be two types of constitutive strains, (*a*) regulator constitutive, and (*b*) operator constitutive (Fig. 15.19).

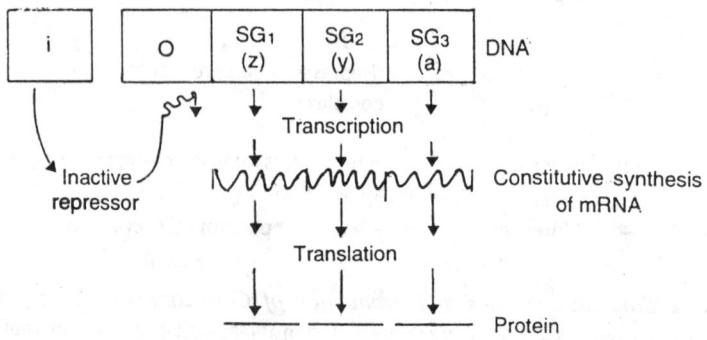

Fig. 15.19: (*a*) Regulator Constitutive

Metabolism of Amino Acids and Protein

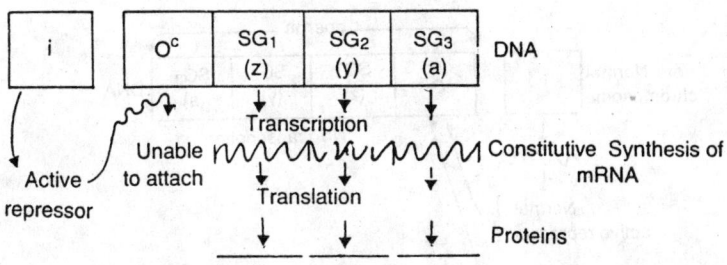

(b) Operator Constitutive

Fig. 15.19: Constitutive strains resulting due to mutations in regulator gene and operator gene.

The Differences between Regulator Mutant and Operator Mutant

The two types of mutants, though have the same affect on protein synthesis, differ in two major respects : (i) in location, (ii) in dominant-recessive relationship with respect to wild type. This dominant-recessive relationship has been demonstrated in special partially diploid cells, which contain two copies of the relevant chromosomal regions. When diploid heterozygotes for regulator constitutive were produced, no protein synthesis was observed suggesting that it is dominant over mutuant (i^c) (Fig. 15.20). In contrast, partial diploid heterozygotes for operator constitutive exhibited constitutive synthesis of proteins suggesting that normal operator gens (O) is recessive to mutant gene, (O^c). (Fig. 15.21).

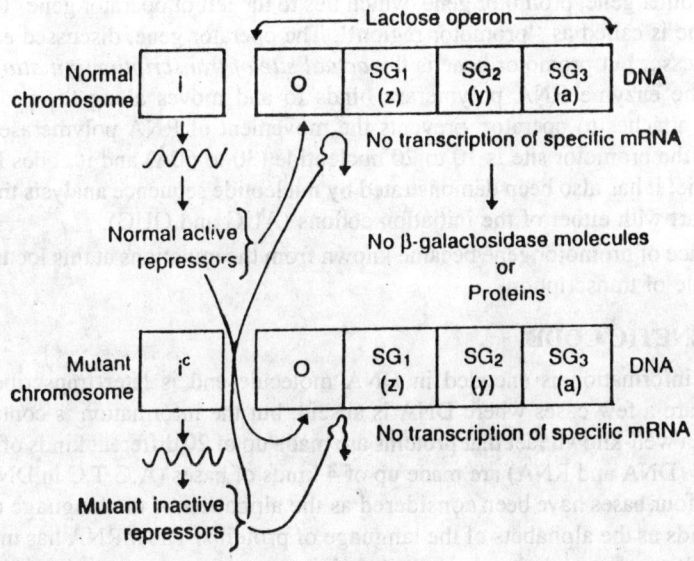

Fig. 15.20: Check on protein synthesis in a partial diploid cell heterozygous for regulator gene (i/i^c). Note, here is dominant over i^c.

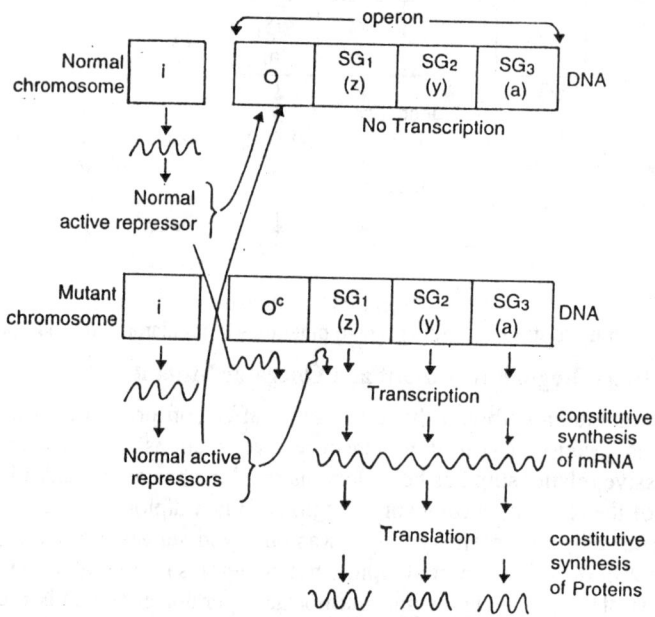

Fig. 15.21: Constitutive protein synthesis in a partial diploid heterozygote for operater gene (O/O^c) Note, here O^c is dominant over O.

Promotor Gene and Promotor Region

There is another gene, promotor gene, which lies to the left of operator gene. The region which contains this gene is called as "promotor region". The operator gene, discussed earlier, is the *site of action* of repressor but promotor gene is the *actual site of transcription initiation*. It is believed, therefore, that the enzyme RNA polymerase binds to and moves along the promotor site. The repressor, when attaches to operator, prevents the movement of RNA polymerase (Fig. 15.22). It is proposed that the promotor site is 10 to 20 nucleotide (30 to 60Å) and it codes for the initiating mRNA nucleotide. It has also been demonstrated by nucleotide sequence analysis that many mRNA chains do not start with either of the initiation codons (AUG and GUG).

The existence of promotor gene became known from the mutations at this locus which affected the maximum rate of transcription.

15.7. THE GENETIC CODE

All the genetic information is encoded in DNA molecule and is later transcribed into mRNA. However, there are a few cases where DNA is absent, but the information is contained in genetic RNA. Now, it is a well-known fact that proteins are made up of 20 different kinds of aminoacids and the nucleic acids (DNA and RNA) are made up of 4 kinds of bases (A,G,T,C in DNA and A,G,U,C in RNA). These four bases have been considered as the alphabets of the language of nucleic acids and 20 aminoacids as the alphabets of the language of proteins. The mRNA has many codons and to read these codons various codes were proposed, but out of those, the "triplet code" is the most acceptable one as it has found support from experimental evidences of both genetic and biochemical nature

The genetic code has been experimentally deciphered and perfected independently by *Marshall Warren Nirenberg, Robert Holley* and *Hargovind Khorana,* for which they were

Metabolism of Amino Acids and Protein

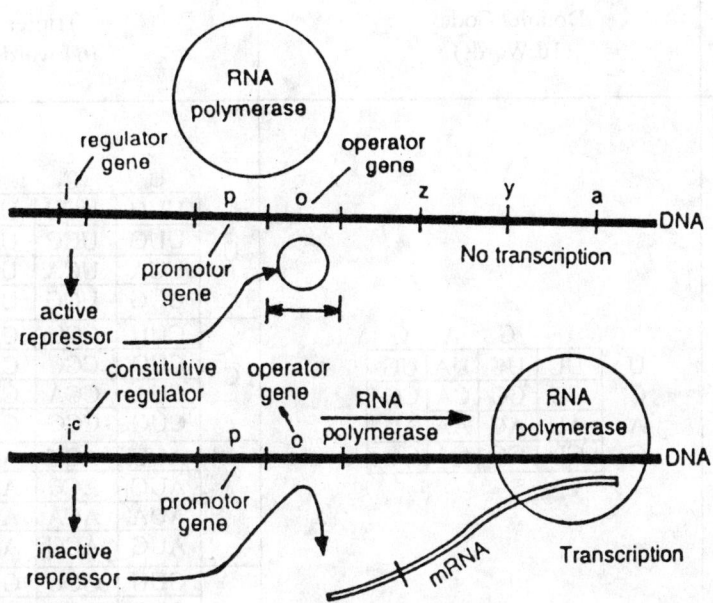

Fig. 15.22. Binding of RNA polymerases the promotor for lactose operon. The figure also shows the relationship between promotor gene, repressor and RNA polymerase. After Waston 1970, Molecular Biology of the gene.

Jointly awarded the Noble Prize for medicine and physiology in 1968. The genetic code dictionary as represented in Table 15.3. shows that three codons, namely UAA, UAG and UGA, are chin *termination codons* (non-sense codons). They do not code for any aminoacid and simply act as signals in the process of chain termination. The codons AUG and GUG help in the initiation of polypeptide and are called as chain initiation codons.

Properties of Genetic Code:- The genetic code possesses the following important properties which have now been proved by definite experimental evidences :

1. The code is triplet : DNA contains four kinds of nucleotides (A.T.G. and C), and proteins are synthesized from 20 different types of amino acids. A basic problem regarding the genetic code was: how many bases of DNA specify one amino acid? In singlet code each base or letter would specify one amino acid. Only 4 of twenty types of amino acids would be coded unambiguously by a singlet code (Table 15.2). In a two letter or doublet code two bases would specify one amino acid. Here 16 (4x4) of the 20 amino acids can be specified, but there would be ambiguous determination of a number of amino acids. A triplet or three-letter code was first suggested by a physicist Gamow in 1954. According to the triplet code three letters or bases specify one amino acid. Thus, 64 (4x4x4) distinct triplets of purine and /or pyrimidine bases determine the 20 amino acids. These triplets have been called codons. Since there are 64 codons and only 20 amino acids, it is obvious that there are many more codons than there are amino acids it is obvious that there are many more codons that there are amino acids, i.e., the code is degenerate. Experimental evidence shows that the code is a triplet one and that 61 of the 64 codons code for individual amino acids during protein synthesis.

Table 15.1 Possible singlet, doublet and triplet codes of m-RNA.

Singlet Code (4 words)	Doublet Code (16 Words)	Triplet (64 words)

Singlet Code:
U
C
A
G

Doublet Code:

	U	C	A	G
U	UU	UC	UA	UG
C	CU	CC	CA	CG
A	AU	AC	AA	AG
G	GU	GC	GA	GG

Triplet Code:

		U	C	A	G	
U		UUU	UCU	UAU	UGU	U
		UUC	UCC	UAC	UGC	C
		UUA	UCA	UAA	UGA	A
		UUG	UCG	UAG	UGG	G
C		CUU	CCU	CAU	CGU	U
		CUC	CCC	CAC	CGC	C
		CUA	CCA	CAA	CGA	A
		CUG	CCG	CAG	CGG	G
A		AUU	ACU	AAU	AGU	U
		AUC	ACC	AAC	AGC	C
		AUA	ACA	AAA	AGA	A
		AUG	ACG	AAG	AGG	G
G		GUU	GCU	GAU	GGU	U
		GUC	GCC	GAC	GGC	C
		GUA	GCA	GAA	GGA	A
		GUG	GCG	GAG	GGG	G

Table 15.2 The maximum possible number of codons in the singlet, doublet and triplet codons

Type of Code	Number of Bases Bases in codon	Number of codons	Ambiguous degenerate
Singlet Code	1	4	Ambiguous
Doublet Code	2	4 × 4 = 16	Ambiguous
Triplet Code	3	4 × 4 × 4 = 64	Degenerate

Thus, each codon is made up of three letter-words and in a triplet code of 64 codons, 20 are used for 20 amino acids, 3 are used as chain termination codons (as signals) and remaining 41 are again used for various amino acids (see Table 15.3)

Evidence Favouring a Triplet Code :- The first experimental evidence supporting the concept of a triplet code was provided by Crick and coworkers 1961 by the study of addition or deletion of single base pairs, in a particular region of DNA of T_4 bacteriophage. It was found that the bacteriophage ceased to perform their normal function after addition or deletion of single base pari, because it completely changes the sequence of the codes. For example, if a functional unit of gene in DNA consists of the following five codons :

        TCA   GGC   TAA   AGT   CGG
         1   2   3   4   5

The addition of a single base 'G' at the end of the first codon would shift all other codons out of order, and correct reading of the sequence will not be possible as shown below :

Metabolism of Amino Acids and Protein

TCA	GGG	CTA	AAG	TCG
1	2	3	4	5

It was also found that one or two base addition failed to produce a functionally normal protein. But there base additions could restore the original sequence. The same would happen as a result of single base deletion. Several such evidences indicate that the codon is a sequence of three nucleotides, and the genetic code is generally called a triplet code.

Wobble Hypothesis :- An important feature of the triplet code is that the first and the second bases in a triplet are more important than the third in identifying the amino acid. Thus, all the codons for a particular amino acid have the same first two letters, and only the third letter can vary. This flexibility in the third nucleotide of a codon will reduce the consequences of simple errors during transcription. For example, it has been shown that the same t-RNA can recognise more than one codons differing only at the third position. This pairing is not very stable and takes place due to wobbling in the base pairing at the third position. This 'wobble hypothesis' was proposed by Crick (1965). The wobbling helps in the economy of the number t-RNA molecules, and different codons meant of the same amino acid can be recognised by the same t-RNA.

Table 15.3 Genetic Code Dictionary

First letter (Base)		Second letter (Base) U	Second letter (Base) C	Second letter (Base) A	Second letter (Base) G	Third letter (Base)
	U	UUU, UUC } Phe; UUA, UUG } Leu	UCU, UCC, UCA, UCG } Ser	UAU, UAC } Tyr; UAA*–Tmn; UAG*–Tmn	UGU, UGC } Cys; UGA*–Tmn; UGG–Tryp	U C A G
	C	CUU, CUC, CUA, CUG } Leu	CCU, CCC, CCA, CCG } Pro	CAU, CAC } His; CAA, CAG } GluN	CGU, CGC, CGA, CGG } Arg	U C A G
	A	AUU, AUC } Ileu; AUA; AUG Mett	ACU, ACC, ACA, ACG } Thr	AAU, AAC } Asp.N; AAA, AAG } Lys	AGU, AGC } Ser; AGA, AGG } Arg	U C A G
	G	GUU, GUC, GUA, GUG } Val	GCU, GCC, GCA, GCG } Ala	GAU, GAC } Asp; GAA, GAG } Glu	GGU, GGC, GGA, GGG } Gly	U C A G

* Tmn-chain termination codons (nonsense codons). Bold lettered codons are chain initiation codons.
Note : The names of amino acids against the codons are given in abbreviated form.

Another notable feature of the genetic code is that the amino acids with similar structural properties tend to have similar codons. Thus, codons for aspartic acid (GAU, GAC) resemble the codons for glutamic acid (GAA, GAG), in their first two letters. Similarly, codons for phenylalanine, tyrosine and tryptophan, all begin with the letter 'U', This feature is considered to reduce the effect of mistakes during translation of genetic information, because one amino acid will be replaced by another with similar property.

2. The code is degenerate : When a particular amino acid is coded by more than one codon, it is called *degenerate* i.e. there are more than one codons for a particular amino acid, e.g., UUU, UUC codons are for phenylalanine and AAA and AAG are for lysine. Thus codons starting with UU specify phenylalanine and codons starting with AA specify lysine. Similarly, codons starting the GC (GCU, GCC, GCA, GCG) specify *alanine,* codons starting with CC (CCU, CCC, CCA, CCG) specify *proline* and codons starting with GU (GUU, GUC, GUA, GUG) specify *valine.* Unequal distribution of amino acids in protein may be due to this variability in number of codons for amino acids.

In genetic code dictionary consists of 64 codons for 20 amino acids, of which 4 codons are signal codons (3 stop codons and one start codon) and 61 codons (including start codon) code for all 20 amino acids. Thus, it is obvious that more than one codons may code for a particular amino acids. The number of codons coding for different amino acids in genetic code dictionary are as follows :

(i) Tryptophan, methionine .... one codon.
(ii) Phenylalanine, tyrosine, cysteine, histidine, glutamine,
    asparagine and lysine .... Two codons.
(iii) Isoleucine .... Three codons.
(iv) Valine, proline, threonine, alanine, glycine .... four codons.
(v) Leucine, serine and arginine .... six codons.

3. The code is non-overlapping: The code is non-overlapping which means that the same letter is not used for two different codons. The overlapping and non-overlapping codons are shown below :

Overlapping   Non-overlapping

Although the code is non-overlapping but *Barell et al* (1976) reported that bacteriophage φ × 174 possesses overlapping genes.

4. The code is commaless : It means, there are no punctuations or comma etc. between two codons, in the genetic code. In other words no codon is reserved for punctuations. This results into continuous coding of amino acid without any interruption. However, the coding process may be interrupted in presence of introns in transcribing DNA. The commaless codons are generally written are follows :

UUU CUU AUU GUU   – codon sequence
Phe Leu Ileu val   – Amino acids

However, the codons in genetic code are also written as -

UUU – CUU – AUU – GUU   – codon sequence
Phe – Leu – Ileu – val   – Amino acids

If only one base in deleted from above codon sequence, (e.g. *c* base from second codon), it will result into change in coding and sequence of amino acids e.g.,

UUU UUA UUGUU – codon sequence with deleted *c* base in second codon
Phe-Leu-Leu ------ – change amino acid sequence.

5. The code is non - ambiguous : It means there is no ambiguity about a particular codon. A particular codon always codes for the same amino acid , wherever it is found.

In an ambiguous code, the same codon can code for two or more different amino acids. For example, UUU codon coding for phenylalanine can code for isoleucine, leucine and serine is presence of streptomycin. In non-ambiguous code, the same codon never codes for two different amino acids. However, there is some ambiguity in genetic code that AUG and GUG both act as initiating codon during protein synthesis and in that case both code for amino acid *methionine*, although GUG code is for *valine.*

A different genetic code is found in some eukaryotic mitochondria, so that in cytoplasm and mitochondria the same codon may code for different amino acids. The genetic code of mitochondria differs from universal genetic code, so this point can not be considered as a point of ambiguity.

6. The code is universal : Although the genetic code has been worked out by using *in vitro* systems of micro-organisms, there is no doubt about its universality, i.e., in all kinds of organisms (micro - or macro-, plants or animals), the same genetic code is used. The universality of genetic code has been demonstraled by *Nirenberg* et al (1967) using aminoacyl tRNAs of *E. coli, xenoplus laevis* (an amphibian) and *guinea pig* (a mammal). When they used purified in RNAs from rebbit reticulocytes for haemoglobin synthesis and injected into frog oocytes, the frog synthesized rabbit haemoglobin by using it own machinery for translation.

New Genetic codes in Mitochondria and Ciliate Protozoans

Genetic code is usually described as universal but during the last few years it has been reported that the genetic code in mitochondria of mammals, *Drosophila, Neurospora* and yeasts differs from the universal code for nuclear DNA. The genetic code in mitochondria not only differs with universal code but also differs in mitochondria of different groups of organisms. For example, codon UGA, which is a stop (termination) codon in universal code for nuclear encoded protein in plants and all other organisms, codes for *tryptophan* in mitochondrial codes in mammals, *Drosophila, Neurospora* and yeasts. The codons AGA and AGG are universal codons for *arginine* also code for fungal and plant mitochondrial DNA (mt DNA) but they act as stop codons in mammalian mt DNA and serine codon in Drosophila mt DNA. Similarly AUA codon which codes for *isoleucine*, codes for methionine in mt DNAs of mammals, Drosophila and yeasts. AUU codon which codes for isoleucine in universal code and in plants, codes for methionine in mt DNAs of all four organisms mentioned above. The codons CUU, CUC, CUA, CUG which code for leucine, codes for threonine in yeasts. The alterations in universal code (for nuclear DNA) in mitochondria of different organisms are in table 15.4.

Table 15.4. Showing alterations in the universal code in mitochondira.

Codon	'universal' code	Mitochondrial codes				
		Mammals	*Drosophila*	Neurospora	Yeasts	Plants
UGA	Stop	Trp*	Trp*	Trp*	Trp*	stop
AGA, AGG	Arg	Stop*	Ser*	Arg	Arg*	Arg
AUA	Ileu	Met*	Met*	Ileu	Met*	Ileu
AUU	Ileu	Met*	Met*	Met*	Met*	Ileu
CUU, CUC CUA, CUG	Leu	Leu*	Leu	Leu	Thr*	Leu

* codes differ fro 'universal' code

Explanation for number of tRNAs used in mitochondrial genetic code.

The tRNA used translate mt mRNAs are entirely derived from mt chromosomes. It was unusually reported in case of mt genetic code that only 24-types of tRNAs could be found. According to crick's wobble hypothesis, at least 32 tRNAs are required for the translation of 61 codons. There are two possibilities to this problem -

(*i*) The mitochondrial genes do not utilize all 61 codons.

(*ii*) Wobble rules might be different for mitochondria.

Actually, second is the case. According to Crick's original wobble rules, at least two different tRNAs are required to translate four codon families. In all these case except arginine codon family, single tRNAs have been found responsible for specifying all four code words and these tRNAs contain a *U* in the wobble position of their anticodons. It seems that mitochondrial ribosome allows these tRNAs to pair with all four members of the codon family. The six mitochondrial tRNAs pairing with normal two codons contain an altered *U* in the wobble position. This modification causes them to conform the normal 'wobble' rules.

Table 15.5. Showing genetic code of yeast mitochondira.

First Base Position (5' end)		Second Base Position								Third Base Position (3' end)
		U		C		A		G		
		codons (5'-3')	anticodons (3'-5')	codons (5'-3')	anticodns (3'-5')	codons (5'-3')	anticodns (3'-5')	codons (5'-3')	anticodns (3'-5')	
	U	UUU, UUC	Phe AAG	UCU, UCC, UCA, UCG	Ser AGU	UAU, UAC	Tyr AUG	UGU, UGC	cys ACG	U, C
		UUA, UUG	Leu AAU*			UGA, UGG	stop	UGA, UGG	Trp ACU*	A, G
	C	CUU, CUC, CUA, CUG	Thr GAU	CCU, CCC, CCA, CCG	Pro GGU	CAU, CAC	His GUG	CGU, CGC, CGA, CGG	Arg GCA[b]	U, C, A, G
						CAA, CAG	Gln GUU*			
	A	AUU, AUC	Ileu UAG	ACU, ACC, ACA, ACG	Thr UGU	AAU, AAC	Asn UUG	AGU, AGC	Ser UCG	U, C
		AUA, AUG	Met UAC[a]			AAA, AAG	Lys UUU*	AGA, AGG	Arg UCU*	A, G
	G	GUU, GUC, GUA, GUG	Val CAU	GCU, GCC, GCA, GCG	Ala CGU	GAU, GAC	Asp CUG	GGU, GGC, GGA, GGG	Gly CCU	U, C, A, G
						GAA, GAG	Glu CUU			

* designates U in 5' position of anticodon (tRNA) that carries the - $CH_2NH_2CH_2COOH$ grouping on 5' position of the pyrimidine.
a Two tRNA are found for methionine.
b Arg tRNA has been reported in yeast mitochondria but its extent of utilization by CGN codons is not clear.

The study of genetic code of yeast mitochondria shows another peculiarity. The comparison of tables 15.3 and 15.5 indicates that the mitochodrial code has many differences in code word meaning. The codons beginning with CU represent *Thr* instead of *Leu*, the *AUA* codon represent *Met* instead of *Ileu*, and the codon UGA represents *Trp* rather than a sto signal.

The main differences betwen universal genetic code and mitochondrial genetic code are given in table 15.6.

Table 15.6

Universal genetic code (U.G.C.)	Mitochondrial genetic code (mt G.C.)
1. It contains 55 anticodons (tRNAs).	1. It contains 22 anticodons (tRNAs)
2. It includes three termination (stop) codons - UAA, UAG, UGA.	2. It includes 4 stop codons - UAA, UAG, AGA, AGG (in mammals) and 2 codons
3. UGA acts as termination codon only	3. UGA codes for tryptophan
4. AGA and AGG code for *arginine*.	4. AGA and AGG are *stop codons* in mammalian mt genetic code but in Drosophila these codons code for *serine*.

Metabolism of Amino Acids and Protein

5. AUA codon codes for *isoleucine*	5. AUA codes for *methionine* in mammalsy, drosophila and yeast.
6. AUU codes for *isoleucine*	6. AUU codes for *methionine* in mt G.C. of all discovered organisms.
7. CUU, CUC, CUA, CUG codons code for *leucine*.	7. These codons code for *threonine* in yeast mitochondrial genetic code.
8. UAG tRNA is aminoacylated by *leucine*.	8. UAG tRNA is aminoacylated by *threonine* in yeast mitochondira.

Species-Specific Rules Regarding Anticodon Codon Pairing

The genetic code differs very little between species. The species show considerable differences in anticodon translation system of tRNA e.g. mitochondrial tRNA system. In these systems the bases in anticodon-codon complex run antiparallel like that of standard double helix pairing and in them the base pairing occurs Watson-Crick like between the first two bases in the codon and the opposing bases in the anticodon segment of the tRNA. However, the rules for pairing for 3' base in codon vary with the species. These rules for codon- anticodon pairing are summarized in table 15.7.

Table 15.7. Rules for anticodon - codon pairing.

Anticodon First Base (5')	Codon Third Base (3')	Examples.
U	U, C, A, G	Mitochondrial code in family boxes.
*U	A, G	Mitochondrial code in two codon-sets.
†U	A	Eukaryotes
‡U	U, A, G	Eubacteria in family boxes.
C	G	All codes
*C	A	Bacteria, isoleucine codon AUA.
G	U, C	All codes
A	U	Rare
I (inosine)	U, C, A,	Eukaryotes, ICG in eubacteria.

*, +, ‡ are modified forms of U. (After Yoko yama et al., 1985)
*C is modified form of C.

The table 15.7 indicates that an unmodified U can pair with any of the four bases (U, C, A, G) present in 3' position of codon (See -family box for mitochondrial genetic code in table 15.7). *Yokoyama et al* (1985) prepared various modified U bases and showed that modified *U can pair with either A or G in two codon sets in mitochondria; modified †U can pair with only A base of codon in eukaryotes; modified ‡U can pair with either U or A or G base of codon in eubacteria. If C base is present at 5' position in anticodon (tRNA), it pairs with G base present at 3' position of codon in all codes. The only known exception to this rule is found in eubacteria, where c is covalently modified in tRNA that recognizes the AUA codon. Thus, a modified *C base at 5' position of anticodon can pair with 3' A in the codon. If G is present at 5' base position in anticodon, it can pair with either 3' U of a codon. This is applicable in all organisms. If 5' base in the anticodon is A, it can pair with only 3' U or C base of codon. However, A is rarely found at 5' of anticodon because at this position in eukaryotes the base A is usually deaminated to another base inosine (I), which possesses broad pairing capacity. The base 5' A can pair only with 3' U, out 5' I can pair with 3' U, C or A codons. In eubacteria, deamination is limited to the conversion of the ACG sequence to an ICG anticodon.

Methods of Cracking of Genetic Code
or
Methods of Cryptoanalysis of Genetic Code

Following methods have been adopted for cracking or deciphering the genetic code :

A. Codon Assignments In Vitro -

M. Nirenberg and H.G. Khorana made following approaches to establish *in vitro* that which triplet codon of the possible 64 codons should code for which of the 20 amino acids.

1. *Codon Assignment with unknown sequences -*

   (i) *Assignment of Codons with unknown sequence of homopolymers :*

   Nirenberg et. al (1961), for the first time, provided the method of deciphering genetic code by the use of unknown sequence of homopolymers. In vitro, they incorporated *radioactive amino acids* in cell free protein synthetic systems having synthetic or artificial mRNA (ribopolynucleotides). A *cell free system for protein synthesis consists of ribosomes, enzymes mRNA and tRNA* etc. A cell free system for protein synthesis was produced by breaking the cell e.g. *E. Coli*, containing protein synthesizing components and separating them by ultracentrifugation. During the experiment, the natural mRNA was degraded replaced by introducing *artificial mRNA*, prepared for each nitrogen base. By adding the enzyme *polynucleotide phosphorylase* extracted from *Micrococcus lysodeikticus* or *Azotobacter vinelandii*. This enzyme was quite different from RNA polymerase as it did not require DNA primer for the synthesis of RNA.

   By above procedure, Nirenbery et al artificially prepared polyuridylic acid (UUUUU.....), polycytidylic acid (CCCCC ....), polyadenylic acid (AAAAA ...) and polyguanylic acid (GGGGG ....) homopolymers from uracil, cytosine, adenine and guanine respectively. All there homopolymers *represent mRNAs*. When they introduced these artificial homopolymers separately in different cell free protein synthesizing systems of *E.coli* containing radioactive amino acids, they found that each artificial mRNA stimulated the synthesis of polypeptides of single kind of amino acids. For example, *poly U* acids stimulated the synthesis of only polypeptide *polyphenylalanine* whose amino acid residues were *phenylalanine*. Therefore, Nirenberg and his coworkers concluded that the triplet *UUU* coded for the amino acid *phenylalanine*. This news that the genetic code had been cracked splashed in the newspapers all over the world in 1961. Subsequently, poly C stimulated the synthesis of *polyproline* and poly A stimulated the synthesis of polylysine. Therefore, triplet CCC was assigned to amino acid *proline* and triplet AAA to amino acid *lysine*. Similar experiments with poly G were not successful, because it attained secondary. Structure and thus could not be attached ribosomes. Later on, the triplet GGG was found to specify *glycine*. Thus, UUU, CCC and AAA codons could be assigned to phenylalanine, proline and lysine amino acids respectively without any difficulty using homopolymers of RNA. GGG sequence was assigned to glycine. In this way 4 codons out of 64 codons were easily accounted for

   (ii) *Codon assignment with unknown sequence of copolymers.*

   In order to gain some insight into the meaning of remaining 60 codons containing more than one kind of nucleotides, Nirenberg ad colleagues continued these experiments by using artificially synthesized random ribopolynucleotides or mRNA (mixed copolymers) containing two or three different nucleotides. By using artificially mRNAs like poly UA, poly UC and poly UG, he cracked the codes for amino acids arginine, alanine, serine, proline, tyrosine, isoleucine, valine, leucine, cysteine, tryptophan, glycine, methionine and glutamic acid .

   When mixed copolymers are prepared, all the possible triplets from each copolymer can be calculated assuming that the bases are incorporated randomly in the molecule. For example, if only A and C are used to synthesize a copolymer *poly AC* from the mixture containing equal proportions of A and C, six triplets (AAA, ACA, ACC, CCC, CCA and CAA) should occur in equal frequency. The use of poly AC resulted into incorporation of six amino acids in polypeptides, i.e., asparagine, glutamine, histidine, lysine, threonine and proline.

In the next step *poly AC* was synthesized from a mixture containing unequal proportions of A and C (more A then C), the ratio of asparagine to histidine increased in the polypeptide. By such experiments the composition of the bases in certain triplets could be deduced. A large number of synthetic copolymers with unknown sequences were used for preparing a dictionary of codon composition for amino acids.

If only A and C are used, the poly AC will consist of eight possible codons namely AAA, AAC, ACA, CAA, CCA, CAC, ACC and CCC. The proportion of these codons can be calculated if the unequal known quantities of A and C are used for the synthesis of *poly AC*. For example, if A : C = 5 : 1 (i.e. 5/6 is A and 1/6 is C) the calculated relative proportions of 8 codons on random basis would be as given in table 15.8.

Table 15.8. Relative proportions of different codons in mRNA on the basis of 5 : 1 ratio of A and C bases.

Base Composition	Codon	Probability	Ratio using min (as 1)	max (as 100)
3A	AAA	5/6 × 5/6 × 5/6 = 125/216	125	100
2A, 1C	AAC	5/6 × 5/6 × 1/6 = 25/216	25	20
	ACA	5/6 × 1/6 × 5/6 = 25/216	25	20
	CAA	1/6 × 5/6 × 5/6 = 25/216	25	20
1A, 2C	CCA	1/6 × 1/6 × 5/6 = 5/216	5	4
	CAC	1/6 × 5/6 × 1/6 = 5/216	5	4
	ACC	5/6 × 1/6 × 1/6 = 5/216	5	4
3C	CCC	1/6 × 1/6 × 1/6 = 1/216	1	0.80

The calculated relative proportions of codons were compared with the proportions of different amino acids present in the polypeptides synthesized by poly AC (Table 15.9).

Table 15.9. codon assignment based on A : C (5 : 1) composition in synthesized mRNA

Base composition	Codons	Amino acids
3A	AAA	Lysine
2A, 1C	AAC	Asparagine
	ACA	Threonine
	CAA	Glutamine
1A, 2C	CCA	Proline
	CAC	Histidine
	ACC	Threonine
3C	CCC	Proline

Initially the codons were assigned on the basis of base composition and not on the basis of base sequences in codons. Hence, by using poly AC six amino acids described above (on equal proportion basis of A and C) were found in a polypeptide of poly AC. However, the successive increase in the ratio of asparagine to histidine was noticed with the increase in ratio of poly A to poly C.

2. Codon Assignment with Known sequences

The next approach was to find out the base sequences in codons of known composition. It was done by following methods :

(i) *Codon assignment through filter binding technique :* This method was used by *nerenberg* and *Leder* in 1964. They found that if a synthetic simple trinucleotides of a known sequence (with known bases at 5' end and 3' end) is added to ribosome and a particular aminoacyl-tRNA, it forms a complex, provided the trinucleotide codes for the amino acid attached to the added aminoacyl-tRNA.

$codon_1$ know (trinucleotide) + Ribosome + AA_1- $tRNA_1 \rightarrow$ Ribosome - $codon_1–AA_1$ – $tRNA_1$

The above reaction indicates that if AA_1, is used with a given $codon_1$, and complex formation takes place, it proves that the used $codon_1$ codes for amino acid AA_1. For example, a trinucleotide GCC formed a complex with ribosome and alanyl-tRNA, it indicated that GCC codon was for alanine and not for any other amino acid.

In another approach, Nirenberg and Leder found that if ribosome - codon - AA -tRNA complex is allowed to pass through nitrocellulose membrane, it is absorbed on the membrane. They took a mixture of one radioactive amino acid, 19 remaining non-radioactive amino acids, codon of known sequence, ribosome and tRNA and passed through nitrocellulose membrane. This complex was also absorbed on the membrane. Now, they detected the radioactivity of the complex on the membrane on the basis of radioactive amino acid. The radioactivity was observed only when radioactive amino acid took part in complex formation, otherwise no radioactivity was observed. They made various samples and tested them using only one radioactive amino acid separately in each sample i.e., each sample contained one separate radioactive aminoacid and other remaining 19 non-radioactive amino acids.

By using above method Nirenbery and Leder cracked 45 codons for amino acids arginine, alanine, cysteine, glutamine, glycine, isoleucine, leucine, methionine, proline, tryptophan, tyrosine, serine and valine.

(ii) *Codon assignment using copolymers of repetitive sequences :*

Khorana used copolymers of repetitive sequences for exact codon assignment. In order to ascertain that which structural isomer of a codon with same empirical formula designates which amino acid, artificial mRNAs of known sequences were employed. Khorana prepared artificial mRNAs by using copolymers of repetitive sequences. For example, when he added copolymer containing two base CU repeatedly $(CU)_n$. it gave rise to copolymer repetitive sequence as CUCUCUCU ..... Theoretically, only two codons are possible from this sequence, i.e., CUC and UCU. Because these two codons are present in alternating sequence, the resultd polypeptide fro this sequence wold have only two amino acids (leucine and serine in alternating sequence. These two amino acids (leucine and serine) can be assigned to the two codons CUC and UCU respectively (Table 15.10)

Table 15.10 codon assignment using copolymers of repetitive sequences of two bases.

Copolymers	codons	Aminoacids	codons
$(CU)n$	CUC/UCU/CUC	leucine/serine	CUC/UCU
$(AC)n$	ACA/CAC/ACA	threonine/histidine	ACA/CAC
$(UG)n$	UGU/GUG/UGU	Cysteine/Valine	UGU/GUG

Similarly the addition of copolymer with three bases (ACU) resulted into sequence as ACU ACU ACU ......) which may give rise to three kinds of homopoly peptides, depending upon from where the reading is started. The codon assignment with this sequence is given in Table 15.11.

Metabolism of Amino Acids and Protein

Table 15.11- Codon assignment, having known sequences, using copolymers of repetitive sequences with three base (ACU)n

Codons	Homopolypeptide	codon assignment
ACU/ACU/ACU/ACU/ACU =Poly (ACU)	(threonine)n	ACU = threonine
A/CUA/CUA/CUA/CUA/CUA = Poly (CUA)	(leucine)n	CUA = leucine
AC/UAC/UAC/UAC/UAC/UAC = Poly (UAC)	(Tyrosine)n	UAC = tyrosine

Using some method Khorana (1968) synthesized a polynucleotide or mRNA containing alternate copolymers such as UGU-GUG-UGU-GUG. When the particular alternating polynucleotide was used as mRNA in the *in vitro* protein-synthesizing system, it resultd into formation of alternating polypeptide (cysteine - valine - cysteine....). Thus, codons UGU and GUG were assigned to aminoacids cysterin and *valine* respectively. This result established that both UGU (cysteine) and UUG (leucine) are different codons though base composition in then 2U, 1G. Similarly, GUG (valine) UGG (tryptophan) and GGU (glycine) are different codons though all the three have base composition 2G, 1U. Thus, UUG is an isomer of UGU, and UGG and GGU are the isomers of GUG.

On the basis of above techniques, a complete genetic code dictionary was propared which is shown in table. The dictionary indicates that the two codons (AUG and GUG) are the *initiation codons* and the three codons (UAA, UAG and UGA) are the *termination* codons.

B. Codon Assignment In VIVO

For codon assignment in vitro, mostly cell free protein synthesing systems were used which have proved their significance in cracking the genetic code but these could not tell whether the genetic code so cracked (deciphered) is also used in living systems of all organisms also (in vivo conditions) or not. three types of approaches were made by molecular biologists to confirm that some code which used in vitro conditions, is also used in vivo. These are :

(*i*) **By replacement of amino acids** : Such studies were made during tryptophan synthetase synthesis in *Ecoli* by *Yanofsky* et al (1963) and hoemoglobin synthesis in man.

(*ii*) **By frame shift mutations** : They were studied by *Terzaghi* et. al (1966) on enzyme lysozyme of the T4 bactersiophages.

(*iii*) **By comparison between DNA or mRNA polynucleotide cryptogram with its corresponding polynucleotide** : Such studies were made by *S. Cory* et al. (1970) during synthesis of

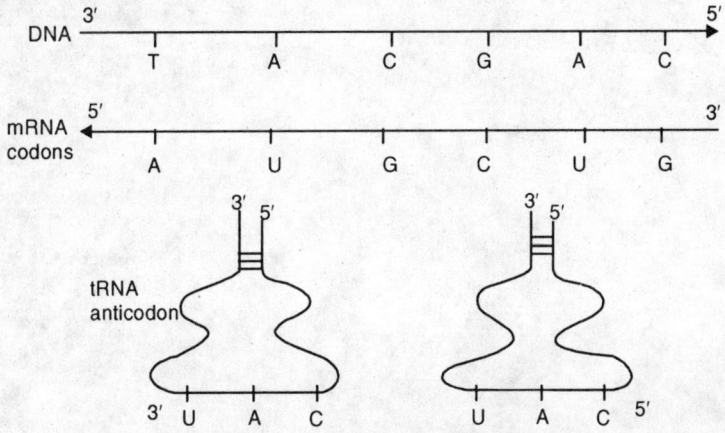

Fig. 15.28 The complementarity and antipolarity of triplets in DNA codons in mRNA and anticodons in tRNA.

coat-protein in R_{17} bacteriophage. They compared the amino acid sequence of R_{17} bacteriophage coat protein with that of nucleotide sequence of R_{17} mRNA regin for coat protein synthesis and found that both show complementarity in bases.

Thus, the studies *in vitro* and *in viva* helped in formulating the genetic code dictionary for various 20 amino acids. The codons for amino acids in dictionary are written on the basis of mRNA script reading in 5' → 3' direction. The corresponding codons in DNA will be complementary to mRNA and will be written in reverse order on the strand in 5' → 3' direction. Similarly, the bases in the corresponding tRNA anticodons will be both complementary and antipolat to mRNA codons.

Pattern to Genetic Code

The critical study of genetic code dictionary (Table) reveals following pattern in genetic code.

(*i*) Out of 64 codons, 61 (including AUG for methionine) correspond to various 20 amino acids correspond to various 20 amino acids.

(*ii*) Four codons act as signals UAA, UAG and UGA are *chain termination codons* which act as *stop signals* or *stop codons* where as single AUG methionie codons is *chain initiation codon* which acts as *start signal* or *start codon* in most of the cases during protein synthesis. Rarely, GUG codon (for valine) also acts as start codon.

(*iii*) Amino acids with similar structural properties tend to have related codons. Therefore, the *aspartic acid* codons (GAU and GAC) are similar to glutomic acid codons (GAA and GAG). Similarly, the codons for aromatic acids like UUU and UUC for *phenylalanine*. UAU and UAC for *tyrosine* and UGG for tryptophan resemble each other because in all these case the codons start with uracil. This characteristic feature of genetic code helps to minimize the consequences of mistakes arising during translation or mutagenic base substitution. Thus, if an amino acid in a protein is replaced erroneously by another amino acid having similar properties, the process of protein synthesis will continue.

(*iv*) For many of the synonym codons specifying the same amino acid, the first two bases of the triplet are constant whereas the third base varies. For example, all codon starting will CC (CCU, CCC, CCA, CCG), specify *proline* all codons starting with GG (GGU, GGC, GGA, GGG) specify *glycine* and all codons starting with AG (AGU, AGC, AGA, AGG) specify *arginine*. This flexibility in third base of codons may help to minimize the errors during transcription. *F. Crick* (1966) has given molecular interpretation for its occurrence. The *Wobble hypothesis* is also based on the above fact.

16
Lipid Metabolism

16.1. INTRODUCTION

Lipids are organic compounds that include fats, waxes, phospholipids, glycolipids and sterols. All of them are present in almost every living cell. Like carbohydrates, they are made up of carbon, hydrogen and oxygen, but are poorer in oxygen and are made up of carbon chains of various lengths. The lipids are insoluble in water, but soluble in organic solvents like ether, alcohol, chloroform and carbon disulphide. The fats containing saturated fatty acids are usually solid at room temperature, whereas, fats containing unsaturated fatty acids are liquid and are called 'oils'.

Fats are esters of the trihydric alcohol, glycerol and fatty acids. In nature, they are synthesized from these two contituents and upon oxidation release a huge amount of energy. Thus, the lipids serve as the prime fuel reserve for metabolism.

16.2. BIOSYNTHESIS OF FATS

As the fats are made up of glycerol and fatty acids, their biosynthesis includes the following three steps:

1. Synthesis of fatty acids,
2. Synthesis of glycerol,
3. Condensation of fatty acids and glycerol.

(1) Synthesis of Fatty Acids

(a) Saturated Fatty Acids: The main pathway of saturated fatty acid synthesis in plants, animals and bacteria is common and takes place through malonyl CoA pathway. In germinating pea-nuts and in slices of maturing pea-nut cotyledons, *Newcomb* and *Stumpf* (1952) using radio-active substrates showed that acetate was the most effective substrate incorporated into long chain fatty acids. The de novo synthesis of fatty acids in higher plants and in bacterial systems is identical. The final product is palmityl-ACP, where ACP is acyl carrier protein and is an essential component of the complete system.

Earlier when the β-oxidation sequence was described by *Knoop* (1904), it was thought that synthesis of fatty acids is just the reverse of β-oxidation, but experiments of *Stansly* and *Beinert* (1954) in which five purified enzymes of β-oxidation and acetyl CoA were used, did not support β-mutiple cendensation theory.

The malonyl CoA pathway of *S. Wakil* (1959) for synthesis of saturated fatty acids runs as under:

$$\text{Acetyl - CoA} + CO_2 \xrightarrow{\text{Acetyl–CoA Carboxylase}} \text{Malonyl-CoA}$$

$$\text{Acetyl-CoA+ACP-SH} \xrightarrow{\substack{\text{Acetyl transacylase} \\ \text{or ENZ (I)}}} \text{Acetyl -S-ACP + CoA}$$

Acetyl-S-ACP + [β-Ketoacyl ACP - synthetase] → Acetyl-S-Enz. (3) + ACP
or Enz. (3)

$$\text{Malonyl-CoA + ACP-SH} \xrightarrow{\text{Malonyl transacylase or Enz. (2)}} \text{Malonyl-S-ACP + CoA}$$

Acetyl-S-Enz. (3) + Malonyl-S-ACP → Acetoacetyl-S-ACP + Enz. (3) + CO_2

$$\text{Acetoacetyl-S-ACP + NADPH + H}^+ \xrightarrow{\text{β-Ketoacyl ACP-reductase}} \text{D(-) - β-hydroxy-butyryl-S-ACP + NADP}^+$$

$$\text{D(-)-β-Hydroxybutyryl-S-ACP} \xrightarrow{\text{Enoyl ACP hydrase}} \Delta^2 \text{ trans crotonyl-S-ACP} + H_2O$$

$$\Delta^2\text{-trans-crotonyl-S-ACP + NADPH + H}^+ \xrightarrow{\text{Enoyl ACP reductase}} \text{Butyryl-S-ACP + NADP}^+$$

Butyryl-S-ACP + Enz. (3) → Butyryl-S-Enz. (3) + ACP

Butyryl-S-Enz. (3) + Malonyl-S-ACP → β-Ketohexanoyl-S-ACP + Enz. (3) + CO_2, etc.

The process is repeated till a long chain fatty acid is formed. The final end-product is generally palmitic acid with 16 C atoms. In this scheme since acetyl-CoA, which is a 2-C fragment, is added the number of carbon atoms in the fatty acid formed is always even.

(b) Un-saturated Fatty acids: There are two possible pathways for the synthesis of unsaturated fatty acids.

(i) Oxidative Desaturation under aerobic conditions.

(ii) Dehydration and Acylation under anaerobic conditions.

(i) *Oxidative Desaturation:* In yeast, vertebrate tissues, fungi, some bacteria and in all higher plants oxidative desaturation of a long-chain saturated acyl-CoA to a mono-unsaturated acyl CoA occurs in presence of a reductant and molecular oxygen.

$$\text{Stearyl CoA} \xrightarrow[O_2]{\text{NADPH}} \text{Olelyl CoA}$$

The enzyme *oxidative desaturase* in yeast, fungi, bacteria and vertebrate tissue is associated with microsomal particles. Cytochrome b_5 as electron carrier couples the reductant NADPH or NADH in presence of O_2 with the desaturase. Under aerobic condition from ^{14}C-acetate ^{14}C-Oleic acid has been shown to be formed. Under anaerobic conditions mainly stearic acid and some palmitic acids are formed. In higher plants stearyl desaturase is not associated with microsomal particle, but is present in the soluble form as stearyl-ACP. It requires ferredoxin and NADPH as electron donor. A more effective reducing system is an ascorbate-dichlorophenol-indophenol (DCIP)—photosystem I (Chloroplast)—ferredoxin system. In the presence of light, electrons derived from ascorbate are channeled through the photosystem I carrier system to ferredoxin. Reduced ferredoxin then in some manner interacts with molecular oxygen, the desaturase and stearyl-ACP leading to the desaturation reaction.

The desaturation of stearyl-ACP is probably the primary route for the synthesis of oleic acid in higher plants. The mechanism of synthesis of linoleic acid [18 : 2 (9, 12)] and a-linoleic acid [18 : 3 (9, 12, 15)] is controversial and somewhat unclear at present.

Lipid Metabolism

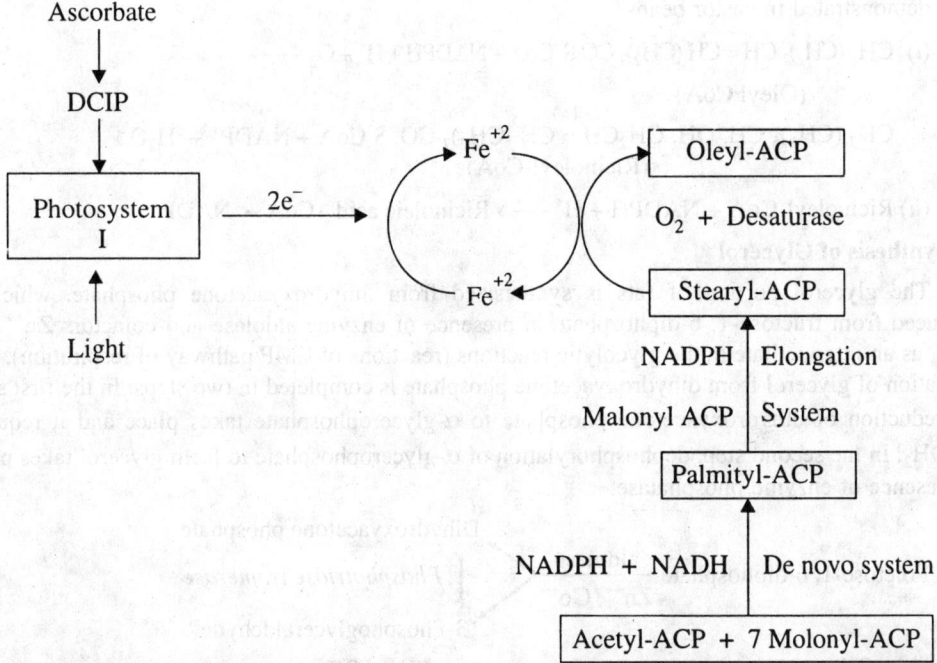

Stumpf et al. (1972) studied the conversion of Oleyl-CoA to linolyl-CoA by microsomal preparation of maturing safflower seeds. The system was (NADH + O_2)-dependent. Photosystem-I ferrodoxin-light reducing system also brought about desaturation.

$$CH_3-(CH_2)_3-\overset{H}{\underset{H}{C}}-\overset{H}{\underset{H}{C}}-HC=CH-(CH_2)_7-\overset{O}{\overset{\|}{C}}-S.CoA \quad \text{Oleyl-CoA}$$

$$\downarrow 2e \rightarrow O_2$$

$$CH_3-(CH_2)_4-HC=CH-HC=CH-(CH_2)_7-\overset{O}{\overset{\|}{C}}-S.CoA \quad \text{Linoleyl-CoA}$$

(ii) **Dehydration and Acylation under anaerobic conditions** : Bloch described a second system which is an-aerobic sobuble complex limited to certain eubacteria. It involves the β-, γ-enoic acid and subsequent addition of malonyl-CoA units to make the final mono-unsaturated fatty acid. Thus, to synthesize oleic acid the following reactions can be considered.

$$10:0 \xrightarrow{C_2} 3\text{-OH-}12:0 \xrightarrow{-H_2O} 12:1(3) \xrightarrow{+3C_2} 18:1(9)$$

| Saturated fatty acid | β-hydroxyl acyl fatty acid | β-γ -enoic acid | Malonyl CoA | Unsaturated fatty acid (Oleic Acid) |

This mechanism is, however, in-operative in higher plants.

(c) **Hydroxy fatty acids** : The hydroxy fatty acid (ricinoleic acid) is synthesized from Oleyl-CoA, NADPH and O_2. The reaction requires ferrous ions for activation. This synthesis has been demonstrated in castor beans.

(i) $CH_3(CH_2)_7 CH=CH(CH_2)_7 CO.S.CoA + NADPH + H^+ + O_2 \longrightarrow$
   (Oleyl CoA)
   $CH_3(CH_2)_5 CH.OH.CH_2CH=CH(CH_2)_7 CO.S CoA + NADP^+ + H_2O$.
   (Ricinoleyl-CoA)

(ii) Ricinoleyl-CoA + NADPH + H$^+$ $\longrightarrow$ Ricinoleic acid + CoA + NADP$^+$

(2) Synthesis of Glycerol

The glycerol portion of fats is synthesized from dihydroxyacetone phosphate, which is produced from fructose-1, 6-diphosphate in presence of enzyme aldolase and cofactors Zn^{++} and Co^{++}, as an intermediate in the glycolytic reactions (reactions of EMP pathway of respiration). The formation of glycerol from dihydroxyacetone phosphate is completed in two steps. In the first step, the reduction of dihydroxyacetone phosphate to α-glycerophosphate takes place and it requires $NADH_2$. In the second step, dephosphorylation of α-glycerophosphate to form glycerol takes place in presence of enzyme phosphatase.

$$\text{Fructose-1, 6-diphosphate} \xrightarrow[Zn^{++}, Co^{++}]{\text{aldolase}} \begin{array}{c} \text{Dihydroxyacetone phosphate} \\ \updownarrow \text{Phosphotriose isomerase} \\ \text{3-Phosphoglyceraldehyde} \end{array}$$

(i)
$$\begin{array}{c} CH_2\text{—}OP \\ | \\ C=O \\ | \\ CH_2OH \end{array} + NADH_2 \xrightarrow{\text{dehydrogenase}} \begin{array}{c} CH_2\text{—}OP \\ | \\ CH\text{—}OH \\ | \\ CH_2\text{—}OH \end{array} + NAD^+$$

Dihydroxyacetone phosphate   α-glycerophosphate.

(ii)
$$\begin{array}{c} CH_2\text{—}OP \\ | \\ CH\text{—}OH \\ | \\ CH_2\text{—}OH \end{array} + H_2O \xrightarrow{\text{Phosphatase}} \begin{array}{c} CH_2OH \\ | \\ CHOH \\ | \\ CH_2OH \end{array} + H_3PO_4$$

α-glycerophosphate   Glycerol

(3) Condensation of Fatty acids and Glycerol or Biosynthesis of Triglycerides

The condensation of glycerol and fatty acids is shown in Fig. 15.1. of chapter "Lipids". The scheme for the biosynthesis of triglycerides (fats) is outlined in Fig. 16.1.

The biosynthesis of triglycerides is completed in many steps (Fig. 16.1). The stages of the reactions are as follows:

(i) The glycerol is phosphorylated to form L-α-glycerophosphate with the help of ATP (Reaction 1, Fig. 16.1.). The reaction is reversible. (Reaction 2).

(ii) L-α-glycerophosphate undergoes condensation with two molecules of acyl-CoA to form L-α-phosphatidic acid. The reaction is catalysed by enzyme acyl transferase (Reaction 3, Fig. 16.1).

(iii) The L-α-phosphatidic acid at this stage undergoes dephosphorylation in presence of enzyme phosphatase to form an α-, β-diglyceride (Reaction 4, Fig. 16.1).

Lipid Metabolism

(*iv*) In the last step, the condensation of one molecule of an acyl-CoA with the free hydroxyl group of the α-, β-diglyceride takes place in presence of enzyme diglyceride acyl transferase to form a triglyceride (Reaction 5, Fig. 16.1.).

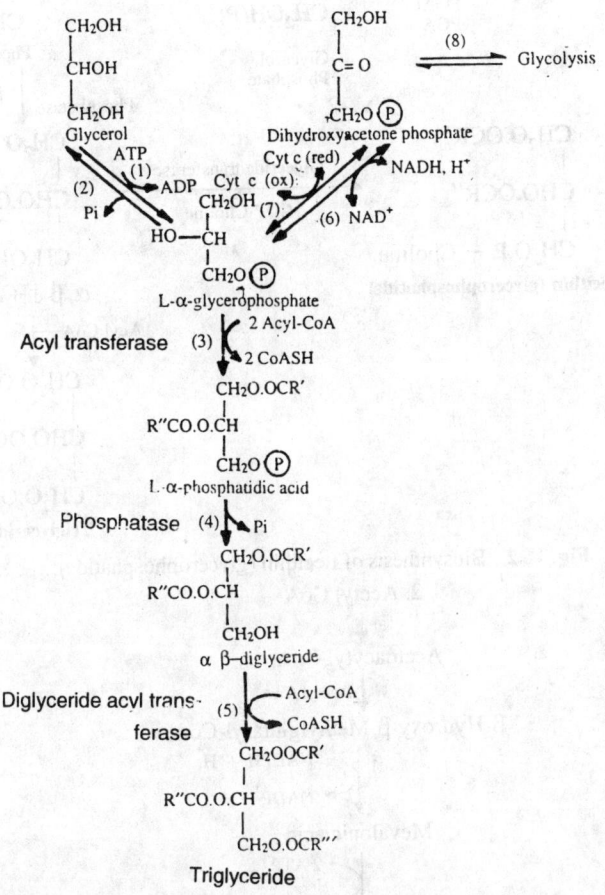

Fig. 16.1 : Biosynthesis of triglycerides.

The glycerol itself is not the starting point of triglyceride synthesis in vivo, rather L-α-glycerophosphate which is formed by diversion of glycolytic pathway at the dihydroxyacetone phosphate stage. The reactions 6, 7 and 8 are the bridge reactions in the formation of L-α-glycerophosphate.

16.3. BIOSYNTHESIS OF LECITHIN (GLYCEROPHOSPHATIDES)

Kennedy (1957) reviewed the ideas about the biosynthesis of phospholipids (glycerophosphatides). Very little work on this line has been carried out with plant materials. Most of the ideas are based on the experiments with animals and yeast. The biosynthesis of phospholipids (glycerophosphatides) suggested by Kennedy (1957) is outlined in Fig. 16.2.

16.4. BIOSYNTHESIS OF CHOLESTEROL

Cholesterol is the most important sterol found in the mammals. It has also been isolated from red algae. The scheme for its biosynthesis is outlined in Fig. 16.3.

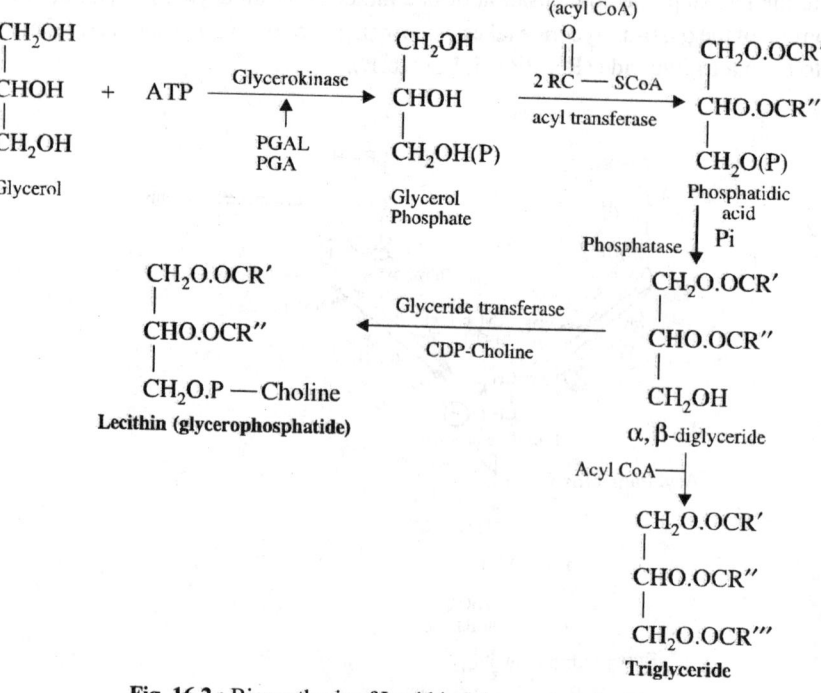

Fig. 16.2 : Biosynthesis of Lecithin (glycerophosphatide).

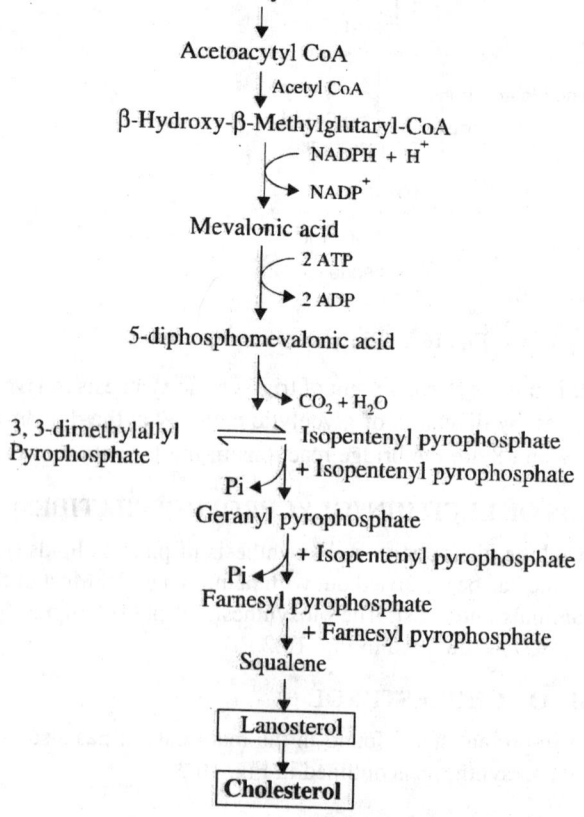

Fig. 16.3 : Biosynthesis of Cholesterol.

16.5. BIOSYNTHESIS OF SPHINGOSINE AND SPHINGOMYELINS (PHOSPHOSPHINGOSIDES)

During the formation of phosphosphingosides or sphingomyelins, the amino alcohols, sphingosine or phytosphingosine, are formed as intermediates. The phytosphingosine (amino alcohol of plant phosphosphingosides) synthesis is very little known. Anyhow, they might have been synthesized in plants similar to that of sphingosine in animals. The biosynthetic scheme for the phosphosphingosides (sphingomyelins) is shown in Fig. 16.4.

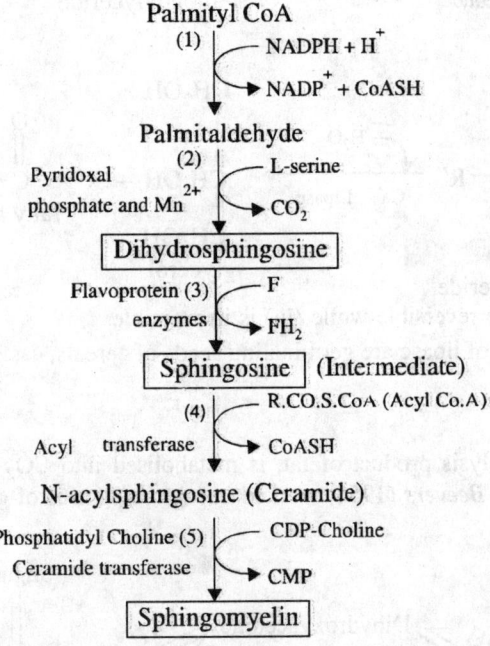

Fig. 16.4 : Biosynthesis of phosphosphingosides (sphingomyelins).

16.6. FAT OXIDATION (CATABOLISM)

Fat oxidation requires three processes:
- (*i*) Hydrolysis of fat into glycerol and fatty acids.
- (*ii*) Metabolism of glycerol.
- (*iii*) Oxidation of fatty acids.

Hydrolysis of fat (Triglyceride) by Lipase

Glycerides are the fatty acid esters of glycerol and the initial step in their catabolism is almost invariably hydrolysis. During germination of the castor oil seeds, the enzyme 'lipase' catalyses this reaction. Triglycerides are hydrolysed stepwise and the process is completed in three steps. The enzyme lipase first attacks the α-carbons and then in the last β-carbon atom. The various steps of hydrolysis of triglycerides are as follows:

(ii)
```
    αCH₂OH                              αCH₂OH
    |     O           H₂O               |      O                O
    βCH.O.C—R''    ⇌                    βCH.O.C—R''  +  R'''—C—OH
    |     O         Ca⁺⁺, Lipase        |                    fatty acid
    αCH₂.O.C—R'''                       αCH₂OH
    α, β⁻, Diglyceride                  β–monoglyceride
```

(iii)
```
    CH₂OH                               CH₂OH
    |      O         –H₂O               |                      O
    CH.O.C—R''    ——————→               CH.OH  +  R''—C—OH
    |              Ca⁺⁺, Lipase         |               fatty acid
    CH₂OH                               CH₂OH
    β⁻, Monoglyceride                   glycerol
```

Steps (*i*) and (*ii*) are reversible while (*iii*) is irreversible.

The richest sources of lipase are germinating seeds of cereals, castor bean, sunflower, lettuce and rape.

Metabolism of Glycerol

Glycerol, the hydrolysis product of fat, is metabolised into CO_2 and H_2O through various steps. *Stumpf* (1955) and *Beevers* (1956) worked out the sequences of glycerol metabolism which are outlined in Fig. 16.5.

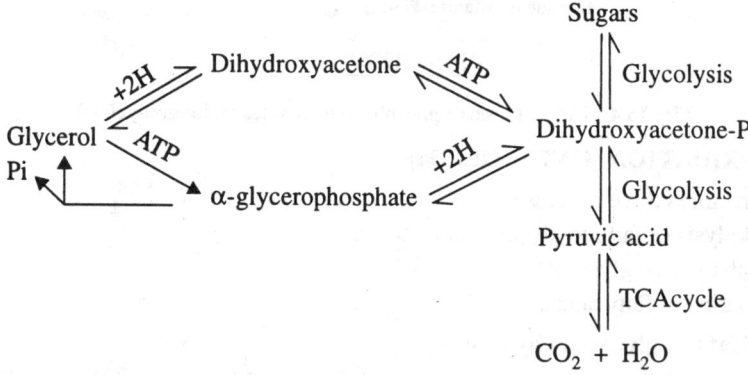

Fig. 16.5 : Metabolism of glycerol.

Reactions 1, 6, 7 and 8 of Fig. 16.1 also indicate the stages of glycerol metabolism. After these steps the glycerol is converted into acetyl CoA which may be oxidized in the TCA cycle to CO_2 and H_2O.

Oxidation of Fatty Acids

The fatty acids are oxidized in three ways :

(*a*) β-Oxidation..
(*b*) α-Oxidation.
(*c*) ω-Oxidation.

(*a*) *β-Oxidation of fatty acids*: The classical experiments on this aspect were carried out by

Lipid Metabolism

Knoop. His experiments demonstrated that fatty acids are degraded by the successive removal of C_2 units after oxidation at the β-carbon atom. The various steps of β-oxidation were studied first in animal tissues, but later on the experiments were also conducted with plant tissues which confirmed this type of oxidation of fatty acids.

β-oxidation spiral requires five enzymes which are located only in mitochondria. The experiments conducted with plant tissues indicate that the plant cell contains two sets of β-oxidation enzymes, one in the mitochondria and the other in the soluble fraction of the cytoplasm. Uptil now, it is not clear how these two β-oxidation sites within the cell are inter-related. The β-oxidation spiral is shown in Fig. 16.6. The various steps of the β-oxidation spiral are as follows.

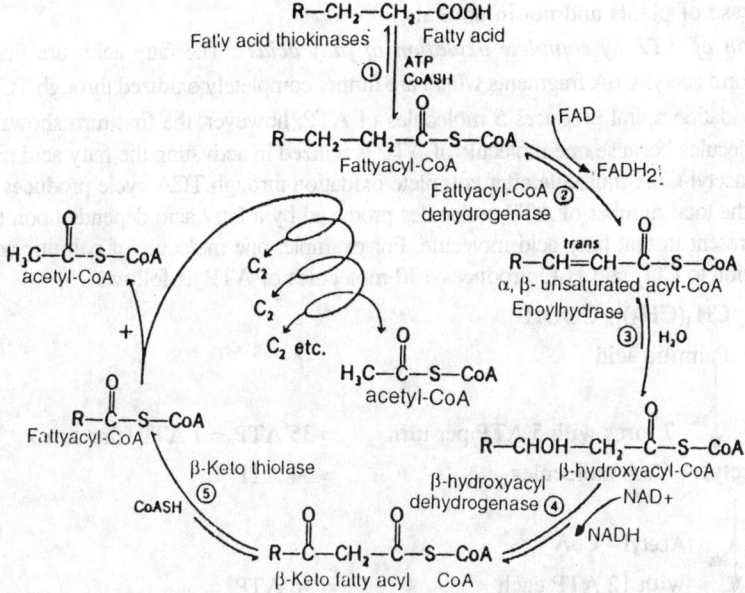

Fig. 16.6 : The β-oxidation spiral of fatty acids.

(i) The first step is the thioester formation of fatty acid with Coenzyme A. This is an endergonic reaction and the energy is supplied by ATP. The enzymes catalysing this reaction are called thiokinases (Thiokinase type of reaction).

(ii) The second step involves the removal of two hydrogen atoms from the α and β-carbon atoms of fatty acyl CoA and the reduction of FAD to $FADH_2$ in presence of enzymes acyl-CoA dehydrogenase-(Acyldehydrogenase type of reaction).

(iii) In the third step, an enzyme enoylhydrase, brings about the stereospecific addition of one molecule of water across the double bond of the α-, β-unsaturated acyl-CoA forming β-hydroxyacyl-CoA (Enoylhydrase type of reaction).

(iv) The fourth step involves the removal of two hydrogen atoms from the β-carbon atom of β-hydroxyacyl-CoA to form corresponding β-Keto derivative and the conversion of NAD^+ to $NADH + H^+$ (β-hydroxyacyl dehydrogenase type of reaction).

(v) The final step involves the thiolytic cleavage of the β-Keto fatty acyl CoA by CoA-SH to form the acetyl CoA and the thioester of the remaining fatty acid chain. This reaction is catalysed by enzyme. β-ketothiolase. The thioester of the remaining fatty acid (fatty acyl-CoA) is shorter by two carbon atoms than when it entered this sequence of reactions (β-Ketothiolase type of reaction).

The resulting fatty acyl-CoA with two carbon atoms less than the starting compound, now re-enters the spiral by combining with coenzyme-A and passes round it. Each β-oxidation spiral removes a C_2 unit (acetyl-CoA) from the fatty acid. This process continues step by step until the whole molecule of fatty acid is degraded.

The end product of β-oxidation of fatty acids with an even number of carbon atom is acetyl-CoA, while fatty acids with odd number of carbon atoms produce propionyl-CoA (C_3) as well as acetyl-CoA (C_2). Thus, in both the cases, the formation of acetyl-CoA is common and is the main end product of β-oxidation. The acetyl-CoA molecules produced by β-oxidation of saturated and unsaturated fatty acids may be completely oxidized to CO_2 and H_2O through TCA cycle. They may also be converted into carbohydrates by glyoxylate cycle. This conversion takes place only in case of plants and not in animals.

Production of ATP by complete oxidation of fatty acids : The fatty acids are first oxidized by β-oxidation to form acetyl-CoA fragments which are further completely oxidized through TCA cycle. Each turn of the β-oxidation spiral produces 5 molecules of ATP; however, the first turn shows a net gain of only 4 ATP molecules because one molecule of ATP is utilized in activating the fatty acid molecule in the first step. Each acetyl-CoA molecule after complete oxidation through TCA cycle produces 12 molecules of ATP. Thus, the total number of ATP molecules produced by a fatty acid depends upon the number of carbon atoms present in that fatty acid molecule. For example, one molecule of palmitic acid (C_{16}) after complete oxidation to CO_2 and H_2O produces 130 molecules of ATP as follows:

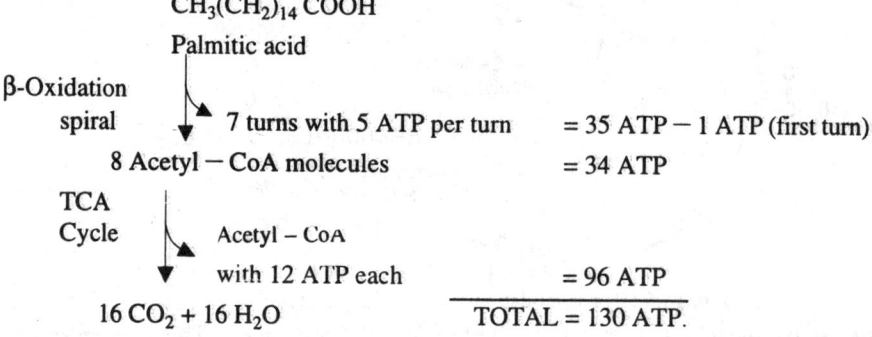

Now, one ATP molecule represents a gain of about 7600 calories of free energy. Thus, the complete oxidation of one molecule of palmitic acid by β-oxidation and TCA cycle will show a gain of 7600 × 130 = 988,000 calories of utilizable energy.

(b) α-Oxidation of fatty acids : *Stumpf* discovered another type of oxidation of fatty acids, known as α-oxidation, in peanut cotyledons. The whole α-oxidation is completed with the help of two enzymes namely, a soluble cytoplasmic *fatty acid peroxidase* and a microsomal *NAD-specific aldehyde dehydrogenase*. The enzyme fatty acid peroxidase catalyses the peroxidative decarboxylation of fatty acids to yield CO_2 and an aldehyde one carbon atom shorter than the acid. The enzyme aldehyde dehydrogenase now catalyses the oxidation of aldehyde (formed by first enzyme) to yield the corresponding acid. This resulted acid is again utilized by the enzyme fatty acid peroxidase as substrate for another turn round the two stage of α-oxidation spiral (Fig. 16.7). The enzymes of α-oxidation are specific for long chain saturated fatty acids. The acids with more than C_{12} (usually with C_{14}, C_{16} and C_{18}) are utilized as substrate in α-oxidation.

It should be noted that the substrates for the α-oxidation are always free acids whereas for the β-oxidation are acyl-CoA derivates. In this respect, the two types of oxidations are quite different.

A slightly different type of α-oxidation was observed by *Hitchcock* and *James* in 1964 in the

Lipid Metabolism

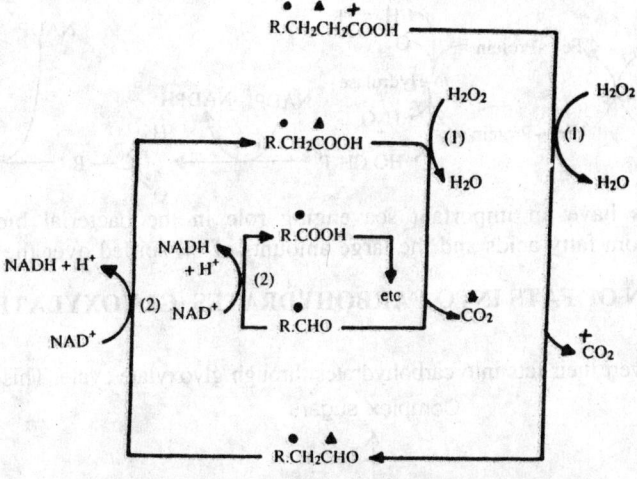

(1) = Fatty acid peroxidose, (2) = Aldehydrogenase.

Fig. 16.7 : The α-oxidation spiral in peanut cotyledons. (after Stumpf, 1952).
required and 2-hydroxy fatty acids and fatty aldehydes are formed as intermediates.

The introduction of imidazole inhibits reaction (1) but not reactions (2) and (3). The reaction (3) requires NAD^+ and in absence of it the aldehyde accumulates.

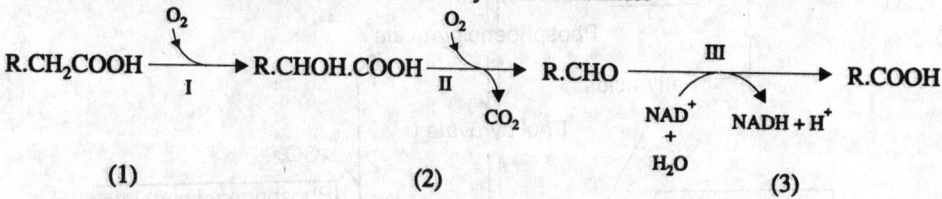

Fig. 16.8 : Leaf α-oxidation (Hitchcock and James, 1964).

The leaf α-oxidation process resembles with that of brain microsomal α-oxidation process discovered by Mead and Levis (1962).

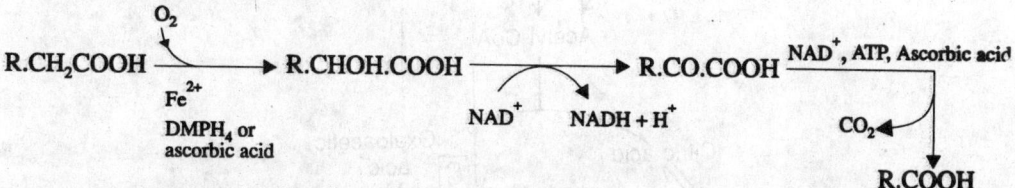

Fig. 16.9 : Brain microsomal α-oxidation system. (After Mead and Levis, 1962).

Significance of α-oxidation

(i) α-oxidation may provide a source of long-chain alcohols by reduction of aldehydes produced in the system.

(ii) The alcohols thus formed may be used in the formation of waxes.

(iii) The oxidation may provide a source of fatty acids with odd number of carbon atoms.

(c) *ω-Oxidation*: A number of aerobic bacteria have been isolated from oil-soaked soil which rapidly degrade hydrocarbons of fatty acids to water-soluble products. The reactions involve an initial hydroxylation of a terminal methyl group to a primary alcohol and subsequent oxidation to a carboxylic acid. Thus, straight chain hydrocarbons are oxidized to fatty acids which in turn are β-oxidized to Acetyl-CoA.

These reactions have an important scavenging role in the bacterial bio-degradation of

$$NADH \rightarrow \text{Reductase (oxidized)} \rightarrow Fe^{+2}\text{-Protein} \rightarrow \begin{array}{c} CH_3-R \\ O_2 \end{array} \omega\text{-Hydrolase}$$

$$NAD^+ \leftarrow \text{Reductase (Reduced)} \leftarrow Fe^{+3}\text{-Protein} \leftarrow \begin{array}{c} H_2O \\ HO.CH_2R \end{array} \xrightarrow{NADP^+ \; NADPH} \underset{O}{\overset{H}{\underset{\|}{C}}} - R \xrightarrow{NADP^+ \; NADPH} \underset{O}{\overset{HO}{\underset{\|}{C}}} - R$$

These reactions have an important scavenging role in the bacterial bio-degradation of detergents derived from fatty acids and the large amounts of oil spilled over the ocean surface.

16.7. CONVERSION OF FATS INTO CARBOHYDRATES (GLYOXYLATE CYCLE) (Fig. 16.10)

Many fatty seeds convert their fats into carbohydrates through glyoxylate cycle. This conversion takes

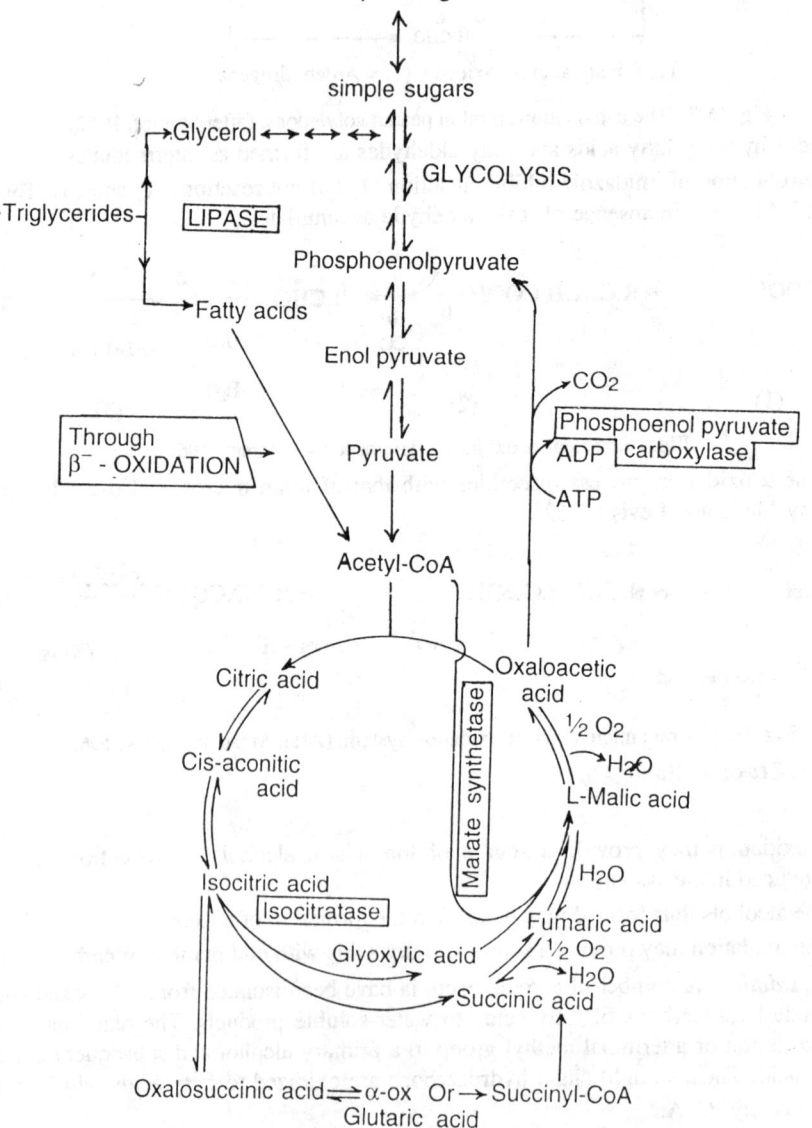

Fig. 16.10 : Conversion of fats into carbohydrates via the glyoxylate cycle (After Kornberg and Kreb's, 1957).

Lipid Metabolism

place as the seeds germinate. The scheme of glyoxylate cycle as suggested by Kornberg and Kreb's (1957) is represented in Fig. 16.10 which shows complete conversion of fats into carbohydrate.

Kornberg and Kreb's suggested that isocitritase and malate synthetase are the two key enzymes of glyoxylate cycle which prevent the entrance of acetyl-CoA into TCA cycle to oxidize into CO_2 and H_2O. The enzyme isocitritase catalyses the reversible aldol fission of isocitrate to succinate and glyoxylate. The second enzyme malate synthetase catalyses the condensation of acetyl-CoA with glyoxylate to form malate.

The succinate, so formed by the fission of isocitrate is now converted into oxaloacetate via fumarate and malate (dicarboxylic acids of TCA cycle). The oxaloacetate does not enter into TCA cycle and is directly converted into phosphoenol pyruvate which in turn can readily be converted into sugars by reversal of the Embden-Meyerhof glycolytic path way (Glycolysis). The conversion of oxaloacetate into phosphoenol pyruvate is catalysed by the enzyme phosphoenol pyruvate carboxylase which requires ATP and removes CO_2.

$$\text{Oxaloacetate} + \text{ATP} \xrightarrow{\text{Phosphoenol Pyruvate carboxylase}} \text{Phosphoenolpyruvate} + CO_2 + \text{ADP}$$

Thus, acetyl-CoA formed either by β-oxidation of fatty acids or from glycerol is utilized in the glyoxylate cycle to form sugars. As soon as the fat stores in the seeds are used up, the activity of two key enzymes of glyoxylate cycle disappears.

Beevers (1967) suggested that glyoxysomes are the intracellular particles which contain specific enzymes (isocitritase, malate synthetase, citrate synthetase, aconitase, malate dehydrogenase, glycollate oxidase and catalase) of glyoxylate cycle and they are present only in those tissues which convert fats into carbohydrates. The TCA cycle enzymes (succinic dehydrogenase, fumarase, NADH oxidase and cytochromes) are completely lacking in these glyoxysomes. He further suggested that acetyl-CoA produced by mitochondrial β-oxidation of fatty acids is converted into succinate through the reaction sequences with the glyoxysome as shown in Fig. 16.11.

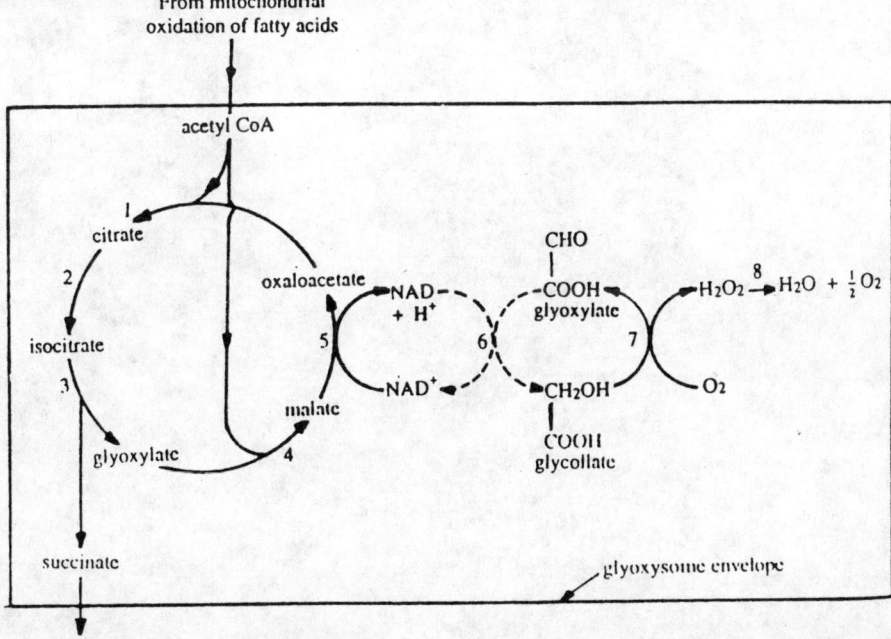

Enzymes of Reactions–1. Citrate synthetase 2. Aconitase 3. Isocitratase 4. Malate synthetase 5. Malate dehydrogenase 6. Glyoxylate reductase 7. Glyoxylate 8. Catalase.

Fig. 16.11 : Proposed reaction sequence within the glyoxysome (Beevers, 1967).

The present evidences indicate that the conversion of succinate into phosphoenolpyruvate and ultimately to sugar takes place in the following sequence.

Succinate $\xrightarrow{(i)}$ fumarate $\xrightarrow{(ii)}$ malate $\xrightarrow{(iii)}$ oxaloacetate $\xrightarrow{(iv)}$ phosphoenol pyruvate $\xrightarrow{\text{Glycolysis}}$ simple sugar.

The reactions (i), (ii), (iii) and (iv) do not take place in the glyoxysomes because of absence of appropriate enzymes.

17

Carbohydrate Metabolism

17.1. INTORDUCTION

During the synthesis of metabolites energy is required and the reactions are called *endergonic*, whereas in their breakdown energy is released and the reactions are called *exergonic*. The growth of plants and other organisms occurs when synthesis exceeds breakdown.

17.2. BIOSYNTHESIS OF CARBOHYDRATES

Monsaccharides

(1) *Synthesis from Carbon Dioxide* : Most of the plants capable of photosynthesis, fix atmospheric CO_2 in presence of RuDP carboxylase through Calvin cycle. This enzyme catalyses the carboxylation of ribulose 1-5-diphosphate to form 3-phosphoglyceric acid. For this reaction the energy is supplied from the sun.

$$\text{Ribulose 1,5-di P} + CO_2 + H_2O \xrightarrow{\text{RuDP carboxylase}} \text{2 mol of 3-Phosphoglyceric acid.}$$

The 3-phosphoglyceric acid is converted into 3-phosphoglyceraldehyde which is then converted into tetrose, pentose, hexose and heptose phosphates. The sugars undergo interconversion and are ultimately utilized in the synthesis of oligo- and polysaccharides.

In tropical grasses and other plants which have been called as C_4 plants, phosphoenolpyruvate carboxylase (PEP carboxylase) catalyses the carboxylation of phosphoenolpyruvate (PEP), to form oxaloacetate. Thus, in tropical grasses PEP acts as a CO_2-acceptor. The enzyme PEP carboxylase is found in the chloroplasts of the mesophyll cells. The oxaloacetate is converted into 3-phosphoglyceric acid and then into other mono- and oligosaccharides by the calvin cycle in the bundle sheath.

Calvin Cycle (See Chapter 10, Fig. 10.14).

(2) *Synthesis by Reversal of the Embden-Meyerhof Glycolytic pathway*: Monosaccharides and their phosphates can be synthesized from pyruvic acid by a reaction sequence which is essentially a reversed glycolytic pathway (see Fig. 11.7). Although, all the reactions of the glycolytic sequence are reversible, the standard free energy change of the overall process is about 50,000 cal/mol which clearly indicates that the equilibrium lies towards the break-down rather than synthesis.

(3) *Biosynthesis of Ascorbic Acid*: The ascorbic acid can be biosynthesized either from D-glucose or from D-galactose. During its formation from D-glucose, C-1 and C-6 of the sugar become C-6 and C-1 of the ascorbic acid. The biosynthesis from D-galactose does not involve this change.

First pathway from D-glucose :

D-Glucose → D-Glucuronic acid. → L-Gulono-γ-lactone →
→ L-Ascorbic acid (In Strawberry fruits)

Second pathway from D-galactose:

D-Galactose → D-Galacturonic acid ester → L-galactono-

$$NADP^+ \rightleftharpoons NADPH + H^+$$

γ-lactone ⟶ L-Ascorbic acid (In Pea seedings and cauliflower florets)

Biosynthesis of UDP-D-Glucose: UDP-D-glucose is a sugar nucleotide and is biosynthesized from -D-glucose in the following three steps in plants:

(1) α-D-glucose + ATP $\underset{}{\overset{Kinase}{\rightleftharpoons}}$ α-D-glucose-6-P + ADP.

(2) α-D-glucose-6-P $\overset{Phosphoglucomutase}{\rightleftharpoons}$ α-D-glucose - 1-P.

(3) α-D-glucose-1-P + UTP $\overset{UTP: a\text{-}D\text{-}glucose\text{-}1\text{-}P\text{-}}{\underset{Uridyltransferse}{\rightleftharpoons}}$ UDP-D-glucose + PPi

(5) *Biosynthesis of Ribose* : The pentose sugar, ribose, is formed from glucose through the hexose monophosphate shunt. The various steps are shown in a sequence.

Glucose ⟶ Glucose-6-P $\overset{Glucose\text{-}6\text{-}P}{\underset{dehydrogenase}{\longrightarrow}}$ 6-Phosphoglucono-γ-lactone $\overset{Lactonase}{\rightleftharpoons}$ 6-Phosphogluconic acid $\rightleftharpoons$

3-Keto-6-Phosphogluconic acid $\overset{Mn^{++}}{\rightleftharpoons}$ D-Ribulose-5-P $\overset{Isomerase}{\rightleftharpoons}$ D-Ribose-5-P

(Intermediate Compound)

The formation of Ribose 5-phosphate is also shown in Fig. 10.14, reaction 10.

Biosynthesis of Oligosaccharides

Disaccharide

Sucrose : The sucrose is synthesized in plants by two reactions involving UDP-D-glucose. The first reaction was studied in wheat germ tissues and sugar-beet leaves where the enzyme *sucrose synthetase* catalyses the transfer of glucose residue of UDP-glucose to fructose with the formation of sucrose and UDP.

UDP - D-Glucose + D-Fructose $\overset{Sucrose}{\underset{synthetase}{\rightleftharpoons}}$ Sucrose + UDP

The second type of reaction was studied in wheat-germ and the chloroplasts from tobacco and sugarcane leaves. Here, the enzyme which catalyses the reaction is *sucrose phosphate synthetase* and the reactants are UDP-D-glucose and D-fructose 6-phosphate. The glucose residue of UDP-D-glucose is transferred to fructose-6-phosphate to form sucrose phosphate. The sucrose-phophate, thus, formed is dephosphorylated to sucrose and Pi in presence of phosphatase enzyme.

UDP-D-Glucose + D-Fructose-6-P $\overset{Sucrose\text{-}P}{\underset{synthetase}{\rightleftharpoons}}$ Sucrose Phosphate + UDP

Sucrose phosphate $\overset{Phosphatase}{\rightleftharpoons}$ Sucrose + Pi.

Doudoroff in *Pseudomonas saccharophila* discovered that sucrose synthesis takes place in the presence of another enzyme *sucrose phosphorylase* which is not functional in higher plants.

Glucose-1-P + D-Fructose $\overset{Sucrose}{\underset{phosphorylase}{\rightleftharpoons}}$ Sucrose + Pi.

Carbohydrate Metabolism

Biosynthesis of Polysaccharides

Polysaccharides are the main constituents. Most of them are synthesized from atmospheric CO_2 through a number of steps as shown in Fig. 17.1. About 75 per cent of the carbon of photosynthetically fixed CO_2 is rapidly incorporated into structural and other polysaccharides. Fig. 17.1 shows the overall picture of carbohydrate biosynthesis.

Starch, a well-known storage polysaccharide, is synthesized from ADP-glucose and UDP-glucose. The structural polysaccharides like xylans and hemicelluloses are synthesized from UDP-xylose, and celluloses from GDP-glucose. The pectic substances, arabans, galacturonan and galactan are formed from UDP-arabinose, UDP-galacturonic acid and UDP-galactose respectively. The conversion of fats into carbohydrates takes place via glyoxylate cycle.

17.3. OXIDATION (CATABOLISM) OF CARBOHYDRATES

Monosaccharides

All green plants synthesize carbohydrates by photosynthesis. At the time of their break-down the cell utilizes the enzymes of the cytoplasm and mitochondria. The process involves a number of steps. The steps that occur in the cytoplasm comprise *glycolysis*; those occurring in the outer lamellae of mitochondria comprise the *Kreb's cycle* or the Citric Acid Cycle and those occurring on the inner lamellae comprise the *electron transport* and *oxidative phosphorylation*. The whole process produces an efficient amount of ATP. The various steps, in which the oxidation of glucose into CO_2 and H_2O takes place, are completed through following several processes:

  (*i*) Glycolysis or Embden-Meyerhof Parnas Pathway.
  (*ii*) Oxidative Decarboxylation.
  (*iii*) Kreb's Cycle or citric acid cycle.
  (*iv*) Respiratory chain or Electron transport system.
  (*v*) Oxidative phosphorylation.
  (*vi*) Direct Oxidation pathway.

Note : For (*i*), (*ii*), (*iii*), (*iv*) and (*v*) consult chapter 11—Respiration.

Direct Oxidation Pathway (Fig. 17.2)

Harry Beevers (1954, 56) pointed out the existance of direct oxidation pathway in plants. This pathway is also known as 'oxidative glycolysis pathway', 'pentose phosphate pathway' and 'hexose monophosphate shunt' (H.M.P.). Simply it is also known as 'pentose shunt'. The direct evidence for the existence of this pathway came through the inhibition of Kreb's cycle by malonic acid and with the radioactive carbon (C^{14}). Various steps of this pathway are as follows :

(1) Glucose is first phosphorylated to glucose-6-phosphate as in EMP pathway (Enzyme-*Hexokinase*).

(2) The glucose-6-phosphate is oxidized to 6-phosphogluconate and NADP is reduced to NADPH. (Enzyme-*Glucose-6-phosphate dehydrogenase*).

(3) The 6-phosphogluconate is converted into a keto-pentose sugar, ribulose-5-phosphate, with the liberation of one molecule of CO_2 for every sugar molecule. The coenzyme NADP is reduced to $NADPH_2$ at this stage (Enzyme-*6-phosphogluconate dehydrogenase*).

(4) Ribulose-5-phosphate is then changed to its aldopentose isomer, ribose-5-phosphate. (Enzyme-*Phosphoribose isomerase*).

(5) The two molecules of ribose-5-phosphate now combine and form one molecule of sedoheptulose and another of glyceraldehyde-3-phosphate. Both of them are phosphate esters. (Enzyme-*Transketolase*).

(6) The 3-carbon atom chain from sedoheptulose is now linked to the 3-carbon chain of 3-phosphoglyceraldehyde with the formation of fructose-6 phosphate molecule. (Enzyme-*Transaldolase*).

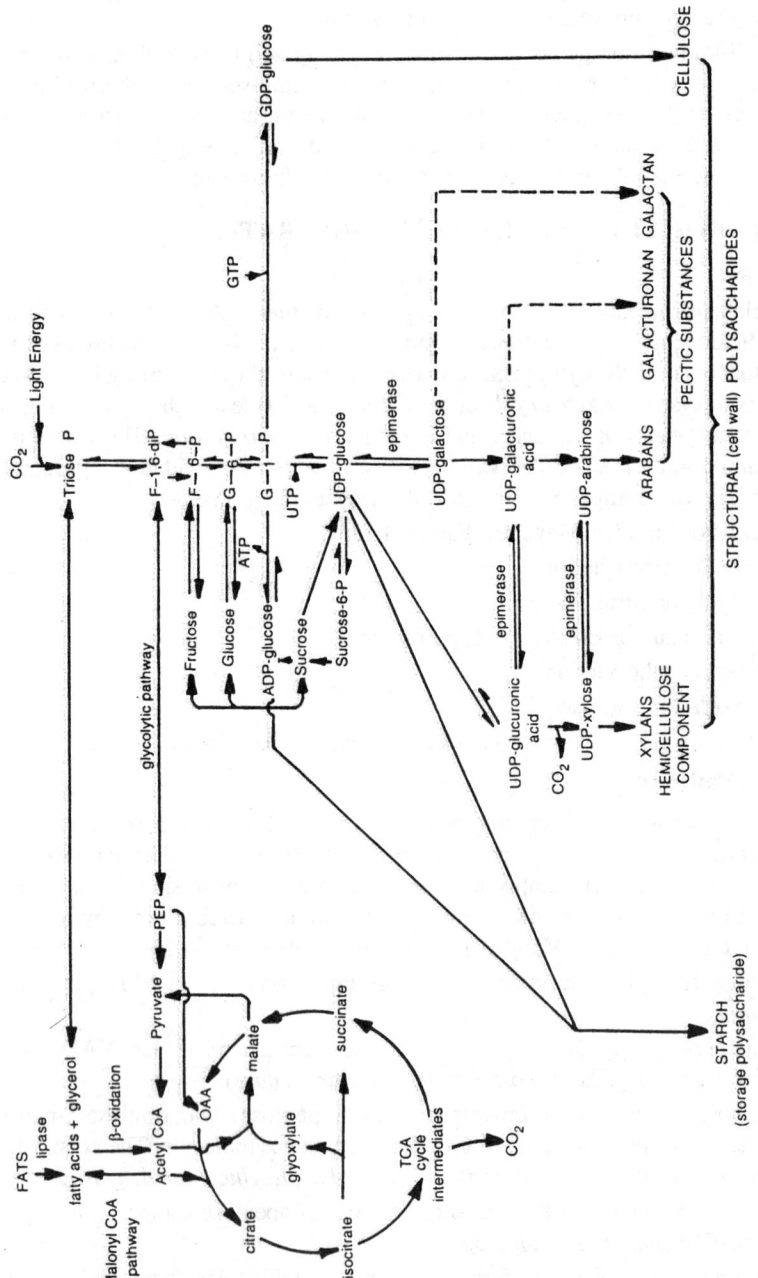

Fig. 17.1: Carbohydrate biosynthesis in plants (as proposed by Mercer and Goodwin, 1972).

Carbohydrate Metabolism

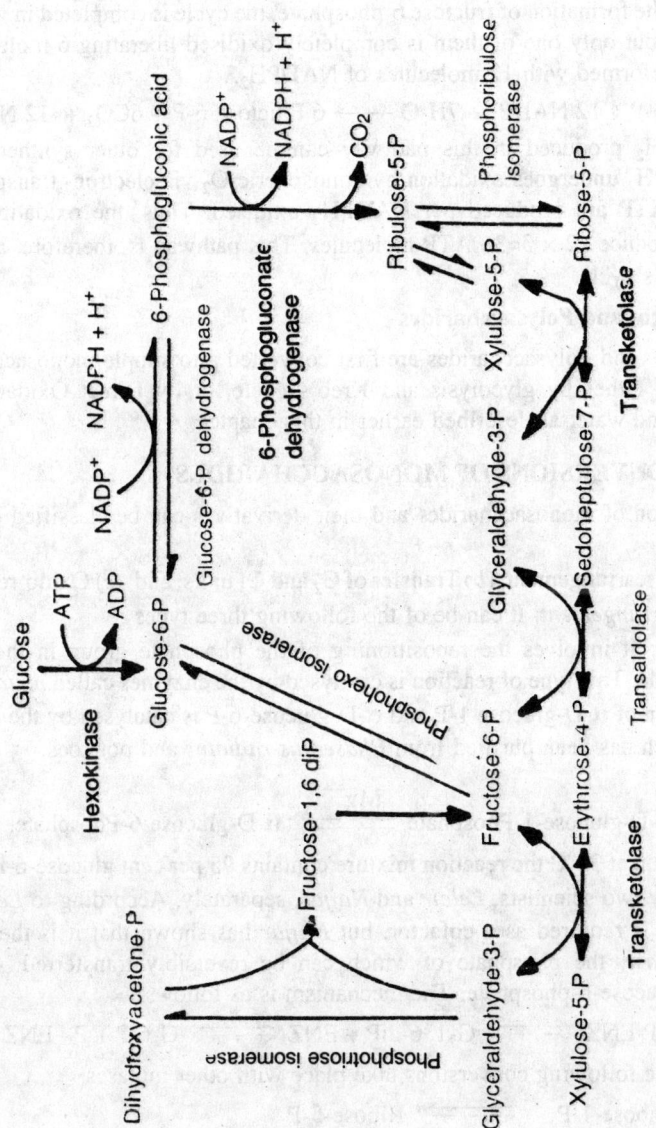

Fig. 17.2: Diagrammatic representation of Direct Oxidation Pathway (Hexose monophosphate shunt)– After Devlin (1969).

(7) A second molecule of fructose 6-phosphate is formed by the linkage of the remaining chain of 4-carbon atom (erythrose 4-P) from the sedoheptulose, with a 2-Carbon fragment of a pentose molecule (xylulose-5-P) along with the formation of 3-phosphoglyceraldehyde. (Enzyme-*Transketolase*).

(8) Two molecules of 3-phosphoglyceraldehyde link to form fructose 1,6-diphosphate which after dephosphorylation form fructose-6-P.

Thus, with the formation of fructose 6-phosphate, the cycle is completed in which 6-molecules of glucose enter, but only one of them is completely oxidised liberating 6-molecules of CO_2 and the rest five are reformed with 12 molecules of $NADPH_2$.

$$6 \text{ Glucose-6-P} + 12 \text{ NADP}^+ + 7H_2O \longrightarrow 6 \text{ Fructose-6-P} + 6CO_2 + 12 \text{ NADPH}_2 + H_3PO_4.$$

The $NADPH_2$ produced in this pathway can be used for other synthetic processes (Fat synthesis). $NADPH_2$ undergoes oxidation by atmospheric O_2 via electron transport system where 3 molecules of ATP are produced per $NADPH_2$ oxidised. Thus, the oxidation of $12 NADPH_2$ molecules will produce $12 \times 3 = 36$ ATP molecules. This pathway is, therefore, almost as efficient as the EMP-Kreb's cycle.

Oxidation of Oligo- and Polysaccharides

All the oligo- and polysaccharides are first converted into simple monosaccharides, and then they are oxidized either by glycolysis and Kreb's cycle, or by Direct Oxidation pathway into carbon-di-oxide and water as described earlier in this chapter.

17.4. INTERCONVERSIONS OF MONOSACCHARIDES

The interconversion of monosaccharides and their derivatives can be classified broadly into three main groups:

(*a*) Internal rearrangement, (*b*) Transfer of C_2 and C_3 units, and (*c*) Oxido-reduction reactions.

Internal rearrangement: It can be of the following three types :

(*i*) *Mutation*: It involves the repositioning of the phosphate group in the monosaccharide phosphate molecule. This type of reaction is catalysed by the enzymes called *mutases*. For example, the interconversion of α-D-glucose 1-P and α-D-glucose-6-P is catalysed by the enzyme *Phosphoglucomutase* which has been purified from *Phaseolus radiatus* and potatoes.

$$\alpha\text{-D-glucose-1-Phosphate} \xrightleftharpoons{Mg^{++}} \alpha\text{-D-glucose-6-Phosphate}.$$

At equilibrium at 30°C the reaction mixture contains 95 per cent glucose-6-P. The mechanism was discovered by two scientists, *Leloir* and *Najjar*, separately. According to *Leloir*, α-D-glucose 1,-6-di-phosphate is required as a cofactor, but *Najjar* has shown that it is the phosphoenzyme, *phosphoglucomutase*, the phosphate of which can be reversibly transferred to either glucose 1-phosphate or glucose-6-phosphate. The mechanism is as follows:

$$\text{G-1-P} + \text{P-ENZ} \rightleftharpoons \text{G-1,6-diP} + \text{ENZ} \rightleftharpoons \text{G-6-P} + \text{P-ENZ}$$

Similarly, the following conversions take place with other mutases.

$$\text{Ribose-1-P} \rightleftharpoons \text{Ribose-5-P}$$
$$\text{Mannose-1-P} \rightleftharpoons \text{Mannose-6-P}$$
$$\text{Galactose-1-P} \rightleftharpoons \text{Glactose-6-P}$$

(*ii*) *Epimerization*: Epimerization forms another class of internal rearrangement. It involves the inversion of the configuration of the H and OH substituents at one of the carbon atoms of the monosaccharide. Most of the reactions of this type are catalysed by the enzymes called *epimerases*. They may utilize monosaccharide-phosphate alone or UDP derivatives as their substrates. The epimerization can be seen in the following reactions :

Carbohydrate Metabolism

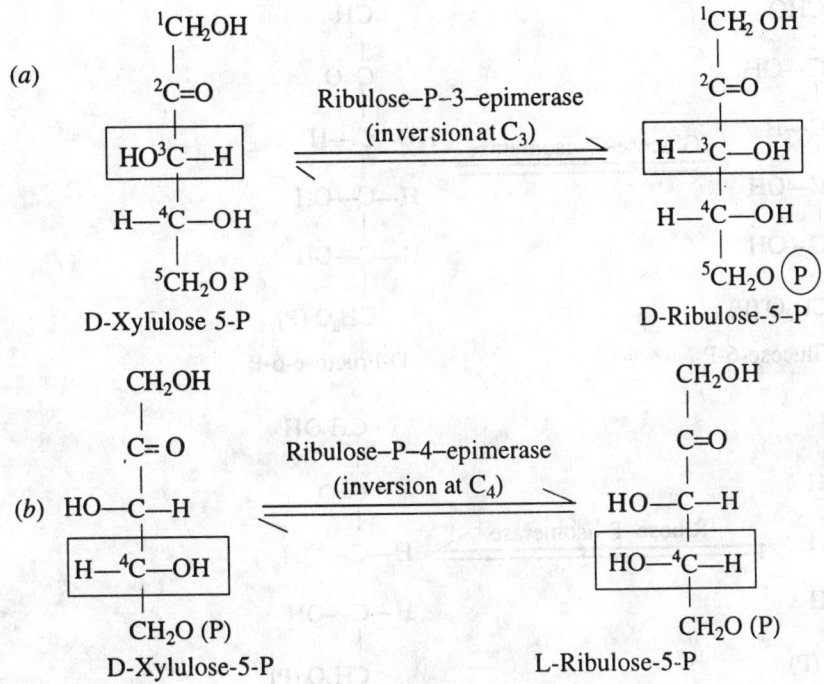

It should be noted that in some of these epimerase-catalysed reactions there is a change from a D-sugar to a L-sugar.

(iii) *Aldose-Ketose interconversions*: In certain biochemical reactions the enzymes called as isomerases catalyse aldose-ketose interconversions.

```
    CHO              CH₂OH
    |                |
    CHOH    ⇌        C=O
    |                |
    R                R
   Aldose          Ketose
```

The conversion of D-glucose-6-phosphate to D-fructose-6-phosphate, D-ribose-5-phosphate to D-ribulose-5-phosphate and glyceraldehyde-3 phosphate to dihydroxy-acetone-phosphate, are examples of this type.

$$\begin{array}{c}\text{CHO}\\|\\\text{H—C—OH}\\|\\\text{HO—C—H}\\|\\\text{H—C—OH}\\|\\\text{H—C—OH}\\|\\\text{CH}_2\text{O (P)}\end{array} \quad\underset{\rightleftharpoons}{\xrightarrow{\text{Glucose–P–isomerase}}}\quad \begin{array}{c}\text{CH}\\|\\\text{C=O}\\|\\\text{HO—C—H}\\|\\\text{H—C—OH}\\|\\\text{H—C—OH}\\|\\\text{CH}_2\text{O (P)}\end{array}$$

D-Glucose-6-P   D-Fructose-6-P

$$\begin{array}{c}\text{CHO}\\|\\\text{H—C—OH}\\|\\\text{H—C—OH}\\|\\\text{H—C—OH}\\|\\\text{CH}_2\text{O (P)}\end{array} \quad\underset{\rightleftharpoons}{\xrightarrow{\text{Ribose–P–isomerase}}}\quad \begin{array}{c}\text{CH}_2\text{OH}\\|\\\text{C=O}\\|\\\text{H—C—OH}\\|\\\text{H—C—OH}\\|\\\text{CH}_2\text{O (P)}\end{array}$$

D-Ribose-5-P   D-Ribulose-5-P

$$\begin{array}{c}\text{CHO}\\|\\\text{H—C—OH}\\|\\\text{CH}_2\text{O (P)}\end{array} \quad\underset{\rightleftharpoons}{\xrightarrow{\text{Triose–P–isomerase}}}\quad \begin{array}{c}\text{CH}_2\text{OH}\\|\\\text{C=O}\\|\\\text{CH}_2\text{O (P)}\end{array}$$

D-Glyceraldehyde-3-P   Dihydroxyacetone-P

Transfer of C_2 and C_3 Units : *Transketolase* and *transaldolase* are two important enzymes which catalyse the transfer of 2-C and 3-C Units from one monosaccharide to another respectively.

(a) *Transketolase* : It catalyses the tranfer of a ketol residue ($CH_2OH-CO-$) from a ketose to aldose. D-Sedohetulose-7-P, D-fructose-6-P, and D-xylulose-5-P are ketol donors and all of them (ketol donors) possess the L-configuration at C_3.

(b) *Transaldolases* : It catalyses the transfer of a dihydroxyactone residue ($CH_2OH.CO.CHOH-$ from a ketose to an aldose.

Oxido-Reductions : (Conversion of aldoses into aldonic and alduronic acids).

The aldoses are oxidized at C_1 and C_2 positions. When oxidation at C_1 occurs aldonic acids are formed. For example, the glucose-6-phosphate is converted into 6-phosphogluconolactone (an aldonic acid) in presence of an enzyme *glucose-6-Phosphate dehydrogenase* and coenzyme $NADP^+$.

When the oxidation occurs at C_6 position of aldoses, alduronic acids are formed. The example of this type is the oxidation of C_6 of UDP-D-Glucose to form UDP-D-glucuronic acid in presence of the enzyme UDP-D-*glucose dehydrogenase* which has been detected from pea seedlings.

Carbohydrate Metabolism

$$\text{Glucose-6-P} \xrightleftharpoons[\text{+NADP}]{\text{G-6-P-dehydrogenase}} \text{6-Phosphogluconolactone} + \text{NADPH} + H^+$$

$$\text{UDP-D-Glucose} + 2NAD^+ \xrightarrow{\text{UDP-D-glucose dehydrogenase}} \text{UDP-D-Glucuronic Acid} + 2NADH + 2H^+$$

Recent studies have shown that in plants glucose can be directly converted into myo-inositol which upon oxidation with plant extracts yields D-glucuronic acid.

$$\text{D-Glucose-6-P} \xrightarrow{-Pi} \text{m-Inositol} \xrightarrow{\text{Oxidation}} \text{D-Glucuronic acid}$$

Conversion of Aldoses into Polyhydric Alcohols

When an aldehyde group of free aldoses is reduced, the conversion of aldoses into polyhydric alcohols takes place. Example of this type of oxido-reduction reaction is the conversion of D-glucuronic acid into L-gulonic acid which is an important step in the biosynthesis of ascorbic acid.

Gluconeogenesis

It takes place in animal tissues. The formation of D-glucose from non-carbohydrate precursors is called gluconeogenesis ("formation of new sugar "). The important precursors are lactate,, pyruvate, glycerol, certain amino acids (alanine cysteine, serine, aspartic acid) and the intermediates of citric acid cycle, Pyruvic acid and oxaloacetic acid). Gluconeogenesis largely occurs in the liver at the time of starvation. Regardless of the precursor, all these compounds must "bypass" following three irreversible three reactions of glycolytic pathway in forming glucose.

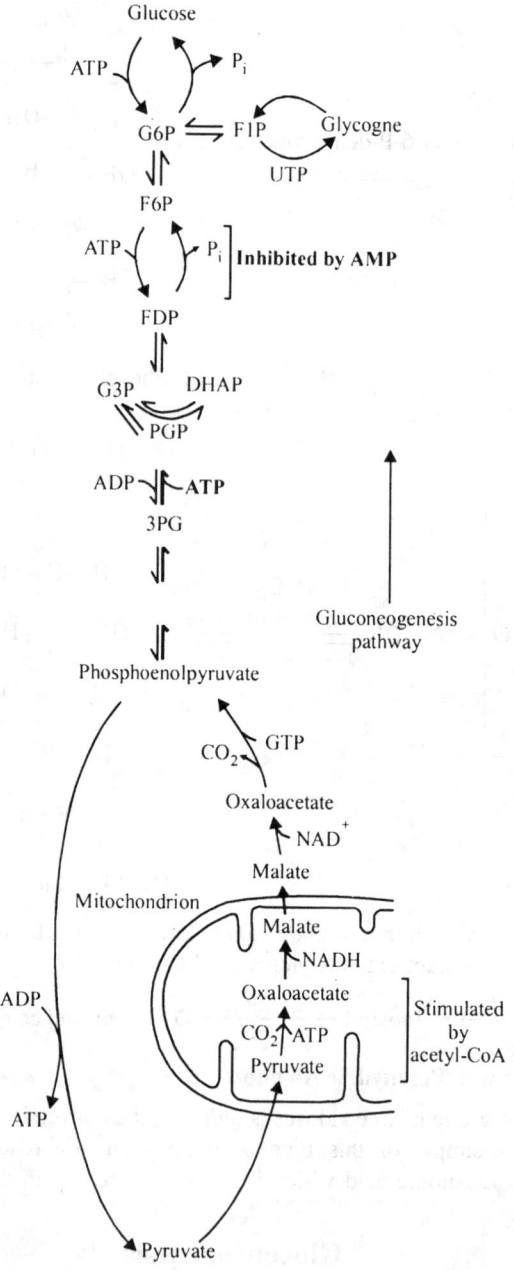

Fig. 17.3. The opposing pathways of glycolysis and gluconeogenesis

Carbohydrate Metabolism

1. Conversion of posphoenol pyruvic acid into pyruvic acid.
2. Conversion of fructose 6-phosphate into fructose 1, 6-di phosphate.
3. Conversion of glucose into glucose -6 phosphate

Bypass of Reaction 1 (Conversion of phosphoenol pyruvic acid into pyruvic acid :.

The conversion of this reaction cannot occur by reversal of the *pyruvate kinase*.

Phosphoenolpyruvic acid + ADP $\longrightarrow$ Pyruvic acid + ATP ($\Delta G = 7.5$ kcal/mol). This reaction has a large negative standard free energy change and has been found irreversible in intact cells. On the other hand, the phosphorylation of pyruvic acid is carried out by a sequence of reactions in cooperation with enzymes in cytosol and the mitochondria of liver cells.

The bypassing of this reaction involves participation of two new enzymes - *Pyruvic carboxylase* and phosphoenal pyruvic (PEP) *carboxykinase,* that catalyse CO_2 fixation reactions during functioning of the kreb's cycle. The enzyme *pyruvic carboxylase is* located in mitochondria and phosphoenol pyruvic carboxykinase in cytosol. The pyruvic carboxylase, a biotin containing enzyme, catalyses the formation of oxaloacetic acid from pyruvic acid. For this purpose, the pyruvic acid produced in the cytosol from phosphoenol pyruvic acid must enter the mitochondria. The $\Delta G'$ of this reaction is quite small, and therefore the reaction in readily reversible.

$$\begin{array}{c} CO_2H \\ | \\ C=O \\ | \\ CH_3 \end{array} + CO_2 + ATP \underset{Mg^{2+}}{\overset{Acetyl\ co\ A}{\rightleftharpoons}} \begin{array}{c} CO_2H \\ | \\ C=O \\ | \\ CO_2H \end{array} + ADP + H_3PO_4 \quad (\Delta G' = 500 \text{ cal at pH7}) \quad \text{(Reaction-1)}$$

Pyruvic acid   Oxaloacetic acid

Pyruvic carboxylase is a regulatory enzyme which remain inactive in absence of acetyl-CoA.

The oxaloaceti acid produced in mitochondria through above reaction is reversibly reduced to malic acid by mitochondrial enzyme, *malic dehydrogenase* at the expense of NADH.

$$NADH + H^+ + \text{Oxaloacetic acid} \rightleftharpoons NAD^+ + \text{Malic acid} \quad \text{(Reaction – 2)}$$
(red)   (ox)

The malic acid now leaves the mitochondria via a special dicarboxylate transport system in the inner mitochondrial membrane to get entry into the cytosol, where malic acid (malate) is reoxidized by cytosolic NAD- linked malate dehydrogenase to form extramitochondrial oxaloacetic acid.

$$\text{Malate} + NAD^+ \longrightarrow \text{Oxaloacetic acid} + NADH + H^+ \quad \text{(Reaction-3)}$$

The oxaloacetic acid, thus, formed is converted in to phosphoenol pyruvic acid in presence of guanosine triphosphate (GTP), a phosphate donor, and enzyme *phosphoenolpyruvic carboxykinase*.

$$\text{Oxaloacetic acid} + GTP \rightleftharpoons \text{Phosphoenol pyruvic acid} + CO_2 + GDP \quad \text{(Reaction -4)}$$

$$\begin{array}{c} COOH \\ | \\ CH \\ | \\ C=O \\ | \\ COOH \end{array} + GTP \xrightarrow{\text{Phosphoenolpyruvic carboxykinase}} \begin{array}{c} CH_2 \quad O^- \\ \parallel \quad | \\ C-O-P-O^- \\ | \quad \parallel \\ COOH \quad O \end{array} + GDP + CO_2$$

oxaloacetic acid   Phosphoenolpyruvate

Written as the some of reactions 1 to 4.

$$\text{Pyruvil acid} + ATP + GTP \longrightarrow \text{Phosphoenolpyruvic acid} + ADP + GDP + Pi$$
$$\Delta G' = +0.2 \text{ K.cal/mol}$$

The phosphorylation of one molecule of pyruvic acid to form phosphoenolpyruvic acid requires 14.8 K cal/mol under standard conditions. The above reaction shows that during phosphorylation of pyruvic acid, two high-energy phosphate groups, one from ATP and other from GTP must be expended. Each phosphate group yields - 7.3 K cal/mol. In contrast, the conversion of phosphoenolpyruvic acid to pyruvic acid during glycolysis generates only one ATP molecule from ADP. The standard free energy change of the reaction leading to synthesis of phosphoenolpyruvic acid is + 0.2 kcal/mol but the actual free energy change under intracellular conditions is highly negative (about - 6.0 k cal). It is thus irreversible reaction.

In several species, PEP carboxykinase is chiefly found in the cytoplasm. This creates complications because the oxaloacetate produced in reaction 1 can not pass through mitochondrial membrane which is permeable to malate. In order for oxaloacetate to made available to cytoplasmic PEP carboxykinase, it is first reduced to malic acid in presence of enzyme malic dehydrogenase in the mitochondria. The malic acid diffuses out from mitochondria into the cytoplasm where it is reoxidised to oxaloacetate by cytoplasmic malic dehydrogenase and converted to phosphoenol pyruvic acid by PEP carboxykinase.

Oxaloacetate + NADH + H⁺ —Mitochondrial malic dehydrogenase→ Malate + NAD⁺

Malate (mitochondrial) —Mitochondrial inner membrane→ Malate (cytoplasmic)

Malate + NAD⁺ —Cytoplasmic malic dehydrogenase→ oxaloacetate + NADH + H⁺

Bypass of Reaction 2 (Conversion of Fructose 1, 6-diphosphate into Fructose 6-phosphate)

The conversion of fructose 6-P into fructose 1, 6-diphosphate by enzyme *phosphofructokinase* in glycolysis is a downhill reaction and it can not participate in uphill process of gluconeogenesis because it is irreversible in intact cells.

Fructose 6-P + ATP —Phospho fructokinase→ Fructose 1, 6 - dip + ADP

This reaction is bypassed by enzyme *fructose diphosphatase* - which causes irreversible hydrolysis of 1-phosphate group to yield fructose - 6 P.

Fructose 1, 6-di P + H_2O —Fructose diphosphates / Mg^{2+}→ Fructose - 6P + Pi ($\Delta G° = 3.9$ kcal)

The enzyme fructose diphosphatase has a molecular weight of about 150,000 and requires Mg^{2+} for activity. It is a regulatory enzyme and is found in a large number of tissues. It accounts for the formation of fructose 6-P from hexose diphosphate and thus provides a means by which glucose or glycogen can subsequently be formed from hexose diphosphate. However, the enzyme diphosphatase is strongly inhibited by the negative modulated AMP.

Bypass of Reaction 3 (conversion of Glucose-6 P into Free glucose)

The dephosphorylation of glucose - 6 P to form free D-glucose does not occur by reversal of hexokinase reaction because it is irreversible. The dephosphorylation or bypass is brought about by enzyme *glucose 60-phosphatase* which catalyses irreversible hydrolytic reaction.

Glucose 6-P + H_2O —Glucose 6-phosphatase / Mg^{2+}→ Glucose + Pi ($\Delta G° = -2.9$ Kcal./mol)

Glucose 6-phosphatase is found associated with endoplasmic reticulm (ER). It has been isolated from microsomal fractions during ultracentrifugation of cell homogenate. The enzyme is found primarily in tissues that can produce free glucose.

The overall reaction (sum of all reactions) for gluconeogenesis leading from pyruvic acid (pyruvate) to free glucose may be written as follows-

2 Pyruvic acid + 4 ATP + 2 GTP + 2 NADH + 2H⁺ + 4H_2O ⟶
Glucose + 2NAD⁺ + 4 ADP + 2GDP + 6 Pi

Carbohydrate Metabolism

The summary of gluconeogenesis starting from pyruvic acid (Pyruvate) in the form of sequencial reactions is given in table 17.1

Table 17.1. Sequential Reactions in Gluconeogenesis Starting from Pyruvate†

Pyruvate + CO_2 + ATP → oxaloacetate + ADP + P_i	× 2
Oxaloacetate + GTP ⇌ phosphoenolpyruvate + CO_2 + GDP	× 2
Phosphoenolpyruvate + H_2O ⇌ 2-phosphoglycerate	× 2
2-Phosphoglycerate ⇌ 3-phosphoglycerate	× 2
3-Phosphoglycerate + ATP ⇌ 3-phosphoglyceroyl phosphate + ADP	× 2
3-Phosphoglyceroyl phosphate + NADH + H^+ → glyceraldehyde 3-phosphate + NAD^+ + P_i	×2
Glyceraldehyde 3-phosphate ⇌ dihydroxyacetone phosphate	
Glyceraldehyde 3-phosphate + dihydroxyacetone phosphate ⇌ fructose 1, 6-diphosphate	
Fructose 1, 6-phosphate + H_2O → fructose 6-phosphate + P_i	
Fructose 6-phosphate ⇌ glucose 6-phosphate	
Glucose 6-phosphate + H_2O → glucose + P_i	
Sum : 2 Pyruvate + 4ATP + 2GTP + 2NADH + $2H^+$ + $4H_2O$ → glucose + $2NAD^+$ + 4ADP + 2GDP + $6P_i$	

18

Photoperiodism and Photomorphogenesis

18.1. INTRODUCTION

Since ancient times the man was sub-cautiously aware of controlling effects of light on plant growth. it was demonstrated earlier that light was essential and the plants could not grow in dark. *Henfrey* proposed that variations in day length with change in latitude controlled plants distribution. *Sachs* (1882) first proposed that flowering is controlled by hormonous stimulus. *Cajlachjan* (1936) and *Moskov* (1936) proposed that the stimulus is mobile in the plants. American scientists *Garner* and *Allard* (1920) concluded that length of the day is a controlling factor of flowering.

Garner and *Allard* (1920) first initiated the principles of photoperiodism. Their experiments with tobacco and soybean etc. indicated that of the various environmental factors, which affect the plant, the length of day is unique in its action on sexual reproduction. They formulated this action in the following broad terms.

"Sexual reproduction can be attained by the plants only when it is exposed to a specifically favourable length of day. Exposure to a length of day unfavourable to reproduction but favourable to growth tends to produce gigantism or indefinite continuation of vegetative development, while exposure to a length of day favourable alike to sexual reproduction and vegetative growth tends to induce the ever bearing type of fruiting."

The term 'photoperiod' was introduced to designate the favourable length of day for each organism, and 'photoperiodism' to response of an organism to the relative length of day and night. Photoperiodism is defined as *"the response of a plant to the relative length of light and dark period"* or "The duration of dark period is much more important than duration of light period".

18.2. CRITICAL DAY LENGTH

The photoperiod required to induce flowering is called *critical day length*. It is 12 hours for Maryland Mammoth tobacco and 15.5 hours for *Xanthium strumarium*.

18.3. CLASSIFICATION OF PLANTS ACCORDING TO PHOTOPERIODIC REACTION

The plants can be grouped into following five classes according to photoperiodic requirements for their flowering.

1. *Short day plants* : The plants requiring a photoperiod of less than 12 hours are called *short day plants*. The common examples are *Xanthium pensylvanicum, Chrysanthemum, Cosmos bipinnatus,* some varieties of soybean, rice, strawberry, *Euphorbia pulcherimma* etc.

2. *Long day plants* : The plants requiring a photoperiod of more than 12 hours are called *long-day plants*. The common examples are *Hyoscyamus, Spinacea oleracea, Hibiscus syriacus,* wheat, some variaties of barley, *Anethum graveolens* etc.

3. *Photo-neutral Plants* : Photoperiod has no effect on the flowering of such plants, *e.g., Lycopersicum esculentum, Capsicum annum* etc.

4. *Intermediate Plants* : These plants require a photoperiod of 12 to 15 hours for flowering. e.g. *Mechania*.

5. *Long short day plants* : These plants require first long photoperiod and then short

photoperiod for flowering, e.g., *Bryophyllum*.

6. *Short long day plants* : These plants require first short photoperiod and then long photoperiod for flowering.

18.4. PHOTOPERIODIC INDUCTION OR PHOTOPERIODIC AFTER EFFECT

The plants, whether short or long-day, require a definite photoperiod to flower. It has been experimentally demonstrated that the light treatment necessary to cause flowering is needed continuously until that event actually occurs. Soybean requires only ten short days to bring about flowering (Garner and Allard). After this the plants could be placed under long-day conditions without any check in flowering process. It was shown later that soybean plants produce flower primordia even when subjected to still shorter periods (only 2 to 4 short day treatments).

This condition in which suitable cycles of light and dark persist in the plant and lead ultimately to flowering under normal unsuitable condition, is called photoperiodic induction or *photo-induction* or *photoperiodic after effect*.

Naylor (1941) demonstrated in *Xanthium* that on increasing the favourable treatment, corresponding increase in number of flower buds takes place. *Borthwick* and *Parker* (1938) demonstrated that there is a close relation between the age of the plant and the effectiveness of photo-induction treatment. *Hamner* and *Bonner* (1938) demonstrated in *Xanthium* and *soybean* that effectiveness of light treatment in flowering depends on the age of the plant. They suggested that young immature leaves play no part in photo-induction and the reproductive growth occurs only when the plant has achieved a certain amount of vegetative growth or maturity.

18.5. CO_2 SUPPLY AND PHOTOPERIODIC INDUCTION

It has been demonstrated by many scientists that even the appropriate light cannot induce flowering in the absence of CO_2. *Parker* and *Borthwick* (1940) demonstrated in soybean that there is a great correlation between the amount of CO_2 and formation of flowers (Table 18.1).

Table 18.1: The effect of CO_2 on flowering in Soybean

Hours of Treament	Normal air	0	2	4	6	8
	CO_2 free air	8	6	4	2	0
	% of flower bearing Plants	0	20	57	80	80

The data in table indicate that in absence of CO_2 no plants could bear flowers. It may be either due to suspension of carbohydrate synthesis or the production of other food materials during this period.

18.6. THE PERCEPTION AND TRANSMISSION OF STIMULUS

The stimulus of day length is received by leaves and not by the growing points where the flowers are produced. *Knot* (1934) and *Cajlachjan* (1936) for the first time realised the importance of leaves in the perception of the photoperiodic stimulus long day and short day plants respectively. Cajlachjan grew *Chrysanthemum* plants under long day conditions and then decapitated each plant so that the buds below the headless top could develop into long potentially flower bearing lateral shoots. All the leaves of the lateral shoots were removed and the leaves of the lower part of the stem were allowed to develop normally. He divided the plants into four groups and varied the regime of light and dark cycles in all the four groups by using light proof cases. The four groups of plants are as follows:

Group A : Entire plant continuously received long day treatment.

Croup B : Lower leafy portion received short day treatment while upper defoliated portion received long day treatment.

Group C : Lower leafy portion received long day treatment while upper defoliated portion received short day treatment.

Group D : Entire plant continuously received short day treatment.

He observed that flowering occurred in those plants where the leaves received short day treatment (Group B and D) but it failed in those plants where the leaves received long day treatment (Group A and C). On the basis of above observations he concluded that :

(i) The first reaction to day length takes place in the leaves. In other words short day stimulus is perceived by the leaves.

(ii) The stimulus perceived by leaves is transmitted to the buds through downward movement in the petiole and then upward movement in the stem to the terminal bud. Cajlachjan called this stimulus as "florigen".

Cajlachjan further investigated the role of light in the perception of the photoperiodic stimulus in a short day plant, *Perilla nankensis*. The results of his experiment are illustrated below. The unshaded portions in leaves represent which receive long days treatment, lightly shaded portions represent which receive short days treatment and black portions represent which receive no light at all (kept in darkness).

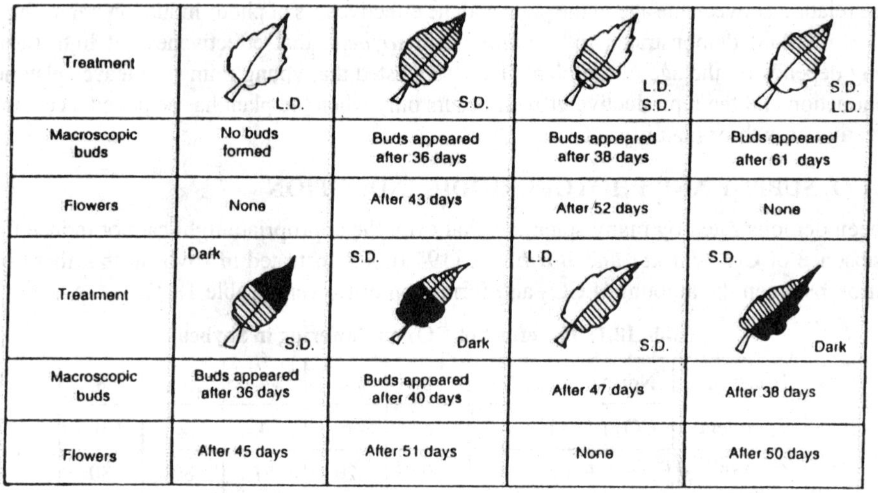

Treatment	L.D.	S.D.	L.D. / S.D.	S.D. / L.D.
Macroscopic buds	No buds formed	Buds appeared after 36 days	Buds appeared after 38 days	Buds appeared after 61 days
Flowers	None	After 43 days	After 52 days	None
Treatment	Dark / S.D.	S.D. / Dark	L.D. / S.D.	S.D. / Dark
Macroscopic buds	Buds appeared after 36 days	Buds appeared after 40 days	After 47 days	After 38 days
Flowers	After 45 days	After 51 days	None	After 50 days

Fig. 18.1: Response of shoots of *Perilla nankensis* under different lengths of illumination of halves of leaves.

The results of his experiment can be briefly written as follows

(i) Flowering did not occur when either whole leaf (LD) or basal half (SD/LD) or lateral half (LD/SD) were subjected to long days treatment.

(ii) Flowering occurred when:

(a) Whole leaf received short days treatment (SD).

(b) The basal half of the leaf received short days treatment (LD/SD).

(c) The basal half of the leaf received short days treatment and apical half received continuous darkness (D/SD).

(d) The apical half of the leaf received short days treatment and the basal half continuous darkness (SD/D).

(e) The lateral half received short days treatment and other lateral half received continuous darkness.

He concluded that in order to induce flowering, the basal portion should not be allowed to subject to long days treatment because it acts as a barrier preventing the transmission of stimulus.

18.7. IMPORTANCE OF DARK PERIOD

Plants under normal conditions are subjected to a twenty-four hours cycle of light and darkness. It has been demonstrated that during flowering the dark period is more important than light period. Short day plants usually flower when critical dark period exceeds while long day plants flower when critical dark period is reduced or light period exceeds the critical value.

The importance of dark period on flowering was first demonstrated by *Hamner* and *Bonner* (1938) in a short day plant, *Xanthium*. He demonstrated that :

(*i*) In *Xanthium*, the dark period is critical and must be continuous. If this dark period is interrupted with a brief exposure of light, (red light of 660-665 mµ wavelength), the plant does not flower.

(*ii*) Maximum inhibition of flowering with red light occurs at about the middle of dark period.

(*iii*) The inhibitory effect of red light can be overcome by a subsequent exposure with far-red light of 730-735 mµ wavelength.

(*iv*) Breaking up the light period with a brief period of darkness had very little effect.

(*v*) Interruption of light period with red light does not have inhibitory effect on flowering.

(*vi*) Prolongation or continuous dark period initiates early flowering.

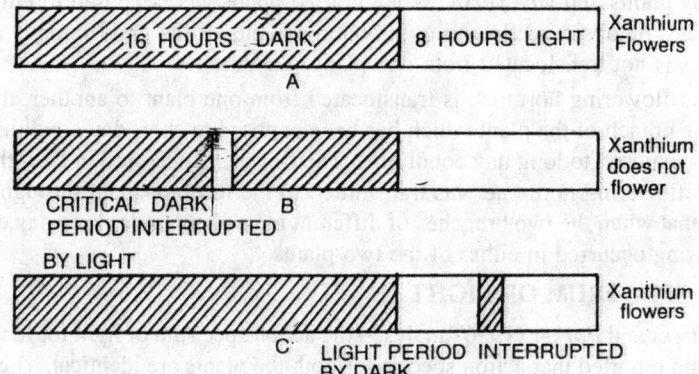

Fig. 18.2: Effect of brief exposure of red light during dark and light periods on flowering in *xanthium*.

The dark period is the critical part of the photoperiodic cycle. This view was supported by Hamner (1940) by performing experiments or *Biloxi soybean*, a short day plant, where flowering could not be induced unless the plant received a dark period excess of 10 hours. There the length of photoperiod did not matter.

18.8. IMPORTANCE OF PHOTOPERIOD

Athough the length of photoperiod has no effect on flower initiation but an increase in length of photoperiod shows an increase in number of floral primordia. *Hamner* (1940) demonstrated that the length of the dark period determines actual initiation of floral primordia and the length of light period determines the number of floral primordia initiated. (Fig. 18.3)

The intensity of light could have an

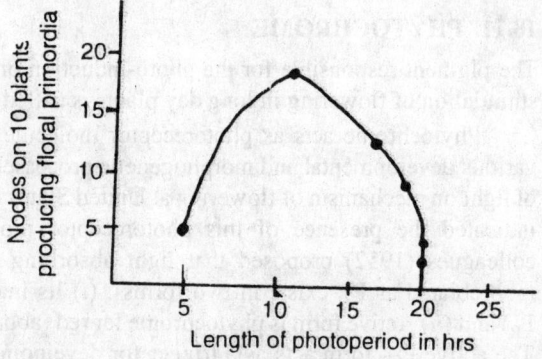

Fig. 18.3: Correlation between length of photoperiod and number of floral primordia on the nodes of plants.

indirect effect, such as controlling the amount of sugar flowing in meristematic regions capable of initiating floral primordia. The light intensity may help in the synthesis of some factors or hormones necessary for flower formation.

18.9 FLOWERING HORMONE : FLORIGEN

Although there are a number of evidences for the existence of flowering hormone but it has not yet been isolated. Therefore, the nature of this hormone is not very much clear. The name "florigen" was proposed to this flowering hormone by Cajlachjan. He advocated that this hormone was produced in the leaves and was easily transported down the petiole and then up the stem to the growing apical tips situated at other parts of the plant where it caused flower buds to develop in place of vegetative ones. Thus, florigen hormone is synthesized in the leaves and is transported to the apical tips of the stem and branches where it results flowering.

Grafting experiments on cocklebur plants (*Xanthium pennsylvanicum*) have been performed by *Zeevaart* (1958) to prove that :

(*i*) Flowering stimulus is similar in long and short day plants both.

(*ii*) Flowering hormone can be translocated from one plant to another.

To study that flowering stimulus is similar in long and short day plants, he grafted long day plants to short day plants and *vice versa*. When grafted plants were exposed to either long day or short day conditions, he observed flowering in both the plants. If the graft union is not formed, the flowering stimulus is not translocated from one plant to another.

To study that flowering hormone is translocated from one plant to another, the grafting was made between one branch of the plant which has been exposed to short day conditions and another branch which was exposed to long day conditions. The flowering occurred in both the plants which indicated that the flowering hormone was transmitted to the receptor plant through graft union. It was also noticed that when the two branches of different plants kept under long day conditions were grafted, no flowering occurred in either of the two plants.

18.10. ACTION SPECTRUM OF LIGHT

Borthwick, Hendricks and *Parker* (1956) analysed the action spectrum of light for long day and short day plants both and reported that action spectra for both the plants are identical. The most effective region of spectrum for photo-induction is about 660-665 nm (red) and much smaller effective region is blue. The phenomenon of photo-induction is due to a pigment called *phytochrome*. It exists in two forms, one red absorbing (P_R) and other far red absorbing (P_{FR}). Both forms are photochemically interconvertible. The P_R form is converted to P_{FR} in presence of red light (660-665 nm) and P_{FR} is converted to P_R in presence of either far red light (730-735 nm) or darkness. The conversion of P_{FR} to P_R in darkness is under thermal control.

18.11. PHYTOCHROME

The pigment responsible for the photo-induction or inhibition of flowering in short day plants and stimulation of flowering in long day plants is called *phytochrome*. It is proteinaceous in nature.

Phytochrome acts as photoreceptor molecule by absorbing light and showing its effects on various developmental and morphogenetic processes. A group of scientists while working on effects of light on mechanism of flowering at United States department of Agriculture, Beltsville, Maryland, indicated the presence of this photoreceptor molecule. *Borthwick*, H.A. *Hendricks*, S.B. and colleagues (1952) proposed that light absorbing pigment which was later called *phytochrome* (abbrebiated as P), exists in two forms : (*i*) Its inactive form is phytochrome red (abbrebiated as P_R) and (*ii*) Active form is phytochrome far red (abbrebiated as P_{FR}). Both forms are interconvertible. The active P_{FR} form acts as a trigger for developmental and morphogenetic processes. It confirms that red light induces photomorphogenetic responses while far red light reverses the red light induced responses. The P_R form can be converted into P_{FR} form in presence of red light (660-665 nm).

Similarly, P_{FR} form can be converted into P_R form in presence of either far red light (730–735 nm) or darkness. The conversion of P_{FR} to P_R in darkness is under thermal control and is slower process. It can be accelarated by reductant dithionite, indicating thereby that $P_R \rightleftharpoons P_{FR}$ interconversions are oxidation reduction processes. This interconversion does not take place directly but involves many intermediate species of phytochrome called I_a, I_b and so on.

$$P_R \begin{array}{c} \nearrow I_a\ 700 \searrow \\ \searrow I_b\ 700 \nearrow \end{array} I_{b1} \rightarrow I_{b2} \rightarrow I_{b3} \rightarrow P_{FR}$$

Differences between P_R and P_{FR} Forms

P_R Form	P_{FR} Form
1. It is an inactive form of phytochrome.	1. It is an active form of phytochrome.
2. Being inactive, it does not show phytochrome mediated responses.	2. Being active, it shows phytochrome mediated responses.
3. It has an absorption maximum in red region (about 680 nm).	3. It has an absorption maximum in far red region (abut 730 nm).
4. It is found diffused throughout the cytosol.	4. It is usually found in discrete areas of cytosol.
5. It is converted into P_{FR} form in presence of red light (660-665 nm).	5. It is converted into P_R form in presence of far red light (730-735 nm).
6. When the extract is centrifuged at 20,000 x g, it remains present in the supernatent.	6. On centrifugation P_{FR} form settles down in the form of pillets.
7. It shows activity in presence of urea, metal ions Cu^{2+}, Co^{2+}, Zn^{2+} etc., and N-ethyl maleimide.	7. It shows comparatively more activity in presence of these chemicals.
8. Its original structure contains many double bonds in pyrrole rings.	8. The P_{FR} form shows rearrangement of double bonds in all four pyrrole rings.

(1) *Occurrence* : Phytochrome is found in green and red algae, desmids, bryophytes, gymnosperms and angiosperms. It has been detected from roots, stems, coleoptiles, hypocotyls, cotyledons, petioles and blades of leaves, vegetative buds, floral tissues, seeds and developing fruits of higher plants. It may also be present at membrane surfaces.

(2) *Physico-Chemical Nature* : Chemically it is made up of a *protein* and a *chromophore*. Protein is composed of structural units called amino acids which form polypeptide chain through peptide linkages. The aminoacids composition and their sequence in different proteins phytochromes detected from different plants usually differ but the proteins show more or less similar size and configuration.

Phytochromes of two different sizes, smaller and larger, with different molecular weights have been isolated and identified so far. The small-sized phytochrome possesses a m.w. of about 60 kDa. and is believed to be produced by the degradation of large-sized phytochrome. It does not exist in vivo. The large-sized natural phytochrome is a dimer. Its each monomer unit has a m.w. of about 120 kDa. The recent researches have shown that large natural phytochrome is about 5 per cent smaller than native molecule (m.w. about 124 kDa) and is produced by the proteolytic degradation of native phytochrome.

Phytochrome is a conjugated protein and chemically it is made up of a water soluble *protein* and a *chromophore*. Due to presence of this chromophore, the phytochrome appears bluish.

Each phytochrome molecule is a dimer, containing two identical monomeric units. Each monomeric unit is made up of one *globular protein monomer* and one *chromophore*. Thus, each phytochrome molecule (dimer) contains two globular protein monomer units and two chromophores.

Globular Protein Monomer : Each protein monomer contains three disulphide (S–S) bonds inside which are not involved in the cohesion of two identical monomer protein units. Each unit is made up of about 1100 amino-acids. About 46% aminoacids including lysine, histidine, arginine, asparagine, serine, glutamic acid, and glutamine are polar and the sequence of aminoacids in a protein monomer is as follows :

Leucine – arginine – alanine – proline – histidine – serine – cysteine – histidine – leucine glutamine – tyrosine.

The aminoacids jointly form a polypetide chain through peptide linkages. The amino acids composition and their sequence in different proteins of phytochromes detected from different plants usually differ but the proteins show more or less similar size and configuration.

Chromophore : It is proposed that each monomer unit contains one chromophore. Because each phytochrome molecule (dimer) has two monomer units, therefore two chromophores will be present in each phytochrome. Each chromophore remain linked to protein monomer through cysteine residue. The basic structure of chromophore is almost similar to the cyanophycean algal pigment δ-phycocyanin.

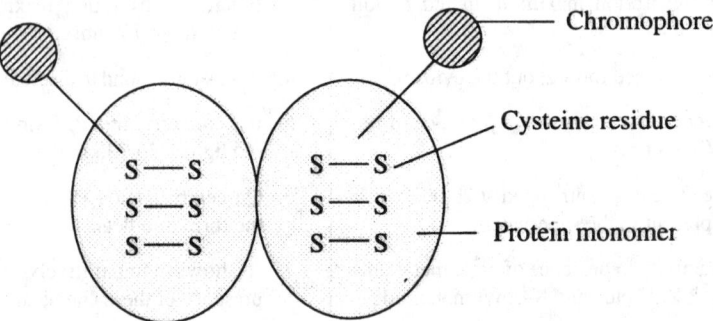

Fig. 18.4: Diagrammatic representation of a phytochrome molecule.

The *chromophore* of phytochrome is made up of four pyrrole rings arranged in a linear row. These rings possess many double bonds which become rearranged as a result of absorption of far-red light. Thus, during inter-conversion of red form (P_R) and far red from (P_{FR}), the redistribution or shifting of double bonds takes place in the pyrrole rings of chromophore. The chemical nature of phytochrome is based on the studies on algal chromoproteins which resemble with phytochrome in absorption spectra. The chromophore remains attached to the protein through thio-ester linkage through cysteine to the C-2 side chain of ring A. Ring B is also involved in the attachment with

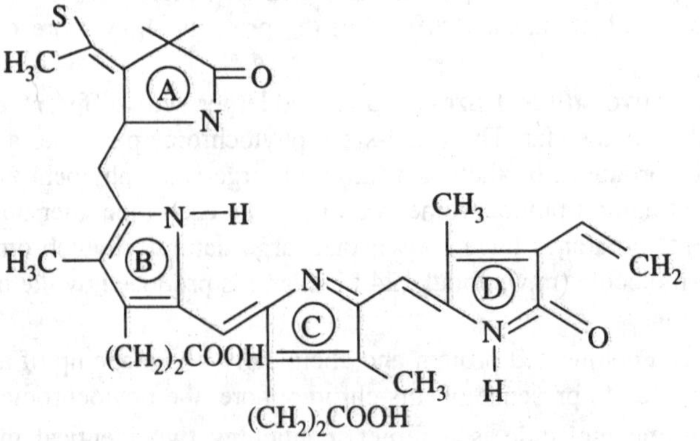

Fig. 18.5: Chemical structure of the chromophore of phytochrome.

protein through its side chain CH_2-CH_2-COOH. In addition to chromophore, the phytochrome also contains about one phosphate per monomer unit whose function is unknown. The light absorbing property of phytochrome is due to chromophore.

Properties: The phytochrome possesses following properties :
- (*i*) Phytochrome exists in two different forms, (*a*) red light (660-665 nm) absorbing form called P_R and (*b*) far-red light (730-735 nm) absorbing form called P_{FR}.
- (*ii*) These two forms are photochemically interconvertible.
- (*iii*) On absorbing red light, P_R form is converted into P_{FR}.
- (*iv*) On absorbing far-red light, the P_{FR} form is converted into P_R form.
- (*v*) The P_{FR} form gradually converted into P_R form in dark.
- (*vi*) Phytochrome is produced in light grown tissues.
- (*vii*) The biologically active P_{FR} form interacts with some substance (X) to form P_{FR}. X complex which results in physiological response.

Mode of Action: It is supposed that during day time P_{FR} form of phytochrome is accumulated in the plants causing inhibition of flowering in short day plants and stimulation in long day plants. During critical dark period in short day plants, P_{FR} form gradually changes into P_R form resulting in flowering. The P_R form is again converted into P_{FR} form in presence of brief red light exposure resulting inhibition of flowering. In short day plants, the inhibitory effect of red light during critical dark period can be reversed by exposing the plant with far red light which causes the conversion of P_{FR} form into P_R form.

In long day plants, the prolongation of critical light period or the reduction in dark period or the interruption of dark period by red light results in further accumulation of P_{FR} form which stimulates flowering.

Isolation : Phytochrome pigment molecule was isolated by *W.L. Butler* and colleagues in 1959 and 1964 and also by physiologists of United States Department of Agriculture, Beltsville.

Since phytochrome is proteinaceous in nature, it can be isolated from the plant seedlings through those standard procedures which are commonly used for isolation of proteins. Following scheme yields poor quality and quantity of phytochrome.

To prepared plant extract add ammonium sulphate → Precipitate of pigments → Purify through column, ion-exchange or gel-exclusion chromatography → Poor quality and quantity of phytochrome.

19
Dormancy of Seeds and Seed Germination

19.1. INTRODUCTION

Most of the viable seeds germinate immediately if placed under suitable conditions necessary for germination (moisture, air, suitable temperature and proper sequence of light and dark) but the seeds of certain plants fail to germinate even if placed under all favourable conditions. The germination of such seeds may be delayed for days, weeks, months or even years. Such seeds are said to be in the state of dormancy. The dormancy may be defined as the condition of seed when it fails to germinate even though the favourable environmental conditions are present.

Dormancy in seeds may be either due to lack of some necessary external environmental factor or as a result of internal causes. *Wareing* (1969) defined dormancy as any phase in the life-cycle of a plant in which active growth is temporarily suspended. According to him the dormancy due to unfavourable environmental conditions is called *imposed dormancy* or *quiescence*, while the dormancy due to conditions within the dormant plant or organ is called *innate dormancy* or *rest*.

Innate dormancy is a condition in which germination or growth fails to occur even though the external environmental conditions are favourable. The fully dormant condition of a seed is a gradual process and is not attained suddenly. During the entire process, there may be following three phases of dormancy.

(i) *Predormancy* or *early rest* : During this phase, the dormant organ has capacity to resume growth by various treatments i.e., capacity of germination or growth is not completely lost. It is called *predormancy*.

(ii) *Full dormancy* or *mid rest* : When a seed or organ becomes completely dormant and germination or growth cannot be induced immediately by changes in environmental conditions, it is called *full dormancy* or *mid-rest*.

(iii) *Post dormancy* or *after rest* : When a dormant seed or organ gradually emerges from full dormancy and in it the germination or growth can be induced by changing environmental conditions, it is called *post-dormancy* or *after-rest*.

The dormancy may be true, relative or secondary.

(i) *True dormancy* : When in a seed or organ, the germination or growth cannot be induced under any set of environmental conditions, it is called *true dormancy*.

(ii) *Relative dormancy* : When in a seed, the germination can be induced under specific conditions even at the time of its deepest dormancy, it is called *relative dormancy*.

(iii) *Secondary dormancy* : When a seed has not fully emerged from dormancy and is again thrown back into full dormancy by certain environmental conditions, e.g., temperature etc., it is called *secondary dormancy*.

19.2. SEED DORMANCY

Crocker (1916) divided seed-dormancy into seed-coat induced and embryo-induced.

1. *Seed-coat induced dormancy*: The dormancy of seeds due to extreme hardness of seed-coat is called *seed-coat induced dormancy*. The hard seed coat checks the entry of water, exchange of gases and expansion of embryo.

(2) *Embryo induced dormancy* : The dormancy of seeds due to rudimentary or complete dormant embryo is called *embryo induced dormancy*. In this case the rudimentary embryo is incompletely developed and requires a rest period for complete development while complete dormant embryo shows dormancy due to physiological conditions.

Other types of dormancy may be:

(3) *Secondary dormancy* : When the seeds become dormant again after breaking the dormancy, it is called *secondary dormancy*. It may be due to combination of different kinds of dormancy in a single seed, *e.g., Xanthium pennsylvanicum.*

(4) *Special type of dormancy* : The failure of seedling development is not always traceable due to dormancy of seed itself. In many of the spring wild plants the germination of seed takes place but the growth is restricted due to establishment of young roots. Sometimes the system of epicotyl fails to germinate. In some cases, the epicotyl may be pushed through the seed-coat but remains dormant. This dormancy is often broken by exposure to low temperature.

19.3. CAUSES OF THE SEED DORMANCY

The dormancy of seeds may be either due to single or a combination of many different factors. As stated earlier, it may be seed coat-induced dormancy or embryo induced dormancy.

Seed-Coat Induced Dormancy

The seed coat of most of the seeds is formed by the integumentary layers of ovules. Chemically it is composed of a complex mixture of polysaccharides, hemicelluloses, fats, wax and proteins. During seed ripening, the chemical components of the seed coat become dehydrated and form a hard and tough protective covering around the embryo. The seed-coat induced dormancy may be due to following causes.

(1) *Water impermeability*: The seed coats of many plant species are completely impermeable to water. This condition is very common in the seeds belonging to families Leguminosae, Malvaceae, Chenopodiaceae, Convolvulaceae, Solanaceae and Nymphaeaceae etc. Here the germination fail to occur until water penetrates through the seed-coat. In many such seeds permeability of coat to water increases slowly in dry stage. The action of micro-organisms like bacteria and fungi also increases the permeability of seed-coats to water and shortens the dormant period of seeds. In many plants the seeds have waxy coating.

(2) *Gas impermeability* : The seed coats of certain seeds are impermeable to gases such as oxygen and carbon-di-oxide. Since oxygen is required for early respiratory activity in germinating seeds, the seeds fail to prolong germination.

In *Xanthium*, two types of seeds are found in a fruit which are not equally dormant under natural conditions. The lower seeds generally germinate in the spring following maturity while upper seeds remain dormant. The dormancy of upper seeds has been demonstrated due to impermeability of seed-coat to oxygen. If oxygen pressure is increased, the seed germination occurs. The oxygen requirement in upper seeds is greater than the lower seeds. Such dormancy of seeds is reduced by storing them for a longer time. This type of dormancy has also been reported in many grasses and some members of Compositae.

(3) *Mechanical resistance of seed-coat to the growth of embryo* : The seeds of some common weeds such as *Alisma, Amaranthus* and *Capsella* etc. have such hard and tough seed-coat that it prevents any appreciable expansion of embryo. Thus, they remain dormant. In the seeds of *Amaranthus,* the water and oxygen penetrate through the seeds readily but enlargement of embryo is limited by mechanical strength of seed coat. As long as the seed coats are saturated with water, the dormancy in *Amaranthus* may persist for about 30 years. However, if the seed-coats become dry and then again become saturated with water, they are no longer able to resist the expansion of embryo. The seed-coats rupture and germination takes place. At above 40°C some germination of seeds also takes place because the seed-coats become less resistant to pressure developed by imbibitional forces in the embryo.

Embryo-induced Dormancy

Embryo-induced dormancy may be of two kinds.

(1) *Rudimentary and poorly developed embryo* : In many plant species like *Anemone nemorosa, Fraxinus excelsior, Ginkgo biloba,* members of Orchidaceae, *Orobancheae* etc., the seed dormancy may be due to immature and rudimentary embryo. In such seeds the embryo does not develop as rapidly as surrounding tissues. Thus, when the seeds are shed, they are still imperfectly developed due to incomplete embryo. The germination of such seeds takes place only after a period of rest (dormancy) during which the further development of embryo is completed.

(2) *Embryo fully developed but unable to resume growth* : In many species, *e.g.,* seeds of apple, peach, Iris, Hemlock, peas, cherry, etc., although the embryos are completely developed in ripe seeds but the seeds fail to germinate even when the environmental conditions for germination are favourable. Dormancy of such seeds is due to physiological conditions of embryo. The embryo of such seeds does not germinate even if the seed-coats are removed. The germination in such seeds can be induced if they are stored in moist, well aerated and low temperature conditions. This process is called *stratification* or *after-ripening*. The ripening involves principally a series of physiological changes in the dormant embryo which gradually convert a dormant embryo into one that resumes growth.

Dormancy due to Specific Light Requirement

The seeds of certain plant species such as *Lactuca sativa, Lythrum salicaria, Nicotiana tabacum* etc. have a specific light requirement for germination. In imbibed *Lactuca sativa* seeds the germination is stimulated by red light of 660 nm wavelength, while it is inhibited by far-red light of 730 nm wavelength. This indicates the involvement of photoreversible pigment, phytochrome, in germination. Light not only affects qualitatively but also quantitatively. The reciprocal increase in germination with increase in intensity and exposures of light duration has been reported in *Rumex* seeds. The germination of certain seeds requires a specific photoperiod, *e.g. Bignonia* requires a photoperiod of 12 or more hours for seed germination. The light sensitive seeds are called *photoblastic*.

Dormancy due to Germination Inhibitors

In many seeds the dormancy occurs due to presence of germination inhibitors in seed-coats, endosperm, embryos or structures surrounding them such as the juice or the pulp of fruits *e.g.* in tomatoes, and in glumes, *e.g.* in Oats. A number of chemical substances such as organic acids, phenolics, tannins, alkaloids, unsaturated lactones, mustard oil, ammonia releasing substances, cyanide releasing substances, indoles and gibberellins etc. have been isolated from the seeds which act as germination inhibitors. Besides, the other natural inhibitors are coumarin, parascorbic acid, ammonia, phthalides, ferulic acids, and abscisin II etc. If these inhibitors are leached out, the germination of seeds takes place.

19.4 METHODS OF BREAKING SEED DORMANCY

The dormancy of seeds can be broken and the dormant seeds can be induced to germinate by one or combination of more than one methods described below.

Scarification

This method is used for breaking dormancy of seeds caused by hard seed coats which become impermeable to water and gases etc. In this method the seed-coat is rendered permeable to water and gases either by mechanical method or chemical treatments. The seed coat becomes soft and weak by this treatment. The method employed in softening or weakening the seed-coat is called *scarification*. When mechanical breaking of seed coat is done at one or more places, it is called *mechanical scarification*. The treatment of seed coat with strong mineral acids or other chemicals is called *chemical scarification*. Mechanical scarification is done by shaking the seeds with sand or by scratching or nicking the seed-coat with knife. Chemical scarification is usually done by dipping the seeds into strong acids like H_2SO_4 or into organic solvents like acetone or alcohol. It can also be done by boiling the seeds in water. Under natural conditions in the soil, certain micro-organisms like bacteria and fungi act upon the seed coat to decompose it but this process requires a lot of time.

Stratification

This method is used to break the dormancy of seeds caused due to condition of embryo. In this process the seeds are exposed to well aerated, moist conditions under low temperature (0° to 10°C) for weeks to months. This treatment is called *stratification* or *after-ripening*. During stratification some chemical changes occur in the immature embryo of seeds which are necessary for seed germinaion. These changes are as follows—

(*i*) The concentrations of nitrogen and phosphorous are shifted to the various parts of the seeds.

(*ii*) Various constituent amino acids, organic acids and enzymes are also shifted.

(*iii*) Cyanogenic glycosides are decomposed.

(*iv*) The concentration of various growth regulators is changed.

Alternating Temperature

In some seeds, *e.g. Poa pratensis*, the seed dormancy is broken by the treatment of an alternating low and high temperatures. The difference between the alternating temperatures should not be more than 10-20°C. This method is beneficial in those seeds in which the dormancy is due to immature embryos. Alternating temperature of 15°C and 25°C is useful in breaking the dormancy of photoblastic seeds like *Rumex crispus*.

Light

The light sensitive seeds are called *photoblastic* which may be of following three types :

(*i*) *Positive photoblastic seeds* : The seeds requiring single exposure of light for germination are called *positive photoblastic seed, e.g., Lactuca sativa*.

(*ii*) *Negative photoblastic seeds* : The seeds requiring complete darkness for germination are called *negative photoblastic seeds*.

(*iii*) *Non-photoblastic seeds* : The seeds requiring either light or darkness for germination are called *non-photoblastic seeds*.

The dormancy of positive photoblastic seeds can be broken by exposing them to red light (660 nm). Far-red light inhibits the seed germination indicating the involvement of photoreversible pigment phytochrome in the process of seed germination. This pigment occurs in two forms, one red absorbing and other far-red absorbing. Both these forms are photochemically interconvertible. The red absorbing form (P_R) is converted into far-red form (P_{FR}) after absorbing the red light. The far-red form absorbs the far-red light and is converted back into red absorbing form of the pigment.

$$P_R \underset{735 \text{ nm}}{\overset{660 \text{ nm}}{\rightleftharpoons}} P_{FR}$$

It is supposed that in positive photoblastic seeds, the far-red absorbing form of the pigment is stimulatory to seed germination while red-absorbing form is inhibitory to seed germination.

Pressure

The seed germination in certain plants like sweet clover (*Melilotus alba*) and alfalfa (*Medicago sativa*) can be greatly improved after being subjected to hydraulic pressure of about 2000 atm. at 18°C for about 5-20 minutes. This pressure changes the permeability of seed coat to water resulting into seed germination.

Growth Regulators

Growth regulators are most widely used to hasten the development of roots or cuttings and to increase the number of roots. Kinetins and gibberellins have been used to induce germination in positively photoblastic seeds like lettuce and tobacco etc. Besides, a number of chemicals such as KNO_3, thiourea and ethylene etc. have also the capacity to induce seed germination.

19.5. ADVANTAGES OF SEED DORMANCY

(i) The dormancy of seeds help the plants of temperate zones to tide over the severe colds.

(ii) The dormancy of seeds due to impermeable seed coats ensures good chances of survival to the plants of tropical regions.

(iii) The dormancy of seeds in cereals is most important to mankind. If these seeds germinate immediately after harvest, they will be quite useless for mankind.

(iv) Dormant seeds and organs in perennial plants resist unfavourable conditions for their development.

(v) The seeds form a measure of the quantity and duration of rainfall, both of which determine the amount of soil-moisture available for plant growth.

19.6. CHANGES DURING SEED GERMINATION

Seed germination starts with the imbibition of water which increases the respiration rate rapidly. In cereals this increase is 0.05 μ l O_2/g/hr at resting level and when water content reaches 40 percent, it becomes 100 μ l O_2/g/hr within a few hours. The increased activity is mainly confined to embryo and the endosperm contains probably sucrose as respiratory substrate. The increase in respiration depends upon the activation of mitochondria which in dormant seeds are little active.

19.7. LIGHT DEPENDENT GERMINATION

In many seeds, the seed germination depends upon light. The seeds contain a photoreceptor pigment called inactive *phytochrome* (P_R) which becomes active (P_{FR}) in red light (660 nm) and can be converted into inactive form (P_R) in presence of far-red (730 nm) light. Thus, both forms are interconvertible and red light induces germination (For details see phytochrome).

Phytochrome mediates several other physiological responses which include control of flowering, regulation of etiolation, opening of plumular hook and interaction with inducers.

19.8. CHANGES IN FOOD RESERVES ETC

As the seed germination proceeds, the food reserves like carbohydrates, lipids and proteins are mobilized to provide building blocks for embryonic development, energy for biosynthetic processes and development in general. The nucleic acids control protein metabolism. The various changes in carbohydrates, lipids, proteins, nucleic acids and inorganic materials etc. are discussed below.

Carbohydrates

The cereal endosperms usually contain starch as carbohydrate reserve which during germination is hydrolysed into maltose by two enzyme: α-amylase and β-amylase. The activity of these enzymes increases rapidly after imbibition. The reason behind is that in the endosperm cells, the α-amylase is synthesized at the time of germination while the β-amylase which remains already present in the cells in latent condition, becomes active.

The imbibition causes secretion of gibberellic acid by embryonic axis into either aleurone layer as in wheat or scutellum as in maize where gibberellic acid (GA) stimulates total synthesis of α-amylase. The mechanism of stimulation is unknown. Abscisic acid (ABA) nullifies the effect of GA. Sometimes other protein synthesis inhibitors also prevent GA stimulated α-amylase synthesis. The addition of extra GA will again induce the α-amylase synthesis.

β-Amylase in ungerminated seeds is found in small quantity and represents a small portion of the total enzyme activity. In wheat, β-amylase enzyme remains linked by disulphide linkage to glutenin. Its activation is due to its release from the glutenin with the formation of active –SH groups.

The maltose produced by the hydrolysis of starch is then converted into glucose by maltase enzyme whose activity is regulated by GA. *R.R.Swain* and *E.E.Dekker* (1966) demostrated that in germinating pea seeds the enzyme α-glucosidase is involved in the hydrolysis of starch and conversion of maltose into glucose. The enzyme acts in combination with amylases. The glucose is absorbed into scutellum, converted into sucrose via UDPG and transferred to embryonic axis.

The above situation is applicable only with the seeds containing carbohydrate reserves in endosperm. The situation in seeds containing reserves in cotyledon is not very clear. For example in pea seeds the α-amylase activity increases after five days of germination and exogenous supply of GA does not stimulate enzyme activity.

Lipids

The seeds with high lipid reserves (triglyceride content) show their loss during germination. At first the lipids are hydrolysed by lipases at neutral pH. The lipases from various species differ in their pH optima, substrate specificity and localization in tissues. The phospholipids are hydrolysed by phosphatidase. These enzymes become active by imbibition. In certain seeds like peanuts and cotton the enzyme lipase is not present but is synthesized de novo and its synthesis is triggered by GA. The hydrolysis of neutral fats yields glycerol and fatty acids. The glycerol is converted into α-glycerophosphate which can be channelled into glycolytic sequence via dihydroxyacetone phosphate.

$$\text{glycerol} \xrightarrow[\text{glycerokinase}]{ATP \quad ADP} \alpha\text{-glycerophosphate} \xrightarrow[\text{glycerophosphate dehydrogenase}]{NAD^+ \quad NADH, H^+} \text{Dihydroxyacetone phosphate}$$

The fatty acids may be either metabolised by β-oxidation or converted into sucrose through glyoxylate cycle (consult chapter on lipid metabolism). The two key enzymes of glyoxylate cycle, *isocitratase* and *malate synthetase*, are absent from mature seeds but are rapidly synthesized during early stages of germination. Their number becomes maximum when the rates of disappearance of fat and appearance of sucrose are maximum. Later on, they disappear when the fat reserves of the seeds are completely used up and photosynthesis starts. The synthesis of these enzymes may be stimulated due to liberation of GA from the embryonic axis.

Proteins

Storage proteins in seeds are hydrolysed during germination into their constituent amino acids and amides through proteolytic enzymes especially peptidases. Germinating seeds contain numerous proteolytic enzymes which may be present either in dry seeds or synthesized during germination. The proteolytic enzymes may be proteinases which cause hydrolysis of large protein molecules or peptidases which hydrolyse smaller polypeptides. They become activated when the seeds imbibe. The peptidases are abundantly present in the aleurone layers of cereal seeds. The activity of peptidase is stimulated by endogenous GA. The amino acids and amides, thus, produced are then translocated into embryonic axis where they are used either in the synthesis of new proteins for developing embryo or to provide energy by oxidation of carbon skeleton after deamination. The ammonia produced in the latter reaction may cause toxic effect which is prevented by fixng it into glutamine and asparagine. The protein synthesis is initiated by formation of different types of RNAs and polysomes in the seedlings. The scutellum in cereal seeds is an important structure which hydrolyse polypeptides and plays a special role in nitrogen metabolism.

Nucleic Acids

In storage tissues of seeds, the nucleic acid content usually does not change during germination. However, the increase in DNA and RNA content of embryonic axis has been recorded in germinating seeds. During early stages of imbibition and in presence of embryo, the DNA and RNA content in monocotyledonous seeds disappears and the same amount appears in embryonic axis. It is proposed that the DNA and RNA are transported as macromolecules where they synthesize new DNA and RNA. However, rapid appearance of RNA in aleurone layers of endosperm has been noticed after 16 hours. This is mainly de novo synthesis and is probably related with active enzyme synthesis in endosperm.

The cells of embryonic axis undergo cell-division and elongation. Before entering into cell-divison the cells prepare themselves for cell-division i.e., synthesize protein and RNA and DNA in major amount at interphase stage. The DNA contents also increase with the increase in cell-number. The increase in nucleic acid contents and protein helps in the formation of mitotic apparatus. During germination the RNA content of mitochondria also increases in castor bean endosperm and corn scutellum.

Many seeds like cotton, wheat, radish and vigna etc. contain long-lived m-RNA of disputed nature and function and others contain cytoplasmic ribonucleoprotein (RNP) granules. According to *John Ingle* and colleagues (1964) the nucleic acid content in maize root and shoot increases rapidly after 48 hours while those of endosperm and scutellum remains unchanged even after 120 hours of germination.

Inorganic Materials

Much of the phosphorus in seeds remain stored in the form of *phytin*, a salt of *inositol hexaphosphate* or *phytic acid*. The phytin also contains other macronutrients like Ca, Mg, K and Mn. It occurs mainly in protein bodies. In most seeds it represents about 80 per cent of the total phosphorus. Various metabolic reactions like phospholipid synthesis and energy generating processes require orthophosphate. During germination, the activity of various phosphatases including *phytase* increases remarkably. The activity of phytase in cereals is highest in scutellum and aleurone layer which causes liberation of sufficient amount of phosphate, Ca, Mg and K for most of the metabolic processes. This mechanism of liberation is unknown and the reaction may be catalysed by some non-specific enzyme. This indicates that during germination : (*i*) the amount of phytin decreases and that of inorganic phosphate increases. (*ii*) Phytin acts as a reserve pool of phosphorus in seeds.

19.9. FACTORS AFFECTING GERMINATION

Seed germination is affected by various factors. Only viable seeds can germinate and the viability period varies from a few days to several years. Lotus seeds (*Nelumbo nucifera*) can survive for more than 400 years. Most of the seeds become non-viable due to biochemical, ultrastructural or morphological changes. Some factors affecting germination are as follows :

(1) *Water* : The water is an important factor which controls and activates various enzymatic activities. It helps the seeds in imbibition causing increase in osmotic effects.

(2) *Atmospheric Composition* : It also affects seed germination. The atmosphere consists of several gases like O_2, CO_2 and N_2 etc. Oxygen increases respiration. The percentage of O_2 for germination usually varies from 8 to 20. Excess of CO_2 concentration decreases germination in general but in a few seeds like *Phleum pratense* show increase in germination. Most of the pollutants lie NO_2, SO_2, O_3, NH_3, H_2S, F and high concentration of ethylene inhibit germination.

(3) *Temperature* : The optimum temperature for seed germination varies according to species. The optimum temperature for most species ranges between 25-35°C. Generally, the germination is

inhibited at very low temperature and very high temperature. A few seeds of certain species require low temperature for germination. Temperature may interact with light and humidity causing germination.

(4) *Light* : The seeds of most cultivated plants germinate equally well in light and dark. Photoblastic seeds can germinate only in presence of light. The light induces phytochrome activity. **Mayer** (1986) proposed the involvement of calcium binding protein, the **calmodulin**, which stimulate the metabolic responses during germination.

Inactive phytochrome (P_R) $\xrightarrow{\text{Red light}}$ Active phytochrome (P_{FR}) $\longrightarrow$ Transport of Ca^{2+} ions $\longrightarrow$ Calmodulin mediation $\longrightarrow$ Primary light reaction $\longrightarrow$ Secondary light reaction $\longrightarrow$ Metabolic reactions $\longrightarrow$ Seed germination.

Scheme showing phytochrome mediated induction of seed-germination

(5) *Soil Conditions* : The soil conditions which induce germination include water holding capacity, aeration of soil, mineral composition, soil-texture, pH of soil and organic matter etc. Saline condition of soil inhibits germination. Other soil conditions also play important role in germination.

(6) *Seed Structure* : Sometimes the seeds contain such structures which inhibit germination. These include hard and impermeable seed-coat; reduced or rudimentary embryo and presence of germination inhibitors etc. Hard seed-coat may be impermeable to water and gases.

20

Vernalization

20.1. INTRODUCTION

It has been noticed that certain plants in addition to an appropriate photoperiod require low temperature treatment during their early stage of life for subsequent flowering in the later stages. Since the word **vernal** pertains to the spring season, it usually implies the ability of such treatment to turn a winter annual or biennial plant into a spring annual or into a spring habit of growth. *Klippart* (1857) for the first time found that winter wheat which is usually sown in winter and flowers in summer, can be converted into spring wheat which is sown in spring and flowers in summer, if slightly germinated seeds are kept at 0°C to 5°C. This conversion of winter variety of wheat into spring variety by low temperature or chilling treatment was termed as *Vernalization* by *Lysenko* (1928). Due to low temperature treatment the period of vegetative growth of the plant is reduced resulting into early flowering.

Chouard (1960) defined vernalization as *"acquisition or acceleration of the ability to flower by a chilling treatment"*.

The phenomenon of vernalization was studied in several winter annuals *e.g.*, Petkus winter rye (*Secale cereale*), some biennials *e.g.*, Henbane (*Hyoscyamus niger*) and also in certain perennials *e.g.*, apples (*Pyrus malus*). The cold requirement for flowering has been recorded in a number of day-neutral plants *e.g.*, *Pyrethrum cinerariaefolium, Saxifraga rotundifolia, Senecio jacobaea* and various species of *Erysimum, Geum* and *Lychnis*. All the above plants require a low temperature treatment for subsequent flowering. However, certain plants such as Petkus winter rye, do not have an absolute cold requirement for flowering. In such cases the cold temperature only shortens the time of flowering.

The effect of cold stimulus on plant is not immediately visible. It is expressed only at a certain later stage in the form of flowering. Thus, vernalization may be considered as an excellent example of the *physiological preconditioning*. Such conditioning is also done through photoperiodism. Many species responding to vernalization are known to be sensitive to photoperiodic stimulations of flowering. *Chrysanthemum* may be induced by short day and low temperature, China aster by short day and high temperature, spinach by long day and high temperature and many cruciferae, beet and winter cereals by long day and low temperature. The interaction of these two environmental signals may be complementary or supplementary. Hence, low temperature may modify the critical day length for flowering (*Lang* and *Melchers*, 1943), enhance the responsiveness to photoperiods without changing the critical day-length required (*Cathey*, 1957) or even abolish a photoperiodic requirement (*Koller* and *Highkin*, 1960).

20.2. PERCEPTION OF THE COLD STIMULUS

The experiments conducted with biennial or perennial plants strongly suggest that the cold stimulus is either perceived by *apical meristems* (*Melchers*, 1937; *Schwabe*, 1954) *e.g.*, Henbane, or by the fragments of *embryos* as in rye (*Purvis*, 1940), or by all *actively dividing cells* of leaves and roots (*Wellensiek*, 1964) *e.g.*, *Lunaria biennis*. Thus, apical meristems, fragments of embryos and actively dividing cells are the potential sites of vernalization.

20.3. MORPHOLOGICAL CHANGES ASSOCIATED WITH SEED VERNALIZATION

Following two morphological changes are associated with seed vernalization:

(i) Extensive development of vascular tissues leading to the growing points (*Chakravarti*; 1950, 1954).

(ii) Much enlargement of embryo at the expense of the endosperm (*Stokes,* 1952).

20.4. INDUCED STATE

Once a meristem is vernalized, the subsequent growth from it acts as vernalized growth (*Schwabe,* 1954) and the induced state seems to be permanent. Certain cereals can be vernalized when moistened and then dried and stored for months to years without damaging the induced condition. However, the induced state is far less stable in many plants.

20.5. PRESENCE OF A FLORAL HORMONE

Most of the physiologists believe that the perception of the cold stimulus results in the formation of a floral hormone which is transmitted to other parts of the plant. This hormone has been named as **vernalin** by *Melchers* (1939) but has not yet been isolated. The cold stimulus can be transmitted from one plant to another by a graft union *e.g.*, Henbane. Here, if a vernalized plant is grafted to an unvernalized plant, the later also shows flowering.

20.6. OTHER CONDITIONS NECESSARY FOR VERNALIZATION

(1) *Age of the plant*: The age of the plant is an important factor for vernalization phenomenon. The age at which a plant is sensitive to vernalization is quite different in different species. In cereals, vernalization can be brought about in the germinating seeds and developing embryos (*Lang,* 1961; *Purvis,* 1961) if they are given low temperature treatment for sufficient time, while in case of biennial variety of henbane (*Hyoscyamus niger*) vernalization is only effective after the plant has reached the rosette stage and has completed at least 10 days of growth. The treatment of the seed or immature embryo in this plant is ineffective.

(2) *Appropriate low temperature and duration of the exposure*: The most effective temperature for vernalization ranges between 0°C to 5°C. Vernalization fails at temperature below –4°C (Hansel, 1953) and is observed up to 14°C in winter rye. Low temperature at about –6°C is completely ineffective. Similarly, at higher temperature from 7°C onwards the response of the plants is decreased. Temperatures at about 12-14°C are almost ineffective in vernalizing the plants.

The duration of vernalization treatment is also important which varies in different species. Minimum duration may be of four days while maximum of three months. The saturation times vary from three weeks for winter wheat to three months for henbane. In certain cases, a long duration of several months is needed for vernalization. This phenomenon was referred to as *oververnalization* by *Salisbury* and *Ross* (1969).

(3) *Oxygen*: The vernalization is an *aerobic process* which requires metabolic energy and it will not proceed in the absence of O_2. A cold treatment is completely ineffective if given in an atmosphere of pure nitrogen and in absence of O_2 (*Gregory* and *Purvis,* 1948). If a little amount (0.2%) of O_2 is given, vernalization occurs.

(4) *Sugar*: By excising and culturing embryo on various media, it has been demonstrated that cold treatments on media without added sugar are not successful (*Gregory* and *de Ropp,* 1938).

(5) *Water*: Sufficient amount of water is also needed for vernalization. Vernalization of dry seeds is not possible and seeds must imbibe at least 50 per cent of water for the adequate vernalization treatment (*Purvis,* 1961).

20.7. MECHANISM OF VERNALIZATION

There are two main theories regarding mechanism of vernalization.

(1) *Phasic Development Theory*: This theory, was proposed by *Lysenko* (1934). The main points of this theory are :
  (i) The growth and development are two different phenomena.
  (ii) The development of an annual seed-plant consists of a series of phases.
  (iii) These phases always proceed in a strict sequence and a subsequent phase cannot start till the preceding one is completed.
  (iv) The phases require different external conditions for their completion.
  (v) Two phases are involved in the development of a plant.
    (a) *Thermophase*: It depends upon temperature.
    (b) *Photophase* : It depends upon light.
The vernalization accelerates the thermophase.

The winter wheat requires low temperature treatment for the completion of the thermophase. After this the photophase starts. Vernalization accelerates the thermophase in winter wheat so that there is an early swing from vegetative to the reproductive phase or flowering.

Although the term vernalization is restricted to cover only the inductive low-temperature stimulation of flowering e.g., beet, cabbage, celery etc., the broader definition will include such non-inductive examples as Brussel's sprouts in which some flowers are produced before the removal of plants from the cold. Vernalization by low temperature treatment can be done in barley, field pea, oats, rice, rye, crimson clover, meadow foxtail, melilotous, red-clover, and white lupin etc., while maize, millet, soybeans, Sudan grass and sorghums require a high temperature treatment of 20-30°C for 5-10 days.

(2) *Hormonal Theory*: The vernalization involves the formation of a floral hormone called **vernalin**. To explain this, several schemes have been proposed by different workers from time to time. The first hormonal theory was proposed by *Lang* and *Melchers* (1947). According to this theory, a precursor **A** is converted into a thermolabile compound **B** during cold treatment. Under normal temperature **B** changes into a stable product **C** (**Vernalin**) which causes flowering. But at higher temperature **B** is converted into **D** and flowering does not take place due to devernalization. (Fig. 20.1).

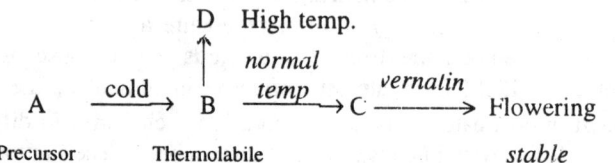

Fig 20.1: Schematic representation of Lang and Melchers hormonal theory.

20.8. DEVERNALIZATION

The effect of low temperature treatment for flowering is nullified if the plants are immediately given high temperature treatment. This phenomenon is called *devernalization*. During devernalization the product, formed during low temperature treatment, is destroyed before it is converted to stable form. The degree of devernalization decreases if the duration of the cold treatment has been longer. However, the devernalized plants can be vernalized again by subsequent cold treatments.

20.9. VERNALIZATION AND GIBBERELLINS

Lang et al. (1957) demonstrated that application of gibberellins can replace the low temperature requirement for vernalization in many plants, e.g., *Henbane*. This plant remains vegetative and retains its rosette habit during first growing season and after passing through the winter period flowers in the next season. The gibberellins cause such plants to flower even during the first year.

It was shown that the endogenous level of gibberellins was enhanced in vernalized plants.

Since the hormone vernalin has not yet been isolated and identified, it is supposed that the product of vernalization which induces flowering could be a particular gibberellin or a mixture of gibberellins.

20.10. IMPORTANCE OF VERNALIZATION
   (i) Vernalization increases the cold resistance of plants.
   (ii) It reduces the vegetative period of development of plants and induces early flowering.
   (iii) It increases the resistance of plants to fungal diseases.
   (iv) It helps in crop-improvement.
   (v) Winter varieties of crop plants can be converted into spring varieties by vernalization.

21

Growth and Metabolism of Growth Hormones

21.1. INTRODUCTION

Growth is an important character of all the living beings. Each living organism increases in size, shape, volume and weight. This process is called *growth*. It requires nutrition. Physiological processes play important roles during the growth of plants and animals. Growth is always caused due to **internal forces**, during which the cells divide, increase in length and finally differentiate into different type of tissues. In animals, the growth stops after a definite stage while in plants it continues during the entire life. In lower plants, it takes place usually in the entire body while in higher plants it takes place mostly at the apex of different organs like root, stem and leaves because the meristems remain situated at this part. They are called *primary meristems*. In the older regions of these organs secondary meristems like cambium and cork-cambium arise which cause secondary growth. The non-living materials like stones and crystals of copper sulphate in its solution also increase in size but this growth always takes place by the deposition of particles of these substances through **external forces**.

21.2. COURSE OF GROWTH

The growth rate of a cell, a plant organ or a whole plant does not remain the same but always varies. This variation in growth is due to irregular changes. In initial stages during the phase of cell formation, the growth rate is quite slow while it increases rapidly during the phase of cell elongation and becomes maximum, later on it slows down during the phase of cell maturation. At the last, a stage comes when the growth rate becomes zero i.e., the growth stops. Thus, the total period from initial to final stage of growth is called *grand period of growth*. The total period of growth can be divided into following stages :

1. *Lag period of growth* : During this period, the growth rate is quite slow because it is the initial stage of growth. In other words, the growth starts from this period.

2. *Log period of growth* : During this period the growth rate is maximum and reaches at the top because at this stage the cell-division and physiological processes are quite fast.

3. *Decline period of growth* : This period comes after log-period. During this stage the growth-rate slows down because the metabolic processes become slow.

4. *Senescence period or steady period* : During this period the growth is almost complete and becomes static. Thus, the growth-rate becomes zero.

If this growth rate is plotted against time, a 'S' shaped curve is obtained which is called *sigmoid curve* or *grand period curve*. Environmental conditions may alter growth rates but not the sigmoid form of the growth curve.

21.3. MEASUREMENT OF GROWTH

The growth of a plant can be measured on the following basis:

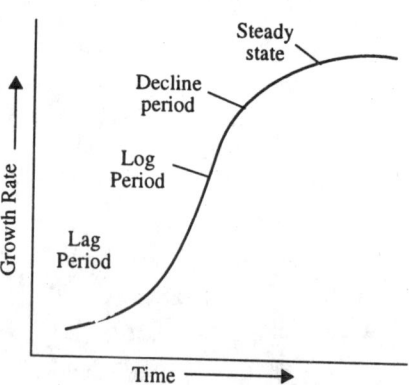

Fig. 21.1: Sigmoid curve.

Growth and Metabolism of Growth Hormones

(*i*) Increase in length or girth as in case of stems and roots.
(*ii*) Increase in weight.
(*iii*) Increase in volume or area as in case of fruits and leaves respectively.
(*iv*) Increase in number of cells per unit area.

When various parts of a plant grow in length, it is called *linear growth*. For its measurement following methods are employed:

(1) *Direct method* (Fig. 21.2): In this method, the length of the growing plant or its organs is measured directly with the help scale after certain intervals.

(2) *By Arc auxanometer* : The growth measuring instrument is called *auxanometer* by which the linear growth of any stem-apex can be easily measured. In an arc auxanometer, there is a stand on which an arched sextant remain tight with the help of two screws. The sextant is made up of three metal strips of which two are linear but perpendicular to each other and are fixed on a pully. The pully moves on an axis. The ends of both the linear strips remain attached with the third arched strip which is graduated. The long pointer moves on the scale. The broad end of pointer is fixed with the pully and it moves along with the pully. A thread is passed over the pully with one end tied to the growing point of the plant and other end carrying a weight to keep the thread stretched. When the plant grows, pully moves resulting in the movement of the pointer to indicate the rate of growth on the arched scale. The reading is taken. Actual increase in length of the stem or plant is then calculated by knowing the diameter of the pully and the length of the pointer. If the diameter of the pully is 4″ and the length of the pointer 20″, growth is magnified ten times on the graduated arc.

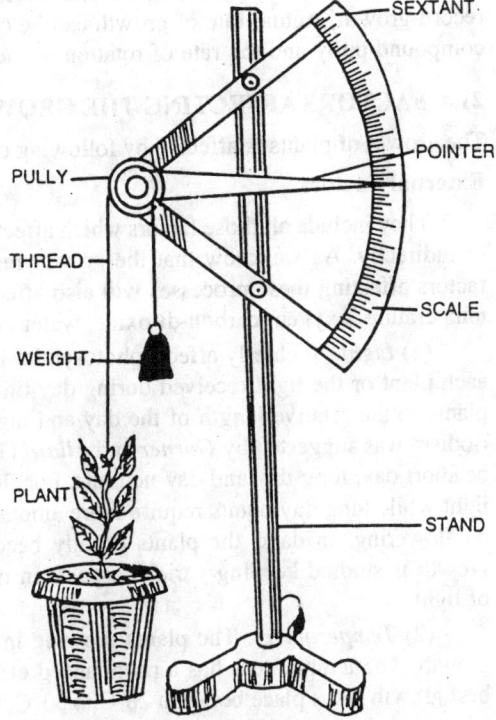

Fig. 21.2 : Arc auxanometer.

(3) *By Pfeffer's drum Auxanometer*: Pfeffer's auxanometer consists of a vertical stand holding a compound pully with a small and a large wheel on a horizontal base. Both wheels have the same axil. A thread is passed over the small wheel, one end of which is connected to the growing point of the plant or stem and the other end to a weight to keep it stretched. Another thread is passed over the large wheel whose both ends are tied with weights to keep it stretched. Near one end of this thread (Fig. 21.3) is attached a fine scratching pointer which remains in touch with a cylindrical drum rotating automatically. The drum is covered by a smoked paper.

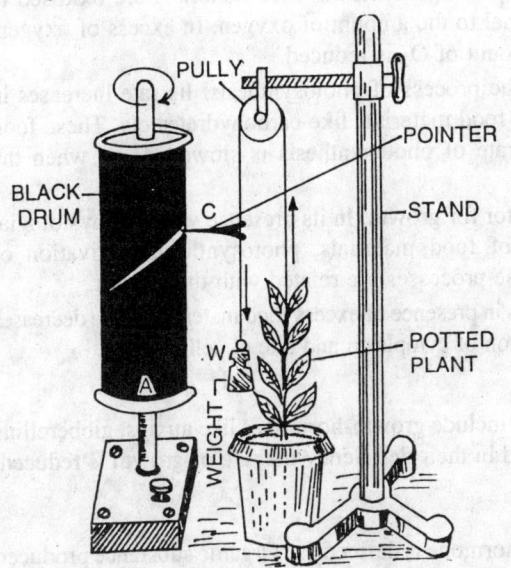

Fig. 21.3 : Pfeffer's auxanometer.

When the plant grows in length, the wheels of the pully move so that the pointer also moves downward continuously making a white scratch mark on the smoked paper of the rotating drum to record growth. Actual rate of growth can be calculated with the help of radii of two wheels of the compound pully and the rate of rotation of the drum.

21.4. FACTORS AFFECTING THE GROWTH OF PLANTS

The growth of plants is affected by following external and internal factors:

External Factors

They include all those factors which affect various physiological activities of the plants directly or indirectly. As we know that the growth results due to various physiological processes, so the factors affecting these processes will also affect growth. Some of these important factors are light, temperature, oxygen, carbon-di-oxide, water and food materials.

(1) *Light* : It chiefly affects photosynthesis and transpiration. The favourable length of day for each plant or the light received during day time by a plant is called *photoperiod*. The response of plants to the relative length of the day and night is termed as *photoperiodism*. The term photoperiodism was suggested by *Garner* and *Allard* (1920). According to the light duration the plants may be short day, long-day and day-neutrals. For flowering, the short day plants require less amount of light while long day plants require more amount of light. Day neutral plants have no effect of light on flowering. In dark, the plants usually become weak, thin and yellow. The effect of light on growth is studied keeping various aspects in mind like light intensity, light quantity and duration of light.

(2) *Temperature* : The plants growing in different regions require different temperatures for growth. The temperature has a pronounced effect on growth. It occurs between 4°C to 45°C. The best growth takes place between 28°C to 33°C. In plants occurring in colder regions the best growth takes place between 35°C to 40°C, while in plants growing in warmer regions it takes place between 38°C to 44°C. Some seeds show germination when they are kept at low temperature. Similarly, some plants require low temperature in addition to the sufficient photoperiod for flowering. When they are given low temperature treatment at the initial stages of life, they show an early flowering later on. Such property of plants is called *Vernalization*. It has been observed in wheat, rye, rice etc.

(3) *Oxygen* : It is necessary for respiration during which the food materials are oxidised to release energy. The growth is directly proportional to the amount of oxygen. In excess of oxygen, the growth is more while it is reduced when amount of O_2 is reduced.

(4) *Carbon-di-oxide:* CO_2 chiefly effects the process of photosynthesis. Its rate increases in excess of CO_2 and results in the manufacture of food materials like carbohydrates etc. These food materials are most important for growth. The rate of photosynthesis is slowed down when the amount of CO_2 is not sufficient.

(5) *Water* : The water is most important factor for growth. In its presence various physiologial activities like water absorption, translocation of food materials, photosynthesis, activation of enzymes and protoplasm etc. take place. All these processes are related with the growth.

(6) *Food materials* : The growth-rate increases in presence of excess food materials while decreases during shortage. They also increase the concentration of cytoplasm and rate of cell-division.

Internal Factors

Internal factors necessary for growth mainly include growth-hormones like auxins, gibberellins and cytokinin etc. They are required in traces and in their deficiency the rate of growth is reduced.

21.5. GROWTH-HORMONES

According to *Pincus* and *Thimann* (1948), a plant hormone is defined as "Organic substance produced naturally in the higher plants, controlling growth or other physiological functions at a site remote

Growth and Metabolism of Growth Hormones

from its place of production and active in minute amounts".

The term hormone was used by *Starling* (1906) for the first time as stimulating substance. The hormones are special type of organic substances related with the growth phenomenon. For this reason they are called by various names like growth-hormones, growth-regulators, phytohormones, growth-factors or growth-substances. Hormones possess following characteristics :

(*i*) They are usually produced at the tips of roots, stems and leaves.

(*ii*) The transfer of hormones from producing part to other parts takes place through phloem.

(*iii*) Their growth promoting action occurs only when they are used in definite quantity and concentration.

(*iv*) They are required in traces.

(*v*) They can be isolated from the plant organs by chemical methods.

(*vi*) Some growth regulating substances are called vitamins which act like hormones.

(*vii*) All hormones are organic in nature.

A number of growth regulating natural and unnatural or synthetic substances are known which can be classified into following categories :

(*i*) Auxins, *e.g.* Indole acetic acid (IAA).

(*ii*) Gibberellins, *e.g.* Gibberellic acid (GA).

(*iii*) Cytokinins, *e.g.* Kinetin and Zeatin.

(*iv*) Dormins, *e.g.* Abscissic and (ABA), Xanthoxin etc.

(*v*) Ethylene, *e.g.* Ethylene.

(*vi*) Flowering hormone, *e.g.* Florigen and Vernalin.

(*vii*) Phenolic substances, *e.g.* Coumarin.

(*viii*) Other natural substances, *e.g.* Vitamins, Phytohormones, Traumatic, substance etc.

(*ix*) Synthetic growth retardants, *e.g.* Phosphon-D, Morphactins.

(*x*) Other synthetic substances, *e.g.* Synthetic auxins and Cytokinins.

Some important growth regulators classified above are described here.

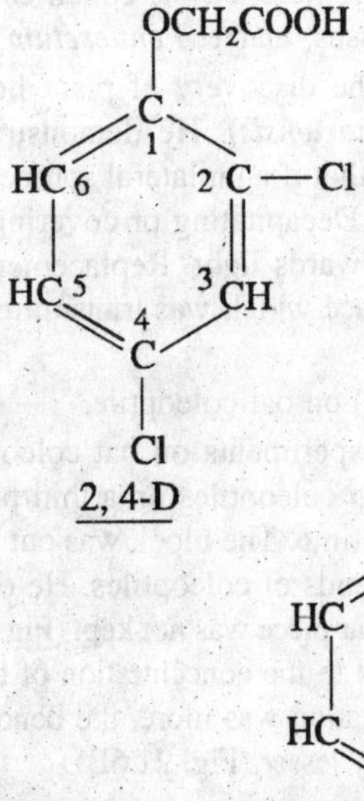

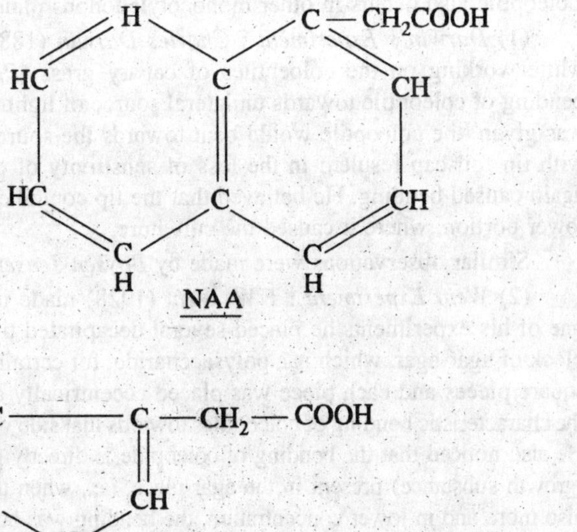

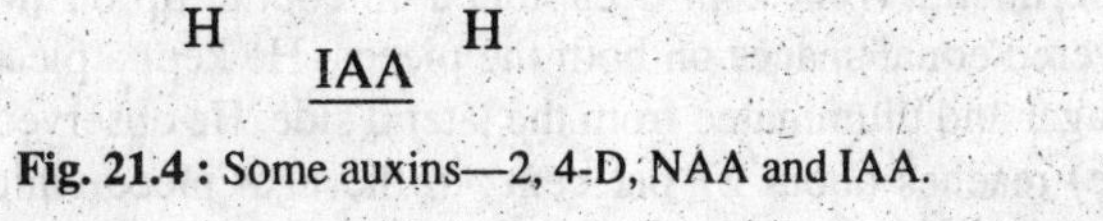

Fig. 21.4 : Some auxins—2, 4-D, NAA and IAA.

Auxins

All those growth regulating organic compounds which are produced at the tips of roots and stems as a result of metabolism and are transported to the region of elongation causing elongation of cells are called *auxins*. Thus, the chief function of auxins is cell-elongation showing growth. Indole acetic acid (IAA), Indole propionic acid (IPA), Indole butyric acid (IBA), Napthalene acetic acid (NAA), Phenylacetic acid (PAA), 2, 4-Dichlorophenoxyacetic acid (2, 4-D), 2, 4, 5-Trichlorophenoxy acetic acid (2, 4, 5-T) are the examples of auxins.

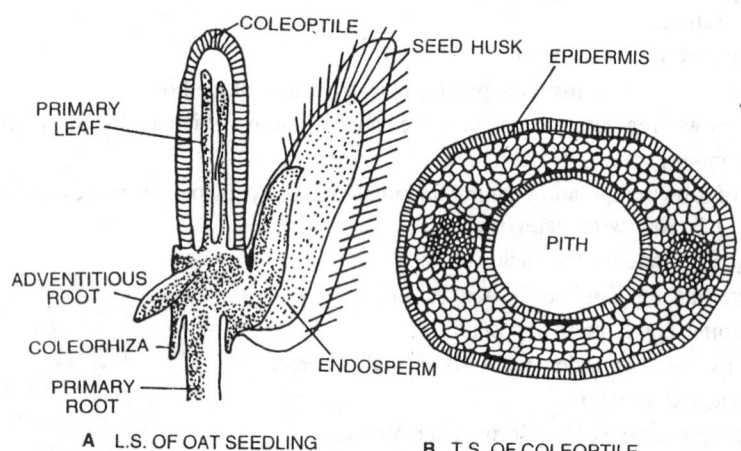

Fig. 21.5 : (A) L.S. of Oat seedling, (B) T.S. of Coleoptile.

A number of scientists have performed experiments on various plants regarding the presence of auxins. Some of the important ones are *Darwin, Boysen-Jensen, Went* and *Pal*. The discovery of auxin was made by the experiments on the seedlings of Oat (*Avena sativa*). The seedlings of oat plant possess a thin protective covering over the newly formed plumule-leaves, called *coleoptile*. Coleoptile also occurs in other monocotyledonous plants like grasses, maize, *Pennesetum* etc.

(1) *Darwin's Experiment* : Charles Darwin (1881) led to the discovery of plant hormones while working on the coleoptiles of canary grass (*Phalaris canariensis*). He demonstrated the bending of coleoptile towards unilateral source of light. He found that if a unilateral source of light was given, the coleoptile would bent towards the source of light. Decapitating or covering the tip with tin foil cap resulted in the loss of sensitivity of the plant towards light. Replacement of tip again caused bending. He believed that the tip contained a substance which was transmitted to the lower portion, where it caused the curvature.

Similar observations were made by *Boysen-Jensen* (1910-13) on oat-coleoptile.

(2) *Went Experiment* : F.W. Went (1928) made important experiments on oat coleoptile. In one of his experiments he placed several decapitated tips of *Avena* coleoptiles on a thin plate like block of agar-agar, which is a polysaccharide, for certain period of time. The block was cut to small square pieces and each piece was placed eccentrically on the cut ends of coleoptiles. He observed the characteristic bending of coleoptiles towards that side where the agar piece was not kept (Fig. 21.6A). He also noticed that the bending of coleoptile is directly proportional to the concentration of hormone (growth substance) present in the agar piece, i.e., when the concentration was more, the bending was also more and in lower concentration, the bending was comparatively lesser (Fig. 21.6B).

In another experiment, Went kept decapitated coleoptile tip on two pieces of agar in such a way that the tip covered equal spaces on both the pieces. He kept a piece of mica sheet in between these two pieces of agar and illuminated from the lateral side. He observed that the growth promoting substance (hormone) reaches about 27 per cent in the agar piece which is present towards light side while about 57 per cent in the agar piece present towards non-illuminated side (Fig. 21.6C). When he performed this experiment in horizontal position, he observed that the flow of hormone was towards lower side due to gravitational force (Fig. 21.6D).

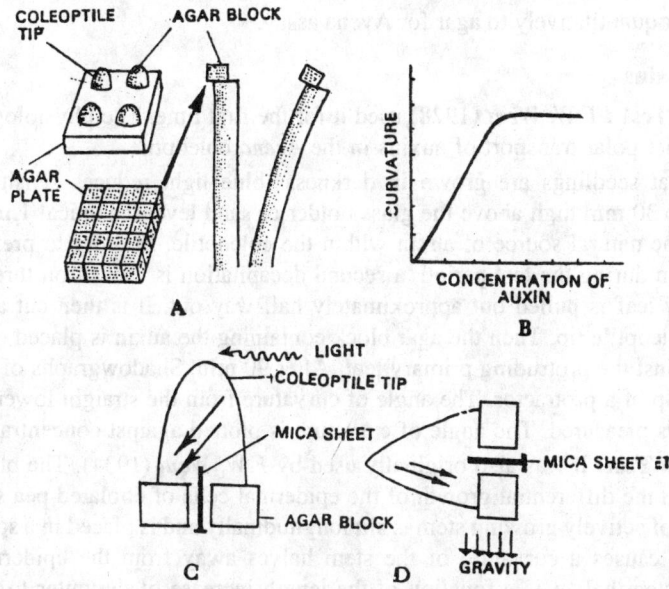

Fig. 21.6 : F.W. Went's experiments.

(3) *Isolation of Auxins* : The chief plant auxin is indole acetic acid (IAA). It is generally known as IAA. It was isolated for the first time by *Salkowski* (1885). Later on, *Kogl, Haagen Smit, Erxleben* (1934) isolated it from corn germ oil and human urine. *Thimann* (1935) isolated heteroauxin from the fungus, **Rhizopus**. In addition to indole acetic acid, its several derivatives like indole 3-pyruvic acid, indole 3-acetonitrile, indole 3-ethanol, indole 3-acetadehyde etc. have also been isolated. They act like auxin after being converted into IAA. Certain synthetic auxins like indole butyric acid (IBA), 2, 4-dihlorophenoxy acetic acid (2-4 D), α- and β-naphthalene acetic acid (NAA) and phenyl acetic acid (PAA) etc. are also known.

Extraction of Auxins

The following two methods are used to extract auxins from the plant materials:
(a) Diffusion
(b) Solvent Extraction

Diffusion

The simplest method to obtain a growth hormone from plant material is by diffusion into agar. The organ to be tested for auxin activity is placed on an agar block (1.5.% agar) for a period of an hour under conditions which do not permit transpiration. The auxin so obtained by diffusion in agar block is bio-assayed by the *Avena* curvature test. Excessive transpiration from the test organ and enzymic destruction of the growth hormone at the cut-end create difficulties in this method.

Solvent Extraction

Thimann (1934) employed chloroform as the solvent for extracting growth hormone from plant tissues. However, *Boysen-Jensen* (1936) found diethyl ether as most satisfactory. Diethyl ether which is peroxide-free and contains 5 per cent water is recommended. *Van Overbeek et al.* (1945) have recommended the following simple technique for obtaining free auxin:

1. Freeze plant material on CO_2 ice or liquid nitrogen.
2. Slice bulky tissues into 2-5 mm slices.
3. Extract with peroxide-free ether at $0°$ C for two and half hour intervals.
4. Combine the ether extracts and reduce volume by evaporation to a few ml.
5. Transfer quantitatively to agar for Avena assay.

Bioassays for Auxins

The Avena Test : *F.W. Went* (1928) used it for the first time. The physiological basis of this test lies in the strict polar transport of auxins in the *Avena* coleoptile.

Dehusked oat seedlings are grown in darkness (blue light reduces sensitivity). When the coleoptile is 15 to 30 mm high above the glass holder or sand level the apical 1 mm is removed in order to cut off the natural source of auxin within the coleoptile. In order to prevent the renewed formation of auxin during the test period, a second decapitation is carried on three hours after the first. The primary leaf is pulled out approximately half-way out. It is then cut about one quarter inch above the coleoptile tip. Then the agar block containing the auxin is placed on one side of the coleoptile tip against the protruding primary leaf. After 90 min. Shadowgraphs of the curvature are taken with the help of a protractor. The angle of curvature from the straight lower region to the tip of the coleoptile is measured. The angle of curvature is plotted against concentration.

The Slit Pea Test: It was also originally used by *F.W. Went* (1934). The physiological basis for this test lies in the differential growth of the epidermal cells of etiolated pea stems in response to auxin. A piece of actively growing stem is slit longitudinally and is placed in a solution containing the auxin, which causes a curvature of the stem halves away from the epidermal side. Such a curvature of the stem halves is a function of the length increase of the outer to inner cells. After the stem segments have been kept for 6-24 hr in the test solutions, the angle formed between, (*a*) the tangent at the point where inward curvature commences, and (*b*) the tangent at the point where inward curvature ceases, is noted. The angle of curvature is plotted against auxin concentration.

Straight Growth Test : The physiological basis for straight growth test is the simple stimulation of straight growth by auxins. Pea internodes of standard length are floated on test solutions for 6-24. hr. The increase in length is measured and is plotted against auxin concentration.

Pea Root Test : The physiological basis for this test is essentially the same as for the other straight growth. The roots are extremely sensitive to auxins. The linear growth is inhibited quantitatively. Pea and cress roots have been used for this test.

Biosynthesis of Indole Auxins

Tryptophan as the precursor of indole auxins : The amino acid tryptophan is a precursor of IAA as it has a close chemical similarity with it and is presumably present in all cells. Free tryptophan levels in leaves are much higher than those of IAA. However, free tryptophan may show compartmentation in tissues.

The evidences of the various pathways of IAA biosynthesis are: (*i*) The presence of interme-

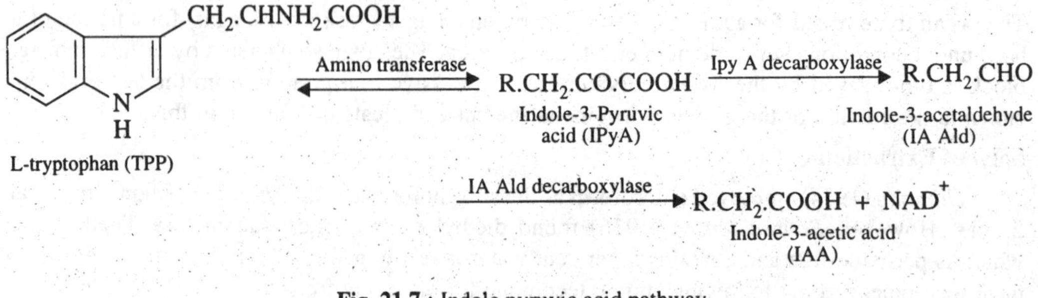

L-tryptophan (TPP) $\xrightarrow{\text{Amino transferase}}$ R.CH$_2$.CO.COOH (Indole-3-Pyruvic acid (IPyA)) $\xrightarrow{\text{Ipy A decarboxylase}}$ R.CH$_2$.CHO (Indole-3-acetaldehyde (IA Ald)) $\xrightarrow{\text{IA Ald decarboxylase}}$ R.CH$_2$.COOH + NAD$^+$ (Indole-3-acetic acid (IAA))

Fig. 21.7 : Indole pyruvic acid pathway.

diates as native compounds, (ii) The biological activity of intermediates, (iii) In vivo interconversion of the intermediates, and (iv) Isolation of the necessary enzyme systems.

Based on these evidences the following pathways have been established.

(1) *The Indole-pyruvic acid Pathway* : Indole-pyruvic acid is unstable and its isolation as a native compound from plant tissues has not been possible. However, there is good evidence for the occurrence of indole-3-acetaldehyde in sterile pea shoots (*Rajgopal,* 1968) and in cucumber seedlings (*W.K. Purves*).

The conversion of (^{14}C) TPP to (^{14}C) IPyA, (14_C) IA Ald and (^{14}C) IAA has been demonstrated in cell-free extracts of mungbean seedlings (*Wightman* and *Cohen*, 1968).

(2) *The Tryptamine Pathways*: Tryptamine occurs sporadically in higher plants. It was first isolated from *Acacia* and has since been found in several other species. However, it is not common in pea, squash, cabbage etc. Tryptamine is active after a lag period in the Avena curvature test. The tryptamine pathway is as under:

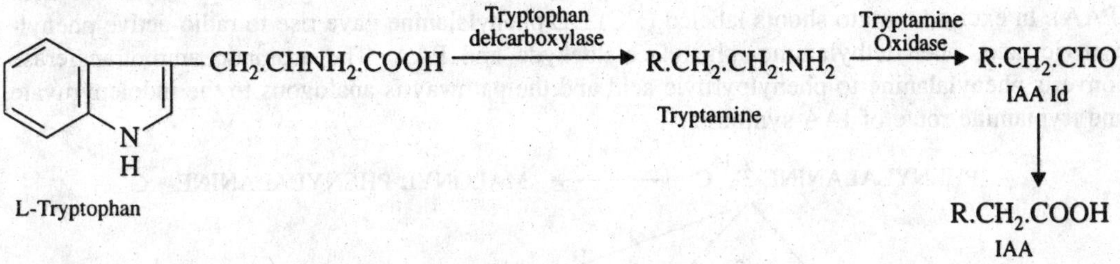

Fig. 21.8 : The tryptamine pathway.

(3) *The Indoleacetaldoxime Pathway* : This pathway is characteristic of the family Cruciferae. The *Brassica* plants probably contain small amounts of indole-3-acetonitrile (IAN) which is also obtained by the breakdown of glucobrassicin by the enzyme myrosinase. Glucohrassicin is active in the oat and wheat coleoptile tests. In vivo experiments have shown that cabbages can convert (^{14}C) TPP to IAOx to IAN and to Glucobrassicin. The enzyme responsible for converting TPP to IAOx in vivo has not yet been isolated, although this reaction can be catalysed in vitro by horse radish peroxidase. Indole-3-acetaldoxime hydrolase converts IAOx to IAN, and Nitrilase converts IAN to IAA.

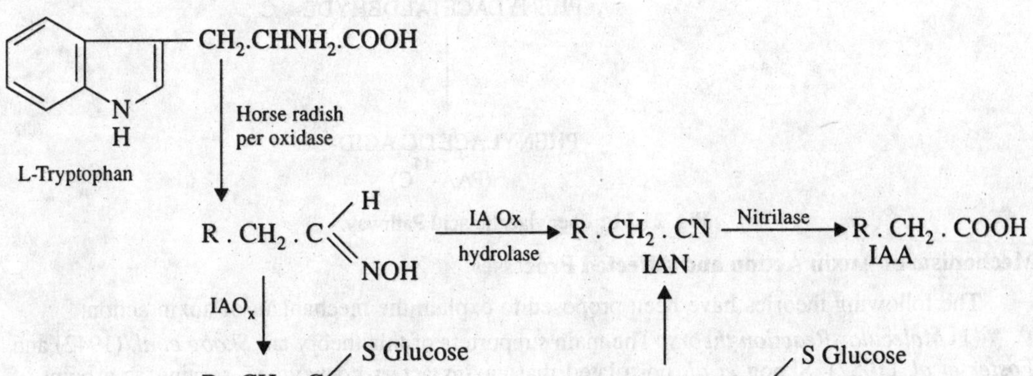

Fig. 21.9 : Indoleacetaldoxime pathway.

(4) *The Tryptophol Pathway*: This pathway is really a modification of the IPyA pathway. It appears to involve the two main intermediates required for the IPyA pathway and additionally

involves the formation of tryptophol (indole-3-ethanol, Tol) as a transitory side-reaction product. In nature tryptophol has been extracted from cucumber shoots.

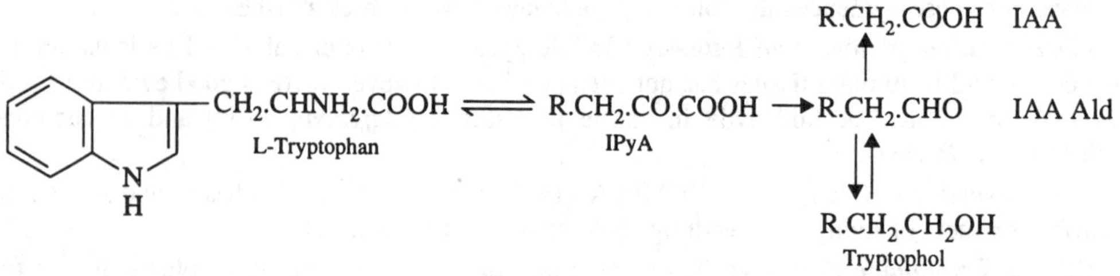

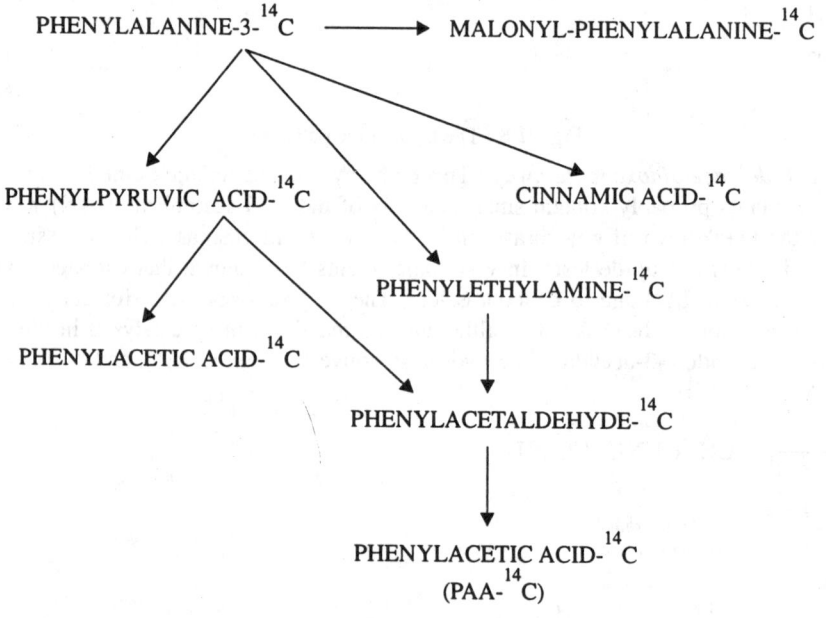

Fig. 21.10 : The tryptophol pathway.

Biosynthesis of Non-indole Auxins

The only non-indole auxin whose biosynthesis has been investigated is phenylacetic acid (PAA). In excised tomato shoots labeled (^{14}C) DL-phenylalanine gave rise to radio-active phenylpyruvic acid, phenylethylamine, phenylacetaldehyde and PAA. The aromatic aminotransferase converts phenylalanine to phenylpyruvic acid and the pathway is analogous to the indolepyruvate and tryptamine route of IAA synthesis.

Fig. 21.11 : Phenylacetic acid Pathway.

Mechanism of Auxin Action and Affected Processes

The following theories have been proposed to explain the mechanism of auxin action:

(1) *Molecular Reaction theory*: The main supporters of this theory are *Skoog et al.* (1942) and *Foster et al.* (1952). Skoog et al. postulated that auxins act as coenzymes, serving as a point of attachment for some substrate on to an enzyme controlling growth.

Foster et al. (1952) proposed a theory of two point attachment in which the auxin reacts with some material in the cell at two positions—(*i*) At some position in the ring, and (*ii*) At the acid group of the side chain.

2. *Auxins affect enzymes* : When growing tissues are treated with an auxin, they show an increased activity of a number of enzymes either directly or indirectly (*Hand*, 1939; *Northen*, 1942; *Bonner*, 1949; *Thimann*, 1951; *Bonner et al.*, 1952).

3. *Auxin affects osmotic pressure* : Auxin affects growth by increasing the Osmotic pressure in the treated tissue thereby increasing the absorption of water (*Czaja*, 1935).

4. *Auxin affects cell-elongation* : Auxin causes cell elongation which may be due to:
   (*i*) Increasing osmotic solutes and pressure of the cells.
   (*ii*) Increasing the permeability of cells to water.
   (*iii*) Increasing the wall synthesis.
   (*iv*) Decrease in wall pressure.
   (*v*) Inducing the synthesis of specific DNA dependent new m-RNA and a specific enzymic protein. The latter brings about an increase in cell plasticity.

5. *Auxin affects nucleic acid metabolism* : Auxin is associated with nucleic acid metabolism. It affects the site of DNA responsible for increasing cell plasticity and its extension (*Coartney et al.*, 1967; *Masuda et al.*, 1967; *Nooden*, 1968).

Functions of Auxins

Auxins perform following important functions :

1. *Cell elongation*: The primary and chief function of auxin in plants is to stimulate the cell elongation in shoot. The growth of cells in length is called *cell-elongation*.

2. *Apical dominance* : The influence of apical bud in suppressing the growth of lateral buds is called *apical dominance*. It is commonly found in many vascular plants especially the tall and sparcely branched ones. In such plants if the terminal bud is intact and growing, the lateral buds remain suppressed. Removal of the apical bud results in the fast growth of the lateral buds. The gardeners usually cut the upper portin (apical buds) of hedges so that the lateral buds and branches could develop fast and the hedge could be compact. The reason for this was explained by *Thimann* and *Skoog* (1934) and *Thimann* (1937). According to them, the auxin is synthesized in the apical meristem from where it is translocated downwards causing inhibition of growth of lateral buds. During downward movement of auxin some correlative inhibitors are synthesized which inhibit the

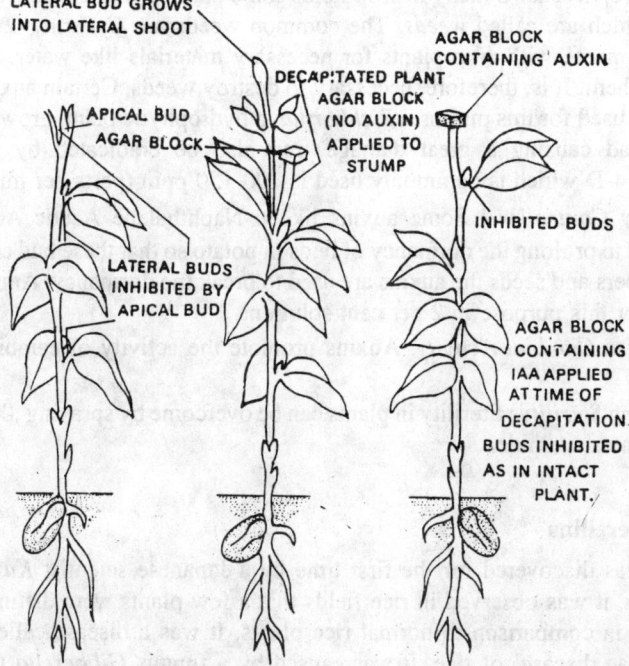

Fig. 21.12 : Demonostration of apical dominance.

growth of lateral buds. When the apical meristem is cut, the effect of auxin is lost causing inititation and growth of lateral buds. If an agar block containing auxin is again placed on the decapitated plant or stump, the growth of lateral buds is again checked. (Fig. 21.12).

3. *Root Initiation*: The same concentration of auxin does not show the same effect on the different parts of a plant. Sometimes the concentration of auxin which increases growth of one organ, reduces or inhibits the growth of other organ, *e.g.* the concentration of auxin which accelerates the growth of stem, reduces the growth of roots but the number of lateral branches in roots is increased. On the basis of increase in number of lateral roots it can be said that auxin helps in root initiation. Usually higher concentration of auxin is used for this purpose.

4. *Prevention of Abscision* : Natural auxins control the falling of fruits, flowers and leaves from the plants. The leaves and fruits fall down from the plants only when an abscision layer is formed between petiole or pedicel or fruit stalk and stem at the point of attachment.

5. *Parthenocarpy*: The formation of seedless fruits without fertilization is called *parthenocarpy*. Although it is common in nature but parthenocarpic fruits can also be induced by auxins. Parthenocarpy has been induced successfully by synthetic auxins (IAA, IPA, NAA, PAA) in fruits like orange, banana, apple, tomato, pineapple, cucumber etc.

6. *Respiration* : Auxin stimulates respiration in many plants and there is a correlation between auxin induced growth and an increased respiration rate.

7. *Callus Formation*: Sometimes the cut parts of plants show fast cell-division resulting into formation of a group of cells at that place to form a knot or tumor. It is called *callus*. The cells of callus are similar, homogeneous and undifferentiated. It also contains meristematic zone whose cells form different organs. If auxin is applied at the cut end of stem or other part, the cells at that place become meristematic and help in the formation of callus.

8. *Shortening of Internodes*: Auxin helps in the reduction of size of branches. It is generally observed in plants like apple and pear where both types small and big branches are found and fruits are born only on small branches.

9. *Eradication of Weeds*: Usually in crop fields some harmful and unwanted plants grow along with crop plants which are called *weeds*. The common weeds are *Cyperus, Oxalis, Cynodon* and *Medicago*. They compete with crop plants for necessary materials like water, light, minerals etc. and cause harm to them. It is, therefore, necessary to destroy weeds. Certain auxins, *e.g.* 2, 4-D and 2, 4, 5-T have been used for this purpose. *Eichhornia*, a hydrophytic plant, growing in water bodies of India and abroad causing a great damage, can also be eradicated by auxins spray. The concentration of 2, 4-D which is commonly used is 100-150 ppm (parts per million).

10. *Dormancy Controller* : Some auxins like α-Naphthalene Acetic Acid (NAA) and its derivatives are used to prolong the dormancy of buds in potato so that these bud could not germinate. In corms, bulbs, tubers and seeds the auxins are used to break the dormancy. Ammonium thiocynate is generally used for this purpose in 2 per cent solution.

11. *To Increase Cambial Activity*: Auxins promote the activity of cambium when supplied from outside.

12. *Overcoming Sterility* : Sterility in plants can be overcome by spraying .001 per cent solution of auxin naphthalene acetamide.

Gibberellins
Discovery of Gibberellins

Gibberellin was discovered for the first time by a Japanese scientist *Kurosawa* in 1926. In Japan and Formosa, it was observed in rice fields that a few plants were distinctly taller, seedless and pale in colour in comparison to normal rice plants. It was a disease called 'foolish seedling disease' or 'Bakanae disease' of rice. It was caused by a fungus *Gibberella fujikuroi* (*Fusarium moniliforme*). *Kurosawa* (1926) proved that if the extract of this fungus is sprayed over the healthy plants of rice, they get infected and produced the symptoms of Bakanae disease. *Yabuta* and *Hayashi* (1939) isolated a crystalline growth promoting substance which they called *gibberellin—A*. Japanese chemists have suggested that this substance is a mixture of several different growth promotors collectively known as gibberellins. About 52 gibberellins have been isolated so far from different type of plants.

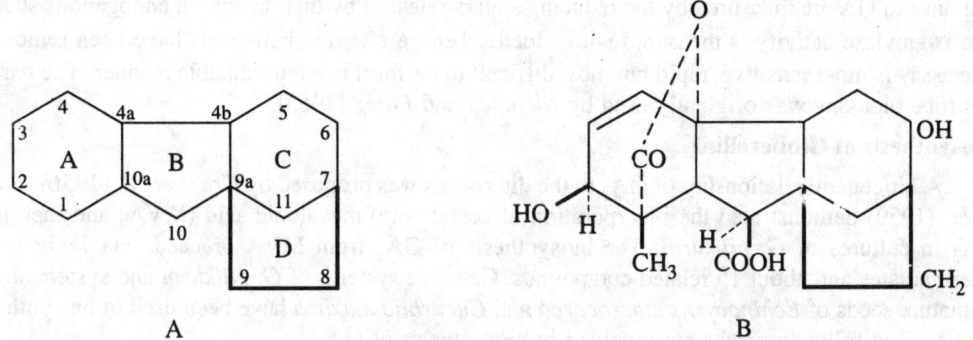

Fig. 21.13 : A—Gibbane skeleton, B—Gibberelic acid.

Structure of Gibberellins

All gibberellins are almost similar in structure. They contain a *gibbane ring* made up of cyclohexane ring and 4-lactone ring. (Fig. 21.13A).

They differ only in minute details *viz*., the number and position of —OH and sometimes —CH_3 and —COOH groups at different carbon atoms of the gibbane ring. GA_3 is generally called gibberellic acid (Fig. 21.13B).

Extraction

For extraction of GAs, the plant material is usually macerated with cold methanol or acetone containing 50 per cent water. After evaporation of the organic solvent and centrifugation, the aqueous residue is partitioned against ethyl acetate at different pH values. The GAs are extracted into ethyl acetate at pH 2.5-3.0, re-extracted into buffer at pH 8.0 and again extracted into ethyl acetate after adjusting pH to 2.5.-3.0. Almost all free GAs may be obtained in the final (acid) ethyl acetate fraction.

Bioassay of Gibberellins

Three types of bioassays are widely used, because of their ease of performance, reliability, sensitivity and range of response. They are:
1. Dwarf seedling bioassay
2. The hypocotyl bioassay
3. Cereal endosperm bioassay

Dwarf Seedling Bioassays

They depend on the increased growth of certain dwarf cultivars in response to applied GA. *Brian* and *Hemming* (1955) used dwarf peas for the first time and applied 5 µ l of the test solution to the apical buds. The log dose-log response curve for GA_3 was linear from 10^{-1} to 10 µ g per plant.

Hypocotyl Bioassays

They depend on the increased elongation of the hypocotyl of certain seedlings in response to GA. The assays are easy to perform, rapid but are less sensitive. *Frankland* and *Wareing* (1960) used lettuce hypocotyl and under optimal conditions the log dose-log responsse curve for GA_3 was linear for 10^{-2} to 10 µ g/ml.

Cereal Endosperm Bioassay

In germinating cereal seeds, GA produced by the embryo stimulates the formation of hydrolytic enzymes in the aleurone layer. The enzymes are secreted into the endosperm where they hydrolyse reserve carbohydrates. For bioassay, the embryos are removed by cutting the seeds transversely and the embryo-less half-seeds are incubated with the test solutions. The hydrolytic enzymes formed in response to GA are measured by the reducing sugars released by their action on endogenous starch. The α-amylase activity is measured with added substrate after the half-seeds have been removed. The assay is most sensitive, rapid but most difficult to perform in a reproducible manner. The barley aleurone bioassay was originally used by *Nicholas* and *Paleg* (1963).

Biosynthesis of Gibberellins

A biogenetic relationship of GA_3 to the diterpenes was proposed by *Cross et. al.* (1956). *Birch et al.* (1959) demonstrated the incorporation of acetate into mevalonic acid (MVA) and then into GA_3 in cultures of *G. fujikuroi*. The biosynthesis of GA_3 from MVA proceeds via 18 or more intermediates and about 15 related compounds. Cell-free systems of *G. fujikuroi* and systems from immature seeds of *Echinocystis macrocarpa* and *Cucurbita maxima* have been used in biosynthetic studies. The following steps are involved in biosynthesis of GA_3:

1. *Formation of Mevalonate (MVA) from Acetate*

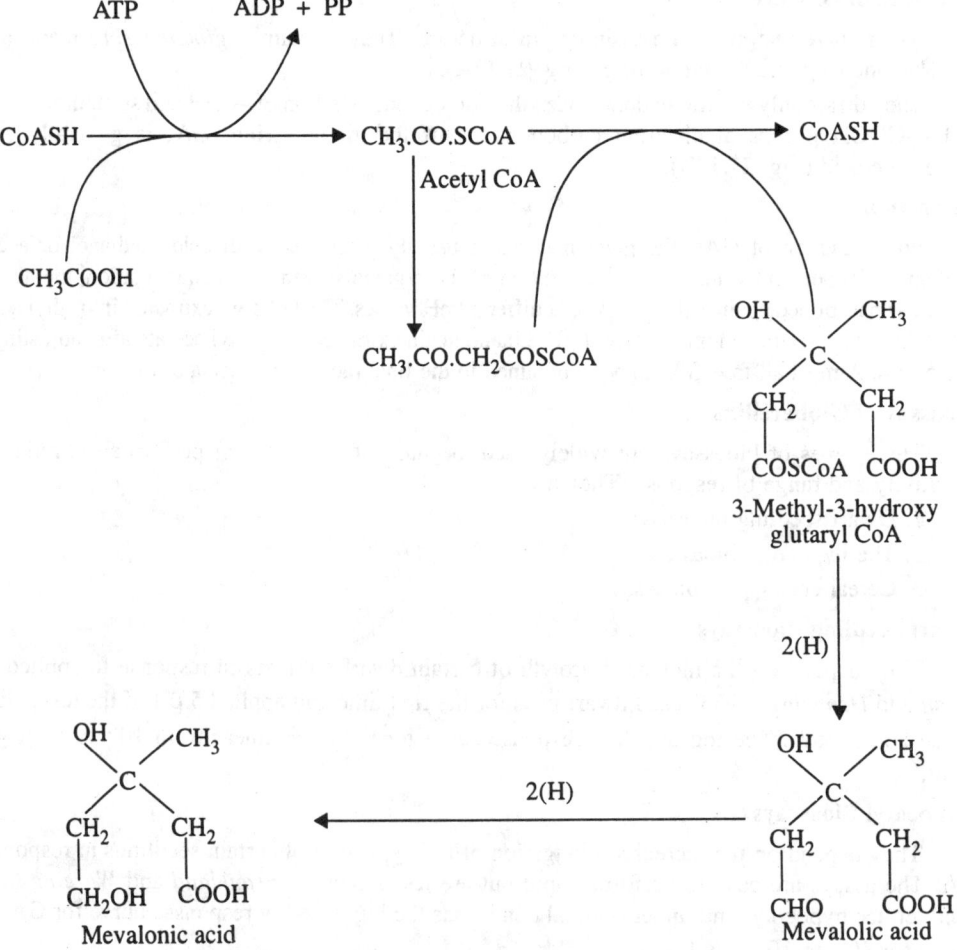

Fig. 21.14 : Formation of mevalonic acid.

2. Formation of Isopentenyl Pyrophosphate (IPP) from Mevalonate

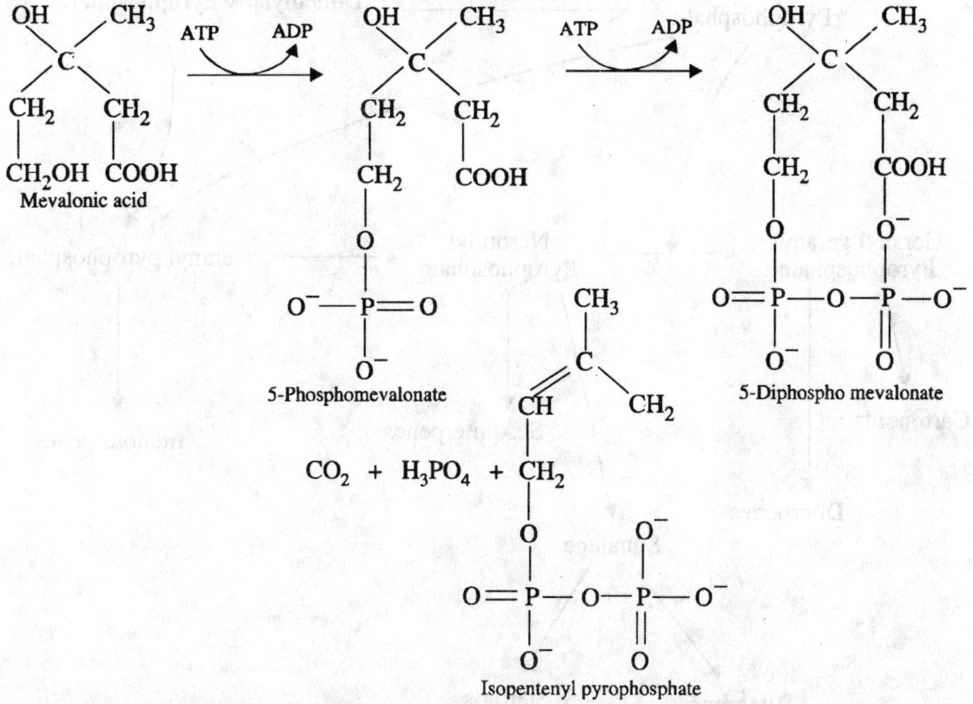

Fig. 21.15 : Formation of Isopentenyl pyrophosphate (IPP).

3. Condensation of Isopentenyl Pyrophosphate to form Geranyl Pyrophosphate.

Isopentenyl pyrophosphate has its isomer, the dimethyl allyl pyrophosphate.

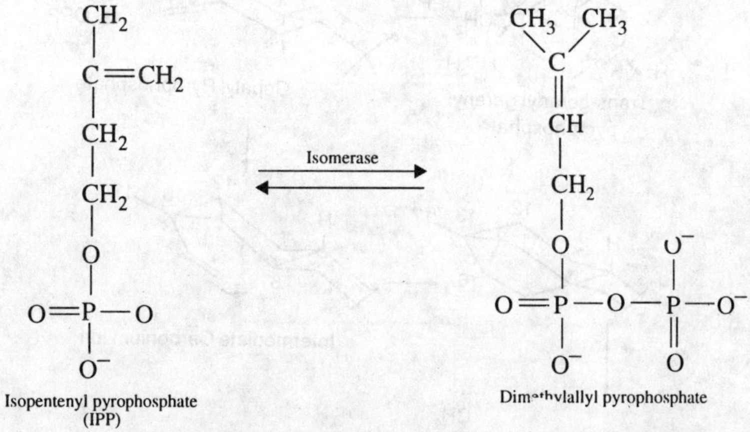

The two isomers condense to geranylgeranyl pyrophosphate by alkylation.

4. Cyclisation of Geranylgeranyl Pyrophosphate

The following scheme (Fig. 21.17) shows the generally accepted mechanism for cyclization of geranylgeranyl pyrophosphate to ent-Kaurene from (I^{14} - C) acetate.

5. Conversion of ent-Kaurene to ent-7 α-hydroxy Kaurenoic acid

ent-Kaurene is oxidised step-wise at C-19 to form ent-Kaurenol, ent-Kaurenal, and ent-Kaurenoic acid. The latter is hydroxylated to ent-7 α-hydroxy Kaurenoic acid. The sequence has

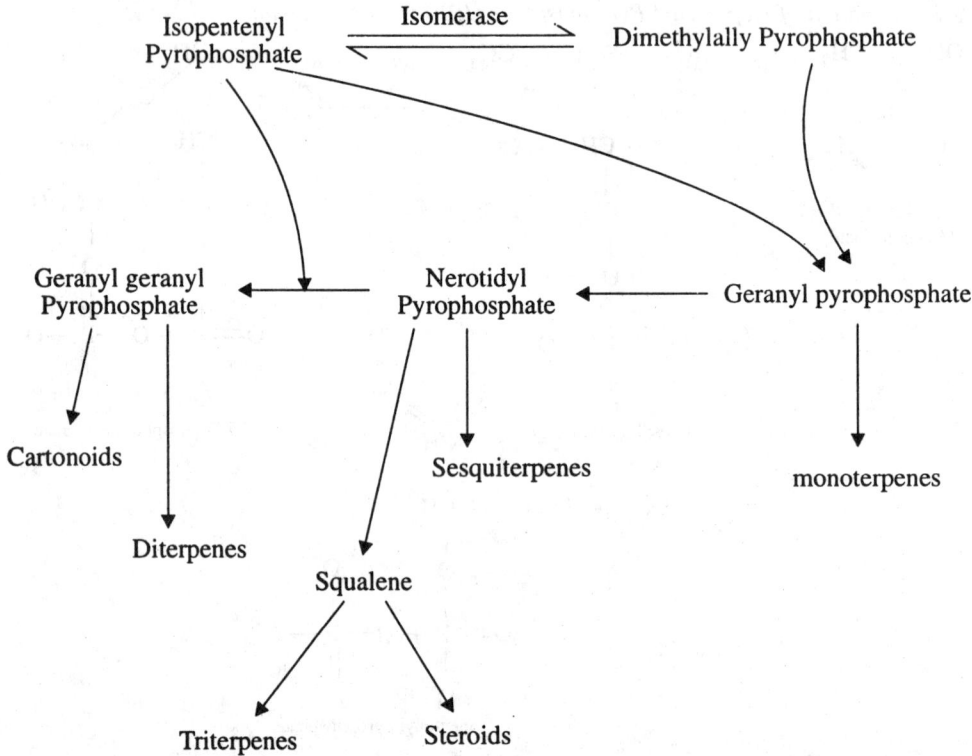

Fig. 21.16: Conversion of isopentenyl pyrophosphate to geranylgeranyl pyrophosphate.

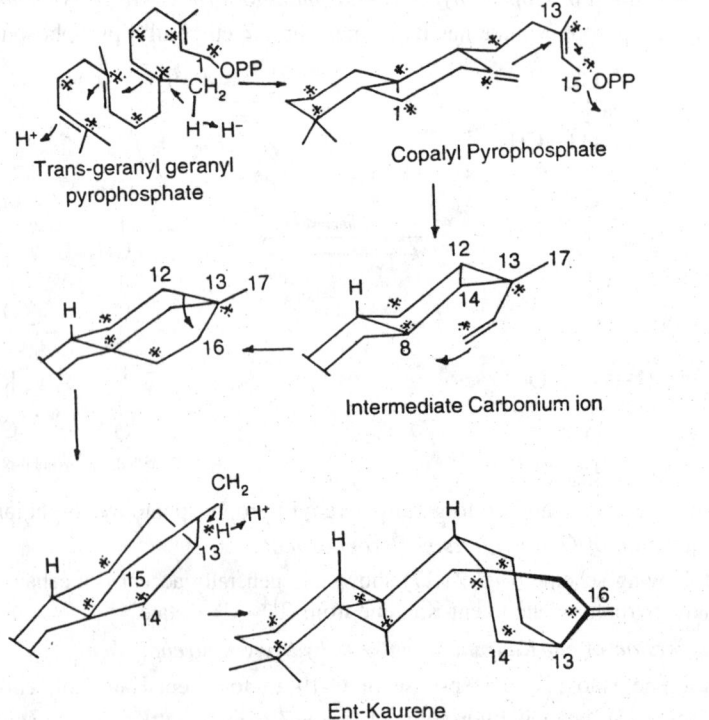

Fig. 21.17: Cyclization of geranylgeranyl pyrophosphate to ent-Kaurene.

been established in *Echinocystis* cell free system and in cultures of *G. fujikuroi*.

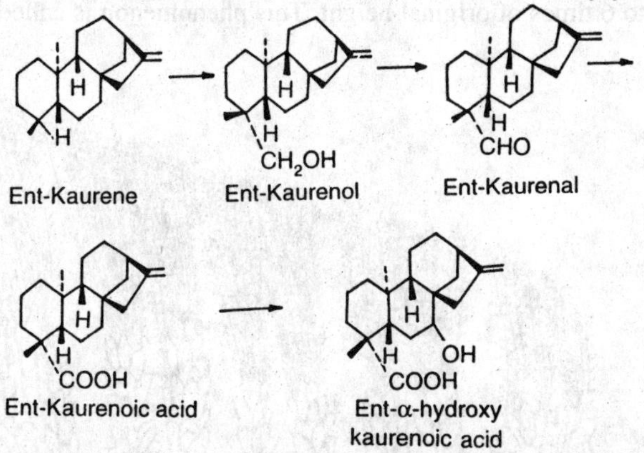

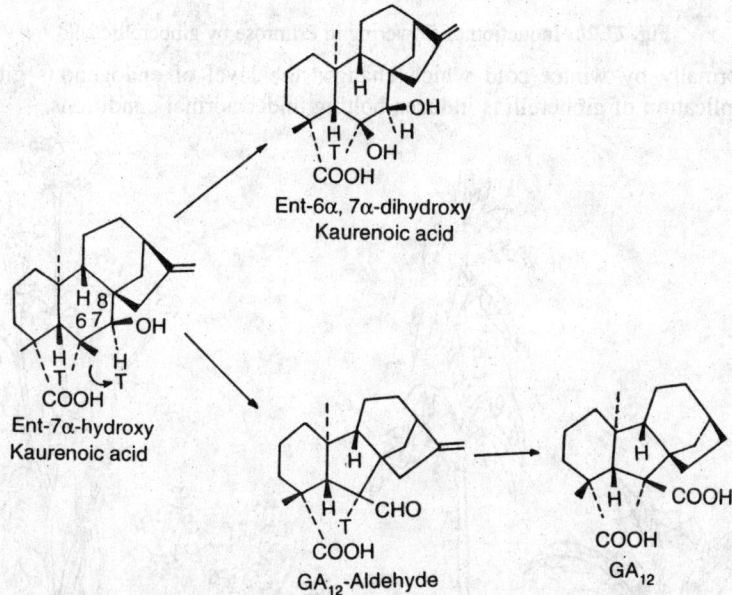

Fig. 21.18: Conversion of ent-Kaurene to ent-7 α-hydroxy Kaurenoic acid.

6. *Contraction of β-ring and β-hydroxylation*

The conversion of ent-7 α-hydroxy kaurenoic acid to GA_{12}-aldehyde involves loss of 6 β-hydrogen, a shift of the 7,8 bond to 6,8-position and loss of a proton from the extruded C-7 (Fig. 21.19).

Fig. 21.19: Conversion of ent- 7 α-hydroxy Kaurenoic acid.

7. *Loss of C-20 to form C-19 GAs (GA_3)*

Loss of one carbon must occur to give rise to C-19 GAs such as GA_3.

Functions of Gibberellins

(*i*) *Cell-elongation*: The plants of dwarf species of pea and maize become very tall in comparison to normal ones due to effect of gibberellin. It is because the cells grow in length. The dwarf character may be either due to absence of endogenous gibberellin or presence of natural inhibitors.

(*ii*) *Bolting* : Certain plants like Henbane and carrot etc. show profuse development of leaves but retardation in growth of internode. It is called 'rosette' type of growth. Just before the reproductive stage, there is a striking stimulation of internode elongation and sometimes the internode elongates 5 to 6 times of original height. This phenomenon is called *bolting*. The bolting

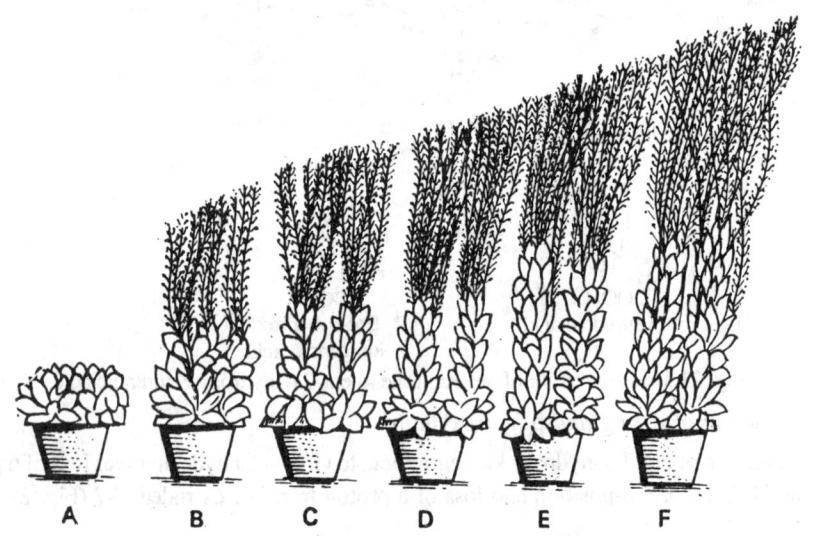

Fig. 21.20: Induction of flowering in Primrose by gibberellic acid.

is induced normally by winter cold which changes the level of endogenous gibberellins. The exogenous application of gibberellins induces bolting under normal conditions.

Fig. 21.21: A-B. Bolting effect in Henbane plant, C-D. Bolting effect in carrot.

(iii) *Parthenocarpy* : Formation of seedless fruits without pollination and fertilization is called *parthenocarpy*. It can be induced in different plants like tomato, guava, orange, banana etc. by gibberellins. In many cases gibberellins show higher activity than the native auxins. In many cases, e.g. pome and stone fruits, the auxin has proven ineffective but gibberellins effective.

(iv) To help in flowering and development of fruits.

(v) To help in the conversion of biennial plants into annual plants.

(vi) To increase the size and number of fruits, *e.g.*, lemon, orange, grapes etc.

(vii) To break the dormancy of seeds, buds and underground organs, *e.g.* bulb, tuber, corm, etc.

(viii) Activation of fermentation during wine formation.

(ix) Increase the size and number of flowers.

(x) Increase the activity of cambium in woody plants.

(xi) de novo synthesis of the enzyme α-amylase.

(xii) Increase the growth of stem growing in light, i.e. revival of light inhibited stem growth.

(xiii) Increase the size of leaves.

Table 21.1 : Differences between Auxins and Gibberellins

	Auxins		Gibberellins
1.	They are mostly found in higher plants.	1.	They are mostly found in fungi and few higher plants.
2.	They are synthesized in the meristematic regions of the plants e.g., in coleoptile, buds, growing tips of roots and leaves.	2.	They are synthesized in fungal mycelia, in immature bean seeds, in pea seedlings, hypocotyls, cereal endosperms.
3.	Chemically they are indole 3-acetic acid (IAA) or its derivatives or chemicals with identical properties like IBA, NAA, IPA etc.	3.	Chemically they are gibberellic acid which are diterpenoid acids derived from the tetracyclic diterpenoid hydrocarbon, ent-Kaur-16-ene or (–)-kaurene, having 20 carbons.
4.	They promote apical dominance.	4.	They show no effect on apical dominance.
5.	They cause growth in dwarf pea stem sections, but show no effect on intact plant.	5.	They cause growth of intact plant, but show no effect on its stem sections.
6.	They show no effect on bolting and flowering.	6.	They promote seed germination and breaking of dormancy.
7.	Auxins do not effect seed germination and breaking of dormancy.	7.	They promote bolting and flowering in non-vernalized biennials and long day plants.
8.	They do not cause de-novo synthesis of hydrolytic enzymes.	8.	They cause de-novo synthesis of hydrolytic enzymes.
9.	Auxins in higher concentrations inhibit growth of roots but initiate formation of new roots.	9.	Gibberellins do not effect growth of roots.
10.	Auxins promote cell-elongation, cell-division and cambial activity.	10.	They promote cell-elongation by synthesizing enzymes and checking the effect of inhibitors.
11.	Auxins like 2, 4-D; 2, 4, 5-T and Dalapon etc. are used as weed killers.	11.	Gibberellins are not used as weed killers.

Cytokinins-Kinetins

The term *cytokinin* was proposed by *Letham* (1963). The group cytokinin includes those hormones which stimulate and help in cell-division and cytokinesis (division of cytoplasm). *Kinetin* and *zeatin* are chief cytokinins. Miller *et al.* (1955) isolated kinetin from yeast DNA by its degradation. Chemically kinetin is 6-furfuryl-amino-purine. The name kinetin was suggested due to its function in helping cell-division. It is richly found in liquid endosperm of coconut (*Cocos nucifera*).

The cytokinin—*Zeatin* was isolated from immature seeds of maize for the first time by *Letham* (1963). Chemically zeatin is 6-(4 hydroxy, 3-methylbut-trans-2-enyl) aminopurine. It was synthesized by *Shaw* and *Wilson* (1964). Zeatin is more active *than* all other known cytokinins because it contains a highly reactive allylic —OH group in its side chain.

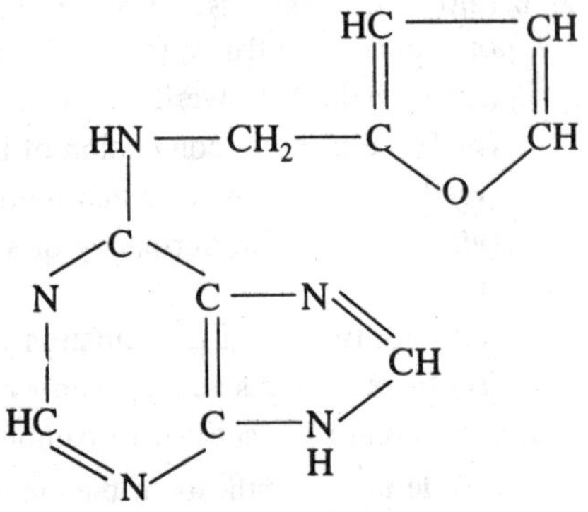

Fig. 21.22: Structure of Kinetin.

Bioassay of Cytokinins

The different bioassays for cytokinins fall into 6 groups:

1. Those which are based on the ability of cytokinins to promote expansion of excised leaf or cotyledon.

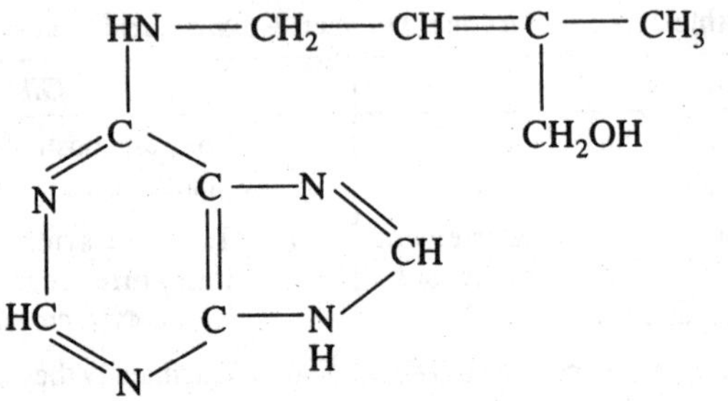

Fig. 21.23: Structure of Zeatin.

2. Those which depend on the promotion of growth of *Lemna*.
3. Those which depend on the promotion of growth of stem and coleoptile sections.
4. Those which depend on the ability of cytokinins to induce cell divisions in tissue cultures.
5. Those which are based on retardation of leaf senescence.
6. Those which depend on pigment formation (chlorophyll and betacyanin respectively).

The tissue culture assays have high sensitivity under sterile conditions and are specific. The soybean callus assay is probably the most popular, because it has a wide concentration range over which a linear relationship between response and concentration is obtained. However they need a long assay time (21-35 days).

The *Amaranthus* betacyanin bioassay is a rapid assay. It has high specificity and sensitivity under sterile condition. The radish cotyledon assay is simpler to perform than the Amaranthus assay. The chlorophyll formation assay in cucumber cotyledons requires a short assay time and exhibits excellent sensitivity and specificity.

Effects of Cytokinins

(*i*) Stimulate cell-division (*Miller et al.* 1956; *Braun & Wood*, 1962).

(*ii*) Induce cell-enlargement (*Miller*, 1956; *Arora et al.*, 1959).

(*iii*) Break dormancy. (*Khan*, 1964).

(*iv*) Cause morphogenetic changes (*Skoog and Miller*, 1957).

(*v*) Counteract the apical dominance of bud (*Wickson* and *Thimann*, 1958; *Sachs* and *Thimann*, 1964).

(*vi*) Increase the rate of protein synthesis (*Osborne*, 1962).

(vii) Delaying of senescence (*Richmond & Lang*, 1957).
(viii) Induce tuber formation (*Palmer* and *Smith*, 1972).

Florigen

The hormones which induce flowering in plants are called **florigen. Gibberellin** and **anthesin** hormone jointly initiate flowering in plants. They are synthesised in the leaves as a result of dark and light phases. From leaves, they are transferred to buds where they perform the function of differentiation of floral organs.

Flowering Mechanism : The modified vegetative shoot buds which have the power of formation of reproductive gametes and seeds are called flowers. Different environmental factors induce flowering. Temperature and light durations specially play an important role during this phenomenon.

(a) *Light effected Photoperiodism*: American scientists *Garner* and *Allard* (1920) recognized that the ratio of day and night light duration effect flowering. It is called *photoperiodism*. According to the light duration, the plants have been classified into following three groups—

(i) *Long day Plants* : They require a light duration of more than 12 hours for flowering, e.g. radish, spring wheat and turnip etc.

(ii) *Short day plants* : They require a light duration of less than 12 hours for flowering, e.g. Soybean and tobacco.

(iii) *Day neutral plants* : The plants in which flowering is not affected by length of day are called *day neutral* plants, e.g. Cotton and Pea.

Some plants require a number of short days followed by long days for flowering. Such plants are called *short long day* plants, e.g. Winter rye and Candytuft. Some plants require a number of long days followed by short days for flowering. Such plants are called *long short day plants, e.g. Bryophyllum* and *Cestrum nocturnum.*

The florigen hormones in leaves and buds are produced by metabolic processes due to photoperiodism which is responsible for differentiation of floral organs. The flowering in different seasons also takes place due to photoperiodism. The flowering in plants can be induced in any season by increasing or decreasing the duration of day or night. Due to this reason, photoperiodism is especially useful in agriculture and gardening. Indian scientist Prof. *S.M. Sirkar* has proved that

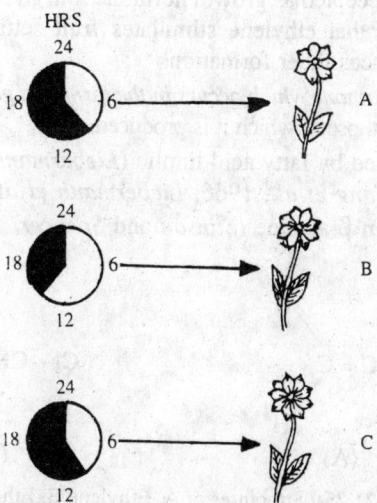

Fig. 21.24: Classification of plants according to photoperiodism. A— Short day plants, B—Long day plans, C—Day-neutral plants.

Aman variety of rice which normally flowers in 140 days can flower in only 50 days if the plants are covered. Gibberellins are also useful in inducing flowering.

(b) *Vernalization*: Russian scientist *Lysenko* (1938) kept wet and germinated seeds of winter wheat artificially at 0°—5°C temperature and felt that they develop the characters of spring wheat which can be sown in spring. These plants produce seeds at the end of summer during the same year. This phenomenon is called *vernalization*. Thus, vernalization can be defined as the phenomenon of low temperature promotion of flowering in plants.

The hormone responsible for vernalization is called *vernalin*. This name was suggested by a German botanist *G. Melchers* (1939). The vernalin is a hypothetical substance and has not been isolated so far. The plants can be protected from natural calamities like fog etc. by vernalization. In Siberia, the crop of wheat is prepared by this process. Melchers also demonstrated that the product of vernalization could be transmitted from a vernalized to an unvernalized *Hyoscyamus* plant through a graft union. For plants growing in temperate regions it is necessary that they should pass through cold conditions for flowering otherwise the flowering in them will not take place.

21.6. GROWTH INHIBITORS

Ethylene

It is a volatile gas present in the atmosphere as a component of smoke and other industrial gases. In plant cells, it is produced as a result of metabolism. Its quantity increases in plants on plucking the flowers. The chemical formula of ethylene is $CH_2 = CH_2$. Chemically it is the most simple among plant hormones. Acetylene and propylene also act like ethylene but ethylene is 60 to 100 times more active than propylene (*Pratt* and *Goeschl*, 1969). They are low molecular weight hydrocarbons containing unsaturated bond.

Neljubow (1901) demonstrated that ethylene gas alters the tropic responses of roots. *Denny* (1924) observed that ethylene gas helps in inducing fruit ripening. *Wallace* (1926) observed that ethylene gas helps in falling of plant leaves. *Vacha* and *Harvey* (1927) proved that ethylene gas can break the dormancy of underground stems, seeds and buds etc. *Zimmermann* and *Hitchcock* (1933) demonstrated that ethylene induces root formation. *Gane* (1934) proved that ethylene gas plays an important role in ripening of fruits.

Although, various effects described above have been proved by various scientists but until 1960 the ethylene gas was not accepted as growth hormone. *Burg* (1962) finally proved it as a plant hormone. *Varga* (1969) found that ethylene stimulates fruit setting. *Palmer* and *Smith* (1972) demonstrated that ethylene induces tuber formation.

Ethylene is the only plant hormone which occurs in the form of a gas. Due to its gaseous state, it can readily diffuse to cells other than those in which it is produced.

Ethylene can be synthesized by fatty acid linolic (*Liebermann* and *Mapson*, 1962, 1964), or from aminoacid methionine (*Yang et al.*, 1966; *Liebermann et al.* 1966); or from fumaric acid (*Burg* and *Burg*, 1965), or from β-alanine (*Stinson* and *Spencer*, 1969) or from isoamyl alcohol (*Lona* and *Raffi*, 1970).

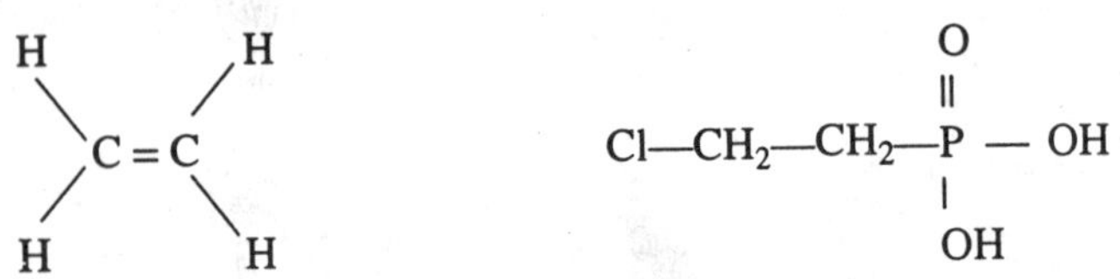

Fig. 21.25: Structures of A. Ethylene B. Ethephone.

Growth and Metabolism of Growth Hormones

Roles of Ethylene

(i) Helps in the ripening of unripe fruits like banana, mango, orange etc. Due to this character it is also called *ripening hormone*.

(ii) Stimulates the formation of abscission layer in leaves, flowers and fruits.

(iii) Inhibits liner growth of stem and increases diameter.

(iv) Inhibits geotropism in stems.

(v) Controls epinasty.

(vi) Increases the number of female flowers and decreases the number of male flowers.

(vii) Accelarates apical dominance.

(viii) Normally reduces flowering in plants but in pineapple it increases flowering.

(ix) Stimulates the formation of new roots.

(x) Induces growth in underground stems and seed germination but after sprouting checks the growth of stem and leaves.

(xi) Regulates the growth of cell-wall.

(xii) Promotes the synthesis of β-1, 3-glucanase.

(xiii) Induces de novo synthesis of peroxidase.

Being volatile in nature, ethylene gas is not used directly. Commercially, it is being used in the form of a substance *ethephone* (2-chloroethyl phosphonic acid) which after hydrolysis yields ethylene gas.

Abscisic Acid (ABA)

Carns and *Addicott* (1965) separated a substance from cotton fruits and named it *abscisin II*. This substance similar to dormin was extracted by *Wareing* and *Cornforth* (1964). Later on, it was confirmed that abscisin II and dormin were the same substance. Therefore, they were jointly

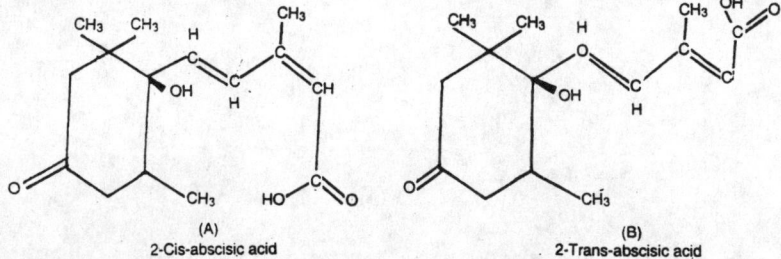

Fig. 21.26 : Structure formulae of abscisic acids.

given the name abscisic acid (ABA). It is synthesized in the leaves from where it is translocated to the stem apex through phloem.

Roles of Abscisic Acid

(i) Helps in abscission of leaves, flowers and fruits by forming abscission layer.

(ii) Induces dormancy in seeds and buds, thus, inhibiting growth of buds and seed germination.

(iii) Helps in reducing transpiration rates by closing stomata.

(iv) Spoils chlorophylls, proteins and nucleic acids of leaves making them yellow.

(v) Helps in the formation of potato tubers.

(vi) Normally, ABA is a growth inhibitor. It also reduces the growth induced by gibberellins.

(vii) Inhibits cell-division and cell-elongation.

(viii) Checks the synthesis of enzyme α-amylase which results in inhibition of seed germination.

Morphactins

They are synthetic growth inhibitors which contain a skeleton *fluorenol* (9-hydroxy-fluorene-9-carboxylic acid). They inhibit or retard the growth of seedlings, seed germination, linear growth of stems and development of leaf-lamina. They also inhibit apical dominance and induce the growth of lateral buds. They check sprouting of buds in rosette type of plants.

In addition to morphactins, chlorocholine chloride (CCC) and maleic hydrazide (MH) are also growth inhibitors.

22

Plant Movements

22.1. INTRODUCTION

The change in the state of either complete plant body or any specific plant part due to influence of environment or internal conditions is called *movement*. It takes place due to sensitivity or irritability of the protoplasm. The changes in the environment or external conditions of the plant which force the plant to response is called *stimulus*. Light, temperature, touch, gravitation etc. are such factors. The alteration in the life processes of the plant produced by stimulus is called *response*. The protoplasm receives the stimulus and shows the response due to its properties of sensitivity and irritability.

The parts of the plants which have the capacity to receive the stimulus are called *perceptive organs* or *perceptive regions* and those parts which show the response against stimulus are called *responsive organs* or *regions*. The period between receiving the stimulus and showing the response is called *reaction time*. The minimum time required for stimulus to produce response is called *presentation time*.

22.2. LAW OF SUMMATION OF STIMULII

Sometimes when the stimulus of low potency does not produce response, the same stimulus is given frequently at short intervals by reducing time and the response is produced. It is called law-of summation of stimulii.

22.3. CLASSIFICATION OF MOVEMENTS (Table 22.1)

The movements in plants are mainly of two types—I. Physical and II. Vital.

(1) *Physical movements* are those which are found in dead parts of the plants and they are not related to any irritability of the protoplasm. They are also called *hygroscopic movements* or *mechanical movements*. Dispersal of spores and seeds, dehiscence of sporangia, bursting of seeds and movement of elaters are the examples of physical or hygroscopic movements.

(2) *Vital movements* are those which are exhibited by the living cells or plants or organs. They are always related to the irritability of the protoplasm. The various type of vital movements can be classified according to the Table 22.1. The vital movements are mainly of two types:

    I. Movements of locomotion.

    II. Movements of curvature or orientation.

I. Movements of Locomotion

These movements include the movement of protoplasm inside the cell or movement of whole unicellular or multicellular plant body as in *Chlamydomonas* and *Volvox* or the movement of free organs of the plants like gametes and zoospores from one place to another place. All such movements are called *movements of locomotion*. These movements are of two types—

    (*a*) Autonomous or Spontaneous or Autogenic movements.

    (*b*) Induced or Paratonic or tactic or taxis movements.

(*a*) **Autonomous Movements**

The movements arising from internal changes or internal stimuli of plant-body are called

Table 22.1: Classification of Plant Movement

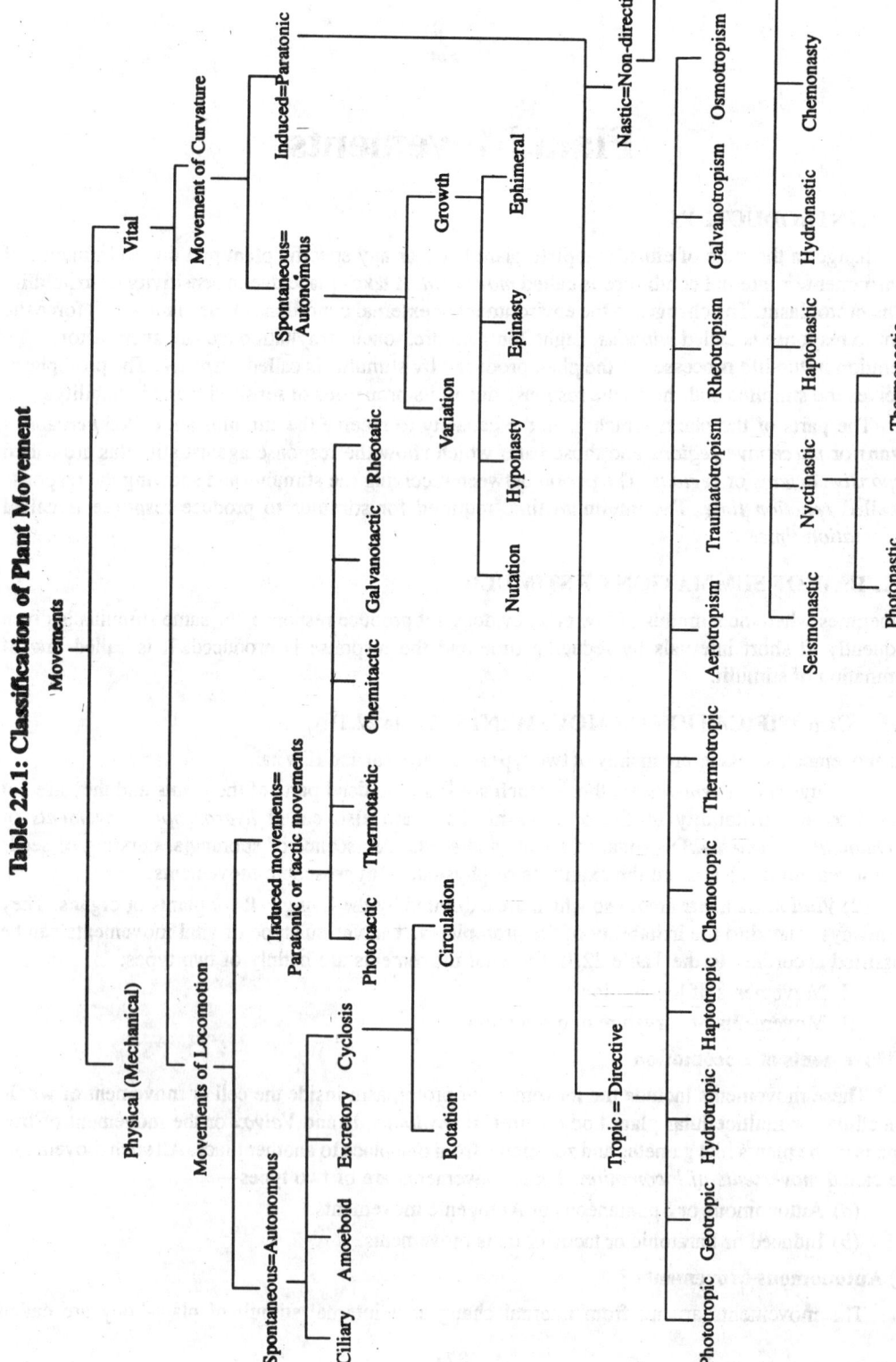

autonomous movements. They are of following types:

(1) *Ciliary movements* : The movement produced due to activity or movement of cilia is called *ciliary movement*. It is found in certain unicellular algae like *Chlamydomonas* or muliticellular algae like *Volvox, Pandorina* and *Eudorina*, zoospores of *ulothrix, spirogyra* and many other algae and in antherozoids of mosses and ferns

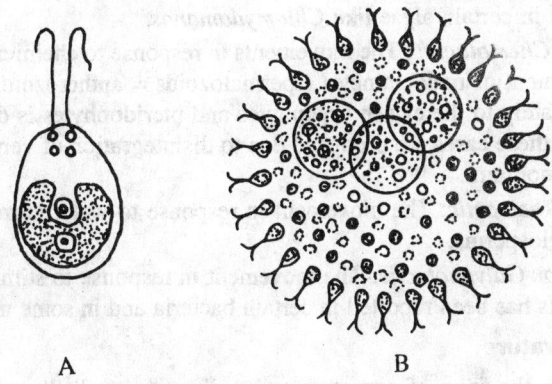

Fig. 22.1: Ciliary movements in (*a*) Chlamydomonas (*b*) Volvox.

(2) *Amoeboid movement*: The movement due to pseudopodia is called *amoeboid movement*. It is found in the members Myxomycetes (slime fungi) particularly in the naked protoplasm which undergo changes in shape resulting in creeping movement of protoplasm.

(3) *Excretory movement* : In certain algae like *Oscillatoria*, the movement in filaments takes place due to influence of excretory products. It is called *excretory* movement.

(4) *Cyclosis* : The movement of protoplasm around the vacuoles is called *cyclosis*. When the movement takes place only in one direction around a single vacuole, it is called *rotation*. It occurs in leaf cells of *Vallisneria, Hydrilla, Elodea* and internodes of *Chara* and *Nitella*. When the movement of protoplasm takes place around many vacuoles in several directions, it is called *circulation*. It occurs in the staminal hair cells of *Tradescantia* and *Rhoeodiscolor*.

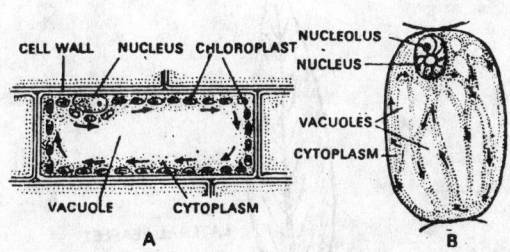

Fig. 22.2: (A) Rotation movement in a leaf cell of *Vallisneria* (B) Circulation movement in the cell of staminal hair of *Tradescantia*.

(*b*) **Induced Movements**

The movements due to external factors or stimulii like light, temperature, chemicals, water or electric currents are called *induced* or *tactic movements*. These movements are directional and may be positive (towards the source of stimulus) or negative (away from the source of stimulus). Depending upon the kind of stimulus, the induced movements may be of the following types.

(1) *Phototaxis = Phototactic* : The movements in response to light are called *phototaxis*. These are found in motile algae, chloroplasts in the cells and phototropic bacteria etc. Some algae move towards diffused light and others against deep sunlight.

(2) *Thermotaxis = Thermotactic* : The movements due to temperature stimulus are called *thermotaxis*. It is found in certain algae like *Chlamydomonas*.

(3) *Chemotaxis = Chemotactic* : The movements in response to chemical substances are called *chemotaxis*. The movement of male gametes (spermatozoids = antherozoids) towards the neck of archegonium and ultimately to the egg in bryophytes and pteridophytes is due to chemotaxis. The chemical substances in these cases are produced due to disintegration of venter canal cell and neck canal cells of the archegonium.

(4) *Rheotaxis or Rheotactic*: The movement in response to water current is called *rheotaxis*. It occurs in aquatic angiosperms.

(5) *Galvanotaxis or Galvanotactic*: The movement in response to stimulus of electric current is called *galvanotaxis*. Is has been reported in certain bacteria and in some members of volvocales.

II. Movements of Curvature

The movements in the form of curvature or bending of attached organs of fixed plants are called *movements of curvature*. These movements are also of two types—

(a) Autonomous movements

(b) Paratonic = induced movements.

(a) **Autonomous movements:** The movements arising from internal changes or internal stimuli of plant body are called *autonomous movements*. They do not require any external stimulus and are of two types :

(1) *Movements of Variation*: They are commonly found in the leaflets of *Desmodium gyrans* (Indian telegraph plant), *Eleiotis soraria*, *Trifolium pratense* and *Oxalis acetocella*. In *Desmodium gyrans* and *Eleiotis soraria* the leaf comprises one large terminal leaflet and two much smaller opposite lateral leaflets (Fig. 22.3). The movement of variation is performed by lateral leaflets. In *Desmodium gyrans* the movement is brought about by rhythmic changes in turgor of a group of cells at the base of leaflet stalk. (Fig. 22.3).

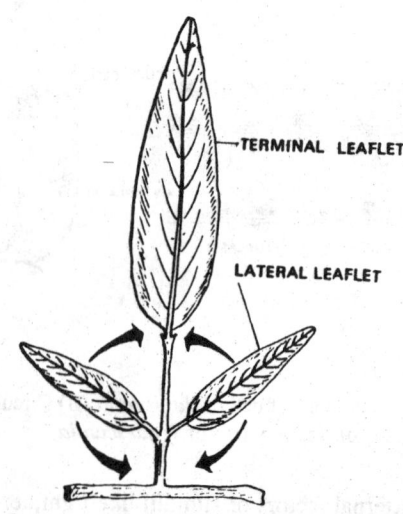

Fig. 22.3: Movement of variation in Indian telegraph plant.

(2) *Movements of Growth*: The movements due to difference in the growth of two surfaces or two regions are called *movements of growth*. They may be of following types:

Nutation : It is commonly found in the stems of climbers of Cucurtibaceae and also in other organs such as leaves, tendrils, roots, stolons, flower stalks, fungal sporangiophores and filamentous algae. (Fig. 22.4).

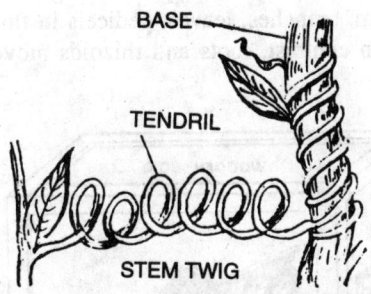

Fig. 22.4: Nutation in the tendril of Cucurbita.

Hyponasty and Epinasty: The petals and sepals of the flowers close and open due to photonasty and thermonasty. When unequal growth on both the surfaces of petals takes place, it results into movement of growth. When growth is more on lower surface, petals show curvature on upperside and ultimately the flower becomes closed. Such type of movement is called *hyponasty*. When the growth is more on upper surface, petals, show curvature on lower side and ultimately the flower opens. Such movement is called *epinasty*. Flowers usually open at high temperature and remain closed at low temperature. (Fig. 22.5).

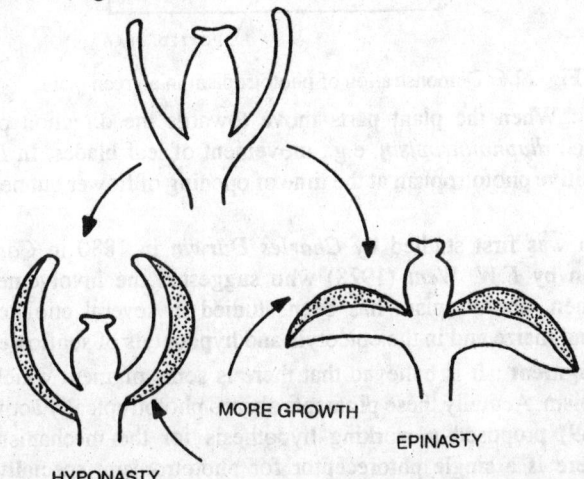

Fig. 22.5: Opening and closing of flower petals due to hyponasty and epinasty.

Ephemeral movements : The opening of flowers and leaves takes place due to unequal growth on two sides of the organ but these movements take place only once and for this reason they are called *ephemeral movements*.

(*b*) **Paratonic = Induced Movements**: The movements arising from some external stimulus are called induced or *paratonic movements*. The stimulus may be unidirectional or diffused. These movements are of two types :

(*i*) Tropic movements = Directive movements.

(*ii*) Nastic movements = Non-directive movements.

(1) Tropic Movements = *Tropisms*: The movements occurring in response to an unidirectional stimulus are called *tropic movements* or *tropisms*. In such movements the stimulus acts on protoplasm from one direction only and the response has direct relation to the direction of stimulus. These movements may be of following types:

(i) **Phototropism** : The tropic movement in response to light stimulus is called *phototropism*. Some parts of the plants like stem, branches, leaves, pedicels in flowers move towards light and are called positive *phototropic*. In contrast, roots and rhizoids move against light and are called *negative phototropic*. (Fig. 22.6).

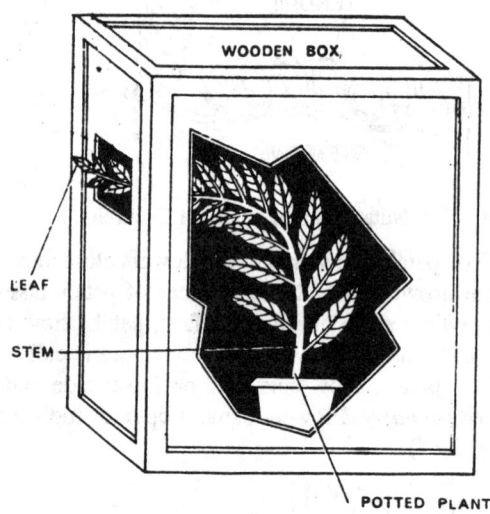

Fig. 22.6: Demonstration of phototropism in a green plant.

Diaphototropism : When the plant parts move towards the direction perpendicular to the ncident light, it is called *diaphototropism*. e.g., movement of leaf blades. In *Lynaria cymbalaria*, the peduncle shows positive phototropism at the time of opening of flower but negative phototropism after fertilization.

The phototropism was first studied by *Charles Darwin* in 1880 in *Canary grass* and *Oat coleoptiles* and later on by *F.W. Went* (1928) who suggested the involvement of auxin in this phenomemon. Since then, phototropism has been studied in several etiolated grass coleoptiles including oats, wheat and maize and in the epicotyls and hypocotyls of sunflower and pea seedlings.

Photoreceptor Pigment : It is believed that there is some pigment which acts as a photoreceptor during phototropism. Acutally these pigments absorb phototropically active light and produce responses. *Curry* (1969) proposed a working hypothesis for the mechanism of phototropism, according to which there is a single phtoreceptor for phototropism, specially a yellow pigment (probably carotenoid) located in plastid or proplastid–like organelles. Light absorption by the photo-receptor initiates a sequence of events (largely unknown) which changes the permeability properties of the cells containing the photoreceptor. These changes in turn affect the basipetal transport of auxin. An asymmetrical light stimulus causes asymmetrical changes in auxin transport, and hence curvature.

Present physiologists believe that two types of yellow pigments, **β-carotene and riboflavin, are the photoreceptors** which are involved in phototropism. Riboflavin or similar flavin compound is probably most effective in phototropic reception. The action spectrum for photoropism indicate that blue region is maximum sensitive while red region is practically ineffective. However, the photo-tropic reaction is extraordinarily sensitive to light of low intensity. For example, itiolated

Avena seedlings curve towards light with an intensity of only 0.00017 m-c, i.e., $\frac{1}{1000}$ to that of full moonlight, but a longer duration of about 43 hours is required to produce a given response at this intensity. The effective photoreceptor pigment concentration required is very small i.e., about 10^{-9}M. In blue region, two peaks can be observed; one at 440 nm and other at 480 nm. These may be correlated with the two pigments carotenes and riboflavin which absorb blue light.

Location of Photoreceptor

Photoreceptor pigment may be located either at the tip of coleoptile as in grasses or in hypocotyl as in the seedlings of dicots. In coleoptiles of grasses, the photosensitive zone is the first 1-5 mm from the tip. The terminal 0.2 mm of the colecoptile is about 6000 times as senstitive to light as an area only 3 mm below the tip. *S. Lam* and *A.C. Leopold* (1966) demonstrated that photoreceptor pigment in sunflower seedling is located in hypocotyl. The cotyledons of seedlings when covered with black paper, still showed positive phototropism although intensity of response was reduced. However, the covering of hypocotyl with black paper prevented phototropic response, even if the cotyledons were exposed to light.

The phototropic reaction can determine the growth habit of plants (Langham, 1941). Bermuda and some other grasses, which grow prostrate under usual field conditions of high-light intensity, become upright in light of low intensity. It is apparent that such species are positively phototropic at low-light intensities and negatively phototropic at high-light intensities.

Role of Auxins in Phototropism

The phototropic response is due to the unequal growth rate of two sides of an organ caused by asymmetrical distribution of auxins. This results into accumulation of more auxin on the darkened side than on the illuminated side. Hence, the darkened side grows more rapidly and curves towards light. According to *Galston* and *Hand* (1949), there may be following reasons for the unequal distribution of auxins:

1. The rate of auxin production may be reduced on the illuminated side.
2. The auxin may be destroyed by the action of light.
3. Transverse or lateral transport of auxins from illuminated to the darkened side.
4. Decreased translocation of auxins out of the tip.
5. Decreased sensitivity to the auxin.

Out of the above possibilities, second and third are considered to be most important because they have been supported by some evidences. The experiment performed by *Went* (1928) may be considered in this regard. He excised the tip of a coleoptile that had been exposed unilaterally to light and placed it upon two agar blocks, separated by a razor blade, in such a way that auxin from the darkened and illuminated sides diffused into different blocks. It was noticed that more auxin diffused out of the darkened half of the coleoptile tip than out of the illuminated half. Thus, one-sided exposure to light would seem to force the transverse transport of auxin.

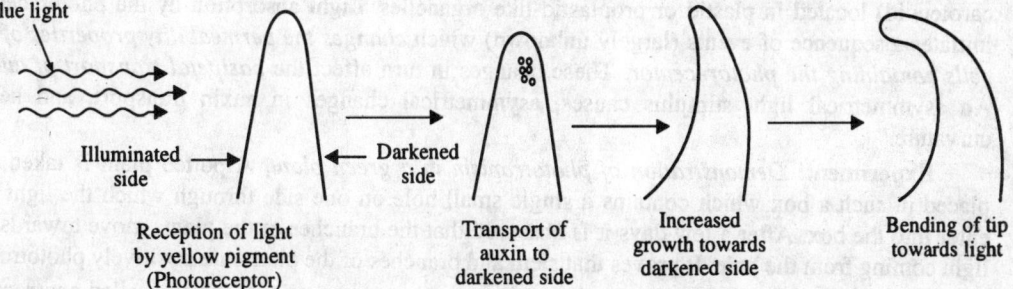

Fig. 22.7: Diagrammatic representation of positive phototropism in coleoptile.

Similar type of auxin transport has also been reported in maize coleoptile and pea seedlings. In maize coleoptile, when the labelled (^{14}C) IAA was exogenously supplied, it moved away from the illuminated side to the darkened side. In pea seedlings, when blue light was illuminated from one side, it caused the transport of 3H-IAA auxin from illuminated side to the darkened side. The accumulated IAA on darkened side caused increased growth on that side resulting into bending of coleoptile or seedlings.

First Positive and the Second Positive Response

The intensity of light has a peculiar effect on phototropic curvature of coleptile. When the blue light of very low intensity (10^{-11} einstein per cm^2) is given to a coleoptile tip, it shows positive phototropism and the phototropic curvature at this intensity of blue light is called *first positive response*. If the intensity of light is increased above 10^{-11} e per cm^2, it reduces curvature and at 10^{-9} e per cm^2 light intensity the curvature is reversed showing negative phototropism. If the intensity of light is incresed further, again a positive phototropic curvature is initiated. It is called *second positive response*. The natural sun light is very intense. So, the positive phototropic responses in nature must be due to second positive response. Both types of positive responses (first and second) are governed by the same photoreceptor pigment.

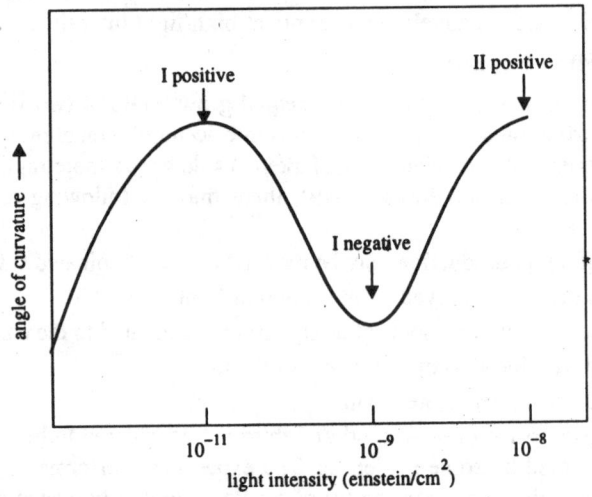

Fig. 22.8: Phototropic response with increasing energies of light.

Curry Hypothesis for Phototropism

A working hypothesis for the mechanism of phototropism, adopted by Curry (1969), is as follows : " there is a *single photoreceptor for phototropism, specifically, a yellow pigment* (probably carotenoid) located in plastid or proplastid-like organelles. Light absorption by the photoreceptor initiates a sequence of events (largely unknown) which *changes the permeability properties of the cells containing the photoreceptor*. These changes in turn affect the *basipetal transport of auxin*. An asymmetrical light stimulus causes, asymmetrical changes in auxin transport, and hence curvature."

Experiment: *Demonstration of phototropism in a green plant,* A potted plant is taken and placed in such a box which contains a single small hole on one side through which the light can enter into the box. After a few days it is observed that the branches of the plants move towards the light coming from the hole. It proves that stem and branches of the plants are positively phototropic.

(ii) **Geotropism :** The tropic movement in response to gravitational force is called *geotropism*. Mostly the stems grow away from the soil i.e., against the gravitational force and are called *negative geotropic*. In contrast, the roots and rhizoids grow towards the gravitational force and are called

positive geotropic. The first demonstration of geotropism was given by *Dodart* (1703). The term geotropism was first coined by *Frank* (1868).

Gravitropism (L. gravitas = heavy) and *barytropism* (Gk. barytes = weight) terms were also used for geotropism. It may be of the following types :

(i) *Positively Orthogeotropic* : When the organ grows in the centre of the earth, *e.g.*, roots and rhizomes.

(ii) *Negatively Orthogeotropic* : When the organ grows vertically away from the earth, *e.g.*, stems and pneumatophores etc.

(iii) *Diageotropic* : When the axis of the organ tends to lie at right angles to the direction of gravitational force, *e.g.*, rhizomes and runners.

(iv) *Plagiogeotropic* : When the axis of the organ makes an angle other than a right angle with the vertical, *e.g.*, side roots of the first order.

Effect of Various Conditions on Geotropism

The geotropic behaviour of an organ can be changed by a range of conditions. The peduncle of *Papaver* bud is positively geotropic but gradually becomes negatively geotropic if the flower opens. The stamens of *Hosta caerulea* are positively geotropic in the dark but become negative by

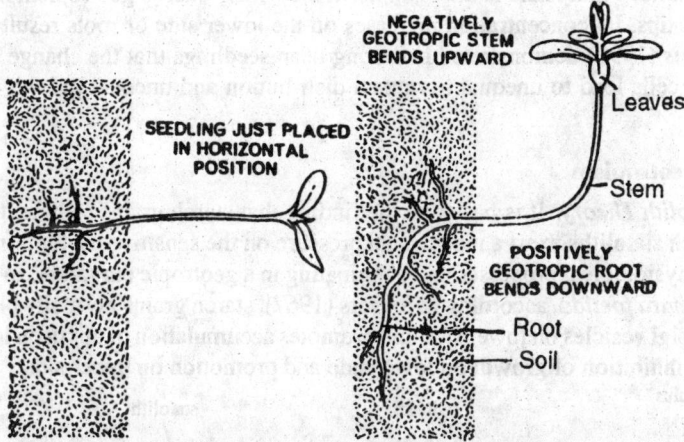

Fig. 22.9: Positive geotropism in roots and negative geotropism in stem. (A) Experiment in the beginning, (B) After a few days of experiment.

geotropic as they are illuminated. The stolons of *Cynodon dactylon, Paspalum vaginatum* and *Stenotaphrum secundatum* are diageotropic in the light but become negatively geotropic when the parent plant is darkened. The rhizomes of *Aegopodium podagraria* are diageotropic in darkness and positively plagiogeotropic after given a short exposure of red light. If the stem tip or primary root is excised, the lateral (leaf) nearest to the wound becomes orthogeotropic.

Steps in Geotropic Response

The geotropic response consists of four steps, namely, *perception* (physical action of the gravity stimulus), *transformation of information* (induction of a metabolic change in the sensitive region), *transmission of information* (transport of auxins from root apex to extending zone) and *reaction* (differential growth on two sides of the organ).

Role of Auxins in Geotropic Responses

Like phototropism, geotropism is due to the asymmertic distribution of auxin in the plant. The total amount of auxin given off by *Avena* coleoptile tips remain unchanged following geotropic stimulation induced by placing the organ in a horizontal position (*Dolk*, 1936) *Navez* and *Robinson*,

(1932) have shown experimentally that more than half of the auxin diffuses out to the lower half of the tip. The shoot turn upwards because its growth is accelerated and the root curves downwards because its growth is inhibited by auxin (*Cholodny*, 1922).

The asymmetric distribution of auxin in horizontal stems may have other influences. In certain varieties of pineapple plants, stems maintained in a vertical position remain vegetative and those in a horizontal position develop flowers (*van Overbeek* and *Cruzado*, 1948). Flowering is initiated due to auxin accumulation on the lower side of the horizontal stem tip.

A role of auxins also exists in the geotropism of stems. The usual geotropic curvatures of sugarcane stem is not evident when immersed in water at 52°C for about 20 min (Brandes and *van Overbeek*, 1948). This hot-water treatment decreases concentration of free auxin in the stem by about 50 per cent. When the lower half of a horizontal sugarcane stem is soaked in a very dilute solution of IAA after the hot-water treatment, geotropic bending takes places. The geotropic reaction occurs at the nodes of sugarcane stems, arising from the formation of new cells on the lower side of the stem as well as from an increase in the size of cells (*Brandes* and *McGuire*, 1951).

Role of Other Hormones in Geotropism

I.D.J. Phillip (1972) has shown the role of gibberellins in the geotropism of stems. Other physiologists believe that ABA is the main hormone which causes geotropism in roots. It is usually present in root tips. Its concentration increases on the lower side of roots resulting into bending of tips. *M.L. Evans* (1990) demonstrated in moong bean seedlings that the change in electric potential of the cortical cells lead to unequal hormonal distribution and unequal growth causing bending of roots.

Theories of Geotropism

1. *Statolith Theory:* It is based on the finding that starch grains sediment under gravity. The starch statoliths exert a mechanical pressure on the sensitive protoplasm, initiating a chain of physiological processes and culminating in a geotropic curvature. In the case of rhizoids of *Chara foetida*, according to Sievers (1967), starch grains normally block the distribution of golgi vesicles on lower side and promotes accumulation of auxin on lower sides resulting into inhibition of growth on lower side and promotion on upper side.

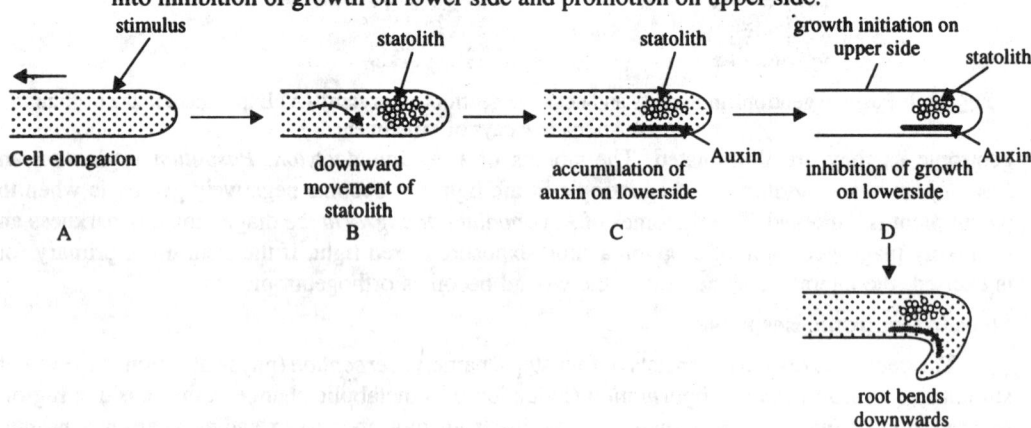

Fig. 22.10: Positive geotropism inroots due to statolith.

2. *Microsome Theory* : It was proposed by *Cholodny* (1922). According to this theory, the microsomes are electrically charged which induce a associated migration of active ions like K^+ and Ca^{2+} leading to differential permeability in plasmamembrane on two sides of a cell and symmetric growth of these two sides.

3. *Micropendulum Theory* : It was proposed by *Larsen* (1957, 1962) according to which the statolith, a gravity sensor, is a asymmetric particle which does not remain free but it remains hinged at one end and shows pendulum like movement in vertical position when the cell is turned.

4. *Geoelectrical Theory* : According to this theory, the gravitational force is changed to an electric charge which ultimately leads to the response. According to *Wilkins* (1966), the development of geoelectric effect is directly linked to lateral differences in auxin concentrators.

When seedlings are placed in a dark box in different directions, the roots always move towards gravitational force and the stems against the gravitational force.

A clinostat is used to abolish the effect of gravitational force in plants. Its structure is shown in Fig. 22.11. No part of the plant could remain constant for longer duration in one direction till the clinostat turns. As a result plant moves horizontally and there is no effect of gravitational force

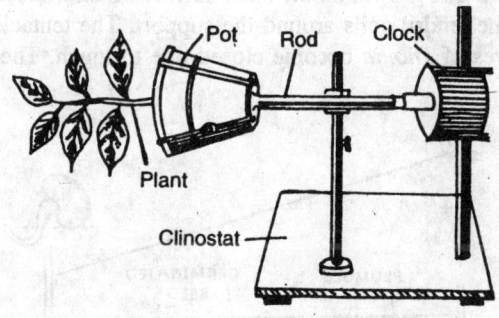

Fig. 22.11: Clinostat which can change the effect of geotropism.

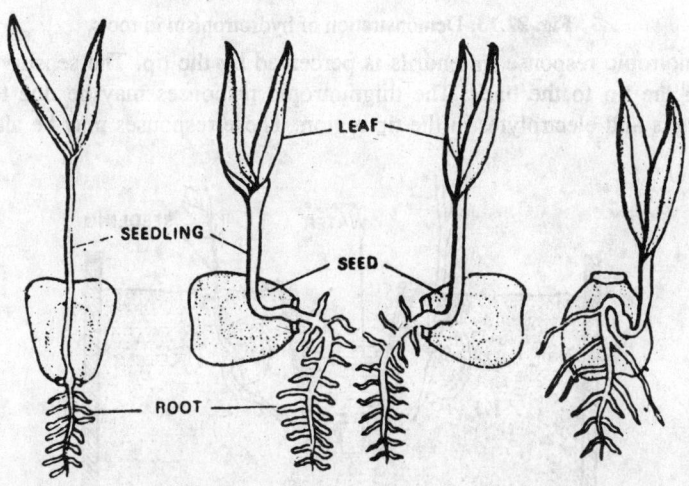

Fig. 22.12: Demonstration of geotropism in the maize seedlings. The experiment shows no effect of seed direction on geotropism.

on the plant. If the rotation of clinostat is stopped, the effect of gravitational force becomes lateral. The stem shows negative geotropism and the roots positive geotropism.

(*iii*) **Hydrotropism:** The tropic movement in response to water or moisture stimulus is called *hydrotropism*. It was first demonstrated by *Johnson* (1829).

Primary and secondary roots of all the plants and rhizoids of bryophytes move towards water and show *positive hydrotropism*. In contrast, stems and branches move against the water and show negative hydrotropism.

For the demonstration of hydrotropism, take a rectangular tub and after filling it with water slightly tilt it. Now place the perforated plate on it and cover the plate with wet wooden powder. Place some seeds on the powder for germination. When seeds germinate, their radicles move towards gravitational force through pores and then turn towards water. It proves that roots show positive hydrotropism and hydrotropism is stronger than geotropism.

(*iv*) **Thigmotropism** : The tropic movement in response to touch stimulus is called *thigmotropism*. The stem tendrils of the members of Cucurbitaceae when come in contact with any solid, they grow fast and encircle around the support and help the plant to climb up.

The part which is not in touch with the support grows fast and the part which touches the support shows less growth. The growth on the outer surface of tendril is always more in comparison to inner surface. Thus, the tendril coils around the support. The tentacles of *Drosera* bent due to touch stimulus. The leaves of *Dionia* become closed due to touch. These are all the examples of thigmotropism.

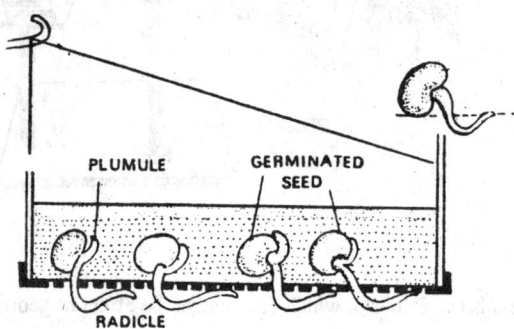

Fig. 22.13: Demonstration of hydrotropism in roots.

The thigmotropic response in tendrils is perceived by the tip. The sensitivity of the tendrils decreases from the tip to the back. The thigmotropic responses may be due to involvement of movement of salts and electrolytes in the tip region. These responses may be also due to changes

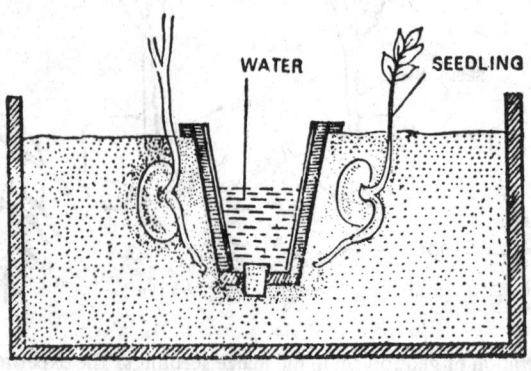

Fig. 22.14: Demonstration of Thigmotropism.

Plant Movements

in ATP and inorganic phosphate upon touch stimulus as in pea tendrils or may be also due to changes in membrane permeability or movement of metabolites or excessive water uptake.

(v) **Chemotropism** : The tropic movement due to stimulus of chemical substances is called *chemotropism*. The growth of pollen tube on the stigma into style is always due to chemotropism. It was investigated by *Pfeffer* (1884).

(vi) **Thermotropism** : The tropic movement in response to temperature change on one side of a plant organ is called *thermotropism*. It has been reported in *Anemone stellata, A. nemorosa* and *Tulipa sylvestris* where peduncles curve towards sun throughout the day. Thermotropism may be regarded as a special type of phototropism. In thermotropism, the stimulus is due to longer wavelengths of infra-red radiation whereas in phototropism the stimulus is due to shorter waves of visible light.

The bending of shoots is due to *negative thermotropism* whereas bending of roots is considered to be a *positive thermotropism*. If the temperature is below the optimum, plant will curve in the direction of the more warmer side and if above the optimum, it will curve towards the colder side.

(vii) **Aerotropism** : The tropic movement due to changes in concentration of a gas particularly O_2 is called *aerotropism*.

(viii) **Traumatotropism** : The tropic movement as a result of wounding of apex is called *traumatotropism*.

(ix) **Rheotropism** : The tropic movement in response to water-current is called *rheotropism* e.g., bending of root tip. Generally, roots of seedings are more sensitive to stimulus of running water than those of more adult plants. The roots of Cruciferae and Gramineae are most sensitive whereas the roots of certain species are insensitive. It has been noticed that the hyphae of *Mucor* and *Phycomyces* are *negatively rheotropic* while those of *Botrytis cinerea* are *positively rheotropic*. Following views have been but forwarded to explain rheotropism :

1. It is a reaction to the stimulus of water pressure.
2. It is a type of hydrotropism.
3. It is a special case of chemotropism.

(x) **Galvanotropism** : The tropic movement in response in electric current is called *galvanotropism*.

(xi) **Osmotropism** : The tropic movement due to osmoregulation in the system is called *osmotropism*.

(2) Nastic Movements

The movements occurring in response to diffused stimulus are called *nastic movements*. In such movements the stimulus acts on protoplasm from all sides and the response has no relation to the direction of stimulus. The stimulus in such movements may be light, temperature, touch or water etc. Nastic movements are of following types:

Seismonastic movements: The nastic movement in response to touch is called *seismonastic movement* e.g., leaflets of *Mimosa pudica*. On touching, the leaflets of *Mimosa* become closed and the leaves hang down. The upper side of the leaflets shrinks causing the upward folding while in petiole the shrinking occurs towards lowerside causing the drooping of whole leaf. The response in this case is so fast that the leaflets start closing only after 0.1 second of stimulation and complete closing of leaves takes place within a few seconds. When one leaflet is stimulated once, the same stimulus moves through the plant upward and downward with a speed of about 40-50 cm per second. Thus, all the leaflets of a plant become closed and petiole hang down within a few seconds.

The seismonastic response is completed in two steps : (i) transfer of stimulus (ii) the reaction in the base of leaflet or leaf. According of *J.C. Bose* the transfer of stimulus to the base of leaf is through a nervous system while *A.L. Houwink* was of the opinion that it was in the form of electrical impulse. Present physiologists believe that transfer of stimulus is due to transfer of action potential in plant tissues. The action potential moves through the parenchyma cells of xylem and phloem with a velocity of about 2 cm per second. *Barbara G. Pickard* and her colleagues noticed that the long distance spread of action potential depends on its movement through an unidentified chemical, called *Ricca's factor*, after its discoverer, *Ubaldo Ricca*.

Thus, the stimulus reaches the base of the leaf or leaflet where it causes the transport of water from the parenchyma cells to the intercellular spaces. It was also explained due to turgidity and flaccidity of parenchyma cells present in the pulvinus.

The upper half portion of the pulvinus is made up of thick membraned cells which are without or with only a few intercellular spaces. The lower half portion of pulvinus is made up of thin walled cells containing large intercellular spaces. Before touching the leaves, the cells of the pulvinus remain in turgid state but as soon as we touch or provide the stimulus, the turgid cells of the pulvinus lose their water into intercellular spaces and become flaccid. Now, the cells of the upper side of the pulvinus exert a pressure on the cells of lower side of pulvinus. Thus, the leaflets become hanged. After sometime the cells again become turgid and the leaflets regain their original position.

The transport of water from parenchyma cells to the intercellular spaces was also explained in the following ways :

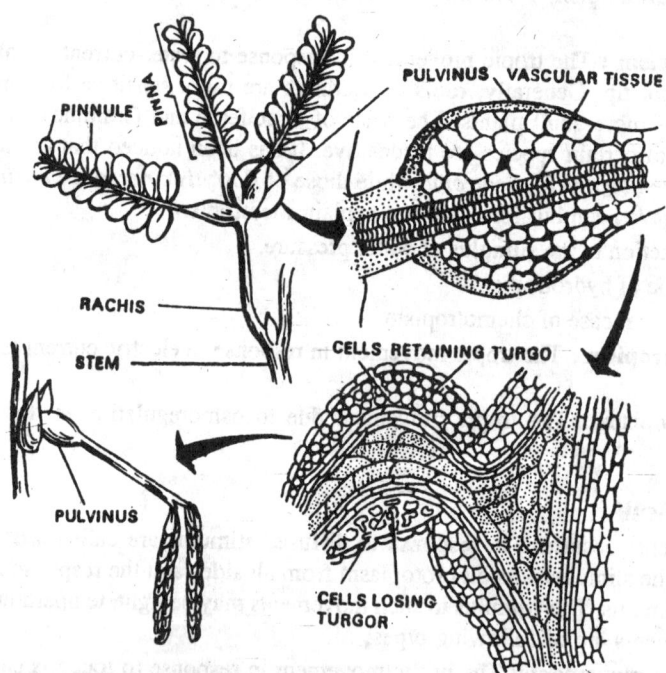

Fig. 22.15: Demonstration of seismonastic movement in Mimosa pudica.

1. The stimulus causes change in the permeability of cells rapidly leading to transport of water.
2. The pulvinus contains large vacuoles filled with water which due to stimulus disappear and lose water.
3. The pulvinus contains certain colloidal proteins which undergo hydration and dehydration due to stimulus, causing change in cell-volume. The energy for such change in cell-volume due to action of water in colloidal proteins and water transport is supplied through ATP.

Nyctinastic or Sleeping movements

The nastic movements in response to light and temperature are called *nyctinastic* or *sleeping movements*.

Plant Movements

The nyctinastic movements are caused by relative changes in cell sizes on opposite sides of the base of the leaflets in a shoot zone called *pulvinus*. The nyctinastic movements depend upon

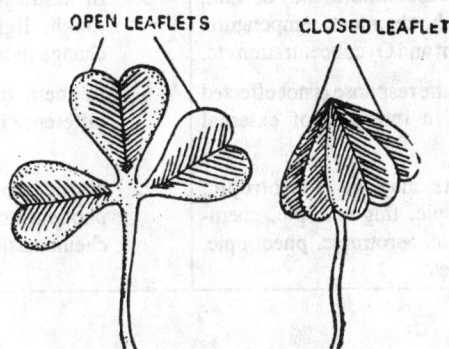

Fig. 22.16: Photonastic movement in the leaves of Oxalis. (A) Leaflet during day time (B) Leaflets during night.

the amount of auxin, transfer of K^+ ions and movement of water. It is believed that during early morning, large amount of auxins are produced, which are transferred mainly to the lowerside of the petiole. The accumulation of auxins causes preferential accumulation of K^+ in this area which ultimately results into translocation of water in the cells. Thus, the leaves become erect. During night, the reverse reaction takes place i.e., the accumulation of auxins and K^+ is reduced very much resulting into loss of water. Thus, the leaves hang down.

Nyctinastic movements may be of following types:

(a) Photonastic movements : The nastic movement in response to light is called *photonastic movement* or *photonasty*. The opening of leaves and flowers in most of the plants during day time and closing during night is an example of photonasty. The leaves of *Oxalis* show such type of movement which is called *sleeping movement*.

(b) Thermonastic movements : The nastic movement in response to temperature is called *thermonastic movement*. The opening of flowers at high temperature and closing at low temperature is an example of thermonasty *e.g.,* in *Crocus* plant.

Haptonastic movement : The nastic movement in response to touch of insects is called *haptonastic*. In *Drosera* and other insectivorous plants the infolding of glandular hairs due to touch of insects is an example of haptonasty.

Hydronasty : The nastic movements due to changes in relative humidity of atmosphere is called *hydronasty*.

Chemonasty : The nastic movement due to chemical stimulus is called *chemonasty*.

Table 22.2: Differences between Tropic and Nastic movements

Tropic Movement	*Nastic Movement*
1. These movements occur in response to an unidirectional stimulus.	1. These movements occur in response to diffused stimulus.
2. In tropic movements the stimulus acts on protoplasm from one direction only and the response is directly related to the direction of stimulus.	2. In nastic movements the stimulus acts on protoplasm from all sides and response has no relation to the direction of stimulus but with the organ.
3. They are movements of curvature caused by unilateral growth i.e., one side of an organ grows faster than other causing curvature.	3. These are also movements of curvature but they can also be caused by reversible turgor changes.

4. They are paratonic movements.	4. They may be autonomic growth movements or paratonic variation movements.
5. In tropic movements the stimulus may be light, gravity, water, touch, chemical, temperature, wound, water current and O_2 concentration etc.	5. In nastic movements the stimulus may be touch, light, temperature, insect touch, change in humidity or chemical etc.
6. In such movements the response is not effected by the difference in intensity of external stimulating factors.	6. In them the response is effected by the difference in the intensity of external factors.
7. Tropic movements may be phototropic, geotropic, hydrotropic, thigmotropic, chemotropic, thermotropic, aerotropic, pheotropic, and traumatotropic etc.	7. Nastic movements may be seismonastic, photonastic, thermonastic, hydronastic and chemonastic etc.

23

Physiological Responses in Plants Growing Under Stress Conditions or Stress Physiology

INTRODUCTION

The physiological processes described in Plant Physiology portion are operated in plants growing under normal or ideal environmental conditions. In nature plants have to face various competitions and adverse climatic and environmental variations. The variations in temperature, water and salt etc. create stress in plant systems. The plants react with such stress conditions and adjust themselves through a number of biochemical and physiological changes. Such adjustments are necessary to overcome such stress situations. Presently the scientists are interested to know that how and what changes in chemical composition and physiology of a plant cell take place during any type of stress condition. The common stress conditions of Northern India are long periods of drought, winter cold, early frost and flood etc. The physiologists in collaboration with plant breeders are working to evolve new resistant varieties to face such stresses.

In recent years the developing countries to make them modernize are setting up various industries. These industries are releasing a number of air pollutants and toxic effluents which are causing another type of stress conditions to the plants. The ozone, CFCs, unsaturated hydrocarbons, metals like lead (Pb), Cadmium, Mercury, Polonium etc, carbon mono-oxide, sulphur oxides released in the atmosphere and used of insecticides, pesticides, herbicides etc. All are affecting the growth of plant parts due to stress. The plants show dwarfism chlorosis, non-flowering and low seed setting etc. The only solution for such stresses is understand the nature of pollution and pollutants and their preventive measures.

What is stress and Stress Physiology?

Although it is very difficult to define stress but *Jacob Levitt* (1972) defined biological stress as "Any change in environmental conditions that might reduce or adversely change a plant growth or development." Stress physiology is that branch of environmental physiology which is concerned with how organisms respond to environmental conditions that deviate significantly from those that are optimal for the organism.

Types of stress

Various types of physiochemical stresses which have to face the plants are given in Table 23.1

Of the various stresses classified in table drought, heat, cold, salinity and frost are most common. Recently excess pollutants and effluents have become more common stresses because of their toxic effects in the environment. Some of the common stresses will be discussed here.

Stress Resistance

In general *stress resistance* is of two types - *avoidance* and *tolerance*.

Avoidance :- When an internal environment is created within the plant to overcome or reduce

Table 23.1

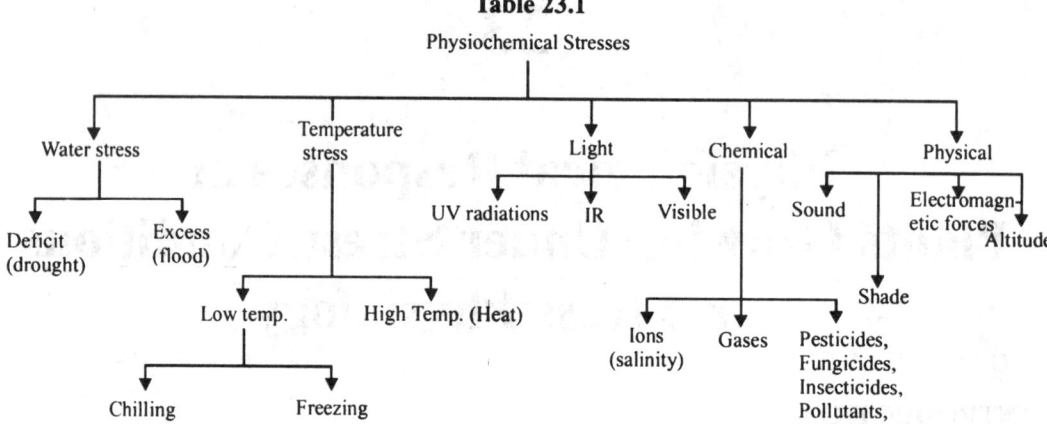

the stress of cells, it is called *avoidance*. For example, a leaf maintains coolness inside, during high atmospheric temperature due to internal process of transpiration. Similarly, succulents conserve water to avoid drought stress.

Tolerance : - When a plant possesses the capacity to withstand the stress, it is called *tolerance*. For example, the mosses and the *resurrection* plant (*selaginella*) can tolerate dessication conditions.

Effects of Stress

If a plant is subjected to any stress condition, it shows reaction which may be *elastic* or *plastic*. When the reaction is temporary and the plant reverts back to its original normal state, this effects is called *elastic*. When the reaction is permanent, (i.e.., the plant is deformed) and the plant does not return back to its original state, this effect is called *plastic*. Thus , plastic effects are always irreversible.

Sometimes the plants exposed to stress conditions become resistant. This state or effect is called *hardening*. In certain cases the effect produced is carried over in subsequent generations. For example - Wheat, pea and bean plants subjected to low temperature become dwarf and this effect is carried over in several generations. The reaction of plants to the stress conditions is a complex process which results into numerous physiological responses including accumulation of proteins, carbohydrates and other stress tolerant organic compounds etc. They maintain osmotic potential without limiting enzyme function.

DROUGHT STRESS

Drought stress is one of the commonest stresses experienced by the plants. *Parker* (1956) defined drought as deficiency of available soil moisture which results in water deficits in plants severe enough to reduce plant growth The injurious effects of drought increase due to factors favouring high rate of transpiration and increasing water deficits such as low humidity, high temperature and wind velocity etc. In general, soil moisture deficit is the most common cause of internal water deficit.

The nature and causes of drought resistance has been controversial (*Levitt,* 1951, *Parker,* 1956 *Stocker* 1956). The present view supports that numerous factors are involved in drought stress including those which post pond dehydration, such as efficiency of absorbing surface and water conducting system, leaf area, leaf structure, stomatal behaviour, osmotic pressure, cell size and protoplasmic, characteristics.

Kinds of Drought stress enduring Plants

Following types of drough stress enduring plants have been observed.

1. Plants which can not endure drought at all and are injured or die as soon as soil moisture

becomes deficient. Few or no woody plants occur in this category.

2. Plants such as succulents which store large amount of water and have very low water loss because of their small surface to volume ratio, thick cuticle and few stomata. The protoplasm of these plants usually, is not resistant to dehydration. Only a few trees belong to this category.

3. The plants which are truely drought reducing. Their protoplasm can be dehydrated without permanent injury, e.g., Ferns, mosses and lichens.

4. Plants with protoplasm having some capacity to endure dehydration combine morphological and anatomical characteristics which reduce rate of water loss and post pone critical internal water deficits. Most of the drought resistant trees fall under this category (*Oppenheimer, 1951*).

Morphological and Anatomical Adaptations.

The plants have developed several morphological and anatomical adaptations which include shiny surface of leaves leathery texture of leaves, deposition of cutin or wax on leaf surfaces, development of hairs, reduction of leaves into scales or needles, development of thick cuticle, sunken stomata, formation of seeds with low water contents, multilayered palisade parenchyma, excessive sclerenchyma, infolding of mesophyll cells, life cycle of short duration, development of deep root system in seedlings etc.

Water molecules perform several functions including maintenance of complex fluids in stable configuration. Dehydration leads to loss of water molecules and disruption of proteins. Water loss causes concentration of solutes leading to high concentration of cell sap and intercellular fluids cause great decrease in the water potential of the fluids. This causes stress on the protoplasm which badly affect the biochemical processes. This effect may be due to water imbalance and change in pH of the cell-sap.

Biochemical Adaptations

Plants develop certain biochemical adaptations to face drought stress. The most important one is the production of hydrophilic substances like high MW proteins, alginic acid, polyhydric alcohol's resistant proteins and other sugars in the protoplasm. Such substances help in retaining and conserving water, reduce water potential of cell sap and protect protoplasm from dessication. The polyhydric alcohols are low MW compounds of hydrophilic nature. The increase is sugars during drought in protoplasm directly lowers the water potential of cell sap which helps in retaining water. It is observed that sugarcane although contains high amounts of sugars but are drought susceptible whereas pineapple contains less amounts of sugars but are drought resistant. This indicates that the water binding capacity in these cases is not related with the sugars but depends upon specific proteins. It is also suggested that certain resistant proteins appear in the cell during drought stress. These proteins are resistant to denaturation and also check the denaturation of other proteins. In general drought resistant plants have smaller cells, less starch and high amount of sugars and nucleic acid contents. Some physiologists are of the opinion that drought resistance is associated with the elastic nature of the protoplasm.

Effects of Drought on certain Physiological Processes

1. On Functioning of Stomata- According to *Ilfin* (1922), a plant exposed to severe drought can not re-establish its normal functions and remains abnormal even though it regains its turgescence. The stomata open slightly or partly and loose their function, or may be killed. It has been observed in *Centurea orientalis* That 8% retained ability to open, 73% were closed and 19% were killed. In other plants upto 45% stomata remained inactive although the leaves appeared to be perfectly normal. It is a known fact that the loss of water by plants stimulates the transformation of sugars to starch and lowers the osmotic pressure (O.P) but when the wilting passes certain limits, the starch in guard cells decomposes. Complete decomposition has been observed in plant species adapted to moist habitats. When the water loss ranges from 15 to 30%, the drought resistant species show decomposition from 50 to 60%.

2. On Carbohydrate Metabolism - Many investigators have observed that starch disappears from wilted leaves (*Neger*, 1915; *Molisch*, 1921; *Montfort*, 1937) following decrease in soil moisture (Woodhams, 1954). According to *Iljin* (1930), *Magness* et. al (1933), *Julander* (1945) and *Eaton* and *Ergle* (1948) the sugars accumulate simultaneously after water is added by irrigation. If the water loss is not too severe, starch formation will become re-established (*Wadleigh et, al,* 1943). Drying stimulates the activity of enzyme *amylase* (*Spocher*, 1939) and *Phosphatase* (*Sisakiran*, 1940) not only in the guard cells but throughout the leaf. The loss of water decomposition inhibits carbon assimilation. The accumulation of carbohydrates in plants depends upon several factors. An increased supply of nitrogen stimulates the utilization of carbohydrates and if sufficient moisture is available, growth and formation of new organs are accelarated. If sufficient moisture is not available, growth is interrupted and accumulation of polysaccharides takes place. The break down of carbohydrates in the leaves may be accompanied by their deposition in the roots (*Granfield*, 1943; *Eaton and Ergle*, 1948). It is not possible to state that water loss includes the breakdown and disappearance of starch in all parts of the plant but it can be said to take place in the leaves of majority of species. When water content was reduced under controlled conditions in *Rumex,* an increase in sugar content was noticed. It was observed that 20% of the water loss increased sugar concentration upto 49 to 76% after several hours.

3. On Photosynthetic Activity – Drought affects photosynthesis firstly, by decreasing the size of stomatal opening which limits CO_2 absorption and secondly by limiting or reducing photosynthetic activity of green tissues. A comparison of photosynthetic rate of fresh plants with wilted ones having closed stomata showed marked differences. The loss of water from 16 to 47% or more caused a decrease of 20% in photosynthetic rate but no correlation should be made between amount of water lost and the rate of photosynthesis. The wilted plant do not carry photosynthesis normally but their capacity to do so is reduced by 35 to 59%. According to *Stalfelt* (1939) photosynthesis is also reduced in plants without stomata such as in mosses and lichens. Stomatal movement is very important factor in photosynthesis. There is a certain relationship between size of aperture, photosynthesis and transpiration. This relationship changes in different habitats. According to *Iljin* (1923) in normal habitats with humid air, mesophytes transpire less than xerophytes. The mesophytes loose 7 to 10 grams of water in assimilating 1 ml CO_2 and xerophytes 19 to 84 grams. The value for mesophytes growing in dry sites are much higher than those of xerophytes.

Reduction in photosynthesis rate in dry soil has been observed in orchard and forest trees. Photosynthesis is depressed in many plants before the onset of wilting. CO_2 enters the leaves not only through the stomata but also through the epidermis when the stomata are closed. A careful observation of plant species having stomata either on one or both the surfaces shows that assimilation is more active through surfaces with open stomata than those without stomata. Photosynthesis is checked completely by a well developed cuticle. The leaves with delicate cuticle are able to assimilate equally well whether the stomata are open or closed.

4. On Respiration - The effect of drought and wilting has been noticed in several plants. In wilting condition due to drought the hydrophytes and mesophytes show increased respiration. The investigations based on seeds are seedling show that a decrease in respiration accompanies a reduction in water content of tissues. In case of leaves and stems the results are different.

The loss of large amount of water by a plant induces the breakdown of polysaccharides into simpler compounds other than sugars. Starch may disappear with accumulation of sugars. The probable products are on the way towards becoming the final product of respiration, i.e.. H_2O and CO_2.

5. On Osmotic Pressure - The first observations on the relation between drought and osmotic pressure (O.P.) were those of *Pringshein* (1906) in pumpkin seedlings. O.P. in plants resistant to drought increases during growth of seedlings at low humidities. *Fitting* (1911) made many measurements of O.P. in plants of Sahara desert. The species which were not protected against transpiration and high osmotic values utilized soil moisture. The theoretical conclusions made by

Fitting although have been criticized by many physiologists but his observations have been confirmed by *Iljin* (1924), *Bartel* (1947) and *Mayer* (1953). The species of temperate zone have an average O.P. of 10 atm and those of Arizona desert 20 atm. Large cells have low osmotic values while small cells have higher osmotic values.

6. On Resistance of Cells to dessication - Presence of large sized cells with large vacuole, expansion of protoplast, delicate nature of cytoplasmic membrane are the chief causes of death of cells subjected to dessication. In presence of above conditions the destruction of cell takes place easily and its resistance is reduced. The plants growing in dry sites usually have smaller cells than plants growing in moist site. The cells of certain xerophytic mosses are much elongated with narrow lumen and their opposite walls are closed appressed. They are adapted to dessication. No change in the shape of such cells takes place when water is lost from them.

In dessicated cells the water is transferred to vacuole. The vacuole imbibes water and increases in its size or volume stretching membrane and enlarging protoplast.

The buds of higher plants are usually resistant to drought due to absence of vacuoles. As the buds develop vacuoles their resistance to drought is decreased.

7. On Growth - Soil moisture often becomes a limiting factor because of its effects on internal water balance which in turn affects the processes controlling growth. *Fritts* (1958) found a high degree of correlation between variations in soil moisture and diameter growth of beech trees during late summer when soil moisture is often low.

The studies of *stanhill* (1957) on 80 plant species dealing with the effects of soil moisture content on growth indicated that in 66 species the growth was affected before soil moisture was lowered to permanent wilting percentage. The growth stopped with the falling of soil moisture content below the permanent wilting percentage. In general, the diameter growth is affected more than height growth by deficit soil moisture. Diameter growth is more sensitive to annual rainfall than height growth (*Tryon et al.* 1957).

SALT STRESS

The water stress caused due to high concentration of salts or solutes in the external medium is called *salt stress*. The salts and solutes highly influence the water potential and in very high concentration cause toxicity. The water stress due to salt concentration causes osmotic imbalance, closure of stomata due to formation of ABA, ion imbalance and toxicity by accumulating ions etc. IT also affects nitrogen metabolism carbohydrate metabolism, growth and action of several enzymes.

When concentration of organic solutes becomes very high in the cytoplasm, it performs two important roles, (i) It maintains osmotic balance and (ii) It protects enzymes of metabolisms essential for life. At high salinity, the quantity of certain amino acids like *proline* and *glycine* and sugars like *sucrose* increases several fold. Proline helps in increasing the solubility of proteins and sucrose helps in protecting isolated chloroplasts against injury. Various plant responses, their adverse effects and adaptations by plants to endure salinity are given in table 23.2.

On the basis of sensitivity to salt concentration, the plants were classified into two main groups (i) halophytes and (ii) glycophytes. The plants growing in saline habitats are called halophytes. They adapt themselves to a high salt concentration in soil by the development of certain features and properties during ontogenic evolution. The plants growing in on-saline habitats are called *glycophytes*. They possess no or limited ability to adapt themselves to salt stress. According to data available the salt concentration in saline soils varies from 0.3 to 20% but most of the halophytes grow in soils containing 2 to 6% salt concentration.

Table 23.2

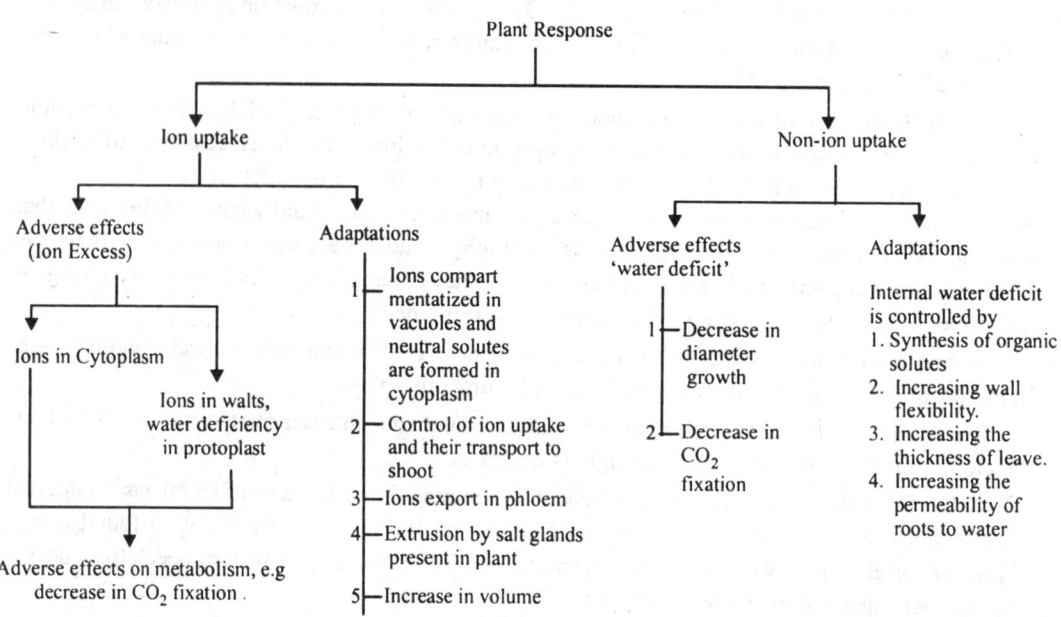

Effects of Salt Stress on Some physiological Processes

1. On Maintenance of Water Potential and Enzyme Activity –

The concentration of salts in external medium causes water stress and osmotic imbalance in protoplast. Excess ions penetrate in the cytoplasm and walls causing water deficiency in protoplast which produces adverse effects such as decrease in CO_2 fixation. The salts can also damage the enzymes that control metabolic processes necessary to continue life. An important adaptation seen in water stressed plants is the appearance and accumulation of certain organic solutes like amino acids (proline and glycine) and sucrose which lower the osmotic potential. Thus, water potential in cell is maintained without limiting enzyme function. Such organic solutes are called *compatible solutes*. They have been reported in the cells of many xerophytes during water stress. The resulting drop in osmotic potential due to newly appeared organic solutes is called *osmotic adjustment* or *osmoregulation*. IT results due to presence of excess dissolved salts in soil solution. The osmotic potential of soil solution becomes negative which causes diffusion of water from the tissues into soil solution. The water stress enduring plants can survive in higher concentration without damaging the metabolic enzymes.

In mesophytic plants, the water stress affects enzyme activities. During water stress certain enzymes like *phenylalanine ammonia lyase* (PAL), which is required for the synthesis of flavonoids and amino acids, and *nitrate reductase* which is required for reduction of nitrate to nitrite during nitrogen metabolism, show decreased activity while enzymes like α-*amylase* and *ribonuclease* show increased activities. It was proposed that such enzymes help in the break down of starch and other complex materials to make the osmotic potential negative of the cell. It is an adaptation to endure salt stress and is called *osmotic adjustment*.

2. On closing and Opening of Stomata –

It is observed that abscisic acid (ABA) in extremely low concentration when applied externally causes the closure of stomata. Further it is reported that during water stress the ABA appears in leaf tissues. In slow drying leaves the formation of ABA before the closure of stomata suggests that during water stress the stomatal closure is mediated by

ABA. In such leaves the loss of water from guard cells is not rapid enough cause closure of stomata. To explain this situation it is proposed that there are two feed back loops which control the opening and closing of stomata. The first feed back loop provides CO_2 for photosynthesis and the second loop protects the tissues against excessive water loss. The first loop causes the opening of stomata and entry of CO_2 while the second loop causes closing of stomata and checking of water loss. The second loop is mediated by ABA produced in adjacent mesophyll cells. First loop works when concentration of CO_2 decreases in the intercellular spaces. The decrease in CO_2 concentration allows the movement of potassium ions (K^+) into guard cell to open stomata for the entry of CO_2 from the atmosphere. This CO_2 is utilized in photosynthesis again lowering CO_2 concentration. The second loop works for the water stress condition. During water stress condition the water potentia is lowered which leads to increase or synthesis of ABA in the water of adjacent mesophyll cells. The ABA moves into the guard cells and simultaneously K^+ ions and water move out from guard cells to the adjacent mesophyll cell. Thus, stomata become closed to check water loss. The two loops interact together. The degree of stomatal response to ABA depends upon CO_2 concentration in the guard cells and response to CO_2 depends upon ABA. If the rate of drying is high, the water is lost from the guard cells directly.

3. On Nitrogen Metabolism -Salinity plays an important role in nitrogen metabolism. IT inhibits the enzyme *nitrate reductase* which is required for reducing nitrate to nitrite during nitrogen assimilation. The enzymes of ammonium assimilation pathway are also inhibited by NaCl concentration above 100 mM. Salinity also directs the transfer of carbon fragments from organic acids to amino acids. This act of transfer is achieved by inhibition of malate dehydrogenase activity, and stimulation of transamination reaction. As stated earlier, salt stress causes an increase in the production of amino acids like proline, hydroxyproline and glycine. Free proline which represents about 10-20% of shoot dry weight acts as compatible solute in balancing cytoplasmic and vacuolar water potential. In some species where accumulation of proline does not occur, this function is performed by accumulation of other amino compounds. They maintain water potential and control nitrogen metabolism.

4. On Mineralisation and Ion Imbalance - Salinity causes mineralisation due to ion imbalance especially by NaCl. Their excess accumulation may cause toxicity. In halophytes on average about 90% sodium is found in shoots and about 20% in leaves, In contrast, the glycophytes contain sodium ions chiefly in the leaves which are released in the medium. The glycophytes growing in salive soil show lower metabolic rates. They also show accumulation of non-nutrient salts in the protoplasm of leaf cells causing mineralisation which affects hydrogen ion(H^+) concentration and may cause toxicity. In acidic medium the effect of anions and in alkaline medium those of cations is noticed.

The penetration of salts into the plants is regulated by the permeability of the tissues. The root tissues are usually impermeable and their salt resistance character is maintained upto a certain limit, above which the salts are absorbed causing poisoning and sometimes death of the plant. At high concentration of salts the plant cells are damaged due to replacement of selective salt absorption by passive absorption.

5. On Accumulation of waste products of Disturbed Metabolism - Salt stress causes disturbances in the normal metabolism. The abnormal or disturbed metabolism produces several toxic substances which accumulate in the cells causing toxicity and drying of cells. In plants, the toxic products of nitrogen metabolism and ammonia, urea, thiourea and amines etc which can not be utilized for the synthesis of new proteins. These toxic substances get accumulated in the cells and cause adverse effects on plants. Some metabolic products act as precursors for the formation of more complex nitrogenous substances. *Gluatamine* and *asparagine* have been considered to show adaptive response to reduce the concentration of ammonia during its metabolism.

During salt stress the degradation of protoplasmic substances and activation of oxidative enzymes result into disturbed metabolism and drying of cells. When the concentration of salt becomes very high in the medium, the salts penetrate and accumulate in the cells resulting into toxic effects.

Salt poisoning is characterized by the presence of necrotic spots or yellowish white patches on leaves showing disintegration of chlorophyll. It is caused due to disturbed metabolism and autolysis.

TEMPERATURE STRESS

Majority of plants have to face temperature stress. It is of two types - (i) Stress due to high temperature and (ii) Stress due to low temperature.

Stress Due to High Temperature – Majority of the plants can endure high temperature stress due to their internal built up. Most of the xerophytes (e.g., *cactii*), thermal algae, lichens and other desert plants survive at a temperature of 70° or more. High temperature causes *heavy water losses and denaturation of enzymes and other proteins*. The denaturation is compensated by increased enzyme production. It is an adaptation to high temp. In some plants a rise in temperature causes reduction or slowing down of some metabolic processes which can be restored by addition of some compounds like *ascorbic acid* and *vitamins*. It is reported that addition of *adenine* increases temperature tolerance in plants. High temperature resistant plants usually possess heat stable enzymes and proteins. Such enzymes may be isoenzymes which have been developed at high temperatures. The heat stable proteins are not affected by high temperatures to denaturation and they have ability to place thermal denatured proteins immediately.

High temperature is responsible for heat injury which results into degradation of proteins, lipids, chemicals, respiratory metabolism and changes in kinetics of metabolism. These changes cause stress injuris like break down of cell products and membranes (membrane injury), chemical injury, toxin induced injury, starvation injury etc. The effects of heat injury are summarised in table 23.3.

Table 23.3.

Stress due to Low temperature – Low temperature stress may be caused due to freezing, frost or chilling. These result into injury.

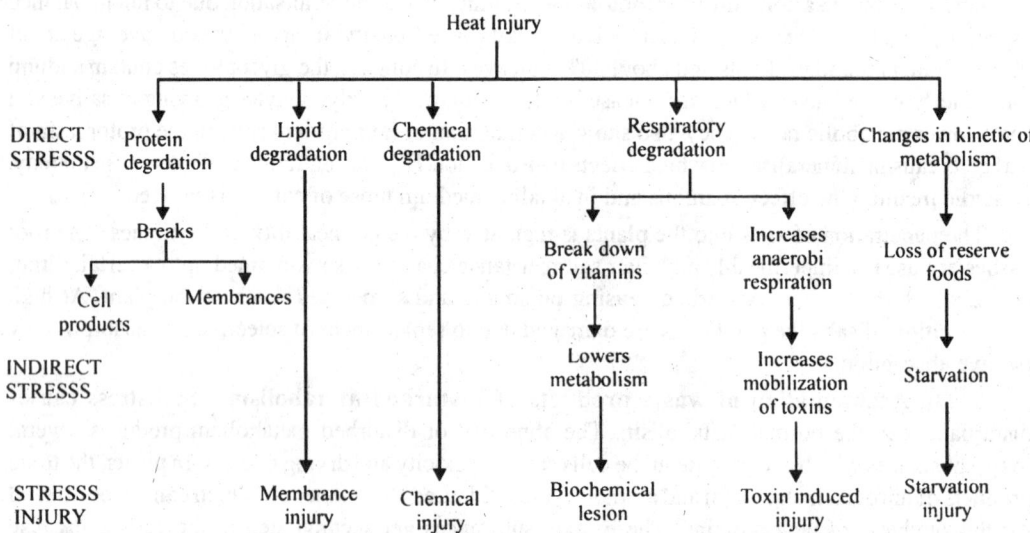

Injury due to Freezing – Most of the plants are capable to resist freezing. The *plants of tropical climate* are chilling-prone. They are sensitive to low temperature of 12-13°C and the temperature between 0–5° causes lethal effect. This indicates that the proteins present in such plants are sensitive to low temperature. The chilling also causes destruction of tissues and organs. On the contrary *alpine* and *arctic* plants are low temperature resistant and do not show damaging effects but their tissues may not undergo water formation in their cells. Certain plant parts like seeds, pollen and embryos can be stored at −190°C. This low temperature is obtained through liquid nitrogen.

Freezing causes damage in plants by two ways - (i) By mechanical effects especially disrupting membrance and cell organisation due to formation ice crystals by increasing water volume. (ii) By producing drought situation due to water loss by ice formation. It is observed that the intercellular water has quite higher potential than the water of cell cytoplasm and vacuole. In the begining the water of intercellular spaces get crystallized to form ice crystals and later on the freezing continues to form large sized crystals. This results into release of water from the cell cytoplasm and its accumulating into intercellular spaces. This creates a type of drought situation where solutes of protoplasts to become concentrated due to precipitation causing abrupt shifting or change in cell pH. If the temperature is further lowered upto - 35 to −40°C, total water of the tissues will be crystallized *Causes of Freezing Resistance.*

The plants of cold regions (alpine and arctic) are freezing resistant. This property is freezing resistant plants is contributed by several factors like increase in concentrations of electrolytes, formation of low temperature resistant proteins, dehydration of compounds, presence of high levels of sugar earlier and synthesis of more sugars later on, formation of ice crystals and allosteric enzymes. Some physiologists believe that shrinkage is caused due to association of protein molecules. The resistant proteins possess more hydrophilic bonds.

Effects of Freezing on some Physiological and Biochemical Processes -

Freezing temperature causes increase in the cell permeability of protoplast, root cells and cell-organelles like chloroplast and mitochondria. Permeability increase in protoplast results into loss of solute and direct injury; in foot cells causes decrease in water uptake resulting into wilting; in chloroplast results into disrupted photosynthesis and ; in mitochondria results into shifting of aerobic respiration into anaerobic respiration which causes ATP deficit and decrease in active ion uptake. During freezing the carbohydrates and proteins remain unaffected but the chloroplast and mitochondrial membranes bound enzymes become inactive. The inactivation affects photophosphorylation in chloroplast and oxidative phosphorylation in mitochondria. Freezing results into uncoupling of phosphorylation of electron transport system producing no or less ATP. It is reported that during freezing, water splitting site in chloroplast near PS II becomes active and PSI almost remains resistant to freezing.

Frost Hardening

When a plant becomes resistant to frost stress, it is known as *frost hardening* and such a plant is called *frost-hardy plant.* Frost hardening is similar to that of deep freezing resistance. Frost stress causes denauration of proteins, increase in sugar concentration, reduction in growth and formation of hydrophobic substances.

Frost injury affects the PS II resulting into reduction in photosynthesis but the carbohydrates and soluble proteins remain unaffected. In frost injured cells membrane bound enzyme systems are damaged and low molecular weight proteins are denatured. The membranes of chloroplasts and mitochondria with play an important role in cellular energy conversion become inactive due to damage of enzymes. This affects photophosphorylation, Oxidative phosphorylation and non-production of ATP.

Causes of Frost Hardening - Several plants growing in colder regions become frost hardy. This character develops due to several factors which are described below -

(*i*) **Due to Photoperiod and dormancy -** In several plants, frost hardening is associated with photoperiod. This conclusion is based upto the fact that most of the plants become maximum frost hardy when they are given the treatment of photoperiod and dormancy.

(*ii*) **Due to specific inhibitors and starvation -** In certain plants frost hardiness seems to be associated with the occurrence of specific metabolic inhibitors and is others specific degree of starvation.

(*iii*) **Inadequate light or shade** -It is observed that excessive shading or inadequate supply of light causes deformity and starvation in plants. To become frost hardy such plants have to adjust their leaf area, blade thickness, chlorophyll contents and number and orientation of chloroplasts in their cells.

(*iv*) **Formation of Low Temperature Resistant Proteins** - The proteins present in frost-prone plants are low temperature susceptible and get denatured easily to low temperature. In frost hardy plants the formation of low temperature resistant proteins takes place similar to that of freezing resistant plants. Although frost hardy plants possess high amounts of sugar in cells but the low temperature further increases this amount of sugars. It is reported that plant tissues may be made frost hardy by placing in sugar solution. The new frost resistant proteins may be synthesized which be resistant to high sugar concentration.

Chilling Injury

Chilling injury is caused with the lowering in temperature due to crystallization of lipids present in cellular membranes of chilling sensitive plants. The crystallization temperature is called *critical temperature* and it is equivalent to the temperature causing chilling injury. The crystallization is determined by the ratio of saturated fatty acids to unsaturated fatty acids. An increase in the proportion of unsaturated fatty acids or sterols maintains the normal functioning of membrances at low temperature.

The membrane lipids normally exist in liquid crystalline state at which the enzymes activity is optimum and the permeability of membrane remains under control below the critical temperature. The change in state of lipids from liquid crystalline to solid increases the permeability of membrane which results into imbalance of solute concentration and ultimately into disturbance of solute balance concentration and ultimately into disturbanced of solute balance enzyme activity. This disturbance in enzyme activity causes the accumulation of metabolites of glycolysis etc because they are not oxidized by mitochondrial enzyme system. Thus, only a tittle ATP is produced. The membrane resumes its normal activity with the increase in temperature as its lipids are again converted the liquid crystalline state.

Sometimes semipolar compounds of the soluble cell constituents like leucine, isoleucine, phenylalanine and valine etc. become toxic to the cell membrane. These compounds have been isolated during freezing of lipids in membrane. It is assumed that such semipolar compounds become distinct during freezing and cause damage to the membrane. Levitt (1972) is of the opinion that chilling injury is caused due to oxidation of -SH group of structural protein to –S–S– (disulphide) bridge. This change is irreversible and results into denaturation of membrane protein.

POLLUTION STRESS

Since the dawn of civilization man is dependent on nature and natural resources. All living organisms including plants need a balanced environment for their proper growth and development. The various components of a balanced environment always occur in a definite proportion. With the population growth, social and technological advancements and huge industrialization in developing countries, the man has increased the exploitation of natural resources at an alarmingly rapid rate with the release of various industrial gases and effluents in the environment (in water and air). The effluents are chemical materials with variable toxicity and are called *pollutants*. When these pollutants are released in the environment, they disturb the proportion of various components of a balanced environment and cause stress to the plants. This is called *pollution stress* which affects the growth and metabolisms of the plants. The effects of some important pollutants on plants are described here.

1. Effects of UV-B Rays on Plants

DNA is most important and fundamental component of all living beings including plants. It is the basic genetic material the chemicals like plastoquinone and plastoquinol involved in photosynthetic light reaction show their peak absorption spectra between 260 and 320 nm. When UV-b rays are absorbed by DNA, it produce mutation, genetic defects and cancer development due to its disorder. Impact of UV on DNA brings about photochemical changes as pyrimidine dimers, 6-4 photoproducts, DNA protein cross links and lesions that can lead to single and double strand breaks - UV rays increase CO_2 concentration and green house effect. In *Nostoc muscorum* UV rays caused reduction in phycocyanin pigments, photosynthesis and nitrogen fixation.

The complex photochemical reactions take place in presence of UV radiations producing toxic and irradiating chemicals like photochemical smog. These chemicals are highly injurious and cause reduction in rate of photosynthesis and primary production in plants. They damage leaves of crop plants like tobacco, peas, pines etc and reduce crop yield.

2. Hydrocarbons

They are released into atmosphere by anaerobic decomposition of organic matter, forest fire, gasoline fueled vehicles, industries and automobile exhausts. When these hydrocarbons combine with nitrogen oxides in the atmosphere under the influence of UV radiations of sunlight, they form ozone, peroxyacetyl nitrate (PAN), aldehydes and other matters which are commonly known as *photochemical smog*. It causes adverse effects on metabolisms.

3. Sulphur Dioxide (SO_2)

Most of the SO_2 (about 75%) is released into atmosphere by burning of coal in thermal power plants and smelting of sulphur containing ores and about 25%, SO_2 is emitted from petroleum refineries and automobiles. The effects of SO_2 on plants are as follows -

1. SO_2 damages cereal crops, coniferous plants of forests, apple and mango orchards.
2. SO_2 causes chlorosis and necrosis of vegetation in as low concentration as 0.032 ppm. chlorosis results from the destruction of chlorophyll which is changed to phaeophytin (Rao and Lc Blane, 1967).

$$SO_2 + H_2O \longrightarrow H_2SO_3 \longrightarrow HSO_3^- + H^+$$
$$\text{Chlorophyll a} + 2H^+ \longrightarrow \text{Phaeophytin a} + Mg^{+2}$$

3. Lichen vegetation is completely destroyed by SO_2. So, lichens are called *Pollution indictors*.
4. SO_2 released in the atmosphere is converted to H_2SO_4 and causes various injuries in the cells like damage of membrane and its permeability, plasmolysis, destruction of chlorophyll and metabolic inhibitions. H_2SO_4 formed by SO_2 and atmospheric moisture or water when precipitates down on earth, it is called *acid rain*. Sometimes HNO_3 also precipitates down along with H_2SO_4. Thus, acid rain may be due to either H_2SO_4 alone or HNO_3 alone or by combination of both the acids. Acid rain causes following effects.

   (*a*) It reduces crop productivity.
   (*b*) Soil becomes less fertile due to leaching away of mineral nutrients like Ca, Mg, K etc by these acids.
   (*c*) Acid rains mobilize heavy metals like Hg, Pb, Cd etc.
   (*d*) Acid rains at low pH release man toxic metals and trace elements which in high concentrations are highly toxic to the plants. The reduce the alkalinity of the medium.
   (*e*) Excessive amount of acids disturbs soil chemistry which affects plant metabolisms.

4. Increased Carbondioxide Concentration

The concentration of CO_2 in the atmosphere is increasing due to burning of fossil fuels (coal, oil, gas etc) in automobiles, industries, thermal power plants, hotmix plants and domestic cooking etc. Each year, more than 18×10^{12} tons of CO_2 is being released from fossil fuels alone. IT is also released during respiration of plant and animals and during volcanic erruptions.

Increase in CO_2 concentration is not only responsible for *green house effect* and *global warming* but it also affects plants in various ways. Some of the important effects are as follows -

1. Increase in CO_2 concentration increases the rate of CO_2 assimilation in leaves as reported in case of wheat, rice, soybean cotton and banana. IT may be due to increased intracellular CO_2 concentration at CO_2 fixation site.
2. The rate of transpiration usually decreases in C_3 plants as compared to C_4 plants which is considered to be due to decrease in stomatal conductance.
3. Increase in CO_2 concentrations also decreases activity of RUBISCO and carbonic anhydrase enzymes and reduction in chlorophyll and total soluble protein contents.
4. It increases the translocation of solutes in leaves.
5. Increased CO_2 concentration lowers the percentage of total nitrogen content and total soluble protein content of plants and seeds but increases the activity of enzyme *nitrate reductase* which acts as a rate limiting enzyme in nitrate assimilation. The symbiotic nitrogen fixation in field grown crops usually increases in increased CO_2 concentration.
6. In general the growth of most C_3 plants increases upto CO_2 concentration of 600 ppm as observed in wheat, rice and *Plantago etc*. The increase in growth may be related to increase in number of tillers and crop yield as in rice or may be in the form of increased leaf area, increased leaf dry weight and increased number of branches, fruits and seeds per plant.
7. Increase CO_2 concentration causes closure of stomata in plants.

5. Industrial Effluents or Discharges

These are the main contributors of water pollution. Industrial effluents from breweries, tanneries, sugar mills, butcheries, textile and dyeing mills, paper and pulp mills, steel and electroplating industries, mining etc containing organic wastes, compounds of heavy metals and metalloids eg. Hg, Pb, Cd, Cu, As, Ni, Zn, Mn, Mo etc, acids, alkalies, cyanides, chromates, thiocyanates, organic solvents etc when disposed off into water cause water pollution and ultimately pollution stress to the plants.

Although the role of heavy metals like Cu, Zn, Mn, Mo etc is very much clear in plants as micronutrient but those of Hg, Pb and Cd is indefinite. The heavy metals cause adverse effects. In higher concentration heavy metals are toxic to plants. They increase permeability of cells, reduce rate of respiration and photosynthesis. They show reduction in chlorophyll content, protein content, nitrogen assimilation and growth in general. Thus, a number of physiological and metabolic disorders can be observed due to accumulation of heavy metals in the medium. It is reported that Pb, Hg and Cd at above 0.1 mM concentration inhibit seed germination and seedling growth in certain crops like wheat, maize, barley etc.

PART II
BIOCHEMISTRY

1
Methods of Biochemical Analysis

1. CHROMATOGRAPHY

1.1 INTRODUCTION

The name chromatography was given by a Russian Botanist, **Michael Tswett** in 1906 to a process in which the solution of green pigments was allowed to percolate down in a packed column of alumina or chalk inside a glass cylinder to give coloured bands, representing the different pigments present. By this method Tswett showed that the chlorophyll is made up of two pigments and it became more clear when petroleum ether was poured in the packed column. The preparation was called as the *chromatogram*.

The concept of chromatography as given by *Tswett* was revised with the advancement in techniques. Now not only the solids, but liquids and the gases are also used to separate the pigments and other colourless compounds. **A.F.P. Martin** has described chromatography as 'the uniform percolation of a fluid through a column of more or less uniformly divided substance which selectively retards, by whatever means, certain components of a fluid.' This may also be defined as **a technique in which the components of a mixture are caused to migrate at different rates through an apparatus which involves equilibration of compounds between a stationary and a mobile phase.**

It, thus, involves separation of components in a fluid by causing the fluid, which is either a liquid solution or a mixture of vapours, to flow over a stationary phase which may be a pure solid powder as in adsorption chromatography, a porous gel or resin as in partition and ion-exchange chromatography, or may be a solid coated with a liquid as in gas chromatography.

Although, the procedure is called chromatography as it was initially used to separate coloured substances, it is also applied to colourless substances that absorb ultraviolet light or fluorescein or that give a colour when treated chemically.

1.2. KINDS OF CHROMATOGRAPHY

In all the chromatographic techniques the phenomenon of adsorption or partition is involved. In adsorption, the binding of a compound, to the surface of the solid phase (stationary phase) takes place, whereas in partition the relative solubility of a compound in two phases, results in the partition of the compound in two phases. Thus, all types of chromatography known so far have been grouped in either of the two categories.

I. Partition Chromatography.
   1. Partition column chromatography.
   2. Paper chromatography
   3. Thin layer chromatography.
   4. Gel filtration

   } Liquid as mobile phase & solid as stationary phase.

		5.	Gas liquid chromatography or vapour phase chromatography.	Gas as mobile & liquid as stationary phase.
II.	Adsorption Chromatography	1.	Adsorption column chromatography.	Organic solvent as mobile phase and some ionic polymer as stationary phase.
		2.	Thin layer chromatography.	
		3.	Ion-exchange chromatography.	

Flow chart of classification

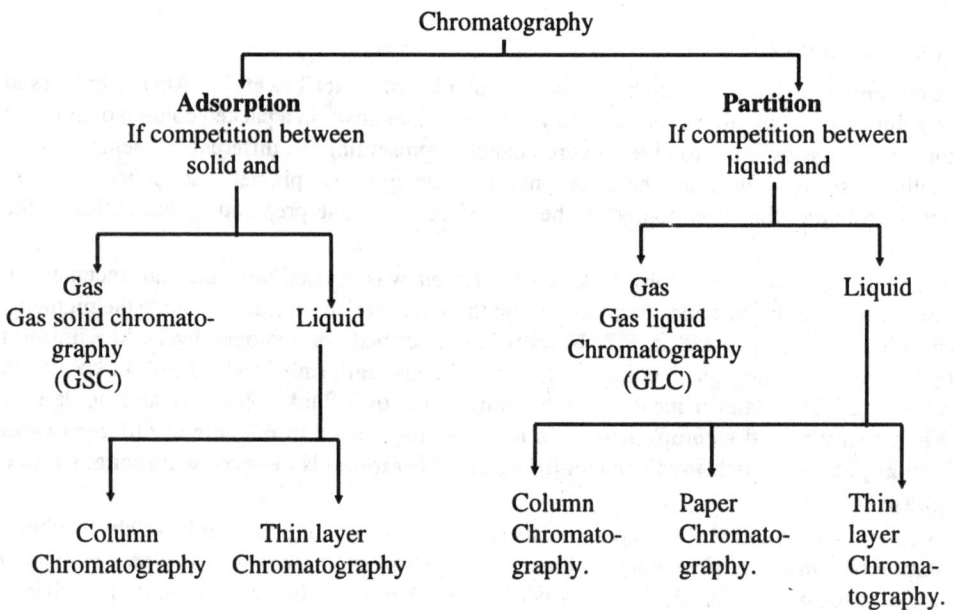

1.3. OUTLINE STEPS IN DIFFERENT KINDS OF CHROMATOGRAPHY

1. Partition column Chromatography

The column is packed with a porous solid of high surface area, e.g., silica gel and cellulose which is coated with water (stationary phase). The components of a mixture are separated by passing an organic solvent (mobile phase) through the column.

2. Paper Chromatography

The dissolved substances are applied as a small spot on a cellulose bound filter paper which is then kept in a container dipped in an organic solvent. The mixtures are partitioned between paper (stationary) and organic solvent (mobile phase).

3. Thin layer Chromatography

The adsorbent (stationary phase) is spread over a glass plate in a thin film of even thickness. The solvent (mobile phase) moves up the plate by capillary action and thus affects separation.

4. Ion-exchange Chromatography

Ionized compounds are separated in aqueous solution (mobile phase) by virtue of their differences in affinity for ionized compounds which are an integral part of the insoluble solid phase (stationary phase).

5. Gas liquid Chromatography

A column is packed with a porous inert solid coated with a thin layer of an involatile liquid as the stationary phase. Components of a mixture are separated by being partitioned between this phase and a gaseous (mobile) phase.

6. Adsorption Column Chromatography

The separation of a mixture is determined by the differential adsorption of the components on an active solid like alumina, silica gel (stationary phase) when an organic solvent (mobile phase) containing them passes over and affects separation.

1.4. CHROMATOGRAPHIC METHODS OF POPULAR USE

1. Adsorption Column Chromatography

It was the earliest method used in the history of chromatography where the phenomenon of adsorption was used in separating and identifying the unknown substances. The name column chromatography was given to it because column of materials were used for adsorption of different substances. By Column chromatography the different coloured components of a mixture are separated and the separated components travel down the column at different rates. By changing receivers the different fractions are collected separately as they leave the bottom of the column. Colourless substances are frequently separated by collecting many small fractions in succession from the column, testing each fraction chemically or by other means, and combining all fractions containing single component.

The essential apparatus for a column chromatography consists of a tube with means for

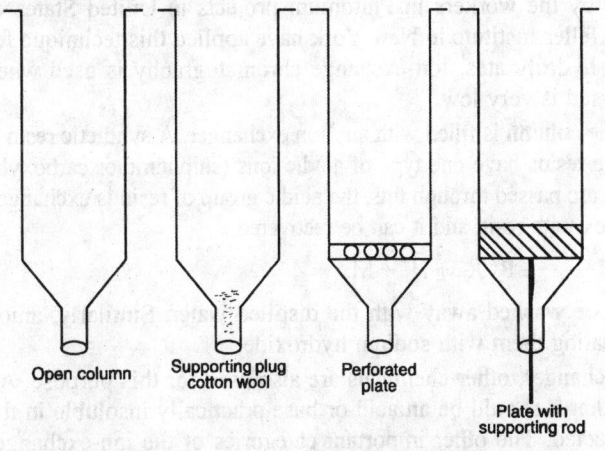

Fig. 1.1A. Types of chromatographic columns.

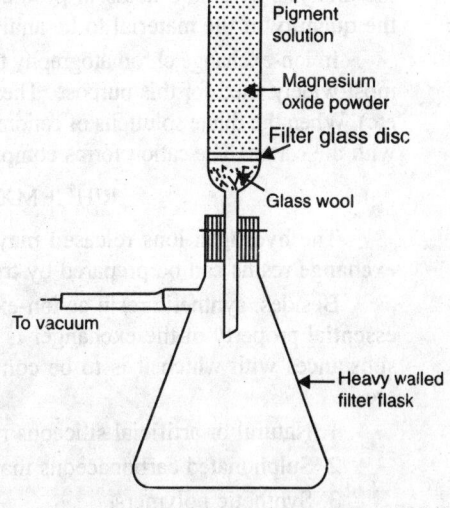

Fig 1.1B: Apparatus for the separation of pigments by column chromatography.

supporting the packed adsorbent and openings for the admission and collection of the mobile phase. The tube can be made up of any chemically resistant material. Usually it is made up of glass.

The criteria for selecting an adsorbent are : (*a*) The adsorbent should be insoluble in the solvent used, and (*b*) It should not react chemically either with the solute or with the solvent. **Deutz** (1951) has described the physical properties of more than 100 compounds which can be used as adsorbents. Sometimes, inert solid diluents are mixed to improve the adsorbing capacity of the substance used. The purified diatomous earth is frequently used as a filter and is commonly known as celite or supercell under trade name.

Various adsorbents which are in common use are usually inert compounds like charcoal, alumina, calcium carbonate, calcium phosphate, cellulose, glass kaolin (clay material), magnesium oxide, magnesium silicate, sucrose etc. Silicagel and kiesulguhr are most commonly used now a days. The individual adsorbent particles should be of uniform size and sufficiently hard for adsorption during the packing of the column. For these purposes, the adsorbents are finely divided before they are packed in the column.

During the selection of solvent, two things are kept in mind : (*a*) the solvent should be pure and stable, (*b*) the recovery of dissolved substance from it should be easy. The most common method of recovery is distillation. However, electrolytic and ion-exchange mechanisms have also been employed for this purpose. Petroleum ether, butanol, acetic acid, water, acetone, chloroform, benzene, phenol, ammonia and many other organic and inorganic solvents are used. Sometimes a mixture of two or more solvents are used in definite proportion in resolving a particular substance. In separating the photosynthetic products of *Scenedesmus*, for example, **Calvin** and **Benson** used the solvents phenol saturated with water and butanol, propionic acid and water. Similarly, in resolving, gibberellin-like substances a mixture of isopropanol : 7N ammonium hydroxide : Water (8 : 1 : 1 by volume) is usually employed. Acetone or organic compounds of this type are frequently used in the chromatography of plant pigments.

Adsorption chromatography is widely used in the separation of aminoacids, lipids, steroids, sugars and similar other compounds of relatively low molecular weight. Purines, pyrimidines and nucleic acids have also been resolved by this method.

2. Ion-Exchange Chromatography

This method is generally used to separate ionic-substances. The first biochemical application was made by **Folin** and **Bell** in 1917, when they determined ammonia in urine. However, the actual procedure of the method was adopted by the workers in Plutonium projects in United States of America. **Stein** and **Moore** of the Rockfeller Institute in New York have applied this technique for the analysis of amino acids in protein hydrolysates. Ion-exchange chromatography is used when the quantity of the material to be analysed is very low.

In ion-exchange chromatography the column is filled with any ion exchanger. A synthetic resin is most widely used for this purpose. These resins have one type of acidic ions (sulphonic or carboxylic etc). When the dilute solutions of cations are passed through this, the acidic group of resin is exchanged with the cation. The cation forms complex with resin and it can be recovered.

$$R'H^+ + MX^+ \longrightarrow R'X^+ + H^+ + M.$$

The hydrogen ions released may be washed away with the distilled water. Similarly, anion exchange resins can be prepared by treating them with sodium hydroxide.

Besides, synthetic resin as ion-exchanger, other chemicals are also used for this purpose. An essential property of the exchanger is that it should be an acid or base practically insoluble in the substances with which it is to be contacted. The other important categories of the ion-exchanger are—

1. Natural or artificial siliceous matter,
2. Sulphonated carbonaceous materials,
3. Synthetic polymers,
4. Derivatives of cellulose and
5. Miscellaneous.

Of these categories only (3) and (4) are of importance in the biochemical field. The synthetic polymers were first discovered by **Adams** and **Halms** of Chemicals Research Laboratory in England in 1934. These were prepared by the condensation or polymerisation of polyhydric phenols with formaldehyde, and their cation exchange properties are due to the presence of free phenolic groups.

Some other commercially used ion-exchangers are : Phenols with active-SO_3H, polysterene with SO_3H, and polymethocrylic acid with-COOH.

The apparatus used for ion exchange chromatography is slightly different from column. A wider column is used for it. When cellulose paper is used as ion-exchanger, the apparatus is the same as in paper chromatography. When resin or other chemicals are used as ion-exchanger, the apparatus is similar to column chromatography.

The method of ion-exchange chromatography has been applied to a wide variety of compounds of biochemical importance, both synthetic as well as of natural importance. Certain ionic complexes have been discovered by **Gabrielson** and **Samuelson** (1953) for the separation of carbonyl compounds. Certain biologically important proteins were separated by **Heinrinch, Dewey** and **Kidder** (1953) by this method. Separation of folic acid and its derivatives was accomplished by **Usdin** and **Porath** (1957). Vitamin thiamine and its esters were separated by **Siliprandi** and **Siliprandi** (1959).

Aminoacids bearing an appropriate charge can be deposited on the column of the proper type of resin by an exchange reaction with the functional groups of resin, while other aminoacids pass through the column. The deposited amino acids can then be displaced from the column by using strong solutions of the proper acids or bases.

3. Partition Chromatography

For this chromatography, credit goes to two British scientists **Martin** and **Synge**, who in the year 1941 originally described their method as liquid-liquid chromatography. They considered that basic mechanism of separation of solutes is their separation between two immiscible liquids. The term 'partition chromatography' was suggested later by **Lester Smith**. In actual procedure the material to be analysed or separated is distributed between two immiscible liquids.

In partition chromatography, the stationary phase consists of a gel which can be penetrated by diffusion. In partition chromatography the gel is saturated with water or another solvent, so acting as a stationary water phase and the separation of components put on the column is mainly due to a partition between the flowing solvent and the stationary solvent.

Theory : Every compound capable of being analysed chromatographically possesses a definite speed of migration in a definite solvent. Thus, long chain aminoacids migrate more rapidly than glycols, monosaccharides quicker than disaccharides, aglycons faster than glycosides. A substance, thus, can be characterized by the speed at which it migrates.

Rf Value : Position of a substance on a chromatograph is characterized by its Rf value (flow ratio) which is the measure of the velocity of migration of the substance. Rf value is defined as the quotient or ratio of the distance of the solute divided by the distance of the solvent from the starting point.

$$Rf = \frac{\text{Distance of the solute from starting point}}{\text{Distance of the solvent from starting point}}$$

It is important to note that the Rf value of a solute is always less than one.

4. Paper Chromatography

In paper chromatography separation is carried out on strips of paper which correspond to the chromatographic column of Tswett. Many grades of filter papers are used in paper chromatography, but the most widely used have been those manufactured by Whatman, Schleicher and Shull, and D. Arehes. Papers woven from glass fibres or from synthetic materials such as nylon, orlon or teflon have also been manufac-

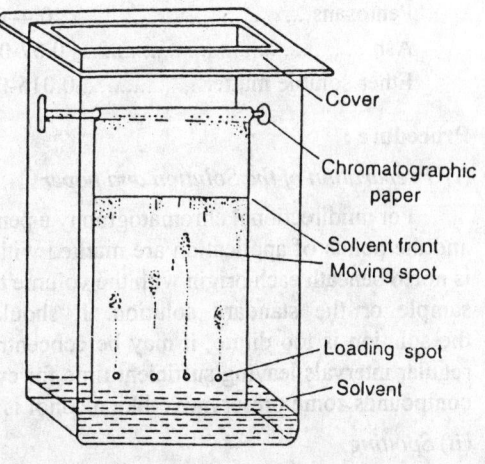

Fig 1.2: Demonstration of ascending paper chromatoatography.

tured for special purposes. Paper impregnated with inorganic adsorbents such as alumina or silicic acid was at one time advocated for the separation of some substances like vitamins, steroids and derivatives of ketones.

Whatman No. 1 is the standard slow running paper. The apparatus used is a cylindrical jar or cubical specimen jar.

Solvents used are mostly immiscible with water or partially miscible. It should be preferably of low vapour pressure. As a rule, previous purification of solvents is more necessary. Phenol, butanol, butanol mixture or a wide variety of organic acids and alcohols are used as solvent. Alcohols, pyridines, carboxylic acids, acetone and tetrahydrofuran are the solvents which are miscible with water. These compounds when employed for the purpose must contain small quantities of water, otherwise they do not migrate at all. Rf values of these substances depend on high degree of the water content. The paper chromatography may be :

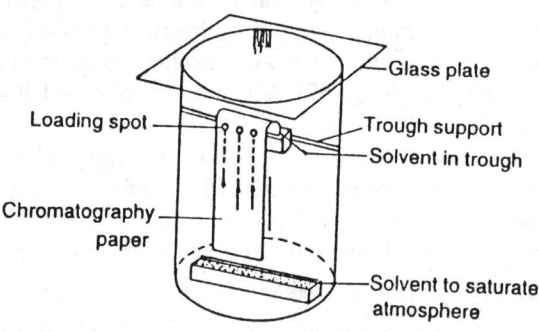

Fig 1.3: Demonstration of descending chromatography.

1. Vertical (*a*) Ascending,

   (*b*) Descending

2. Horizontal

3. Circular

For good separation of components of a mixture, now a days *bidimensional* (or bidirectional) paper chromatography is most widely used. In this technique, after development of the chromatogram with a solvent in one direction the paper is dried, turned right angles (90°) and the process of development is repeated with another solvent.

Chemical composition of *Whatman filter paper no. 1*

α-Cellulose 98–99%

β-Cellulose 0.3–1.0%

Pentosans 0.4–0.8%

Ash 0.07-0.1%

Ether soluble matter 0.015-0.1%

Procedure :

(i) *Preparation of the Solution and paper*

For unidirectional chromatography a pencil line is drawn about 2.5 cm up from the lower edge and the points of application are marked with a cross or dot. The name and code of each solution is noted beneath each origin with the volume to be supplied. Spotting solution is either the unknown sample or the standard solution. It should be about 1% with respect to each component. If the solution is too dilute, it may be concentrated on the paper itself by applying several spots at regular intervals leaving sufficient time for evaporation between each application. In water soluble compounds some preservative like alcohol is mixed with the aqueous solution.

(*ii*) *Spotting*

Spotting is done either by a platinum loop or by a micropipette. In spotting the aqueous

Methods of Biochemical Analysis

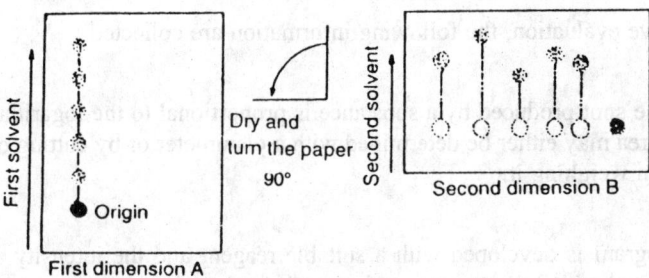

Fig 1.4: Representation of two-dimensional chromatographs.

solutions, a platinum loop of about 2-3 mm in diameter is used. The advantage of this method is that the wire can be cleaned by placing it over the flame for a few seconds. For volatile substances graduated micro-pipettes are used. The pipettes used for taking blood samples are well suited for this purpose. Melting point tubes with their ends cut-off are also of limited use. After some little practice drops may also be applied with a glass-rod. However, there is always a danger of applying too much of solution.

(iii) Running of the Chromatogram

After applying the spot it is dried and then the paper is suspended in the chromatographic chamber vertically for several hours, so that the solvent may travel up to maximum distance. When the solvent front has reached the end of the paper strip, this is cautiously raised up out of the trough and the front is marked.

(iv) Drying of the paper

The chromatogram is hung up on a washing line or clamped in an adjusting drying frame. According to the nature of the solvent, it is either dried in the air at room temperature or placed horizontally in a drying oven. Drying may also be accomplished with a drier which is the best method for preventing damage to the paper.

(v) Developing

If the spots on the chromatogram are not visible, it is necessary to develop them by treating with some suitable reagent. The reagent is either sprayed over or the chromatogram is dipped in that. A few spots are rendered visible by exposing them to gases or vapours. For example, the spots of the fats are seen when the chromatographic paper is exposed to iodine vapours. Aminoacids are visible after treating with ninhydrin solution and sugars with $KMnO_4$ or benzidine solution.

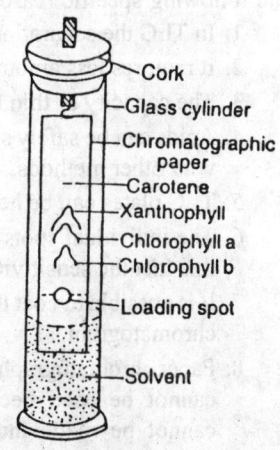

Fig 1.5: Demonstration of separation of plant pigments by paper chromatography.

(vi) Evaluation of Results

Qualitative Evaluation :

Pencil circles are drawn around the spots on the completed chromatogram. The centres of the spots are marked and distances from centre point to starting point are measured off. Division of these lengths by the distance of starting point to solvent front gives Rf values.

Quantitative Evaluation :

For quantitative evaluation, the following information are collected :

(a) Size of the spot

The area of the spot produced by a substance is proportional to the logarithm of the quantity of the solute. The area may either be determined with a planimeter or by cutting out paper covered by the spot and then weighing it.

(b) Photometry

The chromatogram is developed with a suitable reagent and the intensity along the strip is subsequently measured with a photometer and a graph is plotted.

(c) Extraction of the spots

After locating the spots on the chromatogram, the respective portions are cut off and the solute to be estimated is eluted. It is then estimated by a suitable micromethod.

1.4.5. Thin layer Chromatography

Thin layer chromatography (TLC) was introduced recently by **Izmailov** and **Williams.** They achieved some success with adsorbents plated out on a glass surface, rather than packed in the column. In this chromatography a glass plate coated with an adsorbent is used as a substitute of paper. A slurry of adsorbent and a binder (usually silicagel and plaster of Paris) is spread evenly on a glass plate and is dried in a hot air oven at about 105°C for 30 minutes to activate the silicagel.

Spotting, running, drying, developing and evaluation of the results are done in the same way as in paper chromatography.

TLC is considered far superior to paper chromatography and column chromatography because of the following specific reasons :

1. In TLC the separation is sharpest as compared to other two methods.
2. It requires less amount of the substances (0.4 mg) and less time (15-50 minutes).
3. The capacity of thin layer of an adsorbent is higher than that of paper.
4. Acids can be safely sprayed on TLC plates for identification purpose which is not possible with other methods.
5. TLC plates can be heated to higher temperatures without causing any damage.
6. The individual spots are less diffused as compared to paper chromatography. Because of this fact the sensitivity of detection is increased several times.
7. It is possible to coat the plates with a variety of corrosive reagents that would destroy paper chromatogram.
8. Paper chromatography is limited only to cellulose and other media, as alumina and silicagel cannot be used because they cannot be made into suitable sheets. But this difficulty can be overcome in thin layer chromatography where thin layers of these substances are put on plates which are known as **chromatoplates.**

1.5. LAWS OF ABSORPTION

Substances absorb certain wavelengths of visible light and emit the rest. The light that is reflected or transmitted through a substance imparts colour to that sub-

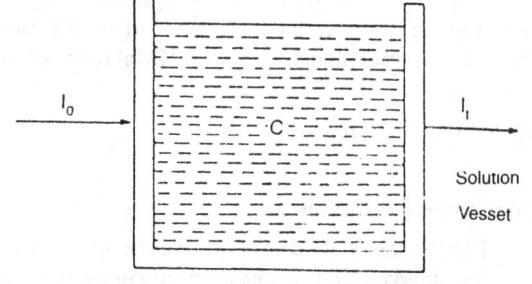

Fig 1.6: Showing transmittance of light through the solution.

Methods of Biochemical Analysis

stance. The intensity of the transmitted light may be compared with that of a standard. The amount of light absorbed differs with the concentration of the substance. The mathematical relationship between absorption of light and concentration of the substance is provided by the following fundamental laws:

1. Beer's law

Beer's law states that **the amount of light absorbed is directly proportional to the concentration of the solute in solution**.

Suppose, the solution of 'C' concentration is filled in a vessel and I_o and I_t are the initial and transmitted light intensities respectively (Fig. 1.6).

The fraction of light transmitted will be I_t/I_o. In general it will be less than I_o, whenever the concentration of the absorber is greater than zero. The amount of light transmitted (I_t/I_o) is related to the concentration and length of the substance. According to the general equation—

$$I_t / I_o = a\ b\ c$$

where,   a, is the absorptivity or absorbance of the sample,

b, is the length of the light through the solution i.e., the thickness of sample, and

c, is the concentration of the sample.

Or

$A = a\ b\ c$ (Where 'A' is the absorbance of the solution).

Mostly, 'a' 'b' are constant i.e., 'A' is directly proportional to the concentration of the substance. This is simply Beer's law. If we plot a graph between 'A' and 'c', we get a linear line but if the concentrations are very high, linear relationship is not obtained. Some of the solutions do not obey Beer's law due to following reasons:

1. In concentrated solutions, physical and chemical changes in the solute particles may occur which may further cause interaction between solute and solvent molecules.
2. The temperature of the absorbing substances may change the absorbance.
3. Extraneous materials may affect the absorbance of the solution. For example, large molecules may cause an excessive amount of light scattering.
4. Instrumental difficulties may also lead to change in Beer's law.

The abov- mentioned reasons may be invalidated to some extent if the following precautions are taken:

(a) Monochromatic light is used,

(b) The system is made up of randomly oriented molecular or atomic alcohols.

2. Lambert's law

This law states that **when a monochromatic light passes through a solution of constant concentration, the light absorbed is directly proportional to the thickness of the solution**.

The combined *Lambert-Beer's* law can be expressed as:

$\log\ I_o/I_t = K.\ C\ b$

where, I_o = Intensity of incident light

I_t = Intensity of emergent light

C = Concentration of solute in moles per litre

b = Length of light path in centimetres

K = molar extinction coefficient.

or $\log I_o/I_t = A$ or OD where A = absorbance of the medium

OD = optical density.

The constant K is referred to as the *molar extinction coefficient* only when the concentration is expressed in moles per litre and light path in centimetres. The constant has the dimensions:

litre mole^{-1} cm^{-1}

$$\text{O.D.} = \log \frac{I_o}{I_t} = \log \frac{100}{\% \text{ Transmittance T}} = 2 - \log T.$$

1.6. COLORIMETRY

Optical methods of biochemical analysis are better than other methods because of their speed, simplicity and precision. For such analysis very small amount of reagents and the biological materials are needed. One further advantage is that the solution listed often remains unchanged and thus it may be recovered.

Of the various optical methods of analysis, the colorimetric and spectrophotometric methods are most widely used in biological laboratories. In colorimetric analysis a comparison is made between the light transmitted through an unknown and a standard solution.

The term colorimeter refers either to an instrument which compares the light transmitted by an unknown and a standard solution or to an instrument capable of measuring directly the amount of light energy absorbed by a solution. The first type is known as a 'colour comparator' and the second type as a 'photometer'. Now, in practice colorimeters with a photocell galvanometer system are employed to measure the intensity of transmitted light. Such photometers are commonly called as 'photoelectric colorimeters'. Colorimeters with photoelectric cell are more sensitive than visual colorimeters.

Photoelectric colorimeters are designated in such a way that the length of the substance remains constant. For this purpose, there is an absorption cell with a fixed holder. Definite indicators are prescribed and transmittance (I_t/I_o) or percentage transmission ($I_t/I_o \times 100$) is calculated. I_o is the initial intensity of light at the time of entering in the solution and I_t is the intensity of light after coming out of the solution or after transmittance. Optical density of the substance is log I_o/I_t. These values are calculated according to the colorimeter used. Absolute values of I_o and I_t are not calculated because first the apparatus is set with the solvent for 100% transmittance. On replacing the blank with the coloured sample, the fraction of the light absorbed or transmitted by the latter is obtained.

By colorimetric analysis only coloured substances or those colourless substances which are capable of producing colour after reacting with suitable reagents, are analysed. The solutions to be analysed should have the following characteristics :

1. The colour of the solution should be sufficiently intense and should not fade rapidly.
2. There should be no effect of temperature and pH on the colour of the solution.
3. The reacting reagent with colourless solution should not itself be coloured to show light absorption.

1.7. SPECTROPHOTOMETRY

The basic principles of spectrophotometry are the same as for colorimetry. In spectrophotometer the spectrum of the coloured substance is analysed. Sometimes, the substances are not coloured but their spectrum differs from that of pure water or pure sample. The spectrophotometry shows the following advantages :

(a) Increased sensitivity,
(b) Increased spectral range,
(c) The presence of monochromatic light source,
(d) Narrowness of the spectral region,
(e) Determination of absorbancy even in very dilute solutions,
(f) Measuring absorption at ultraviolet and infrared regions, and
(g) In the determination of the absorption spectrum of the substance.

Methods of Biochemical Analysis

1. Spectrophotometer

Spectrophotometer (spectrometer and photometer) is an instrument which is used for measuring the distribution of energy in the different wavelengths of spectra. It is employed principally to supply a discrete narrow wavelengths band of radiation from any portion of the spectrum. The range of spectrophotometer is from ultraviolet to infrared. It is equipped with phototubes, with some indicators to measure intensity of radiation emerging from the slit of the monochromator segment of the instrument. The essential components of a spectrophotometer are shown in fig. 1.7.

In general, a spectrophotometer is made up of following essential parts :

1. A source of radiant energy. e.g.,
   (a) Incandescent lamps,
   (b) Hydrogen and Mercury discharge lamps,
   (c) Globars, and
   (d) Nernst glowers.

2. A monochromator, i.e., a device for isolating monochromatic (homogenous) or, more generally, narrow bands of radiant energy from the source.

3. Cell or holders for substances under investigations.

4. A device to receive and measure the radiant flux passing through or reflected from the substances under investigation. The different devices used are—(a) Photoemissive cells, (b) Gas filled photoemissive cells, (c) Photomultipliers, (d) Barrier layer cells, (e) Thermo-couples etc.

A typical instrument, the Beckman' Spectrophotometer, is outlined in Fig. 1.8.

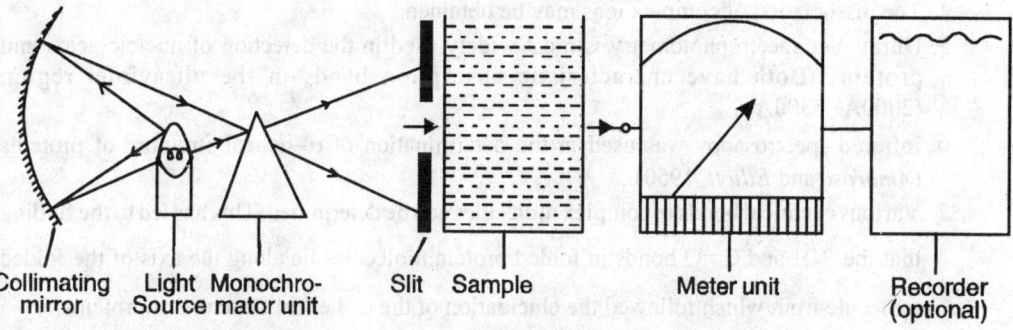

Fig. 1.7. Essential components of a Spectrophotometer.

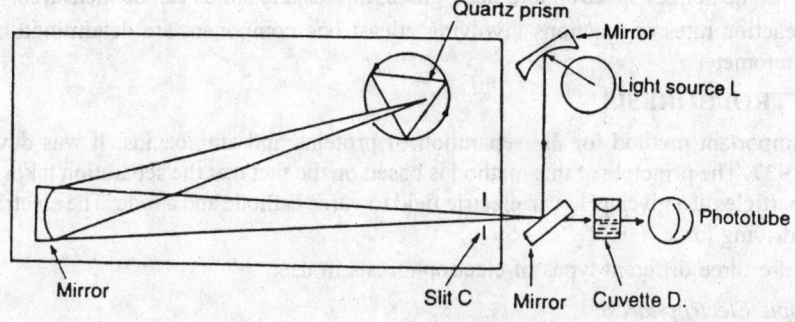

Fig 1.8: Beckman's Spectrophotometer.

Different types of spectrophotometers are available in the market. Some of them have additional absorption vessel compartments for maintaining a constant temperature, and devices that automatically record the optical density of a sample over the whole or part of the spectrum. This operation is referred to as "scanning". Some of these scanning devices can be set to repeat the operation at definite intervals of time. The degree to which the instrument can separate out individual wavelengths is called *resolution*.

2. Quantitative estimation

Quantitative estimation of any solution by spectrophotometer is done in the same way as by colorimeter. It can deliver radiation of a particular wavelength and then the optical density of a layer of solution containing an unknown concentration of a substance which absorbs this wavelength is compared with optical density of a layer of the same thickness of a solution containing a known concentration of the same substance (standard).

$$\text{Conc. of unknown substance} = \frac{\text{Conc. of standard}}{\text{Optical density of standard}} \times \text{Optical density of unknown substance}$$

This equation applies only when the system meets the requirements of Beer's law.

3. Applications

1. Spectrophotometer is used in determining the concentration of the absorbing substances.
2. pH of any solution can be determined.
3. It is used in the determination of ionization constant of weak acids and bases.
4. The dissociation of complex ions may be obtained.
5. Ultraviolet spectrophotometry is most widely used in the detection of nucleic acids and proteins. Both have characteristic absorption bands in the ultraviolet region (2000Å–3300Å)
6. Infrared spectroscopy was used in the determination of α–helical structure of proteins (*Ambrose* and *Elliott*, 1950).
7. Various chemical bonds in complex molecules can be determined. This has led to the finding that the NH and C =O bonds in folded protein molecules lie along the axis of the folded molecule from which followed the elucidation of the α–helical structure of proteins.
8. The progress of many metabolic reactions, particularly those of an oxidative nature, is frequently determined. This utilizes the fact that the most cofactors of biological oxidation have different absorption spectra in the reduced and oxidized state.
9. With the help of spectrophotometer gases, liquids and solids can be measured.
10. Reaction rates of reactions involving atleast one component are determined by spectrophotometer.

1.8. ELECTROPHORESIS

This is an important method for the separation of proteins and aminoacids. It was developed by *Tiselius* in 1937. The principle of this method is based on the fact that the separation takes place only of charged particles at a given pH in an electric field towards cathode and anode. The electric potential here, is the driving force.

There are three different types of electrophoresis in use.

1. *Microscopic Electrophoresis*

In this method, first the solution in a glass tube is placed on the stage of microscope horizontally and then the movement of various particles is watched. This method was used in the separation of microscopic materials like colloidal particles, bacteria, blood cells, protozoa etc.

Methods of Biochemical Analysis

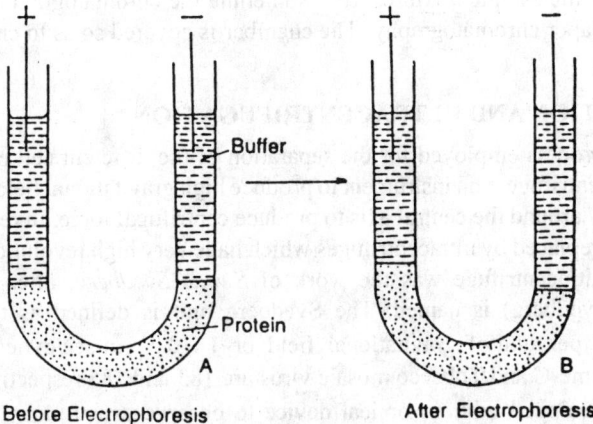

Fig 1.9: Demonstration of moving boundary electrophoresis.

2. Moving Boundary Electrophoresis

This method is applied where the electrical behaviour of molecules and ions in relatively large quantities is to be studied. In this method, the movement of the boundary of a mass of particles is measured. The apparatus is simply made up of U-tube. It is first filled with the substance which is to be studied and then the rest is filled with a buffer solution. Lastly a positive and a negative electrode is introduced into the two different arms of the U-tube and the current is passed. As a result of passing of the current, the boundary moves towards the positive or negative electrode depending upon the particles of the material in study.

3. Zone-Electrophoresis

This method is similar to that of paper chromatography and column chromatography. Here, the movement of the particle takes place on a solid medium or paper. After electrophoresis, the individual constituents occupy definite bands or zones as in chromatography. Zone electrophoresis is also known as *paper electrophoresis or ionography*.

The whole process consists of different steps. At first, a narrow strip of about 3-4 cm in width of whatman paper No. 1 is cut and in the middle of it a drop or a narrow zone of the sample mixture is applied with the help of an applicator and hair drier. Now, it is moistened in the electrolyte and is kept in the electrophoretic chamber containing the buffer solution in such a manner that both the ends are dipped in it. An electric current is now passed through the electrodes for the separation of

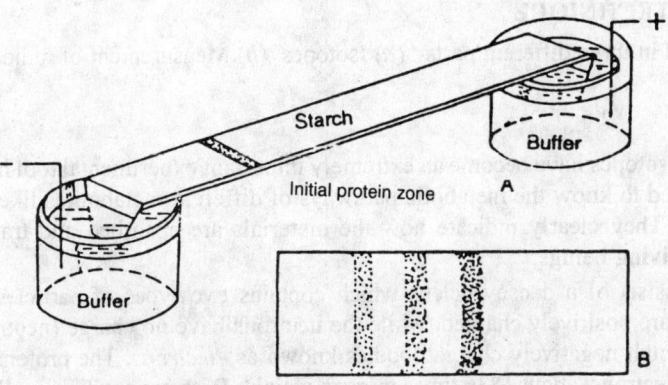

Fig 1.10: Demonstration of zone electrophoresis.

different contituents of the sample mixture. After sometime the chromatogram is developed by the same technique as in paper chromatography. The chamber is covered so as to check evaporation of the solvent.

1.9. CENTRIFUGATION AND ULTRACENTRIFUGATION

Centrifugation is the process employed for the separation of the different cell organelles to a high degree of purity. The centrifuge is an instrument to produce high gravitational field. Here, the samples are rotated horizontally around the central axis to produce centrifugal force. In recent years ordinary centrifuges have been replaced by ultracentrifuges which have very high revolutions per minute. The development of the ultracentrifuge was the work of *Swede Svedberg*, after whom the unit of sedimentation (S = Svedberg) is named. The Svedberg unit is defined as the velocity of the sedimenting molecule per unit of gravitational field or 1×10^{-13} cm/sec/dyne/g. Typical S values calculated for cytochrome-C and Tobacco mosaic virus are 183 and 185 respectively.

Ultracentrifuge equipped with an optical device to observe exactly how fast the molecules sediment is extremely valuable in obtaining data on the molecular weight of protein and establishing the concept that proteins are of discrete molecular weights and shapes. This work has revealed that sizes of proteins vary greatly, with a continuous range in weights between the extremes of approximately 10,000 and 1,000,000. Ultracentrifuge also facilitates in isolation, localization and characterization of enzymes. Ultracentrifuges are used further in the isolation of viruses, ribosomes, nucleic acids and in the determination of molecular weight of nucleic acids.

1.10. X-RAY DIFFRACTION

This technique is based on the diffraction of radiations when they counter small obstacles. If a ray of white light impringes upon a diffraction grating that has 1000 lines per millimeter, it will be diffracted and will show the various bands of spectrum.

To record the diffraction pattern of the material to be analysed, a beam of collimated X-rays is passed through the material and a photographic plate is placed beyond this. On the plate a series of concentric spots or bands appear due to interference between the different diffracted rays. The distance between these spots and the centre of the pattern depends upon the spaces between the regularly repeating units. From diffraction pattern, two things may be clear–(1) greater distance between repeating units will indicate that the angle of diffraction is small, (2) The sharper spots indicate that the spacing is more regular.

X-ray diffraction analysis has been made by a number of workers including **Pauling, Perutz, Kendrew, Wilkins, Franklin** and **Bernal**. The materials which have been analysed are crystals of hemoglobin, DNA, collagen fibre, keratin, muscle and myelin. The X-ray diffraction analysis throws light on the orientation of the molecules, distances between molecules and their atomic organization.

1.11. TRACER TECHNIQUE

It can be described in three different parts : (*a*) Isotopes, (*b*) Measurement of radioactivity, and (*c*) Tracer technique.

(a) Isotopes

Radioactive isotopes have become an extremely important experimental tool in Biochemistry. They are being used to know the metabolic pathways of different metabolites like carbohydrates, fats and proteins. They clearly indicate how the materials are absorbed and transported to the different parts in living beings.

An atom consists of a dense nucleus which contains two types of particles : *Protons* and *neutrons*. Protons are positively charged, while the neutrons have no charge (neutral). Around the nucleus, revolve small negatively charged bodies known as *electrons*. The protons are very large in comparison to electrons, about 1836 times more in weight. Both are equal in number, and thereby

Methods of Biochemical Analysis

create a condition of neutrality, e.g., hydrogen has one of each. Neutrons have the weight about equal to that of protons, thus, adding weight to the atom without altering its charge. Besides, they also play an important role in the stability of atom.

The atom contains a high amount of bond energy and when alteration in its neutron number takes place, an unstable condition arises in which the atom tends to split or give off particles or energy, thereby achieving stability. Such unstable atoms are known as *radioactive isotopes* or *radioisotopes*. The radioisotopes are defined as "the atoms of same element having different atomic weights, but the same atomic number," e.g., the carbon atom usually has six protons and six neutrons in its nucleus and six electrons in the orbit around the nucleus. The carbon atom is called ^{12}C. Thus, it has its atomic number 6 and atomic weight 12. There is another form of carbon atom (radioisotopic form) which has 8 neutrons in place of normal 6 and has a weight of 14 rather than 12. Other known isotopes of carbon are ^{11}C (6P + 5n) and ^{13}C (6P + 7n).

Radioisotopes usually emit radiations (energy) in the form of X-rays, β– or γ-rays. The movement of energy in either particles or wave form through the space is considered as *radiation*. The radiations are of many types. The radiations in the form of high energy atomic particles, which can transfer their kinetic energy to any matter through which they pass, is known as *particulate* or *Corpuscular radiation*. The radiations in the form of high energy short waves, which cause electric and magnetic disturbances affecting the internal structure of matter, is known as *electromagnetic radiation*. The treatment of plants or animals with radiations is called *irradiation*.

The common biologically important radioactive isotopes are ^{14}C, ^{11}C, ^{32}P, ^{35}S, which emit β–, γ– or X-rays. The radioactive isotopes move in transpiration stream as ions of these elements, but being unstable, transmutate to other elements (^{32}P to S; ^{35}S to Cl; ^{11}C to B etc.) when β–particles are emitted. This process is called *transmutation* and it is due to the fact that on emission conversion of a proton to a neutron takes place.

The complete life span of many isotopes is not definite, but half life is definite. Within a certain time period half of the atoms of the unstable isotope will disintegrate. This time period is known as the half-life of an isotope. The half lives varies from a few seconds to several thousand years. The table 1.1 shows the properties of some biologically important radioactive isotopes.

Table 1.1

S.No.	Name of isotope	Atomic number	Atomic weight	Half life	Nature of radiation
1.	Carbon	6	^{11}C	20.35 Minutes	β-rays
2.	Carbon	6	^{14}C	5568 Years	β-rays
3.	Chlorine	17	^{36}Cl	308,000 years	β-rays
4.	Cobalt	27	^{60}Co	5.27 years	β–, γ-rays
5.	Hydrogen	1	^{3}H	12.46 years	β-rays
6.	Iron	26	^{55}Fe	2-7 years	γ-rays
7.	Magnesium	12	^{28}Mg	21.4 hours	β–, γ-rays
8.	Molybdenum	42	^{93}Mo	67 hours	β-rays
9.	Phosphorus	15	^{32}P & ^{33}P	14.5 & 25.4 days	β-rays
10.	Potassium	19	^{42}K	12.4 hours	β-rays
11.	Sodium	11	^{22}Na	2.6 years	β–, γ-rays
12.	Sulphur	16	^{35}S	87.1 days	β-rays
13.	Zinc	30	^{66}Zn	250 days	β-rays

(b) Measurement of radioactivity

There are many methods for the measurement of radioactivity, but some of the important ones are described here.

1. Geiger-Muller Counter (G-M tube)

This is one of the most popular devices for measuring radioisotopes. The instrument is made up of two main parts—a *Geiger Muller-Tube* and a *recording device* (Scaler). The G.M. tube is simple and is filled up with helium and ethanol. It contains two electrodes. One electrode is the tube itself, while the other electrode is a large, round and fine wire stretched in the center. The fine wire electrode is maintained at a high potential (1000-2500 V) with respect to the second tube electrode. The open end of the tube is covered with a thin window of mica or plastic, which is filled up with helium and ethanol. The radioactive material, which is to be measured, is kept beneath this mica window. The radioactive material emits radiations into the tube and the radioactive particles then ionise the gas molecules of helium with the release of free electrons. These free electrons are attracted towards the positive electrode and thus a mild current is established, which can be measured by suitable electric counter (scaler). The result is referred as ionizations or counts per minute.

2. Mass spectroscope

This instrument is applied to determine isotopic contents of electromagnetic field. In this method, the substances to be measured are first changed into gases and then the separation of different isotopes is done by means of an ion collector present in the spectroscope, in which the

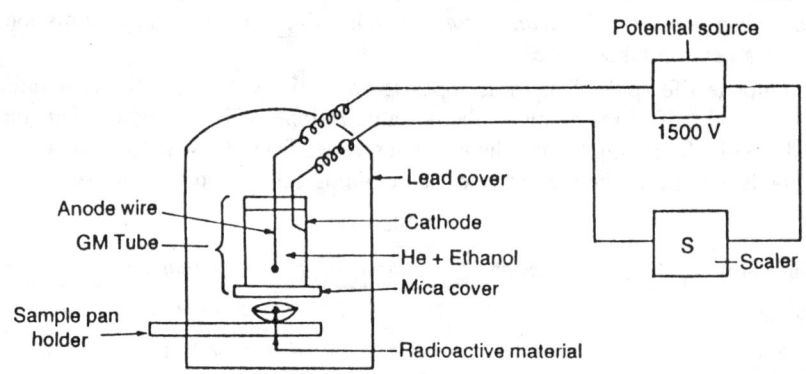

Fig 1.11: Schematic diagram of Geiger-Muller tube & scaler (Geiger Muller counter).

quantity of any particular ion can be estimated by an amount of electric current generated in the collector. The results obtained are calculated in either of the two ways.

(a) It is calculated as the ratio of one isotope to the other. For example, C^{12}/C^{11} in air is 0.011%.

(b) It is calculated on the basis of percentage composition as the oxygen in air.

(c) Tracer Technique

Any element which is a constituent of plant or animal body can be used as a tracer-atom. Usually radioactive isotopes (^{18}O, ^{11}C, ^{13}C, ^{14}C, ^{33}S, ^{34}S, ^{35}S, ^{32}P) are used as tracers in biochemistry. The compounds containing these radioactive isotopes in their molecular structure are said to be labelled. The different tracer atoms which have been used in finding out the biochemical reactions, are Deuterium, Tritium, ^{14}C, ^{15}N, ^{18}O, ^{32}P, Deuterium and tritium isotopes of hydrogen are given as water. ^{15}N isotope of nitrogen has been used as $^{15}NO_3$ or $^{15}NH_4$ in nitrogen metabolism. Similarly,

Methods of Biochemical Analysis

^{14}C has been used as $^{14}CO_2$ gas to trace its role in photosynthesis and fat metabolism.

The isotopes are applied to the plants or fed to the animals for a required time. The material (Plant or animal) is then killed to stop further biochemical reactions. The labelled compounds are then isolated from the tissues and identified by chromatographic or other chemical methods. Their radioactivity is measured by any of the above methods. The ultimate result can be represented by one of the following three ways.

(a) Total isotopic content.

(b) Specific isotopic content (cpm/mg of the substance).

(c) Distribution of the isotope within the molecule of the compound itself.

Application of radioactive materials

There are different methods of application of radioactive materials in plants and animals.

In Plants

In plants, various parts like roots, leaves, flowers and seeds are treated with the solution of required concentration containing the radioisotopes or stable isotopes. In case of root-application, they are dipped directly into the solution, from where the solution ascends through the roots to the various other organs. In case of foliar application, a drop of the isotope is applied to the lower or upper surface of the leaf and is allowed to be absorbed for a specific period of time. In case of flower treatment, the cotton moistened with isotopic solution is used. It is directly put at the tip of inflorescence. In seed treatment, they are first soaked in the solution and then grown in the laboratory conditions. All these treatments indicate that the isotope should be present in the solution or liquid form. Sometimes gaseous forms of the isotopes are also applied as in case of photosynthesis where labelled $^{14}CO_2$ is fed to the plants.

In Animals and Micro-organisms

In case of animals and micro-organisms, the isotope is mixed in their diet and through the mouth it enters into the body of the individual.

1.12. AUTORADIOGRAPHY

Autoradiography is the most widely used technique in biochemical and microbiological research. It is used to find out the exact location and distribution of radioactive materials in the living tissues. In this technique, first the plants are supplied and animals are fed on radioactive material and then after a required time the photograph is taken on X-ray film. This process is called autoradiography and the photograph as autoradiograph. The autoradiograph shows spots because the supplied isotope reaches and distributes itself in the tissues and into the structure of compounds taking part in the metabolism. At the same time it disintegrates and produces radiations. Before taking the photograph, the tissue is killed and dried between two hot copper-plates.

1.13. GAS CHROMATOGRAPHY

Gas chromatography is almost similar to that of column chromatography, except that a gas is used as the mobile phase instead of liquid. The gas chromatography may be Gas-solid chromatography (GSC) or Gas-liquid chromatography (GLC). In both the cases, the separation is carried out in a tubular column made up of glass, metal or teflon. The column is filled with a sorbent which acts as a stationary phase. In GSC, the adsorbents are first finely divided into fine size graded powder and then filled in the column. In GLC, the column wall is either coated with a liquid as fine film or the column is supported by an inert graded porous powder coated with a non-volatile liquid. It. acts as a stationary phase. A gas serving as mobile phase flows continuously through the column. It is known as *carrier gas*. The sample is injected into a stream of carrier gas (e.g., helium) in the vapour form. As the gas passes through the column, it comes in contact with some type of sensor (detector and recorder) which detects changes in gas composition. The different components of the gas sample are sorbed on the stationary phase to different extents depending upon their distribution coefficients.

GSC is based upon selective adsorption on a solid, whereas GLC is based upon the partition between the gas and immobile (stationary) liquid phase.

The basic components of gas chromatographs (for GSC and GLC) are the same and are shown in Fig. 1.12.

Although gas chromatography was initially restricted to such volatile substances as low molecular weight hydrocarbons, alcohols, aldehydes and fatty acids but recently it is being used

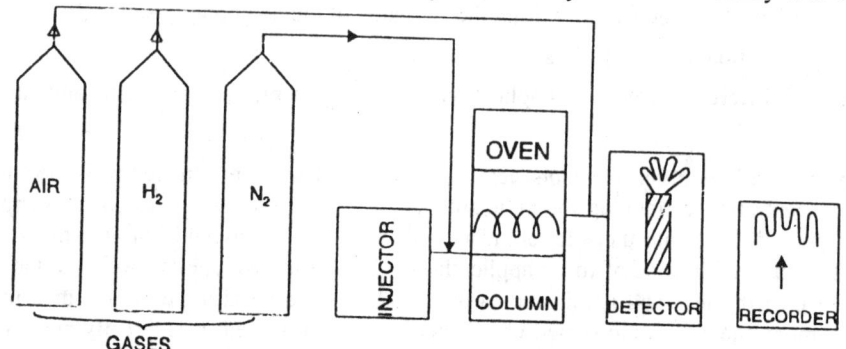

Fig. 1.12. Line diagram of a gas chromatograph.

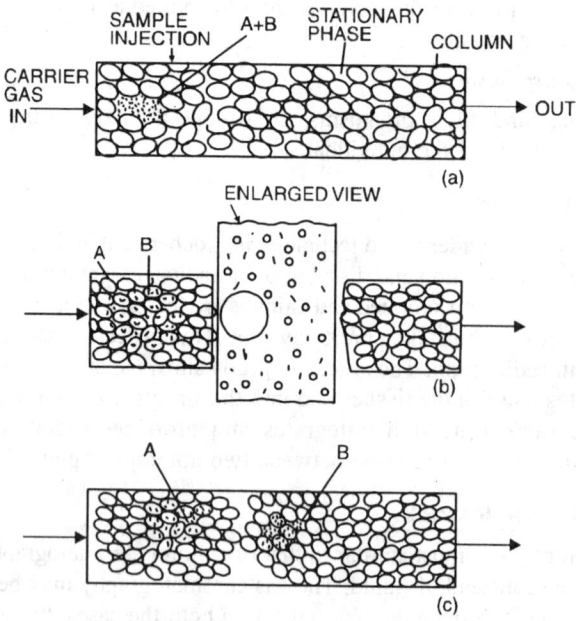

Fig. 1.13. Separation of gas mixture (A and B).

with some success to separate certain thermostable derivatives of sugars and amino acids.

Detecting Devices in Gas Chromatography

Following two detecting devices are generally used in gas chromatography.

1. Thermal Conductivity Cell

This detecting device is based on the principle that heat is conducted away from a hot wire by a gas passing over it. The cell contains a metal block made up of two parts. Two fine coils of

Methods of Biochemical Analysis

wire (C' and C) with a high temperature coefficient of resistance are placed separately in these parts. Suitable electric resistors are inserted in the circuit of each coil to form a wheatstone bridge circuit. When current is passed through the bridge, the connecting wires C' and C are heated. The thermal conductivity of the gas passing over the wire coil finally controls the equilibrium temperature of the wires. If the gas is the same, the wires will have the same temperature and resistance and the bridge will remain balanced. If an effluent gas now passes through C' and only carrier gas passes through C, the temperature of the wire will differ and its resistance will be changed. Thus, the bridge will become unbalanced. The extent of unbalance is measured with a recording potentiometer.

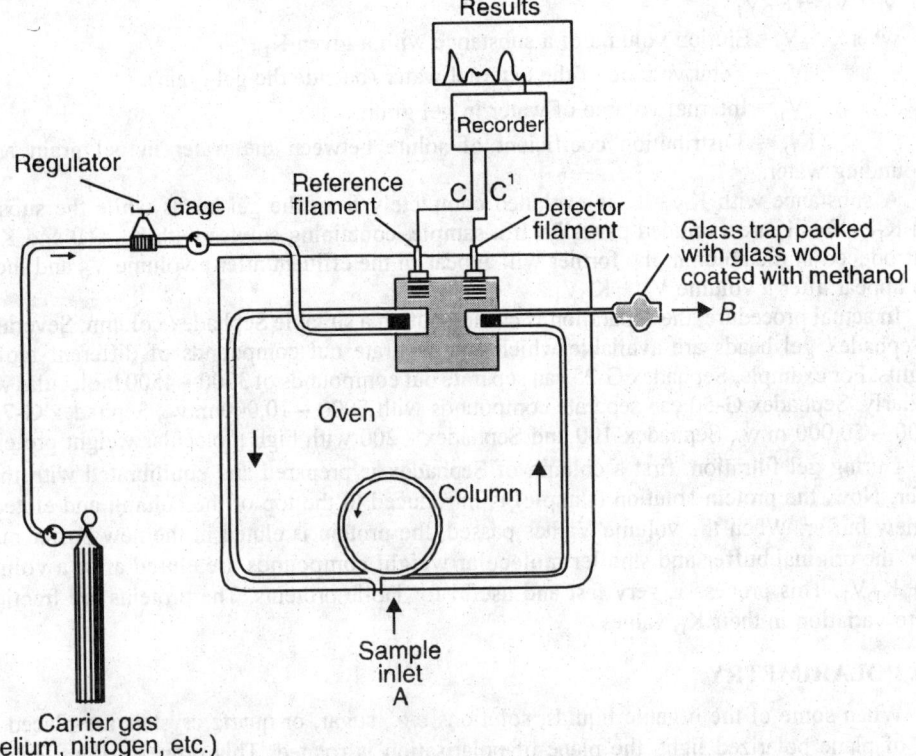

Fig. 1.14. Gas chromatograph with the help of thermal conductivity cell detector.

2. Hydrogen Flame Ionization Detector

It is based on the principle that when organic material is burned in a hydrogen flame, electrons and ions are produced. The negative ions and electrons move in a high voltage field to an anode and produce a very small current, which is changed to a measurable current by appropriate circuit. The electric current is directly proportional to the amount of material burned. These detectors are highly sensitive with wide response but are insensitive to water.

1.14. GEL FILTRATION

It is also known as *three dimensional dialysis, molecular sieve chromatography, molecular exclusion chromatography and molecular sieving*. It is carried out by filtration of a mixture of solutes through a column of gel particles. The internal matrix of column gel can filter only molecules of limited size and compact shape. The technique of separating molecules of different size by passage through a gel column is called **gel-filtration**. The commercial name for the gel is **Sephadex**. The basic principle behind this separation is the property of sephadex to exclude solutes of large molecular size and to be accessible for diffusion to molecules of small dimension.

Dextrans of varying chain length and degree of artificial cross-linkage provide gels with graded ability to take up molecules of various sizes. Desalting of protein solutions is accomplished readily. The salts penetrate the gel and leave the column after the protein (which enters the gel matrix less readily, if at all). By appropriate choice of gel, proteins of different molecular weights can be separated from each other. Through gel filtration, molecules are separated according to their molecular size.

Anionic and cationic groups introduced by chemical incorporation has produced a series of ion-exchange dextran gels, the functions of which are a combination of ion exchange and molecular sieving.

The appearance of a solute in an effluent is expressed by following general equation.

$V = V_0 + K_D V_1$

where, V = Elution volume of a substance with a given K_D.

V_0 = Total volume of the external water (outside the gel grain).

V_1 = Internal volume of water in gel grain.

K_D = Distribution coefficient of solute between the water in gel grain and the surrounding water.

A substance with $K_D = 0$, is excluded completely from the gel beads while the substances with $K_D = 0 - 1$, are excluded partially. If a sample, containing solutes with $K_D = 0$ and $K_D = 1$, is introduced in the column, the former will appear in the effluent after a volume V_0 and the latter with appear after a volume $V_0 + K_D V_1$.

In actual procedure, the separation is carried out on a suitable Sephadex column. Several types of Sephadex gel beads are available which can separate out compounds of different molecular weights. For example, Sephadex G-25 can separate out compounds of 3500 – 4500 molecular weight. Similarly, Sephadex G-50 can separate compounds with 8000 – 10,000 m.w., Sephadex G-75 with 40000 – 50,000 m.w., Sephadex-100 and Sephadex - 200 with high molecular weight proteins.

During gel filtration, first a column of Sephadex is prepared and equilibrated with the new buffer. Now, the protein solution (sample) is introduced to the top of the column and eluted with the new buffer. When the volume V_0 has passed, the protein is eluted in the new buffer medium whi' the original buffer and smaller molecular weight compounds are eluted after a volume of $V_0 + K_D V_1$. This process is very fast and useful for labile proteins. The proteins are fractionated due to variation in their K_D values.

1.15. POLARIMETRY

When some of the organic liquids, solutions e.g., sugar, or quartz crystals are placed in the path of plane polarized light, the plane of polarisation is rotated. This property is called *optical activity* and the substances as *optically active*. The substances which rotate the plane of polarised light towards the right (clockwise) are called *dextro-rotatory* (+) and those which rotate towards the left (anticlockwise) are called *laevorotatory* (—). A mixture of these two varieties in equal proportions remain optically inactive and is called *racemic form*. The deflection in plane of polarised light is due to presence of an asymmetric carbon atom is substances. Such deflections can be measured through polarimetry.

Theory

According to wave theory of light, an ordinary ray of light vibrates in all planes at right angles to the direction of propagation. However, certain optical systems, e.g., Nicol prism, transmit light in only one plane. Such a light which vibrates in only one plane, perpendicular to the path of propagation is called *plane polarised light*.

The magnitude of the rotation depends upon following factors —

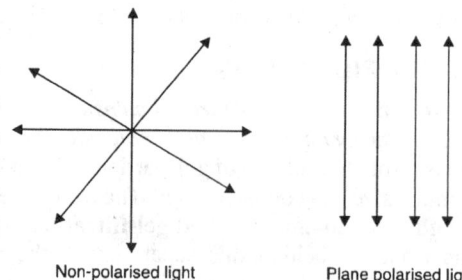

Non-polarised light   Plane polarised light

Fig. 1.15. Plane polarised light.

Methods of Biochemical Analysis

(i) Nature of substance.
(ii) Length of the liquid column (l) through which light passes.
(iii) Concentration of solution.   (iv) Nature of solvent.
(v) Temperature of the solution.   (vi) Wavelength of light used.

The rotatory power of a given sample is usually expressed as *specific rotation*, $[\alpha]_n^t$. The optical rotation by a compound at a given temperature and wavelength is expressed as :

$$A° = [\alpha]_\lambda^t \times C \times l$$

where Å = the observed rotation in degrees.

$[\alpha]_\lambda^t$ = the specific rotation of a compound at a specific temperature (t) and wave length (l) of light.

C = concentration of compound in solution (g/ml)
l = length of path of light through solution in decimeters

Specific rotation $[\alpha]_\lambda^t = \dfrac{A}{C \times l}$

Specific rotation may be defined as the rotation in degrees of a solution having 1g of material in a 10 cm (1 dm) polarimeter tube.

For very dilute solutions in any given non-active solvent, the specific rotation is constant.

The **optical rotation** of the compounds varies according to the wavelength of light used. The variation is called **optical rotatory dispersion**.

The apparatus – Polarimeter

The essential components of a polarimeter are shown in Fig. 1.16. It contains the sodium lamp as light source which emits the light at 589 nm (called *D* line of sodium). This light is filtered through a slit and made parallel through a lens. Then it is passed on to a nicol prism which functions as a polariser (polarising nicol prism). The length of the polarimeter tube (sample tube) is generally 1 dm. The light coming through the polarimeter tube is passed through another nicol prism which functions as an analyser (analysing nicol prism). The entire system (all components) is filtted inside a tube. First, with the blank tube (polarimeter tube without optically active sample), the lens on analyser side is so adjusted that the light intensity seen on the viewer side (eye) becomes minimum. Now, the tube is filled with the optically active compound which rotates the plane of light. The analyser side lens is rotated in order to minimise the light intensity again. This lens is fitted on a scale and the angle of desired rotation can be measured on that scale. By knowing this absorbed rotation in degrees, the specific rotation can be calculated by the formula as mentioned earlier.

If the analysing lens is rotated clockwise, the substance is called *dextro-rotatory* (+). If the lens is rotated counter-clockwise, the substance is called *laevo-rotatory* (−).

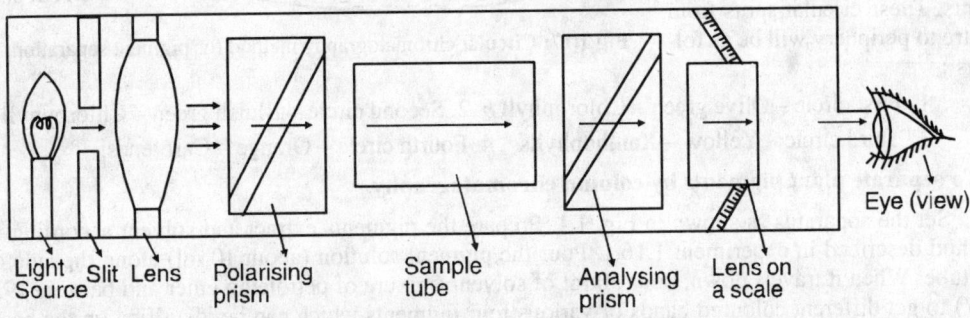

Fig. 1.16. Essential components of a polarimeter

Uses of Polarimetry

1. It is used for identifying unknown compounds.

2. It is used for knowing concentration of known compounds.

3. It is employed for studying an enzymatic reaction where the substrate and product have different optical rotations.

4. Optical rotatory dispersion measurements are utilized in getting information about atomic organisation of a molecule.

5. It is being used for plant control in pharmaceutical industry and sugar industry.

1.16 EXPERIMENTS RELATED TO CHROMATOGRAPHY

1. To separate major plant pigments by strip paper chromatography.

The procedure for paper chromatography is described in the text. The method of extraction of pigments extract and solvent to be used is given below :

To prepare the pigments extract, take about 100 gm of fresh spinach or grass leaves in a pestle and mortar. Crush them well with 40 ml of 80% acetone and a pinch of $CaCO_3$. Stir the mixture occasionally till a deep green colour is obtained. Filter the solution. The filtrate contains the extract of chlorophyll a, chlorophyll b, carotenes and xanthophyll. This filtrate is used for spotting on Whatman's filter paper no. 1.

The solvent used for separation of pigments is a mixture of petroleum ether and benzene in the ratio of 9:1 by volume. The various pigments can be identified on the basis of their colours (See Fig. 1.5).

2. To separate major plant pigments by circular or filter paper chromatography.

Take Whatman's filter paper No. 1 and cut is circular. Again cut this circular paper from the periphery to the centre to form a wick (Fig. 1.17). Now apply a few drops of pigment extract (or acetone extract) to the centre of the disc with the help of micropipette or glass rod or capillary tube or platinum loop and hair drier. Several applications are needed to concentrate the spot. Place this spotted filter paper over a small petrish containing solvent (Petroleum ether and benzene, 9:1). Now bent down the wick in such a manner that it dips or touches the solvent. Cover this paper and small petridish by a large petridish to check evaporation of the solvent and observe.

The solvent will rise through the wick and run on the filter paper centripetally. After about two hours take out the filter paper and dry it with the help of hair drier. Four circular spots will be separated which can be identified on the basis of colour showing presence of four pigments. These circular spots from centre to periphery will be as follows :

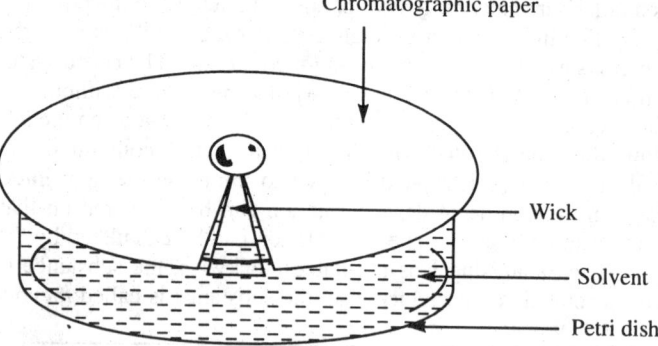

Fig 1.17: Circular chromatography method for pigment separation.

1. First circle – Olive green – Chlorophyll *b* 2. Second circle – Bluish green – Chlorophyll *a*
3. Third circle – Yellow – Xanthophylls 4. Fourth circle – Orange – Carotenes.

3. To separate plant pigments by column chromatography.

Set the apparatus as shown in Fig. 1.1. Prepare the pigments extract and solvent according to method described in experiment 1.16.1. Pour the pigment solution (about 10 ml) along the side of the tube. When it travels down, pour 50 ml of solvent (mixure of petroleum ether and benzene, 9:1 V/V) to get different coloured bands of various four pigments which can be identified on the basis of their colours.

4. To separate the mixture of amino acids by thin layer chromatography mehtod.

Methods of Biochemical Analysis

For this purpose glass plates are coated with adsorbents and then used as paper in paper chromatography. The best adsorbent for the preparation of TLC plates is *silicagel*.

Preparation of the plate: Take a square glass plate of 20 cm × 20 cm. Clean it thoroughly. Now weigh 8 gm silicagel in chemical balance and transfer in pastle and motor. Add 30 ml water and a pinch of gypsum ($CaSO_4$). Mix it thoroughly to form homogeneous solution and transfer it into measuring cylinder. See, the solution should be 30 ml and if it is less, add water to make it 30 ml. Stir it and transfer immediately on the glass plate placed on plain surface. Spread silicagel solution uniform on the plate with the help of clean slide. Leave it for 15 minutes till the gel solidifies. Now keep this plate carefully in an oven at 80°C. When the silicagel coated plate (TLC plate) becomes dry, it can be used for the experiment.

To start the experiment, the TLC plate should be kept again in oven at 120°C for 30 to 40 minutes to activate the silicagel. Cool this plate and use.

Spotting: The spotting should be done carefully with the help of capillary tubes or micropipettes (0.001 to 0.1 ml). Apply few drops of amino acid mixture and dry with the help of 60 W bulb or hair drier from lower side of the plate. Direct dry will damage the plate. Once spotting is done, it can be used like paper.

Solvent: The solvent in this case will be Butanol, Acetic acid and water in the ratio of 4:1:5 (V/V) respectively. Keep this mixutre in a separating funnel, mix thoroughly. Two layers will be separated. The upper layer is butanol saturated with water and lower layer is water saturated with butanol. Discard lower layer and use upper layer as solvent.

Developer: Developer for amino acids is ninhydrin solution. To prepare it weight 200 mg ninhydrin powder and dissolve in 100 ml acetone. Thus, 2% ninhydrin solution is prepared. It is sprayed on the plate, after removing it from the solvent and drying with the help of fine glass sprayer. The spots of amino acids will develop after heating the plate at 80°C. They can be identified by calculating the Rf values and then by comparing with standard ones.

The aminoacid mixture from pea seeds can be prepared by making their power and dissolving it in 80% alcohol. This solution is centrifuged and the supernatent is used for spotting the amino acids.

5. To separate mixture of plant sugars by thin layer chromatography method.

The procedure, preparation of the plate, spotting and solvent etc. are the same as described for separation of amino acids but only the developer is different. For developing the spots of sugars the developers used are $KMnO_4$ solution or benzidine solution or ammonium molybdate solution. The spots develop after heating the sprayed plate at 80°C.

6. Separation of amino acids and sugars by paper chromatography.

The procedure for their separation will be exactly similar as described in thin layer chromatography except that here paper will be used in place of glass plate.

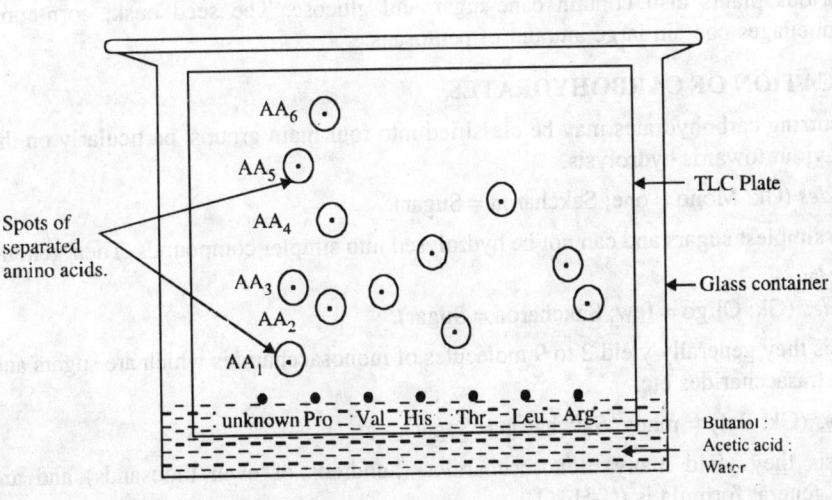

Fig. 1.18: Separation of amino acids by TLC method.

2
Carbohydrates

2.1. INTRODUCTION

The name carbohydrate indicates that they are hydrates of carbon and contain carbon, hydrogen and oxygen. Most of them contain hydrogen and oxygen in the ratio of 2:1. For that reason, the general empirical formula of carbohydrates is given as $[C(H_2O)]_n$ e.g., glucose ($C_6H_{12}O_6$), fructose ($C_6H_{12}O_6$), and sucrose ($C_{12}H_{22}O_{11}$) have the ratio of hydrogen to oxygen as 2:1 as in water. There are certain other sugars like rhamnose ($C_6H_{12}O_5$) and sorbitol ($C_6H_{14}O_6$) where this ratio of hydrogen to oxygen is not like water. There are certain other organic compounds like formaldehyde (H.CHO), acetaldehyde (CH_3.CHO), and lactic acid (CH_3.CHOH.COOH) which contain C, H and O and the ratio of H:O is also the same as in water, but are not carbohydrates.

Thus, *carbohydrates are substances which are either polyhydroxy aldehydes or ketones or are substances that yield polyhydroxy aldehydes or ketones on hydrolysis*, e.g. glucose is a polyhydroxy aldehyde and fructose a polyhydroxy ketone.

The carbohydrates formed in photosynthesis plays an important role in the life of plants and animals. Their molecules store and transport energy that is utilized in various biochemical and physiological processes in the cell. The carbohydrates in a living cell are in a constant flux participating in many enzyme catalysed reactions. This is necessary to convert bond energy into chemical energy for the growth and development of the cell.

A large number of carbohydrates isolated from the plants are components of the cell wall, protoplasm and cell-sap while others accumulate as insoluble storage products.

Carbohydrates occur in grains, tubers, roots, flowers, fruits and in certain other secretions. In grains, tubers and roots, the carbohydrates are starch and cellulose and form the staple food for men. The wood produced by plants is cellulose. The nectar contains cane-sugar and glucose. Fruits and juices of various plants also contain cane-sugar and glucose. The seed husk, corn-cobs, plant-gums and mucilages contain large amount of pentosans.

2.2. CLASSIFICATION OF CARBOHYDRATES

The naturally occurring carbohydrates may be classified into four main groups, particularly on the basis of their behaviour towards hydrolysis.

1. *Monosaccharides* (Gk: Mono = one; Sakcharon = Sugar).

They are the simplest sugars and can not be hydrolysed into simpler compounds. Their general formula is $C_nH_{2n}O_n$.

2. *Oligosaccharides* (Gk: Oligo = few; Sakcharon = Sugar).

On hydrolysis they generally yield 2 to 9 molecules of monosaccharides which are sugars and include di-, tri-, tetrasaccharides etc.

3. *Polysaccharides* (Gk: Poly = many; Sakcharon = Sugar).

On hydrolysis they yield many monosaccharides (hundreds or even thousands) and are non-sugars. Their general formula is $(C_6H_{10}O_5)_x$.

4. Glycosides

They are a kind of oligosaccharides which on hydrolysis yield a carbohydrate, and a non-carbohydrate fragment, which may be a hydroxy-compound or a nitrogen base *e.g.*, Arbutin and Salicin are o-glycosides.

$$\underset{\text{Arbutin}}{C_{12}H_{16}O_7} + H_2O = \underset{\substack{\text{Glucose}\\\text{(Carbohydrate)}}}{C_6H_{12}O_6} + \underset{\substack{\text{Hydroquinone}\\\text{(non-carbohydrate)}}}{C_6H_6O_2}$$

Adenosine gives a pentose sugar and adenine—a nitrogenous base.

1. **Monosaccharides** : They are further classified on the basis of number of carbon atoms present in a molecule *e.g.*
    (*i*) Trioses contain 3-carbon atoms *e.g.* Glyceraldehyde and Dihydroxy acetone.
    (*ii*) Tetroses contain 4-carbon atoms *e.g.* Erythrose and threose.
    (*iii*) Pentoses contain 5-carbon atoms *e.g.* Ribose, Ribulose, Deoxyribose, Arabinose, Xylose, and Xylulose.
    (*iv*) Hexoses contain 6-carbon atoms. *e.g.* Glucose, Fructose, Mannose and Galactose.
    (*v*) Heptoses contain 7-carbon atoms. *e.g.* Sedoheptulose.
Others are Octoses and Nonoses, containing eight and nine carbon atoms respectively.

2. **Oligosaccharides** : May be classified into :
    (*i*) *Disaccharides.* Yield 2-molecules of monosaccharides on hydrolysis. *e.g.* Sucrose, Maltose, Lactose, Melibiose, Cellobiose, Gentiobiose, Trehalose.
    (*ii*) *Trisaccharides.* Yield 3-molecules of monosaccharides on hydrolysis. *e.g.* Gentianose, Reffinose & Melezitose.
    (*iii*) *Tetrasaccharides.* Yield 4-molecules of monosaccharides on hydrolysis *e.g.* Stachyose.

Others are penta and hexasaccharides which yield 5 and 6-molecules of monosaccharides respectively.

3. **Polysaccharides**. Include the following non-sugars.
    (*i*) *Pentosans. e.g.* Araban and Xylan.
    (*ii*) *Hexosans* — They are further classified into-
        (*a*) Glucosans. *e.g.* Starch, Cellulose and Glycogen.
        (*b*) Fructosans. *e.g.* Inulin, Synanthrin, Graminin, etc.
        (*c*) Mannans. *e.g.* Mannane, Mannocellulose.
        (*d*) Galactans. *e.g.* Galactane, Paragalactane.
    (*iii*) *Pectic Compounds. e.g.* Pectic acid, Pectin & Protopectin.
    (*iv*) *Gums.*
    (*v*) *Amino-hexosans. e.g.* Chitin.
    (*vi*) *Mucilages.*

2.3. CHEMISTRY OF MONOSACCHARIDES

In the chemistry of monosaccharides the following two facts are important :
   (1) Isomerism, and
   (2) Ring Structure

1. Isomerism

The term isomer (Gk: *isos* = equal; *meros* = part) was first used by **J.J. Berzelius** (1827) to different compounds with same molecular formula. The isomerisms are of two types.
   (*i*) Structural isomerism,

(*ii*) Stereoisomerism.

Structural Isomerism

The structural isomers have similar molecular formulae, but different structural formulae *e.g.* D-glucose and D-fructose have the same molecular formula i.e., $C_6H_{12}O_6$, but differ in their structural formulae.

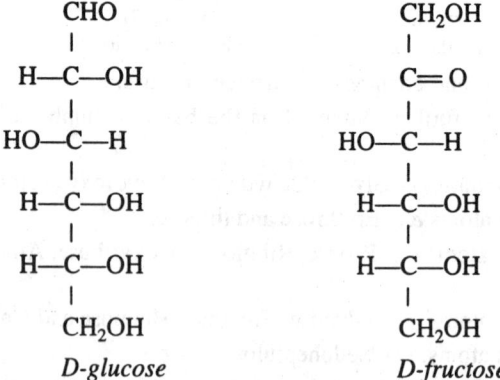

The structural isomers are again of three types:

(*a*) *Chain isomers*: They have different arrangement of carbon atoms generally in the form of a chain or link.

(*b*) *Positional isomers*: A substituent group in two compounds is at different positions, but their chain is the same.

(*c*) *Functional isomers*: Isomers in which the compounds have different functional groups, *e.g.* compounds having formula (C_2H_6O) may be an ethyl alcohol ($CH_3—CH_2OH$) or a dimethyl-ether ($CH_3—O—CH_3$), and D-glucose and D-fructose.

Stereoisomerism

Stereoisomers have the same structural and molecular formula, but differ in spatial arrangement of atoms or groups in the molecule. The arrangement of the groups in different patterns always takes place around the **asymmetric carbon** atoms i.e. carbon atoms in which their 4 valencies are completely satisfied by different kinds of atoms, *e.g.* D-glucose and D-mannose have change in spatial arrangement of hydrogen and hydroxyl groups.

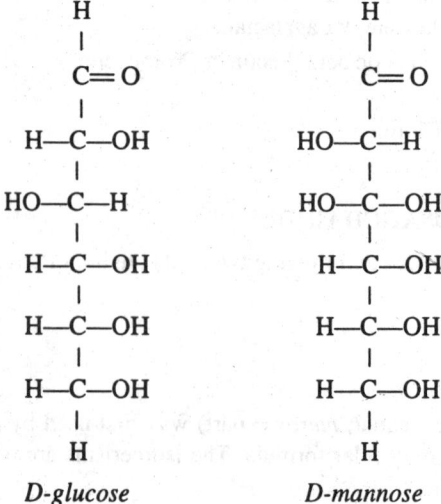

Carbohydrates

'D' molecule is one in which the
  (i) Asymmetric carbon atom is farthest from aldehyde or keto-group, and
  (ii) Adjacent to the terminal –CH_2OH group, the hydroxyl or –OH group is projected to the right side.

In L-forms the –OH group of the same carbon is shown on the left. The L-form of glucose and mannose would be:

```
           H                                    H
           |                                    |
         1 C=O                                1 C=O
           |                                    |
     ┌─────────────┐                      ┌─────────────┐
     │ HO—²C — H   │                      │ H —²C — OH  │
     └─────────────┘                      └─────────────┘
           |                                    |
      H — ³C — OH                         H — ³C — OH
           |                                    |
      HO—⁴C — H                           HO—⁴C — H
           |                                    |
      HO—⁵C — H                           HO—⁵C — H
           |                                    |
      HO—⁶C — H                           HO—⁶C — H
           |                                    |
           H                                    H
       L-glucose                            L-mannose
```

The structural formulae of L-glucose and L-mannose would be the mirror image of D-glucose and D-mannose respectively. This is also known as **opitcal isomerism** which is a kind of stereoisomerism.

In the same way the D-glyceraldehyde and L-glyceraldehyde would be—

```
           H                                    H
           |                                    |
           C=O                                  C=O
           |                                    |
       H — C — OH                         HO — C — H
           |                                    |
          CH₂OH                                CH₂OH
     D-glyceraldehyde                    L-glyceraldehyde
```

2.3.2. Ring Structure

The molecules of sugars may exist in two different ring forms. The two forms differ from each other in their stability and reactivity. The two ring forms are —

Pyranose ring

It is more stable than furanose ring. Here C-1 and C-5 are linked by an oxygen atom, thus

```
              OH
              |
        H — ¹C ─────────┐
              |         |
        H — ²C — OH     |
              |         |
       HO — ³C — H      O
              |         |
        H — ⁴C — OH     |
              |         |
        H — ⁵C ─────────┘
              |
            ⁶CH₂OH
```

D-glucose (Ring structure)
Straight Chain form

D-glucose (Pyranose ring)
Closed-form

forming a large-sized stable ring. This ring structure is exhibited by pentose and hexose sugars which are generally written as closed hexagons or pentagons. The ring may be written either in the form of a straight carbon chain with oxygen atom shared between appropriate carbon atoms or in a closed form. Thus, pyranose ring for the glucose may be written in either way.

Furanose ring

It is less stable. Here C-1 and C-4 are linked by an oxygen atom, thus forming a small-sized less stable ring. This ring structure is exhibited by pentose and hexose sugars. They may be written as closed carbon rings or as chain-carbon rings.

(D-fructose showing ring structure)

(Furanose form of Ribose)
Closed ring.

Each of the hexoses exists in α- and β-forms depending upon the position of the -H and -OH groups to C-1 which is also asymmetric in pyranose and furanose rings. In β-form the -OH group is attached to C-1 on the left side, while in α-form towards the right side.

β-D-Glucose   α-D-Glucose

Hemiacetal Formula

Generally it has been seen that the aldehyde can add hydroxyl compounds to the carbonyl (C=O) bond. When water is added, the hydrate of aldehyde is formed, but if an alcohol is added in place of water, the *hemiacetal* is formed. Further, when other molecules of alcohol are added in the hemiacetal, they form the full acetals and the elimination of water takes place. This reaction is the basis of glycoside formation by carbohydrates.

(i) $R_2C=O$ + HOH → $R_2C(OH)_2$

    Aldehyde   Water   Aldehyde hydrate

(ii) $R_2C=O$ + ROH → $R_2C(OH)(OR)$

    Aldehyde   Alcohol   Hemiacetal

(iii) $\begin{array}{c}H\\R\end{array}C\begin{array}{c}OH\\OR\end{array}$ + R'OH → $\begin{array}{c}H\\R\end{array}C\begin{array}{c}OR\\OR\end{array}$ + H_2O

    Hemiacetal   Alcohol   Full acetal

Carbohydrates

A ring is produced with the formation of hemiacetal. A fives membered ring with one oxygen atom forms **furan** and a six-membered ring with one oxygen atom forms **pyran**. The sugars containing these forms are correspondingly known as *Furanose* or *Pyranose* sugars.

Furan Pyran

On the basis of this ring, the β-D-glucose would be a *glucopyranose* while α-D glucose would be a *glucofuranose*.

Fischer's Projection and Haworth's Projection

The formula written in the chain form with carbon atoms, along with oxygen atom forming a ring was given by *Fischer*. This formula does not show the proper arrangement of molecules in the ring form *e.g.*, in the Fischer's formula the ring form of glucopyranose does not show that—

(*i*) C_6 and its attached groups are trans i.e., they are alternately up and down with respect to hydroxyl groups on carbons 1, 2, and 4.

(*ii*) It does not show the proximity of carbon 1 and 5. Both these things are clear in *Haworth's Projection*.

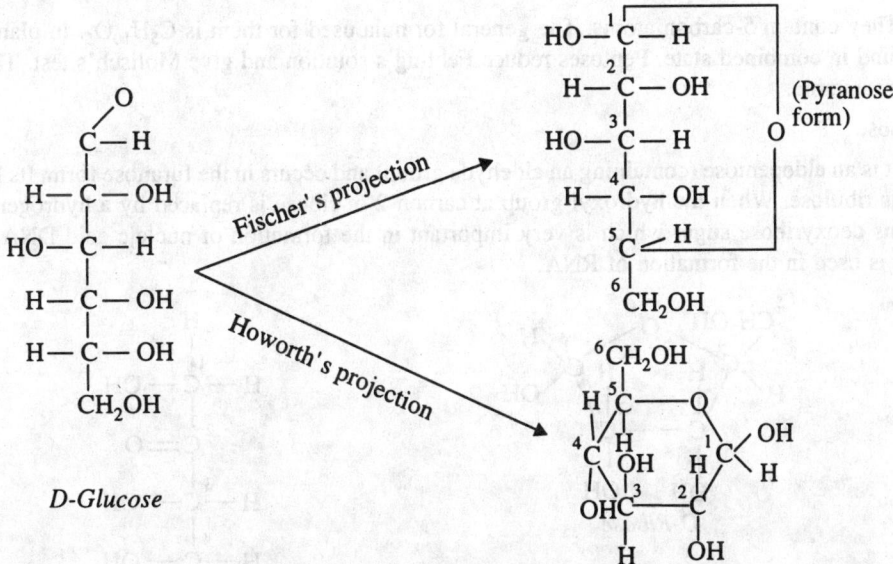

β-D-glucose, (Pyranose) or Glucopyranose

2.4. CLASSIFICATION OF MONOSACCHARIDES

1. Trioses

The simplest compounds having three carbon atoms are known as **trioses** *e.g.*, glyceraldehyde

D-Glyceraldehyde   L-glyceraldehyde   Dihydroxyacetone

and dihydroxyacetone. Glyceraldehyde is an aldotriose (containing-CHO gp), while the dihydroxyacetone is a ketotriose (containing C = Ogp.).

Both are colourless, sweet, crystalline and soluble in water, but insoluble in ether. Both show the properties of an aldehyde and a ketone, and cannot be hydrolysed. They are formed in plants during glycolysis.

2 Tetroses

(i) **Erythrose**. This sugar has four carbon atoms and properties like that of a triose. It is produced in plants in photosynthesis in presence of transketolase from fructose -6-phosphate.

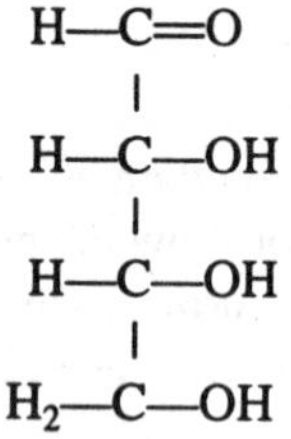

$$\begin{array}{c} H-C=O \\ | \\ H-C-OH \\ | \\ H-C-OH \\ | \\ H_2-C-OH \end{array}$$

Erythrose

3. Pentoses

They contain 5-carbon atoms. The general formula used for them is $C_5H_{10}O_5$. In plants they are found in combined state. Pentoses reduce Fehling's solution and give Molisch's test. They are not fermentable.

1. Ribose

It is an aldopentose (containing an aldehyde group) and occurs in the furanose form. Its ketonic form is ribulose. When the hydroxyl group at carbon-2 of ribose is replaced by a hydrogen atom, it forms deoxyribose sugar which is very important in the formation of nucleic acid DNA, while ribose is used in the formation of RNA.

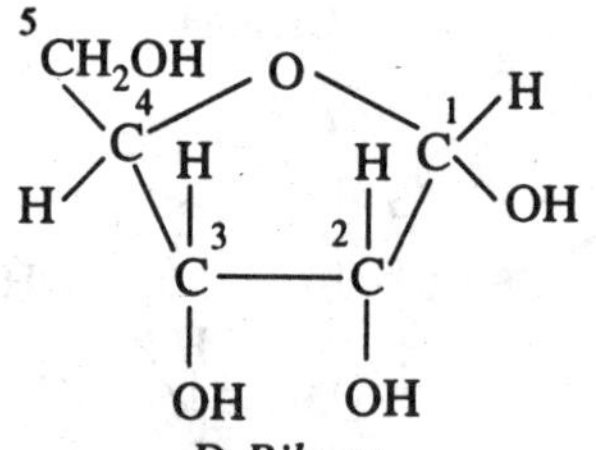

D-Ribose

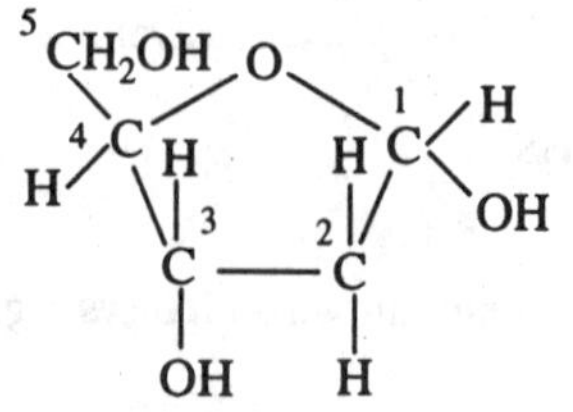

D-2 Deoxyribose
(Here oxygen atom at Carbon no. 2 is lacking)

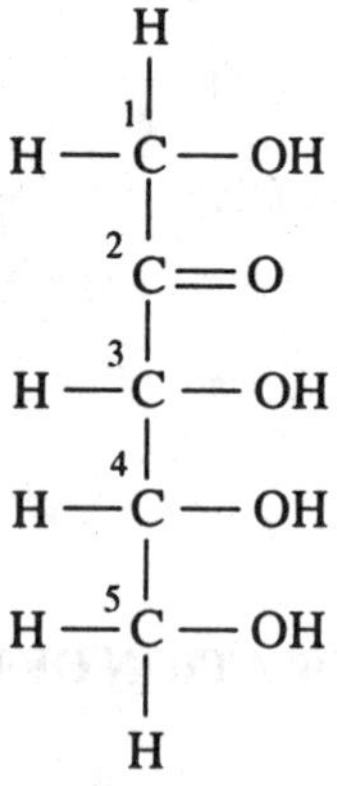

Ribulose (Ketonic form)

2. Arabinose

It is colourless, crystalline and sweet in taste. It can be obtained by the hydrolysis of gum arabic, peach gum and cherry gum. It reduces Fehling's solution and with diphenyl hydrazine forms a characteristic diphenylhydrazone. The melting point of arabinose is 160°C.

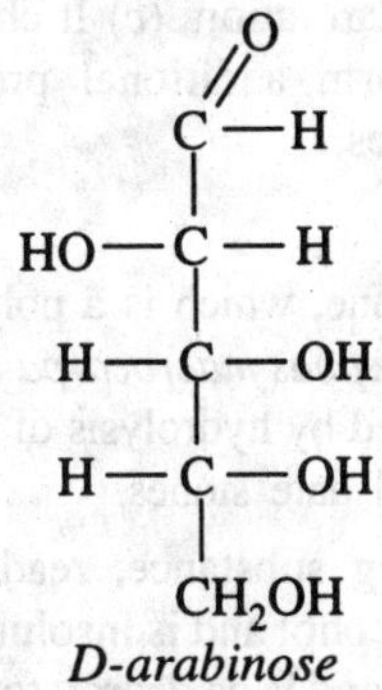

```
      O
      ‖
      C—H
      |
   HO—C—H
      |
    H—C—OH
      |
    H—C—OH
      |
     CH₂OH
   D-arabinose
```

3. Xylose

It is an aldose sugar. Its ketonic form is xylulose and is formed in photosynthesis. It is colourless, crystalline, optically inactive having melting point of 144-145°C. It yields xylonic acid when oxidized with bromine. It is obtained by the hydrolysis of wood gum or xylane and also from maize fruits, straw or from other forms of celluloses.

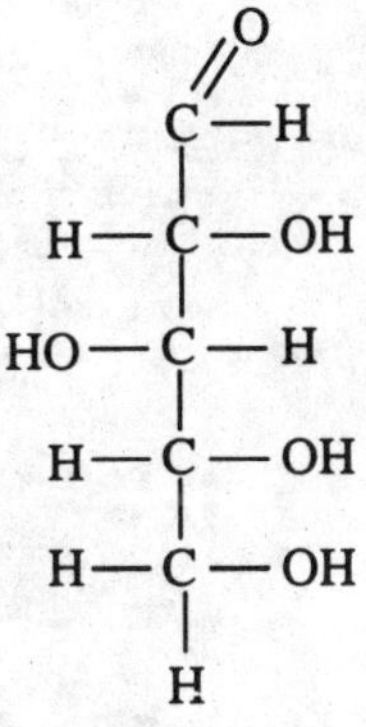

```
      O                          H
      ‖                          |
      C—H                     H—C—OH
      |                          |
    H—C—OH                       C=O
      |                          |
   HO—C—H                     HO—C—H
      |                          |
    H—C—OH                    H—C—OH
      |                          |
    H—C—OH                    H—C—OH
      |                          |
      H                          H
   D-xylose                  D-xylulose
```

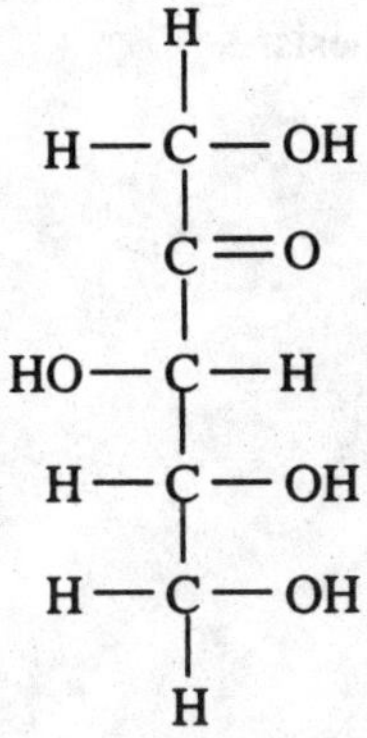

4. Hexoses

These sugars have 6-carbon atoms and have the same properties as pentoses. They may be present either as aldoses or as ketoses. They cannot be hydrolysed. Examples are:

1. Glucose

It is also known as dextrose. It is formed by the hydrolysis of cane-sugar, glucosides and many polysaccharides, such as starch, cellulose etc. It has needle-shaped crystals, which are anhydrous. It crystallizes in the form of plates with one molecule of water. It has an aldehyde group and thus shows the properties of aldehyde.

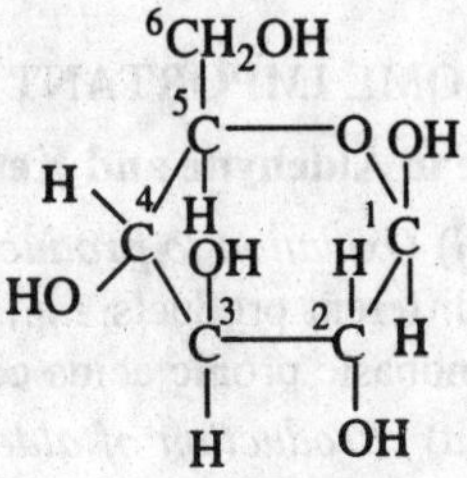

Glucose differs from an aldehyde in :

(a) It shows no addition reaction with NH_3 and $NaHSO_3$, (b) It has properties of -OH groups, (c) It does not give any colour with schiff's reagent, (d) It shows mutarotation.

D-glucose

2. Fructose

It is a ketose sugar having 6-carbon atoms. It is formed in equal quantity with glucose by the hydrolysis of cane-sugar. It is soluble in hot absolute alcohol and ether.

Fructose differs from ketones because, (a) It shows the properties of alcohols or –OH gp, (b) It shows mutarotation, (c) It chars in cold with conc. H_2SO_4, (d) It does not form additional product with $NaHSO_3$, (e) It shows reducing properties.

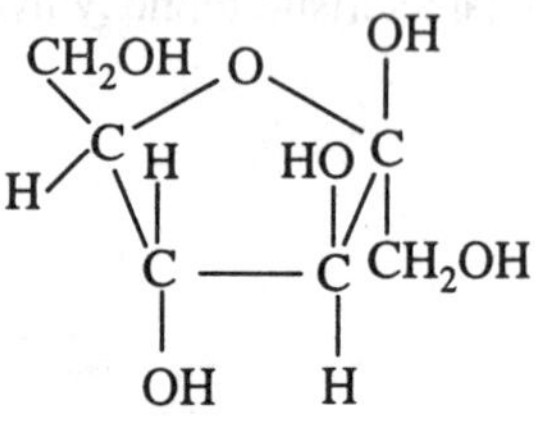

3. Mannose

It is prepared by hydrolysing Mannane, which is a polysaccharide and found in Ivory nuts, the fruits of *Phytelephas macrocarpa* and in salep mucilage (Orchid *Morio*). It is also prepared by hydrolysis of hemi-celluloses contained in peas, coffee beans and date stones.

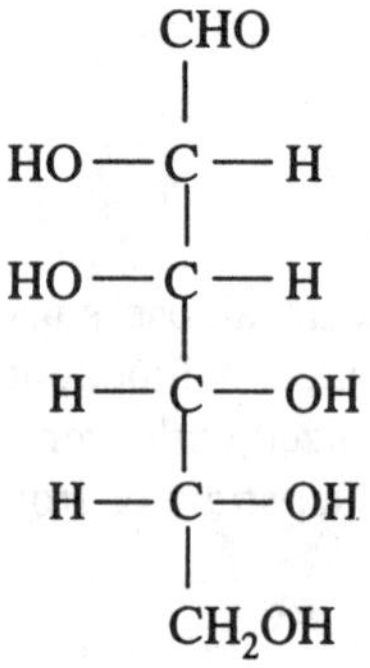

In dry state it is a hard crumbling substance, readily soluble in water, slightly soluble in hot alcohol and is insoluble in ether. It is readily fermentable by yeast and is dextrorotatory. It is detected by phenylhydrazones and reduces fehling's solution.

2.4.5. Heptoses

They are 7-C sugars. Sedoheptulose is a ketose sugar and is formed is photosynthesis.

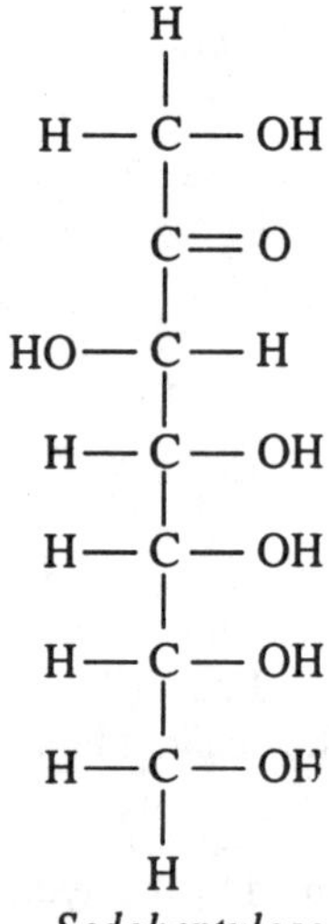

Fructose

Mannose

Sedoheptulose

2.5. SOME IMPORTANT REACTIONS OF MONOSACCHARIDES

1. Due to Aldehyde and Ketone groups

(i) *Oxidation to produce sugar acids*: Monosaccharides on oxidation under proper conditions form different products e.g., aldoses may form monobasic aldonic acids or dibasic saccharic acids or monobasic uronic acids containing the aldehyde group.

(a) *Production of aldonic acid* : Aldoses when oxidised in presence of bromine water, the aldehyde group is converted into carboxyl group and forms the corresponding acids. The hypobromous acid, an oxidizing agent, is formed by the reaction of bromine with water and it oxidises glucose into gluconic acid.

1. $Br_2 + HOH \longrightarrow HOBr + HBr$

Hypobromous acid

Carbohydrates

Similarly, galactose, mannose and arabinose give galactonic, mannonic and arabonic acid respectively. The Ketoses are not readily oxidised by bromine water.

Monosaccharides with Fehling's solution and Tollen's reagent give rise to red precipitate of cuprous oxide and silver mirror respectively and are oxidized into gluconic acid.

$$\begin{array}{c} \boxed{H-C=O} \\ | \\ H-C-OH \\ | \\ HO-C-H \\ | \\ H-C-OH \\ | \\ H-C-OH \\ | \\ CH_2OH \\ \text{D-glucose} \end{array} + HO\boxed{Br} \longrightarrow \begin{array}{c} COOH \\ | \\ H-C-OH \\ | \\ HO-C-H \\ | \\ H-C-OH \\ | \\ H-C-OH \\ | \\ CH_2OH \\ \text{D-gluconic acid} \end{array} + HBr$$

$$\text{Reducing sugar} + 2\,Cu^{++} \underset{\text{Blue}}{\longrightarrow} \text{Oxidized sugar} + 2\,Cu^{+}$$

$$2\,Cu^{+} + 2\,OH^{-} \longrightarrow \underset{\text{yellow}}{2\,CuOH} \longrightarrow \underset{\text{Red}}{Cu_2O} + H_2O$$

$$\underset{\text{Glucose}}{C_6H_{12}O_6} + Ag_2O \longrightarrow \underset{\text{gluconic acid}}{C_6H_{12}O_7} + \underset{\text{silver mirror}}{2\,Ag}$$

$$\begin{array}{c} H-C=O \\ | \\ H-C-OH \\ | \\ HO-C-H \\ | \\ H-C-OH \\ | \\ H-C-OH \\ | \\ CH_2OH \\ \text{D-glucose} \end{array} + 2Cu(OH)_2 \longrightarrow \begin{array}{c} COOH \\ | \\ H-C-OH \\ | \\ HO-C-H \\ | \\ H-C-OH \\ | \\ H-C-OH \\ | \\ CH_2OH \\ \text{D-gluconic acid} \end{array} + Cu_2O + 2H_2O$$

Red precipitate

(b) *Saccharic acids*: The aldehyde and the primary alcohol groups of addoses when oxidized with nitric acid under suitable conditions are converted to carboxyl to form saccharic acids or dibasic sugar acids e.g., D-glucose, D-Mannose and D-galactose are converted into D-glucaric acid (D-gluco-saccharic acid), D-mannaric acid (D-mannosaccharic acid) and D-galactosaccharic acid respectively. These names have been given by adding to saccharic a prefix indicating the sugar from which the acid is derived. Fructose is oxidized to mesotartaric acid and glycollic acid and trihydroxy glutaric acid.

(i)
$$\underset{\text{D-glucose}}{\begin{array}{c} CHO \\ | \\ H-C-OH \\ | \\ HO-C-H \\ | \\ H-C-OH \\ | \\ H-C-OH \\ | \\ CH_2OH \end{array}} \xrightarrow[HNO_3]{3O} \underset{\text{D-glucosaccharic acid}}{\begin{array}{c} COOH \\ | \\ H-C-OH \\ | \\ HO-C-H \\ | \\ H-C-OH \\ | \\ H-C-OH \\ | \\ COOH \end{array}} + H_2O$$

(ii) D-mannose $\xrightarrow[HNO_3]{3\,[O]}$ D-mannosaccharic acid + H_2O

(iii) D-galactose $\xrightarrow[HNO_3]{3\,[O]}$ D-galactosaccharic acid

(iv) $\underset{\text{fructose}}{CH_2OH \cdot (CHOH)_3 \cdot CO \cdot CH_2OH} \xrightarrow[\text{con. } HNO_3]{[O]} \underset{\text{tartaric acid}}{HOOC \cdot (CHOH)_2} \cdot$

$+ \underset{\text{glycollic acid}}{HOOC \cdot CH_2OH} + \underset{\text{trihydroxyglutaric acid}}{COOH \cdot (CHOH)_3 \cdot COOH}$

(c) *Uronic acids*: Those aldoses in which the primary alcohol group is converted to carboxyl, but the aldehyde group is not oxidised, form uronic acid. These are produced under proper conditions and the aldose derivatives are oxidized in presence of activated platinum carbon catalyst by oxygen.

$$\underset{\text{D-glucose}}{\begin{array}{c} CHO \\ | \\ H-C-OH \\ | \\ HO-C-H \\ | \\ H-C-OH \\ | \\ H-C-OH \\ | \\ CH_2OH \end{array}} \xrightarrow{\text{Acetone}} \underset{\substack{\text{1, 2-0 isopropylidine}\\\text{D-glucose}}}{\begin{array}{c} H-C-O \\ | \quad\quad\>\!\!\!\searrow \\ H-C-O \quad C(CH_3)_2 \\ | \quad\quad\>\!\!\!\nearrow \\ HO-C-H \\ | \\ H-C-OH \\ | \\ H-C \\ | \\ CH_2OH \end{array}} \xrightarrow{\text{Oxidized}} \underset{\substack{\text{1, 2-0-isopropylidene}\\\text{D-glucuronic acid}}}{\begin{array}{c} H-C-O \\ | \quad\quad\>\!\!\!\searrow \\ H-C-O \quad C(CH_3)_2 \\ | \quad\quad\>\!\!\!\nearrow \\ HO-C-H \\ | \\ H-C-OH \\ | \\ H-C \\ | \\ COOH \end{array}} \to \underset{\text{D-glucuronic acid}}{\begin{array}{c} CHO \\ | \\ H-C-OH \\ | \\ HO-C-H \\ | \\ H-C-OH \\ | \\ H-C-OH \\ | \\ COOH \end{array}}$$

(ii) *Reduction to produce Sugar Alcohols*

Both aldoses and Ketoses on reduction yield polyhydric or Polyhydroxy alcohols. The reaction may be accomplished either with sodium amalgam and water or in presence of a catalyst by hydrogen under high pressure *e.g.*, the glucose, mannose and fructose on reduction yield Sorbitol, mannitoal and mannitol and sorbitol respectively.

Carbohydrates

(i)
```
H—C=O                    CH₂OH
H—C—OH                   H—C—OH
HO—C—H         2H        HO—C—H
H—C—OH         ──→       H—C—OH
H—C—OH                   H—C—OH
CH₂OH                    CH₂OH
D-glucose                D-sorbitol
```

(ii)
```
H—C=O                    CH₂OH
HO—C—H                   HO—C—H
HO—C—H         2H        HO—C—H
H—C—OH         ──→       H—C—OH
H—C—OH                   H—C—OH
CH₂OH                    CH₂OH
D-mannose                D-mannitol
```

(iii)
```
CH₂OH              CH₂OH                 CH₂OH
C=O                HO—C—H                H—C—OH
HO—C—H             HO—C—H                HO—C—H
H—C—OH   4H        H—C—OH        +       H—C—OH
H—C—OH   ──→       H—C—OH                H—C—OH
CH₂OH              CH₂OH                 CH₂OH
D-fructose         D-mannitol            D-sorbitol
```

Glucose forms n-hexane in presence of stronger reducing agents like red phosphorus and hydrogen iodide.

$$CH_2OH \cdot (CHOH)_4 \cdot CHO \xrightarrow[HI]{Red\ P} CH_3(CH_2)_4 \cdot CH_3$$

glucose   N-hexane

Glyceraldehyde and dihydroxy-acetone on reduction form the trihydroxy alcohol or glycerol.

```
    O
    ‖
    C—H                        CH₂OH
H—C—OH          2H             CHOH
    |           ──→            |
    CH₂OH                      CH₂OH

Glyceraldehyde                 Glycerol
```

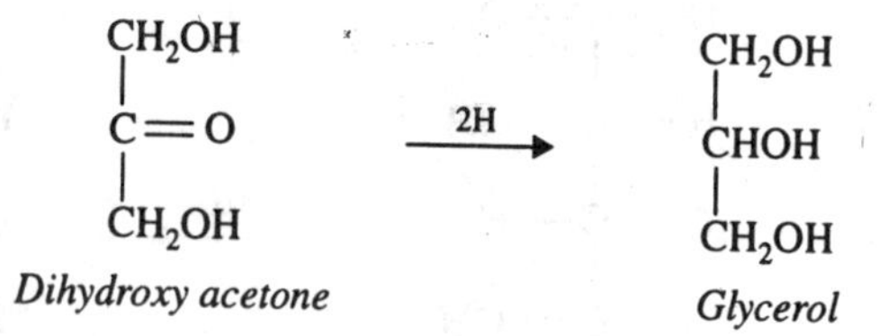

$$\underset{\text{Dihydroxy acetone}}{\begin{array}{c}CH_2OH\\|\\C=O\\|\\CH_2OH\end{array}} \xrightarrow{2H} \underset{\text{Glycerol}}{\begin{array}{c}CH_2OH\\|\\CHOH\\|\\CH_2OH\end{array}}$$

(iii) Reaction with hydrazines to form Hydrazones and Osazones. Monosaccharides react with phenylhydrazine and other substituted hydrazines to form osazones. These contain a free sugar group to form hydrazones and osazones e.g., glucose forms glucose phenylhydrazones and finally glucosazone.

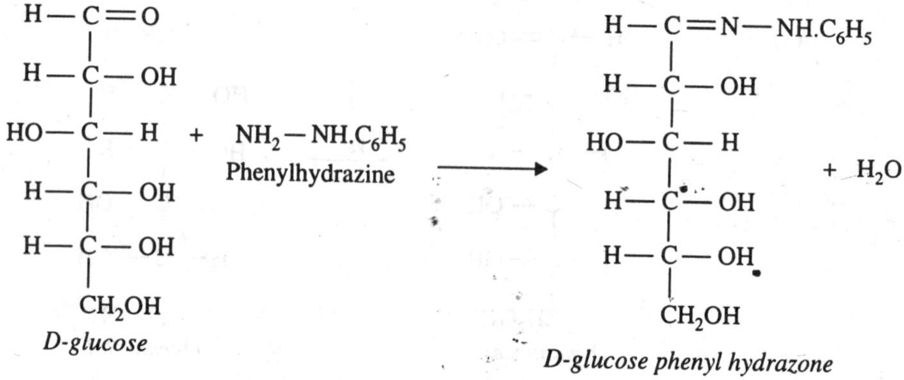

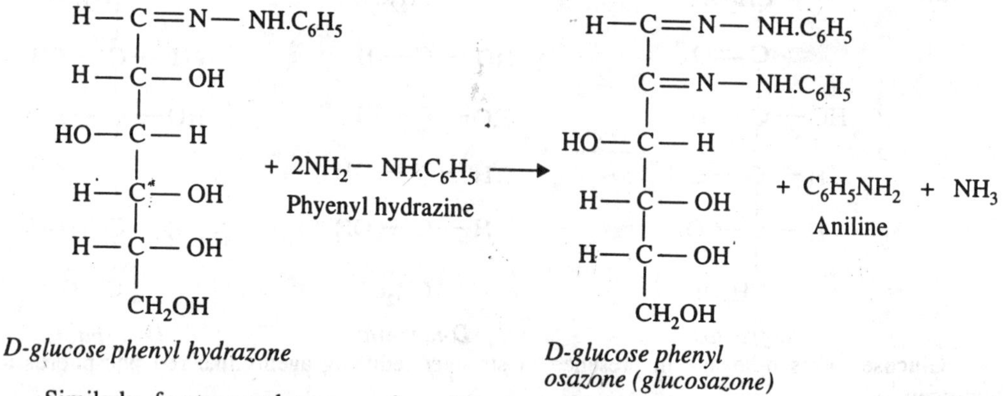

$$\underset{\text{D-glucose}}{\begin{array}{c}H-C=O\\|\\H-C-OH\\|\\HO-C-H\\|\\H-C-OH\\|\\H-C-OH\\|\\CH_2OH\end{array}} + \underset{\text{Phenylhydrazine}}{NH_2-NH.C_6H_5} \longrightarrow \underset{\text{D-glucose phenyl hydrazone}}{\begin{array}{c}H-C=N-NH.C_6H_5\\|\\H-C-OH\\|\\HO-C-H\\|\\H-C-OH\\|\\H-C-OH\\|\\CH_2OH\end{array}} + H_2O$$

$$\underset{\text{D-glucose phenyl hydrazone}}{\begin{array}{c}H-C=N-NH.C_6H_5\\|\\H-C-OH\\|\\HO-C-H\\|\\H-C-OH\\|\\H-C-OH\\|\\CH_2OH\end{array}} + \underset{\text{Phyenyl hydrazine}}{2NH_2-NH.C_6H_5} \longrightarrow \underset{\substack{\text{D-glucose phenyl}\\\text{osazone (glucosazone)}}}{\begin{array}{c}H-C=N-NH.C_6H_5\\|\\C=N-NH.C_6H_5\\|\\HO-C-H\\|\\H-C-OH\\|\\H-C-OH\\|\\CH_2OH\end{array}} + \underset{\text{Aniline}}{C_6H_5NH_2} + NH_3$$

Similarly, fructose and mannose form Osazones of fructose and mannose.

(iv) Reaction with hydrogen cyanide to form cyanohydrins: When HCN is added to an aldehyde group, an asymmetric carbon is formed with two cyanohydrins.

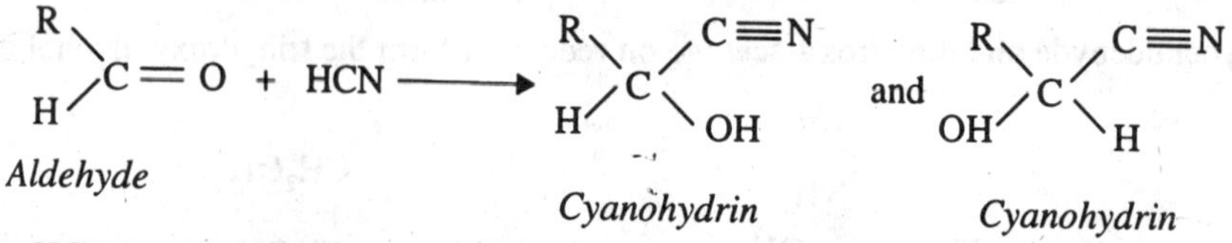

$$\underset{\text{Aldehyde}}{\begin{array}{c}R\\ \diagdown \\ C=O\\ \diagup \\ H\end{array}} + HCN \longrightarrow \underset{\text{Cyanohydrin}}{\begin{array}{c}R\quad C\equiv N\\ \diagdown \diagup \\ C\\ \diagup \diagdown \\ H\quad OH\end{array}} \text{ and } \underset{\text{Cyanohydrin}}{\begin{array}{c}R\quad C\equiv N\\ \diagdown \diagup \\ C\\ \diagup \diagdown \\ OH\quad H\end{array}}$$

When HCN is added to glucose or glyceraldehyde, cyanohydrins of glucose and glyceraldehyde are formed.

Carbohydrates

$$\begin{array}{c} \text{CHO} \\ | \\ \text{H}-\text{C}-\text{OH} \\ | \\ \text{HO}-\text{C}-\text{H} \\ | \\ \text{H}-\text{C}-\text{OH} \\ | \\ \text{H}-\text{C}-\text{OH} \\ | \\ \text{CH}_2\text{OH} \end{array} \quad + \quad \text{HCN} \quad \longrightarrow \quad \begin{array}{c} \text{CH} \stackrel{\text{OH}}{\underset{\text{CN}}{<}} \\ | \\ \text{H}-\text{C}-\text{OH} \\ | \\ \text{HO}-\text{C}-\text{H} \\ | \\ \text{H}-\text{C}-\text{OH} \\ | \\ \text{H}-\text{C}-\text{OH} \\ | \\ \text{CH}_2\text{OH} \end{array}$$

D-glucose   *D-glucose cyanohydrin*

$$\begin{array}{c} \text{H}-\text{C}=\text{O} \\ | \\ \text{H}-\text{C}-\text{OH} \\ | \\ \text{CH}_2\text{OH} \end{array} + 2\text{HCN} \longrightarrow \begin{array}{c} \text{C}\equiv\text{N} \\ | \\ \text{H}-\text{C}-\text{OH} \\ | \\ \text{H}-\text{C}-\text{OH} \\ | \\ \text{CH}_2\text{OH} \end{array} + \begin{array}{c} \text{C}\equiv\text{N} \\ | \\ \text{OH}-\text{C}-\text{H} \\ | \\ \text{H}-\text{C}-\text{OH} \\ | \\ \text{CH}_2\text{OH} \end{array}$$

D-glyceraldehyde   *D-glyeraldehyde cyanohydrin*

(v) Reaction with hydroxylamine to form oxines: Aldoses and ketoses undergo condensation reaction to form *oximes*.

$$\begin{array}{c} \text{H}-\text{C}=\text{O} \\ | \\ \text{H}-\text{C}-\text{OH} \\ | \\ \text{HO}-\text{C}-\text{H} \\ | \\ \text{H}-\text{C}-\text{OH} \\ | \\ \text{H}-\text{C}-\text{OH} \\ | \\ \text{CH}_2\text{OH} \end{array} + \text{H}_2\text{N}-\text{OH} \longrightarrow \begin{array}{c} \text{H}-\text{C}=\text{N}-\text{OH} \\ | \\ \text{H}-\text{C}-\text{OH} \\ | \\ \text{HO}-\text{C}-\text{OH} \\ | \\ \text{H}-\text{C}-\text{OH} \\ | \\ \text{H}-\text{C}-\text{OH} \\ | \\ \text{CH}_2\text{OH} \end{array} + \text{H}_2\text{O}$$

D-glucose   *D-glucose oxime*

(vi) Action of alkalies: Monosaccharides form enol salt in alkaline solution and when two hydroxyl groups are attached to the double bond-carbon atom, such enolic forms of sugar are known as *enediols*.

$$\begin{array}{c} \text{H}-\text{C}=\text{O} \\ | \\ \text{H}-\text{C}-\text{OH} \\ | \\ \text{HO}-\text{C}-\text{H} \\ | \\ \text{H}-\text{C}-\text{OH} \\ | \\ \text{H}-\text{C}-\text{OH} \\ | \\ \text{CH}_2\text{OH} \end{array} \rightleftharpoons \begin{array}{c} \text{H}-\text{C}-\text{OH} \\ || \\ \text{C}-\text{OH} \\ | \\ \text{HO}-\text{C}-\text{H} \\ | \\ \text{H}-\text{C}-\text{OH} \\ | \\ \text{H}-\text{C}-\text{OH} \\ | \\ \text{CH}_2\text{OH} \end{array} + \text{NaOH} \rightleftharpoons \begin{array}{c} \text{H}-\text{C}-\text{ONa} \\ || \\ \text{C}-\text{OH} \\ | \\ \text{HO}-\text{C}-\text{H} \\ | \\ \text{H}-\text{C}-\text{OH} \\ | \\ \text{H}-\text{C}-\text{OH} \\ | \\ \text{CH}_2\text{OH} \end{array} + \text{H}_2\text{O}$$

D-glucose   *1-2 enediol*   *enediol salt*

In the same way enediol salts of fructose and mannose are also formed.

(vii) Action of acids: Monosacharides are stable to hot dilute acids, but when the concentration of acid is increased the molecule is decomposed e.g. glucose and fructose from *Levulinic acid.*

$$C_6H_{12}O_6 \longrightarrow CH_3CO-CH_2CH_2COOH + HCOOH + H_2O$$
glucose or fructose   Levulinic acid   Formic Acid

(viii) Fermentation: The monosaccharides, such as glucose, fructose and mannose are readily fermented by yeast. This process of yeast fermentation is very complex. During this process sugar phosphate & Sugar acid phosphate are formed. Ordinarily this process in the formation of CO_2 and alcohol.

$$C_6H_{12}O_6 \xrightarrow{yeast} 2C_2H_5OH + 2CO_2$$
glucose   ethyl alcohol

2. Reactions of Sugars Due to hydroxyl (–OH) gp

(i) Formation of glycosides: The simple sugars react with alcohols in presence of hydrogen chloride as catalyst and form glycosides. Generally, the hydroxyl group of carbon-I of sugar undergoes reaction under these conditions e.g., glucose or fructose forms glucosides or furctosides respectively.

$$C_6H_{11}O_5 \cdot OH + HOCH_3 \xrightarrow{dry\ HCl} C_6H_{11}O_5 \cdot OCH_3 + H_2O$$
glucose or fructose   methyl glucoside or fructoside

(ii) Formation of esters: When the sugars are treated with an approriate acid anhydride or chloride under proper conditions, their hydroxyl groups are esterified to give esters, such as Sugar acetates, benzoates, propionates etc. e.g., glucose and furctose with acetic-anhydride frm penta-acetyl derivative in presence of fused $ZnCl_2$.

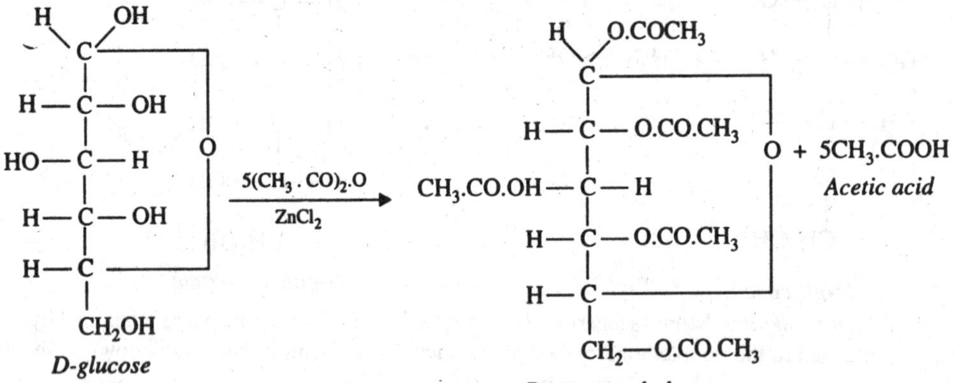

<!-- Structure: D-glucose + 5(CH3.CO)2.O / ZnCl2 → Penta-acetyl glucose or α D-glucopyranose penta acetate + 5 CH3.COOH (Acetic acid) -->

2.6. OTHER SUGAR DERIVATIVES

The more common sugar derivatives which are of biological importance are as follows :

1. Sugar-Phosphates

They are the phosphate derivatives of trioses, tetroses pentoses, hexoses, heptoses etc. and are formed by the process of phosphorylation. Some of them are: Glyceraldehyde-3-Phosphate, Dihydroxyacetone-Phosphate, Erythrose-4-phosphate, Ribose-5-Phosphate, Ribulose-5-phosphate, Xylulose-5-Phosphate, Glucose-6-phosphate, Fructose-6-phosphate etc.

Carbohydrates

2. Amino-Sugars

Amino sugars may be defined as the sugars in which the amino group has been added at the second position in place of hydroxyl group e.g., glucose, mannose and galactose form glucosamine, mannosamine and galactosamine respectively. These amino-sugars occur as N-acetyl derivatives or with an acetyl group attached to the nitrogen.

```
         ¹CHO                          ¹CHO                              O
          |                             |                                ‖
       ²  |                          ²  |                          ¹ C — H    O
      H—C—OH                        H—C—NH₂                         |         ‖
       ³  |                          ³  |                        ² |       
      HO—C—H         NH₃          HO—C—H         CH₃COOH         H—C—NH—C—CH₃
       4  |         ───→          4  |         ───→              ³  |
      HO—C—H        −H₂O          HO—C—H        −H₂O            HO—C—H
       5  |                        5 |                           4  |
      H—C—OH                      H—C—OH                        HO—C—H
       6|                          6|                            5  |
       CH₂OH                       CH₂OH                        H—C—OH
                                                                 6|
       Galactose            Galactosamine or chondrosamine       CH₂OH
                                                              N-acetyl galactosamine
```

```
         ¹CHO                          ¹CHO                              ¹CHO
       ²  |                          ²  |                       O     ²  |
      HO—C—OH                       NH₂—C—H                     ‖       |
       ³  |                          ³  |                    CH₃—C—HN—C—NH
      HO—C—H         NH₃          HO—C—H         CH₃COOH            ³  |
       4  |         ───→          4  |         ───→               HO—C—H
      H—C—H         −H₂O          H—C—H         −H₂O               4  |
       5  |                        5 |                            HO—C—H
      H—C—OH                      H—C—OH                           5  |
       6|                          6|                             H—C—OH
       CH₂OH                       CH₂OH                           6|
                                                                   CH₂OH
       Mannose                    Mannosamine                     Mannosamine
```

```
         ¹CHO                          ¹CHO                              ¹CHO       O
       ²  |                          ²  |                             ²  |         ‖
      H—C—OH                        H—C—NH₂                          H—C—NH—C—CH₃
       ³  |                          ³  |                             ³  |
      HO—C—H         NH₃          HO—C—H         CH₃COOH            HO—C—H
       4  |         ───→          4  |         ───→                 4  |
      HO—C—H        −H₂O          HO—C—H        −H₂O                HO—C—H
       5  |                        5 |                               5  |
      H—C—OH                      H—C—OH                            H—C—OH
       6|                          6|                                6|
       CH₂OH                       CH₂OH                             CH₂OH
       Galactose            Galactosamine or             N-acetyl galactosamine
                                chondrosamine
```

3. Deoxy Sugars

They may be defined as sugars in which the oxygen of the –OH group has been removed, leaving hydrogen. Thus, the -CHOH or –CH₂OH group becomes a –CH₂ or –CH₃ group. The most common example is D-2-deoxyribose and others are, 3-deoxyglyceric aldehyde, 6-deoxy- L-mannose, 6-deoxy-L-galactose (L-fucose).

Structures

D-2, deoxyribose (Present in DNA)
```
H—C=O
H—C—H
H—C—OH
H—C—OH
CH₂OH
```

6-deoxy-L-mannose (L-rhamnose) (Present in many glycosides and di- and trisaccharides)
```
H—C=O
H—C—OH
H—C—OH
HO—C—H
HO—C—H
CH₃
```

6-deoxy-L-galactose (L-fucose) (Present in glycoproteins and Bacterial polysaccharides)
```
H—C=O
HO—C—H
H—C—OH
H—C—OH
HO—C—H
CH₃
```

3-deoxyglyceric aldehyde (Common in the leaves of Poplar)
```
H—C=O
H—C—H
CH₃
```

4. Ascorbic Acid or Vitamin C

This is very interesting sugar acid and is found in plants, especially in green walnuts and citrus juices. It is synthesized from L-sorbose by oxidation with nitric acid.

L-Sorbose →(oxidation HNO₃)→ **2-keto-L-gulonic acid** →(enolized)→ **L-ascorbic (1, 4-lactone) acid**

2.7. REDUCING AND NON-REDUCING SUGARS

Any carbohydrate which is capable of being oxidized and causes the reduction of other substances without having to be hydrolysed first is known as **reducing sugar,** but those which are unable to be oxidized and do not reduce other substances are known as **non-reducing sugars**. Generally, all the free monosaccharides having free aldehyde or α-hydroxy ketonic group are capable of being oxidized. After being oxidized they cause the reduction of the other substances and so are known as reducing sugars. Fehling's solution and Benedict's solution are used to carry out the oxidation e.g.,

Reducing carbohydrate + Fehling's solution ⟶ Oxidized Carbohydrates + 2 Cu⁺
(Deep blue)

$2Cu^+ + 2OH^- \longrightarrow 2CuOH \longrightarrow Cu_2O + H_2O$
(light yellow or green) (Red)

Carbohydrates

$$C_6H_{12}O_6 + 2Cu(OH)_2 \longrightarrow C_6H_{12}O_7 + Cu_2O + H_2O$$
Glucose   Gluconic   (Red)
                                          acid

2.8. CHEMISTRY OF OLIGOSACCHARIDES

This group is composed of disaccharides, trisaccharides, tetrasacccharides, pentasaccharides and hexasaccharides. Di-, tri-, tetra- and so on indicate the number of monosaccharides present in a molecule. In this case also, the carbohydrates which contain hemiacetal (free sugar group) possess α- and β-forms, like those of monosaccharides.

1. Disaccharides

The general molecular formula of a disaccharide is $C_{12}H_{22}O_{11}$. They are mostly sugars and may be reducing or non-reducing. According to the classification some of the important disaccharides are as follows :

(i) *Non-reducing sugars:*
  (a) Sucrose consists of glucose, fructose.
  (b) Trehalose consists of glucose, glucose.

(ii) *Reducing sugars*:
  (a) Maltose consists of glucose, glucose.
  (b) Lactose consists of glucose, galactose.
  (c) Melibiose consists of glucose, galactose.
  (d) Cellobiose consists of glucose, glucose.
  (e) Gentiobiose consists of glucose, glucose.

1. Sucrose

Sucrose is found widely distributed in plants. It occurs especially in plants such as sugar-cane, sugarmaple, sugarbeet, pine apple, sorghum and in little quantities in wheat, Barley, Carrots, maize and in mostly sweet fruits. It is a non-reducing sugar and on hydrolysis yields one molecule of glucose and one molecule of fructose.

$$C_{12}H_{22}O_{11} + H_2O \longrightarrow C_6H_{12}O_6 + C_6H_{12}O_6$$
Sucrose   Glucose   Fructose

It does not give the reactions of the sugar group (Hemiacetal group). The linkage of glucose to fructose in the molecule involves -OH of C-1 of glucose and OH of C-2 of fructose.

Hydrolysis of sucrose is done by enzyme *invertase* or *sucrase* or by dilute acids. The glucose and fructose molecules are produced with a change in optical rotation from positive to negative because D-fructose is more levorotatory than D-glucose which is dextrorotatory. This is known as **inversion**.

$$C_{12}H_{22}O_{11} + H_2O \longrightarrow C_6H_{12}O_6 + C_6H_{12}O_6$$
+66.5°   +52.5°   −92°

α-D-glucopyranosyl   *β-D-fructofuranoside*

Sucrose

Sucrose is directly fermentable by yeast. It crystallizes from water in monoclinic crystals, readily soluble in water, slightly soluble in alcohol and insoluble in ether. It is dextrorotatory and its specific orientation is +66.5°. It is colourless with m.p. 188° and does not reduce Tollen's reagent and Fehling's solution. The reduction takes place only after boiling with dil HCl. In presence of conc. H_2SO_4 it chars with evolution of CO_2 and SO_2.

2. Maltose

Maltose is composed of 2 molecules of glucose and is obtained by the hydrolysis of starch as an intermediate product either by amylase or by diastase. It contains a free sugar group. Maltose is a reducing sugar and gives the reactions of hemiacetal. On hydrolysis it yields two molecules of glucose. So, it is a glucose α-glucoside and generally crystallizes as the β-form. Its structural formula is as follows which shows an α-1, 4 glucoside linkage in between the two glucose units.

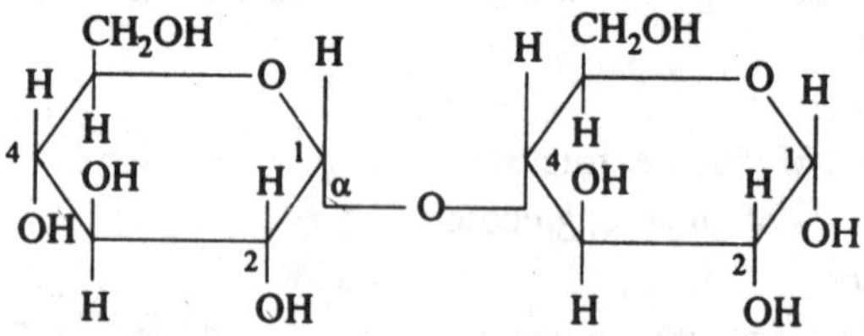

Maltose

Maltose is colourless, crystalline, soluble in water, and insoluble in ether; on crystallization it forms slender, white needles; It is dextrorotatory with specific rotation +137°, and is fermentable by yeast directly. Maltose reduces Fehling's solution, Benedict's solution and Nylander's reagent.

D-glucopyranose group *α-D-glucopyranosyl group*

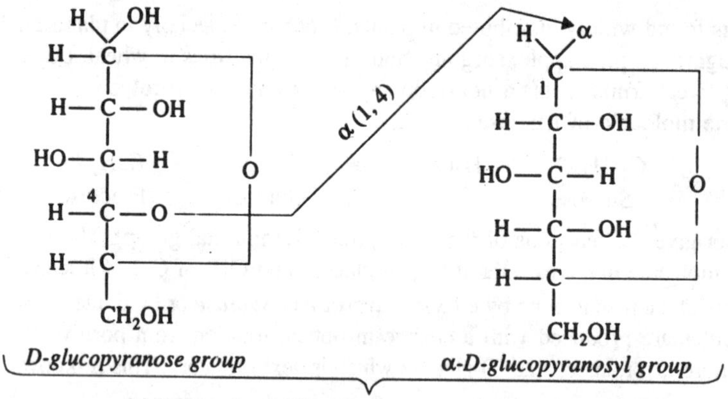

Maltose α-form (Two glucose α-1, 4 glucosidal linkage)

3. Lactose

Lactose is composed of one molecule of glucose and one molecule of galactose and is formed by the mammary glands. The hydrolysis by enzyme *lactase* or by acid yields its constituent monosaccharides. Lactose contains a free sugar group in its structure. Since it is a reducing sugar, it gives the reactions of sugar group viz. forms cyanohydrins, osazones, oximes, reduces fehling's solution, Benedicts solution and Nylander's reagent.

It exists in solution as a mixture of α-, β-and aldehydo-compounds. Its structure shows β-1,4 galactoside linkage.

Carbohydrates

Lactose (α-form) with 1, 4 galactoside linkage

(D-glucopyranose group — β(1,4) — β-D-glucopyranosyl group)

Lactose

4. Cellobiose

Cellobiose is composed of two molecules of glucose and is obtained by the incomplete hydrolysis of cellulose. It is a reducing sugar and gives these reactions of free sugar group. Its structural formula is as follows:

4-O-β-D-glucopyranosyl-D-glucopyranose
Cellobiose (α-form)

2.8.2. Trisaccharides

(i) *Non reducing Sugars :*
   (a) Raffinose consists of fructose, glucose, galactose.
   (b) Gentianose consists of glucose, fructose, glucose.
   (c) Melezitose consists of glucose, fructose, glucose.

(ii) *Reducing Sugars :*
   (a) Rhamnotriose consists of galactose, rhamnose, rhamnose.
   (b) Mannotriose consists of galactose, galactose, glucose.

(i) (a) *Raffinose* : This trisaccharide occurs in barley, cotton seeds, and in sugarbeets. It is concentrated in sugarbeet and molasses and is also found in Eucalyptus, fungi and higher plants.

On hydrolysis, raffinose yields one molecule of fructose, one molecule of glucose and one molecule of galactose. The fructose and glucose are joined by a sucrose linkage and galactose molecule is attached to glucose. The products of hydrolysis depend upon the nature of hydrolysing agents *e.g.*, *Hydrolysis by Maltase* :

$$\underbrace{\text{Fructose–glucose–galactose}}_{\text{Raffinose}} + \text{HOH} \xrightarrow{\text{Maltase}} \text{galactose} + \underbrace{\text{Fructose-glucose}}_{\text{Sucrose}}$$

Hydrolysis by Weak acids :

$$\underbrace{\text{Fructose – glucose – galactose}}_{\text{Raffinose}} \xrightarrow[+ \text{HOH}]{\text{Weak acid}} \text{Fructose} + \underbrace{\text{glucose – galactose}}_{\text{Melibiose}}$$

Hydrolysis by Sucrase :

$$\underbrace{\text{Fructose – glucose – galactose}}_{\text{Raffinose}} \xrightarrow[+ \text{HOH}]{\text{sucrase}} \text{Fructose} + \underbrace{\text{glucose – galactose}}_{\text{Melibiose}}$$

Structure formula of Raffinose

Raffinose is colourless, crystalline, soluble in water and methyl alcohol. It is strongly dextro-rotatory and its specific orientation is *+104.4°* in 10% solution. Raffinose does not give characteristic reaction of sugar group because all the sugar groups of galactose, glucose and fructose are in combind state and not in free state.

(*i*) (*b*) ***Gentianose*** : It occurs in Gentian root and on hydrolysis yields one molecule of fructose and two molecules of glucose.

(*i*) (*c*) ***Melezitose*** : It is found in the sap of Scrub pine, Douglas fir and larch which on hydrolysis yields two molecules of glucose and one molecule of fructose and does not possess any free sugar group.

(*ii*) (*b*) ***Mannotriose*** : It is formed by partial hydrolysis of *stachyose*.

3. Tetrasaccharide

The most common example of tetrasaccharide is *Stachyose*. It is isolated from the rhizomes of *Stachys tuberifera* and from bulbs of garlic and onion. It is a α-D-galactopyranosyl-α-D-galactopyranosyl-α-D-glucopyranosyl- β-D-fructofuranoside.

2.9. CLASSIFICATION AND CHEMISTRY OF POLYSACCHARIDES

Polysaccharides are formed by the combination of many monosaccharides joined together by glycosidic linkages, like oligosaccharides. They are also known as "Glycans" and are again classified into : (1) *Homoglycans*, and (2) *Heteroglycans*. Those polysaccharides which are made up of only one kind of monosaccharides are known on as **Homoglycans**, while those which are made up of two or more kinds of monosaccharides are known as **Heteroglycans**. Polysaccharides are hydrolysed either by enzymes or by mineral acids and are resistant to alkaline hydrolysis. The most common examples of polysaccharides are starches, celluloses, glycogen, chitin and Inulin.

Carbohydrates

1. Starch

The structure of starch includes two chemical substances, viz., amylose and amylopectin and these two substances are again made up of a number of glucose units joined through the glycosidic linkages like maltose i.e., amylose consists of glucose residue with repeating maltose units, while amylopectin consists of glucose units in α-glycosidic linkage. In its structure amylose shows α-1, 4 glucoside linkage, while amylopectin in addition to α-1, 4 glucoside linkage has α-1,6 glucoside linkage in side chains. Thus the chain of amylopectine is branched.

Structure of amylose and amylopectin

The starches are found as reserve food material in potatoes, in seeds, in rhizomes, in grams, in fruits and in pith of plants. The shape of starch grains differ in plants. The general structural formula used for starch is $(C_6H_{10}O_5, H_2O)n$. Starch is slightly soluble in hot water, but insoluble in cold water. It generally gives a characteristic test i.e. **Starch-iodide test**.

In presence of enzyme *amylase*, it is hydrolysed and in presence of enzyme *diastase*, starch is digested i.e. it is converted into sugars.

2. Inulin

This polysaccharide is a poly fructosan i.e., consists of fructose residue in the furanose form and the number of fructose units is less than 100. It generally occurs in the bulbs of onion and garlic, in dahlia, and in the tubers of *Jerusalem artichoke* and chicory.

It is white, crystalline powder, readily soluble in hot water and slightly soluble in cold water and can be easily hydrolysed by acids.

3. Glycogen

This polysaccharide is also known as animal starch. It occurs in animal tissues, especially in liver, and muscles of the bones. The structure of glycogen is similar to that of amylopectin, but glycogen is more branched and of higher molecular weight than amylopectin. In presence of water it forms suspensions

4. Chitin

Those amino-sugars which occur in nature in combination with protein and glucosamine, are called **chitin**. Generally, it is made up of chitibiose, a disaccharide, which on decomposition yields N-acetyl glucosamine. When chitin is hydrolysed by acids, it yields acetic-acid and glucosamine.

Chitin is used as a structural material of the shells of Arthropods, and particularly of crustaceans and insects. It is also important in the formation of lenses of eyes, and in the lining of digestive, respiratory and excretory tracts of these insects.

5. Cellulose

Cellulose is the condensation product of glucose and is the chief constituent of fibrous plants *e.g.*, cotton, flax, ramie etc. It is also found in the cereal straws, woods of plants and in cell walls. The structure of cellulose is similar to that of amylose, but here the glucose units are held together by β-1, 4-glucoside linkage.

6. Agar

It is a polysaccharide obtained from marine algae or seaweeds. It consists of a mixture of D & L-galactoses with some amount of sulphuric acid. In presence of water it swells up, but does not dissolve. Agar is tasteless, odorless and soluble in hot water to form sol which upon cooling sets to a gel.

7. Gum Arabic or Gum Acacia

Vegetable gums and mucilages are the substances which contain hexoses or pentoses or both in the form of glycosidic linkage with a carbohydrate acid group *e.g.*, Uronic acids. Gum arabic on complete hydrolysis yields galactose, arabinose, rhamnose and glucuronic acid. It is used as an adhesive and also in the preparation of pharmaceuticals and in confectionaries.

8. Pectins

Pectins are the substances which form the jellies in presence of sugar and acids. They occur in the pulp of citrus fruits, beets, apples and carrots.

9. Mucopolysaccharides

Those substances which are generally made up of aminosugars or their derivatives and uronic acid units, are known as **mucopolysaccharides**. Some of them are made up of aminoacid and monosaccharide only and Uronic acid is completely absent. The mucopolysaccharides which show the acidic properties are known as **acid mucopolysaccharides** *e.g.*, Hyaluronic acid, chondroitin sulphates and heparin.

All the mucopolysaccharides are generally acetylated because of the presence of hexamine and they are found in combination with protein as mucoids or mucoproteins in tissues.

Table 2.1 : Distinction between Mono-, Oligo- and Polysaccharides

S.No.	Monosaccharides	Oligosaccharides	Polysaccharides
1.	Monosaccharides are simplest sugars *e.g.*, glucose and fructose and are generally reducing sugars.	Oligosaccharides are also sugars, may be reducing or non-reducing *e.g.* Sucrose, and maltose	Polysaccharides are non-sugars *e.g.*, Starch, glycogen, cellulose inulin & pectin.
2.	Monosaccharides contain generally up to 9-Carbon atoms.	Oligosaccharides contain generally 12 to 36-carbon atoms.	Polysaccharides contain more number of carbon atoms.
3.	They contain a carbonyl group (C = O) and show the properties of aldehyde or ketones.	Do not contain the carbonyl group and fail to give the reactions of aldehydic or ketonic group.	Do not contain the carbonyl group and so fail to give the reactions of aldehydic or ketonic group.

4.	They are colourless, crystalline and sweet.	They are also generally colourless, crystalline, and sweet.	These are colourless amorphous and taste-less.
5.	Soluble in water.	Soluble in water.	Insoluble in water.
6.	These are optically active.	Same.	These are optically inactive.
7.	They have free or potential aldehyde or ketonic group.	No.	No.
8.	Monosaccharides cannot be hydrolysed.	Oligosaccharides can be hydrolysed & generally yield 2 to 6 molecules of monosaccharides.	Can be hydrolysed and upon hydrolysis yield many number of monosaccharides.

2.10 SIGNIFICANCE OF CARBOHYDRATES

The carbohydrates are of great importance to plants as well as to animals and human beings.

1. In Plants and Animals

(i) Carbohydrates are the structural materials of the plants for example, cellulose is found in plant fibres and in wood.

(ii) They are widespread and act as a reserve food material as starch in tubers, grains and roots.

(iii) Sucrose is present in the nectar of flowers, in roots, and in fruits. Glucose, fructose and simple sugars are also found in small amount in plants as reserve food material.

(iv) The carbohydrates on oxidation release energy which is utilized by plants for various physiological processes.

2. For Human Beings

They are of great significance for human beings both from biochemical and industrial point of view.

(i) The carbohydrates such as starches and sugars are the main food for human beings. They are easily digestible and are easily oxidised to provide energy for various physiological processes. These are present in cereals.

(ii) Various carbohydrates which are present in seeds such as rice, maize, rye, barley and also in fruits are utilized in the production of alcoholic beverages.

(iii) The carbohydrate derivatives such as glucosides, form important drugs and other medicines for various diseases.

(iv) The carbohydrates, particularly cellulose and its derivatives are used in the production of artificial silk, paper, plastics, cinema films and explosives.

(v) Carbohydrates form other derivatives which are of practical use and are used in the detection of certain chemicals.

(vi) All animal tissues, blood, milk and tissue fluids contain carbohydrates and their derivatives as important constituents, e.g. Blood contains glucose as sugar.

(vii) *Importance of Blood glucose.*

(a) Muscles and other tissues remove glucose from blood and form glycogen which provides energy on oxidation.

(b) Mammary glands form milk sugar i.e. lactose from blood glucose.

(c) Many tissues are formed by combinations of sugars or sugar derivatives and proteins.

2.11 SOME IMPORTANT TESTS FOR CARBOHYDRATES

1. **Fehlings Test :** Mix Fehling solution A and B in equal amount in a test tube. Now add equal volume of glucose solution and boil till a brick red precipitate is produced. The Fehling solution contains cupric sulphate, sodium hydroxide and sodium potassium tartrate. When this solution is mixed with glucose and boiled, the aldehyde group of glucose is oxidized and cupric salt is reduced to cuprous oxide giving red precipitate.

$$C_6H_{12}O_6 + 2Cu(OH)_2 \rightleftharpoons C_6H_{12}O_7 + Cu_2O + H_2O$$
$$\text{blue} \qquad\qquad\qquad\qquad \text{red}$$

2. **Benedict's Test :** Mix Benedict's solution (about 3 ml) and glucose solution (about 1 ml) and warm the mixture gently. A redish-brown colour of cuprous oxide is produced. On warming the colour of mixture turns from blue to green and from green to redish-brown. Benedict's solution contains blue coloured copper sulphate. When alkaline solution of glucose is added to this and warmed, the aldehyde group of glucose is oxidized and blue coloured copper sulphate is changed to green coloured copper hydroxide and ultimately $Cu(OH)_2$ changed to redish brown cuprous oxide.

3. **Molisch Test :** Sugars and their polymers give characteristic colour in presence of strong sulphuric acid and α-naphthol. The sugars in presence of acids undergo dehydration to form furfural and the colour is produced by the condensation of aldehyde and phenol.

4. **Selivanoff Test:** Sugars containing keto group (Ketoses) when heated with HCl and resorcinol produce bright red colour.

2.12. EXPERIMENTS

1. **To test the presence of starch :**

   ke 2 ml of 1% starch solution in a test tube and add a few drop of iodine solution. Blue colour of starch iodide appears confirming the presence of starch. On boiling, the blue colour disappears while on cooling reappears.

2. **To test the presence of non-reducing sugar or sucrose :**

   Prepare sucrose solution by dissolving sucrose in water. Take 2 ml of sucrose solution in a test tube and add 2-3 drops of conc. HCl to hydrolyse sucrose. Boil the mixture and cool. Neutralize the mixture by adding 10% NaOH solution. Test it with litmus paper. The sucrose is hydrolysed into glucose and fructose reducing sugars which can be tested by Fehling's test or any other test descried in the text.

3. **To test the presence of cellulose in gram seeds :**

   Crush a few gram seeds and place in a watch glass. Add a few drops of conc. H_2SO_4 and 2 drops of iondine solution (I–KI). A blue colour appears due to presence of cellulose in the cell walls.

3

Amino acids and Proteins

3.1. WHAT ARE AMINO ACIDS

The amino acids are derivatives of carboxylic acids in which a hydrogen atom in a α-carbon chain is replaced by an amino group (-NH_2). They are represented by a general formula shown below.

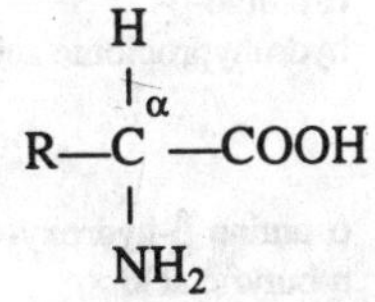

$$R-\overset{H}{\underset{NH_2}{\overset{|\alpha}{C}}}-COOH$$

(where R-represents a great variety of structures).

The amino acids containing this structure are known as **α-amino acids**. If one hydrogen atom of methyl group of acetic acid (CH_3COOH) is replaced by an amino-group (–NH_2), it forms amino-aceticacid –CH_2 (NH_2) COOH, also known as **glycine**.

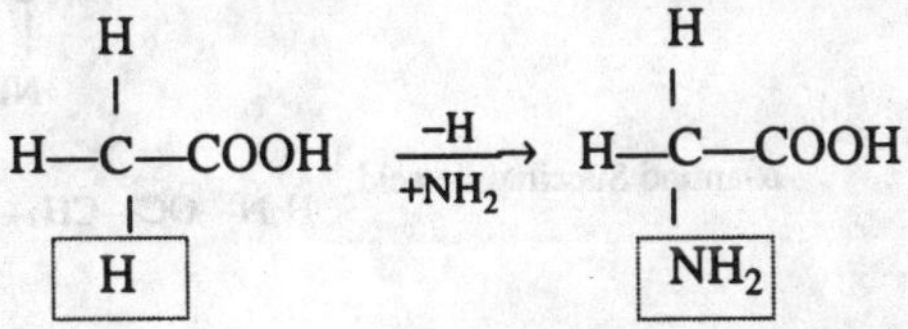

$$H-\overset{H}{\underset{\boxed{H}}{\overset{|}{C}}}-COOH \xrightarrow[+NH_2]{-H} H-\overset{H}{\underset{\boxed{NH_2}}{\overset{|}{C}}}-COOH$$

acetic acid   amino-acetic acid
                             (glycine)

Proteins, on hydrolysis yield a mixture of α -amino acids.

3.2. CLASSIFICATION OF AMINO ACIDS

The protein amino acids are classified into three main classes on the basis of their structure and properties:

(*i*) Aliphatic amino acids.

(*ii*) Aromatic amino acids.

(*iii*) Heterocyclic amino acids.

1. Aliphatic aminoacids

Amino acids having an aliphatic structure are known as **aliphatic amino acids**. They are further sub-divided into following four sub-classes.

1. *Monoamino-monocarboxylic acids (Neutral amino acids)* :

S. No.	Name	Abbreviation	Schematic name	Formula
1.	Glycine	Gly	Aminoacetic acid	NH_2-CH_2-COOH
2.	Alanine	Ala	α-amino propionic acid	$CH_3-\overset{\alpha}{\underset{NH_2}{\overset{|}{CH}}}-COOH$

#	Name	Abbr	Systematic name	Structure
3.	Valine	Val	α-amino isovaleric acid	$(CH_3)_2CH-CH(NH_2)-COOH$
4.	Leucine	Leu	α-amino iso-caproic acid	$(CH_3)_2CH-CH_2-CH(NH_2)-COOH$
5.	Isoleucine	Ileu	α-amino-β-methyl-n-valeric acid	$CH_3-CH_2-\overset{\beta}{C}H(CH_3)-\overset{\alpha}{C}H(NH_2)-COOH$
6.	Serine	Ser	α-amino-β-hydrohypropionic acid	$\overset{\beta}{C}H(OH)-\overset{\alpha}{C}H(NH_2)-COOH$
7.	Theonine	Thr	α-amino-β-hydroxy-n-butyric acid	$CH_3-\overset{\beta}{C}H(OH)-\overset{\alpha}{C}H(NH_2)-COOH$

2. Monoamino-dicarboxylic acids (acidic amino-acids):

#	Name	Abbr	Systematic name	Structure
8.	Aspartic acid	Asp	α-amino Succinic acid	$HOOC-CH_2-\overset{\alpha}{C}H(NH_2)-COOH$
*8.(a)	Asparagine	Asp-NH2	α-amino Succinamic acid	$H_2N-OC-CH_2-\overset{\alpha}{C}H(NH_2)-COOH$
9.	Glutamic acid	Glu	α-amino glutaric acid	$HOOC-CH_2-CH_2-\overset{\alpha}{C}H(NH_2)-COOH$
*9(a)	Glutamine	GLU-NH2	α-amino glutamic acid	$H_2N-OC-CH_2-CH_2-\overset{\alpha}{C}H(NH_2)-COOH$

3. Diamino-mono-carboxylic acid (Basic aminoacids)

#	Name	Abbr	Systematic name	Structure
10.	Arginine	Arg	α-amino-δ-guanidino n-valeric acid	$H_2N-\underset{\parallel}{\overset{NH}{C}}-NH-CH_2-CH_2-CH_2-\overset{\alpha}{C}H(NH_2)-COOH$
11.	Lysine	Lys	α-ε-Diamino n-caproic acid	$H_2N-CH_2-CH_2-CH_2-CH_2-CH(NH_2)-COOH$

4. Sulphur-Containing amino-acids

---

* Asparagine and glutamine are amides of aspartic acid and glutamic acid respectively and occur naturally in proteins.

Amino acids and Proteins

12.	Cysteine	CySH	α-amino-β-mercapto Propionic acid.	$HS-CH_2-CH-COOH$ $\quad\quad\quad\quad\mid$ $\quad\quad\quad\quad NH_2$
13.	Cystine	CYs $\mid$ Cys	Bi-S-(α-aminopropionic acid) β-disulphide or Di [α-amino-β-thio Propionic acid]	$S-CH_2-CH-COOH$ $\mid\quad\quad\quad\quad\mid$ $\mid\quad\quad\quad\quad NH_2$ $S-CH_2-CH-COOH$ $\quad\quad\quad\quad\mid$ $\quad\quad\quad\quad NH_2$
14.	Methionine	Met	α-amino-γ-methyl thio-n-butyric acid	$CH_3-S-CH_2-CH_2-CH-COOH$ $\quad\quad\quad\quad\quad\quad\quad\quad\quad\mid$ $\quad\quad\quad\quad\quad\quad\quad\quad\quad NH_2$

2. Aromatic aminoacids

Aminoacids having aromatic structure formula (benzene derivatives) are known as aromatic amino acids e.g.,

15.	Phenyl alanine	Phe	α-amino-β-phenyl propionic acid	⌬$-CH_2-CH-COOH$ $\quad\quad\quad\quad\quad\mid$ $\quad\quad\quad\quad\quad NH_2$
16.	Tyrosine	Tyr	α-amino- -(p-hydroxy-phenyl) propionic acid	$HO-$⌬$-CH_2-CH-COOH$ $\quad\quad\quad\quad\quad\quad\mid$ $\quad\quad\quad\quad\quad\quad NH_2$

3. Heterocyclic aminoacids

Amino acids having a closed chain ring of carbon with nitrogen are known as hetero-cyclic amino acids. e.g.,

17.	Proline	Pro	Pyrrolidine-α-carboxylic acid	H_2C-CH_2 $\mid\quad\quad\mid$ $H_2C\quad CH-COOH$ $\ \ \ \backslash\ /$ $\ \ \ \ N$ $\ \ \ \ \mid$ $\ \ \ \ H$
18.	Hydroxy-proline	Hypro.	γ-hydroxy pyrrolidine-α-Carboxylic acid.	$HO-HC-CH_2$ $\quad\quad\mid\quad\quad\mid$ $\ \ H_2C\ \ CH-COOH$ $\ \ \ \ \ \backslash\ /$ $\ \ \ \ \ \ N$ $\ \ \ \ \ \ \mid$ $\ \ \ \ \ \ H$
19.	Histidine	His	α-amino-β-unidazole Propionic acid	$CH=C-CH_2-CH-COOH$ $\mid\quad\quad\mid\quad\quad\quad\quad\mid$ $N\quad NH\quad\quad\quad NH_2$ $\ \backslash\ /$ $\ \ CH$
20.	Tryptophan	Try	α-amino -β-indole Propionic acid	(indole ring)$-CH_2-CH-COOH$ $\quad\quad\quad\quad\quad\quad\quad\mid$ $\quad\quad\quad\quad\quad\quad\quad NH_2$

Note : Amino acids 15, 16, 17, 18 are monoamino-monocarboxylic acids and 20 is diamino-monocarboxylic acid.

3.3. Essential and Non-essential Amino acids

Essential amino acids are those which cannot be synthesized in the body in adequate amount and must be supplied from outside as part of food. The presence of these aminoacids is essential for growth of the young and maintenance of the adult. The essential amino acids are :

(i) Lysine (ii) Tryptophan (iii) Histidine (iv) Leucine (v) Phenylalanine (vi) lso-leucine (vii) Threonine (viii) Methionine (ix) Valine and (x) Arginine.

Non-essential amino-acids are those which can be synthesized in the body in adequate amount. They need not be supplied from the outside.

3.4. SEPARATION OF AMINO ACIDS

For the separation of individual amino acids from protein hydrolysates the following methods are employed.

1. Electrophoresis

An electric current is passed into a mixture of amino acids at pH 5.5. The acidic amino acids (negatively charged) migrate to anode, the basic amino acids (positively charged) pass to cathode, while the neutral amino acids remain in the centre.

2. Ion-Exchange

Two types of polymeric exchange resins are available for separation of mixture of amino acids.

(a) *Cation exchangers (acidic).* They are either polysulphonic resins or polycarboxylic resins. A typical cation exchanger is Amberlite IR-100.

$$\text{Resin} - SO_3^- \; Na^+ + \overset{+}{N}H_3R \text{ (Basic amino acids)} \rightleftharpoons \text{Resin} - SO_3^- \; \overset{+}{N}H_3 - R + Na^+$$

(b) *Anion exchangers (basic):* They are the polymers of basic nitrogenous compounds, e.g. Amberlite IR-4.

$$\text{Resin} - \overset{+}{N}H_3 \; OH^- + RCOO^- \text{ (Acidic amino acids)} \rightleftharpoons \text{Resin} - \overset{+}{N}H_3 RCOO^- + OH^-$$

3. Chromatography

A mixture of amino acids can be separated by paper chromatography due of differences in their partition coefficients between water and an organic solvent.

A two-dimensional chromatographic procedure as described earlier is followed to separate aminoacids present in a mixture. Amino acids that have the greatest solubility in the organic solvent move more rapidly. This two-dimensional movement gives better resolution of the amino acids. The spots of aminoacids can be made visible by first drying and then spraying the paper with *nin-hydrin solution.* The Rf value of each aminoacid is calculated. The Rf value of each aminoacid is specific in one solvent system under the same conditions.

3.5. PROTEINS

Proteins are regarded as the most important constituents of all living organisms as they play important role in their life. The important functions of proteins are : (a) They form the structural framework of the cell, (b) Maintain osmotic balance, (c) catalyse biochemical reactions, (d) regulate metabolism, (e) help in storage of some elements, (f) act as oxygen carriers, (g) form the colloidal system in protoplasm, (h) transport lipids as lipoproteins, and (i) act as storage proteins (proteinoplasts).

Amino acids and Proteins

1. Characteristic features of Proteins

(1) Proteins are colloidal, optically active, highly complex, naturally occurring organic compounds of no definite melting point. They have very high molecular weights.

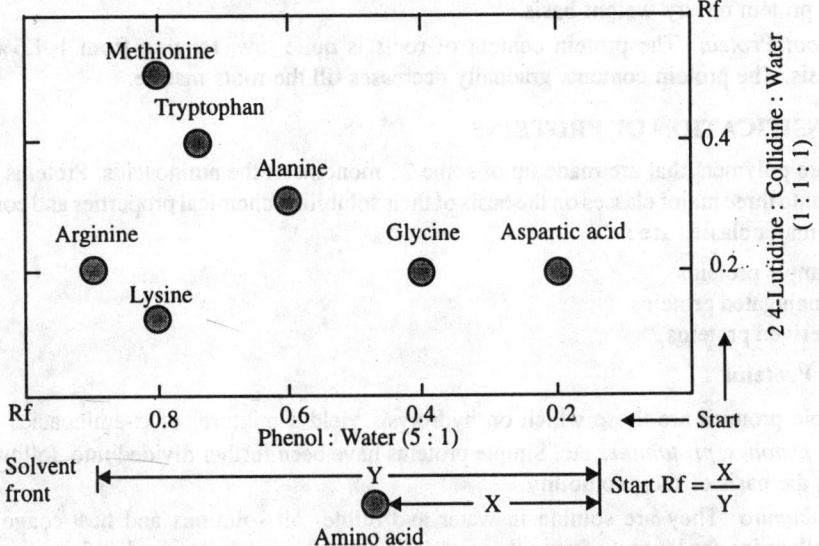

Fig. 3.1: Showing the chromatogram of amino acids.

(2) They are essentially made up of C, H, O, N and S, but may also contain Fe and phosphorus. The percentage of these elements varies in different proteins.
C=50-55%, H=6.5-7.3%, N=15-19%, O=21.9-24%, S=0-2.8%.

(3) Each protein molecule is made up of several amino acids bound together by 'amide bonds' between the α-amino group of one molecule and the carboxyl group of another. These bonds are called **peptide bonds**.

(4) They may contain disulphide linkages, hydrogen bonds and Vander Waal forces.

(5) Proteins are mostly insoluble in water, alcohols and ethers, but are soluble in dilute salt solutions.

(6) On heating they get coagulated, while on hydrolysis (acid, alkaline or enzyme) they yield a mixture of α-amino acids.

2. Protein Content of the Plant

The protein content of plant tissues is variable. It is mainly found in seeds and leaves. Roots contain very small amount.

1. *Seed Proteins* : The seeds like groundnut, soybean, maize, wheat, barley, rice, cotton-seeds, peas and beans etc contain maximum amount of protein which remains stored as reserve form of nitrogen in aleurone grains or protein-bodies. The protein-bodies are single membraned structures of about 1-20 μm in diameter. The percentage of protein in seeds varies from 10 (in maize) to 30 (in ground nut).

When polyacrylamide gel electrophoresis of water extracts of wheat flour is done, it yields 15-17 protein bands. The starch get electrophoresis of wheat glutin yields more than 20 bands. Barley-albumin can be separated into 7 protein fractions and each fraction contains 4-30% carbohydrates. In this case, the sugar xylose is joined to protein through an ester linkage found between C-4 hydroxyl of xylose and carboxyl of L-alanine. It indicates that barley albumin is a mixture of glycoproteins.

2. *Leaf Protein:* Leaf protein is mainly present in chloroplasts but in small amounts it is also present in other cell-organelles and cytoplasm. In chloroplasts, the pigments exist as *Chromoproteins*. The fraction I protein (Ribulose biphosphate carboxylase – oxygenase, RUBISCO) is the major protein of chloroplasts. Little is known about other leaf-proteins. The leaves contain only about 2% protein on dry weight basis.

3. *Root Protein:* The protein content of roots is quite low, ranging from 1–1.5% on fresh weight basis. The protein contents gradually decreases till the roots mature.

3.6. CLASSIFICATION OF PROTEINS

Proteins are polymers that are made up of some 22 monomers, the aminoacids. Proteins have been classified into three major classes on the basis of their solubility, chemical properties and composition. The three major classes are :

1. Simple proteins
2. Conjugated proteins
3. Derived proteins

1. Simple Proteins

Simple proteins are those which on hydrolysis yield a mixture of α-aminoacids only. *e.g.*, *Albumins, histones, protamines* etc. Simple proteins have been further divided into following seven groups on the basis of their solubility.

1. *Albumins*: They are soluble in water and dilute salt solutions and heat-coagulable. The common albumins are *leucosin* from *wheat, ricin* from *castor seeds, legumelin* from *legume seeds*, β-*amylase* from *barley*, and *avalbumin* from egg white.

2. *Globulins*: They are insoluble in water, soluble in dilute salt solutions and are heat coagulable. Globulins are common in seeds as storage proteins, *e.g. arachin* in ground nuts, *legumin* in peas, *tuberin* in potato tuber, *plasma globulins* in blood plasma etc.

3. *Glutelins:* They are insoluble in water, salt solutions and netural solvents but are soluble in dilute acids and bases. These are not heat sensitive and are confined to plants only *e.g. glutenin* from wheat and *oryzenin* in rice.

4. *Prolamines*: They are insoluble in water and absolute alcohol, but soluble in 70-80% aqueous ethyl alcohol. Upon hydrolysis they yield ammonia and proline (an amino-acid) *e.g.*, *Gliadin* in wheat, *Zein* in maize and *hordein* in barley.

5. *Protamines*: They are soluble in water, dilute acids and ammonia, contain more basic amino acids such as lysine and arginine, but lack tryptophan and tyrosine. Sulphur is also absent from these proteins. They do not coagulate easily and are generally found associated with the nucleic acids, *e.g.*, *Salmin* in the sperms of fishes and *clupein* in the herring sperm.

6. *Histones*: They are small in size, soluble in water and do not coagulate easily by heat. Like protamines they are also rich in basic aminoacids such as lysine and arginine. Histones are also found in the nuclei along with the nucleic acids.

7. *Scleroproteins*: They are also known as albuminoids. Scleroproteins are insoluble in water, salt solutions, acids and bases, and in other common solvents. They lack complex amino-acids like tryptophan and are easily digestible, but slowly by gastrointestinal enzymes; Typical scleroproteins are *keratin* from horn, nails, hoofs, and feathers; *collagen* from skin, *tendon* and bone, and *elastin* from ligament.

2. Conjugated Proteins

Conjugated proteins are those which on hydrolysis yield some substance or substances (carbohydrates, nucleic acids, phosphoric acid or lipids) in addition to α-amino acids, *e.g.*, *Casein* from milk, *mucin* from saliva, *vitellin* from egg yolk etc. The additional non amino acids are referred

Amino acids and Proteins

to as prosthetic groups. Conjugated proteins are further classified in accordance with their prosthetic group into the following seven major types :

1. *Nucleoproteins*: Nucleoproteins are characterized by the possession of prosthetic groups known as nucleic acids. Upon hydrolysis they yield amino-acids and a nucleic acid. Nucleoproteins are weakly acidic in nature and are soluble in water. They are found in the nucleus of the cell. Other examples of the nucleoproteins are viruses and ribosmes.

2. *Glycoproteins*: Glycoproteins are characterized by the possession of prosthetic groups known as polysaccharides. Thus, they are the complexes of amino-acids (proteins) and carbohydrates. They are soluble in alkalies and are acidic in nature. They are found in egg white, jelly fish, cell-membranes and saliva.

3. *Lipoproteins*: They are the complexes of lipids and proteins. The common lipids which are found as prosthetic groups are lecithin and cephalin. They are insoluble in water and are found in the membranes, nucleus, and lamellae of the chloroplast. Egg yolk contains the lipoprotein, *lipovitellin*.

4. *Chromoproteins*: As the name indicates, chromoproteins contain pigments as prosthetic groups. The pigments also contain metals like Fe, Cu, Co, Mg, etc. The chlorophyll or green pigment containing proteins are known as **chlorophylloproteins** and iron-porphyrin containing proteins are known as **haemoproteins**. The chromoproteins include *hemoglobins, myoglobins, carotenoid proteins* etc.

5. *Metalloproteins*: They contain no special non-protein substances other than the metallic atoms themselves. Among them are *ferrodoxin* and *ferritin* (iron containing proteins), copper-binding and iron-binding plasma proteins, **carbonic anhydrase** (a zinc protein enzyme), *insulin* (a zinc-containing hormone) and certain metal-activated *peptidase enzymes*.

6. *Lecithoproteins*: They contain phosphorylated fats, i.e., lecithin as the prosthetic group.

7. *Phosphoproteins*: They possess phosphoric acid as prosthetic group. They are insoluble in water and soluble in alkalies. They can also be precipitated by the acids from alkaline solution e.g., *casein* in milk and *vitelline* in egg.

3. Derived Proteins

They are produced from natural proteins by the action of enzymes, acids or alkalies. The derived proteins may be of the size of parental protein or smaller than that. If it is of the parental size, it is known as **primary derived protein** and the small-sized proteins are known as **secondary derived proteins**.

Primary derived proteins: They are of two types:

(i) Metaproteins

(ii) Coagulated proteins.

(i) *Metaproteins* : They are insoluble in water and dilute salt solutions, but soluble in acids and alkalies. They are produced by the hydrolysis of natural proteins, by alkalies or prolonged treatment with acids.

(ii) *Coagulated proteins* : They are formed by heating the natural proteins. They are insoluble in water and are produced by the action of alcohol or heat. e.g., coagulated egg white.

Secondary derived proteins: They are of the following three types :

(i) *Proteoses*: They are produced when hydrolysis is continued even after the formation of metaproteins. They are soluble in water and do not coagulate by heat. They can be easily precipitated with the ammonium sulphate. They show toxic effects when injected in animals, e.g. *Albuminose* from albumin.

(ii) *Peptones*: They are produced by the action of dilute acids like HCl and H_2SO_4 or enzymes on natural proteins. This hydrolysis is known as **graded hydrolysis** as it is continued even after the formation of proteoses. They are soluble in water and do not coagulate by heat. They cannot be salted out by ammonium sulphate or by any other salt. Peptones give biuret test.

(iii) Peptides : They are produced by the graded hydrolysis with HCl or H_2SO_4. They do not give biuret test. They are soluble in water and are not coagulated by heat.

3.7. CHEMICAL TESTS OF AMINO-ACIDS AND PROTEINS

1. Biuret Reaction

Substance containing two or more peptide linkages produce a blue-violet colour with dilute copper sulphate solution in a strong alkali. This is called **biuret reaction**, after the compound biuret, which happens to give the test. The blue-violet colour is devploped due to formation of a complex between the cupric ion and two adjacent peptide chains.

Test: Add 40% NaOH solution to 2 ml of protein solution in a test tube. Now add one drop of 1% $CuSO_4$ solution. A blue violet colour appears due to the formation of a complex between cupric ion and two adjacent peptide chains (Fig 3.2).

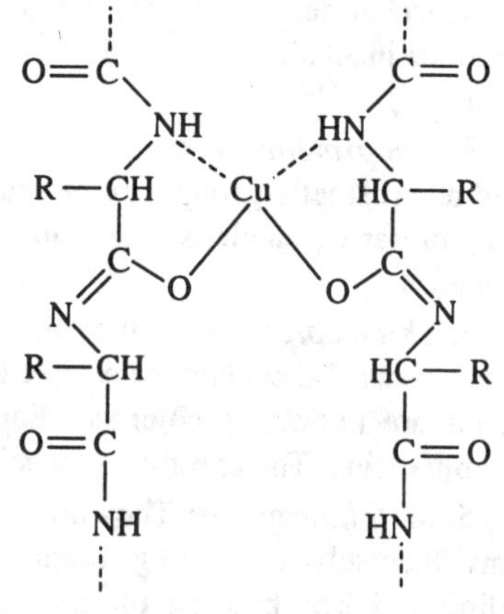

2. Millon's Test

Million's reagent (a mixture of mercuric & mercurous nitrates and nitrites) on heating with compounds containing phenolic groups, such as tyrosine, produces a red colour. The reaction probably involves mercuration and nitration or nitrosation. Tyrosine gives this test in free state as well as combined state.

Fig. 3.2: Biuret Protein Complex.

3. Xanthoproteic test

To 2 ml of protein solution, 1 ml of conc. HNO_3 is added and then heated, then it is cooled. The yellowish colour develops due to formation of nitrocompounds by the action of the nitric acid on the aromatic or heterocyclic rings. Xanthoproteic test comes positive in the presence off either tyrosine or tryptophan. The yellowish nitrocompounds are converted to orange-coloured salts by the alkali like NH_4OH. Human skin also gives this test.

3.7.4. Nitro-prusside test

To 2 ml of protein solution to be tested, 1 ml of 2% sodium notiro-prusside ($Na_2Fe(CN)_5NO$) solution and then 2-3 drops of 40% NH_4OH solution are added. A deep purple-violet or red colour is produced. Free sulph-hydryl groups as in cysteine, some proteins, and glutathions (in its reduced form) give this test. Disulphide groups (as in cystine) give a positive reaction only after reduction to sulph-hydryl groups.

3.7.5. Ninhydrin reaction

Amino acids, peptides or proteins i.e., free amino groups containing compounds give this test. When these are treated with ninhydrin (triketohydrindene hydrate) and heated, they produce deep blue to violet pink or even red colour. Ninhydrin undergoes an oxidation-reduction reaction with free amino groups, oxidatively deaminating them to carbonyl groups and ammonia. The reduced form of ninhydrin then couples with NH_3 and the residual ninhydrin (oxidized ninhydrin) to give rise a blue-violet or even red dye. The reactions are given in Fig 3.3.

(i) Oxidation-reduction reaction.

(ii) Hydrolysis of imino-acid and decarboxylation of α-ketoacid.

(iii) Formation of blue product.

Proline and hydroxyproline are not alpha amino acids and have a secondary amino group in their cyclic structure. Due to lack of alpha-amino group, the proline and hydroxyproline give yellow colour with ninhydrin.

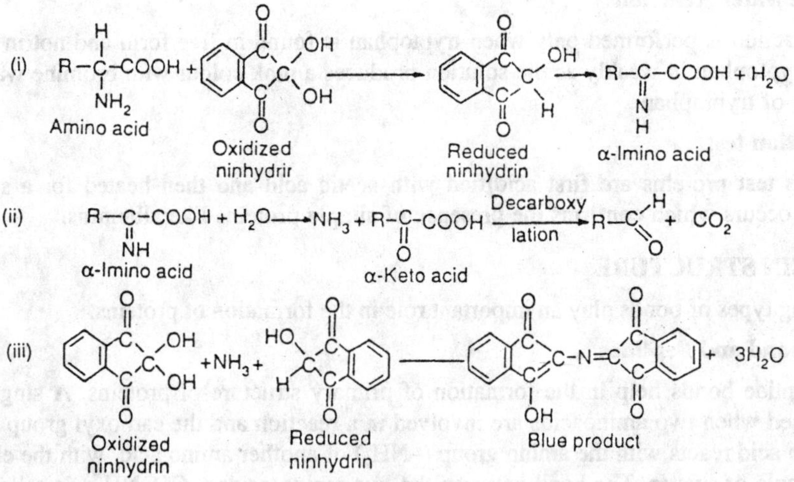

Amino acids and Proteins

Fig. 3.3: Showing ninhydrin reaction.

6. Hopkin's-Cole reaction

This is the characteristic test for those proteins which contain aminoacid, tryptophan. The protein solution is taken in a test tube and a little glyoxylic acid is added. Conc. H_2SO_4 along the side of the test-tube is poured. The appearance of bluish-violet colour confirms the presence of proteins. The colour is due to the interaction of tryptophan and glyoxylic acid, as shown in Fig. 3.4.

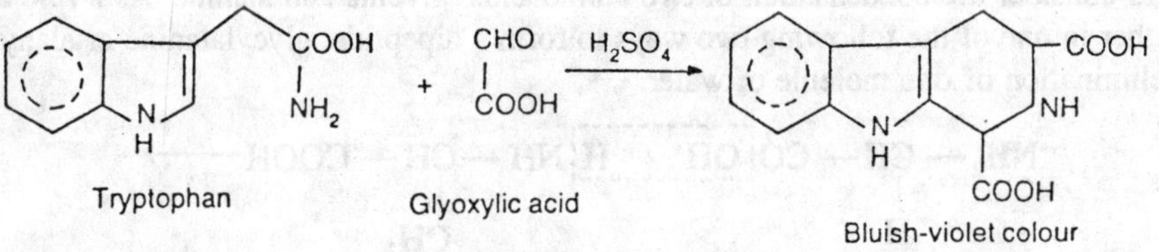

Fig. 3.4: Hopkin's-Cole Reaction.

7. Lead sulphide test

When protein solution is mixed with NaOH and lead acetate-Pb $(CH_3COO)_2$, a black colour is obtained on boiling. This test is due to the presence of either disulphide (-S-S-.) or -SH group.

8. Folin's Phenol test

When proteins containing tyrosine or tryptophan, with alkaline solution are added to Folin's Phenol reagent(a phosphomolybdotungstic acid), a blue colour is obtained due to the reduction of folin's reagent.

9. Pauly reaction

When diazotized sulphanilic acid is added to the alkaline solution of histidine and tyrosine, a red product is obtained. It is due to the coupling of histidine and tyrosine with diazotized sulphanilic acid to form azo dyes.

10. Sakaguchi reaction

When arginine is treated with α-naphthol and sodium hypochloride or hypobromide it develops a red colour and the reaction is known as Sakaguchi reaction.

11. Bromine water Reaction

This reaction is performed only when tryptophan is found in free form and not in combined form. Free tryptophan in weakly acidic solution produces a pink colour with bromine water due to halogenation of tryptophan.

12. Coagulation test

For this test proteins are first acidified with acetic acid and then heated for a short time. Coagulation occurs which confirms the presence of simple proteins like albumins.

3.8. PROTEIN STRUCTURE

The following types of bonds play an important role in the formation of proteins.

1. Peptide bond and Peptides

The peptide bonds help in the formation of primary structure of proteins. A single peptide bond is formed when two aminoacids are involved in a reaction and the carboxyl group (–COOH) of one amino acid reacts with the amino group (–NH$_2$) of another amino acid, with the elimination of one molecule of water. The bond between the two amino acids (–CO–NH–) is called **peptide bond** and the compound formed by the condensation of amino acids is known as a **dipeptide**.

$$R-\underset{NH_2}{\underset{|}{\overset{H}{\overset{|}{C}}}}-CO\boxed{OH\ +\ H}NH-\underset{COOH}{\underset{|}{\overset{H}{\overset{|}{C}}}}-R' \xrightarrow{-H_2O} R-\underset{NH_2}{\underset{|}{\overset{H}{\overset{|}{C}}}}-\boxed{CO-NH}-\underset{COOH}{\underset{|}{\overset{H}{\overset{|}{C}}}}-R'$$

Formation of peptide linkage   dipeptide

Let us consider the condensation of two aminoacids, glycine and alanine. As a rule they will react together in one of the following two ways to form a dipeptide, glycylalanine or alanylglycine with the elimination of one moleule of water.

$$NH_2-\underset{H}{\underset{|}{CH}}-CO\!:\!OH\ +\ H\!:\!NH-\underset{CH_3}{\underset{|}{CH}}-COOH \longrightarrow$$
$$\text{glycine} \qquad\qquad \text{alanine}$$

$$NH_2-\underset{H}{\underset{|}{CH}}-\!:\!CO-NH\!:\!-\underset{CH_3}{\underset{|}{CH}}-COOH\ +\ H_2O$$
Peptide bond   glycylalanine

$$NH_2-\underset{CH_3}{\underset{|}{CH}}-CO\!:\!OH\ +\ H\!:\!NH-\underset{CH_3}{\underset{|}{CH}}-COOH \longrightarrow$$
alanine   glycine

$$NH_2-\underset{CH_3}{\underset{|}{CH}}-\!:\!CO-NH\!:\!-\underset{H}{\underset{|}{CH}}-COOH\ +\ H_2O$$
alanylglycine

Depending upon the number of amino acids involved in a reaction, the compound is known as a **dipeptide** (two amino acids with one peptide bond), a tripeptide (three amino acids with two peptide bonds), a **tetrapeptide** (four amino acids with three peptide bonds) or a polypeptide [many (n......) aminoacids with n-1 peptide bonds] where n = number of amino acids. When a polypeptide

Amino acids and Proteins

chain is formed, one free amino and one free carboxyl group is left at the two different ends. The end having free amino group is referred as **amino terminal** or **N-terminal**, while the end having free carboxyl group is called *carboxy terminal* or *C-terminal*.

2. Disulphide bond

Disulphide bond is also characteristic of the primary structure of proteins. It is a covalent bond and is generally established between cysteine residues as in insulin and in ribonucleases. The disulphide bond may also be established in other sulphur containing amino acids like Cystine and Methionine. When the thiol groups of two cysteine molecules are reversibly oxidized, they form the disulphide compound, cystine, and the linkage established between them is known as **disulphide (—S—S—) linkage**. The disulphide bond may be formed either within the single polypeptide chain (intramolecular) or between the two polypeptide chains (intermolecular).

Disulphide bond helps in stiffening the folded polypeptide chain and also in joining two or more polypeptide chains together.

$$HS-CH_2-CH(NH_2)-COOH$$
$$HS-CH_2-CH(NH_2)-COOH$$
2 molecules of cysteine

$\longrightarrow$

$$S-CH_2-CH(NH_2)-COOH$$
$$S-CH_2-CH(NH_2)-COOH + 2H^+ + 2e^-$$
disulphide bond — cystine

Formation of disulphide bond

3. Hydrogen bonds

Hydrogen bonds are commonly found among the proteins. They are electrostatic in origin and reflect the interaction of the incompletely shielded nucleus of the hydrogen atom—which is a portion of unit positive charge, with the electronic system of another atom. The hydrogen bonds are formed by only electronegative atoms. This bond results by sharing of electrons between hydrogen atom and other electronegative atoms like oxygen. This is an electrovalent bond and has low energy (-1 to -5 K Cal) than most covalent bonds. As a consequence, hydrogen bonds can be broken and reformed with relative ease at ordinary temperatures.

4. Hydrophobic bonds

Hydrophobic bonds arise from the mutual cohesion of non-polar hydrocarbon side chains. In biological systems there are a number of amino acids having side-chains which are of hydrocarbon nature. These are hydrophobic groups in that they do not form hydrogen bonds with water molecules. On the other hand, water molecules have a strong tendency to form hydrogen bonds among themselves. As a result of the hydrogen bonding among water molecules, hydrocarbons are forced out of any water phase in which they may be placed. Similarly, the hydrocarbon side-chains of the various aminoacids tend to be forced together as a result of the hydrogen bonding among water molecules. The hydrophobic bonds are believed to contribute most of the structural stabilization energy for the majority of proteins.

5. Structure of Proteins

The sequential arrangement of the amino acids in a protein molecule is known as the primary structure. When an interaction between polypeptide takes place, it gives rise to a helical type structure known as secondary structure. Further folding and coiling gives rise to the highly specific and complex tertiary and quarternary structures. The elucidation of the structure of proteins has been one of the major triumphs of biochemistry.

Primary Structure

The sequential arrangement of the various aminoacids in a protein (Polypeptide chain) through the peptide bonds is known as the **primary structure**. Each protein molecule consists of one or more polypeptide chains in which the aminoacids are linked by peptide linkages. **Myoglobin**, a protein, consists of only single polypeptide chain, whereas **hemoglobin** molecule consists of four polypeptide chains. The covalent bonds and disulphide (—S—S—) bonds are again characteristic of the primary structure of proteins. The disulphide bond is generally established between cysteine residues, as in *insulin* and *ribonucleases*.

Sanger (1953) for the first time determined the complete amino acid sequence of a protein *Oxinsulin*. He showed that Oxinsulin consists of two polypeptide chains joined together by disulphide bonds. The long chain contained 30 amino acids, while the small only twenty-one. Thus, the total number of aminoacids in insulin as shown by Sanger is 51 (Fig. 3.5). This was a great achievement, although insulin is a relatively small protein molecule. Later on, the number and sequence of amino acids of various enzymes and proteins including *lysozyme, subtilase, ribonuclease* and protein of Tobacco Mosaic Virus (TMV), have been determined. The enzyme lysozyme consists of 129 amino acids, subtilase 276 amino acids, ribonuclease 124 amino acids and TMV polypeptide chain consists of 158 amino acids.

A team of Chinese workers in 1965 synthesized the first protein molecule in vitro was also insulin. **Denkewalter** and **Hirschmann** (1969) and **Merrifield** and **Gutte** (1969) synthesized the ribonuclease separately.

If the protein has only one polypeptide chain, it can have only one free α-amino group (-NH₂ terminal) and one free carboxyl (-C-terminal) group. In the determination of primary structure of protein, it is essential to know what amino acids are at N-terminal and C-terminal ends. The free α-amino group can react with the reagents like dinitrofluorobenzenes to give a dinitrophenyl (DNP) derivative which on hydrolysis yields the yellow coloured DNP-amino acids. These can be isolated chromatographically. If the protein is pure and its molecular weight is known accurately, the number of –NH₂ terminals can be determined by DNP anaylsis. If the protein contains more than one polypeptide chain, it is necessary to split the cross linkages that hold the strands together. Then free ends are treated with the reagents and hydrolysis is made. Finally, aminoacids are separated chromatographically.

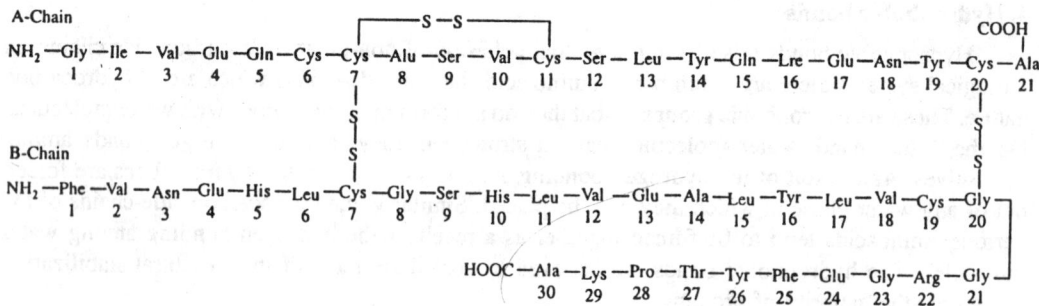

Fig. 3.5: Sequence of aminoacids in the cattle insulin molecule (After Sanger).

Secondary Structure

When the long polypeptide chains in a protein undergo folding, they form the secondary structure or helical structure. The secondary structure is determined by hydrogen bonding between the components of the peptide chain itself. The hydrogen bonds can occur either within one polypeptide chain or between different polypeptide chains of the protein molecule. Thus, the secondary structure of proteins is represented by helical structures which are ultimately formed by the hydrogen bonding between the chains or chain. *Astbury* (1930) suggested that all the proteins are made up of primary chains which can undergo changes in their shape and folding under special conditions. *Astbury* and *Street* (1933) showed that wool or hair fibres when placed in steam can be stretched to twice their original length. The X-ray diffraction analysis of these stretched fibres shows similar patterns to those of silk fibres, but is quite different from those of unstretched wool fibres. They named the unstretched keratin, the α-*form* and the stretched type, the β-*form*.

Fine structure of Proteins

α-Structure: Fine structure of proteins was discovered by *Pauling* and *Corey* (1951) of California Institute of Technology and were awarded Nobel Prize in chemistry for this discovery. They proposed the α-helical conformation of Protein based on theoretical grounds. In searching for the intrinsically most stable configuration of a polypeptide chain they were guided by the following tenets :

1. The maximum possible number of hydrogen bonds between –CO and –NH groups are formed.
2. The peptide (–CO–NH–) group is planar, as is the case in small molecules.
3. The N-H bond for each N-H-.....O hydrogen bond does not deviate by more than 30° from the vector joining the N and O atoms.
4. The orientations about C-N and C-C single bonds are near the potential energy minima for rotation about these bonds.
5. The spatial progression from one residue to the next is constant.

The α-helix proposed by *Pauling* and *Corey* (1951) consists of a single strand twisted about a helical axis. The coiling is maintained chiefly by the hydrogen bonds between each ⟩C=O group and the –NH group of the third peptide residue. They are roughly parallel to the helical axis. The

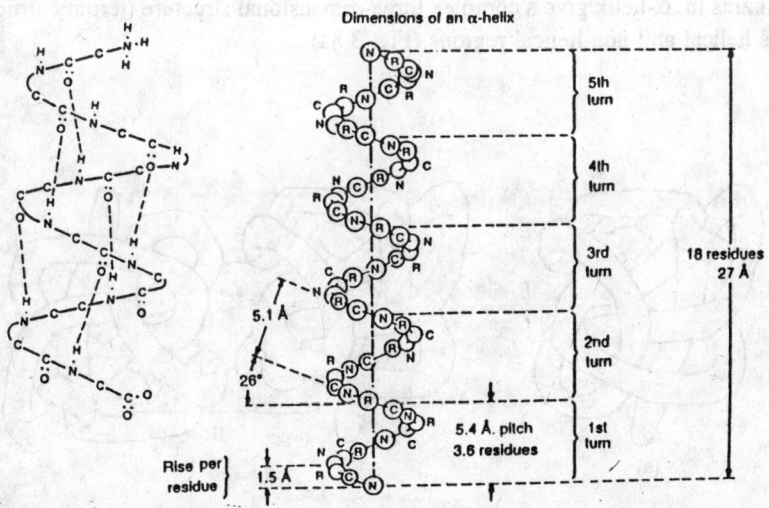

Fig. 3.6: (*a*) α-helix model of protein coiling showing intrachain hydrogen bonding which gives stability to the molecule. (*b*) Representation of a polypeptide chain as in α-helical structure showing dimensions of α -helix.

helix contains 3.6 amino acid residues for each complete turn and each residue rises by 1.5Å. The pitch or spacing between successive turns, is 5.4Å. (Fig. 3.6 a, b).

Ambrose, Elliott and Others (1950) also examined the polypeptide chains in α-forms in the polarizing infra-red spectrometer and showed that in α-form the hydrogen bonds between the NH and C=O groups lie parallel to the long axis of chains and not at right angles as they do in β-forms or β -Structure. They proposed a folding of polypeptide chain by the formation of hydrogen bonds within the chain the α-helix structure has been within the chain. The α-helix structure has been discovered in a great variety of proteins, *e.g.*, myoglobin.

β-*Structure*: *Astbury* and *Street* (1933) proposed β-structure of proteins which was later modified by *Pauling* and *Corey* (Fig. 3.7). The β-structure is represented by parallel zig-zag polypeptide chains which form a pleated sheet-like structure. The hydrogen bonds are formed between NH and C=O groups on the neighbouring chains which stabilize the β-structure of proteins. The side chain attached to the amino acid residues lies above and below the hydrogen bonded sheets. This structure is a stable arrangement where side chains are small and do not cause distortion of the pleated structure. In fibroin, the chains run anti-parallel to each other, *i.e.* the free amino (or carboxyl) groups are at opposite ends of neighbouring chains. β-structure is found in milk and keratin.

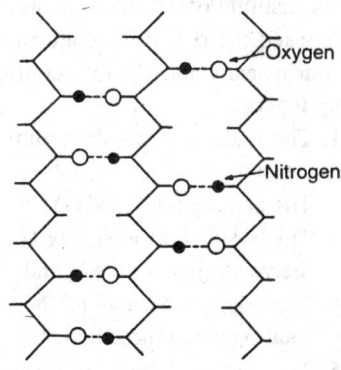

Fig. 3.7: β-structure of proteins.

Tertiary Structure

Very few protein molecules exist as a simple α-helix. Further degrees of folding or coiling of polypeptide chains in α-helix give a complex three-dimensional structure (tertiary structure) which often contains helical and non-helical regions (Fig. 3.8a).

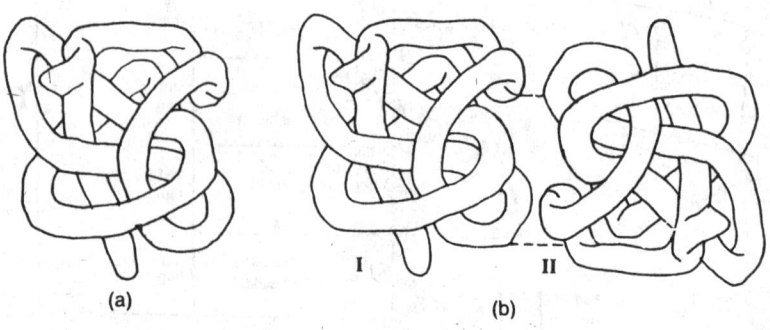

Fig. 3.8. (*a*) Model representing the tertiary structure of proteins.
(*b*) A tetramer of protein units illustrating the quarternary structure.

Amino acids and Proteins

Folding of the α-helix occurs where the amino acid proline is present. Proline has an imino (NH) group instead of an amino group which causes unstability in the α-helix by producing hindrance in the regular internal hydrogen bonding.

Three main types of bonds, ionic, hydrogen and hydrophobic, are responsible for the formation of the tertiary structure of a protein. Dipole-dipole interaction and disulphide linkages are also responsible for the formation of tertiary structure. All types of bonds are shown in Fig 3.9. Ionic bonds are formed between the basic and acidic amino acid residues, while the disulphide bonds are formed between the cysteine residues. Tertiary structure of proteins is probably thermodynamically the most stable and is of much importance, because the enzymatic properties of a protein depend on it.

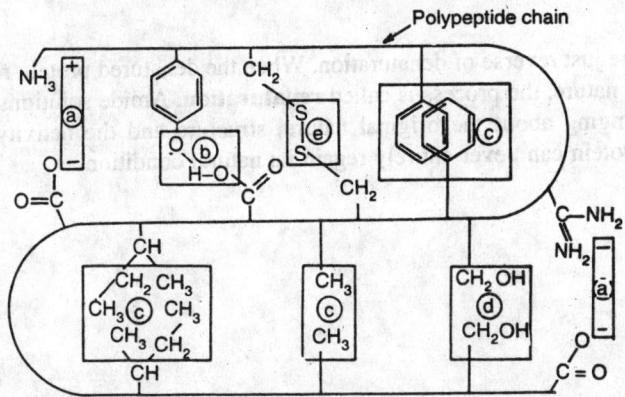

Fig. 3.9. Types of interaction which may contribute to the stabilization of protein structure. (a) Electrostatic bonds (b) hydrogen bonds, (C and D) hydrophobic bonds, (e) disulphide linkage.

Quarternary Structure

This defines the degree of polymerization of a protein unit. The quarternary structure is exhibited by haemoglobin molecule which was determined by *Perutz* and coworkers (1960). They showed that this protein undergoes further organization, being made up of 4-polypeptide chains. This further organization is known as the quarternary structure (Fig. 3.8b). The chains undergo secondary folding, two of the structures consisting of α-chain and other two of β-chain. All 4-chains fit together in a compact tetrahedral arrangement to form a complete haemoglobin molecule. The forces maintaining the quarternary structure are similar to those involved in tertiary structure, but the association of the sub-units, in general is more flexible.

3.9 DENATURATION AND RE-NATURATION OF PROTEINS

1. Denaturation

If a protein is exposed to heat or some other abnormal conditions, its complex structure (secondary and tertiary) is lost without breaking the primary structure. Thus randomly oriented and biologically inactive single polypeptide chains are obtained. This unfolding process of proteins is called **denaturation**. As a result of denaturation the physical and chemical nature of protein alters. The following changes take place :

(a) The linkages of polypeptide chains, mainly of hydrogen bonds, are partly split.
(b) Just after splitting hydrogen bonds reduction and oxidation also take place.
(c) As a result of reduction, disulphide bonds (—S—S—) linking two polypeptide chains are split and the liberation of cysteine-SH radicals takes places.
(d) The quantity of diamino acids increases, while that of monoamino acids decreases.

These effects vary according to the (*i*)pH prevailing during denaturation, (*ii*) to the mutual concentrations of chemcal factors and of proteins, and (*iii*) to the intensity of physical factors.

Putnam (1953) summarized the process of denaturation as follows:
1. Solubility decreases.
2. Biochemical activity disappears.
3. Activity of some radicals present in the structure of proteins increases.
4. Original size and shape of the molecule alters.

Denaturating agents: These may be physical or chemical. The physical agents are heat, surface action, UV light, ultrasound, and high pressure. The chemical agents are organic solvents, acids and alkalies, urea and guanidine, and detergents.

2 Renaturation

This process is the just reverse of denaturation. When the denatured protein recovers its entire biological activity and nature, the process is called **renaturation**. Amide solutions, detergents, and antibodies help in bringing about the original nature, structure and the activity of the protein. However, renatured protein can never entirely regain its natural condition.

4

The Lipids

4.1. INTRODUCTION

The lipids comprise a large group of structurally distinct organic compounds including fats, waxes, phospholipids, glycolipids etc. The lipids are characterized by their solubility in organic solvents such as ether, alcohol, chloroform, carbon disulphide and insolubility in water. From the living tissues, lipids can be extracted as a complex mixture with the help of alcohol and ether. Like carbohydrates, they are composed of carbon, hydrogen, and oxygen, but are poorer in oxygen, and are made up of carbon chains of various lengths. **Lipids are esters of fatty acids and alcohols**.

The lipids are found widely distributed in plant and animal kingdom. In plants, they are found in both vegetative and reproductive parts. In reproductive structures (seeds) they are present in higher amounts, whilst in vegetative structures in only small amounts. In animals, they are found in blood, muscles, brain, skin, membrane and other parts of the body.

4.2. CLASSIFICATION OF LIPIDS

The lipids are classified into the following classes :

1. *Simple Lipids*

   (a) Neutral fats (Triglycerides)

   (b) Waxes

2. *Compound Lipids*

   (a) Phospholipids

   (i) Glycerophosphatides

   (ii) Phosphoinositides

   (iii) Phosphosphingosides.

   (b) Glycolipids

   (c) Lipoproteins

3. *Derived Lipids*

1. Simple Lipids

(a) *Neutral fats* (Triglycerides) : The fats are esters of the trihydric alcohol, glycerol and fatty acids. In nature, three molecules of fatty acids combine with one molecule of glycerol with the elimination of three molecules of water. The enzyme *lipase* catalyses this reaction (Fig 4.1).

In the formation of fat molecule, one, two or all the three hydroxyl groups of glycerol may be esterified with fatty acids forming mono-, di-, and tri-glycerides respectively. In a triglyceride, all the three fatty acid molecules may be identical as in **tripalmitin** or may be different. **All the fats contain glycerol in common, but differ in amount and types of fatty acids**.

The triglycerides which contain saturated fatty acids are solid at room temperature and are called **fats**. Other triglycerides which contain unsaturated fatty acids are liquid at room temperature and are called **oils**. Oils can be converted to solid fats by saturating the fatty acids. This process is utilized in the manufacturing of vegetable ghee from oils.

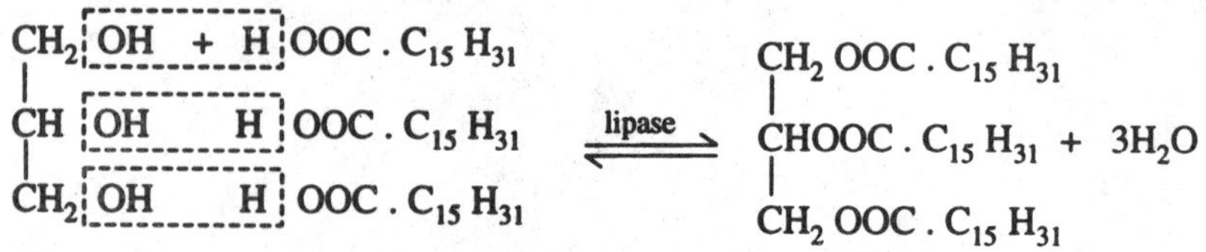

$$\begin{array}{llll}
CH_2\,OH + H\,OOC.C_{15}H_{31} & & CH_2\,OOC.C_{15}H_{31} & \\
CH\,OH H\,OOC.C_{15}H_{31} & \xrightleftharpoons{\text{lipase}} & CHOOC.C_{15}H_{31} & +\ 3H_2O \\
CH_2\,OH H\,OOC.C_{15}H_{31} & & CH_2\,OOC.C_{15}H_{31} & \\
\text{Glycerol} \quad\ \text{Palmitic acid} & & \text{Tripalmitin} & \\
\text{(one molecule)} \ \text{(3-molecules)} & & \text{(fat)} &
\end{array}$$

Fig. 4.1: Formation of fat molecule from glycerol and Palmitic acid.

The acids which are combined with glycerol in the naturally occurring fats are called **fatty acids**. These acids contain even number of carbon atoms linked together in long chains which are in general unbranched. The chain contains from six to about twenty carbon atoms in plant triglycerides and the most frequent number being 16 or 18. The long chain molecule possesses the carboxyl group at one end, while the remainder part consists of carbon and hydrogen atoms only.

Classification of fatty acids: The fatty acids can be classified into following four groups:

1. Saturated fatty acids
2. Unsaturated fatty acids
3. Branched chain fatty acids
4. Cyclic fatty acids.

1. *Saturated fatty acids:* Saturated fatty acids are those which contain only single bonds in their hydrocarbon chain. The general formula for saturated fatty acids is as follows:

$$R-COOH$$

Where R is $CH_3(CH_2)n$, n varies from zero in acetic acid to 24 in cerotic acid and to 86 in mycolic acids. The most commonly occurring saturated fatty acids of the higher plants are palmitic acid (C_{16}) and stearic acid (C_{18}). Other major saturated fatty acids found in plant lipids are shown in Table 4.1.

Table 4.1: Major saturated fatty acids and their sources

S.No.	Name of fatty acid	Structure	Source
1.	Butyric acid	$CH_3(CH_2)_2 COOH$	Butter
2.	Caproic acid	$CH_3(CH_2)_4 COOH$	Butter, palm oil, coconut oil.
3.	Caprylic acid	$CH_3(CH_2)_6 COOH$	Palm oil, coconut oil.
4.	Capric acid	$CH_3(CH_2)_8 COOH$	Palm oil, coconut oil.
5.	Lauric acid	$CH_3(CH_2)_{10} COOH$	Lauraceae, coconut oil palm oil.
6.	Myristic acid	$CH_3(CH_2)_{12} COOH$	Seed fats, butter, coconut oil.
7.	Palmitic acid	$CH_3(CH_2)_{14} COOH$	Plant fats, palm oil, peanut oil.
8.	Stearic acid	$CH_3(CH_2)_{16} COOH$	Plant and animal fats.
9.	Arachidic acid	$CH_3(CH_2)_{18} COOH$	Peanut oil.
10.	Behenic acid	$CH_3(CH_2)_{20} COOH$	Plant fats.
11.	Lignoceric acid	$CH_3(CH_2)_{22} COOH$	Plant lipids.
12.	Cerotic acid	$CH_3(CH_2)_{24} COOH$	Bees wax, wool.

2. *Unsaturated fatty acids:* Unsaturated fatty acids are those which contain one or more double bonds in their hydrocarbon chain. The general formula is as follows:

The Lipids

$$R-CH=CH(CH_2)n-COOH.$$

These acids are also characterized by the presence of cis-trans isomerism at their double bonds. The most common examples of unsaturated fatty acids are shown in table 4.2. Oleic and linoleic acids are found widely distributed among plants.

Table 4.2 : Major unsaturated fatty acids and their sources

S.No.	Name of fatty acid	Structure	Source
1.	Palmitoleic acid	$CH_3(CH_2)_5CH=CH(CH_2)_7 COOH$	Sardine oil.
2.	Oleic acid	$CH_3(CH_2)_7CH=CH(CH_2)_7 COOH$	Olive oil, peanut oil, Linseed oil.
3.	Linoleic acid	$CH_3(CH_2)_4CH=CHCH_2CH=CH(CH_2)_7COOH$	Olive oil, peanut oil, castor oil, Linseed oil, Soyabean oil.
4.	γ-Linolenic acid	$CH_3(CH_2)_4CH=CHCH_2CH=CHCH_2CH=CH(CH_2)_4COOH$	Linseed oil.
5.	Parinaric acid	$CH_3CH_2CH=CHCH=CHCH=CHCH=CH(CH_2)_7COOH$	Plant lipids.
6.	Erucic acid	$CH_3(CH_2)_7CH=CH(CH_2)_{11}COOH$	Rapeseed oil.
7.	Arachidonic acid	$CH_3(CH_2)_4CH=CHCH_2CH=CHCH_2CH=CHCH_2CH=CH(CH_2)_3.COOH$	Pea nut

In all above unsaturated fatty acid the double bonds have a cis-configuration.

3. *Branched Chain fatty acids*: These fatty acids contain odd number of carbon atoms and are usually found in animal fats. The general formula is as follows :

$$CH_3-CH_2-CH-(CH_2)n-COOH.$$
$$|$$
$$CH_3$$

4. *Cyclic fatty acids*: They have been reported from plants and bacteria and are rare in higher animals. The example of cyclic fatty acid is chaulmoogric acid. It has bactericidal action on the leprosy bacillus, *Bacillus leprae* and due to this property it plays an important role in the leprosy therapy. Table 4.3 shows some examples of fatty acids with unusual structures in plant lipids.

Table 4.3 : Fatty acids with unusual structures

S. No.	Name of fatty acid	Structure	Type	Plant source
1.	Tariric	$CH_3(CH_2)_{10}C\equiv C(CH_2)_4COOH$	With acetylenic double bond	*Pieramnia* spp.
2.	Chaulmoogric	$\begin{array}{c}CH=CH\\|\diagdown\\CH_2-CH_2\end{array}CH(CH_2)_{12}.COOH$	With ring structure.	*Hydnocarpus wightiana*.

3.	Ricinoleic	$CH_3(CH_2)_5 CH CH_2CH=CH(CH_2)_7COOH$
		    |
		    OH
			A hydroxyl substituent	*Ricinus* spp.

4.	Juniperic	$HOCH_2(CH_2)_{14} COOH$		Conifers.

5.	Vernolic	$CH_3(CH_2)_4CH-CHCH_2CH=CH(CH_2)_7COOH$
		        \\ / ||
		         O O
			An epoxy substituent	*Veronica anthelmintica*.

| 6. | Licanic | $CH_3(CH_2)_3(CH=CH)_3(CH_2)_4 \overset{O}{\underset{\|}{C}}.CH_2CH_2COOH$ | a keto-substituent | *Licania rigida.* |
| 7. | Japanic | $HOOC(CH_2)_{19}COOH$ | With an additional carboxyl group | *Rhus* spp. |

Properties of fatty acids and fats

Physical Properties

1. Fats and fatty acids are soluble in organic solvents such as petroleum ether, benzene, and choloroform and are insoluble in water.
2. Saturated fatty acids are solid at room temperature, while unsaturated fatty acids are liquid.
3. Unsaturated fatty acids show cis-trans isomerism due to presence of double bonds.
4. They are bad conductors of heat.
5. Saturated glycerides containing fats require high temperature for melting, while unsaturated glycerides containing fats require relatively lower temperature for its melting.

Chemical Properties

1. *Hydrolysis* : Fats undergo hydrolysis when they are treated with mineral acids, the alkalies or fat splitting enzyme *lipase* or hydrolases to yield glycerol and the constituent fatty acids.

$$\begin{array}{l} CH_2O\,|\,OC.C_{15}H_{31} \quad HO\,|\,H \\ | \\ CHO\,|\,OC.C_{15}H_{31} \quad HO\,|\,H \\ | \\ CH_2O\,|\,OC.C_{15}H_{31} \quad HO\,|\,H \end{array} \xrightarrow{\text{Enzyme lipase}} \begin{array}{l} CH_2OH \\ | \\ CHOH \\ | \\ CH_2OH \end{array} + 3C_{15}H_{31}COOH$$

Fat (tripalmitin)   Water   glycerol   Palmitic acid

Hydrolysis by alkalies such as NaOH or KOH leads to the formation of sodium or potassium salts of fatty acids. The salts are known as soaps and process of its formation is known as **saponification**.

$$\begin{array}{l} CH_2\,|\,O.OCR \quad NaOH \\ | \\ CH\,|\,O.OCR + NaOH \\ | \\ CH_2\,|\,O.OCR \quad NaOH \end{array} \longrightarrow \begin{array}{l} CH_2OH \\ | \\ CHOH \\ | \\ CH_2OH \end{array} + 3RCOONa$$

Fat   sodium hydroxide   Glycerol   sodium salt of fatty acid (soap)

2. *Hydrogenation* : Oils containing unsaturated fatty acids can be hydrogenated in presence of high temperature, pressure and finely divided nickel. By this process the oils are converted into solid fats (glycerides of saturated fatty acids).

$$\begin{array}{l} CH_2OOC\,C_{17}H_{33} \\ | \\ CHOOC\,C_{17}H_{33} \\ | \\ CH_2OOC\,C_{17}H_{33} \end{array} \xrightarrow[\text{Ni, } 180°-200°]{3H_2} \begin{array}{l} CH_2OOC\,C_{17}H_{35} \\ | \\ CHOOC\,C_{17}H_{35} \\ | \\ CH_2OOC\,C_{17}H_{35} \end{array}$$

Triolein (m.p. 17°)   Tristearin (m.p. 60°)
(oil)   (solid fat)

The Lipids

This reaction forms the basis of the industrial production of hydrogenated oil (vegetable ghee or margarine).

3. *Hydrogenolysis* : Oils and fats are converted into glycerol and a long chain aliphatic alcohol when excess of hydrogen is passed through them under pressure and in presence of copper-chromium catalyst. This splitting of fat by hydrogen is called **hydrogenolysis**.

$$\begin{array}{c} CH_2OOC\,C_{17}H_{35} \\ | \\ CHCOC\,C_{17}H_{35} \\ | \\ CH_2OOC\,C_{17}H_{35} \end{array} \xrightarrow[200\,atm.]{6H_2,\,Cu-Cr} \begin{array}{c} CH_2OH \\ | \\ CHOH \\ | \\ CH_2OH \end{array} + 3\,C_1H_{35}CH_2OH.$$

Tristearin (fat)   Glycerol   Octadecyl alcohol.

4. *Halogenation* : When unsaturated fatty acids are treated with halogens such as iodine and chlorine, they take up iodine or other halogens at their double bond site. This process of taking of iodine is known as **halogenation** and it is an indication of unsaturation. **Iodine number** is the percentage of iodine absorbed by a fat.

$$CH_3(CH_2)_7CH=CH(CH_2)_7COOH + I_2 \longrightarrow CH_3(CH_2)_7\underset{I}{CH}-\underset{I}{CH}(CH_2)_7COOH$$

oleic acid   Iodine

Halogenated oleic acid

5. *Rancidity* : Oils and fats, on long storage in contact with heat, light, air and moisture, develop an unpleasant odour. Such oils and fats are known as **rancid oils and fats**. The rancidity develops due to certain chemical changes taking place in the fat. The changes include :

(*i*) *Enzymatic hydrolysis* : In presence of enzymes and micro-organisms the fats and oils form bad smelling lower fatty acids.

(*ii*) *Air oxidation of unsaturated fatty acids* : During air oxidation the unsaturated fatty acid portions of fats are oxidized at the site of double bonds into aldehyde and ketones with unpleasant odour.

(*iii*) *β-oxidation of saturated fatty acids* : The saturated fatty acids undergo β-oxidation followed by decarboxylation to form ketones of unpleasant odour.

In all the above cases, hydrolysis of fats is caused by enzyme lipase which is produced by micro-organisms present in them. To check the development of rancidity it is, therefore, essential to protect oils and fats from air, light and moisture during storage. Other compounds like phenol, ascorbic acid and vitamin E (α-tocopherol) are antioxidants and they prevent the rancidity of fats.

6. *Emulsification* : The process of breaking of large-sized fat molecules into smaller ones is known as **emulsification**. In animals, it is brought about by bile juice liberated from liver. Other emulsifying agents are water, soaps, proteins and gums.

Waxes

Waxes are usually a complex mixture of esters of long-chain fatty acids and long-chain alcohols, free long chain fatty acids, free long-chain alcohols and aldehydes and saturated high molecular weight hydrocarbons. In this mixture there is always dominance of esters. The number of carbon atoms in the constitutional fatty acids and alcohols varies between 24 and 36. For example, industrially used Carnauba wax is made up of 75% myricyl (30 C) and cerotate (26 C).

There are certain waxes which possess characteristic odour. This is due to presence of hydroxy acids in the form of lactones in them. For example, the wax Ambretolide extracted from the seeds of *Hibiscus abelmoschus* has a characteristic musk-like smell.

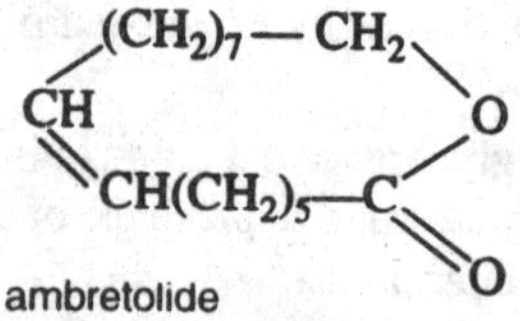

Wax ambretolide (structure with $(CH_2)_7-CH_2$, CH, $CH(CH_2)_5-C$, O, O)

$HOOC(CH_2)_{14}CH_2OH$
Juniperic acid

The waxes from the confiers contain polymers formed by the ester linking of many ω-hydroxy acids, such as juniperic acid with each other.

The waxes are found distributed throughout the entire plant kingdom where they form protective coatings on the stems, leaves, flowers, and fruits of plants. They are epidermal in origin.

2. Compound Lipids

Phospholipids

Structure and Distribution : As the name indicates, phospholipids are those lipids which contain phosphorus. In these, the phosphorus is present in the form of esterified phosphoric acid.

Depending upon the alcoholic component present, the phospholipids are classified into :

(*i*) Phosphoglycerides (glycerol containing)
(*ii*) Phosphoinositides (inositol containing)
(*iii*) Phosphosphingosides [containing amino alcohol phytosphigosine (4-D-hydroxysphinganine) in plants, or sphingosine (4-sphingenine) in animals].

(*i*). *Phosphoglycerides*: This is the largest and most widespread class. It is again subdivided into the following groups :

(a) Phosphatidylcholines (Choline phosphoglycerides)
(b) Phosphatidylethanolamines (Ethanolamine phosphoglycerides).
(c) Phosphatidylserines (Serine phosphoglycerides)
(d) Plasmalogens
(e) Phosphatidylglycerols.

The members of the group (a), (b) and (c) are made up of glycerol, phosphoric acid, fatty acid and a nitrogenous base in the molar ratio of 1 : 1 : 2 : 1. They resemble in their structure, because they are all derivatives of L-α-phosphatidic acid (Fig 4.2 a), but differ in having different nitrogenous bases esterified at the phosphoric acid residue. The nitrogenous bases of groups (a), (b) and (c) are :

(a) Contains *Choline* (Fig. 4.2*b*)
(b) Contains *ethanolamine* (Fig. 4.2*c*)
(c) Contains *serine* (Fig. 4.2*d*)

All of them form internal salts or zwitterions. Certain other compounds of the groups (a), (b) and (c) have also been detected which contain only one esterified fatty acid residue. These are present in small quantities and are known as *Lyso-compounds*, e.g., *Lyso-phosphatidylcholine*.

Phosphatidylcholine and phosphatidylethanolamines are very commonly found in the seeds and vegetative parts (stem and leaves) of higher plants. Phosphatidylserines are less common in higher plants and have been reported from *Scenedesmus* and *Scorzonera* sp.

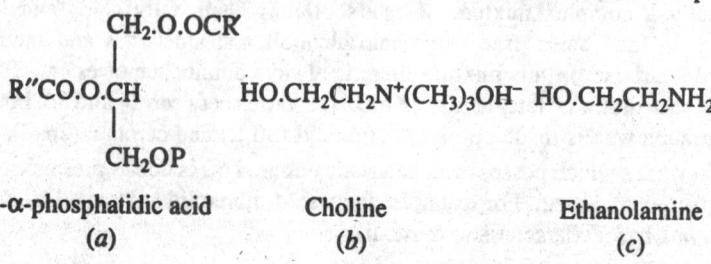

L-α-phosphatidic acid   Choline   Ethanolamine
(a)   (b)   (c)

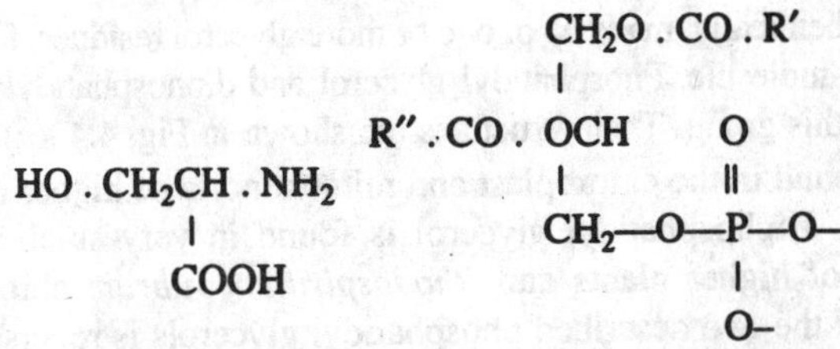

The Lipids

$$\text{HO.CH}_2\text{CH.NH}_2$$
$$|$$
$$\text{COOH}$$

Serine
(d)

$$\text{CH}_2\text{O.CO.R}'$$
$$|$$
$$\text{R''.CO.OCH} \quad \text{O}$$
$$| \quad \quad \|$$
$$\text{CH}_2\text{—O—P—O—R}$$
$$|$$
$$\text{O}^-$$

General strucure
(e)

$R = -\text{CH}_2\text{CH}_2\overset{+}{\text{N}}(\text{CH}_3)_3$: Phosphatidylcholine

$R = -\text{CH}_2\text{CH}_2\overset{+}{\text{NH}}_3$: Phosphatidylethanolamine

$R = -\text{CH}_2\text{CHN}^+\text{H}_3$: Phosphatidylserine
$\quad\quad\quad\quad|$
$\quad\quad\quad\text{COO}^-$

R' & R'' = residues of long chain fatty acids.

Fig. 4.2. (a), (b), (c), (d) and (e) showing the structures of phosphatidic acid, various nitrogenous bases and general structure for the groups (a), (b) and (c).

Plasmalogens are slightly different types of glycerophosphatides. In their structures, the fatty acid residue on the α-carbon atom (c-1) is usually replaced by a long chain acyl group joined to the glycerol by a vinyl ether type linkage as in Choline plasmalogen (Fig. 4.3)

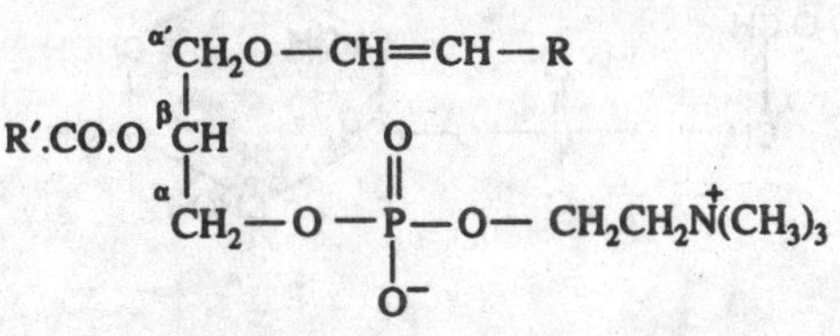

$$^{\alpha''}\text{CH}_2\text{O—CH=CH—R}$$
$$|$$
$$\text{R'.CO.O}^\beta\text{CH} \quad \quad \text{O}$$
$$^\alpha| \quad \quad \quad \|$$
$$\text{CH}_2\text{—O—P—O—CH}_2\text{CH}_2\overset{+}{\text{N}}(\text{CH}_3)_3$$
$$|$$
$$\text{O}^-$$

Fig. 4.3. Structure of Choline plasmalogen.

Choline and ethanolamine plasmalogens are found in the seeds of higher plants, brain and muscles of the animals. They are found only in traces in the liver of animals.

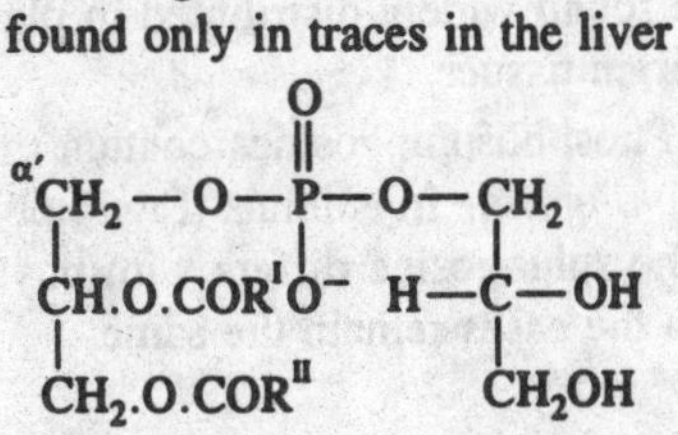

$$\begin{array}{c}
^{\alpha'}\text{CH}_2\text{—O—}\overset{\overset{\displaystyle O}{\|}}{\text{P}}\text{—O—CH}_2 \\
| \quad \quad \quad \quad | \\
\text{CH.O.COR}^\text{I} \quad \text{O}^- \quad \text{H—C—OH} \\
| \quad \quad \quad \quad \quad \quad | \\
\text{CH}_2\text{.O.COR}^\text{II} \quad \quad \quad \text{CH}_2\text{OH}
\end{array}$$

(a)

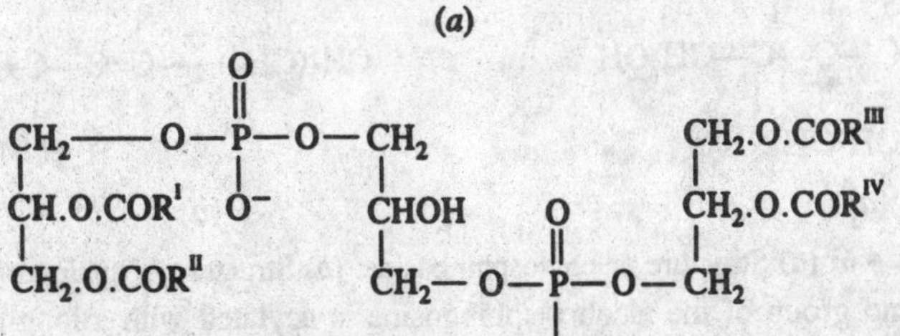

$$\begin{array}{c}
\text{CH}_2\text{——O—}\overset{\overset{\displaystyle O}{\|}}{\text{P}}\text{—O—CH}_2 \quad \quad \quad \text{CH}_2\text{.O.COR}^\text{III} \\
| \quad \quad \quad \quad | \quad \quad \quad \quad \quad \quad \quad \quad | \\
\text{CH.O.COR}^\text{I} \quad \text{O}^- \quad \text{CHOH} \quad \quad \quad \text{CH}_2\text{.O.COR}^\text{IV} \\
| \quad \quad \quad \quad \quad \quad | \quad \quad \text{O} \quad \quad | \\
\text{CH}_2\text{.O.COR}^\text{II} \quad \quad \quad \text{CH}_2\text{—O—}\overset{\|}{\text{P}}\text{—O—CH}_2 \\
\quad \quad \quad \quad \quad \quad \quad \quad \quad \quad \quad | \\
\quad \quad \quad \quad \quad \quad \quad \quad \quad \quad \quad \text{O}^-
\end{array}$$

(b)

Fig. 4.4. (a) Structure of phosphatidyl glycerol. (b) Structure of diphosphatidyl glycerol.

Phosphatidyl glycerol are, chemically, made up of two or more glycerol residues, fatty acid residues and at least one phosphoric acid molecule. Phosphatidyl glycerol and diphosphatidyl glycerol are the two wide occurring members of this group. Their structures are shown in Fig. 4.4. (a) and (b).

Phosphatidyl glycerol is found in the chloroplast and mitochondria of higher plant leaves and in the green alga *Scenedesmus*. Diphosphatidyl glycerol is found in very small amounts in the chloroplasts and mitochondria of higher plants and *Rhodospirillum rubrum* chromatophores. In animal mitochondria the ratio of the two described phosphatidyl glycerols is reversed as compared with that of plant mitochondria.

(ii) Phosphoinositides

Phosphoinositides are made up of glycerol, fatty acid, phosphoric acid and an inactive cyclic hexahydric alcohol meso-inositol. They are further divided into two groups, the mono and the diphosphoinositides. Their structures are shown in Fig. 4.5 (b) and (c). The monophosphoinositides upon hydrolysis yield inositol phosphate, whereas diphosphoinositides yield inositol-m-diphosphate.

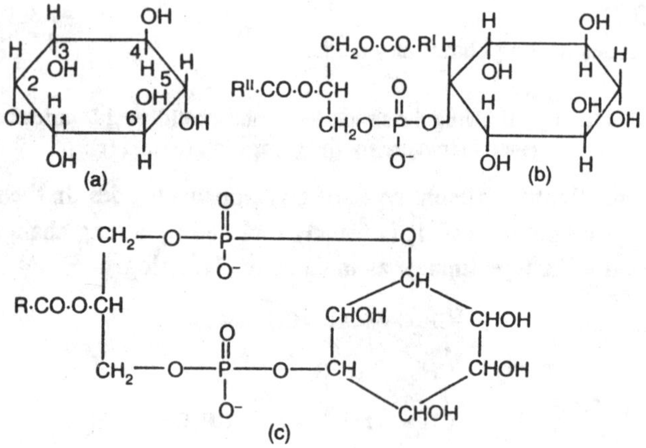

Fig. 4.5. (a) Structure of inactive meso-form of inositol. (b) Structure of monophosphoinositides (c) Structure of diphosphoinositides.

Monophosphoinositides are found widely distributed in plants and animals both, but diphosphoinositides occur only in the brain tissues.

(iii) Phosphosphingosides: Phosphosphingosides contain the constituent alcohol *phytosphingosine* in plants (Fig. 4.6a) and *sphingosine* in animals (Fig. 4.6b). They also possess a phosphoric acid residue in their structure. The sphingosine differs with phytosphingosine in having a double bond, but other structures in both the cases remain the same.

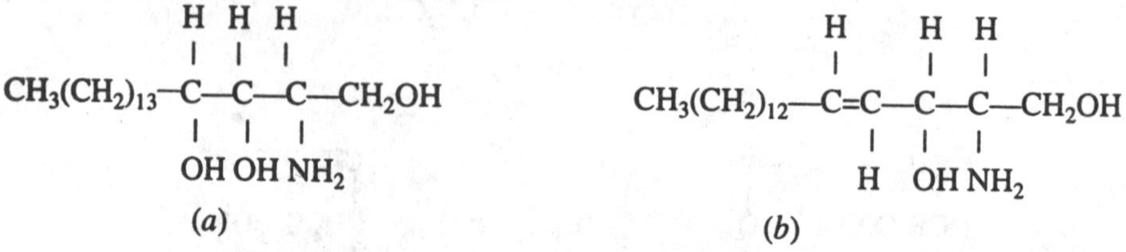

$$CH_3(CH_2)_{13}-\underset{OH}{\underset{|}{\overset{H}{\overset{|}{C}}}}-\underset{OH}{\underset{|}{\overset{H}{\overset{|}{C}}}}-\underset{NH_2}{\underset{|}{\overset{H}{\overset{|}{C}}}}-CH_2OH$$

(a)

$$CH_3(CH_2)_{12}-\overset{H}{\overset{|}{C}}=\underset{H}{\underset{|}{C}}-\underset{OH}{\underset{|}{\overset{H}{\overset{|}{C}}}}-\underset{NH_2}{\underset{|}{\overset{H}{\overset{|}{C}}}}-CH_2OH$$

(b)

Fig. 4.6: (a) Structure of phytosphingosine. (b) Structure of sphingosine.

When the amino group of the alcohol sphingosine is acylated with a long chain fatty acid, the compound **ceramide** is formed. They are formed as intermediates in the biosynthesis of sphingomyelins which are phosphosphingosides.

The Lipids

Sphingomyelins occur in the myelin sheaths of nerves. In their structure the primary alcoholic hydroxyl group at C-1 of the ceramide is esterified with phosphoric acid which in turn is esterified with choline.

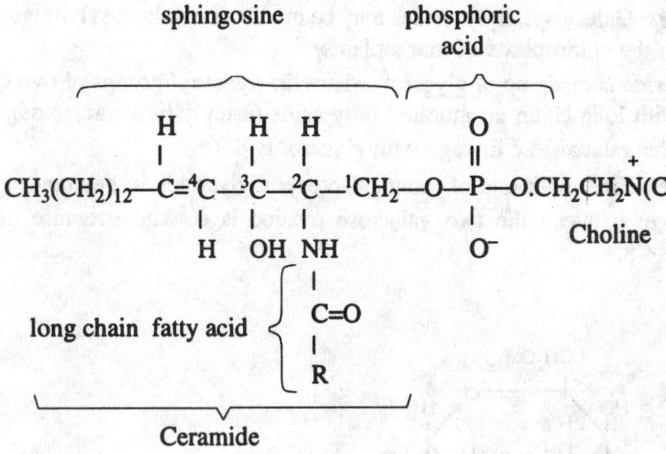

$$\underbrace{CH_3(CH_2)_{12}-\overset{H}{\underset{H}{C}}=\overset{4}{C}-\overset{3}{\underset{OH}{C}}-\overset{2}{\underset{NH}{C}}-\overset{1}{CH_2}}_{\text{sphingosine}}-\underbrace{O-\overset{O}{\underset{O^-}{P}}-}_{\text{phosphoric acid}}O.CH_2CH_2\overset{+}{N}(CH_3)_3 \text{ Choline}$$

long chain fatty acid { C=O, R }

Ceramide

Structure of Sphingomyelin

Glycosphingolipids is another class of phosphosphingosides. The members of this class are made up of phytosphingosine residue in the form of ceramide, phosphoric acid residue, inositol, and several monosaccharide residues. Glycolipids (carbohydrate containing lipids) differ from those of glycosphingolipids in lacking a phosphoric acid residue and are, therefore, not phospholipids. The glycolipids include glycoceramides which are made up of ceramide residue bound by its primary alcoholic hydroxyl group to a carbohydrate. When this carbohydrate is a monosaccharide, the glycoceramide is known as **ceramide oligosaccharide**. The structure of glycosphingolipids extracted from the seeds of soybean and maize is shown in Fig. 4.7. They occur widely in the seeds of plants.

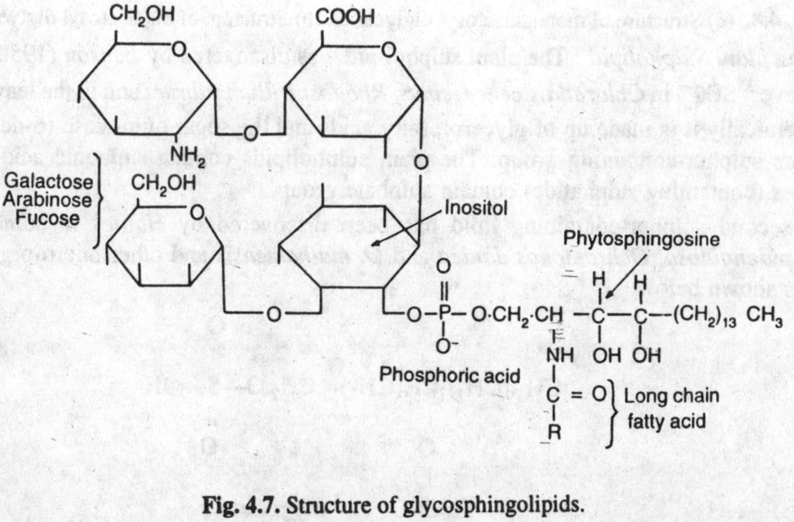

Fig. 4.7. Structure of glycosphingolipids.

Glycolipids

Glycolipids are those lipids which contain carbohydrates. They have been isolated from different plant spp. including soybean, maize, peanut, and wheat. The seeds of sunflower and cotton also contain glycolipids. The chloroplasts of green algae, spinach leaves and other plants contain large amounts of these glycolipids.

Galactosyl diglycerides and sulpholipid form the major glycolipids of plants.

I. *Galactosyl diglycerides*: Galactosyl diglycerides may be mono- or digalactosyl diglycerides. Both have been isolated from the chloroplasts of many plants.

Monogalactosyl diglyceride is made up of glycerol, where the hydroxyl groups of two carbon atoms have been esterified with long chain unsaturated fatty acids (usually linolenic acids), and a monosaccharide galactose. The galactosidic linkage with glycerol is β.

Digalactosyl diglyceride differs with that of monogalactosyl diglyceride in having one more galactose residue. The linkage between the two galactose residue is α. The structure of both diglycerides is shown in Fig. 4.8.

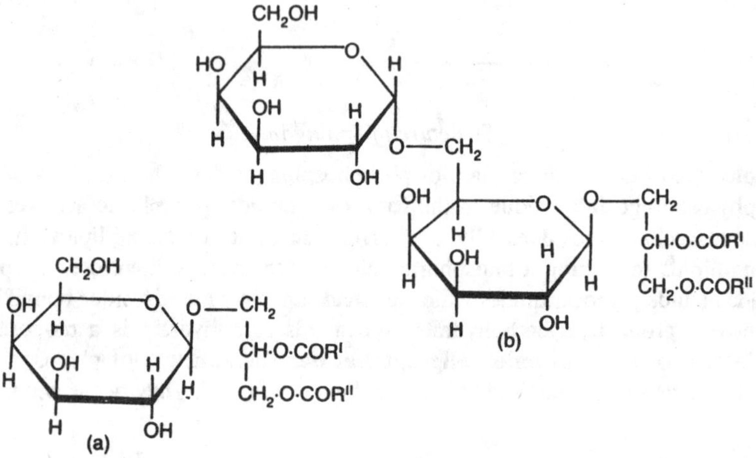

Fig. 4.8. (*a*) Structure of monogalactosyl diglyceride (*b*) Structure of digalactosyl diglyceride

II. *The plant Sulpholipid* : The plant sulpholipid was discovered by *Benson* (1950) by the use of radioactive $^{35}SO_4^-$ in *Chlorella, Scenedesmus, Rhodospirillum rubrum* and in the leaves of higher plants. Chemically it is made up of glycerol, fatty acids and the sugar quinovose (6-deoxyglucose) which bears sulphur-containing group. The plant sulpholipids contain suphonic acid and animal cerebrosides (containing sulphatide) contain sulphate group.

The second sulphur-containing lipid has been discovered by *Haines* in some algae like *Chlorella pyrenoidosa, Ochromonas danica* and *O. malhamensis* and other micro-organisms. The structure is shown below.

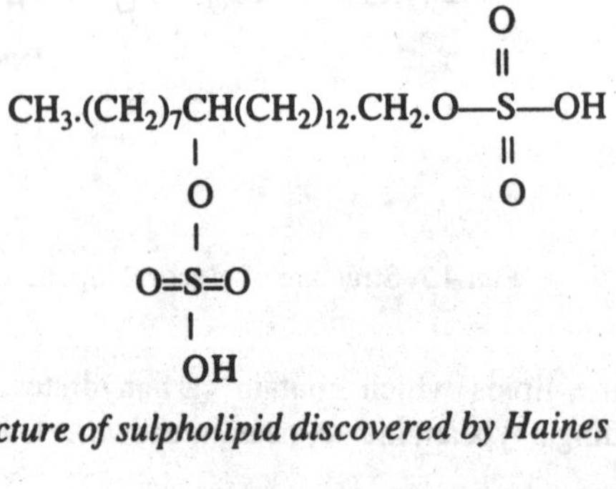

$$CH_3.(CH_2)_7CH(CH_2)_{12}.CH_2.O-\underset{\underset{O}{\|}}{\overset{\overset{O}{\|}}{S}}-OH$$

with substituent $-O-S(=O)(=O)-OH$ on the CH

Structure of sulpholipid discovered by Haines

The Lipids

Lipoproteins

Lipoproteins are complexes which contain lipids and proteins in association. Lipoproteins are most important complexes because in the living systems they are found in the form of soluble liquid drops and build the plasmamembrane and the membranes of various cell-organelles like mitochondria, chloroplast, lysosomes, golgi-complex, endoplasmic reticulum and nucleus. The membranes of these cell-organelles possess a number of enzymes for biochemical reactions.

The lipoproteins of the membranes may be water soluble or fat-soluble. For this reason, they help in the entry and exit of water and fat-soluble compounds and also maintain the concentration and equilibrium for such compounds in the cell. The lipoprotein complexes are affected by agents that influence the purity of proteins such as heat, pH, and other chemicals. Lipoproteins provide the surface for biochemical reactions and energy production and also help in the transportation of lipids and proteins to the various parts of the plant or animal body.

3. Derived Lipids – Steroids

Steroids are manufactured from lipid precursors in the cells. Each steroid molecule possesses a fused four ring (A, B, C, D) skeleton, known as **Cyclopentanoperhydrophenanthrene**. Steroids are the members of the group triterpenoids. **Cholesterol** is the most important and common sterol found in the mammals. It has also been isolated from red algae. The structure is shown in Fig. 4.9. Steroids that possess an —OH group are called **sterols**.

Plant sterols are characterized by having supernumerary side chains. The chains are formed at C-24 and are one or two carbons long. In sterols, at C_{10}, C_{13} and C_{17} the substituents may be present either above or below th plane of the ring system. Usually they are are above and are represented by thick black lines showing the β-configuration. The α-configurations are usually indicated by dotted lines. A very small portion of total sterols in the cells of higher plants is esterified with long chain fatty acids. *Ecdysones* form a new group of plant steroids and are derived from cholesterol. Ecdysones are also known as insect moultin hormones.

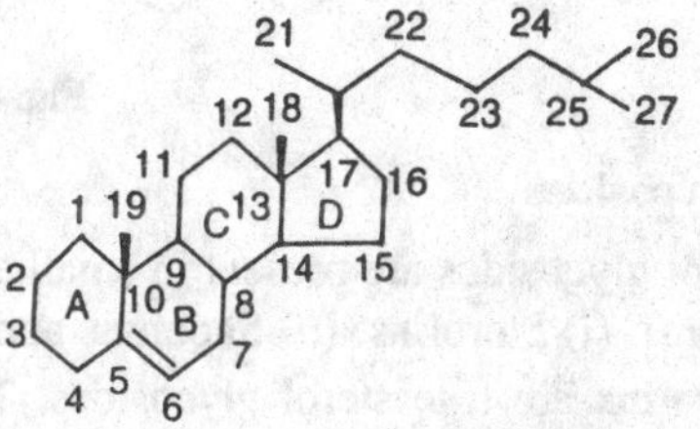

Fig. 4.9: Cholesterol.

Table 4.4: Some Typical Plant Sterols

S.No.	Name	Structure	Source
Sterols			
1.	Stigmasterol (Ethyl group at C-24)	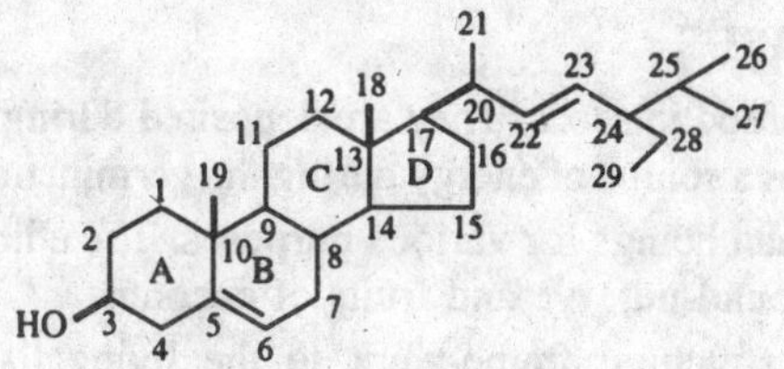	Widely Distributed
2.	Ecdysone (Insect moulting hormone)	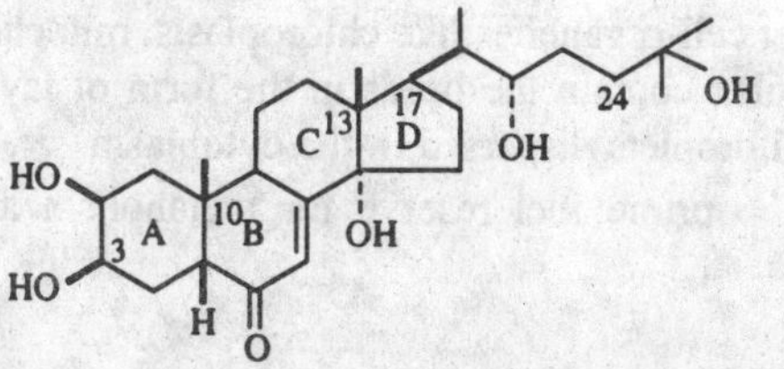	*Podocarpus* spp.

4-α-Methyl Sterol
3. **Lophenol** — Lophocereus Schottii (Cactus)

4,4-Dimethyl Sterols
4. **Cyclolaudenol** (Mythyl group at C-24) — Opium poppy

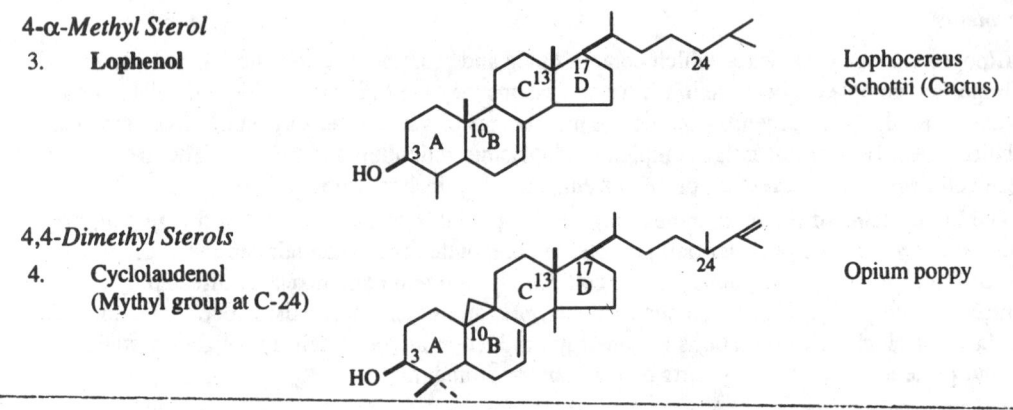

Ergosterol occurs in yeast and fungi (Fig. 4.10).

Fig. 4.10: Ergosterol.

Sterol Glycosides

Sterol glycosides are present in small amount in higher plants. They can be divided into three main groups: (*i*) Sterolins, (*ii*) Saponins and (*iii*) Cardiac glycosides.

Sterolins are true sterol glycosides. The clearly defined example of this type is **ipuranol**, isolated from *Ipomoea purpurea*. Ipuranol is a glycoside of sitosterol.

Saponins are a type of terpenoid glycosides. In low concentrations, they are toxic to animals and cause haemolysis of the red blood cells. The aglycones are called **sapogenins**. Saponins have been isolated from many plants including *Olea europaea*, *Dioscorea* spp., *Soya* spp. and *Digitalis purpurea*.

Cardiac glycosides show poisonous effect in animals caused by their action action on heart muscle. They have been isolated from many plant families e.g., Apocyanaceae, Liliaceae, Ranunculaceae, Moraceae and Scrophulariaceae. Family Scrophulariaceae is commercially important because one of its members, *Digitalis*, yields cardiotonic drugs (digitanides A, B and C).

4.3. IMPORTANCE OF LIPIDS

1. Fats serve as reserve food in seeds. They are deposited during the development of seed and are mobilized and reutilized as a source of energy during the germination and growth of the seedling.

2. Oils are used by human beings for various purposes. The edible oils are obtained from the seeds of mustard, cotton, ground-nut, rye and fruits of coconut.

3. Lipids provide the structural frame-work to the living tissues of plants and animals. Plasmamembrane and other cell-organelles like chloroplasts, mitochondria, endoplasmic reticulum, lysosomes and golgi-complex contain the lipids in the form of layers. In the plant cells they are found in the form of small droplets dispersed in the cytoplasm.

4. Lipids serve as prime fuel reserve for metabolism and provide more energy than carbohydrates and proteins.

The Lipids

5. Fats provide a bridge in filling the gap between water soluble and insoluble phases.

6. Waxes (lipids) give a protective covering on the upper surface of leaves, stems and fruits and this covering also provides resistance to water, insects and bacteria.

7. Certain oils, like castor oil, mustard oil, clove oil and coconut oil are used medicinally.

8. Other oils are used in the preparation of soaps and vegetable ghee.

4.4. TESTS FOR LIPIDS

I. *Physical Tests* : These include determination of melting point, refractive index, viscosity and specific gravity.

II. *Stainability* : Oils and fats give a red colouration when a few drops of dye Sudan III is added in the living tissue.

III. *Chemical Tests* : These include acid value, saponification value, iodine value and Reichert-Meissl value. These values and tests give a good idea about the quality of a given oil or fat.

   (a) *Acid Value* : It is defined as the number of milligrams of potassium hydroxide (KOH) required to neutralise 1 gram of the oil or fat.

   The acid value gives an idea of the amount of free acids present in a given oil or fat.

   (b) *Saponification Value* : It is defined as the number of milligrams of KOH required to neutralise the fatty acids resulting from the complete hydrolysis of 1 gm of the oil or fat.

   Saponification value of a fat or oil gives an idea about its molecular weight. The higher the molecular weight of a fat or oil, the smaller is its saponification value.

   (c) *Iodine Value* : It is defined as the number of grams of iodine taken up by 100 gm of the oil or fat. This value gives an idea about the degree of unsaturation of fat or oil.

   (d) *Reichert-Meissl Value* : It is defined as the number of milliliters of 0.1 N KOH solution required to neutralise the volatile water soluble acids obtained by the hydrolysis of 5 gm of fat. This value is important for testing the purity of butter or ghee. It is usually lower for adulterated ghee than for pure ghee.

5

The Enzymes

5.1. INTRODUCTION

Beadle (1948) defined enzymes as indispensable compounds that play a key role in metabolism by bringing direction and control to the physiological processes of living cells. Any change in enzyme complement of living cells is immediately reflected in change in the physiological and biochemical processes of the cell.

According to *Fairley* and *Kilgour* (1966), biological catalysts are called enzymes, and with no known exception, all enzymes are proteins.

Devlin (1970) also defined the enzymes as organic catalysts which increase the rate of biochemical reactions. All enzymes are proteins, but all proteins are not enzymes.

5.2. OCCURRENCE

Every cell synthesizes its own enzymes with the help of genes. A single gene helps in the synthesis of a single and a specific enzyme (One-gene one-enzyme hypothesis). This hypothesis has been replaced, recently, by one-gene one-polypeptide hypothesis. The synthesized enzymes are then distributed to the different parts of the cell and ultimately to the different parts of the plant or animal body. The enzymes that are produced within the cell for metabolic activities are known as **endoenzymes** and those which act away from the site of synthesis are called **exo-enzymes**. The endoenzymes are produced inside the various cell-organelles like the nucleus, the cytoplasm, mitochondria, chloroplast, lysosomes etc.

5.3. NOMENCLATURE AND CLASSIFICATION OF ENZYMES

Since the enzymes are specific for a particular reaction, they are named according to the substrate on which they act or on the nature of the reaction they catalyse. The most common method for naming them is to suffix-ase at the end of the name of the substrate attacked. Thus, peptide is attacked by *peptidase*, lipid by *lipase*, urea by *urease*, arginine by *arginase* and tyrosine by *tyrosinase*.

A commission of enzymes (International Union of Biochemistry, I.U.B., 1961) has developed a complete, systematic, but rather a complex system of nomenclature and classification.

The commission established a numerical system of classification with the following recommendations :

    (a) Each enzyme has a systematic code number (E.C.) of four digits.
    (b) The first digit of the four figures indicates the main class (See Table 5.1 showing six main classes and their sub-classes and 5.13 key to numbering and classification).
    (c) The second digit indicates the sub-class.
    (d) The third digit indicates the subdivision of the sub-class (sub-subclass).
    (e) The fourth digit designates the serial number of the specific enzyme in the fourth sub-class.
    (f) In the system of nomenclature, enzyme commission recommended both "systematic" and "trivial" names for enzymes.

For example, the code number E.C. 1.1.1.1. stands for the enzyme alcohol dehydrogenase, where:

The Enzymes

E.C. Stands for enzyme commission.

1. Stands for oxidoreductase.

1.1. Stands for the enzyme, which utilizes substrate as-CHOH (alcoholic group).

1.1.1. Stands for those enzymes which utilize NAD as an acceptor.

Similarly, code number E.C. 2.7.1.1. denotes main class 2 (a transferase), sub-class 7 (transfer of phosphorus containing group), sub-sub-class 1 (an alcohol functions as the phosphate acceptor). The fourth digit 1 indicates hexokinase.

Table 5.1: I.U.B. Classification of Enzymes (Condensed)

1. Oxidoreductases. Transfer oxygen, hydrogen or electrons
2. Transferases. Transfer following groups :
   2.1 One carbon
   2.2 Aldehydic or ketonic residues
   2.3 Acyl
   2.4 Glycosyl
   2.5 Alkyl or related
   2.6 Nitrogenous
   2.7 Phosphorus-containing
   2.8 Sulphur-containing
3. Hydrolases. Act on the following bonds :
   3.1 Ester
   3.2 Glycosyl
   3.3 Ether
   3.4 Peptide
   3.5 Other C-N bonds
   3.6 Acid anhydride
   3.7 C-C
   3.8 Halide
   3.9 P-N
4. Lyases. Act on the following bonds :
   4.1 C-C
   4.2 C-O
   4.3 C-N
   4.4 C-S
   4.5 C-halide.
5. Isomerases. Bring about the following types of isomerisations :
   5.1 Racemases and epimerases
   5.2 *Cis-trans* isomerases
   5.3 Intramolecular oxidoreductases
   5.4 Intramolecular transferases
   5.5 Intramolecular lyases
6. Ligases. Join the following bonds :
   6.1 C-O
   6.2 C-S
   6.3 C-N
   6.4 C-C

5.4. MAJOR CLASSES OF ENZYMES

There are six major classes of enzymes based on the type of reaction they catalyse.

1. Oxido reductases

They catalyse the oxidation reductions of CH-OH, CH-CH, C=O, $CH-NH_2$ and CH=NH groups. In these reactions one compound is oxidized and the other is reduced. During oxidation, there is either an addition of O_2 atom, or removal of electrons, e.g.

$$2Mg + O_2 \longrightarrow 2MgO \quad \text{(Oxygen transfer)}$$
$$PbO + H_2 \longrightarrow Pb + H_2O \quad \text{(Hydrogen transfer)}$$
$$Fe^{++} \longrightarrow Fe^{+++} + e^- \quad \text{(Electron transfer)}$$
$$\text{(reduced)} \quad \text{(oxidized)}$$

Some of the important sub-classes are :

1.1 Enzymes acting on CH-OH group as electron donors e.g., 1.1.1.1 Alcohol: NAD oxidoreductase (*Alcohol dehydrogenase*)

$$\text{Alcohol} + NAD^+ \longrightarrow \text{Aldehyde or Ketone} + NAD + H^+$$

1.4 Enzymes acting on the $CH-NH_2$ group as electron donors, e.g., 1.4.1.3 L-Glutamate : NAD(P) oxidoreductase (deaminating).

$$\text{L-Glutamate} + H_2O + NAD(P)^+ \rightarrow \alpha\text{-Ketoglutarate} + NH_4^+ + NAD(P)H + H^+$$

1.9 Enzymes acting on heme groups as electron donors, e.g. 1.9.3.1. Cytochrome C : O_2 oxidoreductase (*Cytochrome oxidase*)

1.11 Enzymes acting on H_2O_2 as electron acceptor e.g. 1.11.1.6. H_2O_2 oxidoreductase (*Catalase*).

Thus, this class includes the enzymes which were previously termed as oxidases or dehydrogenases or peroxidases, e.g., Alcohol dehydrogenase, Cytochrome oxidase and Catalase.

2. Transferases

Transferases catalyse the transfer of a chemical group (aldehyde, ketone, one carbon groups, acyl, alkyl, glycosyl, amino, sulphur or phosphorus containing groups) from one molecule to another. These chemical groups are not present in the free state during the transfer, e.g. transaminases transfer amino group, transmethylases transfer methyl group and transaldolases transfer aldehyde group. The following reaction shows the transfer of an amino group from glutamic-acid to oxaloacetic acid to form aspartic acid in presence of *glutamic-aspartic transminase enzyme*.

```
COOH        COOH                           COOH         COOH
 |           |                              |            |
CH2         CH2                            CH2          CH2
 |     +     |         transaminase         |     +      |
CH2         C=O        ⇌                   CH2          CH.NH2
 |           |                              |            |
CHNH2       COOH                           C=O          COOH
 |                                          |
COOH       Oxalo-acetic                    COOH        Aspartic acid
Glutamic    acid                           α-keto glutaric
acid                                        acid
```

3. Hydrolases

This large group of enzymes brings about hydrolysis of various compounds. These reactions are enhanced in presence of water molecule, as one of the reactants. Since most of the hydrolytic reactions are reversible, they may also be called as synthetic or condensation enzymes. For example, peptidases.

The Enzymes

$$RCO\text{-}OR \xrightleftharpoons{HOH} RCOOH + ROH$$

The following are the sub-groups of hydrolases :

1. Esterases

Esterases catalyse the hydrolysis of esters to alcohols and acids, *e.g.* lipase catalyses the hydrolysis of fats to glycerol and their constituent fatty acids.

Like lipases, phosphatases catalyse the hydrolysis of phosphate esters; carboxylesterases of ethyl butyrate; and acetyl choline esterase of acetylcholine.

$$\begin{array}{c} CH_2OOC\ R \\ | \\ CHOOC\ R \\ | \\ CH_2OOC\ R \\ \text{Fat} \end{array} \xrightleftharpoons[\text{Esterification } -3H_2O]{\text{Hydrolysis}+3H_2O} 3(R.COOH) + \begin{array}{c} CH_2OH \\ | \\ CHOH \\ | \\ CH_2OH \\ \text{Glycerol} \end{array}$$

Glycosylases (Glycosidases)

They attack glycosidic linkages. The sub-group may be divided into two sub-subgroups : (1) first sub-sub-group includes those enzymes which are specific for simple glycosides and oligosaccharides, (2) Second sub-subgroup includes those enzymes which attack the polysaccharides. The enzymes hydrolyzing carbohydrates are sometimes collectively known as "Carbohydrases". The examples of carbohydrases are *invertase, amylase, cellulase, maltase* etc.

Hydrolases acting on ethers

This subgroup includes only one known enzyme, *thioether hydrolase*. The enzyme is found in liver and attacks S-adenosyl-homocysteine.

Peptidases

Peptidases are sometimes called as proteases or proteolytic enzymes. They catalyse the hydrolysis of peptide bonds in proteins and related compounds. Like glycosidases, peptidases are also divided into two sub-subclasses. The enzymes of sub-subclass 'exopeptidases' hydrolysed only terminal peptide bonds, whereas 'endopeptidases can hydrolyse peptide linkages of terminal and interior bond of the peptide chain. Examples of peptidases are, *trypsin, pepsin, erepsin, bromelin* etc.

Hydrolases acting on C-N bonds other than peptide (Amidases and related enzymes)

They hydrolyse the amide linkages. For example, under the influence of urease, urea (the diamide of carbonic acid) is hydrolysed to a molecule of CO_2 and two molecules of NH_3. These products form ammonium carbonate in the solution.

Hydrolases acting on acid anhydride bonds

Hydrolases of this group catalyse the hydrolysis of anhydride bonds. It has been noted that all substrates of enzymes of this class recognized upto this time involve phosphoric acid as one or both of the participants in the susceptible linkages.

Other Hydrolases

A number of enzymes are known which do not fit into the above classes. These include :

(i) *Fumasylacetoacetate hydrolase*, an enzyme acting on a 'C-C' linkage. The enzyme is involved in tyrosine metabolism.

(ii) An alkylhalidases of unknown significance, acting on 'C-halide' bonds.

(iii) Phosphoamidase, acting on 'P-N' bonds. The enzyme attacks phosphocreatine, phosphoarginine, and other phosphoamides.

4. Lyases

Enzymes of this group reversively catalyse the removal of groups from their substrate by mechanisms other than hydrolysis (non-hydrolytically), and a double bond is introduced at the place of removal of the group. Lyases act on C-C, C-O, C-N, C-S, and C-halide bonds. In most of the cases they require a coenzyme. Examples are : fumarase, carboxylase etc. The important subclasses of lyases are:

Dehydratases

They catalyse the reactions with the removal of H_2O, *e.g.*, carbonic anhydrase splits H_2CO_3 to CO_2 and H_2O fumarase : acts on fumarate aconitase converts citrate to cis-aconitate with the removal of water.

Desulfhydrases

These enzymes are analogous to dehydratases except that after catalysing the reaction they remove H_2S instead of H_2O. For example, Cysteine desulfhydrase removes NH_3 and H_2S from cysteine with the formation of pyruvate; Homocysteine desulfhydrase acts on homocysteine and converts it to α-ketobutyrate with the removal of NH_3 and H_2S.

Decarboxylases

These enzymes split off carboxyl groups, e.g., Glutamate decarboxylase, Tyrosine decarboxylase, Pyruvate decarboxylase, Oxaloacetate decarboxylase etc. Most of the decarboxylations of amino acids of bacteria and mammals take place in the presence of coenzyme, pyridoxal phosphate.

Amidine and Ammonia Lyases

This sub-group includes those enzymes which catalyse the removal, of ammonia or substituted ammonia in the free state by means of reactions which are not of the hydrolytic, oxidative, or transfer type, *e.g.*, aspartase, which converts aspartate to fumarate and ammonia.

Aldolases and Ketoacid Lyases

The aldolases show aldol condensation, while ketoacid lyases show related reactions to aldolases, but here the reactions are more complex due to the participation of coenzyme A and in some cases ATP. The examples of aldolases are fructose diphosphate aldolase, pentose aldolase, ketotetrose aldolase etc; and of ketoacid lyases are citrate synthetase, ATP-citrate lyase etc.

5. Isomerases

This class includes those enzymes which catalyse the reactions where intramolecular rearrangements take place in the substrate and optical, geometric or positional isomers are formed, *e.g.*, triosephosphate isomerase, glucose phosphate isomerase, mannose phosphate isomerase etc. Most of the isomerases are involved in the metabolism of carbohydrates. The subclasses of isomerases are as follows:

Racemases and epimerases

These enzymes change L-compounds to D-compounds or α-compounds to β-compounds. The ribulosephosphate 3-epimerase and UDP glucose epimerase catalyse the rearrangement of alcoholic hydroxyl groups in monosaccharides from one steric configuration to the other.

Cis-trans isomerases

These enzymes catalyse the conversion of double bonds from trans- to cis configurations, *e.g.*, Maleylacetoacetate isomerase, retinene isomerase etc.

Intramolecular transferases

These enzymes transfer a variety of groups from one carbon of a compound to another *e.g.*,

The Enzymes

Phosphoglyceromutase which changes 3-phosphoglycerate to 2-phosphoglycerate, and methyl-malonyl-CoA mutase which changes methylmalonyl-CoA to succinyl-CoA.

Intramolecular oxido-reductases

The enzyme of this sub-group catalyse the interconversion of aldoses (aldehydes) and ketoses (ketones) i.e., conversion of aldehyde to ketones or ketones to aldehyde, *e.g.*, triose pentose and hexose-phosphate isomerases. The enzymes are of importance in Carbohydrate metabolism.

6. Ligases

Ligases catalyse the linkage of two molecules coupled with the cleavage of a pyrophosphate bond in Adenosine triphosphate (ATP) or similar compounds. Ligases may form C-O, C-S, C-N, or C-C bonds, *e.g.*, acteyl-CoA synthetase, succinyl CoA synthetase, pyruvate carboxylase, glutamine synthetase, acetyl CoA carboxylase etc.

5.5. ISOENZYMES

Certain enzymes which are formed by genetical changes specially by the processes which form alleles and iso-alleles, are known as **isoenzymes** or **isozymes**. The isoenzymes show very small differences in the molecular structure with that of original enzyme. Physically and chemically, the enzymes and isoenzymes are very similar and they catalyse the same reaction. For example, the enzyme lactic dehydrogenase, which catalyses the reaction of pyruvate to lactate, occurs in five different forms. All forms are known as isoenzymes and can be separated by electrophoresis.

5.6. ISOLATION AND PURIFICATION OF ENZYMES

There are a number of methods to extract enzymes. If specific enzymes are not to be separated then leaves germinated seeds or fungal mats may be used as materials. These materials are first killed by freezing or by adding chloroform or toluene and allowing to stand for a while killed material is quickly dried at low temperature and ground to make powder which is later utilized to extract crude enzymes.

Extraction

Take plant tissue, homogenise in a pestle and mortar in presence of a buffer of specific strength and pH at below 4°C. Add ethylene diamine tetra-acetic acid (EDTA) in the extraction medium to dissolve the membranes and to chelate heavy metals which inhibit enzyme activity. Detergents like **Triton-X** can also be used in place of EDTA. Sometimes disulphide (–S–S–) bonds containing enzymic proteins are oxidized during enzyme extraction causing loss of enzyme structure and function. This difficulty can be overcome by adding thiols like meriaptoethanol or aminoacid cysteine in the extraction medium. Sometimes proteins like casein or bovine serum albumin are also added in the extraction medium to protect enzyme from hydrolytic action of endogenously present proteases which become active during tissue homogenization.

Filtration and Centrifugation

Filter the homogenized tissue extract through musclin cloth or filter paper. Take filtrate in the test tube and centrifuge in cold using a refridgerated centrifuge at about 20,000 × g to remove cell debris and bigger cell-organelles which settle down in the bottom of the test tube. The supernatant obtained is again centrifuged through ultracentrifuge at about 100000 × g to remove all cell-organelles which settle down in the bottom. Decant the cytosolic protein containing supernatant carefully. This supernatant can be used for further purification of cytosolic enzymes.

Precipitation

Like proteins, the enzymes are also highly charged molecules and can be precipitated out by proper charge breaking chemicals. By adding chemicals their charges are neutralized and proteins are precipitated. For this purpose acids, bases or organic solvents can be used. The precipitation of proteins or enzymes by concentrated salt solutions is called *salting out*. The addition of a salt dehydrates the protein molecule, aggregates and precipitated out. The suitable concentration of $(NH_4)_2.SO_4$ used for precipitation of enzymes is between 30-70%. The preciptate is separated out by centrifugation. Now dissolve the precipitate in a buffer of desired pH and ionic strength of dialyse to remove excess of $(NH_4)_2.SO_4$ and other low molecular weight contaminents. Organic solvents like acetone, methanol and ethanol etc. are also used for the precipitation of enzymes. The solvents or precipitants are first cooled upto $-40°C$ before their use and precipitation is done at $0°C$ by adding drop by drop.

Purification

The dialysed enzyme preparation can be purified by chromatographic or electrophoretic techniques. Adsorption column chromatography and affinity chromatography are widely used for this purpose.

5.7. CHEMICAL NATURE OF ENZYMES

They are colloidal and proteinaceous in nature, and exhibit the same properties as shown by proteins such as;

    (i) *Thermolability* i.e. sensitivity to heat.

    (ii) *Hydrolysis* : Enzymes yield a mixture of α amino acids on hydrolysis like proteins.

    (iii) *Substrate specificity* : An enzyme acts only on a specific substrate. For example, diastase acts only on starch whereas catalase acts on H_2O_2.

The enzymes differ from inorganic catalysts in the following points :

1. The enzymes are thermolabile in nature.
2. The enzymes show reaction specificity and substrate specificity.
3. The enzymes have capacity to bring about reactions at low temperature in approximately neutral amounts.

The catalytic power of an enzyme is measured by the "turnover number" which is defined as the number of molecules of substrate converted into products per minute per enzyme molecule. The molecular weight of an enzyme generally lies between 50,000 to 2,00,000 daltons.

Most of the enzymes are soluble in water, glycerol and dilute alcohols. They are insoluble in strong alcohol and alkaline reagents. Protein precipitating agents have the ability to precipitate these enzymes. Such agents are concentrated alcohols, alkalies, sodium chloride and ammonium sulphate. The chemical analysis of purified crystalline enzymes gives values typical of proteins.

The enzymes are least soluble at the isoelectric point. They are more stable in concentrated solutions and are subjected to denaturation by change in the pH or by increase in the temperature of the solution.

Enzymes have been divided chemically into two categories.

1. Simple protein enzymes
2. Complex protein enzymes

The simple protein enzymes contain only simple proteins, whereas complex protein enzymes have a prosthetic group associated with a specific protein. The dissociation of the prosthetic group from the specific protein leads to inactivation of the enzyme. **Apoenzyme** is the term which is used for the protein part of the conjugated or complex enzyme. The apoenzyme in combination with **prosthetic group** (coenzyme) forms the complete enzyme or **holoenzyme**.

$$\text{Holoenzyme} \rightleftharpoons \text{apoenzyme + coenzyme}$$
$$\text{or}$$
$$\text{Complex protein enzyme} \rightleftharpoons \text{protein + Prosthetic group.}$$

The Enzymes

5.8. MODE OF ACTION OF ENZYMES

1. Collision theory

Biological systems with chemicals in dilute aqueous solutions, differ from chemical reactions in which collision takes place between reacting molecules to a greater extent, since there is a greater reaction between water molecules than the reacting molecules in the former. Hydrolysis of sucrose to glucose and fructose is one example of such a biological reaction.

$$C_6H_{11}O_5\text{-}C_6H_{11}O_6 + H_2O \rightleftharpoons C_6H_{12}O_6 + C_6H_{12}O_6$$
$$\text{Surcrose} \qquad\qquad\qquad \text{Glucose} \quad\;\; \text{Fructose}$$

If the collision takes place all water molecules and the glycosidic bond, the sucrose molecules would disappear in a short while; but since, in nature it does not take place easily, the collision of sucrose with water in neutral solution does not result in hydrolysis. The hydrolysis takes place only on acidification. The hydrogen ions here act as catalysts.

Some enzymes need water to break the bond, but other require compounds of low molecular weight to make the enzyme active. The enzymes such as Chymotrypsin, lipases and phosphatases generally catalyse the reaction of the following type :

1. $EH + RX \longrightarrow ER + HX$
   Enzyme substrate   Intermediates

2. $ER + H_2O \longrightarrow EH + ROH$

On summing up : $RX + H_2O \longrightarrow ROH + HX$

Another type of reaction generally catalysed by enzymes such as myokinases, hexokinases, CoA transferases is the one in which the enzyme catalyses in the following manner :

$$E + AB \longrightarrow E \sim A + B$$
$$E \sim A + C \longrightarrow E + AC$$

Here the enzyme (E) first combines with the 'A' part of the compound (AB) and conserves the bond energy. It then reacts with (C) compound to form (AC). Water molecule here does not react with (E ~ A) complex and the energy of the bond is conserved to be used for the synthesis of a new compound (AC).

2. Enzyme-Substrate Complex Theory

This theory was postulated by Michaelis and Menten to explain the mechanism of action of enzymes. The following equations will give the idea of the mechanism.

Enzyme + Substrate $\rightleftharpoons$ Enzyme-substrate complex

Enzyme-substrate complex $\longrightarrow$ Enzyme + Products of reaction.

According to the above equations, the mechanism of action presumes that before hydrolysis the substrate molecule combines with the enzyme and leaves the enzyme surface in the form of reaction products. Stern proposed the enzyme substrate complex as an intermediate which is decomposed rapidly into the enzyme and the product molecules.

The enzymes may be called as temporary reactants because they are not utilized during the completion of the raction and act unlike substrate i.e., the substrate is utilized during the course of reaction. According to Keilin, when the substrate and enzyme come together in the reaction they combine and form a complex known as "enzyme-substrate complex". It is now known that the rate of each reaction is governed by the law of mass action, whether it is catalysed or not. As a proof, the case where the concentration of the substrate varies, but that of the enzyme remains constant will be discussed. These reactions, show an increase in the velocity of the initial reaction when the substrate concentration is increased. The rate of reaction increases until the limiting initial velocity is reached. This condition reaches only when all the active sites on the enzyme molecule get occupied by the substrate molecules. The term "initial reaction velocity" is used without considering the time for the progress of the reaction. A constant known as Michaelis constant is used to measure the relative activity of an enzyme in different conditions.

5.9 DERIVATION OF MICHAELIS CONSTANT

Michaelis and *Menten* in 1913 gave a mathematical expression to an enzyme catalysed reaction. The equation is commonly called the Michaelis equation.

This equation can be derived rather simply with the help of certain assumptions. The first assumption is that the enzymatic process occurs as a stepwise set of reactions which can be shown as :

$$E + S \underset{k_{-1}}{\overset{k_1}{\rightleftharpoons}} ES \overset{k_2}{\rightarrow} E + P$$

(Enzyme) (Substrate)   (Enzyme Substrate Complex)   (Enzyme) (Products)

The second assumption made by *Michaelis* is that the process of reaching equilibrium between E, S, and E S is quite rapid as compared to the rate of reaction of ES to give free enzyme and products.

The letter E, S, ES and P stand for the enzyme, substrate, enzyme-substrate complex and the product respectively. If these reactions are considered from the kinetic point of view, the molar concentration of the reactants and the products are used and the symbols are written in parenthesis.

In describing the kinetics of these reactions in mathematical terms there is no theoretical problem. However, a practical difficulty arises from the fact that the concentration of the enzyme-substrate complex or the concentration of the enzyme cannot be determined. The success of this derivation lies in replacing these immeasurable quantities with those which could be measured experimentally.

According to the law of mass action the rate of appearance of products, v (for velocity), should be proportional to the concentration of the enzyme-substrate complex, E S.

$$v \propto ES$$

or $\qquad v = K_1 \, ES \qquad$ ...(1)

For simplicity the brackets used to indicate molar concentration are omitted.

The maximum reaction possible, V_{max}, will occur at a substrate concentration sufficiently high so that essentially all of the enzyme is bound to substrate. The maximum concentration of ES which is possible is thus equal to the total enzyme concentration.

Thus, $\qquad V_{max} \propto E_{total}$

or $\qquad V_{max} = K_1 \times E_{total} \qquad$ ...(2)

Dividing equation 1 by equation 2, we get

$$\frac{v}{V_{max}} = \frac{ES}{E_{total}} \qquad ...(3)$$

Equation 3 expresses the immeasurable quantities ES and E_{total} in terms of reactions rates that can be experimentally measured.

The equilibrium constant for the dissociation of ES complex can be written as :

$$Km = \frac{E \times S}{ES} \qquad ...(4)$$

The Enzymes

Equation 4 contains the quantity ES with E rather than E_{total}. The total enzyme concentration (E_{total}) comprises of the enzyme which is free (E_{free}) at any instant and the enzyme which is bound to the substrate (ES).

Hence

$$E_{free} = E_{total} - ES \qquad \ldots(5)$$

Substituting the value of E in equation 4 we get,

$$Km = \frac{(E_{total} - ES) \times S}{ES} \qquad \ldots(6)$$

or $\quad Km \cdot ES = E_{total} \times S - ES \times S$

or $\quad Km \cdot ES + ES \cdot S = E_{total} \cdot S$

or $\quad ES(Km + S) = E_{total} \cdot S$

or $$ES = \frac{E_{total} \cdot S}{Km + S}$$

or $$\frac{ES}{E_{total}} = \frac{S}{Km+S} \qquad \ldots(7)$$

Now substituting $\frac{v}{V_{max}}$ for $\frac{ES}{E_{total}}$ of equation 3

we get $$\frac{v}{V_{max}} = \frac{S}{Km+S} \qquad \ldots(8)$$

or $$v = \frac{V_{max} \times S}{Km+S}$$

This expression contains with the exception of Km only easily measurable variables, the reaction rate v at a given substrate concentration and the rate V_{max} achieved at saturation of the enzyme. These measurements then permit the calculation of Km. The evaluation is simplified greatly if the reciprocal of equation 8 is used.

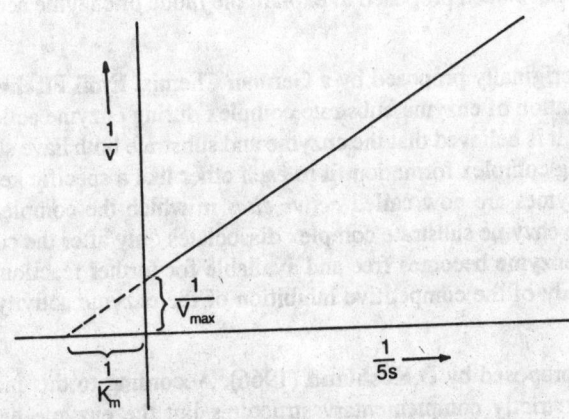

Fig. 5.1: Determination of V_{max} and Km from the Lineweaver Burk plot of enzyme kinetics found experimentally.

$$1/v = \frac{K_m + S}{V_{max} \times S}$$

$$= \frac{K_m}{V_{max}} \cdot 1/S + 1/V_{max} \qquad ...(9)$$

Since $1/v$ and $1/S$ are variables, the equation to the right represents the function of a straight line, $y = ax + b$. By plotting $1/v$ against $1/S$, V_{max} and K_m are found easily from the co-ordinate intercepts (Fig. 5.1).

The Michaelis constant determined in this way (= S at $v = 1/2\ V_{max}$) is a dynamic value. Michaelis constant may also be derived from equation 8 as under:

$$\frac{V_{max}}{v} = \frac{K_m + S}{S}$$

or

$$\frac{V_{max} \cdot S}{v} = K_m + S$$

or

$$K_m = \frac{V_{max} \cdot S}{v} - S$$

or

$$K_m = \left(\frac{V_{max}}{v} - 1\right) \cdot S \qquad ...(10)$$

If $V_{max} = 2v$ in equation 10

Then,

$$K_m = \left(\frac{2v}{v} - 1\right) \cdot S$$

or $K_m = S$

It means that Michaelis constant is numerically equal to the substrate concentration when the reaction proceeds at half its maximum rate, and is expressed in moles/litre.

A high K_m value shows that a high substrate concentration is necessary to attain half saturation or V_{max} and the enzyme has a low affinity for the substrate in question. Michaelis constants usually range between 10^{-3} and 10^{-5} moles/litre.

5.10. MODELS FOR EXPLAINING ENZYME ACTION

Following two models have been proposed to explain the mode of enzyme action:

1. Lock and Key model

This model was originally proposed by a German Chemist **Emil Fischer** in 1898. According to this model, the formation of enzyme-substrate complex during enzyme action is analogous to the fitting of lock and key. It is believed that the enzyme and substrate both have strictly complementary *structures* which during complex formation fit to each other like a specific key in a particular lock. Such structures of enzymes are now called *active sites* in which the complementary structures of substrate get fit in. The enzyme substrate complex dissociates only after the conversion of substrate into products and the enzyme becomes free and available for further reactions. This mechanism is supported from the study of the competitive inhibition of the enzyme activity.

2. Induced fit model

This model was proposed by **D.Koshland** (1966). According to this model, the enzyme and substrate do not have strictly complementary structures but the enzyme has flexible active site structures which are changed according to substrate configuration. Thus, the enzymes show 'induced fit' mechanism during enzyme substrate complex formation. This model can be compared with a type of 'hand and glove'. The same glove can be fit in the hands of many persons. In other words,

The Enzymes

the glove adjusts itself in the hands of those persons who bear it. The conformational changes in the enzyme are induced by the substrate. During the reaction the alteration in the geometry of

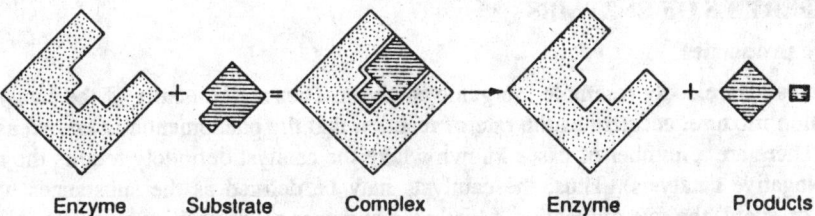

Fig. 5.2: Explaining lock and key mechanism of enzyme action.

enzyme protein takes place and the reaction is carried out by the strain exerted by enzyme–substrate binding forces. In presence of competitive inhibitors, the enzyme becomes inactivated. Induced fit mechanism has been supported through X-ray studies on carboxypeptidases–A and other enzymes.

5.11. ENERGY OF ACTIVATION IN THE MECHANISM OF ENZYME ACTION

Arrhenius believed that all the molecules in a given population do not have the same kinetic energy and the energy constant of the molecules changes continuously. Some molecules acquire more energy by collisions as they collide in their ceaseless motion. The molecules with higher energy are called "activated" and the activation may be achieved by such means as heating, irradiation etc. The activated molecules react with the energy poor ones in comparison to others which are all energy poor ones. This difference in the energy of the molecules is an energy barrier to the reaction. All reactions have this energy barrier and the higher the energy barrier for a molecule, the greater is its stability. The energy needed to induce the molecules to overcome this barrier is called the "activation energy" (Fig. 5.3). Only the activated molecules react chemically and their number is less than one millionth of the sucrose molecules in the solution. With the rise of temperature by 10° the number of activated molecules

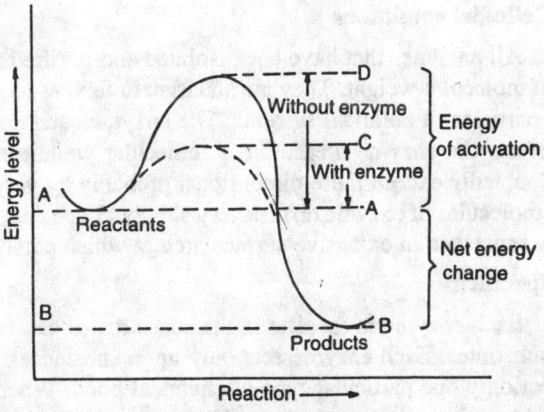

Fig. 5.3: The energy of activation of a biochemical reaction in presence and absence of a specific enzyme.

increase 2 to 3 times explaining the behaviour of the increase in the velocity of the reaction to an equal extent. The activation of the molecules increase the velocity or make, the structure more labile making collision more efficient than with the non-activated molecules. The minimum energy required to bring about this chemical reaction is known as "activation energy".

From the energy point of view reactions are divided into three types :

1. *Exergonic reactions*: These reactions involve the loss of free energy.
2. *Reversible reactions*: No change in the free energy takes place in such reactions so that the reactions come to equilibrium and the product molecules and substrate molecules are present in constant amounts.

3. *Endergonic reactions*: These reactions need supply of energy which is usually possible due to coupling with another exergonic reaction. The synthesis of NADP from ATP and NAD is a good example of endergonic reaction.

5.12. PROPERTIES OF ENZYMES

1. Catalytic properties

Enzymes very closely resemble inorganic catalysts. The small quantity of the catalyst present in the reaction mixture, accelerates the rate of reaction and the phenomenon is known as 'positive catalysis'. There are a number of cases known where the catalyst definitely retards the rate of the reaction (Negative catalysis). Thus, the catalysts may be defined as the substances which alter (accelerate or retard) the rate of reaction. Mostly the enzymes act as positive catalysts. A very small quantity of enzyme can catalyse the transformation of vastly larger quantities of the substrate, without itself (enzyme) being consumed. The enzyme catalase brings about the reduction of H_2O_2 to water and molecular oxygen.

$$2H_2O_2 \xrightarrow{catalase} 2H_2O + O_2$$

Catalase is one of the most efficient enzymes, one molecule of this enzyme being able to catalyse the conversion of 5,000,000 molecules of H_2O_2 per minute when the conditions are favourable (*Sumner* and *Somers*, 1947). Similarly, the enzyme invertase can hydrolyse at least 1,000,000 times its own weight of sucrose. The catalytic power of an enzyme is measured by the 'turnover number' which is defined as the number of molecules of substrate converted into products per minute per molecule of the enzyme.

2. Colloidal conditions

All enzymes that have been isolated and purified are molecules of very large dimensions and high molecular weight. They are so large, in fact, as to fall within the size limits which characterize the particles of colloidal systems. The enzyme catalase has a molecular weight of about 2,48,000, whereas, the enzyme urease has a molecular weight of about 4,83,000. Some enzyme molecules undoubtedly exceed these dimensions, probably by very large margins. Because of their great size, the molecules of enzyme diffuse very slowly. Since enzyme dispersed in water are colloidal systems, they represent an extensive surface area at which catalysed reactions can occur.

3. Specificity

Each enzyme is specific in the sense that it can operate only upon certain substrate or group of substrates. Each enzyme acts only upon substances having a certain molecular pattern and can affect only one particular type of chemical bond. When a number of different compounds possess this bond in common, they can be acted upon by the same enzyme. For example, the enzyme emulsion can hydrolyse any β-glycoside.

Enzyme specificity can also be illustrated by the fact that one and the same substrate under the influence of different enzymes yields different end products. For example, the trisaccharide raffinose in presence of enzyme *sucrase* is hydrolysed into melibiose and fructose, while in presence of enzyme *emulsion* the end-products sucrose and galactose are formed.

Absolute Specificity : Many enzymes apparently act on only a single kind of substrate. For example, the enzyme *urease* can act only upon urea and no other molecule, *invertase* can act upon sucrose and; *cytochrome oxidase* can catalyse only the oxidation of "cytochrome C" by O_2. These all represent the examples of absolute specificity.

4. Reversibility of Action

Like a true catalyst the enzymes can also accelerate the rate of reaction in whichever direction it is taking place, provided suitable sources of energy are available. Usually a single enzyme brings

The Enzymes

about the synthesis and digestion (hydrolysis) of a particular substance. For example, the enzyme *lipase* is required for the synthesis of fats from glycerol and fatty acids. The same enzyme lipase is again required for the digestion of fats. Similarly, enzyme *emulsion* is required for the synthesis and hydrolysis of glycosides. Sometimes two different enzymes are involved in the synthesis and digestion of a particular substance. For example, the enzyme *sucrase* (*invertase*) is required in the hydrolysis of sucrose to glucose and fructose, but its synthesis apparently requires the action of *sucrose phosphorylase*.

5. Heat Sensitivity

The enzymes are proteinaceous in nature. Hence, they are thermolabile, i.e., in a liquid medium they are inactivated at high temperature ($60°-70°$). Thus, unlike catalysts enzymes are inactivated or destroyed at temperatures considerably below the boiling point of water. The destruction of the enzymes in this temperature range is a heat coagulation phenomenon. The enzymes of dry tissues such as seeds and spores can survive temperatures of $100°C$ to $120°C$ or even more, for considerable period without suffering deleterious effects.

The exact temperature at which a given enzyme will be destroyed varies greatly, depending upon conditions prevailing in the medium in which it is dispersed. The pH of the medium has a marked effect upon the heat sensitivity of enzyme. When the temperature is reduced to freezing point or below freezing point, the enzymes become inactivated, but they are not destroyed.

6. Enzyme Inhibitors

There are certain products which inhibit the enzyme activity. These are known as **enzyme inhibitors**. During the reaction, the active sites of the enzymes are filled up with these substances (inhibitors) instead of substrate molecules. Thus, the activity of the enzyme is lost. The inhibitors raise the activation-energy required for the reaction.

5.13. ENZYME INHIBITION

There are two types of enzyme inhibitions : irreversible and reversible. Irreversible inhibition results from the formation of a stable enzyme inhibitor (EI) complex which results in complete inhibition of the enzyme, *e.g.*, inhibition of 'SH enzymes' by iodoacetamide, the inhibition of Xanthine oxidase by CN-, and inhibition of cholinesterase by nerve gases.

Reversible inhibition is of three types competitive inhibition, non-competitive inhibition and uncompetitive inhibition.

1. Competitive Inhibition

This type of inhibition depends on the fact that the inhibitor competes with the true substrate for the "active site" of the enzyme. The inhibition is relieved by increasing the concentration of the substrate.

$$\begin{array}{cc} CH_2COOH & COOH \\ | & \diagup \\ CH_2COOH & CH_2 \\ & \diagdown \\ & COOH \\ I & II \\ \text{(Succinic acid)} & \text{(Malonic acid)} \end{array}$$

The most important example of competitive inhibition is the inhibition of succinic dehydrogenase by malonic acid. The molecular structure of succinic acid is very similar to malonic acid.

Because of this similarity in structure, the enzyme can react with both to form complexes. However, only the enzyme-succinic acid complex decomposes to yield a reaction product.

$$E + I \underset{}{\overset{Ki}{\rightleftharpoons}} EI$$

Where E is the enzyme, I is inhibitor (malonic acid), Ki inhibitor association constant and EI enzyme inhibitor complex.

2. Non-Competitive Inhibition

In non-competitive inhibition the inhibitor reacts with the enzyme to reduce catalytic activity without preventing the formation of enzyme-substrate complex. The affinity of the enzyme for substrate is not reduced, but the maximum velocity of the reaction is reduced, *e.g.* fluoride ions inhibit enolase and thiocyanate ions inhibit fumarase. In both the cases the inhibition cannot be overcome by increasing the concentration of the substrate.

3. Uncompetitive Inhibition

In case of uncompetitive inhibition, the inhibitor is thought to combine with forms of the enzyme, but they do not combine actually with the substrate (*e.g.* the enzyme-substrate complex). These forms of the enzyme are known as 'substrate non-combining' forms and cannot then be converted back into the 'substrate-combining' forms of the enzyme. This type of inhibition is not relieved by increasing the concentration of the substrate. Uncompetitive inhibition is quite common in multi-substrate reactions, but is rare in reactions involving single substrate.

4. Allosteric Inhibition

In the case of allosteric inhibition, the inhibitor, which is structurally quite different from the substrate, is bound at a site other than the active site of the enzyme. This binding of the inhibitor alters the conformation of the enzyme protein, and thereby prevents it from binding to the substrate. Since the inhibitors bind at a site other than the active site of the enzyme they are called **allosteric effectors** or determinants and the sites to which they bind, **allosteric sites** (allows=other). The whole phenomenon is also called as **allosteric effect** or feedback inhibition and it is always reversible. The allosteric inhibition is of great physiological and biochemical importance.

The enzymes whose catalytic activities are controlled allosterically are charaterized by having some of all of the following features :

1. The enzyme may contain multiple binding sites and the binding of an effector at allosteric site facilitates the utilization of the substrate at the active site. When identical molecules are bound at allosteric and active sites, the resulting interaction between these sites is called a homotropic effect. It is usually cooperative (the catalytic activity of the enzyme is stimulated).

2. Sometimes different molecules are bound at allosteric and active sites of the enzyme. The resulting interaction between these sites is called a heterotropic effect. It may be cooperative or antagonistic (the catalytic activity of the enzyme is inhibited). These allosteric effectors (activators and inhibitors) are frequently end-products in metabolic pathways and the enzymes to which they bind is frequently that catalysing the first reaction in the pathway; in this way the rate at which the pathway operates is controlled by positive or negative feedback.

3. When the enzyme is subjected to conformation altering conditions (e.g, high or low pH, heat, high ionic strength, heavy metal ions), its catalytic activity is released from allosteric control.

4. Allosteric enzymes are formed by the aggregation of many subunits. Their purification is difficult because there is a marked tendency for disaggregation to occur. The disaggregating subunits are frequently catalytically inactive.

Monod, Wyman and *Changeux* (1965) proposed a theoretical model to explain the allosteric control. According to them the allosteric enzymes are oligomers made up of identical units, monomers (Protomers). The monomers are linked together in such a way that they all occupy equivalent positions. Each protomer possesses one site for the binding of each molecule of either substrate or allosteric effector which are able to bind with the enzyme. The conformation (three-dimensional structure) of each protomer is dictated by its association with other protomers. The oligomer can exist in at least two states which differ in distribution and energy of inter-protomer bond. When a transition from one state to the other occurs there is a change in the affinity of one (or several) of the binding sites towards the molecule(s) which usually bind at those sites. Thus when the enzyme goes from one state to another, its molecular symmetry is conserved. Table 5.2 shows the properties of some allosteric enzymes.

Table 5.2

	Enzyme	Substrate	Inhibitor	Activator
1.	Biosynthetic L-threonine deaminase (E.Coli K12 and Yeast).	L-Threonine	L-Isoleucine	L-Valine
2.	NAD-isocitric dehydrogenase (N.Crassa).	D-Isocitrate + NAD$^+$	α-Ketoglutarate	Citrate
3.	NAD-isocitric dehydrogenase (yeast).	D-Isocitrate + NAD$^+$		5'-AMP
4.	Glycogen synthetase (yeast)	UDP-glucose		Glucose 6-P.

5.14. FACTORS AFFECTING ENZYME ACTIVITY

The rate of an enzyme-catalysed reaction is influenced by the following environmental factors :

(a) Enzyme concentration.

(b) Substrate concentration.

(c) Temperature.

(d) pH.

(e) Ions (Cations and anions).

(f) Accumulation of end-products.

1 Enzyme Concentration

Enzyme molecules are larger than the substrate molecules. They possess many active sites in which the substrate molecules get attached at the time of reaction. The reaction proceeds until all the active sites of all the enzyme molecules are filled up by substrate molecules. If in the system, the substrate molecules are in relatively larger concentration as compared to the specific enzyme concentration, the enzyme-catalysed reaction will attain the maximum speed. Further addition of the substrate will have no effect on the reaction because all the active sites of the enzyme molecules have been saturated. Now, if the enzyme concentration is increased and the substrate concentration is relatively higher in the system, the rate of reaction will increase. Again increase in the enzyme concentration will increase the rate of reaction, but a time will come when the rate will not increase due to limiting effects of the substrate concentration. Further addition of the substrate and enzyme concentration will again increase the rate of reaction. The effect of enzyme concentration is shown in Fig. 5.4.

2. Substrate concentration

If the enzyme concentration is fixed and the substrate concentration is relatively higher, the velocity of the reaction is increased to its maximum. If the substrate concentration is relatively low at a particular enzyme concentration, the reaction proceeds, but at a slow rate, because the active sites of only few enzyme molecules become saturated with the substrate molecules. Other remaining enzyme molecules undergo no reaction due to shortage of the substrate molecules. An increase in the substrate concentration increases the rate of reaction. If the substrate concentration is increased beyond a limit, when all the active sites of enzyme molecules have become saturated, the reaction will become constant without accelerating its velocity. The effect of substrate concentration on the rate of enzyme-catalysed reaction is shown in Fig. 5.4.

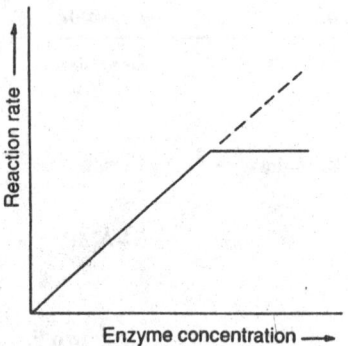

Fig 5.4: Effect of enzyme concentration on the rate of a reaction when substrate concentration is relatively higher.

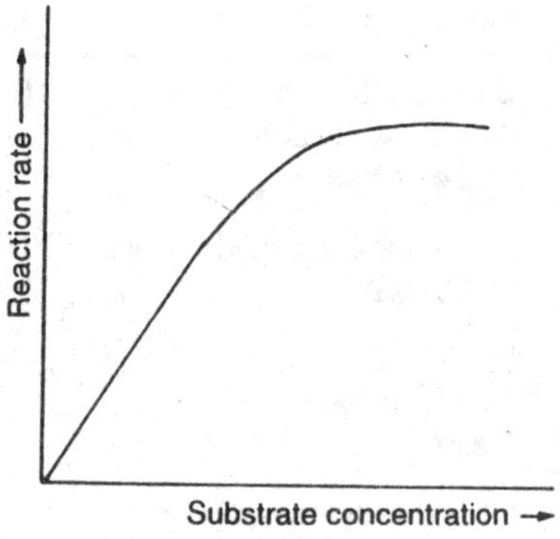

Fig. 5.5: Effect of substrate concentration on the rate of an enzyme catalysed reaction.

3. Temperature

With certain exceptions, the rates of enzyme catalysed reactions are increased as the temperature is raised. Vant Hoff's Q_{10} law states, that with an increase in temperature by 10°C the rate of most of the reactions is doubled. By raising the temperature, the number of activated molecules is increased which ultimately results in increase in velocity of the reaction. The enzyme catalysed reactions show the increase in velocity in between 25°C- 35°C (Fig. 5.6). At 0°C or below 0°C the enzymes become inactivated, but they are not destroyed. At 60°C to 70°C, in a liquid medium the enzymes are inactivated and destroyed. This destruction of enzymes are inactivated and destroyed. This destruction of enzymes at high temperature results in coagulation and denaturation. Thus, as the temperature is raised the reaction rate increases up to a certain limit and above that the enzymes get denatured. The temperature at which the rate of reaction is maximum, is known as "optimum temperature".

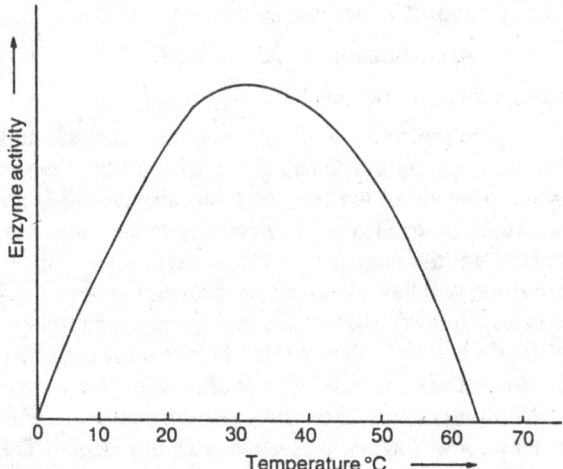

Fig. 5.6: Effect of temperature on enzyme activity.

4. Hydrogen ion Concentration (pH)

Hydrogen ion concentration of an enzyme solution also has a marked effect on its activity. Every enzyme acts at a particular pH. The pH at which the rate of reaction is maximum, is known as "optimum pH". If the pH value is increased or decreased on either side of its optimum range, the rate of reaction, usually decreases. At extremes of pH the enzymes are denatured and inactivated.

The Enzymes

The pH value for the enzyme activity is important because enzyme molecules possess a multiplicity of ionizable groups. The state of these ionizable groups obviously depends on pH. At the time of enzyme-activity some of these groups have to be ionized and others need no ionization. This ionization of certain groups is controlled by a limited range of pH. The optimum pH of certain enzymes at which they show maximum degree of dissociation and maximum activity are given below :

Pepsin	2.0
Invertase	4.5
Cellobiase	5.0
Peroxidase	5.0
Chlorophyllase	5.9
Maltase	7.0
Amylase	7.0
Catalase	7.0
Urease	7.0
Trypsin	8.0
ATP	9.0

The hydrogen ion concentration also alters the ionization, solubility of the substrates, inhibitors, activators and absorbed ions.

5. Effects of ions

Hydrogen ion concentration is the most important factor in the activity of all enzymes, but there are also many other cations like Mg^{2+}, Ca^{2+}, Mn^{2+}, Zn^{2+}, Na^+ or K^+ which play an important role in the activity of certain enzymes. These enzymes in the absence of a particular cation remain inactive. Some enzymes already possess a loosely bound cation in their structure. In other words these enzymes require particular cation for their complete structure. For example, the enzyme phosphopyruvate hydratase contains a loosely bound Mg^{2+} cation without which it is inactive. In other cases the ions may combine with substrate. Anions are of less importance in the enzyme activity. There are only a few instances which show the importance of anions in the enzyme activity. For example, the enzyme salivary amylase requires chloride ions (Cl^-) for its activity.

6. Accumulation of Products

The accumulation of the products of enzyme-catalysed reaction, in most of the cases, retards the rate of the reaction. There may be following three points for this retardation :
   (*i*) The accumulation of products may increase the rate of reverse reaction.
   (*ii*) The enzymes become inactive due to accumulation of these products on the surface of enzyme molecules itself.
   (*iii*) They may also change the pH of the enzyme-solution.

5.15. KEY TO NUMBERING AND CLASSIFICATION OF ENZYMES

Based on recommendations of International Union of Biochemistry (IUB).

1. **OXIDOREDUCTASES**

    1.1 Acting on the CH-OH group of donors
      1.1.1 With NAD or NADP as acceptor
      1.1.2 With a cytochrome as an acceptor
      1.1.3 With O_2 as acceptor
      1.1.99 With other acceptors.

    1.2 Acting on the aldehyde or Ketogroup of donors
      1.2.1 With NAD or NADP as acceptor

1.2.2 With a cytochrome as an acceptor
1.2.3 With O_2 as acceptor
1.2.4 With lipoate as acceptor
1.2.99 With other acceptors.

1.3 Acting on CH=CH group of donors
1.3.1 With NAD or NADP as acceptor
1.3.2 With a cytochrome as acceptor
1.3.3 With O_2 as acceptor
1.3.99 With other acceptors.

1.4 Acting on $CH-NH_2$ group of donors
1.4.1 With NAD or NADP as acceptor
1.4.3 With O_2 as acceptor

1.5 Acting on the C-NH group of donors
1.5.1 With NAD or NADP as acceptor
1.5.3 With O_2 as acceptor

1.6 Acting on $NADH_2$ or $NADPH_2$ as donor
1.6.1 With NAD or NADP as acceptor
1.6.2 With a cytochrome as an acceptor
1.6.4 With a disulphide compound as acceptor
1.6.5 With a quinone or related compound as acceptor
1.6.6 With a nitrogenous group as acceptor
1.6.99 With other acceptors.

1.7 Acting on other nitrogenous compounds as donors
1.7.3 With O_2 as acceptor
1.7.99 With other acceptors

1.8 Acting on sulphur groups of donors
1.8.1 With NAD or NADP as acceptor
1.8.3 With O_2 as acceptor
1.8.4 With a disulphide compound as acceptor
1.8.5 With a quinone or related compound as acceptor
1.8.6 With a nitrogenous group as acceptors.

1.9 Acting on heme groups of donors
1.9.3 With O_2 as acceptor
1.9.6 With a nitrogenous group as acceptor

1.10 Acting on diphenols and related substances as donors.
1.10.3 With O_2 as acceptor

1.11 Acting on H_2O_2 as acceptor

1.98 Enzymes using H_2 as reductant

1.99 Other enzymes using O_2 as oxident
1.99.1 Hydroxylases
1.99.2 Oxygenases.

2. TRANSFERASES

2.1 Transfer one-carbon group
2.1.1 Methyltransferases
2.1.2 Hydroxymethyl-, formyl- and related transferases.
2.1.3 Carboxyl- and Carbamyltransferases

The Enzymes

    2.2 Transferring aldehydic or ketonic residues

    2.3 Acyltransferases
        2.3.1 Acyltransferases
        2.3.2 Aminoacyltransferases

    2.4 Glycosyltransferases
        2.4.1 Hexosyltransferases
        2.4.2 Pentosyltransferases

    2.5 Transferring alkyl or related groups

    2.6 Transferring nitrogenous groups
        2.6.1 Aminotransferases
        2.6.2 Amidinotransferases
        2.6.3 Oximinotransferases

    2.7 Transferring phosphorus-containing groups
        2.7.1 Phosphotransferases with an alcohol group as acceptor
        2.7.2 Phosphotransferases with a carboxyl group as acceptor
        2.7.3 Phosphotransferases with a nitrogenous group as acceptor
        2.7.4 Phosphotransferases with a phospho group as acceptor
        2.7.5 Phosphotransferases, apparently intramolecular
        2.7.6 Pyrophosphotransferases
        2.7.7 Nucleotidyltransferases
        2.7.8 Transferases for other substituted phospho groups.

    2.8 Transferring sulphur-containing groups
        2.8.1 Sulphurtransferases
        2.8.2 Sulphoransferases
        2.8.3 Co A-transferases.

3.   HYDROLASES

    3.1 Acting on ester bonds
        3.1.1 Carboxylic ester hydrolases
        3.1.2 Thiolester hydrolases
        3.1.3 Phosphoric monoester hydrolases
        3.1.4 Phosphoric diester hydrolases
        3.1.5 Triphosphoric monoester hydrolases
        3.1.6 Sulphuric ester hydrolases

    3.2 Acting on glycosyl compounds
        3.2.1 Glycoside hydrolases
        3.2.2 Hydrolysing N-glycosyl compounds
        3.2.3 Hydrolysing S. glycosyl compounds

    3.3 Acting on ether bonds
        3.3.1 Thioether hydrolases

    3.4 Acting on peptide bonds (Peptide hydrolases)
        3.4.1 α-Aminopeptide aminoacidohydrolases
        3.4.2 α-Carboxypeptide aminoacidohydrolases
        3.4.3 Dipeptide hydrolases
        3.4.4 Peptide peptidohydrolases

    3.5 Acting on C-N bonds other than peptide bonds
        3.5.1 In linear amides

3.5.2 In cyclic amides
3.5.3 In linear amidines
3.5.4 In cyclic amidines
3.5.99 In other compounds

3.6 Acting on acid-anhydride bonds
   3.6.1 In phosphoryl-containing anhydrides

3.7 Acting on C-C bonds
   3.7.1 In ketonic substances

3.8 Acting on halide bonds
   3.8.1 In C-halide compounds
   3.8.2 In P-halide compounds

3.9 Acting on P-N bonds

4. LYASES
   4.1 Carbon-Carbon lyases
      4.1.1 Carboxy-lyases
      4.1.2 Aldehyde-lyases
      4.1.3 Ketoacid-lyases
      4.2 Carbon-oxygen lyases
      4.2.1 Hydro-lyases
      4.2.99 Other carbon-oxygen lyases

   4.3 Carbon-nitrogen lyases
      4.3.1 Ammonia-lyases
      4.3.2 Amidine-lyases

   4.4 Carbon-Sulphur lyases

   4.5 Carbon-halide lyases

5. ISOMERASES
   5.1 Racemases and epimerases
      5.1.1 Acting on amino acids and derivatives
      5.1.2 Acting on hydroxyacids and derivatives
      5.1.3 Acting on carbohydrates and derivatives

   5.2 Cis-trans isomerases

   5.3 Intramolecular oxidoreductases
      5.3.1 Interconverting aldolases and ketolases
      5.3.2 Interconverting keto and enol groups
      5.3.3 Transposing C=C bonds

   5.4 Intramolecular transferases
      5.4.1 Transferring acyl groups
      5.4.2 Transferring phosphoryl groups
      5.4.99 Transferring other groups
      5.5 Intramolecular lyases

6. LIGASES
   6.1 Forming C-O bonds
      6.1.1 Amino acid-RNA ligases

   6.2 Forming C-S bonds
      6.2.1 Acid thiol ligases

The Enzymes

6.3 Forming C-N bonds
    6.3.1 Acid-ammonia ligases (amide synthetases)
    6.3.2 Acid-amino acid ligases (Peptide synthetases)
    6.3.3 Cyclo-ligases
    6.3.4 Other C-N ligases
    6.3.5 C-N ligases with glutamine as N-donor
6.4 Forming C-C bonds.

Table 5.4 : Differences between Enzymes and Hormones

	Enzymes		*Hormones*
1.	All the enzymes are proteins.	1.	The hormones may be polypeptides, terpenoids, steroids, phenolic compounds or amines.
2.	Mostly the enzymes perform reactions at the place of origin i.e. in cells where they are produced.	2.	Hormones perform activity at some distance away from the place of origin.
3.	Enzymes are not translocated from one part to another part of the cell or plant part.	3.	Hormones are first synthesized in one part and then translocated to another part where they perform activity. Most of the hormones show polar translocation.
4.	Enzymes are biological catalysts.	4.	Hormones are not catalysts.
5.	They catalyse the biological reactions.	5.	They simply initiate biochemical reactions.
6.	As enzymes are catalysts, at the end of reaction they remain unchanged and can be reutilized.	6.	As hormones are not catalysts, they participate in biological reactions and their chemical composition is changed.

5.16. TESTS FOR SOME ENZYMES

Experiment 1: To test the activity of catalase in plant tissues

Catalase activity can be tested with the help of fresh potato tuber. Catalase is an enzyme which acts on hydrogen peroxide (H_2O_2) and decomposes it to water and molecular oxygen as follows:

$$2H_2O_2 \xrightarrow{catalase} 2H_2O + O_2$$

To test the activity of enzyme catalase cut a thin slice of fresh potato tuber. Take a test tube, pour dilute solution of H_2O_2 (1 ml of 3% H_2O_2 + 30 ml distilled water) in it and place the potato slice in the test tube and observe.

The bubbles of oxygen come out rapidly from the potato slice. It indicates that potato tuber contain enzyme catalase in its tissues. If boiled potato is used in place of fresh potato, the bubbles do not evolve as the activity of catalase enzyme is destroyed during boiling, *i.e.*, the denaturation of enzyme takes place.

Experiments 2: To test activity of enzyme diastase

Diastase activity can be tested with the help of germinating barley seeds. Diastase is an enzyme which acts on starch and converts it into glucose. The starch gives a positive test with iodine solution to form blue-coloured starch iodide but glucose fails to give this starch iodide test.

To prepare diastase solution crush germinating barley seeds with water in the pestle mortar. Filter this with the help of muslin cloth. The filtrate is centrifuged and supernatant is separated which contains enzyme diastase. Take test tube and pour 2 ml of 1% starch solution (boil 1 gm starch in 100 ml water) in it. Now add one or two drops of iodine solution. A blue colour appears showing the formation of starch iodide. Now add 3 ml diastase solution into the tube containing starch iodide. Blue colour disappear on slight heating due to conversion of starch into glucose. Further on adding fehlings solution into it and warming the test tube, red precipitate of cuprous oxide is obtained (It is the positive test of glucose). The above experiment indicates that germinating barley seeds contain diastase which converts starch into glucose. The glucose gives positve test with fehling's solution.

Experiment 3: To test the presence of oxidases in a plant material

Oxidases are enzymes which catalyse the oxidation of their substrates with the help of molecular O_2 serving as an electron acceptor. Thus, they can transfer of hydrogen of their substrates to the molecular O_2 as follows :

$$AH_2 \text{ (substrate)} + O \longrightarrow A + H_2O$$

The presence of oxidases can be tested in potato tuber by the following method :

Take a fresh potato tuber, crush it with water in a pestle mortar and filter to get extract containing oxidases. Now prepare 2% guaiacum solution by dissolving 2 gm Gum guaiacum in 100 ml of absolute alcohol. Take 1 ml of guaiacum solution in a test tube and add 5 ml enzyme extract and observe. Blue colour appears indicating that guaiacum which is a phenolic compound, has been oxidized by enzyme oxidase.

In our daily life we see that when fresh apple is cut and left for sometime, it turns brown. It is also due to enzyme oxidase which catalyses the oxidation of phenolic compounds present in the apple with the help of atmospheric oxygen.

Experiment 4: To test the presence of peroxidases in potato tuber

Peroxidases are enzymes which catalyse the oxidation of their substrates by removing hydrogen which combines with H_2O_2 as follows :

$$AH_2 + H_2O_2 \xrightarrow{\text{peroxidase}} A + 2H_2O$$

Prepare the enzyme extract by crushing fresh potato tuber in water and filtering. Prepare 2% guaiacum solution by dissolving 2 gm gum guaiacum in 100 ml of absolute alcohol. Take 1 ml of guaiacum solution in a test tube and add 5 ml of enzyme extract and immediately 10-15 drops of 3% H_2O_2. Observe the change in colour.

It is observed that the colour of the mixture rapidly turns blue indicating that the phenolic compound guaiacum has been oxidized by enzyme peroxidase in presence of H_2O_2. It shows that the potato tuber contains enzyme peroxidase.

6
Coenzymes

6.1. INTRODUCTION

Certain enzymes in addition to their protein structure have a non-protein group attached to them. The protein part is called the **apoenzyme** and the non-protein part as **prosthetic group** or the **Coenzyme**. The complete enzyme is called the **holoenzyme**.

Holoenzyme = Apoenzyme + Prosthetic group or Coenzyme.

The Coenzymes are small molecular weight organic, dialyzable, thermostable compounds required for the catalytic activity of one or a group of enzymes. Some of the common examples of coenzymes are NAD^+, NADP, FAD, FMN, COA, pyridoxals etc.

In contrast to the organic substance-requiring enzymes, there are enzymes which require certain metal ions for activity. The metal ion is generally referred to as an 'activator'. Some of them are iron, copper, zinc, magnesium, manganese and molybdenum.

6.2. STRUCTURE AND CLASSIFICATION

1. Structure

In the case of coenzymes it is striking that, with the exception of iron-porphyrins, a nucleotide is always present. Thus, the coenzymes are mono or dinucleotides. Many coenzymes, especially those concerned with biological oxidations, contain members of the B-complex group of vitamins such as pyridine, flavin, thiamine, and pyridoxine in their structure. The coenzymes are much smaller molecules than proteins and can be removed from their apoenzymes by dialysis if they are dissociable.

2. Classification

There is no definite classification of the coenzymes, but for convenience they are classified on the basis of the reactions in which they participate as a catalyst. The broad classification is as follows:

I. *Hydrogen Transferring Coenzymes*:

  A. *Pyridine Coenzymes*
   (i) Nicotinamide adenine dinucleotide–NAD.
   (ii) Nicotinamide adenine dinucleotide phosphate–NADP.
   (iii) Nicotinamide mononucleotide–NAM.
  B. *Riboflavin Coenzymes*
   (i) Rioboflavin mononucleotide–FMN.
   (ii) Riboflavin adenine dinucleotide–FAD.
  C. Iron Prophyrin Coenzymes
   (i) Porphin
   (ii) Hematin.
     (a) Catalase

(b) Cytochromes (Cyt. a,b, c, f)

II. *Group Transferring Coenzymes*

A. *Coenzymes acting as decarboxylases and transaminases.*
  (i) Pyridoxal phosphate–PALP.
  (ii) Pyridoxamine phosphate.
  (iii) Pyridoxine phosphate.

B. *Coenzyme for acetylation*
  (i) Coenzyme A – Co A

C. *Coenzymes for oxidative decarboxylation*
  (i) Thiamine pyrophosphate–TPP.

D. *Other group-transferring Coenzymes*
  (i) Adenosine-tri-phosphate–ATP (Phosphate gp is transferred)
  (ii) Adenosine-di-phosphate–ADP (Phosphate gp is transferred)
  (iii) Adenosine-mono-phosphate–AMP (Phosphate gp is transferred)
  (iv) Folic acid (Formyl gp is transferred).
  (v) Uridine-di-phosphate–UDP (Sugar, Uronic acid is transferred)
  (vi) Cytidine-di-phosphate–CDP (Phosphonyl choline group is transferred)
  (vii) Cytidine-tri-phosphate–CTP (Phosphonyl choline group is transferred)
  (viii) Phosphoadenyl suphate–PAPS (Sulphate gp is transferred)

III. *Other Coenzymes*
  (i) Vitamin A—Vit. A
  (ii) Vitamin B—Vit. B
  (iii) Vitamin C—Vit. C
  (iv) Vitamin D—Vit. D
  (v) Vitamin K—Vit. K
  (vi) Lipoic acid
  (vii) Biotin (Helps in CO_2 fixation reactions)
  (viii) Vitamin B_{12} (Metabolism of methyl groups & synthesis of DNA).
   (ix) Glutathione ascorbic acid (oxidation-reduction-reactions)
   (x) Glucose-1-6, di-phosphate (In carbohydrate metabolism).
   (xi) Glyceric acid -2,3-di-phosphate (In carbohydrate metabolism)
   (xii) Glucose-1-phosphate uridine-nucleotide. (In carbohydrate metabolism)

6.3. ACTION OF COENZYMES

The coenzymes usually occur in living cells in low concentrations, and it is of the highest importance that reactions have been studied in vitro. The coenzyme is as essential a reactant as the substrate that is activated by the enzyme. The following equation represents the reaction of the coenzyme with the substrate:

$$\text{Substrate (A) + Coenzyme} \xrightarrow{\text{Enzyme (A)}} \text{Substrate (A) derivative + Coenzyme derivative.}$$

The Coenzyme is regenerated and again becomes available as an essential component for the above reaction.

$$\text{Substrate (B) + Coenzyme derivative} \xrightarrow{\text{Enzyme (B)}} \text{Substrate (B) – derivative + Coenzyme.}$$

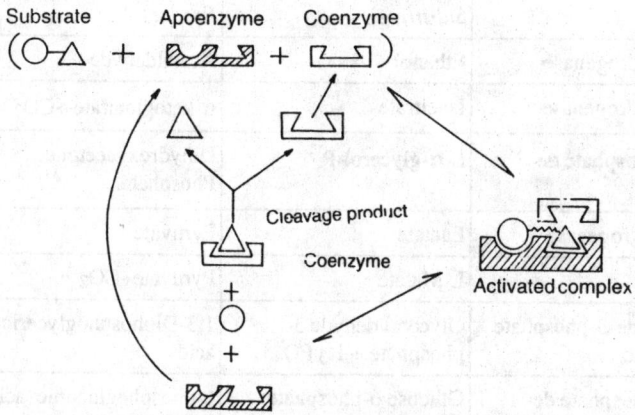

Fig. 6.1a: Function of Coenzymes (after A. Cantarow, & B. Schepartz, 1967).

The function of coenzyme in the enzymatic reaction is thus to assist in the cleavage of the substrate by acting as an acceptor for one of the cleavage products, as shown in Fig. 6.1a and b. Here, the substrate and apoenzyme form a complex in the presence of the coenzyme. When the bond in the substrate becomes activated, one of the cleavage products (usually a small fragment of the entire substrate molecule) is transferred directly to the coenzyme. The coenzyme here acts as a receptor because it has an appropriate receptor site in its structure. What is left of the substrate now dissociates from the apoenzyme. The attached fragment of the substrate in the coenzyme is either liberated as such or is passed on to other enzyme systems for additional changes; in either case the coenzyme is regenerated. Both apoenzyme and coenzyme are then able to repeat the same cycle of events, hence it can be said that both act catalytically. Thus, the coenzyme becomes bound with apoenzyme during the reaction. For example, dehydrogenases utilize either NAD+ or NADP. Their function is to transfer the hydrogen nuclei with two electrons from the substrate, thus oxidizing it:

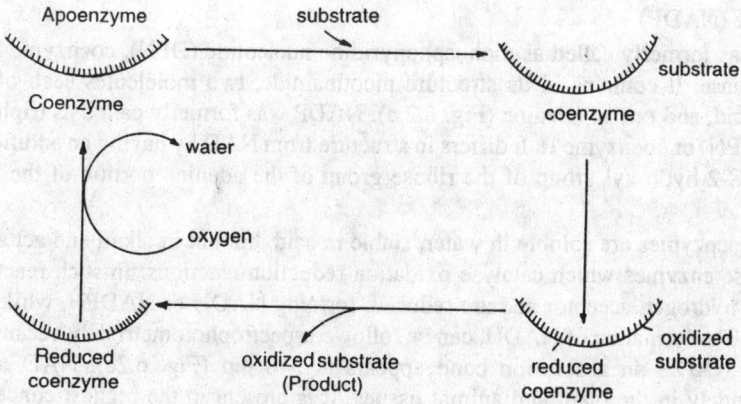

Fig. 6.1b: Cyclic nature of coenzyme action in an oxido-reductase system.

Substrate + NAD^+ + ENZYME $\rightarrow$ Oxidized substrate + NADH + H^+. In the reverse direction the substrate reduced. NADP and NADPH behave in a similar way. In a cell the energy producing catabolic processes require NAD^+ while the synthetic processes, use NADPH.

Table 6.1 : Some reactions Catalysed by Pyridine nucleotide enzymes.

	Enzyme	Substrate	Product	Coenzyme
1.	Alcohol dehydrogenase	Ethanol	Acetaldehyde	NAD^+
2.	Isocitric dehydrogenase	Isocitrate	α-ketoglutarate $+CO_2$	NAD^+
3.	α-glycerol phosphate dehydrogenase	L-α-glycerol-P	Dihydroxyacetone Phosphate.	NAD^+
4.	Lactic dehy-drogenase	Lactate	Pyruvate	NAD^+
5.	Malic enzyme	L-Malate	Pyruvate$+CO_2$	$NADP^+$
6.	Glyceraldehyde-3-phosphate dehydrogenase.	Glyceral dehyde 3-phosphate + $H_3 PO_4$	1,3-Diphosphoglyceric acid	NAD^+
7.	Glucose 6-phosphate dehydrogenase	Glucose 6-phosphate	6-phosphogluconic acid	$NADP^+$
8.	Glutamic dehydrogenase	L-Glutamic-acid	α-ketoglutarate+ NH_3	NAD^+ $NADP^+$
9.	Glutathione reductase	Oxidized glutathione	Reduced glutathione	NADPH
10.	Quinone reductase	P-Benzo-quinone	Hydroquinone	NADH NADPH
11.	Nitrate reductase	Nitrate	Nitrite	NADPH

The prosthetic group acts in an analogous fashion, the only difference being that the acceptor of the substrate fragment remains attached to the surface of the apoenzyme. The coenzymes generally exhibit a much less restricted range of specificity than do the apoenzymes with which they co-operate. It has been shown in the systems of biological oxidations.

6.4 SOME IMPORTANT COENZYMES

1. Nicotinamide Adenine Dinucleotide (NAD) and Nicotinamide Adenine Dinucleotide-Phosphate (NADP)

NAD was formerly called as diphosphopyridine nucleotide (DPN), coenzyme I, cozymase or codehydrogenase. It contains in its structure nicotinamide, two molelcules each of D-ribose and phosphoric acid, and one of adenine (Fig. 6.2 a). NADP was formerly called as triphosphopyridine nucleotide (TPN) or coenzyme II. It differs in structure from NAD by having an additional phosphate esterified at C-2 hydroxyl group of the ribose group of the adenine portion of the molecule (Fig. 6.2 d).

These coenzymes are soluble in water, stable in acid, but not in alkali and act as cofactors for dehydrogenase enzymes which catalyse oxidation-reduction reactions. In such reactions the coenzymes act as hydrogen acceptor and are reduced, forming NADH or NADPH, while the substrates are oxidized. The formation of NADH, can be followed spectrophotometrically because on reduction of NAD^+ to NADH an absorption band appears at 340 nm (Fig. 6.2c). NAD and NADP are distributed widely in the plant and animal tissues. It is present in the highest concentration in the germ and pericarp in cereal grains. Yeast is rich in these coenzymes. Other important sources are liver, kidney, fish, meat, certain nuts, legumes (peas, beans, lentils), coffee, tea, certain green vegetables, wheat, rye etc. Fruits, milk and eggs are generally poor sources.

The role of NAD as a coenzyme is now clear. In the formation of lactic acid NADH is oxidised, while pyruvic acid is reduced. NADH represents a store of hydrogen atoms or reductive energy in the cell, available for certain hydroxylase systems, synthesis of fatty acids and steroids, and reduction

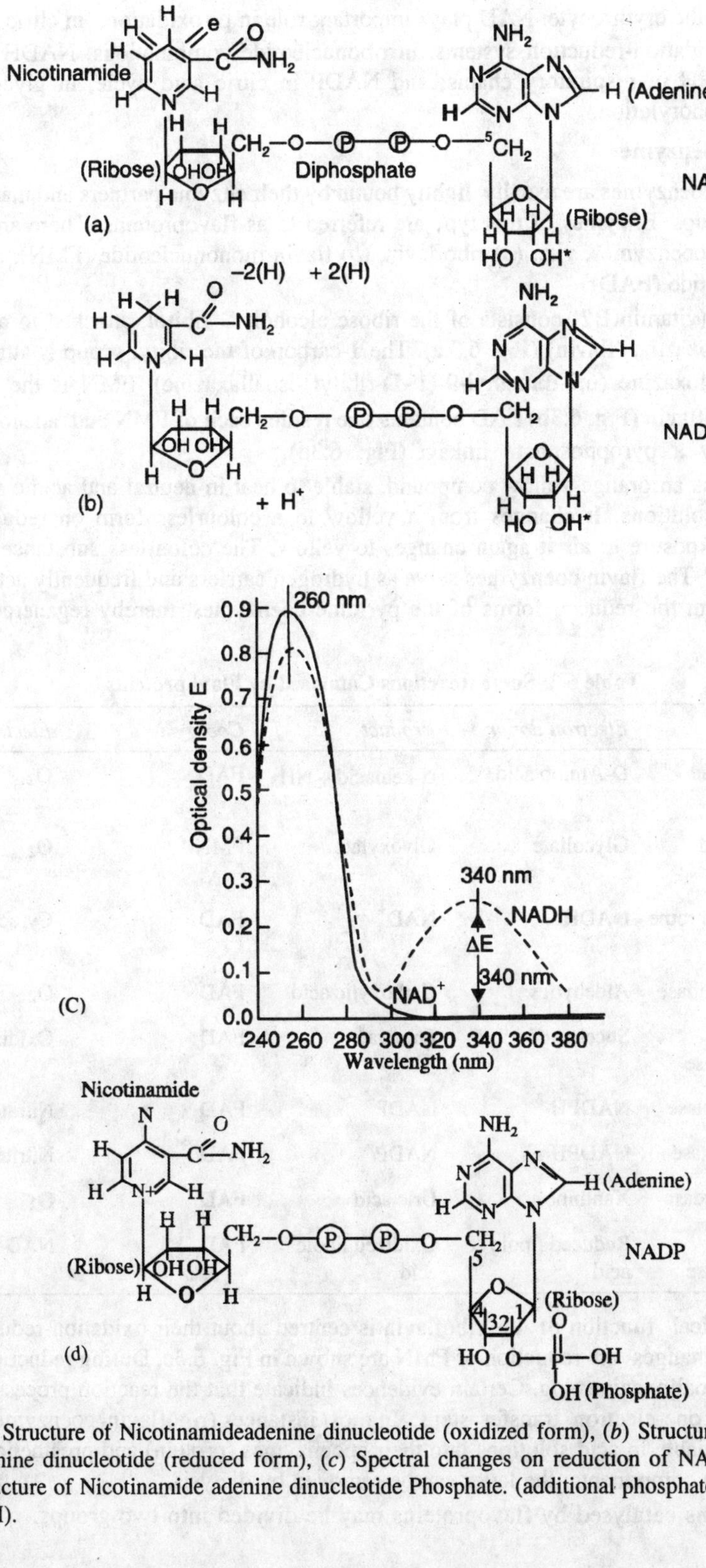

Fig. 6.2: (a) Structure of Nicotinamideadenine dinucleotide (oxidized form), (b) Structure of Nicotinamide adenine dinucleotide (reduced form), (c) Spectral changes on reduction of NAD^+ to NADH, (d) Structure of Nicotinamide adenine dinucleotide Phosphate. (additional phosphate is attached to the -OH).

of glutathione in the erythrocyte. NAD plays important role in β-oxidation, in citric acid cycle, in glycolysis, in oxidation-reduction systems, in ribonucleotide combinations; NADH in fatty acid synthesis; NADPH in respiratory chains; and NADP in citric acid cycle, in glycolysis and in non-cyclic phosphorylation.

2. Riboflavin Coenzyme

The flavin coenzymes are usually lightly bound by their enzyme partners and may be regarded as prosthetic groups. Enzymes of this type are referred to as flavoproteins. There are three most important flavin coenzymes, viz., (a) riboflavin, (b) flavin mononucleotide (FMN), and (c) flavin adenine dinucleotide (FAD).

Riboflavin (vitamin B2) consists of the ribose alcohol, D-ribitol, attached to a heterocyclic substance, isoalloxazine (flavin) (Fig. 6.3.a). The 1-carbon of the ribityl group is attached at the I position of isoalloxazine (6,7-dimethyl-9-(1'-D-ribityl)-isoallaxazine). FMN is the 5' phosphate derivative of riboflavin (Fig. 6.3b) FAD contains one residue each of FMN and adenosine-5'-phosphate, united by a pyrophosphate linkage (Fig. 6.3d).

Riboflavin is an orange-yellow compound, stable to heat in neutral and acidic solutions, but not in alkaline solutions. It changes from a yellow to a colourless form on reduction and on reoxidation by exposure to air it again changes to yellow. The colourless substance is known as 'leuco-riboflavin' The flavin coenzymes serve as hydrogen carriers and frequently act as acceptors for hydrogen from the reduced forms of the pyridine coenzymes, thereby regenerating NAD of NADP.

Table 6.2: Some Reactions Catalysed by Flavoproteins

	Enzyme	Electron donor	Product	Coenzyme	Electron acceptor
1.	D-Amino acid oxidase.	D-Amino acid	α-ketoacid + NH_3	FAD	O_2
2.	Glycolic acid oxidase	Glycollate	Glyoxylate	FMN	O_2
3.	NAD^+ cytochrome C reductase	NADH	NAD^+	FAD	Cytochrome Cox
4.	Aldehyde oxidase	Aldehydes	Carboxylic acid	FAD	O_2
5.	Succinic dehydrogenase	Succinate	Fumarate	FAD	Oxidized dyes
6.	Nitrate reductase	NADPH	$NADP^+$	FAD	Nitrate
7.	Nitrite reductase	NADPH	$NADP^+$	FAD	Nitrite
8.	Xanthine oxidase	Xanthine	Uric acid	FAD	O_2
9.	Lipolyl dehydrogenase	Reduced lipoic acid	Oxidized lipoic acid	FAD	NAD^+

The biological function of the riboflavin is centred about their oxidation-reduction properties. Spectral changes on reduction of FMN are shown in Fig. 6.3c. During reduction, hydrogen is added to the isoalloxazine ring. Certain evidences indicate that the reaction process consists of two consecutive one-electron transfer steps. In most instances (riboflavin coenzymes + Protein) are readily dissociable in acid solutions into their apoenzymes (protein) and prosthetic (coenzyme; flavin nucleotide) components; the latter can be removed by dialysis.

The reactions catalysed by flavoproteins may be divided into two groups.

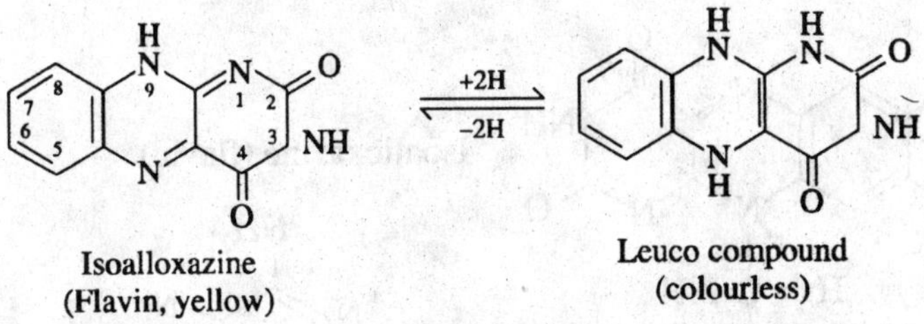

Coenzymes

[Isoalloxazine (Flavin, yellow) ⇌ Leuco compound (colourless), +2H / −2H]

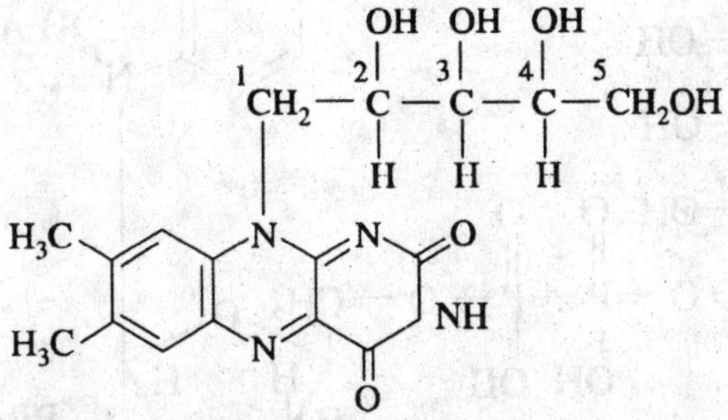

Fig. 6.3a: Riboflavin (6, 7 dimethyl-9-(1D ribityl) isoalloxazine

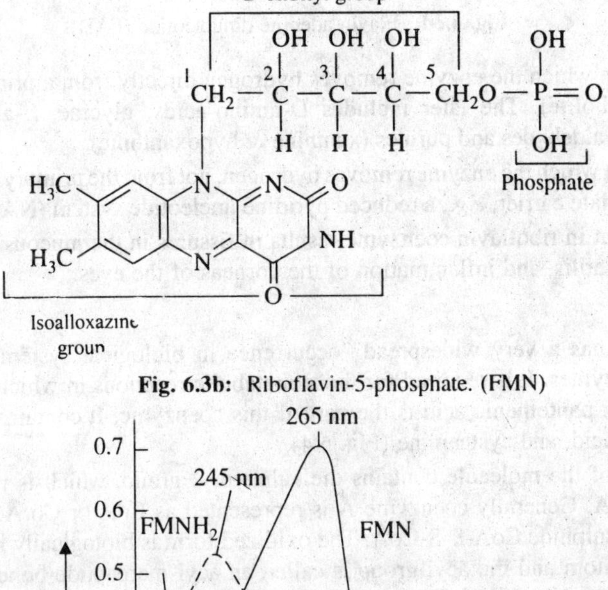

Fig. 6.3b: Riboflavin-5-phosphate. (FMN)

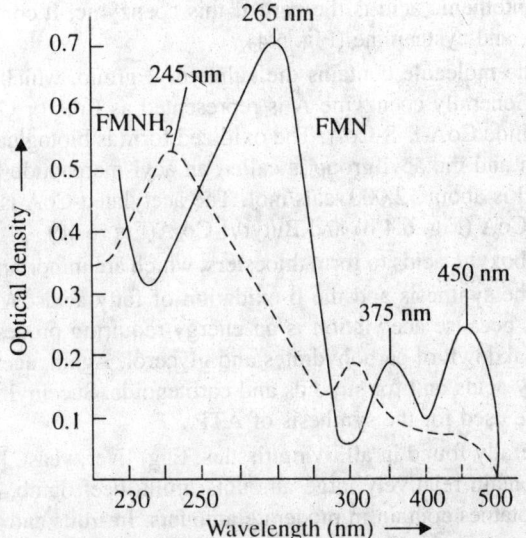

Fig. 6.3c: Spectral changes on reduction of FMN.

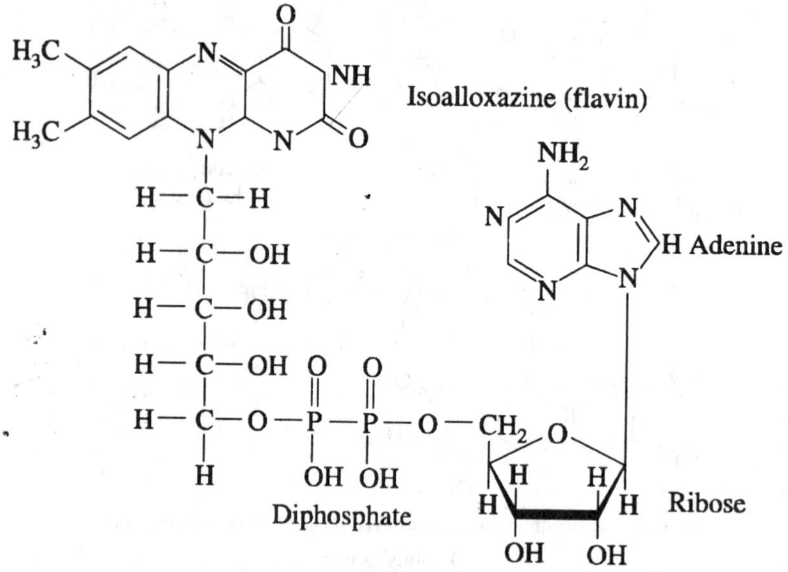

Fig. 6.3d: Flavin adenine dinucleotide (FAD).

1. Reactions in which the enzyme removes hydrogen directly from a primary substrate-(organic metabolite). The later includes D-amino-acids, glycine, L-amino-acids, L-hydroxyacids, aldehydes and purines (xanthine & hypoxanthine).
2. Reactions in which the enzyme removes hydrogen, not from the primary substrate, but from an intermediate carrier, e.g., a reduced pyridine nucleotide system (NADPH, NADH).

The diet deficient in riboflavin coenzyme results in fissures in the mucous membranes of the mouth and face, dermatitis, and inflammation of the corneas of the eyes.

3 Coenzyme-A

This coenzyme has a very widespread occurrence in biological systems and serves as a cofactor for many enzymes. It is required for those metabolic reactions in which organic acids are involved. The vitamin pantothenic acid is the part of this coenzyme. It contains one residue each of ATP, pantothenic acid, and cysteamine (Fig. 6.4).

The active end of the molecule contains the sulphydryl group, which is readily acylated to yield acyl coenzyme A. Generally coenzyme A is represented as CoA or Co A. SH. Co A. SH is readily oxidized to disulphide CoA-S-S-CoA. The oxidized form is biologically inactive. The bond between the sulphur atom and the acyl group is called an acyl mercaptide bond. The free energy on hydrolysis of this bond is about 12,000, cals/mol. The acetylated CoA is Co A-S-CO CH_3. The other forms are Malonyl-CoA (Fig. 6.4 b) and Butyryl-Co A(Fig. 6.4c).

Co A reacts with carboxylic acids to form thioesters, which are important in many biochemical reactions particularly in the synthesis and the β-oxidation of fatty acids. ATP is required for the formation of acetyl-Co A because acetylation is an energy-requiring process Acetyl CoA acts as an intermediate in the breakdown of carbohydrates and glycerol. Again, acetyl CoA is the starting point of synthesis for fatty acids and for steroids and carotenoids. Succinyl Co A is analogous to the acetyl CoA and can be used for the synthesis of ATP.

Coenzyme A is generally found in all living tissues. Egg, liver, yeast, kidney, wheat and rice bran, peanut, and peas contain relatively large amounts; milk, beef, lamb, chicken, pork, certain fish, rye, oat and sweet potatoes contain in moderate amounts. In fruits and vegetables, it is found in very-very small amounts.

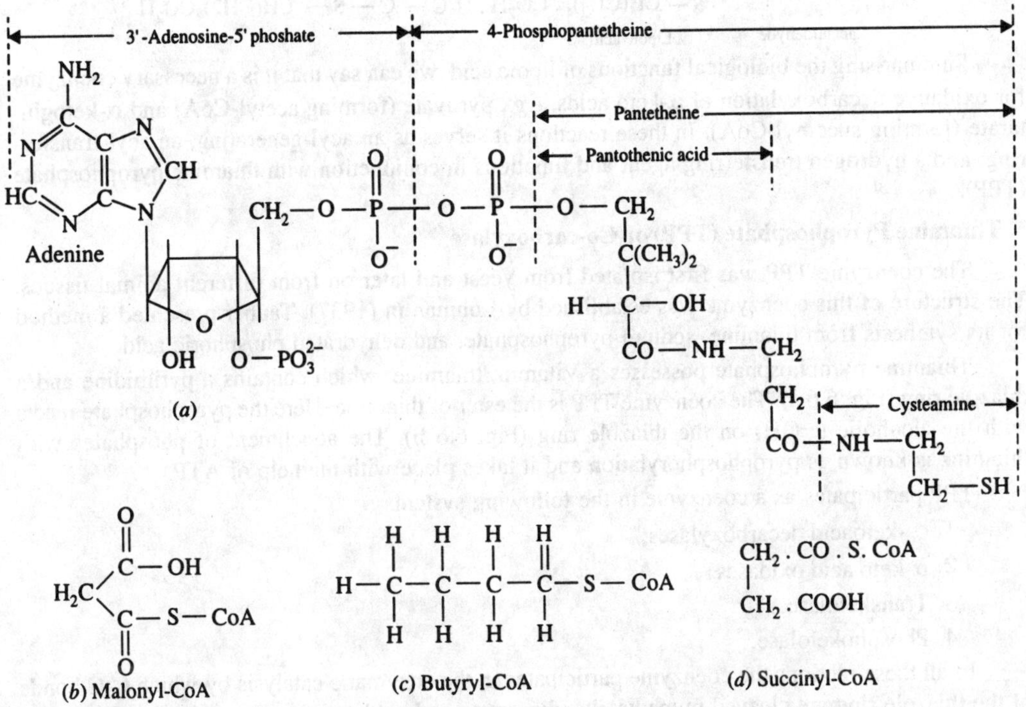

Coenzymes

Fig. 6.4: Structure of Coenzyme A.

(a) [Structure showing 3'-Adenosine-5' phosphate, 4-Phosphopantetheine, Pantetheine, Pantothenic acid, Cysteamine components]

(b) Malonyl-CoA

(c) Butyryl-CoA

(d) Succinyl-CoA

4. Lipoic Acid

Lipoic acid, a bacterial growth factor, occurs in liver and in yeast. The lipoic acid attached to protein can be released by acid, base, or proteolytic hydrolysis. It consists of a five-membered, disulphide-containing ring plus an aliphatic hydrocarbon chain terminating in a carboxyl group (Fig. 6.5). The carboxyl group is usually bound to the enzyme protein in amide or amide like linkage.

Lipoic acid (throtic acid) or 1, 2-Dithiolane-3-Valeric acid

Fig. 6.5: Structure of α-lipoic

The biological role of lipoic acid appears to arise from the presence of an intramolecular disulphide linkage and the reactions which it makes possible. Lipoic acid is a hydrogen transferring coenzyme. Its prime role is in oxidative decarboxylation. It cooperates in the conversion of primary decarboxylation product, an active aldehyde to an active acid. Therefore, it is involved in both dehydrogenation and group transfer. During oxidative-decarboxylation of pyruvic acid the lipoic acid undergoes a reaction with the acetaldehyde unit in which the aldehyde is oxidized to an acid and made into a thioester; at the same time the lipoic acid molecule is reduced to the equivalent of the dithiol form.

In the conversion of pyruvate to acetate and CO_2, thiamin, CoA, and lipoic acid all are involved.

$$H_3C-\underset{acetaldehyde}{\overset{O}{\underset{\|}{C}}-H} + \underset{Lipoic\ acid}{\begin{array}{c}S-CH_2\\|\quad\ \ \ \diagdown CH_2\\S-CH(CH_2)_4 \cdot CO_2H\end{array}} \longrightarrow H_3C-\overset{O}{\underset{\|}{C}}-S-\begin{array}{c}HS-\overset{H_2}{C}\\\quad\ \ \diagdown CH_2\\CH(CH2)_4CO_2H\end{array}$$

Summarising the biological functions of lipoic acid, we can say that it is a necessary coenzyme for oxidative decarboxylation of α-keto acids, e.g., pyruvate (forming acetyl-CoA) and α-ketoglutarate (forming succinyl-CoA). In these reactions it serves as an acyl-generating, an acyl-transferring, and a hydrogen transferring agent and functions in conjunction with thiamine pyrophosphate (TPP).

5. Thiamine Pyrophosphate (TPP) or Co-carboxylase

The coenzyme TPP was first isolated from yeast and later on from different animal tissues. The structure of this coenzyme was established by Lohman in (1937). Tauber presented a method for its synthesis from thiamine, sodium pyrophosphate, and dehydrated phosphoric acid.

Thiamine pyrophosphate possesses a Vitamin, thiamine, which contains a pyrimidine and a thiazole ring (Fig. 6.6 a). The coenzyme TPP is the ester of thiamine. Here the pyrophosphate reacts with the alcoholic (—OH) on the thiazole ring (Fig. 6.6 b). The attachment of phosphates with thiamine is known as pyrophosphorylation and it takes place with the help of ATP.

TPP participates as a coenzyme in the following systems :

1. α-ketoacid decarboxylases.
2. α-keto acid oxidases.
3. Transketolase.
4. Phosphoketolase.

In all these ractions the coenzyme participates in the enzymatic catalysis by which C-C bonds of the thiazole ring are cleaved immediately adjacent to carbonyl groups. The carbanion formed in the reaction is believed to participate in the decarboxylation of α-keto acids. In each of the above

$$-\overset{O}{\underset{\|}{C}}-XH + Y \longrightarrow -\overset{O}{\underset{\|}{C}}-YH + X$$

cases, the over-all reaction is a removal of a group 'X' and replacement by a group 'Y'.

The 'X' groups are invariably of such a structure that they can leave as stable compounds provided that the bond ruptured is cleaved by withdrawal of the bonding electron pair toward the carbonyl and into the thiamine-substrate-enzyme complex. Thus, a hydrogen atom is removed as a proton, a carboxylate group as CO_2, and an alcoholic group as an aldehyde plus a proton.

Thiamine and TPP undergo both oxidation and reduction. Treatment in vitro with mild oxidizing agents (e.g. Potassium ferricyanide) results in the formation of thiochrome (Fig. 6.6 c).

In biological systems the coenzyme TPP plays the important role in oxidative decarboxylation of pyruvic acid and α-ketoglutaric acid and also in transketolase reaction, a step in the phosphogluconate oxidative pathway of carbohydrate metabolism.

(a) Thiamine

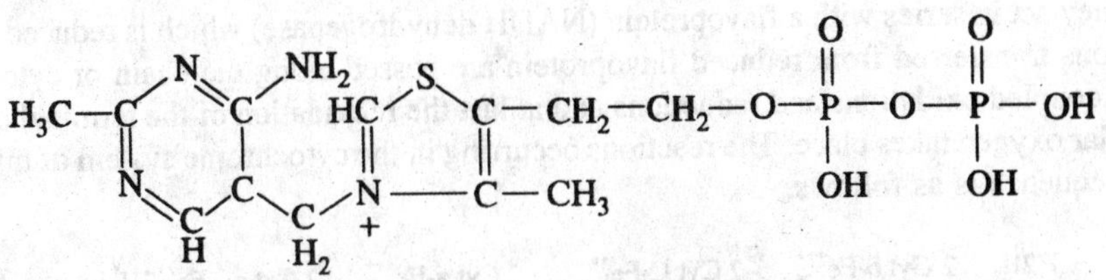

Coenzymes

(b) Thiamine pyrophosphate (cocarboxylase)

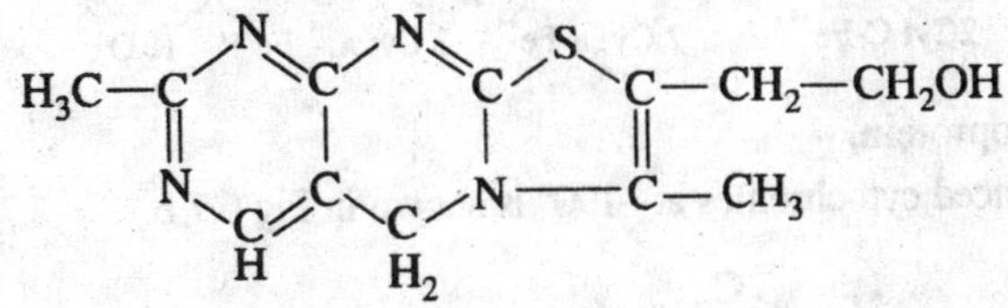

(c) Thiochrome

Fig. 6.6: Thiamine and its derivatives.

6. The Cytochromes

The cytochromes are a group of hemoproteins which have a central collective role in the terminal stages of oxidative metabolism. At present four different types of cytochromes –a, b –, c and f are known which have been isolated from the mitochondria of plant and mammalian cells. The prosthetic groups of these cytochromes are similar, but not identical.

The structure of each cytochrome shows that it contains an atom of iron as a complex with the prosthetic group. The iron atom may either be present in oxidized (Fe^{+++}) or reduced (Fe^{++}) form (Fig. 6.7 a).

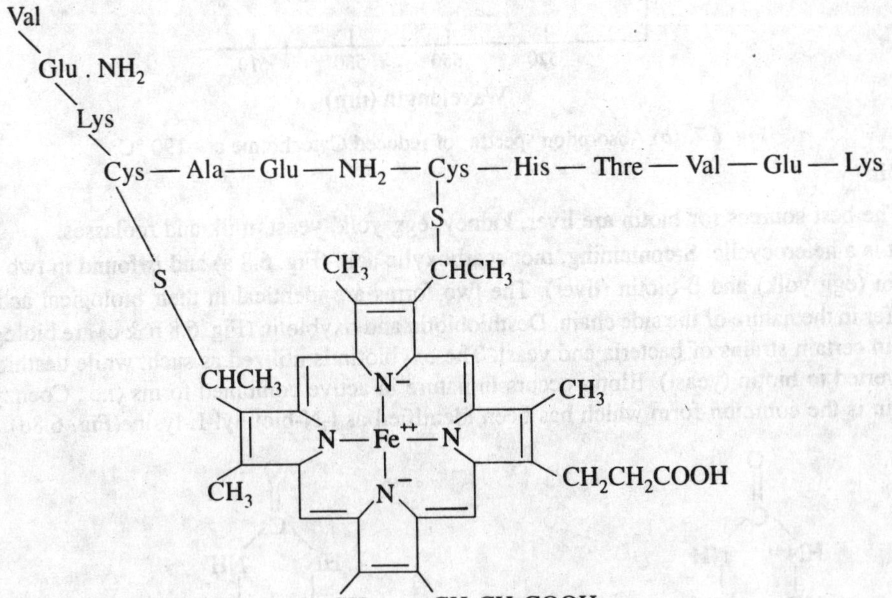

Fig. 6.7. *(a)* Cytochrome C

The cytochromes form the final link in the chain of oxidation-reduction reactions. During the process, the pyridine coenzymes reduced by the operation of TCA cycle are reoxidized by molecular oxygen. They act in series with a flavoprotein (NADH dehydrogenase) which is reduced by NADH. The electrons transferred from reduced flavoprotein are passed along the chain of cytochromes in a series of coupled oxidations and reductions and at last the reoxidation of the terminal cytochrome by molecular oxygen takes place. The reactions occurring in the cytochrome system of mitochondria and their sequence is as follows.

$$F.2H \searrow \nearrow 2\ Cyt\ b\text{-}Fe^{+++} \searrow \nearrow 2\ Cyt\ C\text{-}Fe^{++} \searrow \nearrow 2\ Cyt\ a\text{-}Fe^{+++} \searrow \nearrow 2\ Cyt\ a_3\text{-}Fe^{++}2H^+ \searrow \nearrow \tfrac{1}{2}O_2$$
$$F\ 2H^+ \nearrow \searrow 2\ Cyt\ b\text{-}Fe^{++} \nearrow \searrow 2Cyt\ C\text{-}Fe^{+++} \nearrow \searrow 2\ Cyt\ a\text{-}Fe^{++} \nearrow \searrow 2\ Cyt\ a_3\text{-}Fe^{+++} \nearrow \searrow H_2O$$

Here 'F' represents the flavoprotein.

The absorption spectra of reduced cytochromes at $-190°$ is shown in Fig 6.7 b.

Fig. 6.7. (b) Absorption spectra of reduced Cytochrome at - 190 °C

7. Biotin

The best sources for biotin are liver, kidney, egg yolk, yeast, milk and molasses.

It is a heterocyclic, S-containing, monocarboxylic acid (Fig. 6.8 a) and is found in two forms, α-biotin (egg yolk) and β-biotin (liver). The two forms are identical in their biological activities, but differ in the nature of the side chain. Desthiobiotin and oxybiotin (Fig. 6.8 b & c) are biologically active in certain strains of bacteria and yeast. The oxybiotin is utilized as such, while desthiobiotin is converted to biotin (yeast). Biotin occurs in nature in active combined forms (i.e., Coenzymes). Biocytin is the common form which has been identified as t-N-biotinyl-L-lysine(Fig. 6.8d).

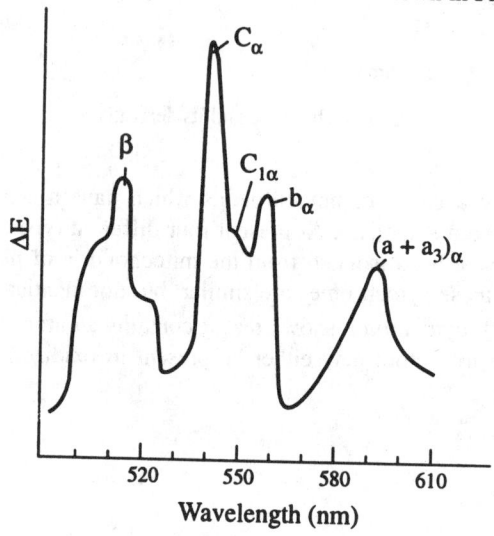

Biotin
(a)

Desthiobiotin
(b)

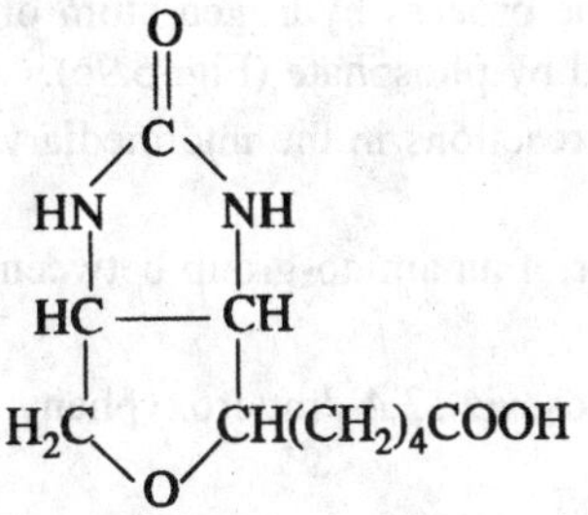

Coenzymes

Oxybiotin (c)

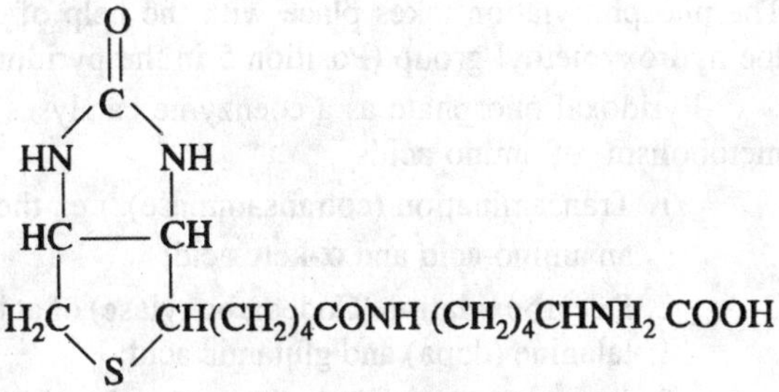

Biocytin (t-N-biotinyl-L-lysine) (d)

Fig. 6.8: Biotin and related Compounds.

Biotin acts as a coenzyme for two types of carboxylation reactions. In both biotin serves as a transient carrier of CO_2. The first type of reactions are those which require or depend on ATP. For example, acetyl-coenzyme A carboxylase reaction. The second type are those which do not need any cofactor such as ATP; example is the methylmalonyl-oxaloacetic transcarboxylase reaction. Thus, biotin acts as a prosthetic group of certain enzymes that catalyse CO_2 transfer reactions, i.e., CO_2 fixation and β-decarboxylation. In mammalian tissues, these include the following.

(a) Acetyl-CoA carboxylase, which catalyses carboxylation of acetyl-CoA to form malonyl-CoA (Fatty acid bio-synthesis);
(b) Propionyl CoA carboxylase, which catalyses transformation of propionyl CoA to methyl-malonyl CoA (formation of succinyl-CoA)
(c) pyruvic acid carboxylase, which catalyses the transformation of pyruvic acid to oxaloacetic acid.
(d) Biotin is involved in the production of acetoacetate from leucine (carboxylation of β-methylcrotonyl CoA).

8. Pyridoxal Phosphate

Pyridoxal phosphate is a coenzyme derived from a group of vitamins, pyridoxine, pyridoxal, and pyridoxamine. The phosphorylated derivative of pyridoxal is known as 'pyridoxal phosphate'.

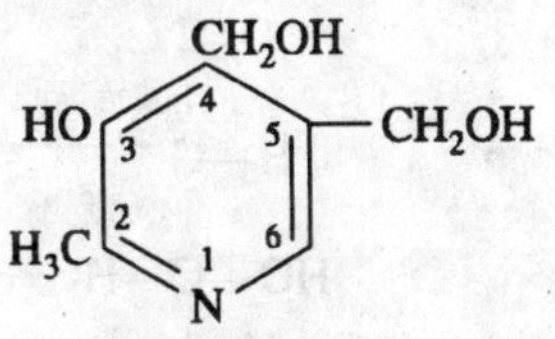

Pyridoxol (a)

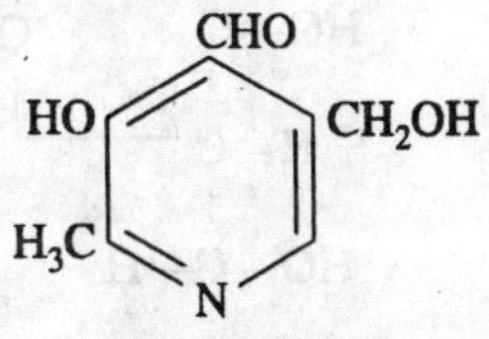

Pyridoxal (b)

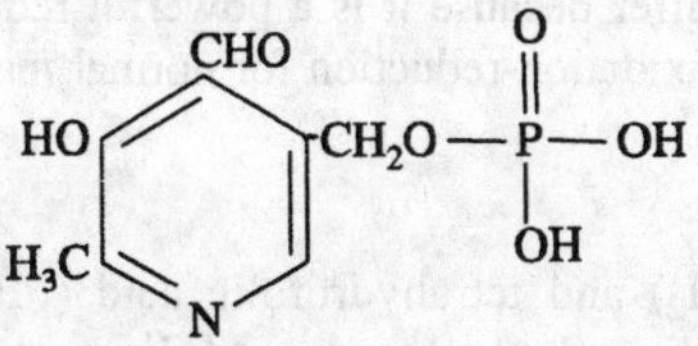

Pyridoxal phosphate (c)

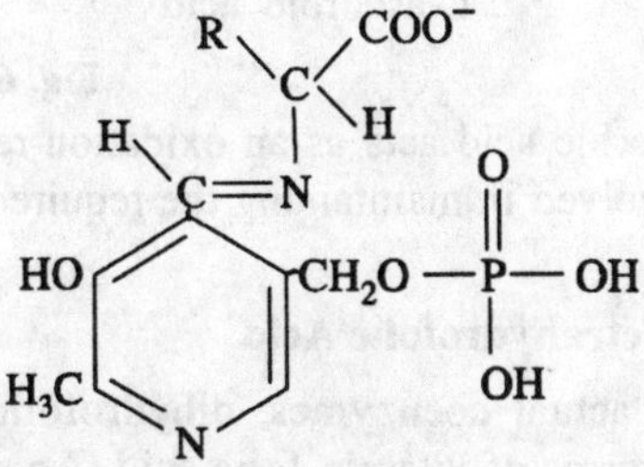

Schiff's base (d)

Fig. 6.9: Structure formulae of Pyridoxol, Pyridoxal, Pyridoxal phosphate and Schiff's base.

The phosphorylation takes place with the help of ATP and during the process hydrogen atom of the hydroxymethyl group (Position 5 in the pyridine ring) is replaced by phosphate (Fig. 6.9c).

Pyridoxal phosphate as a coenzyme catalyses certain important reactions in the intermediary metobolism of amino acids:

1. Transamination (cotransaminase), i.e., the reversible transfer of an amino-group between an amino-acid and α-keto acid;
2. Decarboxylation (Codecarboxylase) of at least two amino acids, viz., 3,4-dihydroxyphenylalanine (dopa) and glutamic acid;
3. Interconversion of glycine and serine, involving also tetra hydrofolic acid as a cofactor;
4. Conversion of 3-hydroxykynurenine to 3-hydroxy-anthranilic acid, a step in the pathway of formation of nicotinic acid from tryptophan;
5. Transsulphurase and thionase reactions, converting homocysteine and cystathionine to cysteine.
6. Transformation of linoleic to arachidonic acid.
7. Synthesis of sphingosine (from serine).

Snell has postulated that aminotransferases, decarboxylases, lipases, and synthetases reactions are catalysed by an intermediate product, Schiff's base (Fig. 6.9 d).

$$\text{Pyridoxal phosphate + amino acid} \longrightarrow \text{Schiff's Base} + H_2O.$$

9. Ascorbic Acid

This acid has been proposed as a coenzyme for the system responsible for the conversion of p-hydroxy phenylpyruvic acid to 2,5 dihydroxy phenyl pyruvic acid in intermediary metabolism of tyrosine. The enzyme (ascorbic acid + protein) apparently operates as a dehydrogenase involving the enediol structure of vitamin. Ascorbic acid (Fig. 6.10) is generally known as Vitamin C.

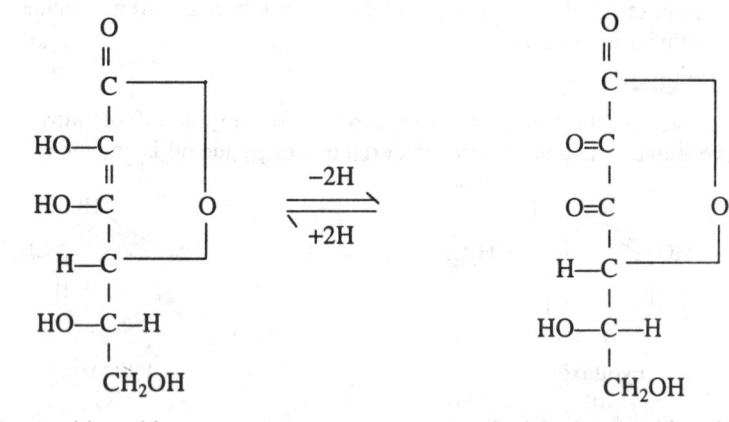

Fig. 6.10 : Ascorbic acid.

Ascorbic acid acts as an oxidation-reduction buffer because it is a powerful reducing agent and is involved in maintaining the required level of oxidation-reduction for normal metabolism in cells.

6.4.10. Tetrahydrofolic Acid

The actual coenzymes, dihydrofolic acid (FH_2) and tetrahydrofolic acid (FH_4), are the reduced forms of vitamin folic acid. An enzyme, folic reductase, reduces folic acid to dihydrofolic acid (FH_2). The other enzyme which reduces the dihydrofolic acid to tetrahydrofolic acid

Coenzymes

is dihydrofolic reductase. The reducing agent leading to the formation of FH_2 is NADPH in both the cases.

The functional role of FH_4 coenzyme is in the transfer of one-carbon fragments. Tetrahydrofolic acid is required in the biosynthesis of purines, pyrimidines and aminoacids. The folic acid and its active formates are known to stimulate the growth of many bacteria. Derivatives of folic acid also play an important role in the formation of normal erythrocytes.

The folic acid coenzymes are found in abundance in green leafy vegetables, yeast, and liver. Other green vegetables, beef, kidney and wheat are also good sources.

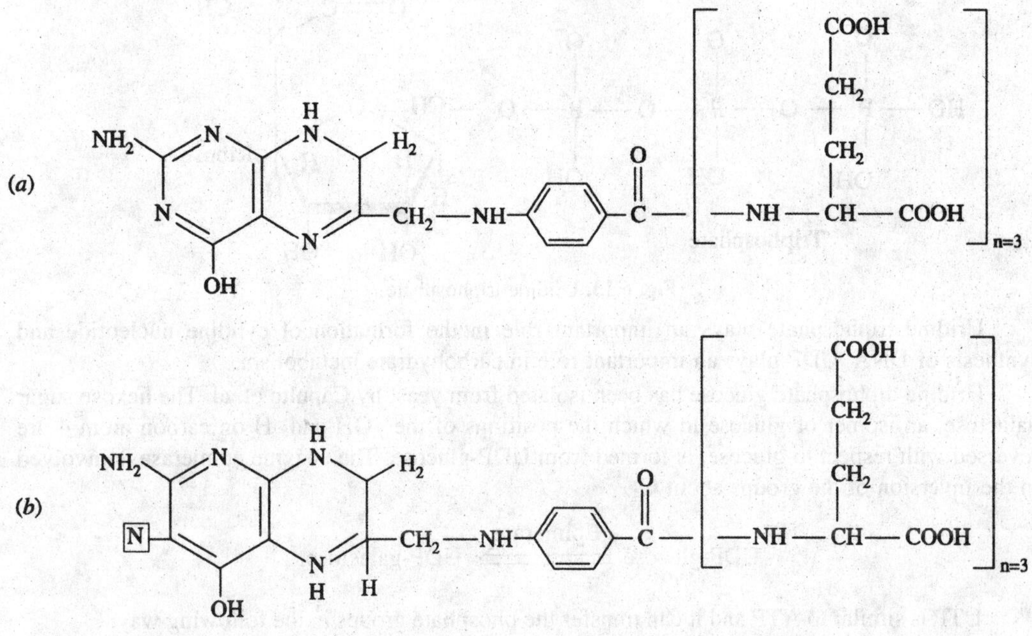

Fig. 6.11: (a) Dihydrofolic acid (FH2) (b) Tetrahydrofolic acid (FH4)

11. Cytidine Di-Phosphate Choline (CDPC)

Cytidine-di-phosphate choline is the coenzyme utilized in the synthesis of lecithins. The synthesis involves the reaction of coenzyme with 1,2-diglyceride to form a lecithin and cytidine-mono-phosphate. Thus, a choline molecule is transferred to the diglyceride molecule.

$$CDPC + 1{,}2\text{-diglyceride} \longrightarrow Lecithin + CMP$$

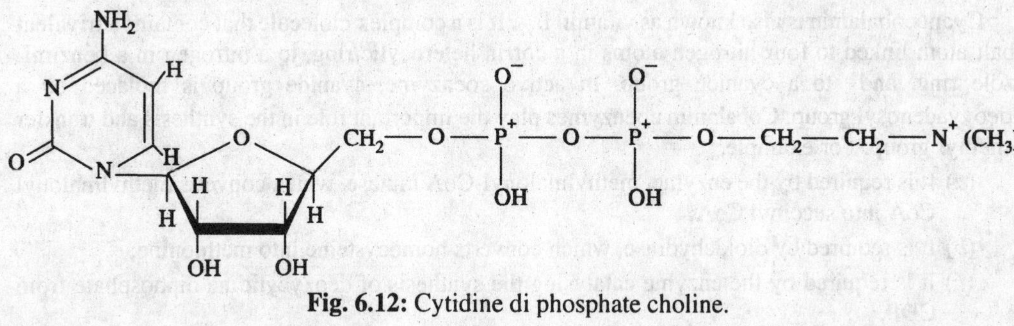

Fig. 6.12: Cytidine di phosphate choline.

12. Uridine Diphosphate (UDP)

Uridine diphosphate is formed from one molecule of uracil, one ribose, and two phosphate molecules. The uridine triphosphate possesses one additional phosphate molecule (Fig. 6.13).

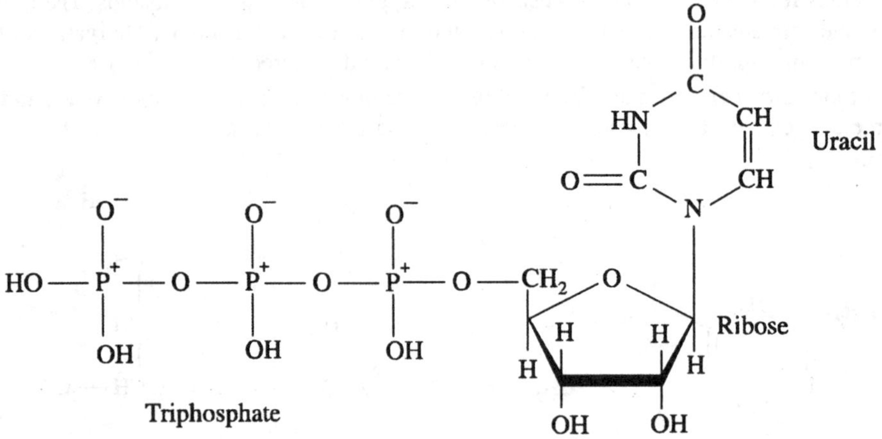

Fig. 6.13. Uridine triphosphate.

Uridine triphosphate plays an important role in the formation of cytidine nucleotide and synthesis of DNA. UDP plays an important role in carbohydrate metabolism.

Uridine diphosphate glucose has been isolated from yeast by Capulto et. al. The hexose sugar galactose, an isomer of glucose in which the positions of the –OH and -H on carbon atom 4 are reversed with respect to glucose, is formed from UDP-glucose. The enzyme epimerase is involved in the inversion of the groups about C_4.

$$\text{UDP-glucose} \xrightleftharpoons{\text{Epimerase}} \text{UDP-galactose.}$$

UTP is similar to ATP and it can transfer the phosphate groups in the following way:

(a) Transfer of orthophosphate group and release of UDP.

(b) Transfer of polyphosphate group and release of UMP.

(c) Transfer of Uridine monophosphate and release of pyrophosphate.

(d) Transfer of Uracil group and release of both ortho-phosphate and pyrophosphate.

UTP contains two and UDP contains one energy rich bond and have high potential for transferring the group.

13. Cyanocobalamin

Cyanocobalamin is also known as vitamin B_{12}. It is a complex molecule that contains a trivalent cobalt atom linked to four nitrogen atoms in a corrin heterocylic ring, to a nitrogen in a benzimidazole ring, and to a cyanide group. In active coenzyme, cyanide group is replaced by a 5'-deoxyadenosyl group. Cobalamin coenzymes play the important role in the synthesis and transfer of methyl group. For example,

(a) It is required by the enzyme, methylmalonyl-CoA mutase, which converts methylmalonyl CoA into succinyl CoA;

(b) It is required by dioldehydrase, which converts homocysteine into methionine;

(c) It is required by the enzyme catalysing the synthesis of deoxycytidine diphosphate from CDP.

The coenzyme is found in kidney, liver, pancrease, heart muscle, egg, milk, meat and fish.

Coenzymes

Table 6.3 : List of some important Coenzymes and Prosthetic groups showing the essential nutritional factors concerned.

S. No.	Coenzyme or prosthetic group	Enzymic and other function	Essential nutritional factors or Vitamins
1.	Nicotinamide-adenine dinucleotide (NAD^+)	As hydrogen acceptor of dehydrogenases	Nicotinic acid
2.	Nicotinamide-adenine dinucleotide-phosphate ($NADP^+$)	As hydrogen acceptor of dehydrogenases	Nicotinic acid
3.	Adenosine-tri-phosphate (ATP)	Transphosphorylation	None
4.	Pyridoxal phosphate	Transaminase, amino-acid decarboxylases, racemases, etc.	Pyridoxine
5.	Thiamine pyrophosphate	Oxidative decarboxylation	Thiamine or Vitamin B_1
6.	Flavin mononucleotide (FMN)	As hydrogen acceptor of dehydrogenases	Riboflavin.
7.	Flavin-adenine dinucleotide (FAD)	As hydrogen acceptor of dehydrogenases	Riboflavin
8.	Coenzyme A(CoA)	Acetyl or other acyl group transfer; fatty acid synthesis and oxidation.	Pantothenic acid
9.	Iron-protoporphyrin	In catalase, peroxidase, cytochromes, haemoglobin	None
10.	Lipoic acid	Oxidative decarboxylation; as hydrogen and acyl acceptor.	Required by micro-organisms.
11.	Tetrahydrofolic-acid	One-carbon transfer	Folic acid
12.	Biotin	CO_2 transfer	Biotin
13.	Cobamide	Group transfer	Cobalamine
14.	Coenzyme Q.	As hydrogen acceptor in electron transfer system	None

14. Coenzyme Q (Ubiquinone)

Coenzyme Q occurs is almost all the plants and animals. Mitochondria of the living plant and animal cells contain the richest proportion of coenzyme Q.

Coenzyme Q is a benzoquinone derivative with a long lipophilic terpenoid side chain (Fig. 6.14 a). The coenzyme plays an important role in the process of electron transport system because it collects electrons from the NADH-flavoprotein complex and the succinate flavoprotein complex and then channels them into the cytochrome system. Many investigators have shown that Ubiquinone

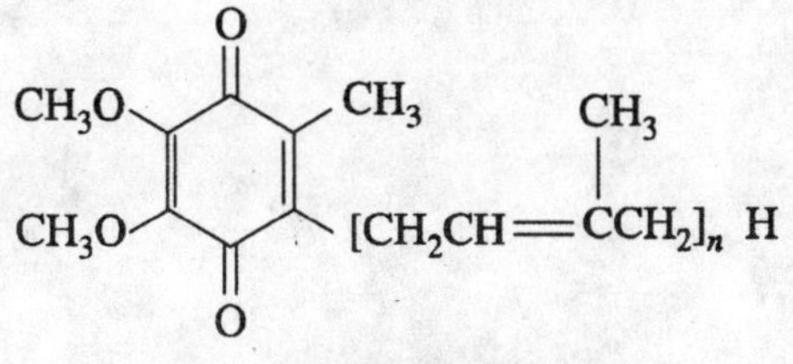

(n = 9 or 10 in higher plants)
(n = 6 to 10 in animals).

Fig. 6.14. (*a*) Structure of Ubiquinone;

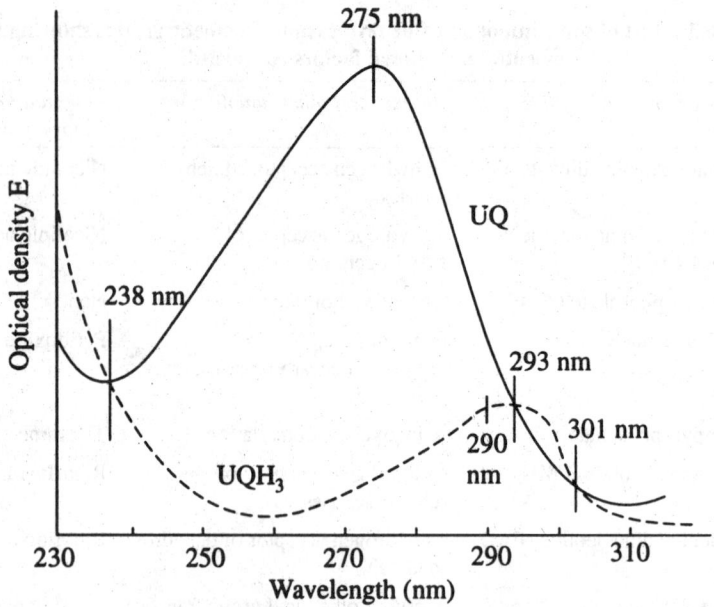

Fig. 16.4 : (b) Spectral changes on reduction of UQ in ethanol.

(UQ) is not on the main pathway of electron transport, but it is in this area that the system changes from a two-electron transport system ($NAD^+ \longrightarrow NADH$, $FAD \longrightarrow FADH_2$) to a one-electron transport system in which the cytochromes are involved. For the oxidation of each molecule of NADH or $FADH_2$, two molecules of each cytochrome are required. The oxidation reduction behaviour of coenzyme Q during electron-transport can be studied spectro photometrically as shown in Fig. 6.14 b.

7
Vitamins

7.1 INTRODUCTION

Vitamins are naturally occurring essential organic constituents of the diet which, in minute amounts, aid in maintaining the normal activities of the tissues. Vitamins are organic compounds of varying complexity that cannot be synthesized in the body. They are essential for normal cellular function in small quantities in diet. Their deficiency in the body results in various diseases. Vitamins differ form other organic foodstuffs in that they do not enter into the tissue structure and do not undergo degradation for purposes of providing energy.

7.2 GENERAL CHARACTERISTICS OF VITAMINS

The vitamins show a number of common characteristics and hence are grouped together
1. The vitamins are required only in small quantities.
2. They cause marked physiological effects.
3. Almost all the vitamins have complex chemical structure except a few, like nicotinic acid and p-aminobenzoic acid.
4. They possess high specificity of function i.e., each vitamin is related to a definite disease which is caused by its deficiency. Its cure is effected only by supply of the same vitamin.
5. They cannot be synthesized by animals and organisms and therefore must be supplied through the agency of food.
6. They have catalytic action.

7.3. VITAMINS AND OTHER RELATED COMPOUNDS

There are some compounds which are related to vitamins. They have no biological activity but are transformed in vivo into the active vitamins. These are called "**Pro-vitamins**". For example, β-carotene is provitamin A and ergosterol is pro-vitamin D_2.

The chemical nature of vitamins led in many instances to the discovery that there are several other structurally related substances which possess similar vitamin activities. Such substances have been termed as "*isotels* and *vitamers*". For example; activated 7-dehydrocholesterol (Vitamin D_3) and activated ergosterol (Vitamin D_2) are isotels, or D vitamins.

7.4. VITAMINS AND HORMONES

Vitamins and hormones are natural organic compounds which possess a marked physiological activity and are characterized by specificity of function. Both are required in small quantities and are essential for maintenance and growth of all animals and organisms.

Vitamins differ from hormones in their chemical structure. The former are supplied to the body chiefly through the food, while the later are produced in the body by the ductless or endocrine glands. For example parathyroid gland produces 'parathormone'; pancreas produces insulin; thyroid gland produces thyrocalcitonin; and testes (interstitial cells) produce testosterone. Recently it has been shown that some members of the vitamin B complex are synthesized in certain animals through the agency of bacteria. The plants and various micro-organisms can very well synthesize the vitamins. The animals eat the plants and get their vitamin supply from them.

Table 7.1: Differences between Vitamins and Hormones

	Vitamins		Hormones
1.	Vitamins are natural compounds with specific complex chemical structure except a few like nicotinic acid and P-aminobenzoic acid.	1.	They are also natural organic compounds with different structures than vitamins.
2.	They are synthesized in various plants and supplied to the animal body chiefly through food.	2.	Plant hormones are synthesized in different parts of the plants or apical meristems, from where they are translocated to other parts. Animal hormones are produced in the body by ductless or endocrine glands.
3.	They have no specific influence on growth.	3.	They have specific influence on growth.
4.	They influence growth and metabolism through direct nutritive effect.	4.	Nutritive effect is not involved in case of hormones.
5.	Each vitamin is related to a difinite disease which is caused due to its deficiency. Its cure is effected only by the supply of that vitamin.	5.	Each hormone is related to the growth of definite organ. They perform and control various affects in the animal body. Plant hormones perform other functions like growth, rooting, cell division etc.
6.	They have catalytic action.	6.	They have no catalytic action.

7.5. NOMENCLATURE AND CLASSIFICATION

Previously the vitamins were named by a letter of alphabet; *e.g.*; Vitamin A, B, C, etc. But, now International Union of Pure and Applied Chemistry (IUPAC) had formulated definite rules for the nomenclature of vitamins, although the terms Vitamin A, Vitamin B, Vitamin C, Vitamin D and so on are still in use. Once the structure of the vitamin had been established, they have generally been renamed; *e.g.*, Vitamin C is known as **ascorbic acid**. Derivatives have been given the name with the suffix accordingly. For example, alcohols and aldehydes, have been given the suffix -ol and -al. IUPAC recommended as far as possible the correct chemical name or a good description to trivial name.

Vitamins were originally classified according to their solubility in water or fats. Thus, they can be classified into the main groups :

A. *Fat-soluble Vitamins* :

Those vitamins which are soluble in fats and fat solvents; *e.g.*, Vitamin A, Vitamin D, Vitamin K and Vitamin E etc.

B. *Water-soluble Vitamins* :

Those vitamins which are soluble in water; *e.g.*, Vitamin B complex and Vitamin C etc.

1. Fat-Soluble Vitamins

A number of experiments with animals have shown that certain lipid compounds are Vitamins. Vitamins A, D, and K are required for human beings and a variety of other animals. A definite need of Vitamin E has been shown for many animals, but not for humans. Vitamin A, D, K, and E are not single compounds, but each of them represents a type of compounds which has a specified physiological effect. Sometimes single vitamin represents a family because it includes a number of other chemical compounds which show the same activity. For example, vitamin A and related compounds show "vitamin A-activity" and similarly "Vitamin D and related compounds show" "Vitamin D Activity".

Vitamins

1. Vitamin-A, or Retinol

Occurrence: Vitamin A, also known as Vitamin A_1, occurs in animal tissues. The most important animal food source of vitamin A are whole milk, butter, and egg yolk–where it occurs free or as esters. The rich sources of vitamin A are Cod-liver oil, Shark-liver oil, the flesh of oily fish, and liver of other animals. The liver of marine fishes contain Vitamin A_1 and those of freshwater fishes A_2.

It is generally believed that carotenoids or provitamin A, are present in both plant and animal tissues, whereas vitamin A occurs in animal tissues only. It is also believed that the carotenoids found in animals are derived from the plants as food source and in animals these carotenoids are converted into vitamin A by some process. The possibility of the formation of carotenoids in animal organism had not been excluded. The chief sources of provitamin A (Carotenoids) are green vegetables like spinach, cabbage, and beet green and yellow coloured vegetables and fruits, *e.g.*, corn, sweet potatoes, carrots, tomatoes, apricots, yellow peaches etc.

Chemistry: Vitamin A is soluble in fats and solvents, insoluble in water. It is stable at high temperatures and unstable in air because it is destroyed by ultraviolet light. The stability of the Vitamin can be produced by the addition of antioxidants like hydroquinone or α-tocopherol (Vitamin E).

Vitamin A is a derivative of certain carotenoids, which are hydrocarbon pigments (yellow, red) widely distributed in nature. The most important carotenoid precursors are α–, β– and γ–carotenes. The carotenoid precursors are known as **provitamins A**. The carotenes are hydrocarbons with the formula $C_{40}H_{56}$. Out of α–, β– and γ–carotenes, β–carotene is most important and active physiologically. The conversion of carotene into Vitamin A takes place in the intestinal tract, but the mechanism in unknown. The monohydroxy β–carotene ($C_{40}H_{55}OH$) is termed as "**Cryptoxanthin**".

The structure of β–carotene (Fig. 7.1) shows that it is a symmetrical molecule, containing two terminal β–ionone rings (A & B), connected by 18 C–chain with 11 conjugated double bonds. The other forms of carotene differ from β–carotene only in the nature of ring B (Fig. 7.1.) Such compounds are subjected to *cis-trans* isomerization.

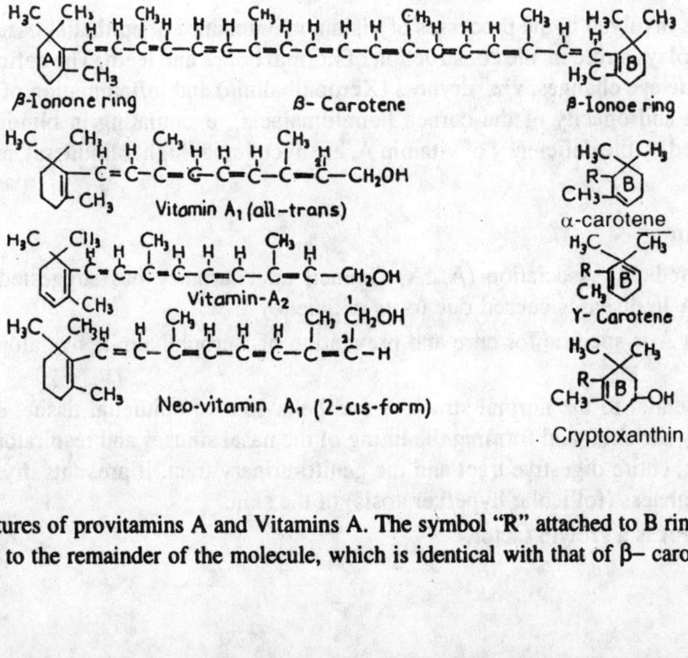

Fig. 7.1: Structures of provitamins A and Vitamins A. The symbol "R" attached to B rings of the provitamin refers to the remainder of the molecule, which is identical with that of β– carotene.

There are following differences in the structure of b-carotene and Vitamin A :

	β– Carotene	Vitamin A
1.	β–Carotene contains two β– ionone rings.	Vitamin A contains only one β– ionone ring.
2.	β–Carotene ($C_{40}H_{56}$) contains 40 C atoms.	Vitamin A ($C_{20}H_{29}OH$) contains 20 C atoms
3.	β–Carotene has 11 conjugated double bonds in the hydrocarbon chain.	Vitamin A has five bonds.
4.	β–Carotene had no alcoholic group.	Vitamin A has a terminal primary alcoholic group.

Two molecules of vitamin A are formed by symmetrical oxidative scission of β–carotene. Similar splitting of α– or γ–carotene, or crypto-xanthin, containing only one β–ionone ring, gives rise to only one molecule of the vitamin.

Isomers of Vitamin A : Vitamin A occurs in nature in different forms. These forms are known as **isomers**. The following isomers of vitamin A are known :

(a) *Vitamin A_1* : The usual form or vitamin A, described above, is also known as vitamin A_1. On treatment with antimony trichloride, the Vitamin A_1 shows an absorption band at 620 mµ. This form of Vitamin is found in the liver of marine fishes and all land animals.

(b) *Vitamin A_2* : Vitamin A_2 is the second isomer of Vitamin A which shows an absorption band at 693 mµ when treated with $SbCl_3$. It has been isolated from natural sources and was synthesized by *Jones et. al.* (1951, 52). It is dehydrovitamin A_1. The most important source of vitamin A_2 is the liver of fresh water fishes. The Vitamin A_2 differs from A_1 in-

  (i) Having an additional unsaturated linkage in the ring, i.e., a double bond between carbons 3 and 4.

  (ii) The biological activity of Vitamin A_2 is approximately 40% of that of Vitamin A_1.

(c) *Neo-Vitamin A_1* : Neo-vitamin A_1 is a stereoisomer of vitamin A_1. The former is a cis-trans-form (2-Cisform), while the later (Vitamin A_1) all *trans*-form. This isomer is biologically active and the activity being about 70 to 80% of vitamin A_1. It has m.p. 59-60° and has been isolated from fish-liver oils by *Robeson et.al.* (1947).

Functions and Effects of Deficiency

Vitamin A is involved in the processes of vision, maintenance of epithelial tissues, and growth in general. It also plays a role in the construction of normal bones and teeth. The deficiency develops rather characteristic eye changes, viz., dryness (Xerophthalmia) and inflammation of the conjuctiva, ulceration, edema and opacity of the cornea (keratomalacia), eventuating in blindness. The other diseases, developed by the deficiency of vitamin A, are nyctalopia (night blindness), and hemeralopia (day blindness).

Deficiency in Man

American Medical Association (AMA) Council on Pharmacy has suggested the following roles of vitamin A in diseases caused due to its deficiency :

(a) Vitamin A is specific for cure and prevention of Xerophthalmia, nyctalopia, and hemeralopia.

(b) It is essential to the normal structure and behaviour of epithelial tissue, *e.g.*, epithelium covering the skin, and forming the lining of the nasal sinuses and respiratory tract, mouth, pharynx, entire digestive tract and the genito-urinary tract. It prevents dryness, scaliness, and roughness (follicular hyperkeratosis) of the skin.

(c) Vitamin A is a growth factor.

(*d*) It regulates the osteoblastic and Osteoclastic activity.

Effects of Excess of Vitamin A

Continued intake of excessive amounts of Vitamin A in children produces roughening of the skin, irritability, coarsening and falling of the hair, headache, loss of weight, vertigo, hyperesthesia, occasionally hepatomegaly, splenomegaly, hyperlipemia, hemorrhages, and certain characteristic skeleton changes.

In Eskimos, large amount of vitamin A resulted in drowsiness, sluggishness, severe headache, vomiting, and peeling of the skin about the mouth and elsewhere.

Vitamin A Requirement : The minimum daily requirement of vitamin A recommended is about 1.5 to 2.0 mg. For pregnant and nursing women it is slightly more, and during lactation period it must be about 3.0-3.2 mg. The children under one year require about 0.5 mg of Vitamin A daily.

2. Vitamin D or Calciferol

Occurrence: Vitamin D generally occurs along with Vitamin A. So, the chief sources of vitamin D are also Cod-liver oil, other fish liver oils, egg yolk, milk and butter. The ergosterol (Provitamin D_2) is found widely distributed in plants and yeast. It has also been found in animal species (snail, earthworm, chicken-egg, milk), perhaps of dietary origin. 7-Dehydro-cholesterol (provitamin D_3) is found in higher animals and man, and is believed to be formed from cholesterol. Mushrooms also contain a little amount of vitamin D.

VItamin D_2 (calciferol) is produced by irradiation of a steroid, ergosterol. Provitamin D_3 (7-dehydrocholesterol) can be converted by irradiation with ultraviolet light to vitamin D_3. The vitamin D content of milk can be increased by, (1) irradiation of the milk, (2) addition of calciferol or of fish liver oil concentrates, and (3) by feeding irradiated yeast to cows.

Chemistry: Vitamin D is used for curing rickets and is therefore known as "antirachitic" vitamin. At least 10 such substances are known which differ only in the hydrocarbon side chain. These antirachitic substances are designated as D_1, D_2, D_3, and so on. All the antirachitic vitamins are formed from corresponding provitamins which are cyclopentanophenanthrene derivatives. The provitamin D_2 (ergosterol) and provitamin D_3 (7-dehydrocholesterol) are most important. The former is found to occur in plant, while the later in animals.

All provitamins D possess following essential structural characteristics :

(1) -OH group at C-3;

(2) Two conjugated double bonds (between C-5 and C-6 and between C-7 and C-8);

(3) a hydrocarbon chain at C-17.

All D vitamins possess the empirical formulae similar to their corresponding provitamins, i.e., they are isomers. The inactive provitamins are converted photochemically into active vitamins by the ultraviolet rays present in sunlight or artificially by ultraviolet irradiations (275 to 300 mμ), or by α, β, γ, and other similar types of radiations. The photochemical activation process results into—

(1) Opening of the ring B between C-9 and C-10;

(2) Conversion of the methyl group (-CH_3) into a methylene (=CH_2) group at C-10; and

(3) Hydrogenation of C-9.

In this way an intramolecular rearrangement takes place without any oxidation.

Vitamin D and its provitamins are soluble in fat and fat-solvents. They are insoluble in water. Vitamin D is quite stable in crystalline form or in vegetable-oil solution; but it must be kept anaerobically in the dark. Vitamin D is resistant to heat and to oxidation (in neutral solution), and is not effected by acids and alkalies.

Functions and effects of deficiency : The chief function of vitamin D is to increase the

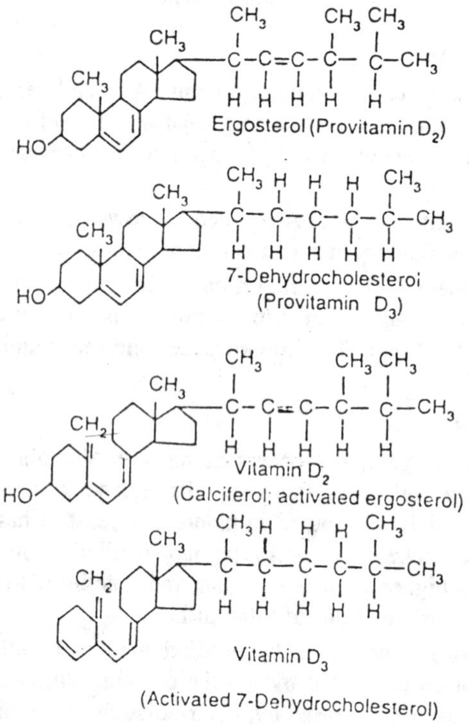

Fig. 7.2: Structure of Vitamin D₂ and D₃ & their provitamins.

availability and retention of calcium and phosphate and their utilization for proper mineralization of the skeleton. It is a well established fact that vitamin D deficiency is responsible for the production of **rickets** (Softening of bones) in children. The bones become soft and pliable leading to deformities. The characteristic skeleton manifestation deficiency in adults is a type of defective mineralization of osteoid tissues termed as "Osteomalacia". In osteomalacia the parts of the bone become softer than the rachitic bone and the ratio of calcium and phosphorous is also changed. The manner in which the adequate supply of vitamin D prevents these symptoms, is unknown. Other functions of the Vitamin D are :

1. Vitamin D promotes absorption of calcium in the intestine.
2. It increases the intestinal absorption of phosphate.
3. Vitamin D regulates the proper growth of the bones.
4. It promotes mineralization of the skeleton in both adults and growing organisms.
5. It controls certain characteristic effects of parathyroid hormone.
6. It lowers the pH in the colon, cecum, and distal ileum. The urinary pH increases simultaneously. This may increase the rate of absorption of calcium.
7. It counteracts the inhibitory effect of calcium ions on the hydrolysis of phytate (inositol hexaphosphate) which produces the rickets. The mechanism of action is unknown.

Effects of excess Vitamin D: Extremely large amounts (500 to 1000 times the normal requirement) of vitamin D produce a number of symptoms. These symptoms include anorexia, thirst, lassitude, constipation, and polyuria, followed later by nausea, vomiting, and diorrhea.

Requirement: The minimum daily requirement of Vitamin D recommended is about 0.025 mg.

3. Vitamin E or Tocopherol

Occurrence: Vitamin E is of wide occurrence in plants as well as in animals. α–Tocopherol is most abundant. The rich natural sources are vegetable fats; *e.g.*, wheat germ oil, cotton seed oil, corn oil, peanut oil, and other seed-germ oils. In considerable amounts it is found in all green plants; *e.g.*, lettuce, and alfalfa. Animal tissues contain relatively small amounts, *e.g.*, fish-liver oil, egg yolk, animal fats, milk of cow and human, and muscles of heart and kidney.

Chemistry: The vitamin E or "tocopherols" possess a hypothetical "tocol" nucleus. Like vitamin A and D, it occurs in more than one form. About seven naturally occurring tocopherols have been identified. Of the several tocopherols (α–, β–, γ–, δ–tocopherol), α–tocopherol is most important because it is most stable, potent, and commercially available. The known tocopherols differ from one another in the number or position or both of methyl groups on the chroman portion (ring) of the "tocol" nucleus (Fig. 7.3). They also contain the phenolic hydroxyl group which permits the formation of esters.

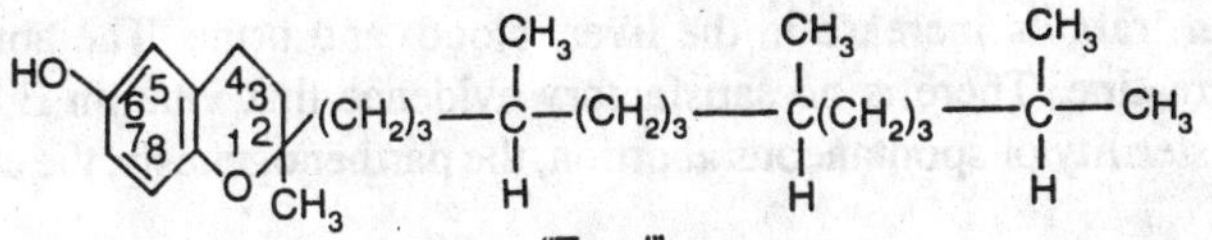

"Tocol"

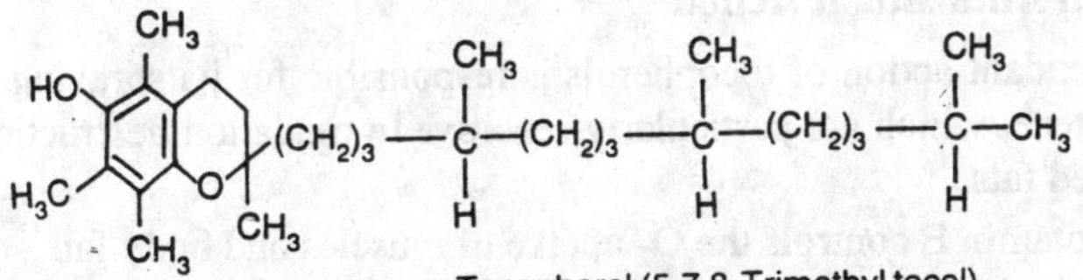

α-Tocopherol (5,7,8-Trimethyl tocol)

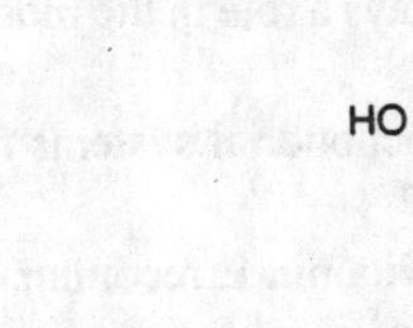

β-Tocopherol
(5,8-Dimethyltocol)

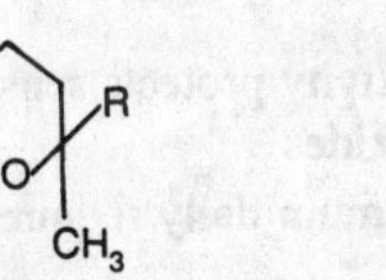

γ-Tocopherol
(7-8-Dimethyltocol)

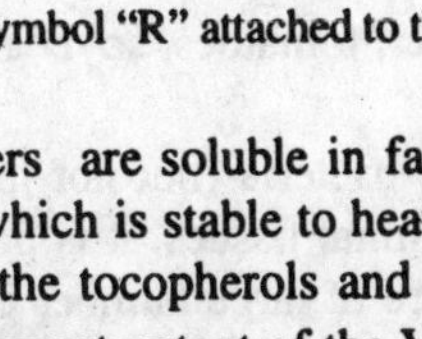

δ-Tocopherol
(8-Methyl tocol)

Fig. 7.3: Vitamin E (tocopherols). The symbol "R" attached to the ring of tocol nucleus refers to the remainder of the molecule.

Free tocopherols and their esters are soluble in fat and fat-solvents and insoluble in water. Vitamin -E is a light yellow liquid which is stable to heat and acids. The antioxidant activity is the most striking chemical property of the tocopherols and is due to the phenolic hydroxyl group at C-6 in the ring. δ–Tocopherol is the most potent of the Vitamin E in antioxidant activity, followed in order by γ–, β–, and α–tocopherols.

Functions and effects of deficiency: Vitamin E was discovered by *Evans* and *Sure* in 1922 independently. With the discovery of its properties, the vitamin was designated as "Fat-soluble vitamin E", and the "antisterility" or "fertility" vitamin.

Most of the information about function and deficiency are based on the experiments with certain rodent species, *e.g.*, rat, rabbit, guinea pig, and chicks. They may be arranged in two categories :

(1) gonadal and reproductive function; (2) muscle metabolism and structure.

The characteristic symptoms of vitamin E deficiency vary with the animal species. In the female rat, vitamin deficiency does not affect the ovary. However, the foetus does not develop normally, dying in uterus and undergoing reabsorption. In male rates and guinea pigs the deficiency or vitamin E results in permanent sterility.

The most common mainfestations of vitamin E deficiency in a number of species (*e.g.*, rat, rabbit, hamster, young quinea pig, lamb, calf and duckling) is muscle dystrophy, a degenerative change in the skeleton muscles, leading to necrosis, edema, inflammation, and fibrosis. This results in either weakness or paralysis. The deficiency of vitamin E in chickens results in vascular abnormalities and in rabbits increase in the liver, blood, and urine. The abnormalities in rabbits are due to loss of creatine. There is no satisfactory evidence that vitamin E deficiency is a factor in the production of sterility or spontaneous abortion, the parthenogenesis, the etiology, and muscular dystrophies in man.

Functions Based on Antioxidant Action

1. The antioxidant action of tocopherols is responsible for its spraying action on vitamin A and carotene, which are particularly sensitive to oxidation destruction in the presence of unsaturated fats.
2. In vitro, vitamin E controls the O_2 uptake of muscles and body fat.
3. Tocopherols prevent the peroxidation and brown pigmentation of the adipose tissues due to highly unsaturated fatty-acids in rats and chicks.
4. In rats, the tocopherols prevent the massive hepatic necrosis, caused by feeding certain brands of yeast.
5. Vitamin E also prevents the liver injury induced by feeding excessive amounts of Cod-liver oil.
6. In chicks, the administration of tocopherols prevents the development of "encephalomalacia (cerebellar disorder)" and an "exudative diathesis".
7. Tocopherolactone, a metabolite of tocopherol, plays a role in the biosynthesis of coenzyme Q (ubiquinone).
8. The tocopherols' activity protects sensitive mitochondrial systems from irreversible inhibition by lipid peroxides.

Requirement: The minimum daily requirement of vitamin E recommended is about 5 mg.

4. Vitamin K

Occurrence: Vitamin K occurs in two well known forms, vitamin K_1 and K_2. Vitamin K_1 is found in plant kingdom and is present chiefly in green leafy tissues, *e.g.*, alfalfa and spinach leaves. It has also been extracted from cauliflower, tomatoes, soybeans, cabbage, Kale, rice bran, and Oat shoots.

Vitamin K_2 is found in most of the bacteria (but not in yeasts, molds, or fungi) where it is produced as a result of metabolism. In animal tissues, it is present in large amounts in putrid fish meal. This large amount is due to presence of large number of bacteria and their luxuriant growth in this material. In small amounts it is also found in milk and yolk of the egg due to the presence of intestinal bacteria.

Vitamins

Chemistry: Vitamins K is the most recently recognized member of the group of fat-soluble vitamins. The term vitamin K was applied to the missing factor which was later (1939) identified as a naphtoquinone. The factor was discovered by *Dam* in 1934 which was responsible for the hemorrhagic disease in chicks (Slowness of blood Clotting). Because this disease was cured by vitamin K it was named as coagulation vitamin (Vitamin K).

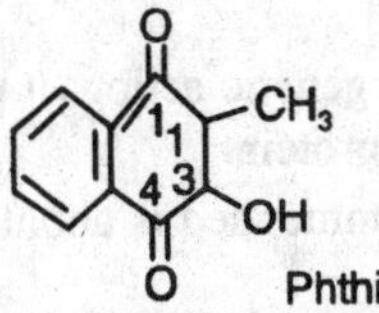

Phthiocol
(2-Methyl-1, 4,-naphthoquinone)

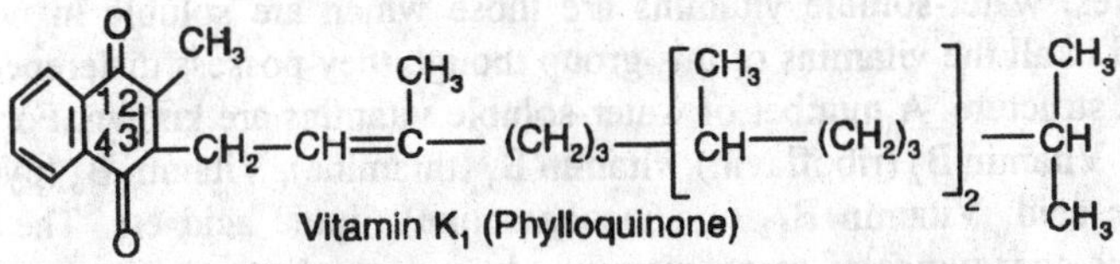

Vitamin K_1 (Phylloquinone)
(2-Methyl-3-phytyl-1, 4-Naphthoquinone)

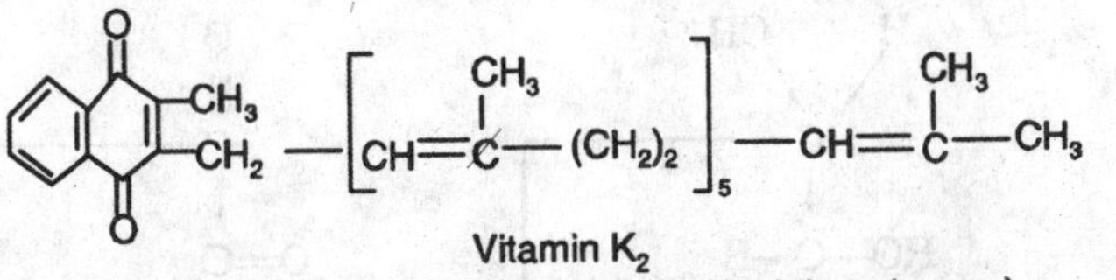

Vitamin K_2
(2-Methyl-3-Difarnesyl-1,4-Naphthoquionone)

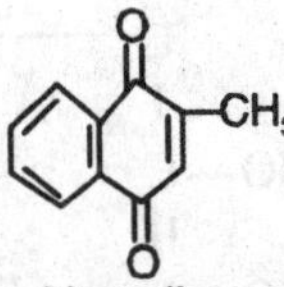

Menadione

Fig. 7.4: Vitamins K and Menadione.

Vitamin K and other several natural and synthetic substances which have the same activity as vitamins K, are **naphthoquinones**. The antihemorrhagic activity of these substances is due to the formation of menadione (2-methyl-1,4-naphthoquinone) which ultimately forms vitamin K_2 by alkylation with digeranyl pyrophosphate.

Vitamin K_1 (phylloquinone) possesses a phytyl chain attached at position 3 of the menadione nucleus (2-methyl-3 phytyl-1, 4-naphthoquinone), while vitamin K_2 (farnoquinone) contains a longer, difarnesyl chain attached at position 3(2-methyl-3-difarnesyl-1, 4-naphthoquinone).

The activity of the vitamin is related to the presence of the methyl group at the 2 position in the quinonoid ring. When the 2-methyl group is substituted by other alkyl radicals or by hydrogen, its activity is decreased.

Vitamin K is insoluble in water and quite soluble in most fat-solvents. It is stable to heat, but is readily destroyed by light, alkali, and alcohol. It also possesses characteristic ultraviolet absorption spectrum which helps in its identification. Due to sensitivity to light, the vitamin is kept in dark bottles. Researches have shown that blood-clotting defect in hemorrhagic diseased persons is due to deficiency in "Prothrombin", a compound essential for blood clotting. Vitamin K is required for the synthesis of this compound.

Functions: The chief functions of vitamin K are the formation of

(1) proconvertin, (2) plasma thromboplastin, (3) Stuart's factor, and (4) prothrombin by hepatic cells.

These functions have been observed in animals and human beings, but the exact mechanism of action is unknown. In plants also it performs some most important functions. Some of them are:

(1) Cell-metabolism, (2) It participates in electron transfer in the oxidative chain, (3) Vitamins K is associated with phosphorylations, (4) It plays an important role in the coupling mechanisms.

Deficiency: The main symptom of vitamin K deficiency is enormous bleeding from minor wounds and slight bruises changing into extensive sub-cutaneous hemorrages. It causes blood-clotting defect i.e., the blood clots vary slowly. The low plasmaprothrombin activity which occurs consistently in the new-born, during the first few days of life, is attributed to vitamin K deficiency. Deficiency utilization may result in hepatocellular damage, the liver cells being unable to synthesize prothrombin despite of an adequate supply of vitamin K.

Recent researches have shown that vitamin K has a genetic action. It plays an important role in the formation of RNA and synthesis of blood-clotting protein.

The daily minimum requirement of vitamin K recommended is about 0.001 mg.

2. Water-Soluble Vitamins

As the name indicates, water-soluble vitamins are those which are soluble in water. This common property is found in all the vitamins of this group though they possess differences in their chemical composition and structure. A number of water-soluble vitamins are known. For example, Vitamin C (ascorbic acid), Vitamin B_2 (riboflavin), vitamin B_1 (thiamine), Vitamin B_6 (pyridoxine), niacin, biotin, pantothenic acid, Vitamin B_{12} (cyanocobalamine), lipoic acid etc. The structure, chemistry and functions, as coenzymes of many vitamins, have been discussed in the chapter on Coenzymes.

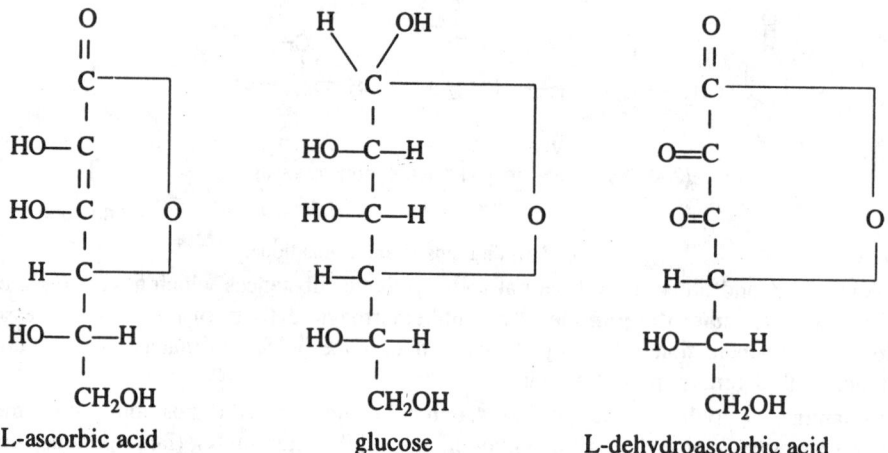

Fig. 7.5: Vitamin C, L-glucose, L-dehydro-ascorbic acid.

1. Vitamin C (Ascorbic acid)

Occurrence: Amla (*Phyllanthus emblica*), citrus fruits and tomatoes are the best sources of vitamin C. Other sources of vitamin C are leafy vegetables, green peas, beans, peppers, potatoes, turnips, and other fruits (oranges, lemons, limes, grapefruits, strawberries, banana etc.). Cow's milk, human milk and liver are the animal sources. Thus, the vitamin C is distributed widely throughout the plant and animal kingdom.

Chemistry: Vitamin C, also known as ascorbic acid, was isolated by *Szent-Gyorgyi* in 1928. It was the first discovered vitamin which was recognized as early as in sixteenth century, but its exact presence, nature, chemical structure, and antiscorbutic properties were later on confirmed in the nineteenth century.

Chemically ascorbic acid is a derivative of monosaccharide, L-glucose. The naturally occurring vitamin C is L-ascorbic acid. Its D-forms are generally inactive as antiscorbutic agents.

Vitamin C is soluble in water and insoluble in fat-solvents. Its most important and prominent chemical property is its strong reducing activity. It is oxidized to dehydroascorbic acid by a number of agents like air, ferricyanide, silver nitrate, $FeCl_3$, H_2O_2, iodine, quinones, methylene blue,

2,6-dichlorophenol, indophenol etc. It is easily destroyed by alkalies, but stable in weak acid solutions. Its sensitivity for oxidation increases in presence of silver and cupric ions.

Functions: The exact biochemical role of vitamin C is not clearly understood : (1) Its sensitivity to reversible oxidation suggests its role in cellular oxidation-reduction reactions and perhaps it serves as a hydrogen transport agent. (2) Vitamin C plays a key role in the biosynthesis of serotonin and especially in the hydroxylation of tryptophan to 5-hydroxytryptophan. (3) It is also involved in the conversion of pteroylglutamic (folic) acid to the active formyl tetrahydrofolic derivative, and in the mobilization of iron from its storage form, ferritin. (4) Vitamin C is required for functional activities of fibroblasts and osteoblasts, and consequently for formation of collagen fibers and mucopolysaccharides of connective tissue, osteoid tissue, dentin, and the intercellular, "cement substance" of the capillaries.

Deficiency: In animals, ascorbic acid deficiency results in a disease, known as "Scurvy". The "Scurvy" results from failure of certain specialized cells, i.e., fibroblasts, osteoblasts, odontoblasts, which promote normal deposition of collagen, osteoid and dentin. The animals deficient of vitamin C assume the "Scurvy position", lying flat with the hind legs extended. The joints become swollen, tender and loose with constant pain in them. The gums start bleeding and teeth become loose which ultimately fall down later on. In this case the formation of ground substance (mucopolysaccharides) is also impaired.

The daily requirement of vitamin C recommended, varies from children to adults to pregnant women. The official recommended minimum daily requirements are : Children, 30 mg. (infants under one year) to 80 mg. (adolescence); adults, 75 mg.; pregnant women, 100 mg.; during lactation, 150 mg.

2. Vitamin B_1 or thiamine

(For structure and function see chapter on Coenzymes)

Occurrence: Thiamine is widely distributed throughout the plant kingdom. In highest concentrations it is found in seeds, but also being present in the leaf, root, stem, and fruit. In cereals, it is concentrated in the outer germ and bran layers *e.g.* rice and wheat. The good sources of vitamin are peas, beans, whole cereal grains, bran, nuts, prunes, gooseberries, killed yeast.

Thiamine is also found in animal tissues, *e.g.*, egg yolk, liver, milk, ham, and pork.

Deficiency: Thiamine deficiency results in a disease known as "beriberi". This is characterized by cardiovascular and neurological manifestations, and, in some cases, edema ("Wet" beriberi).

The daily requirement of vitamin B_1 recommended for children ranges from 0.4 mg for infants to 1.3 mg for Pre-adolescents (10 to 12 years); for adults 1.0 mg to 1.5 mg.

3. Vitamin B_2 (Riboflavin)

(For chemistry, function, and deficiency see chapter on Coenzymes)

Occurrence: Riboflavin is distributed widely in all plants and animal cells. In tissues, it occurs as coenzyme, FMN and FAD. Both are nucleotides. High concentrations of this vitamin occur in yeast and fermenting bacteria. Appreciable amounts are present in liver, kidney, crab meat, whole grains, nuts, milk, eggs, dry beans and peas, meat, and green leafy vegetables.

Requirement: The recommended daily intake is as follows : adults, 1.5 to 1.8 mg. depending on weight; women in the latter half of pregnancy, 2.0 mg; during lactation, 2.5 mg; infants, 0.6 mg; children, 1.0 to 1.8 mg. and 2 to 2.5 mg. during aldolescence.

4. Niacin (Nicotinic Acid)

(For occurrence, Chemistry, and function seechapter on Coenzyme)

Deficiency: Nicotinic acid in plants and animal tissues is found in the form of its coenzyme nucleotides–NAD and NADP. It may also be found in free form. Its deficiency causes defects in skin, gastrointestinal and cerebral troubles. The skin of the persons, feeding on deficient diet of this vitamin, becomes red and later on brown, thickened, and scaly. Gastrointestinal manifestations include : nausea, vomiting, abdominal pain with alternating constipation and diarrhoea, thickening and inflammation of the colon.

Cerebral manifestations include : headache, irritability, forgetfullness, confusion, depression anxiety and other mental symptoms: The general defects are: inadequate growth (children), loss of weight and strength, anemia, dehydration and diarrhoea.

The recommended daily intake is as follows : adults : 17 to 21 mg; children range from 6 mg. for infants to 17 mg. for pre-adolescents (10 to 12 years).

5. Vitamin B_6 (Pyridoxine)

The term 'Vitamin B_6' is used as a group designation for naturally occurring pyridine derivatives possessing B_6 activity. These derivatives are pyridoxol, (-CH$_2$OH) pyridoxal (-CHO), and pyridoxamine (-CH$_2$NH$_2$). It functions as a coenzyme (pyridoxal phosphate) in certain important reactions in the intermediary metabolism of amino-acids-transamination and decarboxylation reactions. The chief sources of vitamin B_6 are liver, egg, meat, and cereal grains. It has also been isolated from rice cover-bran.

Deficiency: The general manifestations of vitamin B_6 deficiency are inadequate growth or failure to maintain weight, anemia, leukopenia, skin lesions, nervous system symptoms. All of these do not occur in every species. In rats its deficiency causes dermatitis (acrodynia).

The daily requirement of vitamin B_6 recommended is about 2.0 mg.

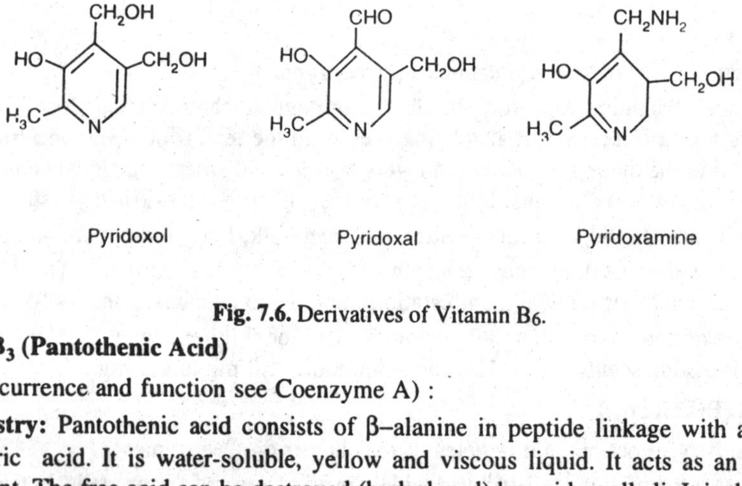

Fig. 7.6. Derivatives of Vitamin B$_6$.

6. Vitamin B_3 (Pantothenic Acid)

(For occurrence and function see Coenzyme A) :

Chemistry: Pantothenic acid consists of β–alanine in peptide linkage with a dihydroxydimethyl butyric acid. It is water-soluble, yellow and viscous liquid. It acts as an oxidizing and reducing agent. The free acid can be destroyed (hydrolyzed) by acid or alkali. It is thermolabile. In nature it is found as coenzyme A which plays an important role in carbohydrate and fat-metabolism.

Deficiency: Pantothenic acid deficient diet when supplied to rats showed poor growth, dermatitis, and graying of hairs, decreased reproductive capacity, scaling of the paws and tail and the loss of hair. It also develops anemia. Hemorrhages occur beneath the skin and in the kidneys and adrenal cortex.

The daily requirement of pantothenic acid is about 3-5 mg.

7. Vitamin H. (Biotin)

(For occurrence, chemistry and function see chapter on Coenzymes)

Deficiency: The chief manifestations of biotin-deficiency are dermatitis (rat, pig, fowl); spectacleeyed appearance in rats; thinning or loss of fur (alopecia) (Mouse pig, and monkey); paralysis (dog, cow, rat); graying of black or brown fur (mouse, monkey). Vitamin H was detected

Vitamins

from all parts of higher plants by Bonner and Bonner in 1948. The daily requirement is about 100 to 300 mg.

8. Vitamin B_{12} (Cobalamine) : See Chapter on Coenzymes.
9. Lipoic Acid: See chapter on Coenzymes.
10. Folic Acid : See chapter on Coenzymes.

Table 7.1 : Various Vitamins; their sources, principal functions and daily requirements; and diseases caused by their deficiency

S.No.	Vitamin	Chemical name	Important Sources of Vitamin	Principal metabolic functions	Disease caused due to Vitamin deficiency	Daily requirement
I FAT-SOLUBLE VITAMINS						
1.	Vitamin $A(A_1)$ A_2	Retinol (all transform) Dehydro-vitamin A1	Whole milk, butter, egg yolk, Cod-liver-oil. Vitamin A_1- livers, of marine fishes. Vitamin A_2- livers of fresh water fishes.	Vitamin A helps in the process of vision; maintain epithelial tissues and growth; Construct normal bones and teeth.	Xerophthalmia, nyctalopia, hemeralopia (day blindness), follicular hyperkeratosis.	Children, 1 year 0.5 mg; 10-12; adults; 1.5-2.0 mg. Women L.P. (lactation period) 3.0–3.2 mg
	Neo-Vitamin A_1	Stereoisomer of vitamin A1 (2-cis-form)	Fish liver oil.			
2.	Vitamin D_2	Calciferol (Ergo).	Widely distributed in plants & yeasts–but also in milk, egg, earthworm, snail.	To increase the availability and retention of calcium and phosphate and their utilization for proper mineralization of the Skeleton; regulates growth of bones and effects of parathyroid hormone; toxic in large amount.	Rickets (softening of bones) in children, and Osteomalacia.	0.025 mg. (general)
	D_3	Cholecalciferol	In animals– Cod-liver oil, egg yolk, milk, and butter.			
3.	Vitamin E	α–Tocopherol	Richest sources are vegetable oils, e.g., Wheat germ oil, Cotton Seed oil, Corn oil etc.	Antioxidant, protects Vitamin A and unsaturated fatty-acids.	Necrosis, edema, fibrosis which results in "Paralysis".	5.00 mg. (general)

#	Vitamin	Chemical Name	Sources	Functions	Deficiency	Daily Requirement
4.	Vitamin K_1	Phylloquinone	Green leafy tissues; e.g., alfalfa and spinach-leaves.	In animals (i) formation of proconvertin (ii) plasma thromboplastin, (iii) Stuart's factor (iv) Prothrombin.	Delayed blood clotting.	0.001 mg. (general).
	K_2	Farnoquinone	Most of the bacteria, putrid fish meal.	In plants (i) Cell metabolism, (ii) electrontransfer in oxidative chain (iii) Phosphorylation (iv) Coupling mechanisms.		

II. WATER-SOLUBLE VITAMINS

#	Vitamin	Chemical Name	Sources	Functions	Deficiency	Daily Requirement
1.	Vitamin B_1	Thiamine	Best sources are seeds of plants, e.g., pea, beans, rice bran, nuts, prunes. In animals-egg, liver, milk, pork, yeast.	Decarboxylation of -Keto acids, Oxidation of ketoacids; transketolation phosphoketolation.	Beriberi (Dry and Wet).	Children 0.4 mg, Adults 1.3 mg, Women L.P. 1–1.5 mg.
2.	Vitamin B_2 Complex.	Riboflavin	Yeast and fermenting bacteria, liver, kidney enriched grains, milk, egg, green leafy vegetables.	Coenzymes serves as hydrogen carriers in redox systems.	Unknown (but according to few dermatitis, inflammation of the cornea).	Children 0.6 mg, adults 1.5-1.8 mg, women L.P. 2.5 mg.
		Niacin (Nicotinic-acid, Niacinamide).	Germ and pericarp in cereal grains, yeast, meat, enriched grains liver, kidney.	Hydrogen acceptor in redox reactions (Coenzyme); oxidation to produce ATP; biosynthesis of fatty acids, steroids; B-oxidation, in TCA cycle, glycolysis.	Pellagra, inadequate growth, anemia, dehydration, diarrhea.	Children 6 mg; adults 17 mg; women L.P. 17-21 mg.
		Folic Acid	Green leafy vegetables, yeast, liver.	Transfer of 1-carbon fragments (formyl), biosynthesis of purines, methionine, choline, pyridines etc.	Megaloblastic anemia.	1-2 mg (Details unknown)

Vitamins

		Pantothenic acid	Egg, liver, yeast, kidney, wheat and rice bran, legumes, milk.	Acylation reactions (acetyl group transfer)-Carbohydrate and fat metabolism.	Burning foot syndrome, dermatitis, greying of hairs in rats, poor growth.	Unknown (approx. 3-5 mg).
3.	Vitamin B_6	Pyridoxine	Liver, egg, meat, cereal grains.	As Coenzyme in amino acid metabolism, transamination, decarboxylation, transulfuration, tryptophansynthestase, amino-acid transport.	Inadequate growth, anemia, leukopenia, skin lesions.	2.00 mg.
4.	Vitamin B_{12}	Cobalamine	Kidney, liver, pancrease, egg, milk, meat, fish.	Synthesis and transfer of methyl groups—Synthesis of methionine, pruines, choline, succinyl CoA.	Pernicious anemia.	0.001 mg.
5.	Vitamin C	Ascorbic acid	Citrus fruits, tomatoes, strawberries, banana, grape fruits.	Serves as hydrogen transport agent, synthesis of serotonin, mycopoly-saccharides, and intracellular cement substance formation, iron absorption.	Scurvy	Children 1 year, 30 mg; adults 10-12 yr, 75 mg; Women-lactation period; 150 mg.
6.	Vitamin H	Biotin	Liver, kidney, egg yolk, yeast, milk and molasses.	Carboxylation and transcarboxylation.	Unknown in man, Dermatitis (rat, pig fowl) paralysis (dog, cow, rat) graying of hairs (mouse, monkey)	0.25 mg.

8

Plant Growth Substances

8.1. INTRODUCTION

The substances which are required in very small amounts for the growth and development of plants are known as **growth regulating substances**. They have been variously called as hormones, phytohormones and growth promoting substances. They are organic compounds which are synthesized at one part of the plant body especially at the tips of stem and leaf and from there exert their effect. The hormones are neither enzymes nor nutrient food materials.

The term "hormone" was used by *Bayliss* and *Starling* in 1904 for the first time which meant 'Chemical-massengers' within organisms. *Pincus* and *Thimann* (1948) defined plant hormone as an organic substance produced naturally in the higher plants, that controlled growth or other physiological functions in minute amounts, at a site remote from its place of production.

Van Overbeek et al. (1954) defined phytohormones as organic compounds, other than nutrients, which in small amounts promote, inhibit or otherwise modify any physiological process in plants. Plant hormones are of the following types:

1. Auxins
2. Gibberellins
3. Cytokinins
4. Abscisic acid
5. Ethylene
6. Morphactins

8.2. AUXINS

Auxins are one of the most important groups of plant growth hormones because they regulate many physiological processes in plants. They are the oldest known among the plant hormones and have been discovered by many scientists including *Charles Darwin* (1880), *Boysen-Jensen* (1910), *Paal* (1919), *Stark* (1921), but practically the credit goes to *F.W. Went* (1928), of the University of Utrecht, Holland, who isolated and confirmed the presence of such hormones from *Avena* coleoptile.

Auxin is found in the meristematic regions of the plants, *e.g.*, in coleoptile tips, in buds, in the growing tips of roots and leaves. Most of our knowledge of auxins is based on experiments done on *Avena* coleoptile.

8.2.1. Chemical nature of Auxins

Chemically the auxin is indole 3-acetic acid (IAA). Several auxins have been isolated in pure form. *Kogl* and *Haagen-smit* (1931) isolated the active compound of molecular weight 328 from human urine which was called as Auxin-A (auxanotriolic acid). Later on in 1934, a similar active compound was isolated from malt and corn-grain oil with molecular weight 372 and was named as Auxin-B (auxanolonic acid). Indole-3-acetic acid is the only true natural auxin of the higher plants. In addition to IAA, there are other auxins with identical properties like Indole butyric acid (IBA), and α-naphthalene acetic acid (NAA), 2,4-dichlorophenoxy acetic acid (2,4-D), 2,4,5-trichlorophenoxyacetic acid (2,4,5-T) etc.

Plant Growth Substances

Sometimes complexes of various types occur in plant cell which include Indole 3-acetyl aspartic acid, a number of glycosides, and glycobrassicin, that occur naturally. The structures of different compounds with auxin activity are shown in Fig. 8.1.

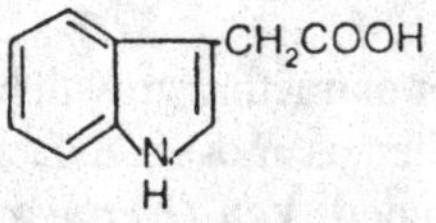

Indole acetic acid (IAA)

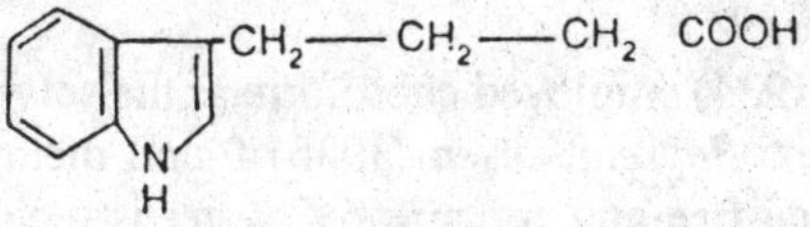

α-Indole butyric acid (IBA)

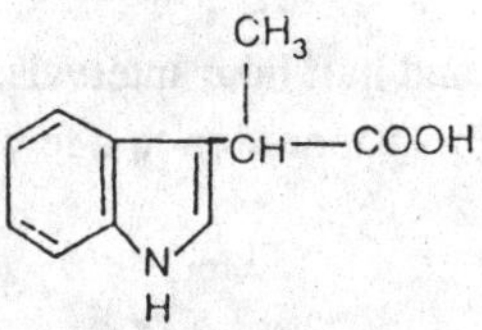

α-Indole propionic acid (IPA)

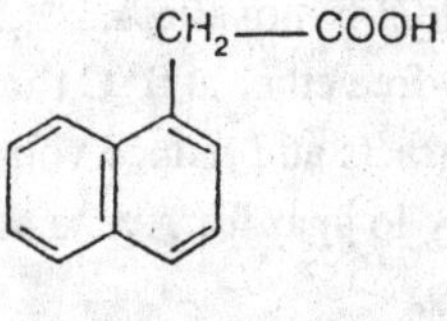

a-Naphthalene acetic acid (α-NAA)

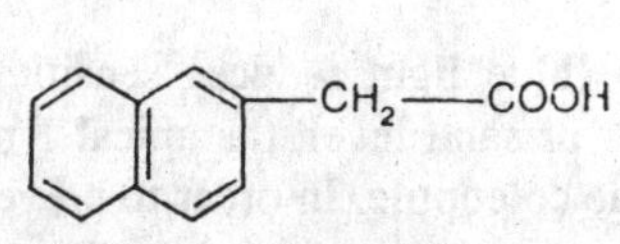

β-Naphthalene acetic acid (β-NAA)

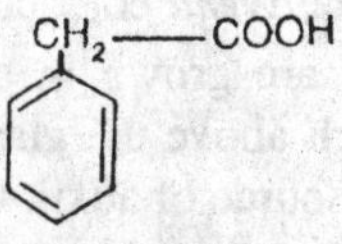

Phenyl acetic acid (PAA)

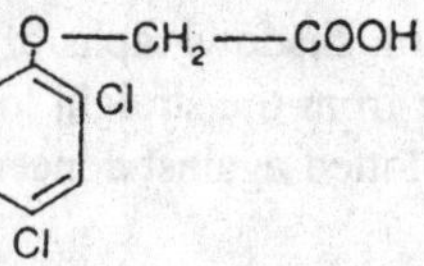

2,4-Dichlorophenoxy acetic acid (2, 4-D)

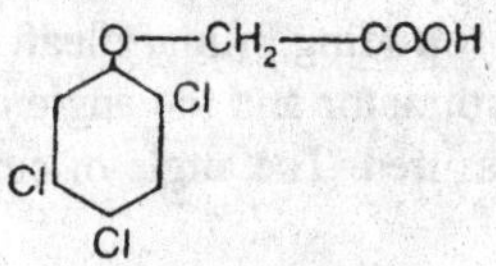

2, 4, 5-Trichlorophenoxy acetic acid (2, 4, 5-T)

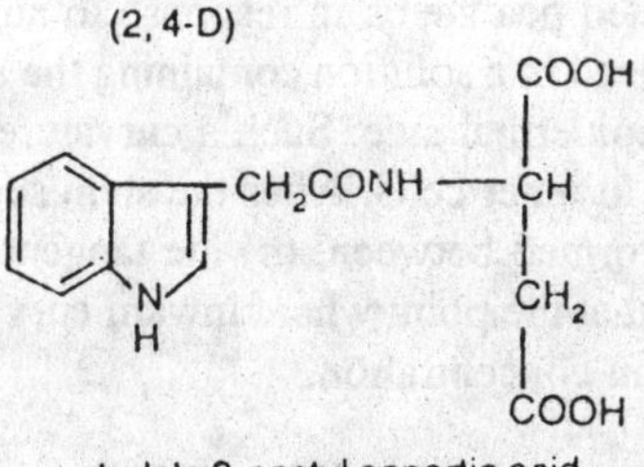

Indole 3-acetyl aspartic acid

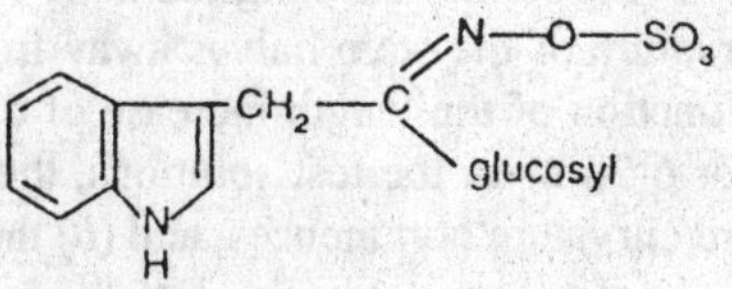

Glucobrassicin

Fig. 8.1: Structure of some Auxins.

8.2.2. Extraction of Auxins

The following two methods are used to extract auxins from the plant materials:
(a) Diffusion
(b) Solvent Extraction

Diffusion:

The simplest method to obtain a growth hormone from plant material is by diffusion into agar. The organ to be tested for auxin activity is placed on an agar block (1.5.% agar) for a period of an hour under conditions which do not permit transpiration. The auxin so obtained by diffusion in agar block is bio-assayed by the *Avena* curvature test. Excessive transpiration from the test organ and enzymic destruction of the growth hormone at the cut-end create difficulties in this method.

Solvent Extraction

Thimann (1934) employed choroform as the solvent for extracting growth hormone from plant tissues. However, *Boysen-Jensen* (1936) found diethyl ether as most satisfactory. Diethyl ether which is peroxide-free and contains 5% water is recommended. *Van Overbeek et al.* (1945) have recommended the following simple technique for obtaining free auxin:

1. Freeze plant material on CO_2 ice or liquid nitrogen.
2. Slice bulky tissues into 2-5 mm slices.
3. Extract with peroxide-free ether at $0°$ C for two and half hour intervals.
4. Combine the ether extracts and reduce volume by evaporation to a few ml.
5. Transfer quantitatively to agar for Avena assay.

8.2.3. Bioassays for Auxins

The Avena Test

F.W. Went (1928) used it for the first time. The physiological basis of this test lies in the strict polar transport of auxins in the *Avena* coleoptile.

Dehusked oat seedlings are grown in darkness (blue light reduces sensitivity). When the coleoptile is 15 to 30 mm high above the glass holder or sand level, the apical 1 mm is removed in order to cut off the natural source of auxin within the coleoptile. In order to prevent the renewed formation of auxin during the test period, a second decapitation is carried on three hours after the first. The primary leaf is pulled out approximately half-way out. It is then cut about one quarter inch above the coleoptile tip. Then the agar block containing the auxin is placed on one side of the coleoptile tip against the protruding primary leaf. After 90 min. shadowgraphs of the curvature are taken with the help of a protractor and the angle of curvature from the straight lower region to the tip of the coleoptile is measured. The angle of curvature is plotted against concentration.

The Slit Pea Test

It was also originally used by *F.W. Went* (1934). The physiological basis for this test lies in the differential growth of the epidermal cells of etiolated pea stems in response to auxin. A piece of actively growing stem is slit longitudinally and is placed in a solution containing the auxin, which causes a curvature of the stem halves away from the epidermal side. Such a curvature of the stem halves is a function of the length increase of the outer to inner cells. After the stem segments have been kept for 6-24 hr in the test solutions, the angle formed between, (*a*) the tangent at the point where inward curvature commences, and (*b*) the tangent at the point where inward curvature ceases, is noted. The angle of curvature is plotted against auxin concentration.

Straight Growth Test

The physiological basis for straight-growth test is the simple stimulation of straight growth by auxins. Pea internodes of standard length are floated on test solutions for 6-24. hr. The increase in length is measured and is plotted against auxin concentration.

Pea Root Test

The physiological basis for this test is essentially the same as for the other straight-growth. The roots are extremely sensitive to auxins. The linear growth is inhibited quantitatively. Pea and cress roots have been used for this test.

8.2.4. Biosynthesis of Indole Auxins

Tryptophan as the precursor of indole auxins : The amino acid tryptophan is a precursor

Plant Growth Substances

of IAA as it has a close chemical similarity with it and is presumably present in all cells. Free tryptophan levels in leaves are much higher than those of IAA. However, free tryptophan may show compartmentation in tissues.

The evidences of the various pathways of IAA biosynthesis are: (*i*) The presence of intermediates as native compounds, (*ii*) The biological activity of intermediates, (*iii*) In vivo interconversion of the intermediates, and (*iv*) isolation of the necessary enzyme systems.

Based on these evidences the following pathways have been established.

1. The Indole-pyruvic acid Pathway

Indole-pyruvic acid is unstable and its isolation as a native compound from plant tissues has not been possible. However, there is good evidence for the occurrence of indole-3-acetaldehyde in sterile pea shoots (*Rajgopal*, 1968) and in cucumber seedlings (*W.K. Purves*).

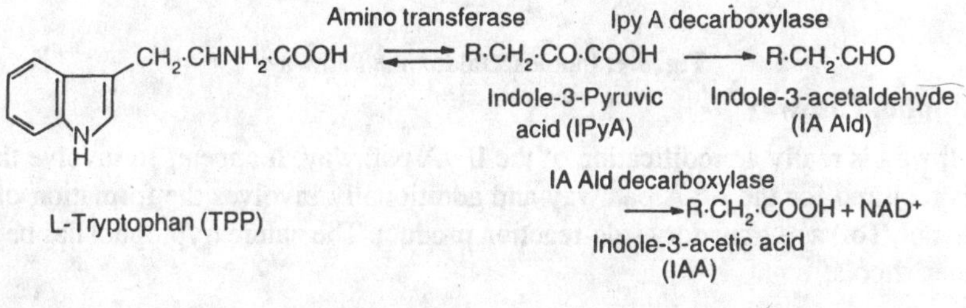

Fig. 8.2: Indole-pyruvic acid Pathway.

The conversion of (^{14}C) TPP to (^{14}C) IPyA, (^{14}C) IA Ald and (^{14}C) IAA has been demonstrated in cell-free extracts of mungbean seedlings (*Wightman* and *Cohen*, 1968).

2. The Tryptamine Pathway

Tryptamine occurs sporadically in higher plants. It was first isolated from *Acacia* and has since been found in several other species. However, it is not common in pea, squash, cabbage etc. Tryptamine is active after a lag period in the Avena curvature test. The tryptamine pathway is as under:

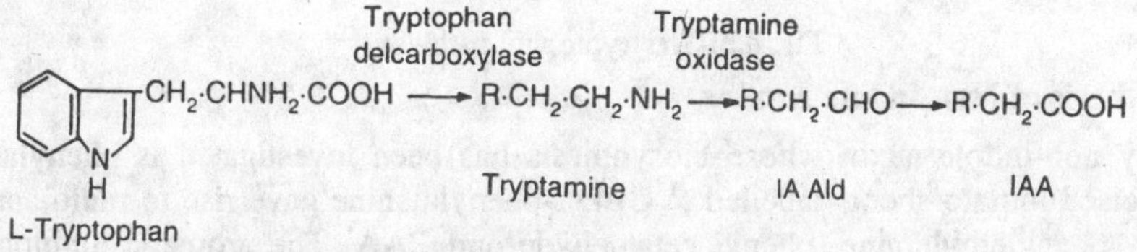

Fig. 8.3: The tryptamine pathway.

3. The Idoleacetaldoxime Pathway

This pathway is characteristic of the family *Cruciferae*. The *Brassica* plants probably contain small amounts of indole-3-acetonitrile (IAN) which is also obtained by the breakdown of glucobrassicin by the enzyme *myrosinase*. Glucobrassicin is active in the oat and wheat coleoptile tests. The vivo experiments have shown that cabbages can convert (^{14}C) TPP to IAOx to IAN and to Glucobrassicin. The enzyme responsible for converting TPP to IAOx in vivo has not yet been isolated, although this reaction can be catalysed in vitro by horse radish peroxidase. *Indole-3-acetaldoxime hydrolase* converts IAOx to IAN, and *nitrilase* converts IAN to IAA.

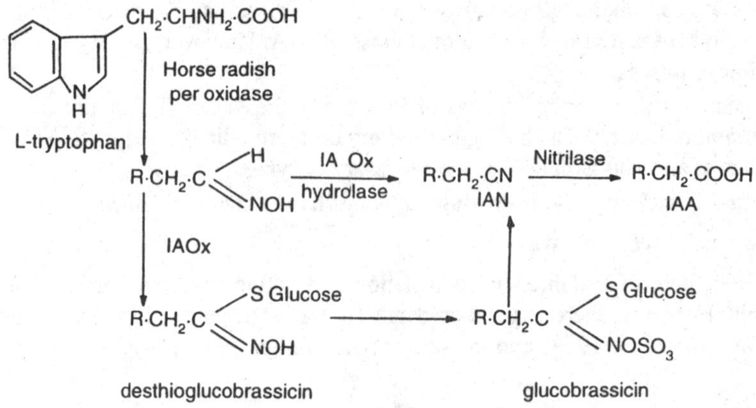

Fig. 8.4: Indoleacetaldoxime Pathway

4. The Tryptophol Pathway

This pathway is really a modification of the IPyA pathway. It appears to involve the two main intermediates required for the IPyA pathway and additionally involves the formation of tryptophol (indole-3-ethanol, Tol) as a transitory side-reaction product. The nature tryptophol has been extracted from cucumber shoots.

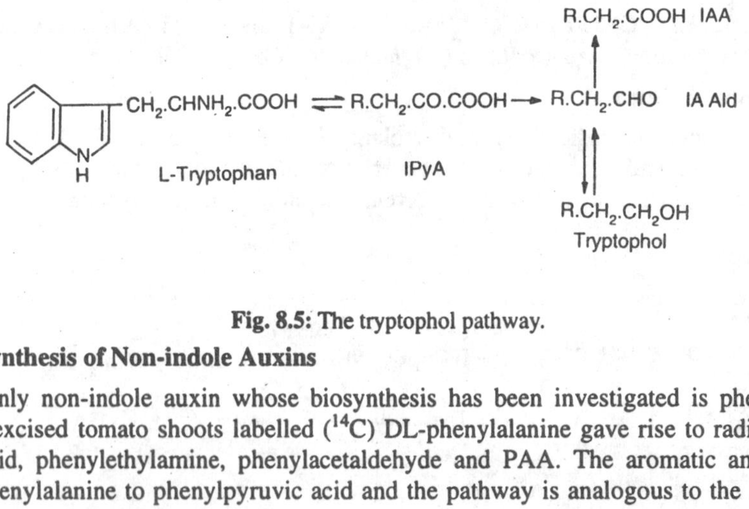

Fig. 8.5: The tryptophol pathway.

8.2.5. Biosynthesis of Non-indole Auxins

The only non-indole auxin whose biosynthesis has been investigated is phenylacetic acid (PAA). In excised tomato shoots labelled (^{14}C) DL-phenylalanine gave rise to radio-active phenlpyruvic acid, phenylethylamine, phenylacetaldehyde and PAA. The aromatic aminotransferase converts phenylalanine to phenylpyruvic acid and the pathway is analogous to the indolepyruvate and tryptamine route of IAA synthesis.

8.2.6. Mechanism of Auxin action

The following theory have been proposed to explain the mechanism of auxin action:

Molecular Reaction theory

The main supporters of this theory are *Skoog et al.* (1942) and *Foster et. al.* (1952). Skoog et al. postulated that auxins act as coenzymes, serving as a point of attachment for some substrate on to an enzyme controlling growth.

Foster et al. (1952) proposed a theory of two-point attachment in which the auxin reacts with

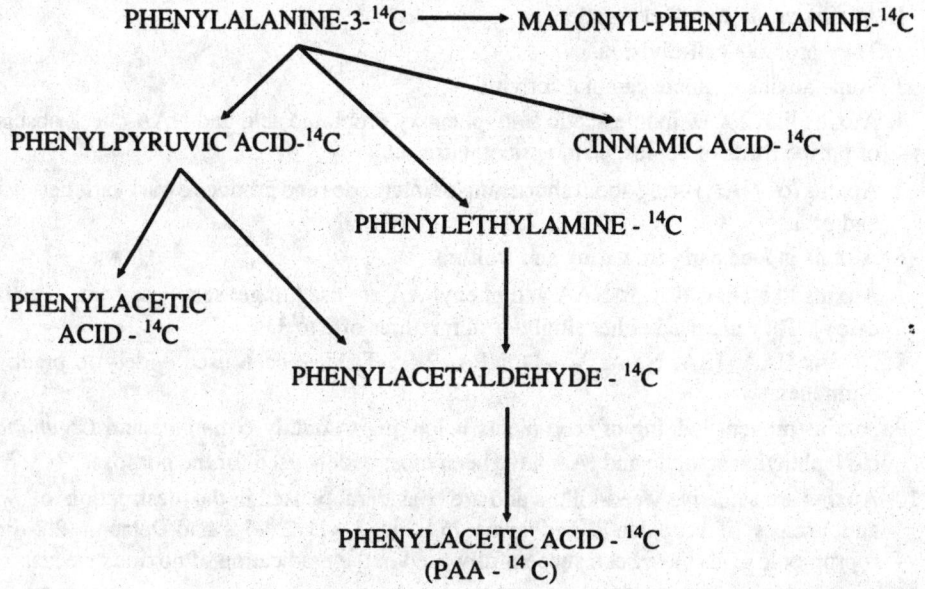

Fig. 8.6: Phenylacetic acid Pathway.

some material in the cell at two positions—(*i*) at some position in the ring, and (*ii*) at the acid group of the side chain.

Affects of Auxin

1. *Auxins affect enzymes*

When growing tissues are treated with an auxin, they show an increased activity of a number of enzymes either directly or indirectly (*Hand*, 1939; *Northern*, 1942; *Bonner*, 1949; *Thimann*, 1951; *Bonner et al.*, 1952).

2. *Auxin affects Osmotic pressure*

Auxin affects growth by increasing the osmotic pressure in the treated tissue thereby increasing the absorption of water (*Czaja*, 1935).

3. *Auxin affects Cell-elongation*

Auxin causes cell elongation which may be due to:
   (*i*) Increasing osmotic solutes and pressure of the cells.
   (*ii*) Increasing the permeability of cells to water.
   (*iii*) Increasing the wall synthesis.
   (*iv*) Decrease in wall pressure.
   (*v*) Inducing the synthesis of specific DNA dependent new m-RNA and a specific enzymic protein. The latter brings about an increase in cell plasticity.

4. *Auxir affects nucleic acid metabolism*

Auxin is associated with nucleic acid metabolism. It affects the site of DNA responsible for increasing cell plasticity and its extension (*Coartney et al.*, 1967; *Masuda et al.*, 1967; *Nooden*, 1968).

8.2.7. Biological Effects of Auxins

Auxins promote growth in the stem and root by promoting cell division and cell elongation. The various important roles played by auxins are:

1. Auxins promote cell elongation.
2. They promote cell-division.
3. Some auxins promote cambial activity.
4. Auxins like 2,4-D, indole acetic acid, phenoxy propionic acid and NAA check abscission of unripe fruits of apples, pear, apricot, citrus etc.
5. Auxins (α–NAA) bring about shortening of internodes and produce dwarf varieties in apple and pear.
6. Auxins induce early flowering and fruiting.
7. Auxins like IAA, IPA, α-NAA and phenyl AA are used to get seed-less fruits (Parthenocarpy). They are used either singly or in mixtures of 2 to 3.
8. Auxins (IAA, IBA, NAA, 2, 4-D, NPA, 2,4,5-T) have been used widely to break seed dormancy.
9. Auxins prevent lodging of crop plants belonging to family *Graminae* and *Leguminosae*. α-Naphthyl acetamide and IAA have been most widely used for the purpose.
10. Auxins are selective weed-killers and are considerably used in the destruction of weeds and grasses of crop lands and lawns. In India, 2,4-D, 2,4,5-T and Dalapon (2,2-dichloropropionic acid) have been successfully used in the eradication of noxious weeds.
11. Auxins like IAA, PAA, IBA and α-NAA, promote early rooting in the cuttings of stems of *Psidium guajava*, *Mangifera indica*, *Ipomea batata* and other plants.
12. Naphthalene acetamide is used in overcoming the sterility in plants.
13. IAA, IBA, IPA and NAA promote fruit setting in *Eriobotrya japonica*. They have been used in other plants belonging to families *Cucurbitaceae*, *Solanaceae*, *Caricaceae*, *Papaveraceae* and *Cactaceae*.
14. IAA also promotes tissue and organ culture.

Auxins are applied in very low concentrations for good results. Higher concentrations inhibit growth and exert toxic effects on plants.

8.3. GIBBERELLINS

The gibberellins, which form another group of growth regulating substances, were not discovered in higher plants, but first became known as metabolites of the fungus, *Gibberella fujikuroi*, that caused overgrowth or the "bakanae" disease of rice. *Yabuta* and *Hayashi* (1939) at the University of Tokyo successfully obtained crystals of the active material from the fungus and named it "gibberellin". These crystals later turned out to be a mixture of compounds. The first pure GA was isolated by *Cross* (1954) and *Borrow et. al.* (1955) in Great Britain closely followed by *Stodola et al.* (1955) in USA and by *Takahashi et. al.* (1955) in Japan.

I
ent-Kaur-16-ene or
(-)-Kaurene

II
ent-Gibberellane
(C_{20}-GA$_s$)

III
ent-20-Nor-Gibberellane
(C_{19} - GAs)

The discovery of GAs in *G. fujikuroi* led to the discovery of GAs in higher plants. The first GA of higher plants was isolated from immature bean seeds by Mac Millan *et al.* (1960).

The GAs are diterpenoid acids derived from the tetracyclic diterpenoid hydrocarbon, ent-Kaur-16-ene or (–)-Kaurene, having 20 carbons. GAs which retain the full 20-Carbon atoms of this precursor are referred to as C_{20}-GAs and have the carbon skeleton II. Others which have lost one carbon atom in course of their biosynthesis are referred to as C_{19}-GAs and have the skeleton III.

The numbering of carbon atoms is in accord with tetracyclic diterpenes. The systematic nomenclature is based on the name "Kaurane" for the structure (IV) and gibberellane for the structure (V). Since the GAs and their precursors are enantiomeric to these structures, they are systematically denoted ent-gibberellanes (C_{20}-GAs), ent-20-norgibberellanes (C_{19}-GAs) and ent-Kauranes respectively.

IV	V	VI
Kaurane	Gibberellane	Gibbane

The gibbane skeleton (VI) used earlier in nomenclature is unrelated to other diterpenoids and is no longer used.

Fiftyone gibberellins have been fully characterized to date and are numbered as GA_1, GA_2, GA_3....GA_{51}. Their structures are shown in Fig. 8.7. Of the 51 GAs, 13 have been found only in *G.fujikuroi*, 29 only in higher plants and 9 have been found in both.

8.3.1. Extraction

For extraction of GAs, the plant material is usually macerated with cold methanol or acetone containing 50% water. After evaporation of the organic solvent and centrifugation, the aqueous residue is partitioned against ethyl acetate at different pH values. The GAs are extracted into ethyl acetate at pH 2.5-3.0, re-extracted into buffer at pH 8.0 and again extracted into ethyl acetate after adjusting pH to 2.5.-3.0. Almost all free GAs may be obtained in the final (acid) ethyl acetate fraction.

8.3.2. Bioassay of Gibberellins

Three types of bioassays are widely used, because of their ease of performance, reliability, sensitivity and range of response. They are:

1. Dwarf seedling bioassay
2. The hypocotyl (bioassay)
3. Cereal endosperm (bioassay)

Dwarf seedling bioassays

They depend on the increased growth of certain dwarf cultivars in response to applied GA. Brian and Hemming (1955) used dwarf peas for the first time and applied 5 µ g of the test solution to the apical buds. The log dose-log response curve for GA_3 was linear from 10^{-1} to 10 µ g per ml.

Hypocotyl Bioassays

They depend on the increased elongation of the hypocotyl of certain seedlings in response to

GA. The assays are easy to perform, rapid, but are less sensitive. *Frankland* and *Wareing* (1960) used lettuce hypocotyl and under optimal conditions the log dose-log responsse curve for GA_3 was linear for 10^{-2} to 10μ g/ml.

Cereal Endosperm Bioassay

In germinating cereal seeds, GA produced by the embryo stimulates the formation of hydrolytic enzymes in the aleurone layer. The enzymes are secreted into the endosperm where they hydrolyze reserve carbohydrates. For bioassay, the embryos are removed by cutting the seeds transversely and the embryo-less half-seeds are incubated with the test solutions. The hydrolytic enzymes formed in response to GA are measured by the reducing sugars released by their action on endogenous starch. The α-amylase activity is measured with added substrate after the half-seeds have been removed. The assay is most sensitive, rapid, but most difficult to perform in a reproducible manner. The barley aleurone bioassay was originally used by *Nicholas* and *Paleg* (1963).

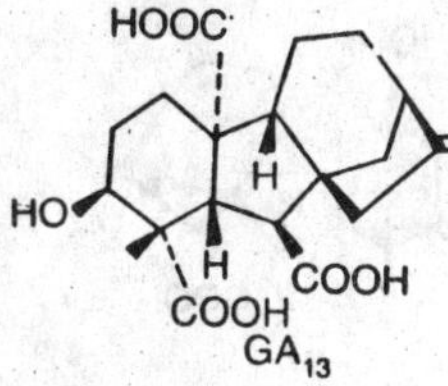

Plant Growth Substances

GA$_{13}$

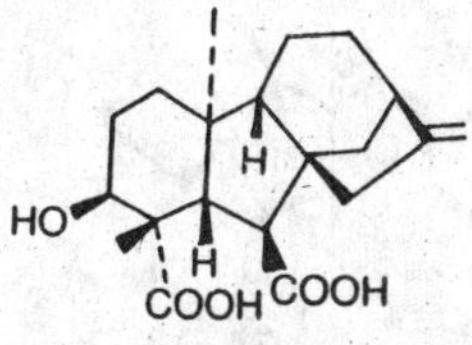

GA$_{14}$

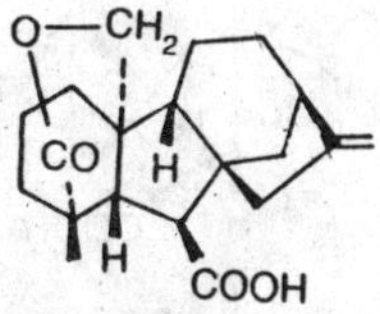

GA$_{15}$

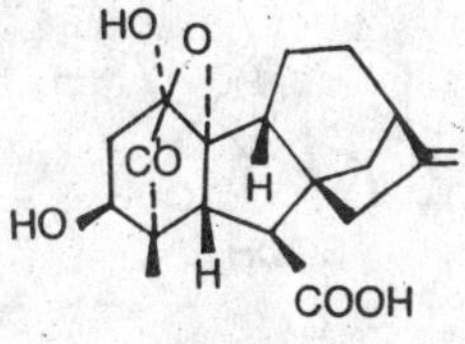

GA$_{16}$

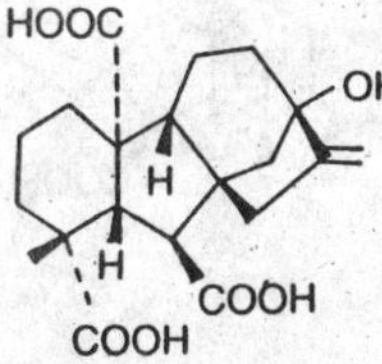

GA$_{17}$

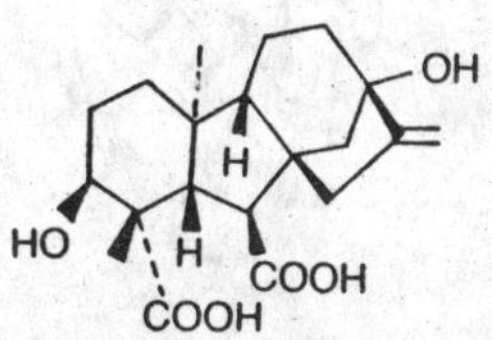

GA$_{18}$

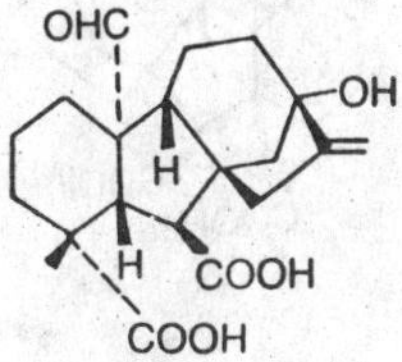

GA$_{19}$

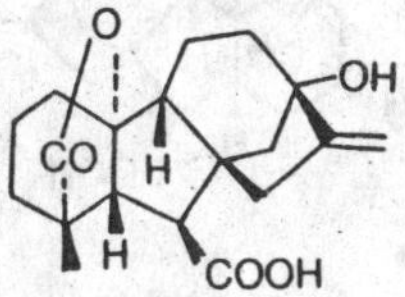

GA$_{20}$

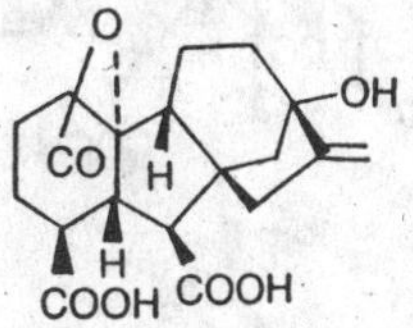

GA$_{21}$

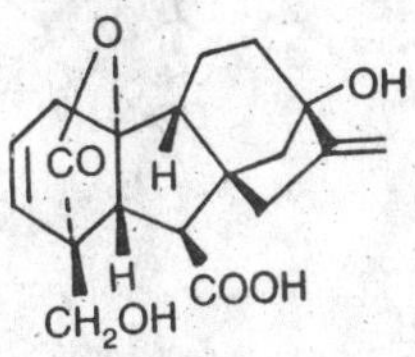

GA$_{22}$

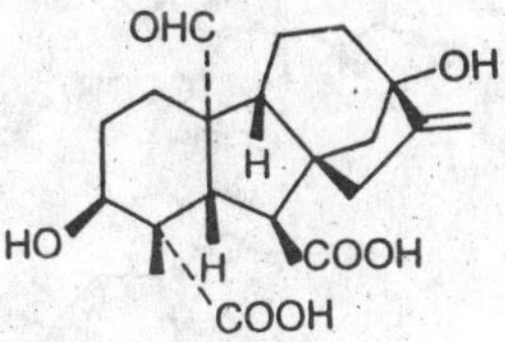

GA$_{23}$

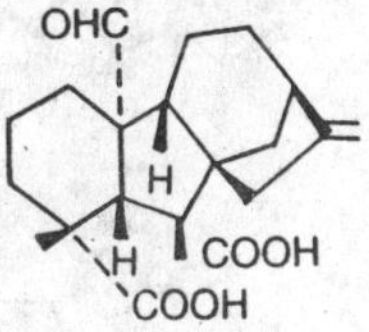

GA$_{24}$

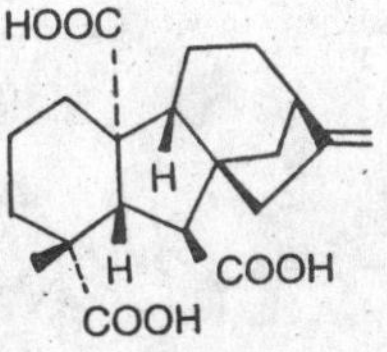

GA$_{25}$

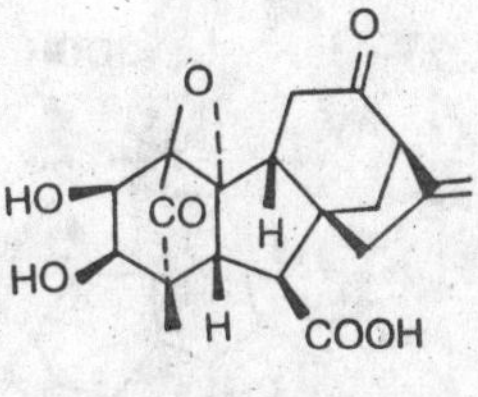

GA$_{26}$

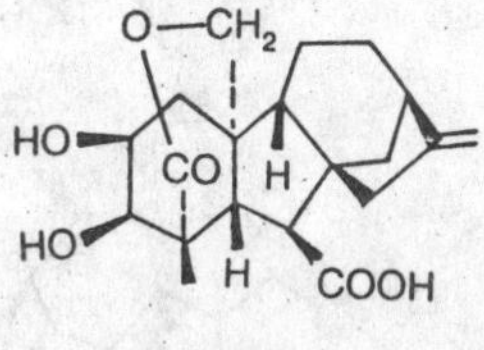

GA$_{27}$

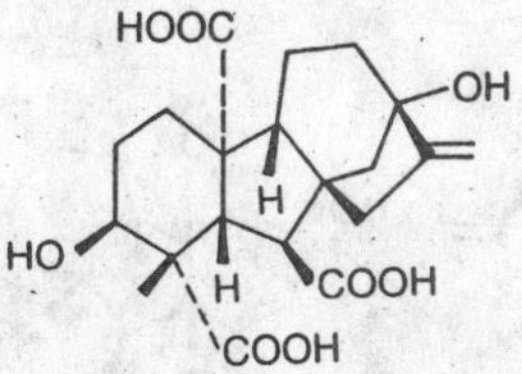

GA$_{28}$

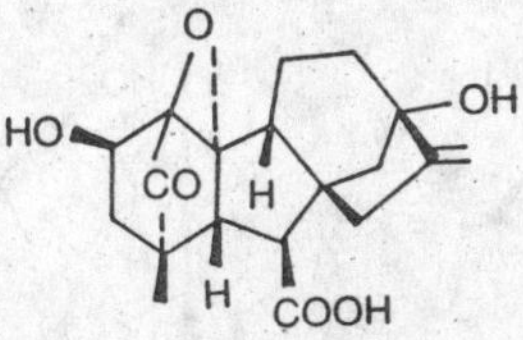

GA$_{29}$

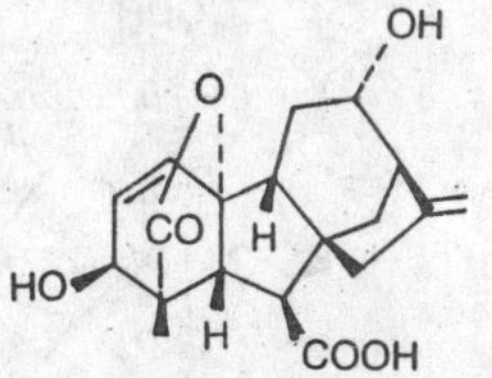

GA$_{30}$

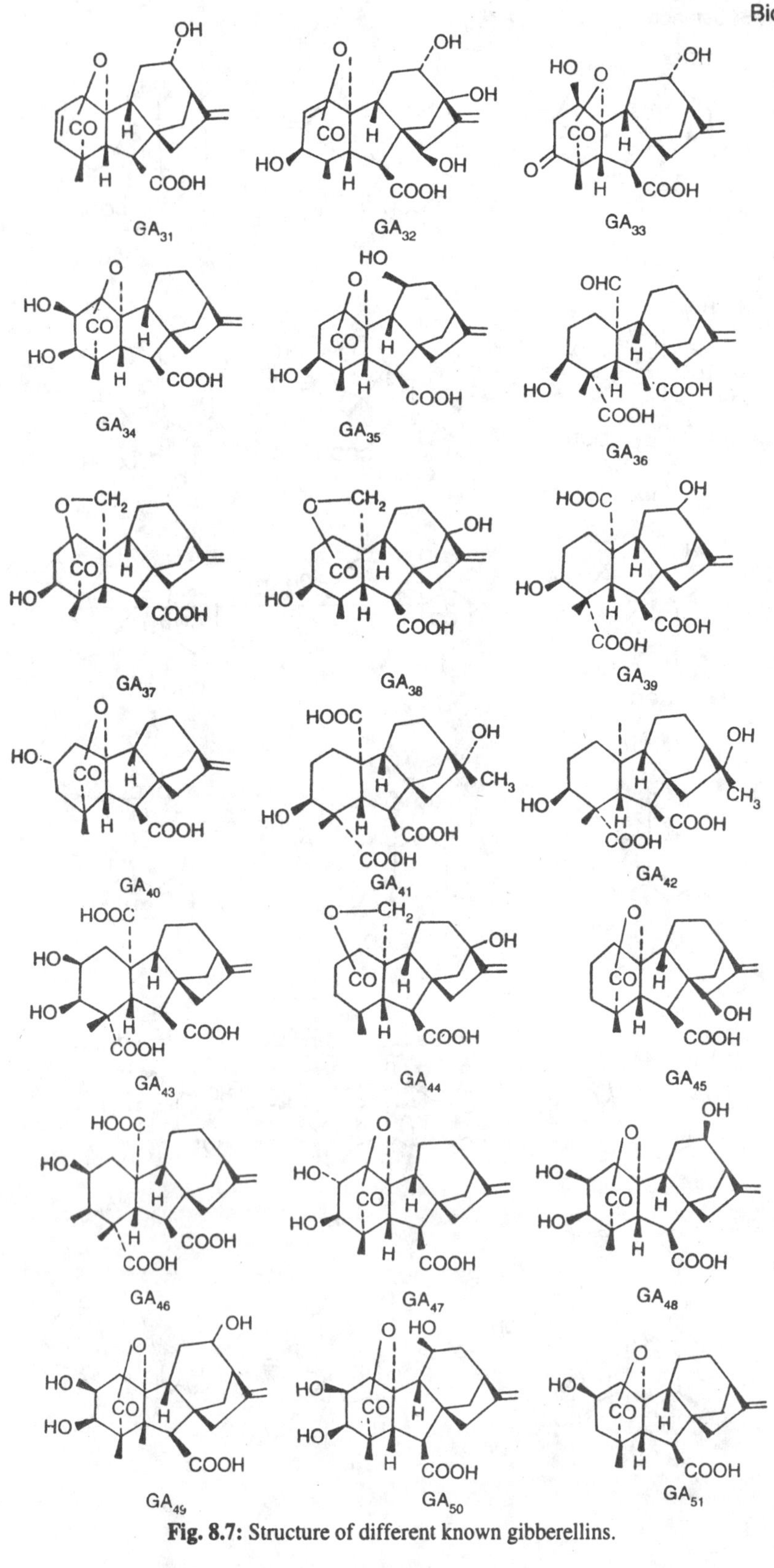

Fig. 8.7: Structure of different known gibberellins.

Plant Growth Substances

8.3.3. Biosynthesis of Gibberellins

A biogenetic relationship of GA_3 to the diterpenes was proposed by *Cross et al.* (1956). *Birch et al.* (1959) demonstrated the incorporation of acetate into mevalonic acid (MVA) and then into GA_3 in cultures of *G. fujikuroi*. The biosynthesis of GA_3 from MVA proceeds via 18 or more intermediates and about 15 related compounds. Cell-free systems of *G. fujikuroi*, and systems from immature seeds of *Echinocystis macrocarpa* and *Cucurbita maxima* have been used in biosynthetic studies. The following steps are involved in biosynthesis of GA_3

1. *Formation of Mevalonate (MVA) from Acetate*

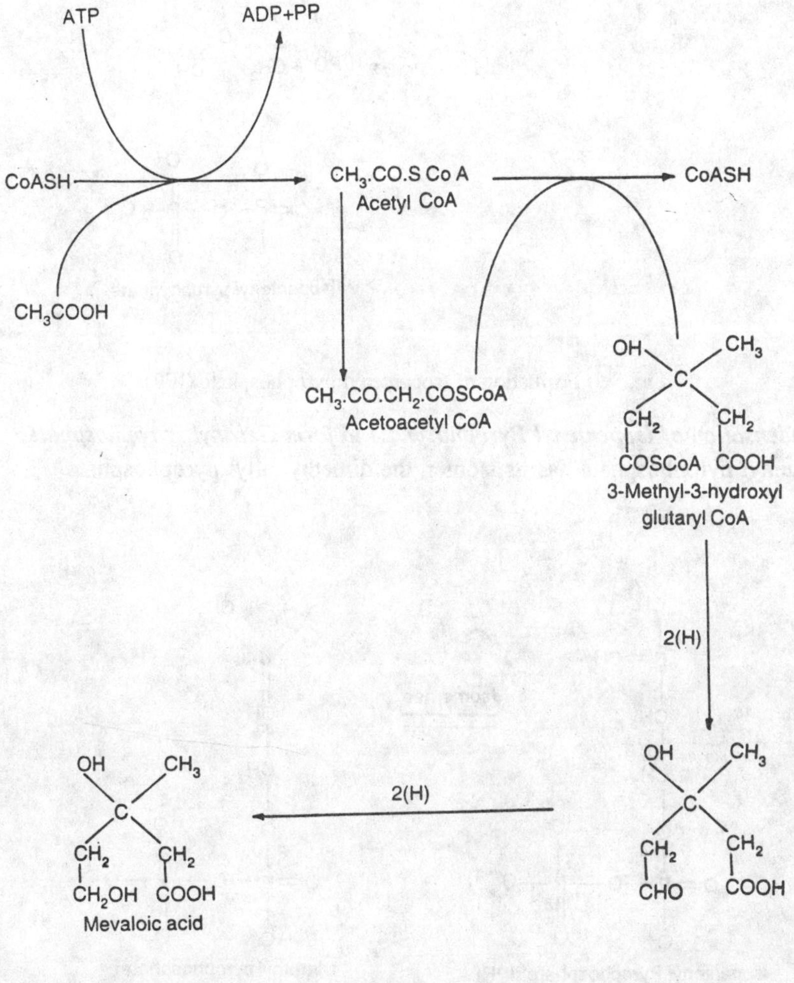

Fig. 8.8: Formation of Mevalonic acid.

2. Formation of Isopentenyl Pyrophosphate (IPP) from Mevalonate

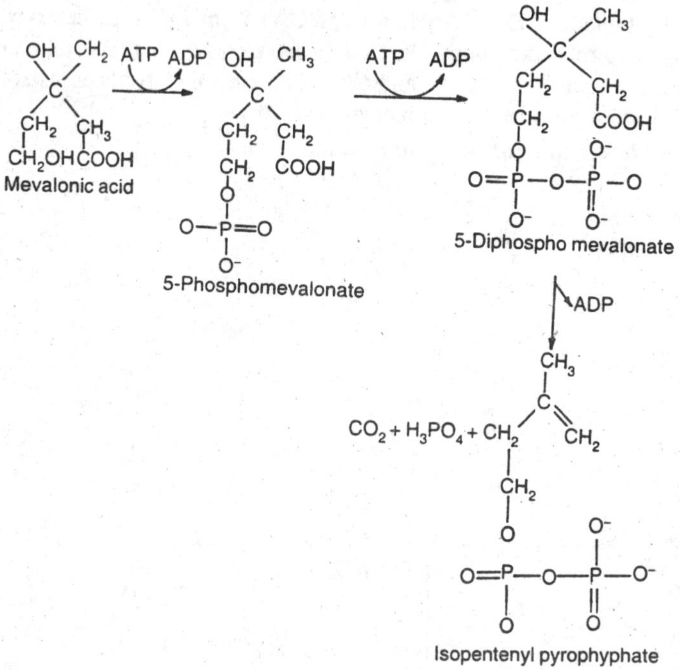

Fig. 8.9: Formation of Isopentenyl pyrophosphate (IPP).

3. Condensation of Isopentenyl Pyrophosphate to form Geranyl Pyrophosphate.

Isopentenyl pyrophosphate has its isomer, the dimethyl allyl pyrophosphate.

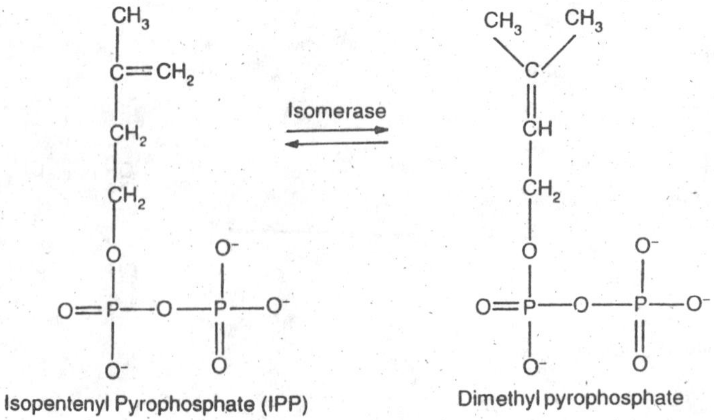

The two isomers condense to geranylgeranyl pyrophosphate by alkylation.

Plant Growth Substances

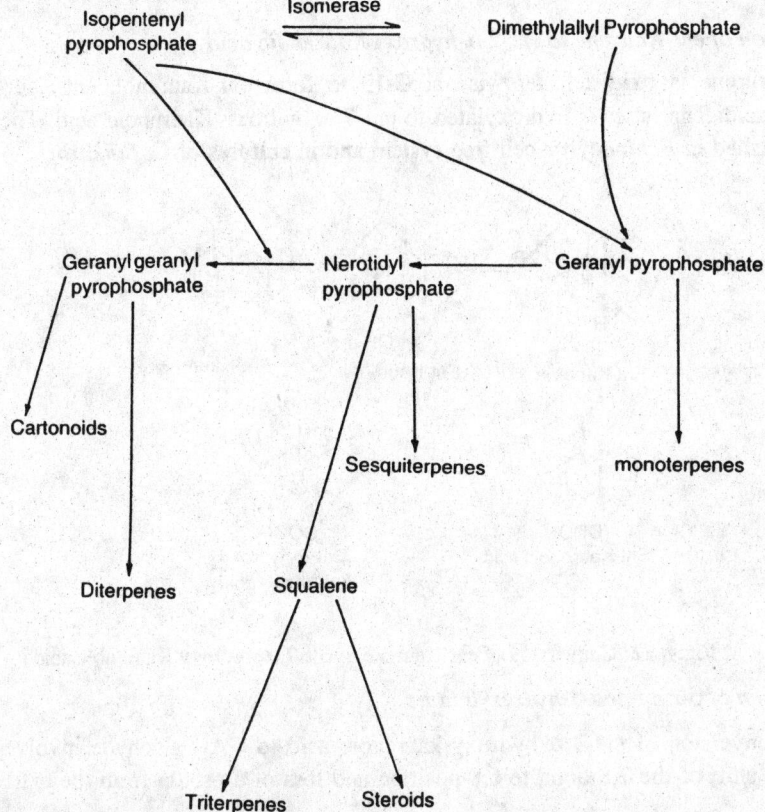

Fig. 8.10: Conversion of Isopentenyl pyrophosphate to geranylgeranyl pyrophosphate.

4. *Cyclization of Geranylgeranyl Pyrophosphate*

The following scheme shows the generally accepted mechanism for cyclization of geranylgeranyl pyrophosphate to ent-Kaurene from ($I^{14} - C$) acetate.

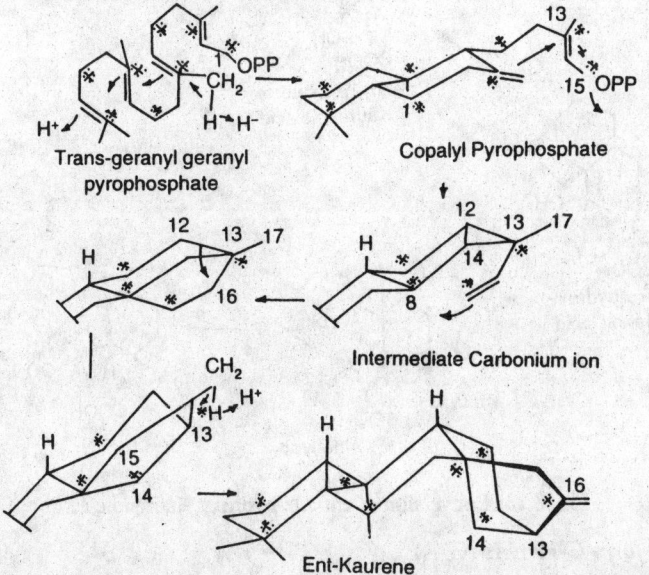

Fig. 8.11: Cyclization of geranylgeranyl pyrophosphate to ent-Kaurene.

5. Conversion of ent-Kaurene to ent-7 α-hydroxy Kaurenoic acid

ent-Kaurene is oxidised step-wise at C-19 to form ent-Kaurenol, ent-Kaurenal, and ent-Kaurenoic acid. The latter is hydroxylated to ent-7 α-hydroxy Kaurenoic acid. The sequence has been established in *Echinocystis* cell free system and in cultures of *G. fujikuroi*.

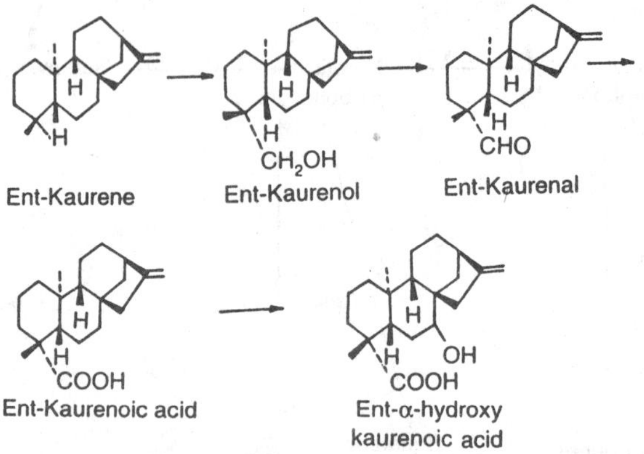

Fig. 8.12: Conversion of ent-Kaurene to ent-7 α-hydroxy Kaurenoic acid.

6. Contraction of β-ring and β-hydroxylation

The conversion of ent-7 α-hydroxy kaurenoic acid to GA_{12}-aldehyde involves loss of 6 β-hydrogen, a shift of the 7,8 bond to 6,8-position and loss of a proton from the extruded C-7 (Fig. 8.13).

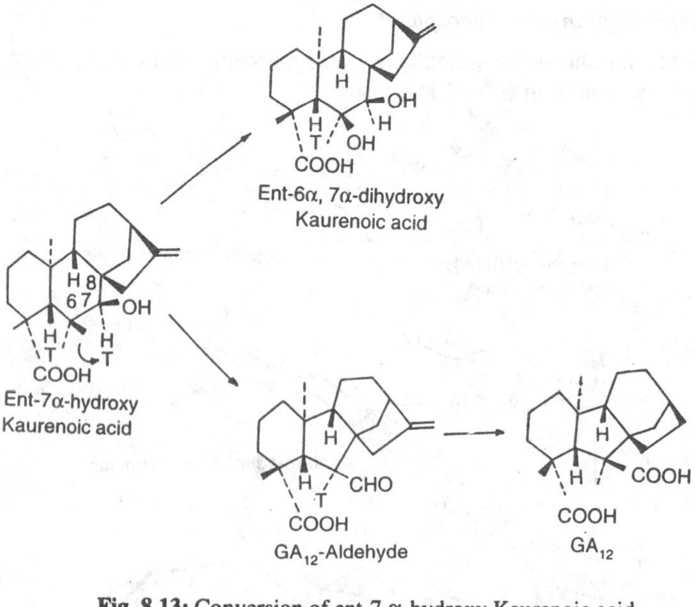

Fig. 8.13: Conversion of ent-7 α-hydroxy Kaurenoic acid.

7. Loss of C-20 to form C-19 GAs (GA_3)

Loss of one carbon must occur to give rise to C-19 GAs such as GA_3.

8.3.4. Physiological Effects of Gibberellins

Gibberellins cause the following physiological effects:

1. Genetic Dwarfism

Gibberellins are able to overcome genetic dwarfism by cell elongation. This effect has been studied in a number of plants including *Pisum sativum, Vicia faba, Phaseolus multiflorus,* tomato, pepper, cucumber and cabbage. There are two views, for the dwarfism in plants. According to one view, it is caused due to lack of enzymes which synthesize gibberellin and according to other view it is due to presence of some natural inhibitor in dwarf plants. The introduction of gibberellin counteracts the effect of this inhibitor.

2. Bolting and Flowering

Gibberellins induced bolting and flowering in many plants including–*Chrysanthemum morifolium, Sesamum indicum* and *Brassica* sp.

3. Light Inhibited Stem Growth

In plants light inhibits the growth of stem. The application of gibberellin increases the plasticity of young cell wall and counteracts this effect of light, as studied in mustard seedlings by *Lockhart* (1961).

4. Parthenocarpy

Gibberellins like auxins, have been used in the production of parthenocarpic fruits. In certain cases, for example, pome and stone fruits, the auxins have proved in-effective and gibberellins effective.

5. Mobilization of Storage Compounds During Germination

Gibberellins induce the enzymatic activity in the endosperm where the mobilization of the stored starch takes place to produce simple sugars for the growth of embryo. It has been shown that at least two enzymes (α-amylase and protease) induced by GA treatment arise through de novo synthesis.

6. Leaf Expansion

In certain plants like pea, bean, tomato, lettuce and cabbage the GA treatment induces the production of broader and elongated leaves.

7. Sex-Expression

It has been shown that spraying of GA produces a marked reduction in the number of male flowers and increase in the number of female flowers in cucumbers.

Table 8.1 : Differences between Auxins and Gibberellins

	Auxins		*Gibberellins*
1.	They are mostly found in higher plants.	1.	They are mostly found in fungi and few higher plants.
2.	They are synthesized in the meristematic regions of the plants e.g., in coleoptile, buds, growing tips of roots and leaves.	2.	They are synthesized in fungal mycelia, in immature bean seeds, in pea seedlings, hypocotyls, cereal endosperms.
3.	Chemically they are indole 3-acetic acid (IAA) or its derivatives or chemicals with identical properties like IBA, NAA, IPA, etc.	3.	Chemically they are gebberellic acids which are diterpenoid acids derived from the tetracyclic diterpenoid hydrocarbon, ent-kaur 16-ene or (–)-kaurene, having 20 carbons.
4.	They promote apical dominance.	4.	They show no effect on apical dominance.
5.	They cause growth in dwarf pea stem sections, but show no effect on intact plant.	5.	They cause growth of intact plant, but show no effect on its stem sections.
6.	They show no effect on bolting and flowering.	6.	They promote bolting and flowering in non-vernalized biennials and long day plants.

	Auxins		*Gibberellins*
7.	Auxins do not effect seed germination and breaking of dormancy.	7.	They promote seed germination and breaking of dormancy.
8.	They do not cause de-novo synthesis of hydrolytic enzymes.	8.	They cause de-novo synthesis of hydrolytic enzymes.
9.	Auxins in higher concentrations inhibit growth of roots but initiate formation of new roots.	9.	Gibberellins do not effect growth of roots.
10.	Auxins promote cell-elongation, cell-division and cambial activity.	10.	They promote cell-elongation by synthesizing enzymes and checking the effect of inhibitors.
11.	Auxins like 2, 4-D; 2, 4, 5-T and Dalapon etc. are used as weed killers.	11.	Gibberellins are not used as weed killers.

8.4. CYTOKININS

Miller et al. (1954) isolated the third growth substance from autoclaved herring sperm DNA. Because of its cell division activity on tobacco pith callus it was called as **kinetin**. Chemically it is a derivative of adenine with a furfuryl group at C-6 and is called as **6-furfurylaminopurine**. The kinetin is formed from deoxyadenosine, a degradation product of DNA.

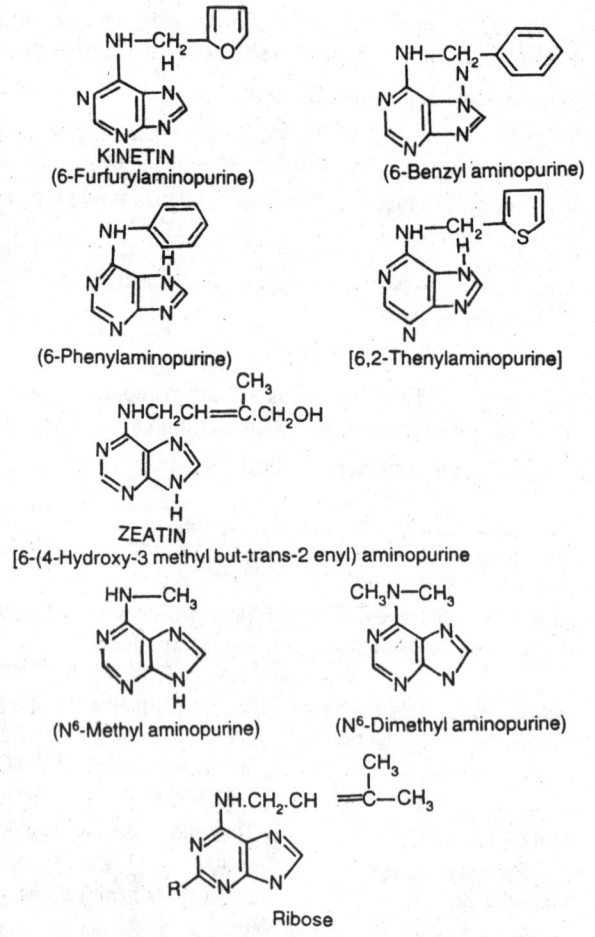

Fig. 8.14: Structure of Kinetin and other naturally occurring and synthesized cytokinins.

Plant Growth Substances

Many analogs of kinetin have been synthesized and grouped under a generic term cytokinins (substances which bring about cytokinesis). All of them show similar activities. *Van Overbeek et al.* (1941) isolated another naturally occurring cytokinin from the endosperm of coconut, *Cocos nucifera*. *Miller* (1961) extracted partially pure cytokinin from immature seeds of maize. *Letham* (1963) successfully extracted the pure crystalline cytokinin from maize seeds and called it as **Zeatin**. Zeatin has been identified as 6-(4-hydroxy-3-methylbut-trans-2-enyl) aminopurine and is more active than other cytokinins. Its high activity is due to the side chain of a very reactive allylic hydroxyl group.

Cytokinins have also been reported from t-RNAs. *Zachau et al.* (1966) isolated two cytokinins, 2-iso-pentenyl adenosine and its methyl thio-derivative, from t-RNA's of serine and tyrosine. They do not occur in other amino acids of higher plants.

8.4.1. Bioassay of Cytokinins

The different bio-assays for cytokinins fall into 6 groups.

1. Those which are based on the ability of cytokinins to promote expansion of excised leaf or cotyledon.
2. Those which depend on the promotion of growth of *Lemna*.
3. Those which depend on the promotion of growth of stem and coleoptile sections.
4. Those which depend on the ability of cytokinins to induce cell divisions in tissue cultures.
5. Those which are based on retardation of leaf senescence.
6. Those which depend on pigment formation (chlorophyll and betacyanin respectively).

The tissue culture assays have high sensitivity under sterile conditions and are specific. The soybean callus assay is probably the most popular, because it has a wide concentration range over which a linear relationship between response and concentration is obtained. However they need a long assay time (21-35 days).

The **Amaranthus betacyanin bioassay** is a rapid assay. It has high specificity and sensitivity under sterile condition. The radish cotyledon assay is simpler to perform than the Amaranthus assay. The chlorophyll formation assay in cucumber cotyledons requires a short assay time and exhibits excellent sensitivity and specificity.

8.4.2. Physiological Effects of Cytokinins

1. *Cell Division* : Cytokinins promote cell division in presence of auxin IAA, especially in tobacco pith culture, carrot root tissue, pea callus and soybean cotyledon. This effect was studied by *Miller et al.* (1956) and *Das* (1956) separately.

2. *Cell Enlargement* :- Cytokinins also induce cell enlargement, an effect usually associated with IAA and gibberellin. Cell-enlargement has been observed in *Phaseolus vulgaris*, pumpkin cotyledons, tobacco pith cultures etc.

3. *Root Inhibition and Growth* : Cytokinins are able both to stimulate and to inhibit root initiation and development.

4. *Breaking of Dormancy* : Cytokinins can break the dormancy of certain light sensitive seeds like lettuce and tobacco.

5. Cytokinins promote the growth of lateral buds in pea stem by neutralizing the effect of natural auxin.

6. By entering into the t-RNA cytokinins can regulate enzyme synthesis in plants.

7. Cytokinins also induce the phenomenon of sex-reversal.

8. They also stimulate RNA synthesis in the elongating zones of onion roots.

8.5. ABSCISIC ACID

Abscisic acid is the most recently discovered plant hormone. *Okhuma et al.* (1965) first isolated it from young cotton fruit. *Addicott and Lych* (1969) also isolated this hormone from cotton fruits and *Sycamore* leaves. The compound, abscisic acid, was previously known as abscisin II and dormin. It is found widely distributed in plants.

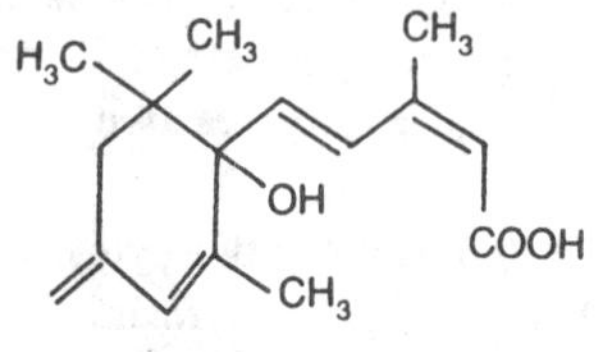

A. Abscisic acid

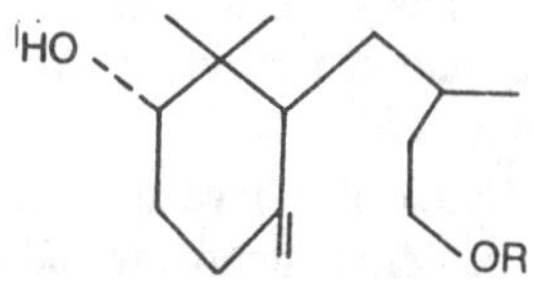

B. Farnesiferol–β

Fig. 8.15: Structure of Abscisic acid and farnesiferol β.

Abscisic acid (Fig. 8.15a) is a sesquiterpene. Another naturally occurring compound similar to abscisic acid is farnesiferol β (Fig. 8.15b). The abscisic acid is biosynthesized from mevalonic acid, but the details are still unknown.

The abscisic acid inhibits action of auxins, gibberellins and cytokinins, so it is also a growth inhibitor.

8.5.1. Physiological Effects of Abscisic Acid (ABA)

1. Abscisic acid accelerates leaf and fruit abscission.
2. It prolongs bud dormancy in woody plants and in tubers.
3. It inhibits germination by prolonging seed dormancy.
4. Abscisic acid inhibits flowering of long-day plants when they are held under short days.
5. Abscisic acid neutralizes the abscission-retarding effect of IAA.
6. It inhibits growth of Oat coleoptiles.
7. In germinating seeds it inhibits enzyme-synthesis stimulated by gibberellic acid.
8. It promotes senescence.
9. It causes loss of chlorophyll and turgor.

8.5.2. Biochemical Responses of Abscisic Acid

1. ABA causes the inhibition of GA_3 induced synthesis of α-amylase and ribonucleases (*Chrispeels & Varner*, (1967).
2. It causes the promotion of the activity of phynylalanine ammonia-lyase, an enzyme which catalyses the conversion of phenylalanine to anti-auxin, transcinnamic acid.
3. ABA decreases the transcription sites coding for RNA synthesis, causing inhibition of RNA synthesis (*Pearson and Wareing*, 1969).
4. It increases the production of nucleases, the enzymes which cause the degradation of DNA and RNA (*Letham*, 1971).
5. It brings about a change in read out pattern of the genome by changing the UMP:GMP base ratio (*Khan and Anojulu*, 1971).

8.6. ETHYLENE

Ethylene is a ripening hormone and is produced in traces in the form of gas by almost all plant tissues. Previous experiments made on this aspect could only demonstrate the liberation of ethylene, but the technique of Gas-chromatography helped in detection of exact amounts of ethylene from different plant tissues like lemons and oranges. Many immature fruits contain only about 0.05 ppm of ethylene which increases to 100-folds during their ripening.

Plant Growth Substances

The ethylene gas is $CH_2=CH_2$ and it is biosynthesized in plants from the amino acid methionine. The methionine is first converted into precursor molecule, methionol either by oxidative deamination or by transamination followed by decarboxylation in presence of enzyme soluble peroxidase and cofactors p-coumaric acid and methanesulphonic acid. C-3 and C-4 of methionine are the source of ethylene.

8.6.1. Physiological effects of Ethylene

1. Ethylene stimulates respiration and ripening in fruits.
2. It stimulates germination in some seeds.
3. It induces auxin-mediated geotropic responses.
4. Ethylene in low concentrations induces resistance to pathogenic infection.
5. Ethylene causes stimulation of enzymes peroxidase and polyphenol oxidase in the tissues.
6. Ethylene induces epinastic movements.
7. It also induces flowering in pineapple and root-hair formation in many other plants.
8. Ethylene stimulates radial growth in stems and roots.
9. Ethylene regulates the level of endogenous auxin by feedback mechanism.
10. Ethylene promotes leaf abscission.

8.7. MORPHACTINS

These growth hormones were recently discovered by *Schneider* (1970) along with Alnusin, Heliangine, Portulal, Xanthoxin and Xanthinin. Morphactins are synthetic growth regulators and affect various physiological and morphogenetic processes.

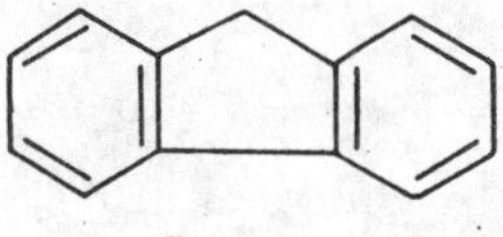

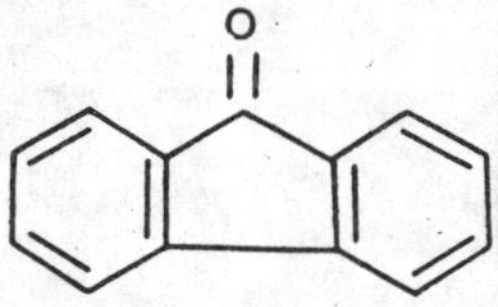

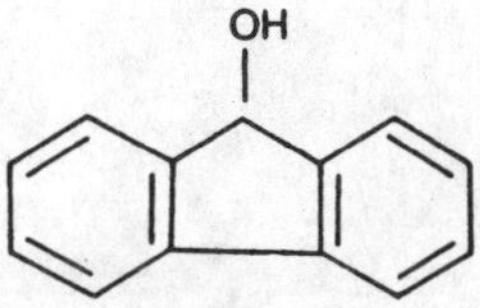

Fig. 8.16: Structure of Fluorene and its derivatives.

Chemically, morphactins are made up of inactive basic skeleton **fluorene** which becomes active by the introduction of -COOH group at 9 position. More active forms contain -Cl at 2 position of the fluorene skeleton in addition to -COOH group at 9 position. In addition to the free acids, their esters and salts are also effective. Some of the important synthesized derivatives are:

1. *Chloroflurenol*-CFL-(2-chloro-9-hydroxy fluorene-9 carboxylic acid).
2. *Chlorofluren*-(2-chloro-fluorene-9-carboxylic acid)
3. *Flurenol* - FL (9-hydroxy fluorene-9-carboxylic acid).
4. *Dichlorflurenol*-DCFL-(2,7-Dichloro-9-hydroxy-florene-9-carboxylic acid)

8.7.1. Physiological Effects of Morphactins

1. Morphactins help in the organization and correlation mechanism of higher plants.
2. They stimulate cell division in the cambium and pericycle of plants and cuttings.
3. They stimulate callus growth.
4. They inhibit mitosis in apical meristems, in apical cells, and in young growing tissues of intact plants.
5. They abolish the polarity phenomenon in plants.
6. Morphactins promote branching.
7. They inhibit lateral root formation.
8. They help in the abolition of geotropism and phototropism.
9. They induce internode shortening in seedlings.

Other effects of morphactins are: induced parthenocarpy; retardation of flowering, fruiting and ripening; retardation of chlorophyll breakdown; abscission; shifting the composition of plants–high sugar contents, low N_2, high amino acids, high protein, high latex yield; checking pathogenic infection and increasing resistance in plants against wilting disease etc. Thus, morphactins show polyvalent action on the growth and development of plants.

8.8. OTHER HORMONES

Other plant hormones are; **Florigen** (flowering hormone), **Vernalin** (Vernalisation hormone), **Anthesins** (Flowering hormone), **Calines** (Formative hormones) and **Traumatic acid** (Wound hormone). Very little is known about them.

9
Nucleic Acids

9.1. INTRODUCTION

The nucleic acids are of considerable importance in biological systems. They are of two types: (1) Ribose nucleic acid, and (2) Deoxyribose nucleic acid. The basic chemical subunits of the nucleic acids are nucleotides. The nucleotides are made up of three components : (*i*) A heterocyclic ring containing nitrogen, known as a *nitrogenous base*, (*ii*) A five carbon *pentose sugar*, and (*iii*) A *phosphate group*. The bases found in nucleic acids are of two kinds—purines and pyrimidines. Adenine and guanine are purine and cytosine, uracil and thymine are pyrimidine bases.

The nucleotides found in nucleic acids are much fewer in number than the α-amino acids. DNA is found in almost all the cells as a major component of chromosomes of the nucleus. In 1962 the presence of chloroplast DNA was reported by *Ris* and *Plaut* from *Chlamydomonas*. Segments of DNA as long as 150 μ have been reported from chloroplasts by *Woodcock* and *Fernandez-Morgan* (1968). In 1963 *M.M.K. Nass* and *S. Nass* reported the presence of DNA from mitochondria. Certain viruses, including many of the bacterial viruses or bacteriophages, are DNA-protein particles. Mostly the plant viruses are RNA-protein particles.

Ribose nucleic acid (RNA) is also of common occurrence in plants as well as animals. It is of three types–(*i*) ribosomal RNA (r-RNA); (*ii*) soluble RNA or transfer RNA (t-RNA) and (*iii*) messenger RNA (m-RNA). Ribosomal-RNA is found in small sub-cellular particles, the ribosomes. RNAs with sendimentation Coefficient value, 5S, 16S and 23S have been reported from 70S ribosomes, while 18S, 28S, 5.8S and 5S r-RNAs have been reported from 80S ribosomes. *t*-RNA is found in free form in the cytoplasm. m-RNA is found in small quantities in association with ribosomes.

9.2. NITROGENOUS BASES

Nitrogenous bases are of common occurrence in nucleic acids and are five in number. They are of two kinds—purines and pyrimidines. Both RNA and DNA contain the two purines adenine and guanine and one pyrimidine cytosine in common. The fourth nitrogenous base found in two nucleic acids is different. RNA contains uracil, whereas, DNA contains thymine. The bases are joined to the pentose sugar by N-C glycosidic bond. For the purines, the **glycosidic bond** is in between the C_1 position of the pentose and the N_9 position of the base. For the pyrimidines the linkage joins the C_1 and N_3 positions.

9.2.1. Pyrimidine bases

Pyrimidine bases are all derivatives of the parent compound pyrimidine which shows a six-membered ring. The derivatives are found in the nucleic acids. Cytosine is found in both types of nucleic acids, uracil in RNA, and thymine and 5-methylcytosine in DNA. In certain strains of **coliphage**, the cytosine in replaced by a fifth pyrimidine, 5-hydroxymethylcytosine.

DNA and RNA differ not only in the pentose structure, but also in the pyrimidine base. Generally, radioactive thymidine is used to label DNA and similarly, radioactive uracil can be used for RNA. The thymine differs from uracil only in having a methyl substituent in the C_5 position.

The structural formulae and the numbering system for the pyrimidine bases of the nucleic acids are shown in Fig. 9.1. The pyrimidine and its compounds are capable of undergoing keto-enol tautomerism.

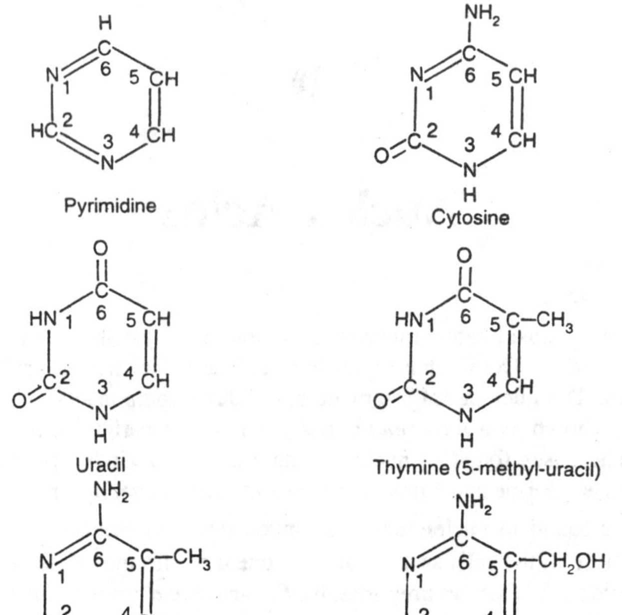

Fig. 9.1: Pyrimidine and its derivatives (Keto form).

9.2.2. Purine Bases

Adenine and **guanine** comprise the purine bases and occur in both DNA and RNA. In RNA a large number of bases are methylated and occur as methyladenine, methylguanine or methylcytosine.

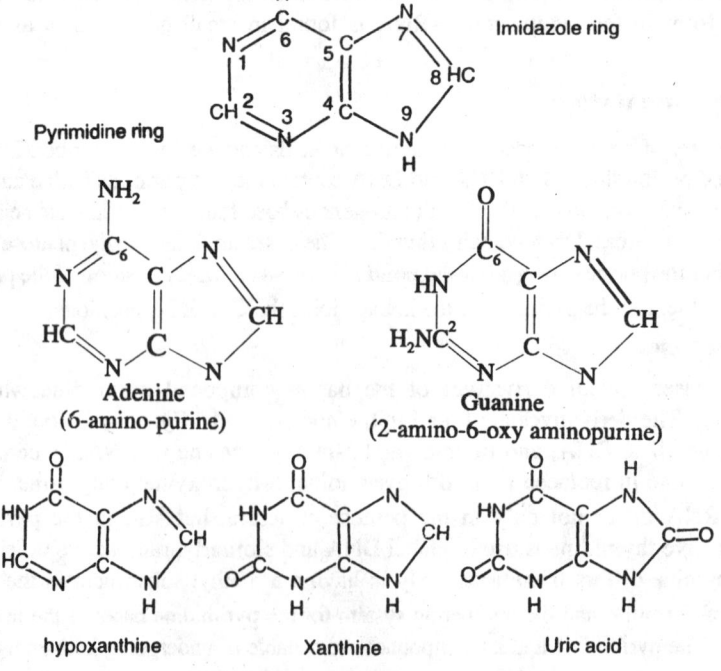

Fig. 9.2: Purine and its derivatives.

A **purine ring** is formed by the fusion of a 6-membered pyrimidine ring and a 5-membered imidazole ring. In a purine the positions are designated by the numbers 1 through 9 (Fig. 9.2). Adenine and guanine also undergo keto-enol tautomerism.

Other naturally occurring purine derivatives are **hypoxanthine, xanthine,** and **uric acid**. Certain minor bases have also been reported from some nucleic acids in small amounts. For example, tRNA contains such methylated bases as 2-methyladenine, 6-methylamino-purine, 6-dimethylamino purine, 1-methylguanine, 6-hydroxy-2-methylaminopurine, 5-methylcytosine etc. These unusual bases comprise less than 5 per cent of the total base content of the t-RNA and very in relative amounts from species to species. It has been confirmed that all nitrogenous bases have double bonds between carbons alternating with single bonds. These bonds can interchange continuously, producing the phenomenon known as **resonance**. This enables the bases to absorb ultraviolet light at 2600Å.

9.3. CHEMISTRY OF STRUCTURE OF BASES

9.3.1 The Purines

It is possible to write alternative tautomeric structures for both adenine and guanine. Thus, adenine and guanine could exist wholly, or in part, as an imino (=NH) derivative or an enol form (Fig. 9.3).

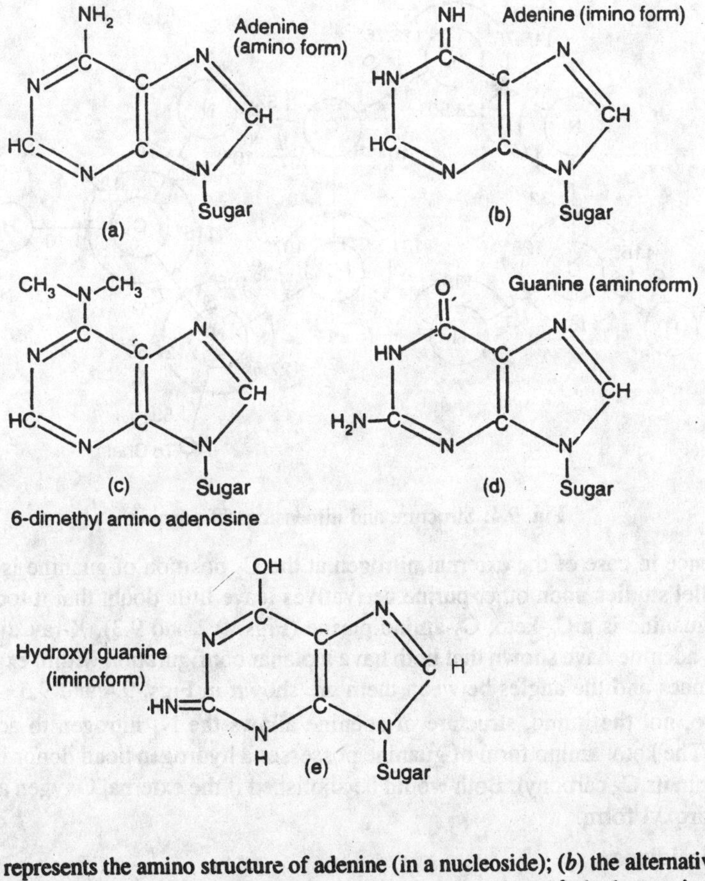

Fig. 9.3: Part (*a*) represents the amino structure of adenine (in a nucleoside); (*b*) the alternative imino structure of adenine (in a nucleoside); (*c*) the 6-dimethylaminoadenosine; (*d*) the ketoamino form of guanine (in a nucleoside); **and** (*e*) the alternative hydroxyl, imino form of guanine.

In actuality, neither possibility appears to be realized. X-ray diffraction studies have indicated that the C_6-O bond length (1.20 Å) of guanine is considerably shortened from the normal value (1.37 Å) for single bond of this type. This is characteristic of and provides strong evidence for, a predominantly double bond character for this linkage, as would be the case for a keto (C=O) group.

The existence of the external nitrogen as a primary amino group ($-NH_2$) in adenine has been confirmed through infra-red spectroscopy. The absorption of light in the infra-red region (>7000 Å or 0.7μ) provides a sensitive and selective means of examining the state of particular bonds or groups. The different arrangement of double bonds for the amino and imino forms of adenine should lead to very different infra-red absorption spectra. Through the use of infra-red spectrum in amino form of adenosine and in its C_6-dimethylamino derivative, it has been confirmed that the ring structure and the nature of C_6-N bond are the same for both compounds and hence adenine is a C_6-amino purine.

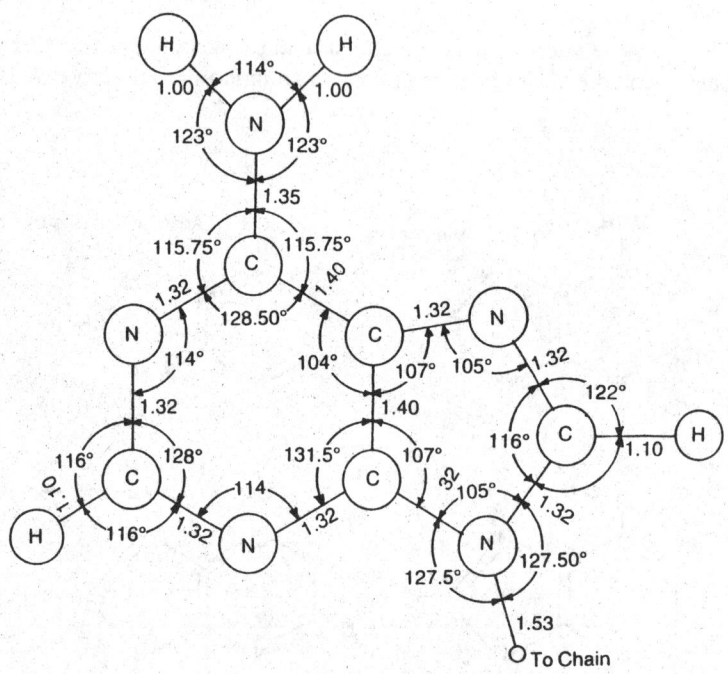

Fig. 9.4: Structure and dimensions of adenine.

The evidence in case of the external nitrogen at the C_2 position of guanine is less conclusive. However, parallel studies upon other purine derivatives leave little doubt that it too is in the amino form and that guanine is a C_6-keto, C_2-amino purine (Figs. 9.2 and 9.3). X-ray diffraction studies on guanine and adenine have shown that both have a planar configuration within experimental error. The bond distances and the angles between them are shown in Figs. 9.4 and 9.5.

The amino, not the imino, structure of adenine allows the N_1 nitrogen to act as a hydrogen bond acceptor. The keto, amino form of guanine possesses a hydrogen bond donor in its N_1 nitrogen and an acceptor in its C_6 carbonyl. Both would be abolished if the external Oxygen at the C_6 position were in the hydroxyl form.

9.3.2. The Pyrimidines

Like purines, the pyrimidines can exist in alternative tautomeric forms. Thus, cytosine could in, principle, exist partially or wholly as a imino derivative, while a hydroxyl form would be a

Nucleic Acids

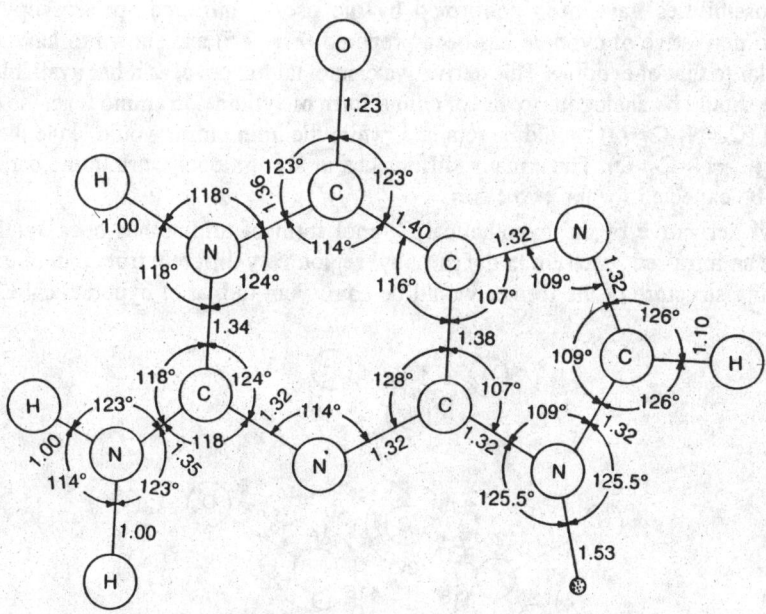

Fig. 9.5: Structure and dimensions of guanine.

possibility for uracil or thymine as shown in Fig. 9.6.

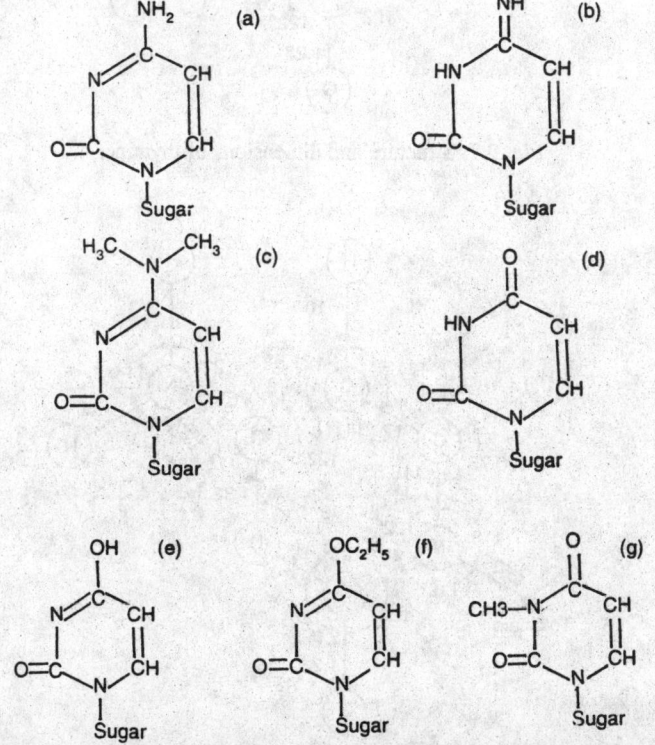

Fig. 9.6: (a) Represents the amino form of cytosine (in a nucleoside); (b) the alternative imino form of cytosine; (c) the C6-dimethylamino cytosine; (d) the diketo form of uracil (in a nucleoside); (e) the alternative hydroxyl form of uracil; (f) the ethyl derivative of hydroxyl form of uracil; and (g) the N_1-methyl uracil.

These possibilities have been confirmed by the use of infra-red spectroscopy. Thus, the dimethylamino derivative of cytidine has been prepared (Fig. 9.6) and shown to have an infra-red spectrum similar to that of cytidine. This derivative cannot tautomerize, as it has available hydrogen, so its structure should be analogous to that of amino form of cytidine. In amino form the conjugation the C_2 oxygen ($C_6=N_1-C_2=O$) should be retained, while the imino form would leave the C_2 oxygen unconjugated ($C_6=N_1-C_2=O$). Thus, major differences in absorption spectra in the carbonyl (6 pe) region would be expected for the two cases.

The ethyl derivative of the hypothetical C_6-enol form of uridine has been synthesized and shown to have an infra-red spectrum in the carbonyl region very different from that of uridine itself. Because the ring structure of the former would be equivalent to that of hypothetical C_6-enol form

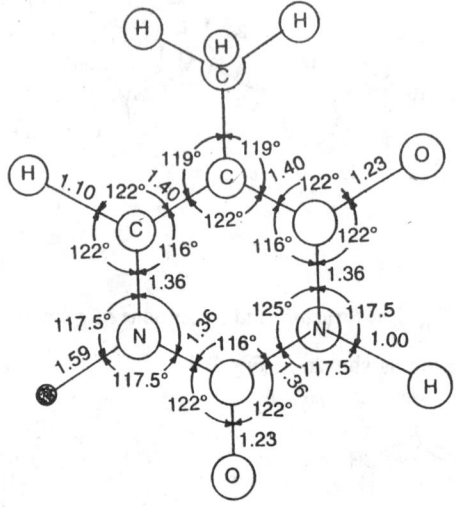

Fig. 9.7: Structure and dimensions of thymine.

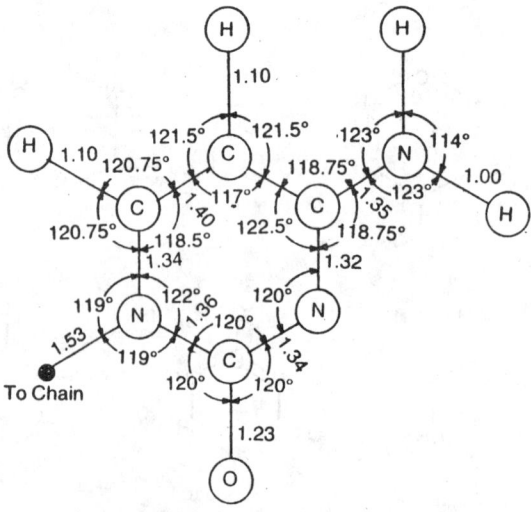

Fig. 9.8: Structure and dimensions of cytosine.

of uridine, this provides evidence counter to the presence of the enol species (Fig. 9.6). In contrast, the N_1-methyl derivative of uridine has been shown to have an infra-red spectrum similar to that of uridine (Fig. 9.6). Since the ring structure of this derivative should be equivalent to that of the diketo form of uracil, it is very probable that this is the prevalent form.

In summary, the prevailing evidence indicates that cytosine may be assigned as amino, keto structure and that uracil and thymine are diketo derivatives. X-ray diffraction studies have indicated that the pyrimidine rings have a planar configuration (Figs. 9.7 and 9.8).

9.4. FREE PURINES AND PYRIMIDINES

Free purines and pyrimidines are found well distributed in plants, but are rare in animal tissues where they are formed by the degradation of nucleic acids. In plants, certain bases have been reported which are not found in nucleic acids. These include the N-methylated bases with important pharmacological properties. The examples of such exceptional bases are theobromine from *Theobroma cacao* and caffeine from *Coffea* spp. Certain end-products of the purine catabolism in mammals like uric acid and allantoin, are also found widely distributed in higher plants. In some species, allantoin and allantoic acid account for the major portion of the soluble nitrogen in the sap, and they are considered to be important storage and transport forms of nitrogen in these species. Their formation in these species from purines has been confirmed, but at what level they are formed (free base, nucleoside or nucleotide) is still unknown.

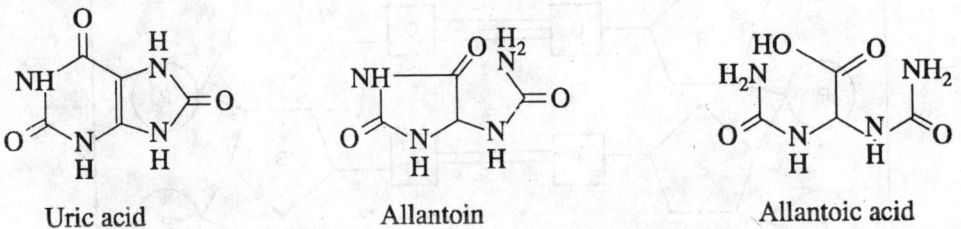

Uric acid   Allantoin   Allantoic acid

9.5. PENTOSE SUGARS

They contain five carbon atoms and are found in a combined state in the plants. They reduce the fehling solution and are not fermentable. Two naturally occurring pentoses ribose and deoxyribose found in RNA and DNA respectively have a pentagonal ring with 5 carbons, two of which (3' and 5') are linked to phosphoric acid and a third one (Carbon C_1) to the base (Fig. 9.9).

Fig. 9.9: Showing the strucuture of different pentose sugars.

The pentose sugar in DNA, as the name implies, is deoxyribose. It is always the OH on the C_1 which is the point of attachment of the base. In deoxyribose sugar the oxygen on the second carbon is lacking. This sugar is responsible for the Feulgen reaction, a very specific reaction for DNA.

9.6. PHOSPHORIC ACID

Nucleic acids are formed by the union of many nucleotides or polynucleotides. The union between two nucleosides is made through phosphoric acid. The nucleoside becomes a nucleotide, with the attachment of a phosphate group. During attachment of the two consecutive nucleosides the phosphate forms a **phosphate ester bond** between the pentoses of the nucleosides. These bonds link carbon 3' of one nucleoside with carbon 5' of the next (Fig. 9.10).

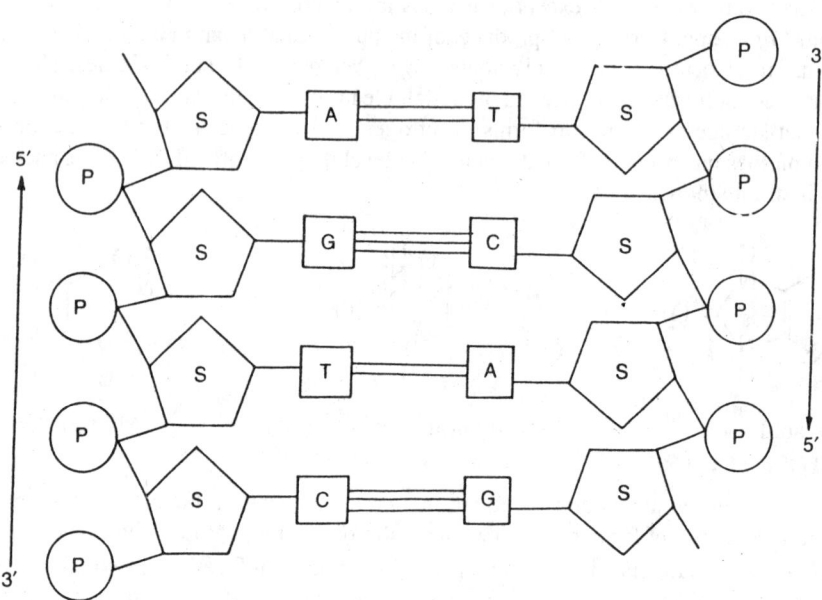

Fig. 9.10: Segment of a DNA molecule, showing two complementary pairs of bases (cytosine-guanine, adenine-thymine) with the hydrogen bonds in between.

In this way the phosphoric acid uses two of its acid groups. The 3rd acid group enables the molecule to form ionic bonds with basic proteins, histones and protamines. This group makes nucleotides highly basophilic. Thus, the phosphoric acid plays an important role in the formation of nucleotides and nucleic acids.

9.7. NUCLEOSIDES

A nucleoside consists of a purine or pyrimidine base and a pentose or deoxypentose component only and is obtained by the hydrolysis of nucleotides, together with inorganic phosphate.

$$\text{Base — Sugar — Phosphate} \xrightarrow{\text{Alkaline hydrolysis}} \text{Base — Sugar} + H_3PO_4$$
$$\text{(Nucleotide)} \qquad\qquad\qquad \text{(Nucleoside)}$$

With five bases only five nucleosides are possible. Thus, adenine combines with ribose to form **adenosine**. Similarly, guanine forms **guanosine**, cytosine forms **cytidine**, thymine forms

Nucleic Acids

thymidine, and uracil forms **uridine**. The ribonucleoside from the hypoxanthine is named as **inosine**. Adenosine, guanosine, cytidine and uridine are the ribonucleosides obtained from RNA. The nucleosides of the DNA are deoxyadenosine, deoxyguanosine, deoxycytidine and thymidine.

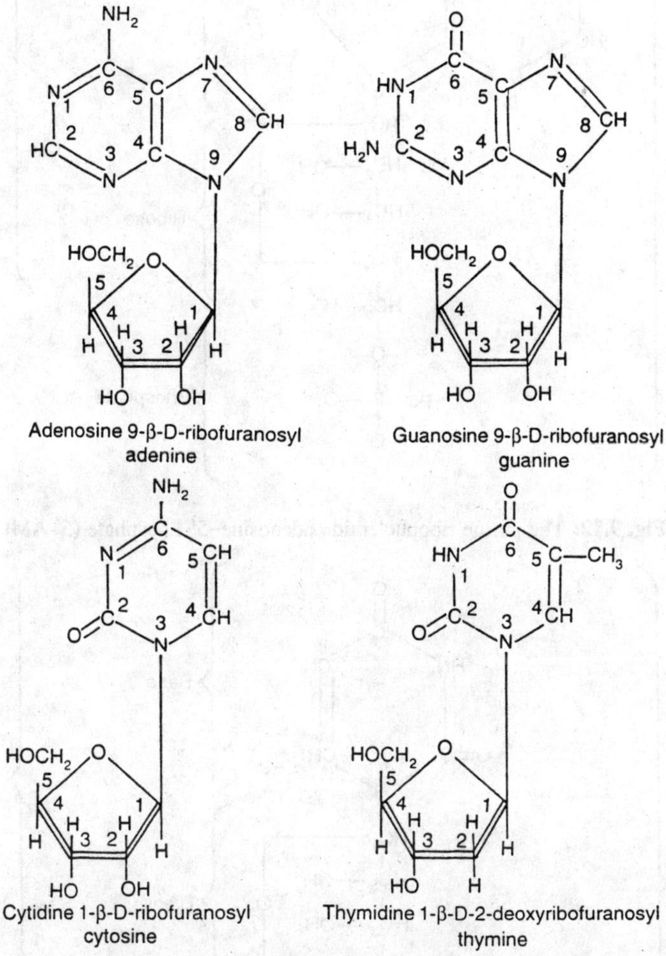

Adenosine 9-β-D-ribofuranosyl adenine

Guanosine 9-β-D-ribofuranosyl guanine

Cytidine 1-β-D-ribofuranosyl cytosine

Thymidine 1-β-D-2-deoxyribofuranosyl thymine

Fig. 9.11: Structure of different nucleosides.

9.7.1. Position of the glycosidic linkage

It has been shown that the purine nucleosides can be easily hydrolysed by acid to free base and sugar. Because the C-C bonds are normally resistant to hydrolysis, this was indicative that the base-sugar linkage was to the N-C type. The synthetic analogs of adenosine and guanosine were prepared and their properties were compared with the naturally occurring adenosine and guanosine. The observations from such experiments indicated that the glycosidic linkage is to the N_9 position of the base (Fig. 9.12). Similar studies for the primidine nucleosides have placed the glycosidic linkage at the N_3 position of the pyrimidine base (Fig. 9.13)

9.8. INTERCONVERSIONS OF THE NUCLEOSIDES

Interconversion of the various nucleosides can be done with the help of nitrous acid. This reaction provides confirmatory evidence for the accepted structures of these compounds.

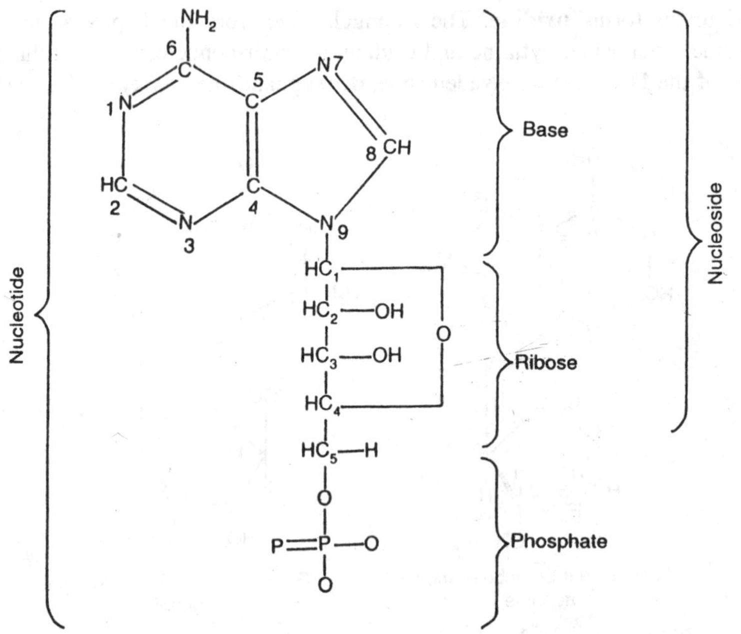

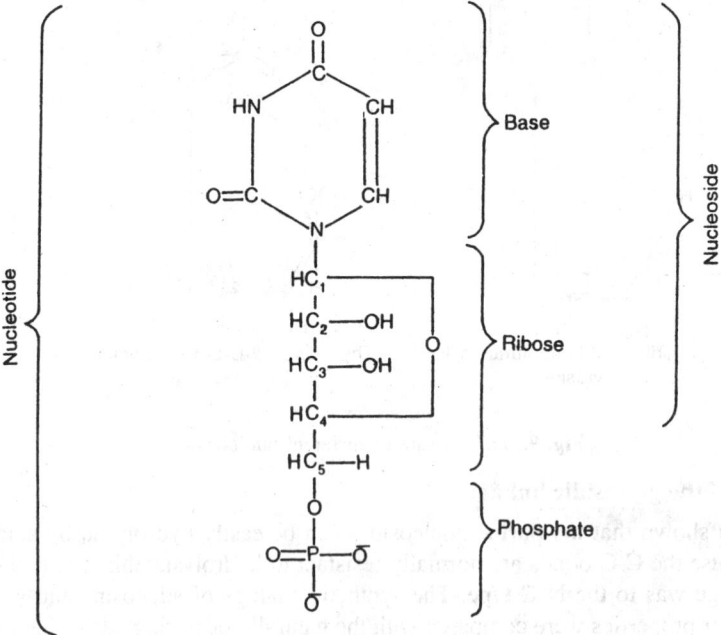

Fig. 9.12: The purine ribonucleotide adenosine–5′ Phosphate (5′-AMP).

Fig. 9.13: The pyrimidine ribonucleotide uridine –5′ Phosphate (5′-UMP).

9.9. THE NUCLEOTIDES

The nucleotides are the phosphate esters of the nucleosides. There, phosphoric acid (H_3PO_4) forms an ester linkage with one of the free hydroxyl groups on the ribose or deoxyribose component. Those nucleotides derived from ribose nucleosides are usually called as **ribonucleotides** and those from deoxyribose nucleosides as **deoxyribonucleotides**. In the past these

Nucleic Acids

terms have commonly been abbreviated to ribostide, ribotide, deoxyribotide and deoxyriboside—but this usage is incorrect.

The ribose portion of a ribonucleoside has three possible positions where the phosphate could be esterified, the 2'-hydroxyl, 3'-hydroxyl, and 5'-hydroxyl, whereas, the deoxyribonucleoside has only the 3'-and 5'-positions available. Since there are three positions available in the ribose of ribonucleosides, three possible nucleoside monophosphates can be formed. Adenosine, for example, can give rise to three monophosphates (adenylic acids), adenosine 5'-phosphate, adenosine 3'-phosphate, and adenosine 2'-phosphate. Nucleotides of all three types have been isolated and identified. Adenosine 5'-phosphate was originally discovered in the free state in muscles and referred to as muscle adenylic acid, while adenosine 3'-phosphate was originally obtained from alkaline hydrolysates of yeast RNA and used to be called yeast-adenylic acid.

In the same way guanosine, cytidine, and uridine can give rise to three guanosine monophosphates (guanylic acids), three cytidine monophosphates (cytidylic acids) and three uridine monophosphates (uridylic acids) respectively. The structure of nucleotides or nucleoside monophosphates are shown in Fig. 9.14.

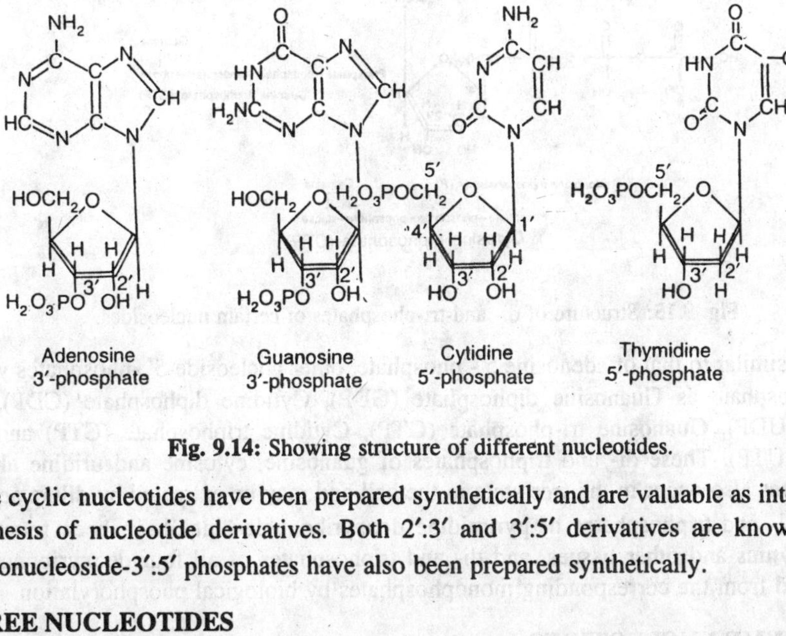

Fig. 9.14: Showing structure of different nucleotides.

The cyclic nucleotides have been prepared synthetically and are valuable as intermediates for the synthesis of nucleotide derivatives. Both 2':3' and 3':5' derivatives are known. The cyclic deoxyribonucleoside-3':5' phosphates have also been prepared synthetically.

9.10. FREE NUCLEOTIDES

Many metabolically important free nucleotides have been isolated from plants. They occur in the form of coenzyme nucleotides. For example, ATP, NAD, FAD, FMN and coenzyme A, ADP-glucose and GDP-glucose are the sugar nucleotides. Another important nucletide derivative, Vitamin B_{12}, is not found in higher plants.

9.11. THE NUCLEOSIDE DI- AND TRIPHOSPHATES

The nucleoside 5'-phosphate may be further phosphorylated at position 5' to yield di- and triphosphates. Thus, adenosine 5'-phosphate (AMP) yields adenosine diphosphate (ADP) and, adenosine triphosphate (ATP). The structure of these compounds has been confirmed by periodate oxidation and by synthesis. The naturally occurring adenosine triphosphate (ATP) and diphosphate (ADP) play an essential role in the conservation and utilization of energy in many biochemical reactions in cells.

Their unique importance depends upon the ability to accept or donate a phosphate group. The ribonucleoside-5'-diphosphates are the substrates for the enzyme polynucleotide phosphorylase, which catalyses their polymerization to polyribonucleotides. The deoxyribonucleoside-5'-triphosphates are the substrates for the enzyme DNA polymerase which is responsible for the biosynthesis of DNA.

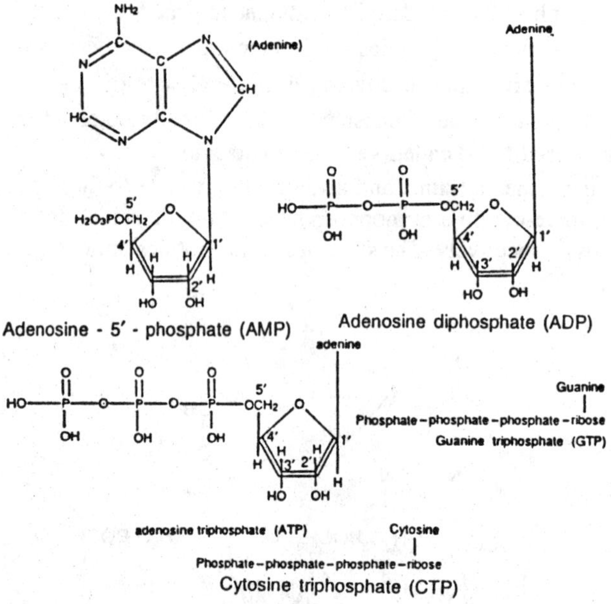

Fig. 9.15: Structure of di- and-tri-phosphates of certain nucleosides.

Further, similar to that of adenosine 5'- phosphate, other nucleoside-5'-phosphates yield such di- and triphosphate as Guanosine diphosphate (GDP), Cytidine diphosphate (CDP), Uridine diphosphate (UDP), Guanosine tri-phosphate (GTP), Cytidine triphosphate (CTP) and Uridine triphosphate (UTP). These di- and triphosphates of guanosine, cytosine and uridine along with monophosphates also occur in the free state in the cell and may be extracted by dilute acid.

Mono, di- and triphosphates of pyrimidine deoxyribonucleosides have been found in acid extracts of thymus and other tissues, and di- and triphosphates of all four deoxyribonucleosides may be formed from the corresponding monophosphates by biological phosphorylation.

9.12. COENZYME NUCLEOTIDES

There are other important nucleotides which do not occur in nucleic acids, and which may contain bases other than purines or pyrimidines. Many important biological compounds also have nucleotide sturctures. They include coenzymes such as nicotinamide nucleotides, flavin-adenine dinucleotide and coenzyme A, which are complex derivatives of AMP. The uridine nucleotide coenzymes play a part in the interconversion of sugars; cytidine triphosphate (CTP) is important in the biosynthesis of phospholipids; and guanosine triphosphate (GTP) is involved in the biosynthesis of proteins. The detailed structures and functions of these various coenzyme nucleotides have been discussed in the chapter on coenzymes.

9.13. PROPERTIES OF THE NUCLEOTIDES

Ionization: Each of the bases occurring in the ribo- and deoxyribo-nucleotides has one ionizable site,

Nucleic Acids

except for guanine, which has two. The bases of adenosine and cytidine are uncharged at neutral and alkaline pH, but bind a proton at acid pH to acquire a cationic charge. The bases of thymidine and uridine are uncharged at neutral and acid pH, but dissociate a hydrogen ion at alkaline pH to become negatively charged (Fig. 9.16). The base of guanosine is uncharged at neutral pH and binds a proton at acid pH and dissociates one at alkaline pH.

Phosphoric acid is tribasic and ionizes in the following three steps :

$$\text{Primary :} \quad H_3PO_4 \rightleftharpoons H^+ + H_2PO_4^-$$
$$\text{Secondary :} \quad H_2PO_4^- \rightleftharpoons H^+ + HPO_4^=$$
$$\text{Tertiary :} \quad HPO_4^= \rightleftharpoons H^+ + PO_4^\equiv$$

The phosphate group of the nucleotides carries a single negative charge at acid pH and a double negative charge at alkaline pH.

$$-C_5'-O-\overset{\overset{O}{\|}}{\underset{\underset{O^-}{|}}{P}}-O^- + H^+ \rightleftharpoons -C_5'-O-\overset{\overset{O}{\|}}{\underset{\underset{O^-}{|}}{P}}-OH$$

Thymidylic acid and uridylic acid are anionic over the entire pH range, while cytidylic acid and the purine nucleotides form anions at alkaline pH and zwitterions at acid pH. Generally, all the bases occurring in the ribo- and deoxyribonucleotides have dissociation constants which may or may not depend upon the state of ionization of the other site.

(2) *Loci of ionization* :—The X-ray diffraction analyses have indicated that in adenine and cytosine the most probable positions for the bound protons are the N_1 nitrogens (Fig. 9.16). In case of guanine the positions is less certain. the ionization of pyrimidine bases results in the loss of the N_1 hydrogen. The negative charge is localized in either of C_2 or C_6 external oxygen (Fig. 9.16).

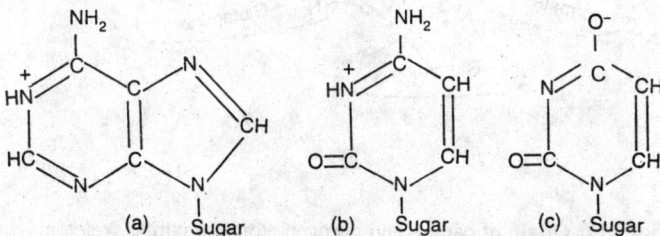

Fig. 9.16: Ionized forms of the nucleoside bases. (*a*) Acid form of Adenine (in adenosine); (*b*) the acid form of cytosine; (*c*) alkaline form of uracil.

(3) *Ultraviolet absorbancy* : The sugar and phosphate groups do not absorb light in the visible or near-ultraviolet wave-length (>220 mμ). However, the presence of bases endows the nucleotides with very intense absorption in the near ultraviolet. The position of maximum absorption is in the range 260-280 mμ, depending upon the base. The molar absorbancies of different nucleotides at pH 7 are as follows :

(*a*) 5′- AMP. The molar absorbancy at 259 mμ (pH 7) is 15.5×10^3.

(*b*) 5′- UMP. The molar absorbancy at 262 mμ (pH 7) is 10.0×10^3.

(*c*) 5′- CMP. The molar absorbancy at 271 mμ (pH 7) is 9.0×10^3.

(*d*) 5′- GMP. The molar absorbancy at 252 mμ (pH 7) is 13.7×10^3.

The molar absorbancy is dependent upon the state of ionization of the base.

9.14. BIOSYNTHESIS OF NUCLEOTIDES

9.14.1. Synthesis of Purines

Purines are synthesized in the cell in the form of their **nucleotide monophosphates**. The synthesis of purines has been discussed by a number of workers including **Hartman** and **Buckmann** (1959), **Warren** (1961), **Schulman** (1961) and **Grav** (1967).

A purine ring (skeleton) is made up of four nitrogen atoms and five carbon atoms (Fig. 9.17). It is proposed that these atoms of the purine ring originated from the following molecules :

1. N_3 and N_9 – from NH_3 after its conversion into amide group of **glutamine**.
2. N_1 – Also from NH_3 but after its conversion into **aspartate**.
3. N_7 – from **glycine** molecule.
4. C_4 and C_5 – from **glycine** molecule
5. C_6 – from CO_2 molecule.
6. C_2 and C_8 – from **active formate**.

Free formate cannot be incorporated as such.

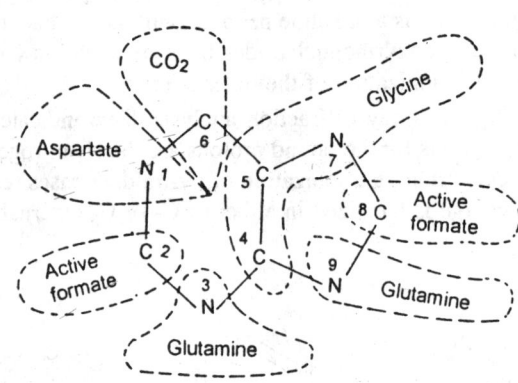

Fig. 9.17: Origin of carbon and nitrogen atoms of purine skeleton.

Biosynthesis of Inosine-5 Monophosphate (IMP)

During the biosynthesis of purine bases e.g., guanine and adenine, first inosine 5-monophosphate (IMP) is to be formed. The formation of IMP is completed in following steps (Fig. 9.18).

1. Ribose-5-phosphate is phosphorylated by ATP to produce 5-phosphoribosyl-1-pyrophosphate (PRPP) in presence of enzyme *phosphorylase*.
2. PRPP combines with glutamine to form 5-phosphoribosyl amine (PRA), glutamate and pyrophosphate group in presence of enzyme *amido phosphoribosyl transferace*. This is an amination reaction where glutamine acts as most efficient donor. NH_2 becomes attached at C_1 of ribose.
3. PRA combines with glycine to produce glycinamide ribonucleotide (GAR). During the reaction amino group (NH_2) of PRA molecule reacts with the carboxyl group of glycine to form an amide linkage between glycine and PRA. Ultimately Glycinamide ribonucleotide is produced with the liberation of one molecule of water.

Nucleic Acids

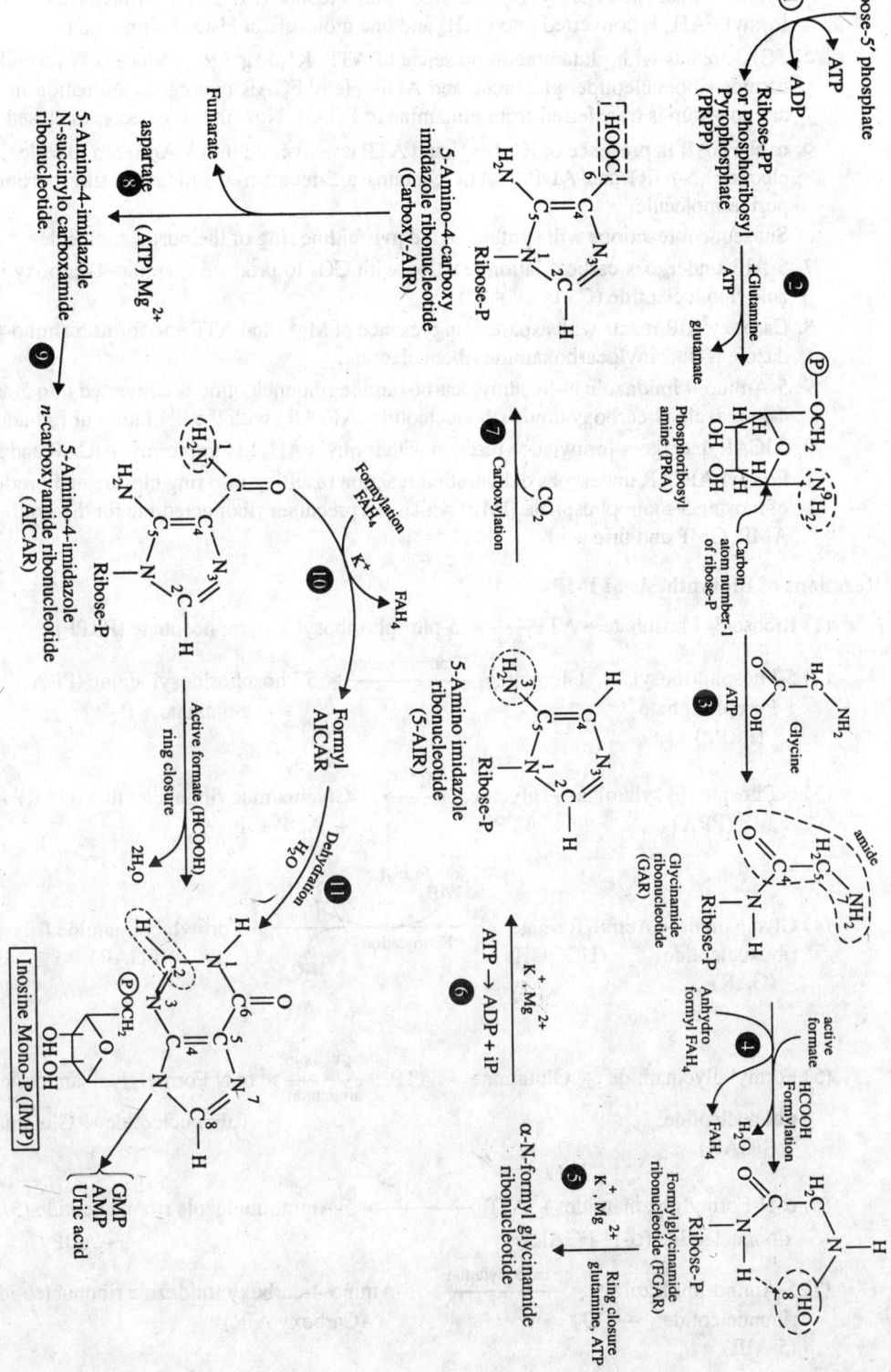

Fig. 9.18: Outline of Purine Synthesis. (Numbers refer to atoms in the final purine molecule).

4. GAR is formylated by the transfer of formyl group from the N_5, N_{10} anhydroformyltetrahydrofolic acid and Formyl glycinamide ribonucleotide (FGAR) is formed. Here Anhydroformyl FAH_4 is converted into FAH_4 and one molecule of H_2O is eliminated.
5. FGAR reacts with glutamine in presence of ATP, K^+, Mg^{2+} to produce α-N formylglycinamide ribonucleotide, glutamate and ADP. Here, FGAR undergoes amination in which amino group is transferred from glutamine to FGAR. Now the ring becomes closed.
6. α-N-FGAR in presence of K^+, Mg^{2+} and ATP is converted into 5-Aminoimidazole ribonucleotide (5-AIR) and ADP. 5-AIR contains a 5-membered imidazole ring portion of a purine-molecule.
Subsequent reactions will synthesise the pyrimidine ring of the purine molecules.
7. 5-AIR undergoes carboxylation reaction with CO_2 to produce 5-amino-4-carboxy imidazole ribonucleotide (Carboxy AIR).
8. Carboxy AIR reacts with aspartate in presence of Mg^{2+} and ATP and forms 5 amino-4-imidazole N-succinylocarboxamide ribonucleotide.
9. 5-Amino-4 imidazole N-Succinylocarboxamide ribonucleotide is converted into 5-amino-4-imidazole N-carboxyamide ribonucleotide (AICAR) with the liberation of fumarate.
10. AICAR undergoes formylation reaction with formyl FAH_4 to form formyl AICAR and FAH_4.
11. Formyl AICAR undergoes dehydration reaction resulting into ring closure and production of inosine 5-monophosphate (IMP) which is a precursor ribonucleotide for the synthesis of AMP, GMP and uric acid.

Reactions of Biosynthesis of IMP

(1) Ribose-5-Phosphate + ATP $\longrightarrow$ 5-phosphoribosyl 1-pyrophosphate (PRPP)

(2) 5-Phosphoribosyl 1-Pyrophosphate (PRPP) + Glutamine $\xrightarrow{\text{Transamination}}$ 5′Phosphoribosyl amine (PRA) + glutamate + P–P

(3) 5′ Phosphorib sylamine (PRA) + glycine + ATP $\xrightarrow{Mg^{2+}}$ Glycinamide ribonucleotide (GAR) + H_2O + ADP + ip

(4) Glycinamide ribonucleotide (GAR) + Active formate (HCOOH) $\xrightarrow[\text{Formylation}]{\text{Anhydro formyl } FAH_4 \quad FAH_4, H_2O}$ Formylglycinamide Ribonucleotide (FGAR)

(5) Formyl glycinamide ribonucleotide (FGAR) + Glutamine + ATP $\xrightarrow[\text{aminaiton}]{k^+, Mg^{2+}}$ α-N Formylglycinamide Ribonucleotide + Glutamate + ADP

(6) α-N-Formylglycinamide ribonucleotide (α-N-FGAR) + ATP $\xrightarrow{k^+, Mg^{2+}}$ 5-Aminoimidazole ribonucleotide (5-AIR) + ADP

(7) 5-Amino-imidazole ribonucleotide (5-AIR) + CO_2 $\xrightarrow{\text{carboxylation}}$ 5-Amino-4-carboxy imidazole ribonucleotide (Carboxy AIR)

Nucleic Acids

(8) 5 Amino-4-Carboxyimidazole ribonucleotide + Asparate + ATP (carboxy AIR) $\xrightarrow{Mg^{2+}}$ 5-Amino-4-imidazole N-Succinylo carboxamide ribonucleotide + ADP

(9) 5 Amino-4-imidazole N-Succinylocarboxyamide ribonucleotide $\longrightarrow$ 5- Amino-4-imidazole N-Carboxyamide ribonucleotide (AICAR) + Fumarate

(10) 5-Amino-4-imidazole N-carboxyamide ribonucleotide (AICAR) $\xrightarrow[K^+]{N^{10}\text{-formyl FAH}_4 \quad FAH_4}$ Formyl AICAR

(11) Formyl AICAR $\xrightarrow[H_2O]{dehydration}$ Inosine 5'-Phosphate (IMP)

Synthesis of Guanosine Monophosphate (GMP) from IMP

The synthesis of GMP from IMP takes place through following steps (Fig. 9.19):

(1) IMP undergoes oxidation via NAD^+ to produce xanthosine monophosphate (XMP) or Xanthylic acid. The purine base of XMP is called **xanthine.**

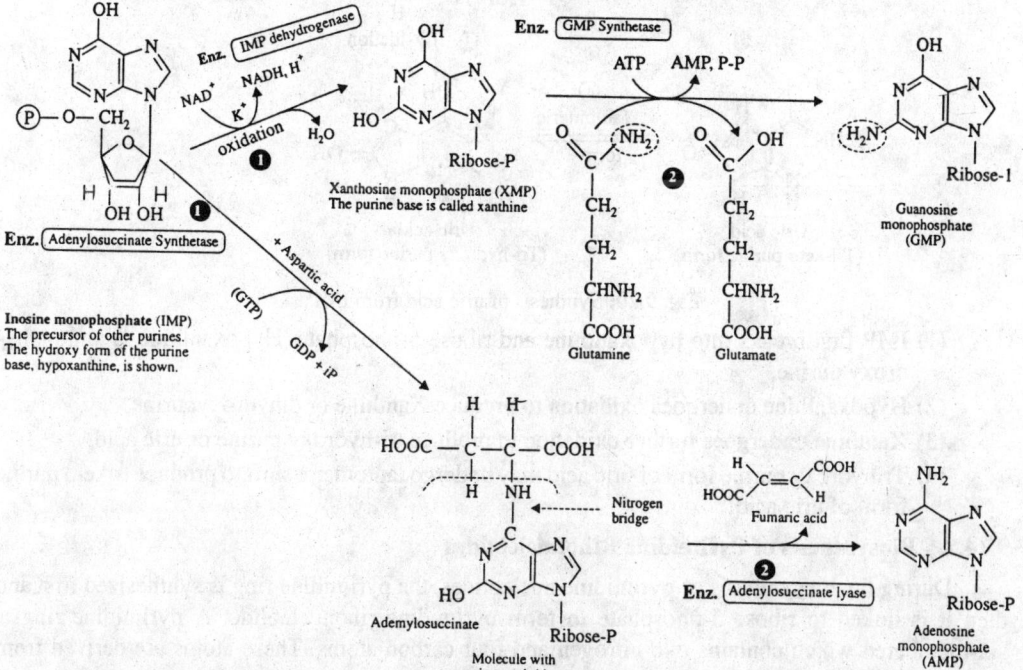

Fig. 9.19: Outline of the Synthesis of GMP and AMP from inosine monophosphate (IMP).

(2) XMP undergoes amination reaction in which XMP reacts with glutamine in presence of ATP and Mg^{2+}. During the reaction amino group (–NH$_2$) of glutamine is transferred to XMP at C_2 position resulting into synthesis of GMP. The glutamine and ATP are converted into glutamic acid and AMP respectively.

Xanthylic acid + ATP + glutamine $\xrightarrow{Mg^{2+}}$ Guanylic acid (GMP) + glutamic acid
(XMP) + AMP + PPi

Synthesis of Adenosine monophosphate (AMP) from IMP

The synthesis of AMP from IMP is outlined in Fig 9.19. The steps are as follows :

(1) IMP undergoes condensation with aspartic acid in presence of GTP to form adenylosuccinate (nic-acid) which contains a nitrogen bridge.

(2) Adenylosuccinic acid subsequently cleaves into adenosine monophosphate (AMP = adenylic acid) and fumaric acid.

Synthesis of Uric acid from IMP

The synthesis of uric acid from IMP takes place in following steps (Fig. 9.20).

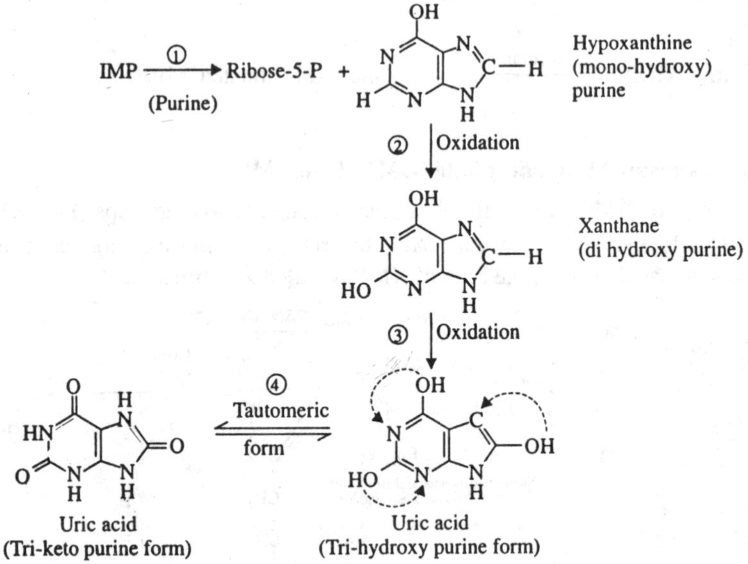

Fig. 9.20: Synthesis of uric acid from IMP.

(1) IMP first breaks into hypoxanthine and ribose-5-phosphate. Hypoxanthine is a monohydroxy purine.

(2) Hypoxanthine undergoes oxidation to produce Xanthine or dihydroxypurine.

(3) Xanthine undergoes further oxidation to produce tri-hydroxy purine or uric acid.

(4) Trihydroxy purine form of uric acid may undergo tautomeric shift to produce triketo purine form of uric acid.

9.14.2. Biosynthesis of Pyrimidine Ribonucleotides

During the biosynthesis of pyrimidine nucleotides, the pyrimidine ring is synthesized first and then it is linked to ribose 5-phosphate to form pyrimidine ribonucleotide. A pyrimidine ring is six-membered which contains two nitrogen and four carbon atoms. These atoms are derived from aspartate, glutamine (NH_3) and CO_2.

Orotic acid is the key substance in the biosynthesis of pyrimidine. From orotic acid first orotidine 5'-monophosphate (OMP) is produced and then from OMP, uridine 5'-monophosphate (UMP) is produced. UMP with glutamine synthesize cytidine 5'-monophosphate (CMP). Thus, UMP is the first pyrimidine nucleotide to be formed. Various steps in the biosynthesis of orotic acid and from Orotic acid to UMP and CMP are shown in fig 9.21.

(1) In first step NH_3 (from glutamine) combines with CO_2 (from bicarbonate) to form carbamyl phosphate in presence of ATP and enzyme *carbamyl phosphate synthetase* also known as *carbamate kinase*.

Nucleic Acids

(2) Carbamyl phosphate condenses with aspartic acid to form carbamyl aspartate (=ureidosuccinic acid) in presence of enzyme *aspartate carbamoyl transferase*.

$$NH_3 + CO_2 \xrightarrow[\text{ADP}]{\text{ATP}} \text{Carbamyl-P}$$

$$\text{Carbamyl-P} + \text{Aspartic acid} \xrightarrow{Pi} \text{Carbamyl aspartate}$$

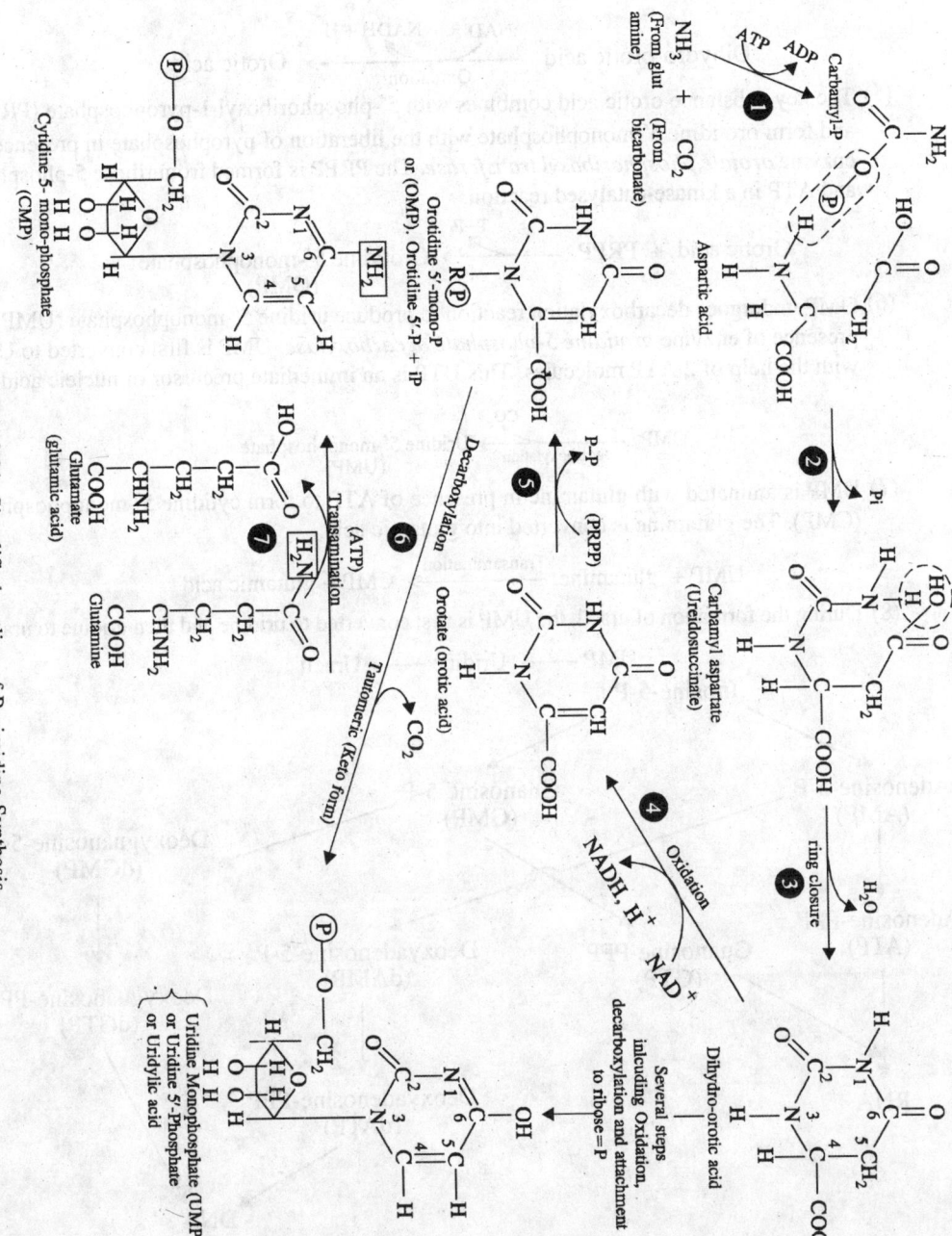

Fig. 9.21: Simplified outline of Pyrimidine Synthesis.

(3) Carbamyl aspartate undergoes dehydration and ring closure to form dihydro-orotic acid in presence of enzyme *dihydro-orotase*.

$$\text{Carbamyl aspartate} \xrightarrow[\text{ring closure}]{H_2O} \text{Dihydro-orotic acid}$$

(4) Dihydro-orotic acid is oxidized with NAD+ coenzyme to orotic acid in presence of enzyme *dihydro-orotate dehydrogenase*

$$\text{Dihydro-orotic acid} \xrightarrow[\text{Oxidation}]{NAD^+ \quad NADH+H^+} \text{Orotic acid}$$

(5) The key substance orotic acid combines with 5'-phosphoribosyl 1-pyrophosphate (PRPP) and form orotidine 5'-monophosphate with the liberation of pyrophosphate in presence of enzyme *orotate phosphoribosyl transferase*. The PRPP is formed from ribose 5-phosphate and ATP in a kinase-catalysed reaction.

$$\text{Orotic acid} + \text{PRPP} \xrightarrow{P-P} \text{Orotidine 5'-monophosphate (OMP)}$$

(6) OMP undergoes decarboxylation reaction to produce uridine 5'-monophosphate (UMP) in presence of enzyme *orotidine 5-phosphate decarboxylase*. UMP is first converted to UTP with the help of 2 ATP molecules. This UTP is an immediate precursor of nucleic acids.

$$\text{OMP} \xrightarrow[\text{decarboxylation}]{CO_2} \text{Uridine 5'-monophosphate (UMP)}$$

(7) UMP is aminated with glutamine in presence of ATP to form cytidine 5'-monophosphate (CMP). The glutamine is converted into glutamic acid.

$$\text{UMP} + \text{glutamine} \xrightarrow{\text{Transamination}} \text{CMP} + \text{glutamic acid}$$

(8) During the formation of uracil, the UMP is first converted to uridine and then uridine to uracil.

$$\text{UMP} \longrightarrow \text{Uridine} \longrightarrow \text{Uracil}$$

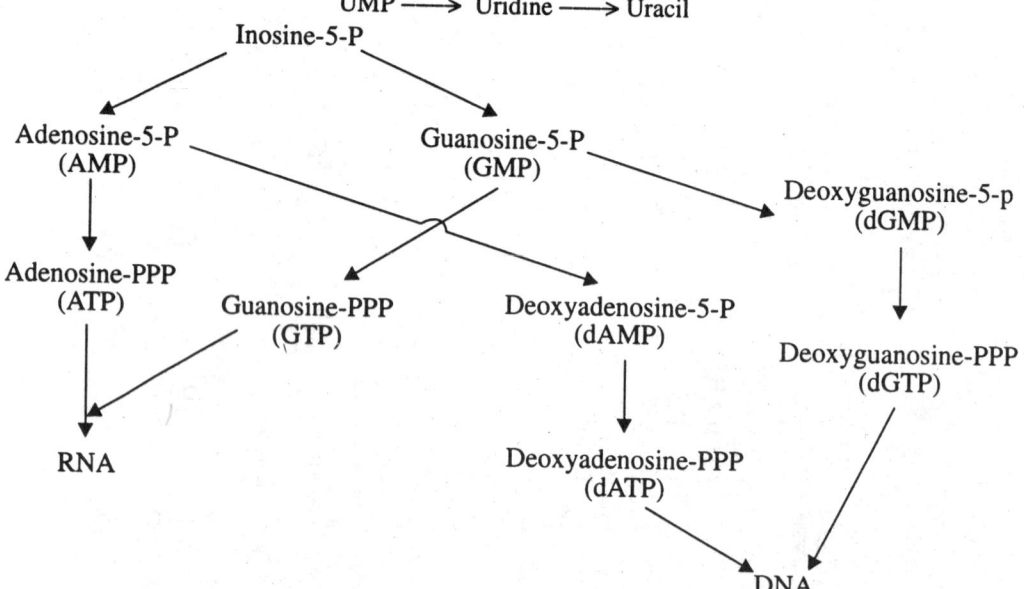

Fig. 9.22 : Outline Scheme for the synthesis of purine nucleotides of RNA and DNA from IMP. M=mono, P=phosphate, d = deoxy, T = tri and A, G= purine bases.

Nucleic Acids

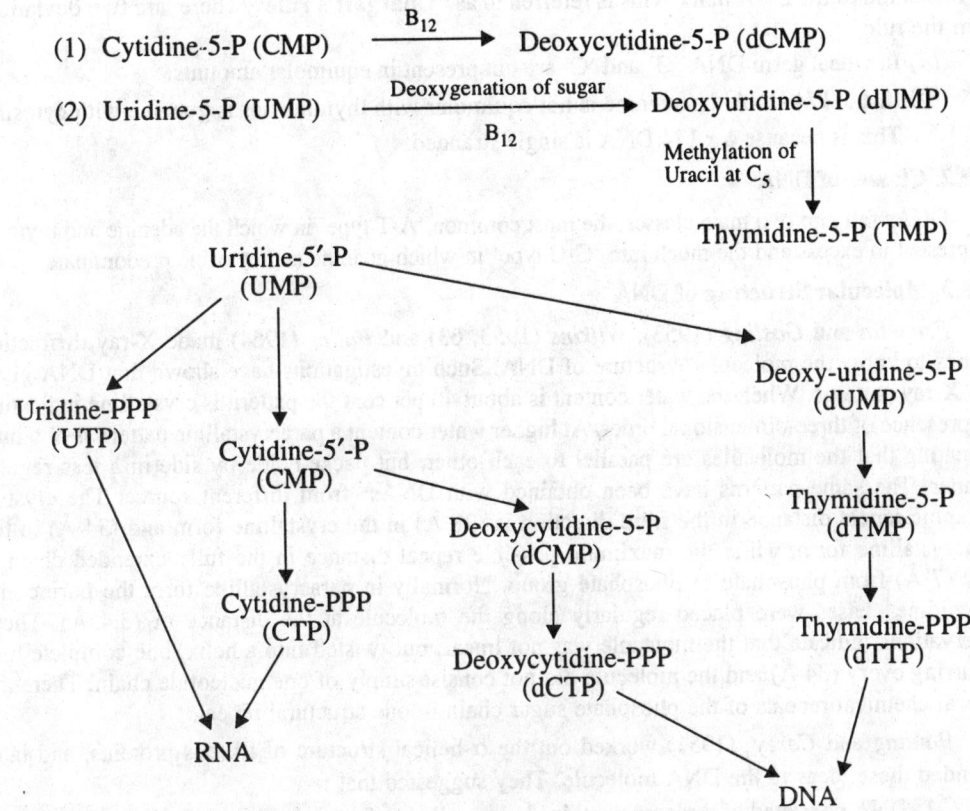

Fig. 9.23 : Outline Scheme for the synthesis of pyrimidine nucleotides of RNA and DNA.

9.15. THE DEOXYRIBONUCLEIC ACID (DNA)

Studies since 1940s have indicated that DNA is the universal genetic material of all forms of life except certain viruses, which have RNA as their genetic material. In the unclei of plant and animal cells the DNA is present in the chromosomes in association with protein molecules, but in mycoplasmas and bacteria it is not found associated with any other molecule. It has attracted the attention of a number of biochemists and biologists as it is directly involved in hereditary mechanisms. Because the DNA has been considered as the life of cell, it has occupied a central place in biochemistry, cytology, genetics, microbiology, molecular biology and in other branches of science. Though most of the details regarding the structure and chemistry of DNA have been worked out in recent years, still the investigations are being made to know more and more about DNA. Generally, DNA has been extracted from the nuclei of animal and plant cells, but in recent years the DNA has also been extracted from mitochondria, plastids and centrioles. In certain cases, for example, *Paramecium, Amoeba proteus,* amphibians and ferns, the DNA molecules are found in cytoplasmic matrix.

9.15.1 The Structure of DNA

The building units of DNA are nucleotides, each one of which is made up of nitrogenous base, a deoxyribose sugar and a phosphoric acid molecule. The nucleotide units are joined together to form a polynucleotide chain.

Chargaff further studied the chemical structure of DNA and showed that DNA contained equal proportions of the large purine bases and the smaller pyrimidines, and even more interesting that adenine and thymine were present in equimolecular proportions and so are cytosine and guanine. This equivalence of 'A' and 'T' and of 'G' and 'C' was of the utmost importance in relation to the formation of the DNA helix. This is referred to as **"Chargaff's rule"**, There are two deviations from the rule :

(a) In wheat germ DNA 'G' and 'C' are not present in equimolar amounts.

(b) In $\phi \times 174$ DNA the adenine is not equimolar with thymine nor is guanine with cytosine. This is because $\phi \times 174$ DNA is single stranded.

9.15.2. Classes of DNA

DNAs fall into two main classes, the most common 'A-T type' in which the adenine and thymine are present in excess and the much rare 'G-C type' in which guanine and cytosine predominate.

9.15.3. Molecular Structure of DNA

Franklin and *Gosling* (1953), *Wilkins* (1953, 63) and *Fuller* (1964) made X-ray diffraction studies to know the molecular structure of DNA. Such investigations have shown that DNA gives two X-ray patterns. When the water content is about 40 per cent the pattern is crystalline indicating the presence of three-dimensional order. At higher water content a paracrystalline pattern is obtained indicating that the molecules are parallel to each other, but packed side by side in a less regular manner. The same patterns have been obtained with DNAs from different source. The crystallographic repeat distance in the fibre direction is (28 Å) in the crystalline form and (34 Å) in the paracrystalline form, while the maximum possible repeat distance in the fully extended chain is only (7 Å) from phosphate to phosphate group. Normally in paracrystalline form the purine and pyrimidine bases were placed regularly along the molecule at the distance of (3.4 Å). These observations indicate that the molecule was not linear, but twisted into a helix, one complete turn occurring every (34 Å) and the molecule did not consist simply of one nucleotide chain. There are several chemical repeats of the phosphate sugar chain in one structural repeat.

Pauling and *Corey* (1951) worked out the α-helical structure of fibrous proteins, and later extended these ideas to the DNA molecule. They suggested that :

1. DNA consisted of three nucleotide chains coiled to form a helix.
2. In such structures, the phosphate groups are oriented towards the inner side and the purine and pyrimidines groups towards outside.

But the modern biochemists working on this aspect suggested that hydrogen bonding between the bases probably played a part in stabilizing the molecular structure. This could only occur if, in contrast to Pauling and Corey's suggestion, the bases on the polynucleotide chains are inwards towards each other.

9.15.4. Watson and Crick Model of DNA

Watson and *Crick* (1953) constructed their famous "doble helix" model of DNA, which explained all the evidence than available, and for which they and Wilkins were later awarded the Noble-Prize. The model is illustrated in Fig. 9.24 a, b. This model is important because it explains the physio-chemical and biological properties of DNA, particularly its duplication in the cell. The essential characteristics of the model are explained as follows :

(a) Each DNA molecule is composed of two long polynucleotide chains that run in opposite directions, forming a double helix around a central axis.

(b) In each chain the deoxyribose sugar units on adjacent nucleotides are linked by phosphate groups to form an outer sugar-phosphate back-bone.

(c) Each nucleoside is disposed in a plane that is perpendicular to that of the polynucleotide chain.

(d) The two chains are held together by hydrogen bonds established between the pair of bases. Further, the purine and pyrimidine bases of the nucleotide units are found inwards.

(e) The pairing of the bases is highly specific. Because there is a fixed distance of 11 Å between two sugar moieties in the opposite nucleotides, one purine base can pair with one pyrimidine

Nucleic Acids

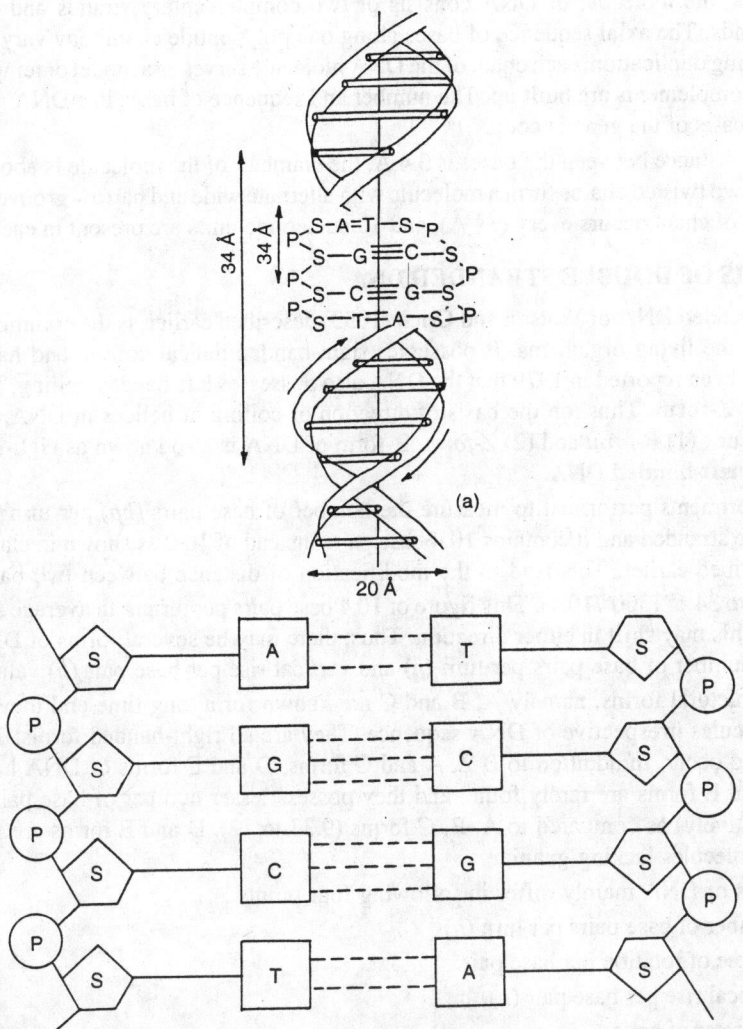

Fig. 9.24: (a) Diagrammatic representation of the DNA molecule as proposed by Watson & Crick.
(b) Diagrammatic representation of part of a hypothetical polynecleotide chain in DNA.

base. Thus, A-T and G-C pairs are the only ones that can be formed. Fig. 9.24b show that two hydrogen bonds are formed between A and T, and three hydrogen bonds are formed, between C and G.

Because of this type of base pairing, the model satisfies *chargaff's* chemical observations that the DNA molecule contains equal number of adenine and thymine bases, and of cytosine and guanine bases. It also satisfies the Wilkins' X-ray diffraction observations.

(f) The two polynucleotide chains are complementary to each other, in that there is complementary relationship between their sequence of bases. Thus, if one chain has a region which goes -adenine -guanine-cytosine-thymine-guanine, then the corresponding region of the complementary chain will go–thymine–cytosine–guanine–adenine–cytosine. This can be shown as follows :

```
    1st Chain   A,  G,  C,  T,  G
                ||  |||  |||  ||  |||
    2nd Chain   T   C   G   A   C
```

Thus, the molecule of DNA consists of two complementary strands and not indentical strands. The axial sequence of bases along one polypeptide chain may vary considerably. During duplication, each chain of the DNA molecule serves as a model or template on which its complements are built up. The number and sequence of bases in a DNA molecule form the basis of the genetic code.

(g) The distance between the bases is 3.4 Å, the diameter of the molecule is about (20 Å), and the two twisted chains form a molecule with alternate wide and narrow grooves. A complete turn of chain occurs every (34 Å), and 10 nucleotide units are present in each turn.

9.16. FORMS OF DOUBLE STRANDED DNA

The double stranded DNA of Watson and Crick (1953) described earlier, is the commonly occurring DNA in most the living organisms. It possesses right-handed helical coiling and has been called **B-form**. It has been reported in 1979 that the DNA also possesses left-handed coiling. This DNA has been called as **Z-form**. Thus, on the basis of direction of coiling of helices in DNA, two forms of DNA may occur : (1) *B-form* and (2) *Z-form*. B-form of DNA is also known as **righ-handed DNA** and Z-form as **left-handed DNA**.

The experiments performed to measure the number of base pairs (*bp*) per turn indicated that DNA is double stranded and it contains 10.4 base pairs instead of 10.0 as shown in classical B-form of DNA described earlier. This lead to the modification of distance between two base bairs from 36° (360°/10) to 34.6° (360°/10.4). This figure of 10.4 base pairs per turn is an average and according to conditions this may shift in either direction. Thus, there may be several forms of DNAs showing variations in number of base pairs per turn (*n*) and vertical rise per base pair (*h*) values.

Three structural forms, namely A, B and C are known for a long time and they are found in all DNA molecules irrespective of DNA sequence. They are all right-handed forms. The transition forms may also occur. In addition to B Z, A and C forms, D and E forms of DNA have also been reported. D and E forms are rarely found and they possess lesser number of base pairs per turn (8 and 7.5 respectively) as compared to A, B, C forms (9.33 to 12). D and E forms are found only in some DNA molecules lacking guanine.

All forms of DNA mainly differ in following four points :

(a) Number of base pairs per turn (*n*)

(b) Degree of rotation per base pair.

(c) Vertical rise per base pair (*h*).

(d) Diameter of helix.

The existence of these forms also requires certain conditions. In their absence the existences of these forms is impossible. The conditins, number of base pairs per turn (*n*), degree of rotation per base pair, vertical rise per base pair (h) and diamter of A, B, C and Z forms of DNAs are shown in the table 9.1.

Table 9.1: Showing characteristic features of A, B, C and Z forms of DNA

DNA Form	Conditions	Base pairs per turn (n)	Rotation per base pair	Vertical rise per base pair	Diamter of helix
A	75% relative humidity; Na^+, K^+ or Cs^+ ions.	11	+32.7° (right-handed)	2.56 Å	23 Å
B	92% relative humidity, low ionic concentration	10	+36.0° (right-handed)	3.38 Å	19 Å
C	66% relative humidty; Lithium ions (Li^+)	9.33	+38.6° (right-handed)	3.32 Å	19 Å
Z	Very high salt concentration	12	−30.0° (left-handed)	3.71 Å	18 Å

Nucleic Acids

Z-DNA or left-handed DNA

It was reported in 1979. This DNA was obtained artificially by synthesizing d $(C-G)_3$ molecules in crystallized form, the Name Z-DNA was given to this because this DNA follows a zig-zag course. It resembles with B-form DNA in certain points and also differs in many other respects. The similarities and dissimilarities between Z-DNA and B-DNA are as follows :

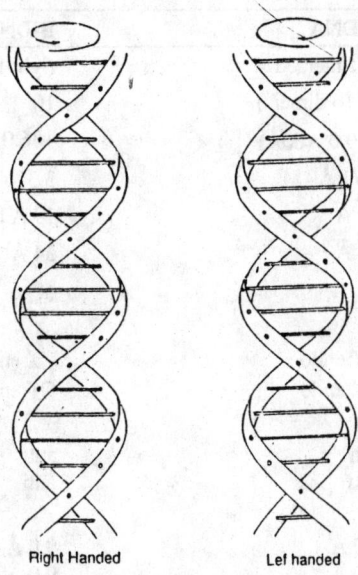

Right Handed   Lef handed

Fig. 9.25: Two helices showing right-handed sense in B-DNA and left-handed sense in Z-DNA.

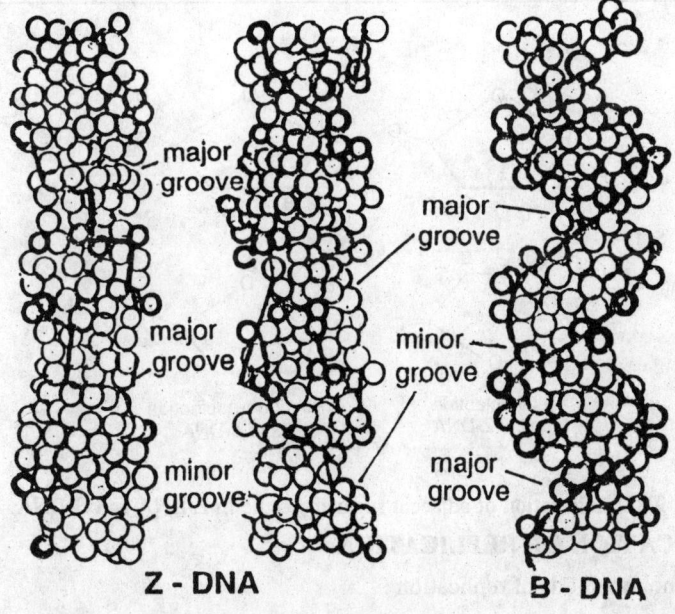

Fig. 9.26: Side views of Z-DNA and B-DNA.

Similarities between Z-DNA and B-DNA

(1) Both forms of DNA are double helical.
(2) Two strands of double helix run antiparallel in both DNAs.
(3) Both forms exhibit G ≡ C pairing.

Table 9.2: Dissimilarities between Z-DNA and B-DNA

	Character	Z-DNA	B-DNA
1.	Helix sense	left-handed	right-handed
2.	Base pairs per turn	12 (6 dimers)	10
3.	Rotation per base pair	−30.0° (60Å)	+36.0
4.	Vertical rise per base pair	3.71 Å	3.38 Å
5.	Helix diameter	18 Å	19 Å or 20 Å
6.	Helix pitch	45 Å	34 Å
7.	Base pair tilt	7°	6°
8.	Sugar pucker (a) deoxyguanosine (b) deoxycytidine	C_3' endo C_2' endo	C_2' endo C_3' endo
9.	Glycoside torsion angle (a) deoxyguanosine (b) deoxycytidine	syn anti	anti anti
10.	Distance of P from axis d G_pC d C_pG	8.0 Å 6.0 Å	9.0 Å 9.0 Å
11.	Phosphate backbone	zig-zag	regular
12.	Orientation of sugar molecules	alternate	not alternate
13.	Repeating unit	dinucleotide	Mononucleotide

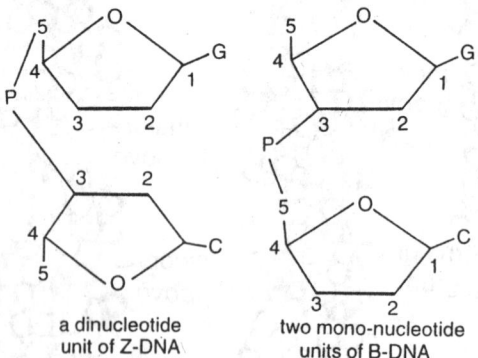

a dinucleotide unit of Z-DNA   two mono-nucleotide units of B-DNA

Fig. 9.27: Orientation of adjacent sugar molecule in Z-DNA and B-DNA.

9.17. DNA DUPLICATION OR REPLICATION

There are two common methods of replication :

9.17.1. Watson and Crick model

Watson and *Crick* (1953) also proposed the method of DNA replication. Their model itself suggests the manner in which DNA molecules are replicated (Fig. 9.28).

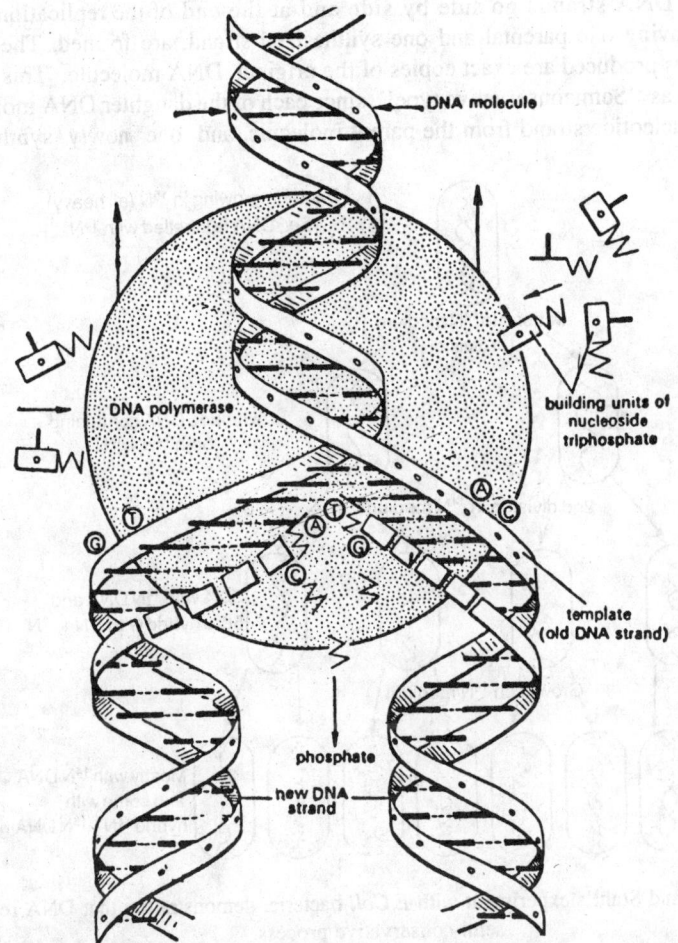

Fig. 9.28: Diagrammatic represenation of the replication of DNA molecule.

According to their hypothesis, during replication the nucleotides of a DNA molecule split by separation of their hydrogen bonds. As a result, the two strands of the DNA molecule unwind and separate completely. The strands, thus, separated are complements of one another. When the separation is completed, the nucleotides of the separated strands attract other complementary nucleotides which are already present in the cell nucleus as nucleotides pool. During the formation of complementary strand of each separated strand, first purine attaches with the pyrimidine and pyrimidine with purine ('A' couples with 'T' and 'G' with 'C' or Vice-Versa). The attachment of bases takes place through hydrogen bonds. After the attachment of bases, the sugar molecules unite with one another by their phosphate components. Thus, the formation of nucleotide and polynucleotide chain takes place. In the final stage, the complementary units link up to form two DNA molecules, each of them is an exact copy of the original molecule. This is known as **DNA replication** or **DNA duplication**.

9.17.2. Meselson and Stahl's Theory

This theory was proposed by *Meselson* and *Stahl* in 1958. According to them, during replication the two strands do not separate completely and synthesize their replica, but instead they start unzipping at one end and simultaneously the unzipped segments start attracting their respective nucleotide pairs. Thus, the unzipping of the original DNA strands and synthesis of fresh DNA strands go side by side and at the end of the replication two daughter DNA molecules, having one parental and one synthesized strand, are formed. The two daughter DNA molecules thus produced are exact copies of the original DNA molecule. This type of DNA replication is called as "Semiconservative type", since each of the daughter DNA molecule consists of one 'Old' polynucleotide strand from the parent molecule, and one newly synthesized strand.

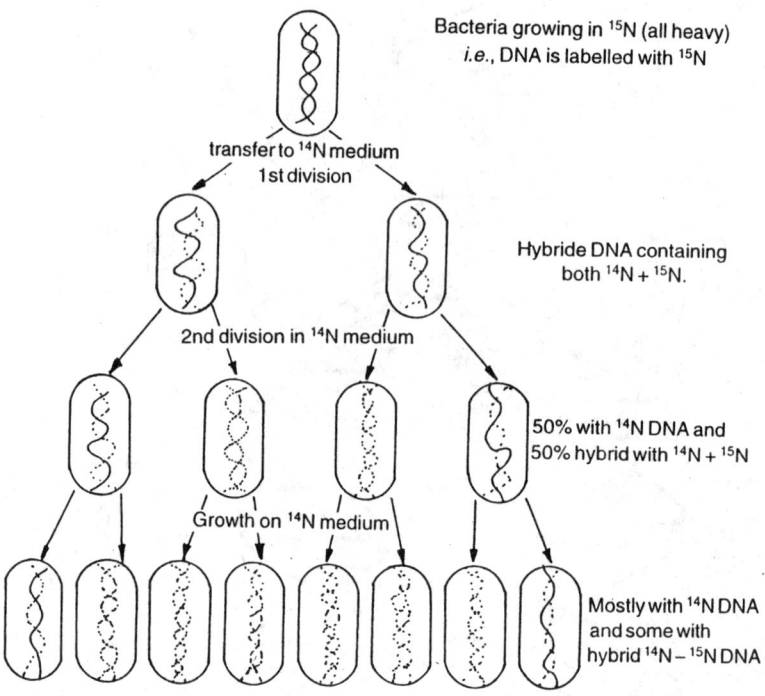

Fig. 9.29: Meselson and Stahl's experiment with *E.Coli* bacteria, demonstrating that DNA replication is a semi-conservative process.

The over-all process is summarised in the following steps :

(a) During replication the DNA molecule first unzips (the two strands move in the opposite direction and start to open).

(b) During unzipping process the hydrogen bonds between the organic bases are broken. As a result two strands start to open.

(c) Attachment of new complementary nucleotides, which are present in the nucleus, takes place.

(d) Two similar DNA molecules are formed. The semi-conservative type of DNA replication was shown by Meselson and Stahl in *E. coli* by the use of ^{15}N isotope of nitrogen which was heavier than ^{14}N. The experiment is shown in Fig. 9.29.

9.18. MECHANISM OF DNA REPLICATION

The complete process of DNA replication involves following steps in *E.coli*.

1. Recognition of the Initiation Point

DNA replication starts at a specific point, called **initiation point** or **origin** where replication fork begins. This is a nucleotide sequence of 100 to 200 base pairs. **Specific initiator proteins** recognise the initiation point on DNA. Such proteins along with DNA directed RNA polymerase initiate the synthesis of RNA primer for the formation of DNA chain. Prokaryotic chromosomes usually possess one initiation point or replication fork whereas eukaryotic chromosomes may possess several (about a thousand) replication forks.

2. Unwinding of DNA

The unwinding proteins bind to the nicked strand of the duplex and open up a loop or bubble, separating the two strands of DNA duplex.

3. RNA Priming

Now, RNA primers are synthesized from the DNA directed RNA polymerase. The priming RNA strands (RNA primers) are complementary to the two stands of DNA and are made up of 50 to 100 nucleotides.

4. Formation of DNA on RNA Primers

The new strands of DNA are synthesized in the 5' – 3' direction from the 3'–5' template DNA by the addition of deoxyribonucleotides to the 3' end of the primer RNA. This addition is achieved by DNA polymerase III (poly III-copol III) in presence of ATP. Once the synthesis of DNA strand

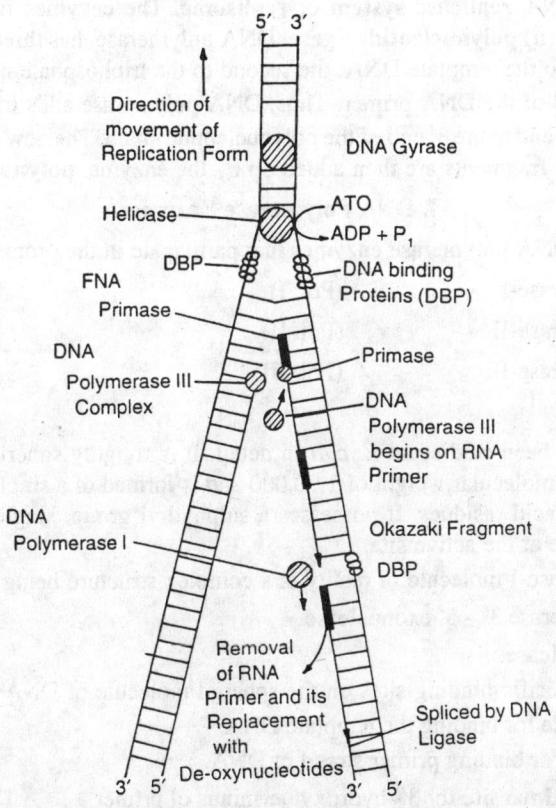

Fig. 9.30: Summary of major steps in DNA replication.

has been intiated, copol III detaches and poly III carries out replication of the DNA strand. The unwinding proteins separate of DNA duplex strands ahead of replication fork.

The **leading strand of DNA** is synthesized in 5′–3′ direction as one piece. The **lagging strand of DNA** is synthesized in its opposite direction in short segments consisting of 1,000 to 2,000 nucleotides. These segments are called **Okazaki fragments.**

5. Excision of RNA Primers

Once a small segment of an Okazaki fragment has been formed, the nucleotides of RNA primer are removed from the 5′ end one by one by the action 5′–3′ exonuclease activity of DNA polymerase-I.

6. Joining of Okazaki Fragments

The gaps left between Okazaki fragments are filled with complementary deoxyribonucleoride residues by DNA polymerase-I. Finally, the adjacent 5′ and 3′ ends are joined by **DNA-ligase.**

In eukaryotic cells, the mechanism of DNA replication is expected to be more complex than in prokaryotes. The replication of eukaryotic DNA begins at multiple points of origin. Their RNA primer is formed of about 10 nucleotides and the Okazaki fragments are much shorter (about 100 to 150 nucleotides).

Enzymes for DNA Synthesis

About 20 or more different proteins and enzymes are required during DNA replication. These collectively form a **DNA replicase system** or **replisome**. The enzymes fall into two types—(*i*) **DNA polymerase** and (*ii*) **polynucleotide ligase.** DNA polymerase has three sites for attachement. One of them attaches to the template DNA, the second to the triphosphate nucleotide and the third one to the 3′—OH end of the DNA primer. Thus, DNA polymerase adds triphosphate nucleotides to primer DNA from 5′ end to the 3′ end of the polynucleotide chain. The new strands are synthesized in fragments and these fragments are then added up by the enzyme, **polynucleotide ligase.**

I. DNA Polymerase Enzymes

There are three DNA polymerase enzymes that participate in the process of DNA replication.

(*i*) DNA polymerase-I   (Pol. I)

(*ii*) DNA polymerase-II   (Pol. II)

(*iii*) DNA polymerase-III   (Pol. III)

(*i*) DNA Polymerase-I

This enzyme has been studied in *E. coli* in detail. It is roughly spherical with a diameter of about 6.5 mm. It has a molecular weight of 1,90,000 and is formed of a single polynucleotide chain of about 1,000 amino acid residues. It possesses a sulphydryl group, single interchain disulphide and one zinc mole cule at the active site.

A DNA polymerase-I molecule in reality is a complex structure being formed of :

(*i*) DNA polymerase 3′ – 5′ exonuclease

(*ii*) 5′–3′ exonuclease.

There are five specific binding sites on the spherial molecule of DNA polymerase-I:

(*i*) **Template site** for binding the template DNA.

(*ii*) **Primer site** for binding primer strand of DNA.

(*iii*) **Primer terminus site** for 3′–hydroxyl terminus of primer.

(*iv*) **5′–triphosphate site** – a locus for incoming deoxyribonucleotide 5′–triphosphate group.

(*v*) **5′–3′ exonuclease site**, a locus for 5′–3′ exonuclease activity situated in the path of growing chain.

Nucleic Acids

DNA polymerase-I was discovered by KORNBERG and his colleagues in 1955, then it was considered to carry out DNA replication. It is now known that DNA polymerase-I performs varied functions (**multifunctional enzyme**).

(1) It catalyses the addition of mononucleotide units (the deoxyribonucleotides) to free 3'–OH end of DNA strand, thus, participating in the repair of DNA molecule. A pure DNA polymerase-I can add about 1000 nucleotide units per minute per molecule of enzyme at 37°C.

(2) It catalyses 3'–5' exonuclease activity and removes nucleotide residues of primer RNA at 3' end.

(3) It also catalyses 2'–3' exonuclease activity.

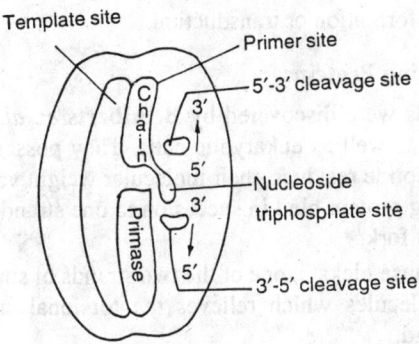

Fig. 9.31: A model of DNA polymerase-I enzyme with different sites.

(ii) DNA Polymerase-II

Its role is not yet fully understood. It is effective only on DNA duplex with gaps and it cannot replicate long strands of DNA.

(iii) DNA Polymerase-III

It was discovered by **T. Kornberg** and **M.L. Gefter** in 1972. It is most active enzyme among all the three polymerases. It is made up of α, β, θ, δ and ε sub-units.

DNA polymerase-III is chiefly responsible for DNA chain elongation. It is complex molecule with molecular weight (MW) about 550,000. the sub-unit β is also known as **copolymerase-III**. It recognizes and binds to the primer strand of parental DNA. The copolymerase III is released just after the polymerase-III binds to the correct initiation point. DNA polymerase-III now helps in elongating the DNA strand in 5'–3' direction by addition of new nucleotide residues to the 3' end of primer strand. Thus, DNA polymerase-III does not initiate replication but helps in elongation of chain.

It also acts as 5' – 3' exonuclease and 3' –

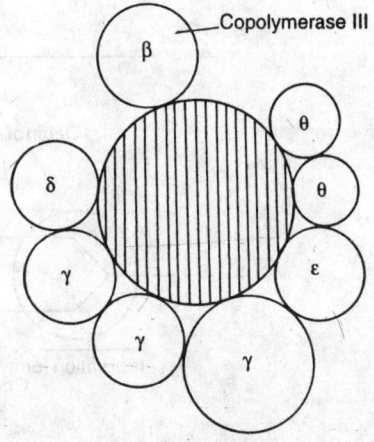

Fig. 9.32: DNA polymerase-III holoenzyme.

5' exonuclease. It can hydrolyse terminal nucleotides from either end of a DNA strand.

II. DNA-Ligase or Polynucleotide Ligase

DNA-ligase enzyme helps in joining or sealing the joints of the two DNA fragments by catalysing the synthesis of a phosphodiester bond between a 3'–OH group at the end of one chain and 5'–phosphate group at the end of other chain. This enzyme was extracted from E. coli. It is made up of single polynucleotide chain with a molecular weight about 77,000. Each E. coli cell contains about 2000 to 4000 copies of this enzyme. DNA ligase enzyme performs following important functions.

1. It helps in joining the DNA fragments during DNA replication.
2. It helps in repairing signle-stranded nicks in duplex DNA molecule.
3. It helps in linking the ends of linear DNA duplexes to form circular DNA.
4. It helps in joining the segments of DNA during recombination which takes place during meiosis, genetic transformation or transduction.

DNA Unwinding and Untwisting Proteins

DNA unwinding proteins were discovered by **B. Alberts** *et. al.* in T_4 phage. A number of them are found in prokaryotic as well as eukaryotic cells. They possess specific binding sites for short segments of about 8 nucleotide residues, their molecular weight varies from 10,000 to 75,000. Several molecules of unwinding protein bind in succesion to one strand of DNA duplex in advance for the formation of replication fork.

The untwisting proteins cause nicks in one of the two strands of supercoiled DNA. This allows some unwinding of DNA molecules which relieves the torsional stress. The nick is released afterwards completing the strand.

DNA Replication is Discontinuous

DNA replication is usually discontinuous and takes place in short steps. DNA polymerase

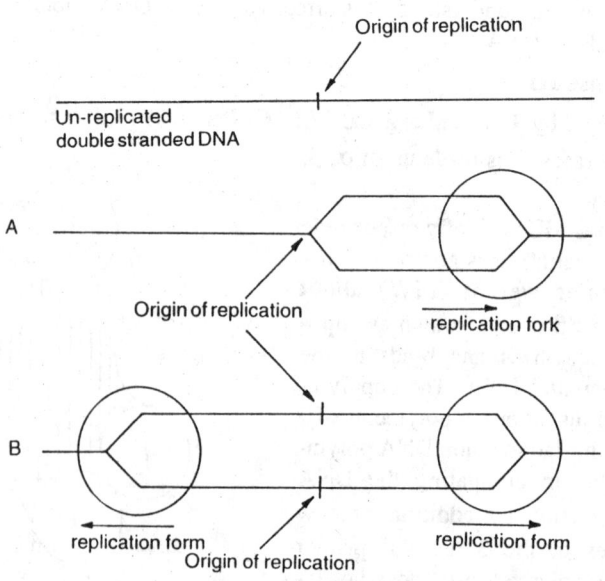

Fig. 9.33: Explains unidirection and bidirectional replications with movement of replication forks (A) unidirectional replication (B) bidirectional replication.

helps in the formation of polynucleotide chain in 5' – 3' direction along the 3' – 5' polynucleotide strand of DNA. It means the same enzyme cannot help in the synthesis of 3'– 5' chain along the 5'–3' polynucleotide strand of parental DNA.

Direction of DNA Replication

Unidirectional replication was suggested by **J. Cairans** but the recent experiments have suggested the bidirectional replication of DNA. During replication of DNA, replication forks are also formed. One replication fork is formed during unidirection replication while two forks are formed during bidirectinal replication.

Evidences in Support of Semiconservative Mode of DNA Replication

1. Meselson and Stahl's Experiment (using N^{15})

M. Meselson and F.W. Stahl (1958) performed an experiment on *E.coli* bacteria which demonstrates the semiconservative type of DNA replication (Fig. 9.29). following two basic principles were considered during the experiment :

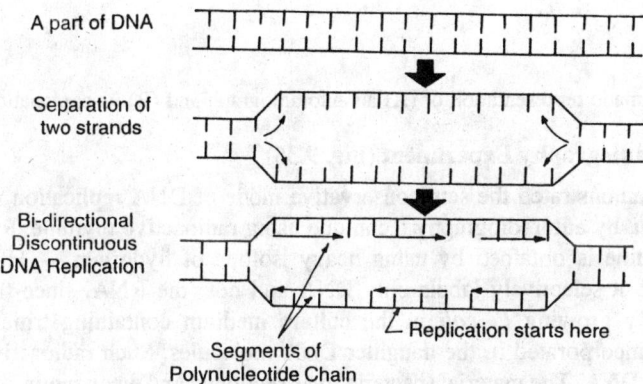

Fig. 9.34: Showing bidirectional DNA replication.

(*i*) They cultured *E. coli* bacteria in a culture medium containing N^{15} isotope of nitrogen. The bacteria replicated in the medium for a number of generations and both the strands of their DNA now contained N^{15} as constituent of purines and pyrimidines. Thus, the DNA of all the bacteria was labelled with heavy nitrogen (N^{15}). These DNA labelled bacteria were transferred in the culture medium containing N^{14} and were allowed to replicate. After first generation of replication, the separated DNA possessed one strand with N^{15} and other with normal N^{14} nitrogen. The heavier N^{15} strand represents the parental strand and the lighter N^{14} strand represents newly synthesized strand. Thus, this, DNA containing one N^{15} and other N^{14} strand is called **hybrid DNA** and the replication is of semiconservative type because it conserves the parental DNA.

(*ii*) The second principle utilized was the preparation of caesium chloride density gradient. This density gradient is based on the fact that when heavy salt solutions are subjected to ultracentrifugation, a continuous density gradient is set up. When a substance having density within the range of this gradient is dissolved in salt solution, this substance will find place at its own level of density. Thus, very light differences in density can be detected.

Based on above principle Meselson and Stahl conducted density gradient experiments. The DNA of the parental generation bacteria was labelled with heavier N^{15} nitrogen and hence bacteria were transferred to normal N^{14} culture medium. In first generation the DNA was lighter than the parental labelled N^{15} DNA. In second generation two bands were observed indicating the presence of DNA with two different densities, as is expected according to semiconservative type of DNA replication. These two bands were of equal intensity in the second generation. The intensity of light density bands gradually increased and the intensity of hybrid density bands gradually decreased during subsequent generations.

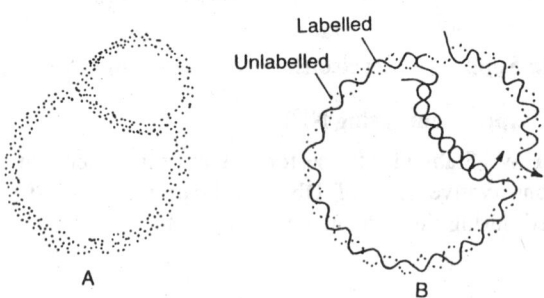

Fig. 9.35: Diagrammatic representation of (A) an autoradiograph and (B) its explanation in terms of DNA.

2. Cairans Autoradiography Experiment (Fig. 9.35)

J. *Carians* demonstrated the semiconservative mode of DNA replication in the chromosome of bacterial. *E. coli*, by autoradiography technique using radioactive thymine. Radioactive thymine or tritiated thymidine is obtained by using heavy isotope of hydorgen — H_3. He used tritiated thymidine because it selectively labels only DNA and not the RNA, since thymine base is not found in RNA. By growing *E. coli* in the culture medium containing tritiated thymidine, the radioactivity was incorporated in the daughter DNA molecules. Such radioactive DNA molecules are called **labelled DNA.** The material (bacteria) was sectioned and when mounted on a photographic emulsion for autoradiography, the labelled DNA exposed the film producing a diffused profile of DNA molecule. The duplicating DNA molecule showed a replicating fork, the point at which two chains become four. After first replication, the radioactivity was incorporated to only one of the strands of DNA and both strands become labelled after second generation. Thus, the photographs showed the presence of tritium. The results of first and second generation showed the semiconservative mode of DNA replication.

3. Taylor's Experiment of *Vicia faba* Root Tips (Fig. 9.36)

Autoradiography was also utilized by J.H. Taylor and his co-workers (1957) to demonstrate the semiconservative mode of DNA replication in the root tip cells of *Vicia faba*. The roots were grown in a medium containing radioactive thymidine, so that radioactivity is incorporated in the DNA of these cells. The chromosomes, thus, become labelled. When these root tips with labelled chromosomes were transferred to the unlabelled medium containing colchicine and studied for radioactivity, in the first generation of duplication both chromatids were labelled. In the second generation of duplication, in each chromosome, one of the two chromatids was found to be labelled. This was considered as showing semi-conservative mode of DNA replication.

Taylor's experiment is important as it demonstrated the semiconservative mode of replication in chromosmes of a higher plant. But as it is known that each chromatid is made up of only one double helical DNA molecule, it is the replication of DNA.

Nucleic Acids

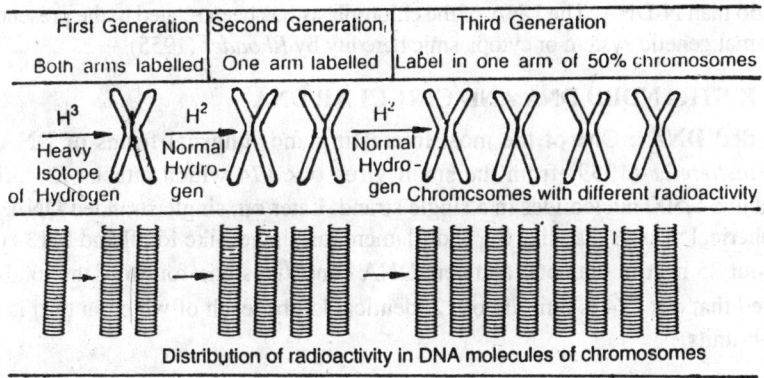

Fig. 9.36: Taylor's experiment on *Vicia faba* root tips using autoradiographic technique.

9.19. MITOCHONDRIAL DNA (Mt-DNA)

The presence of mitochondrial DNA (Mt-DNA) was reported by *M.M.K. Nass* and *S. Nass* (1963) and it was confirmed later on by *Kleinschmidt*. A single mitochondrion may contain one or more molecules of DNA according to its size. The common length which has been reported from many mitochondria is 5 µ. The molecules of Mt-DNA are highly twisted double stranded and circular in shape, and contain displacement (d) loop (*Borst*, 1972). They are similar to bacterial DNA in their circular shape, but there are certain differences in between mitochondrial DNA (Mt-DNA) and nuclear DNA (N-DNA) which are as follows :

1. Mitochondrial DNA(Mt-DNA) is circular in shape with a molecular weight ranging from 220 to 166×10^6.
2. Mt-DNA contains more guanine and cytosine (G-C) contents than the N-DNA (Rabinowitz, 1968).
3. Mt-DNA has higher denaturation temperature than the N-DNA.
4. Mt-DNA is shorter and contains few coded informations than the N-DNA.
5. DNA polymerase of the mitochondria is also different from the N-DNA polymerase.
6. Mt-DNA is devoid of histone proteins.

Recently, it has been established that mitochondria have the ability to conduct protein synthesis. They have the enzyme DNA-polymerase for the duplication of Mt-DNA themselves. During replication, the Mt-DNA behaves as if it is a mitochondrial chromosome. *Suzama* and *Bonner* (1966) and *Breiedenbach et al.* (1967) assumed that the mitochondria are self replicating systems and show continuity through cell generations—as they contain DNA and RNA in them. *Nass* (1969) reported that Mt-DNA codes for some structural proteins of the mitochondria and the structural genes for the Mt-enzymes are essentially localized on the nuclear genome.

9.20. CHLOROPLAST-DNA

It is now generally accepted that a characteristic DNA occurs in chloroplasts of algae and higher plants. *Ris* and *Plaut* (1962) reported the presence of fine filaments of DNA from the chloroplasts of *Chlamydomonas*. The presence of DNA has been confirmed in several other algae and higher plants. Segments of DNA as long as 150 µ have been separated from chloroplasts by *Woodcock* and *Fernandez-Morgan* (1968). The DNA isolated from plastids of *Phaseolus vulgaris* and *Nicotiana tabacum* ranges in size between about 0.25 to 3.25 µm. DNA of the chloroplast resembles closely with the bacterial DNA and differs with the N-DNA. The algal chloroplast DNAs from several species have lower G-C content than the corresponding N-DNA, but chloroplast-DNAs from angiosperms possess higher G-C ratio than N-DNA. The DNA of the chloroplasts has been related to the presence of a special non-chromosomal genetic system or cytoplasmic heredity by *Rhoades* (1955).

9.21. SINGLE-STRANDED DNA AND CIRCULAR DNA

Single Stranded DNA : One of the most interesting and unusual forms of DNA is that first isolated by *Sinsheimer* (1959) from the small virus $\phi \times 174$ which attacks *E. coli*. This DNA molecule contains 5,500 nucleotides in a single strand. Later on, single stranded DNA was reported from other spherical viruses also like S_{13} and filamentous phages like fd., fl and M13 etc. The phage $\phi \times 174$ is about 25 m μ in diameter and here DNA constitutes one-fourth of the total weight. *Hall* (1957) reported that $\phi \times 174$ is built up of 12 identical knobs, each of which in turn is composed of 5-identical sub-units.

9.22. DISTINGUISHING FEATURES BETWEEN NATIVE DNA AND SINGLE-STRANDED DNA (SS-DNA) :

   (*i*) SS-DNA does not show the phenomenon of helix → coil transition on heating.
   (*ii*) In contrast to the normal DNA the ultraviolet absorption of $\phi \times 174$ DNA begins to rise at 20°C and increases rapidly until temperature reaches to 90°C.
   (*iii*) SS-DNA is susceptible to the action of enzymes like hydrolases, proteolytics etc., and various chemicals like formaldehyde; but double stranded DNA is not susceptible because the amino groups of its bases are protected strongly by the hydrogen-bonded structure of the double helix.
   (*iv*) In SS-DNA, after treatment with formaldehyde, some changes in the ultraviolet absorption take place.
   (*v*) Chemical composition shows the single stranded nature of the molecule in this phage.
   (*vi*) The molar proportions of bases of SS-DNA do not show the equivalence of A and T and of G and C as required for double helix formation. Here the base contents are adenine, 1.0; thymine, 1.3; guanine, 1.0; and cyltosine, 0.8.
   (*vii*) When SS-DNA is heated, its density does not change.

9.23. CIRCULAR DNA

Some of the DNA molecules can exist in circular form. In certain viruses, the circular DNA is wrapped in a protein coat. It then forms a filament of two nucleo-protein strands by bringing the opposite sides of the circle together. The circular DNA is found in certain phages and viruses such as *λ-phage, Polyoma virus, $\phi \times 174$ virus, fd virus, T-phages*, etc. When *λ-phage* DNA molecule is isolated from free phage, it is linear in shape. However, by controlled heating and recooling it can be reversely converted to a circular form.

9.24. SINGLE-STRANDED CIRCULAR DNA

The DNA of bacteriophage $\phi \times 174$ has been shown by Sinsheimer as single stranded and in the form of a continuous annular polynucleotide chain of mol. wt. 1.7×10^6. The circular DNA can be converted by a single scission with pancreatic DNase to a linear or acyclic form. At pH 12, this change causes a 10% decrease in sedimentation velocity. Other spherical phage particles like S13 and filamentous phages like fd, contain similar DNA molecules. Some animal viruses such as MVM contain single-stranded DNA of mol. wt. 1.5×10^6, though as yet these have not been shown to be circular.

9.25. DOUBLE-STRANDED CIRCULAR DNA

The double stranded circular DNA is found in *Polyoma virus* (mol. wt. 3×10^6). The two strands of the DNA have been shown to be topologically bonded (interlocked, but not covalently linked). At neutral pH, the DNA of the virus can be separated by sedimentation velocity into three components, I(20S), II (16S) and III (14.5S). *Vinograd* and *Watson* (1965) have shown that 20S can be converted to 16S by single scission with pancreatic DNase and to 14.5 S by a single scission with *E. coli* endonuclease I. Component 20 S is a circular molecule with both strands continuous polynucleotide chains, but there is a deficiency to turns in the *Watson-Crick* double helix. The twisted circular forms produced by deficiency are called **supercoiled forms**. The twists in the supercoiled form can be released by breaking one strand to form the open-circular component (Fig. 9.37 II). Component III (14.5S) is a double-stranded linear structure. Three components of polyoma viral DNA are shown in Fig. 9.37.

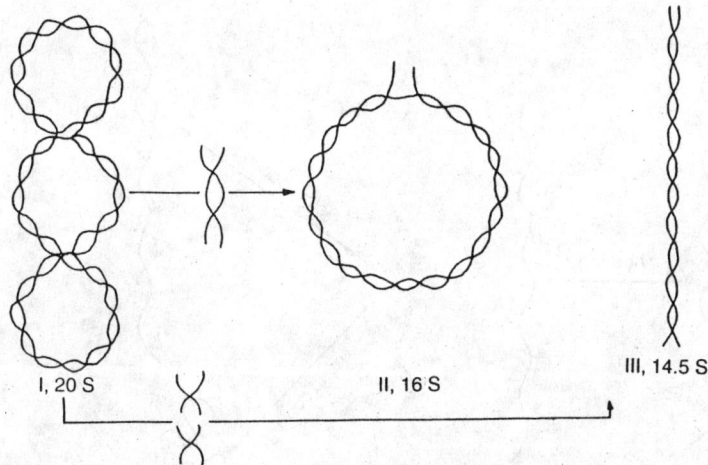

Fig. 9.37: Diagrammatic representation of three components of polyoma DNA at pH 8.

Many viruses contain double-stranded linear DNA, but most of those which have been studied in detail exhibit some additional features such as cohesive ends, terminal repetitions, circular permutations or "nicks".

9.26. THE MOLECULAR WEIGHT OF DNA

The molecular weight of DNA is usually determined by light scattering measurements or by measurements of sedimentation rate or intrinsic viscosity or by a combination of measurements of sedimentation coefficient and intrinsic viscosity. Most convincing molecular weight values come from DNA-containing viruses. For example, the bacterial virus T_2 contains one DNA molecule of MW=1.3×10^8, and the bacterial virus λ contains a single molecule of MW=3.2×10^7. Table 9.3 indicates DNA molecular weights obtained from various sources.

Table 9.1 DNA molecular weights

	Source	Mol. Wt.	Length	No. of nucleotide Pairs	Conformation
1.	*E. coli* chromosome	2.2×10^9	1mm	3×10^6	Cyclic, duplex
2.	*H. influenzae* Chromosome	8×10^8	300 μ	12×10^5	—

3.	*Mycoplasma PPLO* Strain H-39	4×10^8	150 µ	6×10^5	—
4.	Bacteriophage T2 or T4	1.3×10^8	50 µ	2×10^5	Linear, duplex.
5.	Bacteriophage λ	3.2×10^7	13 µ	0.5×10^5	Linear, duplex.
6.	Bacteriophage φ ×174	1.6×10^6	0.6 µ	—	Cyclic, Single stranded.
7.	*Polyoma virus* mouse	3×10^6	1.1 µ	4.6×10^3	Cyclic, duplex.
8.	Mitochondria	9.5×10^6	5 µ	14×10^3	Cyclic, duplex.

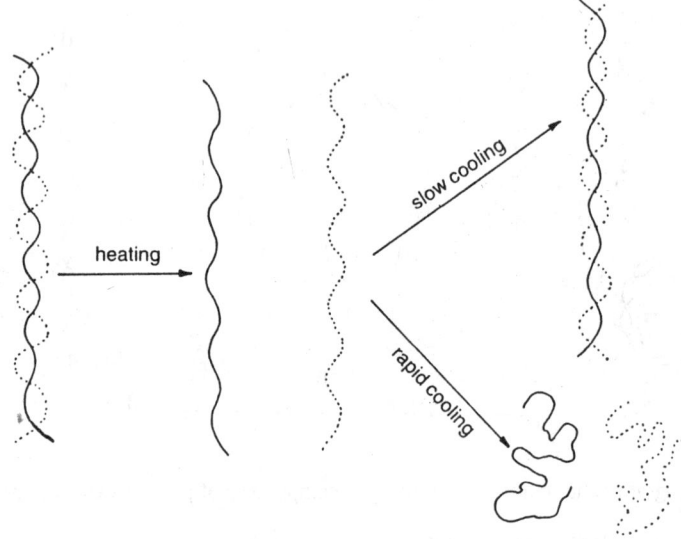

Fig. 9.38: The denaturation of DNA by heat causes strand separation. On slow cooling the strands recombine and on rapid cooling they remain separate.

9.27. DENATURATION & RENATURATION OF DNA

Denaturation of DNA may be defined as the phenomenon of breaking of hydrogen bonds and separation of the two polynucleotide chains. It is done by raising the temperature to near 100 °C. When the denaturation is done by heating or temperature, it is called "thermal denaturation" (Fig. 9.38). The DNA is also denatured by extremes of pH, exposure to low ionic strength, or treatment with such reagents as urea, trichloroacetate, and thiocyanate. Denaturation also known as melting of DNA, is accompanied by an increase in absorbancy at 2600 Å. This is called the "hyperchromatic shift". The melting and hyperchromatic shift vary for the DNAs having different AT/GC ratios because a higher temperature is needed to break the GC pair (three hydrogen bonds) than the AT pair (two hydrogen bonds).

If thermally denatured DNA is cooled slowly, the two complementary strands can meet and reform the normal double helix. Thus, reversal of the denaturation occurs. This process is called as renaturation or "annealing". On the other hand, rapid cooling (quenching) prevents this and preserves the nucleic acid in its denatured form. The process of renaturation is used to produce hybrid nucleic acids.

Nucleic Acids

9.29. HYDROLYSIS OF DNA

When the DNA is hydrolysed under different conditions, three kinds of low molecular weight products are produced. These are a pentose sugar, purines and pyrimidines, and phosphoric acid. The hydrolysis of DNA is shown in the Fig. 9.39.

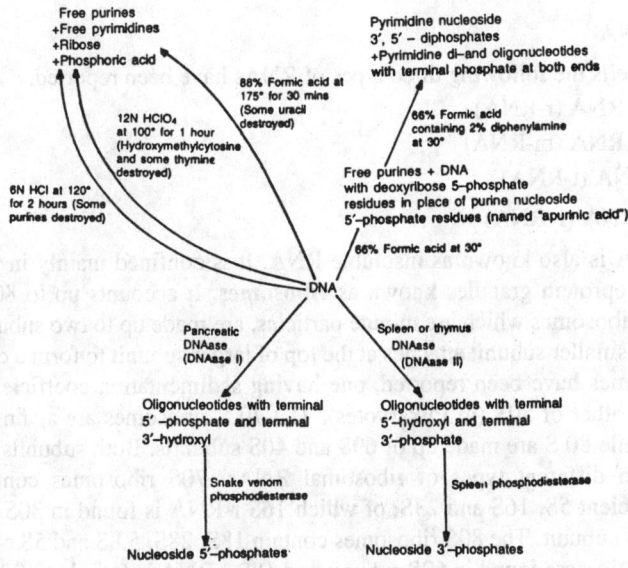

Fig. 9.38: Hydrolysis of DNA.

9.29. RIBONUCLEIC ACID (RNA)

RNA exists chiefy in the cytoplasm of animal and plant cells, but nucleus also contains a very small amount of RNA. The naturally occurring RNA does not show the regularity of base composition as is found in DNA. They also do not show the equivalence of adenine to uracil, or of guanine to cytosine. It is, therefore, not to be expected that the structure of RNA will correspond to a perfect two-stranded helix of the *Watson-Crick* type. There are a number of evidences which show that the RNA is single-stranded and it forms partially helical structure. The helical regions remain separated by the non-helical regions. Thus, the RNA shows loop like structures.

The structure of RNA is closely related with the DNA; like DNA, RNA is also made up of a pentose sugar, organic bases (Purines and Pyrimidines) and a phosphoric acid molecule. The pentose sugar in RNA is **ribose instead of deoxyribose** found in DNA. The pyrimidines in RNA are cytosine and uracil instead of cytosine and thymine which are found in DNA. In this way we can say that RNA is made up of :

(i) Ribose sugar

(ii) Purines-Adenine, Guanine ; Pyrimidines-Cytosine and Uracil.

(iii) Phosphoric acid molecule.

9.30. Differences between RNA and DNA

(i) RNA molecule is much shorter than DNA.

(ii) Mostly RNA is single-stranded, while the DNA is double-stranded.

(iii) The sugar in RNA is ribose, while in DNA it is deoxyribose.

(iv) DNA contains the pyrimidine-thymine, but RNA contains uracil.

(v) RNA is generally confined to the cytoplasm and its inclusions, while DNA is found always in nucleus. In recent years the presence of DNA in mitochondria and chloroplast and of RNA in chromosomes has been reported by many workers.

(vi) The two types of nucleic acids are distinguishable by their stainable reactions. DNA is stainable by methyl green and RNA is stainable by pyromine. Differences no. 3 and 4 are most important.

9.31. TYPES OF RNA

In animal and plant cells the following three types of RNAs have been reported.
(i) Ribosomal RNA (r-RNA)
(ii) Messenger RNA (m-RNA)
(iii) Transfer RNA (t-RNA)

9.31.1. Robosomal RNA (r-RNA)

Ribosomal RNA is also known as insoluble RNA. It is confined mainly in the cytoplasm in the minute ribonucleoprotein granules known as **ribosomes.** It accounts up to 80 per cent of the total cell RNA. The ribosomes which are minute particles, are made up to two subunits, one smaller and other larger. The smaller subunit attaches at the top of larger subunit to form a cap like structure. Two types of ribosomes have been reported, one having sedimentation coefficient value of 70 S (in Prokaryotes) and other of 80S (in Eukaryotes). The 70 S ribosomes are again made up of 50S and 30S subunits, while 80 S are made up of 60S and 40S subunits. Both subunits of the two types of ribosomes contain different types of ribosomal RNAs. 70S ribosomes contain RNAs with sedimentation Coefficient 5S, 16S and 23S, of which 16S r-RNA is found in 30S subunit and 23S and 5S r-RNA in 50S subunit. The 80S ribosomes contain 18S, 28S, 5.8S and 5S r-RNAs, of which 28S, 5.8S and 5S r-RNAs are found in 60S subunit and 18S r-RNA is found in 40S smaller subunit.

The 28S r-RNA and 23S r-RNA have the molecular weight of about 1.7×10^6 daltons and 1.1×10^6 daltons respectively, while 16S and 18S r-RNAs about 6×10^5 and 7×10^5 daltons respectively. 5S r-RNA with mol. wt. 3.2×10^4 was discovered by *Knight* and *Darnell* (1967) and by *Forget & Weissman* (1968). Forget and Weissman reported that 5S r-RNA has a length to 120 nucleotides and it has a clover leaf shape, similar to 4S t-RNA. 5.8S rRNA has mol. wt. 3.5×10^4 daltons.

The role of r-RNS in the process of protein synthesis is still little understood. However, it is supposed that they definitely play a significant role in the process of protein synthesis. The function of 5S r-RNA is still quite unknown.

Processing of r-RNA

The r-RNA is formed first in the nucleolus with the help of nucleolar organizer which contains ribosomal DNA (r-DNA) cistrons. The r-DNA cistrons first form a precursor molecule of 45 S r-RNA which after a number of events is broken into two molecules of 28S and 18S r-RNA and ultimately they reach the cytoplasm from the nucleolus (Fig. 9.40). The 18S r-RNA enters the smaller subunit of ribosome (40S), whereas, 28S rRNA enters the larger subunit (60S) of ribosome.

The 45S r-RNA precursor molecule is cleaved into 32S and 18S r-RNAs through several intermediate steps. The 18S r-RNA is immediately exported to cytoplasm, but 32S r-RNA is again cleaved to form 28S r-RNA. The formation of 28S rRNA from 32S rRNA takes about 40 minutes. After its formation the 28S rRNA remains in the nucleolus for 30 minutes and then it is transferred to the cytoplasm.

Miller (1970) after doing a number of experiments on amphibian nucleoli suggested that each r-DNA cistron coding for a 45S molecule is separated by segments of non-transcribed DNA. It is also believed that there are about 100 RNA polymerases which act on each r-DNA cistron at the same time. Each polymerase transcribes a single 45S RNA.

Nucleic Acids

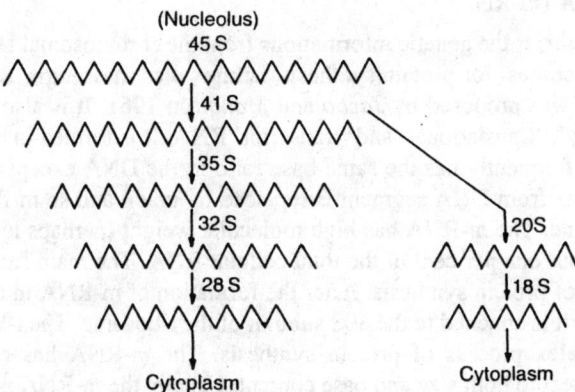

Fig. 9.40: Processing of 45S nucleolar RNA into 28S and 18S r-RNA.

The 5S r-RNA is transcribed outside the nucleolus because the genes forming it are not linked to the nucleolar organizer and are present as anucleolate as shown in Fig. 9.41. Those regions of the DNA that code for messenger RNA are called the **structural genes** and other regions which code for the different ribosomal and transfer RNAs are frequently called the **determinants** for RNA. The 5S rRNA, which has been formed anucleolate, finally enters the 60S ribosomal subunit. Fig. 9.41 summarizes, at the cellular level, the transcription and transport of nuclear RNAs, in the eukaryotic cell and also m-RNA participation in protein synthesis. It also indicates the transcription sites in DNA for 4S and 18-180S m-RNAs.

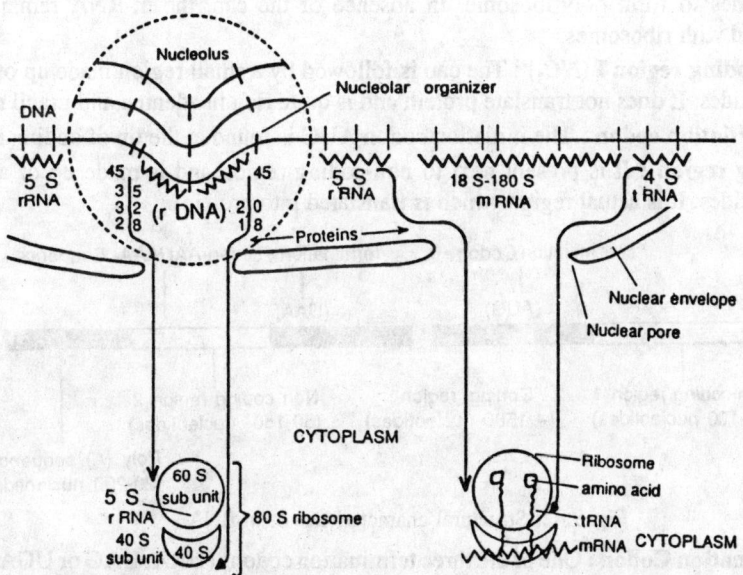

Fig. 9.41: Diagram showing transcription and transport of a nuclear RNA's in a eukaryotic cell and participation of tRNA and mRNA in protein synthesis.

9.31.2. Messenger RNA (m-RNA)

The RNA which carries the genetic informations from the chromosomal DNA to the cytoplasm (particularly to the ribosomes) for protein synthesis, is known as **messenger** RNA or m-RNA. The name messenger RNA was proposed by *Jacob* and *Monod* in 1961. It is also known as 'informational', 'complementary', 'translational' and 'transcript' RNA. It is formed in the nucleus under the influence of DNA and frequently has the same base ratio as the DNA except that in the formation of m-RNA - the thymine from DNA segment is replaced by uracil and so m-RNA will have uracil bases in place of thymine. The m-RNA has high molecular weight (perhaps up to several millions) and represents only about one per cent of the total cellular RNA. The main function of the m-RNA is to direct the process of protein synthesis. After the formation of m-RNA in the nucleus, it moves to the cytoplasm where it is attached to the 30S subunit of the ribosome. The t-RNA and aminoacids also help in this complex process of protein synthesis. The *m*-RNA has a rapid turnover and heterogeneity with respect to both size and base content. Usually the m-RNA is also single stranded like r-RNA, but it has got different sizes. The size of m-RNA is directly related with the size of codons for the protein molecules. Two main types of m-RNA have been discovered so far.

(a) *Monocistronic m-RNA* : These *m*-RNA are those which carry the codes of single cistron of the DNA i.e, codes for one complete protein molecule.

(b) *Polycistronic m-RNA* : It is also known as polygenic m-RNA. These m-RNA are those which carry the codes from several adjacent DNA cistrons and become much longer in size. This type of m-RNA has been recorded during the metabolism of the histidine.

Characteristics of m-RNA

m-RNA possesses following characteristic features :

1. **Cap** : The m-RNA of eukaryotic cells and animal viruses possesses a '*cap*' at 5' end. It is a blocked methylated structure containing several methylated bases like m^7 Gpp Cmp Amp or m^7 Gpp Cmp Cmp Amp. Cmp and Amp contains 2'O methyl ribose. The rate of protein synthesis depends upon the presence of the cap which helps the m-RNA in binding with ribosomes to form polyribosome. In absence of the cap, the m-RNA remains loosely attached with ribosomes.
2. **Non-coding region I (NC_1)** : The cap is followed by a small region made up of 10 to 100 nucleotides. It does not translate protein and is quite rich in adenine and uracil residues.
3. **The initiation codon** : The initiation codon AUG is found at the tip of coding region.
4. **Coding region** : It is present next to non-coding region and is made up of about 1500 nucleotides. It is actual region which is translated into protein.

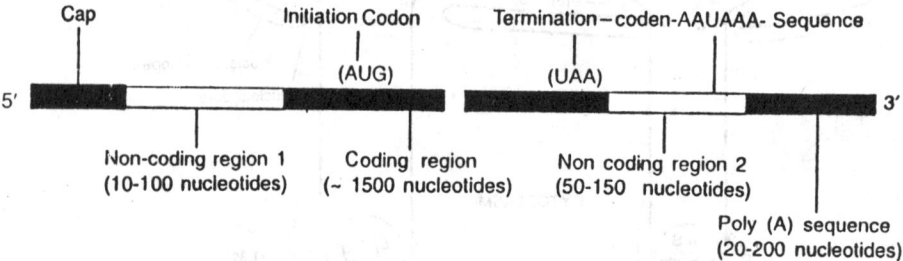

Fig. 9.42: Structural characteristics of m-RNA.

5. **Termination Codon** : One of the three termination codons (UAA, UAG or UGA) is always present at the end of coding region which helps in polypetide chain termination.
6. **Non-coding region II (NC_2)** : It is the present next to termination code. It is made up of 50-150 nucleotides and is not involved in translation of protein. NC_2 contains AAUAAA sequence.

Nucleic Acids

7. Poly (A) sequence : The 3' end of m-RNA contains a polyadenylate or Poly(A) sequence which is added in the nuclues before its entry into the cytoplasm. Initially this sequence contains 200-250 nucleotides but with the lapse of time nucleotide number is reduced.

9.31.3. Transfer RNA (t-RNA)

Transfer RNA is also known as **soluble RNA (s-RNA)** and **acceptor RNA**. It is found free in the cytoplasm and accounts for 15 to 20 per cent of the total cellular RNA. These contain 75 to 80 nucleotides and have a molecular weight of about 25,000 daltons which is much lower than r-RNA. The t-RNA functions in the transportation of amino acids from the cytoplasm to the ribosomes. Each tRNA is specific for each amino acid. Thus, there are about twenty or twenty-two different types of t-RNAs which help in the transfer of the same number of aminoacids from the cytoplasm to the ribosomes. The t-RNA differs from other RNAs (r-RNA or m-RNA) in having unusual bases which are mostly methylated derivatives of more common bases, e.g., methylated purines, pseudo-uridines, methyl cytosine, methylaminopurine etc.

All t-RNAs possess almost similar molecular weights (25,000), but different nucleotide sequences. The sedimentation cofficient of t-RNA is 4S. The nucleotide sequence within the RNA molecule is known as primary structure and when within the single-stranded RNA the base-pairing

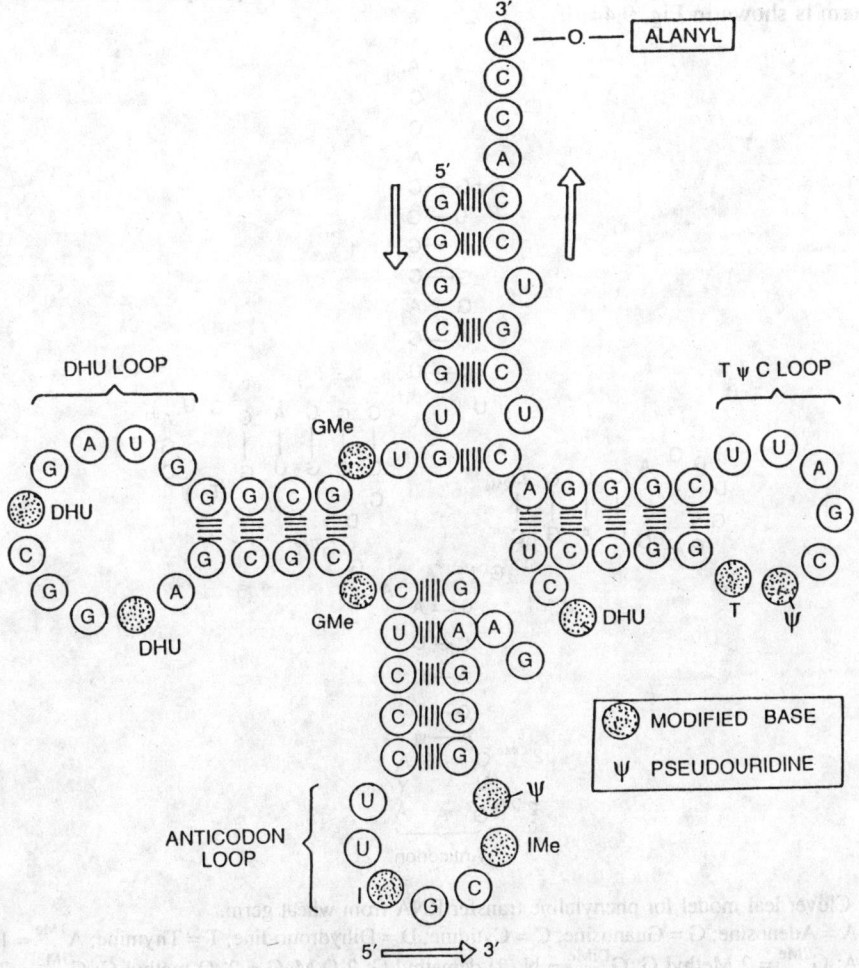

Fig. 9.43: Structure of yeast alanyl t-RNA.

takes places, it gives rise to the secondary structure of the RNA. Previously, it was supposed that the single strand of RNA twists to form a hair pin like structure (helical structure) which has mostly paired bases. The unpaired bases which are generally found in the centre are known as **anti-codons**. The terminal end has also exposed unpaired bases (trinucleotide sequence) which consists of two cytidylic residues followed by an adenosine in all t-RNAs. Later on, *Lake* and *Beeman* (1967) suggested that each t-RNA consists of three folds of double helix to give it the shape of the clover leaf. This clover leaf pattern is the characteristic of all t-RNA's.

9.31.4. Clover leaf model of t-RNA : (Two dimensional structure)

The detailed structure of t-RNA is most important because, (1) It recognises the amino acid through the enzyme *amino acid RNA synthetase* (*AA-RNA synthetase*) (2) It recognises the appropriate base sequences on m-RNA.

The detailed study of t-RNA was easily worked out because they are comparatively smaller than other types of RNAs. The first t-RNA which was worked out in detail is alanyl t-RNA of yeast. The studies were made by *Holley* and colleagues for which he received Nobel Prize along with *Khorana* and *Nirenberg* in 1968. This information was later used by Khorana and his colleagues for the synthesis of the gene for this t-RNA. The detailed structure of yeast alanyl t-RNA as worked out by *Robert Holley* is shown in Fig. 9.43 and the structure of phenylalanine transfer RNA of wheat germ is shown in Fig. 9.44.

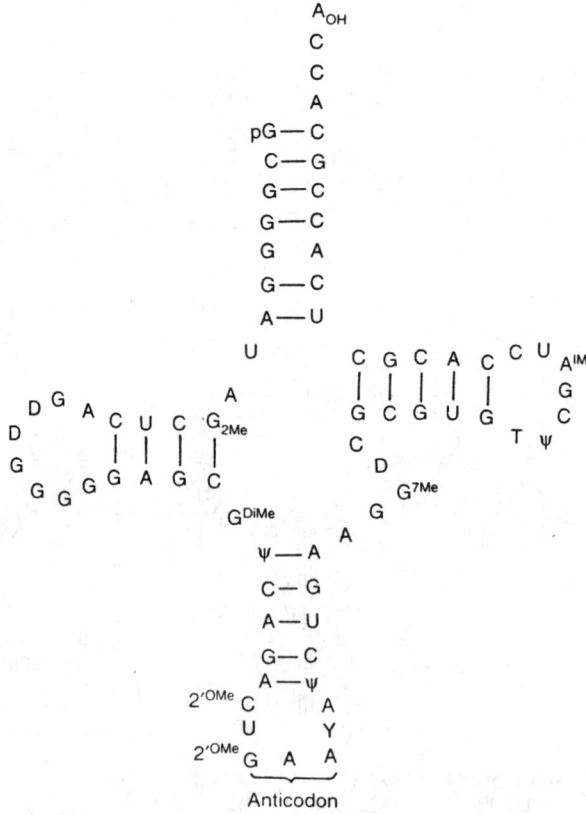

Fig. 9.44: Clover leaf model for phenylaline transfer RNA from wheat germ.

A = Adenosine; G = Guanosine; C = Cytidine; D = Dihydrouridine; T = Thymine; A^{1Me} = 1-Methyl A; G^{2Me} = 2 Methyl G; G^{DiMe} = N (2) demethyl G; 2'O MeG = 2'-O-methyl G; G^{7Me} = 7-Methyl G; Ψ = Pseudouridine; y = modified A; P = 5' phosphate; OH = 3' hydroxyl.

On the basis of the detailed structures of several t-RNAs worked out, the following general conclusions may be drawn :

(i) Each t-RNA has the two ends, 3'-end and 5'-end.

(ii) The 3'-end has the base sequence 'CCA' and 5'-end has 'G'-nucleotide (guanine residue).

(iii) The amino acid is always attached at 3'-end only and the sequence 'CCA' remains unpaired.

(iv) Each t-RNA contains a large number of unusual nucleosides e.g. pseudouridine (ψ), Inosine (I), dihydrouridine (DHU) etc. The unusual bases are formed by the process of methylation of normal bases. The phenylalanine t-RNA of wheat germ (Fig. 9.44) shows the methylated bases like A^{1M}, $G7^{Me}$, $G^{2O'Me}$, G^{2Me}, G^{DiMe}, $C^{2,OMe}$ etc.

(v) The ratios of A:U and G:C are near unity which, suggests the formation of DNA like double helical segments (secondary structure).

(vi) In these double helical regions of RNA, G:C base pairs are more common than A:U.

(vii) Each t-RNA possesses a tertiary structure and Mg^{++} ion concentration plays an important role in its stabilization. Therefore, although each t-RNA is a polynucleotide, it resembles with enzymes in its structure and function. The details of the tertiary sturctre of t-RNA are still unknown.

(viii) The intermediates portion of a t-RNA seems to be invariably folded in a 'clover leaf' pattern with mostly three or sometimes more double helical regions. Each of these arms has a loop. The first and the third loop (side loops) are known as **aminoacyl synthetase** (DHU loop) and **ribosomal binding** (GTψC loop) loops respectively. The DHU loop usually contains 8-12 bases, whereas GTψC loop only 7 bases (fixed). The 'anticodon' is usually present in the loop on the second helical region. The loop is known as **anticodon loop** and is made up of 7 bases. The anticodon recognizes the appropriate codon triplet in the m-RNA. It is found at the centre of the loop and is different for each t-RNA.

9.31.5. Three-dimensional structure of t-RNA

The Three-Dimensional Structure (TDS) of t-RNA was studied with the help of x-ray crystallography. About a dozen models have been proposed by many workers to understand the structural and functional relationship of t-RNA, but the latest and most acceptable model was

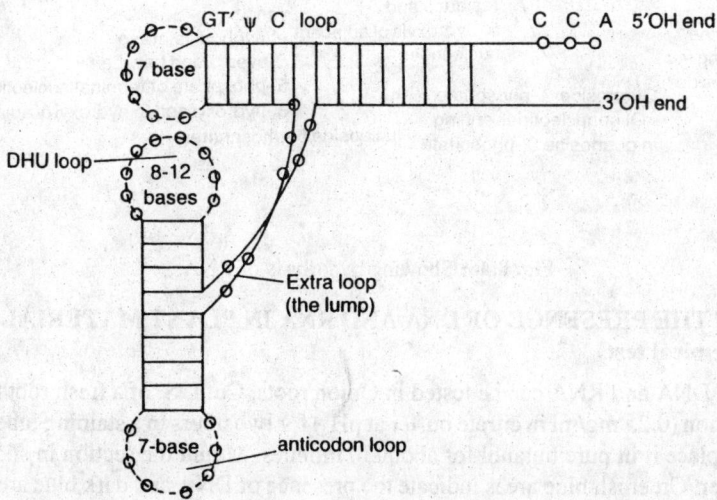

Fig. 9.45: A model for three-dimensional structure of t-RNA (as proposed by *Kim*, 1973).

propsed by S.H. Kim in 1973. He suggested that TDS of tRNA is like letter 'L' with a thickness of about 20Å as shown in Fig. 9.27. This model also confirms its derivation from the two-dimensional clover leaf model. The extra arm which varies in different t-RNA molecules can be extended without distorting TDS conformation. The CCA stem projects out and can take different orientations.

9.32. HYDROLYSIS OR DEGRADATION OF RNA

Like DNA, RNA can also be hydrolysed under different conditions (Fig. 9.46). The acid hydrolysis yields the different components which actually form it. The various components are free purines and pyrimidines, ribose sugar and phosphoric acid.

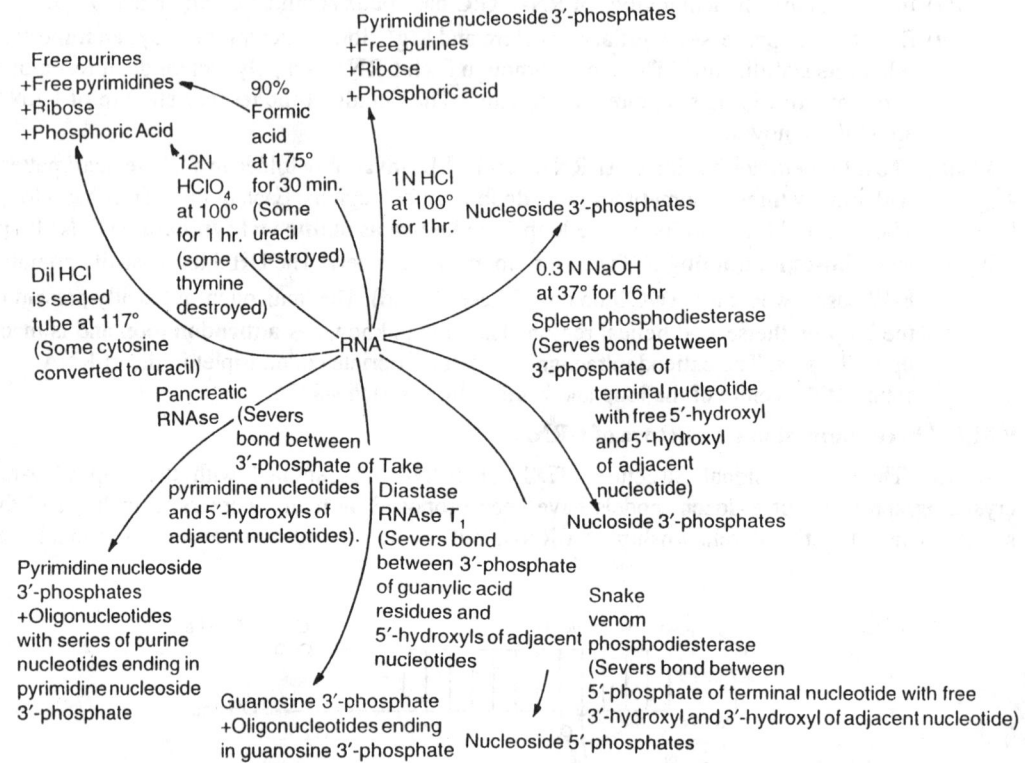

Fig. 9.46: Showing bydorlysis of RNA.

9.33. TO TEST THE PRESENCE OF DNA AND RNA IN PLANT MATERIAL (Microchemical test)

The presence of DNA and RNA can be tested in Onion roots. Cut L.S. of a fresh root tip and place it in Azure B solution (0.25 mg/ml in citrate buffer at pH_4) for two hours in a staining tube. Now remove the section and place it in pure butanol for about 30 minutes. Mount the section in slide and observe under high power. Greenish blue areas indicate the presence of DNA and dark blue areas indicate the presence of RNA.

10

Introduction to Bioenergetics

10.1 ENERGY

Energy is defined as the capacity to do work, which is the product of a given force acting through a given distance.

$$\text{Work} = \text{Force} \times \text{Distance}.$$

Energy may be of different kinds, such as, potential, kinetic, electrical, radiant etc. Of these, radiant energy is most important to a biochemist. The radiant energy of the sun is utilized by green plants to 'fix' atmospheric CO_2 and to convert it into complex carbon compounds through the formation of covalent linkages. Thus, an external source of energy is required to join atoms together to form stable molecules. The energy stored in such compounds is used to drive the mechanical, osmotic, chemical and electrical work in plants and animals.

The reactions which require an external source of energy are termed as 'endergonic' and reactions which liberate energy are termed as 'exergonic'. The energy is liberated by the cleavage of energy-rich bonds of carbon compounds.

10.1.1. Free Energy

The free energy may be defined as the total energy present in the substrate which can be liberated by any means. This quantity cannot be measured experimentally. However, the energy liberated or consumed in a reaction which can be used in or supplied to another system is termed as the "free energy change" ($\Delta G'$) of the reaction and it can be measured experimentally. In an exergonic reaction $\Delta G'$ is always negative because there is an overall loss of free energy. For example, in the conversion of A + B into C + D, if there is a loss of free energy, the reaction is called exergonic.

$$A + B \longrightarrow C + D + \text{'Free Energy'} \qquad \ldots(1)$$

The reverse reaction (reaction 2) must be endergonic with $\Delta G'$ being positive because C + D takes up energy in order to be converted into A + B.

$$C + D + \text{'Free Energy'} \longrightarrow A + B \qquad \ldots(2)$$

During the occurrence of these two reactions, a stage comes when $\Delta G'$ becomes zero and the reaction is said to be at equilibrium stage.

$$A + B \rightleftharpoons C + D$$
$$\Delta G' = 0$$

Table 10.1 shows the $\Delta G'$ values of number of biologically important compounds.

10.2. ENERGETIC COUPLING

In the living systems, the energy released or made available in an exergonic reaction is utilized to drive other endergonic reactions and is thereby made to do work. The energy transfer can only occur between two reactions which share a common substrate. The process by which this energy transfer between two reactions takes place is known as 'Energetic Coupling' or 'Coupling of reactions'. Thus, for energetic coupling it is necessary that the two reactions, exergonic and endergonic, should occur simultaneously and also the $\Delta G'$ (free energy change) value of exergonic reaction must be greater than endergonic reaction. The $\Delta G'$ is always negative for exergonic reaction and positive for endergonic reaction.

Table. 10.1: The free energy of hydrolysis ($\Delta G'$) of a number of biologically important compounds.

	Compound	Temperature (°C)	pH	$\Delta G°$ (Kcal/mole)
1.	Glucose 6-P	25	7.0	−3.3
2.	Glucose 1-P	25	7.0	−5.0
3.	ATP (ADP + Pi)	37	7.0	−7.6
4.	1, 3 diphosphoglyceric acid.	25	6.9	−12.0
5.	Phosphoenol pyruvate	25	8.5	−12.0
6.	Acetyl-CoA	25	7.0	−7.7

A very common energy yielding reaction in biochemistry is the hydrolysis of adenosine triphosphate (ATP) to give ADP and Pi. The $\Delta G'$ for this reaction is about 7600 calories/mole.

$$ATP \longrightarrow ADP + Pi + 7600 \text{ calories.}$$

The energy released by the hydrolysis of ATP is used to drive other endergonic reactions. Thus, the energy of one reaction is coupled with another reaction. Most of the phosphate anhydrides undergo hydrolysis to yield large negative free energy change. The formation and hydrolysis of this type of bond provide a principal means for storing and releasing chemical energy.

In glycolysis the following reactions make a considerable contribution to the negative $\Delta G'$.

$$\text{Glucose + ATP} \underset{\text{Hexokinase}}{\overset{Mg^{2+}}{\rightleftharpoons}} \text{Glucose–6–(P) + ADP} \quad \text{...(1)}$$

$$\begin{cases} \Delta G' = -4600 \text{cal/mole} \\ K = 2.3 \times 10^3 \end{cases} |pH7|$$

$$\text{Fructose-6-(P)+ATP} \underset{\text{Phosphofructokinase}}{\overset{Mg^{2+}}{\rightleftharpoons}} \text{Fructose 1, 6–di (P) + ADP}$$

$$\begin{cases} \Delta G' = -4200 \text{cal/mole} \\ K = 1.2 \times 10^3 \end{cases} |pH7| \quad \text{...(2)}$$

$$\text{Phosphoenol pyruvate + ADP} \underset{\text{Pyruvatekinase}}{\overset{\text{Pyruvatekinase}}{\rightleftharpoons}} \text{Pryruvate + ATP}$$

$$\begin{cases} \Delta G' = -5200 \text{ cal/mole} \\ K = 6.5 \times 10^3 \end{cases} |pH7| \quad \text{...(3)}$$

Reaction (1) is in fact the sum of two coupled reactions; One is the esterification of glucose by phosphoric acid to give glucose-6-phosphate, a process requiring an expenditure of 3000 cal/mole

Introduction to Bioenergetics

($\Delta G' = +3000$ cal/mole glucose) and the other is the hydrolysis of ATP to yield ADP and Pi, an exergonic reaction producing 7600 cal/mole ($\Delta G' = -7600$ cal/mole ATP). The net free energy change of this coupled reaction is, therefore, -4600 cal/mole of glucose-6-phosphate formed. In other words, it can be said that 4600 calories are lost as heat from the system during the production of glucose-6-phosphate from glucose. This extra-expenditure of energy makes the reaction (1) very difficult to reverse. The whole process of coupling of this reaction may be summarized as follows:

$$ATP + H_2O \xrightarrow{\text{Hydrolysis}} ADP + H_3PO_4 \; (\Delta G' = -7600 \text{ cal/mole}) \quad —(i)$$

$$Glucose + H_3PO_4 \xrightarrow{\text{Esterification}} Glucose\; 6\text{-}(P) + H_2O \; (\Delta G' = +3000 \text{ cal/mole}) \quad —(ii)$$

Sum of (i) & (ii):   Glucose + ATP $\longleftrightarrow$ Glucose 6-(P) + ADP ($\Delta G' = -4600$ cal)

A similar type of explanation can be given to reaction (2) where fructose-6-phosphate is phosphorylated to fructose 1, 6-diphosphate. In this reaction there is a net loss of 4200 cal/mole of energy. This reaction is performed by the following two reactions.

$$ATP + H_2O \rightarrow ADP + H_3PO_4 \; (\Delta G' = -7600 \text{ cal/mole}) \quad \ldots (i)$$

$$\text{Fructose-6-(P)} + H_3PO_4 \rightarrow \text{Fructose 1,6-di-(P)} + H_2O \; (\Delta G' = +3400 \text{ cal/mole}) \quad \ldots (ii)$$

Sum of (i) & (ii)   Fructose-6-(P)+ATP $\rightarrow$ Fructose 1,6-di (P) + ADP ($\Delta G' = -4200$ cal)

In case of reaction (3) the explanation is essentially the same, but there the synthesis of ATP takes place rather than break down. The phosphoenolpyruvate is dephosphorylated to pyruvate with the loss of 5200 cal/mole of energy. The reaction is completed in the following two steps:

$$\text{Phosphoenol pyruvate} + H_2O \longrightarrow \text{Pyruvate} + H_3PO_4 \; (\Delta G' = -12{,}800 \text{ cal/mole}) \quad —(i)$$
$$ADP + H_3PO_4 \longrightarrow ATP + H_2O \; (\Delta G' = +7600 \text{ cal/mol}) \quad —(ii)$$

Sum of (i) & (ii)   Phosphoenol pyruvate + ADP $\longrightarrow$ Pyruvate + ATP ($\Delta G' = -5200$ cal/mole)

The dephosphorylation of phosphoenolpyruvate to give pyruvate is a strongly exergonic reaction yielding 12,800 cal/mole of PEP ($\Delta G' = -12{,}800$ cal/mole). This reaction couples with another reaction showing phosphorylation of ADP. It is an endergonic reaction which requires the input of only 7600 cal/mole of ADP ($\Delta G' = +7600$ cal/mole). Thus, the complete reaction shows the loss of 5200 cal/mole in the form of heat.

10.3. ENERGY RICH COMPOUNDS

10.3.1. Adenosine Tri-phosphate (ATP)

ATP is a compound which acts as a link between cellular endergonic and exergonic processes. This is the triphosphate ester of adenine ribonucleoside and is made up of a purine base, adenine; a pentose sugar, ribose; and three molecules of phosphoric acid. The combination of adenine and ribose sugar is known as its nucleoside, **Adenosine** (adenine ribonucleoside). Adenosine monophosphate (AMP) and Adenosine Diphosphate (ADP) are the mono- and diphosphate esters of adenine ribonucleoside respectively. In all these three compounds (ATP, ADP and AMP) the CH_2OH group of ribose forms an **ester link** with the phosphate group of phosphoric acid, H_3PO_4 as shown in Fig. 10.1).

ATP is hydrolysed in presence of suitable enzyme, or by dilute mineral acids, or alkalies. As a result of hydrolysis the terminal phosphate group is removed, leaving ADP. The release of standard free energy for this reaction is about -7600 calories. The hydroylsis of phosphate group in ADP results in AMP and the release of about –6500 calories energy takes place. The hydrolysis of third phosphate in AMP releases only small amount, about –2200 calories, of energy.

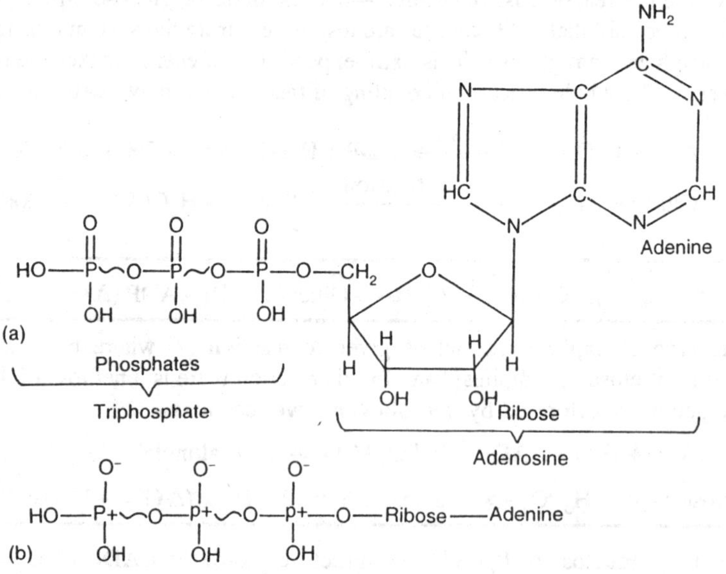

Fig. 10.1: (a) and (b) Structures of adenosine triphosphate (ATP).

$\left.\begin{array}{l}\text{Adenine}\\\text{riboncleoside}\\\text{(Adenosine)}\end{array}\right|$ ATP + H_2O → ADP + H_3PO_4 ($\Delta G' = -7600$ cal/mole).
ADP + H_2O → AMP + H_3PO_4 ($\Delta G' = -6500$ cal/mole).
AMP + H_2O → Adenine ribonucleoside (Adenosine) + H_3PO_4 ($\Delta G' = -2200$ cal/mole).

The energy released by the hydrolysis of ATP is utilized in various physiological functions, like motility, contraction, growth, maintenance, active transport, bioluminescence, locomotion, cell division, nervous transmission, and biosynthesis of carbohydrates, fats, proteins etc.

10.3.2. Causes of Energy Richness of ATP

There are two main reasons behind this energy richness of ATP.

1. In each of the phosphate groups of ATP the oxygen atom, because of its tendency to acquire electrons, assumes a negative charge (Fig. 10.1.b) which induces a positive charge on the neighbouring phosphorus atoms. This means that in ATP molecule energy is required to overcome the electrostatic repulsion of like positive charges on the phosphorous atoms and to hold the molecule together. Thus, the sufficient amount of energy can remove this electrostatic repulsion and is stored in these phosphates and when one phosphate group is removed by hydrolysis this stored energy in released.

2. Several resonance forms exist for both the reactants and the products of hydrolysis. The more resonance forms produce greater stability. ATP possesses lesser number of resonance forms than ADP and for this reason ATP molecule is less stable than ADP, which is the hydrolysis product of ATP.

Similar changes take place during the hydrolysis of second and third phosphate groups in ADP and AMP respectively. AMP has the maximum number of resonance forms and is most stable among ATP, ADP and AMP. It shows the lowest free energy change.

10.4. OTHER ENERGY RICH COMPOUNDS

Besides, ATP, a number of other energy rich compounds have been discovered which play important

Introduction to Bioenergetics

role in cellular metabolism. For example, Guanosine-triphosphate (GTP), Uridine-triphosphate (UTP), Cytidine-triphosphate (CTP), Guanidinium phosphate, Phosphoenol pyruvate, Acetyl CoA etc.

UTP, GTP and CTP are the triphosphate esters of uridine, guanine and cytosine ribonucleosides respectively. Their energy richness can be explained similar to that of ATP showing less number of resonance forms of the reactant and more number of products. In guanidinium phosphate, there are no obvious electrostatic repulsion forces; and no obvious ionization or tautomerization processes. The creatine phosphate possesses only 12 resonance forms, whereas, its hydrolytic product has 18 resonance forms. This resonance-stabilization of the product of reaction clearly explains the energy-richness of creatine phosphate.

The Acetyl-CoA exhibits a $\Delta G'$ of hydrolysis of $-8,000$ cal/mole. In this thioester of acetyl coenzyme A the energy richness is due to ionization of product.

$$CH_3-\underset{\underset{Acetyl-CoA}{}}{\overset{\overset{O}{\|}}{C}}-S-CoA + H_2O \longrightarrow CH_3-\overset{\overset{O}{\|}}{C}-O^- + CoA-SH + H^+$$

The hydrolysis of acetyl coenzyme A results in the formation of acetic acid and CoA-SH. The acetic acid thus formed ionizes with a decrease in free energy of about -8000 cal/mole. The thiol group of CoA (CoA-SH) will not ionize at pH7.

Phosphoenolpyruvate contains a phosphate group and upon hydrolysis this compound yields a large amount of free energy ($\Delta G' = -12,800$ cal/mole).

$$\underset{\underset{\text{(enol-unstable form)}}{\text{Phosphoenol pyruvate (PEP)}}}{\overset{COOH}{\underset{CH_2}{\overset{|}{\underset{\|}{C}}-O-PO_3H_2}}} + H_2O \longrightarrow \underset{\underset{\text{(keto-stable form)}}{\text{Pyruvate}}}{\overset{COOH}{\underset{CH_3}{\overset{|}{\underset{|}{C=O}}}}} + H_3PO_4 \quad (\Delta G' = -12,800 \text{ Cal/mole})$$

There are two reasons for the energy richness of this compound Phosphoenol pyruvate (PEP).

(*i*) *Isomerization of the hydrolysis product* : In the phosphate form of pyruvate (PEP), the pyruvic acid is held in the unstable enol form, while on hydrolysis it isomerizes to a much more stable keto form. The enol-form has greater free energy, whereas, the ketoform has lower free energy.

(*ii*) *Resonance-stabilization of the Product* : The enol form of pyruvate has less number of resonance forms showing unstability, whereas, the keto-form possesses many resonance forms and greater stability.

The examples of energy-rich-compounds cited above clearly indicate that the large decrease in free energy occurs during hydrolysis because the products are significantly more stable than the reactants. The following 4 important factors contribute to this stability :

(*i*) Bond strain in the reactant caused by electrostatic repulsion (e.g., ATP, GTP, UTP, and Creatine phosphate).

(*ii*) Stabilization of the produce by ionization (e.g., Acetyl CoA).

(*iii*) Stabilization of the products by isomerization (e.g., phosphoenolpyruvate).

(*iv*) Stabilization of the products by resonance, (e.g., ATP, Phosphoenolpyruvate, GTP and UTP etc.).

10.5. LAWS OF THERMODYNAMICS

Thermodynamics is that branch of science which deals with the transformation of energy and includes all kinds of interconversions.

10.5.1. First Law of Thermodynamics

This law is essentially the law of conservation of energy and states, that "The total energy of a system plus its surroundings is constant and independent of any transformation which the system may undergo. Energy can neither be created nor destroyed by any means. Thus, whenever, energy in one form disappears, an equal amount of energy in some other form must appears".

Let a quantity of heat 'q' be absorbed by a system whose internal energy in the given state-A is E_1. As a result of absorption of heat, let the system do work equal to 'w' on its surroundings and let its final energy in the second state-B be E_2.

Increase in the internal energy of the system = $E_2 - E_1 = \Delta E$

According to the first law of thermodynamics,

$$\Delta E + w = q$$

$$\therefore \quad \Delta E = q - w$$

In cyclic processes, where the initial and final states of the system are the same, $E_1 = E_2$, and in such cases the ΔE being zero.

$$q - w = 0$$
$$\text{or} \quad q = w$$

The work done is equal to the heat absorbed.

10.5.2. Applications

All such processes where the transformation of one form of energy into another takes place, may be considered as the points of applications of the first law of thermodynamics. Some of the processes and steps as points of applications are given below :

1. The process of photosynthesis is the most important and common example of this type where the radiant energy of sun is converted into chemical energy. The process involves the reception of light quantum (Photon) by chlorophyll and its conversion into chemical energy takes place through many intermediate steps.

2. In cell only a part of energy liberated from the foodstuff is dissipated as heat, the rest is recovered as new chemical energy. The energy liberated in exergonic reactions is used in different cellular functions. Here, the chemical energy is converted into mechanical energy. The various processes which require this conversion of energy are—

   (*i*) Synthesis of carbohydrates, proteins and fats.

   (*ii*) Cell-division, cyclosis or cytoplasmic streaming indicate the mechanical work;

   (*iii*) Initiation of active transport against an osmotic or ion gradient.

   (*iv*) Maintenance of membrane potentials as in nerve conduction, transmission or to produce electric discharges (e.g., in electric fish).

3. Cell permeability includes a series of mechanisms that require energy. These mechanisms are generally described as active transport which indicates that a certain amount of work must be done in order for the molecules or ions to penetrate. ATP, which is mainly produced by oxidative phosphorylation in mitochondria, is generally used as the source of energy.

4. *Chemosynthesis* : There is a small group of bacteria that is able to obtain energy from inorganic molecules and the process is called, chemosynthesis. For example, the bacterium Nitrobacter oxidizes nitrites to nitrates ($NO_2^- + \frac{1}{2}O_2 \longrightarrow NO_3^-$). Some bacteria transform ferrous into ferric oxides and others oxidize -SH_2 to sulphate.

5. *Formation of vitamins* : The light energy helps in the photochemical activation of certain provitamins, like provitamin-A and provitamin-D. The light energy is converted into chemical energy and ultimately the production of vitamins takes place.

Introduction to Bioenergetics

6. In growth period 'q' is always greater than 'w'. Therefore when ΔE increases the child grows.

10.6. ENTROPY

Clausius in 1850 introduced a new function of the state of the system to serve as a reliable criteria of spontaneity (irreversible) and designated it by the symbol 'S', called **entropy**. As the entropy of a system is difficult to define directly, it is found more convenient to define it numerically by the equation,

$$dS = dQ/T$$

where, dQ is the heat absorbed or transferred reversibly by a system at constant temperature (T) and dS is the entropy change for an infinitesimal reversible process. Usually, the entropy is called the measurement of the disorder of the system.

Like ΔE, ΔS is dependent only on the state of the system and can be calculated if the substance is brought reversibly from one state to the other. It is independent of the path taken. The entropy is expressed in calories/degree. This is known as entropy unit.

10.6.1. Physical significance of Entropy

The definition of entropy change embodied in the expression, $dS = dQ/T$, though very useful in making entropy calculations and understanding physiochemical behaviour of the systems, is almost meaningless as a definition. There are two aspects that serve to give to entropy a definite physical significance.

(a) *Entropy is a degree of 'disorder' or 'randomness' of the molecule of a system* : Entropy is directly proportional to the disorder of the system. It means, for ordered systems, the entropy is low and for disordered, it is quite high. The spontaneous processes are accompanied by increase in entropy as well as increase in disorder of the system.

(b) *Entropy is a function of thermodynamic probability* : The state of equilibrium is the state of maximum probability. Any system, therefore, when left to itself, tends to go to a state of equilibrium.

10.6.2. The Concept of Entropy in living Systems :

In living systems, there are two types of processes, anabolic and catabolic. In anabolic processes the molecules are placed in definite order showing low entropy value whereas, in catabolic processes the ordered molecules are displaced giving maximum disorder and entropy value to the system.

Synthetic processes include the synthesis of carbohydrates, fats, proteins and nucleic-acids. In all these compounds the molecules of carbon, hydrogen, Oxygen and nitrogen are placed orderly and they show very low entropy values. In proteins; the aminoacid molecules are arranged in a definite sequence with the help of different type of bondings. Similarly, in nucleic acids the sugar molecules, phosphoric acid and organic bases (purines and pyrimidines) join in a definite order to give nucleosides and nucleotides which ultimately form the strands of the DNA or RNA molecules. The complete structure of proteins and nucleic acids show a low entropy value. When, these proteins and nucleic acid are hydrolysed, they yield their constituent aminoacids and nucleosides or nucleotides respectively. On degrading these molecules, the entropy of the system increases.

When carbohydrates, fats and proteins are oxidized into CO_2 and H_2O, the entropy value again increases. The various processes which help in the oxidation (increase in disorder) of such compounds are glycolysis, Kreb's cycle, α– and β–Oxidation of fatty acids and glyoxylate cycle. During these processes the compounds with high carbon number are degraded to the compounds of low carbon number and ultimately to CO_2 and H_2O.

10.7. CHEMICAL EQUILIBRIUM

In a chemical reaction, when the participating substances are left alone, they will react spontaneously until a position of equilibrium is reached. At this equilibrium, the observable concentrations of the reactants and products no longer change with time. The stage is known as 'chemical equilibrium' and is described by an equilibrium constant. For a general chemical reaction in which 'x' moles of 'A' and 'y' moles of 'B' are transformed into 'P' moles of 'C' and 'Q' moles of 'D'.

$$x A + y B \rightleftharpoons p C + q D$$

The equilibrium constant is given by

$$K = \frac{[C]^p [D]^Q}{[A]^x [B]^y}$$

The bracket letters [A] [B] [C] [D] represent properties of these substances called 'activities'. For dilute aqueous solutions, the activities of solutes are usually approximated by their molar concentrations and the activity of the solvent is usually taken as unity.

If the equilibrium constant is very small, then only a small quantity of reactants (A & B) would be converted into products (C & D) spontaneously. If the equilibrium constant is very large, then a large proportion of reactants would be converted into products spontaneously. A catalyst cannot change the equilibrium position of a reaction. It can only enhance the rate of the approach to equilibrium. The problem to face the cell is to make certain reactions proceed to completion even though their equilibrium constants may be small, that is, there must be a means to drive reactions in a direction which is away from equilibrium.

If a reaction is being carried out at constant temperature and pressure, the change in quantity is called 'free energy' and it represents the maximum useful work which it is possible to obtain from the reaction. The chemical equilibrium is found only in reversible reactions.

10.8. THERMODYNAMIC EQUILIBRIUM, DYNAMIC EQUILIBRIUM AND STEADY STATE

A system is said to be in thermodynamic equilibrium when any of its observable properties such as temperature, pressure, volume etc. do not change with time. For thermodynamic studies, a system must be in three types of equilibria which must exist spontaneously. The system must be in thermal equilibrium i.e., its temperature should not change with time, in chemical equilibrium i.e., its chemical composition should not change with time, and finally in mechanical equilibrium, *i.e.*, there should not be any movement of particles of the constituents of the system in itself and in between itself and surroundings. The systems in which diffusion or fast chemical reactions are taking place, are not in thermodynamic equilibrium and as such cannot be studied by thermodynamic methods.

Dynamic equilibrium occurs in reversible reactions. In a biochemical reaction, the reactants are transformed into products in forward direction and at the same time the products are converted into reactants in backward direction reversively. A stage comes when the rate of forward reaction becomes equal to the rate of backward reaction. The process of conversion of reactants into products and products into reactants does not stop but continues. Such a reaction is dynamic and the chemical equilibrium is known as 'dynamic equilibrium' and the state is known as 'steady state'.

The respiratory chain is a good example of dynamic equilibrium. It depends on the steady state concentrations of various redox systems. Other biochemical reactions of glycolysis, Kreb's cycle. Glyoxylate cycle, α– and β–Oxidation cycles which are reversible can be taken as examples of this type of equilibrium.

11
Biological Oxidation and Reduction

11.1 INTRODUCTION

Biological oxidations and reductions are phenomena which are found in living systems. The idea of oxidation was given at the end of the eighteenth century by Priestley, Lavoisier and others. Their classical researches showed that during respiration, animals take up oxygen from the air and give off carbon-di-oxide in exchange. They compared the phenomenon of respiration with that of chemical process of combustion which is almost similar. The oxygen taken from the air is used to oxidize the different food materials present inside the body into carbon-di-oxide and water. As a result of oxidation a large amount of energy is released which is utilized in a number of physiological processes. The oxidation of food material can also be compared with the oxidation of coal in a steam engine, but the two processes are not analogous. The oxidation of coal shows the conversion of chemical energy into heat energy and ultimately heat energy into mechanical energy. Here only small amount of heat energy is converted into mechanical energy. This conversion of heat energy into mechanical is possible only when there is great difference in the temperature. But in biological systems, the temperature differences are very small. So the living systems adopt some other methods of oxidation of food materials and liberation of energy for various physiological processes. In these methods the heat is not involved as an intermediate.

Lavoisier thought that 'biological oxidation' is the phenomenon related only with the lungs. The lungs utilize air and release CO_2 in exchange, but this view of *Lavoisier* was soon discarded when it was realized that the lungs serve only as a mechanism for exchange of gases between blood and atmosphere. In 1880, it was recognized that most tissues take up oxygen from the blood and release carbon-di-oxide into it. Later on, the uptake of oxygen and the release of carbon-di-oxide was demonstrated in a number of plants and many micro-organisms. Among plants and micro-organisms it was noticed that most nutrients and metabolites are not spontaneously oxidized by molecular oxygen. This problem created an idea of catalytic mechanism in the minds of various biologists and ultimately the concept of enzymes and enzymic mechanisms gradually emerged.

11.2 DEFINITION OF OXIDATION AND REDUCTION

The term oxidation was used in the past to describe the reaction in which a substance combined with oxygen, e.g.,

$$2Mg + O_2 \longrightarrow 2MgO.$$

and the reduction was used to describe that reaction in which oxygen was removed i.e., a reverse process of oxidation, e.g.,

$$PbO + H_2 \longrightarrow Pb + H_2O.$$

Later on, the term oxidation was used to cover reactions such as the change of ferrous ion into ferric ion.

$$\underset{\text{(Reduced)}}{Fe^{2+}} \rightleftharpoons \underset{\text{(oxidized)}}{Fe^{3+}} + e^-$$

This reaction is clearly an oxidation reaction which shows the change of FeO into Fe_2O_3 i.e, the change in corresponding oxides.

A much broader definition for oxidation and reduction includes the loss or gain of electrons. An **oxidizing agent** is one which accepts the electrons. The reduction, which is the reverse of oxidation, can be defined as the process which involves a gain of electrons, and the substance which has the characteristic to give up electrons to another substance is called as a **reducing agent**.

The two processes are complementary. Oxidation is always accompanied by reduction. In other words, they are simply two aspects of the same change–the transfer of electrons from one substance to another. For this reason, they have been given the name oxidation-reduction reactions or 'redox-reactions' or 'redox-systems'.

Thus, a compound is said to be oxidised if it has:

(a) lost one or more electrons.

(b) lost some atom, such as hydrogen, which has a relatively weak attraction for electrons, or

(c) combined with an atom such as chlorine or oxygen which have relatively strong attraction for electrons.

The reduction is the reverse of this process.

The biologically important oxidation and reduction reactions belong to the following three categories :

(a) loss of one or more electrons, for example

$$Fe^{2+} \xrightleftharpoons{-e} Fe^{3+}$$

or (b) loss of one or more atoms of hydrogen, for example

$$CH_3CH_2OH \xrightarrow{-2H} CH_3CHO$$

or (c) addition of one or more atoms of oxygen, for example

$$CH_3CHO \xrightarrow{+O} CH_3COOH$$

(b) and (c) can be again written in the form showing loss of electrons.

$$CH_3CH_2OH \xrightarrow{-2H^+,-2e} CH_3CHO$$

$$CH_3CHO \xrightarrow{+H_2O,-2H^+,-2e^-} CH_3COOH$$

Thus, on the basis of loss of electrons or addition of electrons, the biological oxidations and reductions can be written as follows :

$$C\ red \xrightarrow{-ne} C\ ox\ (oxidation)$$

$$C\ ox \xrightarrow{+ne} C\ red\ (Reduction)$$

Where C-represents the compound taking part in the reaction; Cred-represents the reduced form of the compound; Cox-represents the oxidized form; ne-represents the number of electrons; and '+' and '–' represent the addition of loss of electrons respectively.

Most of the chemical reactions do not involve free electrons or atoms. In these cases, a compound is oxidized only when an equivalent amount of another compound is simultaneously reduced. Thus, they can be represented as follows :

```
     C red          D ox          (C and D = Compounds
   Oxidation      reduction        taking part in the reactions)
     C ox           D red
```

Biological Oxidation and Reduction

This can be illustrated by taking the example of passing the hydrogen over heated oxide of Copper. In this reaction the hydrogen is oxidized to water and the copper-oxide is reduced to copper.

$$\begin{array}{c} H_2 \\ \text{oxidized} \\ H_2O \end{array} \bowtie \begin{array}{c} CuO \\ \text{reduced} \\ Cu \end{array}$$

11.3 REDOX-REACTIONS IN BIOLOGICAL SYSTEMS

Redox reactions are of considerable importance in the living systems. There are many important reactions involving electron transfer in the living cells. One very important process is the oxidation of glucose which yields a considerable amount of energy for various physiological processes. The overall reaction representing the oxidation of glucose may be written as :

$$C_6H_{12}O_6 + 6O_2 \longrightarrow 6CO_2 + 6H_2O \ (\Delta F = -685,000 \text{ cal}).$$

ΔF represents the energy released in the reaction.

This reaction is the reverse of photosynthesis, in which CO_2 and H_2O are converted into glucose in presence of chlorophyll and sunlight. The light energy releases electrons which then take part in a series of reactions. Thus, the process of photosynthesis is completed in a series of gradual steps. Similarly, in the oxidation of glucose a series of gradual steps are involved which ultimately produce a large amount of energy. It shows that the overall reaction is highly exergonic. The energy made available is released in small packets at various stages in the series of reactions.

Certain enzymes and coenzymes play an extremely important role in this process. These compounds can exist in an oxidized or a reduced form and so can act as a means of transferring electrons. The most important among coenzymes, is NAD (nicotinamide adenine dinucleotide) or NADP (nicotinamide adenine dinucleotide phosphate). Their reduced forms NADH and NADPH, take part in the respiratory chain in mitochondria producing a series of coupled oxidation-reduction reactions. In such reactions the coenzymes NAD or NADP act as hydrogen acceptors and are reduced to NADH or NADPH, while the substrate is oxidized. Thus :

$$\begin{array}{c} AH_2 \\ \text{oxidized} \\ A \end{array} \bowtie \begin{array}{c} NAD \\ \text{reduced} \\ NADH_2 \end{array} \quad \text{or} \quad \begin{array}{c} AH_2 \\ \\ A \end{array} \bowtie \begin{array}{c} NADP \\ \\ NADPH_2 \end{array}$$

where 'A' represents the substrate.

When the coenzyme acts as an oxidizing agent or "hydrogen acceptor" as shown above, the pyridine ring of the molecule undergoes reduction as shown below :

$$AH_2 + \underset{\underset{R}{\overset{+}{N}}}{\bigcirc} \rightleftharpoons A + \underset{\underset{H\ R}{\overset{+}{N}}}{\overset{H\ \ H}{\bigcirc}} \quad \text{(reduced form)}$$

The reduced form of coenzyme at physiological pH ionizes and the liberation of one atom of hydrogen takes place.

$$\underset{\underset{H\ R}{\overset{+}{N}}}{\overset{H\ \ H}{\bigcirc}} \xrightleftharpoons[]{\text{Ionization}} \underset{\underset{R}{\overset{+}{N}}}{\overset{H\ \ H}{\bigcirc}} + H^+$$

Reduced form

The reaction may be rewritten more accurately as follows :

Such type of reactions, i.e., the transfer of two hydrogen atoms from the substrate to NAD$^+$ or NADP$^+$, are catalysed by special group of enzymes known as **dehydrogenases**. These dehydrogenases are highly specific for substrates and coenzymes. For example, in the conversion of malic acid into oxaloacetic acid the **malate dehydrogenase** enzyme catalyses the reaction.

Oxidation-Reduction Reactions Catalysed by NAD or NADP in Kreb's Cycle

1. In the conversion of isocitric acid into oxalosuccinic acid, isocitric acid is oxidised and NAD$^+$ is reduced. The reaction is catalysed by the enzyme *isocitric dehydrogenase*.

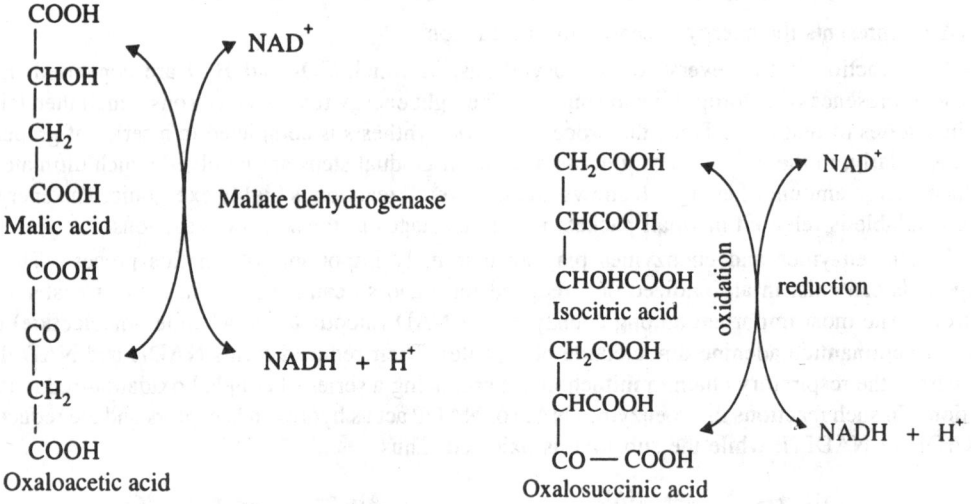

2. During oxidation of lipoic, NAD$^+$ is converted into NADH. This reaction occurs when α-ketoglutaric acid is converted into succinyl CoA. TPP and CoASH coenzymes are also involved.

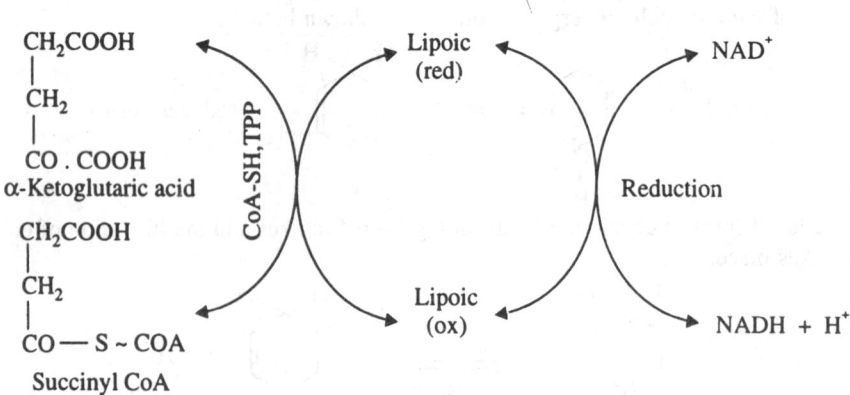

3. In the conversion of malic acid into oxaloacetic acid, the *malate dehydrogenase* catalyses the reaction. Here malic acid is oxidized and NAD$^+$ is reduced.

Biological Oxidation and Reduction

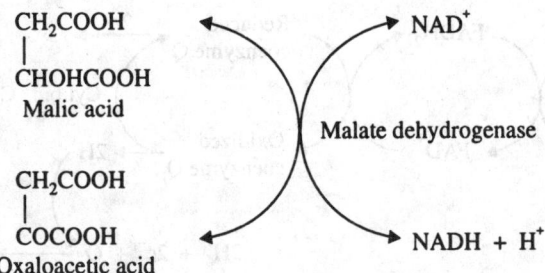

Reaction Catalysed by NAD in Glycolysis

NAD plays an important role in glycolysis. 3-Phosphoglyceraldehyde is oxidized with the release of two electrons (2 e⁻) and two protons ($2H^+$) which, however, reduce NAD to NADH. As a result of oxidation of 3-phosphoglyceraldehyde the energy is released which is trapped and used up in attaching one inorganic phosphate (iP) to the first carbon position of 3-phosphoglyceraldehyde. Thus, the phosphorylation and conversion of 3-phosphoglyceraldehyde into 1,3-diphosphoglyceric acid takes place. The reaction is brought about by the enzyme *phosphoglyceraldehyde dehydrogenase*.

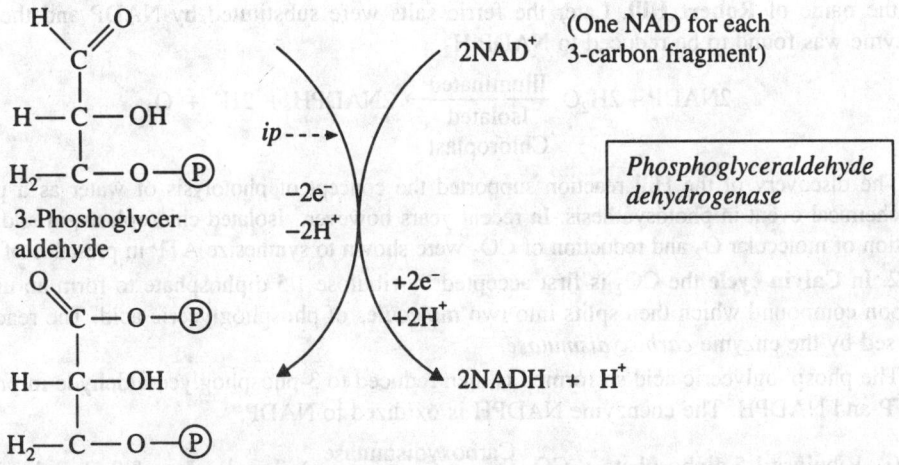

NAD in Electron Transport System

In the initial stages of electron transport the substrate is oxidized by loss of two electrons and two protons, taken up by NAD^+ which is reduced to NADH.

NAD^+ is regenerated when it loses two electrons and two protons to FAD to form $FADH_2$ which is again oxidized to FAD by loss of 2 electrons and 2 protons. The electrons enter into the chain to reduce and oxidize the coenzyme Q, cytochrome b, cytochrome C, cytochrome a and cytochrome a_3 and finally are accepted by one atom of oxygen to form water.

```
   NAD⁺ ←――  FADH₂ ←――  Reduced     ←―2e⁻―→
    )           )       coenzyme Q  )
    (           (           )       (       Cyt.b,... Cyt C₁C₂.... Cytₐ, ...a₃
    )           )           (       )
   NADH₂ ――→   FAD⁺ ――→  Oxidized   ――→ 2H⁺
                         coenzyme Q
```

$$2H^+ + 2e^- + O \longrightarrow H_2O$$

Oxidation-reduction reactions in Photosynthesis Catalysed by NAD or NADP

1. Hill reactions: Hill (1939) separated the light and dark reactions of photosynthesis and found that isolated chloroplasts illuminated in absence of CO_2, but in the presence of exogenously supplied hydrogen acceptors, such as ferric salts, evolved molecular O_2.

$$4Fe^{3+} + 2H_2O \xrightarrow[\text{Isolated Chloroplasts}]{\text{Illuminated}} 4Fe^{2+} + 4H^+ + O_2$$

This property of chloroplast to reduce hydrogen acceptor has been named as **Hill reaction**, after the name of **Robert Hill**. Later the ferric salts were substituted by NADP and the added coenzyme was found to be reduced to $NADP\,H_2$.

$$2NADP + 2H_2O \xrightarrow[\text{Isolated Chloroplast}]{\text{Illuminated}} 2NADPH + 2H^+ + O_2$$

The discovery of the Hill reaction supported the concept of photolysis of water as a primary photochemical event in photosynthesis. In recent years however, isolated chloroplasts, in addition to evolution of molecular O_2 and reduction of CO_2 were shown to synthesize ATP in presence of light.

2. In Calvin cycle the CO_2 is first accepted by ribulose 1,5 diphosphate to form an unstable 6-carbon compound which then splits into two molecules of phosphoglyceric acid. The reaction is catalysed by the enzyme *carboxydismutase*.

The phosphoglyceric acid so formed is then reduced to 3-phosphoglyceraldehyde in presence of ATP and NADPH. The coenzyme NADPH is oxidized to NADP.

(i) Ribulose 1,5 diphosphate + CO_2 $\xrightarrow{\text{Carboxydismutase}}$ 2 molecules of 3 phosphoglyceric acid.

(ii) 3-Phosphoglyceric acid (PGA) + NADPH + H^+ + ATP $\xrightarrow{\text{Phosphoglyceric acid dehydrogenase}}$
3-Phosphoglyceraldehyde + $NADP^+$ + ADP + iP.

Other Redox Reactions Catalysed by NAD

(1) In alcoholic fermentation pyruvic acid is converted into ethyl alcohol and CO_2 is liberated. In this process, aldehyde is formed as an intermediate and the process is completed in the following two steps.

(a) Pyruvic acid $\xrightarrow{\text{Pyruvic decarboxylase}}$ Acetaldehyde.

(b) Acetaldehyde + NADH $\xrightarrow{\text{Alcohol dehydrogenase}}$ $C_2H_5OH + CO_2 + NAD^+$.

The second step shows the reduction of acetaldehyde to C_2H_5OH and oxidation of NADH to NAD^+. Thus, two processes, reduction and oxidation occur simultaneously.

Biological Oxidation and Reduction

(2) During the break-down of pyruvic acid into lactic acid, oxidation and reduction also take place. This is the common process in animal tissues. The reaction takes place in the presence of enzyme *lactic-dehydrogenase*.

$$\text{Pyruvic acid} \xrightarrow[\text{Reduction}]{} \text{Lactic acid} \quad \text{NADH} + \text{H}^+ \xrightarrow[\text{Oxidation}]{\text{Lactic dehydrogenase}} \text{NAD}^+$$

(3) During the synthesis of fatty acids acetoacetyl CoA (5-C compound) is reduced to butyric acid (4-C compound) in presence of $NADH_2$.

Acetoacetyl CoA + NADH + H$^+$ ⟶ Butyryl CoA + NAD$^+$

Butyryl CoA + NADH + H$^+$ ⟶ Butyric acid + CoA + NAD$^+$

(4) In β-oxidation, NAD$^+$ also plays an important role. Here, β-hydroxyacyl dehydrogenase type of reaction results in the loss of hydrogen ions which convert NAD$^+$ to NADH.

$$\text{R--CH(OH)--CH}_2\text{--}\underset{\underset{O}{\|}}{C}\text{--CoA} + \text{NAD}^+ \xrightarrow[\text{dehydrogenase}]{\beta\text{--hydroxyacyl}} \text{R--}\underset{\underset{O}{\|}}{C}\text{--CH}_2\text{--}\underset{\underset{O}{\|}}{C}\text{--CoA} + \text{NADH}$$

(5) In nitrogen metabolism, there are many steps where NADH or NADPH play important role in oxidation and reduction.

(a) *Evans* and *Nason* (1954) demonstrated the reduction of nitrates in presence of *reductase* enzyme which contains FAD as prosthetic group. NADH and NADPH act as hydrogen donors. Molybdenum serves as an electron carrier.

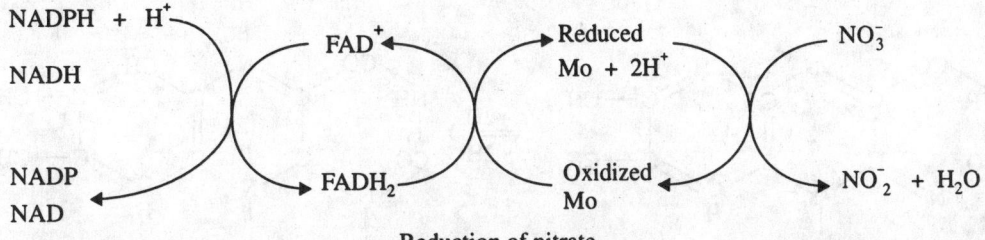

Reduction of nitrate

(b) The reduction of nitrite was demonstrated by *Vaneco* and *Varner* (1955) by photochemical splitting of water, but *Nicholas* (1957) demonstrated in *Neurospora crassa* the reduction of hyponitrite into hydroxylamine in presence of NADH. He further showed that NADH requires iron and copper for its activity.

$$2HNO_2 + 2H_2O \xrightarrow{\text{Light}} 2NH_3 + 3O_2.$$

$$\text{Hyponitrite} + \text{NADH} \xrightarrow{Cu^{2+}, Fe^{3+}} \underset{(NH_2OH)}{\text{Hydroxylamine}} + \text{NAD}^+.$$

(c) In a similar process adopted by *Nicholas* (1957), *Tucker* and *Nason* demonstrated reduction of hydroxyl amine into NH_3 in *Neurospora* and higher plants in presence of NADH.

$$NH_2OH + NADH + H^+ \xrightarrow[\text{reductase, } Mn^{+2}]{\text{Hydroxylamine}} NH_3 + NAD^+ + H_2O.$$

(6) The disulphide form of glutathione is reduced by NADPH and glutathione reductase. Glutathione is oxidized by the system of cytochrome C and *cytochrome oxidase* as well as by the *glutathione peroxidase*.

$$2\ GSH \underset{+2H}{\overset{-2H}{\rightleftharpoons}} G-S-S-G$$

The reduced and oxidized forms of glutathione form reversible oxidation-reduction systems.

(7) The reduced and oxidized forms of ascorbic acid also form reversible oxidation-reduction systems.

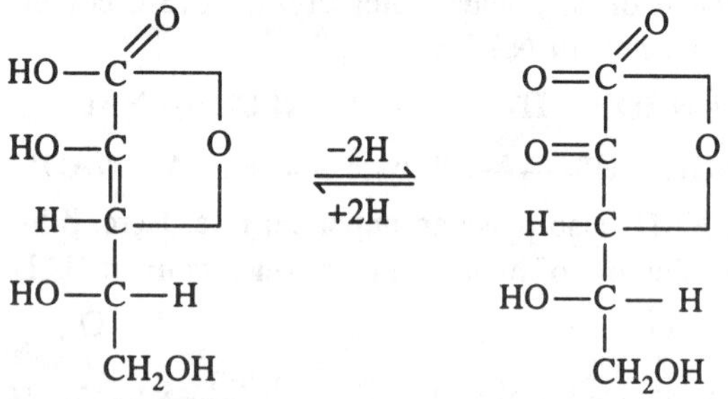

Flavoproteins and FAD enzyme in oxidation-reduction reactions

The reactions described above depend upon the availability of NAD^+ or $NADP^+$. Their total amount is very small in the tissues and gets exhausted in a few reactions. Hence, it is necessary to regenerate them. Flavoproteins bring about their regeneration. Flavoproteins are yellow coloured enzymes and contain FMN (flavin mononucleotide) or FAD (flavin adenine dinucleotide) as the prosthetic groups. They catalyse the oxidation-reduction reactions, in which the dimethylisoalloxazine moiety of the prosthetic group undegoes reversible reduction and oxidation as follows:

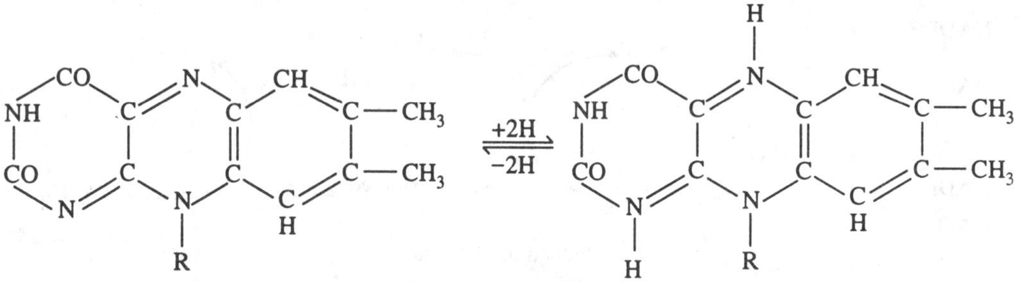

For easy understanding they can also be represented in brief as follows:

$$FP \underset{-2H}{\overset{+2H}{\rightleftharpoons}} FPH_2$$

An enzyme of this group, *NADH dehydrogenase*, appears to reoxidize NADH.

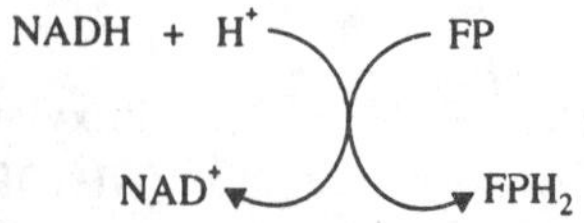

NADH + H⁺ ⟶ FP
NAD⁺ ⟵ FPH₂

The flavoproteins just like NAD^+ or $NADP^+$, are present in very small quantities in the cell. Their reduced forms are reoxidized by the cytochrome systems.

Cytochrome System in Oxidation-reduction reactions :

For structure of different kinds of cytochromes and their role in oxidation-reduction reactions consult the chapter on coenzymes.

Biological Oxidation and Reduction

Lipoic Acid and Other Coenzymes in redox reactions :

Consult the chapter on coenzymes.

11.4. OXIDATION-REDUCTION POTENTIAL OR REDOX-POTENTIAL AND ITS MEASUREMENT

The redox potential of a compound is quantitative measurement of its ability to gain or lose electrons (i.e., its electron affinity). It is expressed as a numerical value which may be positive or negative. On the basis of oxidation-reduction potential, the substances may be arranged in a series, known as **Redox-series**.

Redox potential is determined by measuring electrode potential of various systems compared to that of hydrogen, which is taken as an arbitrary standard. The hydrogen electrode has a potential of zero when an inert metal electrode dips into a solution at pH 0.

The potential of any oxidation-reduction system can be determined if the system can accept of donate electrons reversibly at a metal electrode. In order to determine the redox potential of a given system (=half cell) it must be suitably connected to the standard hydrogen electrode (=standard half cell) and the potential difference between the two half cells (=complete electrical cell) is measured by means of a potentiometer directly and is expressed in volts. A constant potential difference is maintained at the time of equilibrium, provided the conditions are unchanged.

A scale of electrode potentials is set up by measuring the redox potentials of half cells relative to that of the standard hydrogen electrode. This scale is called the **hydrogen scale**. On this scale a half cell which is more strongly reducing (i.e., has a lesser affinity for electrons) than the standard hydrogen electrode has a negative redox potential (E' is –ve); if such a half cell were joined to standard hydrogen electrode, electrons would flow from that cell to the standard hydrogen electrode. Conversely a half cell which is more strongly oxidizing (i.e., has greater affinity for electrons) than the standard hydrogen electrode has a positive redox potential (E' is +ve); if such a cell were joined to the standard hydrogen electrode, electrons would flow from the standard hydrogen electrode to that half cell.

"The potential set up when an electrode is in contact with a molar solution of its own ions at $25°$ and one atmospheric pressure, is defined as **standard electrode potential** of the element ($E'o$)". This value is of great use in oxidation-reduction reactions, since it describes the relative affinities for electrons of various systems, and is then known as **standard redox-potential**. Usually any half cell fulfilling the standard requirements (i.e. molar solution kept at $25°$ and 1 atmosphere pressure) on connecting with standard hydrogen electrode gives the standard redox potential of that half cell. The standard redox potential is designated as $E'o$. The actual redox potential E of any half cell can be calculated by the following equation.

$$E = E'_o + \frac{RT}{nF} \log \frac{[\text{oxidant}]}{[\text{reductant}]}$$

Where,

E = Actual redox potential
$E'o$ = Standard redox potential
R = Gas constant (8.314 absolute joules mole^{-1} degree^{-1})
T = absolute temperature
F = Faraday (96,497 coulombs/g equiv.)
n = number of electrons per grams equivalent of reactant transferred in the reaction.

Knowing the $E'o$ and the concentrations of the oxidant and reductant (or even the ratio of their concentrations) one can calculate, using the above equation, the actual redox potential of a given half cell.

The following Table 11.1 shows E'_o values of some biologically important redox systems.

Table 11.1 : E'_o Values of some Biologically Important Redox Systems

Redox system (half cell)	Temp. (°C)	E'_o (volts) (pH7)
1. Pyruvate/Acetate + CO_2	25	–0.699
2. α-ketoglutarate/succinate + CO_2	25	–0.673
3. Acetate/acetaldehyde	25	–0.600
4. H_2/H^+	25	–0.413
5. H_2/H^+	30	–0.420
6. Malate/pyruvate + CO_2	25	–0.330
7. NADH/NAD^+	25	–0.320
8. $FMNH_2$/FMN	30	–0.219
9. $FADH_2$/FAD^+	30	–0.219
10. Ethanol/acetaldehyde	25	–0.197
11. Lactate/pyruvate	25	–0.185
12. Malate/oxaloacetate	25	–0.166
13. Cytochrome b_5 (Fe^2/Fe^{3+})	25	–0.120
14. Cytochrome b_6 (Fe^{2+}/Fe^{3+})	25	–0.06
15. Cytochrome b (Fe^{2+}/Fe^{3+})	25	–0.04
16. Succinate/fumarate	25	+0.031
17. Ubiquinone (red/ox)	25	+0.10
18. Ascorbate/dehydroascorbate	25	+0.166 (pH4)
19. Cytochrome c(Fe^{2+}/Fe^{3+})	25	+0.25
20. Cytochrome a (Fe^{2+}/Fe^{3+})	25	+0.29
21. Cytochrome f(Fe^{2+}/Fe^{3+})	25	+0.365
22. 1/2 O_2/OH^-	25	+0.817

12

Secondary Metabolites

Secondary metabolites are the groups of natural products such as terpenes and terpenoids, alkaloids, flavonoids, certain fat soluble vitamins, pigments etc. which are produced in plants in unplanned manner.

TERPENOIDS

Structure of Terpenoids

All terpenoids in their structure conain a basic branched 5-carbon unit, the **isopentane unit**.

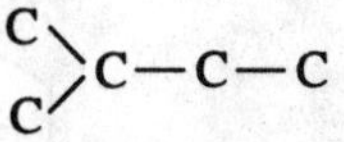

$$\begin{array}{c} C \\ \diagdown \\ C-C-C \\ \diagup \\ C \end{array}$$

ISOPENTANE UNIT

Depending upon the number of isopentane units present, the terpenoids may be of following types:

Table 12.1 : Types of Terpenoids.

	Type	Number of carbon atoms	Example
1.	Hemiterpene	C–5	Isovaleraldehyde
2.	Monoterpene	C–10	Geraniol
3.	Sesquiterpene	C–15	Farnesol
4.	Diterpene	C–20	Geranylgeraniol
5.	Triterpene	C–30	Squalene
6.	Tetraterpene	C–40	Phytoene
7.	Polyterpene	C ~ 4000	Rubber

The structure, type and source of several terpenes are given in Table 12.2, 12.3, 12.4, 12.5, 12.6 and 12.7.

Table 12.2 : Some Hemiterpenes

Name	Structure	Source
Isoamyl alcohol	$(CH_3)_2CHCH_2CH_2OH$	as ester in essential oils, e.g. *Mentha*
Isovaleraldehyde	$(CH_3)_2CHCH_2CHO$	essential oil of *Eucalyptus*
Senecioic acid (ββ-Dimethylacrylic acid)	$(CH_3)_2C=CHCOOH$	*Senecio kaempferi*
β-Furoic acid	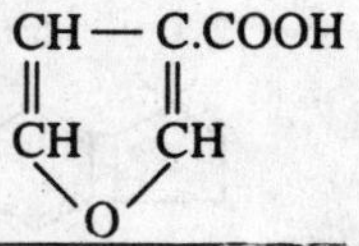 $\begin{array}{c} CH-C.COOH \\ \parallel \parallel \\ CH CH \\ \diagdown \diagup \\ O \end{array}$	*Phaseolus multiformis*

Table 12.3 : Some Typical Monoterpenes

Type	Name	Structure	Source
Open chain	Myrcene		Rhus spp.
Monocyclic	Limonene		Citrus spp.
Bicyclic	α-Pinene		Pinus spp.
Tricyclic	Teresantalol	—CH$_2$OH	Santalum album

Table 12.4 : Some Typical Sesquiterpenes

Type	Name	Structure	Source
Open chain	Farnesol		Widely distributed
Monocyclic	γ-Bisabolene		Widely distributed
Bicyclic (a) Naphthelenic type	α-Cadinene		Cedrus spp.
(b) Spiro type	β-Vetivone		Vetivera zizaniodes
Unusual monocyclic	Humulene		Humulus lupulus
Unusual bicyclic	Cuparene*		Cupressus spp.

* Note the terpenoid-derived aromatic ring

Secondary Metabolites

Table 12.5: Some Typical Diterpenes

Type	Name	Structure	Source
Acyclic	Phytol		Combined in chlorophylls
Monocyclic	α-Camphorene		*Cinnamomum camphora*
Dicyclic	Agathic acid		*Agathis alba*
Tricyclic	Abietic acid		*Pinus palustris*
Tetracyclic	Phyllocladene		*Phyllocladene trichomanoides*

Table 12.6: Some Typical Triterpenes

Type	Name	Structure	Source
Acyclic	Squalene		Widely distributed in traces
Tetracyclic (dimethyl sterols)	Cycloartenol		Potato leaves; widely distributed in traces
Pentacyclic	β-Amyrin		Pea seedlings; resins, latex and waxes of many plants

Table 12.7 : Some Typical Tetraterpenes (Carotenoids)

Type	Name	Structure	Major Source
Carotenes (Hydrocarbons)			
Acyclic	Lycopene		Tamato fruit
Monocyclic	γ-Carotene		Widely distributed in trace amount.
Bicyclic	β-Carotene		All green tissues
Xanthophylls (oxygenated carotenes)			
Hydroxy derivatives	Lutein (3.3'-dihydroxy α-carotene)		All green tissues
Keto derivatives	Canthaxanthin (4, 4'-diketo-β-carotene)		Some flower petals
Epoxides	Violaxanthin (5, 6, 5', 4'- diepoxy zeaxanthin)		All green leaves *Viola* spp. flowers
Furanoid derivatives	Flavoxanthin (5,8-epoxylutein)		Flower petals

Secondary Metabolites

1. **Mono, Sesqui and Diterpenes :** The structure, type and source of these terpenes are given in Table 12.3, 12.4 and 12.5. The hormones abscisic acid and the gibberellins are the derivatives of a sesquiterpene and diterpene respectively. For detail consult chapter on "Metabolism of growth hormones".
2. **Triterpenoids :** Consult chapters on "*Lipids*" and "*Lipid metabolism* (Derived lipids)".
3. **Tetraterpenes :** Carotenoids are the only naturally occurring tetraterpenes. They are found in photosynthetic tissues in chloroplast lamellae. Slight variations in the structure of carotenoids occur in higher plants but much variations occur in carotenoids present in fruits, flowers petals, algae of different classes and photosynthetic bacteria.
   For details on carotenoid structure, functions and biosynthesis consult chapter on "Photosynthesis" and "Vitamins".
4. **Terpenoid Glycosides :** Consult chapter on '*Lipids*'.
5. **Terpene Alkaloids:** The terpenoid compounds containing nitrogen as a part of a heterocyclic ring system are called **terpene alkaloids**. They are found in plants like *Solanum, Actinidia* and *Nupher* etc.
6. **Tropones and Tropolones :** Tropones are terpenoids having seven-membered carbocyclic ring system conjugated with a keto group whereas tropolones are the derivatives of tropones having an extra hydroxy group at C-2. Tropolones are found in higher plants, e.g., **eucarvones** in *Asarum sieboldi* and **thujaplicinol** in *Cupressus pygmaea*. Tropolones occurring in fungi are probably formed by ring expansion of a benzenoid precursor.

A tropone   A tropolone (Thujaplicinol)

7. **Bitter Principles :** The bitter taste in higher plants is mostly due to presence of terpenoids like saponins and cardiac glycosides. In cucurbitaceae the bitter principles are derived from **lanosterol** or cycloartenol with a modified side chain at C-17. Citrus fruits contain **limonin**. The most intense bitter principle known is a diterpene **amarogentin. Stevioside,** a diterpene glycoside is the sweetest natural product. The **aglycone** is steviol.
8. **Long Chain Terpene Alcohols :** Previously it was supposed that long chain terpene alcohols with 20 carbon atoms occur naturally but they are always found combined with non-terpenoid residue like plastoquinone. Tobacco leaves contain terpene alcohol **solanesol** with C-45 atoms. Isoprene alcohols with C-30 to C-110 have been discovered in a number of plant tissues.
9. **Rubber :** It is a polyisoprenoid produced in the latex of *Hevea* and other angiosperms. The rubber obtained from *Hevea* sp is a cis-polymer containing about 3000-6000 isoprene residues. Gutta is a trans-polymer produced by Guayule plant.
10. **Mixed terpenoids :** The terpenoid compounds containing a terpenoid component and another non-terpenoid compound other than a sugar or a fatty acid are called **mixed terpenoids.** *Chlorophylls, Vitamin* K_1 *(phylloquinone),* α-*tocopherol quinone* (all with C-20 isoprenoid side chain); *ubiquinone* with C-50 isoprenoid side chain and *plastoquinone* of higher plants with C-45 isoprenoid side chain are all mixed terpenoids.
    For the details of above mixed terpenoids consult chapter on 'photosynthesis', 'vitamins' and 'coenzymes'.

Biosynthesis of Terpenoids

All terpenoid are synthesized from isoprene unit called **isopentenyl pyrophosphate** (IPP) which ultimately is synthesized from three molecules of acetyl-CoA Via mevalonic acid as outlined in following scheme :

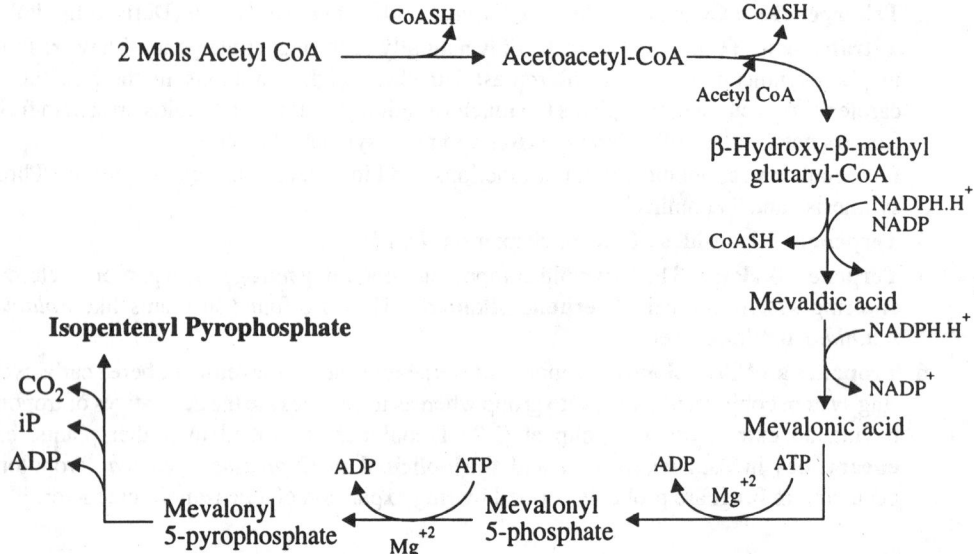

Fig. 12.1: Showing formation of isopentenyl pyrophosphate from acetyl CoA.

The isopentenyl pyrophosphate synthesized through above scheme is converted into geranyl pyrophosphate which is then utilized in the synthesis of different types of terpenoids as shown in following figure :

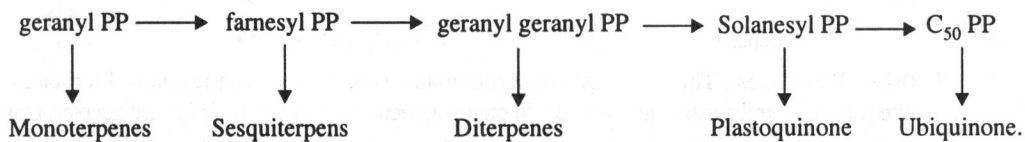

Fig. 12.2: Showing synthesis of different terpenoids from geranyl-PP.

Functions of Terpenoids in Higher Plants

The important functions of different terpenoids are as follows :

1. The monoterpenes show no biological function but their penetrating odour helps in attracting or repelling insects during pollination.
2. The abscisic acid, a sesquiterpene, helps in the regulation of seed dormancy and leaf abscission.
3. In sweet potato the *sesquiterpene ipomeamarone* shows anti-fungal action against the pathogen *Ceratocystis fimbriala.*
4. Sterols are used as structural units of cell-organelles, e.g., cholesterol in animal cells.
5. Carotenoids are accessory pigments and help in light reaction of photosynthesis.
6. In photosynthetic bacteria, carotenoids protect the cells from photosensitization by light absorbed by chlorophylls.
7. Carotenoids act as photoreceptor for response during phototropism.
8. The terpenoids like chlorophylls, plastoquinones, ubiquinones and tocopherylquinones

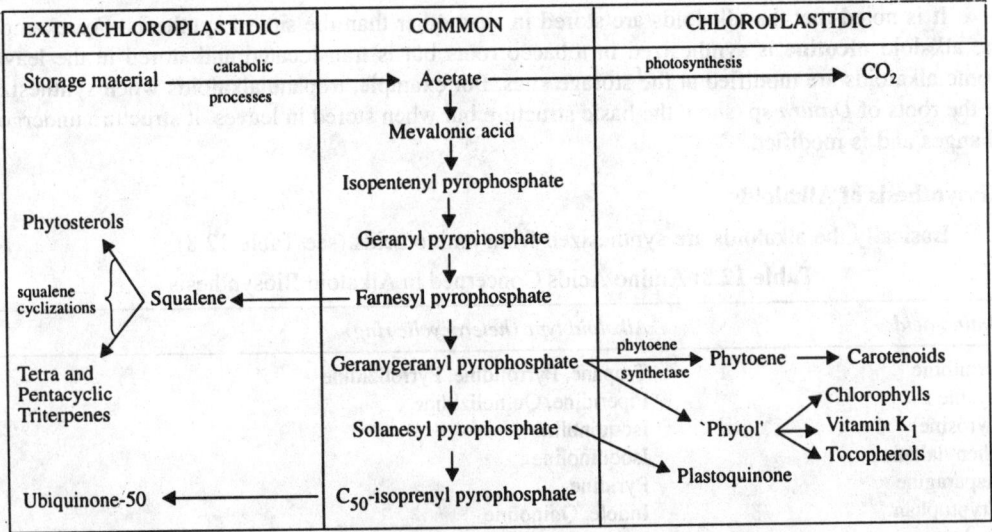

Fig. 12.3: Gives a general and simplified idea of terpenoid biosynthesis in higher plants.

perform their specific functions (consult chapter on Vitamins, Coenzymes and Photosynthesis).

9. The ubiquinones contain long terpenoid side chains and are important components of electron transport system (ETS).
10. For functions of phylloquinone (Vitamin K_1) and tocopherolquinone, consult chapter on 'Vitamins'.

Alkaloids

Alkaloids represent another group of secondary metabolites parallel to terpenes or terpenoids. About 2500 alkaloids have been isolated so far. All alkaloids frequently contain nitrogen in a heterocyclic ring and are basic in nature. They are usually found in plants as salts of organic acids and exhibit important pharmacological properties. The alkaloids may be true alkaloids, protoalkaloids or pseudoalkaloids.

1. **True alkaloids :** The alkaloids containing true heterocyclic rings are called *true alkaloids* and they are classified according to the ring system present in the molecule. The common examples of true alkaloids are *nicotine, atropine, reserpine, quinine, actinidine* and *atisine* etc.
2. **Protoalkaloids :** The alkaloids which do not contain heterocyclic rings are called *protoalkaloids* and they are amines. e.g., *hordenine* found in germinating barley (Hordeum disticha).
3. **Pseudoalkaloids :** Those alkaloids which are not derived directly from amino acids are called pseudo-alkaloids. e.g., terpenoid containing alkaloids.

Distribution and Localization

Alkaloids are poorly distributed in pteriodophytes and gymnosperms and unevenly distributed in angiosperms. They are mainly found in dicotyledonous plants of Centrospermae, Magnoliales, Ranunculales, Papaveraceae, Leguminosae, Papilionaceae and Rutaceae. Among mocotyledonous families, they are found in Liliacea and Gramineae.

The alkaloids are chiefly produced by four types of plant tissue : (*i*) meristematic tissues, (*ii*) epidermal and hypodermal cells, (*iii*) Cells of bundle sheath and (*iv*) latex vessels. They accumulate in the vacuoles and hence do not appear in young cells until they become vacuolated.

It is noted that the alkaloids are stored in sites other than the site of synthesis. For example, the alkaloid **nicotine** is synthesized in tobacco roots but is translocated and stored in the leaves. Some alkaloids are modified at the storage sites. For example, tropane alkaloids when synthesized in the roots of *Datura* sp. show the basic structure but when stored in leaves, it structure undergoes changes and is modified.

Biosynthesis of Alkaloids

Basically the alkaloids are synthesized from amino acids (see Table 12.8).

Table 12.8: Amino Acids Concerned in Alkaloid Biosynthesis

Amino acid	Alkaloid type (heterocyclic ring)
Ornithine	Tropane, Pyrrolidine, Pyrrolizidine
Lysine	Piperidine, Quinolizidine
Tyrosine	Isoquinoline
Phenylalanine	Isoquinoline
Asparagine	Pyridine
Tryptophan	Indole, Quinoline
Proline	Pyrroline
Glutamic acid	Pyrroline
Histidine (?)	Imidazole

1. Biosynthesis of Protoalkaloids

A typical protoalkaloid **hordenine** is synthesized from amino acid *tyrosine* or *phenylalanine* (Fig. 12.4). The methyl groups present in hordenine originate from *methionine* by the process of methylation in presence of enzyme *tyramine* methylpherase.

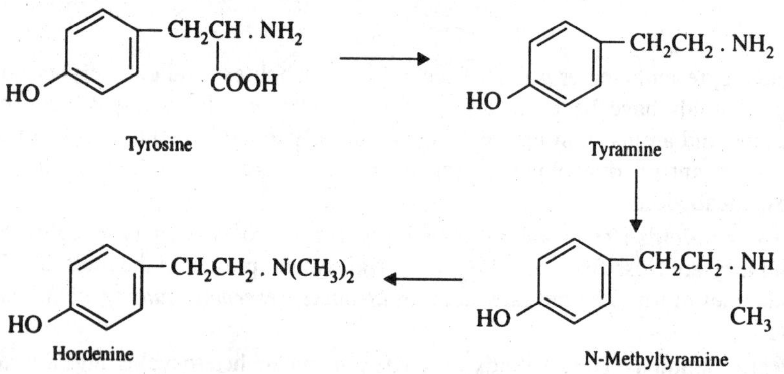

Fig. 12.4: Biosynthesis of hordenine.

2. Biosynthesis of True Alkaloids

The true alkaloids may be derivatives of (*i*) Pyrrolidine, (*ii*) Pyrrolizidine, (*iii*) Pyridine, (*iv*) Piperidine, (*v*) Quinolizidine, (*vi*) Isoquinoline, (*vii*) Indole, (*viii*) Quinoline, and (*ix*) Imidazole. The biosynthesis of only a few important alkaloids from their derivatives is described here.

(*i*) **Pyrrolidine derivatives :** e.g., hyoscyamine alkaloid.

The pyrrolidine ring is a part of tropane skeleton of tropane alkaloids like hyoscyamine. It is synthesized through amino acid *ornithine* via an intermediate pyrrolidine 5-carboxylic acid (Fig. 12.5).

(*ii*) **Pyrrolizidine derivatives :** e.g., *Retronecine* alkaloid.

It is synthesized from two ornithine molecules. The alkaloid contains pyrrolizidine nucleus made up of two fused pyrrolidine nuclei.

(*iii*) **Pyridine dervatives** : e.g., Nicotine

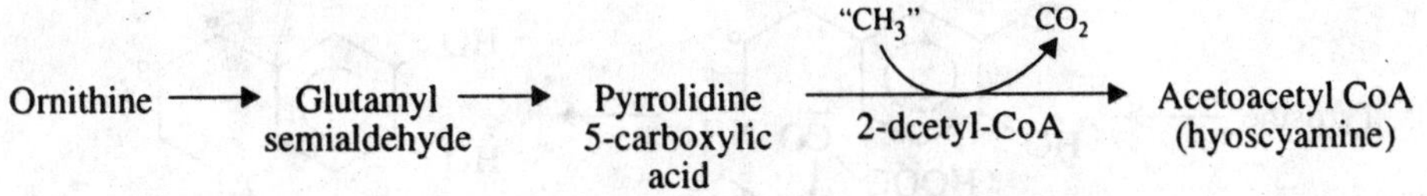

Ornithine ⟶ Glutamyl semialdehyde ⟶ Pyrrolidine 5-carboxylic acid ⟶ "CH₃" / CO_2 / 2-dcetyl-CoA ⟶ Acetoacetyl CoA (hyoscyamine)

Fig. 12.5: Biosynthesis of hyocyamine

The alkaloid nicotine in higher plants, anaerobic yeast and bacteria is synthesized from nicotinic acid which in turn is formed from L-*aspartic acid* and glycerol. But in aminals and aerobic fungi the nicotine is synthesized from *tryptophan*.

(*iv*) **Isoquinoline derivatives** : e.g., *Morphine* group of alkaloids.

Glycerol + Nicotinic acid → Nicotine

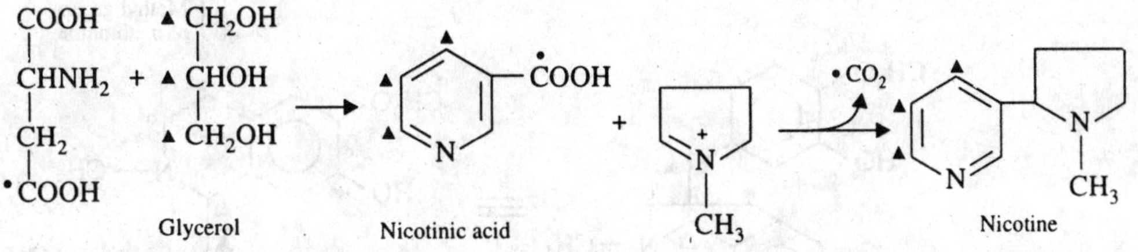

Fig. 12.6: Biosynthesis of Nicotine in higher plants, anaerobic yeast and bacteria.

Tryptophan ⟶ N-formyl kynurenine ⟶ Anthranilic acid ⟶ 3 hydroxy anthranilic acid

Nicotine ↑ Nicotinic acid ⇌ Nicotinic acid mononucleotide ⟵ Phosphoribosyl Pyrophosphate ⟵ Quinolinic acid / CO_2

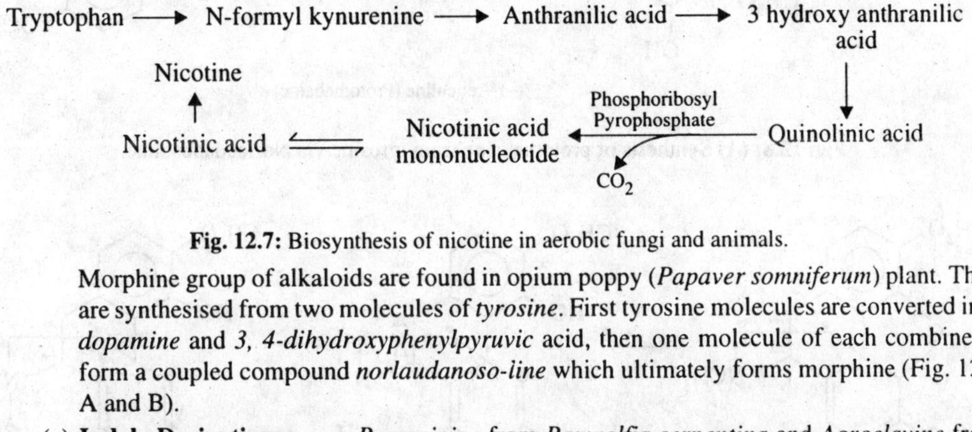

Fig. 12.7: Biosynthesis of nicotine in aerobic fungi and animals.

Morphine group of alkaloids are found in opium poppy (*Papaver somniferum*) plant. They are synthesised from two molecules of *tyrosine*. First tyrosine molecules are converted into *dopamine* and *3, 4-dihydroxyphenylpyruvic* acid, then one molecule of each combine to form a coupled compound *norlaudanoso-line* which ultimately forms morphine (Fig. 12.8 A and B).

(*v*) **Indole Derivatives** : e.g., *Reserpinine* from *Rauwolfia serpentina* and *Agroclavine* from *claviceps* sp. (ergot).

The indole ring of *reserpinine* and *agroclavine* alkaloids arises from amino acid *tryptophan*. The tryptophan first yields *ajmaline*. The remaining part of the molecule arises from a monoterpene via mevalonic acid (MVA). This part is indicated by bold lines (Fig. 12.97A and B).

The ergot alkaloids also arise from tryptophan but here a hemiterpene residue probably dimentylallyl pyrophosphate is involved which after condensation forming 4-dimethylallyl tryptophan is converted into *agroclavine* (Fig. 12.10). Agroclavine provides the basic skeleton for the synthesis of other ergot alkaloids.

(*vi*) **Quinoline and Imidazole derivatives**

Quinoline derivative alkaloids like *Cinchonine* and *Quinine* are synthesized from *tryptophan* via an indole alkaloid like corynantheine. The imidazole alkaloids like *pilocarpine* are probably synthesized from *histidine*. Nothing more is known about the biosynthesis of imidazole alkaloids.

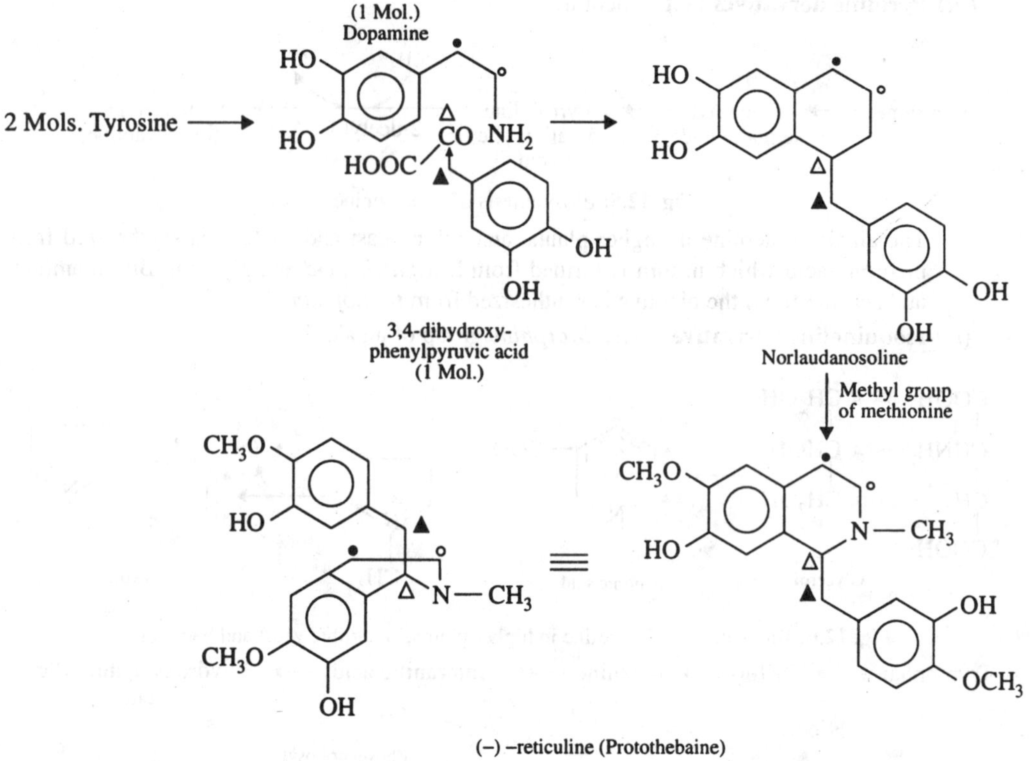

Fig. 12.8: (*A*) Synthesis of protothebaine from tyrosine via Norlaudanosoline.

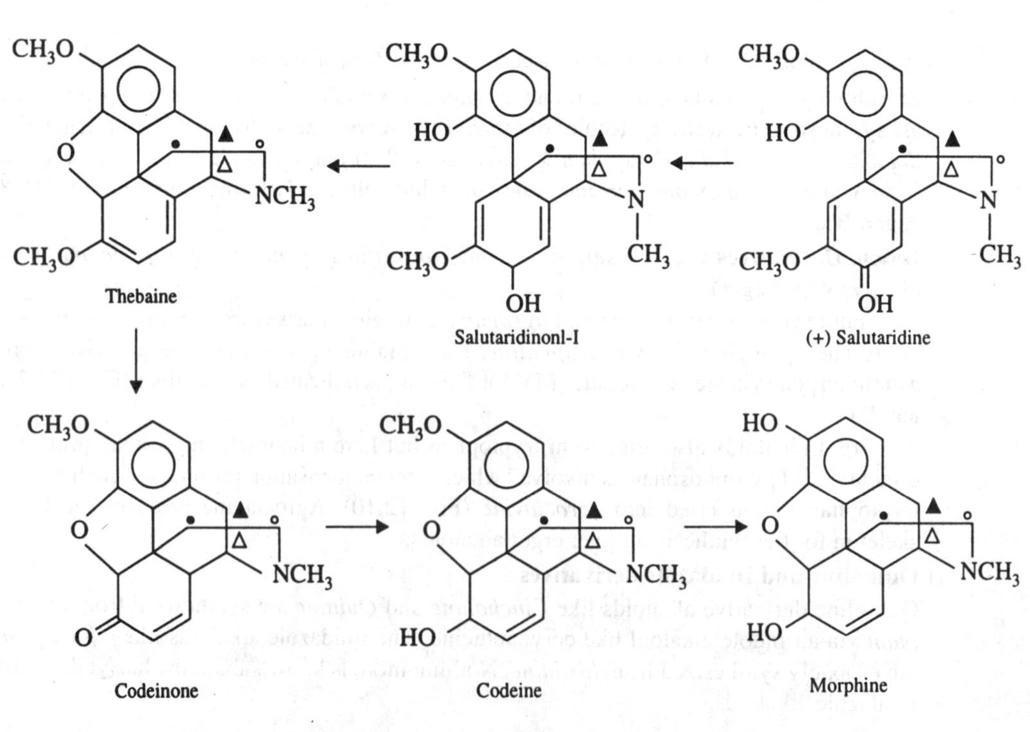

Fig. 12.8: (*B*) Synthesis of morphine fromprotothebaine via Solutaridine.

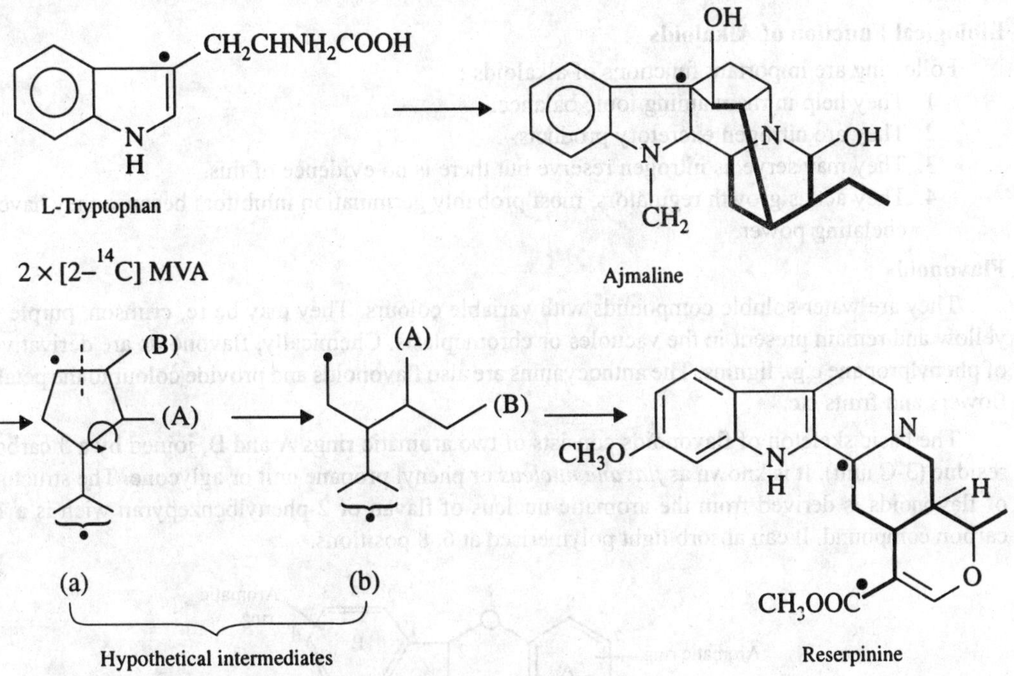

Fig. 12.9: A and B. Biosynthesis of reserpine.

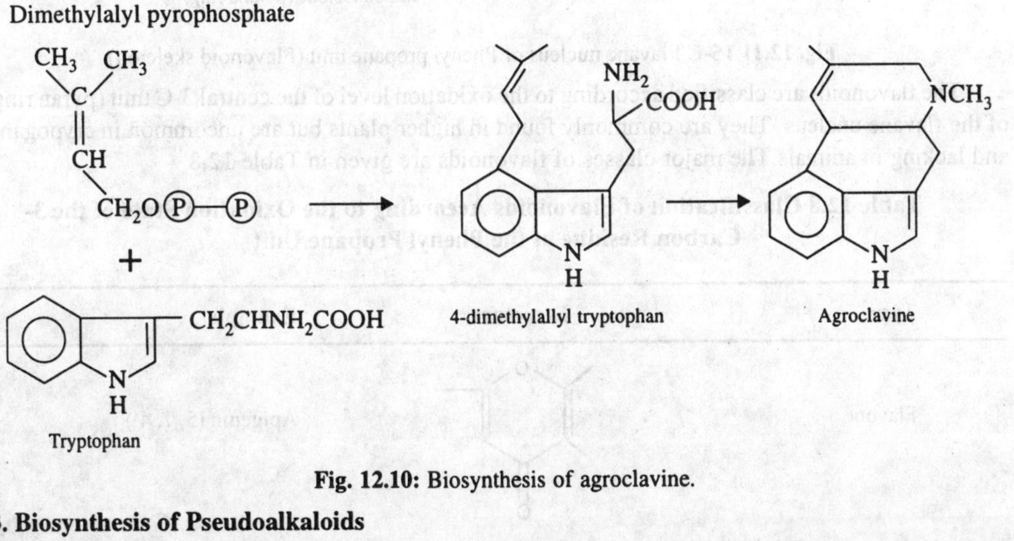

Fig. 12.10: Biosynthesis of agroclavine.

3. Biosynthesis of Pseudoalkaloids

Pseudoalkaloids include terpenoid derivatives which are usually derived from triterpenes. Two types of sterol alkaloids are found in nature (*i*) those which are C-27 compounds derived from cholesterol or related compounds : (*ii*) alkaloids containing C-21 pregnane skeleton. It represents cholestane system where six carbon atoms of the side-chain have been removed. C-27 alkaloids exist as glycosidesin the genus *Solanum*. The alkaoids **tomatine** present in tomatoes is a glycoside of aglycone **tomatidine** whereas solanidine is the aglycone of solanine. These compounds arise from mevalonate via cholesterol. Unlike other alkaloids pseudoalkaloids are synthesized in leaves rather than in the roots and additional changes are produced following amination and oxidation of the C-8 side chain of cholesterol. The examples of C-21 alkaloids are **holarrhimine** and **conarrhimine**. These alkaloids exist in nature as esters.

Biological Function of Alkaloids

Following are important functions of alkaloids :
1. They help in maintaining ionic balance.
2. They are nitrogen excretory products.
3. They may serve as nitrogen reserve but there is no evidence of this.
4. They act as growth regulators, most probably germination inhibitors because they have chelating power.

Flavonoids

They are water-soluble compounds with variable colours. They may be re, crimson, purple or yellow and remain present in the vacuoles or chromoplasts. Chemically, flavonoids are derivatives of phenylpropane e.g., lignins. The anthocyanins are also flavonoids and provide colour to the petals, flowers and fruits etc.

The basic skeleton of flavonoids consists of two aromatic rings A and B, joined by a 3 carbon residue (3-C unit). It is known as *flavane nucleus* or phenyl propane unit or aglycone. The structure of flavonoids is derived from the aromatic nucleus of flavan or 2-phenylbenzepyran wich is a 15 carbon compound. It can absorb light polymerised at 6, 8 positions.

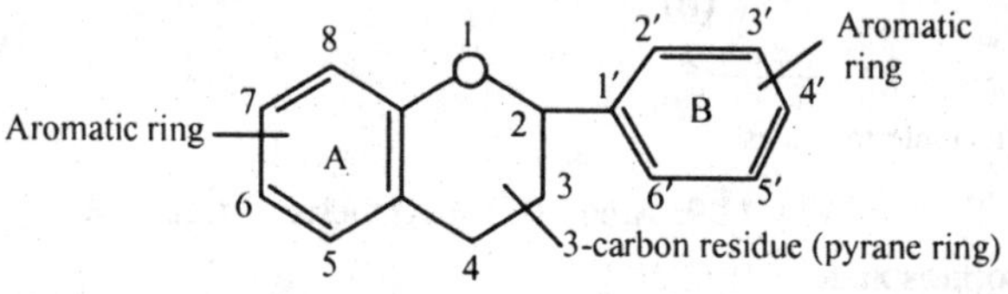

Fig. 12.11 15-C Flavane nucleus or Phenyl propane unit (Flavonoid skeleton)

The flavonoids are classified according to the oxidation level of the central 3-C unit (pyran ring) of the flavane unclues. They are commonly found in higher plants but are uncommon in crypogams and lacking in animals. The major classes of flavonoids are given in Table 12.8

Table 12.8 Classification of Flavonoids According to the Oxidation State of the 3-Carbon Residue of the Phenyl Propane Unit

Type	Structure	Example
Flavone	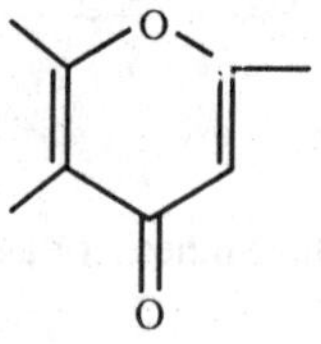	Apigenin (5, 7, 4′)
Flavonol	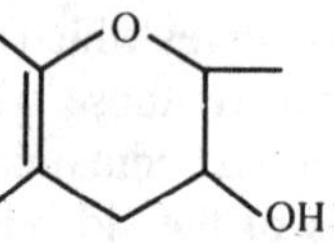	Catechin (5, 7, 3′, 4′)
Flavandiol	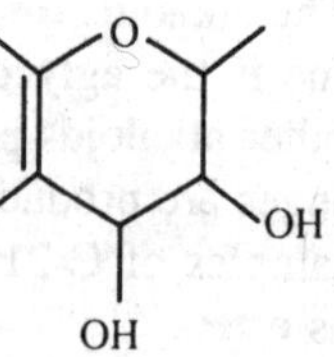	Leucocyanidin (5, 7, 3′, 4′)

Secondary Metabolites

Flavanone	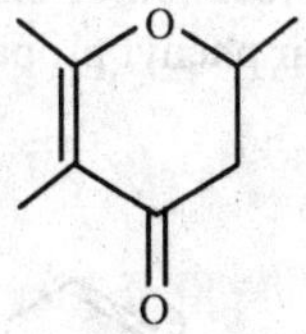	Butin (5, 3', 4')
Dihydroflavonol	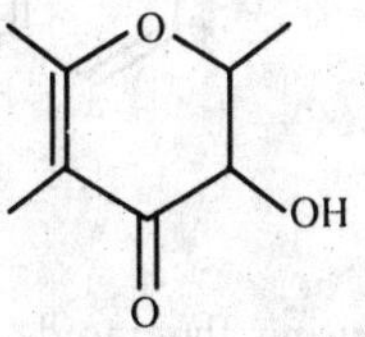	Taxifolin (5, 7, 3', 4')
Flavonol	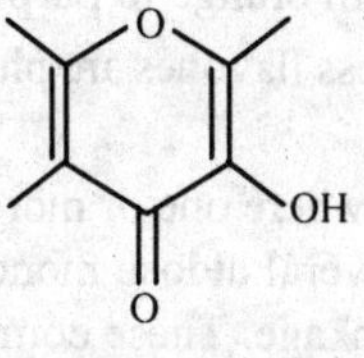	Myricetin (5, 7, 3', 4', 5')
Anthocyanidin	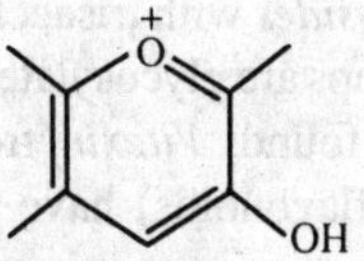	Pelargonidin (5, 7, 4')
Isoflavone	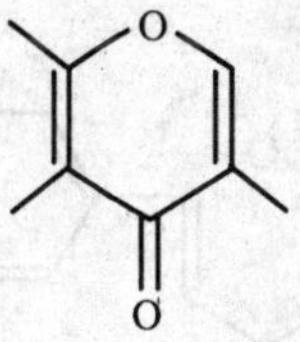	Genistein (7, 7, 4')
Chalcones	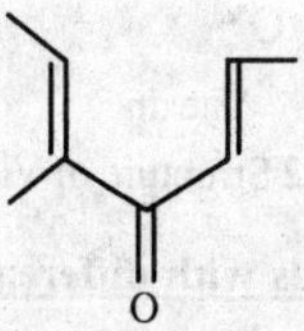	Butein (3, 4, 2', 4')
Dihydrochalcones	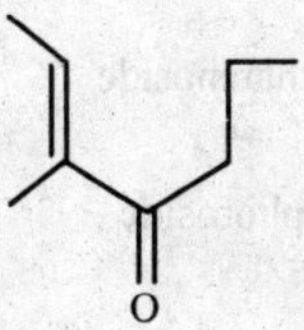	Phoretin (4, 2', 4', 6')
Aurones	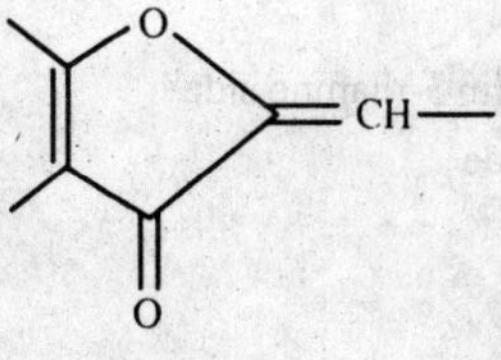	Sulphuretin (6, 3', 4')

* The numbers in brackets indicate the location of the additonal hydroxyl groups; except for aurones and chalcones the numbering is as in phenyl propane unit. The numbering in aurones and chalcones is as follows-

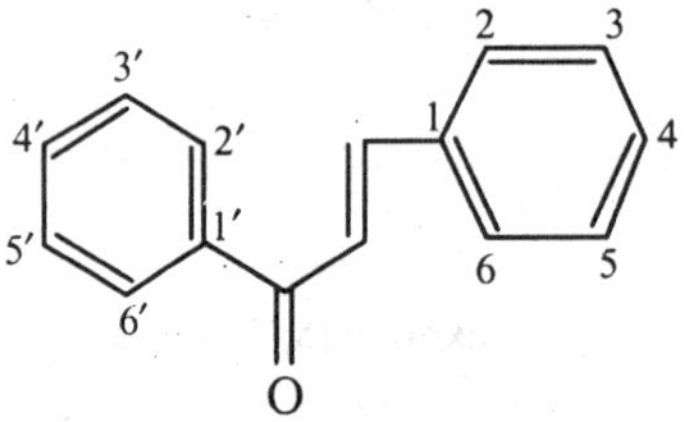

Chalcones 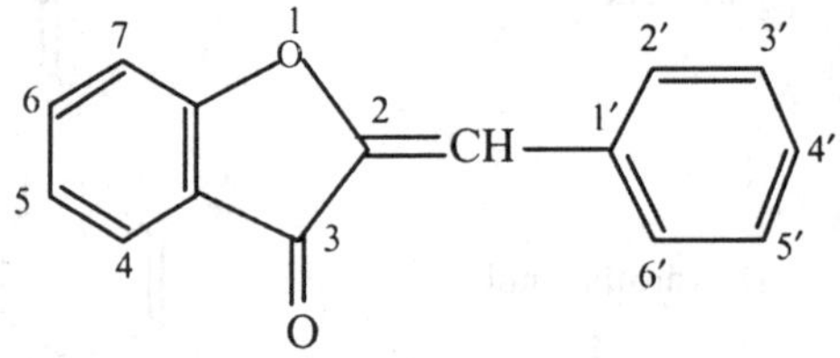   Aurones

Flovone is the simplest naturally occuring flavonoid. It is found in the leaves and stems of *Primula* species. It gives colour ranges from orange to purple blue in association with carotenoids.

Anthocyanidins, flavonols and colourless flavones are physiologically most important flavonoids among prescribed in the table 12.8.

Flavonoids exist in *vivo* as glycosides where one or more hydroxyl groups of flavonoids remain joined by a semi-acetal link to a sugar. Several aldose monosaccharides have been detected which are found linked at C-1 generally by a β-linkage. These compounds are called *monosides*. Similarly *biosides* with disaccharide residues and *triosides* with trisaccharide residues have also been detected. When more than one phenolic hydroxyl groups are glycosylated, they yield *dimonosides, trimonosides, dibiosides* etc. Acylated sugars are also found, *Vitexin* from *Vitex* wood (where c-glycoside is encountered and *negretin* (O-methylated flavonoids) have also been isolated. Various flavonoids with different sugar residues and their sources are given in table 12.9

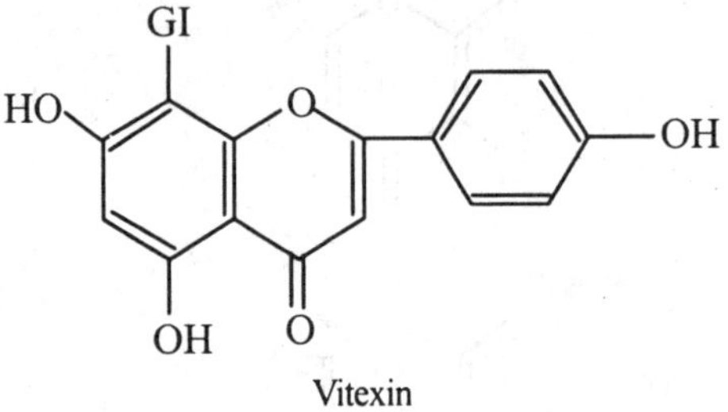

Vitexin

Fig. 12.12 Structure of Vitexin

Table 12.9. Flavonoids with different Sugar Residues

Sugar	Compound	Occurrence
Monoside		
L-Rhamnose	Delphinidin 3-rhamnoside	Sweet pea flowers *Lathyrus odoratus*
Dioside		
Sophorose	Quercetin 3-sophoroside	*Solanum chacoense*
Trioside		
2 G-Glucosylrutinoside	Quercetin 3-glucosyl-2-G rutinoside	*Solanum chacoense*
Dimonoside		
L-Rhamnose	Delphinidin 3-rhamno side	Sweet pea flowers
D-Glucose	5 glucoiside	
Acylated dioside		
Coumaroylrutinoside	Negretein	*Solanum tuberosum* (tubers)

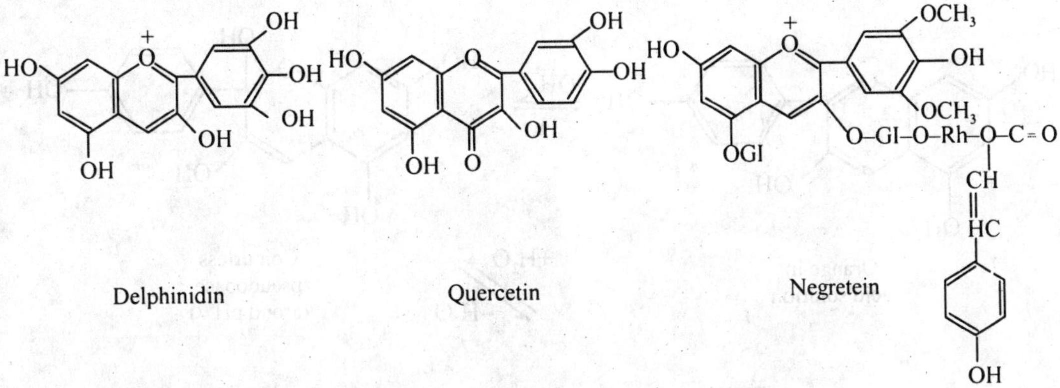

Secondary Metabolites

Fig. 12.13 Structures of Delphinidin Quercetin and Negretein.

Anthocyanidins

They contribue to red, pink, mauve and blue colours in higher plants especially to the flower petals, tubers, fruits and variegated leaves. Their concentration varies from 0.01% to 15% on dry weight basis. Anthiocyanidins are anthocyanin aglycones (sugar free anthocyanins). The chief naturally occuring anthocyanidins are *pelargonidin, cyanidin* and *delphinidin*. Structurally, they differ in *hydroxyl substitution* pattern in ring B. Pelargonidin is a monohydroxyl derivative while cyanidin and dephinidin are di- and trihydroxyl derivatives respectively.

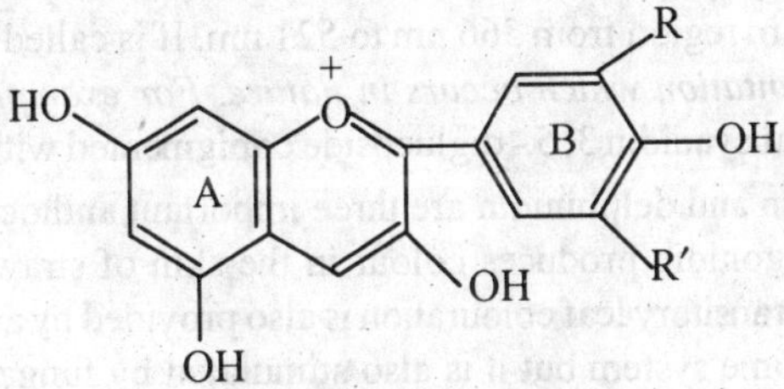

Delphinidin, R = R' = OH
Cyanidin, R = OH, R' = H
Pelargonidin, R = R' = H

Anthocyanins have ironic character, due to which their intensity and colour depends upon pH. In acidic medium their colour varies from orange - red (due to pelargonidin) to mauve (due to delphinidin) depending on the number of free hydroxyl groups. In neutral medium colourless pseudobases are formed. At above $pH_{7.0}$ blue anhydrobases are formed. At very high pH ionization of phenolic groups takes place which results into irreversible changes.

Pelargonidin with one -OH group at B-ring provides shades varying from pink, scarlet and orange. Cyanidin with two – OH groups on B-ring provides crimson and magenta shades and delphinidin with 3 –OH groups on B–ring provides mauve and blue shades. The minor effects on colouration are exerted due to *methylation and variation in glucosylation pattern* but rarely the same pigment gives rise to two totally different flower colours. For example, the pigment cyanidin produces red colour in roses and the blue in corn flowers. In corn flowers, the controlling factor seems to be metal chelation. The protocyanin is cyanin complexed with ferric iron and aluminium and it is also combined with peptide and polysaccharide material to form chelation complex. Such complex is not formed in roses.

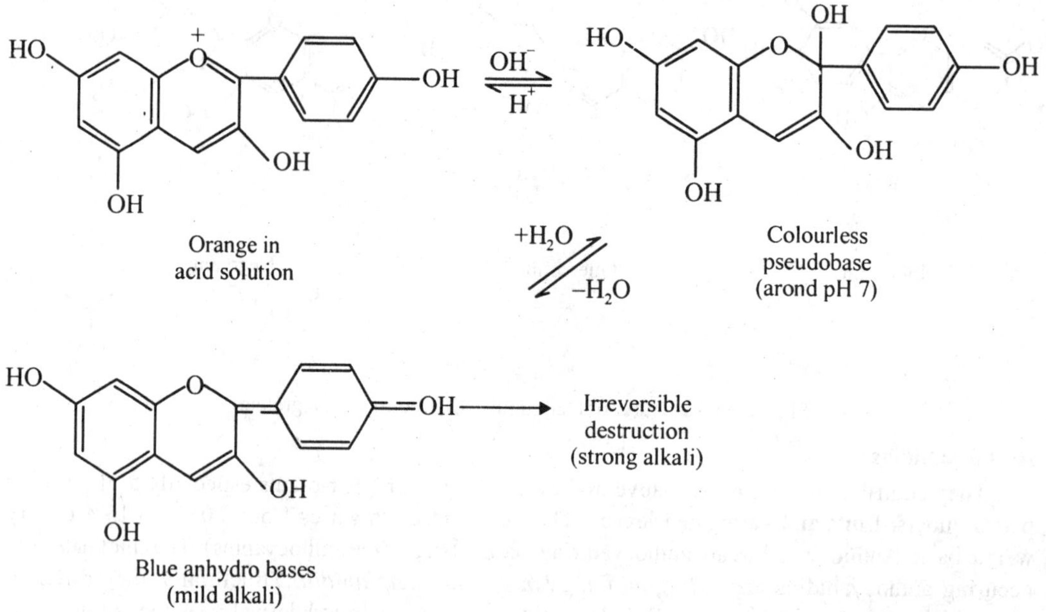

Fig. 12.14. Effect of pH on anthocyanins due to their ionic character.

Anthocyanidin solutions in presence of flavones produce blue tint due to shifting of absorption band in the visible spectrum region from 566 nm to 521 nm. If is called *hypsochromatic shift* and the phenomenon as *co-pigmentation* which occurs in nature. For example, the mauve or deep purple roses (blue) possess normal cyanidin 3, 5 - diglucoside copigmented with huge amouns of gallotannin.

Pelargonidin, cyanidin and delphinidin are three important anthocyanidins wihich contribute to the colour of fruits. Pelargonidin produces colour in the skin of strawberry, cyanidin in apple and dephinidin in grape skin. Transitory leaf colouration is also provided by anthocyanidins. Their synthesis is controlled by phytochrome system but it is also stimulated by fungal infection, wounding and by applying excess growth regulators. In *Begonia* and *Coleus* the synthesis is under genetic control. Red colour in autumn leaves is due to cyanidin 3-glucoside, yellow colour is due to carotenoids and brown due to tannins.

Flavonols

Flavonols contribute yellow colour to the flowers. They contain additional hydroxyl groups at position 6 or 8. The important flovanols corresponding to three main anthocyanidins are *Kaempferol, quercetagetin* and *aurensidin*. Kaempferol is most common and essentially colourless. Quercetagetin is yellow (eg. Primrose) and contains an extra hydroxyl at C-6. Other yellow flavonols are *chalkones* and *aurones*. Aurensidin is 6-glucoside of aurone present in *Antirrhinum majus* flowers. The yellow colour of most flowers is due to carotenoids but flavonols also combine with them. e.g., in *Calendula* spp. Flavonols have also been reported frequently from leaves of the plants.

Quercetagetin is physiologically active in vitro and affect IAA oxidase system in plants. It plays role indirectly in seed germination and plant growth in vivo. In flavonols hydroxyl group is introduced in 3-position of the central ring of flavones. White colour of the flowers is due to flavonol glycosides.

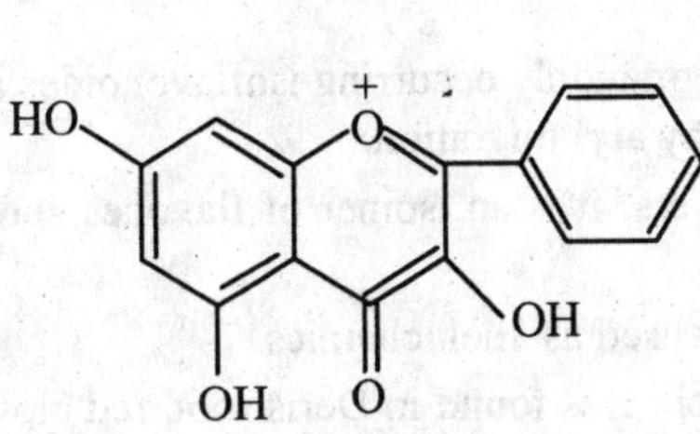

Secondary Metabolites

Figures: Kaempterol, 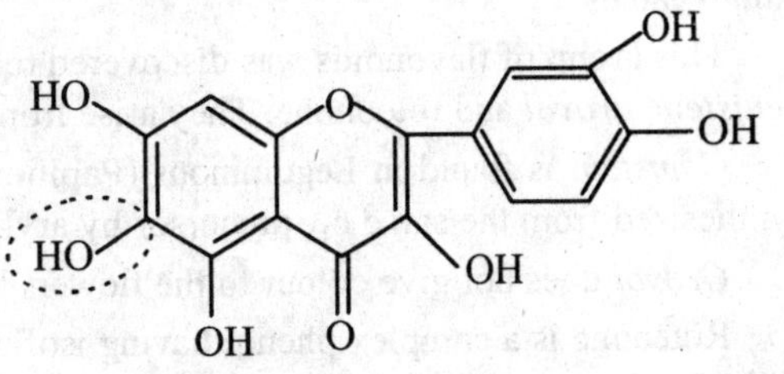 Quercetagetin, 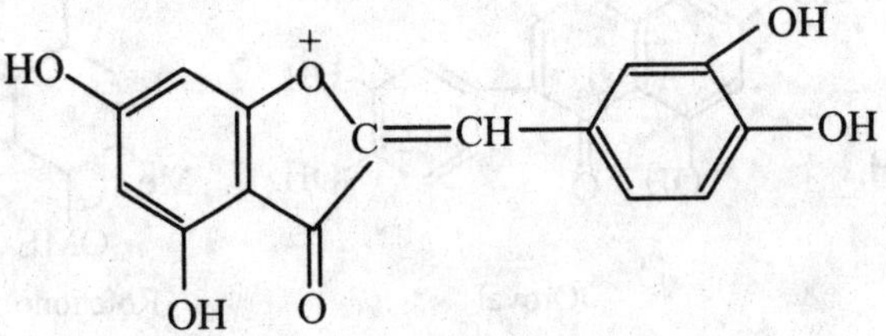 Aurensidin

Fig. 12.15. Structures of some flavonols

Flavones

Majority of flowers contain colourless flavones in their petals. They conribute cream, ivory and white colours to the flowers of herbaceous plants of families scrophulariaceae, Labiatae, Compositae (Asteraceae), unbelliferae and Cruciferae.

Flavones lack 3-hydroxyl group present in flavonols and anthocyanidins. The colourless flavones absorb radiations in UV region of spectrum which can be seen by bees and other insects. They confer a cream or ivory translucent appearance. Due to this character, a true albino mutant of *Antirrhinum majus* without flavones can be easily distinguished from species producing ivory flowers. The commonly occurring flavones are *Apigenin* and *luteolin*. Flavones give yellow to ivory colour to the flowers in association with anthocyanidins.

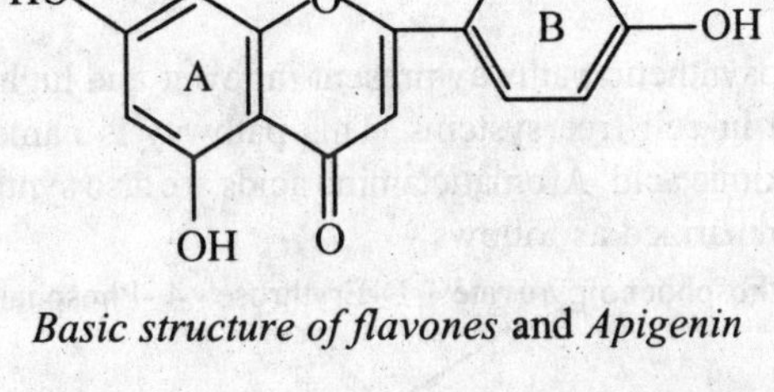

Basic structure of flavones and Apigenin

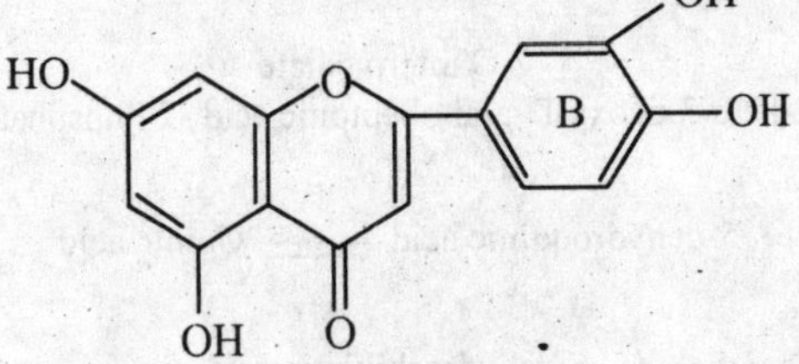

Luteolin

Isoflovonoids

This group of flavonoids was discovered recently. The commonly occurring isoflavonoides are *genistein, orovol* and *roteonone*. They arise from chalcones by aryl migration.

Genistein is found in Leguminous (Papilionaceae) flowers. It is an isomer of flavones and is synthesized from the same c_{15} precursor by aryl migration.

Orovol does not give colour to the flowers but is simply used as allelochemics.

Rotenone is a complex phenol having isoflavone skeleton. It is found in Deris root, red clover and *Pisum sativum* etc. as phytoalexin and pisatin etc. It is used as insecticide.

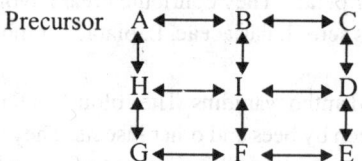

Basic structure of Isoflavonoids (Genistein)   Oroval   Rotenone

Biosynthesis of Flavonoids

All flavonoids can be synthesized if only one precursor is present. It is known as *grid system*.

```
Precursor  A ⇌ B ⇌ C
           ↕   ↕   ↕
           H ⇌ I ⇌ D
           ↕   ↕   ↕
           G ⇌ F ⇌ E
```

As flavonoids contain two aromatic rings A and B, they are derived from different pathways. Ring A is derived from acetate via malonyl– CoA and Ring B 3– C side chain (c atoms 2, 3 and 4 of heterocycle) are derived from p-hydroxycinnamic acid via shikimic acid, a route for phenylpropane building units of lignin biosynthesis.

Shikimic Acid Pathway

It is most important biosynthetic pathway present in lower and higher organisms. The enzymes for this pathway are found in cell free systems. This pathway is named because of formation of intermediate compound shikimic acid. Aromatic amino acids are also synthesized by this intermediate. The pathway in brief is summarized as follows -

$$\text{Phosphoenolpyruvate + D-Erythrose - 4 -Phosphate}$$
$$\downarrow$$
$$\text{7-C intermediate}$$
$$\text{(2-keto-3 deoxy-D-araboheptonic acid -7 Phosphate)}$$
$$\downarrow$$
$$\text{5-dehydroquinic acid} \rightleftharpoons \text{Quinic acid}$$
$$\downarrow$$
$$\text{5-dehydroshikimic acid}$$
$$\downarrow$$

Secondary Metabolites

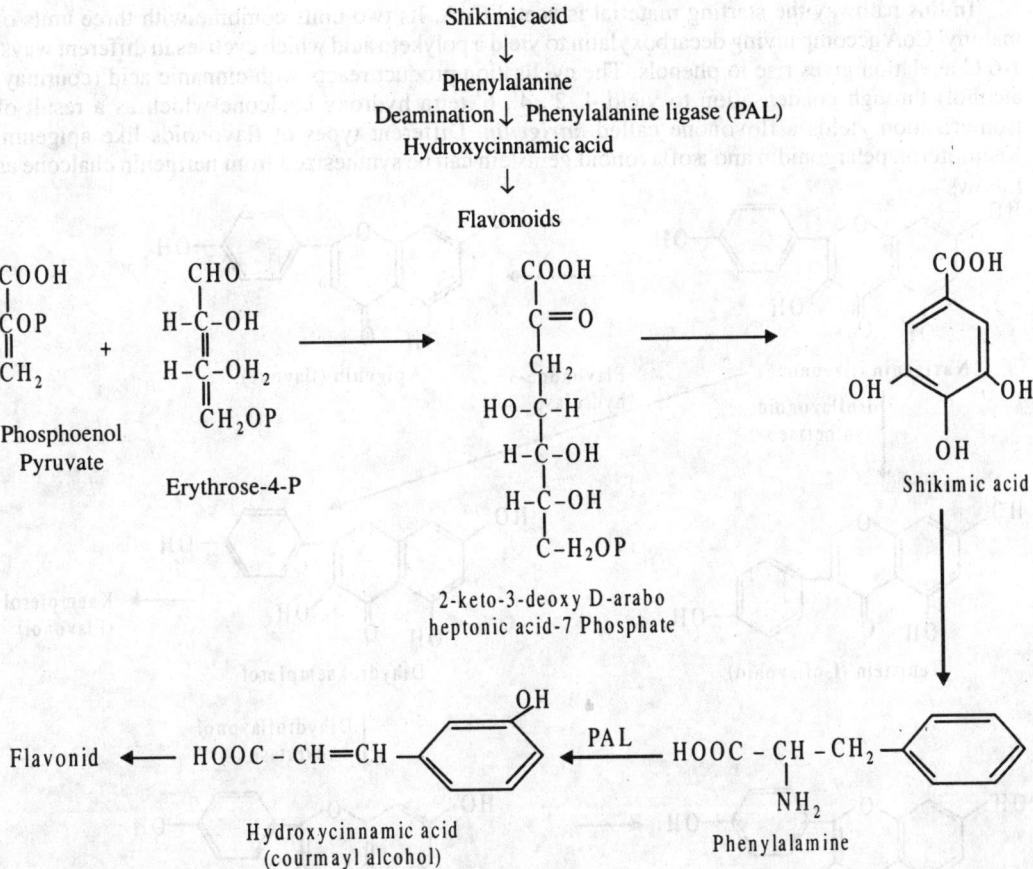

Acetate-Malonate Pathway

The Pathway is almost similar to that of fatty acids synthesis pathway. The pathway is briefly summarized as follows

2 units COOH | CH$_2$.CO.COA (Malonyl CoA) + 3 Units CH$_3$.CO.CoA (Acetyl CoA) ⟶ CH$_3$.CO.CH$_2$CO.CH$_2$.CO.CH$_2$.COOH (Polyketo acid)

In this pathway the starting material is acetyl CoA. Its two units combine with three units of malonyl CoA accompanying decarboxylatin to yield a polyketo acid which cyclises in different ways. 1-6 C acylation gives rise to phenols. The cyclisation product reacts with cinnanic acid (courmayl alcohol) through condensation to yield 4, 2′, 4′, 6′ tetra hydroxy chalcone which as a result of isomerization yields a flovonone called *narigenin*. Different types of flavonoids like apigenin, Kaempferol, pelargonidin and isoflavonoid genistein can be synthesized from narigenin chalcone as follows –

Narigenin (flavonone) → **Apigenin (flavone)** (Flavonone-3 hydrolase)

Narigenin → **Genistein (Isoflavonid)** (Isoflavonoid synthetase)

→ **Dihydrokaempferol** → **Kaempferol (Flavonol)**

Dihydrokaempferol → **Leucopelargonidin** (Dihydroflavonol 4-revertase) → **Pelargonidin (Anthocyanidin)**

Functions of Flavonoids

1. In general, when flavonoids are applied exogenously, they inhibit growth.
2. They retard cell-division and cell enlargement.
3. Certain compounds act as analoges of growth hormones. They affect dormancy, growth and other physiological functions.
4. Flavonoids attract insects and animals and help in pollination
5. Flavonoids can alter certain enzyme systems, i.e. can stimulate or inhibit.
6. Flavonids can provide every colour of spectrum except green to the plant parts and flowers.
7. Anthocyanidins are natural indicators.
8. They help in the dispersal of fruits.
9. Kaempferols stimulate IAA oxidase which causes increased distribution of IAA and decrease in growth rate.
10. Flavonoids play a role in tendril contraction by inhibiting activity of enzymes.
11. Quercetin glycosides can regulate the ATPase – ATP system.
12. Isoflavonoids possess toxic effects, e.g., rotenoids are toxic to insects and fishes, hence are used as insecticides.
13. Genistein, a isoflavonoid, possesses oesterogenic activity.
14. Isoflavonoids in plants produce phyto-allexin activity i.e., they repel the invading pathogenic organisms.
15. Anthocyanidins contribute to red, pink, mauve and blue colours; flavonols to yellow colour Flavones to cream, ivory and white colours; and isoflavoxoids to chemical properties.

PART III

BIOTECHNOLOGY

1
GENETIC ENGINEERING

Introduction

Most of the recent advances have been made in the field of cell biology and molecular biology. These advancements initiated the hope for the development of new technology through which the genotypes of harmful organisms at DNA level can be made correct and improved. The method of artificial synthesis of new genes and their subsequent transplantation in the genome of an organism or method of correcting the defective genes in an organism by molecular techniques formed a discipline called *Genetic Engineering* (**Pair,** 1974: **Bodmer** and **Cavalliofsforza,** 1976). According to **Iingle** (1982) the genetic engineering can be defined *"as the process whereby a foreign sequence of DNA is inserted into the genetic make up of a cell by in vitro techniques, and is subsequently expressed by that cell."*

Through genetic engineering the manipulation of genes of an organism can be done according to our will. Of course, there are a lot of technical difficulties which are to be overcome. However, the advent of genetic engineering has open up exciting possibilities which include (*i*) the application of a relatively quick and accurate method of incorporating specific and desirable genes into an organism and (*ii*) the transfer of genes between sexually incompatible species.

Genetic engineering involves the addition, deletion or repair of a segment of the genetic material, thereby altering the phenotype of an organism. The changed phenotype can be made use of in a variety of ways. Once a perfect control over genetic manipulation is achieved and the genetic materials of several thousands of living objects can be exchanged, replaced or repaired at will, the man will become the master of his own destiny.

Traditionally, breeding method has been used for the exchange of genes between two individuals. This process has several limitations. The crosses can be established only between the related varieties, while interspecific and intergeneric crosses or matings are not successful. This difficulty can be overcome by the protoplast isolation and fusion techniques. Other difficulties can be overcome by other methods. The improved techniques of chemical synthesis of polynucleotides, purification of genes and DNA segments, formation of recombinant DNA and insertion, transfer, replication, integration and expression of heterologous DNA are being widely used for engineering of genetic materials of various organisms.

Steps involved in Gene Transfer

The following are important steps which should be kept in mind while performing an experiment on genetic transfer —

(1) Isolation and identification of the gene or DNA segment that is to be transferred (foreign DNA).

(2) Purification of gene.

(3) Transformation of isolated and purified gene (foreign DNA) into the cell genome or its transfer into the desired cell through bacteriophages or plasmids or any vector.

(4) Manipulation and control of introduced genes.

RECOMBINANT DNA TECHNOLOGY

Recombinant DNA is that DNA which is formed by combination of foreign DNA and plasmid DNA or foreign DNA with phage DNA or foreign DNA with cosmid DNA. In other words, the DNA formed by the combination of foreign DNA and any vector DNA is called **recombinant DNA**. Recombinant DNA molecules are produced by inserting a foreign DNA segment into the DNA molecule of a vector. As the recombinant DNA contains two or more types of DNAs, it is also called **hybrid DNA** and **chimeric DNA**. These recombinant or chimeric DNA molecules are the actual means of gene cloning.

The recombinant DNA is one of the latest arrivals in the list of genetic achievements. Recombinant DNA hold out great promises in terms of reducing human sufferings and improving upon the quality of life. For this purpose it is regarded as the greatest event after the discovery of antibiotics. It has also taken us one step nearer to the geneticist's dream of genetic engineering.

Tools Involved in Recombinant DNA Technology

The following tools are involved in recombinant DNA technology:

I. Enzymes

Following enzymes are used in recombinant DNA technology —

(1) Exonucleases – These are those enzymes which act upon DNA and delete base pairs of either 5' end or 3' end of ss DNA. Sometimes they delete base pairs from DNA at single stranded nicks or near gaps in double stranded DNA (ds DNA).

(2) Endonucleases – These are those enzymes which act upon DNA and cleave them at any point except the ends but they always involve only one strand of DNA.

(3) Restriction enzymes (restriction endonucleases) – One of the important discoveries in last few decades which has contributed greatly to the genetic engineering and to the development of recombinant DNA, is the discovery of *restriction enzymes. They are a special class of enzymes which recognize and cleave the DNA at specific sites* (see table 1.1 and 1.2). These are called **'biological scissor'**. The restriction enzymes recognize only short sequences of double stranded target DNA for cleavage. Different restriction enzymes recognize different but specific sequences of 4 to 8 base pairs length. These specific nucleotide sequences are called *recognition sites* or *target sites*. Restriction enzymes cleave the DNA only at the regions where *palindromic sequences* are found. These sequences read the same on both the strands in 5'–3' direction.

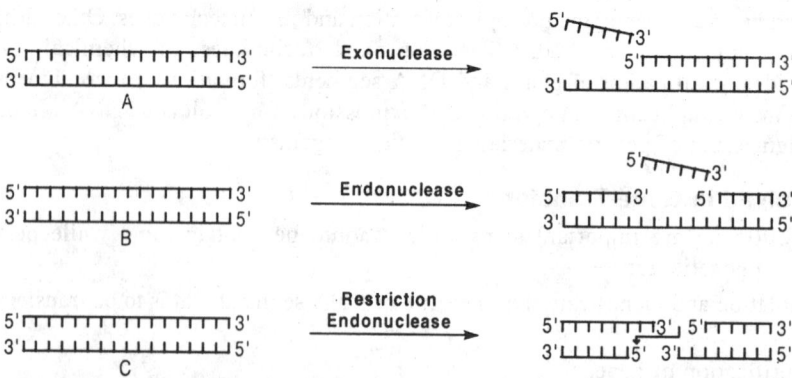

Fig. 1.1. Action of DNA cleaving enzymes: A-Exonuclease: B-Endonuclease: C-Restriction endonuclease.

Genetic Engineering

Certain restriction enzymes are unifunctional *i.e.* they are either involved in restriction or in modification. In such cases, both these functions are performed by two separate enzymes. Other type of restriction enzymes are bifunctional *i.e.*, they are involved in restriction and modification both. In such cases, both functions are performed by the single and the same enzyme, although the restriction and modification sites differ in position.

Werner Arber of Basle University, Switzerland (1962) discovered that certain bacteria contain restriction enzymes which are able to cut DNA into small fragments. For this discovery Professor **Arber** shared 1978 Nobel Prize for medicine with **Nathan** and **Smith**. **Hamilton Smith** and his associates isolated a restriction enzyme from *Haemophilus influenzae* (a bacterium) and called it Hind II, which was able to cut viral DNA at specific sites, producing fragments with 5' and 3' terminals (5'-pPupApCp and pGpTpPY-3'). This showed that DNA is double stranded and each strand runs in opposite direction. Later, these enzymes were used to map DNA sequences, to isolate required genes, to analyze chromosome structure and to create new kinds of DNA molecules.

The restriction enzymes act as a fine chemical scalpel to break the long chain of DNA molecule into smaller units. Even though the role of these enzymes is not fully understood, cells may employ them as defence against foreign DNA. About 150 restriction endonucleases have been discovered (Table 1.2) which recognize more than 40 different nucleotide sequences. However, some restriction enzymes are non-specific and do not cleave DNA at specific sites.

Classes of Restriction Enzymes

There are following classes of restriction enzymes —

(1) Type I enzymes – Type I restriction enzymes perform restriction and modification, both functions. These enzymes break the DNA molecule at random sites, *i.e.*, they are non-specific in cleavage. They possess highest molecular weight (about 3,00,000 daltons) and require ATP, Mg^{2+} and adenosyl methionine as co-factors. They contain three non-identical sub-units, R, M and S (*R* for restriction, *M* for methylation and *S* for specific target site). The recognition site in them is bipartite 4 asymmetrical, *e.g.* $TGAN_8 TGCT$. The cleavage site in them is usually found less than 100 (> 100) bp away from recognition site. The examples of type I restriction enzymes are *Eco B* and *Eco K*. These enzymes make even breaks in DNA strands usually opposite to each other.

(2) Type II enzymes – Type II enzymes perform either restriction or modification function only. These restriction enzymes attack at specific sites only. They make nicks in the opposite strands of DNA, a few base pairs apart to produce staggered breaks. One of the type II enzymes is the Restriction endonuclease-I (Eco R-I), derived from the bacteria *Escherichia coli*. This enzyme recognizes the base sequence G↓ AATTC on the DNA strand and cleaves at the site indicated by arrow. The most important aspect about this cleavage, discovered by **J. Merts** and **R. Davis** of Standford university, U.S.A. (1972), is that the cleavage fragments carry short, single stranded extensions at each end. These ends are sticky and can easily join with other fragments carrying complementary sticky ends. Thus, any two Eco R-I produced DNA fragments can be rejoined. This has made possible the powerful technique of recombinant DNA. These enzymes have been used largely for sequence analysis of DNA and for cloning and amplifying DNA. A few type II enzymes generate blunt ends.

Majority of type II restriction enzymes type II cleave the target DNA having 6 bp length or target site. The enzymes to cut 4 bp target sites are used only when frequent cuts are required to get small DNA fragements and the enzymes to cut 8 bp target sites are used when rare cuts are required to get long DNA fragments. Majority of enzymes cleave only unmethylated target sites but only a few enzymes can cleave both methylated as well as unmethylated target sites. Type II enzymes require *endonuclease* and *methylase* enzymes separately for both the functions. These

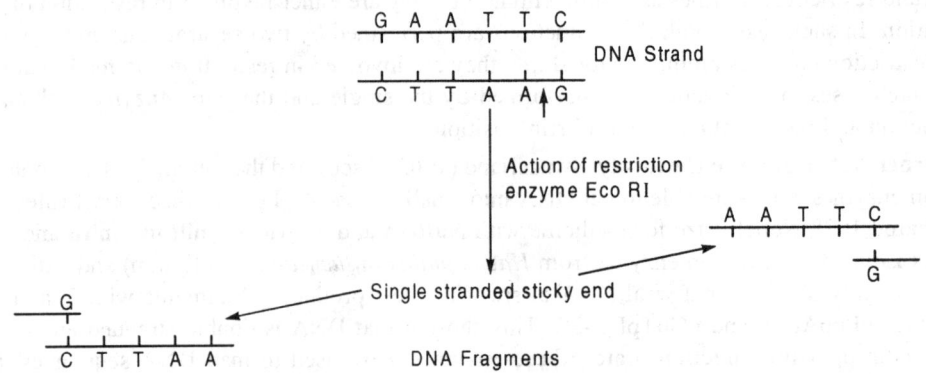

Fig. 1.2. Showing that restriction enzymes can cut long DNA molecule into several small pieces which contain single stranded sticky ends.

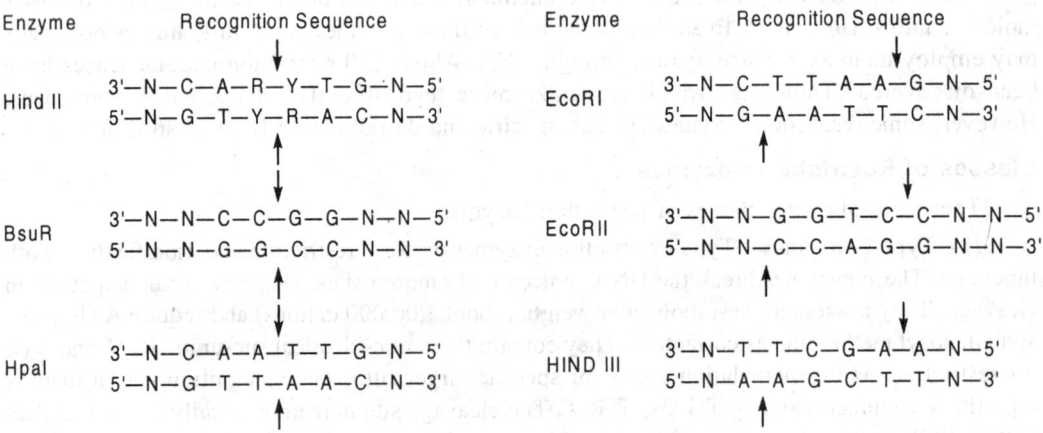

Fig. 1.3. Cleavage sites for some restriction endonucleases (Enzymes of left side cause even breaks while right side, staggered breaks).

enzymes recognize palindromic and short sequences of 4–8 bp. Here, the cleavage and recognition sites are the same or found situated close to each other. They do not require ATP during restriction or methylation.

Besides the above characteristic, the type II enzymes possess MW from 20,000 to 1,00,000 daltons and require Mg^{2+} only as co-factor. They contain two identical sub-units.

(3) Type III enzymes – These are bifunctional *i.e.*, the same enzyme performs the functions of restriction and methylation both. They are made up of two subunits-R and MS (R for restriction and MS for methylation and recognition) and their restriction sites possess asymmetrical sequence of 5-7 bp. The cleavage site in them is found 24-26 bp downstream of recognition site. They possess comparatively higher M.W. The restriction and methylation reactions occur simultaneously which require ATP. The examples of type III enzymes are Eco PI and Eco 15.

The DNA fragments prepared by restriction endonucleases can be rejoined in a desired manner to produce unique DNA sequences. When the unique DNA is introduced in a bacterial cell, it can propagate by replication. Since transformation of bacteria is not possible by introducing DNA fragments, a simple receptor chromosome or a vector is used, with which DNA fragment is joined before a bacterium is infected. In this way copies of newly constructed DNA sequence may be propagated.

Genetic Engineering

Modification By Methylation

The cell protects its own DNA against the action of restriction enzymes by methylating certain bases in the nucleotide sequence recognized by the enzyme. Methylation is done by a methylating enzyme, *restriction methylase*, contained in the same cell that methylates newly synthesized DNA at the endonucleases-sensitive sites. The process is called **modification**. The bases that can be methylated are adenine and cytosine (adenine to 6-methyl amino purine, and cytosine to 5-methyl cytosine). Methylation protects the host cell DNA against endonuclease activity and it distinguishes genes in different states of functioning. The foreign DNA is not methylated, hence it can be cleaved with endonuclease.

Table 1.1. Target Sites and Products of Digestion of Some Restriction Enzymes

Name	Source	Target Sequence	Product
Eco RI	*E. coli* RY 13	$5'G{\downarrow}AATTC3'$ $3'CTTAA{\uparrow}G5'$	$G^5AATTC^{3'}$ $3'CTTAA^5G$
Hind III	*Haemophilus influenzae* Rd	$5'A{\downarrow}AGCTT3'$ $3'TTCGA{\uparrow}A5'$	$A^5AGCTT^{3'}$ $3'TTCGA^5A$
Msp I	*Moraxella spp.*	$5'C{\downarrow}C\ GG3'$ $3'GG\ C{\uparrow}C5'$	$C^5CGG^{3'}$ $3'GGC^5C$
Sma I	*Serratia marcescens* Sb	$5'CCC{\downarrow}GGG3'$ $3'GGG{\uparrow}CCC5'$	$5'CCC\ GGG^{3'}$ $3'GGG\ CCC^{5'}$
Hae II	*Haemophilus aegyptius*	$5'GG{\downarrow}CC3'$ $3'CC{\uparrow}GG5'$	$5'GG\ CC^{3'}$ $3'CC\ GG^{5'}$
Hae III	*Haemophilus aegyptius*	$5'GG{\downarrow}CC3'$ $3'CC{\uparrow}GG5'$	$5'GG\ CC^{3'}$ $3'CC\ GG^{5'}$
Pst I	*Providencia stuartii* 164	$5'CTGCA{\downarrow}G3'$ $3'G{\uparrow}ACGTC5'$	$5'CTGCA^{3'}G$ $G^{3'}ACGTC^{5'}$
Sst I	*Streptomyces standford*	$5'GAGCT{\downarrow}C3'$ $3'C{\uparrow}TCGAG5'$	$5'GAGCT^{3'}C$ $C^{3'}TCGAG^{5'}$
Hsu I	*Haemophilus suis*	$5'A{\downarrow}AGCTT3'$ $3'TTCGA{\uparrow}A5'$	$A^5AGCTT^{3'}$ $3'TTCGA^5A$
Sal I	*Streptomyces albus*	$5'G{\downarrow}TC\ GAC3'$ $3'CAG\ CT{\uparrow}G5'$	$G^5TCGAC^{3'}$ $3'CAGCT^5G$

Table 1.2. Various Restriction Endonucleases and Source

Enzyme	Source	Target Sequence 5' - 3'
Aca I	*Acinetobacter calcoaceticus*	$GT{\downarrow}(^C_A)(^T_G)AC$
Acy I	*Anabaena cylindrica*	$GPu{\downarrow}CGPyC$
Alu I	*Arthobacter luteus*	$AG{\downarrow}CT$
Aos I	*Anabaena oscillaroides*	$TGC{\downarrow}GCA$
Aos II	*Anabaena oscillaroides*	$GPu{\downarrow}CGPyC$

Enzyme	Source	Target Sequence 5' - 3'
Apy I	*Arthrobacter pyridinolis*	CC↓(^A_T)GG
Asu I	*Anabaena subcylindrica*	G↓GNCC
Asu II	*Anabaena subcylindrica*	TT↓CGAA
Ava I	*Anabaena variabilis*	C↓PyCGPuG
Ava II	*Anabaena variabilis*	C↓G(^A_T)CC
Bal I	*Brevibacterium albidum*	TGG↓CCA
BamH I	*Bacillus amyloliquefaciens* H	G↓GATCC
BamH II	*Bacillus amyloliquefaciens*	GAA(N)$_2$↓(N$_3$)TTC
BamNx	*Bacillus amyloliquefaciens* N	G↓G(^A_T)CC
Bbr I	*Bifidobacterium brave*	GGCGC↓C
Bca I	*Bacillus caldolyticus*	T↓GATCA
Bgl I	*Bacillus globigii*	GCC(N)$_4$↓NGGC
Bgl II	*Bacillus globigii*	A↓GATCT
Blu I	*Brevibacterium luteum*	C↓TCGAG
Bpe I	*Bordetella pertussis*	A↓AGCTT
Bst I	*Bacillus sterothermophilus* 1503-4R	G↓GATCC
BstE II	*Bacillus sterothermophilus* ET	G↓GTNACC
BstN I	*Bacillus sterothermophilus* N	CC↓(^A_T)GG
BstP I	*Bacillus stearothermophilus*	G↓GTNACC
BsuR I	*Bacillus subtilis* R	GG↓CC
Bvu I	*Bacillus vulgatis*	GPuGCPy↓C
Cau II	*Chloroflexus aurantiacus*	CC↓(^G_C)GG
Cla I	*Caryophanon latum* L	AT↓CGAT
Cla I	*Caryophanon latum*	GG↓CC
Dde I	*Desulfovibrio desulfuricans* Norway strain	C↓TNAG
Dpn I	*Diplococcus pneumoniae*	GA↓TC
Ecl I	*Enterobacter cloacae*	G↓GTNACC
EcoR I	*Escherichia coli* RY 13	G↓AATTC
EcoR I	*Escherichia coli* RY 13	PuPuA↓TPyPy
EcoR II	*Escherichia coli* R 245	CC($^A_↓$)GG
FnuA I	*Fusobacterium nucleatum* A	G↓ANTC
FnuC I	*Fusobacterium nucleatum* C	↓GATC
FnuD I	*Fusobacterium nucleatum* D	GG↓CC
FnuD II	*Fusobacterium nucleatum* D	CG↓CG
FnuD III	*Fusobacterium nucleatum* D	GCG↓C
FnuE I	*Fusobacterium nucleatum* E	↓GATC
Fnu4H I	*Fusobacterium nucleatum* 4H	G↓CNGC
FspA I	*Flavobacterium species*	G↓GTNACC

Enzyme	Source	Target Sequence 5' - 3'
Gdi I	*Gluconobacter dioxyacetonicus*	AGG↓CCT
Gdi II	*Gluconobacter dioxyacetonicus*	Py↓GGCCG
Hae I	*Haemophilus aegyptius*	(^A_T)GG↓CC(^A_T)
Hae II	*Haemophilus aegyptius*	PuGCGC↓Py
Hae III	*Haemophilus aegyptius*	GG↓CC
Hap II	*Haemophilus aphrophilus*	C↓CGG
HgiA I	*Herpetosiphon giganteus* Hpg[5]	G(^A_T)GC(^A_T)↓C
HgiB I	*Herpetosiphon giganteus* Hpg[5]	G↓G(^A_T)CC
HgiC I	*Herpetosiphon giganteus* Hpg[9]	G↓GPyPuCC
HgiD II	*Herpetosiphon giganteus* Hpa[2]	G↓TCGAC
HgiE I	*Herpetosiphon giganteus* Hpg[24]	G↓G(^A_T)CC
Hha I	*Haemophilus haemolyticus*	GCG↓C
Hind II	*Haemophilus influenza* R_c	GTPy↓PuAC
Hind III	*Haemophilus influenza* R_d	A↓AGCTT
Hinf I	*Haemophilus influenza* R_f	G↓ANTC
Hpa I	*Haemophilus parainfluenzae*	GTT↓AAC
Hsu I	*Haemophilus suis*	A↓AGCTT
Kpn I	*Klebsiella pneumoniae*	GGTAC↓C
Mbo I	*Moraxella bovis*	↓GATC
Mbo II	*Moraxella bovis*	GAAGA(N)$_8$↓ CTTCT(N)$_7$↓
Mla I	*Mastigoladus laminosus*	TT↓CGAA
Mno I	*Moraxella nonliquefaciens*	C↓CGG
Pst I	*Providencia stuartii* 164	CTGCA↓G
Pvu II	*Proteus vulgaris*	CAG↓CTG
Rsp I	*Rhodopseudomonas sphaeroides*	GT↓AC
Rru I	*Rhodospirillium rubrum*	AGT↓ACT
Rru II	*Rhodospirillium rubrum*	CC↓(^A_T)GG
Sac I	*Streptomyces achromogenes*	GAGCT↓C
Sac II	*Streptomyces achromogenes*	CCGC↓GG
Sal I	*Streptomyces albus* G	G↓TCGAC
Sau3A	*Staphylococcus aureus* 3A	↓GATC
Sau96 I	*Staphylococcus aureus* PS96	G↓GNCC
Sfa I	*Streptococcus faecalis var. zymogenes*	GG↓CC
Sla I	*Streptomyces lavendulae*	C↓TCGAG
Sma I	*Serratia marcescens* S_b	CCC↓GGG
Sph I	*Streptomyces phaeochromogens*	GCATG↓C

Enzyme	Source	Target Sequence 5' - 3'
Sst I	*Streptomyces stanford*	GAGCT↓C
Sst II	*Streptomyces stanford*	CCGC↓GG
Taq I	*Thermus aquaticus* YTI	T↓CGA
Tac I	*Thermoplasma acidophilum*	CG↓CG
TthHB8	*Thermus thermophilus* HB8	T↓CGA
Xba I	*Xanthomonas badrii*	T↓CTAGA
Xho I	*Xanthomonas holcicola*	C↓TCGAG
Xho II	*Xanthomonas holcicola*	Pu↓GATCPy
Xma I	*Xanthomonas malvacearum*	C↓CCGGG
Xma III	*Xanthomonas malvacearum*	C↓GGCCG
Xpa I	*Xanthomonas papavericola*	C↓TCGAG

Nomenclature of Restriction Enzymes

The following principles of nomenclature are used naming a particular restriction enzyme.

(1) Each restriction enzyme is named by a three letter abbreviation, written in italic. It identifies its origin from an organism. The first letter is written in *capital* which indicates the name of the genus (bacteria or any other organism) and other two letters are written in small which indicate the name of its species. For example restriction enzyme *Eco* indicates its origin from *Escherichia coli*, and *Hin* indicates its origin from *Haemophilus influenzae*.

(2) Roman numerals (I, II, III etc) are added in the last to distinguish several enzymes with same origin. For example *Eco* I, *Hin* I and *Hpa* I.

(3) The name of the identified strain or type for the genus is written as a subscript *e.g.*, *Eco* K stands for *E. coli* strain K, *Eco* B stands for *E. coli* strain B. and *Hind* for *H. influenzae* strain Rd.

(4) In certain cases where restriction and modification systems are genetically specified by a plasmid or a virus, the extrachromosomal element is written as a subscript *e.g.*, Eco Rl, Eco Pl.

(5) If a strain possesses many restriction and modification systems, these are identified by Roman numerals *e.g. Hind* I, *Hind* II and *Hind* III for *H. influenzae* strain Rd. and *Bam* II and *Bam* HII for *Bacillus amyloliquefaciens* strain H.

OTHER ENZYMES

1. DNA Ligases

They are also known as *polynucleotide ligases*. These are those enzymes which help in sealing the nicks between adjacent nucleotides in a double stranded DNA molecule or in joining the two DNA fragments by catalysing the synthesis of a *phosphodiester bond* between a 3'-OH group at the end of one chain or fragment and 5'-phosphate group at the end of other chain or fragment. The DNA ligases are found in both prokaryotes and eukaryotes. The role of DNA ligase in sealing the cohesive ends of DNA during production of recombinant DNA was first demonstrated by **Mertz** and **Davis** (1972). DNA ligase seals single stranded nicks in DNA which has 5' phosphate → 3'OH termini.

There are two most extensively studied DNA ligases — (*i*) *DNA ligase* from *E. coli* and (*ii*) *DNA ligase* from T_4 phage.

Genetic Engineering

Although both *E. coli* and T_4 phage encoded DNA ligases share commonly the property of sealing nicks that have 5'-phosphate and 3'OH termini, they differ in their cofactor requirements. The *E. coli* DNA ligase requires nicotinamide adenine dinucleotide (NAD^+) as cofactor while the T_4 DNA ligase requires adenosine triphosphate (ATP) as cofactor. Both the enzymes have a-NH_2 group on lysine residue. In both the cases the cofactors (NAD^+ and ATP) split and form an *enzyme adenosine monophosphate complex* (AMP-complex). This complex binds to the nick utilizing free 5'-phosphate and 3'-OH termini and forms a *phosphodiester bond* which seals the nick. The enzyme and AMP are released (Lehman, 1974). The *E. coli* DNA ligase joins the cohesive ends (single stranded sticky ends) of DNA fragments produced by restriction enzymes while T_4 phage DNA ligase joins the blunt ends of DNA fragments produced by restriction enzymes.

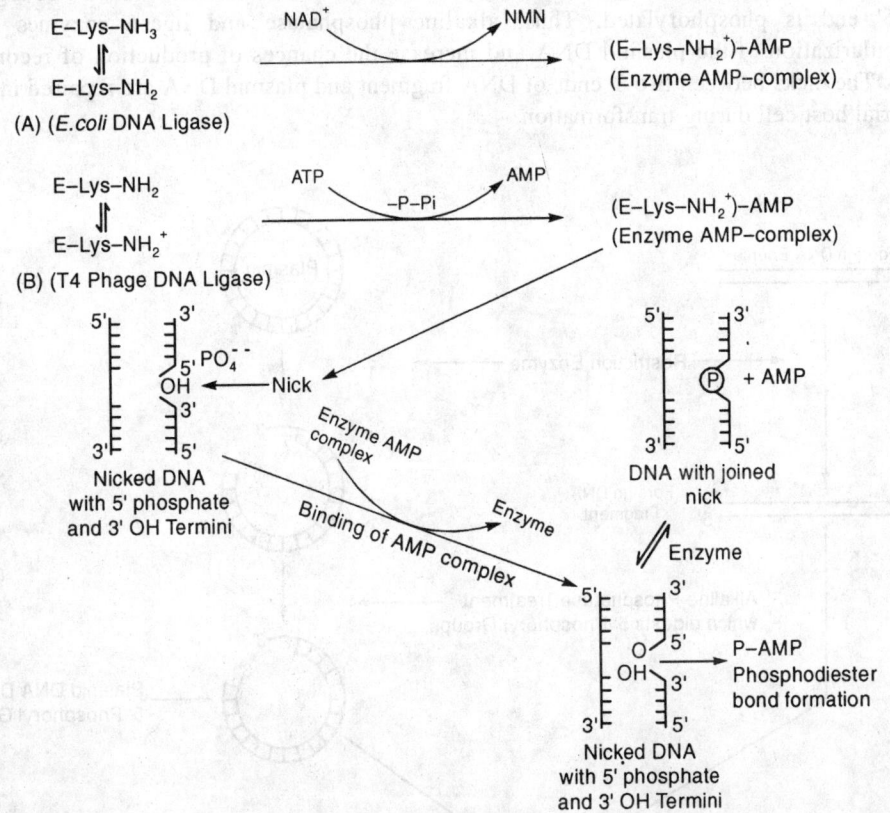

Fig. 1.4. (a) Mechanism of DNA ligase enzyme (Diagrammatic)
P = Phosphate, E = enzyme, AMP = Adenosine monophosphate.

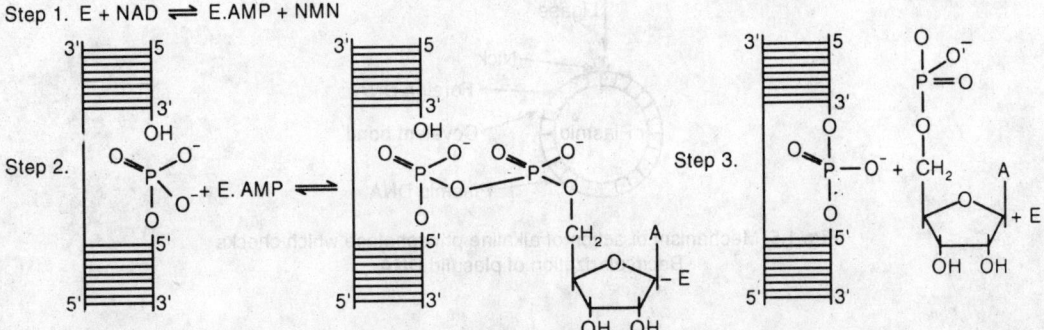

Fig. 1.4. (b) Mechanism of DNA ligase enzyme (structurally).

2. Alkaline Phosphatase

Alkaline phosphatase is the enzyme which is used to check the rejoining of cohesive single stranded ends of the restriction enzyme treated plasmid DNA, otherwise the ends of the same plasmid DNA molecule will join and recircularization of the plasmid DNA will take place. Thus, foreign DNA will not join the plasmid DNA to form recombinant or chimeric DNA. When the restriction enzyme treated plasmid is treated with *alkaline phosphatase* enzyme, the enzyme digests the terminal 5' phosphoryl groups. The restriction enzyme treated foreign DNA fragment is not treated with alkaline phosphatase. Therefore the 5' ends of foreign DNA fragment covalently join to 3' ends of the plasmid. The hybrid DNA obtained possesses a *nick with 3' and 5' hydroxy ends*. The DNA ligase will only join 3' and 5' ends of recombinant DNA together if the 5' end is phosphorylated. Thus, alkaline phosphatase and ligase enzymes prevent recircularization of the plasmid DNA and increase the chances of production of recombinant DNA. The nicks between two 3' ends of DNA fragment and plasmid DNA are repaired inside the bacterial host cell during transformation.

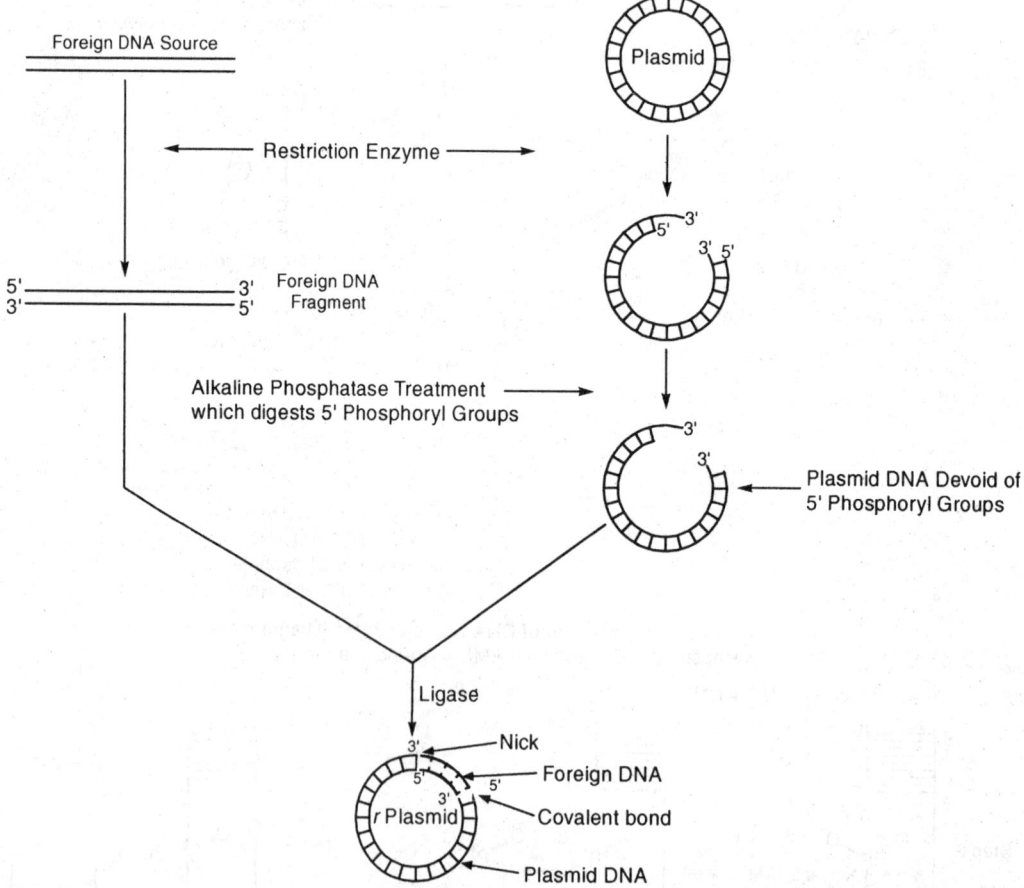

Fig. 1.5. Mechanism of action of alkaline phosphatase which checks Recircularization of plasmid **DNA.**

Genetic Engineering 13

3. S_1 Nuclease — This is the enzyme which degrades the single stranded DNA (ss DNA) or single strand of double stranded DNA (ds DNA) with cohesive ends. The action of S_1 nuclease results into conversion of cohesive ends into blunt ends.

4. DNA Polymerase — (Consult chapter 14 – Nucleic acid metabolism of Physiology part.

5. Reverse transcriptase — This enzyme synthesizes complementary DNA (c DNA) from *m*RNA template. It was discovered by **H. Temin** and **D. Baltimore** (1970). For details, consult chapter 15 on 'Protein Metabolism'.

II. Foreign DNA

It is also known as **passenger DNA**. A foreign DNA is a piece or fragment of that DNA molecule which contains the desired gene sequence. It is first isolated enzymatically and then purified and cloned. The identification of the desired gene on a genome is necessary. It can be done either before or after gene cloning. The cloned foreign DNA fragment expresses normally as it expresses in parental cell. Thus, foreign DNA can be obtained from any source depending upon the aim and scope of cloning experiment.

It is more difficult to identify and characterise DNA sequences on its genome than mRNA. Therefore, purified *m*RNAs are usually used for such purposes. Separate procedures are used to isolate and identify DNA sequences which may be cDNA or directly clonable DNA fragments. If *m*RNA is used and its gene product is not well characterized, the procedure of gene cloning becomes more difficult. An average cell contains about 1-2% *m*RNA of the total cytoplasmic RNAs content. This small amount of *m*RNA carries transcripts for coding various types of proteins. If the amount of *m*RNA is quite low, it causes difficulties in isolation of cDNA clones. The foreign DNA can be isolated and cloned directly from clonable DNA fragments of donor organisms by using restriction endonucleases. They can also be obtained from gene bank for experimental purposes.

III. Vectors or Cloning Vehicles

Vectors are the carriers or vehicles of genetic engineering through which a DNA fragment is introduced permanently into a cell. A vector should possess the following characteristics —

(1) It should be small and capable of autonomous replication in a suitable bacterial host.

(2) It should possess substrate sites for restriction endonucleases so that it is able to integrate foreign DNA segments.

(3) It should show some special properties like drug resistance.

(4) It should have a marker so that necessary prerequisites for the cloning of recombinant DNA molecules are produced. The marker actually permits recognition by complementation of an auxotrophic strain.

(5) It should be distinguishable from a receptor or recombinant phenotypes.

The functions of a vector are replication as well as gene expression.

Types of Vectors

The vectors may be —

(A) Natural vectors : These are different types of naturally occurring plasmids, bacteriophages and viruses.

(1) *Plasmid as vectors.* What are plasmids? The *self replicating circular DNA duplex molecules found freely floating in the bacterial cytoplasm are called plasmids*. They are **extrachromosomal** and **autonomous** elements. They can replicate independently and are maintained in a characteristic number of copies in a bacterial cell. Plasmids in the majority of cases are not the essential constituents of a cell. However, they are important because they contain genes for *fertility, for resistance to antibiotics* and *heavy metals* and for *production of bacteriocins* and *toxins* etc., which are not found in bacterial chromosomes. Due to self-replicating character and these extra properties which are not found in bacterial chromosomes, plasmids have become an important tool in genetic engineering as cloning vehicles.

Two types of plasmids have been recognized —

(*i*) **Single copy plasmids** — They are present as one plasmid DNA per host chromosome.

(*ii*) **Multicopy plasmids** — They are found as 10–20 genomes per cell. There are certain other type of plasmids which contain upto 1000 copies per cell. Such plasmids are maintained under **relaxed replication** control mechanism.

Plasmids may be *self–transmissible* or *non–self–transmissible*. Self transmissible plasmids can be easily transmitted to many bacterial cells. The examples of such plasmids are **RK2** and **RK4**. Non-self-transmissible plasmids are transmitted from one cell to another during conjunction along with other plasmids present in the cell, *e.g.*, pSC_{101} and $ColE_1$.

Isolation of Bacterial Plasmid

Isolation of bacterial plasmid involves the following steps —

(*i*) **Treatment of bacterial cell with suitable detergent** — It causes breaking of cell wall and solubilization of cell membrane, resulting into release of inner content of the cell having plasmid DNA, chromosomal DNA and other molecules. The entire inner content or mixture of these molecules is known as *lysate*.

(*ii*) **Treatment of lysate with potassium acetate or acetic acid solution** — It causes precipitation of chromosomal DNA and other protein molecules.

(*iii*) **Removal of precipitate from lysate** — It is done through centrifuge by rotating it at high rpm (round per minute). The precipitate settles down at the bottom and a clear supernatant of lysate can be seen in the centrifuge tube. The lysate contains plasmid DNA, contaminated RNA, remaining proteins and chromosomal DNA debris.

(*iv*) **Digestion of RNA** — It is done by treating the above lysates mixture [obtained in step (*iii*)] with RNAase.

(*v*) **Removal of contaminated proteins and RNAase** — For this purpose, the lysate containing contaminated proteins, RNAase and plasmid DNA is mixed with phenol in separating funnel and left for some time. It causes the separation of two layers–upper phenol and lower water layer. The phenol layer contains contaminated proteins and RNAase and the water layer contains plasmid DNA and some genomic DNA.

(*vi*) **Removal of phenol layer** — It can be done through stopper of the funnel.

(*vii*) **Removal of genomic DNA from water layer** – The genomic DNA can be removed from the water layer as precipitate when alcohol is added. Now, clear water contains plasmid DNA which can be used for specific purposes at any time.

Experimental procedure for the Formation of Hybrid Plasmid

During the experiment, first the bacterial cells are opened up, plasmids are separated, purified and treated with the restriction enzyme *e.g. Eco RI*. The restriction enzyme cuts it open at a single site and converts the circular plasmid into a linear molecule with sticky ends. The DNA of a donor e.g., mammalian cell or cell of any other organism, is similarly treated with the same restriction enzyme to produce fragments with sticky ends. This isolated fragment is called **foreign DNA.**

The plasmid is now used as a vector to carry a piece of donor DNA into the bacterial cell. When both, plasmid DNA and donor DNA, are mixed under suitable conditions, the fragment of donor DNA with sticky ends may join with the sticky ends of the linear molecule of the plasmid

Genetic Engineering

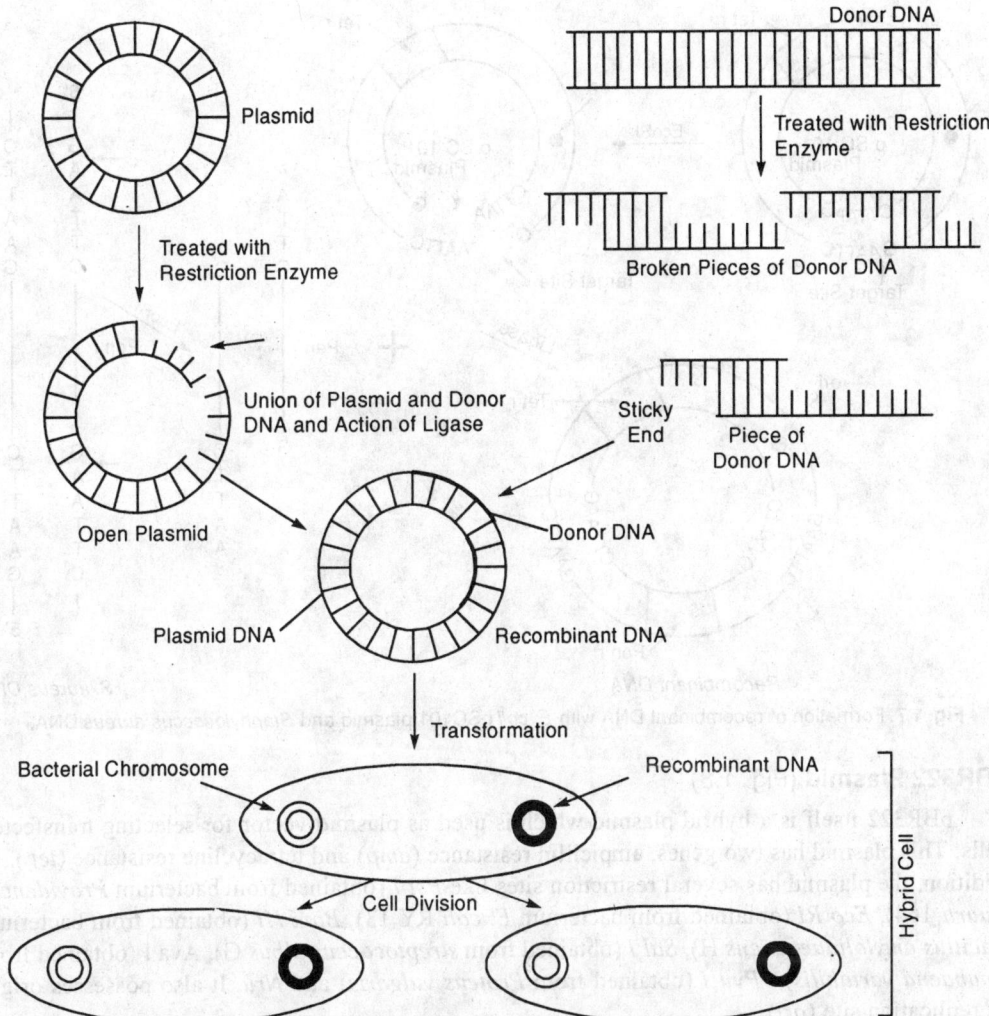

Fig. 1.6. Steps in Recombinant DNA by using plasmid as vector.

DNA. The enzyme DNA ligase may be used to seal the two open ends of the plasmid DNA to make it again circular molecule. Thus, a piece of donor DNA gets incorporated into a plasmid producing a **hybrid plasmid or chimeric plasmid**. This is also called *recombinant* DNA. It can be propagated indefinitely if inserted inside the host bacterium.

An example of *E. coli* plasmid, called *pSC 101*, can be considered here. It possesses tectracycline resistant gene(Tet^r) and a single site for the endonuclease *Eco R-I*, producing staggered cuts. DNA fragment from any other source made with restriction enzyme *E.CoRI* can be inserted into the plasmid. The foreign DNA will replicate with plasmid DNA. Another example of plasmid from *Staphylococcus aureus* may be considered. This plasmid possesses gene for penicillin resistance. The plasmid was treated with *Eco R-I* enzyme and the fragments were cloned in pSC101 *E. coli* cells (resistant for tetracycline). Thus, recombinant DNA was produced which was resistant to both penicillin and tetracycline. (Fig. 1.7).

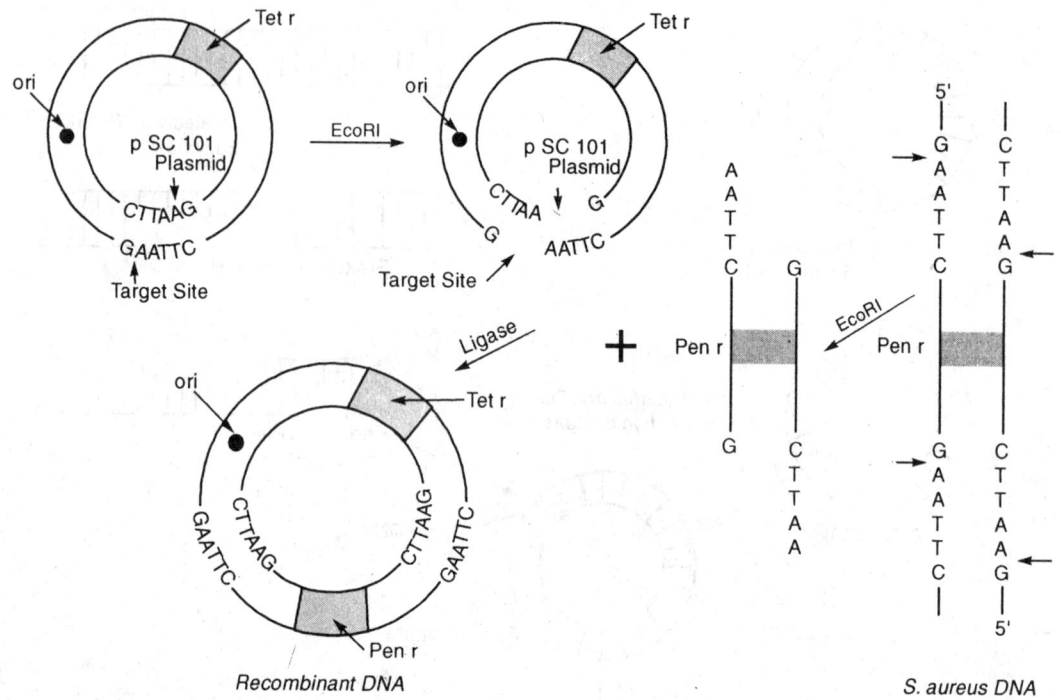

Fig. 1.7. Formation of recombinant DNA with *E. coli* pSC101 plasmid and *Staphylococcus aureus* DNA.

pBR322 Plasmid (Fig. 1.8)

pBR322 itself is a hybrid plasmid which is used as plasmid vector for selecting transfected cells. This plasmid has two genes, ampicillin resistance (*amp*) and tetracycline resistance (*tet*^{r}). In addition, the plasmid has several restriction sites like *Pst I* (obtained from bacterium *Providencia stuarti* 164), *Eco RI* (obtained from bacterium *E. coli* RY 13), *Bam HI* (obtained from bacterium *Bacillus amyloliquefaciens* H), *Sal I* (obtained from *streptococcus albus* G), Ava I (obtained from *Anabaena variabilis*), *Pvu I* (obtained from *Proteus vulgaris*) and *Nru*. It also possesses origin of replication site (*ori*).

From pBR322 plasmid, the hybrid plasmid can be constructed by treating the plasmid and the foreign DNA with *Pst I* restriction enzyme and mixing the two DNA s' together in the presence of DNA ligase. At first foreign DNA is treated with *Pst I* restriction enzyme to produce fragments with staggered ends. Now, pBR322 plasmid is also treated with the same *Pst I* enzyme to produce staggered ends. The restriction fragments of foreign DNA are mixed with treated plasmid at low temperature to permit annealing between the two DNAs. Finally, the annealed fragments are ligated with enzyme *DNA ligase*. The end product will contain some of the original pBR322 fragment and inserted foreign DNA.

When this mixture is used in transfection, (*i*) most of the cells will not be transfected, (*ii*) some will be transfected with desired hybrid plasmids and (*iii*) some will be transfected with pBR322. These three types of cells will differ in their drug resistance properties. Normal cells are killed by tetracycline or ampicillin. The transfected cells with DNA inserted in the plasmid are tetracycline resistant but ampicillin sensitive, since the insert has disrupted the *tet*^{r} gene. The cells containing the desired plasmids can be distinguished from those containing pBR322 by replica plating. The plasmid pBR322 can be digested at any restriction site using appropriate restriction enzyme.

Genetic Engineering

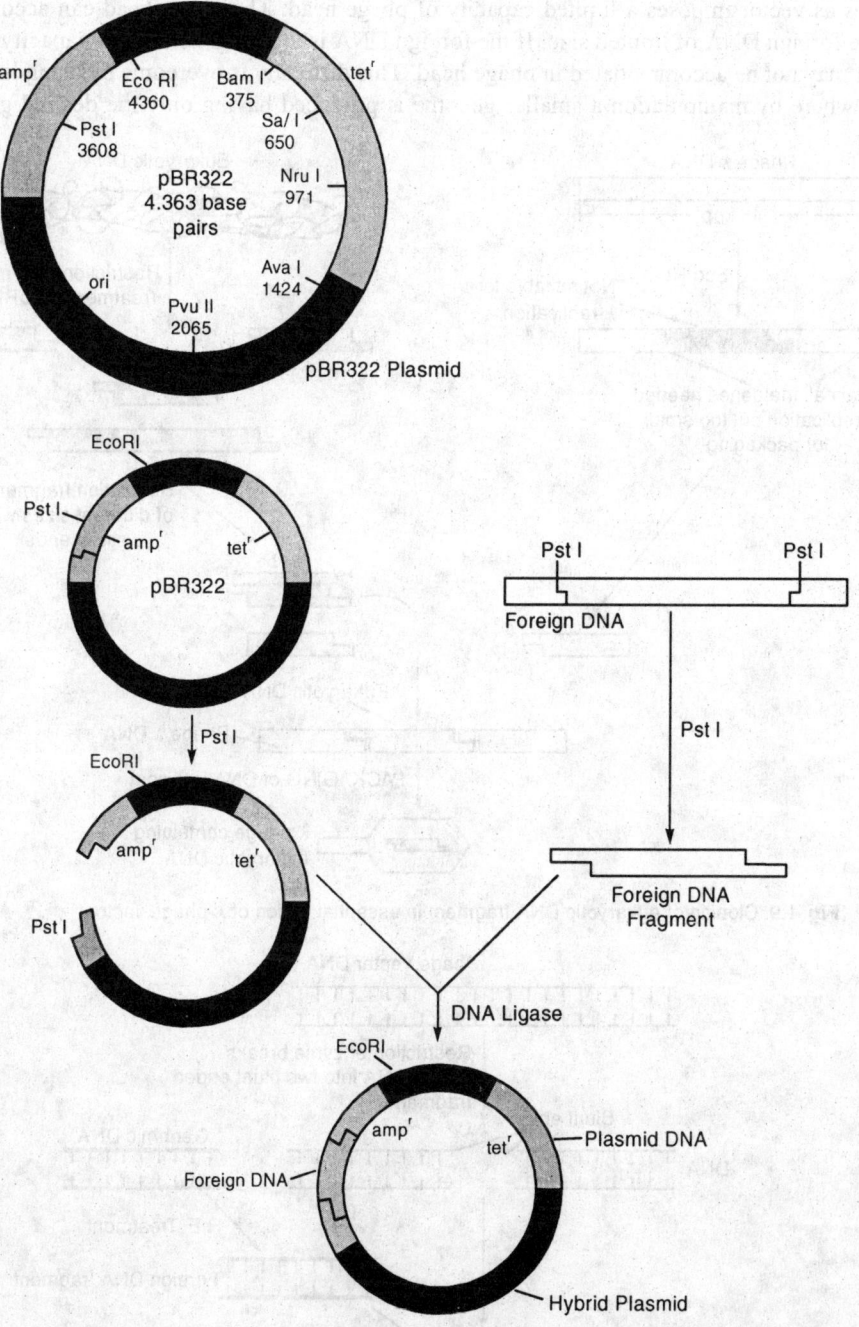

FIG. 1.8. Structure of pBR322 plasmid and construction of hybrid plasmid.

(2) Phages as Vectors

Besides plasmids, phage DNAs can also be used as vector. Usually the phage DNA is a linear molecule which can be cut by single cleavage by restriction endonucleases to produce fragments. The foreign DNA get inserted in between the two fragments of phage DNA to produce a chimeric phage or hybrid phage (recombinant DNA). The chimeric phage can be isolated after allowing it to infect bacteria and collecting progeny particles after a lytic cycle. The use of phage particles as vector imposes a limited capacity of phage head. The phage head can accommodate only the foreign DNA of limited size. If the foreign DNA is too long, *i.e.*, out of capacity of phage head, it may not be accommodated in phage head. This difficulty is overcome by using lambda (λ) phage, where by manipulation a smaller genome is produced having only the desired genes.

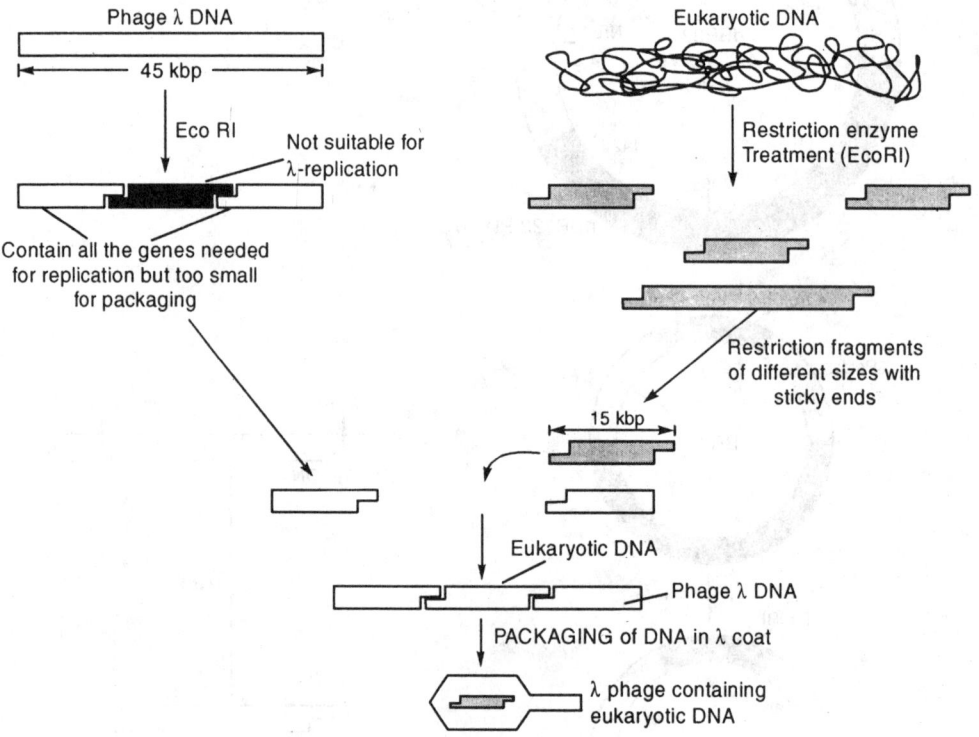

Fig. 1.9. Cloning of eukaryotic DNA fragment in essential region of λ-phage vector.

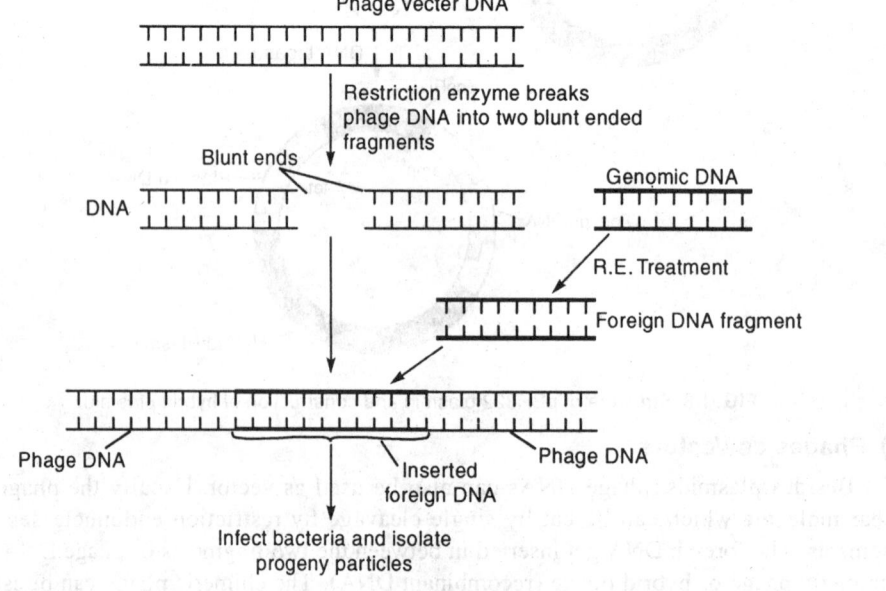

FIG. 1.10. Cloning of foreign dna in a non-essential region of phage vector.

Genetic Engineering

The lambda phage DNA can be inserted into *E. coli* DNA at a specific site between galactose (*gal*) and biotin (*bio*) operons. The circular-phage DNA is cleaved at another point and spliced to cleaved host DNA with the help of *integrase* enzyme. *E coli* DNA containing the phage DNA (recombinant DNA) becomes lysogenic and after a few generations the integrated provirus is induced by chemical treatment to become lytic phage.

Insertion Phage Vectors and Replacement Phage Vectors

These are the derivatives, wild-type phage. The vectors which have a single target site in their DNA at which foreign DNA can be inserted to result chimeric DNA are called ***insertion phage vectors*** and those vectors which have a pair of sites in their DNA fragment which can be removed and replaced by foreign DNA are called ***replacement phage vectors.***

The vector derivatives of insertion and replacement both types, have been produced. Most of them have been constructed by using restriction enzymes like *Eco RI, Bam II* and *Hind III*. The application of above vectors can be extended to other restriction enzymes by using linker molecules.

To accommodate large sized foreign DNA in phage head, a fragment of phage DNA that does not carry essential phage genes is removed to increase the space within the phage DNA and is replaced by foreign DNA.

(3) Viruses as Vectors

In eukaryotes, *Simian virus* 40 (*SV* 40) has been used as vector. A number of other plant and animal viruses have also been used as vectors for various purposes. The viruses have been used for introducing foreign genes into cells, for amplification of genes and for gene expression in host cells. The following are the examples of plant and animal viruses which have been extensively used as vectors.

(*a*) **Plant viruses** — Tobacco Mosaic Virus (TMV), Cauliflower Mosaic Virus (CMV) and Gemini viruses.

(*b*) **Animal viruses** — Adenoviruses, Retroviruses, Herpesvirus, Papovaviruses and Baculovirus.

Simian Virus 40 (SV40)

*SV*40 is a monkey virus that was accidently discovered in kidney-cell cultures from wild monkeys used in the production of polio virus vaccines. It contains a single molecule of double stranded DNA (ds DNA). The virus is without any envelope and contains genome of about 5.2 kb length. *SV* 40 replicates in host cell nucleus and uses host cell enzymes for viral DNA and mRNA synthesis. The DNA of this virus enters the cell nucleus where cellular enzymes are diverted to replication of the viral DNA and its transcription into viral mRNA. This virus has been used in researches for mRNA synthesis and regulation, DNA replication and cell transformation. The known hosts for *SV40* are mostly primates.

SV40 Proteins

*SV*40 virus synthesises two early proteins called '*T*' and '*t*'. T was originally called *T antigen* because it was first demonstrated by immunofluorescence using serum from animals having virus induced tumors. The weight of *T antigen* proteins present in cell nucleus is 90-Kilodaltons (KDa). They remain tightly bound to DNA and play an important role in viral DNA transcription and replication during lytic cycle. The *t* proteins are cytoplasmic and their weight is 20 KDa.

SV40 Replication Origin

It has been observed that a 65 base pair(bp) region in *SV40* chromosome is sufficient to promote DNA replication both in animal cells and in *vitro* conditions. Three segments of the *SV*

40 origins are required for activity. It has been demonstrated during testing of origins containing specific mutations. Mammalian proteins and plasmids carrying *SV40* replication origin have been used to study molecular mechanisms of mammalian DNA replication.

The replication origins in SV40 DNA show the following three important features —

(1) Replication origins are unique DNA segments that contain multiple short repeated sequences.

(2) These short repeats are recognised by multimeric origin-binding proteins which play a key role in assembling the replication enzymes at the origin site.

(3) The origin region is usually flanked by DNA sequences rich in adenine (A) and thymine (T) which facilitate unwinding of DNA duplex. Unwinding is necessary to generate single stranded regions that can serve as templates for DNA synthesis.

Replication of SV40 DNA by Eukaryotic Proteins (Fig. 1.11)

The replication of *SV40* DNA is initiated at a unique location on DNA by interaction of a virus-encoded, site-specific DNA binding protein called **T antigen** (Step I). This protein is multifunctional and it locally unwinds DNA duplex through its helicase enzyme activity. The unwinding of duplex at the *SV40* origin also requires ATP and replication factor A(RFA)–(Step 2). In contrast to replication of *E. coli* DNA, during SV40 DNA replication two distinct *polymerases*, α and δ, appear and function at growing fork by catalysing elongation of the *leading* and *lagging strands*. **Polymerase** α (Pol α), which is tightly associated with a *primase* is thought to form lagging strand, *i.e.*, it forms the 5' end of the leading strand and then is displaced from the template. **Polymerase** δ (Pol δ) is largely responsible for the synthesis of leading strand.

Initially, the synthesis of leading strand is carried out by Pol α and Pol δ in a sequential manner. When the primase-Pol α complex binds to the unwound temperate strands, the primase synthesises RNA primers and Pol α elongates and stretches these primers. The activity of Pol α is stimulated by replication factor C (RFC)–(step 3). Now **proliferating cell nuclear antigen** (PCNA) binds at the primer template terminus and displaces Pol α. Thus, leading strand synthesis is interrupted—(step 4). Now Pol α binds to PCNA AT 3' end of the growing strand. The binding of Pol δ with PCNA increases the processivity of the enzyme which helps in continuous synthesis of the leading strand. As the unwinding of the DNA duplex progresses quite away from the origin, the primase-Pol α complex associates with the unwound temperate strands downstream from the leading strand primers. The synthesis of lagging strand then is carried out by the combined action of Pol α and Pol δ AND RFC, while synthesis of leading strand on the other side of the origin also proceeds—(step 5). Finally, eukaryotic topoisomerases play an important role in relieving torsional stress induced by growing fork movement and in separating the two daugher chromosomes.

(B) Reconstituted Vectors

(1) Cosmids as Vectors — As stated earlier, the cosmids are plasmid particles which contain specific DNA sequences inserted at **cos** sites. They were first developed by **Collins** and **Hohn** (1978). These sites come from lambda phage and enable the DNA to get packed in lambda particles. The DNA fragments are inserted into these cosmid vectors and these vectors are packaged into phage particles *in vitro* similar to that of λ-phage cloning. It permits their purification. The cosmids replicate in the host bacteria similar to that of plasmids. They do not carry the genes for lytic development. The main advantage of cosmids as vector is that through these, quite large sized genomic DNA fragments can be cloned.

Construction of a Cosmid (Fig 1.12)

A vector containing a λ *cos site e.g.*, pJB8, can be used for the construction of a cosmid. The *pBJ8* vector is a 5.1-kb plasmid containing of a λ *cos* site, an *amp^r selectable marker*, an *origin of replication* (ori) and several possible *cloning sites* like **Bam HI** (for enzyme Bam HI, obtained

Genetic Engineering

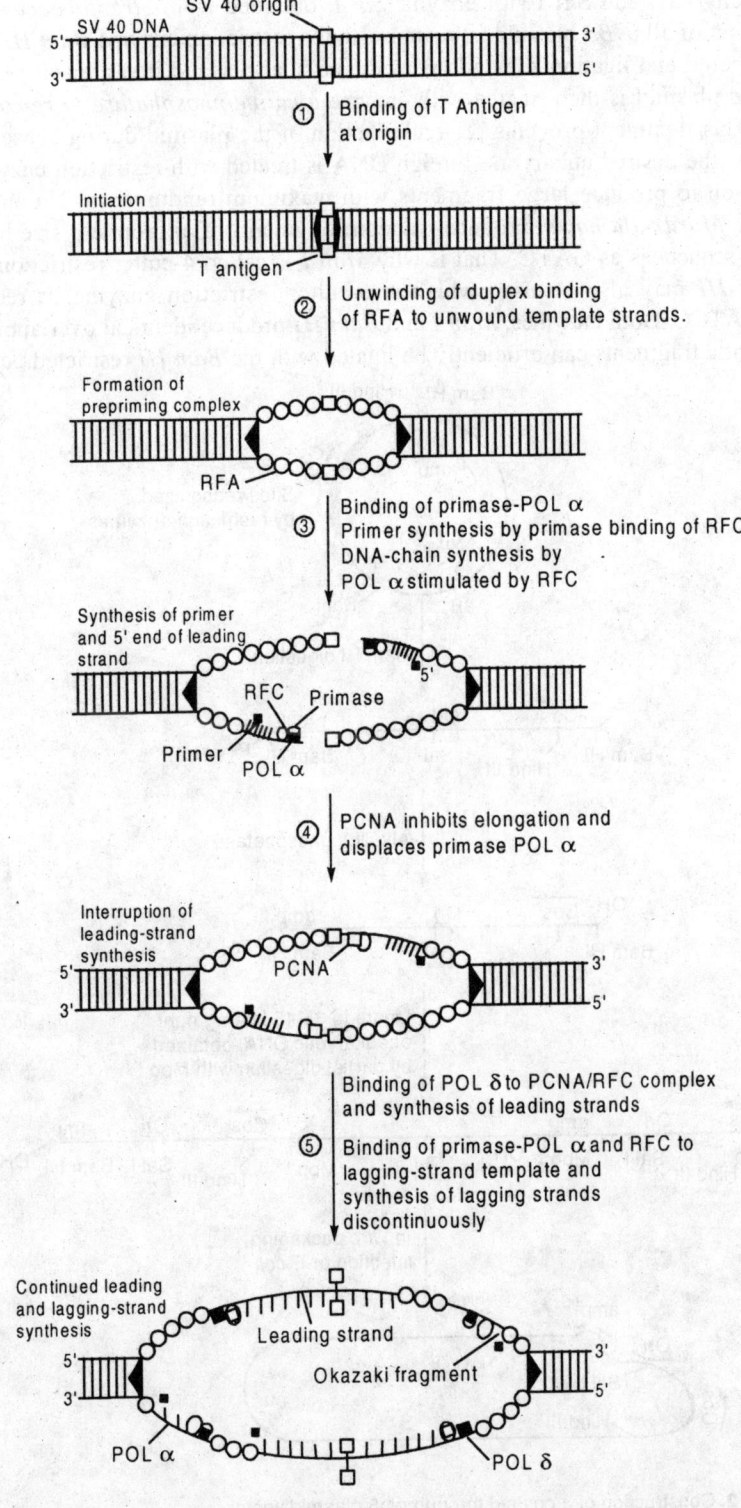

Fig. 1.11. *In Vitro* replication of SV40 DNA by eukaryotic enzymes.

from ***Bacillus amyloliquefaciens* H**), **Hind III** (for enzyme *Hind III*, obtained from *B. amyloliquefaciens* H) and **Sal I** (for enzyme *Sal I*, obtained from *Streptococcus albus* G). To construct it, first of all *pJB8* plasmid is treated with the restriction enzyme ***Bam HI*** which attacks ***Bam HI*** staggered end ligation site and opens the circular plasmid into a linear plasmid (step I). The linearized plasmid is then treated with enzyme *alkaline phosphatase to remove 5'* terminal phosphates. This treatment prevents recircularization of the plasmid during subsequent ligation (step 2). Now, the desired eukaryotic foreign DNA is treated with restriction enzyme *Mbo I* for limited duration to produce large fragments with maximum randomness. The *Mbo I* enzyme, obtained from ***Moraxella bovis***, produces random fragments of appropriate size by recognizing the four base sequences as GATC. That is why ***Mob I*** is called 4-cutter restriction enzyme. The enzyme ***Bam HI*** may also be used which is 6-cutter restriction enzyme. It recognizes base sequence GGATCC. Both enzymes, ***Mob I*** and ***Bam HI*** produce identical overlapping fragments. Thus, eukaryotic fragments can efficiently be ligated with the *Bam HI* restricted cosmid (step 3).

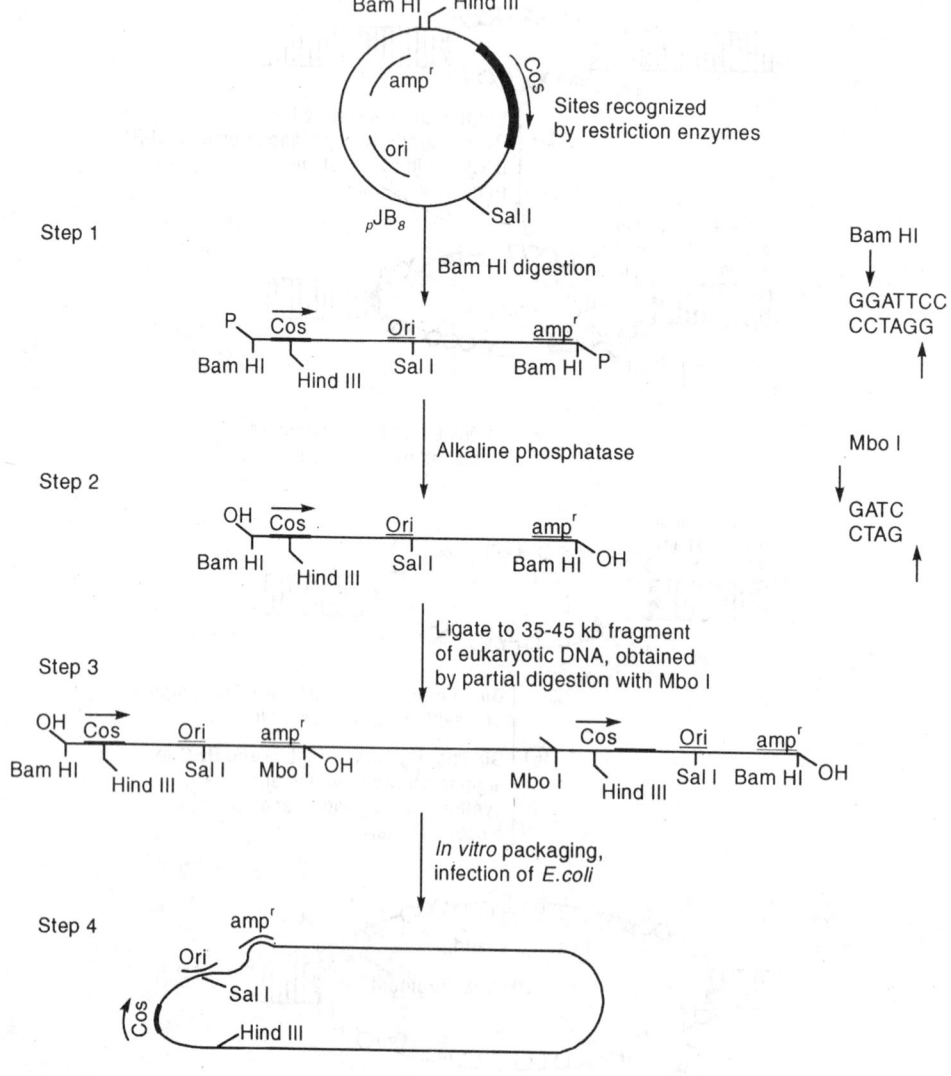

FIG. 1.12. Construction of a cosmid through pjb8 plasmid vector.

Genetic Engineering

Usually fragments of eukaryotic DNA with 35-45 kb are obtained by partial digestion with **MboI** which are ligated with linear, alkaline treated plasmid. After ligation a large concatamer with two *cos* sites is resulted which is packaged *in vitro* into λ particle and introduced into a suitable *E. coli* strain (step 4). The transformants can be isolated through amp^r selectable marker. The transfected structure readily circularizes.

(2) Phasmids

According to **Brenner** *e.t. al.*, (1982), the phasmids are a type of plasmid vectors which contain a fragment of phage DNA at its *att* site. Thus, the phasmids possess the characters of plasmid vector as well as phage vector both. The phasmid may be inserted into a λ-phage DNA in the same way by which phage DNA is inserted into the bacterial genome (chromosome) during lysogenic phase of life cycle.

The insertion of plasmid into λ-phage is done to achieve a specific site responsible for recombinational insertion of the phage into bacterial chromosome during lysogenic cycle. This insertion of plasmid into λ-phage genome is reversible and referred to as *lifting* the plasmid. It generates a phage genome containing *att* site and one or more plasmid molecules. These new genetic recombinations are *phasmids*. They contain functional *ori* genes of plasmids and gene of λ-phage. The plasmids can propagate in suitable *E. coli* strains, similar to those of plasmid and phage vectors. They are released when reversal of lifting process takes place. Phasmids are useful because they can be stored for long time and can be utilized as a cloning vector. The commonly used phasmid is **pBluescrip II KS**. It is about 2961 base pairs long and is derived from *pUC* 19. This is an expression vector whose cloning site is flanked by T_3 and T_7 promoters which are read in opposite directions.

IV. Linkers

These are short pieces of double stranded DNA which have sites for the action of one or more restriction enzymes. The linkers can be ligated to the blunt ends of DNA molecules. When

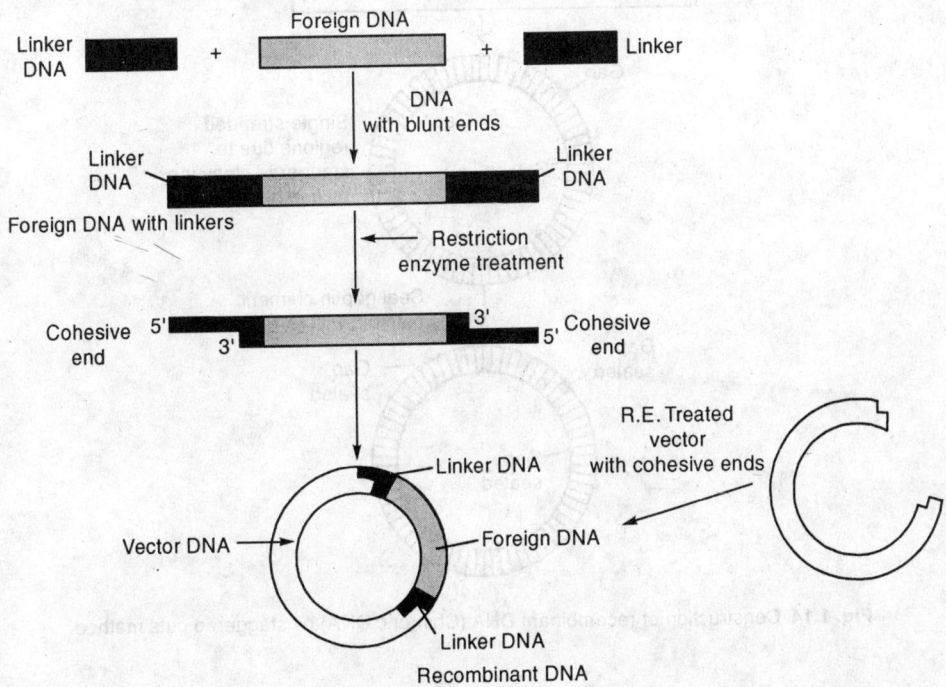

Fig. 1.13. Production of restriction sites on DNA by linkers.

linkers are treated with restriction enzymes, they produce cohesive ended fragments which can be incorporated into any cloning vector (plasmid, phage etc.) after treating with the same restriction enzymes. The examples of linkers are *Eco RI*-linkers and *Sal* I-linkers.

TECHNIQUES INVOLVED IN RECOMBINANT DNA TECHNOLOGY

Following techniques are involved in the construction of recombinant DNA —

1. Palindromes and Staggered Cuts Method (Fig 1.14)

A palindromic sequence is that in which both the strands of a DNA possess the same sequence of bases when read in 5' to 3' direction. They represent an inverse repeat causing

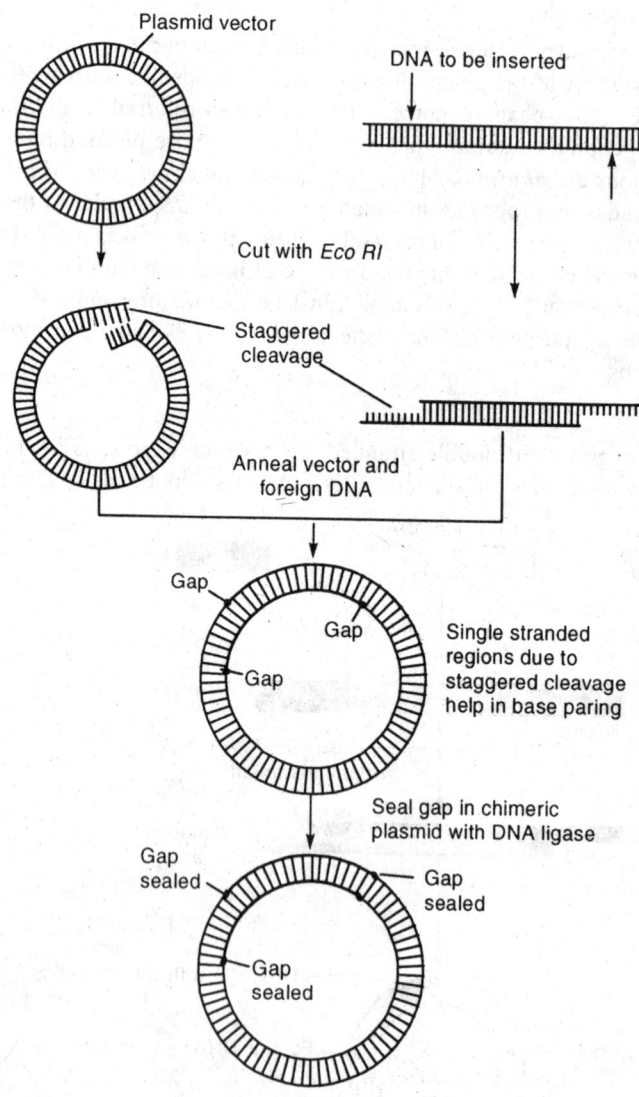

Fig. 1.14. Construction of recombinant DNA (Chimeric DNA) by staggered cuts method.

Genetic Engineering

complementarily between bases. *Palindromes in mRNA cause formation of hairpins*. A palindrome can be represented as follows —

$$5'\ G\ G\ T\ A\ C\ C\ 3'$$
$$3'\ C\ C\ A\ T\ G\ G\ 5'$$

The production of recombinant DNA by staggered cuts method includes the following steps—

(1) The vector DNA containing palindromic sequence is treated with restriction enzyme (*Eco*RI) to cause staggered cuts which produce complementary single stranded sticky ends.

(2) A foreign DNA to be inserted is similarly treated with the same restriction enzyme to cause staggered cuts producing similar single stranded sticky ends having same sequences.

(3) Treated vector DNA is mixed with treated foreign DNA. This mixing causes annealing of two types of DNA producing a recombinant DNA or chimeric DNA. If plasmid DNA is used as vector it is called *chimeric plasmid*.

(4) Finally chimeric plasmid or recombinant DNA is treated with enzyme *DNA ligase* which seals the cut ends of the two types of DNA molecules by joining the bonds.

The advantage of this technique is that it regenerates two restriction sites in the recombinant DNA which help in easy retrieval of foreign DNA segment from the cloned copies of recombinant DNA by cleaving again with the same restriction enzyme.

Besides, the above technique has the following disadvantages—

(1) The two cleaved ends of a vector DNA or of a foreign DNA produced by the restriction enzyme may rejoin before the formation of recombinant DNA.

(2) Sometimes more than one copy of a foreign DNA may join end to end before getting inserted. Thus, expected recombinant DNA will not be produced.

(3) Sometimes only a part of the desired segment is inserted in the vector DNA because the recognition site in the DNA to be cloned (foreign DNA) may not lie in a convenient position.

2. By addition of Poly dA at 3' ends of the Vector and Poly dT at the 3' ends of the DNA Clone (Fig 1.15)

This method includes the following steps —

(1) Vector DNA is cut at the desired position without producing sticky ends.

(2) Similarly foreign DNA to be cloned is also cut at desired positions without producing sticky ends.

(3) Poly dA is added at both the 3' cut ends of the vector DNA using precursor dATP with the help of enzyme *terminal transferase*.

(4) Poly dT is added at both the 3' cut ends of the foreign DNA using precursor dTTP with the help of enzyme *terminal transferase*.

(5) The vector DNA with poly dA 3' ends is mixed with foreign DNA with poly dT 3' ends which results into annealing of poly dA with poly dT tails.

(6) Finally, mixed vector and foreign DNA (cloned) molecules are treated with enzyme *DNA ligase* which results into sealing of cut ends. Thus, recombinant DNA is constructed.

The advantage of this technique is that there is no chance of reannealing between the two cut ends of vector DNA or foreign or between the cut ends of two or more molecules of the foreign DNA to be cloned.

The disadvantage of this technique is that the retrieval of cloned DNA, here, becomes very difficult because the recognition site of enzyme is lost due to insertion of poly dA and poly dT at 3' ends of vector DNA and foreign DNA respectively.

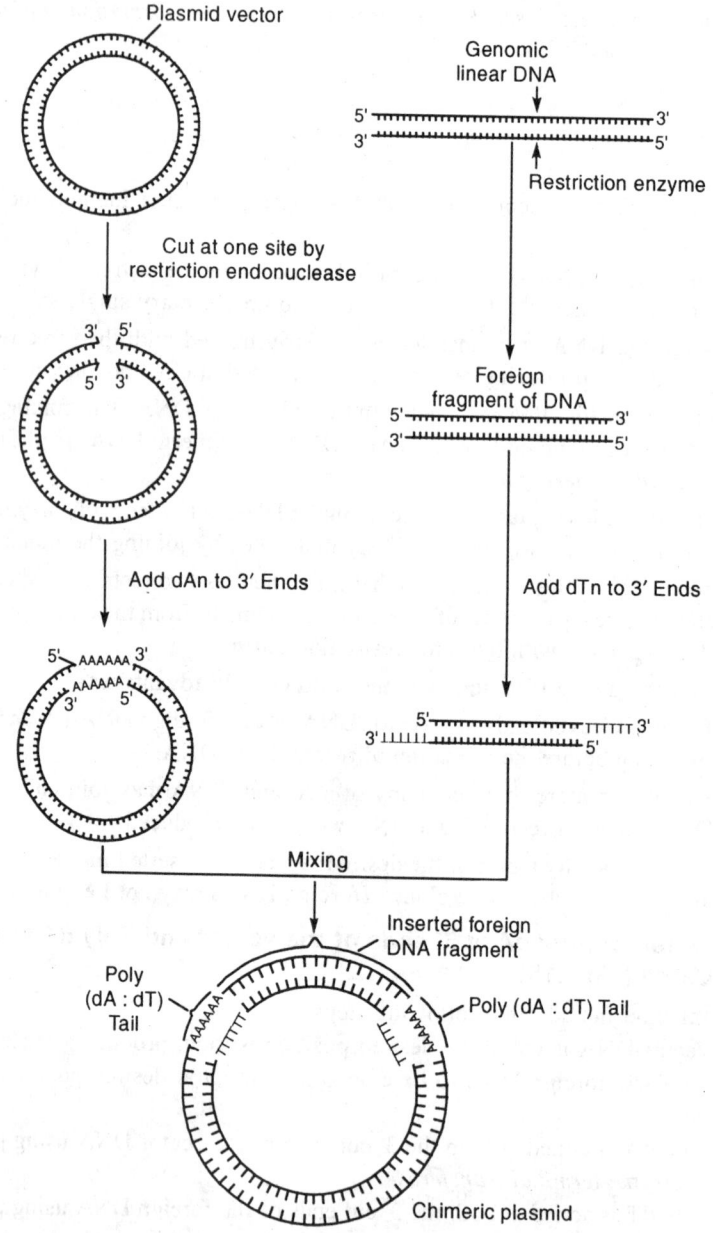

Fig. 1.15. Construction of recombinant DNA through plasmid vector by poly dA and Poly dT tailing technique.

3. Blunt end ligation by T4 DNA Ligase

This method includes the following steps —

(1) Vector DNA (plasmid DNA) is treated with restriction enzyme to cut it open at a particular place producing blunt ends.

(2) The foreign DNA is also treated with the same restriction enzyme to cut it at the same place, as in vector DNA, producing two blunt ends.

Genetic Engineering

(3) Vector DNA with blunt ends is mixed with blunt ended foreign DNA for joining the blunt ends of both the DNAs irrespective of the sequence present at the broken ends.

(4) Ligation of blunt ends is performed by enzyme *T4 DNA ligase*. Thus, recombinant DNA or chimeric DNA is constructed.

This method is beneficial because (*i*) the DNA to be cloned can be inserted at a desired site and (*ii*) it can be easily retrieved whenever required.

The disadvantage of this method is that any two broken ends may join including those belonging to the same DNA molecule.

CLONING VECTORS

These have already been described earlier under heading "tools involved in recombinant DNA technology".

cDNA LIBRARIES AND THEIR CONSTRUCTION

The complementary DNA (cDNA) is the DNA which is derived from mRNA. If the clones are derived from mRNA and carried by vectors, they are called **cDNA clones.** A group of cloned cDNAs form **cDNA library.** The following methods are used to construct cDNA libraries.

1. DNA Libraries from *m*RNA Copies

The Shotgun technique allows random insertion of foreign DNA fragment, hence the chances of implanting a desired gene is rare. Therefore, an indirect method is used for isolation of genes from eukaryotic genome. In this method, the *m*RNA is first transcribed from the gene and then DNA is obtained from this *m*RNA. Highly radioactive probe of RNA or DNA is used for this purpose. This probe can be hybridized with desired gene which can be identified through autoradiography.

2. Reverse Transcriptase Method (Fig 1.16)

In this method, first a specific *m*-RNA is obtained which is then utilized in the synthesis of duplex DNA (double stranded DNA). According to central dogma, normally the information flows as DNA→RNA→Protein, but **Temin** and **Baltimore** (1960) using enzyme *Reverse transcriptase* demonstrated that information flow is also possible from RNA to DNA (RNA-directed DNA Polymerase). Here, the information flows as *m*RNA→*c*DNA→*ds*DNA. This method includes the following steps.

(1) Purification of *m*RNA with a desired nucleotide sequence.

(2) Transcription of *m*RNA into complementary single strand of DNA by enzyme *reverse transcriptase*. Thus, the DNA formed is called *complementary DNA* (cDNA)

(3) Synthesis of *m*RNA–*c*DNA hybrid molecule by base pairing in between them. The enzyme *reverse transcriptase* engages *m*-RNA in 5' to 3' direction during this process.

(4) Removal of *m*-RNA primer from the hybrid molecule by alkaline treatment, leaving *c*DNA.

(5) Conversion of ss complementary DNA (ss *c*DNA) into a duplex DNA (*ds* DNA) by using *DNA polymerase* I of *E. Coli*. The duplex DNA contains hairpin loop at one end.

(6) Cutting of hairpin loop by enzyme S_1 *nuclease* and formation of a conventional duplex DNA. This DNA contains poly (dA) and poly (dT) tails to provide a free 3' end for extension.

(7) Splicing of *ds*DNA to the vector DNA.

(8) Cloning of *ds*DNA and production of multiple copies of *c*DNA which possess the *m*RNA sequence. Thus, *c*DNA library is formed.

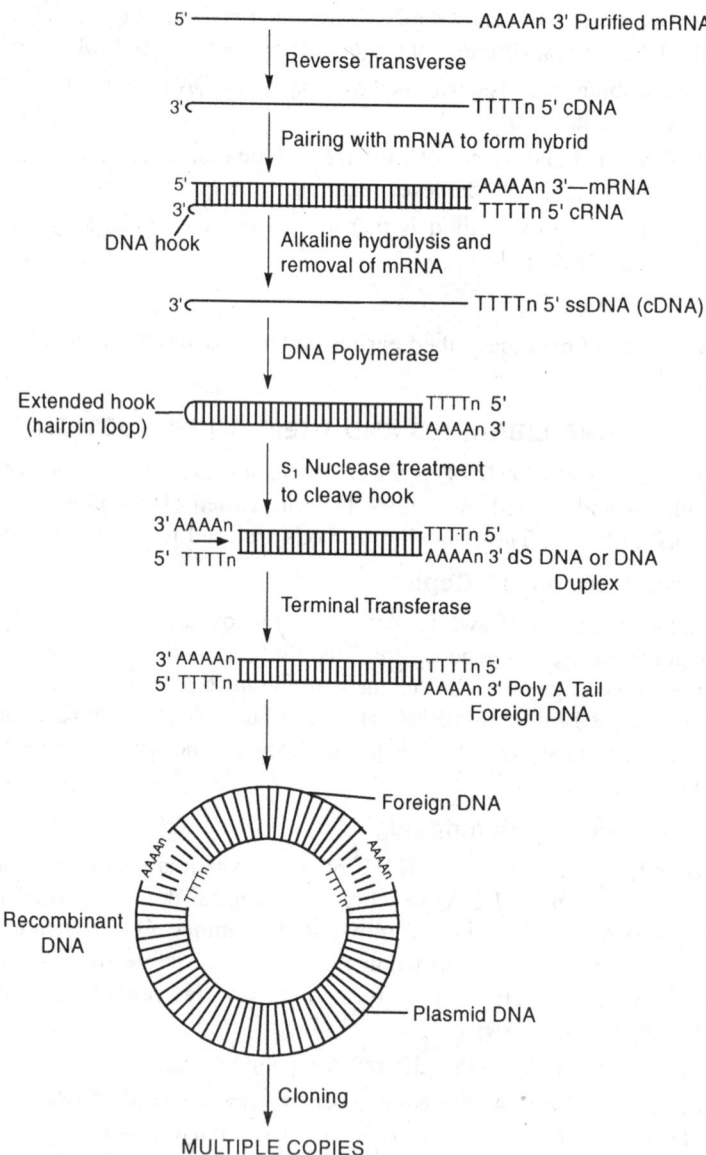

FIG. 1.16. Isolation of a specific gene by reverse transcriptase method and its Splicing to the vector dna to produce multiple copies.

GENOMIC LIBRARIES AND THEIR CONSTRUCTION

When the clones are derived directly from the genomic DNA and are carried by vectors, they are called *genomic clones*.

The cloned sets of fragments are called *genomic library*. For isolation of one or more related genes from a genome, special probes are needed which help in the identification of this gene sequence from the mixture of fragments. A series of clones carrying DNA derived from the genomic DNA directly can be isolated. The following methods can be used to construct genomic libraries–

Genetic Engineering

1. Genomic Library by Shotgun Technique

This method involves cutting of the entire genome into fragments of suitable size by restriction endonucleases. These fragments are then inserted in the DNA of a suitable cloning vector to produce a population of chimeric or recombinant DNA. It is followed by the insertion of recombinants into a host cell, generally *E. coli*. Now, each cell containing recombinant DNA is cultured separately to get cloned sets or fragments. These sets will form genomic library.

Once such a library is available, these clones can be stored indefinitely with either a phage or plasmid vector and can be retrieved with a suitable probe whenever needed for a variety of purposes, mainly for identification or isolation of a gene.

2. Hybridization Method

This method is used for isolation of genes that exist in multiple copies. This method is completed in following steps —

(1) Isolation of the entire DNA from the organism.

(2) Conversion of double stranded DNA (*ds* DNA) into single stranded DNA (*ss* DNA) by heat treatment or alkaline digestion.

(3) Production of DNA-RNA hybrid by mixing *ss*DNA and *m*RNA (transcribed by the gene.)

(4) Separation of *ss*DNA which contains original nucleotide sequence for transcription of *m*RNA strands.

(5) Conversion of *ss* DNA into *ds* DNA by the action of *DNA polymerase 1*.

METHODS TO PICK UP CORRECT DESIRED CLONE FROM A LIBRARY

A library can contain thousands or even more than ten thousands of various kinds of cloned DNAs. To pick up a particular clone (DNA) from the library is a difficult and time taking task. All the cloned DNA*s* may not have selective markers e.g., amp^r or tet^r genes for resistance to antibiotics ampicillin and tetracycline respectively. The selective markers help in the screening of colonies of cloned DNA*s*, plasmids or phasmids etc that have specific DNA inserts. The colonies of plasmids or phages can be screened by following methods —

(1) **Colony hybridization** — This method was developed by **Grunstein** and **Hogness**. Colony hybridization is most recently used method for screening of colonies for such plasmids or phages which contain specific DNA inserts. In this method, specific *radioactive probes* containing some complementary sequences to those of desired DNA are used. The colonies to

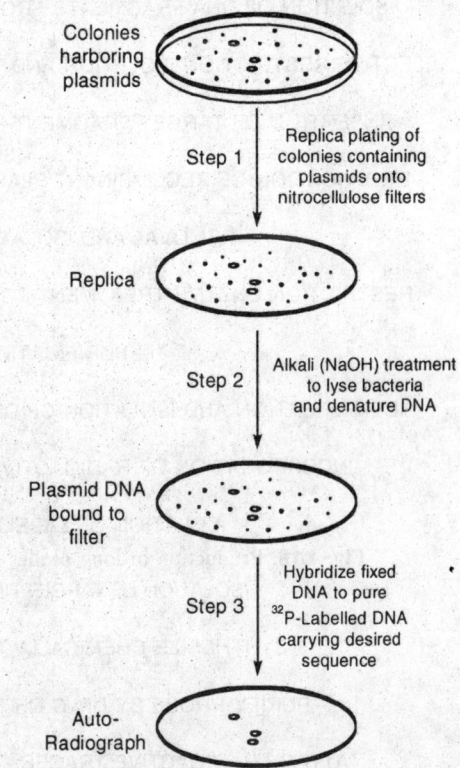

Fig. 1.17. Colony hybridization method for identifying bacterial clones harboring a plasmid containing a specific DNA.

be screened are first grown on Agar Petri plates. A replica of each plate is made on another Agar plate which is stored for reference. A replica is also prepared on a *nitrocellulose filter*. The colonies grown on the filter are lysed and the contents are treated with NaOH which causes denaturation of DNA present in the contents. The filter paper is heated so that the denatured DNA is fixed on the filter at each site where a colony was located. The DNA on the filter is then hybridized with a radioactively labeled DNA probe having complementary bases to the specific DNA sequence to be selected. The presence of hybridized probe at sites occupied by DNA, derived from colonies that include desired DNA fragment, is detected by autoradiography on film. The colony with hybridized DNA (colony DNA + DNA probe) can be picked up from the reference plate which contains a viable bacterial colony at a corresponding location.

Summary of Colony Hybridization Technique

The summary steps of isolation of a gene sequence (particular desired clone) from the genomic library by colony hybridization technique are as follows —

(1) Lysis of bacterial colonies carrying chimeric DNA on nitrocellulose filters.

(2) Denaturation of their DNA in *situ*.

(3) Fixation of DNA on the filter paper.

(4) Hybridization of filter paper with radioactive proves containing desired nucleotide sequences.

(5) Appearance of colonies carrying this sequence as dark spots after autoradiography.

(6) Recovery of original chimeric DNA from original reference library to be used for further experiments.

ISOLATION OF DNA FRAGMENTS FROM TARGET DNA MOLECULE
↓
SELECTIVE RESTRICTION ISOLATION AND TESTING OF TARGET FRAGMENT
↓
INSERTION OF TARGET FRAGMENT INTO A VECTOR (PLASMID)
↓
PROPAGATION OF RECOMBINANT PLASMID INTO HOST BACTERIUM
↓
REMOVAL AND ISOLATION OF DNA
↓
RESTRICTION ENZYME TREATMENT TO PRODUCE DNA FRAGMENTS
↓
PURIFICATION
↓
IDENTIFICATION AND ISOLATION OF DESIRED PROBE FRAGMENTS
↓
INCORPORATION OF RADIOACTIVE TRACER INTO PROBE
↓
ISOLATION OF LABELLED PROBE

Fig. 1.18. Production of long probes (schematic representation).

ISOLATION OF TARGET DNA SEQUENCE
↓
SYNTHESIZE CHEMICALLY MODIFIED PROBE
↓
PURIFY PROBE BY HPLC OR ELECTROPHORESIS
↓
ATTACH RADIOACTIVE TRACER OR REPORTER GROUPS
↓
ISOLATE LABELLED PROBE

Fig. 1.19. Schematic representation for production of short probes.

Genetic Engineering

DNA Probes

DNA probes are small pieces of radioactively lebeled DNA which help in detecting the presence of a particular gene or a long DNA sequence in most of the living systems. Thus, they provide means for diagnosis of infectious diseases, identification of food contaminant, isolation of genes and for several microbiological tests. The use of monoclonal antibodies can be replaced by such DNA probes.

Production of DNA Probes

DNA probes can be produced by any of the following three methods —

(i) By the action of purified biological enzymes on any DNA template.

(ii) With the help of automated DNA synthesizers. These probes contain specific sequence and are only 20 to 40 nucleotides long.

(iii) By inserting the DNA probe in viral or phage DNA (M_{13}) and then by introducing this phage in any bacterium where they multiply producing several copies of DNA probe.

Method of DNA Probe Assays

DNA probe assays consist of the following steps —

(i) Treatment of testing sample with detergents and enzymes to separate non-DNA components.

(ii) Denaturation of DNA. It is done at low pH.

(iii) Binding of single stranded DNA on filters and exposing it to vast excess of several DNA probes. It results in the hybridization of only one probe with single stranded DNA..

(iv) Separation of hybridized probe by using fluorescense and dyes etc.

(2) **Single plaque hybridization** — This method was developed by **Benton** and **Davis** to screen bacteriophage λ recombinant DNA clones in *situ*. The number of phage plaques that can be placed on a single petri plate is much larger than the number of individual bacterial colonies that can be placed on the same size plate (about 10,000 phage plaques versus about 200 bacterial colonies). Therefore, this method is useful especially for the screening of large libraries of eukaryotic DNA for genes which are present at a low frequency, *i.e.*, single copy genes.

(3) **By antibodies directed against the gene encoded protein** — This method is used in the primary screening of cDNA libraries. Antibodies directed against the gene encoded proteins are utilized in such cases where DNA probes are not available.

GENE CLONING TECHNIQUE

(a) **Cloning organisms** — Several strains of *E. coli.* such as HB 101, h 303 and RR_1 have been used as cloning organisms. *E. coli.* has been used largely as a cloning organism for most recombinant studies. **Fink et. al.,** introduced a bacterial gene for *Leucine* production into a yeast strain that was deficient in *Leucine* which showed an achievement of transplantation of gene from a lower organism to a higher organism. Yeasts are safer than *E. coli.* for genetic engineering as they may not prove to be pathogenic in man. The cloning organisms lack restriction enzymes and are also deficient in modification enzymes to protect DNA from degradation.

Some *E. coli.* strains that bud off mini cells have been found most desirable in transcription and translation of gene to be studied. The mini cells contain plasmids and other components required for gene expression, but lack chromosomal DNA. Since there is no interference from other DNA fragments, the mini cells provide better cloning organisms and in them the expression of recombinant DNA can be easily studied.

The genomes of eukaryotic organisms are made up of thousands of genes. The main purpose of cloning technique is to isolate the desired gene. For this purpose a specific probe is needed to isolate the particular sequence of bases or gene.

(b) Isolation of desired genetic material — It involves the successful extraction and purification of DNA from various sources. **Backwitch** and his colleagues of U.S.A. developed the extraction technique and used it in the isolation and purification of some genes of the *lactose operon* from bacterium *E. coli*.

We can isolate and clone single copy of a gene or a DNA segment into an indefinite number of copies which are all identical. If a foreign DNA is inserted in the DNA of a vector (vehicle or carrier), it will replicate with the parental DNA. The vectors which are used may be a bacteriophage or a plasmid. They reproduce in their own style even after the insertion of foreign DNA. This technique of isolation of single copy of gene and its transfer in other bacterium is called **gene cloning**.

(c) Gene cloning — What are clones ? They are descendants produced vegetatively or by apomixis from a single plant, asexually or by parthenogenesis from a single animal, by division from a single cell. The members of a clone are of the same genetic constitution except that mutation occurs amongst them. The gene clones are also descendants of the same single gene. In other words, the gene clone will be the copy of the particular gene. **Gene cloning** is the method of isolation of copies of a particular gene.

The cloning of gene (DNA) is possible only with the help of another replicable DNA molecule. This other replicable DNA molecule belongs to a vector which may be a plasmid or a bacteriophage or a derived cosmid. The cosmids are plasmid particles in which specific DNA sequence for *cos* sites is inserted. The cloning vector should always possess a site at which foreign DNA can be inserted without disrupting any essential function. A specific enzyme is also used to cause a single break in the vector DNA. At this point only, the foreign DNA is inserted.

(i) Gene cloning in bacteria — Gene cloning in bacteria is performed through vectors. As we know, the bacteria are of two types — (*i*) gram negative e.g., *E. coli, Rhizobium* and *Pseudomonas* and (*ii*) gram positive e.g., *Streptomyces sp.* and *Bacillus subtilis*. The plasmid vectors for these two groups differ because a particular plasmid can not maintain stability in both groups of bacteria. In gram negative bacteria the gene cloning can be performed by plasmids, cosmids and bacteriophages while in gram positive bacteria it is performed only by plasmids.

(ii) Gene cloning in Eukaryotes — Gene cloning in eukaryotes is difficult as they contain a well developed nucleus and split genes. Split genes are the silent genes or interrupted genes or intervening sequence of DNA. They are not represented in *m*RNA transcribed from the gene for protein synthesis. Some genes never possess a continuous sequence of nucleotides but always remain interrupted by some intervening sequence. Such genes with intervening sequence are called **split genes, silent genes or Junk DNA**.

Gene cloning in eukaryotes has been done in yeast, mouse and some higher plants. The yeast cells contain a plasmid called **2μ DNA**. It acts as a cloning vehicle. In mouse cells, special animal viruses called **Simian virus** 40 (*SV 40*), were used as cloning vehicle. In *SV 40* viruses, β *globin gene* could be integrated which could be transcribed and translated in mouse kidney cells. Vectors derived from *SV 40* DNA, which are capable of integrating cloned human DNA sequences back into chromosomes of human cells, have already been developed.

Among plants T-DNA (DNA transferred from *Ti* plasmid of *Agrobacterium tumefaciens* or from *Ri* plasmid of *A. rhizogenes*) and CMV (cauliflower mosaic virus) DNA are the best known vectors. T-DNA is very potential vector for cloning experiments with plants. CMV is of only restricted use due to its narrow range. The nitrogen fixing (*nif*) genes of *Klebsiella pneumoniae* are expressed in a mutant which lacks the ability to fix atmospheric nitrogen. They can be utilized in plants.

SOME IMPORTANT TECHNOLOGIES IN GENETIC ENGINEERING

1. Hybridoma Technology and Production of Monoclonal Antibodies

Hybridoma technology was developed by two German born immunologists, **G. F. Koehler** and **C. Milstein** in 1975. For this discovery Koehler and Milstein in 1984 shared the Nobel Prize alongwith **Neils Jerne**, for physiology and medicine. To produce a monoclonal antibody (MAB) from an immunized animal, first of all lymphocytes are isolated, then cultured in *vitro* and there after cloned out when they start producing antibody of interest. During this process, two difficulties were observed — (*i*) Culturing of normal lymphocytes which was not very successful and (*ii*) Providing an antibody producing line for continuous synthesis of antibody which could result into unlimited supply. Koehler and Milstein got success to overcome these difficulties. They demonstrated that somatic hybridization technique could be utilized to produce hybridoma lines that continuously secrete specific antibodies.

George Koehler and Milstein (1975) made the fusion of a single lymphocyte cell (β-lymphocyte) and a single bone marrow tumor cell (cancerous β-lymphocyte) to generate hybridoma cells. The two types of cells were isolated from different sources and they had entirely different characters. The β-lymphocyte cell was isolated from the spleen of mouse and it was immunized with red blood cell of sheep which has the ability to produce normal antibody, while bone marrow tumor cell or myeloma cell (cancerous β-lymphocyte cell) was isolated from bone-marrow which has the ability to multiply indefinitely. As a result of fusion, a hybrid cell was produced called **hybridoma**. Since, hybridoma cell was the fusion product of two cell lines, β-lymphocyte and myeloma β-lymphocyte, it showed the properties of both the parental cells *i.e.*, the hybridoma cell showed the normal antibody producing property inherited from β-lymphocyte and indefinitely multiplying property inherited from myeloma β-lymphocyte. Through this hybridoma cell, hybridoma lines were produced as a result of cell culture or continuous multiplication. These lines were called ***hybridomas*** and the entire technique to produce these hybridomas have been called as *hybridoma technology*. This technology has been utilized in the production of monoclonal antibodies.

Steps Involved in Hybridoma Technology

Hybridoma technology is related with the generation of hybridoma lines which are utilized commercially in the production of desired monoclonal antibodies for specific purposes. The hybridoma technology includes the following main steps —

(1) Immunization of source animal with antigen — The animal, from which the β-lymphocytes are to be isolated, is first immunized with an appropriate antigen. For immunization, the antigen is usually injected into the animal either subcutaneously or into the peritoneal cavity alongwith an stimulant to stimulate the immune system. The immunization of animal is done several times with a gap of a few days which helps in increasing the stimulation of β-lymphocyte cells responding to antigen. The final dose of antigen is given intravenously three days before killing the animal. The intravenously injected antigen ensures high dose in the animal. After three days of final dose of antigen, the immune-stimulated cells grow rapidly which help in their selection and isolation.

(2) Killing of animal and isolation of lymphocytes : Three days after giving the final dose of antigen intravenously, the animal is killed and its spleen is removed aseptically. Now, the spleen is disrupted gently to release the spleen fluid containing lymphocytes and red blood cells. The lymphocytes are separated from red blood cells by density gradient centrifugation method and finally washed.

(3) Selection of myeloma cell line for myeloma β-lymphocytes — A precaution must be taken during this step because any myeloma cell can not be allowed to fuse with normal β-

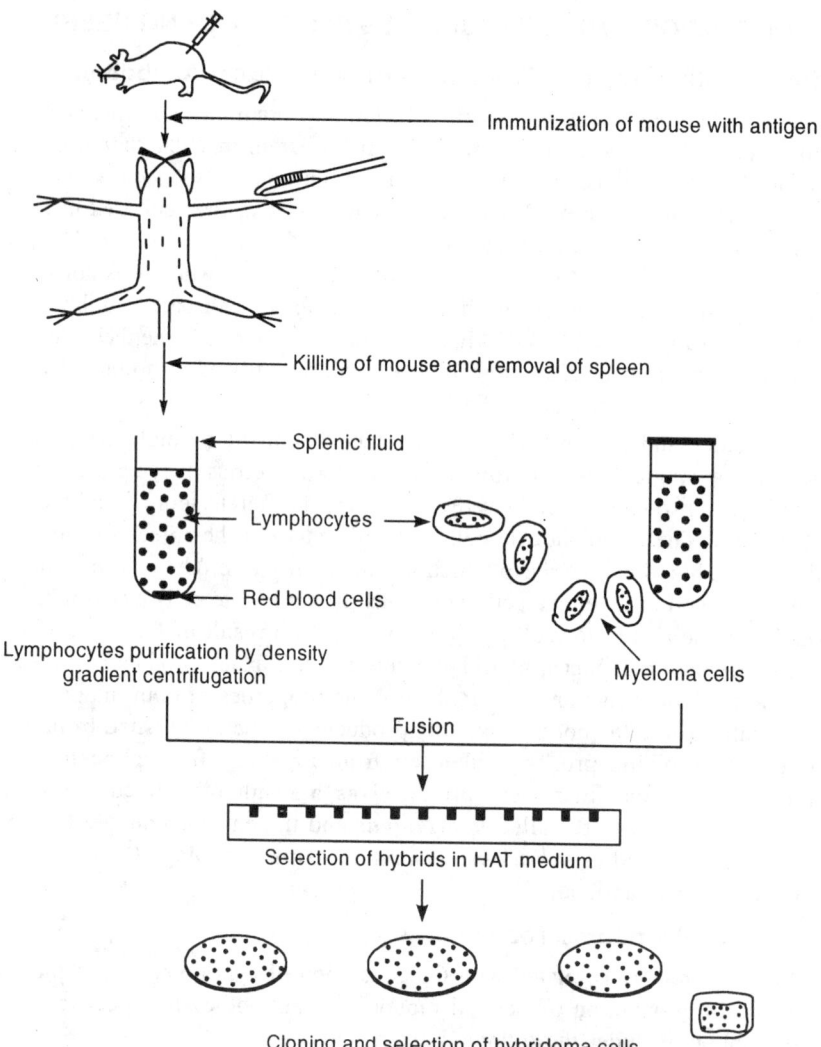

Fig. 1.20. Steps involved in hybridoma technology.

lymphocyte. Only such myeloma cell line should be used which itself does not synthesize any antibody, otherwise hybridoma cell line will produce a mixture of antibodies, produced by myeloma cell (myeloma β-lymphocyte) and normal β-lymphocyte cell both. It should be noted that in most cases of multiple myeloma, there is overproduction of a particular type of antibody. There are a number of cases where a particular myeloma cell line multiplies continuously without producing any antibody, *e.g.,* Hypoxanthine Phosphoribosyl Transferase (HPRT) negative myeloma cell line.

What is HPRT-Negative Myeloma Cell Line ?

When myeloma lymphocytes are grown in **β-azaguanine solution**, some of them are killed and others survive. The surviving myeloma lymphocytes are, thus, resistant to β-azaguanine. This property is due to defect in enzyme, *hypoxanthine phosphoribosyl transferase (HPRT),* present in these resistant lymphocytes. Due to this resistant property against β-azaguanine, such myeloma lymphocytes are called ***HPRT-negative myeloma cell line.*** This test is done before any myeloma lymphocyte is fused with normal lymphocyte so that desired monoclonal antibody could be produced from hybridoma cells.

Genetic Engineering

(4) Fusion of isolated β-lymphocytes and myeloma β-lymphocytes — The washed β-lymphocytes are mixed together with myeloma β-lymphocytes of HPRT-negative cells line to allow fusion. Polyethylene glycol (**PEG**), a strong fusion promoting chemical, is used for this purpose. It is added in the mixture of both the type of cell lines. Because PEG is cytotoxic in nature, it is added in appropriate amount and the cells are placed in it only for a short duration. It causes the fusion of only a few cells. Now the cells are collected and washed thoroughly to make them PEG free. The washed cells comprise a mixture of hybrid (hybridoma) cells, unfused β-lymphocytes and unfused myeloma β-lymphocytes.

(5) Isolation of hybridoma cells from the mixture — The isolation of hybridoma cells from the mixture of hybridoma cells, unfused β-lymphocytes and unfused myeloma β-lymphocytes is done by using a mixture of hypoxanthine, aminopterin and thymidine. All these three chemicals jointly form the growth medium, called **HAT-medium**. When mixture of above three type of cells is grown in HAT-medium, only hybridoma cells grow successfully but unfused β-lymphocytes and myeloma β-lymphocytes fail to grow.

(6) Isolation of monoclonal antibody producing hybridoma cell — This is the final step where a single monoclonal antibody producing hybridoma cells are isolated from the mixture of hybridoma cells, selected through HAT-medium. If all the hybridoma cells isolated through HAT-medium are grown together,

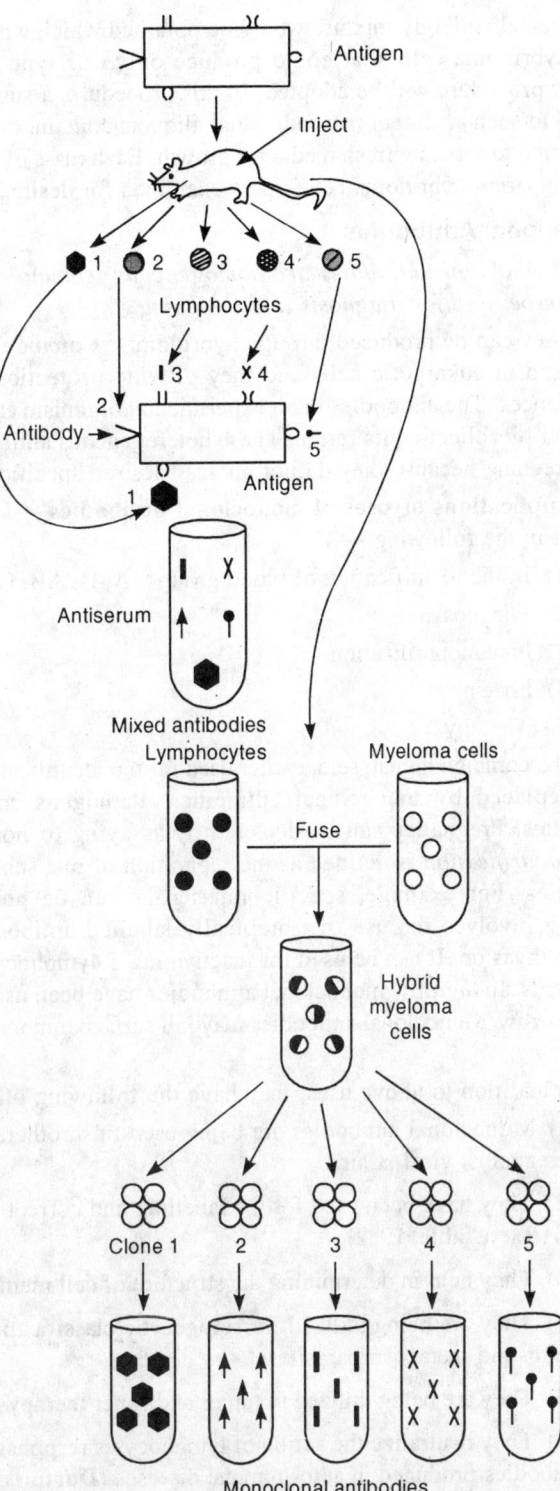

Fig. 1.21. Various steps in production of monoclonal antibodies.

a polyclonal antibody mixture would be obtained which will be of no use. As our aim is to isolate such hybridoma cells that could produce only one type of antibody (monoclonal), a slightly lengthy procedure will be adopted. In this procedure, a suspension of isolated hybridoma cells is diluted to such an extent that individual aliquots contain, on average, only one cell. Such cells are transferred to separate fresh media for growth. Each mass of hybridoma cells (clone) produced from a single parent hybridoma cell is now examined for desired monoclonal antibody.

Monoclonal Antibodies

Monoclonal antibodies are homogeneous immunological reagents of defined specificity and can be used for diagnosis and screening.

They can be produced through hybridomas. Commonly occurring heterogeneous antibodies are found in eukaryotic cells and they provide protection against diseases and other external disturbances. The antibodies in an experimental organism can be produced by injecting an antigen and later by collecting its serum. These heterogeneous antibodies can not be utilized for diagnosis and screening because they do not possess desired specificity.

Applications or uses of monoclonal antibodies — The monoclonal antibodies have been utilized in the following —

(1) In the identification of blood groups (A, B, AB, O).

(2) Diagnosis.

(3) Immunopurification.

(4) Imaging.

(5) Therapy.

The common human sera, earlier used for the identification of A, B, O, AB blood groups, have been replaced by monoclonal antibodies. Pathogens can be detected by the use of such antibodies. Pregnancy can be detected by assaying of hormones with monoclonal antibodies. *Immunopurification* is defined as the separation of one substance from a mixture of very similar molecules. For example, specific interferons can be purified using monoclonal antibodies. *Imaging* involves the use of isotopically labelled antibody to provide an image of an organ without invasion. It can be used for inactivating T-lymphocytes responsible for rejection of organ transplants. In *therapy,* monoclonal antibodies have been used for the removal of tumor cells from bone marrow. Monoclonal antibodies may kill surface tumors and may neutralize the effects of toxic drugs.

In addition to above uses, they have the following other applications —

(1) Monoclonal antibodies are being used in serotherapy and in the preparation of specific vaccines against viral strains.

(2) They have been used for the labelling and correct identification of specialized cells like neurons (**Barnstable**, 1982).

(3) They help in determining the structure of cell membranes (**Milstein**, 1981).

(4) They are being utilized for serogenetic classification of infectious micro-organisms and protozoans and metazoan parasites.

(5) They are being utilized in tumor and caner therapy (**Bomen** and **Fathman**. 1981).

(6) They neutralize the action of lymphocytes responsible for rejection of grafts and destroy autoantibodies produced in auto-immune diseases (**Durum** *et. al.,* 1985).

(7) They help in purification of those enzymes which can not be purified by biochemical methods (**Golfri** and **Milstein**, 1981).

Genetic Engineering

(8) They are being utilized in testing disease susceptibility, drug resistance, sexually transmitted diseases, drug monitoring and genetic screening etc.

2. Blotting Techniques

(1) Southern Blotting — The transfer of DNA fragments from gel to nitrocellulose paper is called *Southern blotting*. It was developed by a British scientist **Dr. E.M. Southern** and named after him in 1975. This technique is analogous to produce xerox copy of a printed page. In this technique, the **agarose gel** containing DNA fragments is placed on a filter paper soaked in **sodium saline citrate** (SSC) solution which is generally used in the preparation of DNA solutions. Now, **nitrocellulose paper strip** is placed over the gel. It is further covered with dry filter paper sheets. Finally a weight is placed over sheets. The SSC solution passes from filter paper to nitrocellulose paper through gel and carries DNA fragments from gel with it. Thus, most of the DNA fragments get transferred from gel to nitrocellulose paper and a few DNA fragments may also pass to the dry filter paper sheets as waste. Its figure is given under DNA finger printing heading.

(2) Northern Blotting — This technique was used to get bands of *mRNA*. Initially, the Southern blotting technique (used for DNA transfer from agarose gel to nitrocellulose paper) was applied for blotting of *mRNA* but it was not successful. To overcome this difficulty another technique was used where *mRNA* fragments were not transferred on to a nitrocellulose paper but they were transferred on to a chemically reactive paper, prepared by *diazotization of aminobenzyloxymethyl paper*. This technique in contrast to southern blotting was named as *northern blotting* although it is not related to the name of any scientist as is in the case of southern blotting.

Later, it was shown that *mRNA* bands can be blotted directly on to nitrocellulose paper. The paper-blotted bands can be hybridized with a labelled DNA or RNA probe. The single stranded regions of the probe are removed by nuclease enzymes like *mungbean nuclease* or *S-I nuclease* to make quantitative estimation of hybridized *mRNA*.

(3) Western blotting — This technique is used to detect the proteins of a particular specificity. The translated proteins can be identified by this technique when a transferred gene expresses in transformed cells forming such proteins.

The translated proteins are first extracted and then subjected to *polyacrylamide gel electrophoresis (PGE)* and then transferred on to nitrocellulose paper. The probing on nitrocellulose paper is done by specific labelled antibodies which do not hybridize with protein as in case of southern or northern blotting but bind with them. An antibody may be labelled with ^{125}I. Now an autoradiograph is prepared. If radioactive label is not used, bound antibody may be detected by a second antibody tagged with an enzyme.

3. DNA Finger Printing (DNA Profiling) and Southern Blotting

Principle of DNA Finger Printing — The technique of DNA finger printing also known as *DNA profiling* was developed by **Professor Alec Jeffreys of Leicester university** in 1984. In India, DNA fingerprinting tests are carried out at centre for **cell and Molecular Biology** (CCMB), Hyderabad. The paternity dispute cases in India are referred to CCMB for DNA evidence. The first such test on DNA fingerprinting was applied in 1989 to settle down a disputed paternity case in Madras. Monika Lewinsky (a affair case of Bill Clinton) and Madumita murder case have been solved through DNA fingerprinting technique. The technique is based on the facts that —

(1) DNA is unique to each individual.

(2) Half of the DNA of each individual is derived from his mother and the other half from his father.

(3) No two persons in this world have identical DNA except indentical twins.

What can be known from DNA Finger Printing ?

DNA finger printing has been hailed as the greatest single break through in forensic science this century. This technique has been utilized in the analysis of disputed parentage, to catch a rapist where a man rapes female, murders her and then bolts away from the scene or any other murderer in murder case. Disputed parentage can be confirmed by analysing and comparing the DNA from the blood samples of child, father and mother. A rapist can be caught by analysing the DNA from seminal fluid (spermatozoa) from the body of victim and then matching this DNA with the DNA taken from all suspects. Similarly a murderer can be identified by analysing the DNA from blood spots or rooted hairs obtained from the spot of murder and then matching this DNA with the DNA taken from all suspects. Thus, blood samples, seminal fluid, blood spots and rooted hairs are chiefly utilized as tools to analyse DNA to solve above problems.

Amount of DNA analysed in DNA Finger Printing

The only problem before 1984 was that there was no way one could 'see' and 'match' the DNA of different individuals. Alec Jeffreys made it possible for the first time in 1984. The DNA which is a double helical structure is found in almost all human beings. Each human cell, except the red blood corpuscles which do not contain a nucleus, contains about 5 picograms (5×10^{-12} g) of DNA. An average human being contains about 50 trillion cells (when the power of a million is multiplied itself by 3 times), so every human being contains almost 250 pg of DNA which is absolutely unique to him. Out of this, DNA finger printing is done from as little as micrograms.

Method of DNA Finger Printing or to See DNA

In DNA finger printing actually restriction fragment length polymorphs (RFLPs) are analysed and compared with suspects.

It involves several steps —

(1) Extraction of DNA from sample tissue — First of all, a sample of tissue from which the DNA finger print is to be prepared is taken. This could be the blood of disputed child or

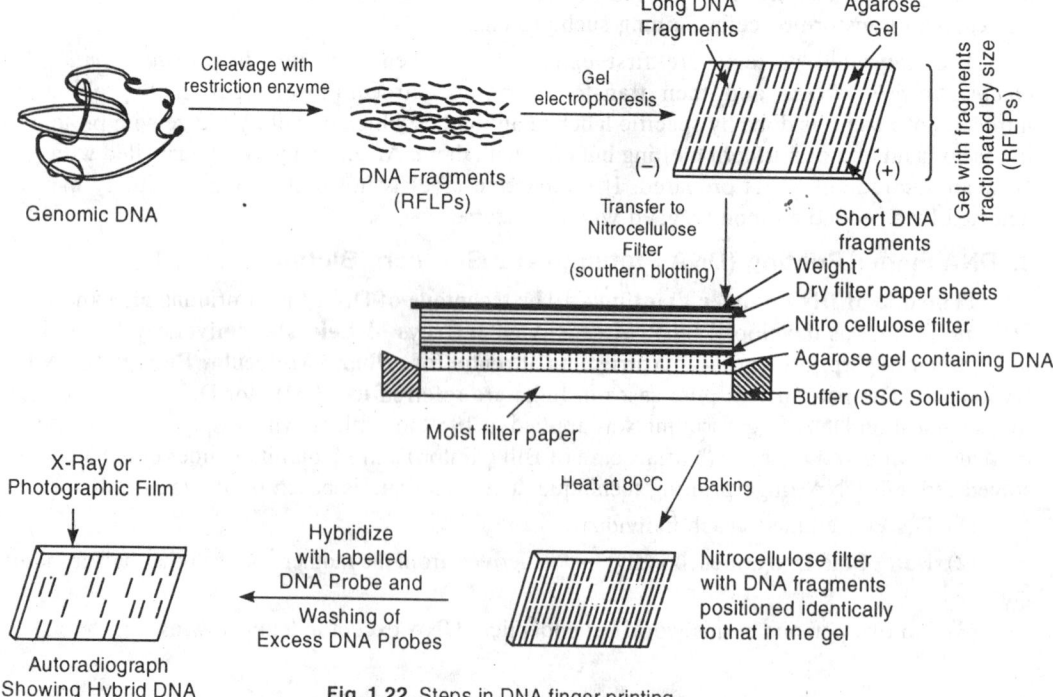

Fig. 1.22. Steps in DNA finger printing.

Genetic Engineering

seminal fluid from the body of the raped woman or a scalp hair from the assailant. From these samples, the genomic DNA is extracted using suitable techniques.

(2) **Production of DNA fragments** — The extracted DNA is treated with restriction enzymes to cut and produce fragments of unequal length (RFLPs). Restriction enzymes are special group of enzymes, normally present in bacterial cells, which cut the DNA only at specific sites. Since each person has a different arrangement of bases in his DNA, the resulting cuts at specific sites by restriction enzymes result into fragments of DNA of unequal length in each person. Since the DNA in each cell of a person comprises as many as 6 billion base pairs, the probability of any two people getting identical fragments is almost zero. This is the basic principle behind DNA finger printing. The above resulted RFLP fragments, are not visible with the help of naked eyes or by microscopic examination. To make these fragments visible a number of following steps are undertaken.

(3) **Separation of DNA fragments** — The separation of DNA fragments (RFLPs) is done by gel electrophoresis technique in the form of invisible bands on gel. During electrophoresis the fragments are put over agarose gel and an electric current of high voltage is applied. Since the fragments have weak negative charge, they begin to move. The larger and heavier fragments move slower while the smaller and lighter fragments move faster. After sometime all the fragments get separated according to their lengths. Since the fragments produced from the DNA of each person are different in length, this fragment or band pattern is also different in each individual. These fragments are still invisible.

(4) **Transfer of DNA fragment from gel to nitrocellulose paper or Southern blotting** — The transfer of DNA fragments from gel to nitrocellulose paper is done by the technique known as *Southern blotting*. It was developed by a British scientist **Dr. E.M. Southern** and named after him in 1975. This technique is analogous to produce xerox copy of a printed page. In southern blotting technique, the agarose gel, containing invisible DNA fragments, is placed on a filter paper soaked in sodium saline citrate (SSC) solution which is also used in the preparation of DNA solutions. Now, a nitrocellulose strip is placed over the gel. It is further covered with dry filter paper sheets and a weight is placed over sheets.

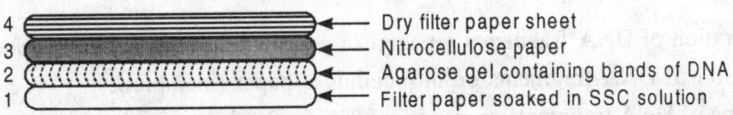

Fig. 1.23. Position of gel and nitrocellulose paper etc.

The SSC solution passes from filter paper to nitrocellulose paper through gel and carries DNA fragments from gel with it. Thus, most of the DNA fragments get transferred from gel to nitrocellulose paper but a few DNA fragments may also pass to the dry filter paper sheets as waste. The nitrocellulose paper is baked at 80°C to fix DNA fragments.

(5) **Preparation of Radioactive DNA probe** — These DNA probes are special type of DNAs having definite sequence of bases and are used in the identification of specific genes or sequence of bases of other DNAs. A DNA probe can be compared with a dedicated sticker which sticks only to a specific DNA fragment like sticker at a fixed position or place on a paper. Each laboratory in the world has developed its own DNA probes. This is an area which is jealously guarded by each laboratory. The probes developed by Professor Alec Jeffreys are also patented and are thus not available to India. The DNA probes in India were developed by **Lalji Singh** from the DNA of the females of the banded krait (*Bungarus fasciatus*), a poisonous Indian snake. He named his own probes as **Bkm** probes (from the initials of banded krait minor satellite). This is a highclass contribution of **Lalji Singh** to India.

(6) Hybridization of DNA fragments with DNA probes on nitrocellulose paper — Hybridization of DNA fragments on nitrocellulose paper is performed by prepared radioactive DNA probes. For this purpose radioactive DNA probes in liquid medium are incubated in the nitrocellulose paper containing DNA fragments. The DNA probes look like a liquid which stick to specific DNA fragments only forming hybridized DNA molecules. Excess DNA probes are then washed off. Now the nitrocellulose paper contains separated normal DNA fragments according to their size alongwith hybridized DNA fragments (DNA fragments + DNA probes). The DNA fragments are however still invisible.

(7) Detection of hybrid DNA fragments — Hybridized DNA fragments can be detected through autoradiography. For this purpose an X-ray plate is placed next to or in front of nitrocellulose paper and is left for 2 to 3 days. The radioactive rays from the DNA probes cause chemical reaction in the X-ray plate just like light causes a chemical reaction on a photographic plate. When the X-ray plate is developed after three days, we get the DNA finger print of RFLPs.

Determination of Results

For determination of parentage of a child, we make use of the fact that half of the DNA of a child comes from his mother and the other half from his father. The half of the bands in the child's DNA finger print should correspond with those of its mother and the other half with those of its father. If some of the bands in a child's finger print do not correspond with either the alleged mother or the alleged father, it is certain that they are not its real parents.

In case of rape and murder, a DNA finger print is prepared from the recovered semen and another print from the suspect. If the bands match exactly, it is 100% proof that the semen indeed came from the suspect. If the bands do not match, we can say with 100% confirmation that the suspect is not the criminal.

Summary of DNA Finger Printing

The DNA finger printing can be summarized in the following steps —

(1) Digestion of DNA sample with restriction enzymes to form its fragments of different size.

(2) Separation of DNA fragments on agarose gel plate through gel electrophoresis.

(3) Transfer of DNA fragments on nitrocellulose paper (membrane).

(4) Baking of DNA fragments on paper at 80°C for 2-3 hrs to fix the fragments permanently on the nitrocellulose paper.

(5) Hybridization of DNA fragments with labelled DNA or RNA probe.

(6) Exposing of nitrocellulose paper with hybridized DNAs on X-ray film to get autoradiograph. This technique is used to get DNA bands.

4. Polymerase Chain Reaction (PCR) Technique

Polymerase chain reaction (PCR) technique is used to obtain millions of copies of a DNA segment of choice. It was developed by **Kary Mullis** at Cetus Corporation in 1985 and for this discovery of PCR, he shared Nobel Prize in 1993 in Chemistry alongwith **Michael Smith** for site directed mutagenesis, PCR is now popularly known as *Peoples Choice Reaction*. The enzyme commonly used in PCR technique is a thermostable DNA polymerase, called *Taq DNA polymerase* (obtained from a thermophilic bacterium, *Thermlus aquaticus*). Vent polymerase is another thermostable enzyme which is used in PCR technology. It is obtained from bacterium, *Thermococcus litoralis*.

Initially PCR technology was used in the amplification of DNA but now-a-days it is also being used in the amplification of RNA. During RNA amplification *rTth DNA polymerase* is used instead of *Taq polymerase*. This enzyme first transcribes RNA to DNA and then amplifies the

Genetic Engineering

DNA. Due to this reason cellular RNA and RNA viruses can be studied even if they are present in small quantities. The PCR technique has now been automated and is carried out by a specially designed machine. It can carry out 25 cycles and amplify DNA 10^5 times in 75 minutes.

Steps Involved in PCR Technology (Method) — The following steps are involved during DNA amplification through PCR technology —

(1) Melting of Target DNA — Target DNA is that DNA which is to be amplified. Its melting is done through *thermal denaturation*. The target DNA with bases between 100 and 5000 is used for amplification. Thermal denaturation of target DNA is done by heating it around 94°C for 15 seconds. It results into separation of both the complementary strands. It is step I and this process is called *melting of target DNA*.

(2) Addition and annealing of primers — Oligonucleotide primers are added to the denatured strands of DNA for annealing.

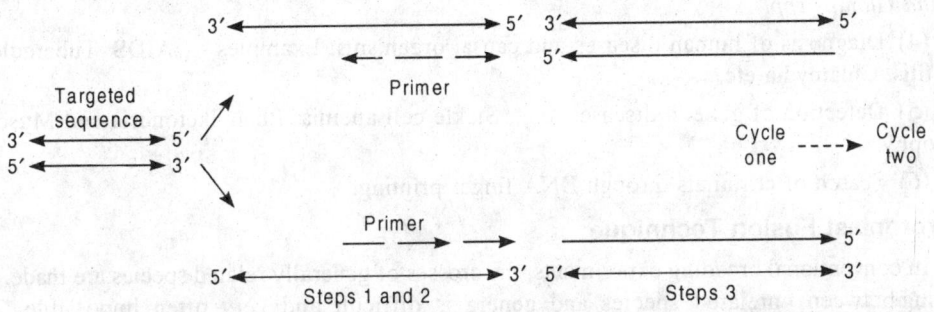

Fig. 1.24. Steps in PCR technique. Cycle-2 follows the steps of cycle one.

Such primers are added in excess and the temperature is lowered down from 94°C to about 68°C for one minute. It results into formation of hydrogen bonds between primers and DNA strands. Thus annealing of two oligonucleotide primers to the two separated strands of DNA (one primer to each DNA strand) takes place. It is step 2.

(3) Primer extension — The third step involves the addition of four nucleoside triphosphates *i.e.*, dATP, dGTP, dCTP and dTTP and a thermostable DNA polymerase (*Taq DNA polymerase* or *Vent polymerase*) in the reaction mixture. The DNA polymerase enzyme accelerates the polymerization process of primers. This results into extension of primers at 68°C. The extended primers synthesize the copies of target DNA. Thus, step 3 is completed.

All the above three steps constitute one cycle and after the completion of this cycle or three steps, the targeted sequences are copied on both the strands. Ultimately four strands are produced. Now, this three step cycle is repeated which forms eight copies from 4 strands. Similarly, third cycle will produce 16 strands and fourth cycle 32 strands. The cycle is usually repeated about 50 times and it is supposed theoretically that 20 cycles of three steps each will produce about one million copies and 30 cycles about one billion copies of the target DNA sequence. In each cycle, the newly synthesized DNA strands serve as targets for subsequent DNA synthesis.

Quantity of Target DNA and Primers Required in PCR technique

In PCR technology the required concentration of target DNA is about 10^{-20} to 10^{-15} M or 1 to 10^5 DNA copies per ml. The required quantity of primers in PCR is about 10-100 picomoles.

Applications of the PCR Technology

PCR technology has been applied extensively in the area of molecular biology, biotechnology and medicines for the following purposes —

(1) For the amplification of DNA and RNA.

(2) Determination of orientation and location of restriction fragments related to one another.

(3) Diagnosis of plant diseases and pathogens. A number of plant pathogens like viroids, viruses, bacteria, mycoplasmas and fungi from various hosts or environmental samples have been detected by using PCR technology. The examples *of* various pathogens detected by PCR are as follows —

(*a*) **Viroids** — The viroids associated with apple, pear, grapes and citrus etc.

(*b*) **Viruses** — *TMV, Cauliflower mosaic virus, Bean yellow mosaic virus, Potyviruses and Plum pox virus.*

(*c*) **Bacteria** — *Agrobacterium tumifaciens, Rhozibium leguminosarum, Pseudomonas solanacearum and Xanthomonas campesiris etc.*

(*d*) **Fungi** — *Phytophthora spp., Colletotrichum gloeosporioides, Laccaria spp, Verticillium spp and Glomus spp.*

(4) Diagnosis of human diseases and causal organisms. Examples — AIDS. Tuberculosis, Hepatitis, Chlamydia etc.

(5) Detection of genetic diseases. *e.g.*, Sickle cell anemia, Phenylketonuria and Muscular dystrophy.

(6) Search of criminals through DNA finger printing.

5. Protoplast Fusion Technique

In conventional breeding experiments, the crosses of generally related species are made. The crossing between unrelated species and genera is difficult and very often impossible. This difficulty has been overcome by the discovery of a new technique of *protoplast fusion.*

The cell wall of bacteria, fungi, yeast and other plant cells can be removed under proper conditions without affecting the integrity and viability of the cell. When the cell-wall of any cell is removed, an unit surrounded by the plasmamembrane is obtained. When this unit is kept in a hypertonic media, it becomes spherical in shape and is called *protoplast*. Thus, the spherical protoplasmic structure surrounded by the plasmamembrane in hypertonic media will be called *protoplast*. When these protoplasts are provided suitable conditions, they can resynthesise their cell walls reverting to their original form. When two different types of protoplasts are mixed under laboratory conditions, they fuse to form a hybrid protoplast. Such phenomenon is called *protoplast fusion.*

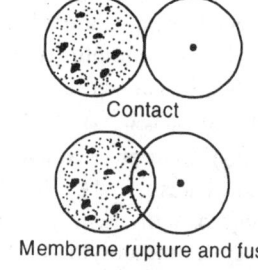
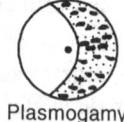

Fig. 2.25. Protoplast fusion and production of somatic hybrid cell.

Polyethylene Glycol (PEG)

Kao and **Michayluk** (1974) reported that polyethylene glycol (PEG) is a very active chemical which can be used for the induction of protoplast fusion. Since then it has been used successfully to induce fusion of protoplasts in a number of microbial systems.

PEG preparations with mean molecular weights between 1000-6000 are suitable for cell fusion. Outside of this range they have an unacceptably high toxicity. The optimum concentration of PEG is normally between 30-50% and the time for which cells are exposed to PEG at this concentration is usually short (1 minute). It minimizes the membrane damaging effects of the fusing agents. The

Genetic Engineering

cells are then gently resuspended, diluted and washed to remove the traces of PEG. The entire operation is conducted at 37°C.

Formation of Recombinant Micro-organisms Through Protoplast Fusion

During the formation of recombinant micro-organisms through protoplast fusion, following three processes take place in the fused parental protoplasts (**Alfoldi**, 1981).

(*i*) Cytoplasmic interaction; (*ii*) DNA-DNA interaction, and (*iii*) reversion to microbial form.

When the plasmamembranes of the two parental protoplasts are disrupted at the union point, the contents of the protoplasts mix freely with each other showing cytoplasmic interaction, DNA-DNA interaction and formation of recombinant micro-organism. In fused protoplasts, there may be interaction between whole genomes via complementation or recombination or both. Special care should be taken to maintain the fine structure and integrity of each of cytoplasm.

Examples of Protoplast Fusion

Protoplast fusion was reported for the first time by **Harris** and **Occo-Watkin** (1965) of Oxford University. They showed that the cells of mouse and man can be made to fuse to form hybrid cells. Their work indicated that the cells from different animal species could be fused together to form viable hybrids. Similarly, the cells from different plant species could also be fused together. Since then protoplast fusion has been reported from different animals but so far no hybrid animal could be produced by this technique because hybrid cells failed to differentiate under culture conditions. Among plants, two well known interspecific hybrids have been reported by protoplast fusion technique. One hybrid was produced by fusing the protoplast from leaf cells of *Nicotiana glauca* and *Nicotiana longsdorfii*. The hybrid cell was cultured on a selective medium for the formation of callus and its differentiation into a new mature plant. This fusion was achieved by **Carlson et. al.,** (1972). The other plant hybrid by protoplast fusion technique was developed from two different species of *Petunia* by **Power et. al.,** (1976).

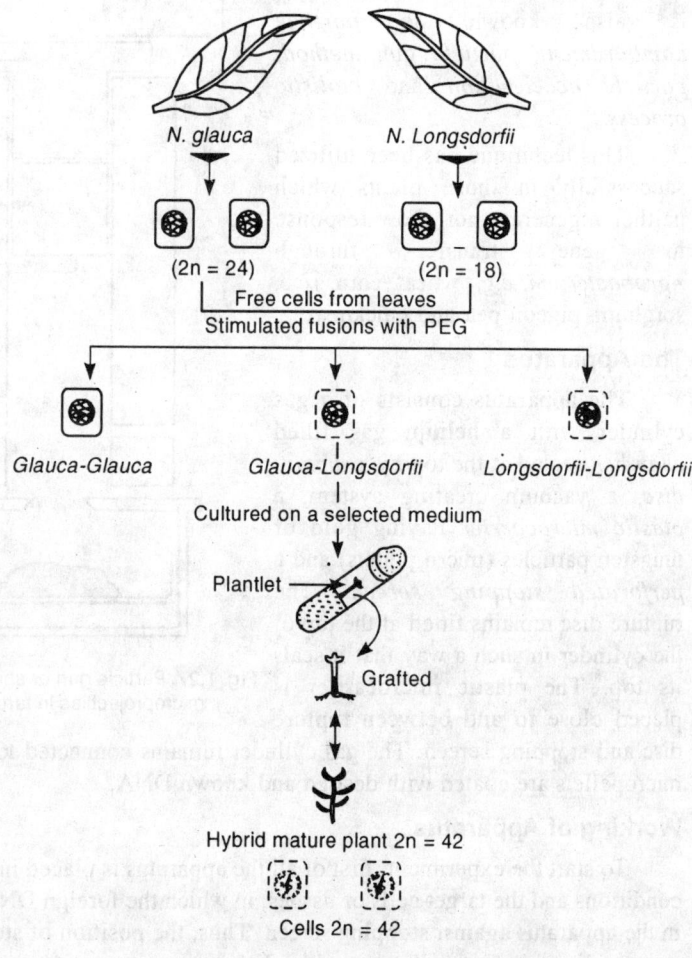

Fig. 1.26. Protoplast fusion in *Nicotiana* species.

Recently, the fusion of protoplast from plants cells and animal cells jointly has also been reported. Cocking *et. al.,* (1975) reported the successful fusion of yeast protoplasts with hen erythrocytes. **Dudits** *et. al.,* (1976) have reported the fusion of protoplast from human cells with carrot cells. It will be interesting if the plant components of these hybrids bring about differentiation.

Protoplast fusion technique has opened the possibility of overcoming the sexual barriers and of mixing and reasoning the genetic elements of hitherto sexually isolated organisms. It offers many exciting vistas for the genetic manipulation of plants. Hybridization through protoplast fusion is known as **parasexual hybridization** because it does not involve fusion.

6. Techniques of Gene (DNA) Transfer or Transformation techniques

(1) Microprojectile Bombardment or Particle Bombardment Gun or Biolistic Method — The foreign DNA can be directly delivered to the target cells at a very high speed through this method.

It was developed by professor *Stanford* and his colleagues of Cornell university, USA in 1987. This technique is also known as *particle bombardment, particle gun method, particle acceleration* and *biolistic process.*

This technique has been utilized successfully in those plants which neither regenerate nor show response to gene transfer through *Agrobacterium, e.g.,* wheat, corn, rice, sorghum, pigeon pea and chickpea.

The Apparatus

The apparatus consists of a gas cylinder with a helium gas filled *chamber,* sealed at the top by a *rupture disc,* a vacuum creating system, a *plastic microcarrier* having gold or tungsten particles (micro pellets) and a *perforated stopping screen.* The rupture disc remains fitted at the top of the cylinder in such a way that it seals its top. The plastic microcarrier is placed close to and between rupture disc and stopping screen. The gas cylinder remains connected to a vacuum creating system. The micropellets are coated with desired and known DNA.

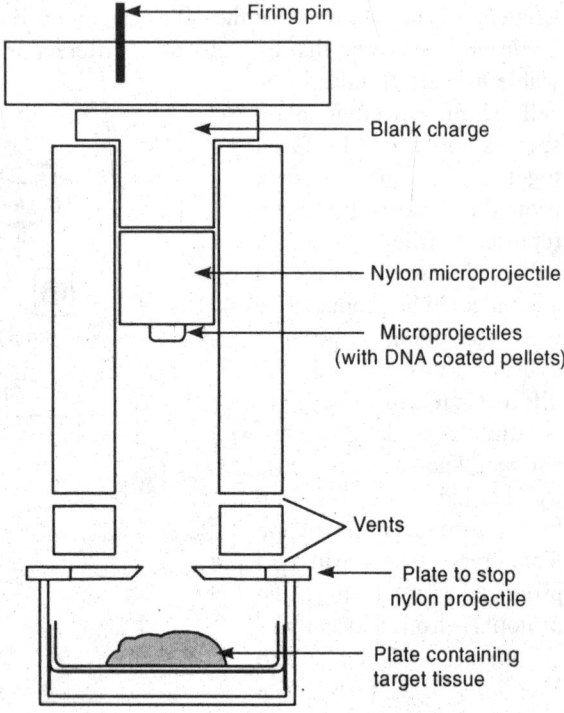

Fig. 1.27. Particle gun or shotgun for delivering DNA coated microprojectiles in target cells through firing pin.

Working of Apparatus

To start the experiment, first of all the apparatus is placed in Laminar flow to maintain sterile conditions and the target cells or tissues, in which the foreign DNA is to be transferred, are placed in the apparatus against stopping screen. Thus, the position of stopping screen will be in between target cells and micropellet assembly. Now vacuum creating system is started and helium gas is flown in the cylinder at high compressor (about 130 kg cm^2 pressure) velocity. When the pressure of the cylinder exceeds the bursting point of rupture disc, it gets ruptured and helium gas comes out of the rupture disc with high pressure waves which pushes the plastic microcarrier containing DNA coated micropellets with a shock. The stopping screen being perforated allows the micropellets to pass through and deliver DNA of the pellets into target cells or tissues. The DNA of micropellets is delived to target cells with a pressure. The transformed cells or tissues are regenerated on to nutrient medium and are screened (selected) through culture media containing antibiotics or herbicide. The selected plants, developed from such tissues, are tested for expression of desired foreign DNA.

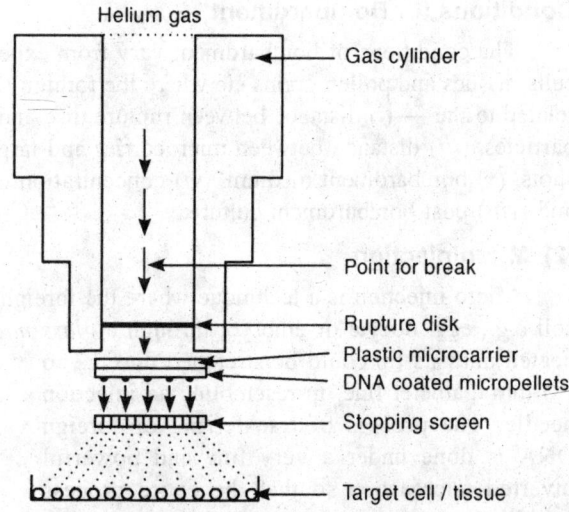

Fig. 1.28. Microprojectile or Particle bombardment.

The working of both the types of guns (shown in Fig. 1.27 and 1.28) is similar.

The successful delivery of foreign DNA through microprojectile bombardment had been achieved in the *epidermal tissues* of onion (*Allium cepa*), *scutellar tissues* of maize and leaf and cell culture of several crops (**Peters, 1993**). Successful transformation by this technique has also been achieved in bacterial cells, algae, fungi, cell-organelles, fruitfly embryos and human cells. **Ramaiah and Skinner (1997)** have got success in producing transgenic *Alfalfa* plants through direct delivery of DNA into pollen grains by microprojectile bombardment technique (Fig. 1.29). The scientists of Plant transformation group at ICGEB, New Delhi, (1998) have achieved success in transforming interferon gamma gene into chloroplasts of maize and tobacco etc. through this technique.

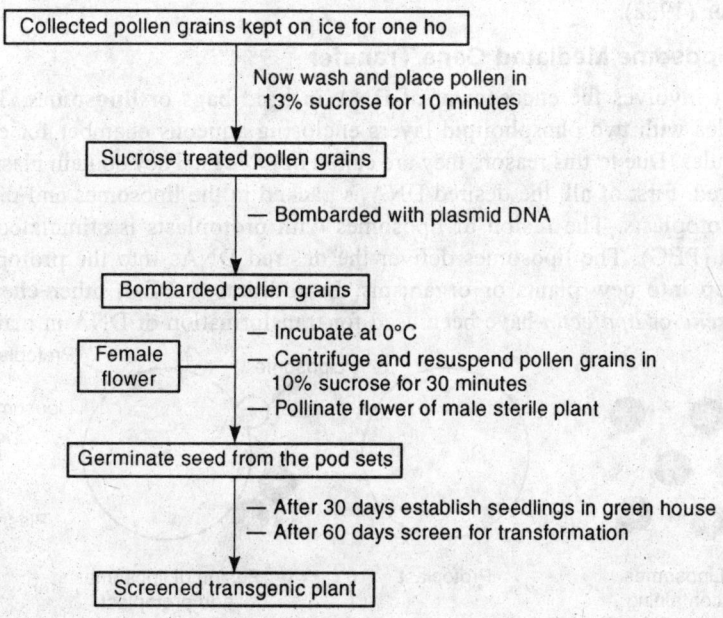

Fig. 1.29. Outline of pollen transformation to get transgenic Alfalfa plants.

Conditions for Bombardment

The conditions of bombardment vary from experiment to experiment and depend upon the cells, tissues and pollen grains etc where the foreign DNA is to be introduced. The conditions are related to the — (*i*) distance between rupture disc and microcarrier, (*ii*) size of gold or tungusten particles, (*iii*) distance between microcarrier and target material, (*iv*) density of the particles or shots, (*v*) bombardment medium, (*vi*) concentration of DNA, (*vii*) helium pressure and vacuum, and (*viii*) post bombardment culture.

(2) Microinjection

Micro injection is a technique where the foreign DNA is mostly transferred into the animal cell *e.g.*, egg, oocyte or embryo, through a *glass micropipette*. The one end of micropipette is heated until its tip could be stretched into 0.3 to 0.5 mm diameter fine tip resembling an injection needle. The process of transferring of foreign DNA is done under a very fine and powerful inverted microscope so that the entire process could be seen.

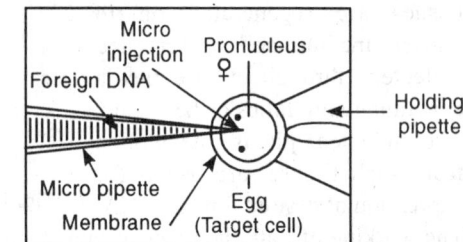

Fig. 1.30. Microinjection technique for foreign DNA transfer in a target cell.

To inject a microinjection of foreign DNA in an animal target cell, to be microinjected, is placed in a container and then it is sucked gently through a holding *pipette* so that it could attach at the tip of the pipette. Now, the tip of holding pipette containing target cell is injected through membrane of the target cell by a prepared microinjection needle having foreign DNA. When the content of the needle is released into the cytoplasm of the target cell, empty needle is taken out.

The technique of microinjection has been utilized for foreign DNA or gene transfer in *Xenopus* oocytes by **Wickens** and **Laskey** (1981), in *Drosophila* embryo by **Rubin** and **Spradling** (1982) and in fertilized eggs and embryos of mouse by **R.D. Palmiter** and **R. L. Brinter** (1982).

(3) Liposome Mediated Gene Transfer

It involves the encasement of DNA in lipid bags or **liposomes**. The liposomes are tiny particles with two phospholipid layers enclosing aqueous chamber for entraping water soluble molecules. Due to this reason, they are called *lipid bags*. They contain plasmids and are artificially prepared. First of all, the desired DNA is packed in the liposomes and then they are mixed with the protoplasts. The fusion of liposomes with protoplasts is stimulated through polyethylene glycol (PEG). The liposomes deliver the desired DNAs into the protoplasts which ultimately develop into new plants or organisms. In addition to PEG, other chemicals like *polycation polybrene* or *lipofectin* have been used for transformation of DNA in maize protoplasts.

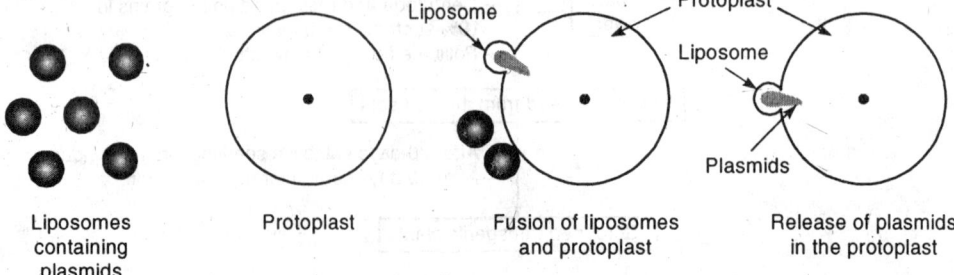

Fig. 1.31. Liposome mediated gene transfer.

Genetic Engineering

Cationic liposome and *polycation mediated DNA delivery* are new protoplast transformation methods. These are considered to be better than other methods of transformation because of the following advantages.

(1) These methods protect DNA/RNA from nuclease digestion.
(2) Cationic liposomes have low cell toxicity.
(3) Encapsulation of nucleic acids makes them more stable during storage.
(4) It showed high degree of reproducibility.
(5) This method is applicable to wide range of cell protoplasts.

(4) Electroporation

It is a technique in which foreign DNA is transferred into the fragile cells using short alternate pulses of high voltage electric current of about 1 MHz or 1-3KV for 10-100 μ seconds. When electric pulses are applied to the protoplasts kept in buffer solution, the pulses through electric field induce the formation of large pores in the cell membrane and through these pores large foreign DNA molecules enter into the protoplasts *i.e.*, transformation of foreign DNA takes place. The technique is optimized by using appropriate electric field strength. The optimum field strength depends upon the following —

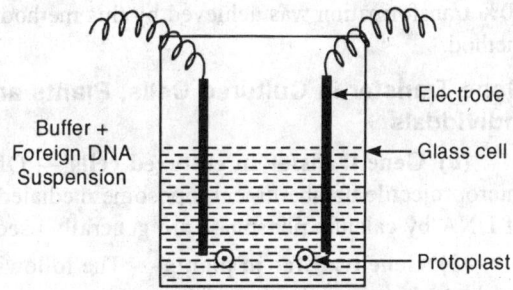

Fig. 1.32. Side view of glass cell with electrodes used for electroporation.

(1) Pulse length of electric current.
(2) Composition and temperature of buffer solution.
(3) Concentration of foreign DNA in the suspension.
(4) Density of protoplasts.
(5) Size of protoplasts.

The transformed protoplasts are cultured which develop *microcalli*. These calli when plated on solid medium containing selective marker, *e.g.*, kanamycin, they show the presence of transformed cells after 37-45 days. The efficiency of this method is as high as 0.5%.

(5) Ultrasonication

This method of gene transfer was used by the scientists of Biotechnology Research centre. Beijing (China), in plants like wheat, sugarbeet and tobacco. They cultured explants and sonicated them with marker genes containing plasmid DNA for transformation. The transformed explants when transferred to selective medium, they produced the shoots successfully. The calli in control and unsonicated experiment did not grow on selective medium and died after sometime. The success of this method was about 22%. In sugar beet and tobacco, mild sonication (20 KHz ultrasound) was used to facilitate the uptake of a *Chloramphenicol acetyl transferase* (CAT) gene in protoplasts. No transgenic plant could be produced by this technique.

(6) Coprecipitation of DNA with Calcium Phosphate

In this method, the DNA of cultured cells is precipitated with the help of calcium phosphate solution and then may be used directly in transformation.

(7) Use of DNA Complexes

The DNA complexes have polycations. These can be used in gene transfer to cultured cells through liposomes.

(8) Laser Microbeam

Weber and co-workers (1988) used an ultraviolet (UV) laser microbeam for the transformation of DNA into plant cells and chloroplasts. A 343 nm beam (wavelength of UV is 200 to 400 nm) was directed through an adjustable attenuator into the optical path of an inverted microscope. The focus of the laser beam was adjusted so that it was identical with that of the objective lens. The laser beam was targeted by focussing on a specimen in the microscope. This laser beam made holes in that part of the cell which was in focus. Laser micropuncture of the cell wall and the plasma membrane allowed uptake (entry) of plasmid DNA into cells. *Brassica napus* (rape seed) cells and mircrospores have been used for transformation by this technique. This technique has also been used to transfer genes into isolated, chloroplasts and chloroplasts of intact protoplasts. 20% transformation was achieved by this method but fertile plants are yet to be produced by this method.

Gene Transfer in Cultured Cells, Plants and Animals for Production of Transgenic Individuals

(*a*) **Gene transfer in cultured cells** — Of the above described gene transfer techniques, microprojectile bombardment, liposome mediated gene transfer, electroporation and precipitation of DNA by calcium phosphate are generally used for gene transfer in cultured cells.

(*b*) **Gene transfer in plants** — The following methods have been utilized for gene transfer in plants during production of transgenic plants..

(*i*) Through *Agrobacterium Ti* and *Ri plasmids*.

(*ii*) Microprojectile bombardment.

(*iii*) Polyethylene glycol (PEG) mediated gene transfer.

(*iv*) Microinjection method.

(*v*) Liposome mediated gene transfer.

(*vi*) Electroporation method.

(*vii*) Laser microbeam method.

(*c*) **Gene transfer in animals or transfection in animals** — The transfection or gene transfer in animals can be achieved at cellular level for a variety of purposes. The transfection methods may be employed in cultured cells (described above) or in fertilized eggs or embryos. The transfection of fertilized egg may involve transfer of whole nucleus, whole chromosomes or their fragment, or DNA segments.

(*i*) **The transfer of whole nucleus** — This technique is described in chapter 40.

(*ii*) **Transfer of whole chromosomes** — During this technique, the chromosomes are first isolated from metaphase cells by hypotonic lysis. They may also be fractioned by density centrifugation method or flow cytophotometry method. The individual chromosomes or their fragments thus obtained are incubated with eggs (whole cells) for incorporation of chromosomes or fragments into nucleus.

(*iii*) **Transfer of DNA segments into fertilized egg** — It is achieved generally by microinjection method, described earlier.

TRANSPOSABLE ELEMENTS

These are accessory genetic elements found both in prokaryotes and eukaryotes. They represent sequences which are capable of moving from one site to another in a genome. The transposable elements are of three types — (*i*) Insertion sequences or IS elements, (*ii*) Transposons, and (*iii*) Retroelements.

Genetic Engineering

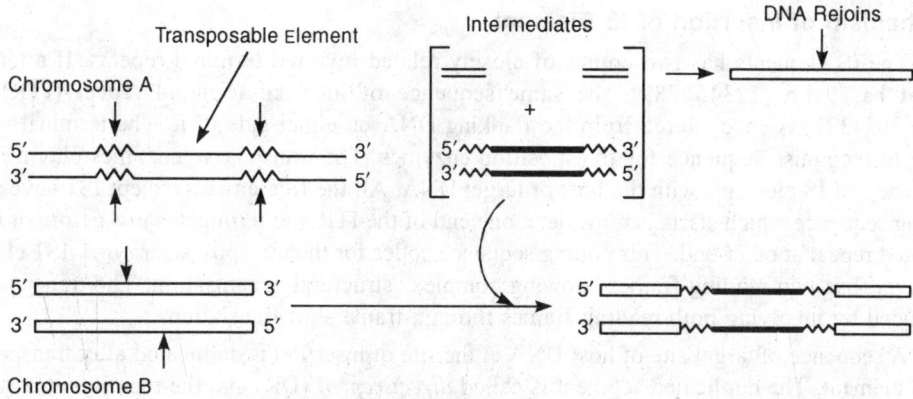

FIG. 1.33. Movement of a transposable element from one chromosomal site to another. Arrows indicate the site for action of enzyme and cutting of dNA stands.

(i) **IS elements** — These are simplest transposable elements which were first discovered as spontaneous insertions in some bacterial operons. The IS elements were first detected in *gal* operon of *E. coli* which synthesized *epimerase, transferase* and *kinase* enzymes required for galactose synthesis. These insertions in *gal* operon, caused inactivation of gene and did not allow the transcription and translation of this gene and other genes of the operon. Thus, the insertion of IS element caused mutations which have been called *polar mutations*. Since these mutations do not revert back and different mutagens do not affect the frequency of reversion, these are neither deletions nor frameshift or point mutations.

The insertion elements are simplest transposable elements because *(i)* They do not contain any genetic information except the DNA sequences necessary for transposition *(ii)* They also lack host gene, *(iii)* They are capable of moving within the genome through illegitimate recombination (moving to unrelated DNA sequences). The insertion sequences were designated by the prefix IS followed by numbers of each type, *e.g.*, IS1, IS2, IS3, IS4 etc. These IS elements are the normal constituents of the chromosome and plasmids. A typical *E. coli* strain may contain more than 10 copies of any IS elements.

Each IS element codes only for protein required for its own transposition. IS elements differ in size of sequences but resemble in their organization. IS elements are about 1000 bp long and contain an inverted terminal repeat of about 10-40 bp length at the end. The characteristics of insertion sequences are given in table 1.3.

Table 1.3 : The Characteristics of insertion sequences

IS elements	Occurrence	Length (bp)	Inverted terminal repeat (bp)	Direct repeat (bp)	No. of copies
IS1	*E. coli* chromosome	768	23	9	6-10
IS2	*E. coli* chromosome	1327	41	5	4-13
IS3	*E. coli* chromosome	1400	38	3.4	5-6
IS4	*E. coli* K12 chromosome	1428	18	11-12	1-2
IS5	*E. coli* K12 chromosome	1195	16	4	10-11
IS50 R	Tn5 on P^{JR67}	1534	9	9	—
IS101	p^{SC101}	201	37	—	—
IS 903	Tn 903 on R6	1057	18	9	—
IS10R	Tn 10 on R100, P^{SM14}	1329	22	9	—

Mechanism of Insertion of IS Element

The IS elements has two copies of closely related inverted terminal repeats. If a terminal repeat has 9 bp (123456789), the same sequence of inverted terminal repeat (ITR) of 9 bp(987654321) is encountered from the flanking DNA on either side of it. The terminal repeats serve to recognise sequence for transposition enzymes. The transposase enzymes play a role in the fusion of IS elements with the host or target DNA. All the IS elements except IS1 have a long coding sequence which starts just inside at one end of the ITR and terminates just before or within inverted repeat at other end. This coding sequence codes for the transposase enzyme. IS1 element, however, has two reading frames showing complex structural organisation. The transposase is produced by involving both reading frames through frame shift translation.

A sequence of target site of host DNA at the site of insertion is duplicated after transposition of IS element. The duplicated segment is called *direct repeat* (DR) and the repetition takes place in the same orientation. The duplication sequence of target site can be recognised if one compares the IS element before and after its insertion. After transposition one copy of target sequence or direct repeat can be observed at both the ends of transposon.

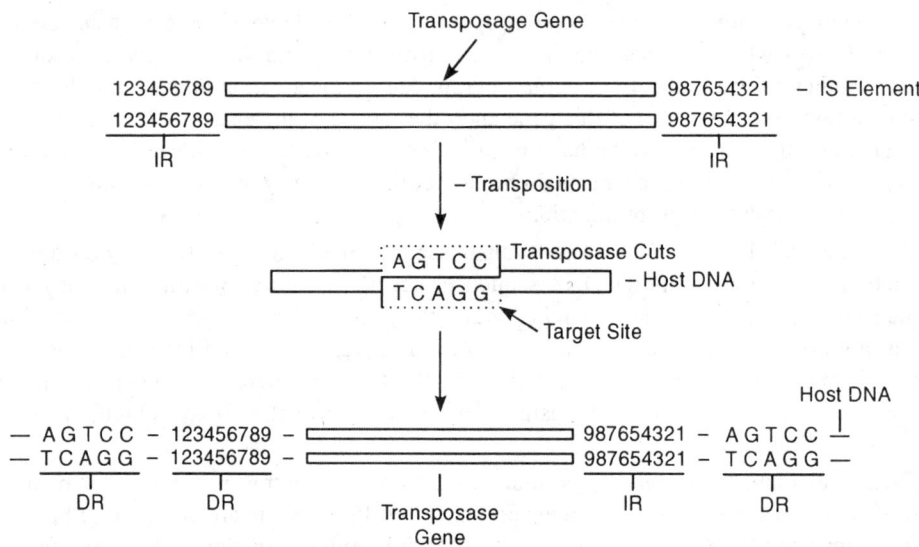

FIG. 1.34. Diagrammatic presentation of insertion sequence (is element) and the mechanism Of its insertion (transposition) into hostchromosome; ir, inverted repeats; dr, Direct repeats of target dna.

A transposition, however, results into different sequences of direct repeats but the length of IS elements remains almost constant, *i.e.* 9 bp. The IS elements can be inserted in host DNA at different sites. The frequency of transposition in different elements may vary from 10^{-5} to 10^{-4} per element per generation. The IS elements may differ in number of bases but each element contains atleast two apparent coding sequences, one initiation and other termination.

Mechanism of Induction of Polar Mutations

As stated earlier, the mutations caused by IS elements are called *polar mutations*. They were studied by using lambda (λ) phage which was able to pick up *gal* region when inserted in this region. λd *gal^m* was isolated from a mutant *gal^m* and its DNA was utilized for the synthesis of radioactive RNA. Certain fragments of radioactive RNA hybridized with *gal^m* but not with *gal^+*. This suggested that an extra piece of DNA was present in *gal^m*. These specific fragments also showed hybridization with DNA from other polar mutants, indicating that same extra piece of

Genetic Engineering

DNA is found inserted in all polar mutants. DNA from λ *d gal*^{m} was hybridized with that from λ *gal*$^+$ and extra piece of DNA with 768 bp could be located in the form of a loop under electron microscope. It was designated as IS1. Several such IS elements were detected from *E. coli* chromosomes or plasmids (See Table. 1.3).

(ii) Transposons

The transposable elements containing additional genetic proportion for encoding genetic information, like drug resistance, not related to process of transposition are called *transposons*. They may or may not be flanked by IS elements. They may occupy different sites on the main DNA molecule of the bacterial chromosome or plasmid. They have also been discovered in prokaryotes and eukaryotes both. They were initially discovered in maize and were described as *controlling elements* due to their controlling property on gene expression. The term *transposon* was first used by *R.W. Hedges* and *A. E. Jacob* in 1974, although the transposons (jumping elements) were first reported by Mc Clintock Barbara in 1940s in maize which affected gene expression. These movable genetic elements have been variously named as *movable genes*, *transposons*, *jumping genes* and *transposable elements*.

Transposons in Prokaryotes

Most of the prokaryotic transposons have been reported from *E. coli* strains. They contain a mobile DNa segment or genetic element alongwith with gene (s) for drug resistance. *Hedges* and

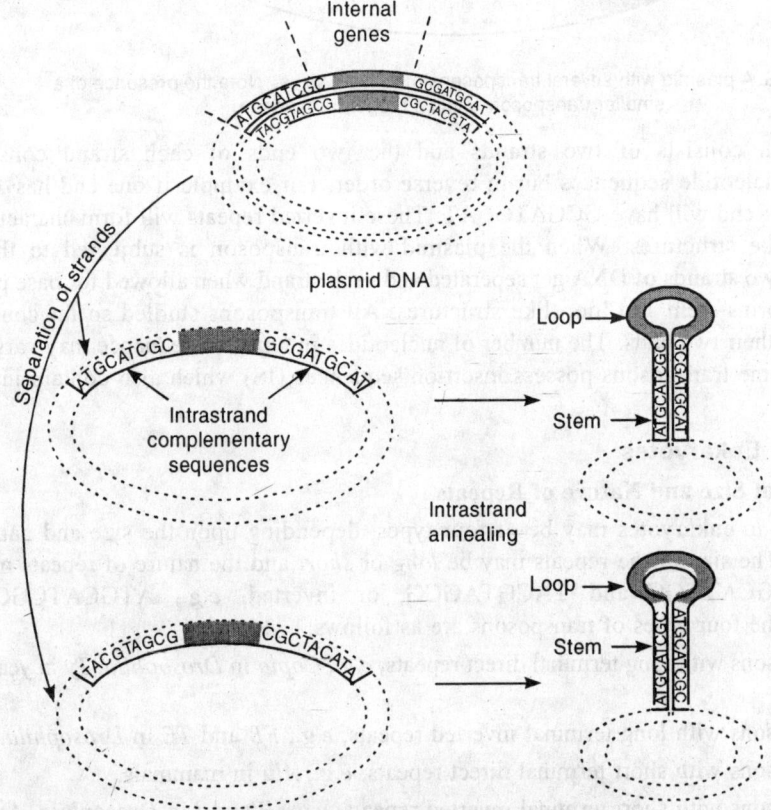

Fig. 1.35. Presence of inverted repeats at the two ends of each strand of a transposon in a plasmid and the formation of characteristic stem and loop in each strand on strand separation due to denaturation. Stem and loop formation is an evidence for inverted repeats.

Jacob (1974) reported that transfer of antibiotic resistance from one plasmid to another is accompanied by an increase in the size of recipient plasmid or DNA molecule. The resistance from recipient plasmid may be transferred to another plasmid with the similar increase in size. This indicates that there is involvement of transfer of DNA segment (transposon). This transposon can occupy different sites in the genome and can be transposed between two bacterial chromosomes or two plasmids or between a bacterial chromosome and plasmid. The transposition of transposon may also take place in bacteria containing mutant gene for recombination, suggesting that these transfers do not involve normal recombination process.

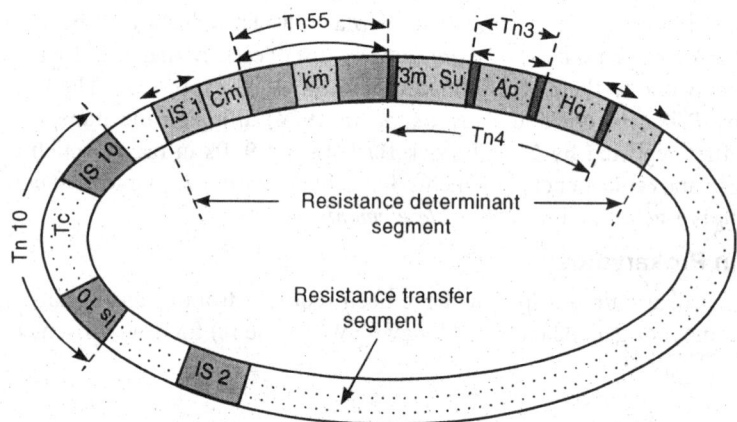

Fig. 1.36. A plasmid with several transposons of varying sizes. Note the presence of a smaller transposon within a bigger transposon.

A transposon consists of two strands and the two ends of each strand consist of complementary nucleotide sequences but in reverse order. For example if one end has ATGC ATC GC, the other end will have GCGATGCAT. These inverted repeats will form characteristic stem and loop like structures. When the plasmid with transposon is subjected to thermal denaturation, the two strands of DNA get seperated and each strand when allowed for base pairing among itself, it forms stem and loop like structures. All transposons studied so far contained inverse repeats at their two ends. The number of nucleotides in repeated sequence may vary from a few to 1400. Some transposons possess insertion sequences (IS) which also contain inverted repeats.

Transposons in Eukaryotes

A. On the Basis of Size and Nature of Repeats

Transposons in eukaryotes may be of four types, depending upon the size and nature of terminal repeats. The size of the repeats may be *long* or *short* and the nature of repeats may be *direct*, *e.g.*, ATGCATCGC and TACGTAGCG, or inverted, *e.g.*, ATGCATCGC and GCGATGCAT. The four types of transposons are as follows:

(*i*) Transposons with long terminal direct repeats, *e.g.*, *Copia* in *Drosophila*, *Ty* in yeast and *IAP* in mice.

(*ii*) Transposons with long terminal inverted repeats, *e.g.*, *FB* and *TE* in *Drosophila*.

(*iii*) Transposons with short terminal direct repeats, *e.g.*, *Alu* in mammals.

(*iv*) Transposons with short terminal inverted repeats, *e.g.*, *P* and *I* in *Drosophila*, Ac/ds in maize, *Tam* 1 in snapdragon and *Tc* 1 in *Coenorhabditis*.

B. On the Basis of Mechanism of Transposition

On above basis, transposons have been grouped into following two classes:

(*i*) **Class I** – This includes those transposons where transposition takes place by reverse transcription of a RNA intermediate (DNA—RNA—DNA), *e.g.*, *Copia* like elements, *Ty*, *LI*, *F* and *IAP* elements.

(*ii*) **Class II** – This includes those transposons where transposition takes place directly, *i.e.*, from DNA to DNA, *e.g.*, *P*, *Ac*, *Tam*, *Tc*1, *Spm/En* elements.

Transposable Elements (TE) in *Drosophila*

About 10% of the genome in *Drosophila* contains *TE*. The important *TE* of Drosophila are *Copia* like elements, fold back (FB) elements, paternal contributing (*P*) and inducer (*I*) elements.

Copia like elements – They possess following properties —

Size = 5 to 8.7 kb
Number = 10 to 80
Long direct terminal repeat (LTR) — 268 to 512 bp.
Short imperfect inverted repeat (SIR) — 0 to 18 bp.

Insertion of these *copia* like elements cause mutations, *e.g.*, white-apricot (w^a) mutation for eye colour in Drosophila. Properties of *Copia* like elements are given in table 1.4.

Table 1.4

Elements	*Number/genome*	*Length (Kb)*	*LTR-length (bp)*	*SIR length (bp)*
Copia	60	5.0	276	13/17
B104	80	8.7	429	3
17.6	40	6.9	512	0
412	40	7.5	481	8/10
297	30	7.0	412/415	3
mdg 3	15	5.5	283/286	15/18
mdg4/gypsy	10	7.3	479	4/5

(*ii*) **Fold Back (FB) elements** — The length of these elements varies from a few hundred to a few thousand base pairs. They carry long inverted terminal repeats. The repeats may be adjoining as in FB1 and FB5 elements or may be separate as in FB3, FB4, FB7, FB8 and FB9 elements. In both the cases, they may be folded back to form stem and loop structures. They cause mutations either by insertion or by their effect on gene expression. FB elements are also found flanked with other class of transposons called TE. The example of such transposons are TE1, TE 28, TE77 and TE 98. TE1 was first identified due to transposition of white eye locus from X–chromosome to second chromosome of *Drosophila*.

(*iii*) **Paternal (P) and Inducer (I) elements** — These elements cause hybrid dysgenesis in *Drosophila melanogaster* in certain combinations during crossing of different strains like P (paternal contributing), M (maternal contributing), I (inducer) and R (reactive). *Hybrid dysgenesis* is a phenomenon responsible for sterility, high frequency of mutation and chromosomal aberrations and non-disjunctions of sex-chromosomes.

Ty elements in Yeast — *Ty* 1 element in Yeast genome is represented by 35 copies. The ends of these elements contain direct repeats which are different from inverted repeats (IR) of bacteria. The direct repeats of *Ty* elements are called delta (δ) and they are represented by 100 copies per genome. *Ty* elements generate 5 bp long repeated sequence during transposition and produce mutation due to insertion. The transposition in *Ty* elements takes place through an RNA intermediate (DNA–RNA–DNA).

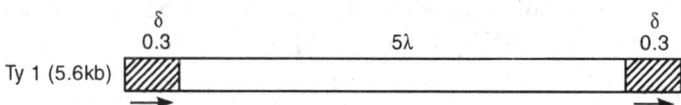

Fig. 1.37. Structure of *Ty* element in yeast genome with direct repeats of Δ sequences at ends.

Controlling Elements in Corn

Controlling elements were originally discovered in maize by *M. Rhoades* and *Barbara Mc Clintock*. These elements were responsible for change in gene expression. Their location on maize chromosome is specific for breakage and reunion of chromosomes which result into change in chromosome structure (chromosomal aberration)

A number of families of controlling elements have been discovered so far. Some of them have *one element system*, e.g., Dotted (*Dt*), Mutator (*Mu*), Modulator (*Mp*), Bergamo (*Bg*), and ubiqnitous (*Uq*) etc, and others have *two element* systems, e.g., *Ac-Ds* system and *Spm-dSpm* sysem. It has been observed that one element system may be converted into two elements system. The number, type and location of these controlling elements are characteristic for each strain of maize.

In each two elements system, the elements have been grouped into two classes – (*i*) Autonomous elements and (*ii*) Non-autonomous elements. The autonomous elements are stable and possess the ability to excise and transpose, rendering sites of their insertion mutable. Non-autonomous elements are unstable and unable to transpose. They can be derived from autonomous elements of the same family. Their unstability is caused when an autonomous element of the same family is present elsewhere in the genome. When a non-autonomous element is complemented by an autonomous element of the same family, the non-autonomous element may show the characteristics of the autonomous element along with transposable ability. The relationship between autonomous and non-autonomous elements of a family is quite specific because a non-autonomous element of one family can not be cross-inactivated by the autonomous element of any other family. The organisation of controlling elements is similar to those of bacterial transposons.

In each two element system, a family consists of a single autonomous element associated with different non-autonomous elements. These elements possess characteristic inverted terminal repeats and short direct repeats in the adjoining target DNA.

In maize, *dotted* (*Dt*) *element* was discovered by *Marcus Rhoades* in 1938 and activator-dissociation (Ac-Ds) system was discovered by *Barbara Mc Clintock* in 1950. Another system, suppressor-mutator (Spm), was discovered in *Nicotiana*. A different Spm system, called defective suppressor-mutator (dSpm) was discovered in maize.

(iii) Retroelements

These are also transposable elements. Retroelements represent those nucleic acid sequences which originate either fully or partly, or propagate through reverse transcription (RNA→DNA). They may be of two types — (*i*) viral retroelements, and (*ii*) non-viral retroelements. On the basis of structural and biological criteria, viral retroelements have been further classified into two subclasses (*a*) retroviruses and (*b*) pararetroviruses and non-viral retroelements into four sub-classes: (I) Retrotransposons, (II) Retroposons, (III) Retrons and (IV) Retrosequences. The characteristics of two types of retroelements are given in table 1.5.

The viral retroelements are infectious and possess an extracellular phase. They are not very important as transposable elements. The characteristics of sub-classes of non-viral retroelements are as follows:

(I) Retrotransposons — They resemble transposons except they have RNA origin.

Table 1.5. Characteristics of retroelements.

(A) Viral retroelements

    (I) **Retroviruses** (RNA in virions) or plus strand viruses *e.g.*, Oncornaviruses, Lentiviruses; Spumaviruses

    (II) **Pararetroviruses** (DNA in virions) *e.g.* Caulimoviruses, Hepadnaviruses

(B) Non-Viral retroelements

	(I) Retrotransposons	(II) Retroposons
LTR	+	−
RT	+	+
Integrase	+	+
Examples	Ty, Copia	SINEs (*Alu*), LINEs (mitochondrial introns & plasmids)

	(III) Retrons	(IV) Retrosequence
LTR	−	−
RT	+	−
Integrase	−	−
Examples	ms DNA (multicopy single stranded DNA)	cDNA genes (pseudogenes)

    (*i*) Retrotransposons have RNA origin.
    (*ii*) They have long terminal repeat (LTR) sequences for transposition.
    (*iii*) They have primer binding site (PBS).
    (*iv*) They have α promoter and processing or polyadenylation signal.
        Examples — Ty, Copia.

(II) Retroposons — They lack
    (*i*) LTR sequences for transposition.
    (*ii*) PBS site
    (*iii*) A promoter and processing signal.
        Examples – SINEs (Alu), LINEs.

(III) Retrons – They contain –
    (*i*) Multicopy single stranded DNA (msDNA) of mycobacteria and *E. coli*.
    (*ii*) Sometimes additional sequences from other prokaryotes.
    (*iii*) Unconventional mechanism of reverse transcription
        Example – msDNA

(IV) Retrosequences
    (*i*) They represent cDNA genes or pseudogenes.
    (*ii*) The retrosequences end in poly (A) tails.
    (*iii*) They are flanked by short repeat sequences.
        Example – cDNA genes (Pseudogenes).

Mechanism of Transposition

The information about mechanism of transposition came through the analysis of DNA sequences and their union with target DNA. The mechanism has been studied extensively in bacterial transposon Tn 3. The movement of the transposon takes place only when transposase enzyme recognises and cleaves at either 5' or 3' of both ends of transposon and causes a staggered

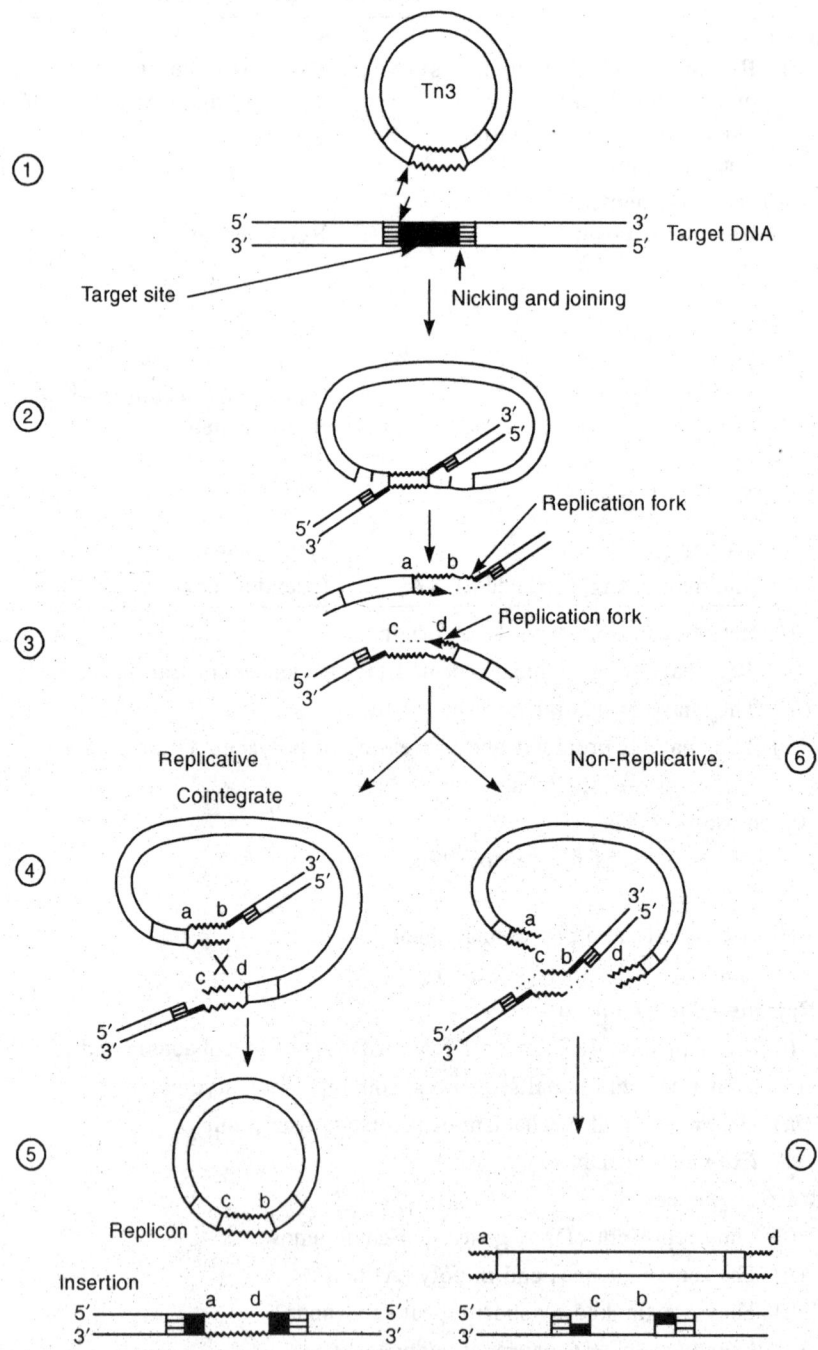

Fig. 1.38. A model to explain transposition mechanism of transposon Tn3.

cut at the target site. A duplication of 3-12 bases in target DNA takes place at the insertion site, depending upon the type of transposon. One copy remains at each end of the transposon sequence.

When both the ends of a transposon get attached to the target site on other DNA, two replication forks are generated immediately. The onward process is carried out either by

Genetic Engineering

replicative method or by non-replicative method. *In replicative method*, the transposon replicates and the replicated DNA joines to the flanking sequences to form a cointegrate. The cointegrate is resolved by the genetic exchange between the two copies of transposon resulting in a simple insertion and regeneration of donor replicon.

In *non-replicative method*, simple insertions are generated without formation of cointegrate. During the process, repair synthesis occurs at the prime termini in the target DNA and the displaced single strand that attaches the transposon to the donor replicon, is broken. Thus, formation of a simple insertion takes place.

Two enzymes, *transposase* coded by *tnp* A and *resolvase* coded by *tnp*R are needed for transposition. Transposase performs the functions of recognising the ends of transposon and connecting them to the target site, whereas resolvase functions to provide a site-specific recombination.

TECHNIQUES OF DNA SEQUENCING AND GENE MAPPING

DNA SEQUENCING

Determination of nucleotide or base sequence of a DNA molecule or fragment is called *DNA sequencing*. It is applicable to those DNA fragments which are only one to few hundred base pair (bp) long. Initial efforts on sequencing of nucleic acids were confined to such RNAs which could be readily isolated in pure form. Tyrosine tRNA was selected for such sequencing. Robert Holly isolated enough tyrosine tRNA in pure form from about 100 pounds of yeast to carry out sequence analysis. The sequencing of RNA is no longer done directly. In fact, the sequencing of RNA has been replaced by sequencing of cDNA which is obtained by transcription of RNA to DNA. This technique requires enzymes (restriction enzymes), electrophoretic techniques, PCR technique (to get large quantities of DNA) and sequencing procedures (Maxam and Gilbert, and enzymatic approach)

Methods – Two methods have been developed for DNA sequencing. One method was developed by *Walter Gilbert* and *Alan Maxam*. It involves cleavage of pre-existing DNA using chemicals (chemical approach). The second method was developed by *Fred Sanger* which involves premature termination of newly synthesized DNA using enzymes (enzymatic approach). Both Gilbert and Sanger received Nobel Prizes for their work on DNA sequencing.

Maxam and Gilbert Procedure (Chemical Approach)

For sequencing of DNA, either single or double stranded DNA molecules can be used. Maxam and Gilbert procedure includes following steps —

1. First step is the ^{32}P labelling of desired specific fragment of DNA at its 5' end with *polynucleotide kinase* enzyme or at its 3' end with *deoxynucleotidyl* transferase enzyme.

In 5' end labelling, the enzyme transfers the $^{32}PO_4^-$ group from $Y^{32}P$-ATP to the 5' hydroxyl end of deoxyribonucleotide chain. If double stranded is used, both 5' ends become labelled and this DNA molecule as such can not be sequenced directly. The complementary strands or the end of doubly labelled DNA need separation.

2. Second step is the separation of complementary 5' end labelled strands of DNA.

The separation can be done by treatment of the DNA with a restriction enzyme that cleaves the DNA into two unequal fragments that can be separated by gel electrophoresis, or by denaturation followed by gel electrophoretic separation of individual strands. The unequal fragments of DNA or complementary strands (both with 5' end labelled) are sequenced separately.

3. Step 3 is the treatment of single and labelled double or single stranded DNAs with a chemical reagent that specifically reacts with one of the four basis.

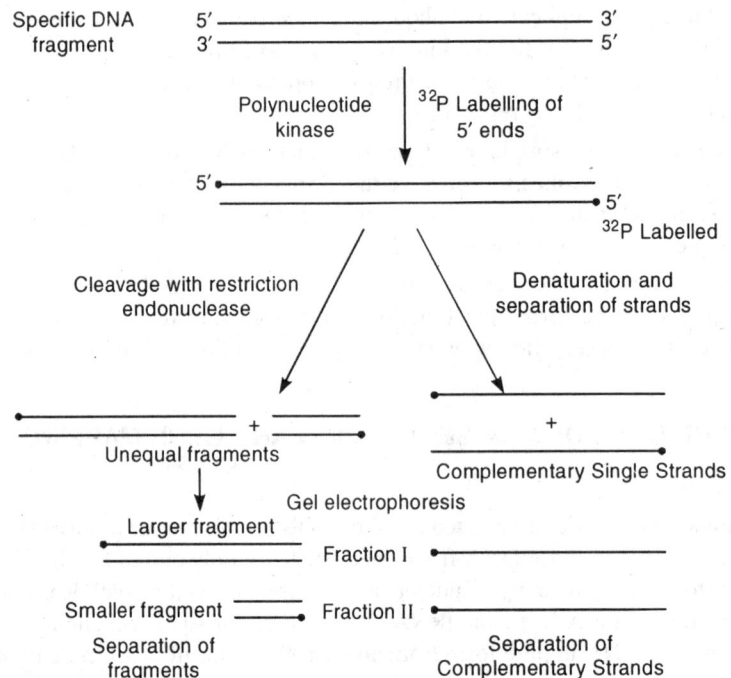

Fig. 1.39. Representation of the production of single-end-labelled DNA molecules stands for nucleotide sequencing through chemical cleavage procedure.

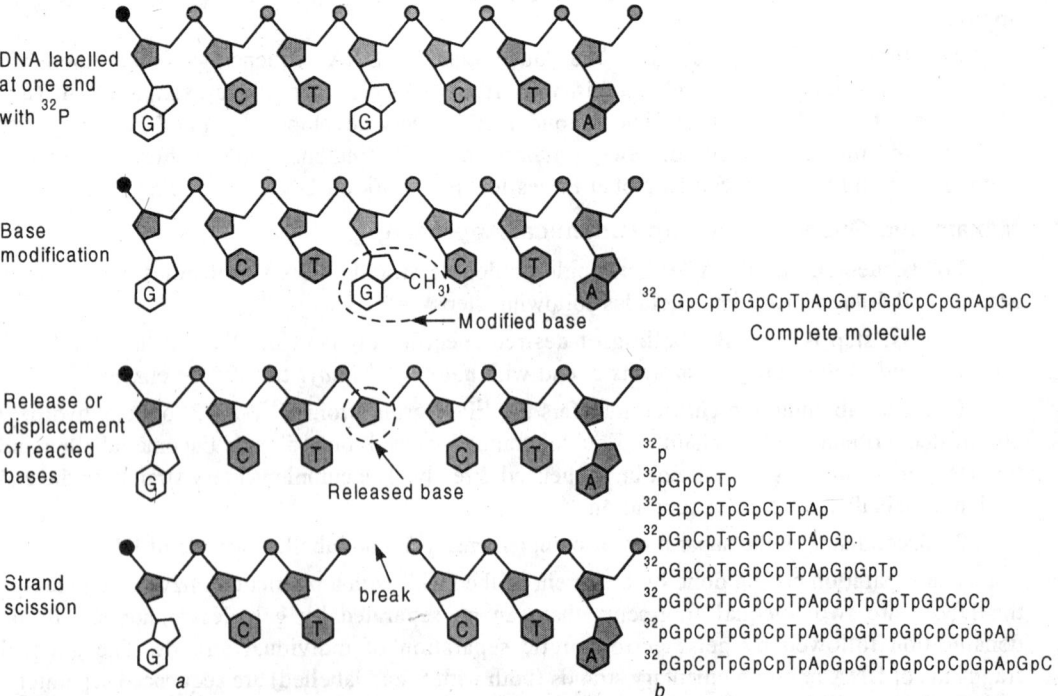

Fig. 1.40. Sequencing end-labelled DNA by limited, base-specific chemical cleavage. (a) A sequence of three reactions leading to strand cleavage at a guanine. If the entire DNA is subjected to this reaction sequence for a limited time, the result is (b) a family of labelled fragments that will be cleaved at different guanine sites.

Genetic Engineering

The reaction is carried out only for a short duration so that only a few residues in the DNA could react with the reagent. Thus, modification of the bases takes place. The modified base introduces a linkage amenable to backbone cleavage by subsequent chemical treatment. All the bases in a given type in DNA are equally susceptible to modification. The net result is a family of products labelled at 5' end with ^{32}p and terminating at the point of cleavage.

Four chemical reactions are used which cleave DNA preferentially at four places viz., at guanines (G > A), adenines (A > G), cytosine alone (C), and cytosines and thymines in equal amount (C + T).

4. Resolving of products of four reactions by gel electrophoresis according to size and production of autoradiogram.

When the products of four reactions have been resolved according to size by gel electrophoresis and their autoradiograms have been prepared, the DNA sequence can be read from the pattern of radioactive bands on autoradiogram.

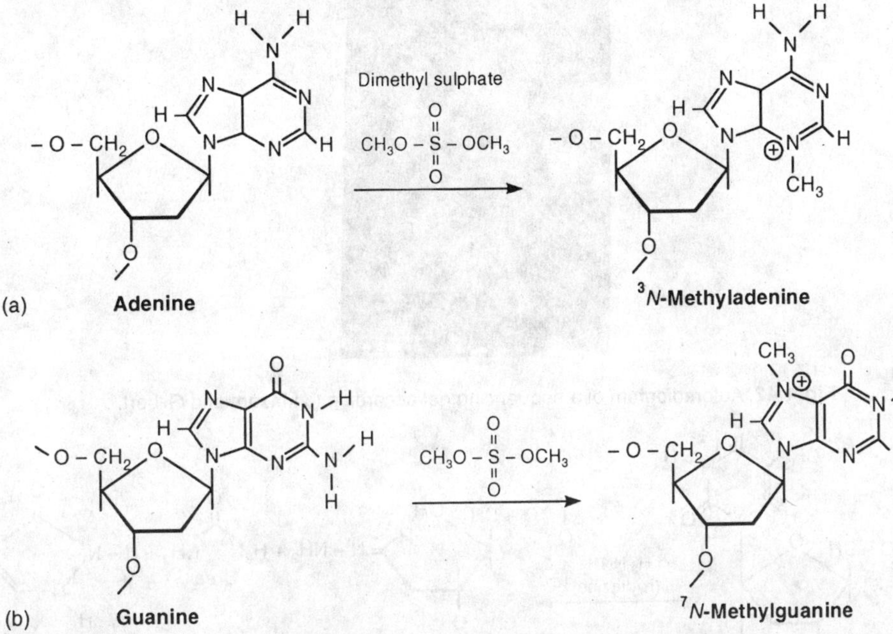

Fig. 1.41. Treatment of DNA with dimethylsulphate, resulting in methylation at N-3 in adenine (a) and at N-7 in guanine (b) Guanine is about five times more reactive than adenine.

An aliquot of the DNA is treated with *dimethylsulphate* for purine specific reaction. The dimethylsulphate methylates the guanines in DNA at the N-7 position and the adenines at the N-3 position (Fig. 1.42). The glycosidic bond of a methylated purine is cleaved on heating at neutral pH, leaving the sugar free. The sugar from the neighbouring phosphate groups can then be cleaved by alkali at 90°C. When the resulting end-labelled fragments are resolved on a gel, the autoradiogram show a pattern of dark and light bands (G > A lane). When methylated DNA is treated with acid, it gives rise to an adenine-enhanced cleavage, releasing methylated adenine preferentially (A > G lane).

Other aliquots are treated with hydrazine which attacks cytosine and thymine basis (Fig 1.43). After a partial reaction in aqueous *hydrazine*, the phosphate backbone of the DNA is cleaved with 0.5 M piperidine. The final gel pattern contains bands of similar intensity owing to the cleavages at cytosines and thymines (C + T lane). However, if 2m Nacl is included in the hydrazine reaction,

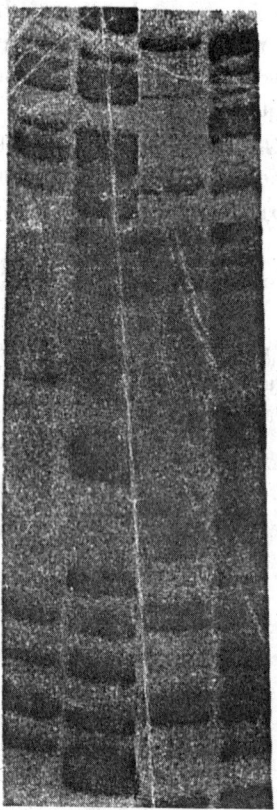

Fig. 1.42. Autoradiogram of a sequencing gel according to Maxam and Gilbert.

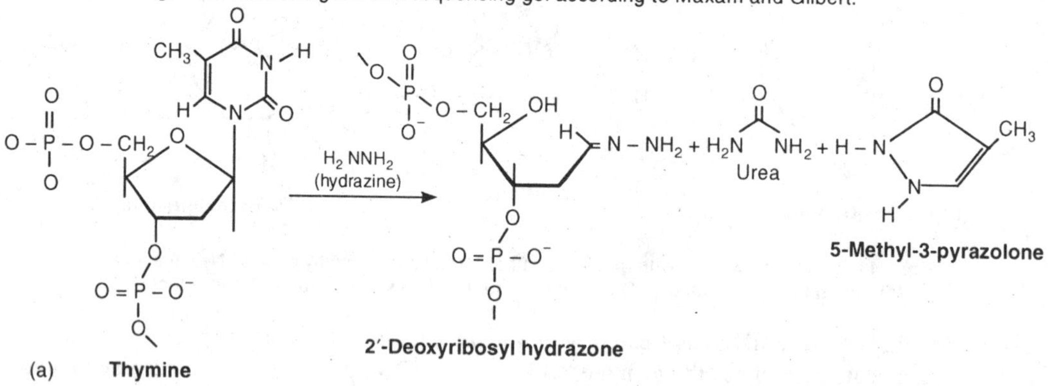

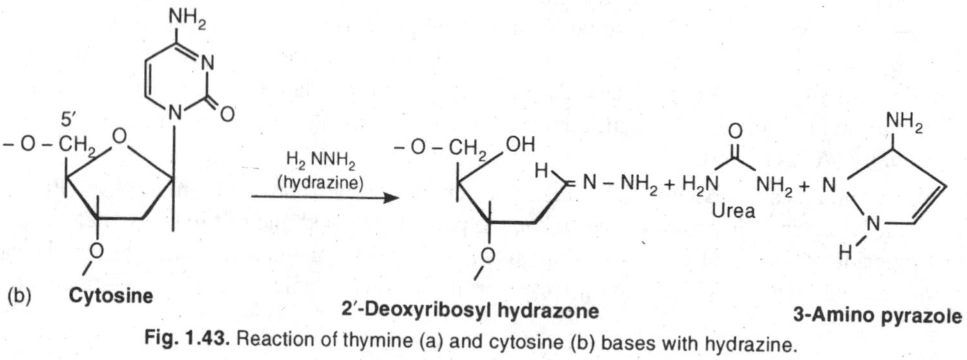

Fig. 1.43. Reaction of thymine (a) and cytosine (b) bases with hydrazine.

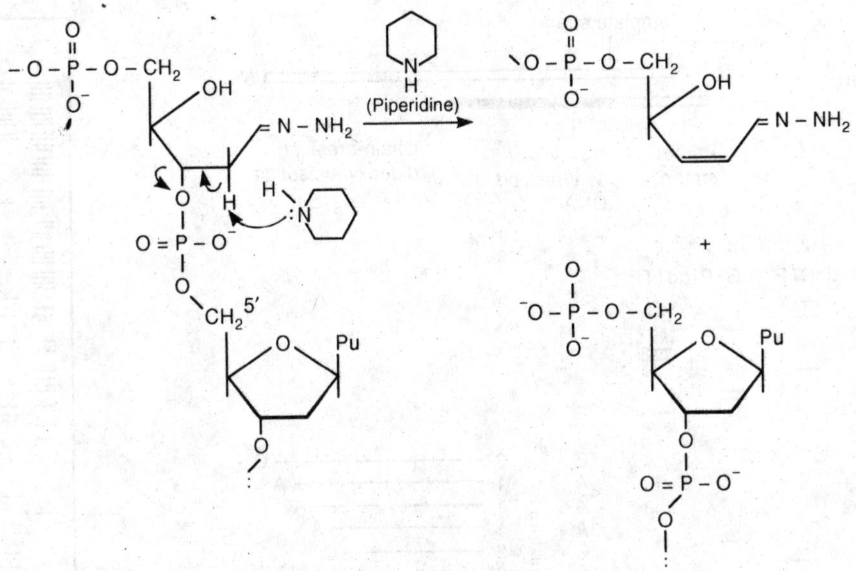

Genetic Engineering

Fig. 1.44. Cleavage of the hydrazine derivative of the cytosine and the thymine in the presence of piperidine.

the reaction of thymine bases is suppressed . Then the piperidine breakage produces bands only from cytosine (C lane). Consequently, if the results with and without added hydrazine are compared, C's and T's in in the sequence can be distinguished.

Sanger Procedure (Enzymatic Approach)

Sanger procedure is based on employing chain terminating di-deoxy-nucleoside triphosphates to produce a continuous series of fragments in the reactions catalysed by polymerase. Di-deoxynucleoside triphosphates resemble deoxynucleoside triphosphates except that they lack a 3'-OH group. They can add to a growing chain during polymerization, but they can not be added onto and therefore serve as chain terminators.

In DNA sequencing, DNA polymerase is used to make a complementary copy of a primed single stranded DNA fragment. By choosing an appropriate primer, the region of the nucleic acid (DNA) that is copied can be predetermined.

In Sanger procedure for DNA sequencing, the DNA fragment to be sequenced is first denatured and the complementary strands are separated through gel electrophoresis. Out of the two complementary strands, one or both (in separate) experiments is used as a template for DNA replication catalysed by larger subunit of *E. coli* DNA polymerase I. The larger subunit of polymerase I is called *Klenow fragment*. The smaller subunit of *E. coli* DNA polymerase I is not used because of its 5'→3' exonuclease activity.

The DNA replication reaction system should necessarily have at least one of the four nucleotide radioactive so that autoradiographic development of bands after gel electrophoresis could be produced. Since a free 3' OH is essential for DNA polymerase I to catalyse DNA replication, a small primer with free 3'OH group is provided to template strand for proceeding DNA replication.

Four separate reaction mixtures are taken for the replication of four separated DNA strands to be sequenced (one mixture for each DNA strand). In one of the reaction mixtures, 2', 3'dideoxy adenosine triphosphate (ddATP) is added in low concentration (about 1/100) than normal dATP in the system. ddATP acts as a terminator for newly synthesized DNA strand because ddATP lacks a free 3'OH group which checks further addition of nucleotide to the newly synthesized strand. At the concentration used here, ddATP would terminate the newly synthesized

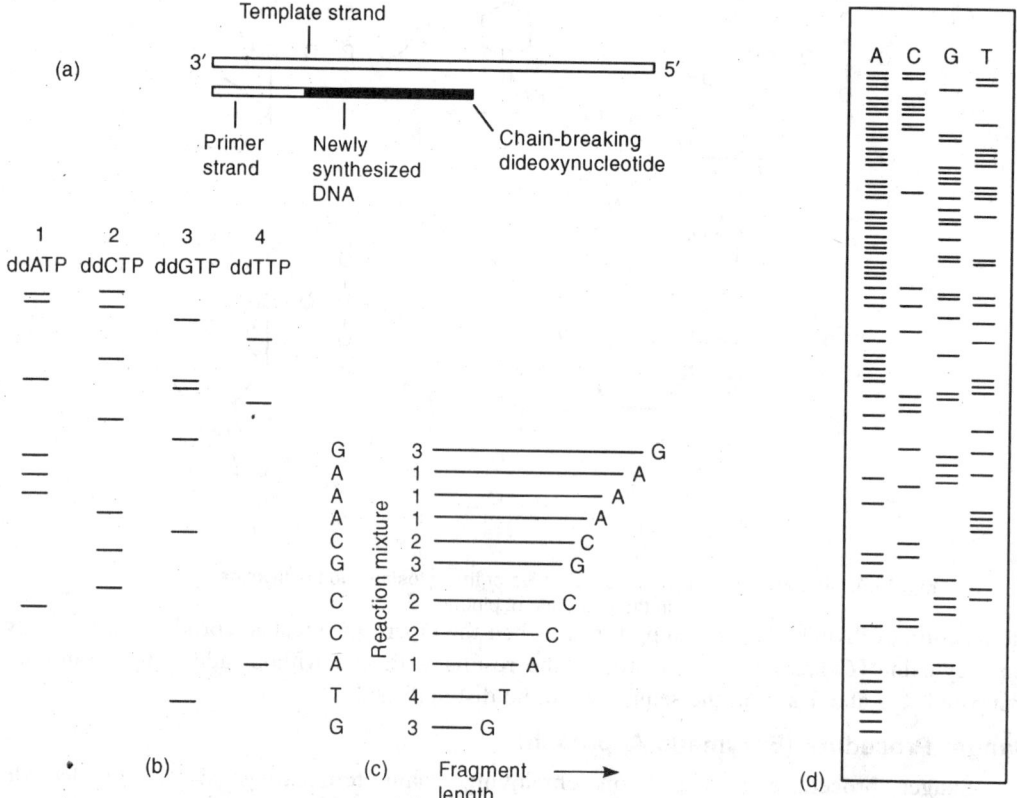

Fig. 1.45. Sanger's dideoxynucleoside method of DNA sequencing (a) selected template and primer which synthesizes new DNA but terminates by chain terminator. (b) Separated bands on gel. (c) Smallest labelled fragment moves the fastest and appears at the bottom of gel; (d) Autoradiograph showing sequenced bands.

DNA (polynucleotide chain) at any one of the 4 possible sites where adenine is to be incorporated in the new strand.

In remaining other three reaction mixtures, having same DNA fragment (used in first mixture), 2', 3' ddCTP (dideoxy cytidine triphosphate), and 2', 3' dd TTP (dideoxy-thymidine triphosphate) 2', 3', ddGTP (dideoxy guanosine triphosphate) are added separately (one in each mixture) as chain terminator to terminate the polynucleotide chains at any one of all the positions where C, G or T, respectively, are to be incorporated in the new chain.

The partially synthesized chains (DNA strands), due to chain termination, from all the four reaction mixtures are separated from the template strand by denaturation and are subjected to gel electrophoresis separately. This helps in separating the newly synthesized strands according to their size. The gel plate with separated strands is placed in front of X-ray film to develop bands for about 24-48 hrs (autoradiography). Thus, an autoradiograph is prepared. The fastest moving strands (fragments) will be smallest one, and each subsequent band will be one nucleotide longer than the previous one. Therefore, by comparing the developed bands of four gels, the nucleotide sequence of the DNA fragment can be determined. The position of a band in the gel from a reaction mixture will indicate the position of a base of which 2', 3'-dideoxynucleotide triphosphate (dd X TP) was used as chain terminator in that mixture.

Summary of Sanger Procedure

1. Denaturation of DNA to be sequenced and separation of complementary strands by gel electrophoresis.

Genetic Engineering

2. Radioactivation of nucleotides A, C, G, T, separately for developing autoradiograph.
3. Replication of DNA by using larger submit of *E. coli* polymerase I and by adding small primer with free 3'OH group.
4. Preparation of four separate reaction mixtures for replication of separated four strand to be sequenced.
5. Addition of 2', 3'-dideoxynucleotide triphosphates (ddATP, ddCTP, ddGTP, ddTTP) separately (one in each reaction mixture) for chain termination.
6. From all reaction mixtures, separation of partially synthesized (new) chains from DNA template by denaturation.
7. Electrophoretic separation of newly or partially synthesized chains according to their size.
8. Preparation of autoradiograph by placing gel plate against x-ray film and development of bands.
9. Comparing of gel bands and determination of nucleotide sequence of DNA.

CHROMOSOME WALKING — A Genetic mapping technique

The technique of identification of genomic fragments with overlapping sequences through hybridization of cloned fragments with sub-cloned ends of all the fragments to prepre large chromosomal regions or complete genome or restriction map is called *chromosome walking*.

Special radio active cDNA probes are utilized for the identification of gene sequences in a genomic library. The original cDNA probe can detect only the members of the genomic library that contain sequences homologous to those present in the probe. The genomic library contains a large number of clones (genes or DNA fragments) obtained by the fragmentation of the large gene or genome. When a original cDNA probe is added or mixed in a mixture of clones carrying a part of single large gene or genome, it hybridizes and selects a clone of interest (clone 1). This hybridized clone of the large gene is converted into radioactive probe and it is used to locate the second member (clone 2) of the library that contain overlapping sequence of the gene.

When the genome is partially digested, it gives rise to partial digests. Different genomes obtained from a large number of cells after partial digestion may result into numerous fragments having overlapping sequences. It happens because the sites cleaved in different genomes of the same organism are random and differ from one another. Since none of these fragments have the entire sequence represented in the probe, overlapping sequence may be utilized to reconstruct and characterize the original genomic sequence.

The technique of chromosome walking includes following steps —

1. The first step in chromosome walking is the selection of a clone (DNA fragment) of interest or clone containing known gene, called clone 1, by a radioactive cDNA probe. This clone provides the starting point of "chromosome walk". It also provides a point of reference in the genetic map.
2. Preparation of a restriction map of clone 1 and isolation and sub-cloning of a small fragment from one end of clone 1. This sub-cloned end of clone 1 is now used as a probe to select and identify clone (s) 2 in the genomic library having complementary sequence of this probe. For this purpose sub-cloned end of clone 1 is hybridized with other clones of the library and clone (s) 2 is selected.
3. Preparation of restriction map of clone 2 and sub-cloning of a fragment from one end of clone 2. The sub-cloned end of clone 2 is now used as a probe to select and identify clone 3 in the genomic library having complementary sequence of this probe (obtained by sub-cloning of clone 2). To get clone 3, the probe obtained by subcloning of hybridized clone 2 is hybridized with other clones of the library.

4. Preparation of a restriction map of clone 3 and sub-cloning of a fragment from the end of clone 3. The sub-cloned end of clone 3 is now used as a probe to select and identify clone (s)4 in the genomic library having complementary sequence of this probe (obtained by subcloning of clone 3). To get clone 4, the probe obtained by sub-cloning of hybridized clone 3 is hybridized with other clones of the library.

5. The above process of step 3 or 4 is repeated till we get all the clones having overlapping sequences or regions of a large gene or a large genome. The process ends when we reach the end of the chromosome.

Chromosome walking technique has been used to map regions of the chromosome having about 1000 Kb.

When other end of clone 1 is used as a probe to identify overlapping clones, the chromosome walking takes place in opposite direction till the otherend of the chromosome is reached. Actually one end of each new clone should be complementary to the probe which is being used to identify it.

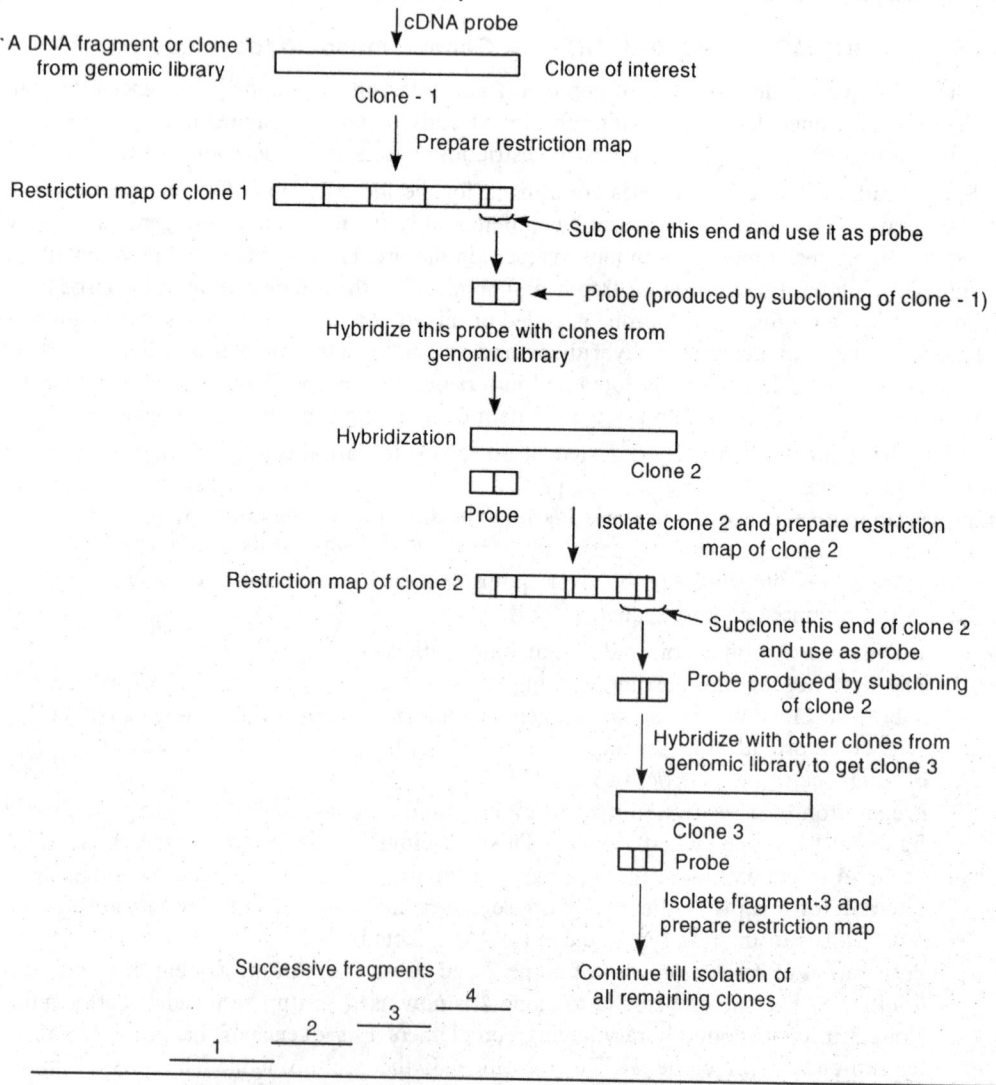

Fig. 1.46. A schematic representation of the technique of chromosome walking.

2

BIOTECHNOLOGY

INTRODUCTION AND FUNCTIONAL DEFINITION

Biotechnology is a new branch which shows a link between the biological sciences, physical sciences and technological achievements. The term "biotechnology" came into use in the mid 1970s, superceding gradually the term "bioengineering". The later term was used to describe both *"biomedical engineering"* which refers to design and manufacture of products like heart valves, body scanners, artificial hips etc., and *"biochemical engineering"* which refers to chemical engineering processes using biological substances or organisms.

Recently biotechnology has been defined by the Organisation for Economic Co-operation and Development (OECD), 1981 as *"the application of scientific and engineering principles to the processing of materials by biological agents to provide goods and services"*. Biological agents included a wide range of biological substances like enzymes, whole cells or multicellular organisms.

Another frequently quoted definition is, *"The application of biological organisms, systems or processes to manufacturing and service industry"*. Biotechnology has been utilized in manufacturing of various products in industries. in food processing and industries. in improving agriculture and in producing various medicines, controlling pollution and health hazards etc. All these are related to human welfare.

Biotechnology deals with the study of technological application of the capacities of microbes and cultured tissue cells. Through biotechnology micro-organisms can be exploited for the production of industrially important biochemicals including proteins, polysaccharides, organic acids, biosolvents, biodetergents and also hormones, enzymes, antibiotics and antibodies etc. This recent branch of biotechnology developed as a result of integrated use of biochemistry, microbiology, chemical engineering and molecular biology. Recent advances in these branches contributed a lot to biotechnology. Thus, biotechnology includes all industrial processes mediated by living organisms at some step or the other. The products of biotechnology are being utilized widely. for the human welfare.

Uses of Biotechnology

The production of vinegar from molasses, formation of yoghurt and cheese from milk, alcoholic fermentation through micro-organisms and production of antibiotics etc are the common examples of the use of biotechnology.

Biotechnology is being utilized now a day for the following purposes —

(1) Production of new pharmaceuticals by fermentation.

(2) Production of enzymes. hormones and antibodies through human gene deroved products.

(3) Production of proteins, organic acids, polysaccharides, biosolvents and biodetergents through micro-organisms.

(4) Production of alternative sources of renewable energy.

(5) Production of antibiotics
(6) Improvement of crops.
(7) Improvement of animals.
(8) Diagnosis and curing diseases.
(9) Production of vitamins
(10) Synthesis of a variety of chemicals.

Some Foreign Biotechnology Companies

Some of the important foreign biotechnology companies are —
(1) Genentech Inc. (U.S.A)
(2) Cetus Corporation (U.S.A)
(3) Hybritech (U.S.A.)
(4) Biogen (U.S.A.)
(5) Biogen (Switzerland)
(6) International centre for genetic engineering and biotechnology (United Nations).

These companies and centres are involved in industrial revolution. A number of biotechnologists and Nobel Laureates like **Carey, Boyer, Hall** and **Burke** are presently associated with the above companies. Gene technology is being utilized in these biotechnology companies.

Cetus — It is a leading bioindustrial company of U.S.A. which is conducting the following biotechnological programmes-
(1) Automated bioscreening.
(2) Genetical improvement of pharmaceutical micro-organisms.
(3) Engineering of a series of organisms for specific industrial use.
(4) Developing immobilized cell and enzyme systems for chemical process industries.
(5) Improved production of vitamin B_{12}.
(6) Manufacturing fructose from inexpensive forms of glucose.
(7) Bioprocessing alkenes to valuable oxides and glycols.
(8) Production of ethanol by continuous fermentation.
(9) Upgrading hydrocarbons microbiologically.
(10) Production of xanthan-gum in oil fields for enhanced crude oil recovery.
(11) Production of human insulin microbiologically.
(12) Production of human interferons microbiologically.
(13) Developing a vaccine to prevent colibacillosis.
(14) Production of monoclonal antibody for organ transplant tissue typing.
(15) Production of diagnostic kits for toxoplasmosis identification.

Biotechnology Boards, Institutes and Centres in India

Various biotechnology boards, institutes and centres have been established in India. Some of them are —
(1) National Biotechnology Board (NBTB)
(2) Centre for cellular and Molecular Biology (CCMB), Hyderabad.
(3) International Centre for Genetic Engineering and Biotechnology (ICGEB), New Delhi.
(4) Indian Agricultural Research Institute (IARI), New Delhi
(5) National Dairy Research Institute (NDRI), Karnal.
(6) Indian Veterinary Research Institute (IVRI), Izatnagar (U.P.)
(7) Council of Scientific and Industrial Research (CSIR), New Delhi.

Biotechnology

(8) Indian Council of Agricultural Research (ICAR), New Delhi.
(9) The Central Plantation Crops Research Institute (ICPCRI), Kasargod.
(10) National Botanical Research Institute (NBRI), . Lucknow.
(11) Central Drug Research Institure (CDRI), Lucknow.
(12) Institute of Genomics and Integrative Biology (IGIB), Delhi.
(13) Central Food Technological Research Institute (CFTRI), Mysore.
(14) Central Institute Medicinal and Aromatic Plants (CIMAP), Lucknow.
(15) Institute of Himalayan Bioresource Technology (IHBT), Palampur.
(16) Indian Institute of Chemical Biology (IICB), Calcutta
(17) Indian Institute of Chemical Technology (IICT), Hyderabad.
(18) Indian Institute of Microbial Technology (IMTECH), Chandigarh.
(19) Industrial Toxicology Research Centre (ITRC), Lucknow.
(20) National Chemical Laboratory (NCL), Pune.
(21) National Environmental Engineering Research Institute (NEERI), Nagpur,
(22) National Institute of Oceanography (NIO), Goa.
(23) National Institute of Immunology, New Delhi.
(24) National Centre for Plant Genome Research, New Delhi
(25) National Centre for Cell Sciences, Pune.
(26) Institute of Life Sciences, Bhuvaneshwar.
(27) Centre for DNA Fingerprinting and Diagnostic, Hyderabad.
(28) Institute of Bioresources and Sustainable Development (IBSD), Imphal, Manipur.
(29) National Brain Research Centre, Gurgaon.
(30) Indian Institute of Petroleum (IIP), Dehradun.

National Biotechnology Board — It was established under the department of Science and Technology, Government of India, on the recommendations of science Advisory Committee to Parliament. This board is co-ordinating and encouraging researches in the field of biotechnology. The NBT has decided to compile a list of biotechnologists in India and abroad working in the fields of — (*i*) genetic engineering, (*ii*) photosynthesis, (*iii*) tissue culture, (*iv*) enzyme engineering, (*v*) alcoholic fermentation and (*vi*) immunotechnology.

Techniques of Biotechnology

The following important techniques are being utilized in biotechnology —

(1) Genetic engineering.
(2) Development of monoclonal antibodies and hybridomas.
(3) Development of DNA probes.
(4) Tissue culture.
(5) Protoplast fusion.
(6) Gene synthesis.
(7) Gene isolation, modification and insertion in existing genomes and vectors.
(8) Reverse transcriptase.
(9) Uptake of free DNA and DNA injection in eukaryotes.
(10) Synthesis of peptides are vaccines.

Most of the techniques have been discussed in chapter 1.

Basic Aspects of Plant Tissue Culture

Plant tissue culture is a common term which is generally used for *vitro* culture of plant cells, tissues and organs. In strict sense, this term is applied for *in vitro* cultivation of plant cells in an unorganised mass i.e., callus cultures. Callus represents an unorganised mass of plant cell capable of cell-division and growth *in vitro*. The term *cell culture* is used for *in vitro* suspension cultures, i.e., culture of single cell or a small group of plant cells. The term plant tissue culture is applied for both callus and suspension cultures.

Plant tissue culture technique is generally used in such cases where conventional breeding methods fail to achieve the target and to save the time, labour and production cost of the products. This technique is being used now a days in inter — and intraspecific crosses, micropropagation, to produce somaclonal variation and encapsulated seeds etc. Conventional breeding methods take about 4-7 years to produce pure lines, haploids and polyploids etc. This duration can be easily reduced through plant tissue culture techniques. The plant tissue culture techniques may be supplemented with genetic engineering to get quick results.

Brief History

1. *Haberlandt* (1902) — Reported culture of isolated single palisade cells from leaves in Knop's salt solution enriched with sucrose.
2. *Hanning* (1904) — Cultured embryo for the first time *in vitro* of certain crucifers.
3. *Gautheret, white* and *Nobecourt* (1939) — Established callus cultures from cambium tissue.
4. *White* (1932) — Developed a tissue culture medium for, *in vitro*, culturing.
5. *Skoog* (1944) — Reported shoot bud differentiation from tobacco pith tissue's cultured *in vitro*.
6. *Rao* (1956) — Cultured ovary of *Phlox drummondii*.
7. *Skoog* and *Miller* (1957) — Proposed that root - shoot differentiation is regulated by auxin - cytokinin ratio.
8. *Braun* (1959) — Regenerated first plant from a mature plant cell.
9. *Reinert* (1958) and *Steward* (1959) — Independently reported the development of somatic embryos for the first time from carrot tissues.
10. *Abraham* and *Ram Chandran* (1960) — Cultured embryos of *Colocasia*.
11. *Nirmala Maheswari* (1961) — First cultured ovary and ovules of *Iberis amara*.
12. *Rangaswami* (1961) — Developed modified White's medium for tissue culture and cultured nucellar tissue of *Citrus microcarpa*.
13. *Maheshwari* and *Baldev* (1962) — Cultured embryo of *Cuscuta reflexa*.
14. *Murashige* and *Skoog* (1962) — Developed a culture medium, called M.S. medium, for tissue and organ culture.
15. *Johri* and *Bajaj* (1963) Cultured embryo of *Dendrophthoe falcata*.
16. *Sabharwal* (1963) developed proembryo from callus of *Citrus aurantifolium* and *C. reticulata*.
17. *Guha* and *Maheshwari* (1964) Cultured anthers of *Datura stramonium* and *D. innoxia* and obtained embryoids.
18. *Johri* and *Bhojwani* (1965) Developed roots and shoots from the callus of *Exocarpus cupressiformis*.
19. *Guha* and *Maheshwari* (1966) Produced complete androgenic haploid plants by pollen culture technique using pollens of *Datura innoxia*.

20. *Sahgal* (1968) Developed plant of *Anethum graveoleus* from its callus has grown the plant upto flowering stage.
21. *Mohan Ram* (1968) Cultured floral buds of *kalanchae*.
22. *Mitra* and *Chaturved* (1969) Developed fruiting plants from leaf tissue culture of *Rauwolfia serpentina*.
23. *Narayana Swamy* and *Chandy* (1971) Developed plantlets of *Datura metal* through tissue culture technique.
24. *Johri* (1971) Developed tissue culture technique of micropropagation of orchids.
25. *Nitsch* (1972) Demonstrated anther culture technique.
26. *San Noeum* (1977) First reported gynogenesis *in vitro*.
27. *Sunderland* and *Roberts* (1977) Developed float culture technique.
28. *De lautour* et al (1978) developed embryo-transplant technique for culturing embryo.
29. *Rashid* and *Reinert* (1981) Studied pollen dimorphism and its role in androgenesis.
30. *Yang* et al (1986) reported synergid-apogamy.
31. *Redenbaugh* (1993) developed technique of producing androgenic haploids from pollen culture of *Datura innoxia*.

Requirements For *in Vitro* Cultures

For tissue culture, following basic things are required.

1. A Tissue Culture Laboratory

A well furnished, spaceous and planned laboratory is required. The laboratory must have good space and facilities for

(*i*) Washing, drying and storage of vessels.

(*ii*) Preparation of nutrient medium, sterilization, cleaning and storage of media and other supplies.

(*iii*) Aseptic conditions for working with living or plant materials.

(*iv*) Controlled environmental conditions for maintaining cultures.

(*v*) Observation and evaluation of cultures.

(*a*) **Washing and Storage Facilities** — These include large sink with cold and hot running water, distillation apparatus, washing machine, drier, cleaning brushes, physical and chemical balances for weighing chemicals and glassware.

(*b*) **Media Preparation Room** — This room must have sufficient space for preparation of media. The space should be provided for bench for chemicals, culture vessels, labware and equipments like hot plates, centrifuge, water-bath, burners, oven, autoclave, refrigerator, pH meter, balances, stirrers and culture vessels etc.

This room should have facility to control temperature (usually at $25° \pm 2°$) with the help of airconditioners and room heaters). It is necessary to have a generator set for providing power to the media room in case of electricity failure. Culture racks should be fitted with 1000 lux light or fluorescent tubes. Borosilicate or Pyrex glasswares and vessels should be preferred for culture work.

The culturing of tissue is usually done in the 25×150 mm or larger rimless tubes, flasks and petridishes. Wide mouth bottles or milk bottles are especially used for micropropagation work. Suspension cultures are maintained in long neck culture flasks. The culture tubes and flasks containing sterilized nutrient medium after inoculation with plant material are plugged with non-absorbent sterile cotton or metal caps. Other laboratory accessories needed are-dissecting microscope, high quality compound microscope, air filter etc.

2. Maintenance of Aseptic Environment

Aseptic environment is maintained to avoid contamination of fungi and bacteria present in air. The contaminants produce toxic metabolites which inhibit the growth of cultured plant tissues. Aseptic environment is usually maintained by sterilizing glassware, instruments, culture room, nutrient media and plant materials.

The sterilization of laboratory material and culture room materials is achieved by one of the following methods:

(*i*) Dry heat, (*ii*) Flame sterilization, (*iii*) Wiping with 70% ethanol, (*iv*) Filter sterilization, (*v*) Autoclaving, and Surface sterilization.

Table 2.1 summarises various methods used for sterilization of specific materials.

Table 2.1

Sterilization Method	Materials sterilized
(*i*) Dry heat (160-180°C for 3 hr)	Empty glasswares like culture vessels, pipettes etc, and certain plasticwares (Teflon FEP); instruments like scalpels, forceps, needles etc.
(*ii*) Flame sterilization	Instruments like scalpels, forceps etc; mouths of culture vessels.
(*iii*) Wiping with 70% Ethanol	Hands of the operator, platform of laminar flow cabinets.
(*iv*) Filter sterilization	
(*a*) Liquid (membrane filter of 0.45 µm or smaller pore size)	Heat labile compounds like GA_3, Zeatin, ABA, Urea, enzymes, certain vitamins.
(*b*) Air (HEPA Filter)	Air blown through laminar flow cabinets.
(*v*) Autoclaving (121°C at 15 p.s.i. for 15-40 min	Media, culture vessels (glasswares and several plasticwares), other glass and plasticwares, contaminated cultures.
(*vi*) Surface sterilization (using one of several sterilizing agents)	All plant materials to be cultured.

Sterilization by *dry heat* is done in an oven at 160-180°C for 3 hours. In *flame Sterilization*, the instruments are dipped in 95% alcohol first and then flamed. Platform of laminar flow and hands of operator are sterilized by *wiping* then with 70% alcohol and then drying. The treatment of plant materials under experimentation with one of the sterilizing agents like 2% sodium hypochlorite, 0.1 to 1% mercuric chloride, 1% silver nitrate, 9-10% calcium hypochlorite, 10-12% Hydrogen per oxide, or 1-2% bromine water etc is called *surface sterilization*. This treatment inactivates microbes present on the surface of plant material. The duration of treatment varies according to plant material from 15 to 30 minutes. Usually soft explants (excised piece or tissue or organ used for culture) like ovule, embryo, pollen grains etc. are surface sterilized alongwith the organ containing them.

The *explants* may be obtained from the field or excised from the seeds aseptically. In such cases, the seeds are first disinfected by the following process:

Autoclaving is done to sterilize nutrient media, glass wares, plasticwares and contaminated cultures. Autoclaving requires 121°C temperature at 15 ponds per square inch (psi) for 15-40 minutes. This variation in time depends upon the volume of the nutrient media (1 to 200 ml — 15 seconds; 200-1000 ml - 30 sec; 1000 - 2000 ml - 40 sec).

Biotechnology

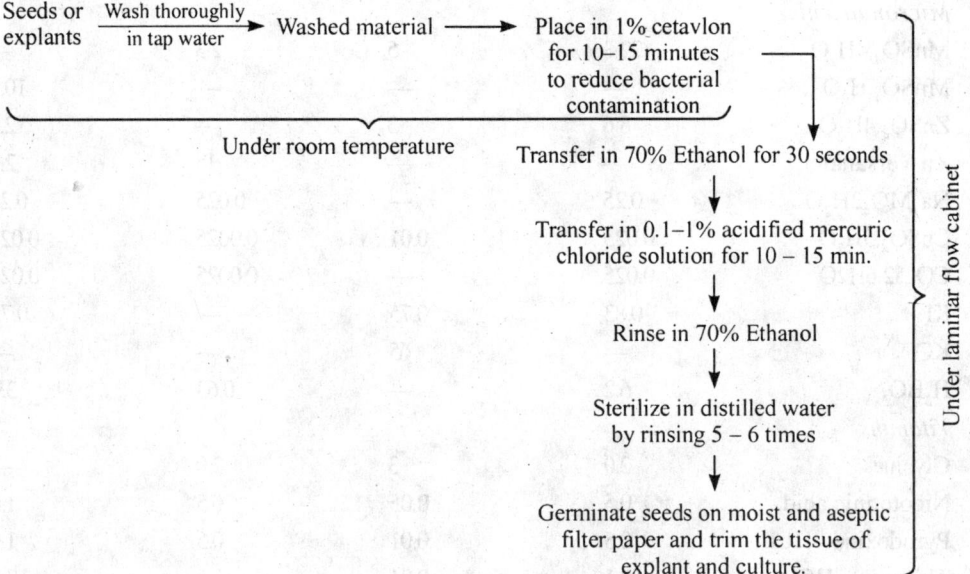

Fig. 2.1. Schematic representation of surface sterilization.

3. *Nutrient Media - Composition and Preparation*

Nutrient media provide nutrition to the culturing cells. The cells absorb nutrients through cell membrane for rapid proliferation or cell-division to increase their number. All nutrient, media for tissue culture are usually synthetic or of chemically defined compounds (Table 2.2) and only a few of them contain complex organic substances, *e.g.*, potato extract, as their normal constituents. Different nutrient media are used for different type of culture experiments because a single medium never suits for all types of tissue cultures.

In general, the nutrient media contain inorganic minerals, growth hormones, vitamins, carbon source, amino acids, solidifying agents and organic constituents.

Table 2.2. Composition of nutrient media required for in *vitro* culture.

	Chemical Constituents	MS medium* (mg/l)	White's medium (mg/l)	ER medium* (mg/l)	B5 medium* (mg/l)
1.	*Macronutrients*				
	NH_4NO_3	1650	—	1200	—
	KNO_3	1900	80	1900	2500
	$CaCl_2 \cdot 2H_2O$	440	—	440	150
	$MgSO_4 \cdot 6H_2O$	370	750	370	250
	KH_2PO_4	170	—	340	—
	$(NH_4)_2SO_4$	—	—	—	134
	Na_2SO_4	—	200	—	—
	NaH_2PO_4	—	19	—	150
	$Ca(NO_3)_2 \cdot 4H_2O$	—	300	—	—
2.	*Iron*				
	Na_2EDTA	37.3	—	37.3	37.3
	$FeSO_4 \cdot 7H_2O$	27.8	—	27.8	27.8

3.	**Micronutrients**				
	$MnSO_4.4H_2O$	22.3	5	2.3	—
	$MnSO_4.H_2O$	—	—	—	10.0
	$ZnSO_4.4H_2O$	8.6	3	—	2.0
	Zn versanate	—	—	15	2.0
	$Na_2MO_4 2H_2O$	0.25	—	0.025	0.25
	$CuSO_4.5H_2O$	0.025	0.01	0.0025	0.025
	$COCl_2.6H_2O$	0.025	—	0.0025	0.025
	KI	0.83	0.75	—	0.75
	KCl	—	65	—	—
	H_3BO_3	6.2	—	0.63	3.0
4.	**Vitamins**				
	Glycine	2.0	3	2.0	—
	Nicototinic acid	0.5	0.05	0.5	1.0
	Pyrodoxine - HCl	0.5	0.01	0.5	1.0
	Thiamine - HCl	0.1	0.01	0.5	10.0
5.	**Cytokinin**				
	Kinetin	0.04-10.0	—	0.02	0.1
	Myo-inositol	100.0	—	—	100.0
	IAA	1.0-30.0	as per need	—	—
	NAA	—	as per need	1.0	—
	2, 4-D	—	—	—	0.1-1.0
6.	**Carbon source**				
	Sucrose (g)	30.0	20	40.0	20.0
7.	**Medium**				
	PH	5.7	5.8	5.8	5.5

* M S, Murashige and Skoog (1962); E R, Eriksson (1965); B5, Gamborg et al. (1968); P.R. White (1963).

(*a*) *Inorganic minerals* — They includes salts of *macronutrients* like nitrogen, phosphorus, potassium, sulphur, calcium and magnesium etc, and *micronutrients* like manganese, zinc, copper, iron, sodium, molybdenum, boron and chloride. First of all a stock solution is prepared in advance and then it is added in the medium according to requirement. To make the iron soluble in the medium, the stock solution of iron is prepared in chelated form as the sodium salt of ferric ethylenediamine tetracetate. It is an iron - EDTA complex and it becomes easily available to the medium even at pH more than 5.8.

(*b*) *Growth Hormones* — Cytokinins, kinetins, myo-inositol, indole acetic acid (IAA), Indole 3-butyric acid (IBA), naphthalene acetic acid (NAA), 2, 4-dichlorophenoxy acetic acid (2, 4-D) and naphthoxy acetic acid (NOA) etc. are commonly used growth hormones.

Cytokinins like kinetin (furfurylamino purine), Zeatin, benzylaminopurine (BAP), isopentenyl adenine (2-ip) and thidiazuron are commonly used to promote cell-division, growth, regeneration of shoots, induction of somatic embryo and axillary buds. The range of concentration used in culture is 0.1–3 mg/litre.

Auxins like IAA, IBA, NAA, NOA and 2, 4-D etc. stimulate cell-division, callus growth, rooting and shoot elongation. Auxin concentration used in cultures is 0.1-3 mg/l.

GA_3 has been exclusively used among gibberellins which promotes shoot elongation and embryo germination in 0.1 -1 mg/l concentration.

For the structure and functions of various growth hormones, consult chapter 21 of physiology portion and chapter 8 of Biochemistry portion of this book.

(c) **Vitamins** — Vitamins B_1, niacin, B_2, B_6, C, B_{12} and H are used for plant tissue cultures. For their structure and function, consult chapter 7 in Biochemistry portion of this book.

(d) **Carbon Source** — The organic compounds are used as a source of carbon. Their concentration varies from 20 to 30 g/litre. Sucrose and D-glucose are commonly used for carbon source but glycerol and myoinosital are also used as chief source. Other organic compounds used are peptone, coconut milk, yeast extract, malt extract and tomato juice etc.

(e) **Amino acids** — They provide nitrogen in addition to inorganic salts. The most widely used amino acids are L-aspartic acid, L-asparagine, L-glutamic acid, L-glutamine and L-arginine. For structure of these amino acid, consult chapter 3 in Biochemistry section.

(f) **Solidifying agents** — Agar, a polysaccharide obtained from a red alga — *Gelidium amansü*, is most commonly used as solidifying or gelling agent. Agar gels do not react with constituents of media and are not digested by tissue enzymes.

(g) **pH** — The pH affects the uptake of ions. The optimum pH required for the growth and development of cultured tissue ranges between 5.00 and 6.00. It should be maintained before sterilization of the nutrient medium.

4. Callus and Suspension Cultures

During culturing of explants in nutrient medium, many of its cells especially parenchyma and often collenchyma undergo division, maturation and differentiation. Thus, these cells get converted into meristematic cells. This phenomenon is called *dedifferentiation*. It involves enhancement of RNA and protein synthesis needed for cellular activity. The cells divide and form unorganised mass of cells consisting of fibres and vascular elements etc. Initially, the cell-mass is formed at cut ends of explant but later on it spreads throughout the explant.

(a) **Callus culture** — An unorganised mass of cells formed by cultured cells or tissues on an agar gelled medium is called *callus*. The callus cultures are needed to be subcultured every 3–5 weeks due to cell-growth, depletion of nutrients and drying of medium.

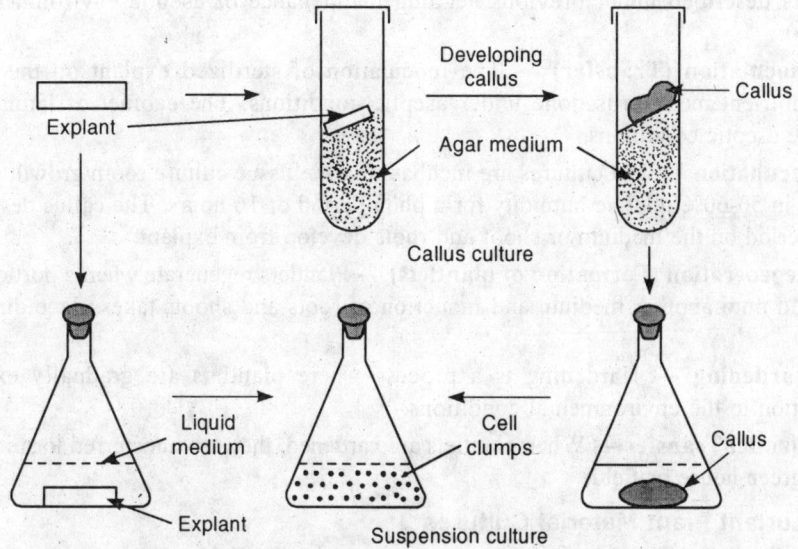

Fig. 2.2. Initiation of callus and suspension cultures *in vitro*.

(b) Suspension cultures — A suspension of single cells or cell clumps of few or many cells produced in a liquid medium during culture of cells and tissue is called *suspension culture*. These cultures need continuous agitation to facilitate aeration and dissociation of cell — clumps into smaller pieces. The suspension cultures grow faster than callus cultures and so need to be sub-cultured every week. The suspension cultures are of following three types:

(*i*) *Batch cultures* — These are those cultures where the same medium and all the cells produced are retained in the culture vessel (tube or flask). The cell number or biomass of batch culture shows a typical sigmoid curve of growth.

(*ii*) *Continuous cultures* — In such cultures, the cell-population is maintained in a steady state by regular replacement of a portion of the used or spent medium by fresh medium. Continuous cultures may be of two types - closed and open. *In closed continuous cultures*, the cells are separated from the used medium and the entire medium is taken out and replaced and then the cells are added back in the fresh medium for the further increasement of cells-biomass. In open continuous cultures, the cells and some volume of the used medium is taken out from the cultures and then equal volume of fresh medium is added alongwith cells.

(*iii*) *Immobilized cell cultures* — In immobilized cell cultures, the plant cells or tissues are encapsulated in suitable materials like agarose or calcium alginate gels, or entrapped in membranes or steel screens and then are packed in a column. The liquid medium is allowed to run continously in the column.

Methods of Plant Tissue Culture — Cell, Tissue and Organ Basic Steps

The basic steps for regeneration of complete plants from an explant or cells are as follows:

(*i*) **Preparation of Suitable Nutrient Medium** — Suitable/nutrient medium is prepared according to the need of culture-experiment and transferred into the flask or petriplate or culture tube. It is autoclaved at 121°C temperature and 15 psi (pond per square inch) for 30 minutes. The hormones and vitamins are sterilized using membrane filter of 0.45 µm or less pore size.

(*ii*) **Selection of Explants** — Any excised part of plant to be used in tissue culture are *explants*. They may be leaf and stem pieces, axillary buds, root and shoot tips, anther, ovary and endosperm etc. During selection of explants, it is necessary that they should be young and healthy.

(*iii*) **Sterilization of explants** — Sterilization of explants can be done by using any of the disinfectants described under previous heading maintenance of aseptic environment (surface sterilization).

(*iv*) **Inoculation (Transfer)** — The inoculation of sterilized explant on the surface of solidified nutrient medium is done under aseptic conditions. The cabinet of laminar airflow provides the aseptic conditions.

(*v*) **Incubation** — The cultures are incubated in the tissue culture room/growth chamber at 25° ± 2°C in 50-60% relative humidity for a photoperiod of 16 hours. The callus develops after a defined period on the medium or shoot and roots develop from explant.

(*vi*) **Regeneration (Formation of plantlets)** — Plantlets regenerate when a portion of callus is transferred onto another medium and induction of roots and shoots takes place directly from explant.

(*vii*) **Hardening** — Hardening is a process where plantlets are gradually exposed for acclimatization to the environmental conditions.

(*viii*) **Plantlet Transfer** — When plantlets are hardened, they are transferred to suitable place, i.e., either green house or field.

Some Important Plant Material Cultures

Different types of cultures are produced by using various plant materials. The important ones are explant culture, callus culture, suspension culture, protoplast culture and organ culture.

Biotechnology

1. Explant Culture

Different types of seed plants (herb, shrubs and trees) although possess similar morphological units, *i.e.*, root, stem and leaves, but they differ in internal organisation of tissues. Of the various types of tissues, parenchyma has been considered as most versatile for culturing expriments because its cells have greater capacity to divide and grow. So, the presence of parenchyma is must in explants. Parenchyma from stems, rhizomes, tubers and roots is easily accessible and it shows better results in culture conditions *in vitro*. Its cells also possess an identical morphogenetic potential.

Explant cultures are the cultures of any excised part of a plant. It may be a young and healthy piece of stem or leaf, axillary bud, node, root and shoot tips, anther, ovary, cotyledon or endosperm. Explant cultures are generally used for induction of callus or regeneration of plant.

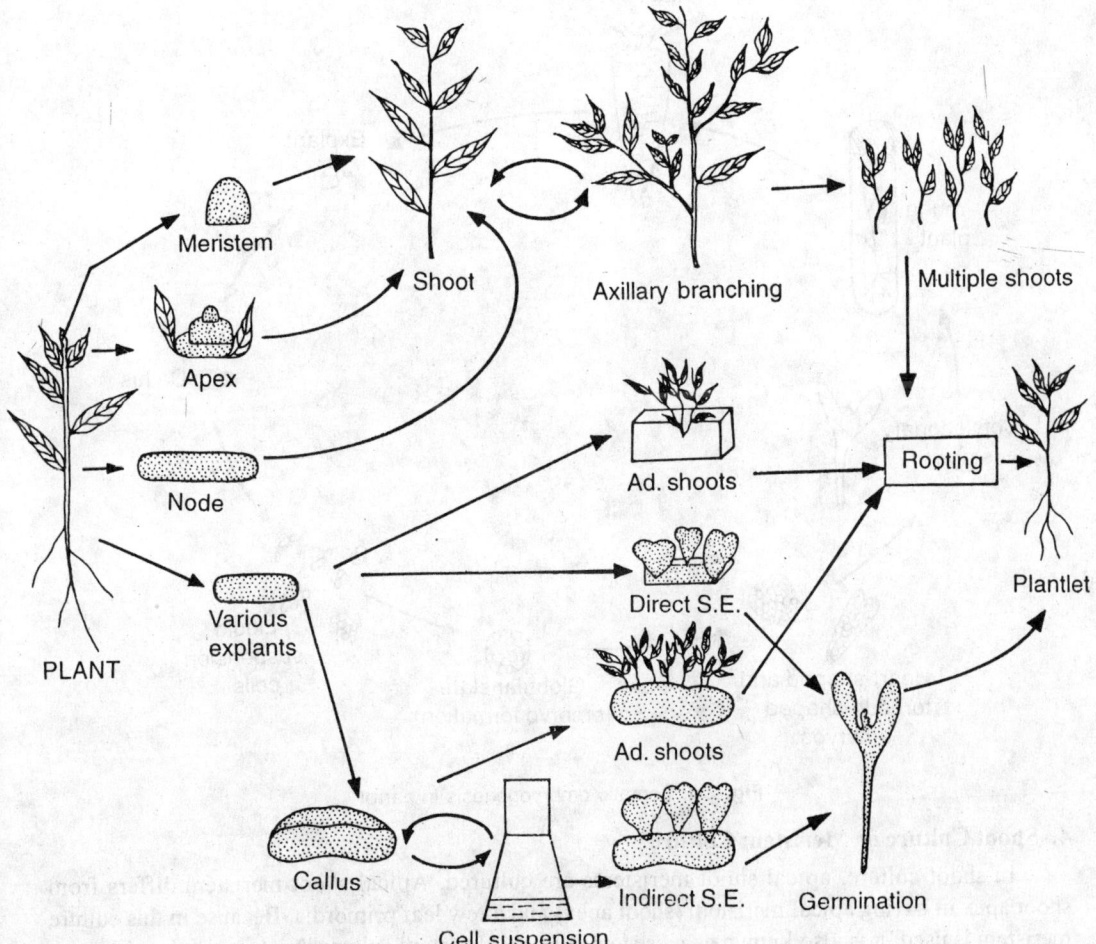

Fig. 2.3. Different methods of regeneration from explant and callus/cell cultures.

2. Callus Cultures

It has been described earlier under heading "Callus and suspension cultures". Callus has been artificially developed through tissue culture techniques. During callus culture a 2-5 mm sterilized piece of explant (preferably root or stem or tuber) is transferred into nutrient medium and inoculated at 25-28°C in an alternate light and dark period of 12 hours. The abnormal growth of callus has potentiality to develop normal roots, shoots and embryoids and ultimately into a plant. The callus cultures are needed to be subcultured every 3-5 weeks due to cell growth, depletion of nutrients and drying of medium.

Organogenesis : The development of organs like root, stem and leaves (but not embryo) induced in plant tissue culture is called organogenesis. It starts due to stimulation of chemicals of the medium especially growth hormones (auxins, cytokinins) endogenous compounds produced by the culture and substances carried over from the original explant.

3. Root Culture

Usually excised roots are cultured in liquid medium. Initially tomato roots were cultured by *White* (1934). Later on, roots have been cultured both from gymnosperms and angiosperm species. Somatic embryogenesis in carrot is shown in fig.2.4.

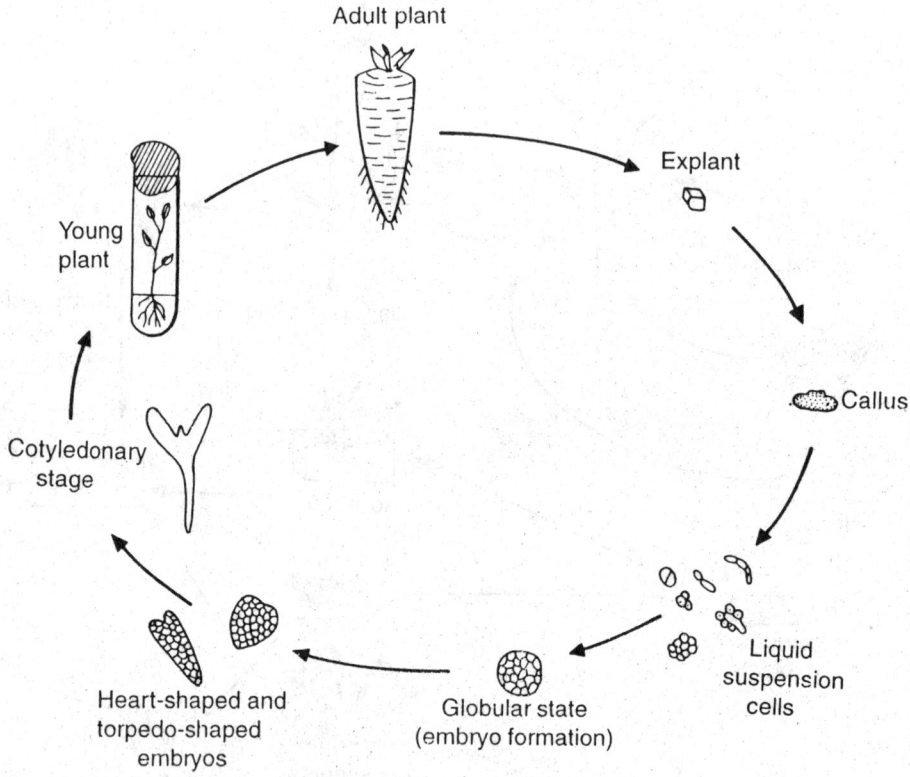

Fig. 2.4. Somatic embryogenesis in carrot.

4. Shoot Culture or Meristem Culture

In shoot culture, apical shoot meristems are cultured. Apical shoot meristem differs from shoot apex in having apical meristem (shoot apex) and a few leaf primordia. Because in this culture meristem is used, it is also known as *meristem culture*. Meristem culture involves the development of already existing shoot meristem which ultimately develops adventitious roots from developed shoot. In this case, a new shoot meristem is not regenerated.

Explant — Apical shoot meristem remains situated in the shoot tip beyond the youngest leaf or first leaf primoridium. It is about 250 µm in length and 100 µm in diameter. A completely developed *shoot tip* of about 100-500 µm contains 1-3 leaf primordia in addition to apical meristem.

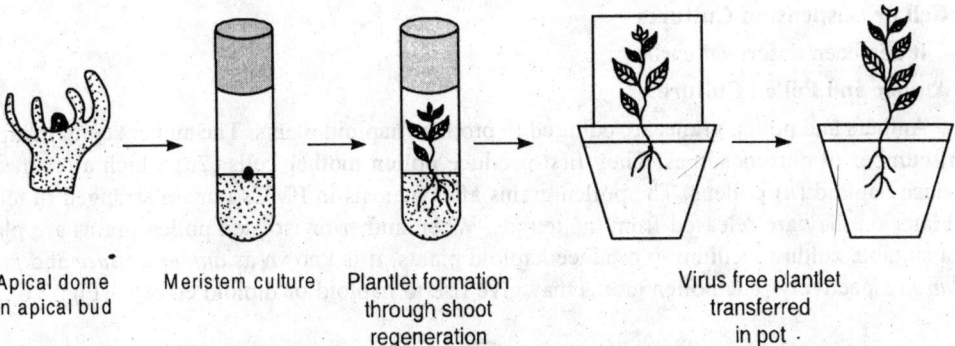

Fig. 2.5. Representation of isolation and meristem culture and regeneration of virus free plantlets.

In meristem culture, the size of shoot tip (explant) varies from 1 µm to 10 mm or more in size according to the objective. In such cultures generally shoot tips are used. Generally, explants taken from actively growing plants in the beginning of growing season are most suitable.

Culture Medium — Generally, MS medium has been used for most of the plant species and it showed satisfactory results. In some species, high concentration of MS medium is not necessary and it shows deleterious or toxic effects, *e.g.*, blueberry. Agar gelled medium and liquid medium both may be used according to convenience.

The *growth regulator* (GR) requirements depend on the stage of culture process, viz., (*i*) Culture initiation, (*ii*) shoot multiplication, (*iii*) rooting of shoot, and transfer of plantlets to soil or pots.

Culture initiation includes surface sterilization of explants and their establishment *in vitro*. During this stage the contamination is detected, eliminated and controlled. When explant is heavily contaminated, a suitable antibiotic or fungicide is added in the medium which is basically GR free. The growth of explant in this medium may or may not occur. For *shoot multiplication*, the explant is transferred into another medium containing growth regulators after 2-3 weeks. Generally, cytokinin in (BAP or 2-iP) alone, or in combination with auxins like NAA, IBA and IAA, is used. Their concentrations depend upon the explant species. Such GR combinations are used which could promote optimum rates of shoot multiplication with minimum risk of production of adventitious shoot buds.

Environmental Conditions — During above two stages, the cultures should be kept at 25°C and 1000 lux light. Sometime 3000 lux light is required. During rooting, the light requirement increases from 3000 to 10,000 lux.

Browning of Medium — Sometimes the nutrient medium becomes brown due to oxidation of phenolic compounds produced by the surfaces of explants. This problem usually arises when explant is excised from mature and woody tree species. This problem can be overcome either by subculturing the explant or using liquid medium or adding antioxidants like ascorbic acid, cystein HCl or citric acid, or by adding activated charcoal as adsorbent or by culturing the explant in dark conditions.

Rooting of Shoots — Rooting of shoot requires low concentration of salts (about 1/4 or 1/2) and reduced sugar levels (1g/l) in the MS medium. In rooting GRs also play important role. Generally auxins (NAA or IBA in 0.1-1 mg/l) are required for rooting but in certain cases, *e.g.*, strawberry and *Narcissus*, GR-free medium is required.

Transfer of Plantlets — When the roots in the plantlets are about 0.5 to 1 cm long, the plantlets are transferred from rooting medium to the soil or pots containing nutrient mixed soil. The plantlets should be kept at about 90% humidity and low light intensities initially. The plants are finally exposed to green house conditions.

5. Cell or Suspension Cultures

It has been described earlier.

6. Anther and Pollen Culture

Anthers and pollen grains are cultured to produce haploid plants. The anthers contain diploid ($2n$) number of chromosomes. They first produce pollen mother cells ($2n$) which after meiosis produce haploid (n) pollens. The pollen grains after meiosis in PMCs remain arranged in tetrads and later on they are released from the tetrads. When anther or isolated pollen grains are placed on a suitable culture medium to produce haploid plants, it is known as *anther culture* and *pollen culture* respectively. The pollen grains may give rise to haploid or diploid embryo (Fig. 2.6).

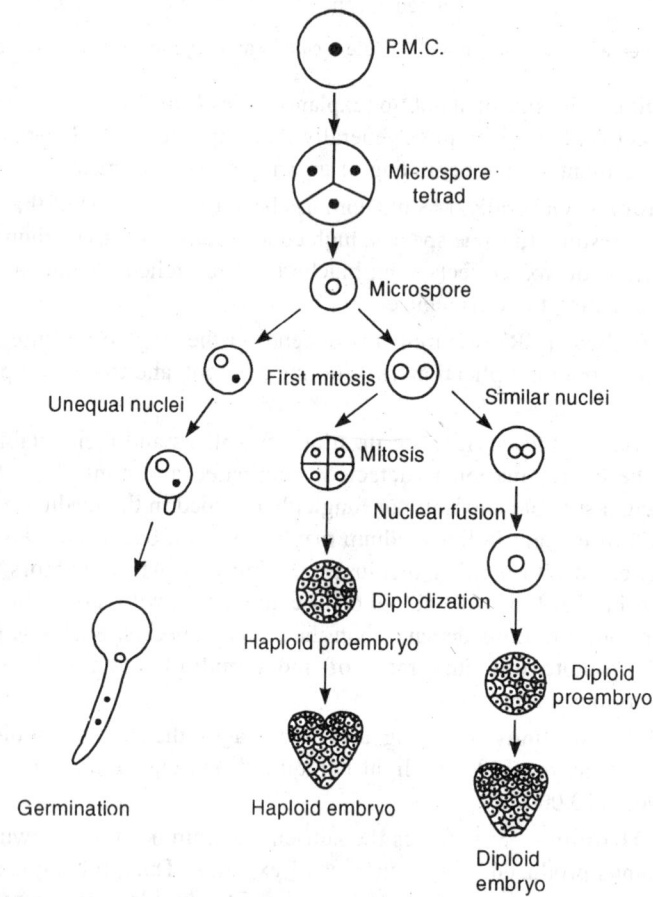

Fig. 2.6. Microsporogenesis and development of haploid or diploid embryo or normal germination of pollen grain.

Culture Technique *in Vitro*

The anthers should be taken from such plants which have been grown in the field or in pots under controlled conditions of humidity, light and temperature. These conditions differ from species to species. In practice, the flower buds are collected at appropriate development stage, surface sterilized and then their anthers are excised and placed horizontally on culture medium. During excising, the injury to anthers should be avoided because it may induce callus formation from anther walls. It is better to take out pollen grains from the anthers first and then these should be cultured on suitable culture medium.

Direct and Indirect Androgenesis

In some plant species e.g., *Brassica campestris*, *B. napus*, Species of *Nicotiana* and *Petunia axillaris* etc., the pollen grains of cultured anthers directly produced somatic embryos (SEs). It is called *direct androgenesis* or *pollen derived embryogenesis*. In other plants, like *Oryza vulgare* (rice), *Triticum* (wheat), *Hordeum vulgare* (barley), *Lycopersicum* (tomato) and *Triticale* etc., the pollen grains first produced callus and the plantlets were regenerated from this callus under suitable culture conditions. It is called *indirect androgenesis*. In direct androgenesis the plants produced haploid embryos whereas in indirect androgenesis, the ploidy level varied.

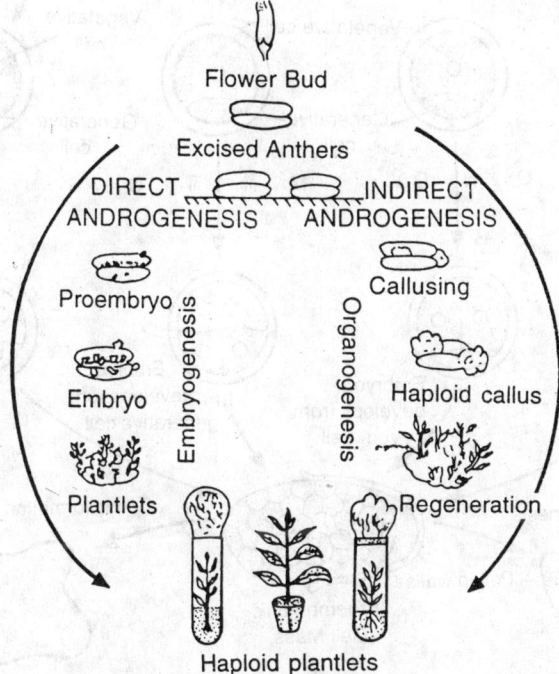

Fig. 2.7. Anther culture and androgenesis leading to production of haploid plants

Pathways of Embryo or Haploid Plant Development

There are following four pathways of embryo or haploid plant development —

(*i*) **Pathway I** — The uniunucleate pollen grain divides symmetrically to form two equal sized daughter cells (vegetative cell and generative cells). Both these cells develop into embryo by further divisions e.g., *Datura innoxia*.

(*ii*) **Pathway II** — The pollen grain divides asymmetrically to form larger vegetative and smaller generative cell. The generative cell degenerates and vegetative cell after successive divisions forms embryo or callus. e.g., *Datura metal*, *Nicotiana tobacum*, *Triticum vulgare*, *Hordeum vulgare*, *Lycopersicum* and *Capsicum* sp, and *Triticale*.

(*iii*) **Pathway III** — The pollen grain divides asymmetrically into a larger vegetative cell and a smaller generative cell. The vegetative cell either does not divide or divide to a limited extent to form a suspensor. The generative cell alone undergoes successive divisions to form embryo. e.g., *Hyoscyamus niger*.

(*iv*) **Pathway IV** — The pollen grain divides asymmetrically to form a larger vegetative and a smaller generative cell. Both these cells divide repeatedly to form embryo. Thus, both the cells are involved in embryo formation.

Different pathways of embryonial mass formation are shown in Fig. 2.8.

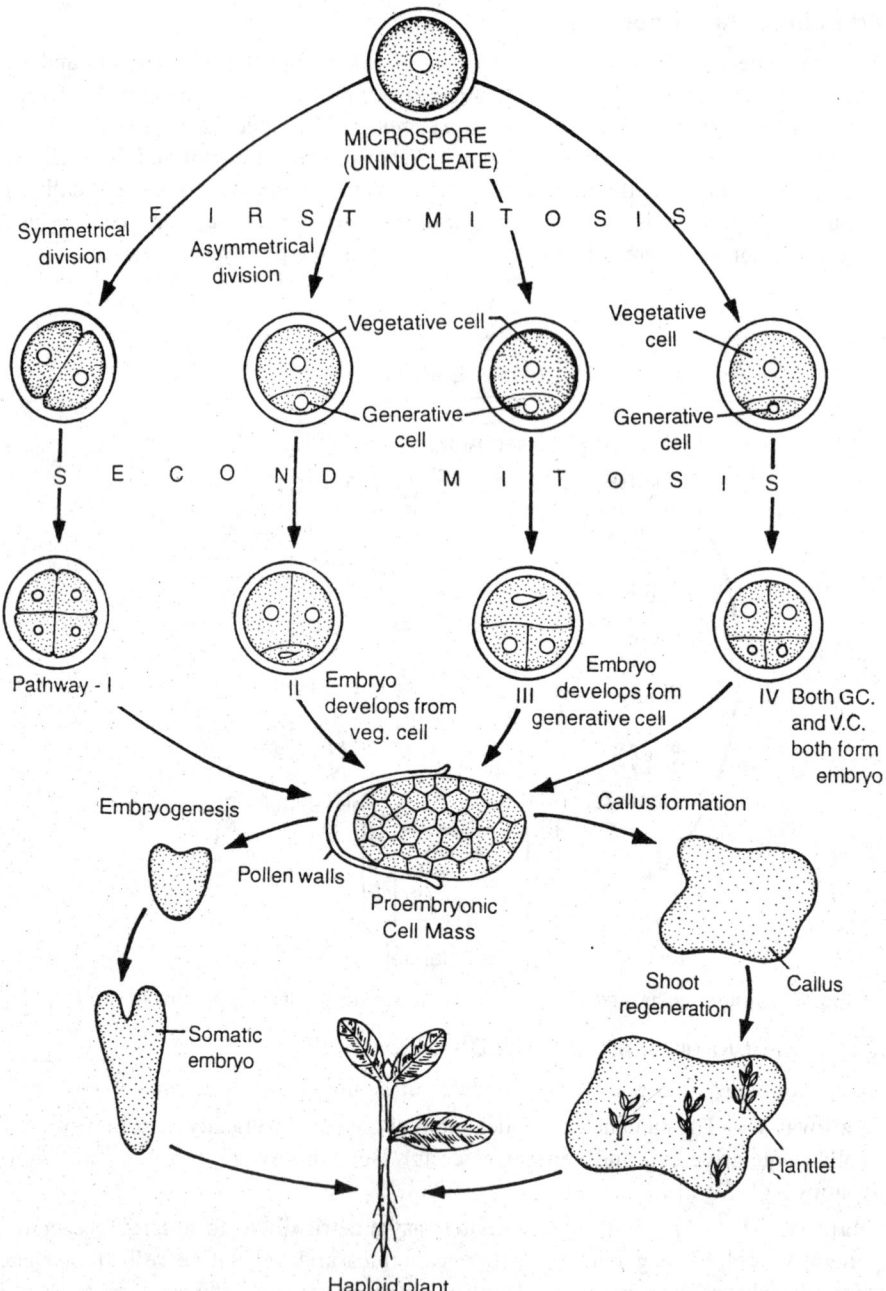

Fig. 2.8. Different pathways of androgenesis and embryonic cell mass formation.

Pollen Dimorphism — The pollen grains of many plant species, *e.g.*, wheat, barley, and tobacco etc., show *dimorphism*. When the pollen grains are stained with acetocarmine nuclear stain, most of the larger pollen grains (99.3%) become deeply stained and remaining a few (0.7%) smaller remain lightly stained. The latter (light stained) pollen grains are called *S-pollen grains* and these actually respond during anther culture. Their number may be increased by giving chilling treatment.

Culture Medium

Generally *Murashige* and *Skoog* (M.S.) and *Linsmaer* and *Skoog* (L.S.) media supplemented with sucrose are used in tissue culture. Sucrose from 3-6% is essential for anther cultures which plays a key role in induction of embryogenesis. Other constituents of the medium help in the post-induction of embryo development. N_6 and Potato-2 media are used for cereals. When White's and Heller's medium is used, it is supplemented with coconut milk because this medium contains low concentrations of salts. Thus, medium requirement varies in different plant species. It is related with the genotype, age of donor plant and anther, and growing conditions of the donor plant. General requirement of sucrose is 2-3%.

Growth Regulators (GR)

Various growth regulators like auxins (IAA, IBA, NAA) are used in embryo cultures in *Hyoscyamus niger*. IAA in 0.1 mg/l gave best results. 2, 4-D (2 mg/l) enhances callus formation, *e.g.*, cytokinins may reduce the number of pollen grains producing embryos. It may be due to interference with cell division in induced pollen grains. Sometimes a mixture of auxins and cytokinins is used. In solanaceous plants, pollen embryogenesis does not require any growth regulators but low concentrations of auxins, cytokinins or even GA_3 may be useful. Wheat anthers cultured in 2, 4-D containing medium produced callus while those anthers cultured in coconut milk supplemented medium gave rise to embryos.

Culture Environment

Anther cultures are generally maintained in alternating periods of light (12-18 hr, 5000-10,000 lux m^2) at 28°C and darkness (12 hr.) at 22°C. When the seedlings (from embryos) or shoots (from callus) become 3 – 5 cm long, they are first transferred to root producing medium and then to soil. The optimum temperature may vary. It is 25°C for tobacco embryos.

7. Embryo Culture

Embryo culture has also been done for the production of haploid plants. It is useful in such cases where embryo fails to develop due to degeneration of embryonic tissues. It is also used for

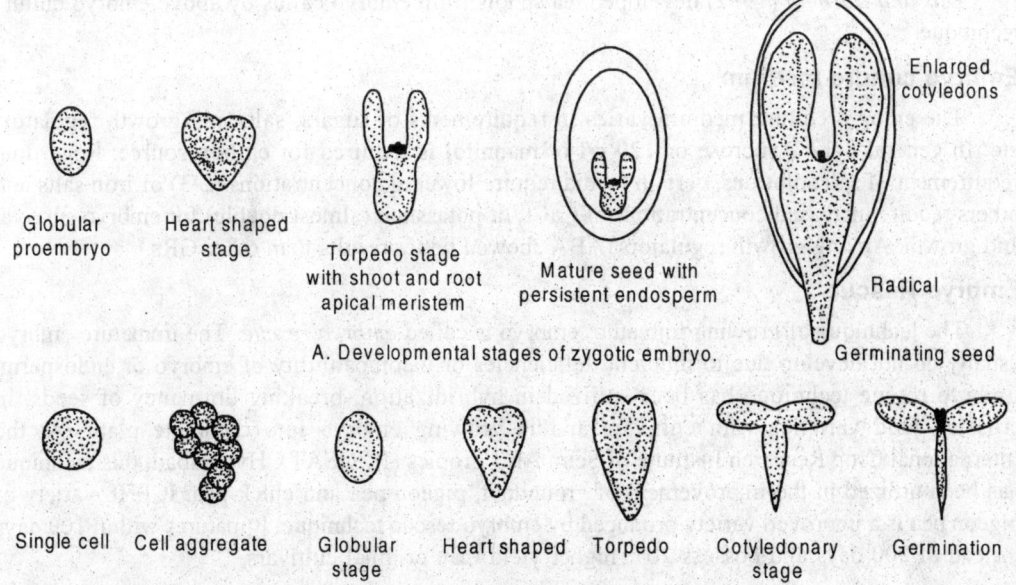

A. Developmental stages of zygotic embryo.

B. Various stages of somatic embryo development in cultures.

Fig. 2.9. Development of zygotic and somatic embryos.

the rediscovery of plants from distinct crosses. *Hannig* (1904) for the first time cultured embryos of *Raphanus* and *Cochlearia* under aseptic conditions. Since then extensive studies on embryo culture have been done by Kundson (1922), Dietrich (1925), Laibach (1929), Norstog (1965), and Das and Barman (1992) etc. The technique of embryo culture has been utilized in the production of haploid barley, Orchid propagation and in the breeding of those species which show dormancy.

Culture Technique

It includes following steps —

1. Selection of healthy and mature fruits from the plant and washing them thoroughly in running water for about an hour.
2. Surface sterilization with any sterilizing agent for 15 min to an hour and then rinsing of seeds 5-6 times in sterilized distilled water.
3. Breaking of seeds aseptically and isolation of embryo.
4. Culturing of embryo in callus proliferation culture supplemented with sugar, vitamins and growth regulators.
5. Incubation of culture at 22-25°C temperature under a photoperiod of 16 h in 2000 lux luminous light. Datura requires a temperature of 35°C for incubation.

   The embryo starts swelling on callus proliferation culture medium after two weeks of inoculation. A distinct callus growth can be observed after 4 weeks.
6. Transferring of callus on shoot regeneration medium after 8 weeks of inoculation. Within 4 weeks of transfer in this medium the callus turns green and develops into soft spongy tissue. Some of these tissue differentiate into small embryo like structures called *embryoids*.
7. Sub-culturings of embryoids onto shoot regeneration medium. It results into production of cluster of budlets in embryoids. The budlets develop into shoots and produce 2-3 leaf primordia within 12 weeks. Each primordial shoot is separated and sub-cultured onto fresh medium until it develops a mature shoot.

Das and Barman (1992) developed tea shoots from embryo callus by above embryo culture technique.

Embryo culture medium

The embryo culture medium varies in requirements of sugars, salts and growth regulators etc. In general 8-12% sucrose or 120 g/l of mannitol is required for carbon source. Regarding requirement of salt solutions, certain media require lowered concentrations (2/3) of iron salts and others require increased concentrations of $CaCl_2$ or potassium (almost double) for embryo survival and growth. Among growth regulators, ABA showed better results than other GRs.

Embryo Rescue

The technique of growing immature embryo is called *embryo rescue*. The immature embryo usually do not develop due to inherent deficiencies or incompatibility of embryo or endosperm. Embryo rescue technique has been utilized in hybridization, breaking dormancy of seeds, in crossing wild varieties with cultivars and in growing embryo into complete plant. At the International Crop Research Institute of Semi Arid Tropics (ICRISAT), Hyderabad, this technique has been utilized in the improvement of groundnut, pigeon pea and chick-pea. ICPH8 variety of pigeon pea is a improved variety produced by embryo rescue technique. It matures within 100 days instead of 200 days and possess 20% higher yield than original cultivars.

Embryo-nurse endosperm Technique

When very young embryos are placed on to or implanted into developing *in vitro* cultured endosperms for their culturing, the technique is known as *embryo-nurse endosperm technique*. It has been utilized to get such hybrids which could not be obtained by usual interspecific or intergeneric crossing. Hybrids from *Hordeumn* × *Secale*, *Trifolium* species, and *Gossypium arboreum* × *G. hirsutum* have been obtained by this technique.

8. Ovary Culture

When unfertilized ovaries are cultured to obtain haploid plants from egg cell or other haploid cells of the embryo sac, it is known as *ovary* culture. The process involved in ovary culture is called *gynogenesis*. It was first reported by *San Noem* (1976) in barley. Later on, gynogenesis has been reported in a number of species in wheat, maize, rice, tobacco, petunia, sunflower, sugarbeet, rubber and *Gerbera* etc.

The ovary culture usually requires a cold pretreatment, *e.g.*, 24-48 hrs at 4°C in sunflower and 24 hr at 7°C in rice, of the inflorescence.

Culture requirement

Ovary culture medium requires either no growth regulators (GRs), *e.g.*, sunflower, or 2-methyl-4-chlorophenoxyacetic acid (MCPA) GR in low concentration for induction of somatic calli or somatic embryos or in higher concentration (0.125 - 0.5 mg/*l*) for gynogenesis. For carbon source, 12% or less sucrose is needed in the medium.

Haploid plants generally originate from egg cell. It is called *in vitro parthenogenesis*. In some cases, they originate from synergids or antipodals. It is called *in vitro apogamy*. Ovary culture has certain limitations, so it is rarely used.

Cryopreservation

The preservation of cells, tissues and organs in liquid nitrogen at $-196°C$ (ultra low temperature) is called *cryopreservation* and the science or study pertaining to this activity is known as *cryobiology*.

The technique of cryopreservation involves four steps, viz., (*i*) Freezing, (*ii*) storage, (*iii*) thawing, and (*iv*) reculture.

CELLULAR TOTIPOTENCY, DIFFERENTIATION AND MORPHOGENESIS

Totipotency is the ability of a living somatic plant cell (with protoplasm) to develop into a complete plant. The totipotent cell performs all the functions of development which are characteristic of zygote. *Haberlandt* (1902) developed initially the *in vitro* techniques to demonstrate the totipotency of plant cells. He cultured isolated single palisade cells from the leaves in Knop's solution enriched with sucrose. The cells remained alive for one month, showed increase in size and accumulation of starch but failed to divide. The demonstration of totipotency led to the development of techniques for cultivation of plant cells under defined conditions. The contributions made by *R.J. Gautheret* in France and *P.R. White* in U.S.A. is important in this direction.

Cellular totipotency was first demonstrated by *Steward* et. al. (1964) using phloem cells of carrot (*Daucus carota*) with the help of tissue culture technique. They developed complete plant from a single phloem cell. This technique is being used, now a days, for multiplying rare and endangered plants through micro-propagation. *Orchids* and *Gladiolii* are being cultivated on large scales through micropropagation where the principle of cellular totipotency is involved. The technique is also widely used in the multiplication of Chrysanthemum, *Dioscorea floribunda*, *Crotons*, *Coleus* and Carnation plants.

Micropropagation

Murashige (1974) established micropropagation technique. It is also known as clonal propagation *in vitro* culture. Clonal propagation is simply the multiplication of genetically identical copies of a plant by asexual or vegetative method. In this technique because minute sized propagules are used in culture, it has been named as *micropropagation*. Micropropagation technique has following advantages —

    (*i*) Rapid multiplication within a short period and space.

    (*ii*) Production of large number of plants from a single explant.

    (*iii*) Multiplication of sexually derived sterile hybrids.

    (*iv*) Being independent of seasonal variation.

    (*v*) Showing uniform multiplication of superior clones.

According to *Murashige* (1974), the procedure of development of micropropagation involves following three stages —

    Stage I — establishment of explant aseptically.

    Stage II — multiplication of propagules by repeated sub-cultures of specific nutrient medium.

    Stage III — rooting and hardening of plantlets and planting into soil.

According to *Fossard* (1987), the micropropagation involves following four stages —

    Stage I — Selection of a suitable explant and inoculation into nutrient medium.

    Stage II — Multiplication and growth of shoots on the medium which takes about 60 days, followed by repeated sub-culture. At this stage rootless cultures are obtained. In fast multiplying cultures, rootless multishooted cultures also develop.

    Stage III — This stage includes following three substages.

    (*i*) **Microcutting stage (MC)** — The culture of stage II is transferred to multishoot inducing medium to get longer shoots for microcutting. These cuttings are sub-cultured on rooting medium enriched with auxin.

    (*ii*) **Culture stage (Intermediate)** — At this stage, stage II multishooted cultures are divided into individual shoots which are induced to form apical dominant shoots and roots.

    (*iii*) **Multishoots Culture Stage** — Small clumps of culture from stage II multishoot culture are transferred to apical dominant inducing medium to get bushy plant. Such plants are preferred for several ornamental horticultural plants.

    Stage IV — At this stage, the plantlets are aseptically removed from the culture environment and planted to natural and harsh environment of soil.

At stage IV, it is necessary to develop acclimatization capacity in plantlets before removing them from the culture environment (test tube). This can be done by (*i*) induction to develop few normal and functional leaves, (*ii*) inducing functional roots, and (*iii*) exposing *in vitro* cultures to harsh environment before two weeks of planting the plantlets.

According to Fossard (1987) the potting with peat alone, vermiculite alone, loam and peat, peat-perlite-vermiculite-ash, perlite-pulverized pine bark-peat-river sand and perlite-vermiculite sand are successful for planting out stage.

During micropropagation, the number of stages may be reduced by taking certain precautions of culture media and pretreatments. In most cases, the micropropagation is achieved by placing sterilized shoot tips or axillary buds into a suitable culture medium. Therefore, the regions of young plants with shoot tips and axillary buds have been considered as most suitable explants for micropropagation.

Micropropagation technique has been utilized as important tool in the cloning of useful tree plants like *Eucalyptus, Santalum, Tectona, phoenix, Poplar and Willows* etc. Orchids, Gladiolus,

Biotechnology

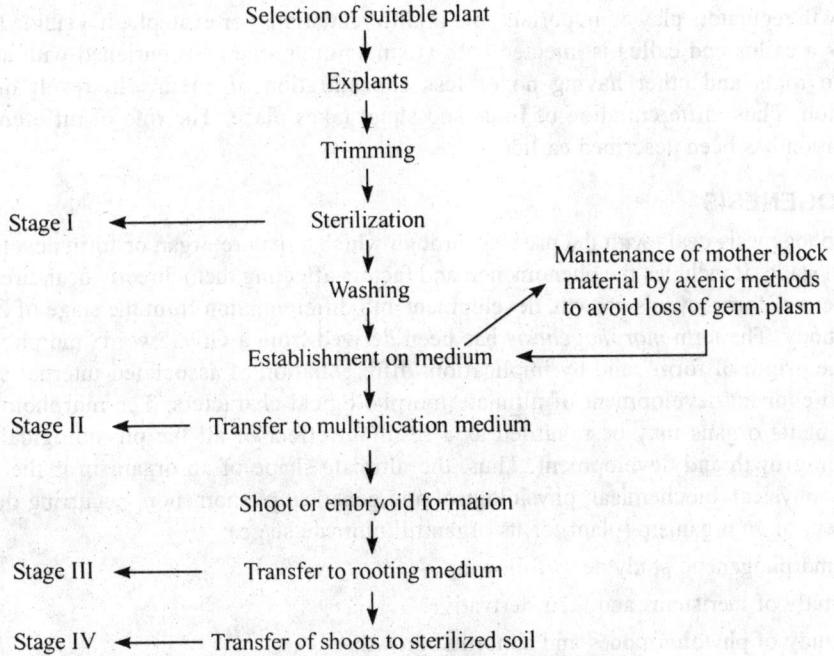

Fig. 2.10. Summary of steps involved in micropropagation.

Allium, Asparagus, Brassica, Arachis, Glycine, Solanum, Zea mays, Phaseolus, Dianthus, Ferns and *Hevea* etc. have also been micropropagated *in vitro*.

Uses of Micropropagation Technique

1. It acts as potential system for storage of germplasm.
2. It helps in the production of disease free varieties.
3. It helps in continue multiplication of desired plants species throughout the year irrespective of season.
4. It reduces production cost of hybrid seeds.
5. It helps in bulking of newly selected cultivars.
6. It produces identical diploid clones of plant species.
7. It helps in apogamy.
8. It helps in the production of such plant which show long dormancy of seeds or do not produce viable seeds.

Differentiation

Differentiation term is often used as synonym for *regeneration*. *Regeneration* deals the development of organised structures like roots, shoots, somatic embryos and flower buds etc. from cultured cells or tissues. These events are also described under the term *Organogenesis*. Root regeneration occurs frequently but it is only useful when attached with shoot and embryo germination. Only shoot and somatic embryo germination result into complete plants. The differentiation chiefly deals with the development of different cells types, e.g., vascular tissues etc. and *Cyto-differentiation*. It is therefore suggested that the terms like shoot differentiation and somatic embryo differentiation etc. should be used instead of term differentiation alone. The differentiation may occur directly from explant or from callus.

Growth regulators play an important role in differentiation. For example, if a single totipotent cell forms a callus and callus is injected with auxin solution, the cells enriched with auxin will result into roots and other having no or less concentration of auxin will result into shoot regeneration. Thus, differentiation of roots and shoot takes place. The role of different GRs in differentiation has been described earlier.

MORPHOGENESIS

Morphogenesis deals with the process through which a mature organ or form develops from embryonal stage. It includes the phenomenon and factors affecting them directly or indirectly. The development of form reveals growth, development and differentiation from the stage of Zygote to a mature body. The term *morphogenesis* has been derived from a Greek word "morphogenous", means "the origin of form" and by implication, differentiation of associated internal structural features to connect development of ultimate morphological characters. The morphology of an organism or its organs may be regarded as a resultant effect of all the physiological process involved in growth and development. Thus, the ultimate shape of an organism is the result of various biophysical, biochemical, physiological and genetical phenomenon, occurring during the development of an organism (plant) or its organ till ultimate stage.

The morphogenetic study deals following —

1. Study of meristems and their derivatives.
2. Study of phytohormones and abnormal growth.
3. Study of organography, polarity, differentiation and regeneration.
4. Study of symmetry, correlation, phyletic causal morphology and development.
5. Study of environmental effects on plant development.

BIOLOGY OF AGROBACTERIUM

Agrobacterium Plasmids or *Ti* and *Ri* Plasmids (Vectors) for Gene Delivery

Agrobacterium is a soil borne bacterium which has been utilized by scientists for gene transfer in plants and to produce transgenic plants. The two species of this bacterium, *Agrobacterium tumifaciens* and *Agrobacterium rhizogenes,* are important because *A. tumifaciens* contains a large *Ti* (Tumor inducing) **plasmid** which ultimately produces crown gall disease and *A. rhizogenes* contains hairy root disease causing *Ri* **plasmid. It is important to note, that** *Agrobacterium tumifacienes* infects only dicotyledonous plants and not monocotyledonous plants (cereals).

Ti-**plasmids** — They are found in *A. tumifaciens. Ti* plasmids are quite large sized ranging between 180–250 kb. It contains a specific T-DNA region of about 23 to 25 kb. As this plasmid is quite large and can not be transferred in plants directly, its only T-DNA region is transferred into host plant cells. The left side of the T-DNA is called **TL-T-DNA** and the right side as **TR-T-DNA**.

When **T-DNA region** of *Ti* plasmid is transferred to host plant cells, the genes in the T-DNA region are expressed in the host cells. In addition to T-DNA region, the plasmid also contains another region, called **vir (virulence) region**, which is essential for successful transformation. **Vir region** consists of 7 or possibly 8 loci labelled as A-G. If eighth locus is present, it is called **O-locus**. A and G are regulatory genes. It has been evidenced since mutation in these loci prevent expression of any of the *vir* genes. The main locus called 'A' is continuously expressed but the expression of other loci is induced by the presence of host.

The *Ti* plasmids have been classified into three types depending upon the opine types (*octapine, nopaline and agropine*). The opines are unusual derivatives of one or more aminoacids, encoded by DNA. They are neigther found normally in plants nor required by plants. The opines

Biotechnology

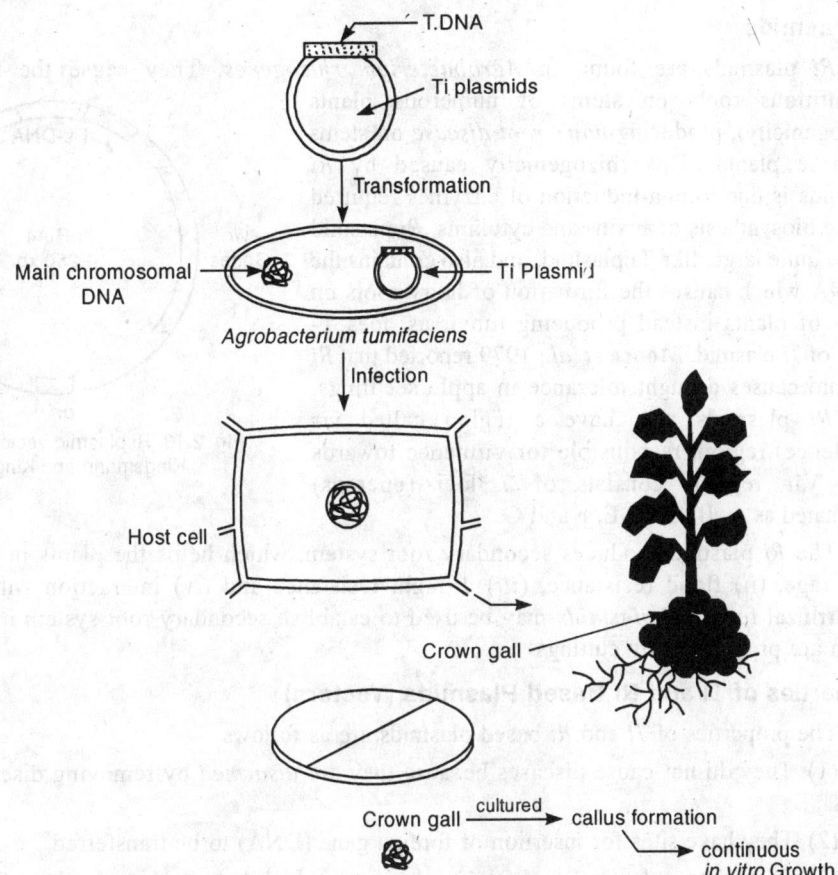

Fig. 2.11. *Agrobacterium tumifaciens* containing Ti plasmid. They are natural genetic engineers.

provide carbon and nitrogen to the bacterium for its growth and multiplication. The *Ti* plasmid also contains *tra* (transfer) gene for transfer of T-DNA, *one* (oncogene) for producing cancer, *ori* (origin) gene for origin of replication. *inc* (incompatibility) gene for incompatibility of plants and *arc* genes, for arginine catabolism.

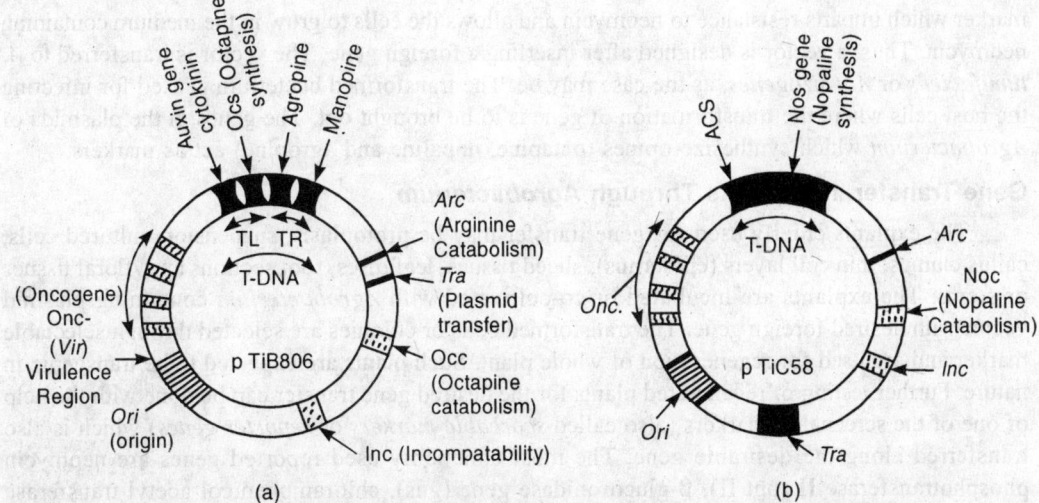

Fig. 2.12. Ti plasmids-a map of Octapine plasmid (A) and Nopaline plasmid (B).

Ri Plasmids

Ri plasmids are found in *Agrobacterium rhizogenes*. They cause the formation of adventitious roots on stems of numerous plants (rhizogenicity), producing *hairy root disease* on stems of these plants. This rhizogenicity caused by *Ri* plasmids is due to non-induction of enzymes required for the biosynthesis of auxins and cytokinis. *Ri* plasmid is also quite large, like Ti plasmid, and also contains the T-DNA which causes the formation of hairy roots on stems of plants instead producing tumor as does T-DNA of *Ti* plasmid. **Moore** *et. al.*, 1979 reported that *Ri* plasmid causes drought tolerance in apple seedlings. The *Ri* plasmids also have a region called *vir* (virulence) region, responsible for virulence towards host. **Vir** regions consists of 7 loci (operons) designated as A, B, C, D, E, F and G.

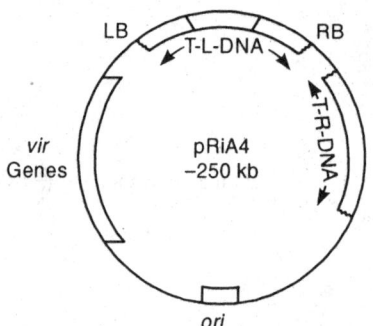

Fig. 2.13. Ri plasmid vector system (after Kingsmann and Kingsmann.

The *Ri* plasmid produces secondary root system, which helps the plants in — (*i*) better anchorage, (*ii*) flood resistance, (*iii*) drought resistance and (*iv*) interaction with soil-borne mycorrhizal fungi. **Ri plasmids** may be used to establish secondary root system in such plants which are propagated by cuttings.

Properties of Ti and Ri Based Plasmids (Vectors)

The properties of *Ti* and *Ri* based plasmids are as follows —

(1) They do not cause diseases because they are *disarmed* by removing diseases causing genes.

(2) They have sites for insertion of foreign gene (DNA) to be transferred.

(3) They carry *selective markers i.e.,* the genes which help in the selection of transformed cells.

Marker Genes of Agrobacterium

The selective marker acts as a powerful promoter, *e.g.*, 35S marker derived from Cauliflower mosaic virus (CaMV) acts for high level of expression and *polyadenylation signal* acts for adding poly A to *mRNA*. *Neomycinphosphotransferase* II (npt II) is the most commonly used selectable marker which imparts resistance to neomycin and allows the cells to grow in the medium containing neomycin. Thus, a vector is designed after inserting a foreign gene. The vector is transferred to *A. tumifaciens* or *A. rhizogenes*, as the case may be. The transformed bacterium is used for infecting the host cells where the transformation of gene is to be brought out. The genes in the plasmids of *Agrobacterium* which synthesize opines (octapine, nopaline and agropine) act as markers.

Gene Transfer in Explants Through *Agrobacterium*

The explants chiefly used for gene transfer may be protoplasts, suspension cultured cells, callus clumps, thin cell layers (epidermis), sliced tissues, leaf discs, root sections and /floral tissues or stems. The explants are incubated or co-cultivated with *Agrobacterium* containing plasmid vector with desired foreign gene. The transformed cells or colonies are selected through selectable marker and are used for regeneration of whole plant. Such plants are expected to be transgenic in nature. Further testing of regenerated plants for the desired gene transfer can be done with the help of one of the screenable markers (also called *scoreable markers* or *reporter genes*) which is also transferred along the desirable gene. The most commonly used reported genes are neomycin phosphotransferase II (npt II), β-glucoronidase gene (gus), chloramphenicol acetyl transferase (cat) and luciferase gene (lux).

Biotechnology

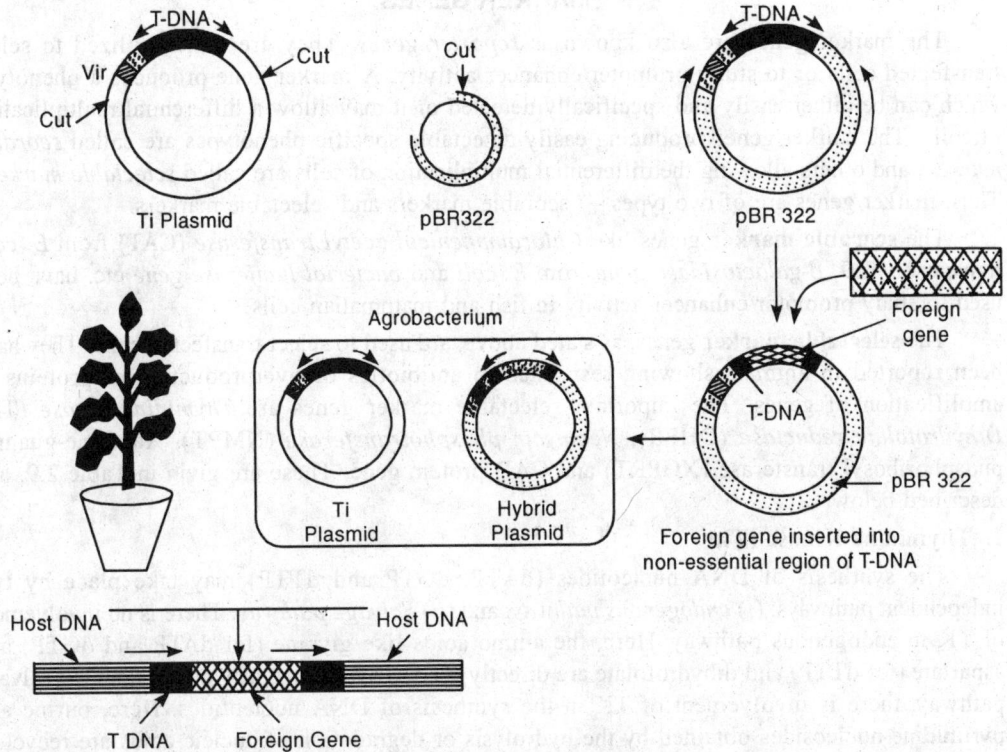

Fig. 2.14. Steps in transfer of a foreign gene into plant using *A. tumifaciens*.

Mechanism of T-DNA Transfer

The successful transformation of a host requires several steps. These includes — (*i*) recognition of a suitable host by the *Agrobacterium*, (*ii*) its attraction to and invasion of the host cells, (*iii*) excision of the T-DNA, (*iv*) integration of T-DNA into a nuclear chromosome, (*v*) presence of *vir* region to help in infection and transformation process and (*vi*) reporter genes for testing.

A. tumifaciens, a soil borne bacterium, infects dicotyledonous plants at the level of soil surface. For infection, wounding of host cell is a necessary pre-requisite through which the bacterial plasmid enters and establishes into plant cells. The bacterial cell wall secretes **lipopolysaccharide** which helps in their attachement with the plant cell wall at the region containing **polygalacturonic acid**. The wounded cell walls of host plant secrete *acetosyringone*, a low molecular weight phenolic compound, which induces the virulence (*vir*) gene of *Ti* plasmids. *Vir* genes encode an enzyme which nicks the double stranded T-DNA (ds T-DNA) on the same strand at two points and produces single stranded DNA (ss DNA) molecule. Usually a cut is made on the TR-T-DNA (region of right side) and a single stranded T-DNA fragment of the 5'-3' direction is generated which is carried into other cells of the plant.

The T-DNA fragment of *Ti* plasmid integrates tightly with the host plant DNA. The sequencing of nucleic acid at the junction between host plant DNA and T-DNA has been done. T-DNA region in plasmids is flanked by a direct repeat of 25 bp. There is a gene *ops* which encodes enzyme for the synthesis of opines in transformed cells. The opines help in proliferation of bacterial cells. TL-T-DNA encodes genes for two enzymes, required for the -biosynthesis of auxin and cytokinin (**Ream, 1989**). This results into tumor formation. The tumors contain colonized bacteria.

MARKER GENES

The marker genes are also known as *reporter genes*. They are either utilized to select transfected cells or to study promoter/enhancer activity. A marker gene produces a phenotype which can be either easily and specifically detected or it may allow a differential multiplication of cells. The marker genes producing easily detectable specific phenotypes are called *scorable markers* and others allowing the differential multiplication of cells are called *selectable markers*. Thus, marker genes are of two types — scorable markers and selectable markers.

The **scorable marker** genes like *Chloramphenicol acetyl transferase* (CAT) from *E. coli*. transposon Tn9, *β-galactosidase gene* from *E. coli* and *bacterial luciferase gene* etc. have been used to study promotor/enhancer activity in fish and mammalian cells.

The **selectable marker** genes, as stated above, are used to select transfected cells. They have been reported in animals showing resistance to antibiotics or overproduction of proteins or amplification of genes. The important selectable marker genes are *Thymidine kinase* (Tk), *Dihydrofolate reductase* (DHFR), *Neomycin phosphotransferase* (NMPT), Xanthine-guanine phosphoribosyl transferase (XGPRT) and CAD protein gene. These are givin in Table 2.9. and described below.

1. Thymidine Kinase (Tk)

The synthesis of DNA nuclcotides (dATP, dGTP and dTTP) may take place by two independent pathways: (*i*) *endogenous pathway*. and (*ii*) *Salvage pathway*. There is no involvement of TK in endogenous pathway. Here, the amino acids like glycine (for dATP and dGTP) and aspartate (for dTTP) and dihydrofolate are directly used to synthesize new nucleotides. In salvage pathway, there is involvement of TK in the synthesis of DNA nucleotides. Here, purine and pyrimidine nucleosides obtained by the hydrolysis or degradation of nucleic acids are recycled. The enzyme thymidine kinase (TK) helps in the phosphorylation of thymidine into thymidine mono-phosphate (dTMP) which is subsequently phosphorylated into thymidine triphosphate (dTTP).

The thymidine kinase deficient (TK$^-$) cells when are placed in HAT medium (containing hypoxanthine, thymidine and a drug aminopterin), they are killed because — (*i*) Aminopterin blocks the endogenous pathway of nucleotide synthesis by inhibiting the enzyme *dihydrofolate reductase* (DHFR). This enzyme catalyses the first reaction involving dihydrofolate (*ii*) TK$^-$ cells are unable to utilize the thymidine of HAT medium for nucleotide synthesis. Due to above reasons, TK$^-$ cells die of nucleotide starvation on HAT medium. The medium acts as an efficient selection agent for TK$^+$ cells.

The thymidine kinase gene acts as a selectable marker gene only when TK$^-$ host cells are used for transfection. When transfected cells are cultured on HAT medium, only TK$^+$ cells survive and multiply but TK$^-$ cells do not survive and die. The TK$^-$ cells show limitation as it restricts the application of this marker. The selectable markers applicable to non-mutant animal cell lines have been developed. These are called dominant selectable markers.

One of the best systems available for selection involves the TK gene of herpes simplex virus. TK$^-$ mutants have been isolated for a number of cell types including mouse, rat and human. The TK$^-$ cells can be transformed to TK$^+$ cells by plasmid DNA containing herpes simplex virus TK gene.

Table 2.3. A list of marker genes used in gene transfers in animal cells.

	Marker gene	Source	Uses/Characteristics
	Scorable marker genes		
1.	Chloramphenical acetyl transferase (CAT)	*E. coli* transposon Tn9	Used in fish, mammalian cells.
2.	β-galactosidase	*E. coli*	Used in fish
3.	Luciferase (*lux*)	Bacteria	Used in fish
	Selectable marker genes		
1.	Thymidine kinase (TK)	Herpes simplex virus (HSV)	Used in Tk⁻ cells; HSV gene promoter is weak in mammalian cells
2.	Dihydrofolate reductase (DHFR)	1. Chinese hamster ovary (CHO) cell line 'A29'	Resistance to methotrexate due to overproduction of DHFR; DHFR gene is amplified during integration
		2. Bacteria	DHFR intrinsically resistant to methotrexate
3.	Xanthine-guanine phosphoribosyltransferase (XGPRT)	*E. coli*	Resistance to mycophenolic acid
4.	Neomycin phospho transferase	Bacterial transposons Tn5 and Tn601	Resistance to antibiotic G418
5.	CAD protein gene	Syrian hamster gene cloned in *E. coli*	Overproduction of CAD protein confers resistance to N-phosphonacetyl-L-aspartate (PALA). CAD gene is also amplified during integration.

2. Dihydrofolate Reductase (DHFR) – Methotrexate Resistance

Methotrexate (*Mtx*) is a strong inhibitor of enzyme dihydrofolate reductase which is involved in the endogenous synthesis of DNA nucleotides (dATP, dGTP and dTTP). The Mtx-resistant cell lines show three important features — (*i*) reduced Mtx uptake, (*ii*) Over production of dihydrofolate reductase (DHFR), and (*iii*) Changed DHFR having reduced affinity for Mtx. It is reported that the gene, producing cell line with overproducing DHFR, is highly amplified. It represents several copies per genome in comparison to normal and the number of copies may be upto 1000 per cell. The gene may either be found integrated to the genome or in the free form as an extrachromosomal element. The latter condition represents *double minute chromosome*.

A dihydrofolate reductase gene was isolated from chinese hamster ovary cell line 'A29' and was named as A29DHFR. It overproduces an altered DHFR and shows resistance to methotrexate (Mtx) due to overproduction of DHFR. The DHFR gene is amplified during integration. A29DHFR encoding sequence was cloned in pBR322 plasmid and was used to transfect Mtx sensitive cells. The transfected cells were selected for Mtx-resistance. Some selected lines survived upto 40 µg/ml Mtx and contained upto 50 copies of DHFR encoding sequence used for transfection. During transfection, DHFR gene undergoes amplification. The DNA sequence which undergoes amplification is called *amplicon* and the DNA sequence liked the DHFR gene becomes a part of DHFR amplicon. This property can be utilized to obtain insertion of multiple copies of a gene in the genome of cell.

The DHFR obtained from bacteria is resistant to methotrexate. This gene for DHFR was fused with SV40 promoter. The integrated gene expressed in mouse cells making them methotrexate-resistant.

Methotrexate, trimethoprim and a number of related compounds inhibit the reduction of dihydrofolate to tetrahydrofolate, a reaction catalysed by dihydrofolate reductase. These inhibitors are structural analogs of folic acid and bind at the catalytic site of dihydrofolate reductase. This enzyme is involved at one step in thymidylate synthesis.

3. Xanthine – guanine Phosphoribosyltransferase (XGPRT) – Mycophenolic Acid Resistance

The XGPRT enzyme obtained from *E. coli* is analogous to mammalian enzyme hypoxanthine - guanine phosphoribosyltransferase (HGPRT). Both XGPRT and HGPRT enzymes convert hypoxanthine into inosine monophosphate (IMP) and IMP into guanosine monophosphate (GMP). These enzymes can also convert guanine into GMP directly. XGPRT enzyme can also convert xanthine into xanthine monophosphate (XMP) and ultimately into GMP.

The mycophenolic acid can inhibit HGPRT catalysed conversion of IMP to XMP. The normal mammalian cells are sensitive to mycophenolic acid and this sensitivity is greatly enhanced in presence of aminopterin which blocks endogenous purine synthesis.

The isolated and cloned bacterial XPGRT gene acts as a dominant selectable marker on a culture medium containing *mycophenolic acid*, *adenine* and *xanthine*. The normal mammalian cells with HGPRT do not survive because HGPRT enzyme is unable to utilize xanthine for GMP production whereas mammalian cells producing bacterial XGPRT can however, utilize xanthine of the culture medium for GMP-production. Thus, mammalian cells producing XGPRT are able to survive and proliferate and normal mammalian cells producing HGPRT but not XGPRT do not survive or die in the medium.

4. Neomycin Phosphotransferase – G418 Resistance

Aminoglycoside antibiotic G418 normally kills neomycin phosphotransferase containing normal mammalian cells through inhibition of protein synthesis but cells with neomycin phosphotransferase encoded by bacterial transposons Tn5 and Tn601 show resistance to G418 and are not killed. Various vectors like plasmids and cosmids have been constructed by introducing neomycin phosphotransferase gene with transposons and are being used for transfection of various mammalian cell lines. The transfected cells are selected on culture medium containing G418 antibiotic which acts as a powerful selection agent.

5. CDA Protein — Phosphonacetyl-L-Aspartate (PALA) Resistance

CDA protein is a multi functional enzyme which catalyses the first three reactions of uridine synthesis. One of these three reactions is inhibited in presence of phosphonacetyl-L-aspartate (PALA). Normal mammalian cells are sensitive to PALA but mammalian cells or cell lines encoded with CDA gene show resistance to high concentrations of PALA. Such PALA-resistant cells lines cause overproduction of CDA protein due to amplification of CDA gene. The CDA gene was isolated from Syrian hamster and cloned in *E. coli*. It served as a dominant marker gene by converting normal mammalian cells into PALA resistant.

From the literature cited above for marker genes, it becomes clear that the marker genes are of two types — scorable marker genes and selectable marker genes. The scorable marker genes are usually used in the studies of promoter/enhancer functions and are occasionally utilized in the selection of transfected cells. The selective marker genes are ordinarily used for the selection of transfected cells. They have been developed for this purpose. The TK gene can be used only in case of TK⁻ host cells. DHFR, XGPRT, Neomycin phospho transferase and CDA protein genes act as dominant selectable markers and are effectively used in the selection of transfected/transgenic cultured mammalian cells.

The marker genes are ordinarily not included in the constructed vectors, used for the production of transgenic mammals. Such mammals are identified on the basis of presence of transgene DNA sequence of its mRNA or the protein or enzyme encoded by the marker gene.

SALIENT ACHIEVEMENTS IN CROP BIOTECHNOLOGY

Techniques of biotechnology have been widely applied for the improvement of various crop plants. Recombinant technology and gene cloning technology have contributed a lot in this direction. As we know that both these techniques require vectors, *Ti* and *Ri* plasmids have been used for this purpose,

For crop improvement, biotechnology has been applied for the following purposes —

(1) For transferring nitrogen fixing (*nif*) genes in crop plants through plasmids, especially *Ti* and *Ri* plasmids.
(2) For increasing photosynthetic efficiency through recombinant DNA-technology by increasing atmospheric CO_2 fixing enzyme, *Ribulose Bisphosphate Carboxylase* (Rubisco).
(3) To eliminate or reduce the competing reactions of photorespiration.
(4) For improving the nutritional value of seed protein.
(5) Production of bacterial biofertilizers.
(6) Production of cyanobacterial biofertilizers.
(7) Production of microbial insecticides.
(8) Production of herbicide-resistant plants.
(9) Production of viral disease resistant plants.
(10) Production of insect pest-resistant plants.
(11) Production of seed genotypes to yield high-protein grains and resistance to heat, moisture and diseases.
(12) Transferring various desired genes.

Genetic Engineering in Cloning of Nitrogen Fixation (*nif*) Genes

Nitrogen is most important for all the living beings as it is a constituent of proteins, nucleic acids and other essential molecules in all organisms. We get nitrogen through leguminous plants which is ultimately absorbed from the soil. The crop fields also need large quantities of nitrogen which they get from nitrogen fixing bacteria or externally supplied fertilizers. The atmospheric air contains about 78% nitrogen by volume. This nitrogen is fixed either by free living nitrogen fixing bacteria like *Klebsiella Pneumoniae* or by symbiotic bacteria like *Rhizobium leguminoserum*. Some of these bacteria possess *nif genes.*

Currently, plasmids are proving to have a bright future in genetic engineering of nitrogen fixation as these play a role in the synthesis of large quantities of *nitrogenase* needed for nitrogen fixation. The *nif* region in *Klebsiella* consists of a DNA segment consisting of seven operons including atleast 17 genes clustered together. The cloning of *nif* region was achieved in the form of plasmid called PWK 120, which confers nitrogen fixing capacity to *E. coli*. Similarly, *nif* region of *Rhizobium* can also be cloned in *E. coli*. Biotechnologists are trying to change the genetic system of cereals so that they may have *nif* genes for fixing atmospheric nitrogen. This is being tried through some plant viruses as vectors. The nitrogen fixing bacteria contain an enzyme called *nitrogenase*. Genetic engineering is being used in constructing nitrogen fixing strains and to obtain more efficient *Rhizobium*.

Cannon and **coworkers** from England have recently constructed amplifiable plasmids carrying 14 out of 18 *Klebsiella pneumoniae* nitrogen fixation *nif* genes, including those that code for the

Mo-Fe and Fe proteins of nitrogenase. **Gene amplification** is a very useful procedure concerned with the 'amplification' of genes fused with the bacterial plasmids. The possibility of enzymatically inserting *nif* DNA into a fast replicating plasmid from *E. coli* (Col E 1) might increase the synthesis of nitrogenase and consequently, the production of NH_4^+.

The nitrogen fixation genes (*nif* genes) have been introduced into cereal plants (Table 2.4). Workers at the University of Sussex have assembled a bacterial plasmid that includes all 17 of the known *nif* genes from the nitrogen-fixing bacterium *Klebsiella pneumoniae*. When the plasmid was transferred to *E. coli* which is normally incapable of nitrogen fixation, it became a nitrogen-fixing microorganism.

Table 2.4. Selected Nitrogen Fixing Microbes Useful for *nif* Gene Transfer Programme.

Donor	*Receptor*	*Expression*
Klebsiella pneumoniae Strain M 5al	*E. coli*	Successful
	Klebsiella aerogenes	Successful
	Salmonella typhimurium	Successful
	Serratia marcescens	Successful
	Erwinia herbiocola	Successful
Rhizobium trifolii Strain T_1	*Agrobacterium tumefaciens*	Successful
Rhizobium trifolii Strain T_1K	*Klebsiella aerogenes*	Successful
	E. coli	Successful
Klebsiella pneumoniae	Yeast	Unsuccessful

Even more promising have been the recent successes of groups of investigators at Cornell University, the Pasteur Institute and the University of Paris. They introduced the 17 *nif* genes of *K. pneumoniae* into yeast (Fig. 2.15). Yeasts, being eukaryotic against bacteria which are prokaryotic, are closely related to the higher plants than to bacteria. Hence the introduction of *nif* genes into yeast cells marks the crossing of a significant biological barrier.

Nevertheless, the yeast cells carrying the *nif* genes were not able to express the inserted DNA; they were not able to fix nitrogen from the atmosphere. The failure illustrates the complexity associated with the genetic engineering of biological functions embodied in more than one gene. The transferred DNA must first transcribed correctly into RNA by the yeast.

Correct transcription cannot be assumed as a matter of course, because the yeast must correctly interpret the bacterial signals to start and stop the transcription. The RNA must then be exported from the nucleus and recognized by the ribosomes as a messenger RNA for translation into protein. The 17 proteins that express the *nif* genes must then function together in the foreign cytoplasm of the yeast cell. There may be impediments to such functioning. For example, the nitrogenase molecule has many iron atoms in its structure. Enough iron is evidently available in nitrogen-fixing bacteria, but it is not certain whether the heavy demand for iron can be met in a plant cell without jeopardizing the synthesis of other enzymes essential to the plant.

(A) Transfer of *nif* genes and production of biofertilizers etc. — The nitrogen fixing apparatus of micro-organisms contains about 17 nitrogen fixing (*nif*) genes which are linked in one region of bacterial DNA. This region is cut and inserted into another organism. Thus, through genetic engineering *nif* genes can be transferred from existing nitrogen-fixing bacteria to those bacteria which are found in the roots of important crop plants but can not fix nitrogen. The *nif* genes can directly be transferred to crop plants so that the transformed crop plants could fix their

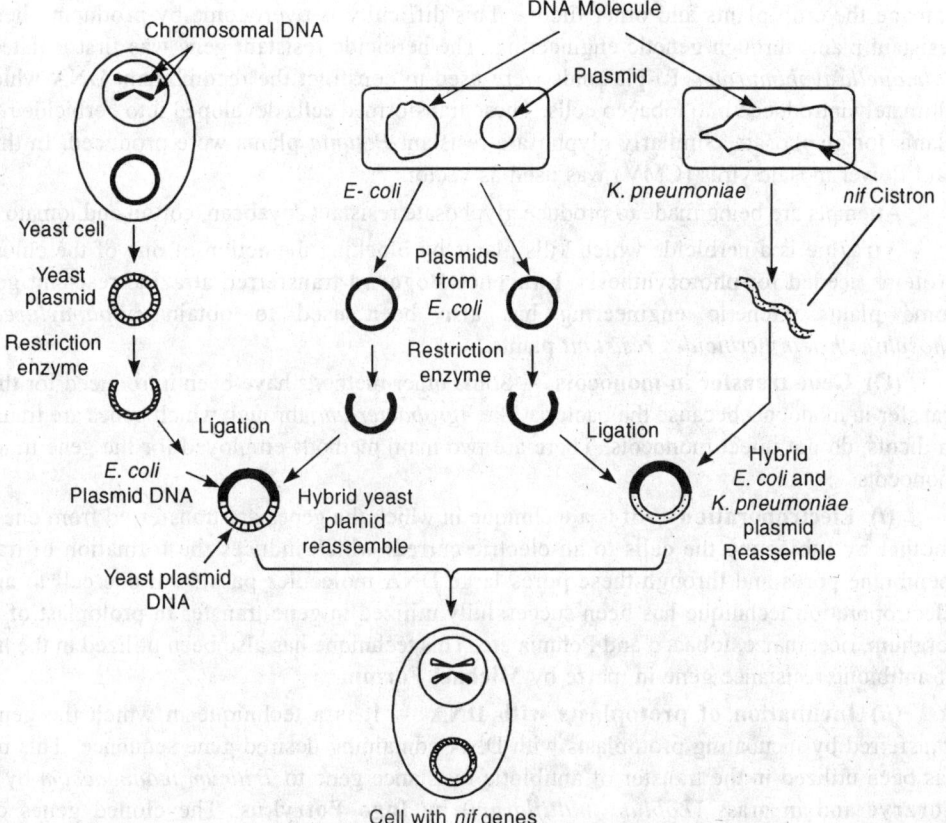

Fig. 2.15. Transfer of nif gene in yeast cell through genetic engineering.

own nitrogen directly from the air. *Agrobacterium-plasmids* (*Ti-* and *Ri*-plasmids) and virus DNA vectors are being employed as cloning vectors for gene transfer in higher plants.

Tissue culture and protoplast fusion techniques can also transfer the genes from one bacterium to other bacterium or plants and can produce new combinations of cereals, pulses and vegetables which show higher yield.

The protoplast fusion is made through *liposomes*. First the desired genes are packed in the liposomes and then are mixed with the protoplasts. The liposomes deliver the desired genes into the protoplast which ultimately develop into new plants or organisms.

For increasing the photosynthetic rates and efficiency of atmospheric CO_2 fixing enzyme (*Ribulose biphosphate carboxylase*), site-directed mutagenesis is used.

Rhizobial biofertilizers and *Azotobacter biofertilizers* are the chief bacterial biofertilizers. For a number of objectives described above for crop improvement, culture protoplast culture, pollen and anther culture techniques are being tried at cellular level. It should be kept in mind that during such excercises the modified cell should be able to develop successfully into complete plant. The plant regeneration from protoplasts has been reported in rice.

(B) Gene transfer in dicots-Herbicide resistant plants : Glyphosate and **Atrazine** are the most widely used herbicides. *Glyphosate* also known as *glyphosphate* (*phoshonomethyl glycine*) is used to control annual and biennial hedges, grasses and broad leaved weed species. It is absorbed by foliar tissues and roots and is then translocated to various plant organs. Thus, it can damage the crop plants and other dicots. This difficulty is overocome by producing herbicide resistant plants through genetic engineering. The herbicide resistant gene was first isolated from *Salmonella typhimurium*. Ri-plasmids were used to construct the recombinant DNA which was ultimately introduced into tobacco cells. These transformed cells developed into herbicide resistant plants for glyphosate. Similarly glyphosate-resistant *Petunia* plants were produced. In this case cauliflower mosaic virus (CMV) was used as vector.

Attempts are being made to produce glyphosate resistant soyabean, cotton and tomato plants.

Atrazine is a herbicide which kills plants by blocking the action of one of the chloroplast proteins needed for photosynthesis. **Laurence Bogorad** transferred atrazine resistant genes in some plants. Genetic engineering has also been used to obtain *sulphonylurea* and *phosphinothricin herbicides resistant* plants.

(C) Gene transfer in monocots — Some other methods have been introduced for the gene transfer in monocots because the bacteria like *Agrobacterium*, through which genes are transferred in dicots, do not infect monocots. There are two main methods employed for the gene transfer in monocots.

(i) Electroporation — It is a technique in which the genes are transferred from one cell to another by subjecting the cells to an electric current which induces the formation of transient membrane pores and through these pores large DNA molecules pass from one cell to another. Electroporation technique has been successfully utilized in gene transfer in protoplast of wheat, sorghum, rice, maize, tobacco and Petunia etc. This technique has also been utilized in the transfer of antibiotic resistance gene in maize by **Michael Formm**.

(ii) Incubation of protoplasts with DNA — It is a technique in which the genes are transferred by incubating protoplasts with DNA containing desired gene sequence. This method has been utilized in the transfer of antibiotic resistance gene to *Triticum monococcum* by **Horst Lorzrye** and in grass (*Lobium multiflorum*) by **Ingo Potrykus**. The cloned genes can be transferred only through the protoplasts as they allow the entry of DNA molecules and also can regenerate into a new plant or organism. The regeneration of rice plants from protoplasts has been reported in recent years.

(D) Transgenic plants — Various techniques utilized in gene transfer (described earlier) permit the introduction of foreign genes or altered form of an endogeneous gene into an organism. These techniques do not replace the endogenous gene but they result into integration of additional copies of the introduced foreign gene. Such introduced genes are called *transgenes* and the organisms carrying them are called *transgenic*. When foreign genes are integrated into endogeneous gene of plant cells, they result into plants called *transgenic plants*. The table 2.5 shows certain transgenic plants which have been produced by any of the methods described earlier.

Table 2.5. A list of higher plants where transgenic plants have been produced, using different methods.

(*i*) **Herbaceous dicotyledons** —

    (1) *Actinidia chinensis* (kiwi)

    (2) *Apium graveolens* (celery)

    (3) *Arabidopsis thaliana*

    (4) *Armoracia* sp. (horse radish)

    (5) *Beta vulgaris* (sugarbeet)

    (6) *Brassica napus* (oilseed rape; canola)

    (7) *Brassica oleracea* (cauliflower)

    (8) *Brassica oleracea* var. capitata (cabbage)

(9) *Brassica rapa* syn. *B. campestris*
(10) *Carica papaya* (papaya)
(11) *Chrysanthemum* sp. (chrysanthemum)
(12) *Cucumis melo* (muskmelon)
(13) *Cucumis sativus* (cucumber)
(14) *Daucus carota* (carrot)
(15) *Dianthus caryophyllus* (carnation)
(16) *Digitalis purpurea* (foxglove)
(17) *Fragaria* sp. (strawberry)
(18) *Glycine max* (soyabean)
(19) *Glycorrhiza glabra* (licorice)
(20) *Gossypium hirsutum* (cotton)
(21) *Helianthus annuus* (sunflower)
(22) *Ipomoea batata* (sweet potato)
(23) *Ipomoea purpurea* (morning glory)
(24) *Lactuca sativa* (lettuce)
(25) *Linum usitatissimum* (flax)
(26) *Lotus corniculatum* (lotus)
(27) *Lycopersicum esculentum* (tomato)
(28) *Medicago sativa* (alfalfa)
(29) *M. varia.*
(30) *Nicotiana tabacum* (tobacco)
(31) *N. plumbaginifolia* (wild tobacco)
(32) *Petunia hybrida* (petunia)
(33) *Pisum sativum* (pea)
(34) *Rosa* sp (rose)
(35) *Solanum melongena* (egg plant)
(36) *Solanum tuberosum* (potato)
(37) *Vaccinium macrocarpon* (cranberry)
(38) *Vigna aconitifolia*
(39) *Vitis vinifera* (grape)

(*ii*) **Woody dicotyledons**
(40) *Azadirachta indica* (neem)
(41) *Juglans regia* (walnut plant)
(42) *Malus sylvestris* (Apple)
(43) *Populus* sp. (Poplar),
(44) *Pyrus communis* (Pear)

(*iii*) **Monocotyledons**
(45) *Asparagus* sp.
(46) *Avena sativa* (oat)
(47) *Dactylis glomerata*
(48) *Festuca arundinacea*

(49) *Oryza sativa* (rice)
(50) *Secale cereale* (rye)
(51) *Triticum aestivum* (wheat)
(52) *Zea mays* (corn)

(iv) **Gymnosperms**
(53) *Picea glauca* (white spruce)

(i) Herbicide Resistance

Herbicides normally effect processes like photosynthesis or biosynthesis of essential amino acids. Some of the herbicide resistant gene transferred cases are given in table 2.6.

Table 2.6

Active Principle of herbicide	*Inhibited Pathway*	*Target Product*	*Use*	*Basis of resistance*
1. Amino acid biosynthesis inhibitor				
(i) Phosphionothricin	Glutamine biosynthesis	Glutamine synthetase (GS)	Broad spectrum	Gene amplification; detoxification; bar gene
(ii) Sulphonyl urea and imidazolinones	Branched chain amino acids.	ALS	Selected crops	Mutant ALS gene
(iii) Glyphosate	Aromatic amino acid biosynthesis	EPSPS	Broad Spectrum	Over expression of EPSPS gene; bacterial aro A gene.
2. Photosynthesis inhibitors				
(i) Atrazine	Photosystem II	QB	Selected crops	Mutant psbA gene; GST gene detoxification.
(ii) Bromoxynil	Photosynthesis	—	Selected crops	bxn gene; detoxification.

(ii) Insect Resistance

(1) Insecticides — Insect pests are commonly being controlled by using pesticides and insecticides. Transgenic plants with insect resistance have been produced through biotechnology by transferring insect resistance gene from various sources. The common sources which have been frequently used are —

(i) *Bt*-2 gene encoding *Bt*-toxin, derived from *Bacillus thuringiensis*.
(ii) Cowpea Trypsin inhibitor gene (CpTi), derived from *Vigna unguiculata*.
(iii) Genes for other *secondary metabolites*, derived from various legumes.

(i) *Bt*-toxins produced by *Bt*-2 gene of *Bacillus thuringiensis* are used as biological insecticide. The insecticidal property of bacteria is due to a protein **delta endotoxin** synthesized during sporulation. This is very costly toxin and is instable in crystalline form in the field.

Bt-2 gene from *B. thuringiensis* has been transferred to tobacco, tomato and cotton through plasmid mediated transformation. Transgenic plants produced *Bt*-toxins. The plants showed resistance to *Manducta sextra*, a pest of tobacco. When the larvae of this insect were fed on

Biotechnology

bacterial toxins, they showed 75-100% mortality. Transgenic cotton plants with *Bt* gene have been produced in India by Maharashtra Hybrid Seeds Company (MAHYCO), Jalna.

(*ii*) **Protease inhibitors or Trypsin inhibitors** — *Vigna unguiculata* (cowpea) trypsin inhibitor (*CpTi*) is responsible for resistance to storage pests (Bruchid beetle). *CpTi* is toxic in a number of insects but cowpea seed with high *CpTi* are not toxic to human beings. During transformation of *CpTi* gene, it was joined with CaMV35S promoter and one or more marker gene. The transformation of *CpTi* gene in tobacco plant has been achieved without decreasing its yield.

(*iii*) **Viral Resistance**

(1) **Cross protection** — It is observed that previous infection by a wild strain gives protection against a virulent strain. The following viruses have been used in case of tomato, potato and citrus.

In Tomato — Tomato-mosaic virus.

In Potato — Potato spindle tuber viriod.

In Citrus — Citrus tristeza virus.

Transformation in tobacco, tomato and potato has been achieved for cross protection by using broad spectrum of plant viruses.

(2) **Virus coat gene or capsid protein** — Capsid protein gene from TMV was transferred to tobacco and it showed stable expression.

(3) **Nucleocapsid protein gene from tomato spotted wild virus (TSWV)** — This gene has been isolated from infected tomato. TSWV is a pathogen for various plants like tomato, tobacco, lettuce, pepper, groundnut and chrysanthemum etc. It has tightly associated RNA with nucleocapsid protein. The transformed plants showed resistant to TSWV.

Bacterial and fungal resistance — It has been achieved by using suitable genes. Table 2.7 shows transgenic plants for bacterial and fungal pathogens.

Table 2.7

S. No.	Pathogen	Disease	Resistant gene	Source of gene	Transgenic crop
1.	*Pseudomans syringae*	Wild fire	Acetyl transferase	—	Tobacco
2.	*Rhizoctonia solani*	—	Chitinase	Bean	Tobacco
3.	*Alternaria langipes*	Brown spot	Chitinase	Soil bacteria	Tobacco
4.	*Phytophthora infestans*	Late blight	Osmotin	Potato	Potato

Disadvantages or Potential Hazards of Genetic Engineering and Biotechnology

Although the world of recombinant DNA technology has undoubtedly tremendous promises, but like all technologies involving manipulation of nature and playing with the environment, it has a negative side also which indeed is quite horifying. There are several negative aspects of this technique.

(1) There is a positive danger that manipulation of genes might, by accident, result in origin of new kinds of diseases and organisms containing fatal genetic elements.

(2) The undesired diseases and organisms, produced through genetic engineering, may escape and contaminate the entire earth.

(3) Many drugs including antibiotics may become ineffective if the bacteria acquires resistance due to uncontrolled recombinant DNAs.

(4) Genetic manipulation of plant will have extensive genetic erosion and destruction of germplasms.

(5) Sometimes a microbe with drug resistance might transform into a pathogenic form which create a new disease for mankind.

(6) Sometimes this technology may produce monsters or dangerous materials.

(7) Accidently unnatural micro-organisms may be produced whose behaviour in the natural environment may be revengeful and destructive, thus, causing biological warfare.

(8) Sometimes highly toxic substances and chemicals may be produced which could destroy the whole population of a city.

(9) It may disrupt the nature's balance.

(10) *E. coli.* when colonizes in the human gut or biosphere, disturbs the physiological and ecological balance.

(11) Some scientists may use this technique with bad intension to produce unsocial elements.

All these potential hazards make it necessary that genetic engineering and biotechnology should be governed by social and ethical regulations in the interest of human health and society. The hazards of this technology were assessed by an International Committee, *Ashby committee*, in 1974. This committee imposed certain restrictions and devised a moral code so that risks could be minimized. The code prepared by the committee not only applies to the application of bio engineering but also to the direction of research in this field.

AUTHOR INDEX

A

A.C. Leopold, 377
A.L. Lehninger, 34
Addicott 148, 369
Agarwal, 116
Allard, 328, 329, 350, 367
Allen, 25, 115
Allison and Mollucci, 30
Allison, 29, 30
Altmann, 17, 32
Ambrose and Easty, 19, 26, 30
Antonie Lavoisier, 137
Antony van Leeuwenhock, 2
Arnon, 114, 117
Arora, et. al., 366
Atkins, 76

B

Bach, 184
Bade, 36
Bager and Hamilton, 41
Baker, 25
Barer Joseph and Meek, 22
Beaufay and BerTher, 30
Beeman, 268
Beevers, 308, 313
Beggins, 40
Beinert, 301
Bell and Muhlethaler, 36
Benda, 32
Bennet-Clark, 118, 123
Bensley, 34
Benson, 11
Benson, 160
Bernhard, 44
Bishop, 7
Blanchard, 105
Boehm, 83
Bonner, 329, 331, 356
Borthwick, 329, 332
Bowen, 25
Boyer, 208, 209
Boysen-Jensen, 353
Brachet, 14, 19
Branton and Park, 40
Braun & Wood, 366

Brawerman and Eisenstadt, 42
Breidenbach et. al., 35, 185
Brian, 359
Brich et al., 360
Brown and Wilson, 115
Brown, 17
Bucher, 5
Buckmann, 249
Burg, 354
Burr, 163
Burstrom, 121
Butschilli, 17
Buvat, 22

C

C. Nagh and C. Cramer, 11
C.R. Slack, 139, 163
Cajal, 27
Cajlachjan, 328, 329, 332
Calvin and Benson, 137
Calvin, 160
Camillo Golgi, 24
Carnahan et al., 241
Cains, 369
Caspersson, 19
Cathry, 344
Chakravarti, 345
Chatt et al., 243
Cholodny, 380
Chouard, 344
Christensen, 19, 20, 23
Christian Huygens, 144
Clark, et. al., 41
Claude, 23, 36, 42
Coartney, 357
Cohen, 355
Comb and Zehavi Willner, 44
Cowdry, 26
Crick, 259
Cross et al., 360
Curry, 378
Czaja, 357

D

D.W. Fawcett, 26
Danielli and Davson, 11
Darnell, 226

Darwin, 352, 376
De Bary, 96, 235
De Duve, 27, 28, 29, 30
De Saussure, 103, 184
De Vries, 42, 126
De-Bary, 237
Decker, 229
Delvin, 118
Denny, 368
De-Robertis, 13, 24, 27
Diller, 105
Dodart, 379
Dolk, 379
Dutrochet and Vierdot, 59
Dutrochet, 137

E

E.E. Dekker, 341
Eaton, 111
Edwards and Lewis, 15
Efraim Racker and Colleages, 206
El-Sharkawy et al., 231
Eltinge, 114
Embden, 185, 186
Emerson, 149
Ernster, 114
Esau, 23, 25
Essner and Novikoff, 22, 23
Evans, 380
Eyster, et. al., 114

F

F. Jacob, 283
F.F. Blackman, 137, 175
Fajer et al., 153
Fawcett and Ito, 19
Fawcett, 22, 23
Flemming, 17, 31, 32
Florendo, 44
Fontana, 51
Forget, 266
Foster, 356
Frank, 379
Frankland, 359
Frederick and Newcomb, 31
Frey-Wyssling and

Muhlethaler, 2
Fritz, 93

G

Galileo, 1
Galston, 377
Gane, 368
Garner, 328, 329, 350, 367
Gatenby, 24
Gauch, 105
George, 24
Gimmler and Avron, 158
Godlewski, 82
Goeschl, 368
Goldacre, 114
Granier, 19
Grav, 249
Gregory, 345
Grew and Malpighi, 2
Grim, 115

H

Haas, 114
Hackett, 185
Hall, et al., 111
Halliman Murty and Grant, 46
Hammer, 329, 331
Hand, 356, 377
Hansel, 345
Hanstein, 42
Harder, 175
Hargovind Khorana, 269, 288
Hartman, 249
Hartt, 163
Harvey, 368
Hayashi, 359
Hemming, 359
Hendricks, 332
Henfrey, 328
Hepler and Newcomb, 8
Heslop and Harrison, 23
Hew and Hortkov, 229
Hewit et al., 116, 117
Highkin, 344
Hill and Bendall, 137
Hirschler, 24
Hitchock, 368
Hober, 118, 120

Hoflet, 42
Hogeboom, et al., 32
Holley, 269
Hoover, 175
Hope, 118
Hopkins, 114, 118, 120
Howell, 40
Hugo Von Mohl, 15
Huxley and Zubay, 43
Hylmo, 118, 120

I

Ingen-Housz, 184

J

J. Hammerling, 51
J. Monod, 283
J.C. Bose, 82, 378
J.Q. Plowe, 11
Jacob, 267
James Clark Maxwell, 144
Jan Ingen Housz, 137
Jean Senebier, 137
Jenny and Overstreet, 119
Jensen, 1, 7
Jensen, 175
John Ingle, 342
Johnson, 382
Jolly, 83
Jones and Fawcett, 23, 24
Joseph Priestly, 137

K

Kalyansundaram, 115
Kavanau, 11
Keilin, 32, 82, 115
Kessel, 22
Khan, 366
Kingsbury, 32
Kisaki, 229
Klein and Carnoy, 17
Kleinschmidt, 35, 185
Klippart, 344
Knight and Darnell, 44, 190, 262
Knight, 226
Knoll and Ruska, 2
Knoop, 301
Knot, 329
Kogl, 353

Kollen, 344
Kollicker, 31
Korenberg, 313
Korn, 11
Kortschak, 163
Kostychev, 186
Kozloskwi, 84
Kramen, 118, 119
Kramer, 76, 78, 84, 118, 120
Kreb, 185, 194, 313
Kurosawa, 358
Kylin and Hylmo, 118, 120

L

Lake, 268
Lakes, 42
Lam, 377
Lang, 344, 345, 346, 367
Langham, 377
Lapetina, 13
Larsen, 381
Latham, 365, 366
Leach, 185
Lehninger, 11, 34, 36
Leibig, 103, 137, 175
Leloir, 320
Levis, 311
Lewis and Lewis, 32
Liebermann, 368
Lindberg, 114
Lloyd, 90
Loewy and Siekevitz, 1
Loftfield, 90
Lona, 368
Lopushmsky, 120
Luck, 24, 36
Lundegardh, 118, 120, 121
Lysenko, 344, 346, 368
Lyttleton, 41

M

M. Traube, 59
M.D. Hatch, 139, 163
M.H. Burgos, 25
M.M.K. Nass, 35
Mac Munn, 184
Maddy, 13
Majelis Young, 185
Malavolte, 112

Index

Mann, 115
Mapson, 368
Marinozzi, 44
Masson and Phillis, 126
Masuda, 357
Matile, 27
Maul, 22
Maurice Traube, 58
Max Plank, 144
Max Schultze, 15
Mazia and Ruby, 13
McDermott, 84
McElroy et al., 114
Mead, 311
Melchers, 344, 345, 346, 368
Meselson, 259
Mayerhof, 185, 187
Michaelis, 32
Miller, 266, 365, 366
Mondrinakis, 40
Monod, 267
Monro, 45
Morrie, 25
Moskov, 328
Mullard et al., 185
Muller et al., 150
Munch, 79, 126

N

Najjar, 320
Nanninga, 43, 44
Nason, 114
Nason, 234
Nass, 35, 151, 185
Naumann, 153
Navez, 379
Naylor, 329
Nelson, 153
Newcomb, 301
Nicholas Theodore de Saussure, 137
Nicholas, 360
Nirenberg, 269, 288
Noggle, 93
Noll, 41
Nollet, 59
Nomura, 45
Nooden, 357

Northcote, 4, 10, 23, 26, 27
Northen, 356
Novikoff, 28, 29, 36, 118
Nowinski and Saez, 24, 27

O

Osborne, 366
Overstreet, 118

P

Palade, 21, 22, 23, 42
Paleg, 360
Palmer, 367, 368
Parat and Painleve, 24
Park, 40
Parker, 329, 332
Parnas, 185, 186
Parsons, 34
Pasteur, 184
Perner, 31
Peter Mitchell, 139, 209
Peters Kelly and Dembitzen, 24
Pfeffer, 58, 184, 377, 383
Phillip, 380
Pickett Heaps, 27
Pierre Favard and Juniper, 26
Pincus, 350
Piper, 114
Pollister, 33
Pon, 40
Porter and Mochado, 8, 24
Porter and Poruni, 24
Porter et al., 19, 24
Possingham, 115, 116, 117
Pratt, 368
Priestley, 76
Purvis, 344, 345, 355

R

R.R. Swain, 341
Racker, 34
Raffi, 368
Rajgopal, 355
Rega, 13
Rhoades and Carvalho, 37
Rhoades, 41, 186
Richmond, 367
Rigard, 32
Robert Brown, 50

Robert Holley, 288
Robert Hooke, 1, 2
Robert M. Dowben, 20
Robert, Hill, 149
Roberts, 26
Robertson, 11, 21, 36, 118, 120, 122
Robinson, 379
Rose and Pomera, 19
Ross, 331
Rothschild, 22, 24, 226
Ruben, Randall and Kamen, 137, 148
Ruinen, 238

S

S. Nass, 35
S. Wakil, 301
S.M. Sirkar, 367
Sabatini, 27
Sachs, 83, 137, 184, 319
Sagar, 36
Salisbury, 345
Salkowski, 353
Saraswati Devi, 115
Saucer, 40
Sayre, 90
Scheide and Lin, 11
Schneider et al., 241
Schulman, 249
Schwabe, 344, 345
Searth, 90
Seikevitz and Palade, 22, 23
Semadeni, 31
Setterfield and Bayley, 7, 264
Shaw, 366
Sievers, 380
Simard, 44
Singer and Nicolson, 11, 12
Sir Issac Newton, 144
Sjostrand, 19
Skoog, et al., 356, 357, 366
Slater, 207, 209
Smith, 367, 368
Soto, 13
Spencer, 368
Stahl, 259
Stansly, 301

Starling, 350
Steinberg and Jeffery, 115
Stephan Hales, 81
Steward, 48, 91, 120
Stiles, 116
Stinton, 368
Stoffler and Wittman, 42
Stokes, 345
Strasburger, 82
Street, 94
Stumpf, 301, 308
Stutz and Noll, 44
Stutz, 41
Suzam and Boawer, 35, 185

T

Thimann, 76, 78, 130, 348, 350, 353, 356, 357, 362, 366
Thirery, 19
Thorneley et al., 243
Thread Gold, 36
Thut, 84
Tolbert et al., 30, 31, 227

Tolbert, 229
Trui, 115
Turner, 118, 120

U

Unger, 82

V

Vacha, 368
Van Overbeek et. al., 353, 380
Vanden Honert, 118
Varga, 368
Vassel, 115
Velami's, 114
Vernon and Ke, 117
Von Mohl, 90

W

W.L. Butler, 335
Walker, 25
Wallace, 36, 368
Warburg, 137, 144
Ward and Ward, 21, 22
Wareing, 336, 357, 369

Warren, 249
Watson, 42, 259
Wightman, 355
Weissman, 45, 266
Wellensick, 344
Went, 352, 354, 376, 377
White, 112
Wickson, 371
Wilkins, 386
Williams, 114
Willstatler, 41
Wilson, 17, 371
Woodcock and Fernandez Morgan, 41, 151
Woodward, 103
Worley, 25
Worlkoff, 176

Y

Yabuta, 364
Yang, 373

Z

Zeevart, 332
Zelitch, 92, 228, 231

Zimmermann, 368

SUBJECT INDEX

A

Abscisic acid, 147, 369
Abscision, 356
Absorption, 71-84
Absorption, spectrum, 148
Acetaldehyde, 193
Acetic acid, 216
Acetobactor, 216
Acetyl CoA, 194, 195
Acetyl lipoic acid, 194
Acetylene propylene, 368
Acidi lactici, 216
Acidification, 168
Aconitase, 195
Aconitic acid, 195
Action spectrum, 148, 177
Active absorption, 76, 120
Active transport, 14
Adenosine, 213
Adsorbents, 4
Adsorption, column chromatography,
Aerobes, 186

Aerobic oxidation of pyruvic acid, 192
Aerobic oxidation, 192
Aerobic respiration, 185
Aerotropism, 383
Aespirator, 224
Agar, 42
Agroclavine, 225
AICAR, 251
AIR, 251
Alanyl glycine, 55
Alanyl tRNA, 193
Albumins, 51
Alcoholic fermentation, 216
Aliphatic amino acids, 49
    non-essential amino acids, 52
Alkaloid, 225
Amides, 52
Amidine lyases, 82
Amino acid, 49
    biosynthesis, 271
    classification, 49

a-amino acids, 49
Amino sugar, 39
Ammonification, 231
Ammonifying bacteria, 231
Amoeboid movement, 373
Amphoteric, 277
Amyloplast, 37
Anabaena, 238
Anabolic, 226
Anaerobes, 186
Anaerobic respiration, 186, 217
Antagonastic function, 109
Antenna complex, 153
Anthesin, 367
Antitranspirants, 93
Apical dominance, 357
Apoenzyme, 84
Apoplast, 79
Arachis hypogea, 51
Aromatic amino acids, 386
Arun spadix, 212
Ascent of sap, 81
Associative nitrogen fixation, 236

Index

ATP synthetase, 206
ATP, 199, 213, 214
Autolysis, 30
Automatically irrigated culture, 106
Autonomous movement, 371, 374
Autophagy, 29
Autoradiography, 18
Autotrophic, 129
Autotrophs, 129
Auxanometer, 349
    arc, 349
    Pfeffer's 349

Auxin, 3, 352-358
    action, 356
    bioassay, 354
    biosynthesis, 354
    chemical nature, 344
    extraction 353
Auxins in geotropsim, 379
Auxins in phototropism, 377
Avena coleoptile, 352
Avena sativa, 352, 354
Avena test, 354
Azotobacter paspali, 342
Azotobacter, 237
α-structure, 61

B

Bacilus, 216
Bacterial photosynthesis, 169
Bacterio Viridin, 38
Bacterium, 216
Bacteroids, 238
Bakane disease, 359
Bakers yeast, 226
Barley, 217
Barytropism, 379
Beckman's spectrophotometer, 12
Beer's Law, 9
Beijerenckia, 237
Benedicts test, 48
Betamers, 115
Biochemical machines, 31
Biological oxidation, 205

Bioluminiscence, 213
Biosynthesis of auxins 354
    functions, 357
    mechanism of action, 356
    non-indol auxin, 356
    pathways of indole auxin, 354
Biosynthesis of cholestrol, 305
Biosynthesis of fats, 301
Biosynthesis of lecithin, 305
Biosynthesis of monosaccharides, 315
Biosynthesis of nucleotides, 249
Biosynthesis of polysaccharides, 317
Biosynthesis of proteins, 277
Biosynthesis of proto alkaloids 226
Biosynthesis of pseudoalhaloids-229
Biosynthesis of ribose, 316
Biosynthesis of sucrose, 316
Biosynthesis of triglyceride, 304
Biosynthesis of true alkaloids-226
Biosynthesis of U.D.P., Dglucose, 316
Biosynthesisis of polysaccharides, 317
Biotin, 113, 130
Bitter principles, 223
Biuret reaction, 57
Blackman's belljar method, 98
Blackmans Law of Limiting Factor, 175
Bleeding, 97
Blue green algae, 235
Bolting, 363, 364
Bordered pit, 9
Boron, 115
Branched chain fatty acid, 68
Bromine water reaction, 58
Brownian movement, 17
Bryophyllum, 220
Burmeda gras, 376
Butyric acid, 214
β-thiogalactoside, 283

β-galactosidase, 283
β-oxidation, 308
β-structure, 59

C

C_4 plants, 166
Cajanus cajan, 235
Calciferol, 123
Calcium carbonate, 49
Calcium nitrate, 234
Calcium oxalate, 50
Calcium, 112
Callus, 358
Calmodulin, 343
Calvin cycle, 161
CAM cycle, 167
Canary grass, 376
Capillary gravitational water, 71
Cappillary force theory, 83
Carbohydrate Continued
    Heptoses, 32
    Pentoses, 30
    Tetroses, 30
Carbohydrate metabolism, 315, 323
Carbohydrates, 24
Carbon, 110
Carbonic acid, 165
Carotenoid, 38, 142, 218
Carrier concept, 120
Catabolic, 226
Catabolism, 307
Causes of energy richness in A.T.P., 200, 213
Causes of seed dormancy, 337
Caustic, 338
Cell division, 213
Cell wall, 3
Cell, 32
Centrosome, 18
Chain termination, 277
Changes during seed germination, 326
    in carbohydrate, 340
    in lipids, 341
    in proteins, 341
Chargoff's rule, 253
Chemical coupling hypothesis 208

Chemiosmotic hypothesis, 210
Chemistry of ATP synthetase, 206
Chemistry of bases, 246
Chemosynthesis, 173
Chemotaxis, 374
Chemotropism, 382
Chitin, 42
Chlorine, 117
Chloro choline chloride (C.CC), 370
Chlorobium, 235
Chlorophyll, 139
Chloroplast DNA, 185
Chloroplast, 37, 38, 42, 227
Chromatophore, 38
Chromocentre, 51
Chromophore, 333, 334
Chromoplast, 37
Cicer arietinum, 235
Ciliary movement, 259
Cillia flagella, 18
Citrate synthetase, 196
Citric acid cycle, 194
Citric acid, 214
Clinostat, 380
Clostridium, 216
Clover leaf model, 268
CO_2 Fixation, 173
Coagulation test, 58
Cobalamin, 131
Cobalt chloride method, 98
Coenzyme A, 108
Coenzyme nucleotides, 166
Coenzyme Q, 117
Coenzyme, 101-118
    Action, 102
    Structure and classification, 101
Cohesive force, 83
Coliophage, 176
Colloidal clay culture, 106
Colorimetry, 9
Competitive inhibition, 91
Complementary codons, 280
Condensation of fatty acid, 304
Condensation, 304
Conformational coupling hypothesis, 308

Conjugated proteins, 368
Controlling genes, 286
Copper, 116
Core compex, 153
Core enzymes, 278
Co-repressure, 283
Corpuscular theory, 145
Corpuscular radiation, 14
Coupling factor, 154
Cristae, 214
Critical element, 117
Cryptoxanthin, 117
Crystalline theory, 83
Curry hypothesis, 376
Cuticular transpiration, 85
Cyanide resistant respiration (CCR), 212
Cyanocobalamine, 112
Cyclic electron transport system, 156
Cyclic fatty acids, 67
Cyclic photophosphorylation, 156
Cyclic photophosphorylative photosynthetic bacteria, 172
Cyclosis, 213, 373
Cynothydrins, 36
Cystolith, 49
Cyt b_6 F complex, 154
Cytidine, 115
Cytochrome C, 111
Cytochrome pump theory, 121
Cytochromes, 111
Cytokinin bioassay and effects, 366
Cytokinins, 365, 367
Cytoplasm, 146, 147

D

Daily periodicity, 87
Dark period, 331
Dark reaction, 160
Deacidification, 168
Decarboxylases, 82
Decarboxylation, 193
Decline period of growth, 348
Deficient nutrient solution, 105
Dehydratase, 82
Denaturation, 63

Denitrification, 241
Deoxyribose, 39
Deplasmolysis, 67
Derived lipids, 75
Derived protein, 55
Descending chromatography, 7
Desulfhydrases, 82
Devernalization, 346
Diaphototropism, 376
Dictyosomes, 27
Difference between aerobic and anaerobic respiration, 187
Difference between auxin and gibberellin, 365
Difference between oxidative and photophosphorylation, 214
Difference between photo and dark respiration, 230
Difference between respiration and photosynthesis, 228
Diffusion hypothesis, 126
Diffusion Pressure Deficit, 55, 66
Diffusion Pressure gradient, 55
Diffusion pressure of liquid, 55
Diffusion, 14, 54, 353
Dihydrolipoly dehydrogenase, 193
Dihydroxy acetone phosphate, 193
Dinitrogen, 234
Dinitrophenyl, 60
Dionea, 136
Dipeptide, 58
Diphosphoglyceric acid, 191
Direct oxidation pathway, 317
Disaccharide, 41
Disulphide Linkage, 59
Diterpenes, 221
DNA duplication, 259
DNA ligase, 264
DNA polymerase, 262
DNA, 257, 176
    chloroplast DNA, 189
    circular, 190
    classes, 176
    denaturation, 192

Index

enzymes, 78-100
evidence experiments, 187
forms, 178
hydrolysis, 193
mechanism of replication, 182, 260
mitochondrial, 189
molecular structure, 176
molecular weight, 191
renatur ration, 192
replication, 180
single stranded, 190
structure, 257, 175
unwinding proteins, 186
watson crick model, 176
DNA, 257, 176
DNP analysis, 60
Donnan equilibrium, 119
Dormancy, 336
Drip culture, 106
Drosera, 134
Dynamic equilibrium, 204

E

E.M.P. Pathway, 188
Ecdysones, 75
Ectotrophic, 134
Einstein's law of photo chemical equibalance, 146
Elaioplast, 37
Electric potential, 275
Electromagnetic radiation, 14
Electromagnetic wave theory, 145
Electron transport particles, 34
Electron transport system (E.T.S.), 153, 198
Electroosmotic theory, 78
Electrophoresis, 12
Electrostatic, 213
Elementary particles, 34
Embryo dormancy, 338
Emeiocytosis, 15
Emerson's effect, 150
Emulsification, 69
Endergonic, 197, 89
Endocytosis, 15
Endolases, 192

Endoplasmic Reticulum, 19, 23
Endosmosis, 59
Endotrophic, 133
Energetic coupling, 201
Enzyme substrate complex theory, 85
Enzymes, 78-100
chemical nature, 84
factors affecting activity, 93
induction, 270
inhibition, 91
mode of action, 85
nomenclature and classification, 78
properties, 90
repression, 283
Ephemeral movements, 375
Epimerases, 82
Epimerization, 320
Epinasty, 375
Epiphytes, 129
Ergastoplasm, 19
Essential oil, 49
Essential element, 103
Ester, 35
Esterases, 81
Ethephone, 369
Ethyl alcohol, 214
Ethylene, 148, 368
ETS, 199-203
Eukaryotic cell, 2
Excretory materials, 49
Excretory movement, 373
Exergonic, 89, 202
Exopeptidases, 81
Exosmosis, 59
Experiments related with inhibition, 68, 70
Experiments related with respiration, 223, 226
Extraction of auxin, 353
Extraction 359

F

Factors affecting germination, 342
Factors affecting respiration rate, 223

Factors affecting transpiration, 93
FAD, 212, 106
Farmers potometer, 100
Fat soluble vitamin, 120
Fat, 217, 65
Fatty acid, 63
Fehling test, 48
Fermentation, 214, 216
Fermentation, 225
Ferrous, 114
Flagella, 18
Flavonoids, 230
Florigen, 330, 332, 362
Fluid mosaic model, 12
Fluorenol, 370
Follish seedling disease, 358
Formylation, 281
Free nucleotides, 248
Fructose, 21
Fucoxanthin, 39
Fumarase, 198
Fumaric acid, 198
Functions of alkaloids - 230
Furanase ring 28

G

G_3P DHAP shuttle, 204
Galactosyl diglycerides, 74
Galvanotaxis, 374
Galvanotropism, 383
Ganong, 99
GAR, 251
Gas chromatography, 317
Genetic code, 289
Gentianose, 393
Geoelectrical theory, 380
Geotropism, 378, 381
Geranyl pyrophosphate, 361
Gibbane ring, 359
Gibberella fuji kuroi, 359
Gibberellins, 349
bioassay, 359
biosynthesis, 360
Gibberellins, 358
Discovery, 358
Functions, 363
Globulins, 54
Gluconeogenesis, 314

Glucose, 31
Glutelins, 54
Glycerophosphatide, 65
Glycine, 49
Glycogen, 45
Glycolate metabolism theory, 92
Glycolate oxidase inhibitor, 231
Glycolate oxidase, 230
Glycolic acid, 230
Glycolipids, 73
Glycolysis, 187, 192
Glycorprotein, 55
Glycosidase, 81
Glycosidic linkage, 159
Glyoxylate cycle, 302
Glyoxysomes, 30
Godlewaski theory, 82
Golgi complex, 24
    chemical composition, 26
    enzymes, 26
    functions, 26
    history, 24
    morphology, 25
    ultrastructure, 26
Grana, 39
Gravel culture, 106
Gravitational water, 79
Gravitropism, 379
Green sulphur bacteria, 169
Growth hormones, 350
Guanosine triphosphate (GTP), 194
Guard cells, 86, 87, 90
Gums, 42, 49
Guttation, 96

H

Halogenation, 69
Harmonal theory, 346
Hatch slack cycle, 164
Helianthus, 227
Hemiacetal, 28
Heteroglycans, 44
Heterophagosomes, 28
Heterotrophic, 129
Hill's reaction, 149
Histones, 54
Holoenzymes, 84
Homoglycans, 44

Homoserine, 237
Hopkin's col reaction 57
Hormones, 96, 115, 116
Howorth's Col projection, 29
Humus, 231
Hydathodes, 96
Hydration, 196
Hydrogen atoms, 205
Hydrogen bonds, 59
Hydrogen ion pool, 202
Hydrogen peroxide, 228
Hydrogen, 110
Hydrogen flame ionization 19
Hydrogenase, 238, 239
Hydrogenation, 68
Hydrogenolysis, 69
Hydrolases, 80
Hydrolysis of fat, 68
Hydroponics, 104
Hygroscopic movement, 371
Hydrotropism, 381
Hydroxy fatty acids, 303
Hydroxy pyruvate, 343
Hydroxylamine, 233
Hygroscopic coefficient, 73
Hygroscopic water, 72
Hyoscyamus niger, 368
Hyperchromatic shift, 192
Hypertonic, 57
Hyponasty, 375
Hyponitrite, 233
Hypoxanthine, 156
Hypotonic, 57

I

Illuminated, 229
Imbibition theory, 82
Imbibition, 68
Imidazole, 247
Imposed dormancy, 336
Inulin, 48
Induced fit model, 88
Inducible system, 284
Initiation point, 178
Innate dormancy, 336
Inorganic materials, 49
Inorganic phosphate, 206
Inosine, 249
Insectivorous plants, 134

Intercellular spaces, 10
Interconversion of monosaccharides, 320
Intermediate plants, 328
Intermediate product, 217
Intramolecular oxido reductases, 83
Intussusception, 7
Inulin, 45
Iodine value, 77
Ion exchange chromatography, 5
Ion exchange, 118
Ion exchanger, 5
Irradiation, 14
Iso enzyme, 83
Isoamyl alcohol, 368
Isocitric acid, 196
Isocitric dehydrogenase, 196
Isoelectric point, 277
Isolation and purification of enzyme, 83
Isomerase, 82
Isopentanyl pyrophosphate, 347
Isotel, 119
Isotonic solution, 57
Isotopes, 18, 229
IUB, 79

K

Kaurenoic acid, 361, 363
Kinetin, 365
Kinetosomes, 18
Klebisilla pneumoae, 241
Kreb's cycle, 194
α-ketoglutaric acid, 196

L

Lactici acidi, 216
Lactose, 42
Lag period of growth, 348
Lag period, 348
Latex, 49
Laws of thermo dynamics, 205
Lecithin, 57
Lectins, 239
Leg haemoglobin, 240

Index

Lenticular transpiration, 85
Leucoplast, 36
Leucoriboflavin, 107
Lichens, 133
Ligases, 83
Light, 223
Lime water, 225
Line Weaver Burk plot, 87
Linseed, 217
Lipase, 66, 68
Lipid, 65
Lipoate dehydrogenease, 194
Lipoic acid 109
Lipoprotein, 75
Lack and key model, 88
Lacomotion, 213
Loctic acid, 214
Long day plant, 328
Loops of t-RNA, 269
Lyases, 82
Lysosome, 27
L-α-glycerophosphate, 304
L-α-phosphatidic acid, 304

M

Macro nutrients, 104
Macroelement, 104
Magnesium, 113
Major elements, 104
Maleic hydrazide, 370
Malic acid, 198, 199
Malic dehydrogenase, 198
Maltose, 42
Manganese, 114
Mannose, 32
Mannotrioe, 44
Mass flow, 120
Mass spectroscope, 15
Matric potental, 64
Mechanism of oxidative phosphorylation, 207
Mechanism of transpiration, 88
Mechanism of water absorption, 76
Messelson and Stahl theory, 261, 181
Messenger RNA (m-RNA), 267
Metalloprotein, 55
Methanobacterium, 237
Methionine, 281, 368

Methylated bases, 269
Mevalonic acid (MVA), 360
Michaelis constant, 86
Micro nutrient, 104
Microbodies, 30
Micropendulum theory, 380
Microsome theory, 380
Millons test, 56
Minor elements, 104
Mitochondria, 31, 227
    chamber, 34
    chemical composition, 34
    cristae, 33
    DNA, 35
    enzymes, 34
    functions, 34
    membranes, 34
    origin, 36
    ribosomes, 36
    Mixed texpenoids, 223
MOA shuttle, 202
Moisture content, 103
Moisture percent, 103
Moisture, equivalent, 73
Molar extinction, coefficient, 9
Molecular oxygen, 203
Molitch test, 48
Molybdenum, 116, 235, 241
Molybdoflavo protein, 234
Monocistronic m-RNA, 267
Monosaccharides, 25
Monoterpenes, 220
Morphactins, 144, 370
Morphine, 224
Movements of curvature, 374
Movements of growth, 374
Movements, 371
Munch's hypothesis, 126
Mycorrhiza, 133
Myeloid bodies, 21
Myrcagale, 235

N

NAA, 358
NAD, 104
    in Kerb's cycle, 208
    in photo synthesis, 210
NADH shuttle system, 203
Nastic movements, 375, 383
N-C glycosidic bond, 16-65, 243

Nectar, 49
Negative phototropism, 376
Nepenthes, 135
Nephthoquinone, 127
Niacin, 129
Nicotiana, 227
Nicotinamide, 105, 200
Nicotinic acid, 129
Nif. Genes, 241
Ninhydrin reaction, 56
Ninhydrin, 56
Nitella, 227
Nitrate reductase, 234
Nitrate reduction, 234
Nitrification, 233
Nitrifying bacteria. 237
Nitrite reductase, 235
Nitro prusside, test, 56
Nitrococcus, 233
Nitrogen cycle, 245
Nitrogen fixation, 236
    Biochemistry, 241
    Biological, 236
Nitrogen in soil, 233
Nitrogen significance, 233
Nitrogen, 110
Nitrogenase, 240, 241
Nitrogenous bases, 246, 155
Nitrosomonas, 233
Nodule formation, 238, 240
Non biological, 236
Non cyclic phosphorylation in photosynthetic bacteria, 169
Non cyclic photophosphorylation, 153
Non cytochrome terminal pathway, 212
Non symbiotic nitrogen fixation, 237
Non-competitive inhibition, 92
Non-cycle ETS, 153
Non-essential elements, 103
Non-sense codons, 282, 155
Nucleic acid, 342, 155
Nucleoplasm, 51
Nucleoprotein, 55
Nucleoside, 248, 162
Nucleotides, 164
Nucleus, 50
Nutation, 375

Nutrient solution, 104
Nutrient vapour bath, 107
Nutrition, 103
Nyctinastic movements, 384

O

Oat coleoptile, 376
Oat, 217
Oil, 62
Okazaki fragment, 180, 264
Oligosaccharide, 25
Opening and closing of stomata, 89
Operator constitutive mutant, 287
Operon model, 284
Organic material, 49
Ornithophilous, 131
Orotic acid, 254
Oryza, 227
Osmosis, 14, 57
    pressure, 14
    types, 59
Osmotic pressure, 60
Osmotic theory, 78
Osmotropism, 383
Outer and apparent free space theory, 118
Oxidation of carbohydrate, 320
Oxidation of extramitochondrial NADH, 203
Oxidation of oligo and polysaccharide, 320
Oxidation of protein, 290
Oxidation Reduction Potential, 211
Oxidative decarboxylation, 192
Oxidative phosphorylation, 204
Oxidising agent, 206
Oxido reduction, 322
Oxidoreductases, 80
Oxinsulin, 66
Oxygen, 110

P

Pantothenic acid, 130
Parasites, 130
Parasponia, 235
Paratonic induced movement, 373, 374

Paratonic movment, 374
Parthenocarpy, 349, 365
Partition chromatography, 6
Paspalam notatum, 236
Passive absorption, 76, 118
Paully reaction, 57
Pea root test, 354
Pectin, 46
Pentose sugar, 247, 161
Peptidases, 81
Peptide bond, 58
Perceptive organ, 371
Permanent cells, 222
Permeability, 56
Peroxysomes, 227
Petunia, 227
Pfeffer's auxanometer, 349
P_{FR}, 332, 339
Phaeoplast, 37
Phagocytosis, 15
Phasic development theory, 346
Phosphate ester bond, 248
Phosphoenol pyruvate, 201, 305
Phosphoglucomutase, 320
Phosphoglyceromutase, 191
Phosphoglyceraldehyde, 191
Phosphoglycerokinase, 191
Phosphoglyceromutase, 191
Phosphoinositide, 72
Phospholipids, 70
Phosphoric acid,
Phosphopyuvic kinase, 192
Phosphoric acid, 248
Phosphorus, 111
Phosphorylation, 154
Photoneutral plants, 328
Photo odixation of H_2O, 151
Photoblastic, 338, 339
Photoelectric calorimeter, 11
Photo-induction, 329, 332
Photometer, 10
Photometry, 10
Photoperiodism, 367, 328
Photorecepter pigments, 376
Photoreceptors, 376
Photorespiration,
    Biochemistry, 229
    Site, 229

Photosynthetic inhibitors, 229
Photosynthetic quotient PQ, 221
Photosystem I, 150
Photosystem, II, 150
Phototaxis, 374
    hototropism, 376
    role of auxins, 377
Phycobillins, 142
Phyllopshere, 236
Physical theories, 82
Physiological preconditioning, 344
Phytin, 342
Phytochrome, 332
Plant ash, 103
Plasma membrane, 11, 15
Plasmalogens, 71
Plasmodesmata, 7
Plasmolysis, 67
Plastids, 36
Polarimeter, 20
Polarimetry, 20
Polycistronic m-RNA, 267
Polyribosome, 46
Polysaccharides, 25
Positive phototropic, 376
Positive response, 379
Potassium, 110
P_R, 318, 325
Presentation time, 371
Pressure potential, 64
Prmotor gene, 288
Processing of RNA, 266
Prokaryotic cells, 2
Prolamines, 54
Promotor region, 288
Properties of fatty acids, 68
Prosthetic group, 84
Protamines, 54
Protein contents, 53
Protein structure, 58
Proteins, 52
Proton transport theory, 92
Protons, 15
Protoplasm, 15
Protoplasmic streaming theory, 126
Purine bases, 247
    Biosynthesis of GMP,

Index

AMP and
    Uric acid, 252
    Biosynthesis of IMP, 251
Purines, 247
Purple sulphur bacteria, 169
Pyranose ring 27
Pyridoxal phosphate, 113
Pyridoxine, 13
Pyrimidine biosynthesis, 254
Pyrimidines, 246
Pyruvate dehydrogenase, 194

Q

Quantum theory, 145
Quiescence, 336

R

Racemases, 82
Radiation, 14
Radio isotopes, 14
Raffinose, 43
Rancidity, 69
Reaction time, 371
Red drop Emersons enhancement effect, 150
Redox potential, 217
    measurement, 217
    sereis, 217
Redox reactions in biological systems, 211
Reducing agent, 40
Reducing, non-reducing sugars, 40
Reduction, 205
Regulatory genes, 286
Renaturation, 64
Repressible system, 286
Reserpine, 225
Reserve material, 47
Resonance, 153
Respiration, 184
    changes associated, 185
    history, 184
    types, 185
Respiratory inhibitors, 212
Respiratory quotient (RQ), 217
Respiratory substrate, 186
Responsive organ, 371

Retinol, 121
Rf. value, 16
Rheotaxis, 374
Rheotropism, 383
Rhizobium species, 241
Rhodoplast, 38
Riboflavin, 106
Riboprotein, 45
Ribose, 30
Ribosomal RNA, 265
Ribosome, 42
    Function, 45
    RNA, 44
    Ultra structure, 43
Ricca's factor, 370
Ricket, 120
Ripening hormone, 369
RNA priming, 179, 261
RNA, 189
    cloverleaf model, 198
    hydrolysis, 200
    Kim model, 199
    messenger, 196
    ribosomal, 194
    types, 194
    transfer, 197
RNA, 265
Role of ethylene, 369
Root pressure theory, 82
Rospiratory substrate, 187
Rotation, 373
Rubber, 223
RUDP carboxylase, 227, 315
RUDP oxygenase, 315
RUDP, 315
Running, water, 72

S

Saccharic acid, 33
Sacrification, 338
Sakaguchi reaction, 57
Salting out, 81
Sand culture, 106
Sapoonin, 76
Saponification, 68
Saprophyte, 132
Saturated fatty acids, 66
Scaning, 11
Scenedesmus, 5
Scleroproteins, 54

Scurvy disease, 129
Secondary metabolites, 219
Secretion theory, 78
Secretory material, 48
Seed coat dormancy, 337
Seed dormancy, 336
Selivanoff test, 48
Semiconservative replication, 261
Senescence, 348, 365
Serine, 229
Sesquiterpenes, 220
Short day plant, 329
Sigma factor, 278
Sigmoid curve, 348
Significance of photorespiration, 231
Simple protein, 54
Sirohaem, 233
Slit pea test, 354
Slope culture, 106
Soil water, 71
Solute potential, 64
Soxhlet type culture, 106
Special feature in plants, 94
Spectrophotometer, 10
Spectrophotometry, 10
Spectroscopy, 11
Sphingosine, 307
Sphingomyeins, 307
Standard electrode potential, 217
Redox potential 217
Starch glucose interconversion theory, 91
Starch, 45
Statolith theory, 380
Steady state, 208
Sterlie root medium, 107
Steroids, 75
Sterol glycoside, 76
Sterolins, 76
Stimulus, 365
Stomata, 86
    types, 87
Stomatal movement, 87
Stomatal transpiration, 86
Straight growth test, 354
Stratification, 338
Streaming, 16
Stress physiology, 373

Substrate phosphorylation, 204
Succinic thiokinase, 197
Sucrase, 41
Sucrose, 41
Suction pressure, 95, 62
Sulpholipid, 75
Sulphur, 110
Svedberg, 14
Symbiosis, 237
Symbionts, 133
Symplast, 79
Synthesis of AMP, 252
Synthesis of fatty acids, 301
Synthesis of glycerol, 304
Synthesis of GMP, 252
Synthesis of uric acid, 252
Synthetic ion exchange materials, 106

T

Termination codons, 289
Terpene alcohols, 277
Terpene alkaloids 223
Terpenoid glycasides, 223
Tetraterpenes, 222
Theory of photosynthesis in
Thermal conductivity cell guard cells, 90
Thermal conductivity cell, 18
Thermotaxis, 374
Thermodynamic equilibrium, 208
Thermotropism, 382
Thiamine pyrophosphate, 110
Thiamine, 106
Thigmotropism, 381
Thin layer chromatography, 9
Thiokinases, 309
Tocopherol, 225
Tracer technique, 15, 17

Transaldolase, 318
Transcription, 278
Transfer RNA (t-RNA), 268
Transferase, 80, 281
Transformylase, 280
Transketolase, 318, 322
Translation, 279
Translocation, 282
Transmutation, 15
Transpiration pull, 84
Transpiration, 85
Traumatotropism, 383
Triglycerides, 65
Tripalmitin, 68
Triterpenes, 222
Tropic movements, 376
Tryptamine pathway, 355
Tryptophan, 354
Tryptophol pathway, 355
Turgor pressure, 61
Turnover number, 81
Types of seed dormancy, 336

U

U.V. spectroscopy, 12
Ubiquinone, 117
Ultracentrifugation, 13
Ultracentrifuge, 13
Unit membrane Model, 12
Unsaturated fatty acids, 66
Unwinding and Untwisting proteins, 182, 261
Unwinding, 262
Uric acid, 172
Uridine diphosphate, 116
Uronic acid, 34
Utricularia, 134

V

Vacuole, 50

Vernalin, 345, 369
Vernalization, 344
Vital force theory, 82
Vitamins, 119-133
    nomenclature and classification, 120
        Vanadium, 117
        Vital theory, 82

W

Wall pressure, 61
Water potential concept, 63
Water soluble vitamins, 128
Watson and Crick Model, 259
Wave theory, 145
Waxes, 69
Weeds eradication, 358
Went experiment, 352
Whatman paper, 1, 7
Wilting coefficient, 73
Wilting, 73
W oxidation, 301

X

Xanthine, 156
Xanthium pennsylvanicum, 332
Xanthoprotein test, 56
X-ray diffraction, 14
Xylan, 318
Xylose, 23
f × 174, 257

Z

Z-DNA, 175
Zeatin, 366
Zinc, 115
Zone electrophoresis, 13
Zwitter ions, 277

Attention: Students

We request you, for your frank assessment, regarding some of the aspects of the book, given as under:

03A 202 A Textbook of Plant Physiology, Biochemistry and Biotechnology
   S.K. Verma, Mohit Verma   *Reprint 2019*

Please fill up the given space in neat capital letters. Add additional sheet(s) if the space provided is not sufficient, and if so required.

(i) What topic(s) of your syllabus that are important from your examination point of view are not covered in the book?
....
....
....
....

(ii) What are the chapters and/or topics, wherein the treatment of the subject-matter is not systematic or organised or updated?
....
....
....
....
....

(iii) Have you come across misprints/mistakes/factual inaccuracies in the book? Please specify the chapters, topics and the page numbers.
....
....
....
....
....

(iv) Name top three books on the same subject (in order of your preference - 1, 2, 3) that you have found/heard better than the present book? Please specify in terms of quality (in all aspects).
   1 ....
   ....
   2 ....
   ....
   3 ....
   ....

(v) Further suggestions and comments for the improvement of the book:
....
....
....
....

Other Details:

(i) Who recommended you the book? (Please tick in the box near the option relevant to you.)
☐ Teacher ☐ Friends ☐ Bookseller

(ii) Name of the recommending teacher, his designation and address:
....
....
....

(iii) Name and address of the bookseller you purchased the book from:
....
....
....

(iv) Name and address of your institution (Please mention the University or Board, as the case may be)
....
....
....

(v) Your name and complete postal address:
....
....
....

(vi) Write your preferences of our publications (1, 2, 3) you would like to have
....
....

The best assessment will be awarded half-yearly. The award will be in the form of our publications, as decided by the Editorial Board, amounting to Rs. 300 (total).

Please mail the filled up coupon at your earliest to:
Editorial Department
S. CHAND & COMPANY LTD.
Post Box No. 5733, Ram Nagar,
New Delhi 110 055